CHILTON®

ASIAN
DIAGNOSTIC SERVICE
2006 EDITION
VOLUME I
Acura
Honda

THOMSON

DELMAR LEARNING

Australia • Canada • Mexico • Singapore • Spain • United Kingdom • United States

CHILTON®
ASIAN
DIAGNOSTIC SERVICE
2006 Edition
Volume I
Acura & Honda

**Vice President,
Technology Professional Business Unit:**
Gregory L. Clayton

**Publisher,
Technology Professional Business Unit:**
David Koontz

Director of Marketing:
Beth A. Lutz

Production Director:
Patty Stephan

Editorial Assistant:
Rebecca Rokitowski

Production Manager:
Andrew Crouth

Marketing Manager:
Brian McGrath

Marketing Coordinator:
Jennifer Stall

Publishing Coordinator:
Paula Baillie

Sr. Content Project Manager:
Elizabeth C. Hough

Managing Editor:
Terry Blomquist

Editor:
Tim Crain

Graphical Designer:
Melinda Possinger

NOTICE TO THE READER

TABLE OF CONTENTS

SECTIONS

USING THIS INFORMATION

Organization

To find where a particular model section or procedure is located, look in the Table of Contents. Main topics are listed with the page number on which they may be found. Following the main topics is a listing of all of the subjects within the section and their page numbers.

Manufacturer and Model Coverage

This product covers 1996-2006 Asian models that are produced in sufficient quantities to warrant coverage, and which have technical content available from the vehicle manufacturers before our publication date. Although this information is as complete as possible at the time of publication, some manufacturers may make changes which cannot be included here. While striving for total accuracy, the publisher cannot assume responsibility for any errors, changes, or omissions that may occur in the compilation of this data.

Part Numbers & Special Tools

Part numbers and special tools are recommended by the publisher and vehicle manufacturer to perform specific jobs. Before substituting any part or tool for the one recommended, you must be completely satisfied that neither your personal safety, nor the performance of the vehicle will be endangered.

ACKNOWLEDGEMENT

The publisher would like to express appreciation to the following vehicle manufacturers for their assistance in producing this publication. No further reproduction or distribution of the material in this manual is allowed without the expressed written permission of the vehicle manufacturers and the publisher. American Honda Motor Co., including Acura and Honda Divisions.

PRECAUTIONS

Before servicing any vehicle, please be sure to read all of the following precautions, which deal with personal safety, prevention of component damage, and important points to take into consideration when servicing a motor vehicle:

• Always wear safety glasses or goggles when drilling, cutting, grinding or prying.

• Steel-toed work shoes should be worn when working with heavy parts. Pockets should not be used for carrying tools. A slip or fall can drive a screwdriver into your body.

• Work surfaces, including tools and the floor should be kept clean of grease, oil or other slippery material.

• When working around moving parts, don't wear loose clothing. Long hair should be tied back under a hat or cap, or in a hair net.

• Always use tools only for the purpose for which they were designed. Never pry with a screwdriver.

• Keep a fire extinguisher and first aid kit handy.

• Always properly support the vehicle with approved stands or lift.

• Always have adequate ventilation when working with chemicals or hazardous material.

• Carbon monoxide is colorless, odorless and dangerous. If it is necessary to operate the engine with vehicle in a closed area such as a garage, always use an exhaust collector to vent the exhaust gases outside the closed area.

• When draining coolant, keep in mind that small children and some pets are attracted by ethylene glycol antifreeze, and are quite likely to drink any left in an open container, or in puddles on the ground. This will prove fatal in sufficient quantity. Always drain the coolant into a sealable container.

• To avoid personal injury, do not remove the coolant pressure relief cap while the engine is operating or hot. The cooling system is under pressure; steam and hot liquid can come out forcefully when the cap is loosened slightly. Failure to follow these instructions may result in personal injury. The coolant must be recovered in a suitable, clean container for reuse. If the coolant is contaminated it must be recycled or disposed of correctly.

• When carrying out maintenance on the starting system be aware that heavy gauge leads are connected directly to the battery. Make sure the protective caps are in place when maintenance is completed. Failure to follow these instructions may result in personal injury.

• Do not remove any part of the engine emission control system. Operating the engine without the engine emission control system will reduce fuel economy and engine ventilation. This will weaken engine performance and shorten engine life. It is also a violation of Federal law.

• Due to environmental concerns, when the air conditioning system is drained, the refrigerant must be collected using refrigerant recovery/recycling equipment. Federal law requires that refrigerant be recovered into appropriate recovery equipment and the process be conducted by qualified technicians who have been certified by an approved organization, such as MACS, ASI, etc. Use of a recovery machine dedicated to the appropriate refrigerant is necessary to reduce the possibility of oil and refrigerant incompatibility concerns. Refer to the instructions provided by the equipment manufacturer when removing refrigerant from or charging the air conditioning system.

• Always disconnect the battery ground when working on or around the electrical system.

• Batteries contain sulfuric acid. Avoid contact with skin, eyes, or clothing. Also, shield your eyes when working near batteries to protect against possible splashing of the acid solution. In case of acid contact with skin or eyes, flush immediately with water for a minimum of 15 minutes and get prompt medical attention. If acid is swallowed, call a physician immediately. Failure to follow these instructions may result in personal injury.

• Batteries normally produce explosive gases. Therefore, do not allow flames, sparks or lighted substances to come near the battery. When charging or working near a battery, always shield your face and protect your eyes. Always provide ventilation. Failure to follow these instructions may result in personal injury.

• When lifting a battery, excessive pressure on the end walls could cause acid to spew through the vent caps, resulting in personal injury, damage to the vehicle or battery. Lift with a battery carrier or with your hands on opposite corners. Failure to follow these instructions may result in personal injury.

• Observe all applicable safety precautions when working around fuel. Whenever servicing the fuel system, always work in a well-ventilated area. Do not allow fuel spray or vapors to come in contact with a spark, open flame, or excessive heat (a hot drop light, for example). Keep a dry chemical fire extinguisher near the work area.
Always keep fuel in a container specifically designed for fuel storage; also, always properly seal fuel containers to avoid the possibility of fire or explosion. Do not smoke or carry lighted tobacco or open flame of any type when
working on or near any fuel related components.

• Fuel injection systems often remain pressurized, even after the engine has been turned OFF. The fuel system pressure must be relieved before disconnecting any fuel lines. Failure to do so may result in fire and/or personal injury.

• The evaporative emissions system contains fuel vapor and condensed fuel vapor. Although not present in large quantities, it still presents the danger of explosion or fire. Disconnect the battery ground cable from the battery to minimize

the possibility of an electrical spark occurring, possibly causing a fire or explosion if fuel vapor or liquid fuel is present in the area. Failure to follow these instructions can result in personal injury.

• The EPA warns that prolonged contact with used engine oil may cause a number of skin disorders, including cancer! You should make every effort to minimize your exposure to used engine oil. Protective gloves should be worn when changing oil. Wash your hands and any other exposed skin
areas as soon as possible after exposure to used engine oil. Soap and water, or waterless hand cleaner should be used.

• Some vehicles are equipped with an air bag system, often referred to as a Supplemental Restraint System (SRS) or Supplemental Inflatable Restraint (SIR) system. The system must be
disabled before performing service on or around system components, steering column, instrument panel components, wiring and sensors. Failure to follow safety and disabling procedures could result in accidental air bag deployment, possible personal injury and unnecessary system repairs.

• Always wear safety goggles when working with, or around, the air bag system. When carrying a non-deployed air bag, be sure the bag and trim cover are pointed away from your body. When placing a non-deployed air bag on a work surface, always face the bag and trim cover upward, away from the
surface. This will reduce the motion of the module if it is accidentally deployed.

• Electronic modules are sensitive to electrical charges. The ABS module can be damaged if exposed to these charges.

• Brake pads and shoes may contain asbestos, which has been determined to be a cancer-causing agent. Never clean brake surfaces with compressed air. Avoid inhaling brake dust. Clean all brake surfaces with a commercially available brake cleaning fluid.

• When replacing brake pads, shoes, discs or drums, replace them as complete axle sets.

• When servicing drum brakes, disassemble and assemble one side at a time, leaving the remaining side intact for reference.

• Brake fluid often contains polyglycol ethers and polyglycols. Avoid contact with the eyes and wash your hands thoroughly after handling brake fluid. If you do get brake fluid in your eyes, flush your eyes with clean, running water for 15 minutes. If eye irritation persists, or if you have taken brake fluid internally, immediately seek medical assistance.

• Clean, high quality brake fluid from a sealed container is essential to the safe and proper operation of the brake system. You should always buy the correct type of brake fluid for your vehicle. If the brake fluid becomes contaminated, completely flush the system with new fluid. Never reuse any brake fluid. Any brake fluid that is removed from the system should be discarded. Also, do not allow any brake fluid to come in contact with a painted or plastic surface; it will damage the paint.

• Never operate the engine without the proper amount and type of engine oil; doing so will result in severe engine damage.

• Timing belt maintenance is extremely important! Many models utilize an interference-type, non-freewheeling engine. If the timing belt breaks, the valves in the cylinder head may strike the pistons, causing potentially serious (also time-consuming and expensive) engine damage.

• Disconnecting the negative battery cable on some vehicles may interfere with the functions of the on-board computer system(s) and may require the computer to undergo a relearning process once the negative battery cable is reconnected.

• Steering and suspension fasteners are critical parts because they affect performance of vital components and systems and their failure can result in major service expense. They must be replaced with the same grade or part number or an equivalent part if replacement is necessary. Do not use a replacement part of lesser quality or substitute design. Torque values must be used as specified during reassembly to ensure proper retention of these parts.

INTRODUCTION TO OBD SYSTEMS

1

Table of Contents

INTRODUCTION TO OBD

Contents

Notes & Cautions

Before servicing any vehicle, please be sure to read all of the following precautions, which deal with personal safety, prevention of component damage, and important points to take into consideration when servicing a motor vehicle:

- Observe all applicable safety precautions when working around fuel. Whenever servicing the fuel system, always work in a well-ventilated area. Do NOT allow fuel spray or vapors to come in contact with a spark, open flame, or excessive heat (a hot drop light, for example). Keep a dry chemical fire extinguisher near the work area. Always keep fuel in a container specifically designed for fuel storage; also, always properly seal fuel containers to avoid the possibility of fire or explosion. Refer to the additional fuel system precautions that follow.
- Fuel injection systems often remain pressurized, even after the engine has been turned OFF. The fuel system pressure must be relieved before disconnecting any fuel lines. Failure to do so may result in fire and/or personal injury.
- Brake fluid often contains Polyglycol Ethers and Polyglycols. Avoid contact with the eyes and wash your hands thoroughly after handling brake fluid. If you do get brake fluid in your eyes, flush your eyes with clean, running water for 15 minutes. If eye irritation persists, or if you have taken brake fluid internally, IMMEDIATELY seek medical assistance.
- The EPA warns that prolonged contact with used engine oil may cause a number of skin disorders, including cancer. You should make every effort to minimize your exposure to used engine oil. Protective gloves should be worn when changing oil. Wash your hands and any other exposed skin areas as soon as possible after exposure to used engine oil. Soap and water, or waterless hand cleaner should be used.
- The air bag system must be disabled (negative battery cable disconnected and/or air bag system main fuse removed) for at least 30 seconds before performing service on or around system components, steering column, instrument panel components, wiring and sensors. Failure to follow safety and disabling procedures could result in accidental air bag deployment, possible personal injury and unnecessary system repairs.
- Always wear safety goggles when working with, or around, the air bag system. When carrying a non-deployed air bag, be sure the bag and trim cover are pointed away from your body. When placing a non-deployed air bag on a work surface, always face the bag and trim cover upward, away from the surface. This will reduce the motion of the module if it is accidentally deployed. Refer to the additional air bag system precautions later in this section.
- Disconnecting the negative battery cable on some vehicles may interfere with the functions of the on-board computer system(s) and may require the computer to undergo a relearning process once the negative battery cable is reconnected.
- It is critically important to observe all instructions regarding ground disconnects, ignition switch positions, etc., in each diagnostic routine provided. Ignoring these instructions can result in false readings, damage to electronic components or circuits, or personal injury.

Preliminary Diagnostics

HISTORY OF OBD SYSTEMS

Starting in 1978, several vehicle manufacturers introduced a new type of control for several vehicle systems and computer control of engine management systems. These computer-controlled systems included programs to test for problems in the engine mechanical area, electrical fault identification and tests to help diagnose the computer

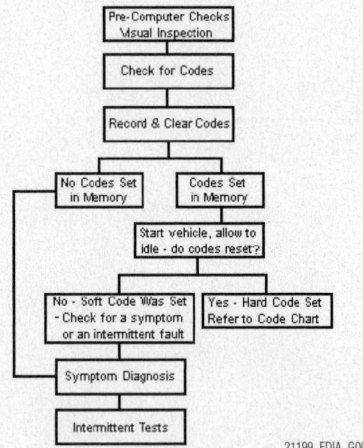

OBD I DIAGNOSTIC FLOWCHART

21199_FDIA_G001

Fig. 1 OBD I diagnostic flow chart

OBD I SYSTEM DIAGNOSTICS

One of the most important things to understand about the automotive repair industry is the fact that you have to continually learn new systems and new diagnostic routines (the test procedures designed to isolate a problem on a vehicle system). For OBD I and II systems, a diagnostic routine can be defined as a procedure (a series of steps) that you follow to find the cause of a problem, make a repair and then verify the problem is fixed.

CHANGES IN DIAGNOSTIC ROUTINES

In some cases, a new Engine Control system may be similar to an earlier system, but it can have more indepth control of vehicle emissions, input and output devices and it may include a diagnostic "monitor" embedded in the engine controller designed to run a thorough set of emission control system tests.

OBD I Diagnostic Flowchart

See Figure 1.

The OBD I Diagnostic Flowchart on this page can be used to find the cause of problems related to Engine Control system trouble codes or driveability symptoms detected on OBD I systems. It includes a step-by-step procedure to use to repair these systems. Compare this flowchart with the one used on OBD II systems.

The steps in this flow chart should be followed as described (from top to bottom).

- Do the Pre-Computer Checks.

- Check for any trouble codes stored in memory.

- Read the trouble codes - If trouble codes are set, record them and then clear the codes.

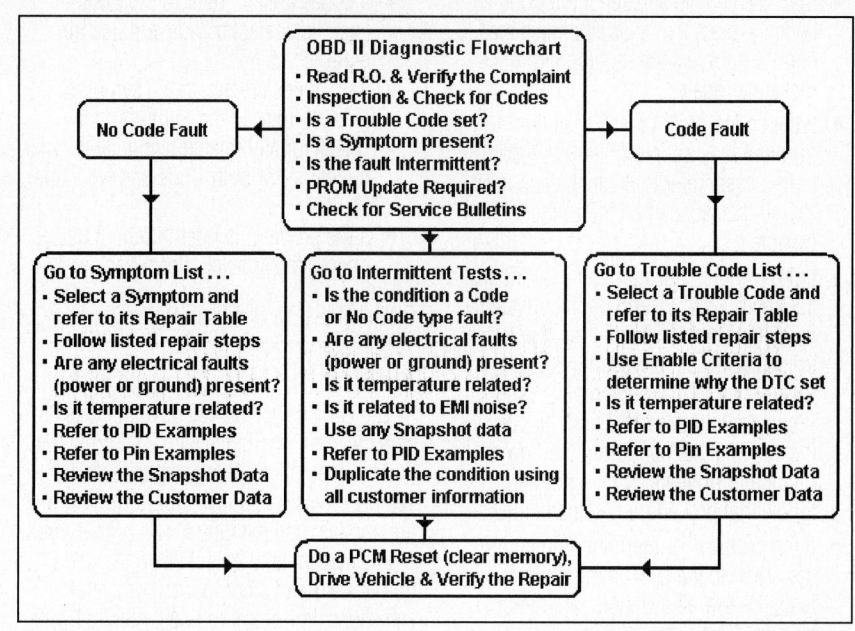

Fig. 2 OBD II diagnostic flow chart

21199_FDIA_G002

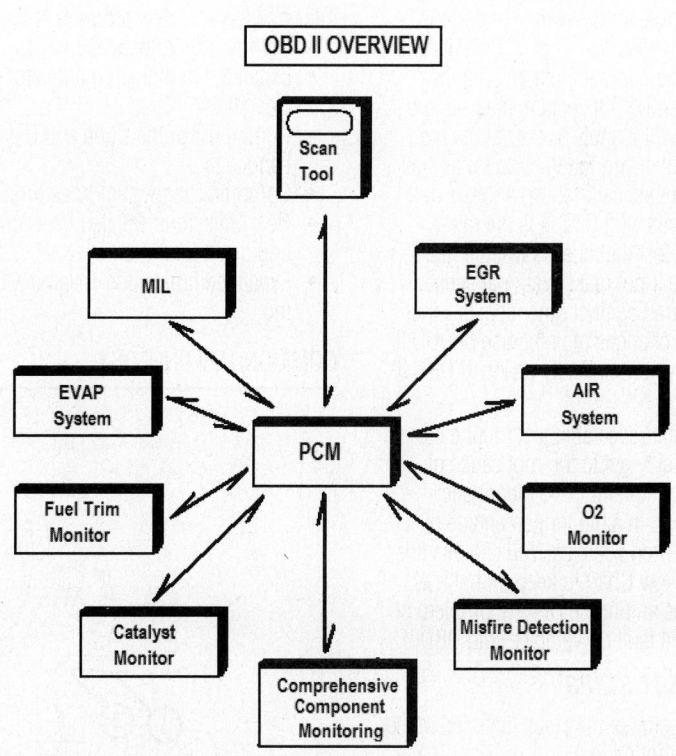

Fig. 3 PCM inputs and outputs

21199_FDIA_G003

- Start the vehicle and see if the trouble code(s) reset. If they do, then use the correct trouble code repair chart to make the repair.
- If the codes do not reset, than the problem may be intermittent in nature. In this case, refer to the test steps used to find the cause of an intermittent fault (wiggle test).
- In no trouble codes are found at the initial check, then determine if a driveability symptom is present. If so, then refer to the approriate driveability symptom repair chart to make the repair. If the first symptom chart does not isolate the cause of the condition, then go on to another driveability symptom and follow that procedure to conclusion.
- If the problem is intermittent in nature, then refer to the special intermittent tests. Follow all available intermittent tests to determine the cause of this type of fault (usually an electrical connection problem).

OBD II System Diagnostics

See Figure 2.

The diagnostic approach used in OBD II systems is more complex than that of the one for OBD I systems. This complexity will effect how you approach diagnosing the vehicle. On an OBD II system, the onboard diagnostics will identify sensor faults (i.e., open, shorted or grounded circuits) as well as those that lose calibration. Another new test that arrived with OBD II is the rationality test (a test that checks whether the value for one input makes rational sense when compared against other sensor input values). The changes plus the use of OBD II Monitors have dramatically changed OBD II diagnostics.

The use of a repeatable test routine can help you quickly get to the root cause of a customer complaint, save diagnostic time and result in a higher percentage of properly repaired vehicles. You can use this Diagnostic Flow Chart to keep on track as you diagnose an Engine Control problem or a base engine fault on vehicles with OBD II.

FLOW CHART STEPS

Here are some of the steps included in the Diagnostic Routine:

- Review the repair order and verify the customer complaint as described
- Perform a Visual Inspection of underhood or engine related items
- If the engine will not start, refer to No Start Tests

- If codes are set, refer to the trouble code list, select a code and use the repair chart
- If no codes are set, and a symptom is present, refer to the Symptom List
- Check for any related technical service bulletins (for both Code and No Code Faults)
- If the problem is intermittent in nature, refer to the special Intermittent Tests

OBD II SYSTEM OVERVIEW

See Figure 3.

The OBD II system was developed as a step toward compliance with California and Federal regulations that set standards for vehicle emission control monitoring for all automotive manufacturers. The primary goal of this system is to detect when the degradation or failure of a component or system will cause emissions to rise by 50%. Every manufacturer must meet OBD II standards by the 1996 model year. Some manufacturers began programs that were OBD II mandated as early as 1992, but most manufacturers began an OBD II phase-in period starting in 1994.

The changes to On-Board Diagnostics influenced by this new program include:

- Common Diagnostic Connector
- Expanded Malfunction Indicator Light Operation
- Common Trouble Code and Diagnostic Language
- Common Diagnostic Procedures
- New Emissions-Related Procedures, Logic and Sensors
- Expanded Emissions-Related Monitoring

COMMON TERMINOLOGY

OBD II introduces common terms, connectors, diagnostic language and new emissions-related monitoring procedures.

The most important benefit of OBD II is that all vehicles will have a common data output system with a common connector. This allows equipment Scan Tool manufacturers to read data from every vehicle and pull codes with common names and similar descriptions of fault conditions. In the future, emissions testing will require the use of an OBD II certifiable Scan Tool.

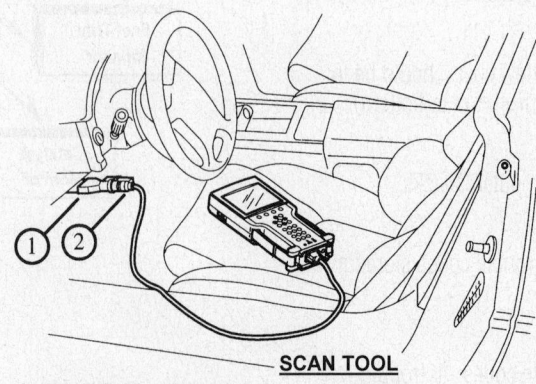

SCAN TOOL

1. DLC Cable Connection
2. SAE 16/19 Pin Adapter

Fig. 4 Typical scan tool hook up

21199_FDIA_G004

Diagnostic Tools & Circuit Testing

HAND TOOLS & METER OPERATION

To effectively use this or any diagnostic information, you should have a solid understanding of how to operate required tools and test equipment.

SCAN TOOLS

See Figure 4.

Vehicle manufacturers designed their computers to have an accessible data line where a diagnostic tester could retrieve data on sensors and the status of operation for components.

These testers became known in the automotive repair industry as "Scan Tools" because they scanned the data on the computers and provided information for the technician.

The Scan Tool is your basic tool link into the on-board electronic control system of the vehicle. Scan Tools are equipped with, or have separate software cards, for each OEM needed to be diagnosed. In this case, always secure a scan tool that has the latest OEM-specific diagnostic software included. Spend some time in the scan tool user's manual to ensure you know how to properly operate the tool and how to select the necessary programs required for full and proper diagnostics.

MALFUNCTION INDICATOR LAMP

Emission regulations require that a Malfunction Indicator Lamp (MIL) be illuminated when an emissions related fault is detected and that a Diagnostic Trouble Code be stored in the vehicle controller (PCM) memory.

When the MIL is illuminated, it is an indication of a problem within one of the electronic components or circuits. When the scan tool is attached to the Data Link Connector (DLC) in the vehicle, it can access the DTCs. In some situations, without the use of a scan tool, the MIL can be activated to flash a series of long and short flashes, which correspond to the numbering of the DTC.

OBD II guidelines define when an emissions-related fault will cause the MIL to activate and set a Diagnostic Trouble Code (DTC). There are some DTCs that will not cause the MIL to illuminate. OBD II guidelines determine how quickly the onboard diagnostics must be able to identify a fault, set the trouble code in memory and activate the MIL (lamp).

ELECTRONIC CONTROLS

You should have a basic knowledge of electronic controls when performing test procedures to keep from making an incorrect

IDENTIFYING THE PROBLEM

2

Table of Contents

Problem Identification

INTRODUCTION

System Control Modules

See Figures 1 and 2.

Before attempting diagnosis of the Electronic Engine Control system, familiarize yourself with the basics of how the system is designed to operate. It consists of a central processing unit: Powertrain Control Module (PCM), Engine Control Module (ECM), Transmission Control Module (TCM) and/or the Body Control Module (BCM). These units are the "heart" of the electronic control systems on the vehicle. In some cases, these units are integral with one another, and on some applications, they are separate. As you get deeper into actual diagnostic testing, you will find out which units are used on the vehicle you are testing.

The PCM is a digital computer that contains a microprocessor. The PCM receives input signals from various sensors and switches that are referred to as PCM inputs. Based on these inputs, the PCM adjusts various engine and vehicle operations through devices that are referred to as PCM outputs. Examples of the input and output devices are shown in the graphic.

Powertrain Subsystems

A key to the diagnosis of the PCM and its subsystems is to determine which subsystems are on a vehicle. Examples of typical subsystems are:

- Cranking & Charging System
- Emission Control Systems
- Engine Cooling System
- Engine Air/Fuel Controls
- Exhaust System
- Ignition System
- Speed Control System
- Transaxle Controls

WHERE TO BEGIN

See Figure 3.

Diagnosis of engine performance or drivability problems on a vehicle with an onboard computer requires that you have a logical plan on how to approach the problem. The "Six Step Test Procedure" is designed to provide a uniform approach to repair any problems that occur in one or more of the vehicle subsystems.

The diagnostic flow built into this test procedure has been field-tested for several years at dealerships - it is the starting point when a repair is required!

It should be noted that a commonly overlooked part of the "Problem Resolution" step is to check for any related Technical Service Bulletins.

Six-Step Test Procedure

The steps outlined as follows were defined to help you determine how to perform a proper diagnosis. Refer to the flow chart that outlines the Six Step Test Procedure as needed. The recommended steps include:

Verify The Complaint & Check For TSBs

To verify the customer complaint, the technician should understand the normal operation of the system. Conduct a thorough visual and operational inspection, review the service history, detect unusual sounds or odors, and gather diagnostic trouble code (DTC) information resources to achieve an effective repair.

PCM INPUTS & OUTPUTS
(OBD II)

Conditions Sensed	Systems Controlled
INPUT SENSORS • Battery Temperature • Camshaft Position • Crankshaft Position • Engine Coolant Temp. • Fuel Level • Intake Air Temperature • Knock Detection • Manifold Air Pressure • Oxygen Content • Throttle Position **INPUT SIGNALS** • A/C Select Switch • Battery Voltage • Brake Switch • Leak Detection Pump Sw. • Ignition Switch • Park Neutral Switch • Power Steering Press. Sw. • SCI Receive • Speed Control Switches	**OUTPUT DEVICES** • A/C WOT Relay • Auto Shutdown Relay • Charge Indicator Lamp • Data Link Connector • EVAP Purge Solenoid • EGR Solenoid • Fuel Injectors • Fuel Pump Relay • Generator Field • Idle Air Control Solenoid • Ignition Coils • Leak Detection Solenoid • MIL or C/E Lamp • Radiator Fan Relay • SCI Transmit • Speed Control Solenoids • Tachometer • TCC Solenoid

PCM

Fig. 1 An example of OBD II input and output devices

21199_FDIA_G005

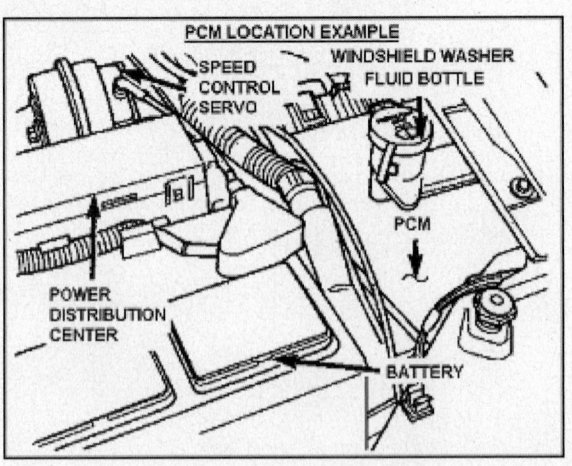

Fig. 2 Typical PCM location

21199_FDIA_G006

This check should include videos, newsletters, and any other information in the form of TSBs or Dealer Service Bulletins. Analyze the complaint and then use the recommended Six Step Test Procedure. Utilize the wiring diagrams and theory of operation articles. Combine your own knowledge with efficient use of the available service information.

Verify the cause of any related symptoms that may or may not be supported by one or more trouble codes. There are various checks that can be performed to Engine Controls that will help verify the cause of a related symptom. This step helps to lead you in an organized diagnostic approach.

Check For Trouble Codes Or Symptoms

Determine if the problem is a Code or a No Code Fault. Then refer to the appropriate published service diagnostic information to make the repair.

Problem Resolution & Repair

Once the problem component or circuit has been properly identified and verified using published diagnostic procedures, make any needed repairs or replacement to restore the vehicle to proper working order. If the condition has set a DTC, follow the designated repair chart to make an effective repair. If there is not a DTC set, but you can determine specific symptoms that are evident during the failure, select the symptom from the symptom tables and follow the diagnostic paths or suggestions to complete the repair or refer to the applicable component or system in service information.

If the vehicle does not set a DTC and has only intermittent operating failures or concerns, to resolve an intermittent fault, perform the following steps:

- Observe trouble codes, DTC modes and freeze frame data.
- Evaluate the symptoms and conditions described by the customer.
- Use a check sheet to identify the circuit or electrical system component.
- Many Aftermarket Scan Tools and Lab Scopes have data capturing features.

PCM Reset

It is a good idea, prior to tracing any faults, to clear the DTCs, attempt to replicate the condition and see if the same DTC resets. Also, once any repairs are made, it will be necessary to clear the DTC(s) - PCM Reset - to ensure the repair has totally resolved the problem. For procedures on PCM Reset, see DIAGNOSTIC TROUBLE CODES.

Repair Verification

Once a repair is completed, the next step is to verify the vehicle operates properly and that the original symptom was corrected. Verification Tests, related to specific DTC diagnostic steps, can be used to verify a repair.

Base Engine Tests

To determine that an engine is mechanically sound, certain tests need to be performed to verify that the correct A/F mixture enters the engine, is compressed, ignited, burnt, and then discharged out of the exhaust system. These tests can be used to help determine the mechanical condition of the engine.

To diagnose an engine-related complaint, compare the results of the Compression, Cylinder Balance, Engine Cylinder Leakage (not included) and Engine Vacuum Tests.

Engine Compression Test

The Engine Compression Test is used to determine if each cylinder is contributing its equal share of power. The compression readings of all the cylinders are recorded and then compared to each other and to the manufacturer's specification (if available).

Cylinders that have low compression readings have lost their ability to seal. It this type of problem exists, the location of the compression leak must be identified. The leak can be in any of these areas: piston, head gasket, spark plugs, and exhaust or intake valves.

The results of this test can be used to determine the overall condition of the engine and to identify any problem cylinders as well as the most likely cause of the problem.

✳✳ CAUTION

Prior to starting this procedure, set the parking brake, place the gear selector in P/N and block the drive wheels for safety. The battery must be fully charged.

COMPRESSION TEST PROCEDURE

1. Allow the engine to run until it is fully warmed up.

2. Remove the spark plugs and disable the Ignition system and the Fuel system for safety. Disconnecting the CKP sensor harness connector will disable both fuel and ignition (except on NGC vehicles).

3. Carefully block the throttle to the wide-open position.

4. Insert the compression gauge into the cylinder and tighten it firmly by hand.

5. Use a remote starter switch or ignition key and crank the engine for 3-5 complete engine cycles. If the test is interrupted for any reason, release the gauge pressure and retest. Repeat this test procedure on all cylinders and record the readings.

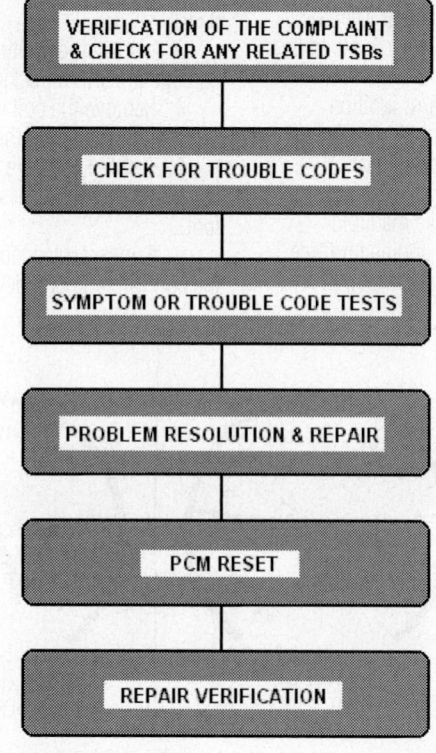

SIX STEP TROUBLESHOOTING PROCEDURE

VERIFICATION OF THE COMPLAINT & CHECK FOR ANY RELATED TSBs

CHECK FOR TROUBLE CODES

SYMPTOM OR TROUBLE CODE TESTS

PROBLEM RESOLUTION & REPAIR

PCM RESET

REPAIR VERIFICATION

Fig. 3 Six-step diagnostic procedure

21199_FDIA_G007

The lowest cylinder compression reading should not be less than 70% of the highest cylinder compression reading and no cylinder should read less than 100 psi.

EVALUATING THE TEST RESULTS

To determine why an individual cylinder has a low compression reading, insert a small amount of engine oil (3 squirts) into the suspect cylinder. Reinstall the compression gauge and retest the cylinder and record the reading. Review the explanations that follow.

Reading is higher - If the reading is higher at this point, oil inserted into the cylinder helped to seal the piston rings against the cylinder walls. Look for worn piston rings.

Reading did not change - If the reading didn't change, the most likely cause of the low cylinder compression reading is the head gasket or valves.

Low readings on companion cylinders - If low compression readings were recorded from cylinders located next to each other, the most likely cause is a blown head gasket.

Readings are higher than normal - If the compression readings are higher than normal, excessive carbon may have collected on the pistons and in the exhaust areas. One way to remove the carbon is with an approved brand of "Top Engine Cleaner."

➡ **Always clean spark plug threads and seat with a spark plug thread chaser and seat cleaning tool prior to reinstallation. Use anti-seize compound on aluminum heads.**

Engine Vacuum Tests

An engine vacuum test can be used to determine if each cylinder is contributing an equal share of power. Engine vacuum, defined as any pressure lower than atmospheric pressure, is produced in each cylinder during the intake stroke. If each cyl-inder produces an equal amount of vacuum, the measured vacuum in the intake manifold will be even during engine cranking, at idle speed, and at off-idle speeds.

Engine vacuum is measured with a vacuum gauge calibrated to show the difference between engine vacuum (the lack of pressure in the intake manifold) and atmospheric pressure. Vacuum gauge measurements are usually shown in inches of Mercury (in. Hg).

➡ **In the tests described in this article, connect the vacuum gauge to an intake manifold vacuum source at a point below the throttle plate on the throttle body.**

ENGINE CRANKING VACUUM TEST PROCEDURE

The Engine Cranking Vacuum Test can be used to verify that low engine vacuum is not the cause of a No Start, Hard Start, Starts and Dies or Rough Idle condition (symptom).

The vacuum gauge needle fluctuations that occur during engine cranking are indications of individual cylinder problems. If a cylinder produces less than normal engine vacuum, the needle will respond by fluctuating between a steady high reading (from normal cylinders) and a lower reading (from the faulty cylinder). If more than one cylinder has a low vacuum reading, the needle will fluctuate very rapidly.

1. Prior to starting this test, set the parking brake, place the gearshift in P/N and block the drive wheels for safety. Then block the PCV valve and disable the idle air control device.

2. Disable the fuel and/or ignition system to prevent the vehicle from starting during the test (while it is cranking).

3. Close the throttle plate and connect a vacuum gauge to an intake manifold vacuum source. Crank the engine for three seconds (do this step at least twice).

The test results will vary due to engine design characteristics, the type of PCV valve and the position of the AIS or IAC motor and throttle plate. However, the engine vacuum should be steady between 1.0–4.0 in. Hg during normal cranking.

ENGINE RUNNING VACUUM TEST PROCEDURE

See Figure 4.

1. Allow the engine to run until fully warmed up. Connect a vacuum gauge to a clean intake manifold source. Connect a tachometer or Scan Tool to read engine speed.

2. Start the engine and let the idle speed stabilize. Raise the engine speed rapidly to just over 2000 rpm. Repeat the test (3) times. Compare the idle and cruise readings.

EVALUATING THE TEST RESULTS

If the engine wear is even, the gauge should read over 16 in. Hg and be steady. Test results can vary due to engine design and the altitude above or below sea level.

Ignition System Tests–Distributor

This next section provides an overview of ignition tests with examples of Engine Analyzer patterns for a Distributor Ignition System.

PRELIMINARY INSPECTION

1. Perform these checks prior to connecting the Engine Analyzer:

2. Check the battery condition (verify that it can sustain a cranking voltage of 9.6v).

3. Inspect the ignition coil for signs of damage or carbon tracking at the coil tower.

4. Remove the coil wire and check for signs of corrosion on the wire or tower.

5. Test the coil wire resistance with a DVOM (it should be less than 7 k/ohm per foot).

6. Connect a low output spark tester to the coil wire and engine ground. Verify that

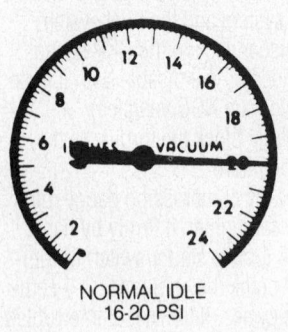
NORMAL IDLE
16-20 PSI

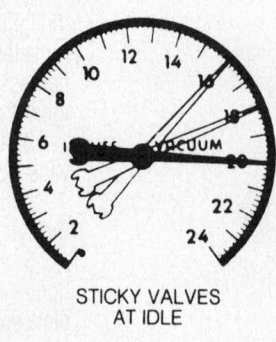
STICKY VALVES
AT IDLE

INCORRECT
MIXTURE
AT IDLE

LATE TIMING OR
INTAKE LEAK
AT IDLE

21199_FDIA_G006

Fig. 4 Engine running vacuum test

the ignition coil can sustain adequate spark output while cranking for 3-6 seconds.

7. Connect the Engine Analyzer to the Ignition System, and choose Parade display. Run the engine at 2000 RPM, and note the display patterns, looking for any abnormalities.

Ignition System Tests–Distributorless

Perform the following checks prior to connecting the Engine Analyzer:

1. Check the battery condition (verify that it can sustain a cranking voltage of 9.6v).

2. Inspect the ignition coils for signs of damage or carbon tracking at the coil towers.

3. Remove the secondary ignition wires and check for signs of corrosion.

4. Test the plug wire resistance with a DVOM (specification varies from 15-30 k/ohm).

5. Connect a low output spark tester to a plug wire and to engine ground. Verify that the ignition coil can sustain adequate spark output for 3-6 seconds.

SECONDARY IGNITION SYSTEM SCOPE PATTERNS (V6 ENGINE)

See Figure 5.

1. Connect the Engine Analyzer to the ignition system.

2. Turn the scope selector to view the "Parade Display" of the ignition secondary.

3. Start the engine in Park or Neutral and slowly increase the engine speed from idle to 2000 rpm.

4. Compare actual display to the examples in the illustration.

Symptom Diagnosis

To determine whether vehicle problems are identified by a set Diagnostic Trouble Code, you will first have to connect a proper scan tool to the Data Link Connector and retrieve any set codes. See DIAGNOSTIC TROUBLE CODES for information on retrieving and reading codes.

If no codes are set, the problem must be diagnosed using only vehicle operating symptoms. A complete set of "No Code" symptoms is found in the SYMPTOM DIAGNOSIS (NO CODES).

DO NOT attempt to diagnose driveability symptoms without having a logical plan to use to determine which engine control system is the cause of the symptom - this plan should include a way to determine which systems do NOT have a problem! Remember, there are 2 kinds of NO CODE conditions:

• Symptom diagnosis, in which a continuous problem exists, but no DTC is set as a result. Therefore, only the operating symptoms of the vehicle can be used to pinpoint the root cause of the problem.

• Intermittent problem diagnosis, in which the problem does not occur all the time and does not set any DTCs.

• Both of these NO CODE conditions are covered in the SYMPTOM DIAGNOSIS.

Accessing Components & Circuits

See Figures 6 and 7.

Every vehicle and every diagnostic situation is different. It is a good idea to first determine the best diagnostic path to follow using flow charts, wiring diagrams, TSBs, etc. Part of choosing steps is to determine how time-consuming and effective each step will be. It may be easy to access a component or circuit in one vehicle, but difficult in

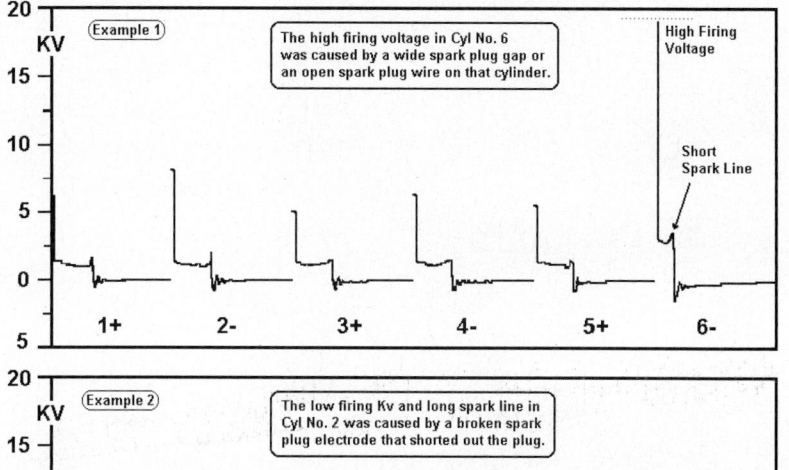

The high firing voltage in Cyl No. 6 was caused by a wide spark plug gap or an open spark plug wire on that cylinder.

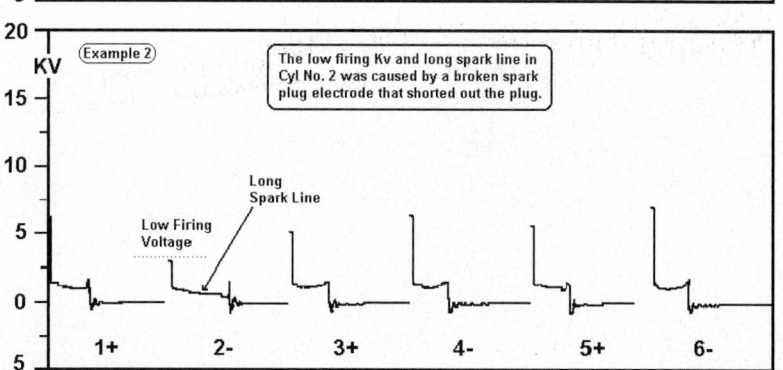

The low firing Kv and long spark line in Cyl No. 2 was caused by a broken spark plug electrode that shorted out the plug.

Fig. 5 Secondary ignition system (V6 engine)

21199_FDIA_G009

21199_FDIA_G010

Fig. 6 Circuits located at the back of the PCM connector

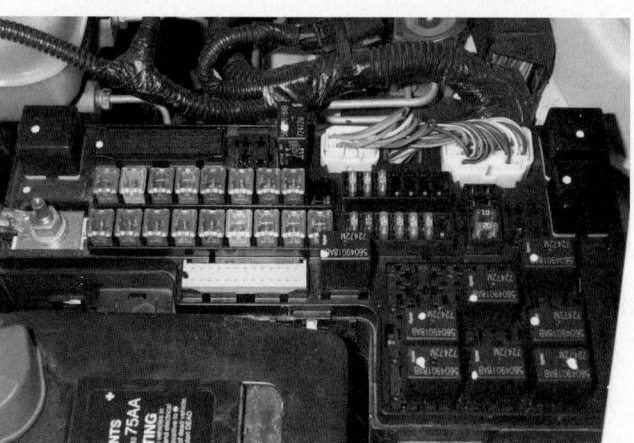

21199_FDIA_G010

Fig. 7 Typical underhood fuse block

another. Many circuits are integrated into a large harness and are difficult to test. Many components are inaccessible without disassembly of unrelated systems.

In the graphic, you will note that the protective covers have been removed from the PCM connectors, and any circuit can be easily identified and back probed. In other cases, PCM access is difficult, and it may be easier to access circuits at the component side of the harness.

Another important point to remember is that any circuit or component controlled by a relay or fused circuit can be monitored from the appropriate fuse box.

There is generally more than one of each type of relay or fuse. Therefore, swapping a suspect relay from another system may be more efficient than testing the relay itself. Relays and fuses may also be removed and replaced with fused jumper wires for testing circuits. Jumper wires can also provide a loop for inductive amperage tests.

Choosing the easiest way has its limitations, however. Remember that an appropriate signal on a PCM controlled circuit at an actuator means that the signal at the PCM is also good. However, a sensor signal at the sensor does not necessarily mean that the PCM is receiving the same signal. Think about the direction flow through a circuit, and not just what signal is appropriate, to save time without making costly assumptions.

INTRODUCTION TO OBD DIAGNOSTIC SYSTEMS

3

Table of Contents

OBD Systems

The California Air Resources Board (CARB) began regulating On-Board Diagnostic (OBD) systems for vehicles sold in California beginning with the 1988 model year. The initial requirements, known as OBD I, required the identification of the likely area of a fault with regard to the fuel metering system, EGR system, emission-related components and the PCM. Implementation of this new vehicle emission control monitoring regulation was done in several phases.

OBD I SYSTEMS

A Malfunction Indicator Lamp (MIL) labeled Check Engine Lamp or Service Engine Soon was required to illuminate and alert the driver of a fault, and the need to service the emission controls. A Diagnostic Trouble Code (DTC) was required to assist in identifying the system or component associated with the fault. If the fault that caused the MIL goes away, the MIL will go out and the code associated with the fault will disappear after a predetermined number of ignition cycles.

Following extensive research, CARB determined that by the time an Emission System component failed and caused the MIL to illuminate, that the vehicle could have emitted excess emissions over a long period of time. CARB also concluded that semi-annual or annual tailpipe tests were not catching enough of the vehicles with Emission Control systems operating at less than normal efficiency.

To take advantage of improvements in vehicle manufacturer adaptive and failsafe strategies, CARB developed new requirements designed to monitor the performance of Emission Control components, as well as to detect circuit and component hard faults. The new diagnostics were designed to operate under normal driving conditions, and the results of its tests would be viewable without any special equipment.

OBD II SYSTEMS

Beginning in the 1994 model year, both CARB and the EPA mandated Enhanced OBD systems, commonly known as OBD II. The objectives of OBD II were to improve air quality by reducing high in-use emissions caused by emission-related faults, reduce the time between the occurrence of a fault and its detection and repair, and assist in the diagnosis and repair of an emissions-related fault.

Differences Between OBD I & OBD II

As with OBD I, if an emission related problem is detected on a vehicle with OBD II, the MIL is activated and a code is set. However, that is the only real similarity between these systems. OBD II procedures that define emissions component and system tests, code clearing and drive cycles are more comprehensive than tests in the OBD I system.

Powertrain Control Module

The PCM in the OBD II system monitors almost all Emission Control systems that affect tailpipe or evaporative emissions. In most cases, the fault must be detected before tailpipe emissions exceed 1.5 times applicable 50K or 100K-mile FTP standards. If a component exceeds emission levels or fails to operate within the design specifications, the MIL is illuminated and a code is stored within two OBD II drive cycles.

The OBD II test runs continuously or once per trip (it depends on the driving mode requirement). Tests are run once per drive cycle during specific drive patterns called trips. Codes are stored in the PCM memory when a fault is first detected. In most cases, the MIL is turned on after two trips with a fault present. If the MIL is "on", it will go off after three consecutive trips if the same fault does not reappear. If the same fault is not detected after 40 engine warmup periods, the code will be erased (Fuel and Misfire faults require 80 warmup cycles).

OBD II Standardization

OBD II diagnostics require the use of a standardized Diagnostic Link Connector (DLC), standard communication protocol and messages, and standardized trouble codes and terminology. Examples of this standardization are Freeze Frame Data and I/M Readiness Monitors.

Changes in MIL Operation

An important change for OBD II involves when to activate the MIL. The MIL must be activated by at least the second trip if vehicle emissions could exceed 1.5 times the FTP standard. If any single component or system failure would allow the emissions to exceed this level, the MIL is activated and a related code is stored in the PCM.

1994 OBD II Phase-In Systems

Starting in 1994 some manufacturers began to "phase-in" the OBD II system on certain vehicles. The OBD II "phase-in" system on these vehicles included the use of a Misfire Monitor that operated with a "lower threshold" Misfire Detection system designed to monitor misfires without setting any codes. In addition, the EVAP Monitor was not operational on these vehicles.

1996 & Later OBD II Systems

By the 1996 model year, all California passenger cars and trucks up to 14,000 lb. GVWR, and all Federal passenger cars and trucks up to 8,600 lb. GWVR were required to comply with the CARB-OBD II or EPA OBD requirements. The requirements applied to diesel and gasoline vehicles, and were phased in on alternative-fuel vehicles.

Diagnostic Test Modes

The "test mode" messages available on a Scan Tool are listed below:
- Mode $01: Used to display Powertrain Data (PID data)
- Mode $02: Used to display any stored Freeze Frame data
- Mode $03: Used to request any trouble codes stored in memory
- Mode $04: Used to request that any trouble codes be cleared
- Mode $05: Used to monitor the Oxygen sensor test results
- Mode $06: Used to monitor Non-Continuous Monitor test results
- Mode $07: Used to monitor the Continuous Monitor test results
- Mode $08: Used to request control of a special test (EVAP Leak)
- Mode $09: Used to request vehicle information (INFO MENU)

Onboard Diagnostics

See Figure 1.

The Diagnostic Repair Chart should be used as follows:
- Trouble Code Diagnosis - Refer to the Code List or electronic media for a repair chart for a particular trouble code.
- Driveability Symptoms - Refer to the Driveability Symptom List in manuals or in electronic media.
- Intermittent Faults - Refer to the Intermittent Test Procedures.
- OBD II Drive Cycles - Refer to the Comprehensive Component Monitor or a Main Monitor drive cycle article.

OBD SYSTEM TERMINOLOGY

It is very important that service technicians understand terminology related to OBD II test procedures. Several of the essential OBD II terms and definitions are explained in the following text.

Two-Trip Detection

Frequently, an emission system or component must fail a Monitor test more

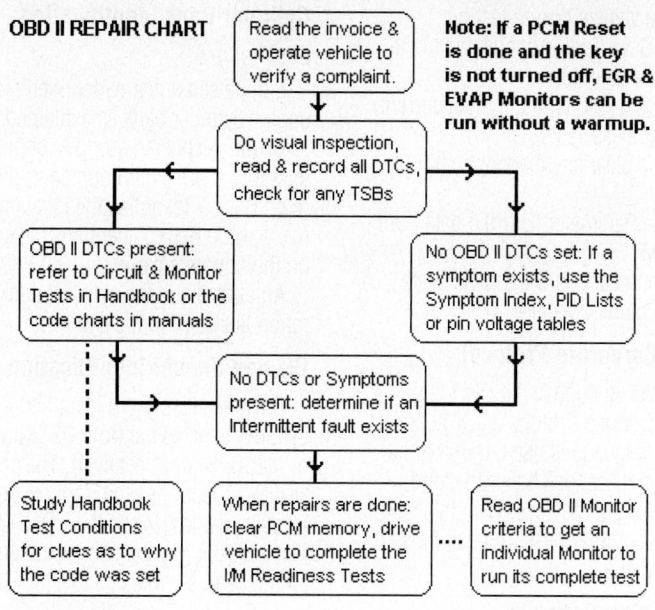

OBD II REPAIR CHART

Read the invoice & operate vehicle to verify a complaint.

Note: If a PCM Reset is done and the key is not turned off, EGR & EVAP Monitors can be run without a warmup.

Do visual inspection, read & record all DTCs, check for any TSBs

OBD II DTCs present: refer to Circuit & Monitor Tests in Handbook or the code charts in manuals

No OBD II DTCs set: If a symptom exists, use the Symptom Index, PID Lists or pin voltage tables

No DTCs or Symptoms present: determine if an Intermittent fault exists

Study Handbook Test Conditions for clues as to why the code was set

When repairs are done: clear PCM memory, drive vehicle to complete the I/M Readiness Tests

Read OBD II Monitor criteria to get an individual Monitor to run its complete test

21199_FDIA_G012

Fig. 1 OBD II repair chart

than once before the MIL is activated. In these cases, the first time an OBD II Monitor detects a fault during any drive cycle it sets a pending code in the PCM memory.

A pending code, which is read by selecting DDL from the Scan Tool menu, appears when Memory or Continuous codes are read. In order for a pending code to cause the MIL to activate, the original fault must be repeated under similar conditions.

This is a critical issue to understand as a pending code could remain in the PCM for a long time before the conditions that caused the code to set reappear. This type of OBD II trouble code logic is frequently referred to as the "Two-Trip Detection Logic".

➡ **Codes related to a Misfire fault and Fuel Trim can cause the PCM to activate the MIL after one trip because these codes are related to critical emission systems that could cause emissions to exceed the federally mandated limits.**

Similar Conditions

If a pending code is set because of a Misfire or Fuel System Monitor fault, the vehicle must meet similar conditions for a second trip before the code matures the PCM activates the MIL and stores the code in memory. Refer to prior Note for exceptions to this rule. The meaning of similar conditions is important when attempting to repair a fault detected by a Misfire or Fuel System Monitor.

To achieve similar conditions, the vehicle must reach the following engine running conditions simultaneously:

- Engine speed must be within 375 RPM of the speed when the trouble code set.
- Engine load must be within 10% of the engine load when the trouble code set.
- Engine warmup state must match a previous cold or warm state.

Summary—Similar conditions are defined as conditions that match the conditions

recorded in Freeze Frame when the fault was first detected and the trouble code was set in the PCM memory.

OBD II Warmup Cycle

See Figure 2.

The meaning of the expression warmup cycle is important. Once the fault that caused an OBD II trouble code to set is gone and the MIL is turned off, the PCM will not erase that code until after 40 warmup cycles. This is the purpose of the warmup cycle: To help clear stored codes.

However, trouble codes related to a Fuel system or Misfire fault require that 80 warmup cycles occur without the fault reappearing before codes related to these monitors will be erased from the PCM memory.

➡ **A warmup cycle is defined as vehicle operation (after an engine off and cool-down period) when the engine temperature rises to at least 40°F and reaches at least 160°F.**

Malfunction Indicator Lamp

If the PCM detects an emission related component or system fault for two consecutive drive cycles on OBD II systems, the MIL is turned on and a trouble code is stored. The MIL is turned off if three consecutive drive cycles occur without the same fault being detected.

Most trouble codes related to a MIL are erased from memory after 40 warmup periods if the same fault is not repeated. The MIL can be turned off after a repair by using the Scan Tool PCM Reset function.

Freeze Frame Data

See Figure 3.

The term Freeze Frame is used to describe the engine conditions that are recorded in PCM memory at the time a Monitor detects an emissions related fault. These conditions include fuel control state, spark timing, engine speed and load.

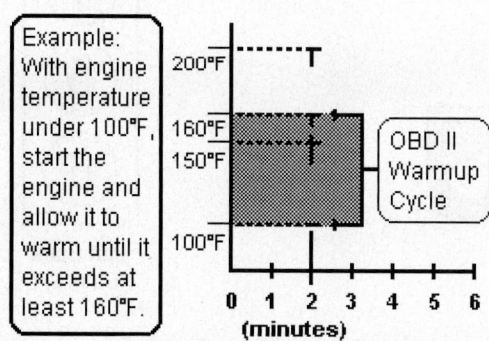

Example: With engine temperature under 100°F, start the engine and allow it to warm until it exceeds at least 160°F.

200°F
160°F
150°F
100°F

OBD II Warmup Cycle

0 1 2 3 4 5 6
(minutes)

21199_FDIA_G013

Fig. 2 OBD II warmup cycle

SCAN TOOL DISPLAY

Freeze Frame Data	
Fuel Sys Status	OL
Load Value	14%
ECT Deg F	+175°F
SHRTFT Adapt	+1.5%
MAP "Hg	18.1"
Engine RPM	750
DTC Priority	01

21199_FDIA_G014

Fig. 3 Scan tool freeze frame

Freeze Frame data is recorded when a system fails the first time for two-trip type faults. The Freeze Frame Data will only be overwritten by a different fault with a "higher emission priority."

Diagnostic Trouble Codes

The OBD II system uses a Diagnostic Trouble Code (DTC) identification system established by the Society of Automotive Engineers (SAE) and the EPA. The first letter of a DTC is used to identify the type of computer system that has failed as shown below:

- The letter 'P' indicates a Powertrain related device
- The letter 'C' indicates a Chassis related device
- The letter 'B' indicates a Body related device
- The letter 'U' indicates a Data Link or Network device code.

The first DTC number indicates a generic (P0xxx) or manufacturer (P1xxx) type code. A list of trouble codes is included.

The number in the hundreds position indicates the specific vehicle system or subgroup that failed (i.e., P0300 for a Misfire code, P0400 for an emission system code, etc.).

Data Link Connector

See Figure 4.

Vehicles equipped with OBD II use a standardized Data Link Connector (DLC). It is typically located between the left end of the instrument panel and 12 inches past vehicle centerline. The connector is mounted out of sight from vehicle passengers, but should be easy to see from outside by a technician in a kneeling position (door open). However, not all of the connectors are located in this exact area.

The DLC is rectangular in design and capable of accommodating up to 16 terminals. It has keying features to allow easy connection to the Scan Tool. Both the DLC and Scan Tool have latching features used to ensure that the Scan Tool will remain connected to the vehicle during testing.

Once the Scan Tool is connected to the DLC, it can be used to:

- Display the results of the most current I/M Readiness Tests
- Read and clear any diagnostic trouble codes
- Read the Parameter ID (PID) data from the PCM
- Perform Enhanced Diagnostic Tests (manufacturer specific)

Standard Corporate Protocol

On vehicles equipped with OBD II, a Standard Corporate Protocol (SCP) communication language is used to exchange bi-directional messages between stand-alone modules and devices. With this type of system, two or more messages can be sent over one circuit.

OBD II Monitor Software

The Diagnostic Executive contains software designed to allow the PCM to organize and prioritize the Main Monitor tests and procedures, and to record and display test results and diagnostic trouble codes.

The functions controlled by this software include:

- To control the diagnostic system so the vehicle continues to operate in a normal manner during testing.
- To ensure the OBD II Monitors run during the first two sample periods of the Federal Test Procedure.
- To ensure that all OBD II Monitors and their related tests are sequenced so that required inputs (enable criteria) for a particular Monitor are present prior to running that particular Monitor.
- To sequence the running of the Monitors to eliminate the possibility of different Monitor tests interfering with each other or upsetting normal vehicle operation.
- To provide a Scan Tool interface by coordinating the operation of special tests or data requests.

Cylinder Bank Identification

See Figure 5.

Engine sensors are identified on each engine cylinder bank as explained next.

Bank—A specific group of engine cylinders that share a common control sensor (e.g., Bank 1 identifies the location of Cyl. No. 1 while Bank 2 identifies the cylinders on the opposite bank).

An example of the cylinder bank configuration is shown in the Graphic.

Oxygen Sensor Identification

Oxygen sensors are identified in each cylinder bank as the front O2S (pre-catalyst) or rear O2S (post-catalyst). The acronym HO2S-11 identifies the front oxygen sensor located (Bank 1) while the HO2S-21 identifies the front oxygen sensor in Bank 2 of the engine, and so on.

OBD II Monitor Test Results

Generally, when an OBD II Monitor runs and fails a particular test during a trip, a pending code is set. If the same Monitor detects a fault for two consecutive trips, the MIL is activated and a code is set in PCM memory. The results of a particular Monitor test indicate that an emission system or component failed: NOT the circuit that failed!

To determine where the fault is located; follow the correct code repair chart, symptom diagnosis or intermittent test. The code and symptom repair charts are the most efficient way to repair an OBD II system.

➡ **Two important pieces of information that can help speed up a diagnosis are code conditions (including all enable criteria), and the parameter information (PID) stored in the Freeze Frame at the time a trouble code is set and stored in memory.**

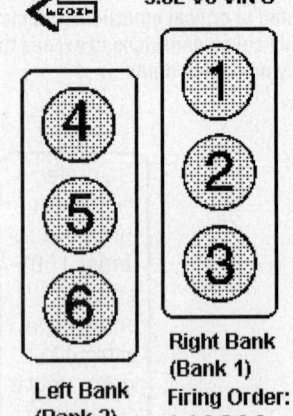

Fig. 5 Typical cylinder bank identification (V6 engine)

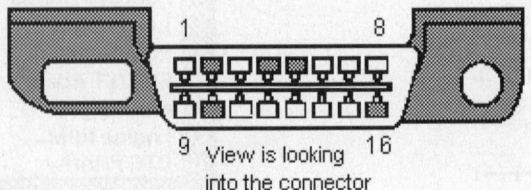

DATA LINK CONNECTOR

View is looking into the connector

Courtesy of Ford Motor Co.

21199_FDIA_G015

Fig. 4 Typical data link connector

Adaptive Fuel Control Strategy

The PCM incorporates an Adaptive Fuel Control Strategy that includes an adaptive fuel control table stored to compensate for normal changes in fuel system devices due to age or engine wear.

During closed loop operation, the Fuel System Monitor has two methods of attempting to maintain an ideal A/F ratio of 14:7 to 1 (they are referred to as short term fuel trim and long term fuel trim).

➡ **If a fuel injector, fuel pressure regulator or oxygen sensor is replaced the, memory in the PCM should be cleared by a PCM Reset step so that the PCM will not use a previously learned strategy.**

Short Term Fuel Trim

Short term fuel trim (SHRTFT) is an engine operating parameter that indicates the amount of short term fuel adjustment made by the PCM to compensate for operating conditions that vary from the ideal A/F ratio condition. A SHRTFT number that is negative (-15%) means that the HO2S is indicating a richer than normal condition to the PCM, and that the PCM is attempting to lean the A/F mixture. If the A/F ratio conditions are near ideal, the SHRTFT number will be close to 0%.

Long Term Fuel Trim

Long term fuel trim (LONGFT) is an engine parameter that indicates the amount of long term fuel adjustment made by the PCM to correct for operating conditions that vary from ideal A/F ratios. A LONGFT number that is positive (+15%) means that the HO2S is indicating a leaner than normal condition, and that it is attempting to add more fuel to the A/F mixture. If A/F ratio conditions are near ideal, the LONGFT number will be close to 0%. The PCM adjusts the LONGFT in a range from -35 to +35%. The values are in percentage on a Scan Tool.

Enable Criteria

The term enable criteria describe the conditions necessary for any of the OBD II Monitors to run their diagnostic tests. Each Monitor has specific conditions that must be met before it will run its test.

Enable criteria information can be different for each vehicle and engine type. Examples of trouble code conditions for DTC P0460 and P1168 are shown below:

Code information includes any of the following examples:

- Air Conditioning Status
- BARO, ECT, IAT, TFT, TP and Vehicle Speed sensors
- Camshaft (CMP) and Crankshaft (CKP) sensors
- Canister Purge (duty cycle) and Ignition Control Module Signals
- Short (SHRTFT) and Long Term (LONGFT) Fuel Trim Values
- Transmission Shift Solenoid On/Off Status

Drive Cycle

The term drive cycle has been used to describe a drive pattern used to verify that a trouble code, driveability symptom or intermittent fault had been fixed. With OBD II systems, this term is used to describe a vehicle drive pattern that would allow all the OBD II Monitors to initiate and run their diagnostic tests. For OBD II purposes, a minimum drive cycle includes an engine startup with continued vehicle operation that exceeds the amount of time required to enter closed loop fuel control.

OBD II Trip

The term OBD II Trip describes a method of driving the vehicle so that one or more of the following OBD II Monitors complete their tests:

- Comprehensive Component Monitor (completes anytime in a trip)
- Fuel System Monitor (completes anytime during a trip)
- EGR System Monitor (completes after accomplishing a specific idle and acceleration period)
- Oxygen Sensor Monitor (completes after accomplishing a specific steady state cruise speed for a certain amount of time)

OBD II Drive Cycle

The ambient or inlet air temperature must be from 40-100°F to initiate the OBD II drive cycle. Allow the engine to warm to 130°F prior to starting the test.

Connect the Scan Tool prior to beginning the drive cycle. Some tools are designed to emit a three-pulse beep when all of the OBD II Monitors complete their tests.

➡ **The IAT PID must be from 50-100°F to start the drive cycle. If it is less than 50°F at any time during the highway part of the drive cycle, the EVAP Monitor may not complete. The engine should reach 130°F before starting before attempting to verify an EVAP system fault. Disengage the PTO before proceeding (PTO PID will show OFF) if applicable. For the EVAP Running Loss system, verify FLI PID is at 15-85%. Some Monitors require very specific idle and acceleration steps.**

Drive Cycle Procedure

The primary intention of the OBD II drive cycle is to clear a specific DTC. The drive cycle can also be used to assist in identifying any OBD II concerns present through total Monitor testing. Perform all of the Vehicle Preparation steps.

Connect a Scan Tool and have an assistant watch the Scan Tool I/M Readiness Status to determine when the Catalyst, EGR, EVAP, Fuel System, O2 Sensor, Secondary AIR and Misfire Monitors complete.

OBD II SYSTEM MONITORS

Comprehensive Component Monitor

OBD II regulations require that all emission related circuits and components controlled by the PCM that could affect emissions are monitored for circuit continuity and out-of-range faults. The Comprehensive Component Monitor (CCM) consists of four different monitoring strategies: two for inputs and two for output signals. The CCM is a two trip Monitor for emission faults on most vehicles.

Input Strategies

One input strategy is used to check devices with analog inputs for opens, shorts, or out-of-range values. The CCM accomplishes this task by monitoring A/D converter input voltages. The analog inputs monitored include the ECT, IAT, MAF, TP and Transmission Range Sensors signals.

DTC	Trouble Code Title & Conditions
	EVAP System Small Leak Conditions: Cold startup, engine running at off-idle conditions, then the PCM detected a small leak (a leak of more than 0.040") in the EVAP system.
	FRP Sensor in Range but Low Conditions: Engine running, then the PCM detected that the FRP sensor signal was out-of-range low. Scan Tool Tip: Monitor the FRP PID for a value below 80 psi (551 kPa).

A second input strategy is used to check devices with digital and frequency inputs by performing rationality checks. The PCM uses other sensor readings and calculations to determine if a sensor or switch reading is correct under existing conditions. Some tests run continuously, some only after actuation.

Output Strategies

An Output State Monitor in the PCM checks outputs for opens or shorts by observing the control voltage level of the related device. The control voltage is low with it on, and high with the device off.

IAC Motor Test

The PCM monitors the IAC system in order to "learn" the closed loop correlation it needs to reposition the IAC solenoid (a rationality check).

Catalyst Efficiency Monitor

The Catalyst Monitor is a PCM diagnostic run once per drive cycle that uses the downstream heated Oxygen Sensor (HO2S-12) to determine if a catalyst falls below a minimum level of effectiveness in its ability to control exhaust emissions. The PCM uses a program to determine the catalyst efficiency based on the oxygen storage capacity of the catalytic converter.

Catalyst Monitor Operation

See Figure 6.

The Catalyst Monitor is a diagnostic that tests the oxygen storage capacity of the catalyst. The PCM determines the capacity by comparing the switching frequency of the rear oxygen sensor to the switching frequency of the front oxygen sensor. If the catalyst is okay, the switching frequency of the rear oxygen sensor will be much slower than the frequency of the front oxygen sensor.

However, as the catalyst efficiency deteriorates its ability to store oxygen declines. This deterioration causes the rear oxygen sensor to switch more rapidly. If the PCM detects the switching frequency of the rear oxygen sensor is approaching the frequency of the front oxygen sensor, the test fails and a pending code is set. If the PCM detects a fault on consecutive trips (from two to six consecutive trips) the MIL is activated, and a trouble code is stored in the PCM memory.

The Catalyst Monitor runs after startup once a specified time has elapsed and the vehicle is in closed loop. The amount of time is subject to each PCM calibration. Certain inputs (enable criteria) from various engine sensors (i.e., CKP, ECT, IAT, TPS and VSS) are required before the Catalyst Monitor can run.

Once the Catalyst Monitor is activated, closed loop fuel control is temporarily transferred from the front oxygen sensor to the rear oxygen sensor. During the test, the Monitor analyzes the switching frequency of both sensors to determine if a catalyst has degraded.

Catalyst Efficiency Monitor

CATALYST TEST–STEADY STATE CATALYST EFFICIENCY TEST

The PCM transfers the input for closed loop fuel control from the front HO2S-11 to the rear HO2S-21 during this test. The PCM measures the output frequency of the rear HO2S. This "test frequency" indicates the current oxygen storage capacity of the converter. The slower the frequency of the test result, the higher the efficiency of the converter.

CATALYST TEST–CALIBRATED FREQUENCY TEST

In Part 2 of the test a second frequency is calculated based on engine speed and load. This frequency serves as a high limit threshold for the test frequency. If the PCM detects the test frequency is less than the calibrated frequency the catalyst passes the test. If the frequency is too high, the converter or system has failed (a pending code is set).

The sequence of counting the front and rear O2S switches continues until the drive cycle completes. The ratio of total HO2S-21 switches to the total of the HO2S-11 switches is calculated. If the switch ratio is over the stored threshold, the catalyst has failed and a code is set.

CATALYTIC MONITOR REPAIR VERIFICATION TRIP

See Figure 7.

Start the engine, and drive in stop and go traffic for over 20 minutes. (Ambient air temperature must be over 50°F to run this test). Drive at speeds from 25-40 mph (6 times) and then at cruise for five minutes.

POSSIBLE CAUSES OF A CATALYST EFFICIENCY FAULT

- Base Engine faults (engine mechanical)
- Exhaust leaks or contaminated fuel

EGR System Monitor

The EGR System Monitor is a PCM diagnostic run once per trip that monitors EGR system component functionality and components for faults that could cause vehicle tailpipe levels to exceed 1.5 times the FTP Standard. A series of sequenced tests is used to test the system.

HO2S-12 WAVEFORM EXAMPLES

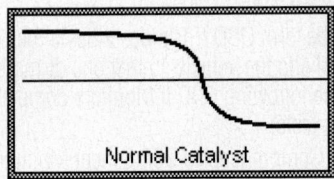
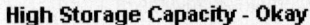

Normal Catalyst

High Storage Capacity - Okay

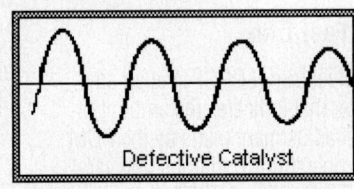

Defective Catalyst

Low Storage Capacity - Not Okay

21199_FDIA_G019

Fig. 6 Typical rear oxygen sensor waveform

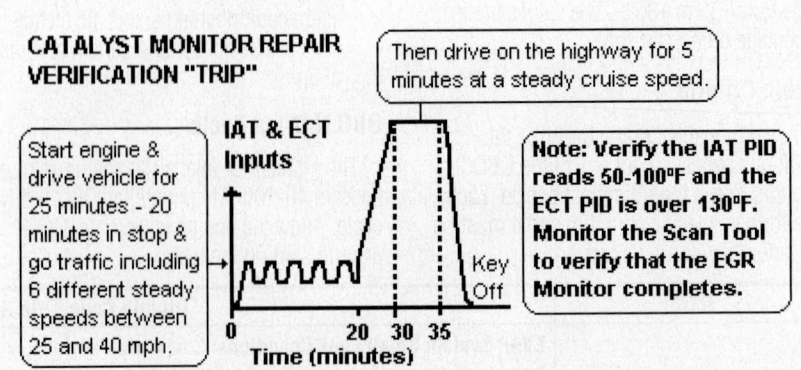

CATALYST MONITOR REPAIR VERIFICATION "TRIP"

Then drive on the highway for 5 minutes at a steady cruise speed.

Start engine & drive vehicle for 25 minutes - 20 minutes in stop & go traffic including 6 different steady speeds between 25 and 40 mph.

IAT & ECT Inputs

Note: Verify the IAT PID reads 50-100°F and the ECT PID is over 130°F. Monitor the Scan Tool to verify that the EGR Monitor completes.

Key Off

0   20 30 35
Time (minutes)

21199_FDIA_G020

Fig. 7 Typical catalyst monitor trip

Possible Causes of an EGR System Failure

See Figure 8.

• Leaks or disconnects in upstream or downstream vacuum hoses
• Damaged DPFE or EGR EVP sensor
• Plugged or restricted DPFE or EGR VP sensor or orifice assembly

Evap System Monitor

The EVAP System Monitor is a PCM diagnostic run once per trip that monitors the EVAP system in order to detect a loss of system integrity or leaks in the system (anywhere from 0.020" to 0.040" in diameter).

Possible Causes of an EVAP System Failure

• Cracks, leaks or disconnected hoses in the fuel vapor lines, components or plastic connectors or lines
• Backed-out or loose connectors to the Canister Purge solenoid
• Fuel filler cap (gas cap) loose or missing
• PCM has failed

On-Board Refueling Vapor Recovery System

An On-Board Refueling Vapor Recovery (ORVR) system is used on late model vehicles to recover fuel vapors during vehicle refueling.

SYSTEM OPERATION

The operation of the ORVR system during refueling is described next:

• The fuel filler pipe forms a seal to stop vapors from escaping the fuel tank while liquid is entering the tank (liquid in the 1" diameter tube blocks fuel vapor from rushing back up the fuel filler pipe).
• The fuel vapor control valve controls the flow of vapors out of the tank (it closes when the liquid level reaches a height associated with the fuel tank usable capacity). The fuel vapor control valve:

   a. Limits the total amount of fuel dispensed into the fuel tank.

   b. Prevents liquid gasoline from exiting the fuel tank when submerged (and also when tipped well beyond a horizontal plane as part of the vehicle rollover protection in an accident).

   c. Minimizes vapor flow resistance in a refueling condition.

• Fuel vapor tubing connects the fuel vapor control valve to the EVAP canister. This routes the fuel tank vapors (that are displaced by the incoming fuel) to the canister.

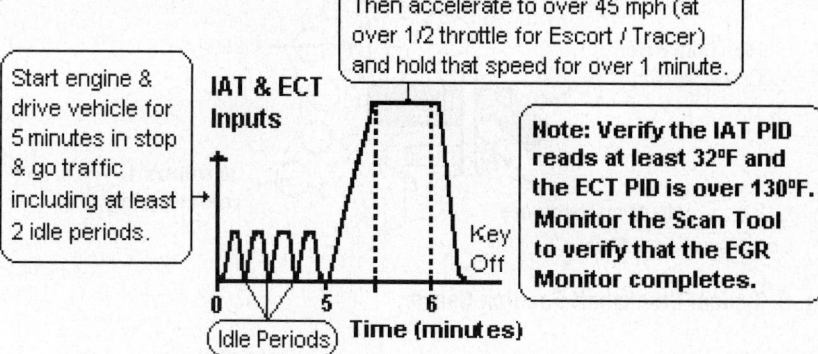

EGR MONITOR REPAIR VERIFICATION "TRIP"

Start engine & drive vehicle for 5 minutes in stop & go traffic including at least 2 idle periods.

Then accelerate to over 45 mph (at over 1/2 throttle for Escort / Tracer) and hold that speed for over 1 minute.

Note: Verify the IAT PID reads at least 32°F and the ECT PID is over 130°F. Monitor the Scan Tool to verify that the EGR Monitor completes.

IAT & ECT Inputs

Key Off

Idle Periods — Time (minutes)

21199_FDIA_G022

Fig. 8 Typical EGR monitor

• A check valve in the bottom of the pipe prevents any liquid from rushing back up the fuel filler pipe during liquid flow variations associated with the filler nozzle shut-off.
• Between refueling events, the charcoal canister is purged with fresh air so that it may be used again to store vapors accumulated during engine soak periods or subsequent refueling events. The vapors drawn from the canister are consumed in the engine.

Evap Monitor Test Conditions

The PCM allows canister purge to occur when the engine is warm, at wide open or part throttle (as long as the engine is not overheated). The engine can be in open or closed loop fuel control during purging.

Fuel System Monitor

The Fuel System Monitor is a PCM diagnostic that monitors the Adaptive Fuel Control system. The PCM uses adaptive fuel tables that are updated constantly and stored in long term memory (KAM) to compensate for wear and aging in the fuel system components.

FUEL SYSTEM MONITOR OPERATION

Once the PCM determines all the enable criteria has been are met (ECT, IAT and MAF PIDs in range and closed loop enabled), the PCM uses its adaptive strategy to "learn" changes needed to correct a Fuel system that is biased either rich or lean. The PCM accomplishes this task by monitoring Short Term and Long Term fuel trim in closed loop mode.

LONG AND SHORT TERM FUEL TRIM

Short Term fuel trim is a PCM parameter identification (PID) used to indicate Short Term fuel adjustments. This parameter is expressed as a percentage and its range

of authority is from -10% to +10%. Once the engine enters closed loop, if the PCM receives a HO2S signal that indicates the A/F mixture is richer than desired, it moves the SHRTFT command to a more negative range to correct for the rich condition.

If the PCM detects the SHRTFT is adjusting for a rich condition for too long a time, the PCM will "learn" this fact, and move LONGFT into a negative range to compensate so that SHRTFT can return to a value close to 0%. Once a change occurs to LONGFT or SHRTFT, the PCM adds a correction factor to the injector pulsewidth calculation to adjust for variations. If the change is too large, the PCM will detect a fault.

➡ **If a fuel injector, fuel pressure regulator, etc. is replaced, clear the KAM and then drive the vehicle through the Fuel System Monitor drive pattern to reset the fuel control table in the PCM.**

Misfire Detection Monitor

The Misfire Monitor is a PCM diagnostic that continuously monitors for engine misfires under all engine positive load and speed conditions (accelerating, cruising and idling). The Misfire Monitor detects misfires caused by fuel, ignition or mechanical misfire conditions. If a misfire is detected, engine conditions present at the time of the fault are written to the Freeze Frame Data. These conditions overwrite existing data.

Misfire Monitor Operation

See Figure 9.

The Misfire Monitor is designed to measure the amount of power that each cylinder contributes to the engine. The amount of contribution is calculated based upon measurements determined by crankshaft acceleration (TDC of compression stroke to

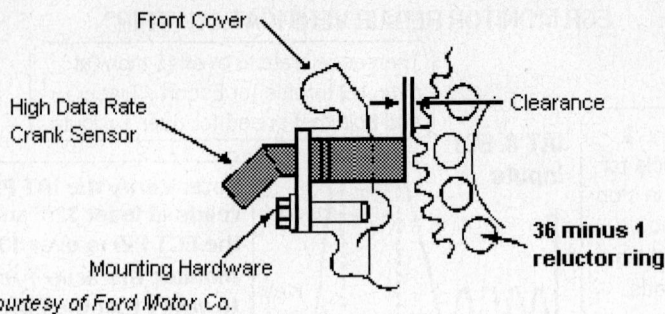

CRANKSHAFT POSITION SENSOR EXAMPLE

Front Cover

High Data Rate Crank Sensor

Clearance

36 minus 1 reluctor ring

Mounting Hardware

Courtesy of Ford Motor Co.

21199_FDIA_G031

Fig. 9 Typical Crankshaft Position Sensor

BDC of the power stroke) for each cylinder. This calculation requires accurate measurement of the crankshaft angle. Crankshaft angle measurement is determined using a low data rate system on 4-Cyl engines. The high data rate system is used to determine crankshaft angle on all other engines.

Catalyst Damaging Misfire (One-Trip Detection)

If the PCM detects a Catalyst Damaging Misfire, the MIL will flash once per second within 200 engine revolutions from the point where misfire is detected. The MIL will stop flashing and remain on if the engine stops misfiring in a manner that could damage the catalyst.

High Emissions Misfire (Two-Trip Detection)

A High Emissions Misfire is set if a misfire condition is present that could cause the tailpipe emissions to exceed the FTP emissions standard by 1.5 times. If this fault is detected for two consecutive trips under similar engine speed, load and temperature conditions, the MIL is activated. It is also activated if a misfire is detected under similar conditions for two non-consecutive trips that are not 80 trips apart.

State Emissions Failure Misfire (Two-Trip Detection)

A State Emissions Failure Misfire is set if the misfire is sufficient to cause the vehicle to fail a State Inspection or Maintenance (I/M) Test. This fault is determined by identifying misfire percentages that would cause a "durability demonstration vehicle" to fail an Inspection Maintenance (I/M) Test. If the Misfire Monitor detects the fault for two consecutive trips with the engine at similar engine speed, load and temperature conditions, the MIL is activated and a code is set. The MIL is also activated if this type of misfire is detected under similar conditions for two non-consecutive trips of not more than 80 trips apart.

➡ **Some vehicles set Misfire codes because of an early version of OBD II hardware and software. If a misfire code is set and the cause of the fault is not found, clear the code and retest. Search the TSB list for possible answers or contact the dealer.**

Misfire Detection

See Figure 10.

The Misfire Monitor uses the CKP sensor signals to detect an engine misfire. The amount of contribution is calculated based upon measurements determined by crankshaft acceleration from each cylinder's power stroke.

The PCM performs various calculations to detect individual cylinder acceleration rates. If acceleration for a cylinder deviates beyond the average variation of acceleration for all cylinders, a misfire is detected.

Faults detected by the Misfire Monitor:
- Engine mechanical faults, restricted intake or exhaust system
- Dirty or faulty fuel injectors, loose or damaged injector connectors
- The vehicle has been run low on fuel or run until it ran out of fuel

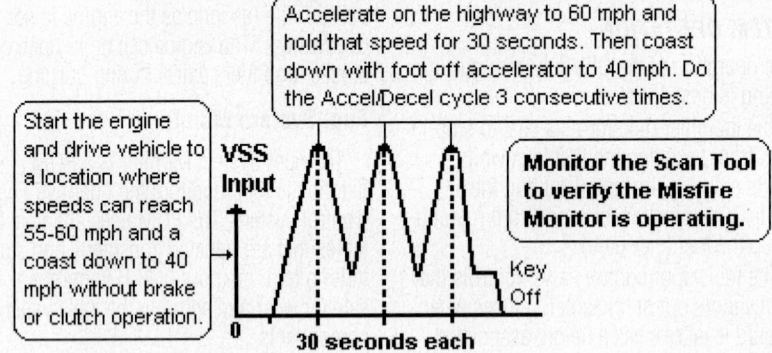

MISFIRE MONITOR REPAIR VERIFICATION "TRIP"

Accelerate on the highway to 60 mph and hold that speed for 30 seconds. Then coast down with foot off accelerator to 40 mph. Do the Accel/Decel cycle 3 consecutive times.

Start the engine and drive vehicle to a location where speeds can reach 55-60 mph and a coast down to 40 mph without brake or clutch operation.

VSS Input

Monitor the Scan Tool to verify the Misfire Monitor is operating.

Key Off

0

30 seconds each

21199_FDIA_G032

Fig. 10 Misfire Detection Monitor

Oxygen Sensor Monitor

The Oxygen Sensor Monitor is a PCM diagnostic designed to monitor the front and rear oxygen sensor for faults or deterioration that could cause tailpipe emissions to exceed 1.5 times the FTP standard. The front oxygen sensor voltage and response time are also monitored.

HO2S Monitor Operation

Fuel System and Misfire Monitors must be run and complete before the PCM will start the HO2S Monitor. Additionally, parts of the HO2S Sensor Monitor are enabled during the KOER Self-Test. The HO2S Monitor is run during each drive cycle after the CKP, ECT, IAT and MAF sensor signals are within a predetermined range.

Fixed Frequency Closed Loop Test

See Figure 11.

The HO2S Monitor constantly monitors the sensor voltage and frequency. The PCM detects a high voltage condition by comparing the HO2S signal to a preset level.

FIXED FREQUENCY TEST

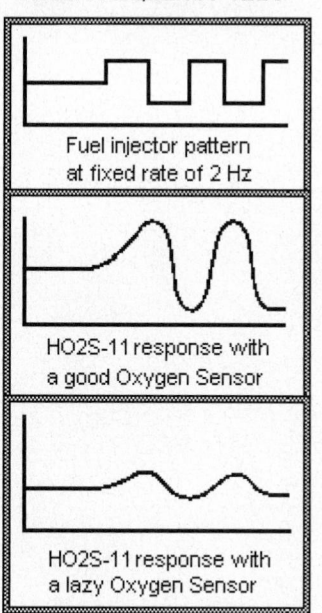

Fuel injector pattern at fixed rate of 2 Hz

HO2S-11 response with a good Oxygen Sensor

HO2S-11 response with a lazy Oxygen Sensor

21199_FDIA_G033

Fig. 11 Fixed Frequency Test

A Fixed Frequency Closed Loop Test is used to check the HO2S voltage and frequency. A sample of the HO2S signal is checked to determine if the sensor is capable of switching properly or has a slow response time (referred to as a lazy sensor).

Oxygen Sensor Heater Monitor

The Oxygen Sensor Heater Monitor is a PCM diagnostic designed to monitor the Oxygen Sensor Heater and its related circuits for faults.

OXYGEN SENSOR HEATER MONITOR OPERATION

The Oxygen Sensor Heater Monitor performs its task by detecting whether the proper amount of O2 sensor voltage change occurred as the HO2S Heater is turned from "on" to "off" with the engine in closed loop. The time it takes for the HO2S-11 and HO2S-12 signal to switch (the response time) is constantly monitored by the Oxygen Sensor Monitor. Once the Oxygen Sensor Heater Monitor is enabled, if the switch time for the HO2S-11 or HO2S-12 signal is too long, the PCM fails the test, the MIL is activated and a trouble code is set.

➡ **Response time is defined as the amount of time it takes for a HO2S signal to switch from Rich to Lean, and then Lean to Rich.**

FRONT AND REAR OXYGEN SENSOR HEATER OPERATION

Both upstream and downstream Oxygen sensors are used on the OBD II system. These sensors are designed with additional protection around the ceramic core to protect them from condensation that could crack them if the heater is turned on with condensation present.

The HO2S heaters are not turned on until the ECT sensor signal indicates that the engine is warm. The delay period can last for as long as 5 minutes from startup. The delay allows any condensation in the Exhaust system to evaporate.

Faults detected by the HO2S or HO2S Heater Monitor:

- A fault in the HO2S, the HO2S heater or its related circuits
- A fault in the HO2S connectors (look for moisture tracking)
- A defective Power Control Module

Air Injection System Monitor

The Air Injection System Monitor is an OBD diagnostic controlled by the PCM that monitors the Air Injection (AIR) system. The Oxygen Sensor Monitor must run and complete before the PCM will run this test. The PCM enables this test during AIR system operation after certain engine conditions are met and these enable criteria are met:

- Crankshaft Position sensor signal must be present
- ECT and IAT sensor input signals must be within limits

AIR MONITOR–ELECTRIC PUMP DESIGN

The AIR Monitor consists of these Solid State Monitor tests:

- A check of the Solid State relay for electrical faults.
- A check of the secondary side of the relay for electrical faults.
- A test to determine if the AIR system can inject additional air.

AIR MONITOR–MECHANICAL PUMP DESIGN

The AIR Monitor for the mechanical (belt-driven air pump) design uses two Output State Monitor configurations to perform two different circuit tests. One test is used to check for faults in the Secondary Air Bypass (AIRB) solenoid circuit. The normal function of the AIRB solenoid and valve assembly is to dump air into the atmosphere.

A second test is used to check for electrical faults in the Secondary Air Divert (AIRD) solenoid. The normal function of the AIRD solenoid and valve assembly is to direct the air either upstream or downstream.

FUNCTIONAL CHECK

See Figure 12.

An AIR system functional check is done at startup with the AIR pump on or during a hot idle period if the startup part of the test was not performed. A flow test is included that uses the HO2S signal to indicate the presence of extra air injected into the exhaust stream.

Diagnostic Trouble Codes

In the Diagnostic Trouble Code charts for the specific manufacturers you will see the following terms in the left column of the chart:

1. 1T–This means the code was activated when the PCM recognized the problem the first time it occurred.

2. 2T–This means the code was activated when the PCM recognized the problem and set the code after it occurred two times.

3. CCM–This means that the code and system affected is an emission related device and has a Comprehensive Component Monitor (CCM) tracking it.

4. MIL: Yes–This means that the Malfunction Indicator Light will be displayed.

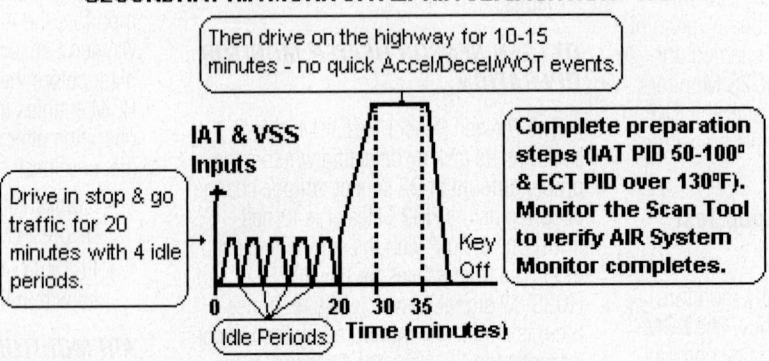

SECONDARY AIR MONITOR REPAIR VERIFICATION "TRIP"

Then drive on the highway for 10-15 minutes - no quick Accel/Decel/WOT events.

IAT & VSS Inputs

Drive in stop & go traffic for 20 minutes with 4 idle periods.

Complete preparation steps (IAT PID 50-100° & ECT PID over 130°F). Monitor the Scan Tool to verify AIR System Monitor completes.

Key Off

Idle Periods

Time (minutes)

21199_FDIA_G036

Fig. 12 Secondary AIR monitor

SYMPTOM DIAGNOSIS (NO CODES)

4

Table of Contents

What To Do When There Are No DTCs

Do not attempt to diagnose a Drivability Symptoms without having a logical plan to use to determine which Engine Control system is the cause of the symptom - this plan should include a way to determine which systems do not have a problem! Drivability symptom diagnosis is a part of an organized approach to problem solving and repair.

DRIVABILITY SYMPTOM INDEX TABLE

To use this list, locate the symptom that matches a particular problem and refer to the areas to test. The items listed under each symptom may not apply to all models, engines or vehicle systems. The repair steps indicate what vehicle component or system to test.

➡ The Drivability Symptoms in this list are intended to be generic. While they apply to most vehicles, some vehicles may not have all of the components listed. Refer to other Chilton repair information and electronic media for specific tests.

Symptom Test Table

Symptom Description	Suggested Areas to Test
Test 1 - No Start, Hard Start Condition • No Crank • Hard Start, Long Crank, Erratic Crank • Stall After Start • No Start, Normal Crank • No Start, MIL is off (if the VREF shorts to ground)	- Check battery, battery circuits to starter - Check for a damaged flywheel, engine compression, base timing and minimum air rate - Check for a failed fuel pump relay - Check for distributor rotor "punch-through" - Check for a faulty ignition control module (ICM) - Check for a VREF circuit shorted to ground - Check SKIM (security system) with a Scan Tool
Test 2 - Rough Idle or Stalls Condition • Low or slow idle speed • Fast idle speed • Hunting or rolling idle speed • Slow return to idle speed • Stalls or almost stalls	- Check for engine vacuum leaks - Check the condition of the PCV valve and lines - Check for excessive carbon buildup - Check for a restricted exhaust - Check base idle speed, check for low fuel pressure - Check the throttle linkage for sticking or binding
Test 3 - Runs Rough Condition • At idle speed • During acceleration • At cruise speed • During deceleration	- Check for engine vacuum leaks at intake manifold - Check condition of ignition secondary components - Check base timing and idle speed settings - Check for low or high fuel pressure - Check for dirty, leaking or shorted fuel injectors - Check for excessive carbon buildup on valves
Test 4 - Cuts-out, Misses Condition • At idle speed • During acceleration • At cruise speed • During deceleration	- Check for engine vacuum leaks at intake manifold - Check condition of ignition secondary components - Check that spark timing advance is available - Check for low or high fuel pressure - Check for dirty, leaking or shorted fuel injectors - Check for excessive carbon buildup on valves
Test 5 - Bucks, Jerks Condition • During acceleration • At cruise speed • During deceleration	- Check for engine vacuum leaks at intake manifold - Check condition of ignition secondary components - Check that spark timing advance is available - Check for low or high fuel pressure - Check for dirty, leaking or shorted fuel injectors - Check operation of the TCC solenoid, brake switch

Symptom Diagnosis Test 1 — No Start, Hard Start Condition

➡ If there is no spark output or fuel pressure available, check for a failed fuel pump relay, no power to the PCM, or loss of the ignition reference signal to the PCM.

PRELIMINARY CHECKS

Prior to starting this symptom test routine, inspect these underhood items:
- Check battery charge and condition, starter current draw.
- Verify the starter relay operation and that the engine cranks (turns over).
- Verify the check engine light (MIL) operation - if it does not activate, check the PCM power and ground circuits, and check for 5v supply at the MAP or TP sensor.
- Check Air Intake system for restrictions (inspect air inlet tubes, air filter for dirt, etc.).
- Check the status of the Smart Key Immobilizer System (SKIM) with the Scan Tool.

Test 1 Chart

Step	Action	Yes	No
1	**Step Description: No Start Condition Only** » Check battery cables, state of charge. » If the engine does not rotate, inspect for a locked engine (hydrostatic lockup condition). » Does the engine crank normally?	Go to Step 2.	Repair the fault in the battery, starter, or Base Engine. Retest for the symptom when all repairs are done.
2	**Step Description: Check the Fuel System** » Verify that the pump operates at key on. » Check the fuel pump relay operation. If the relay does not operate, check for blown fuse. » Inspect pump for a leak-down condition » Test fuel pressure, volume and quality. » Test the operation of the fuel regulator. » Are there any faults in the Fuel system?	Make needed repairs.	Go to Step 3.
3	**Step Description: Check the Ignition System** » Inspect ignition secondary components for damage (look for rotor "punch-through"). » Inspect the coils for signs of spark leakage at coil towers or primary connections. » Check the spark output with a spark tester. » Test Ignition system with an engine analyzer. » Are there any faults in the Ignition system?	Make repairs to the Ignition system. Then retest the symptom.	Go to Step 4.
4	**Step Description: Check the Exhaust System** » Check Exhaust system for leaks or damage. » Check the Exhaust system for a restriction using the Vacuum or Pressure Gauge Test (e.g., exhaust backpressure reading should not exceed 1.5 psi at cruise speeds). » Are there any faults in the Exhaust system?	Make repairs to the Exhaust system. Then retest the symptom.	Go to Step 5.
5	**Step Description: Check the MAP Sensor** » Disconnect the MAP sensor and attempt to start the engine. » Does the engine start and run normally?	Replace the MAP sensor. Retest for the symptom when repairs are completed.	Go to Step 6.
6	**Step Description: Check for a Hot Engine** » Check for signs of an engine overheating condition related to a Hard Start Symptom. » Does the engine appear to be overheated?	Make the repairs to correct the hot engine and then retest for the symptom when done.	Go to Step 7.
7	**Step Description: Check ECT Sensor PID** » Connect a Scan Tool and turn the key to on. » Read the ECT sensor (compare to chart). » Has the ECT sensor shifted out of range?	Replace the ECT sensor. Then retest for the symptom when all repairs are completed.	Go to Step 8.
8	**Step Description: Check the PCV System** » Inspect the PCV system components for broken parts or loose connections. » Test the operation of the PCV valve. » Are there any faults in the PCV system?	Repair the PCV system. Refer to the PCV system tests. Retest the symptom when all repairs are done.	Go to Step 9.
9	**Step Description: Check the EVAP System** » nspect for damaged or disconnected EVAP system components. » Inspect for a fuel saturated charcoal canister. » Are there any faults in the EVAP system?	Refer to the EVAP system tests. Retest for the symptom when all repairs are completed.	Go to Step 10.
10	**Step Description: Test the Base Engine** » Check the engine compression. » Test valve timing and timing chain condition. » Check for a worn camshaft or valve train. » Check for any large intake manifold leaks. » Are there any faults in the Base Engine?	Repair the Base Engine. Refer to the Base Engine Tests. Retest symptom when done.	Return to Step 2 to repeat the test steps in this series to locate and repair the "No Start, Hard Start" condition.

Symptom Diagnosis Test 2 — Rough, Low or High Idle Speed Condition

➡ **If the vehicle has a rough idle and the base timing, idle speed and the IAC (or AIS) motor operates properly, check the engine for excessive carbon buildup.**

PRELIMINARY CHECKS

Prior to starting this symptom test routine, inspect these underhood items:
- All related vacuum lines for proper routing and integrity.
- All related electrical connectors and wiring harnesses for faults (Wiggle Test).
- Check the throttle linkage for a sticking or binding condition.
- Air Intake system for restrictions (air inlet tubes, dirty air filter, etc.).
- Search for any technical service bulletins related to this symptom.
- Turn the key to off. Unplug the MAP sensor connection and restart the engine to recheck for the idle concern. If the condition is gone, replace the MAP sensor.

Test 2 Chart

Step	Action	Yes	No
1	**Step Description: Verify the rough idle or stall** » Does the engine have a warm engine rough idle, low idle or high idle condition in P or N?	Go to Step 2.	Fault is intermittent. Return to the Symptom List and select another fault.
2	**Step Description: Verify idle speed & timing** » Verify the base timing is within specifications » Verify that the base idle speed is set properly » Are the timing and idle speed set properly?	Go to Step 3.	Set the base idle speed and timing to the specifications and then retest for the symptom.
3	**Step Description: Check AIS / IAC Operation** » Check the AIS or IAC motor operation » Inspect the AIS/IAC housing in throttle body for restricted passages. Clean as needed. » Set the parking brake, block the drive wheels and turn the A/C off. Install the Scan Tool. » IAC Motor Tester - Turn the key off and then connect the IAC tester to the IAC valve. » Start the engine and use the IAC tester to extend and retract the IAC valve. » ATM Test - Start the engine. Use the tool to change the speed from min-idle to 1500 rpm. » Did the idle speed change as commanded?	Install an Aftermarket Noid light and check the operation of the PCM and AIS or IAC motor circuits. Check the motor for signs of open or shorted circuits. Replace the IAC motor or PCM as needed or make repairs to the IAC motor wiring. If all are okay, go to Step 4.	If the AIS/IAC motor passages are clean and engine speed did not change as described when the AIS/IAC motor was extended and retracted, replace the AIS/IAC motor. Then retest for the condition.
4	**Step Description: Check/compare PID values** » Connect Scan Tool & turn off all accessories. » Start the engine and allow it to fully warmup. » Monitor all related PIDs on the Scan Tool. » Verify the P/N switch input in gear and Park. » Check the O2S operation with a Lab Scope. » Are all PIDs within normal range?	Go to Step 5. Note: An IAC motor count of over 80 indicates the pintle is extended and an IAC count of (0) indicates the pintle is retracted.	One or more of the PIDs are out of range when compared to "known good" values. Make repairs to the system that is out of range, then retest for the symptom.

5	**Step Description: Check the Ignition System** » Inspect the coils for signs of spark leakage at coil towers or primary connections. » Check the spark output with a spark tester. » Test Ignition system with an engine analyzer. » Were any faults found in the Ignition system?	Make repairs as needed	Go to Step 6.
6	**Step Description: Check the Fuel System** » Inspect the Fuel delivery system for leaks. » Test the fuel pressure, quality and volume. » Test the operation of the pressure regulator. » Were any faults found in the Fuel system?	Make repairs as needed	Go to Step 7.
7	**Step Description: Check the Exhaust System** » Check Exhaust system for leaks or damage. » Check the Exhaust system for a restriction using the Vacuum or Pressure Gauge Test (e.g., exhaust backpressure reading should not exceed 1.5 psi at cruise speeds). » Were any faults found in Exhaust System?	Make repairs to the Exhaust system. Then retest the symptom.	Go to Step 8.
8	**Step Description: Check the PCV System** » Inspect the PCV system components for broken parts or loose connections. » Test the operation of the PCV valve. » Were any faults found in the PCV system?	Make repairs to the PCV system. Refer to the PCV system tests. Then retest for the condition.	Go to Step 9.
9	**Step Description: Check the EVAP System** » Inspect for damaged or disconnected EVAP system components or a saturated canister. » Were any faults found in the EVAP system?	Make repairs to EVAP system. Retest for the condition.	Go to Step 10.
10	**Step Description: Check the Base Engine** » Test the engine compression. » Test valve timing and timing chain condition. » Check for a worn camshaft or valve train. » Check for any large intake manifold leaks. » Were any faults found in the Base Engine?	Make repairs as needed to the Base Engine. Refer to the Base Engine tests. Then retest for the condition when repairs are completed.	Go to Step 2 and repeat the tests from the beginning to locate and repair the cause of the "Rough, Low or High Idle Speed" condition.

Symptom Diagnosis Test 3 — Runs Rough Condition

PRELIMINARY CHECKS

Prior to starting this symptom test routine, inspect these underhood items:

- All related vacuum lines for proper routing and integrity
- Air Intake system for restrictions (air inlet tubes, dirty air filter, etc.)
- Search for any technical service bulletins related to this symptom.

Test 3 Chart

Step	Action	Yes	No
1	**Step Description: Verify engine runs rough** » Start the engine and allow it to idle in P or N. » Does the engine run rough when warm in Park or Neutral position?	Check for any stored codes. If codes are set, repair codes and retest. If no codes are set, go to Step 3.	Go to Step 2.
2	**Step Description: Condition does not exist!** » Inspect various underhood items that could cause an intermittent Runs Rough condition (i.e., dirt in the throttle body, vacuum leaks, IAC motor connections, etc.). » Were any problems located in this step?	Correct the problems. Do a PCM reset and engine "idle relearn" procedure. Then verify the "runs rough" condition is repaired.	The problem is not present at this time. It may be an intermittent problem.
3	**Step Description: Check/compare PID values** » Connect a Scan Tool to the test connector. » Turn off all accessories. » Start the engine and allow it to fully warmup. » Monitor all related PIDs on the Scan Tool. » Were all PIDs within their normal range?	Go to Step 4. Note: The IAC motor should read from 5-50 counts. Check the LONGFT reading for a large shift into the negative range (due to a rich condition).	One or more of the PIDs are out of range when compared to "known good" values. Make repairs to the system that is out of range, then retest for the symptom.
4	**Step Description: Check the Ignition System** » Inspect the coils for signs of spark leakage at coil towers or primary connections. » Check the spark output with a spark tester. » Test Ignition system with an engine analyzer. » Were any faults found in the Ignition system?	Make repairs as needed	Go to Step 5.
5	**Step Description: Check the Fuel System** » Inspect the Fuel delivery system for leaks. » Test the fuel pressure, quality and volume. » Test the operation of the pressure regulator. » Were any faults found in the Fuel system?	Make repairs as needed	Go to Step 6.
6	**Step Description: Check the Exhaust System** » Check Exhaust system for leaks or damage. » Check the Exhaust system for a restriction using the Vacuum or Pressure Gauge Test (e.g., exhaust backpressure reading should not exceed 1.5 psi at cruise speeds). » Were any faults found in Exhaust System?	Make repairs to the Exhaust system. Then retest the symptom.	Go to Step 7.
7	**Step Description: Check the PCV System** » Inspect the PCV system components for broken parts or loose connections. » Test the operation of the PCV valve. » Were any faults found in the PCV system?	Make repairs to the PCV system. Refer to the PCV system tests. Then retest for the condition.	Go to Step 9.
8	**Step Description: Check the EVAP System** » Inspect for damaged or disconnected EVAP system components or a saturated canister. » Were any faults found in the EVAP system?	Make repairs to EVAP system. Retest for the condition.	Go to Step 10.
9	**Step Description: Check Engine Condition** » Test the engine compression. » Test valve timing and timing chain condition. » Check for a worn camshaft or valve train. » Check for any large intake manifold leaks. » Were any faults found in the Base Engine?	Make repairs as needed to the Base Engine. Refer to the Base Engine tests. Then retest for the condition when repairs are completed.	Return to Step 2 and repeat the tests from the beginning to locate and repair the cause of the "Runs Rough" condition.

Symptom Diagnosis Test 4 — Cuts-out or Misses Condition

PRELIMINARY CHECKS

Prior to starting this symptom test routine, inspect these underhood items:
- All related vacuum lines for proper routing and integrity
- Search for any technical service bulletins related to this symptom.

Test 4 Chart

Step	Action	Yes	No
1	**Step Description: Verify Cuts-out condition** » Start the engine and attempt to verify the Cuts-out or misses condition. » Does the engine have a cuts-out condition?	Check for any stored codes. If codes are set, repair codes and retest. If no codes are set, go to Step 3.	Go to Step 2.
2	**Step Description: Condition does not exist!** » Inspect various underhood items that could cause an intermittent Cuts-out condition (i.e., EVAP, Fuel or Ignition system components). » Were any problems located in this step?	Correct the problems. Do a PCM reset and "Fuel Trim Relearn" procedure. Then verify condition is repaired.	The problem is not present at this time. It may be an intermittent problem.
3	**Step Description: Check/compare PID values** » Connect a Scan Tool to the test connector. » Turn off all accessories. » Start the engine and allow it to fully warmup. » Monitor all related PIDs on the Scan Tool (i.e., ECT IAC Counts and LONGFT at idle). » Were all PIDs within their normal range?	Go to Step 4. Note: The IAC motor should be from 5-50 counts. Watch fuel trim (%) for a large shift into the negative (-) range (due to a rich condition).	One or more of the PIDs are out of range when compared to "known good" values. Make repairs to the system that is out of range, then retest for the symptom.
4	**Step Description: Check the Ignition System** » Inspect the coils for signs of spark leakage at coil towers or primary connections. » Check the spark output with a spark tester. » Test Ignition system with an engine analyzer. » Were any faults found in the Ignition system?	Make repairs as needed	Go to Step 5.
5	**Step Description: Check the Fuel System** » Inspect the Fuel delivery system for leaks. » Test the fuel pressure, quality and volume. » Test the operation of the pressure regulator. » Were any faults found in the Fuel system?	Make repairs as needed	Go to Step 6.
6	**Step Description: Check the Exhaust System** » Check Exhaust system for leaks or damage. » Check the Exhaust system for a restriction using the Vacuum or Pressure Gauge Test (e.g., exhaust backpressure reading should not exceed 1.5 psi at cruise speeds). » Were any faults found in Exhaust System?	Make repairs to the Exhaust system. Then retest the symptom.	Go to Step 7.
7	**Step Description: Check the PCV System** » Inspect the PCV system components for broken parts or loose connections. » Test the operation of the PCV valve. » Were any faults found in the PCV system?	Make repairs to the PCV system. Then retest for the condition.	Go to Step 8.
8	**Step Description: Check the EVAP System** » Inspect for damaged or disconnected EVAP system components » Check for a saturated EVAP canister. » Were any faults found in the EVAP system?	Make repairs to EVAP system. Retest for the condition.	Go to Step 9.
9	**Step Description: Check the AIR system** » Inspect AIR system for broken parts, leaking valves or disconnected hoses. » Test the operation of Secondary AIR system. » Were any faults found in the AIR system?	Make repairs as needed. Refer to the Secondary AIR system tests. Retest for the condition.	Go to Step 10.
10	**Step Description: Check Engine Condition** » Test the engine compression. » Test valve timing and timing chain condition. » Check for a worn camshaft or valve train. » Check for any large intake manifold leaks. » Were any faults found in the Base Engine?	Make repairs as needed to the Base Engine. Refer to the Base Engine tests. Then retest for the condition when repairs are completed.	Go to Step 2 and repeat the tests from the beginning to locate and repair the cause of the "Cuts Out or Misses" condition.

Symptom Diagnosis Test 5 — Surge Condition

PRELIMINARY CHECKS

1. Discuss how the operation of the torque converter clutch (TCC) or air conditioning compressor can affect the "feel" of the vehicle during normal operation. Refer to the information in the Owner's Manual to explain how these devices normally operate.
2. Search for any technical service bulletins related to this symptom.

Test 5 Chart

Step	Action	Yes	No
1	**Step Description: Verify the surge condition** » Drive the vehicle and attempt to verify that the vehicle surges at cruise speeds. » Does the engine have a surge condition?	Check for any stored codes. If codes are set, repair codes and retest. If no codes are set, go to Step 3.	Go to Step 2.
2	**Step Description: Condition does not exist!** » Inspect various underhood items that could cause an intermittent surge condition (check for leaks in the MAP sensor vacuum lines). » Were any problems located in this step?	Correct the problems. Do a PCM reset and "Fuel Trim Relearn" procedure. Then verify condition is repaired.	The problem is not present at this time. It may be an intermittent problem.
3	**Step Description: Check/compare PID values** » Connect a Scan Tool to the test connector. » IStart the engine and allow it to fully warmup. » Monitor all related PIDs on Scan Tool (HO2S switching, LONGFT, and the TCC operation) » Compare VSS PID reading to speedometer. » Were all PIDs within their normal range?	Go to Step 4. Note: Verify that the front HO2S responds quickly to throttle changes. Check for silicon contamination on the front HO2S (this can cause a rich A/F signal).	One or more of the PIDs are out of range when compared to "known good" values. Make repairs to the system that is out of range, then retest for the symptom.
4	**Step Description: Check the Ignition System** » Inspect the coils for signs of spark leakage at coil towers or primary connections. » Check the spark output with a spark tester. » Test Ignition system with an engine analyzer. » Were any faults found in the Ignition system?	Make repairs as needed	Go to Step 5.
5	**Step Description: Check the Fuel System** » Inspect the Fuel delivery system for leaks. » Test the fuel pressure, quality and volume. » Test the operation of the pressure regulator. » Were any faults found in the Fuel system?	Make repairs as needed	Go to Step 6.
6	**Step Description: Check the Exhaust System** » Check Exhaust system for leaks or damage. » Check the Exhaust system for a restriction using the Vacuum or Pressure Gauge Test (e.g., exhaust backpressure reading should not exceed 1.5 psi at cruise speeds). » Were any faults found in Exhaust System?	Make repairs to the Exhaust system. Then retest the symptom.	Return to Step 2 and repeat the tests from the beginning to locate and repair the cause of the "Surge" condition.

INTERMITTENT TESTS

Many trouble code repair charts end with a result that reads "Fault Not Present at this Time." What this expression means is that the conditions that were present when a code set or drivability symptom occurred are no longer there or were not met. In effect, the problem was present at least once, but is not present at this time. However, it is likely to return in the future, so it should be diagnosed and repaired if at all possible.

One way to find an intermittent problem is to gather the information that was present when the problem occurred. In the case of a Code Fault, this can be done in two ways: by capturing the data in Snapshot or Movie mode or by driver observations.

The PCM has to detect the fault for a specific period of time before a trouble code will set. While intermittent problems may appear to be occasional in nature, they usually occur under specific conditions. Therefore, you should identify and duplicate these conditions. Since intermittent faults are difficult to duplicate, a logical routine (checklist) must be followed when attempting to find the faulty component, system or circuit. The tests on the next page can be used to help find the cause of an intermittent fault.

Some intermittent faults occur due to a loose connection, wiring problem or warped circuit board. An intermittent fault can also be caused by poor test techniques that cause damage to the male or female ends of a connector.

Test for Loose Connectors

To test for a loose or damaged connection, take the male end of a connector from another wiring harness and carefully push it into the "suspect" female terminal to verify that the opening is tight. There should be some resistance felt as the male connector is inserted in the terminal connection.

The Wiggle Test

See Figures 1 and 2.

A wiggle test can be used to locate the cause of some intermittent faults. The sensor, switch or the PCM wiring can be back-probed, as shown, while the test is done.

During testing, move or wiggle the suspect device, connector or wiring while watching for a change.

If the DVOM has a Min/Max record mode, use this mode during the test.

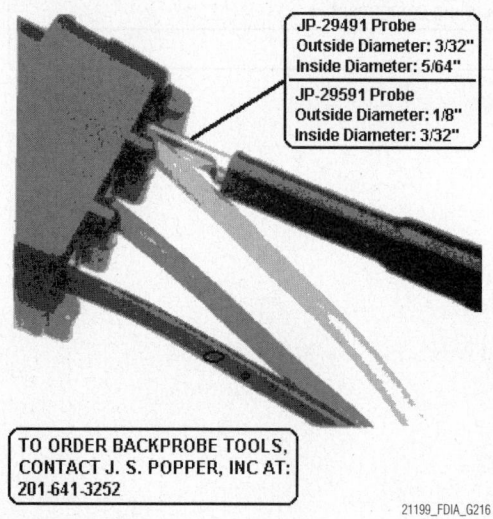

JP-29491 Probe
Outside Diameter: 3/32"
Inside Diameter: 5/64"

JP-29591 Probe
Outside Diameter: 1/8"
Inside Diameter: 3/32"

TO ORDER BACKPROBE TOOLS, CONTACT J. S. POPPER, INC AT: 201-641-3252

21199_FDIA_G216

Fig. 1 Backprobing a connector

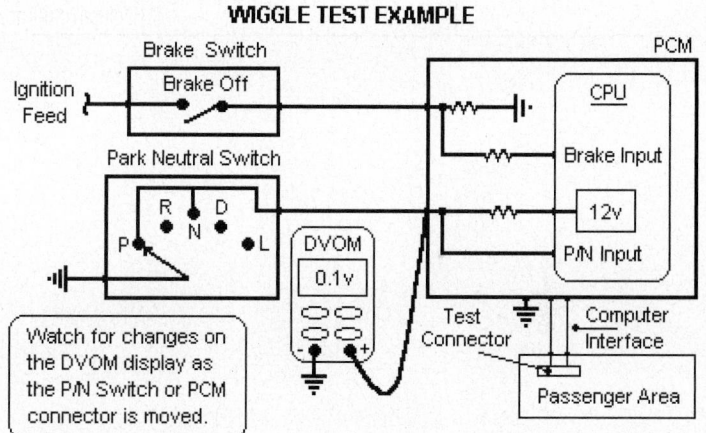

WIGGLE TEST EXAMPLE

Watch for changes on the DVOM display as the P/N Switch or PCM connector is moved.

21199_FDIA_G216

Fig. 2 Wiggle Test Example

Diagnosis And Testing - Vehicle Does Not Fill

CONDITION	POSSIBLE CAUSES	CORRECTION
Pre-Mature Nozzle Shut-Off	Defective fuel tank assembly components.	Fill tube improperly installed (sump)
		Fill tube hose pinched.
		Check valve stuck shut.
		Control valve stuck shut.
	Defective vapor/vent components.	Vent line from control valve to canister pinched.
		Vent line from canister to vent filter pinched.
		Canister vent valve failure (requires double failure, plugged to NVLD and atmosphere).
		Leak detection pump failed closed.
		Leak detection pump filter plugged.
	On-Board diagnostics evaporative system leak test just conducted.	Canister vent valve vent plugged to atmosphere.
		Engine still running when attempting to fill (System designed not to fill).
	Defective fill nozzle.	Try another nozzle.
Fuel Spits Out Of Filler Tube.	During fill.	See Pre-Mature Shut-Off.
	At conclusion of fill.	Defective fuel handling component. (Check valve stuck open).
		Defective vapor/vent handling component.
		Defective fill nozzle.

ACURA
DIAGNOSTIC TROUBLE CODES

TABLE OF CONTENTS

DIAGNOSTIC TROUBLE CODES

OBD II Vehicle Applications

ACURA

2.2CL
1997
2.2L I4 SOHC SMFI VTEC (F22B1) . . . . VIN YA1

2.3CL
1997–1998
2.3L I4 SOHC MFI VTEC (F23A1). VIN YA3

3.0CL
1997–1999
3.0L V6 SOHC SMFI VTEC (J30A1) . . . . VIN YA2

3.2CL
1996–2003
3.2L V6 MFI (C32A6) . . . . VIN UA3
3.2L V6 SMFI 24v (C32A6) . . . . VIN UA3
3.2L V6 SMFI 24v (J32A1) . . . . VIN UA5
3.2L V6 SOHC SMFI VTEC (J32A1) . . . . VIN YA4

2.5TL
1996–1998
2.5L I5 SMFI 20v (G25A4) . . . . VIN UA2

3.2TL
1996–2003
3.2L V6 MFI (C32A6) . . . . VIN UA3
3.2L V6 SMFI 24v (J32A1) . . . . VIN UA5

TL
2004–2006
3.2L V6 SMFI 24v (J32A1) . . . . VIN UA5

3.5RL
1996–2004
3.5L V6 MFI (C35A1) . . . . VIN KA9
3.5L V6 SMFI 24v (C35A1) . . . . VIN KA9

RL
2005–2006
3.5L V6 MFI (C35A1) . . . . VIN KB1

Integra
1996–2001
1.8L I4 DOHC SMFI VTEC (B18C1, B18C5) . . . . VIN DC2
1.8L I4 DOHC SMFI (B18B1) . . . . VIN DC4
1.8L I4 DOHC SMFI (B18B1) . . . . VIN DB7
1.8L I4 DOHC SMFI VTEC (B18C1) . . . . VIN DB8

MDX
2001–2006
3.5L V6 MFI (J35A3) . . . . VIN YD1

NSX
1996–2005
3.0L V6 MFI (C30A1) . . . . VIN NA1
3.2L V6 MFI (C32B1) . . . . VIN NA2

RSX
2002–2006
2.0L I4 MFI (K20A3, K20A2) . . . . VIN DC5

SLX
1996–1999
3.2L V6 SOHC MFI (6VD1) . . . . VIN V
3.5L V6 SOHC SMFI (6VE1) . . . . VIN X

OBD II Trouble Code List (P0xxx Codes)

DTC	Trouble Code Title
DTC: P0010 **1T CCM, MIL: YES** **2002, 2003, 2004, 2005, 2006** **Models:** RSX **Engines:** All **Transmissions:** A/T, M/T	**Variable Valve Timing Control Oil Control Solenoid Valve Circuit Malfunction** Key on or engine running; and the PCM detected an unexpected voltage condition on the VVT Oil Control Solenoid control signal. **Possible Causes:** • VVT oil control solenoid control circuit is open • VVT oil control solenoid control circuit is shorted to ground • VVT oil control solenoid control circuit is damaged or has failed • PCM has failed
DTC: P0011 **2T CCM, MIL: YES** **2002, 2003, 2004, 2005, 2006** **Models:** RSX **Engines:** All **Transmissions:** A/T, M/T	**Variable Valve Timing Control System Malfunction** Engine started, vehicle driven through a hard acceleration period, then returned back to idle speed, and the PCM detected a problem in the VVT System operation. **Possible Causes:** • VVT oil control solenoid valve is stuck • VVT oil control solenoid is damaged or has failed • PCM has failed
DTC: P0097 **2T CCM, MIL: YES** **2003, 2004, 2005, 2006** **Models:** MDX **Engines:** All **Transmissions:** A/T, M/T	**Intake Air Temperature Sensor 2 Circuit Low Input** Key on or engine running; and the PCM detected the Intake Air Temperature (IAT) Sensor 2 signal indicated less than 0.08v (Scan Tool reads 356°F) during the CCM test. **Possible Causes:** • IAT2 signal shorted to chassis ground • IAT2 signal shorted to sensor ground circuit • IAT2 has an internal failure (it is shorted) or has failed • PCM has failed
DTC: P0098 **2T CCM, MIL: YES** **2003, 2004, 2005, 2006** **Models:** MDX **Engines:** All **Transmissions:** A/T, M/T	**Intake Air Temperature Sensor 2 Circuit High Input** Key on or engine running; and the PCM detected the Intake Air Temperature (IAT) Sensor 2 signal indicated more than 4.90v (Scan Tool reads -40°F) during the CCM test. **Possible Causes:** • IAT2 signal shorted to VREF or system power • IAT2 signal circuit is open • IAT2 ground circuit is open • IAT2 has an internal failure (it is open) • PCM has failed
DTC: P0101 **2T CCM, MIL: YES** **1996, 1997, 1998, 1999** **Models:** SLX **Engines:** 3.2L VIN V, 3.5L VIN X **Transmissions:** A/T	**Mass Airflow Sensor Signal Range/Performance** DTC P0106, P0107, P0108, P0121, P0122 and P0123 not set, system voltage at 11-16v, engine running at a stable idle speed, throttle angle stable (±1%), Calculated airflow from 25-40 g/sec, conditions met for 1 second, and the PCM detected a MAF sensor frequency that was significantly higher or lower than a "predicted" MAF airflow based on throttle position and engine speed for 12.5 seconds over a 25 second period during the CCM Rationality test. **Possible Causes:** • Air leaks after the MAF sensor, or in the EGR or PCV system • Engine oil cap missing, engine oil dipstick not fully seated • MAF sensor is contaminated, dirty or out-of-calibration • MAF sensor ground circuit has high resistance • MAF minimum airflow rate to low at idle or during deceleration • MAP or TP sensor signal skewed, stuck or out of calibration • High signal interference (i.e., electrical noise from the ignition) • PCM has failed
DTC: P0102 **1T CCM, MIL: YES** **1996, 1997, 1998, 1999** **Models:** SLX **Engines:** 3.2L VIN V, 3.5L VIN X **Transmissions:** A/T	**Mass Airflow Sensor Circuit Low Frequency** Engine started, engine speed over 500 rpm for 10 seconds, system voltage over 11.5v, the PCM detected the MAF sensor frequency was less than 1000 Hz for a total of 50% of the last 100 samples in the CCM Rationality test (a sample is taken every cylinder event). **Possible Causes:** • MAF sensor signal is shorted to ground • MAF sensor power circuit is open • MAF sensor is contaminated, dirty or is damaged • PCM has failed

DTC	Trouble Code Title, Conditions & Possible Causes
DTC: P0103 **1T CCM, MIL: YES** **1996, 1997, 1998, 1999** **Models:** SLX **Engines:** 3.2L VIN V, 3.5L VIN X **Transmissions:** A/T	**Mass Airflow Sensor Circuit High Frequency** Engine started, engine speed over 500 rpm for 10 seconds, system voltage over 11.5v, the PCM detected the MAF sensor frequency was more than 10,000 Hz for a total of 50% of the last 200 samples in the CCM Rationality test (a sample is taken every cylinder event). **Possible Causes:** • RFI or EMI interference from the Generator or Ignition system • RFI or EMI interference from the an Ignition system component • MAF sensor is contaminated, dirty or is damaged • PCM has failed
DTC: P0106 **2T CCM, MIL: YES** **1996, 1997, 1998, 1999** **Models:** SLX **Engines:** 3.2L VIN V, 3.5L VIN X **Transmissions:** A/T	**Manifold Air Pressure Sensor Signal Range/Performance** DTC P0121, P0122 and P0123 not set, engine speed stable (±100 rpm), throttle angle stable (±1%), IAC counts steady (±10 counts), EGR flow stable (±4%), no change in the A/C clutch, PSPS, Brake switch or TCC status, conditions met for 1 second, and the PCM detected the Actual MAP sensor value varied more than 10 kPa from the Expected MAP value for 10 seconds over a 20 second period. **Possible Causes:** • MAP sensor circuit open or shorted to ground (intermittent) • MAP sensor source vacuum line is leaking or restricted • MAP sensor source vacuum line is plugged at intake manifold • MAP sensor is damaged, out-of-calibration or has failed • PCM has failed
DTC: P0107 **2T CCM, MIL: YES** **1995, 1996, 1997, 1998, 1999,** **2000, 2001, 2002, 2003, 2004,** **2005** **Models:** NSX, NSX-T **Engines:** All **Transmissions:** A/T, M/T	**Manifold Air Pressure Sensor Circuit Low Input** Engine started, engine running in closed loop, and the PCM detected the MAP sensor indicated a value of 0.0" Hg during the CCM test. **Note: The key on, engine off MAP sensor input should be near 2.9v.** **Possible Causes:** • MAP sensor 5-volt power circuit open or shorted to ground • MAP Sensor signal circuit is shorted to ground • MAP Sensor is damaged or has failed • PCM has failed
DTC: P0107 **1T CCM, MIL: YES** **1996, 1997, 1998, 1999, 2000,** **2001, 2002, 2003, 2004, 2005,** **2006** **Models:** 2.2CL, 2.3CL, 3.0CL, 3.2CL, 3.2TL, TL, 3.5RL, RL, Integra, MDX, RSX **Engines:** All **Transmissions:** A/T, M/T	**Manifold Air Pressure Sensor Circuit Low Input** Engine started, engine running in closed loop at idle speed, and the PCM detected the MAP sensor indicated a value of 0.0" Hg during the CCM test. **Note: The key on, engine off MAP sensor input should be near 2.9v.** **Possible Causes:** • MAP sensor 5-volt power circuit open or shorted to ground • MAP Sensor signal circuit is shorted to ground • MAP Sensor is damaged or has failed • PCM has failed
DTC: P0107 **1T CCM, MIL: YES** **1996, 1997, 1998, 1999** **Models:** SLX **Engines:** 3.2L VIN V, 3.5L VIN X **Transmissions:** A/T	**Manifold Air Pressure Sensor Circuit Low Input** DTC P0121, P0122 and P0123 not set, engine started, system voltage from 11-16v, then with the engine speed below 1000 rpm and throttle angle over 1%, or with the engine speed over 1000 rpm and throttle angle over 2%, the PCM detected the MAP sensor was less than 0.04v (11 kPa) for 10 seconds over a 20 second period. **Possible Causes:** • MAP sensor circuit shorted to ground between sensor and PCM • MAP sensor power circuit is open or shorted to ground • MAP sensor is damaged or has failed • PCM has failed
DTC: P0108 **2T CCM, MIL: YES** **1995, 1996, 1997, 1998, 1999,** **2000, 2001, 2002, 2003, 2004,** **2005** **Models:** NSX, NSX-T **Engines:** All **Transmissions:** A/T, M/T	**Manifold Air Pressure Sensor Circuit High Input** Engine started, engine running in closed loop at idle speed, and the PCM detected the MAP sensor indicated a value of 29.9" Hg during the CCM test. **Note: The key on, engine off MAP sensor input should be near 2.9v.** **Possible Causes:** • MAP sensor signal circuit is open, or the ground circuit is open • MAP sensor signal circuit shorted to 5v VREF or system power • MAP sensor is damaged (due to an open circuit) or has failed • PCM has failed

DTC	Trouble Code Title, Conditions & Possible Causes
DTC: P0108 **1T CCM, MIL: YES** **1996, 1997, 1998, 1999, 2000, 2001, 2002, 2003, 2004, 2005, 2006** **Models:** 2.2CL, 2.3CL, 3.0CL, 3.2CL, 3.2TL, TL, 3.5RL, RL, Integra, MDX, RSX **Engines:** All **Transmissions:** A/T, M/T	**Manifold Air Pressure Sensor Circuit High Input** Engine started, engine running in closed loop at idle speed, and the PCM detected the MAP sensor indicated a value of 29.9" Hg during the CCM test. **Note: The key on, engine off MAP sensor input should be near 2.9v.** **Possible Causes:** • MAP sensor signal circuit is open, or the ground circuit is open • MAP sensor signal circuit shorted to 5v VREF or system power • MAP sensor is damaged (due to an open circuit) or has failed • PCM has failed
DTC: P0108 **1T CCM, MIL: YES** **1996, 1997, 1998, 1999** **Models:** SLX **Engines:** 3.2L VIN V, 3.5L VIN X **Transmissions:** A/T	**Manifold Air Pressure Sensor Circuit High Input** DTC P0121, P0122 and P0123 not set, engine started, engine speed less than 1000 rpm and the throttle angle less than 3%, or with engine speed more than 1000 rpm and the throttle angle less than 10%, the PCM detected the MAP sensor was more than 4.40v (90 kPa) for 10 seconds over a 16 second period during the test. **Possible Causes:** • MAP sensor circuit is open between the sensor and the PCM • MAP sensor signal circuit is shorted to VREF or system power • MAP sensor ground circuit is open between sensor and PCM • MAP sensor is damaged or has failed • PCM has failed
DTC: P0111 **2T CCM, MIL: YES** **1997, 1998** **Models:** 2.5TL **Engines:** All **Transmissions:** A/T, M/T	**Intake Air Temperature Sensor Range/Performance** Engine started, engine runtime over 10 minutes, and the PCM detected too large a change in the IAT sensor signal in too short a time during the CCM Rationality test. **Possible Causes:** • IAT sensor ground circuit has high resistance • IAT sensor signal circuit has high resistance • IAT sensor is damaged or has failed • PCM has failed
DTC: P0111 **2T CCM, MIL: YES** **1995, 1996, 1997, 1998, 1999, 2000, 2001, 2002, 2003, 2004, 2005** **Models:** NSX, NSX-T **Engines:** All **Transmissions:** A/T, M/T	**Intake Air Temperature Sensor Range/Performance** Engine started, engine runtime over 10 minutes, and the PCM detected too large a change in the IAT sensor signal in too short a time during the CCM test. **Possible Causes:** • IAT sensor ground circuit has high resistance • IAT sensor signal circuit has high resistance • IAT sensor is damaged or has failed • PCM has failed
DTC: P0111 **2T CCM, MIL: YES** **1996, 1997, 1998** **Models:** Integra, 2.2CL, 2.3CL, 3.2TL, **3.5RL** **Engines:** All **Transmissions:** A/T, M/T	**Intake Air Temperature Sensor Range/Performance** Engine started, engine runtime over 10 minutes, and the PCM detected too large a change in the IAT sensor signal in too short a time during the CCM test. **Possible Causes:** • IAT sensor ground circuit has high resistance • IAT sensor signal circuit has high resistance • IAT sensor is damaged or has failed • PCM has failed
DTC: P0112 **2T CCM, MIL: YES** **1995, 1996, 1997, 1998, 1999, 2000, 2001, 2002, 2003, 2004, 2005** **Models:** NSX, NSX-T **Engines:** All **Transmissions:** A/T, M/T	**Intake Air Temperature Sensor Circuit Low Input** Key on or engine running; and the PCM detected the Intake Air Temperature (IAT) sensor signal indicated less than 0.1v (Scan Tool reads 302°F) during the CCM test. **Possible Causes:** • IAT sensor signal shorted to chassis ground • IAT sensor signal shorted to sensor ground circuit • IAT sensor has an internal failure (it is shorted) or has failed • PCM has failed
DTC: P0112 **2T CCM, MIL: YES** **1995, 1996, 1997, 1998, 1999, 2000, 2001, 2002, 2003, 2004, 2005, 2006** **Models:** 2.5TL, 3.2TL, TL **Engines:** All **Transmissions:** A/T, M/T	**Intake Air Temperature Sensor Circuit Low Input** Key on or engine running; and the PCM detected the Intake Air Temperature (IAT) sensor signal indicated less than 0.1v (Scan Tool reads 302°F) during the CCM test. **Possible Causes:** • IAT sensor signal shorted to chassis ground • IAT sensor signal shorted to sensor ground circuit • IAT sensor has an internal failure (it is shorted) or has failed • PCM has failed

DTC	Trouble Code Title, Conditions & Possible Causes
DTC: P0112 **1T CCM, MIL: YES** **1996, 1997, 1998, 1999, 2000, 2001, 2002, 2003, 2004, 2005, 2006** **Models:** 2.2CL, 2.3CL, 3.0CL, 3.2CL, 3.5RL, RL, Integra, MDX, RSX **Engines:** All **Transmissions:** A/T, M/T	**Intake Air Temperature Sensor Circuit Low Input** Key on or engine running; and the PCM detected the Intake Air Temperature (IAT) sensor signal indicated less than 0.1v (Scan Tool reads 302°F) during the CCM test. **Possible Causes:** • IAT sensor signal shorted to chassis ground • IAT sensor signal shorted to sensor ground circuit • IAT sensor has an internal failure (it is shorted) or has failed • PCM has failed
DTC: P0112 **1T CCM, MIL: YES** **1996, 1997, 1998, 1999** **Models:** SLX **Engines:** 3.2L VIN V, 3.5L VIN X **Transmissions:** A/T	**Intake Air Temperature Sensor Circuit Low Input** DTC P0502 not set, engine started, engine runtime over 2 minutes, vehicle speed over 30 mph, and the PCM detected the IAT sensor indicated less than 0.10v (Scan Tool reads 298°F) for 12.5 seconds over a 20 second period during the CCM test. **Possible Causes:** • IAT sensor circuit shorted to ground between sensor and PCM • IAT sensor is damaged, out-of-calibration or has failed • PCM has failed
DTC: P0113 **2T CCM, MIL: YES** **1995, 1996, 1997, 1998, 1999, 2000, 2001, 2002, 2003, 2004, 2005** **Models:** NSX, NSX-T **Engines:** All **Transmissions:** A/T, M/T	**Intake Air Temperature Sensor Circuit High Input** Key on or engine running; and the PCM detected the Intake Air Temperature (IAT) sensor signal indicated more than 4.90v (Scan Tool reads -4°F) during the CCM test. **Possible Causes:** • IAT sensor signal shorted to VREF or system power • IAT sensor signal circuit is open • IAT sensor ground circuit is open • Sensor has an internal failure (it is open) • PCM has failed
DTC: P0113 **2T CCM, MIL: YES** **1995, 1996, 1997, 1998, 1999, 2000, 2001, 2002, 2003, 2004, 2005, 2006** **Models:** 2.5TL, 3.2TL, TL **Engines:** All **Transmissions:** A/T, M/T	**Intake Air Temperature Sensor Circuit High Input** Key on or engine running; and the PCM detected the Intake Air Temperature (IAT) sensor signal indicated more than 4.90v (Scan Tool reads -4°F) during the CCM test. **Possible Causes:** • IAT sensor signal shorted to VREF or system power • IAT sensor signal circuit is open • IAT sensor ground circuit is open • Sensor has an internal failure (it is open) • PCM has failed
DTC: P0113 **1T CCM, MIL: YES** **1996, 1997, 1998, 1999, 2000, 2001, 2002, 2003, 2004, 2005, 2006** **Models:** 2.2CL, 2.3CL, 3.0CL, 3.2CL, 3.5RL, RL, Integra, MDX, RSX **Engines:** All **Transmissions:** A/T, M/T	**Intake Air Temperature Sensor Circuit High Input** Key on or engine running; and the PCM detected the Intake Air Temperature (IAT) sensor signal indicated more than 4.90v (Scan Tool reads -4°F) during the CCM test. **Possible Causes:** • IAT sensor signal shorted to VREF or system power • IAT sensor signal circuit is open • IAT sensor ground circuit is open • Sensor has an internal failure (it is open) • PCM has failed
DTC: P0113 **1T CCM, MIL: YES** **1996, 1997, 1998, 1999** **Models:** SLX **Engines:** 3.2L VIN V, 3.5L VIN X **Transmissions:** A/T	**Intake Air Temperature Sensor Circuit High Input** DTC P0502 not set, engine started, engine runtime over 4 minutes, vehicle speed over 20 mph, ECT sensor more than 140°F, MAF sensor less than 20 g/sec, and the PCM detected the IAT sensor indicate more than 4.90v (Scan Tool reads -38°F) for 12.5 second over a 25 second period during the CCM test. **Possible Causes:** • IAT sensor signal circuit is open between the sensor and PCM • IAT sensor ground circuit is open between the sensor and PCM • IAT sensor is damaged, out-of-calibration or has failed • PCM has failed

DTC	Trouble Code Title, Conditions & Possible Causes
DTC: P0116 **2T CCM, MIL: YES** **1995, 1996, 1997, 1998, 1999,** **2000, 2001, 2002, 2003, 2004,** **2005** **Models:** NSX, NSX-T **Engines:** All **Transmissions:** A/T, M/T	**Engine Coolant Temperature Sensor Range/Performance** DTC P0116 and P0117 not set, Engine started, and the PCM detected too much change in the ECT sensor signal in too short a period of time during the CCM Rationality test. **Note: The ECT sensor should read 0.47v-0.78v at hot idle speed.** **Possible Causes:** • ECT sensor ground circuit is open (fault may be intermittent) • ECT sensor signal circuit is open (fault may be intermittent) • ECT sensor is damaged or has failed (fault may be intermittent) • PCM has failed
DTC: P0116 **2T CCM, MIL: YES** **1995, 1996, 1997, 1998, 1999,** **2000, 2001, 2002, 2003, 2004,** **2005, 2006** **Models:** 2.5TL, 3.2TL, TL **Engines:** All **Transmissions:** A/T, M/T	**Engine Coolant Temperature Sensor Range/Performance** DTC P0116 and P0117 not set, Engine started, and the PCM detected too much change in the ECT sensor signal in too short a period of time during the CCM Rationality test. **Note: The ECT sensor should read 0.47v-0.78v at hot idle speed.** **Possible Causes:** • ECT sensor ground circuit is open (fault may be intermittent) • ECT sensor signal circuit is open (fault may be intermittent) • ECT sensor is damaged or has failed (fault may be intermittent) • PCM has failed
DTC: P0116 **1T CCM, MIL: YES** **1996, 1997, 1998, 1999, 2000,** **2001, 2002, 2003, 2004, 2005,** **2006** **Models:** 2.2CL, 2.3CL, 3.0CL, 3.2CL, 3.5RL, RL, Integra, MDX, RSX **Engines:** All **Transmissions:** A/T, M/T	**Engine Coolant Temperature Sensor Range/Performance** DTC P0116 and P0117 not set, Engine started, and the PCM detected too much change in the ECT sensor signal in too short a period of time during the CCM Rationality test. **Note: The ECT sensor should read 0.47v-0.78v at hot idle speed.** **Possible Causes:** • ECT sensor ground circuit is open (fault may be intermittent) • ECT sensor signal circuit is open (fault may be intermittent) • ECT sensor is damaged or has failed (fault may be intermittent) • PCM has failed
DTC: P0117 **2T CCM, MIL: YES** **1995, 1996, 1997, 1998, 1999,** **2000, 2001, 2002, 2003, 2004,** **2005** **Models:** NSX, NSX-T **Engines:** All **Transmissions:** A/T, M/T	**Engine Coolant Temperature Sensor Circuit Low Input** Key on or engine running; and the PCM detected the Engine Coolant Temperature (ECT) sensor indicated less than 0.10v (Scan Tool reads more than 302°F) during the CCM test. **Note: The ECT sensor should read 0.47v-0.78v at hot idle speed.** **Possible Causes:** • ECT sensor signal is shorted to chassis ground • ECT sensor signal is shorted to sensor ground circuit • ECT sensor is damaged or has failed (it is shorted internally) • PCM has failed
DTC: P0117 **2T CCM, MIL: YES** **1995, 1996, 1997, 1998, 1999,** **2000, 2001, 2002, 2003, 2004,** **2005, 2006** **Models:** 2.5TL, 3.2TL, TL **Engines:** All **Transmissions:** A/T, M/T	**Engine Coolant Temperature Sensor Circuit Low Input** Key on or engine running; and the PCM detected the Engine Coolant Temperature (ECT) sensor indicated less than 0.10v (Scan Tool reads more than 302°F) during the CCM test. **Note: The ECT sensor should read 0.47v-0.78v at hot idle speed.** **Possible Causes:** • ECT sensor signal is shorted to chassis ground • ECT sensor signal is shorted to sensor ground circuit • ECT sensor is damaged or has failed (it is shorted internally) • PCM has failed
DTC: P0117 **1T CCM, MIL: YES** **1996, 1997, 1998, 1999, 2000,** **2001, 2002, 2003, 2004, 2005,** **2006** **Models:** 2.2CL, 2.3CL, 3.0CL, 3.2CL, 3.5RL, RL, Integra, MDX, RSX **Engines:** All **Transmissions:** A/T, M/T	**Engine Coolant Temperature Sensor Circuit Low Input** Key on or engine running; and the PCM detected the Engine Coolant Temperature (ECT) sensor indicated less than 0.10v (Scan Tool reads more than 302°F) during the CCM test. **Note: The ECT sensor should read 0.47v-0.78v at hot idle speed.** **Possible Causes:** • ECT sensor signal is shorted to chassis ground • ECT sensor signal is shorted to sensor ground circuit • ECT sensor is damaged or has failed (it is shorted internally) • PCM has failed

DTC	Trouble Code Title, Conditions & Possible Causes
DTC: P0117 **1T CCM, MIL: YES** **1996, 1997, 1998, 1999** **Models:** SLX **Engines:** 3.2L VIN V, 3.5L VIN X **Transmissions:** A/T	**Engine Coolant Temperature Sensor Circuit Low Input** Engine started, engine runtime over 1 minute, and the PCM detected the ECT sensor indicated less than 0.10v (Scan Tool reads 302°F) for 6.25 seconds for 50 seconds over a 100 second period during the CCM test. **Possible Causes:** • ECT sensor circuit shorted to ground between sensor and PCM • ECT sensor is damaged, out-of-calibration or has failed • PCM has failed
DTC: P0118 **2T CCM, MIL: YES** **1995, 1996, 1997, 1998, 1999,** **2000, 2001, 2002, 2003, 2004,** **2005** **Models:** NSX, NSX-T **Engines:** All **Transmissions:** A/T, M/T	**Engine Coolant Temperature Sensor Circuit High Input** Key on or engine running; and the PCM detected the Engine Coolant Temperature (ECT) sensor indicated more than 4.90v (Scan Tool reads less than -4°F) during the CCM test. **Note: The ECT sensor should read 0.47v-0.78v at hot idle speed.** **Possible Causes:** • ECT sensor signal circuit is open between the sensor and PCM • ECT sensor signal circuit is shorted to VREF or system power • ECT sensor signal circuit is open between sensor and the PCM • ECT sensor is damaged or has failed (it is open internally) • PCM has failed
DTC: P0118 **2T CCM, MIL: YES** **1995, 1996, 1997, 1998, 1999,** **2000, 2001, 2002, 2003, 2004,** **2005, 2006** **Models:** 2.5TL, 3.2TL, TL **Engines:** All **Transmissions:** A/T, M/T	**Engine Coolant Temperature Sensor Circuit High Input** Key on or engine running; and the PCM detected the Engine Coolant Temperature (ECT) sensor indicated more than 4.90v (Scan Tool reads less than -4°F) during the CCM test. **Note: The ECT sensor should read 0.47v-0.78v at hot idle speed.** **Possible Causes:** • ECT sensor signal circuit is open between the sensor and PCM • ECT sensor signal circuit is shorted to VREF or system power • ECT sensor signal circuit is open between sensor and the PCM • ECT sensor is damaged or has failed (it is open internally) • PCM has failed
DTC: P0118 **1T CCM, MIL: YES** **1996, 1997, 1998, 1999, 2000,** **2001, 2002, 2003, 2004, 2005,** **2006** **Models:** 2.2CL, 2.3CL, 3.0CL, 3.2CL, 3.5RL, RL, Integra, MDX, RSX **Engines:** All **Transmissions:** A/T, M/T	**Engine Coolant Temperature Sensor Circuit High Input** Key on or engine running; and the PCM detected the Engine Coolant Temperature (ECT) sensor indicated more than 4.90v (Scan Tool reads less than -4°F) during the CCM test. **Note: The ECT sensor should read 0.47v-0.78v at hot idle speed.** **Possible Causes:** • ECT sensor signal circuit is open between the sensor and PCM • ECT sensor signal circuit is shorted to VREF or system power • ECT sensor signal circuit is open between sensor and the PCM • ECT sensor is damaged or has failed (it is open internally) • PCM has failed
DTC: P0118 **1T CCM, MIL: YES** **1996, 1997, 1998, 1999** **Models:** SLX **Engines:** 3.2L VIN V, 3.5L VIN X **Transmissions:** A/T	**Engine Coolant Temperature Sensor Circuit High Input** Engine started, engine runtime over 90 seconds, and the PCM detected the ECT sensor was more than 4.90v (Scan Tool reads -38°F) for 50 seconds over a 100 second period during the CCM test. **Possible Causes:** • ECT sensor signal circuit is open between the sensor and PCM • ECT sensor ground circuit open between the sensor and PCM • ECT sensor is damaged, out-of-calibration or has failed • PCM has failed
DTC: P0121 **2T CCM, MIL: YES** **1996, 1997, 1998, 1999** **Models:** SLX **Engines:** 3.2L VIN V, 3.5 VIN X **Transmissions:** A/T	**Throttle Position Sensor Range/Performance** DTC P0121, P0122, P0123 and P1122 not set, engine started, MAP sensor less than 55 kPa, throttle angle stable (±1%), and the PCM detected the Actual throttle angle was not close to the Predicted throttle angle for 12.5 seconds over a 25 second period in the test. **Possible Causes:** • TP sensor signal circuit is open to the PCM (intermittent fault) • TP sensor ground circuit is open (an intermittent fault) • MAP sensor damaged or out-of-calibration • Throttle body is damaged or throttle linkage is bent or binding • TP sensor is damaged or has failed

DTC	Trouble Code Title, Conditions & Possible Causes
DTC: P0122 2T CCM, MIL: YES 1995, 1996, 1997, 1998, 1999, 2000, 2001, 2002, 2003, 2004, 2005 Models: NSX, NSX-T Engines: All Transmissions: A/T, M/T	**Throttle Position Sensor Circuit Low Input** Engine started, engine running in closed loop conditions, and the PCM detected the closed throttle TP signal was less than 0.16v (Scan Tool reads less than 10% open) during the CCM test. **Possible Causes:** • TP sensor signal circuit is shorted to ground • TP sensor VREF circuit is open • TP sensor VREF circuit is shorted to ground • TP sensor is damaged (it may be shorted internally) • PCM has failed
DTC: P0122 2T CCM, MIL: YES 1995, 1996, 1997, 1998, 1999, 2000, 2001, 2002, 2003, 2004, 2005, 2006 Models: 2.5TL, 3.2TL, TL Engines: All Transmissions: A/T, M/T	**Throttle Position Sensor Circuit Low Input** Engine started, engine running in closed loop conditions, and the PCM detected the closed throttle TP signal was less than 0.16v (Scan Tool reads less than 10% open) during the CCM test. **Possible Causes:** • TP sensor signal circuit is shorted to ground • TP sensor VREF circuit is open • TP sensor VREF circuit is shorted to ground • TP sensor is damaged (it may be shorted internally) • PCM has failed
DTC: P0122 1T CCM, MIL: YES 1996, 1997, 1998, 1999, 2000, 2001, 2002, 2003, 2004, 2005, 2006 Models: 2.2CL, 2.3CL, 3.0CL, 3.2CL, 3.5RL, RL, Integra, MDX, RSX Engines: All Transmissions: A/T, M/T	**Throttle Position Sensor Circuit Low Input** Engine started, engine running in closed loop at idle speed, and the PCM detected the TP signal indicated less than 0.16v (Scan Tool reads less than 10%) during the CCM test. **Possible Causes:** • TP sensor signal circuit is shorted to ground • TP sensor VREF circuit is open • TP sensor VREF circuit is shorted to ground • TP sensor is damaged (it may be shorted internally) • PCM has failed
DTC: P0122 1T CCM, MIL: YES 1996, 1997, 1998, 1999 Models: SLX Engines: 3.2L VIN V, 3.5L VIN X Transmissions: A/T	**Throttle Position Sensor Circuit Low Input** Key on or engine running; and the PCM detected the TP sensor indicated less than 0.22v for 0.78 seconds over a 1.5 second period. **Possible Causes:** • TP sensor signal circuit is shorted to sensor or chassis ground • TP sensor VREF circuit is open between sensor and the PCM • TP sensor is damaged or has failed • PCM has failed
DTC: P0123 2T CCM, MIL: YES 1995, 1996, 1997, 1998, 1999, 2000, 2001, 2002, 2003, 2004, 2005 Models: NSX, NSX-T Engines: All Transmissions: A/T, M/T	**Throttle Position Sensor Circuit High Input** Engine started, and after a momentary WOT condition, the PCM detected the TP signal indicated over 4.60v (Scan Tool over 90%). **Possible Causes:** • TP sensor signal circuit is shorted to VREF • TP sensor signal circuit is shorted to system power • TP sensor ground circuit is open between sensor and the PCM • TP sensor is damaged or has failed • PCM has failed
DTC: P0123 2T CCM, MIL: YES 1995, 1996, 1997, 1998, 1999, 2000, 2001, 2002, 2003, 2004, 2005, 2006 Models: 2.5TL, 3.2TL, TL Engines: All Transmissions: A/T, M/T	**Throttle Position Sensor Circuit High Input** Engine started, engine running under a momentary WOT condition, and the PCM detected TP signal was more than 4.60v (Scan Tool reads more than 90%) during the CCM test. **Possible Causes:** • TP sensor signal circuit is shorted to VREF • TP sensor signal circuit is shorted to system power • TP sensor ground circuit is open between sensor and the PCM • TP sensor is damaged or has failed • PCM has failed
DTC: P0123 1T CCM, MIL: YES 1996, 1997, 1998, 1999, 2000, 2001, 2002, 2003, 2004, 2005, 2006 Models: 2.2CL, 2.3CL, 3.0CL, 3.2CL, 3.5RL, RL, Integra, MDX, RSX Engines: All Transmissions: A/T, M/T	**Throttle Position Sensor Circuit High Input** Engine started, engine running under a momentary WOT condition, and the PCM detected the TP signal was more than 4.60v (Scan Tool reads more than 90%) during the CCM test. **Possible Causes:** • TP sensor signal circuit is shorted to VREF • TP sensor signal circuit is shorted to system power • TP sensor ground circuit is open between sensor and the PCM • TP sensor is damaged or has failed • PCM has failed

DTC	Trouble Code Title, Conditions & Possible Causes
DTC: P0123 **2T CCM, MIL: YES** **1996, 1997, 1998, 1999** **Models:** SLX **Engines:** 3.2L VIN V, 3.5L VIN X **Transmissions:** A/T	**Throttle Position Sensor Circuit High Input** Key on or engine running; and the PCM detected the Throttle Position (TP) sensor was more than 4.88v for 0.78 seconds out of a 1.5 second period during the CCM test. **Possible Causes:** • TP sensor signal circuit is open between the sensor and PCM • TP sensor signal circuit is shorted to VREF or system power • TP sensor ground circuit is open between the sensor and PCM • TP sensor is damaged or has failed • PCM has failed
DTC: P0125 **1T ECT, MIL: YES** **2003, 2004, 2005, 2006** **Models:** MDX **Engines:** All **Transmissions:** A/T, M/T	**Engine Coolant Temperature Sensor Slow Response** Engine started, engine running at road load for 10 minutes, and the PCM detected the ECT sensor input did not reach the correct value. **Possible Causes:** • Inspect for low coolant level or for an incorrect coolant mixture • ECT sensor signal circuit is open or shorted (intermittent fault) • ECT sensor is damaged or has failed • PCM has failed
DTC: P0125 **2T ECT, MIL: YES** **1996, 1997, 1998, 1999** **Models:** SLX **Engines:** 3.2L VIN V, 3.5L VIN X **Transmissions:** A/T	**Insufficient Coolant Temperature For Closed Loop** DTC P0112, P0113, P0117, P0118, P1111, P1112, P1114 and P1115 not set, cold engine startup (ECT sensor from 14-82°F), IAT sensor from 17-50°F during testing, and the PCM detected the ECT signal did not reach 84°F after 20 minutes of driving; or with the IAT sensor more than 50°F at startup, the PCM detected the ECT sensor indicated less than 84°F after 2 minutes during the CCM test. **Note: This must fail at least 20 times to set this trouble code.** **Possible Causes:** • Inspect for low coolant level or an incorrect coolant mixture • Check the operation of the thermostat (it may be stuck open) • ECT sensor is damaged or out-of-calibration (it is "skewed") • ECT sensor signal circuit has high resistance • ECT sensor has failed • PCM has failed
DTC: P0128 **2T ECT, MIL: YES** **2000, 2001, 2002, 2003, 2004, 2005, 2006** **Models:** 3.2CL, 3.2TL, TL **Engines:** All **Transmissions:** A/T, M/T	**Cooling System Malfunction** DTC P0107, P0108, P0112, P0113, P0116, P0117, P0118, P0335, P0336, P0300, P0301-P0306, P0335, P0336, P0401, P0505, P1106, P0117, P1108, P1129, P1259 and P1519, engine started, engine running at road load for 10 minutes, and the PCM detected the ECT sensor input did not reach the correct closed loop value. **Possible Causes:** • Inspect for low coolant level or for an incorrect coolant mixture • Check the operation of the thermostat (it may be stuck open) • TSB 01-016 (9/10/01) contains a repair procedure for this code
DTC: P0128 **2T ECT, MIL: YES** **2001, 2002, 2003, 2004, 2005, 2006** **Models:** MDX, RSX **Engines:** All **Transmissions:** A/T, M/T	**Cooling System Malfunction** DTC P0107, P0108, P0112, P0113, P0116, P0117, P0118, P0335, P0336, P0300, P0301-P0306, P0335, P0336, P0401, P0505, P1106, P0117, P1108, P1129, P1259 and P1519, engine started, engine running at road load for 10 minutes, and the PCM detected the ECT sensor input did not reach the correct closed loop value. **Possible Causes:** • Inspect for low coolant level or for an incorrect coolant mixture • Check the operation of the thermostat (it may be stuck open) • TSB 02-009 (3/11/02) contains a repair procedure for this code
DTC: P0130 **2T CCM, MIL: YES** **1995, 1996, 1997, 1998** **Models:** 2.5TL **Engines:** All **Transmissions:** A/T, M/T	**HO2S-11 (Bank 1 Sensor 1) Circuit Malfunction** Engine started, engine running in closed loop at cruise speed, and then back to idle speed, and the PCM detected an unexpected "low" or "high" voltage condition on the HO2S circuit during the CCM test. **Possible Causes:** • HO2S signal circuit is open or it is shorted to ground • HO2S signal circuit is shorted to system power (B+) • HO2S may be contaminated or may have failed • PCM has failed

DTC	Trouble Code Title, Conditions & Possible Causes
DTC: P0131 **2T CCM, MIL: YES** **1995, 1996, 1997, 1998, 1999,** **2000, 2001, 2002, 2003, 2004,** **2005** **Models:** NSX, NSX-T **Engines:** All **Transmissions:** A/T, M/T	**HO2S-11 (Bank 1 Sensor 1) Circuit Low Input** Engine running started, vehicle driven in 4th gear at cruise speed, and then back to idle speed, and the PCM detected the HO2S signal was fixed at less than 0.50v. **Possible Causes:** • HO2S signal circuit is open or it is shorted to ground • HO2S may be contaminated or may have failed • Fuel supply system is too lean (fuel filter is clogged or dirty) • PCM has failed
DTC: P0131 **2T CCM, MIL: YES** **1996, 1997, 1998, 1999, 2000,** **2001, 2002, 2003, 2004, 2005,** **2006** **Models:** 3.2TL, TL **Engines:** All **Transmissions:** A/T, M/T	**HO2S-11 (Bank 1 Sensor 1) Circuit Low Input** Engine started, vehicle driven in 4th gear at cruise speed, and the PCM detected the HO2S signal was fixed at less than 0.50v. **Possible Causes:** • HO2S signal circuit is open or it is shorted to ground • HO2S may be contaminated or may have failed • Fuel supply system is too lean (fuel filter is clogged or dirty) • PCM has failed
DTC: P0131 **1T CCM, MIL: YES** **1996, 1997, 1998, 1999, 2000,** **2001, 2002, 2003, 2004, 2005,** **2006** **Models:** 2.2CL, 2.3CL, 3.0CL, 3.2CL, 3.5RL, RL, Integra, MDX **Engines:** All **Transmissions:** A/T, M/T	**HO2S-11 (Bank 1 Sensor 1) Circuit Low Input** Engine started, vehicle driven in closed loop in 4th gear at cruise speed, and the PCM detected the HO2S signal was fixed at less than 0.50v (the actual value is stored in the PCM memory). **Possible Causes:** • HO2S signal circuit is open or it is shorted to ground • HO2S may be contaminated or may have failed • Fuel supply system is too lean (fuel filter is clogged or dirty) • PCM has failed
DTC: P0131 **1T CCM, MIL: YES** **1996, 1997, 1998, 1999** **Models:** SLX **Engines:** 3.2L VIN V, 3.5L VIN X **Transmissions:** A/T	**HO2S-11 (Bank 1 Sensor 1) Circuit Low Input** DTC P0106, P0107, P0108, P0112, P0113, P0117, P0118, P0121, P0122, P0123, P0171, P0172 and P0300, P0301-P0306 not set, engine running in closed loop with the A/F ratio from 14.5-14.8:1, ECT sensor more than 140°F, throttle angle from 3-19% for 5 seconds, and the PCM detected the HO2S-11 signal was less than 26 mv for 77 seconds over a 90 second period during the CCM test. **Possible Causes:** • HO2S signal circuit is open • HO2S signal circuit is shorted to ground • HO2S element is water or fuel contaminated, or it has failed • PCM has failed
DTC: P0132 **2T CCM, MIL: YES** **1995, 1996, 1997, 1998, 1999,** **2000, 2001, 2002, 2003, 2004,** **2005** **Models:** NSX, NSX-T **Engines:** All **Transmissions:** A/T, M/T	**HO2S-11 (Bank 1 Sensor 1) Circuit High Input** Engine started, vehicle driven in closed loop in D4 or D5 position at cruise speed, and the PCM detected the HO2S signal was fixed at more than 0.90v (the actual value is stored in the PCM memory). **Possible Causes:** • HO2S signal tracking (wet/oily) in connector causing a short between the signal circuit and heater power circuit • HO2S signal circuit is open, or the ground circuit is open • HO2S heater supply circuit is open • PCM has failed
DTC: P0132 **1T CCM, MIL: YES** **1996, 1997, 1998, 1999, 2000,** **2001, 2002, 2003, 2004, 2005,** **2006** **Models:** 2.2CL, 2.3CL, 3.0CL, 3.2CL, 3.2TL, TL, 3.5RL, RL, MDX **Engines:** All **Transmissions:** A/T, M/T	**HO2S-11 (Bank 1 Sensor 1) Circuit High Input** Engine started, vehicle driven in D4 or D5 position in closed loop at cruise speed, and the PCM detected the HO2S signal was fixed at more than 0.90v (the actual value is stored in the PCM memory). **Possible Causes:** • HO2S signal tracking (wet/oily) in connector causing a short between the signal circuit and heater power circuit • HO2S signal circuit is open, or the ground circuit is open • HO2S heater supply circuit is open • PCM has failed

DTC	Trouble Code Title, Conditions & Possible Causes
DTC: P0132 **1T CCM, MIL: YES** **1996, 1997, 1998, 1999, 2000, 2001** **Models:** Integra **Engines:** All **Transmissions:** A/T, M/T	**HO2S-11 (Bank 1 Sensor 1) Circuit High Input** Engine started, vehicle driven in D4 or D5 position, or in 4th gear in closed loop at cruise speed, and the PCM detected the HO2S signal was fixed at more than 0.90v (the actual value is stored in the PCM). **Possible Causes:** • HO2S signal tracking (wet/oily) in connector causing a short between the signal circuit and the heater power circuit (B+) • HO2S ground circuit is open between the sensor and ground • HO2S heater supply circuit is open • PCM has failed • TSB 01-025 (10/29/01) contains a repair procedure for this code
DTC: P0132 **1T CCM, MIL: YES** **1996, 1997, 1998, 1999** **Models:** SLX **Engines:** 3.2L VIN V, 3.5L VIN X **Transmissions:** A/T	**HO2S-11 (Bank 1 Sensor 1) Circuit High Input** DTC P0106, P0107, P0108, P0112, P0113, P0117, P0118, P0121, P0122, P0123, P0171, P0172 and P0300, P0301-P0306 not set, engine running in closed loop with the A/F ratio from 14.5-14.8:1, throttle angle from 3-19% for 5 seconds, the PCM detected the HO2S-11 signal was less than 952 mv for 77 seconds over a 90 second period; or the HO2S-11 signal was more than 500 mv during Decel Fuel Cutoff mode for 3 seconds during the CCM test. **Possible Causes:** • HO2S signal circuit shorted to system power (oil in connector) • HO2S has water or fuel contamination • HO2S is damaged or it has failed • PCM has failed
DTC: P0133 **2T O2S1, MIL: YES** **1995, 1996, 1997, 1998, 1999, 2000, 2001, 2002, 2003, 2004, 2005** **Models:** NSX, NSX-T **Engines:** All **Transmissions:** A/T, M/T	**HO2S-11 (Bank 1 Sensor 1) Circuit Slow Response** Engine started, vehicle driven in closed loop in D4/D5 position at over 55 mph at steady speed, and the PCM detected the HO2S response time to switch between 300-600 mv was too slow, or that the rich to lean or lean to rich switch time was too slow. **Possible Causes:** • Exhaust leak present in the exhaust manifold or exhaust pipes • O2S element fuel contamination • O2S element has deteriorated • PCM has failed
DTC: P0133 **2T O2S1, MIL: YES** **1995, 1996, 1997, 1998** **Models:** 2.5TL **Engines:** All **Transmissions:** A/T, M/T	**HO2S-11 (Bank 1 Sensor 1) Circuit Slow Response** Engine started, vehicle driven in closed loop in 4th gear at over 55 mph at steady speed, and the PCM detected the HO2S response time to switch between 300-600 mv was too slow, or that the rich to lean or lean to rich switch time was too slow. **Possible Causes:** • Exhaust leak present in the exhaust manifold or exhaust pipes • O2S element fuel contamination • O2S element has deteriorated • PCM has failed
DTC: P0133 **2T O2S1, MIL: YES** **1996, 1997, 1998, 1999, 2000, 2001, 2002, 2003, 2004, 2005, 2006** **Models:** 2.2CL, 2.3CL, 3.0CL, 3.2CL, 3.2TL, TL, 3.5RL, RL, Integra, MDX **Engines:** All **Transmissions:** A/T, M/T	**HO2S-11 (Bank 1 Sensor 1) Circuit Slow Response** Engine started, vehicle driven in closed loop in 4th or 6th gear at over 55 mph at steady speed, and the PCM detected the HO2S response time to switch between 300-600 mv was too slow, or that the rich to lean or lean to rich switch time was too slow. **Possible Causes:** • Exhaust leak present in the exhaust manifold or exhaust pipes • O2S element fuel contamination • O2S element has deteriorated • PCM has failed
DTC: P0133 **2T O2S1, MIL: YES** **2003, 2004, 2005, 2006** **Models:** MDX **Engines:** All **Transmissions:** A/T, M/T	**Air Fuel Sensor (Bank 1 Sensor 1) Circuit Slow Response** Engine started, ECT sensor input more than 158°F, vehicle driven at an engine speed from 1200-2250 rpm at a speed over 48 mph, and the PCM detected the AFS1 signal response time was too slow. **Possible Causes:** • Exhaust leak present in the exhaust manifold or exhaust pipes • AFS1 element is loose at its mounting location • AFS1 element is contaminated or deteriorated, or it has failed • PCM has failed

DTC	Trouble Code Title, Conditions & Possible Causes
DTC: P0133 **2T O2S2, MIL: YES** **1996, 1997, 1998, 1999** **Models:** SLX **Engines:** 3.2L VIN V, 3.5L VIN X **Transmissions:** A/T	**HO2S-11 (Bank 1 Sensor 1) Slow Response** DTC P0106, P0107, P0108, P0112, P0113, P0117, P0118, P0121, P0122, P0123, P0171, P0172 and P0300, P0301-P0306 not set, engine runtime 1 minute in closed loop, ECT sensor more than 122°F, engine speed from 1500-3000 rpm, MAF sensor from 9-42 g/sec, Purge duty cycle over 1%, conditions met for 3 seconds, then 90 seconds after entering closed loop, the PCM detected the HO2S-11 lean-to-rich average transition time was over 94 ms, or the rich-to-lean average transition time was over 105 ms during the test. **Possible Causes:** • Exhaust leak present in the exhaust manifold or exhaust pipes • HO2S element has fuel contamination • HO2S element has deteriorated • PCM has failed
DTC: P0134 **2T O2S HTR1, MIL: YES** **2003, 2004, 2005, 2006** **Models:** MDX **Engines:** All **Transmissions:** A/T, M/T	**Air Fuel Sensor (Bank 1 Sensor 1) Heater Malfunction** Engine started, ECT sensor input more than 158°F, and the PCM detected a problem in the AFS1 Heater circuit operation in the test. **Possible Causes:** • AFS1 heater power circuit is open • AFS1 heater control circuit is open or shorted to power • AFS1 (heater portion) is damaged or has failed • PCM has failed
DTC: P0134 **2T O2S1, MIL: YES** **2002, 2003, 2004, 2005, 2006** **Models:** RSX **Engines:** All **Transmissions:** A/T, M/T	**Air Fuel Sensor (Bank 1 Sensor 1) Circuit Stuck Lean** Engine started, vehicle driven in closed loop in 4th or 6th gear at over 55 mph at steady speed, and the PCM detected the AFS1 signal indicated that it was stuck in a "lean" air/fuel ratio condition. **Possible Causes:** • Exhaust leak present in the exhaust manifold or exhaust pipes • AFS1 element is loose at its mounting location • AFS1 element is contaminated or deteriorated, or it has failed • PCM has failed
DTC: P0134 **1T O2S2, MIL: YES** **1996, 1997, 1998, 1999** **Models:** SLX **Engines:** 3.2L VIN V, 3.5L VIN X **Transmissions:** A/T	**HO2S-11 (Bank 1 Sensor 1) Insufficient Activity Detected** DTC P0106, P0107, P0108, P0112, P0113, P0117, P0118, P0121, P0122, P0123, P0171, P0172 and P0300, P0301-306 not set, system voltage from 11-16v, engine runtime over 40 seconds, then after the PCM determined the Oxygen Sensor Heater test passed, it detected the HO2S-11 signal remained from 400-500 mv for 77 seconds over a 90 second period in the Oxygen Sensor Monitor test. **Possible Causes:** • Exhaust leak present in exhaust manifold or exhaust pipes • HO2S element has fuel contamination or has deteriorated • HO2S signal circuit or the ground circuit has high resistance • HO2S heater element has failed, or the heater circuit is open • PCM has failed
DTC: P0135 **1T O2S HTR1, MIL: YES** **1995, 1996, 1997, 1998, 1999, 2000, 2001, 2002, 2003, 2004, 2005** **Models:** NSX, NSX-T **Engines:** All **Transmissions:** A/T, M/T	**HO2S-11 (Bank 1 Sensor 1) Heater Circuit Malfunction** Engine started, engine runtime over 80 seconds, and the PCM detected an unexpected voltage condition on the HO2S heater circuit during the CCM test. **Possible Causes:** • Main relay power circuit to the heater is open (intermittent) • O2S heater ground circuit is open • O2S heater element has high resistance or an open circuit • O2S heater element has a shorted condition • PCM has failed
DTC: P0135 **1T O2S HTR1, MIL: YES** **2001, 2002** **Models:** MDX **Engines:** All **Transmissions:** A/T, M/T	**HO2S-11 (Bank 1 Sensor 1) Heater Circuit Malfunction** Engine started, engine runtime over 80 seconds, and the PCM detected an unexpected voltage condition on the HO2S heater circuit during the CCM test. **Possible Causes:** • HO2S heater power circuit is open (check the underhood fuse) • O2S heater ground circuit is open • O2S heater element has an open or shorted condition • PCM has failed
DTC: P0135 **2T O2S HTR1, MIL: YES** **2003, 2004, 2005, 2006** **Models:** MDX **Engines:** All **Transmissions:** A/T, M/T	**Air Fuel Sensor (Bank1 Sensor 1) Heater Circuit Malfunction** Engine started, ECT sensor input more than 158°F, and the PCM detected a problem in the AFS1 Heater circuit operation in the test. **Possible Causes:** • AFS1 heater power circuit is open (check the LAFHT 15A fuse) • AFS1 heater ground circuit is open • AFS1 heater element has high resistance or an open circuit • AFS1 heater element has a shorted condition • PCM has failed

DTC	Trouble Code Title, Conditions & Possible Causes
DTC: P0135 **1T O2S HTR1, MIL: YES** **1995, 1996, 1997, 1998, 1999,** **2000, 2001, 2002, 2003, 2004,** **2005, 2006** **Models:** 2.5TL, 3.2TL, TL **Engines:** All **Transmissions:** A/T, M/T	**HO2S-11 (Bank 1 Sensor 1) Heater Circuit Malfunction** Engine started, engine runtime over 80 seconds, and the PCM detected an unexpected voltage condition on the HO2S heater circuit during the CCM test. **Possible Causes:** • Main relay output (power) circuit to the heater is open • O2S heater ground circuit is open • O2S heater element has high resistance or an open circuit • O2S heater element has a shorted condition • PCM has failed
DTC: P0135 **1T O2S HTR1, MIL: YES** **1996, 1997, 1998, 1999, 2000,** **2001, 2002, 2003, 2004, 2005,** **2006** **Models:** 2.2CL, 2.3CL, 3.0CL, 3.2CL, 3.5RL, RL, Integra, MDX **Engines:** All **Transmissions:** A/T, M/T	**HO2S-11 (Bank 1 Sensor 1) Heater Circuit Malfunction** Engine started, engine runtime over 80 seconds, and the PCM detected an unexpected voltage condition on the HO2S heater circuit during the CCM test. **Possible Causes:** • Main relay output (power) circuit to the heater is open • O2S heater ground circuit is open • O2S heater element has high resistance or an open circuit • O2S heater element has a shorted condition • PCM has failed
DTC: P0135 **2T O2S HTR2, MIL: YES** **1996, 1997, 1998, 1999** **Models:** SLX **Engines:** 3.2L VIN V, 3.5L VIN X **Transmissions:** A/T	**HO2S-11 (Bank 1 Sensor 1) Heater Circuit Malfunction** DTC P0131 and P0132 not set, ECT and IAT sensors less than 90°F and within 14°F at startup, engine running, system voltage from 11-16v, then with the average Calculated airflow less than 15 g/sec during the test period, the PCM detected the HO2S-11 signal did not vary more than 150 mv from the bias voltage of 400 to 500 mv for up to 150 seconds during the Oxygen Sensor Heater Monitor test. **Possible Causes:** • HO2S heater power circuit is open (check the 20A heater fuse) • HO2S heater ground circuit is open • HO2S heater element has high resistance or has failed • PCM has failed
DTC: P0137 **1T CCM, MIL: YES** **1995, 1996, 1997, 1998, 1999,** **2000, 2001, 2002, 2003, 2004,** **2005** **Models:** NSX, NSX-T **Engines:** All **Transmissions:** A/T, M/T	**HO2S-12 (Bank 1 Sensor 2) Circuit Low Input** Engine started, vehicle driven in closed loop in D4/D5 position at cruise speed, and the PCM detected the HO2S signal was fixed at less than 0.30v. Note: The actual value where the code sets is in the PCM memory. **Possible Causes:** • HO2S signal circuit is open • HO2S signal circuit is shorted to ground • HO2S ground circuit is open • HO2S may be contaminated or may have failed • PCM has failed
DTC: P0137 **1T CCM, MIL: YES** **1996, 1997, 1998, 1999, 2000,** **2001, 2002, 2003, 2004, 2005,** **2006** **Models:** 2.2CL, 2.3CL, 3.0CL, 3.2CL, 3.2TL, TL, 3.5RL, RL, Integra, MDX, RSX **Engines:** All **Transmissions:** A/T, M/T	**HO2S-12 (Bank 1 Sensor 2) Circuit Low Input** Engine started, vehicle driven in 4th gear at cruise speed, and the PCM detected the HO2S signal was fixed at less than 0.30v. Note: The actual value where the code sets is in the PCM memory. **Possible Causes:** • HO2S signal circuit is open • HO2S signal circuit is shorted to ground • HO2S ground circuit is open • HO2S may be contaminated or may have failed • PCM has failed
DTC: P0137 **1T CCM, MIL: YES** **1996, 1997, 1998, 1999** **Models:** SLX **Engines:** 3.2L VIN V, 3.5L VIN X **Transmissions:** A/T	**HO2S-12 (Bank 1 Sensor 2) Circuit Low Input** DTC P0106, P0107, P0108, P0112, P0113, P0117, P0118, P0121, P0122, P0123, P0171, P0172 and P0300, P0301-P0306 not set, engine running in closed loop with the A/F ratio from 14.5-14.8:1, ECT sensor more than 140°F, throttle angle from 3-19% for 5 seconds, the PCM detected the HO2S-12 signal was less than 26 mv for 106 seconds over a 125 second period during the CCM test. **Possible Causes:** • HO2S signal circuit is open or shorted to ground • HO2S has water or fuel contamination • HO2S element is damaged or it has failed • PCM has failed

DTC	Trouble Code Title, Conditions & Possible Causes
DTC: P0138 **1T CCM, MIL: YES** **1995, 1996, 1997, 1998, 1999, 2000, 2001, 2002, 2003, 2004, 2005** **Models:** NSX, NSX-T **Engines:** All **Transmissions:** A/T, M/T	**HO2S-12 (Bank 1 Sensor 2) Circuit High Input** Engine started, vehicle driven in closed loop in D4/D5 position at cruise speed, and the PCM detected the HO2S signal was fixed at more than 0.60v. Note: The actual value where the code sets is in the PCM memory. **Possible Causes:** • HO2S signal tracking (wet/oily) in connector causing a short between the signal circuit and heater power circuit • HO2S signal circuit is shorted to system power • HO2S heater supply circuit is open • PCM has failed
DTC: P0138 **1T CCM, MIL: YES** **1996, 1997, 1998, 1999, 2000, 2001, 2002, 2003, 2004, 2005, 2006** **Models:** 2.2CL, 2.3CL, 3.0CL, 3.2CL, 3.2TL, TL, 3.5RL, RL, Integra, MDX, RSX **Engines:** All **Transmissions:** A/T, M/T	**HO2S-12 (Bank 1 Sensor 2) Circuit High Input** Engine started, vehicle driven in closed loop in 4th gear at cruise speed, and the PCM detected the HO2S signal was fixed at more than 0.60v. Note: The actual value where the code sets is in the PCM memory. **Possible Causes:** • HO2S signal tracking (wet/oily) in connector causing a short between the signal circuit and heater power circuit • HO2S signal circuit is shorted to system power • HO2S heater supply circuit is open • PCM has failed
DTC: P0138 **1T CCM, MIL: YES** **1996, 1997, 1998, 1999** **Models:** SLX **Engines:** 3.2L VIN V, 3.5L VIN X **Transmissions:** A/T	**HO2S-12 (Bank 1 Sensor 2) Circuit High Input** DTC P0106, P0107, P0108, P0112, P0113, P0117, P0118, P0121, P0122, P0123, P0171, P0172 and P0300, P0301-P0306 not set, engine running in closed loop with the A/F ratio command at 14.5-14.8:1, throttle angle from 3-19% for 5 seconds, the PCM detected the HO2S-12 signal was less than 952 mv for 106 seconds over a 125 second period; or the HO2S-12 signal was more than 500 mv during Decel Fuel Cutoff mode for 3 seconds during the CCM test. **Possible Causes:** • HO2S signal circuit shorted to system power (oil in connector) • HO2S has water or fuel contamination • HO2S is damaged or it has failed • PCM has failed
DTC: P0139 **2T O2S1, MIL: YES** **1995, 1996, 1997, 1998, 1999, 2000, 2001, 2002, 2003, 2004, 2005** **Models:** NSX, NSX-T **Engines:** All **Transmissions:** A/T, M/T	**HO2S-12 (Bank 1 Sensor 2) Circuit Slow Response** Engine started, vehicle driven in closed loop in D4/D5 position at over 55 mph at steady speed, and the PCM detected the HO2S response time to switch between 300-600 mv was too slow, or that the rich to lean or lean to rich switch time was too slow. **Possible Causes:** • Exhaust leak present in the exhaust manifold or exhaust pipes • HO2S element fuel contamination • HO2S element has deteriorated • PCM has failed
DTC: P0139 **2T O2S1, MIL: YES** **1995, 1996, 1997, 1998** **Models:** 2.5TL **Engines:** All **Transmissions:** A/T, M/T	**HO2S-12 (Bank 1 Sensor 2) Circuit Slow Response** Engine started, vehicle driven in closed loop in D4 position at over 55 mph at steady speed, and the PCM detected the HO2S response time to switch between 300-600 mv was too slow, or that the rich to lean or lean to rich switch time was too slow. **Possible Causes:** • Exhaust leak present in the exhaust manifold or exhaust pipes • HO2S element fuel contamination • HO2S element has deteriorated • PCM has failed
DTC: P0139 **2T O2S1, MIL: YES** **1996, 1997, 1998, 1999, 2000, 2001, 2002, 2003, 2004, 2005, 2006** **Models:** 2.2CL, 2.3CL, 3.0CL, 3.2CL, 3.2TL, TL, 3.5RL, RL, Integra, MDX, RSX **Engines:** All **Transmissions:** A/T, M/T	**HO2S-12 (Bank 1 Sensor 2) Circuit Slow Response** Engine started, vehicle driven in closed loop in 4th or 6th gear at over 55 mph at steady speed, and the PCM detected the HO2S response time to switch between 300-600 mv was too slow, or that the rich to lean or lean to rich switch time was too slow. **Possible Causes:** • Exhaust leak present in the exhaust manifold or exhaust pipes • HO2S element fuel contamination • HO2S element has deteriorated • PCM has failed

DTC	Trouble Code Title, Conditions & Possible Causes
DTC: P0140 **1T O2S2, MIL: YES** **1996, 1997, 1998, 1999** **Models:** SLX **Engines:** 3.2L VIN V, 3.5L VIN X **Transmissions:** A/T	**HO2S-12 (Bank 1 Sensor 2) Insufficient Activity Detected** DTC P0106, P0107, P0108, P0112, P0113, P0117, P0118, P0121, P0122, P0123, P0171, P0172 and P0300, P0301-306 not set, system voltage from 11-16v, engine runtime over 40 seconds, then after the PCM determined the Oxygen Sensor Heater test passed, it detected the HO2S-12 signal remained from 426-474 mv for 105 seconds of a 125 second period in the Oxygen Sensor Monitor test. **Possible Causes:** • Exhaust leak present in exhaust manifold or exhaust pipes • HO2S element has fuel contamination or has deteriorated • HO2S signal circuit or the ground circuit has high resistance • HO2S heater element has failed, or the heater circuit is open • PCM has failed
DTC: P0141 **1T O2S HTR1, MIL: YES** **1995, 1996, 1997, 1998, 1999,** **2000, 2001, 2002, 2003, 2004,** **2005** **Models:** NSX, NSX-T **Engines:** All **Transmissions:** A/T, M/T	**HO2S-12 (Bank 1 Sensor 2) Heater Circuit Malfunction** Engine started, engine runtime over 80 seconds, and the PCM detected an unexpected voltage condition on the HO2S heater circuit during the CCM test. **Possible Causes:** • Main relay output (power) circuit to the heater is open • O2S heater ground circuit is open • O2S heater element has high resistance or an open circuit • O2S heater element has a shorted condition • PCM has failed
DTC: P0141 **1T O2S HTR1, MIL: YES** **1995, 1996, 1997, 1998** **Models:** 2.5TL **Engines:** All **Transmissions:** A/T, M/T	**HO2S-12 (Bank 1 Sensor 2) Heater Circuit Malfunction** Engine started, engine runtime over 80 seconds, and the PCM detected an unexpected voltage condition on the HO2S heater circuit during the CCM test. **Possible Causes:** • Main relay output (power) circuit to the heater is open • O2S heater ground circuit is open • O2S heater element has high resistance or an open circuit • O2S heater element has a shorted condition • PCM has failed
DTC: P0141 **2T O2S HTR1, MIL: YES** **1996, 1997, 1998, 1999, 2000,** **2001, 2002, 2003, 2004, 2005,** **2006** **Models:** 2.2CL, 2.3CL, 3.0CL, 3.2CL, 3.2TL, TL, 3.5RL, RL, Integra, MDX, RSX **Engines:** All **Transmissions:** A/T, M/T	**HO2S-12 (Bank 1 Sensor 2) Heater Circuit Malfunction** Engine started, engine runtime over 80 seconds, and the PCM detected an unexpected voltage condition on the HO2S heater circuit during the CCM test. **Possible Causes:** • Main relay output (power) circuit to the heater is open • O2S heater ground circuit is open • O2S heater element has high resistance or an open circuit • O2S heater element has a shorted condition • PCM has failed
DTC: P0141 **2T O2S HTR2, MIL: YES** **1996, 1997, 1998, 1999** **Models:** SLX **Engines:** 3.2L VIN V, 3.5L VIN X **Transmissions:** A/T	**HO2S-12 (Bank 1 Sensor 2) Heater Circuit Malfunction** DTC P0151 and P0152 not set, ECT and IAT sensors less than 90°F and within 11°F at startup, engine running, system voltage from 10-16v, then with the average Calculated airflow less than 23 g/sec during the test period, the PCM detected the HO2S-12 signal did not vary more than 150 mv from the bias voltage of 400 to 500 mv for up to 300 seconds during the Oxygen Sensor Heater Monitor test. **Possible Causes:** • HO2S heater power circuit is open (check the 20A heater fuse) • HO2S heater ground circuit is open • HO2S heater element has high resistance or has failed • PCM has failed
DTC: P0151 **1T CCM, MIL: YES** **1995, 1996, 1997, 1998, 1999,** **2000, 2001, 2002, 2003, 2004,** **2005** **Models:** NSX, NSX-T **Engines:** All **Transmissions:** A/T, M/T	**HO2S-21 (Bank 2 Sensor 1) Circuit Low Input** Engine started, engine running in closed loop at cruise speed in D4/D5 position, and the PCM detected the HO2S signal was fixed at less than 0.10v during the CCM test. Note: The actual value where the code sets is in the PCM memory. **Possible Causes:** • HO2S signal circuit is open • HO2S signal circuit is shorted to ground • HO2S ground circuit is open • HO2S may be contaminated or may have failed • PCM has failed

DTC	Trouble Code Title, Conditions & Possible Causes
DTC: P0151 **1T CCM, MIL: YES** **1996, 1997, 1998** **Models:** 3.2TL **Engines:** All **Transmissions:** A/T, M/T	**HO2S-21 (Bank 2 Sensor 1) Circuit Low Input** Engine started, vehicle driven in closed loop in D4 position at cruise speed, and the PCM detected the HO2S signal was fixed at less than 0.10v. Note: The actual value where the code sets is in the PCM memory. **Possible Causes:** • HO2S signal circuit is open • HO2S signal circuit is shorted to ground • HO2S ground circuit is open • HO2S may be contaminated or may have failed • PCM has failed
DTC: P0151 **1T CCM, MIL: YES** **1996, 1997, 1998, 1999** **Models:** SLX **Engines:** 3.2L VIN V, 3.5L VIN X **Transmissions:** A/T	**HO2S-21 (Bank 2 Sensor 1) Circuit Low Input** DTC P0106, P0107, P0108, P0112, P0113, P0117, P0118, P0121, P0122, P0123, P0171, P0172 and P0300, P0301-P0306 not set, engine running in closed loop with the A/F ratio from 14.5-14.8:1, ECT sensor more than 140°F, throttle angle from 3-19% for 5 seconds, the PCM detected the HO2S-21 signal was less than 22 mv for 77 seconds over a 90 second period during the CCM test. **Possible Causes:** • HO2S signal circuit is open or shorted to ground • HO2S has water or fuel contamination • HO2S is damaged or it has failed • PCM has failed
DTC: P01521T CCM, MIL: YES **1995, 1996, 1997, 1998, 1999,** **2000, 2001, 2002, 2003, 2004,** **2005** **Models:** NSX, NSX-T **Engines:** All **Transmissions:** A/T, M/T	**HO2S-21 (Bank 2 Sensor 1) Circuit High Input** Engine started, vehicle driven in closed loop in D4/D5 position at cruise speed, and the PCM detected the HO2S signal was fixed at more than 0.90v. Note: The actual value where the code sets is in the PCM memory. **Possible Causes:** • HO2S signal tracking (wet/oily) in connector causing a short between the signal circuit and heater power circuit • HO2S signal circuit is open, or the ground circuit is open • HO2S heater supply circuit is open • PCM has failed
DTC: P0152 **1T CCM, MIL: YES** **1996, 1997, 1998** **Models:** 3.2TL **Engines:** All **Transmissions:** A/T, M/T	**HO2S-21 (Bank 2 Sensor 1) Circuit High Input** Engine started, vehicle driven in closed loop in D4 position at cruise speed, and the PCM detected the HO2S signal was fixed at more than 0.90v. Note: The actual value where the code sets is in the PCM memory. **Possible Causes:** • HO2S signal tracking (wet/oily) in connector causing a short between the signal circuit and heater power circuit • HO2S signal circuit is open, or the ground circuit is open • HO2S heater supply circuit is open • PCM has failed
DTC: P0152 **1T CCM, MIL: YES** **1996, 1997, 1998, 1999** **Models:** SLX **Engines:** 3.2L VIN V, 3.5L VIN X **Transmissions:** A/T	**HO2S-21 (Bank 2 Sensor 1) Circuit High Input** DTC P0106, P0107, P0108, P0112, P0113, P0117, P0118, P0121, P0122, P0123, P0171, P0172 and P0300, P0301-P0306 not set, engine running in closed loop with the A/F ratio command at 14.5-14.8:1, throttle angle from 3-19% for 5 seconds, the PCM detected the HO2S-21 signal was less than 952 mv for 77 seconds over a 90 second period; or the HO2S-21 signal was more than 500 mv during Decel Fuel Cutoff mode for 3 seconds during the CCM test. **Possible Causes:** • HO2S signal circuit shorted to system power (oil in connector) • HO2S has water or fuel contamination • HO2S is damaged or it has failed • PCM has failed
DTC: P0153 **1T O2S1, MIL: YES** **1995, 1996, 1997, 1998, 1999,** **2000, 2001, 2002, 2003, 2004,** **2005** **Models:** NSX, NSX-T **Engines:** All **Transmissions:** A/T, M/T	**HO2S-21 (Bank 2 Sensor 1) Circuit Slow Response** Engine started, vehicle driven in closed loop in D4/D5 position at 55 mph at steady speed, and the PCM detected the HO2S response time to switch between 300-600 mv was too slow, or that the rich to lean or lean to rich switch time was too slow. **Possible Causes:** • Exhaust leak present in the exhaust manifold or exhaust pipes • HO2S element fuel contamination • HO2S element has deteriorated • PCM has failed
DTC: P0153 **1T O2S1, MIL: YES** **1996, 1997, 1998** **Models:** 3.2TL **Engines:** All **Transmissions:** A/T, M/T	**HO2S-21 (Bank 2 Sensor 1) Circuit Slow Response** Engine started, vehicle driven in closed loop in D4 position at 55 mph at steady speed, and the PCM detected the HO2S response time to switch between 300-600 mv was too slow, or that the rich to lean or lean to rich switch time was too slow. **Possible Causes:** • Exhaust leak present in the exhaust manifold or exhaust pipes • HO2S element fuel contamination • HO2S element has deteriorated • PCM has failed

DTC	Trouble Code Title, Conditions & Possible Causes
DTC: P0153 **2T O2S1, MIL: YES** **2003, 2004, 2005, 2006** **Models:** MDX **Engines:** All **Transmissions:** A/T, M/T	**Air Fuel Sensor (Bank 2 Sensor 1) Circuit Slow Response** Engine started, ECT sensor signal more than 158°F, vehicle driven at an engine speed from 1200-2250 rpm at a speed over 48 mph, and the PCM detected the AFS2 signal response time was too slow. **Possible Causes:** • Exhaust leak present in the exhaust manifold or exhaust pipes • AFS2 element is loose at its mounting location • AFS2 element is contaminated or deteriorated, or it has failed • PCM has failed
DTC: P0153 **2T O2S2, MIL: YES** **1996, 1997, 1998, 1999** **Models:** SLX **Engines:** 3.2L VIN V, 3.5L VIN X **Transmissions:** A/T	**HO2S-21 (Bank 2 Sensor 1) Slow Response** DTC P0106, P0107, P0108, P0112, P0113, P0117, P0118, P0121, P0122, P0123, P0171, P0172 and P0300, P0301-P0306 not set, engine runtime 1 minute in closed loop, ECT sensor more than 122°F, engine speed from 1500-3000 rpm, MAF sensor from 9-42 g/sec, Purge duty cycle over 1%, conditions met for 3 seconds, then 90 seconds after entering closed loop, the PCM detected the HO2S-21 lean-to-rich average transition time was over 94 ms, or the rich-to-lean average transition time was over 105 ms during the Oxygen Sensor Monitor test. **Possible Causes:** • Exhaust leak present in the exhaust manifold or exhaust pipes • HO2S element has fuel contamination • HO2S element has deteriorated • PCM has failed
DTC: P0154 **2T O2S HTR1, MIL: YES** **2003, 2004, 2005, 2006** **Models:** MDX **Engines:** All **Transmissions:** A/T, M/T	**Air Fuel Sensor (Bank 2 Sensor 1) Heater Malfunction** Engine started, ECT sensor signal more than 158°F, and the PCM detected a problem in the AFS2 Heater circuit operation in the test. **Possible Causes:** • AFS2 heater power circuit is open • AFS2 heater control circuit is open or shorted to power • AFS2 (heater portion) is damaged or has failed • PCM has failed
DTC: P0154 **1T O2S2, MIL: YES** **1996, 1997, 1998, 1999** **Models:** SLX **Engines:** 3.2L VIN V, 3.5L VIN X **Transmissions:** A/T	**HO2S-21 (Bank 2 Sensor 1) Insufficient Activity Detected** DTC P0106, P0107, P0108, P0112, P0113, P0117, P0118, P0121, P0122, P0123, P0171, P0172 and P0300, P0301-306 not set, system voltage from 11-16v, engine runtime over 40 seconds, then after the PCM determined the Oxygen Sensor Heater test passed, it detected the HO2S-21 signal remained from 400-500 mv for 77 seconds over a 90 second period in the Oxygen Sensor Monitor test. **Possible Causes:** • Exhaust leak present in exhaust manifold or exhaust pipes • HO2S element has fuel contamination or has deteriorated • HO2S signal circuit or the ground circuit has high resistance • HO2S heater element has failed, or the heater circuit is open • PCM has failed
DTC: P0155 **1T O2S HTR1, MIL: YES** **1995, 1996, 1997, 1998, 1999, 2000, 2001, 2002, 2003, 2004, 2005** **Models:** NSX, NSX-T **Engines:** All **Transmissions:** A/T, M/T	**HO2S-21 (Bank 2 Sensor 1) Heater Circuit Malfunction** Engine started, engine runtime over 80 seconds, and the PCM detected an unexpected voltage condition on the HO2S heater circuit during the CCM test. **Possible Causes:** • Main relay output (power) circuit to the heater is open • O2S heater ground circuit is open • O2S heater element has high resistance or an open circuit • O2S heater element has a shorted condition • PCM has failed
DTC: P0155 **1T O2S HTR1, MIL: YES** **1996, 1997, 1998** **Models:** 3.2TL **Engines:** All **Transmissions:** A/T, M/T	**HO2S-21 (Bank 2 Sensor 1) Heater Circuit Malfunction** Engine started, engine runtime over 80 seconds, and the PCM detected an unexpected voltage condition on the HO2S heater circuit during the CCM test. **Possible Causes:** • Main relay output (power) circuit to the heater is open • O2S heater ground circuit is open • O2S heater element has high resistance or an open circuit • O2S heater element has a shorted condition • PCM has failed
DTC: P0155 **2T O2S HTR1, MIL: YES** **2003, 2004, 2005, 2006** **Models:** MDX **Engines:** All **Transmissions:** A/T, M/T	**Air Fuel Sensor (Bank1 Sensor 1) Heater Circuit Malfunction** Engine started, ECT sensor input more than 158°F, and the PCM detected a problem in the AFS1 Heater circuit operation in the test. **Possible Causes:** • AFS1 heater power circuit is open (check the LAFHT 15A fuse) • AFS1 heater ground circuit is open • AFS1 heater element has high resistance or an open circuit • AFS1 heater element has a shorted condition • PCM has failed

DTC	Trouble Code Title, Conditions & Possible Causes
DTC: P0155 **2T O2S HTR2, MIL: YES** **1996, 1997, 1998, 1999** **Models:** SLX **Engines:** 3.2L VIN V, 3.5L VIN X **Transmissions:** A/T	**HO2S-21 (Bank 2 Sensor 1) Heater Circuit Malfunction** DTC P0151, P0152, P0153 and P0154 not set, ECT and IAT sensor signals less than 90ºF, and within 11ºF at startup, system voltage from 11-16v, throttle angle under 40%, average Calculated airflow less than 18 g/sec in the sample period, and the PCM detected the HO2S-21 signal did not vary more than 150 mv from the bias voltage (400-500 mv) for too long a period (maximum time is 120 seconds). **Possible Causes:** • HO2S power circuit is open (from the O2S heater fuse) • HO2S heater ground circuit is open • HO2S heater element has high resistance • HO2S heater element has failed (open or shorted) • PCM has failed
DTC: P0157 **1T CCM, MIL: YES** **2003, 2004, 2005, 2006** **Models:** MDX **Engines:** All **Transmissions:** A/T, M/T	**HO2S-22 (Bank 2 Sensor 2) Circuit Low Input** Engine started, vehicle driven in closed loop in D4/D5 position at cruise speed, and the PCM detected the HO2S signal was fixed at less than 0.30v (actual value where the code sets is in its memory). **Possible Causes:** • HO2S signal circuit is open or shorted to ground • HO2S ground circuit is open • HO2S may be contaminated or may have failed • PCM has failed
DTC: P0157 **1T CCM, MIL: YES** **1995, 1996, 1997, 1998, 1999,** **2000, 2001, 2002, 2003, 2004,** **2005** **Models:** NSX, NSX-T **Engines:** All **Transmissions:** A/T, M/T	**HO2S-22 (Bank 2 Sensor 2) Circuit Low Input** Engine started, vehicle driven in closed loop in D4/D5 position at cruise speed, and the PCM detected the HO2S signal was fixed at less than 0.30v (actual value where the code sets is in its memory). **Possible Causes:** • HO2S signal circuit is open or shorted to ground • HO2S ground circuit is open • HO2S may be contaminated or may have failed • PCM has failed
DTC: P0157 **1T CCM, MIL: YES** **1996, 1997, 1998, 1999** **Models:** SLX **Engines:** 3.2L VIN V, 3.5L VIN X **Transmissions:** A/T	**HO2S-22 (Bank 2 Sensor 2) Circuit Low Input** DTC P0106, P0107, P0108, P0112, P0113, P0117, P0118, P0121, P0122, P0123, P0171, P0172 and P0300, P0301-P0306 not set, ECT sensor signal more than 140ºF, engine running in closed loop with the A/F ratio at 14.5-14.8:1, throttle angle from 3-19%, and the PCM detected the HO2S-22 signal was less than 26 mv for 106 seconds out of a 125 second period; or that it was more than 400 mv in Power Enrichment Mode during the test. **Possible Causes:** • HO2S signal circuit is open or shorted to ground • HO2S has water or fuel contamination • HO2S is damaged or it has failed • PCM has failed
DTC: P0158 **1T CCM, MIL: YES** **2003, 2004, 2005, 2006** **Models:** MDX **Engines:** All **Transmissions:** A/T, M/T	**HO2S-22 (Bank 2 Sensor 2) Circuit High Input** Engine started, vehicle driven in closed loop at cruise speed, and the PCM detected the HO2S signal was fixed at more than 0.60v. **Possible Causes:** • HO2S signal tracking (wet/oily) in connector causing a short between the signal circuit and heater power circuit • HO2S signal circuit is open, or the ground circuit is open • HO2S heater supply circuit is open • PCM has failed
DTC: P0158 **1T CCM, MIL: YES** **1995, 1996, 1997, 1998, 1999,** **2000, 2001, 2002, 2003, 2004,** **2005** **Models:** NSX, NSX-T **Engines:** All **Transmissions:** A/T, M/T	**HO2S-22 (Bank 2 Sensor 2) Circuit High Input** Engine started, vehicle driven in closed loop in D4/D5 position at cruise speed, and the PCM detected the HO2S signal was fixed at more than 0.60v **Note: The actual value where the code sets is in the PCM memory.** **Possible Causes:** • HO2S signal tracking (wet/oily) in connector causing a short between the signal circuit and heater power circuit • HO2S signal circuit is open, or the ground circuit is open • HO2S heater supply circuit is open • PCM has failed

DTC	Trouble Code Title, Conditions & Possible Causes
DTC: P0158 **1T CCM, MIL: YES** **1996, 1997, 1998, 1999** **Models:** SLX **Engines:** 3.2L VIN V, 3.5L VIN X **Transmissions:** A/T	**HO2S-22 (Bank 2 Sensor 2) Circuit High Input** DTC P0106, P0107, P0108, P0112, P0113, P0117, P0118, P0121, P0122, P0123, P0171, P0172 and P0300, P0301-P0306 not set, engine running in closed loop with the A/F ratio command at 14.5-14.8:1, throttle angle from 3-19% for 5 seconds, the PCM detected the HO2S-22 signal was less than 952 mv for 106 seconds over a 125 second period; or the HO2S-12 signal was more than 500 mv during Decel Fuel Cutoff mode for 3 seconds during the CCM test. **Possible Causes:** • HO2S signal circuit shorted to system power (oil in connector) • HO2S has water or fuel contamination • HO2S is damaged or it has failed • PCM has failed
DTC: P0159 **1T O2S1, MIL: YES** **2003, 2004, 2005, 2006** **Models:** MDX **Engines:** All **Transmissions:** A/T, M/T	**HO2S-22 (Bank 2 Sensor 2) Circuit Slow Response** Engine started, vehicle driven in closed loop at over 55 mph at cruise speed, and the PCM detected the HO2S response time to switch from rich to lean or from lean to rich switch time was too slow. **Possible Causes:** • Exhaust leak present in the exhaust manifold or exhaust pipes • HO2S element fuel contamination • HO2S element has deteriorated • PCM has failed
DTC: P0159 **2T O2S1, MIL: YES** **1995, 1996, 1997, 1998, 1999,** **2000, 2001, 2002, 2003, 2004,** **2005** **Models:** NSX, NSX-T **Engines:** All **Transmissions:** A/T, M/T	**HO2S-22 (Bank 2 Sensor 2) Circuit Slow Response** Engine started, vehicle driven in closed loop in D4/D5 position at over 55 mph at steady speed, and the PCM detected the HO2S response time to switch between 300-600 mv was too slow, or that the rich to lean or lean to rich switch time was too slow. **Possible Causes:** • Exhaust leak present in the exhaust manifold or exhaust pipes • HO2S element fuel contamination • HO2S element has deteriorated • PCM has failed
DTC: P0160 **1T O2S2, MIL: YES** **1996, 1997, 1998, 1999** **Models:** SLX **Engines:** 3.2L VIN V, 3.5L VIN X **Transmissions:** A/T	**HO2S-22 (Bank 2 Sensor 2) Insufficient Activity Detected** DTC P0106, P0107, P0108, P0112, P0113, P0117, P0118, P0121, P0122, P0123, P0171, P0172 and P0300, P0301-P0306 not set, system voltage from 11-16v, engine runtime over 40 seconds, then after the PCM determined the Oxygen Sensor Heater test passed, it detected the HO2S-22 signal remained from 426-474 mv for 105 seconds of a 125 second period in the Oxygen Sensor Monitor test. **Possible Causes:** • Exhaust leak present in exhaust manifold or exhaust pipes • HO2S element has fuel contamination or has deteriorated • HO2S signal circuit or the ground circuit has high resistance • PCM has failed
DTC: P0161 **1T O2S HTR1, MIL: YES** **2003, 2004, 2005, 2006** **Models:** MDX **Engines:** All **Transmissions:** A/T, M/T	**HO2S-22 (Bank 2 Sensor 2) Heater Circuit Malfunction** Engine started, engine runtime 80 seconds, and the PCM detected an unexpected voltage condition on the HO2S heater circuit. **Possible Causes:** • Main relay power circuit to the heater is open (intermittent) • HO2S heater ground circuit is open • HO2S heater element has high resistance or has failed • PCM has failed
DTC: P0161 **1T O2S HTR1, MIL: YES** **1995, 1996, 1997, 1998, 1999,** **2000, 2001, 2002, 2003, 2004,** **2005** **Models:** NSX, NSX-T **Engines:** All **Transmissions:** A/T, M/T	**HO2S-22 (Bank 2 Sensor 2) Heater Circuit Malfunction** Engine started, engine runtime over 80 seconds, and the PCM detected an unexpected voltage condition on the HO2S heater circuit during the CCM test. **Possible Causes:** • Main relay output (power) circuit to the heater is open • HO2S heater ground circuit is open • HO2S heater element has high resistance or an open circuit • HO2S heater element has a shorted condition • PCM has failed
DTC: P0161 **2T O2S HTR2, MIL: YES** **1996, 1997, 1998, 1999** **Models:** SLX **Engines:** 3.2L VIN V, 3.5L VIN X **Transmissions:** A/T	**HO2S-22 (Bank 2 Sensor 2) Heater Circuit Malfunction** DTC P0157 and P0158 not set, ECT and IAT sensor less than 90°F, and within 11°F at startup, system voltage at 11-16v, average Calculated airflow during the test less than 23 g/sec, and the PCM detected the HO2S-22 signal changed less than 150 mv from the bias voltage of 400-500 mv for up to 300 seconds during the test. **Possible Causes:** • HO2S power circuit is open (from the O2S heater fuse) • HO2S heater element has failed (it may be open or shorted) • PCM has failed

DTC	Trouble Code Title, Conditions & Possible Causes
DTC: P0171 **2T FUEL, MIL: YES** **1995, 1996, 1997, 1998, 1999, 2000, 2001, 2002, 2003, 2004, 2005** **Models:** NSX, NSX-T **Engines:** All **Transmissions:** A/T, M/T	**Fuel System Too Lean (Bank 1)** DTC P0107, P0108, P0135, P0137, P0138, P0141, P0151, P0152 and P0155 not set, engine running in closed loop, and the PCM detected the LONGFT value exceeded the calibrated lean limit value. **Possible Causes:** • Air leaks in intake manifold, exhaust pipes or exhaust manifold • One or more injectors restricted or pressure regulator has failed • Air is being drawn in from leaks in gaskets or other seals • O2S element is deteriorated or has failed • A "fuel control" sensor is out of calibration (ECT, IAT or MAP) • PCM has failed
DTC: P0171 **2T FUEL, MIL: YES** **1995, 1996, 1997, 1998** **Models:** 2.5TL **Engines:** All **Transmissions:** A/T, M/T	**Fuel System Too Lean (Bank 1)** DTC P0107, P0108, P0135, P0137, P0138, P0141, P1128, P1129 and P1259 not set, engine running in closed loop, and the PCM detected the LONGFT value exceeded the calibrated lean limit value. **Possible Causes:** • Air leaks in intake manifold, exhaust pipes or exhaust manifold • One or more injectors restricted or pressure regulator has failed • Air is being drawn in from leaks in gaskets or other seals • O2S element is deteriorated or has failed • A "fuel control" sensor is out of calibration (ECT, IAT or MAP) • PCM has failed
DTC: P0171 **2T FUEL, MIL: YES** **1996, 1997, 1998, 1999, 2000, 2001, 2002, 2003, 2004, 2005, 2006** **Models:** 2.2CL, 2.3CL, 3.0CL, 3.2CL, 3.2TL, TL, 3.5RL, RL, Integra, MDX, RSX **Engines:** All **Transmissions:** A/T, M/T	**Fuel System Too Lean (Bank 1)** DTC P0107, P0108, P0135, P0137, P0138, P0141, P1128, P1129 and P1259 not set, engine running in closed loop, and the PCM detected the LONGFT value exceeded the calibrated lean limit value. **Possible Causes:** • Air leaks in intake manifold, exhaust pipes or exhaust manifold • One or more injectors restricted or pressure regulator has failed • Air is being drawn in from leaks in gaskets or other seals • O2S element is deteriorated or has failed • A "fuel control" sensor is out of calibration (ECT, IAT or MAP) • PCM has failed
DTC: P0171 **2T FUEL, MIL: YES** **1996, 1997, 1998, 1999** **Models:** SLX **Engines:** 3.2L VIN V, 3.5L VIN X **Transmissions:** A/T	**Fuel System Too Lean (Bank 1)** DTC P0106, P0107, P0108, P0112, P0113, P0117, P0118, P0121, P0122, P0123, P0131, P0132, P0133, P0134, P0135, P0137, P0138, P0201-206, P0300, P0301=P0306, P0401, P0502, P0503, P0506, P0507, P1406 and P1441 not set, engine running in closed loop, system voltage from 11-16v, BARO sensor over 72.5 kPa, ECT sensor from 77-212°F, IAT sensor from -40 to 248°F, MAP sensor from 24-99 kPa, throttle angle less than 95%, VSS under 85 mph, engine speed from 400-6000 rpm, MAF sensor from 2-20 g/sec, Purge duty cycle over 0%, and the PCM detected the average of the Long Term fuel trim values was more than +20%. **Possible Causes:** • Air leaks after the MAF sensor, or in the EGR or PCV system • Base engine "mechanical" fault affecting one or more cylinders • Exhaust leaks before or near where the front HO2S is mounted • Fuel control sensor is out of calibration (i.e., ECT, IAT or MAP) • Fuel delivery system supplying too little fuel during cruise or idle periods (e.g., faulty fuel pump or dirty, restricted fuel filter) • Fuel injector (one or more) dirty or pressure regulator has failed • HO2S is contaminated, deteriorated or it has failed • Vehicle driven low on fuel or until it ran out of fuel
DTC: P0172 **2T FUEL, MIL: YES** **1995, 1996, 1997, 1998, 1999, 2000, 2001, 2002, 2003, 2004, 2005** **Models:** NSX, NSX-T **Engines:** All **Transmissions:** A/T, M/T	**Fuel System Too Rich (Bank 1)** DTC P0107, P0108, P0135, P0137, P0138, P0141, P0151, P0152 and P0155 not set, engine running in closed loop, and the PCM detected the LONGFT value exceeded the calibrated rich limit. Note: A high MAP sensor signal at idle can cause this code to set. **Possible Causes:** • Base engine fault (i.e., cam timing incorrect, oil level too high) • Engine oil overfill condition • EVAP vapor recovery system has failed (pulling vacuum) • HO2S element may be contaminated with water or alcohol • Leaking/contaminated fuel injector(s) or fuel pressure regulator • Partial engine misfire condition is present

DTC	Trouble Code Title, Conditions & Possible Causes
DTC: P0172 **2T FUEL, MIL: YES** **1995, 1996, 1997, 1998** **Models:** 2.5TL **Engines:** All **Transmissions:** A/T, M/T	**Fuel System Too Rich (Bank 1)** DTC P0107, P0108, P0135, P0137, P0138, P0141, P1128, P1129 and P1259 not set, engine running in closed loop, and the PCM detected the LONGFT value exceeded the calibrated rich limit. Note: A high MAP sensor signal at idle can cause this code to set. **Possible Causes:** • Base engine fault (i.e., cam timing incorrect, oil level too high) • EVAP vapor recovery system has failed (pulling vacuum) • HO2S element may be contaminated with water or alcohol • Leaking/contaminated fuel injector(s) or fuel pressure regulator • Partial engine misfire condition is present
DTC: P0172 **2T FUEL, MIL: YES** **1996, 1997, 1998, 1999, 2000, 2001, 2002, 2003, 2004, 2005, 2006** **Models:** 2.2CL, 2.3CL, 3.0CL, 3.2CL, 3.2TL, TL, 3.5RL, RL, Integra, MDX, RSX **Engines:** All **Transmissions:** A/T, M/T	**Fuel System Too Rich (Bank 1)** DTC P0107, P0108, P0135, P0137, P0138, P0141, P1128, P1129 and P1259 not set, engine running in closed loop, and the PCM detected the LONGFT value exceeded the calibrated rich limit. Note: A high MAP sensor signal at idle can cause this code to set. **Possible Causes:** • Base engine fault (i.e., cam timing incorrect, oil level too high) • EVAP vapor recovery system has failed (pulling vacuum) • HO2S element may be contaminated with water or alcohol • Leaking/contaminated fuel injector(s) or fuel pressure regulator • Partial engine misfire condition is present
DTC: P0172 **2T FUEL, MIL: YES** **1996, 1997, 1998, 1999** **Models:** SLX **Engines:** 3.2L VIN V, 3.5L VIN X **Transmissions:** A/T	**Fuel System Too Rich (Bank 1)** DTC P0106, P0107, P0108, P0112, P0113, P0117, P0118, P0121, P0122, P0123, P0131, P0132, P0133, P0134, P0135, P0137, P0138, P0201-206, P0300, P0301=P0306, P0401, P0502, P0503, P0506, P0507, P1406 and P1441 not set, engine running in closed loop, BARO sensor over 72.5 kPa, ECT sensor from 77-212°F, IAT sensor from -40 to 248°F, MAP sensor from 24-99 kPa, throttle angle less than 95%, VSS under 85 mph, engine speed from 400-6000 rpm, MAF sensor from 2-20 g/sec, Purge duty cycle over 0%, and the PCM detected the average of the Long Term fuel trim values was more than -14% during the Fuel System Monitor test. **Possible Causes:** • Base engine "mechanical" fault affecting one or more cylinders • EVAP system component has failed or canister fuel saturated • Fuel control sensor is out of calibration (i.e., ECT, IAT or MAP) • Fuel delivery system supplying too much fuel during cruise or idle periods (e.g., faulty fuel pump, or faulty pressure regulator) • Fuel injector(s) is leaking or stuck partially open (one or more) • HO2S is contaminated, deteriorated or it has failed
DTC: P0174 **2T FUEL, MIL: YES** **1995, 1996, 1997, 1998, 1999, 2000, 2001, 2002, 2003, 2004, 2005** **Models:** NSX, NSX-T **Engines:** All **Transmissions:** A/T, M/T	**Fuel System Too Lean (Bank 2)** DTC P0107, P0108, P0135, P0137, P0138, P0141, P1128, P1129 and P1259 not set, engine running in closed loop, and the PCM detected the LONGFT value exceeded a calibrated lean limit value. **Possible Causes:** • Air leaks in intake manifold, exhaust pipes or exhaust manifold • One or more injectors restricted or pressure regulator has failed • Air is being drawn in from leaks in gaskets or other seals • O2S element is deteriorated or has failed • A "fuel control" sensor is out of calibration (ECT, IAT or MAP)
DTC: P0174 **2T FUEL, MIL: YES** **1996, 1997, 1998** **Models:** 3.2TL **Engines:** All **Transmissions:** A/T, M/T	**Fuel System Too Lean (Bank 2)** DTC P0107, P0108, P0135, P0137, P0138, P0141, P1128, P1129 and P1259 not set, engine running in closed loop, and the PCM detected the LONGFT value exceeded a calibrated lean limit value. **Possible Causes:** • Air leaks in intake manifold, exhaust pipes or exhaust manifold • One or more injectors restricted or pressure regulator has failed • Air is being drawn in from leaks in gaskets or other seals • O2S element is deteriorated or has failed • A "fuel control" sensor is out of calibration (ECT, IAT or MAP)
DTC: P0174 **2T FUEL, MIL: YES** **2003, 2004, 2005, 2006** **Models:** MDX **Engines:** All **Transmissions:** A/T, M/T	**Fuel System Too Lean (Bank 2)** DTC P0107, P0108, P0133-P0141, P0153-P0161, P0401-P0406, P2251-P2255, P2A00, P2A03 and P2279 not set, engine running in closed loop, and the PCM detected a very lean Air/Fuel condition. **Possible Causes:** • Air leaks in intake manifold, exhaust pipes or exhaust manifold • One or more injectors restricted or pressure regulator has failed • Air is being drawn in from leaks in gaskets or other seals • Fuel control sensor is out of calibration (ECT, IAT or MAP)

DTC	Trouble Code Title, Conditions & Possible Causes
DTC: P0174 **2T FUEL, MIL: YES** **1996, 1997, 1998, 1999** **Models:** SLX **Engines:** 3.2L VIN V, 3.5L VIN X **Transmissions:** A/T	**Fuel System Too Lean (Bank 2)** DTC P0106, P0107, P0108, P0112, P0113, P0117, P0118, P0121, P0122, P0123, P0131, P0132, P0133, P0134, P0135, P0137, P0138, P0201-206, P0300, P0301=P0306, P0401, P0502, P0503, P0506, P0507, P1406 and P1441 not set, engine running in closed loop, system voltage from 11-16v, BARO sensor over 72.5 kPa, ECT sensor from 77-212°F, IAT sensor from -40 to 248°F, MAP sensor from 24-99 kPa, throttle angle steady at less than 95%, VSS under 85 mph, engine speed from 400-6000 rpm, MAF sensor from 2-20 g/sec, Purge duty cycle over 0%, and the PCM detected the average of the Long Term fuel trim values was more than +20%. **Possible Causes:** • Air leaks after the MAF sensor, or in the EGR or PCV system • Base engine "mechanical" fault affecting one or more cylinders • Exhaust leaks before or near where the front HO2S is mounted • Fuel control sensor is out of calibration (i.e., ECT, IAT or MAP) • Fuel delivery system supplying too little fuel during cruise or idle periods (e.g., faulty fuel pump or dirty, restricted fuel filter) • Fuel injector (one or more) dirty or pressure regulator has failed • HO2S is contaminated, deteriorated or it has failed • Vehicle driven low on fuel or until it ran out of fuel
DTC: P0175 **2T FUEL, MIL: YES** **1995, 1996, 1997, 1998, 1999,** **2000, 2001, 2002, 2003, 2004,** **2005** **Models:** NSX, NSX-T **Engines:** All **Transmissions:** A/T, M/T	**Fuel System Too Rich (Bank 2)** DTC P0107, P0108, P0135, P0137, P0138, P0141, P1128, P1129 and P1259 not set, engine running in closed loop, and the PCM detected the LONGFT value exceeded the calibrated rich limit. Note: A high MAP sensor signal at idle can cause this code to set. **Possible Causes:** • Base engine fault (i.e., cam timing incorrect, oil level too high) • EVAP vapor recovery system has failed (pulling vacuum) • HO2S element may be contaminated with water or alcohol • Leaking/contaminated fuel injector(s) or fuel pressure regulator
DTC: P0175 **2T FUEL, MIL: YES** **2003, 2004, 2005, 2006** **Models:** MDX **Engines:** All **Transmissions:** A/T, M/T	**Fuel System Too Lean (Bank 2)** DTC P0107, P0108, P0133-P0141, P0153-P0161, P0401-P0406, P2251-P2255, P2A00, P2A03 and P2279 not set, engine running in closed loop, and the PCM detected a very lean Air/Fuel condition. **Possible Causes:** • Air leaks in intake manifold, exhaust pipes or exhaust manifold • One or more injectors restricted or pressure regulator has failed • Air is being drawn in from leaks in gaskets or other seals • Fuel control sensor is out of calibration (ECT, IAT or MAP)
DTC: P0175 **2T FUEL, MIL: YES** **1996, 1997, 1998, 1999** **Models:** SLX **Engines:** 3.2L VIN V, 3.5 VIN X **Transmissions:** A/T	**Fuel System Too Rich (Bank 2)** DTC P0106, P0107, P0108, P0112, P0113, P0117, P0118, P0121, P0122, P0123, P0131, P0132, P0133, P0134, P0135, P0137, P0138, P0201-206, P0300, P0301=P0306, P0401, P0502, P0503, P0506, P0507, P1406 and P1441 not set, engine running in closed loop, BARO sensor over 72.5 kPa, ECT sensor from 77-212°F, IAT sensor from -40 to 248°F, MAP sensor from 24-99 kPa, throttle angle less than 95%, VSS under 85 mph, engine speed from 400-6000 rpm, MAF sensor from 2-20 g/sec, Purge duty cycle over 0%, and the PCM detected the average of the Long Term fuel trim values was more than -14% during the Fuel System Monitor test. **Possible Causes:** • Base engine "mechanical" fault affecting one or more cylinders • EVAP system component has failed or canister fuel saturated • Fuel control sensor is out of calibration (i.e., ECT, IAT or MAP) • Fuel delivery system supplying too much fuel during cruise or idle periods (e.g., faulty fuel pump, or faulty pressure regulator) • Fuel injector(s) is leaking or stuck partially open (one or more) • HO2S is contaminated, deteriorated or it has failed
DTC: P0201 **1T CCM, MIL: YES** **2003, 2004, 2005, 2006** **Models:** MDX **Engines:** All **Transmissions:** A/T, M/T	**Fuel Injector No. 1 Circuit Malfunction** Engine started, and the PCM detected the identified fuel injector control circuit signal was more than the upper limit, or that it was less than the lower limit, or that no control signal was present. **Possible Causes:** • Fuel injector 1 control circuit is open or shorted to ground • Fuel injector 1 is damaged or has failed • Main relay power supply circuit to the injector is open • Injector "driver" circuit in the PCM is damaged or has failed
DTC: P0201 **1T CCM, MIL: YES** **1996, 1997, 1998, 1999** **Models:** SLX **Engines:** 3.2L VIN V, 3.5L VIN X **Transmissions:** A/T	**Fuel Injector 1 Circuit Malfunction** Engine started, system voltage over 9v, and PCM detected the injector voltage for Cylinder 1 did not equal the system voltage with the injector commanded "off", or that the injector voltage did not equal zero (0) volts with the injector commanded "on". **Possible Causes:** • Fuel injector control circuit is open or shorted to ground • Fuel injector power circuit is open between injector and relay • Fuel Injector has failed • PCM has failed (injector driver circuit may be open or shorted)

DTC	Trouble Code Title, Conditions & Possible Causes
DTC: P0202 **1T CCM, MIL: YES** **2003, 2004, 2005, 2006** **Models:** MDX **Engines:** All **Transmissions:** A/T, M/T	**Fuel Injector No. 2 Circuit Malfunction** Engine started, and the PCM detected the identified fuel injector control circuit signal was more than the upper limit, or that it was less than the lower limit, or that no control signal was present. **Possible Causes:** • Fuel injector 2 control circuit is open or shorted to ground • Fuel injector 2 is damaged or has failed • Main relay power supply circuit to the injector is open • Injector "driver" circuit in the PCM is damaged or has failed
DTC: P0202 **1T CCM, MIL: YES** **1996, 1997, 1998, 1999** **Models:** SLX **Engines:** 3.2L VIN V, 3.5L VIN X **Transmissions:** A/T	**Fuel Injector 2 Circuit Malfunction** Engine started, system voltage over 9v, and PCM detected the injector voltage for Cylinder 2 did not equal the system voltage with the injector commanded "off", or that the injector voltage did not equal zero (0) volts with the injector commanded "on". **Possible Causes:** • Fuel injector control circuit is open or shorted to ground • Fuel injector power circuit open between injector and ECM fuse • Fuel Injector has failed • PCM has failed (injector driver circuit may be open or shorted)
DTC: P0203 **1T CCM, MIL: YES** **2003, 2004, 2005, 2006** **Models:** MDX **Engines:** All **Transmissions:** A/T, M/T	**Fuel Injector No. 3 Circuit Malfunction** Engine started, and the PCM detected the identified fuel injector control circuit signal was more than the upper limit, or that it was less than the lower limit, or that no control signal was present. **Possible Causes:** • Fuel injector 3 control circuit is open or shorted to ground • Fuel injector 3 is damaged or has failed • Main relay power supply circuit to the injector is open • Injector "driver" circuit in the PCM is damaged or has failed
DTC: P0203 **1T CCM, MIL: YES** **1996, 1997, 1998, 1999** **Models:** SLX **Engines:** 3.2L VIN V, 3.5L VIN X **Transmissions:** A/T	**Fuel Injector 3 Circuit Malfunction** Engine started, system voltage over 9v, and PCM detected the injector voltage for Cylinder 3 did not equal the system voltage with the injector commanded "off", or that the injector voltage did not equal zero (0) volts with the injector commanded "on". **Possible Causes:** • Fuel injector control circuit is open or shorted to ground • Fuel injector power circuit open between injector and ECM fuse • Fuel Injector has failed • PCM has failed (injector driver circuit may be open or shorted)
DTC: P0204 **1T CCM, MIL: YES** **2003, 2004, 2005, 2006** **Models:** MDX **Engines:** All **Transmissions:** A/T, M/T	**Fuel Injector No. 4 Circuit Malfunction** Engine started, and the PCM detected the identified fuel injector control circuit signal was more than the upper limit, or that it was less than the lower limit, or that no control signal was present. **Possible Causes:** • Fuel injector 4 control circuit is open or shorted to ground • Fuel injector 4 is damaged or has failed • Main relay power supply circuit to the injector is open • Injector "driver" circuit in the PCM is damaged or has failed
DTC: P0204 **1T CCM, MIL: YES** **1996, 1997, 1998, 1999** **Models:** SLX **Engines:** 3.2L VIN V, 3.5L VIN X **Transmissions:** A/T	**Fuel Injector 4 Circuit Malfunction** Engine started, system voltage over 9v, and PCM detected the injector voltage for Cylinder 4 did not equal the system voltage with the injector commanded "off", or that the injector voltage did not equal zero (0) volts with the injector commanded "on". **Possible Causes:** • Fuel injector control circuit is open or shorted to ground • Fuel injector power circuit open between injector and ECM fuse • Fuel Injector has failed • PCM has failed (injector driver circuit may be open or shorted)
DTC: P0205 **1T CCM, MIL: YES** **2003, 2004, 2005, 2006** **Models:** MDX **Engines:** All **Transmissions:** A/T, M/T	**Fuel Injector No. 5 Circuit Malfunction** Engine started, and the PCM detected the identified fuel injector control circuit signal was more than the upper limit, or that it was less than the lower limit, or that no control signal was present. **Possible Causes:** • Fuel injector 5 control circuit is open or shorted to ground • Fuel injector 5 is damaged or has failed • Main relay power supply circuit to the injector is open • Injector "driver" circuit in the PCM is damaged or has failed

DTC	Trouble Code Title, Conditions & Possible Causes

DTC: P0205
1T CCM, MIL: YES
1996, 1997, 1998, 1999
Models: SLX
Engines: 3.2L VIN V, 3.5L VIN X
Transmissions: A/T
	Fuel Injector 5 Circuit Malfunction
	Engine started, system voltage over 9v, and PCM detected the injector voltage for Cylinder 5 did not equal the system voltage with the injector commanded "off", or that the injector voltage did not equal zero (0) volts with the injector commanded "on".
	Possible Causes:
	  • Fuel injector control circuit is open or shorted to ground
	  • Fuel injector power circuit open between injector and ECM fuse
	  • Fuel Injector has failed
	  • PCM has failed (injector driver circuit may be open or shorted)

DTC: P0206
1T CCM, MIL: YES
2003, 2004, 2005, 2006
Models: MDX
Engines: All
Transmissions: A/T, M/T
	Fuel Injector No. 6 Circuit Malfunction
	Engine started, and the PCM detected the identified fuel injector control circuit signal was more than the upper limit, or that it was less than the lower limit, or that no control signal was present.
	Possible Causes:
	  • Fuel injector 6 control circuit is open or shorted to ground
	  • Fuel injector 6 is damaged or has failed
	  • Main relay power supply circuit to the injector is open
	  • Injector "driver" circuit in the PCM is damaged or has failed

DTC: P0206
1T CCM, MIL: YES
1996, 1997, 1998, 1999
Models: SLX
Engines: 3.2L VIN V, 3.5L VIN X
Transmissions: A/T
	Fuel Injector 6 Circuit Malfunction
	Engine started, system voltage over 9v, and PCM detected the injector voltage for Cylinder 6 did not equal the system voltage with the injector commanded "off", or that the injector voltage did not equal zero (0) volts with the injector commanded "on".
	Possible Causes:
	  • Fuel injector control circuit is open or shorted to ground
	  • Fuel injector power circuit open between injector and ECM fuse
	  • Fuel Injector has failed
	  • PCM has failed (injector driver circuit may be open or shorted)

DTC: P0218
1T CCM, MIL: NO
1996, 1997, 1998, 1999
Models: SLX
Engines: All
Transmissions: A/T
	Transmission Fluid Over-Temperature Malfunction
	DTC P0712 and P0713 not set, and the PCM detected the Transmission Fluid Temperature (TFT) sensor signal indicated more than 284°F, condition met for 21 seconds.
	Note: The ATF LAMP is ON if the TFT sensor signal exceeds 293°F.
	Possible Causes:
	  • TFT sensor signal circuit is shorted to ground
	  • TFT sensor is out-of-calibration (skewed), or it has failed
	  • Torque converter stator is damaged or has failed
	  • PCM has failed

DTC: P0222
1T CCM, MIL: YES
2003, 2004, 2005, 2006
Models: MDX
Engines: All
Transmissions: A/T, M/T
	Throttle Position Sensor 2 Circuit Low Input
	Engine started, engine running in closed loop conditions, and the PCM detected the closed throttle TP2 signal was less than 0.16v (Scan Tool reads less than 10% open) during the CCM test.
	Possible Causes:
	  • TP2 sensor signal circuit is shorted to ground
	  • TP2 sensor VREF circuit is open or shorted to ground
	  • TP2 sensor is damaged (it may be shorted internally)
	  • PCM has failed

DTC: P0223
1T CCM, MIL: YES
2003, 2004, 2005, 2006
Models: MDX
Engines: All
Transmissions: A/T, M/T
	Throttle Position Sensor 2 Circuit High Input
	Engine started, and after a momentary WOT condition, the PCM detected the TP signal indicated over 4.60v (Scan Tool over 90%).
	Possible Causes:
	  • TP2 sensor signal circuit is shorted to VREF or system power
	  • TP2 sensor ground circuit is open between sensor and PCM
	  • TP2 sensor is damaged or has failed
	  • PCM has failed

DTC: P0300
2T MISFIRE, MIL: YES
1995, 1996, 1997, 1998, 1999,
2000, 2001, 2002, 2003, 2004,
2005
Models: NSX, NSX-T
Engines: All
Transmissions: A/T, M/T
	Multiple Misfire Detected
	DTC P0107, P0108, P0131, P0132, P0171, P0172, P1128, P0335, P0336, P0505, P1128, P1129, P1259, P1361, P1362, P1366, P1367 and P1519 not set, engine running under positive torque conditions, and the PCM detected a misfire in 2 or more cylinders.
	Note: If the misfire is severe, the MIL will flash on/off on the 1st trip!
	Possible Causes:
	  • Base engine mechanical fault affecting more than one cylinder
	  • CKP or CMP sensor problem affecting more than one cylinder
	  • Fuel system problem affecting more than one cylinder
	  • Ignition system problem affecting more than one cylinder

DTC	Trouble Code Title, Conditions & Possible Causes
DTC: P0300 **2T MISFIRE, MIL: YES** **1995, 1996, 1997, 1998** **Models:** 2.5TL **Engines:** All **Transmissions:** A/T, M/T	**Multiple Misfire Detected** DTC P0101-P0103, P0171, P0172, P0401, P1102, P1103, P1361, P1362, P1381, P1382, P1491, P1498 and P1508 not set, engine running under positive torque conditions, and the PCM detected a misfire in 2 or more cylinders (MIL will flash if the misfire is severe). **Possible Causes:** • Base engine mechanical fault affecting more than one cylinder • CKP or CMP sensor problem affecting more than one cylinder • Fuel system problem affecting more than one cylinder • Ignition system problem affecting more than one cylinder
DTC: P0300 **2T MISFIRE, MIL: YES** **1996, 1997, 1998, 1999, 2000, 2001, 2002, 2003, 2004, 2005, 2006** **Models:** 2.2CL, 2.3CL, 3.0CL, 3.2CL, 3.2TL, TL, 3.5RL, RL, Integra, MDX, RSX **Engines:** All **Transmissions:** A/T, M/T	**Multiple Misfire Detected** DTC P0101-P0103, P0171, P0172, P0401, P1102, P1103, P1361, P1362, P1381, P1382, P1491, P1498 and P1508 not set, engine running under positive torque conditions, and the PCM detected a misfire in only one cylinder (MIL will flash if the misfire is severe). **Possible Causes:** • Base engine mechanical fault affecting more than 1 cylinder • CKP or CMP sensor problem affecting more than one cylinder • Fuel system problem affecting more than one cylinder • Ignition system problem affecting more than one cylinder
DTC: P0300 **2T MISFIRE, MIL: YES** **1996, 1997, 1998, 1999** **Models:** SLX **Engines:** 3.2L VIN V, 3.5L VIN X **Transmissions:** A/T	**Random Misfire Detected** DTC P0101, P0102, P0103, P0106, P0107, P0108, P0117, P0118, P0121, P0122, P0123, P0336, P0341, P0342, P0502 and P0503 not set, ECT sensor from 20-248°F, system voltage from 11-16v, engine speed from 800-5500 rpm, throttle angle stable (± 3%), and the PCM detected a crankshaft speed variation in one or more cylinders characteristic of a misfire condition during the Misfire Monitor test. Note: If the misfire is severe, the MIL will flash on/off on the 1st trip! **Possible Causes:** • Base engine mechanical fault that affects one or more cylinders • Fuel metering fault that affects more than one cylinder • Fuel pressure too low or too high, fuel supply contaminated • EVAP system problem or the EVAP canister is fuel saturated • EGR valve is stuck open or the PCV system has a vacuum leak • IC control circuit is shorted to ground (an intermittent fault) • MAF sensor contamination (it can cause a very lean condition)
DTC: P0301 **2T MISFIRE, MIL: YES** **1995, 1996, 1997, 1998, 1999, 2000, 2001, 2002, 2003, 2004, 2005** **Models:** NSX, NSX-T **Engines:** All **Transmissions:** A/T, M/T	**Cylinder 1 Misfire Detected** DTC P0107, P0108, P0131, P0132, P0171, P0172, P1128, P0335, P0336, P0505, P1128, P1129, P1259, P1361, P1362, P1366, P1367 and P1519 not set, engine running under positive torque conditions, and the PCM detected a misfire condition in one cylinder. Note: If the misfire is severe, the MIL will flash on/off on the 1st trip! **Possible Causes:** • Base engine (mechanical) problem affecting only one cylinder • Fuel system problem affecting only one cylinder • Ignition problem (i.e., coil or spark plug) affecting one cylinder
DTC: P0301 **2T MISFIRE, MIL: YES** **1995, 1996, 1997, 1998** **Models:** 2.5TL **Engines:** All **Transmissions:** A/T, M/T	**Cylinder 1 Misfire Detected** DTC P0101-P0103, P0171, P0172, P0401, P1102, P1103, P1361, P1362, P1381, P1382, P1491, P1498 and P1508 not set, engine running under positive torque conditions, and the PCM detected a misfire in only one cylinder (MIL will flash if the misfire is severe). **Possible Causes:** • Base engine (mechanical) problem affecting only one cylinder • Fuel system problem affecting only one cylinder • Ignition problem (i.e., coil or spark plug) affecting one cylinder
DTC: P0301 **2T MISFIRE, MIL: YES** **1996, 1997, 1998, 1999, 2000, 2001, 2002, 2003, 2004, 2005, 2006** **Models:** 2.2CL, 2.3CL, 3.0CL, 3.2CL, 3.2TL, TL, 3.5RL, RL, Integra, MDX, RSX **Engines:** All **Transmissions:** A/T, M/T	**Cylinder 1 Misfire Detected** DTC P0107, P0108, P0131, P0132, P0171, P0172, P1128, P0335, P0336, P0505, P1128, P1129, P1259, P1361, P1362, P1366, P1367 and P1519 not set, engine running under positive torque conditions, and the PCM detected a misfire condition in one cylinder. Note: If the misfire is severe, the MIL will flash on/off on the 1st trip! **Possible Causes:** • Air leak in the intake manifold, or in the EGR or PCM system • Base engine (mechanical) problem affecting only one cylinder • Fuel system problem affecting only one cylinder • Ignition problem (i.e., coil or spark plug) affecting one cylinder

DTC	Trouble Code Title, Conditions & Possible Causes
DTC: P0301 **2T MISFIRE, MIL: YES** **1996, 1997, 1998, 1999** **Models:** SLX **Engines:** 3.2L VIN V, 3.5L VIN X **Transmissions:** A/T	**Cylinder 1 Misfire Detected** DTC P0101, P0102, P0103, P0106, P0107, P0108, P0117, P0118, P0121, P0122, P0123, P0336, P0341, P0342, P0502 and P0503 not set, ECT sensor from 20-248°F, system voltage from 11-16v, engine speed from 800-5500 rpm, throttle angle stable (± 3%), and the PCM detected a crankshaft speed variation in one cylinder characteristic of a misfire condition during the Misfire Diagnostic Monitor test. Note: If the misfire is severe, the MIL will flash on/off on the 1st trip! **Possible Causes:** • Base engine mechanical fault that affects only one cylinder • Fuel metering fault that affects only one cylinder • IC control circuit is shorted to ground on only one cylinder • Ignition system fault (a coil) that affects only one cylinder
DTC: P0302 **2T MISFIRE, MIL: YES** **1995, 1996, 1997, 1998, 1999,** **2000, 2001, 2002, 2003, 2004,** **2005** **Models:** NSX, NSX-T **Engines:** All **Transmissions:** A/T, M/T	**Cylinder 2 Misfire Detected** DTC P0107, P0108, P0131, P0132, P0171, P0172, P1128, P0335, P0336, P0505, P1128, P1129, P1259, P1361, P1362, P1366, P1367 and P1519 not set, engine running under positive torque conditions, and the PCM detected a misfire condition in one cylinder. **Note: If the misfire is severe, the MIL will flash on/off on the 1st trip!** **Possible Causes:** • Base engine (mechanical) problem affecting only one cylinder • Fuel system problem affecting only one cylinder • Ignition problem (i.e., coil or spark plug) affecting one cylinder
DTC: P0302 **2T MISFIRE, MIL: YES** **1995, 1996, 1997, 1998** **Models:** 2.5TL **Engines:** All **Transmissions:** A/T, M/T	**Cylinder 2 Misfire Detected** DTC P0101-P0103, P0171, P0172, P0401, P1102, P1103, P1361, P1362, P1381, P1382, P1491, P1498 and P1508 not set, engine running under positive torque conditions, and the PCM detected a misfire in only one cylinder (MIL will flash if the misfire is severe). **Possible Causes:** • Base engine (mechanical) problem affecting only one cylinder • Fuel system problem affecting only one cylinder • Ignition problem (i.e., coil or spark plug) affecting one cylinder
DTC: P0302 **2T MISFIRE, MIL: YES** **1996, 1997, 1998, 1999, 2000,** **2001, 2002, 2003, 2004, 2005,** **2006** **Models:** 2.2CL, 2.3CL, 3.0CL, 3.2CL, 3.2TL, TL, 3.5RL, RL, Integra, MDX, RSX **Engines:** All **Transmissions:** A/T, M/T	**Cylinder 2 Misfire Detected** DTC P0107, P0108, P0131, P0132, P0171, P0172, P1128, P0335, P0336, P0505, P1128, P1129, P1259, P1361, P1362, P1366, P1367 and P1519 not set, engine running under positive torque conditions, and the PCM detected a misfire condition in one cylinder. **Note: If the misfire is severe, the MIL will flash on/off on the 1st trip!** **Possible Causes:** • Air leak in the intake manifold, or in the EGR or PCM system • Base engine (mechanical) problem affecting only one cylinder • Fuel system problem affecting only one cylinder • Ignition problem (i.e., coil or spark plug) affecting one cylinder
DTC: P0302 **2T MISFIRE, MIL: YES** **1996, 1997, 1998, 1999** **Models:** SLX **Engines:** 3.2L VIN V, 3.5L VIN X **Transmissions:** A/T	**Cylinder 2 Misfire Detected** DTC P0101, P0102, P0103, P0106, P0107, P0108, P0117, P0118, P0121, P0122, P0123, P0336, P0341, P0342, P0502 and P0503 not set, ECT sensor from 20-248°F, system voltage from 11-16v, engine speed from 800-5500 rpm, throttle angle stable (± 3%), and the PCM detected a crankshaft speed variation in one cylinder characteristic of a misfire condition during the Misfire Diagnostic Monitor test. **Note: If the misfire is severe, the MIL will flash on/off on the 1st trip!** **Possible Causes:** • Base engine mechanical fault that affects only one cylinder • Fuel metering fault that affects only one cylinder • IC control circuit is shorted to ground on only one cylinder • Ignition system fault (a coil) that affects only one cylinder
DTC: P0303 **2T MISFIRE, MIL: YES** **1995, 1996, 1997, 1998, 1999,** **2000, 2001, 2002, 2003, 2004,** **2005** **Models:** NSX, NSX-T **Engines:** All **Transmissions:** A/T, M/T	**Cylinder 3 Misfire Detected** DTC P0107, P0108, P0131, P0132, P0171, P0172, P1128, P0335, P0336, P0505, P1128, P1129, P1259, P1361, P1362, P1366, P1367 and P1519 not set, engine running under positive torque conditions, and the PCM detected a misfire condition in one cylinder. **Note: If the misfire is severe, the MIL will flash on/off on the 1st trip!** **Possible Causes:** • Base engine (mechanical) problem affecting only one cylinder • Fuel system problem affecting only one cylinder • Ignition problem (i.e., coil or spark plug) affecting one cylinder

DTC	Trouble Code Title, Conditions & Possible Causes
DTC: P0303 **2T MISFIRE, MIL: YES** **1995, 1996, 1997, 1998** **Models:** 2.5TL **Engines:** All **Transmissions:** A/T, M/T	**Cylinder 3 Misfire Detected** DTC P0101-P0103, P0171, P0172, P0401, P1102, P1103, P1361, P1362, P1381, P1382, P1491, P1498 and P1508 not set, engine running under positive torque conditions, and the PCM detected a misfire in only one cylinder (MIL will flash if the misfire is severe). **Possible Causes:** • Base engine (mechanical) problem affecting only one cylinder • Fuel system problem affecting only one cylinder • Ignition problem (i.e., coil or spark plug) affecting one cylinder
DTC: P0303 **2T MISFIRE, MIL: YES** **1996, 1997, 1998, 1999, 2000,** **2001, 2002, 2003, 2004, 2005,** **2006** **Models:** 2.2CL, 2.3CL, 3.0CL, 3.2CL, 3.2TL, TL, 3.5RL, RL, Integra, MDX, RSX **Engines:** All **Transmissions:** A/T, M/T	**Cylinder 3 Misfire Detected** DTC P0107, P0108, P0131, P0132, P0171, P0172, P1128, P0335, P0336, P0505, P1128, P1129, P1259, P1361, P1362, P1366, P1367 and P1519 not set, engine running under positive torque conditions, and the PCM detected a misfire condition in one cylinder. **Note: If the misfire is severe, the MIL will flash on/off on the 1st trip!** **Possible Causes:** • Air leak in the intake manifold, or in the EGR or PCM system • Base engine (mechanical) problem affecting only one cylinder • Fuel system problem affecting only one cylinder • Ignition problem (i.e., coil or spark plug) affecting one cylinder
DTC: P0303 **2T MISFIRE, MIL: YES** **1996, 1997, 1998, 1999** **Models:** SLX **Engines:** 3.2L VIN V, 3.5L VIN X **Transmissions:** A/T	**Cylinder 3 Misfire Detected** DTC P0101, P0102, P0103, P0106, P0107, P0108, P0117, P0118, P0121, P0122, P0123, P0336, P0341, P0342, P0502 and P0503 not set, ECT sensor from 20-248°F, system voltage from 11-16v, engine speed from 800-5500 rpm, throttle angle stable (± 3%), and the PCM detected a crankshaft speed variation in one cylinder characteristic of a misfire condition during the Misfire Diagnostic Monitor test. Note: If the misfire is severe, the MIL will flash on/off on the 1st trip! **Possible Causes:** • Base engine mechanical fault that affects only one cylinder • Fuel metering fault that affects only one cylinder • IC control circuit is shorted to ground on only one cylinder • Ignition system fault (a coil) that affects only one cylinder
DTC: P0304 **2T MISFIRE, MIL: YES** **1995, 1996, 1997, 1998, 1999,** **2000, 2001, 2002, 2003, 2004,** **2005** **Models:** NSX, NSX-T **Engines:** All **Transmissions:** A/T, M/T	**Cylinder 4 Misfire Detected** DTC P0107, P0108, P0131, P0132, P0171, P0172, P1128, P0335, P0336, P0505, P1128, P1129, P1259, P1361, P1362, P1366, P1367 and P1519 not set, engine running under positive torque conditions, and the PCM detected a misfire condition in one cylinder. **Note: If the misfire is severe, the MIL will flash on/off on the 1st trip!** **Possible Causes:** • Base engine (mechanical) problem affecting only one cylinder • Fuel system problem affecting only one cylinder • Ignition problem (i.e., coil or spark plug) affecting one cylinder
DTC: P0304 **2T MISFIRE, MIL: YES** **1995, 1996, 1997, 1998** **Models:** 2.5TL **Engines:** All **Transmissions:** A/T, M/T	**Cylinder 4 Misfire Detected** DTC P0101-P0103, P0171, P0172, P0401, P1102, P1103, P1361, P1362, P1381, P1382, P1491, P1498 and P1508 not set, engine running under positive torque conditions, and the PCM detected a misfire in only one cylinder (MIL will flash if the misfire is severe). **Possible Causes:** • Base engine (mechanical) problem affecting only one cylinder • Fuel system problem affecting only one cylinder • Ignition problem (i.e., coil or spark plug) affecting one cylinder
DTC: P0304 **2T MISFIRE, MIL: YES** **1996, 1997, 1998, 1999, 2000,** **2001, 2002, 2003, 2004, 2005,** **2006** **Models:** 2.2CL, 2.3CL, 3.0CL, 3.2CL, 3.2TL, TL, 3.5RL, RL, Integra, MDX, RSX **Engines:** All **Transmissions:** A/T, M/T	**Cylinder 4 Misfire Detected** DTC P0107, P0108, P0131, P0132, P0171, P0172, P1128, P0335, P0336, P0505, P1128, P1129, P1259, P1361, P1362, P1366, P1367 and P1519 not set, engine running under positive torque conditions, and the PCM detected a misfire condition in one cylinder. **Note: If the misfire is severe, the MIL will flash on/off on the 1st trip!** **Possible Causes:** • Air leak in the intake manifold, or in the EGR or PCM system • Base engine (mechanical) problem affecting only one cylinder • Fuel system problem affecting only one cylinder • Ignition problem (i.e., coil or spark plug) affecting one cylinder

DTC	Trouble Code Title, Conditions & Possible Causes
DTC: P0304 **2T MISFIRE, MIL: YES** **1996, 1997, 1998, 1999** **Models:** SLX **Engines:** 3.2L VIN V, 3.5L VIN X **Transmissions:** A/T	**Cylinder 4 Misfire Detected** DTC P0101, P0102, P0103, P0106, P0107, P0108, P0117, P0118, P0121, P0122, P0123, P0336, P0341, P0342, P0502 and P0503 not set, ECT sensor from 20-248°F, system voltage from 11-16v, engine speed from 800-5500 rpm, throttle angle stable (± 3%), and the PCM detected a crankshaft speed variation in one cylinder characteristic of a misfire condition during the Misfire Diagnostic Monitor test. Note: If the misfire is severe, the MIL will flash on/off on the 1st trip! **Possible Causes:** • Base engine mechanical fault that affects only one cylinder • Fuel metering fault that affects only one cylinder • IC control circuit is shorted to ground on only one cylinder • Ignition system fault (a coil) that affects only one cylinder
DTC: P0305 **2T MISFIRE, MIL: YES** **1995, 1996, 1997, 1998, 1999,** **2000, 2001, 2002, 2003, 2004,** **2005** **Models:** NSX, NSX-T **Engines:** All **Transmissions:** A/T, M/T	**Cylinder 5 Misfire Detected** DTC P0101, P0102, P0103, P0171, P0172, P0401, P1102, P1103, P1361, P1362, P1381, P1382, P1491, P1498 and P1508 not set, engine running under positive torque conditions, and the PCM detected a misfire in 2 or more cylinders. **Note: If the misfire is severe, the MIL will flash on/off on the 1st trip!** **Possible Causes:** • Base engine (mechanical) problem affecting only one cylinder • Fuel system problem affecting only one cylinder • Ignition problem (i.e., coil or spark plug) affecting one cylinder
DTC: P0305 **2T MISFIRE, MIL: YES** **1995, 1996, 1997, 1998** **Models:** 2.5TL **Engines:** All **Transmissions:** A/T, M/T	**Cylinder 5 Misfire Detected** DTC P0101-P0103, P0171, P0172, P0401, P1102, P1103, P1361, P1362, P1381, P1382, P1491, P1498 and P1508 not set, engine running under positive torque conditions, and the PCM detected a misfire in only one cylinder (MIL will flash if the misfire is severe). **Possible Causes:** • Base engine (mechanical) problem affecting only one cylinder • Fuel control system problem affecting only one cylinder • Ignition problem (i.e., coil or spark plug) affecting one cylinder
DTC: P0305 **2T MISFIRE, MIL: YES** **1996, 1997, 1998, 1999, 2000,** **2001, 2002, 2003, 2004, 2005,** **2006** **Models:** 3.0CL, 3.2CL, 3.2TL, TL, 3.5RL, RL, MDX **Engines:** All **Transmissions:** A/T, M/T	**Cylinder 5 Misfire Detected** DTC P0107, P0108, P0131, P0132, P0171, P0172, P1128, P0335, P0336, P0505, P1128, P1129, P1259, P1361, P1362, P1366, P1367 and P1519 not set, engine running under positive torque conditions, and the PCM detected a misfire condition in one cylinder. **Note: If the misfire is severe, the MIL will flash on/off on the 1st trip!** **Possible Causes:** • Base engine (mechanical) problem affecting only one cylinder • Fuel system problem affecting only one cylinder • Ignition problem (i.e., coil or spark plug) affecting one cylinder
DTC: P0305 **2T MISFIRE, MIL: YES** **1996, 1997, 1998, 1999** **Models:** SLX **Engines:** 3.2L VIN V, 3.5L VIN X **Transmissions:** A/T	**Cylinder 5 Misfire Detected** DTC P0101, P0102, P0103, P0106, P0107, P0108, P0117, P0118, P0121, P0122, P0123, P0336, P0341, P0342, P0502 and P0503 not set, ECT sensor from 20-248°F, system voltage from 11-16v, engine speed from 800-5500 rpm, throttle angle stable (± 3%), and the PCM detected a crankshaft speed variation in one cylinder characteristic of a misfire condition during the Misfire Diagnostic Monitor test. **Note: If the misfire is severe, the MIL will flash on/off on the 1st trip!** **Possible Causes:** • Base engine mechanical fault that affects only one cylinder • Fuel metering fault that affects only one cylinder • IC control circuit is shorted to ground on only one cylinder • Ignition system fault (a coil) that affects only one cylinder
DTC: P0306 **2T MISFIRE, MIL: YES** **1995, 1996, 1997, 1998, 1999,** **2000, 2001, 2002, 2003, 2004,** **2005** **Models:** NSX, NSX-T **Engines:** All **Transmissions:** A/T, M/T	**Cylinder 6 Misfire Detected** DTC P0101, P0102, P0103, P0171, P0172, P0401, P1102, P1103, P1361, P1362, P1381, P1382, P1491, P1498 and P1508 not set, engine running under positive torque conditions, and the PCM detected a misfire in 2 or more cylinders. **Note: If the misfire is severe, the MIL will flash on/off on the 1st trip!** **Possible Causes:** • Base engine (mechanical) problem affecting only one cylinder • Fuel system problem affecting only one cylinder • Ignition problem (i.e., coil or spark plug) affecting one cylinder

DTC	Trouble Code Title, Conditions & Possible Causes
DTC: P0306 **2T MISFIRE, MIL: YES** **1996, 1997, 1998, 1999, 2000, 2001, 2002, 2003, 2004, 2005, 2006** **Models:** 3.0CL, 3.2CL, 3.2TL, TL, 3.5RL, RL, MDX **Engines:** All **Transmissions:** A/T, M/T	**Cylinder 6 Misfire Detected** DTC P0107, P0108, P0131, P0132, P0171, P0172, P1128, P0335, P0336, P0505, P1128, P1129, P1259, P1361, P1362, P1366, P1367 and P1519 not set, engine running under positive torque conditions, and the PCM detected a misfire condition in one cylinder. **Note: If the misfire is severe, the MIL will flash on/off on the 1st trip!** **Possible Causes:** • Air leak in the intake manifold, or in the EGR or PCM system • Base engine (mechanical) problem affecting only one cylinder • Fuel system problem affecting only one cylinder • Ignition problem (i.e., coil or spark plug) affecting one cylinder
DTC: P0306 **2T MISFIRE, MIL: YES** **1996, 1997, 1998, 1999** **Models:** SLX **Engines:** 3.2L VIN V, 3.5L VIN X **Transmissions:** A/T	**Cylinder 6 Misfire Detected** DTC P0101, P0102, P0103, P0106, P0107, P0108, P0117, P0118, P0121, P0122, P0123, P0336, P0341, P0342, P0502 and P0503 not set, ECT sensor from 20-248°F, system voltage from 11-16v, engine speed from 800-5500 rpm, throttle angle stable (± 3%), and the PCM detected a crankshaft speed variation in one cylinder characteristic of a misfire condition during the Misfire Diagnostic Monitor test. Note: If the misfire is severe, the MIL will flash on/off on the 1st trip! **Possible Causes:** • Base engine mechanical fault that affects only one cylinder • Fuel metering fault that affects only one cylinder • IC control circuit is shorted to ground on only one cylinder • Ignition system fault (a coil) that affects only one cylinder
DTC: P0325 **1T CCM, MIL: YES** **1995, 1996, 1997, 1998, 1999, 2000, 2001, 2002, 2003, 2004, 2005** **Models:** NSX, NSX-T **Engines:** All **Transmissions:** A/T, M/T	**Knock Sensor 1 Circuit Malfunction** Engine started, engine running for 1 minute, and the PCM detected an unexpected voltage condition on the Bank 1 Knock sensor circuit. **Possible Causes:** • Knock sensor signal circuit is open (rear bank of engine) • Knock sensor signal circuit is grounded (rear bank of engine) • Knock sensor not tightened properly • Knock sensor damaged or has failed (it may be open internally) • PCM has failed
DTC: P0325 **1T CCM, MIL: YES** **1996, 1997, 1998, 1999, 2000, 2001, 2002, 2003, 2004, 2005, 2006** **Models:** 2.3CL, 3.2CL, 3.2TL, TL, 3.5RL, RL, Integra, MDX, RSX **Engines:** All **Transmissions:** A/T, M/T	**Knock Sensor 1 Circuit Malfunction** Engine started, engine running for 1 minute, and the PCM detected an unexpected voltage condition on the Bank 1 Knock sensor circuit. **Possible Causes:** • Knock sensor signal circuit is open (rear bank of engine) • Knock sensor signal circuit is grounded (rear bank of engine) • Knock sensor not tightened properly • Knock sensor damaged or has failed (it may be open internally) • PCM has failed
DTC: P0325 **2T CCM, MIL: YES** **1996, 1997, 1998, 1999** **Models:** SLX **Engines:** 3.2L VIN V, 3.5 VIN X **Transmissions:** A/T	**Knock Sensor Module Range/Performance** DTC P0327 not set, engine started, system voltage from 11-16v, engine runtime over 120 seconds, and the PCM detected the Knock Sensor (KS) signal was present for over 5 seconds in the CCM test. **Possible Causes:** • KS signal circuit is open or shorted to ground • KS signal circuit is shorted to VREF or system power (B+) • Knock Sensor is damaged or has failed • PCM has failed
DTC: P0327 **1T CCM, MIL: YES** **1995, 1996, 1997, 1998** **Models:** 2.5TL **Engines:** All **Transmissions:** A/T, M/T	**Knock Sensor Front Circuit Low Input** Engine started, engine running for 1 minute, and the PCM detected an unexpected low voltage on the Front Knock Sensor circuit. **Possible Causes:** • Knock sensor signal circuit is grounded (front bank of engine) • Knock sensor not tightened properly • Knock sensor is damaged or has failed (it may be shorted internally) • PCM has failed

DTC	Trouble Code Title, Conditions & Possible Causes
DTC: P0327 **2T CCM, MIL: YES** **1996, 1997, 1998, 1999**	**Knock Sensor Circuit Low Input** Engine started, engine runtime over 10 seconds, system voltage from 11-16v, ECT sensor more than 140°F, engine speed from 2000-4000 rpm, throttle angle over 5%, and the PCM detected the Knock Sensor (KS) signal indicated less than 0.20v, or indicated ore than 4.8v for over 15 seconds during the CCM test. **Possible Causes:** • KS signal circuit is open or shorted to ground • KS signal circuit is shorted to VREF or system power (B+) • Knock Sensor is damaged or has failed • PCM has failed
DTC: P0328 **1T CCM, MIL: YES** **1995, 1996, 1997, 1998** **Models:** 2.5TL **Engines:** All **Transmissions:** A/T, M/T	**Knock Sensor Front Circuit High Input** Engine started, engine running for 1 minute, and the PCM detected an unexpected high input on the Front Knock Sensor circuit. **Possible Causes:** • Knock sensor signal circuit is open (front bank of engine) • Knock sensor signal circuit is shorted to VREF or system power • Knock sensor damaged or has failed (it may be open internally) • PCM has failed
DTC: P0330 **1T CCM, MIL: YES** **1995, 1996, 1997, 1998, 1999,** **2000, 2001, 2002, 2003, 2004,** **2005, 2006** **Models:** NSX, NSX-T, 3.5RL, RL **Engines:** All **Transmissions:** A/T, M/T	**Knock Sensor 2 Circuit Malfunction** Engine started, engine running for 1 minute, and the PCM detected an unexpected voltage condition on the Bank2 Knock Sensor circuit. **Possible Causes:** • Knock sensor signal circuit is open (front bank of engine) • Knock sensor signal circuit is grounded (front bank of engine) • Knock sensor not tightened properly • Knock sensor damaged or has failed (it may be open internally) • PCM has failed
DTC: P0330 **1T CCM, MIL: YES** **1996, 1997, 1998** **Models:** 3.2TL **Engines:** All **Transmissions:** A/T, M/T	**Knock Sensor 2 Circuit Malfunction** Engine running for 1 minute, and the PCM detected an unexpected condition on the Knock Sensor (KS) 2 circuit during the CCM test. **Possible Causes:** • Knock sensor signal circuit is open (front bank of engine) • Knock sensor signal circuit is grounded (front bank of engine) • Knock sensor not tightened properly • Knock sensor damaged or has failed (it may be open internally) • PCM has failed
DTC: P0332 **1T CCM, MIL: YES** **1995, 1996, 1997, 1998** **Models:** 2.5TL **Engines:** All **Transmissions:** A/T, M/T	**Knock Sensor Rear Circuit Low Input** Engine started, engine running for 1 minute, and the PCM detected an unexpected low input on the Rear Knock Sensor circuit. **Possible Causes:** • Knock sensor signal circuit is grounded (rear bank of engine) • Knock sensor not tightened properly • Knock sensor damaged or has failed (it may be shorted) • PCM has failed
DTC: P0333 **1T CCM, MIL: YES** **1995, 1996, 1997, 1998** **Models:** 2.5TL **Engines:** All **Transmissions:** A/T, M/T	**Knock Sensor Rear Circuit High Input** Engine started, engine running for 1 minute, and the PCM detected an unexpected high voltage on the Rear Knock Sensor circuit. **Possible Causes:** • Knock sensor signal circuit is open (rear bank of engine) • Knock sensor signal circuit is shorted to VREF or system power • Knock sensor damaged or has failed (it may be open internally) • PCM has failed
DTC: P0335 **1T CCM, MIL: YES** **1995, 1996, 1997, 1998, 1999,** **2000, 2001, 2002, 2003, 2004,** **2005** **Models:** NSX, NSX-T **Engines:** All **Transmissions:** A/T, M/T	**CKP Sensor 1 Circuit Malfunction (No Signal)** Engine cranking or running; and the PCM did not detect any signals from the Crankshaft Position (CKP) Sensor 1 during the test. Note: The engine will crank for a longer period of time, may buck or jerk, but it will start and run without the CKP sensor signal present. **Possible Causes:** • CKP1 sensor signal circuit is open • CKP1 sensor is shorted to ground • CKP1 sensor is damaged or has failed • PCM has failed

DTC	Trouble Code Title, Conditions & Possible Causes
DTC: P0335 **1T CCM, MIL: YES** **1995, 1996, 1997, 1998** **Models:** 2.5TL **Engines:** All **Transmissions:** A/T, M/T	**CKP Sensor 1 Circuit Malfunction (No Signal)** Engine cranking or running; and the PCM did not detect any signals from the Crankshaft Position (CKP) Sensor 1 during the test. Note: The engine will crank for a longer period of time, may buck or jerk, but it will start and run without the CKP sensor signal present. **Possible Causes:** • CKP1 sensor signal circuit is open or shorted to ground • CKP1 sensor is damaged or has failed • PCM has failed
DTC: P0335 **1T CCM, MIL: YES** **1996, 1997, 1998, 1999, 2000, 2001, 2002, 2003, 2004, 2005, 2006** **Models:** 2.2CL, 2.3CL, 3.0CL, 3.2CL, 3.2TL, TL, 3.5RL, RL, Integra, MDX, RSX **Engines:** All **Transmissions:** A/T, M/T	**CKP Sensor 1 Circuit Malfunction (No Signal)** Engine cranking or running; and the PCM did not detect any signals from the Crankshaft Position (CKP) Sensor 1 during the test. Note: The engine will crank for a longer period of time, may buck or jerk, but it will start and run without the CKP sensor signal present. **Possible Causes:** • CKP1 sensor signal circuit is open • CKP1 sensor is shorted to ground • CKP1 sensor is damaged or has failed • PCM has failed
DTC: P0336 **1T CCM, MIL: YES** **1995, 1996, 1997, 1998, 1999, 2000, 2001, 2002, 2003, 2004, 2005** **Models:** NSX, NSX-T **Engines:** All **Transmissions:** A/T, M/T	**CKP Sensor 1 Circuit Intermittent Signal** Engine started, and the PCM detected the CKP Sensor 1 signal was interrupted during the CCM test. Note: This trouble code is usually caused by an intermittent fault. **Possible Causes:** • CKP1 sensor signal circuit is open or shorted to ground • CKP1 sensor is damaged or has failed • Engine timing belt has a skipped teeth condition • PCM has failed
DTC: P0336 **1T CCM, MIL: YES** **1995, 1996, 1997, 1998** **Models:** 2.5TL **Engines:** All **Transmissions:** A/T, M/T	**CKP Sensor 1 Circuit Intermittent Signal** Engine started, and the PCM detected the CKP Sensor 1 signal was interrupted during the CCM test. Note: This trouble code is usually caused by an intermittent fault. **Possible Causes:** • CKP1 sensor signal circuit is open or shorted to ground • CKP1 sensor is damaged or has failed • Engine timing belt has a skipped teeth condition • PCM has failed
DTC: P0336 **1T CCM, MIL: YES** **1996, 1997, 1998, 1999, 2000, 2001, 2002, 2003, 2004, 2005, 2006** **Models:** 2.2CL, 2.3CL, 3.0CL, 3.2CL, 3.2TL, TL, 3.5RL, RL, Integra, RSX **Engines:** All **Transmissions:** A/T, M/T	**CKP Sensor 1 Circuit Intermittent Signal** Engine started, and the PCM detected the CKP Sensor 1 signal was interrupted during the CCM test. Note: This trouble code is usually caused by an intermittent fault. **Possible Causes:** • CKP1 sensor signal circuit is open (intermittent) • CKP1 sensor is shorted to ground (intermittent) • CKP1 sensor is damaged or has failed (intermittent fault) • PCM has failed
DTC: P0336 **1T CCM, MIL: YES** **2001, 2002** **Models:** MDX **Engines:** All **Transmissions:** A/T, M/T	**CKP Sensor 1 Circuit Intermittent Signal** Engine started, and the PCM detected an unexpected interruption in the CKP Sensor 1 signal in the CCM test. **Possible Causes:** • CKP1 sensor signal circuit is open (intermittent) • CKP1 sensor is shorted to ground (intermittent) • CKP1 sensor is damaged or has failed (intermittent fault) • PCM has failed
DTC: P0336 **2T CCM, MIL: YES** **1996, 1997, 1998, 1999** **Models:** SLX **Engines:** 3.2L VIN V, 3.5L VIN X **Transmissions:** A/T	**Crankshaft Position 58X Sensor Circuit Malfunction** Engine started, and the PCM detected extra or missing pulses between consecutive Crankshaft Position (CKP) 58X sensor signals during 10 out of 100 revolutions during the CCM test. **Possible Causes:** • CKP sensor 58X signal circuit is open or shorted to ground • CKP sensor ground circuit is open or has high resistance • CKP sensor power circuit is open between the sensor and PCM • CKP sensor is damaged, or the reluctor wheel is damaged • PCM has failed (the Ignition module function is inside the PCM)

DTC	Trouble Code Title, Conditions & Possible Causes
DTC: P0337 **2T CCM, MIL: YES** **1996, 1997, 1998, 1999** **Models:** SLX **Engines:** 3.2L VIN V, 3.5L VIN X **Transmissions:** A/T	**Crankshaft Position 58X Sensor Circuit Low Input** DTC P0341 and P0342 not set, engine started, and the PCM did not detect any Crankshaft Position (CKP) 58X sensor pulses present between two (2) CMP sensor pulses, or it did not detect any CKP sensor pulses within 8 CMP sensor pulses during the CCM test. **Possible Causes:** • CKP sensor 58X signal circuit is open or shorted to ground • CKP sensor ground circuit is open or has high resistance • CKP sensor power circuit is open between the sensor and PCM • CKP sensor is damaged, or the reluctor wheel is damaged • PCM has failed (the Ignition module function is inside the PCM)
DTC: P0339 **1T CCM, MIL: YES** **2003, 2004, 2005, 2006** **Models:** MDX **Engines:** All **Transmissions:** A/T, M/T	**CKP Sensor 'A' Circuit Intermittent Signal** Engine started, and the PCM detected an unexpected interruption in the CKP Sensor 'A' signal during testing. **Possible Causes:** • CKP 'A' sensor signal circuit is open (intermittent) • CKP 'A' sensor is shorted to ground (intermittent) • CKP 'A' sensor is damaged or has failed (intermittent fault) • PCM has failed
DTC: P0340 **1T CCM, MIL: YES** **2002, 2003, 2004, 2005, 2006** **Models:** MDX, RSX **Engines:** All **Transmissions:** A/T, M/T	**Camshaft Position Sensor Circuit No Signal** Engine started, CKP sensor signals received, and the PCM detected an intermittent loss of the CMP sensor signal. **Possible Causes:** • CMP sensor signal circuit is open (intermittent fault) • CMP sensor signal circuit is shorted to ground (Intermittent) • CMP sensor is damaged or has failed (intermittent fault) • PCM has failed
DTC: P0340 **1T CCM, MIL: YES** **2003** **Models:** 3.2CL **Engines:** All **Transmissions:** M/T	**Camshaft Position Sensor (TDC) 'A' Circuit Intermittent** Engine started, CKP sensor signals received, and the PCM detected an intermittent loss of the CMP sensor 'A' signal during the CCM test. **Possible Causes:** • CMP sensor signal 'A' circuit is open (intermittent fault) • CMP sensor signal 'A' circuit is shorted to ground (Intermittent) • CMP sensor 'A' is damaged or has failed (intermittent fault) • PCM has failed
DTC: P0341 **1T CCM, MIL: YES** **2003** **Models:** 3.2CL **Engines:** All **Transmissions:** M/T	**Camshaft Position Sensor (TDC) 'A' Circuit No Signal** Engine started, CKP sensor signals received, and the PCM detected a loss of the CMP sensor 'A' signal during the CCM test. **Possible Causes:** • CMP sensor signal 'A' circuit is open • CMP sensor signal 'A' circuit is shorted to ground • CMP sensor 'A' is damaged or has failed • PCM has failed
DTC: P0341 **1T CCM, MIL: YES** **2002, 2003, 2004, 2005, 2006** **Models:** RSX **Engines:** All **Transmissions:** A/T, M/T	**Variable Valve Timing Control Phase Gap Malfunction** Engine started, and the PCM detected a malfunction in the Variable Valve Timing phase gap measurement during the CCM test. **Possible Causes:** • VVT component is damaged or has failed • VVT control solenoid is damaged or has failed • PCM has failed • TSB 02-035 (11/25/02) contains a repair procedure for the code
DTC: P0341 **2T CCM, MIL: YES** **1996, 1997, 1998, 1999** **Models:** SLX **Engines:** 3.2L VIN V, 3.5 VIN X **Transmissions:** A/T	**Camshaft Position Sensor Range/Performance** Engine started, engine running with CMP sensor pulses received, and the PCM detected an incorrect number of CMP signals were received during 10 tests over a 100-test sample period (that lasts 15.6 ms) during the CCM Rationality test. **Note: If a CKP sensor code is also set, check the common ground circuit between the CKP and CMP sensors for an open condition. If a fuel injector code is also set, check the power feed circuit as it connects to the fuel injectors and to the CMP sensor.** **Possible Causes:** • CMP sensor circuit is open or shorted to ground (intermittent) • CMP sensor ground circuit is open (an intermittent fault) • CMP sensor is damaged or has failed

DTC	Trouble Code Title, Conditions & Possible Causes
DTC: P0342 **2T CCM, MIL: YES** **1996, 1997, 1998, 1999** **Models:** SLX **Engines:** 3.2L VIN V, 3.5 VIN X **Transmissions:** A/T	**Camshaft Position Sensor Circuit Low Input** Engine started, and PCM did not detect a Camshaft Position (CMP) sensor pulse at least once for every six (6) rotations of the crankshaft in a 10 second period during the CCM test. **Possible Causes:** • CMP sensor signal circuit is open • CMP sensor signal circuit is shorted to ground • CMP sensor power circuit open between the sensor and PCM • CMP sensor ground circuit is open • CMP sensor is damaged or has failed
DTC: P0344 **1T CCM, MIL: YES** **2002, 2003, 2004, 2005, 2006** **Models:** MDX, RSX **Engines:** All **Transmissions:** A/T, M/T	**Camshaft Position Sensor Circuit Intermittent Signal** Engine started, and the PCM detected an unexpected interruption of the CMP sensor signal during testing. **Possible Causes:** • CMP Sensor 1 is damaged or has failed • CMP Sensor 1 signal circuit is open or shorted to ground • Engine timing belt has a skipped teeth condition • PCM has failed
DTC: P0351 **1T CCM, MIL: YES** **1996, 1997, 1998, 1999** **Models:** SLX **Engines:** 3.2L VIN V, 3.5L VIN X **Transmissions:** A/T	**Ignition Control Module Circuit 1 Malfunction** Engine running with CKP 58X signals received, and the PCM detected the IC output signal did not equal 5v with the output commanded "on", or it did not equal 0v with the output commanded "off" in 20 tests over a 40 sample period during the CCM test. **Possible Causes:** • EST signal circuit is open between module and coil or the PCM • EST signal circuit shorted between module and coil or the PCM • ION module is damaged or has failed • PCM has failed
DTC: P0352 **1T CCM, MIL: YES** **1996, 1997, 1998, 1999** **Models:** SLX **Engines:** 3.2L VIN V, 3.5L VIN X **Transmissions:** A/T	**Ignition Control Module Circuit 2 Malfunction** Engine running with CKP 58X signals received, and the PCM detected the IC output signal did not equal 5v with the output commanded "on", or it did not equal 0v with the output commanded "off" in 20 tests over a 40 sample period during the CCM test. **Possible Causes:** • EST signal circuit is open between module and coil or the PCM • EST signal circuit shorted between module and coil or the PCM • ION module is damaged or has failed • PCM has failed
DTC: P0353 **1T CCM, MIL: YES** **1996, 1997, 1998, 1999** **Models:** SLX **Engines:** 3.2L VIN V, 3.5L VIN X **Transmissions:** A/T	**Ignition Control Module Circuit 3 Malfunction** Engine running with CKP 58X signals received, and the PCM detected the IC output signal did not equal 5v with the output commanded "on", or it did not equal 0v with the output commanded "off" in 20 tests over a 40 sample period during the CCM test. **Possible Causes:** • EST signal circuit is open between module and coil or the PCM • EST signal circuit shorted between module and coil or the PCM • ION module is damaged or has failed • PCM has failed
DTC: P0354 **1T CCM, MIL: YES** **1996, 1997, 1998, 1999** **Models:** Amigo, Rodeo, Rodeo Sport **Engines:** 3.2L VIN V, 3.5 VIN X **Transmissions:** A/T	**Ignition Control Module Circuit 4 Malfunction** Engine running with CKP 58X signals received, and the PCM detected the IC output signal did not equal 5v with the output commanded "on", or it did not equal 0v with the output commanded "off" in 20 tests over a 40 sample period during the CCM test. **Possible Causes:** • EST signal circuit is open between module and coil or the PCM • EST signal circuit shorted between module and coil or the PCM • ION module is damaged or has failed • PCM has failed
DTC: P0355 **1T CCM, MIL: YES** **1996, 1997, 1998, 1999** **Models:** SLX **Engines:** 3.2L VIN V, 3.5L VIN X **Transmissions:** A/T	**Ignition Control Module Circuit 5 Malfunction** Engine running with CKP 58X signals received, and the PCM detected the IC output signal did not equal 5v with the output commanded "on", or it did not equal 0v with the output commanded "off" in 20 tests over a 40 sample period during the CCM test. **Possible Causes:** • EST signal circuit is open between module and coil or the PCM • EST signal circuit shorted between module and coil or the PCM • ION module is damaged or has failed • PCM has failed

DTC	Trouble Code Title, Conditions & Possible Causes
DTC: P0356 **1T CCM, MIL: YES** **1996, 1997, 1998, 1999** **Models:** SLX **Engines:** 3.2L VIN V, 3.5L VIN X **Transmissions:** A/T	**Ignition Control Module Circuit 6 Malfunction** Engine running with CKP 58X signals received, and the PCM detected the IC output signal did not equal 5v with the output commanded "on", or it did not equal 0v with the output commanded "off" in 20 tests over a 40 sample period during the CCM test. **Possible Causes:** • EST signal circuit is open between module and coil or the PCM • EST signal circuit shorted between module and coil or the PCM • ION module is damaged or has failed • PCM has failed
DTC: P0365 **1T CCM, MIL: YES** **2003** **Models:** 3.2CL **Engines:** All **Transmissions:** M/T	**Camshaft Position Sensor (TDC) 'B' Circuit Malfunction** Engine started, CKP sensor signals received, and the PCM detected an intermittent loss of the CMP sensor 'B' signal during the CCM test. **Possible Causes:** • CMP sensor signal 'B' circuit is open • CMP sensor signal 'B' circuit is shorted to ground • CMP sensor 'B' is damaged or has failed • PCM has failed
DTC: P0366 **1T CCM, MIL: YES** **2003** **Models:** 3.2CL **Engines:** All **Transmissions:** M/T	**Camshaft Position Sensor (TDC) 'B' Circuit Intermittent Signal** Engine started, CKP sensor signals received, and the PCM detected a loss of the CMP sensor 'B' signal during the CCM test. **Possible Causes:** • CMP sensor signal 'B' circuit is open (intermittent fault) • CMP sensor signal 'B' circuit is shorted to ground (Intermittent) • CMP sensor 'B' is damaged or has failed (intermittent fault) • PCM has failed
DTC: P0385 **1T CCM, MIL: YES** **2003** **Models:** 3.2CL **Engines:** All **Transmissions:** M/T	**Camshaft Position Sensor 'B' Circuit Malfunction** Engine started, CKP sensor signals received, and the PCM detected a loss of the CMP sensor 'B' signal during the CCM test. **Possible Causes:** • CMP sensor signal 'B' circuit is open • CMP sensor signal 'B' circuit is shorted to ground . • CMP sensor 'B' is damaged or has failed • PCM has failed
DTC: P0385 **1T CCM, MIL: YES** **2003, 2004, 2005, 2006** **Models:** MDX **Engines:** All **Transmissions:** A/T, M/T	**Camshaft Position Sensor 'B' Circuit Malfunction** Engine started, CKP sensor signals received, and the PCM detected a loss of the CMP sensor 'B' signal during testing. **Possible Causes:** • CMP sensor 'B' circuit is open • CMP sensor 'B' circuit is shorted to ground • CMP sensor 'B' is damaged or has failed • PCM has failed
DTC: P0386 **1T CCM, MIL: YES** **2003** **Models:** 3.2CL **Engines:** All **Transmissions:** M/T	**Camshaft Position Sensor 'B' Circuit Range/Performance** Engine started, CKP sensor signals received, and the PCM detected an intermittent loss of the CMP sensor 'B' signal during the CCM test. **Possible Causes:** • CMP sensor signal 'B' circuit is open (intermittent fault) • CMP sensor signal 'B' circuit is shorted to ground (Intermittent) • CMP sensor 'B' is damaged or has failed • PCM has failed
DTC: P0389 **1T CCM, MIL: YES** **2003, 2004, 2005, 2006** **Models:** MDX **Engines:** All **Transmissions:** A/T, M/T	**Camshaft Position Sensor 'B' Circuit Intermittent Signal** Engine started, CKP sensor signals received, and the PCM detected an intermittent loss of the CMP Sensor 'B' signal. **Possible Causes:** • CMP sensor signal 'B' circuit is open (intermittent fault) • CMP sensor signal 'B' circuit is shorted to ground (Intermittent) • CMP sensor 'B' is damaged or has failed (intermittent fault) • PCM has failed
DTC: P0401 **2T EGR1, MIL: YES** **1995, 1996, 1997, 1998, 1999,** **2000, 2001, 2002, 2003, 2004,** **2005** **Models:** NSX, NSX-T **Engines:** All **Transmissions:** A/T, M/T	**EGR System Insufficient Flow Detected** Cold engine startup (ECT sensor less than 76°F at startup), engine running in closed loop at 40-55 mph for 2 minutes in high gear followed by a deceleration period back to 35 mph with the throttle closed, and the PCM detected a signal from the EGR position sensor that indicated insufficient EGR flow during the test. **Possible Causes:** • EGR valve source vacuum supply line open or restricted • EGR exhaust manifold passages are clogged or restricted • EGR valve assembly or solenoid valve damaged or has failed • PCM has failed

DTC	Trouble Code Title, Conditions & Possible Causes
DTC: P0401 **2T EGR1, MIL: YES** **1995, 1996, 1997, 1998** **Models:** 2.5TL **Engines:** All **Transmissions:** A/T, M/T	**EGR System Insufficient Flow Detected** Cold engine startup (ECT sensor less than 76°F at startup), engine running in closed loop at 40-55 mph for 2 minutes in high gear, followed by a deceleration period back to 35 mph with the throttle closed, and the PCM detected a signal from the EGR position sensor that indicated insufficient EGR flow during the test. **Possible Causes:** • EGR valve source vacuum supply line open or restricted • EGR exhaust manifold passages are clogged or restricted • EGR valve assembly or solenoid valve damaged or has failed • PCM has failed
DTC: P0401 **2T EGR1, MIL: YES** **1996, 1997, 1998, 1999, 2000, 2001, 2002, 2003, 2004, 2005, 2006** **Models:** 2.2CL, 2.3CL, 3.0CL, 3.2CL, 3.2TL, TL, 3.5RL, RL, MDX **Engines:** All **Transmissions:** A/T, M/T	**EGR System Insufficient Flow Detected** Cold engine startup (ECT sensor less than 76°F at startup), engine running in closed loop at 40-55 mph for 5 minutes in high gear followed by a deceleration period back to 35 mph for 5 seconds with the throttle closed, and the PCM detected a signal from the EGR position sensor that indicated insufficient EGR flow during the test. **Possible Causes:** • EGR valve source vacuum supply line open or restricted • EGR exhaust manifold passages are clogged or restricted • EGR valve assembly or solenoid valve damaged or has failed • PCM has failed
DTC: P0401 **1T CCM, MIL: YES** **1996, 1997, 1998, 1999** **Models:** SLX **Engines:** 3.2L VIN V, 3.5L VIN X **Transmissions:** A/T	**Insufficient EGR System Flow Detected** No ECT, EGR Pintle Position, EVAP, IAC, IAT, MAP, Misfire, TP or VSS codes set, system voltage from 11-16v, ECT sensor more than 140°F, BARO sensor over 75 kPa, IAC position stable (±10 counts), A/C Clutch and TCC status unchanged, and VSS over 15 mph, then with the throttle closed (TP angle under 1%), EGR duty cycle under 1%, MAP sensor from 10-40 kPa (±2 kPa), engine speed from 1100-2000 rpm, the PCM detected the compensated MAP sensor signal indicated a value from 10.3-49.8 kPa during the EGR System test. **Possible Causes:** • Linear EGR valve "low" circuit is open or shorted to ground • Linear EGR valve "low" circuit is shorted to system power (B+) • EGR valve VREF (5-volt) is open between sensor and the PCM • EGR valve feedback circuit is open or shorted to ground • EGR valve is stuck closed, or partially open during the test • EGR exhaust flow path may be restricted • EGR valve is damaged, or has failed • PCM has failed
DTC: P0402 **2T EGR2, MIL: YES** **1998, 1999** **Models:** SLX **Engines:** 3.5L VIN X **Transmissions:** A/T	**Excessive EGR System Excessive Flow Detected** Engine started, IAT sensor more than 38°F, engine running, system voltage at 11-16v, and the PCM detected the EGR position sensor signal indicated more than 21% over a 625 ms period during the EGR System flow test right after engine startup. **Possible Causes:** • Linear EGR valve control circuit is shorted to ground • EGR valve is stuck partially open during the initial startup test • EGR valve is damaged, or has excessive carbon buildup • PCM has failed
DTC: P0403 **2T CCM, MIL: YES** **2003, 2004, 2005, 2006** **Models:** MDX **Engines:** All **Transmissions:** A/T, M/T	**EGR Solenoid Control Circuit Malfunction** Key on or engine running; and the PCM detected an unexpected voltage condition on the EGR solenoid after it was cycled from "on" to "off". The solenoid resistance is 6.3-6.7 ohms. **Possible Causes:** • EGR solenoid control circuit open or shorted to ground • EGR solenoid control circuit shorted to VREF or system power • EGR solenoid is damaged or has failed • PCM has failed (the EGR solenoid driver may be open/shorted)
DTC: P0404 **2T CCM, MIL: YES** **2003, 2004, 2005, 2006** **Models:** MDX **Engines:** All **Transmissions:** A/T, M/T	**EGR Solenoid Control Circuit Range/Performance** Engine started, engine running at cruise speed in closed loop, and the PCM did not detect the correct amount of change in the EGR valve position sensor signal after the valve was cycled "on" to "off". **Possible Causes:** • EGR valve source vacuum supply line clogged (remove EGR valve plate to gain access to the manifold port that is clogged) • EGR valve or exhaust manifold passages are restricted • EGR solenoid connector is disconnected or damaged

DTC	Trouble Code Title, Conditions & Possible Causes
DTC: P0404 **2T EGR2, MIL: YES** **1998, 1999** **Models:** SLX **Engines:** 3.5L VIN X **Transmissions:** A/T	**EGR Pintle Position Sensor Range/Performance** Engine started, IAT sensor more than 38ºF, engine speed less than 600 rpm, system voltage at 11-16v, then with the Desired EGR position at over 0%, the PCM detected the difference between the Actual and Desired EGR position was more than 15% for over 15 seconds. This fault must occur 3 times in a single trip to set a code. **Possible Causes:** • Linear EGR valve control circuit is shorted to ground • EGR valve is stuck partially open during the initial startup test • EGR valve is damaged, or has excessive carbon buildup • PCM has failed
DTC: P0405 **2T EGR2, MIL: YES** **1998, 1999, 2000, 2001, 2002** **Models:** SLX **Engines:** 3.2L VIN V, 3.5L VIN X **Transmissions:** A/T	**EGR Pintle Position Sensor Circuit Low Input** Key on or engine running; system voltage from 11-16v, IAT sensor more than 140ºF, and the PCM detected the EGR position sensor indicated less than 0.10v for 10 seconds during the CCM test. **Possible Causes:** • Linear EGR valve control circuit is shorted to ground • EGR valve is stuck partially open during the initial startup test • EGR valve is damaged, or has excessive carbon buildup • PCM has failed
DTC: P0406 **2T CCM, MIL: YES** **2003, 2004, 2005, 2006** **Models:** MDX **Engines:** All **Transmissions:** A/T, M/T	**EGR Valve Position Sensor Circuit High Input** Key on or engine running; and the PCM detected an unexpected "high" voltage condition (more than 4.88v) on the EGR valve position sensor signal circuit during the CCM test. **Possible Causes:** • EGR valve position sensor circuit is open • EGR valve position sensor circuit is shorted to VREF or power • EGR valve position sensor is damaged or has failed • PCM has failed
DTC: P0406 **1T CCM, MIL: YES** **1998, 1999** **Models:** SLX **Engines:** 3.5L VIN X **Transmissions:** A/T	**EGR Pintle Position Sensor Circuit High Input** Engine started, system voltage from 11-16v, IAT sensor more than 41ºF, and the PCM detected the EGR position sensor indicated more than 4.80v for 10 seconds during the CCM test. **Possible Causes:** • EGR feedback signal circuit is open between sensor and PCM • EGR sensor is damaged or has failed • PCM has failed
DTC: P0420 **3T CAT1, MIL: YES** **1995, 1996, 1997, 1998, 1999, 2000, 2001, 2002, 2003, 2004, 2005** **Models:** NSX, NSX-T **Engines:** All **Transmissions:** A/T, M/T	**Catalyst Efficiency Below Thresholds (Bank 1)** DTC P0137, P0138, P0139, P0141, P0157, P0158, P0159 and P0161 not set, vehicle driven to a speed of 40-55 mph for 2 minutes in closed loop, followed by a deceleration period to 35 mph with the throttle closed, and the PCM detected excessive activity in the Catalyst oxygen sensor (Bank 1) during the Catalyst Monitor test. **Possible Causes:** • Air leaks in at the exhaust manifold or exhaust pipes • Catalytic converter damaged or has failed (deteriorated) • Front HO2S is more aged than the rear HO2S (HO2S is lazy) • PCM has failed
DTC: P0420 **3T CAT1, MIL: YES** **1995, 1996, 1997, 1998** **Models:** 2.5TL **Engines:** All **Transmissions:** A/T, M/T	**Catalyst Efficiency Below Thresholds (Bank 1)** DTC P0137, P0138 and P0141 not set, engine running in closed loop at 40-55 mph for 2 minutes, followed by a deceleration period to 35 mph at closed throttle, and the PCM detected excessive activity in the Catalyst oxygen sensor (Bank 1) in the Catalyst Monitor test. **Possible Causes:** • Air leaks in at the exhaust manifold or exhaust pipes • Catalytic converter damaged or has failed (deteriorated) • Front HO2S is more aged than the rear HO2S (HO2S is lazy) • PCM has failed
DTC: P0420 **3T CAT1, MIL: YES** **1996, 1997, 1998, 1999, 2000, 2001, 2002, 2003, 2004, 2005, 2006** **Models:** 2.2CL, 2.3CL, 3.0CL, 3.2CL, 3.2TL, TL, 3.5RL, RL, Integra, MDX, RSX **Engines:** All **Transmissions:** A/T, M/T	**Catalyst Efficiency Below Thresholds (Bank 1)** DTC P0137, P0138 and P0141 not set, engine running in closed loop at 40-55 mph for 2 minutes, followed by a deceleration period to 35 mph at closed throttle, and the PCM detected excessive activity in the Catalyst oxygen sensor (Bank 1) in the Catalyst Monitor test. **Possible Causes:** • Air leaks in at the exhaust manifold or exhaust pipes • Catalytic converter damaged or has failed (deteriorated) • Front HO2S is more aged than the rear HO2S (HO2S is lazy) • PCM has failed

DTC	Trouble Code Title, Conditions & Possible Causes
DTC: P0420 **1T CAT2, MIL: YES** **1996, 1997, 1998, 1999** **Models:** SLX **Engines:** 3.2L VIN V, 3.5L VIN X **Transmissions:** A/T	**Catalyst Efficiency Below Normal (Bank 1)** DTC P0106, P0107, P0108, P0112, P0113, P0117, P0118, P0121, P0122, P0123, P0131, P0132, P0133, P0134, P0137, P0138, P0140, P0141, P0171, P0172, P0300, P0301-P0306, P0336, P0341, P0342, P0401, P0502, P0506 and P0507 not set, vehicle driven to a speed of 16-75 mph at less than 3500 rpm in closed loop for 2-5 minutes, ECT sensor more than 140°F, MAF sensor from 8-50 g/sec, engine load less than 99% (±8%), predicted Catalyst temperature over 750°F, and the PCM detected the catalyst oxygen storage capacity was below an acceptable threshold during the test. **Possible Causes:** • Air leaks at the exhaust manifold or in the exhaust pipes • Front HO2S or rear HO2S is contaminated with fuel or moisture • Front HO2S older (aged) than the rear HO2S (HO2S is lazy) • Front HO2S and/or the rear HO2S is loose in the mounting hole • Catalytic converter is damaged or has failed
DTC: P0430 **3T CAT1, MIL: YES** **1995, 1996, 1997, 1998, 1999, 2000, 2001, 2002, 2003, 2004, 2005** **Models:** NSX, NSX-T **Engines:** All **Transmissions:** A/T, M/T	**Catalyst Efficiency Below Thresholds (Bank 2)** DTC P0155 and P0161 not set, engine started, vehicle driven in closed loop at 40-55 mph for 2 minutes, followed by a deceleration period to 35 mph with the throttle closed, and the PCM detected excessive activity in the rear HO2S-22 in the Catalyst Monitor test. **Possible Causes:** • Air leaks in at the exhaust manifold or exhaust pipes • Catalytic converter is damaged or has failed (deteriorated) • Front HO2S is more aged than the rear HO2S (HO2S is lazy) • PCM has failed
DTC: P0430 **3T CAT1, MIL: YES** **2003, 2004, 2005, 2006** **Models:** MDX **Engines:** All **Transmissions:** A/T, M/T	**Catalyst Efficiency Below Thresholds (Bank 2)** DTC P0157, P0158 and P0161 not set, engine started, ECT sensor signal more than 158°F, vehicle driven in closed loop at 50-55 mph for 2 minutes, and the PCM detected excessive activity in the rear HO2S-22 during the Catalyst Monitor test. **Possible Causes:** • Air leaks in at the exhaust manifold or exhaust pipes • Catalytic converter is damaged or has failed (deteriorated) • Front HO2S is more aged than the rear HO2S (HO2S is lazy) • Fuel quality is poor or fuel is contaminated • PCM has failed
DTC: P0430 **1T CAT2, MIL: YES** **1996, 1997, 1998, 1999** **Models:** SLX **Engines:** 3.2L VIN V, 3.5L VIN X **Transmissions:** A/T	**Catalyst Efficiency Below Normal (Bank 2)** DTC P0106, P0107, P0108, P0112, P0113, P0117, P0118, P0121, P0122, P0123, P0131, P0132, P0133, P0134, P0137, P0138, P0140, P0141, P0171, P0172, P0300, P0301-P0306, P0336, P0341, P0342, P0401, P0502, P0506 and P0507 not set, engine speed less than 3500 rpm in closed loop, ECT sensor more than 140°F, MAF sensor from 8-50 g/sec, engine load less than 99% (±8%), predicted Catalyst temperature over 750°F, vehicle speed from 16-75 mph, and the PCM detected the catalyst oxygen storage capacity was below an acceptable threshold during the Catalyst test. **Possible Causes:** • Air leaks at the exhaust manifold or in the exhaust pipes • Front HO2S or rear HO2S is contaminated with fuel or moisture • Front HO2S older (aged) than the rear HO2S (HO2S is lazy) • Front HO2S and/or the rear HO2S is loose in the mounting hole • Catalytic converter is damaged or has failed
DTC: P0440 **2T EVAP1, MIL: YES** **1997, 1998** **Models:** 2.5TL **Engines:** All **Transmissions:** A/T, M/T	**EVAP System Leak Detected** Cold engine startup (ECT sensor less than 154°F and IAT sensor more than 14°F at startup), engine runtime 3-5 minutes, then an acceleration period to 50-60 mph with the throttle steady, and the PCM detected a leak somewhere in the EVAP system during the EVAP Leak Test. **Possible Causes:** • Fuel filler cap is loose or missing, or the fuel tank has a leak • EVAP bypass solenoid valve or two-way valve vacuum hose is loose/disconnected, or the valve(s) is damaged or has failed • EVAP purge control solenoid valve is damaged or has failed • EVAP purge control solenoid circuit open or shorted to ground • EVAP control canister vent shut valve is damaged or has failed

DTC	Trouble Code Title, Conditions & Possible Causes
DTC: P0440 **2T EVAP2, MIL: YES** **1996, 1997** **Models:** SLX **Engines:** 3.2L VIN V **Transmissions:** A/T	**EVAP System Performance** DTC P0106, P0107, P0108, P0112, P0113, P0117, P0118, P0121, P0122, P0123, P1640 ad P1650 not set, ECT and IAT sensors less than 90ºF and within 13ºF at startup, ECT sensor over 39ºF at startup, IAT sensor over 4ºF at startup, system voltage from 11-16v, BARO sensor over 75 kPa, throttle angle from 7-30%, fuel level from 15-85% with minimum fuel slosh, vehicle speed under 75 mph, and the PCM determined it was unable to achieve or maintain vacuum in the system for 60-180 seconds during the EVAP Monitor leak test. **Possible Causes:** • Charcoal canister is loaded with fuel or moisture • ECT, IAT, MAP, VSS or TP sensor signals out-of-calibration • Fuel filler cap loose, cross-threaded, incorrect part or damaged • Fuel tank pressure sensor is damaged or has failed • Fuel tank or fuel tank sender assembly 'O' ring is leaking • Fuel tank vapor line(s) block, damaged or disconnected • Purge or Vent solenoid control circuit open or shorted to ground • Purge or Vent solenoid power circuit is open (check the fuse)
DTC: P0440 **2T EVAP2, MIL: YES** **1998, 1999** **Models:** SLX **Engines:** 3.5L VIN X **Transmissions:** A/T	**EVAP System Performance** DTC P0106, P0107, P0108, P0112, P0113, P0117, P0118, P0121, P0122, P0123, P1640 ad P1650 not set, ECT and IAT sensors less than 90ºF and within 13ºF at startup, ECT sensor over 39ºF at startup, IAT sensor over 4ºF at startup, system voltage from 11-16v, BARO sensor over 75 kPa, throttle angle from 7-30%, fuel level from 15-85% with minimum fuel slosh, vehicle speed under 75 mph, and the PCM determined it was unable to achieve or maintain vacuum in the system for 60-180 seconds during the EVAP Monitor leak test. **Possible Causes:** • Charcoal canister is loaded with fuel or moisture • ECT, IAT, MAP, VSS or TP sensor signals out-of-calibration • Fuel filler cap loose, cross-threaded, incorrect part or damaged • Fuel tank pressure sensor is damaged or has failed • Fuel tank or fuel tank sender assembly 'O' ring is leaking • Fuel tank vapor line(s) block, damaged or disconnected • Purge or Vent solenoid control circuit open or shorted to ground • Purge or Vent solenoid power circuit is open (check the fuse)
DTC: P0441 **2T EVAP1, MIL: YES** **1995, 1996, 1997** **Models:** NSX, NSX-T **Engines:** All **Transmissions:** A/T, M/T	**EVAP System Incorrect Purge Flow** Cold engine startup (ECT sensor less than 154ºF and IAT sensor more than 14ºF at startup), engine runtime from 3-5 minutes, followed by an acceleration period to 50-60 mph at steady throttle, and the PCM did not detect enough purge flow through the EVAP system during the EVAP Monitor test. **Possible Causes:** • EVAP purge control valve hose loose or purge port is plugged • EVAP purge control solenoid is damaged • EVAP purge control solenoid circuit open or shorted to ground • EVAP purge flow switch is damaged or the circuit is open
DTC: P0441 **2T EVAP1, MIL: YES** **1995, 1996, 1997, 1998** **Models:** 3.2TL **Engines:** All **Transmissions:** A/T, M/T	**EVAP System Incorrect Purge Flow** Cold engine startup (ECT sensor less than 154ºF and IAT sensor more than 14ºF at startup), engine runtime from 3-5 minutes, followed by an acceleration period to 50-60 mph at steady throttle, and the PCM detected a lack of purge flow in the system in the test. **Possible Causes:** • EVAP purge control valve hose loose or purge port is plugged • EVAP purge control solenoid is damaged • EVAP purge control solenoid circuit open or shorted to ground • EVAP purge flow switch is damaged or the circuit is open
DTC: P0441 **2T EVAP1, MIL: YES** **1995, 1996** **Models:** 2.5TL **Engines:** All **Transmissions:** A/T, M/T	**EVAP System Incorrect Purge Flow** Cold engine startup (ECT sensor less than 154ºF and IAT sensor more than 14ºF at startup), engine runtime from 3-5 minutes, followed by an acceleration period to 50-60 mph at steady throttle, and the PCM detected a lack of purge flow in the system in the test. **Possible Causes:** • EVAP purge control valve hose loose or purge port is plugged • EVAP purge control solenoid is damaged • EVAP purge control solenoid circuit open or shorted to ground • EVAP purge flow switch is damaged or the circuit is open
DTC: P0441 **2T EVAP1, MIL: YES** **1996, 1997, 1998** **Models:** 2.2CL, 3.5RL, Integra **Engines:** All **Transmissions:** A/T, M/T	**EVAP System Incorrect Purge Flow** Cold engine startup (ECT sensor less than 154ºF and IAT sensor more than 14ºF at startup), engine runtime from 3-5 minutes, followed by an acceleration period to 50-60 mph at steady throttle, and the PCM detected a lack of purge flow in the system in the test. **Possible Causes:** • EVAP purge control diaphragm valve hose loose/disconnected • EVAP purge cutoff or purge control solenoid valve is damaged • EVAP purge control diaphragm valve is damaged or has failed

DTC	Trouble Code Title, Conditions & Possible Causes
DTC: P0441 **2T EVAP2, MIL: YES** **1996, 1997** **Models:** SLX **Engines:** 3.2L VIN V **Transmissions:** A/T	**EVAP System No Flow During Purge Detected** DTC P0106, P0107, P0108, P0112, P0113, P0117, P0118, P0121, P0122, P0123 and P1442 not set, ECT and IAT sensors more than 41°F and within 45°F at startup, ECT sensor less than 158°F during testing, BARO sensor over 85 kPa, Calculated manifold pressure over 10 kPa, throttle angle over 14%, engine speed from 800-6000 rpm, system voltage from 11-16v, Purge duty cycle over 95%, and the PCM detected the EVAP vacuum switch indicated a "closed" position (with Purge enabled) for 3 seconds in the EVAP Purge test. **Possible Causes:** • Charcoal canister is damaged, clogged or restricted • Purge solenoid circuit is open or shored to ground (intermittent) • Purge valve vacuum line is clogged, restricted or disconnected • Purge vacuum switch is damaged or has failed • PCM has failed
DTC: P0442 **2T EVAP2, MIL: YES** **1999** **Models:** SLX **Engines:** 3.5L VIN X **Transmissions:** A/T	**EVAP System Small Leak (0.040") Detected** DTC P0106, P0107, P0108, P0112, P0113, P0121, P0122, P0123, P0440, P1650 and P1650 not set, ECT and IAT sensors less than 90°F and within 13°F at startup, ECT sensor over 39°F at startup, IAT sensor over 4°F at startup, system voltage from 11-16v, BARO sensor over 75 kPa, throttle angle from 7-30%, fuel level from 15-85% with minimum fuel slosh, vehicle speed under 75 mph, and the PCM detected a vacuum decaying condition characteristic of a small leak (0.040") in the system during the EVAP Monitor leak test. **Possible Causes:** • Charcoal canister is loaded with fuel or moisture • ECT, IAT, MAP, VSS or TP sensor signals out-of-calibration • Fuel filler cap loose, cross-threaded, incorrect part or damaged • Fuel tank pressure sensor is damaged or has failed • Fuel tank or fuel tank sender assembly 'O' ring is leaking • Fuel tank vapor line(s) block, damaged or disconnected • Purge or Vent solenoid control circuit open or shorted to ground • Purge or Vent solenoid power circuit is open (check the fuse)
DTC: P0443 **1T CCM, MIL: YES** **2003, 2004, 2005, 2006** **Models:** MDX **Engines:** All **Transmissions:** A/T, M/T	**EVAP Canister Purge Valve Circuit Malfunction** Engine started, and PCM detected an unexpected voltage condition on the EVAP canister purge solenoid after it was cycled from "on" to "off". The solenoid resistance is 26-28 ohms. **Possible Causes:** • EGR solenoid control circuit open or shorted to ground • EGR solenoid control circuit shorted to VREF or system power • EGR solenoid is damaged or has failed • PCM has failed (the EGR solenoid driver may be open/shorted)
DTC: P0446 **1T EVAP2, MIL: YES** **1999** **Models:** SLX **Engines:** 3.5L VIN X **Transmissions:** A/T	**EVAP Vent Control System Performance** DTC P0106, P0107, P0108, P0112, P0113, P0121, P0122, P0123, P1640 and P1650 not set, ECT and IAT sensors from 39-86°F at startup, ECT signal within 12°F of the IAT signal at startup, and the IAT signal within 2°F of the ECT signal at startup, system voltage from 11-16v, BARO sensor more than 72 kPa, fuel tank level from 12-87%, Purge duty cycle over 50%, and the PCM detected the FTP sensor did no indicate close to -10" H2O under normal purge conditions with the Canister Vent solenoid "open", or it detected the FTP sensor did not indicate about -1.5 to +1.5" H2O at key "on". **Possible Causes:** • Charcoal canister is clogged, plugged or restricted • EVAP vent control solenoid control circuit is shorted to ground • EVAP vent control solenoid hose is bent, kinked or plugged • EVAP vent control solenoid is damaged or has failed • PCM has failed (the solenoid driver circuit may be shorted)
DTC: P0451 **1T CCM, MIL: YES** **2001, 2002, 2003, 2004, 2005** **Models:** NSX, NSX-T **Engines:** All **Transmissions:** A/T, M/T	**Fuel Tank Pressure Sensor Range/Performance** Engine started, and the PCM detected the fuel tank pressure (FTP) sensor signal was less than the allowable range stored in the PCM memory (a calibrated range adjusted to current conditions). **Note: The FTP sensor PID should be near 2.5v with the fuel cap off.** **Possible Causes:** • FTP sensor vacuum lines loose, damaged or disconnected • Fuel tank pressure sensor is damaged or has failed • PCM has failed
DTC: P0451 **2T CCM, MIL: YES** **2001, 2002, 2003, 2004, 2005, 2006** **Models:** 3.2CL, 3.2TL, TL, 3.5RL, RL, Integra, MDX, RSX **Engines:** All **Transmissions:** A/T, M/T	**Fuel Tank Pressure Sensor Circuit Range/Performance** Engine started, and the PCM detected the fuel tank pressure (FTP) sensor signal was less than the allowable range stored in the PCM memory (i.e., a calibrated range adjusted to current conditions). **Note: The FTP sensor PID should be near 2.5v with the fuel cap off.** **Possible Causes:** • FTP sensor vacuum lines loose, damaged or disconnected • Fuel tank pressure sensor is damaged or has failed • PCM has failed

DTC	Trouble Code Title, Conditions & Possible Causes

DTC: P0452
1T CCM, MIL: YES
1997, 1998, 1999, 2000, 2001,
2002, 2003, 2004, 2005
Models: NSX, NSX-T
Engines: All
Transmissions: A/T, M/T
	Fuel Tank Pressure Sensor Circuit Low Input
	Key on or engine running; and the PCM detected the fuel tank pressure (FTP) sensor signal was less than 0.16v during the test.
	Note: The FTP sensor PID should be near 2.5v with the fuel cap off.
	Possible Causes:
	    • FTP sensor signal circuit is shorted to ground
	    • FTP sensor vacuum lines loose, damaged or disconnected
	    • Fuel tank pressure sensor is damaged or has failed
	    • PCM has failed

DTC: P0452
1T CCM, MIL: YES
1997, 1998, 1999, 2000, 2001,
2002, 2003, 2004, 2005, 2006
Models: 2.5TL, 3.2TL, TL
Engines: All
Transmissions: A/T, M/T
	Fuel Tank Pressure Sensor Circuit Low Input
	Key on or engine running; and the PCM detected the fuel tank pressure (FTP) sensor signal was less than 0.16v during the test.
	Note: The FTP sensor PID should be near 2.5v with the fuel cap off.
	Possible Causes:
	    • FTP sensor signal circuit is shorted to ground
	    • FTP sensor vacuum lines loose, damaged or disconnected
	    • Fuel tank pressure sensor is damaged or has failed
	    • PCM has failed

DTC: P0452
1T CCM, MIL: YES
1997, 1998, 1999, 2000, 2001,
2002, 2003, 2004, 2005, 2006
Models: 2.3CL, 3.0CL, 3.2CL,
Integra, MDX, RSX
Engines: All
Transmissions: A/T, M/T
	Fuel Tank Pressure Sensor Circuit Low Input
	Key on or engine running; and the PCM detected the fuel tank pressure (FTP) sensor signal was less than 0.16v during the test.
	Note: The FTP sensor PID should be near 2.5v with the fuel cap off.
	Possible Causes:
	    • FTP sensor signal circuit is shorted to ground
	    • FTP sensor vacuum lines loose, damaged or disconnected
	    • Fuel tank pressure sensor is damaged or has failed
	    • PCM has failed

DTC: P0452
1T CCM, MIL: YES
1996, 1997, 1998, 1999, 2000,
2001, 2002, 2003, 2004, 2005,
2006
Models: 3.5RL, RL
Engines: All
Transmissions: A/T, M/T
	Fuel Tank Pressure Sensor Circuit Low Input
	Key on or engine running; and the PCM detected the fuel tank pressure (FTP) sensor signal was less than 0.16v during the test.
	Note: The FTP sensor PID should be near 2.5v with the fuel cap off.
	Possible Causes:
	    • FTP sensor signal circuit is shorted to ground
	    • FTP sensor vacuum lines loose, damaged or disconnected
	    • Fuel tank pressure sensor is damaged or has failed
	    • PCM has failed

DTC: P0452
1T CCM, MIL: YES
1998, 1999
Models: SLX
Engines: 3.5L VIN X
Transmissions: A/T
	Fuel Tank Pressure Sensor Circuit Low Input
	Key on or engine running; and the PCM detected the Fuel Tank Pressure (FTP) sensor signal was less than 0.20v for 12.5 seconds with 100 test failures occurring in 200 test samples in the CCM test.
	Possible Causes:
	    • FTP sensor signal circuit is shorted to ground
	    • FTP sensor power circuit is open between sensor and the PCM
	    • FTP sensor is damaged or has failed
	    • PCM has failed

DTC: P0453
1T CCM, MIL: YES
1997, 1998, 1999, 2000, 2001,
2002, 2003, 2004, 2005
Models: NSX, NSX-T
Engines: All
Transmissions: A/T, M/T
	Fuel Tank Pressure Sensor Circuit High Input
	Key on or engine running; and the PCM detected the fuel tank pressure (FTP) sensor signal was more than 4.90v during the test.
	Note: The FTP sensor PID should be near 2.5v with the fuel cap off.
	Possible Causes:
	    • FTP sensor signal circuit is shorted to VREF or power (B+)
	    • FTP sensor ground circuit is open
	    • Fuel tank pressure sensor is damaged or has failed
	    • PCM has failed

DTC: P0453
1T CCM, MIL: YES
1997, 1998, 1999, 2000, 2001,
2002, 2003, 2004, 2005, 2006
Models: 2.5TL, 3.2TL, TL
Engines: All
Transmissions: A/T, M/T
	Fuel Tank Pressure Sensor Circuit High Input
	Key on or engine running; and the PCM detected the fuel tank pressure (FTP) sensor signal was more than 4.90v during the test.
	Note: The FTP sensor PID should be near 2.5v with the fuel cap off.
	Possible Causes:
	    • FTP sensor signal circuit is shorted to VREF or power (B+)
	    • FTP sensor ground circuit is open
	    • FTP sensor vacuum lines loose, damaged or disconnected
	    • Fuel tank pressure sensor is damaged or has failed
	    • PCM has failed

DTC	Trouble Code Title, Conditions & Possible Causes
DTC: P0453 **1T CCM, MIL: YES** 1997, 1998, 1999, 2000, 2001, 2002, 2003, 2004, 2005, 2006 **Models:** 2.3CL, 3.0CL, 3.2CL, Integra, MDX, RSX **Engines:** All **Transmissions:** A/T, M/T	**Fuel Tank Pressure Sensor Circuit Low Input** Key on or engine running; and the PCM detected the fuel tank pressure (FTP) sensor signal was less than 0.16v during the test. **Note: The FTP sensor PID should be near 2.5v with the fuel cap off.** **Possible Causes:** • FTP sensor signal circuit is shorted to ground • FTP sensor vacuum lines loose, damaged or disconnected • Fuel tank pressure sensor is damaged or has failed • PCM has failed
DTC: P0453 **1T CCM, MIL: YES** 1996, 1997, 1998, 1999, 2000, 2001, 2002, 2003, 2004, 2005, 2006 **Models:** 3.5RL, RL **Engines:** All **Transmissions:** A/T, M/T	**Fuel Tank Pressure Sensor Circuit High Input** Key on or engine running; and the PCM detected the fuel tank pressure (FTP) sensor signal was more than 4.90v during the test. **Note: The FTP sensor PID should be near 2.5v with the fuel cap off.** **Possible Causes:** • FTP sensor signal circuit is shorted to VREF or power (B+) • FTP sensor ground circuit is open • FTP sensor vacuum lines loose, damaged or disconnected • Fuel tank pressure sensor is damaged or has failed
DTC: P0453 **1T CCM, MIL: YES** 1998, 1999 **Models:** SLX **Engines:** 3.5L VIN X **Transmissions:** A/T	**Fuel Tank Pressure Sensor Circuit High Input** Key on or engine running; and the PCM detected the Fuel Tank Pressure (FTP) sensor signal was more than 4.90v for 12.5 seconds with 100 test failures occurring in 200 test samples in the CCM test. **Possible Causes:** • FTP sensor signal circuit is shorted to ground • FTP sensor power circuit is open between sensor and the PCM • FTP sensor is damaged or has failed • PCM has failed
DTC: P0455 **2T EVAP1, MIL: YES** 2003, 2004, 2005, 2006 **Models:** MDX **Engines:** All **Transmissions:** A/T, M/T	**EVAP System Very Large Leak (0.080") Detected** IAT sensor signal from 32-86°F at startup, ECT signal over 154°F during testing, engine runtime over 10 minutes, TP sensor signal from 1-4v, vehicle driven to a speed of over 20 mph at an engine speed over 1200 rpm, then with both the purge and vent solenoids closed, the PCM detected a large amount of change in the fuel tank pressure during the EVAP leak test. **Possible Causes:** • Fuel tank cap damaged, loose or the wrong part number • Fuel tank leaks at the fuel fill pipe or at the fuel tank seals • Fuel vapor control valve is damaged or has failed • Fuel tank vapor recirculation valve or vapor tube is damaged • Fuel tank vapor control vent tube is damaged or has failed
DTC: P0456 **2T EVAP1, MIL: YES** 2003, 2004, 2005, 2006 **Models:** MDX **Engines:** All **Transmissions:** A/T, M/T	**EVAP System Very Small Leak (0.020") Detected** IAT sensor signal from 32-86°F at startup, vehicle driven at over 5 mph for over 2 minutes, then with ECT sensor signal more than 154°F and the EVAP Control and Vent solenoids enabled, the PCM detected the fuel tank pressure was incorrect due to a leak in the fuel tank area during the EVAP Monitor Leak Test. **Possible Causes:** • Fuel tank cap damaged, loose or the wrong part number • Fuel tank leaks at the fuel fill pipe or at the fuel tank seals • Fuel vapor control valve is damaged or has failed • Fuel tank vapor recirculation valve or vapor tube is damaged • Fuel tank vapor control vent tube is damaged or has failed
DTC: P0457 **2T EVAP1, MIL: YES** 2003, 2004, 2005, 2006 **Models:** MDX **Engines:** All **Transmissions:** A/T, M/T	**EVAP System Leak Detected (Fuel Cap Loose Or Off)** Engine started, engine running at a speed of more than 5 mph for over 1 minute, and the PCM detected a fuel tank pressure value that indicated a leak present the fuel tank area due to a missing fuel cap. **Possible Causes:** • Fuel tank cap damaged, loose or the wrong part number • Fuel tank leaks at the fuel fill pipe or at the fuel tank seals • Fuel vapor control valve is damaged or has failed • Fuel tank vapor recirculation valve or vapor tube is damaged • Fuel tank vapor control vent tube is damaged or has failed
DTC: P0461 **2T CCM, MIL: YES** 2003, 2004, 2005, 2006 **Models:** MDX **Engines:** All **Transmissions:** A/T, M/T	**Fuel Level Sensor Circuit Range/Performance** Key on or engine running; and the PCM detected an unexpected high or low voltage condition on the Fuel Gauge Unit circuit during the CCM test. **Possible Causes:** • Fuel gauge circuit is open, shorted to ground or to power • Fuel gauge unit is damaged or has failed • PCM has failed

DTC	Trouble Code Title, Conditions & Possible Causes
DTC: P0462 **1T CCM, MIL: YES** **2003, 2004, 2005, 2006** **Models:** MDX **Engines:** All **Transmissions:** A/T, M/T	**Fuel Level Sensor Circuit Low Input** Key on or engine running; and the PCM detected an unexpected low voltage condition (less than 0.05v) on the Fuel Gauge Unit circuit. **Possible Causes:** • Fuel gauge circuit is open, shorted to ground • Fuel gauge unit is damaged or has failed • PCM has failed
DTC: P0463 **1T CCM, MIL: YES** **2003, 2004, 2005, 2006** **Models:** MDX **Engines:** All **Transmissions:** A/T, M/T	**Fuel Level Sensor Circuit High Input** Key on or engine running; and PCM detected an unexpected high voltage condition (less than 0.05v) on the Fuel Gauge Unit circuit. **Possible Causes:** • Fuel gauge circuit is shorted to power • Fuel gauge unit is damaged or has failed • PCM has failed
DTC: P0496 **2T EVAP1, MIL: YES** **2003, 2004, 2005, 2006** **Models:** MDX **Engines:** All **Transmissions:** A/T, M/T	**EVAP Canister System High Purge Flow** IAT sensor signal from 32-86°F at startup, vehicle driven at over 5 mph for over 2 minutes, then with the EVAP canister purge solenoid "on", the PCM detected the fuel tank pressure was incorrect due to a high purge flow condition. The EVAP Function test can also be used. **Possible Causes:** • EVAP canister purge valve has a poor connection • EVAP canister purge valve is damaged or has failed • EVAP canister vent valve has a poor connection • Fuel tank pressure sensor has a poor connection
DTC: P0497 **2T EVAP1, MIL: YES** **2003, 2004, 2005, 2006** **Models:** MDX **Engines:** All **Transmissions:** A/T, M/T	**EVAP Canister System Low Purge Flow** IAT sensor signal from 32-86°F at engine startup, vehicle driven at over 5 mph for over 2 minutes, then with the EVAP Control solenoid enabled, the PCM detected the fuel tank pressure was incorrect due to a low purge flow condition. The EVAP Function test can be used. **Possible Causes:** • Check for leaks or restrictions in the canister purge vacuum line between the canister purge valve and the in take manifold • Check the vacuum line from canister to the solenoid for leaks • EVAP canister purge valve is damaged or has failed • Check the FTP, purge solenoid and vent solenoid connections
DTC: P0498 **1T CCM, MIL: YES** **2003, 2004, 2005, 2006** **Models:** MDX **Engines:** All **Transmissions:** A/T, M/T	**EVAP Canister Vent Shut Valve Control Circuit Low Voltage** Key on or engine running; and the PCM detected an unexpected low voltage condition on the EVAP canister vent shut valve control circuit during the CCM test. The solenoid resistance is 25-30 ohms at 68°F. **Possible Causes:** • EVAP canister vent shut valve control circuit shorted to ground • EVAP canister vent shut valve power circuit is open (test fuse) • EVAP canister vent control circuit is open • EVAP canister vent control circuit is damaged or has failed • PCM has failed
DTC: P0499 **1T CCM, MIL: YES** **2003, 2004, 2005, 2006** **Models:** MDX **Engines:** All **Transmissions:** A/T, M/T	**EVAP Canister Vent Shut Valve Control Circuit High Voltage** Key on or engine running; and the PCM detected an unexpected high voltage condition on the EVAP canister vent shut valve control circuit during testing. The solenoid resistance is 25-30 ohms at 68°F. **Possible Causes:** • Check for loose connections at the EVAP canister vent shut valve (i.e., check the control and ground circuit connections) • Check for a loose EVAP vent shut valve connection at the PCM • PCM has failed
DTC: P0500 **1T CCM, MIL: YES** **1995, 1996, 1997, 1998, 1999, 2000, 2001, 2002, 2003, 2004, 2005** **Models:** NSX, NSX-T **Engines:** All **Transmissions:** A/T, M/T	**Vehicle Speed Sensor Circuit Low Input** Engine started, then accelerated to 4000 rpm in 2nd gear, followed by a deceleration period to 1500 rpm with the throttle closed, and the PCM did not detect any VSS signals during the CCM test. **Note: The VSS signal should pulse from 0-5v as the vehicle moves.** **Possible Causes:** • VSS signal circuit is open or shorted to ground • VSS signal circuit is shorted to VREF or system power (B+) • VSS is damaged or has failed

DTC	Trouble Code Title, Conditions & Possible Causes
DTC: P0500 **1T CCM, MIL: YES** **1996, 1997, 1998, 1999, 2000, 2001** Models: Integra Engines: All Transmissions: A/T, M/T	**Vehicle Speed Sensor Circuit Low Input** Engine started, then accelerated to 4000 rpm in 2nd gear, followed by a deceleration period to 1500 rpm with the throttle closed, and the PCM did not detect any VSS signals during the CCM test. **Note: The VSS signal should pulse from 0-5v as the vehicle moves.** **Possible Causes:** • VSS signal circuit is open or shorted to ground • VSS signal circuit is shorted to VREF or system power (B+) • VSS is damaged or has failed • PCM is damaged
DTC: P0500 **1T CCM, MIL: YES** **2002, 2003, 2004, 2005, 2006** Models: RSX Engines: All Transmissions: A/T, M/T	**Vehicle Speed Sensor Circuit Low Input** Engine started, vehicle driven at cruise speed, followed by a deceleration period with the throttle closed, and the PCM did not detect any VSS signals during the CCM test. **Possible Causes:** • VSS signal circuit is open or shorted to ground • VSS signal circuit is shorted to VREF or system power (B+) • VSS is damaged or has failed
DTC: P0500 **1T CCM, MIL: YES** **1997, 1998, 1999** Models: 2.2CL, 2.3CL, 3.0L CL, 3.5L RL Engines: All Transmissions: A/T, M/T	**Vehicle Speed Sensor Circuit Low Input** Engine started, then the vehicle was accelerated to 4000 rpm in 2nd gear, followed by a deceleration period to 1500 rpm at closed throttle and the PCM did not detect any VSS signal during the CCM test. **Note: The VSS signal should pulse from 0-5v as the vehicle moves.** **Possible Causes:** • VSS signal circuit is open or shorted to ground • VSS signal circuit is shorted to VREF or system power (B+) • VSS is damaged or has failed
DTC: P0501 **1T CCM, MIL: YES** **1997** Models: 2.2CL Engines: All Transmissions: A/T	**Vehicle Speed Sensor Circuit Low Input** Engine started, vehicle driven at cruise speed under road load conditions, and the PCM detected a "low" VSS signal during the test. **Note: The VSS signal should pulse from 0-5v as the vehicle moves.** **Possible Causes:** • VSS signal circuit is open or shorted to ground • VSS signal circuit is shorted to VREF or system power (B+) • VSS is damaged or has failed
DTC: P0502 **1T CCM, MIL: YES** **1996, 1997, 1998, 1999** Models: SLX Engines: 3.2L VIN V, 3.5L VIN X Transmissions: A/T	**Vehicle Speed Sensor Circuit Low Input** Engine started, engine speed from 1800-2500, system voltage from 11-16v, ECT sensor more than 140°F, throttle angle from 10-40%, engine load over 50%, and the PCM did not detect any VSS signals for 12.5 seconds over a 15 second period during the CCM test. **Possible Causes:** • VSS signal circuit is open or shorted to ground • VSS power circuit is open (check the Meter 10A fuse) • VSS ground circuit is open between the sensor and ground • VSS is damaged or has failed, or the VSS rotor is cracked • PCM has failed
DTC: P0505 **1T CCM, MIL: YES** **1997, 1998, 1999, 2000, 2001, 2002, 2003, 2004, 2005** Models: NSX, NSX-T Engines: All Transmissions: A/T, M/T	**Idle Speed Control System** DTC P1519 not set, engine started, engine running at hot idle, and the PCM detected the Actual idle speed and the Target idle speed were too far apart during the CCM Rationality test. **Possible Causes:** • IAC valve circuit open, shorted to ground or shorted to power • IAC valve is damaged or has failed • Throttle body is dirty or full of sludge • PCM has failed
DTC: P0505 **1T CCM, MIL: YES** **1995, 1996, 1997, 1998** Models: 2.5TL Engines: All Transmissions: A/T, M/T	**Idle Speed Control System** DTC P1519 not set, engine started, engine running at hot idle, and the PCM detected the Actual idle speed and the Target idle speed were too far apart during the CCM Rationality test. **Possible Causes:** • IAC valve circuit open, shorted to ground or to power (B+) • IAC valve is damaged or has failed • Throttle body is dirty or full of sludge • PCM has failed

DTC	Trouble Code Title, Conditions & Possible Causes
DTC: P0505 **1T CCM, MIL: YES** **1997, 1998, 1999, 2000, 2001, 2002, 2003, 2004, 2005, 2006** **Models:** 2.2CL, 2.3CL, 3.0CL, 3.2CL, 3.2TL, TL, 3.5RL, RL, Integra, MDX, RSX **Engines:** All **Transmissions:** A/T, M/T	**Idle Speed Control System** DTC P1519 not set, engine started, engine running at hot idle, and the PCM detected the Actual idle speed and the Target idle speed were too far apart during the CCM Rationality test. **Possible Causes:** • IAC valve circuit open, shorted to ground or to power (B+) • IAC valve is damaged or has failed • Fast idle thermo valve is damaged or has failed (some models) • Throttle body is dirty or full of sludge (clean and then retest) • PCM has failed • TSB 03-006 (3/17/03) contains a repair procedure for this code
DTC: P0506 **1T CCM, MIL: YES** **2003, 2004, 2005, 2006** **Models:** MDX **Engines:** All **Transmissions:** A/T, M/T	**Idle Speed Control System Lower Than Expected** Engine started, ECT sensor input over 158°F, vehicle not moving, Fuel Trim from 0.73 to 1.47, and the PCM detected the Actual idle speed was more than 100 rpm lower than the Target idle speed. **Possible Causes:** • IAC valve may be damaged or have failed • Throttle body bore/plate dirty or full of sludge (clean and retest) • Throttle body is damaged • PCM is damaged or has failed
DTC: P0506 **1T CCM, MIL: YES** **1996, 1997, 1998, 1999** **Models:** SLX **Engines:** 3.2L VIN V, 3.5L VIN X **Transmissions:** A/T	**Idle Air Control System Low RPM** DTC P0106, P0107, P0108, P0112, P0113, P0117, P0118, P0121, P0122, P0123, P0125, P0131, P0132, P0133, P0134, P0200-206, P0300, P0301-P306, P0335, P0341, P0342, P0404, P0405, P0440, P0442, P0446, P0452, P0453, P0502, P0507, P0601, P0602, P0705, P1133, P1404 and P1441 not set, engine runtime over 125 seconds, system voltage from 11-16v, ECT sensor more than 122°F, IAT sensor more than -40°F, MAP sensor less than 40 kPa, Purge duty cycle over 10%, BARO sensor over 75 kPa, vehicle speed less than 2 mph with the throttle closed, and the PCM detected the Actual speed was 100-200 rpm below the Desired idle speed for 10 seconds based on the current engine coolant temperature. **Possible Causes:** • High resistance between the IAC 'A' high or low control circuits • Short to ground between the IAC 'B' high or low control circuits • IAC valve is damaged, dirty, sticking or has failed • The throttle plate is carbon fouled (it may need to be cleaned)
DTC: P0507 **1T CCM, MIL: YES** **2003, 2004, 2005, 2006** **Models:** MDX **Engines:** All **Transmissions:** A/T, M/T	**Idle Speed Control System Higher Than Expected** Engine started, ECT sensor input over 158°F, vehicle not moving, Fuel Trim from 0.73 to 1.47, and the PCM detected the Actual idle speed was more than 100 rpm higher than the Target idle speed. **Possible Causes:** • IAC valve may be damaged or have failed • Inspect for any air intake system leaks in the engine or hoses • Throttle body bore/plate dirty or full of sludge (clean and retest) • PCM is damaged or has failed
DTC: P0507 **1T CCM, MIL: YES** **1996, 1997, 1998, 1999** **Models:** SLX **Engines:** 3.2L VIN V, 3.5L VIN X **Transmissions:** A/T	**Idle Air Control System High RPM** DTC P0106, P0107, P0108, P0112, P0113, P0117, P0118, P0121, P0122, P0123, P0125, P0131, P0132, P0133, P0134, P0201-206, P0300, P0301-P306, P0335, P0341, P0342, P0404, P0405, P0440, P0442, P0446, P0452, P0453, P0502, P0507, P0601, P0602, P0705, P1133, P1404 and P1441 not set, engine runtime over 125 seconds, system voltage from 11-16v, ECT sensor more than 122°F, IAT sensor more than -40°F, MAP sensor less than 40 kPa, Purge duty cycle over 10%, BARO sensor over 75 kPa, vehicle speed under 2 mph with the throttle closed, and the PCM detected the Actual speed was 100-200 rpm above the Desired idle speed for 10 seconds based on the current engine coolant temperature. **Possible Causes:** • High resistance between the IAC 'A' high or low control circuits • Short to ground between the IAC 'B' high or low control circuits • IAC valve is damaged, dirty, sticking or has failed • The throttle plate is carbon fouled (it may need to be cleaned)
DTC: P0560 **1T CCM, MIL: YES** **2000, 2001, 2002, 2003, 2004, 2005** **Models:** NSX, NSX-T **Engines:** All **Transmissions:** A/T, M/T	**PCM Backup Circuit Low Voltage** Key on or engine running; and the PCM detected a low voltage condition on the PCM Backup circuit. **Note: This circuit is connected to the Backup/Radio 7.5 amp fuse.** **Possible Causes:** • PCM backup circuit is open • PCM backup circuit is shorted to ground • PCM backup circuit has high resistance • PCM has failed

DTC	Trouble Code Title, Conditions & Possible Causes
DTC: P0560 **1T CCM, MIL: YES** 1996, 1997, 1998, 1999, 2000, 2001, 2002, 2003, 2004, 2005, 2006 **Models:** 3.5RL, RL **Engines:** All **Transmissions:** A/T, M/T	**PCM Backup Circuit Low Voltage** Key on or engine running; and the PCM detected a low voltage condition on the PCM Backup circuit. **Note: This circuit is connected to the Backup/Radio 7.5 amp fuse.** **Possible Causes:** • PCM backup circuit is open or shorted to ground • PCM backup circuit has high resistance • PCM has failed
DTC: P0560 **1T CCM, MIL: NO** 1996, 1997, 1998 **Models:** SLX **Engines:** 3.2L VIN V, 3.5L VIN X **Transmissions:** A/T	**System Voltage Malfunction** Engine started, engine speed over 1000 rpm, TFT sensor more than 302°F, and the PCM detected the system voltage was under 10v, or with the TFT sensor signal less than -40°F, the PCM detected the system voltage was less than 7.3v, or it detected the system voltage was over 16v for 2 seconds under any operating conditions. **Possible Causes:** • Check the drive belt for excessive wear and the proper tension • Check for high resistance at the battery connections or at the starter solenoid connection that connects to PCM power circuit • Check the generator output and the battery condition
DTC: P0562 **1T CCM, MIL: NO** 1999 **Models:** SLX **Engines:** 3.5L VIN X **Transmissions:** A/T	**System Voltage Low Input** Engine started, and the PCM detected the system voltage was less than 11.5v for 15 minutes during the CCM test. **Note: For additional help with this code, view the Failure Records.** **Possible Causes:** • Check the drive belt for excessive wear and the proper tension • Check for high resistance at the battery connections or at the starter solenoid connection that connects to PCM power circuit • Check the generator output and the battery condition
DTC: P0563 **1T CCM, MIL: YES** 2002, 2003, 2004, 2005, 2006 **Models:** MDX, RSX **Engines:** All **Transmissions:** A/T, M/T	**PCM Power Source Circuit Unexpected Voltage** Key on or engine running; and the PCM detected an unexpected loss of voltage condition on the PCM power source circuit during the CCM test. **Possible Causes:** • PCM power source circuit is open or shorted to ground • PCM power source circuit has high resistance • PCM has failed
DTC: P0563 **1T CCM, MIL: NO** 1999 **Models:** SLX **Engines:** 3.5L VIN X **Transmissions:** A/T	**System Voltage High Input** Engine started, and the PCM detected the system voltage was more than 16v for 15 minutes during the CCM test. **Note: For additional help with this code, view the Failure Records.** **Possible Causes:** • Check the drive belt for excessive wear and the proper tension • Check for high resistance at the battery connections or at the starter solenoid connection that connects to PCM power circuit • Check the generator output and the battery condition
DTC: P0600 **1T PCM, MIL: YES** 2002, 2003, 2004, 2005, 2006 **Models:** RSX **Engines:** All **Transmissions:** A/T, M/T	**Serial Communication Link Circuit Malfunction** Key on or engine running; and the PCM detected an unexpected voltage condition on the Serial Communication Link circuit. **Possible Causes:** • PCM serial communication link circuit is open • PCM serial communication link circuit shorted to ground • PCM serial communication link circuit shorted to system power • PCM has failed
DTC: P0601 **1T PCM, MIL: NO** 1996, 1997, 1998, 1999 **Models:** SLX **Engines:** All **Transmissions:** A/T	**PCM Internal Check Sum Error** Key on, and the PCM detected a check sum error had occurred. **Note: For additional help with this code, view the Failure Records.** **Possible Causes:** • The contents of the EEPROM have changed • PCM must be replaced to repair this trouble code
DTC: P0603 **1T PCM, MIL: YES** 2003, 2004, 2005, 2006 **Models:** MDX **Engines:** All **Transmissions:** A/T, M/T	**PCM Internal Module Keep Alive Memory Error** Key on or engine running; and the PCM detected a problem in the Keep Alive Memory (KAM) portion of its internal interface. **Possible Causes:** • Clear codes and recheck for this trouble code. If it resets, update the PCM to the latest software. Then recheck for the same trouble code. If it resets, substitute a known good PCM to determine if the PCM has failed and is causing the code.

DTC	Trouble Code Title, Conditions & Possible Causes
DTC: P0606 **1T PCM, MIL: YES** **2003, 2004, 2005, 2006** **Models:** MDX **Engines:** All **Transmissions:** A/T, M/T	**PCM Processor Error** Key on or engine running; and the PCM detected a problem in its internal processor. **Possible Causes:** • Clear codes and recheck for this trouble code. If it resets, update the PCM to the latest software. Then recheck for the same trouble code. If it resets, substitute a known good PCM to determine if the PCM has failed and is causing the code.
DTC: P0627 **1T CCM, MIL: YES** **2003, 2004, 2005, 2006** **Models:** MDX **Engines:** All **Transmissions:** A/T, M/T	**PGM-FI Main Relay 2 (Fuel Pump) Circuit Malfunction** Key on or engine running; and the PCM detected an unexpected voltage condition on the PGM-FI main relay control circuit. **Possible Causes:** • PGM-FI main relay control circuit is open or shorted to ground • PGM-FI main relay power circuit is open (check the fuse) • PGM-FI main relay is damaged or has failed • PCM is damaged or has failed
DTC: P0641 **1T CCM, MIL: YES** **2003, 2004, 2005, 2006** **Models:** MDX **Engines:** All **Transmissions:** A/T, M/T	**PCM Sensor Reference Voltage 'A' Circuit Malfunction** Key on or engine running; and the PCM detected an unexpected low voltage condition on the Sensor Reference Voltage 'A' circuit. **Possible Causes:** • One or more of these related sensors or circuits (APP1, APP2, MAP, or Counter Shaft Speed sensor) circuits might be shorted to ground. Disconnect one sensor at a time, clear the codes and then recheck for the same trouble code to determine which sensor is causing this trouble code to set. • PCM is damaged or has failed
DTC: P0651 **1T CCM, MIL: YES** **2003, 2004, 2005, 2006** **Models:** MDX **Engines:** All **Transmissions:** A/T, M/T	**PCM Sensor Reference Voltage 'B' Circuit Malfunction** Key on or engine running; and the PCM detected an unexpected low voltage condition on the Sensor Reference Voltage 'B' circuit. **Possible Causes:** • One or more of these related sensors or circuits (EGR Valve Position, Mainshaft Speed, FTP or the TAC module) circuits might be shorted to ground. Disconnect one sensor at a time, clear the codes and then recheck for the same trouble code to determine which sensor is causing this trouble code to set. • PCM is damaged or has failed
DTC: P0657 **1T CCM, MIL: YES** **2003, 2004, 2005, 2006** **Models:** MDX **Engines:** All **Transmissions:** A/T, M/T	**Air Fuel Sensor Power Relay Circuit Malfunction** Key on or engine running; and the PCM detected an unexpected voltage condition on the Air Fuel Sensor power relay control circuit. **Possible Causes:** • AFS power relay control circuit is open or shorted to ground • AFS relay power circuit is open (check the No. 9 LAF HT fuse) • AFS power relay is damaged or has failed • PCM is damaged or has failed
DTC: P0661 **1T CCM, MIL: YES** **2002, 2003, 2004, 2005, 2006** **Models:** RSX **Engines:** All **Transmissions:** A/T, M/T	**Intake Manifold Runner Control Valve Position Sensor Circuit Low Input** Key on or engine running; and the PCM detected an unexpected low voltage condition on the IMRC Valve Position Sensor circuit. **Possible Causes:** • IMRC valve position sensor circuit shorted to sensor ground • IMRC valve position sensor circuit is shorted to chassis ground • PCM has failed
DTC: P0662 **1T CCM, MIL: YES** **2002, 2003, 2004, 2005, 2006** **Models:** RSX **Engines:** All **Transmissions:** A/T, M/T	**Intake Manifold Runner Control Valve Position Sensor Circuit High Input** Key on or engine running; and the PCM detected an unexpected high voltage condition on the IMRC Valve Position Sensor circuit. **Possible Causes:** • IMRC valve position ground circuit is open • IMRC valve position sensor circuit shorted to VREF • IMRC valve position sensor circuit is shorted to system power • PCM has failed
DTC: P0685 **1T PCM, MIL: YES** **2003, 2004, 2005, 2006** **Models:** MDX **Engines:** All **Transmissions:** A/T, M/T	**PCM Power Control Circuit Malfunction** Key on or engine running; and the PCM detected an unexpected voltage condition on the Air Fuel Sensor power relay control circuit. **Possible Causes:** • Clear codes and recheck for this trouble code. If it resets, update the PCM to the latest software. Then recheck for the same trouble code. If it resets, substitute a known good PCM to determine if the PCM has failed and is causing the code.
DTC: P0700 **1T CCM, MIL: YES** **1995, 1996, 1997, 1998** **Models:** 2.5TL **Engines:** All **Transmissions:** A/T	**Automatic Transaxle System Malfunction** Engine started, vehicle driven to over 30 mph for several minutes, and the PCM detected a fault in the Automatic Transaxle system. **Note: DTC P0700 sets along with several other TCM trouble codes.** **Possible Causes:** • Check for other A/T related trouble codes, and then refer to the Possible Causes for these trouble codes for more information.

DTC	Trouble Code Title, Conditions & Possible Causes
DTC: P0700 **1T CCM, MIL: YES** **1996, 1997, 1998, 1999, 2000, 2001, 2002, 2003, 2004, 2005, 2006** **Models:** 2.2CL, 2.3CL, 3.0CL, 3.2CL, 3.2TL, TL, Integra **Engines:** All **Transmissions:** A/T	**Automatic Transaxle System Malfunction** Engine started, vehicle driven to over 30 mph for several minutes, and the PCM detected a fault in the Automatic Transaxle system. **Note: DTC P0700 sets along with several other TCM trouble codes.** **Possible Causes:** • Check for other A/T related trouble codes, and then refer to the Possible Causes for these trouble codes for more information.
DTC: P0700 **1T CCM, MIL: YES** **1995, 1996, 1997, 1998, 1999, 2000, 2001, 2002, 2003, 2004, 2005** **Models:** NSX, NSX-T **Engines:** All **Transmissions:** A/T	**Automatic Transaxle System Malfunction** Engine started, vehicle driven to over 30 mph for several minutes, and the PCM detected a fault in the Automatic Transaxle system. This trouble code (DTC P0700) sets along with several other TCM trouble codes. The D4 Lamp will activate along with the MIL Lamp when this trouble code is set on this model. **Possible Causes:** • Check for other A/T related trouble codes, and then refer to the Possible Causes for these trouble codes for more information.
DTC: P0700 **1T CCM, MIL: YES** **1996, 1997, 1998, 1999, 2000, 2001, 2002, 2003, 2004, 2005, 2006** **Models:** 3.5RL, RL **Engines:** All **Transmissions:** A/T	**Automatic Transaxle System Malfunction** Engine started, vehicle driven to over 30 mph for several minutes, and the PCM detected a fault in the Automatic Transaxle system. This trouble code (DTC P0700) sets along with several other TCM trouble codes. The D4 indicator will activate along with the MIL indicator when this trouble code is set on this model. **Possible Causes:** • Check for other A/T related trouble codes, and then refer to the Possible Causes for these trouble codes for more information.

DTC: P0705
1T CCM, MIL: NO
1996, 1997, 1998, 1999
Models: SLX
Engines: 3.2L VIN V, 3.5L VIN X
Transmissions: A/T
	Transmission Range Switch Illegal Position Malfunction
	Engine started, then driven to a speed of over 8 mph, and the PCM detected "illegal" TR Range or Mode switch signals for 5 seconds.
	Note: For additional help with this code, view the Failure Records.
	Possible Causes:
	   • TR range switch signal is open
	   • TR range switch signal shorted to another switch position signal
	   • TR range switch is damaged or has failed
	   • PCM has failed

DTC: P0706
1T CCM, MIL: NO
1996, 1997, 1998, 1999
Models: SLX
Engines: 3.2L VIN V, 3.5L VIN X
Transmissions: A/T
	Transmission Range Switch Circuit Performance
	DTC P0122, P0123, P0722 and P0723 not set, engine started, then driven with the output speed over 3200 rpm, and the PCM detected the TR Switch indicated Reverse position, or with the Output speed under 3000 rpm and the throttle angle over 20%, it detected the TR switch indicated Park or Neutral position for 4 seconds in the test.
	Note: For additional help with this code, view the Failure Records.
	Possible Causes:
	   • TR switch signal is open
	   • TR switch signal shorted to another switch position signal
	   • TR switch is damaged or has failed
	   • PCM has failed

DTC: P0710
1T CCM, MIL: YES
2001, 2002, 2003, 2004, 2005, 2006
Models: 2.3CL, 3.0CL, 3.2CL, 3.2TL, TL, 3.5RL, RL, Integra, MDX, RSX
Engines: All
Transmissions: A/T
	Automatic Transmission Fluid Circuit Malfunction
	Key on or engine running; and the PCM detected an unexpected voltage condition on the ATF sensor signal circuit during the test.
	Possible Causes:
	   • ATF sensor signal is open between the sensor and the PCM
	   • ATF sensor signal is shorted to sensor or chassis ground
	   • ATF sensor connector is disconnected or damaged
	   • ATF sensor is damaged or has failed
	   • PCM has failed

DTC	Trouble Code Title, Conditions & Possible Causes
DTC: P0711 **1T CCM, MIL: NO** **1998, 1999** **Models:** SLX **Engines:** 3.5L VIN X **Transmissions:** A/T	**Transmission Fluid Temperature Sensor Performance** DTC P0722, P0723 and P1870 not set, engine started, system voltage from 11-16v, TFT sensor from -40°F to 69.8°F at startup, ECT sensor more than 150°F and has changed more than 90°F since startup, vehicle speed over 5 mph with the TCC slip speed over 120 rpm for 410 seconds, and the PCM detected the TFT sensor changed less than 2 counts since startup, or that its delta change was over 36°F at least 14 times during a 7 second period. **Possible Causes:** • TFT signal or ground circuit has a high resistance condition • TFT sensor is out-of-calibration (it may be skewed) • TFT sensor is damaged or has failed • PCM has failed
DTC: P0712 **1T CCM, MIL: NO** **1996, 1997, 1998, 1999** **Models:** SLX **Engines:** All **Transmissions:** A/T	**Transmission Fluid Temperature Sensor Low Input** Engine started, system voltage from 11-16v, and the PCM detected the TFT sensor was less than 0.40v for 20 seconds during the test. **Note: For additional help with this code, view the Failure Records.** **Possible Causes:** • TFT sensor signal circuit is shorted to sensor ground • TFT sensor signal circuit is shorted to chassis ground • TFT sensor is damaged (it may be shorted internally) • PCM has failed
DTC: P0713 **1T CCM, MIL: NO** **1996, 1997, 1998, 1999** **Models:** SLX **Engines:** All **Transmissions:** A/T	**Transmission Fluid Temperature Sensor High Input** Engine started, system voltage from 11-16v, and the PCM detected the TFT sensor signal was more than 4.92v for 409 seconds. **Possible Causes:** • TFT sensor signal circuit is open between the sensor and PCM • TFT sensor ground circuit is open between sensor and ground • TFT sensor signal circuit is shorted to VREF or system power • TFT sensor is damaged (it may be open internally) • PCM has failed
DTC: P0715 **1T CCM, MIL: YES** **1995, 1996, 1997, 1998** **Models:** 2.5TL **Engines:** All **Transmissions:** A/T	**A/T Mainshaft Speed Sensor Circuit Malfunction** Engine running with VSS inputs received, and the PCM detected an unexpected voltage condition on the Mainshaft speed sensor circuit. **Possible Causes:** • Mainshaft speed sensor circuit is open or shorted to ground • Mainshaft speed sensor circuit is shorted to VREF or power • Mainshaft speed sensor connector is disconnected or damaged • Mainshaft speed sensor is damaged or has failed • PCM has failed
DTC: P0715 **1T CCM, MIL: YES** **1995, 1996, 1997, 1998, 1999,** **2000, 2001, 2002, 2003, 2004,** **2005** **Models:** NSX, NSX-T **Engines:** All **Transmissions:** A/T	**A/T Mainshaft Speed Sensor Circuit Malfunction** Engine running with VSS inputs received, and the PCM detected an unexpected voltage condition on the Mainshaft speed sensor circuit. **Possible Causes:** • Mainshaft speed sensor circuit is open or shorted to ground • Mainshaft speed sensor circuit is shorted to VREF or power • Mainshaft speed sensor connector is disconnected or damaged • Mainshaft speed sensor is damaged or has failed • PCM has failed
DTC: P0715 **1T CCM, MIL: YES** **1996, 1997, 1998, 1999, 2000,** **2001, 2002, 2003, 2004, 2005,** **2006** **Models:** 2.3CL, 3.0CL, 3.2CL, 3.2TL, TL, 3.5RL, RL, Integra, MDX, RSX **Engines:** All **Transmissions:** A/T	**A/T Mainshaft Speed Sensor Circuit Malfunction** Engine running with VSS inputs received, and the PCM detected an unexpected voltage condition on the Mainshaft speed sensor circuit. **Possible Causes:** • Mainshaft speed sensor circuit is open or shorted to ground • Mainshaft speed sensor circuit is shorted to VREF or power • Mainshaft speed sensor connector is disconnected or damaged • Mainshaft speed sensor is damaged or has failed • PCM has failed

DTC	Trouble Code Title, Conditions & Possible Causes
DTC: P0719 **1T CCM, MIL: NO** **1996, 1997, 1998, 1999** **Models:** SLX **Engines:** All **Transmissions:** A/T	**TCC Brake Switch Circuit Low Input** DTC P0502 and P0503 not set, engine started, then driven to a speed under 5 mph, then driven from 5-20 mph for 4 seconds, then back to a speed over 20 mph for 6 seconds, and the PCM detected an "open" Brake switch condition for 15 minutes without it changing its status, conditions occurred at least 7 times during the CCM test. **Possible Causes:** • TCC brake switch signal circuit is open or shorted to ground • TCC brake switch is damaged (it may be open internally) • PCM has failed
DTC: P0720 **1T CCM, MIL: YES** **1995, 1996, 1997, 1998** **Models:** 2.5TL **Engines:** All **Transmissions:** A/T	**A/T Countershaft Speed Sensor Circuit Malfunction** Engine running with VSS inputs received, and the PCM detected an unexpected voltage on the Countershaft Speed Sensor circuit. **Possible Causes:** • Countershaft speed sensor circuit is open or shorted to ground • Countershaft speed sensor circuit is shorted to VREF or power • Countershaft speed sensor connector is disconnected or damaged • Countershaft speed sensor is damaged or has failed • PCM has failed
DTC: P0720 **1T CCM, MIL: YES** **1995, 1996, 1997, 1998, 1999,** **2000, 2001, 2002, 2003, 2004,** **2005** **Models:** NSX, NSX-T **Engines:** All **Transmissions:** A/T	**A/T Countershaft Speed Sensor Circuit Malfunction** Engine running with VSS inputs received, and the PCM detected an unexpected voltage on the Countershaft Speed Sensor circuit. **Possible Causes:** • Countershaft speed sensor circuit is open or shorted to ground • Countershaft speed sensor circuit is shorted to VREF or power • Countershaft speed sensor connector is disconnected or damaged • Countershaft speed sensor is damaged or has failed • PCM has failed
DTC: P0720 **1T CCM, MIL: YES** **1996, 1997, 1998, 1999, 2000,** **2001, 2002, 2003, 2004, 2005,** **2006** **Models:** 2.2CL, 2.3CL, 3.0CL, 3.2CL, 3.2TL, TL, 3.5RL, RL, Integra, MDX, RSX **Engines:** All **Transmissions:** A/T	**A/T Countershaft Speed Sensor Circuit Malfunction** Engine started, vehicle driven to over 30 mph for several minutes, and the PCM detected an unexpected voltage condition on the Countershaft Speed Sensor circuit during the CCM test. **Possible Causes:** • Countershaft speed sensor circuit is open or shorted to ground • Countershaft speed sensor circuit is shorted to VREF or power • Countershaft speed sensor connector is disconnected or damaged • Countershaft speed sensor is damaged or has failed • PCM has failed
DTC: P0722 **1T CCM, MIL: NO** **1996, 1997, 1998, 1999** **Models:** SLX **Engines:** All **Transmissions:** A/T	**Output Speed Sensor Circuit Low Input** DTC P0106, P0107, P0108, P0122, P0123, P1106 and P1107 not set, TR switch indicating other than Park or Neutral position, throttle angle over 10%, engine vacuum from 0-70 kPa, engine speed from 3000-5000 rpm, and the PCM detected the Output Speed Sensor (OSS) signal indicated zero (0) rpm, condition met for 5 seconds. **Possible Causes:** • OSS (+) signal circuit is open or shorted to ground • OSS (-) signal circuit is open or shorted to ground • OSS is damaged or has failed • PCM has failed
DTC: P0723 **1T CCM, MIL: NO** **1996, 1997, 1998, 1999** **Models:** SLX **Engines:** All **Transmissions:** A/T	**Output Speed Sensor Circuit Malfunction** Engine started, TR switch indicating other than Park or Neutral position, throttle angle over 10%, engine vacuum from 0-70 kPa, engine speed from 3000-5000 rpm, and the PCM detected an interruption in the Output Speed Sensor (OSS) circuit in the test. **Possible Causes:** • OSS (+) signal circuit open or shorted to ground (intermittent) • OSS (-) signal circuit open or shorted to ground (intermittent) • OSS is damaged or has failed (an intermittent fault) • PCM has failed
DTC: P0724 **1T CCM, MIL: NO** **1996, 1997, 1998, 1999** **Models:** SLX **Engines:** All **Transmissions:** A/T	**TCC Brake Switch Circuit Low (Stuck Off)** DTC P0722 and P0723 not set, engine started, then driven to a speed of over 5 mph, then driven to over 20 mph for 5 seconds, then back to a speed of 5-20 mph for 4 seconds, and the PCM detected a "closed" Brake switch condition for 15 minutes without it changing its status, conditions occurred at least 7 times during the CCM test. **Possible Causes:** • TCC brake switch signal circuit is shorted to system power (B+) • TCC brake switch is damaged (it may be shorted internally) • PCM has failed

DTC	Trouble Code Title, Conditions & Possible Causes
DTC: P0725 **1T CCM, MIL: YES** **1995, 1996, 1997, 1998** **Models:** 2.5TL **Engines:** All **Transmissions:** A/T	**Automatic Transaxle System Malfunction** Engine started, vehicle driven to over 30 mph for several minutes, and the PCM detected a fault in the Automatic Transaxle system. **Note: This trouble code sets along with several other Automatic Transaxle related trouble codes.** **Possible Causes:** • Refer to the repair instructions in a transmission repair manual or the information in other electronic media to repair this code.
DTC: P0725 **1T CCM, MIL: YES** **1995, 1996, 1997, 1998, 1999,** **2000** **Models:** NSX, NSX-T **Engines:** All **Transmissions:** A/T	**Automatic Transaxle System Malfunction** Engine started, vehicle driven to over 30 mph for several minutes, and the PCM detected a fault in the Automatic Transaxle system. **Note: This trouble code sets along with several other Automatic Transaxle related trouble codes.** **Possible Causes:** • Refer to the repair instructions in a transmission repair manual or the information in other electronic media to repair this code.
DTC: P0725 **1T CCM, MIL: YES** **1996, 1997, 1998, 1999** **Models:** Integra **Engines:** All **Transmissions:** A/T	**Automatic Transaxle System Malfunction** Engine started, vehicle driven to over 30 mph for several minutes, and the PCM detected a fault in the Automatic Transaxle system. **Note: This trouble code sets along with several other Automatic Transaxle related trouble codes.** **Possible Causes:** • One or more A/T clutch pressure control switches is damaged • Refer to the repair instructions in a transmission repair manual or the information in other electronic media to repair this code.
DTC: P0730 **1T CCM, MIL: YES** **1995, 1996, 1997, 1998** **Models:** 2.5TL **Engines:** All **Transmissions:** A/T	**Automatic Transaxle Shift Control System** No other A/T trouble codes set, engine started, vehicle driven at cruise speed with VSS inputs received, and the PCM detected the lockup clutch did not lock or unlock correctly during the CCM test. **Possible Causes:** • One or more A/T clutch pressure control switches is damaged • Refer to the repair instructions in a transmission repair manual or the information in other electronic media to repair this code.
DTC: P0730 **1T CCM, MIL: YES** **1995, 1996, 1997, 1998, 1999,** **2000, 2001, 2002, 2003, 2004,** **2005** **Models:** NSX, NSX-T **Engines:** All **Transmissions:** A/T	**Automatic Transaxle Shift Control System** No other A/T trouble codes set, engine started, vehicle driven at cruise speed with VSS inputs received, and the PCM detected the lockup clutch did not lock or unlock correctly during the CCM test. **Possible Causes:** • One or more A/T clutch pressure control switches is damaged • Refer to the repair instructions in a transmission repair manual or the information in other electronic media to repair this code.
DTC: P0730 **1T CCM, MIL: YES** **1996, 1997, 1998, 1999, 2000,** **2001, 2002, 2003, 2004, 2005,** **2006** **Models:** 2.2CL, 2.3CL, 3.0CL, 3.2CL, 3.2TL, TL, 3.5RL, RL, Integra, MDX, RSX **Engines:** All **Transmissions:** A/T	**Automatic Transaxle Shift Control System** No other A/T trouble codes set, engine started, vehicle driven at cruise speed with VSS inputs received, and the PCM detected the lockup clutch did not lock or unlock correctly during the CCM test. **Possible Causes:** • One or more A/T clutch pressure control switches is damaged • Refer to the repair instructions in a transmission repair manual or the information in other electronic media to repair this code.
DTC: P0730 **1T CCM, MIL: NO** **1996, 1997, 1998, 1999** **Models:** SLX **Engines:** All **Transmissions:** A/T	**Transmission Incorrect Gear Ratio** DTC P0722 and P0723 not set, gear selector not in Park, Neutral or Reverse position, engine speed more than 3500 rpm, 3 seconds have passed since an Upshift event, and the PCM detected a slip value over 753 rpm in 1st gear, a slip value over 713 rpm in 2nd gear, a slip value over 694 rpm in 3rd gear, or a slip value over 685 rpm in 4th gear, condition met for 5.5 seconds during the CCM test. **Possible Causes:** • OSS signal is open or shorted to ground (an intermittent fault) • OSS is out-of-calibration (i.e., the tire size or rear axle ratio) • OSS is damaged or has failed • PCM has failed
DTC: P0740 **1T CCM, MIL: YES** **1995, 1996, 1997, 1998, 1999,** **2000, 2001, 2002, 2003, 2004,** **2005** **Models:** NSX, NSX-T **Engines:** All **Transmissions:** A/T	**Automatic Transaxle Lockup Clutch System Malfunction** No other A/T trouble codes set, engine started, vehicle driven at cruise speed with VSS inputs received, and the PCM detected the lockup clutch did not engage or disengage correctly during the test. **Possible Causes:** • Possible problem in the transmission or torque converter • Refer to the repair instructions in a transmission repair manual or the information in other electronic media to repair this code.

DTC	Trouble Code Title, Conditions & Possible Causes

DTC: P0740
1T CCM, MIL: YES
1995, 1996, 1997, 1998
Models: 2.5TL
Engines: All
Transmissions: A/T
	Automatic Transaxle Lockup Clutch System Malfunction
	No other A/T trouble codes set, engine started, vehicle driven at cruise speed with VSS inputs received, and the PCM detected the lockup clutch did not engage or disengage correctly during the test.
	Possible Causes:
	  • Possible problem in the transmission or torque converter
	  • Refer to the repair instructions in a transmission repair manual or the information in other electronic media to repair this code.

DTC: P0740
1T CCM, MIL: YES
1996, 1997, 1998, 1999, 2000, 2001, 2002, 2003, 2004, 2005, 2006
Models: 2.2CL, 2.3CL, 3.0CL, 3.2CL, 3.2TL, TL, 3.5RL, RL, Integra, MDX, RSX
Engines: All
Transmissions: A/T
	Automatic Transaxle Lockup Clutch System Malfunction
	No other A/T trouble codes set, engine started, vehicle driven at cruise speed with VSS inputs received, and the PCM detected the lockup clutch did not engage or disengage correctly during the test.
	Possible Causes:
	  • Possible problem in the transmission or torque converter
	  • Refer to the repair instructions in a transmission repair manual or the information in other electronic media to repair this code.

DTC: P0742
1T CCM, MIL: NO
1996, 1997, 1998, 1999
Models: SLX
Engines: All
Transmissions: A/T
	Torque Converter Clutch Circuit Malfunction (Stuck On)
	DTC P0107, P0108, P0122, P0123, P0502, P0503, P0740, P1810 and P1860 not set, engine started engine speed over 450 rpm for 5 seconds, throttle angle from 15-60%, not in Fuel Cutoff mode, engine speed from 1000-3000 rpm, engine vacuum from 0-105 kPa, engine torque from 40-400 lbs, speed ratio from 0.65-1.25, VSS indicating 20-65 mph, commanded gear not in 1st, TCC is "off", gear range is D4 with no change for 6 seconds, and the PCM detected the TCC slip speed was from -40 to +30 rpm for 5 seconds.
	Possible Causes:
	  • TCC assembly has failed (it may be mechanically stuck "on")
	  • Internal transmission concerns exist

DTC: P0745
1T CCM, MIL: YES
2002, 2003, 2004, 2005, 2006
Models: RSX
Engines: All
Transmissions: A/T
	A/T Clutch Pressure Control Solenoid 'A' Circuit Malfunction
	Engine started, vehicle driven at cruise speed, and the PCM detected an unexpected voltage condition on the A/T Clutch Pressure control solenoid 'A' circuit during the CCM test period.
	Possible Causes:
	  • Clutch pressure control solenoid 'A' circuit is open
	  • Clutch pressure control solenoid 'A' connector is disconnected
	  • Clutch pressure control solenoid 'A' is damaged or has failed
	  • Clear codes and recheck for this trouble code. If it resets, update the PCM to the latest software. Then recheck for the same trouble code. If it resets, substitute a known good PCM to determine if the PCM has failed and is causing the code.

DTC: P0748
1T CCM, MIL: YES
2002, 2003, 2004, 2005, 2006
	A/T Clutch Pressure Control Solenoid 'A' Circuit Malfunction
	Engine started, vehicle driven at cruise speed, and the PCM detected an unexpected voltage condition on the A/T Clutch Pressure control solenoid 'A' circuit during the CCM test period.
	Possible Causes:
	  • Clutch pressure control solenoid 'A' circuit is shorted to ground
	  • Clutch pressure control solenoid 'A' is damaged or has failed
	  • Clear codes and recheck for this trouble code. If it resets, update the PCM to the latest software. Then recheck for the same trouble code. If it resets, substitute a known good PCM to determine if the PCM has failed and is causing the code.

DTC: P0748
1T CCM, MIL: NO
1996, 1997, 1998, 1999
Models: SLX
Engines: All
Transmissions: A/T
	A/T Pressure Control Solenoid Circuit Malfunction
	Engine started, and the PCM detected the difference between the Actual and Commanded Pressure Control solenoid (PCS) current level was more than 200 mA during the CCM test.
	Note: For additional help with this code, view the Failure Records.
	Possible Causes:
	  • Pressure control solenoid control circuit is shorted to ground
	  • Pressure control solenoid is damaged or has failed
	  • PCM has failed

DTC: P0750
1T CCM, MIL: YES
2002, 2003, 2004, 2005, 2006
Models: RSX
Engines: All
Transmissions: A/T
	A/T Shift Solenoid Valve 'A' Circuit Malfunction
	Engine started, vehicle driven at cruise speed, and the PCM detected an unexpected voltage condition on the A/T Shift Solenoid Valve 'A' control circuit during the CCM test period.
	Possible Causes:
	  • A/T shift solenoid valve 'A' control circuit open
	  • AT shift solenoid valve 'A' is damaged or has failed
	  • Clear codes and recheck for this trouble code. If it resets, update the PCM to the latest software. Then recheck for the same trouble code. If it resets, substitute a known good PCM to determine if the PCM has failed and is causing the code.

DTC	Trouble Code Title, Conditions & Possible Causes
DTC: P0751 **1T CCM, MIL: NO** **1996, 1997, 1998, 1999** **Models:** SLX **Engines:** All **Transmissions:** A/T	**Shift Solenoid 'A' Performance Without Input Speed** DTC P0122, P0123, P0722, P0723, P0742, P0753, P0758 and P1860 not set, vehicle driven in D4 Gear at over 6.25 mph, TFT sensor signal at 68-257ºF, then during a 1-2 Shift, TP angle at 10-60% (± 3%), VSS at 11-31 mph, the PCM detected the engine speed in 2nd Gear was 100 rpm more than it was in 1st Gear (1); or during a 2-3 Shift, TP angle at 13-60% (± 5%), VSS at 20-45 mph, the engine speed in 3rd Gear was 64 rpm less than it was in 2nd Gear (2); or during a 3-4 Shift, TP at 7-60% (± 5%), VSS at 25-87 mph, the engine speed in 4th gear was 60 rpm more than it was in 3rd Gear (3); or while in 4th Gear, TP angle at 13-60% (± 5%), speed ratio at 0.85 to 1.2, the TCC slip speed was 100-2000 rpm for 3 seconds (4); or while in 4th Gear with TCC "on", speed ratio at 0.5-0.85, the TCC slip speed was -50 to +500 for 3 seconds (5). **Note: This code is set if the conditions in (1), (2), (3) or (4) are met, or if the conditions in (1), (2), (3) or (5) are met twice in a row.** **Possible Causes:** • Shift solenoid 'A' is damaged or has failed mechanically (on) • Other internal transmission concerns can cause this problem
DTC: P0753 **1T CCM, MIL: YES** **1995, 1996, 1997, 1998** **Models:** 2.5TL **Engines:** All **Transmissions:** A/T	**A/T Lockup Solenoid 'A' Circuit Malfunction** Engine started, engine running at cruise speed with VSS inputs, and the PCM detected an unexpected voltage condition on the Solenoid Valve 'A' circuit during the CCM test. **Possible Causes:** • A/T Solenoid 'A' control circuit is open or shorted to ground • A/T Solenoid 'A' control circuit is shorted to system power • A/T Solenoid 'A' connector is disconnected or damaged • A/T Solenoid 'A' is damaged or has failed • TCM or PCM has failed
DTC: P0753 **1T CCM, MIL: YES** **1995, 1996, 1997, 1998, 1999, 2000, 2001, 2002, 2003, 2004, 2005** **Models:** NSX, NSX-T **Engines:** All **Transmissions:** A/T	**A/T Lockup Solenoid 'A' Circuit Malfunction** Engine started, engine running at cruise speed with VSS inputs, and the PCM detected an unexpected voltage condition on the Solenoid Valve 'A' circuit during the CCM test. **Possible Causes:** • A/T Solenoid 'A' control circuit is shorted to ground or to power • A/T Solenoid 'A' connector is disconnected or damaged • A/T Solenoid 'A' is damaged or has failed • TCM or PCM has failed
DTC: P0753 **1T CCM, MIL: YES** **1996, 1997, 1998, 1999, 2000, 2001, 2002, 2003, 2004, 2005, 2006** **Models:** 2.2CL, 2.3CL, 3.0CL, 3.2CL, 3.2TL, TL, 3.5RL, RL, Integra, MDX, RSX **Engines:** All **Transmissions:** A/T	**A/T Lockup Solenoid 'A' Circuit Malfunction** Engine started, engine running at cruise speed with VSS inputs, and the PCM detected an unexpected voltage condition on the Solenoid Valve 'A' circuit during the CCM test. **Possible Causes:** • A/T Solenoid 'A' control circuit is open or shorted to ground • A/T Solenoid 'A' control circuit is shorted to system power • A/T Solenoid 'A' connector is disconnected or damaged • A/T Solenoid 'A' is damaged or has failed • TCM or PCM has failed
DTC: P0753 **1T CCM, MIL: NO** **1996, 1997, 1998, 1999** **Models:** SLX **Engines:** All **Transmissions:** A/T	**Shift Solenoid 'A' Circuit Malfunction** Engine started, then with Shift Solenoid 'A' (SSA) commanded "on", and PCM detected the solenoid control signal was 12v, or with SSA commanded "off", the solenoid control signal was near 0v, either condition met for 0.84 to 1.0 seconds during the test. **Possible Causes:** • SSA control circuit is open, shorted to ground or to power (B+) • SSA power circuit (from the PCM) is open • SSA is damaged or has failed • PCM has failed
DTC: P0756 **1T CCM, MIL: NO** **1996, 1997, 1998, 1999** **Models:** SLX **Engines:** All **Transmissions:** A/T	**Shift Solenoid 'B' Performance Without Input Speed** DTC P0122, P0123, P705, P706, P0722, P0723, P0742, P0753, P0758, P1106, P1107 and P1860 not set, vehicle driven in D4 Gear to over 6.25 mph at under 8000 rpm, MAP at 0-70 kPa, TFT signal at 68-257ºF, the TCC "off", TP angle over 4%, 1st Gear "on", speed ratio at 0.5-2.65, TSS signal at 320-2000 rpm, the TCC slip speed was -200 to -4000 rpm for 1.8 seconds (1), or during a 2-3 Shift, TP angle at 10-60% (± 5%), VSS at 20-45 mph, the 3rd Gear speed was 64 rpm less than it was in 2nd gear (2), or during a 3-4 shift, TP angle at 7-60% (± 5%), VSS at 25-87 mph, the 4th Gear speed was 60 rpm less than it was in 3rd Gear (3); or in 4th Gear, TP angle at 13-60, speed ratio at 0.5-1.20, the TCC slip speed was 100-2000 rpm (4), or in 4th Gear, TP angle at 13-60, speed ratio at 0.5-0.85, the TCC slip speed was -50 to -500 rpm (5). **Note: This code is set if the conditions in (1), (3) or (4) are met twice (stuck on), or if the conditions in (1) and (3) are met twice (stuck off).** **Possible Causes:** • Shift solenoid 'B' is damaged or has failed mechanically (on) • Other internal transmission concerns can cause this fault

DTC	Trouble Code Title, Conditions & Possible Causes
DTC: P0758 **1T CCM, MIL: YES** **1995, 1996, 1997, 1998, 1999, 2000, 2001, 2002, 2003, 2004, 2005** **Models:** NSX, NSX-T **Engines:** All **Transmissions:** A/T	**A/T Lockup Solenoid 'B' Circuit Malfunction** Engine started, vehicle driven at cruise speed, and the PCM detected an unexpected voltage condition on the Solenoid Valve 'B' circuit during the CCM test. **Possible Causes:** • A/T Solenoid 'B' control circuit is open • A/T Solenoid 'B' control circuit is shorted to ground or to power • A/T Solenoid 'B' connector is disconnected or damaged • A/T Solenoid 'B' is damaged or has failed • TCM or PCM has failed
DTC: P0758 **1T CCM, MIL: YES** **1995, 1996, 1997, 1998** **Models:** 2.5TL **Engines:** All **Transmissions:** A/T	**A/T Lockup Solenoid 'B' Circuit Malfunction** Engine started, vehicle driven at cruise speed, and the PCM detected an unexpected voltage condition on the A/T Solenoid Valve 'B' circuit during the CCM test. **Possible Causes:** • A/T Solenoid 'B' control circuit is open • A/T Solenoid 'B' circuit is shorted to ground or to power • A/T Solenoid 'B' connector is disconnected or damaged • A/T Solenoid 'B' is damaged or has failed • TCM or PCM has failed
DTC: P0758 **1T CCM, MIL: YES** **1996, 1997, 1998, 1999, 2000, 2001, 2002, 2003, 2004, 2005, 2006** **Models:** 2.2CL, 2.3CL, 3.0CL, 3.2CL, 3.2TL, TL, 3.5RL, RL, Integra, MDX, RSX **Engines:** All **Transmissions:** A/T	**A/T Lockup Solenoid 'B' Circuit Malfunction** Engine started, engine running at cruise speed, and the PCM detected an unexpected voltage condition on the A/t Solenoid Valve 'B' circuit during the CCM test. **Possible Causes:** • A/T Solenoid 'B' circuit is open, shorted to ground or to power • A/T Solenoid 'B' control circuit is shorted to system power • A/T Solenoid 'B' connector is disconnected or damaged • A/T Solenoid 'B' is damaged or has failed • TCM or PCM has failed
DTC: P0758 **1T CCM, MIL: NO** **1996, 1997, 1998, 1999** **Models:** SLX **Engines:** All **Transmissions:** A/T	**Shift Solenoid 'B' Circuit Malfunction** Engine started, then with Shift Solenoid 'B' (SSB) commanded "on", and PCM detected the solenoid control signal was 12v, or with SSB commanded "off", the solenoid control signal was near 0v, either condition met for 0.84 to 1.0 seconds during the test. **Possible Causes:** • SSB control circuit is open, shorted to ground or to power (B+) • SSB power circuit (from the PCM) is open • SSB is damaged or has failed • PCM has failed
DTC: P0763 **1T CCM, MIL: YES** **2001, 2002, 2003, 2004, 2005, 2006** **Models:** 3.2CL, 3.2TL, TL, 3.5RL, RL, MDX, RSX **Engines:** All **Transmissions:** A/T	**A/T Lockup Solenoid 'C' Circuit Malfunction** Engine started, engine running at cruise speed, and the PCM detected an unexpected voltage condition on the A/T Solenoid Valve 'C' circuit during the CCM test. **Possible Causes:** • A/T Solenoid 'C' circuit is open, shorted to ground or to power • A/T Solenoid 'C' connector is disconnected or damaged • A/T Solenoid 'C' is damaged or has failed • TCM or PCM has failed
DTC: P0768 **1T CCM, MIL: YES** **2001, 2002, 2003, 2004, 2005** **Models:** NSX, NSX-T, RSX **Engines:** All **Transmissions:** A/T	**A/T Lockup Solenoid 'D' Circuit Malfunction** Engine started, engine running at cruise speed, and the PCM detected an unexpected voltage condition on the A/T Solenoid Valve 'D' circuit during the CCM test. **Possible Causes:** • A/T Solenoid 'D' circuit is open, shorted to ground or to power • A/T Solenoid 'D' connector is disconnected or damaged • A/T Solenoid 'D' is damaged or has failed • TCM or PCM has failed
DTC: P0773 **1T CCM, MIL: YES** **2002, 2003, 2004, 2005, 2006** **Models:** RSX **Engines:** All **Transmissions:** A/T	**A/T Lockup Solenoid 'E' Circuit Malfunction** Engine started, engine running at cruise speed, and the PCM detected an unexpected voltage condition on the A/T Solenoid Valve 'E' circuit during the CCM test. **Possible Causes:** • A/T Solenoid 'E' circuit is open, shorted to ground or to power • A/T Solenoid 'E' connector is disconnected or damaged • A/T Solenoid 'E' is damaged or has failed • TCM or PCM has failed

DTC	Trouble Code Title, Conditions & Possible Causes
DTC: P0775 **1T CCM, MIL: YES** **2002, 2003, 2004, 2005, 2006** **Models:** RSX **Engines:** All **Transmissions:** A/T	**A/T Clutch Pressure Control Solenoid 'B' Circuit Malfunction** Engine started, engine running at 25-40 mph for 10 seconds followed by a deceleration to a complete stop, and the PCM detected invalid operation of the clutch pressure control solenoid 'B'. **Possible Causes:** • Clutch pressure control solenoid 'B' is damaged or has failed • Automatic transmission may also me damaged or have failed
DTC: P0778 **1T CCM, MIL: YES** **2002, 2003, 2004, 2005, 2006** **Models:** RSX **Engines:** All **Transmissions:** A/T	**A/T Clutch Pressure Control Solenoid 'B' Circuit Malfunction** Engine started, engine running at 25-40 mph for 10 seconds followed by a deceleration to a complete stop, and the PCM detected an unexpected voltage condition on the A/T Solenoid Valve 'B' circuit during the CCM test. **Possible Causes:** • Clutch pressure control solenoid 'B' circuit is open or shorted • Clutch pressure control solenoid 'B' is damaged or has failed • TCM or PCM has failed
DTC: P0780 **1T CCM, MIL: YES** **2001, 2002, 2003, 2004, 2005, 2006** **Models:** 3.2CL, 3.2TL, TL, 3.5RL, RL, MDX, RSX **Engines:** All **Transmissions:** A/T	**Automatic Transaxle System Malfunction** Engine started, vehicle driven at cruise speed for several minutes, and the PCM detected an Automatic Transaxle fault. Note: This trouble code sets with along with several TCM related trouble codes. **Possible Causes:** • Refer to the repair instructions in a transmission repair manual or the information in other electronic media to repair this code.
DTC: P0795 **1T CCM, MIL: YES** **2002, 2003, 2004, 2005, 2006** **Models:** RSX **Engines:** All **Transmissions:** A/T	**A/T Clutch Pressure Control Solenoid 'C' Circuit Malfunction** Engine started, vehicle driven at cruise speed, and the PCM detected an unexpected voltage condition on the A/T Clutch Pressure Control solenoid valve 'C' circuit during the CCM test. **Possible Causes:** • Clutch pressure control solenoid 'C' circuit is open • Clutch pressure control solenoid 'C' connector is disconnected • Clutch pressure control solenoid 'C" is damaged or has failed • Clear codes and recheck for this trouble code. If it resets, update the PCM to the latest software. Then recheck for the same trouble code. If it resets, substitute a known good PCM to determine if the PCM has failed and is causing the code.
DTC: P0798 **1T CCM, MIL: YES** **2002, 2003, 2004, 2005, 2006** **Models:** RSX **Engines:** All **Transmissions:** A/T	**A/T Clutch Pressure Control Solenoid 'C' Circuit Malfunction** Engine started, vehicle driven at cruise speed, and the PCM detected an unexpected voltage condition on the A/T Clutch Pressure Control solenoid valve 'C' circuit during the CCM test. **Possible Causes:** • Clutch pressure control solenoid 'C' circuit is shorted to ground • Clutch pressure control solenoid 'C" is damaged or has failed • Clear codes and recheck for this trouble code. If it resets, update the PCM to the latest software. Then recheck for the same trouble code. If it resets, substitute a known good PCM to determine if the PCM has failed and is causing the code.
DTC: P0840 **1T CCM, MIL: YES** **2002, 2003, 2004, 2005, 2006** **Models:** RSX **Engines:** All **Transmissions:** A/T	**A/T Clutch Pressure Switch No. 2 Circuit Malfunction** Engine started, vehicle driven at cruise speed, and the PCM detected an unexpected voltage condition on the A/T Clutch pressure switch No. 2 circuit during the CCM test period. **Possible Causes:** • Clutch Pressure Switch No. 2 circuit is open • Clutch Pressure Switch No. 2 circuit is shorted to ground • Clutch Pressure Switch connector is disconnected or damaged • Clear codes and recheck for this trouble code. If it resets, update the PCM to the latest software. Then recheck for the same trouble code. If it resets, substitute a known good PCM to determine if the PCM has failed and is causing the code.
DTC: P0845 **1T CCM, MIL: YES** **2002, 2003, 2004, 2005, 2006** **Models:** RSX **Engines:** All **Transmissions:** A/T	**A/T Clutch Pressure Switch No. 3 Circuit Malfunction** Engine started, vehicle driven at cruise speed, and the PCM detected an unexpected voltage condition on the A/T Clutch pressure switch No. 3 circuit during the CCM test period. **Possible Causes:** • Clutch Pressure Switch No. 3 circuit is open • Clutch Pressure Switch No. 3 circuit is shorted to ground • Clutch Pressure Switch connector is disconnected or damaged • Clear codes and recheck for this trouble code. If it resets, update the PCM to the latest software. Then recheck for the same trouble code. If it resets, substitute a known good PCM to determine if the PCM has failed and is causing the code.

OBD II Trouble Code List (P1xxx Codes)

DTC	Trouble Code Title
DTC: P1077 **1T CCM, MIL: YES** **2002, 2003, 2004, 2005, 2006** **Models:** MDX, RSX **Engines:** All **Transmissions:** A/T, M/T	**Intake Manifold Runner Control System (Low RPM) Malfunction** Engine started, vehicle driven at cruise speed and then back to idle speed, and the PCM detected a problem in the IMRC system at low engine speed during the CCM test. If P0651 is also set, repair it first. **Possible Causes:** • IMRC component failure affecting low engine speed operation • IMRC solenoid is damaged or stuck at low engine speed • PCM has failed
DTC: P1078 **1T CCM, MIL: YES** **2002, 2003, 2004, 2005, 2006** **Models:** MDX, RSX **Engines:** All **Transmissions:** A/T, M/T	**Intake Manifold Runner Control System (High RPM) Malfunction** Engine started, vehicle driven at cruise speed and then back to idle speed, and the PCM detected a problem in the IMRC system at high engine speed during the CCM test. **Possible Causes:** • IMRC component failure affecting high engine speed operation • IMRC solenoid is damaged or stuck at high engine speed • PCM has failed
DTC: P1102 **1T CCM, MIL: YES** **1997, 1998** **Models:** 2.5TL **Engines:** All **Transmissions:** A/T	**Mass Airflow Sensor Signal Lower Than Expected** Engine started, vehicle driven at cruise speed and then back to idle speed, and the PCM detected the MAF sensor signal indicated less than 11.5 g/sec at some point during the CCM Rationality test. **Possible Causes:** • MAF sensor signal circuit is open (intermittent fault) • MAF sensor signal circuit shorted to ground (intermittent fault) • MAF sensor is damaged or dirty (it may be shorted internally) • BARO sensor signal circuit to the TCM is open or grounded • TCM or the PCM has failed
DTC: P1103 **1T CCM, MIL: YES** **1997, 1998** **Models:** 2.5TL **Engines:** All **Transmissions:** A/T	**Mass Airflow Sensor Signal Higher Than Expected** Engine started, vehicle driven at cruise speed and then back to idle speed, and the PCM detected the MAF sensor signal indicated more than 16.5 g/sec at some point during the CCM Rationality test. **Possible Causes:** • MAF sensor signal circuit is open (intermittent fault) • MAF sensor signal circuit shorted to ground (intermittent fault) • MAF sensor is damaged or dirty (it may be shorted internally) • BARO sensor signal circuit to the TCM is open or grounded • TCM or the PCM has failed
DTC: P1106 **1T CCM, MIL: YES** **1997, 1998, 1999, 2000, 2001,** **2002, 2003, 2004, 2005, 2006** **Models:** 2.3CL, 2.5TL, 3.0CL, 3.2TL, TL, 3.5RL, RL, Integra, NSX, NSX-T, MDX, RSX **Engines:** All **Transmissions:** A/T	**BARO Pressure Sensor Circuit Range/Performance** Engine started, engine running in 4th gear and accelerated under wide-open-throttle conditions, and the PCM detected the BARO sensor signal did not change enough under these test conditions. **Possible Causes:** • BARO sensor signal circuit is open or shorted to ground • BARO sensor ground circuit has high resistance • BARO sensor is damaged or it may be out of calibration • PCM has failed
DTC: P1106 **1T CCM, MIL: NO** **1996, 1997, 1998, 1999** **Models:** SLX **Engines:** 3.2L VIN V, 3.5L VIN X **Transmissions:** A/T	**MAP Sensor Circuit High Input (Intermittent)** DTC P0121, P0122, P0123 not set, engine runtime 10 seconds, engine speed less than 1000 rpm with throttle angle less than 3%, or the engine speed is more than 1000 rpm with the throttle angle less than 10%, and the PCM detected an unexpected high value (over 80 kPa) on the MAP sensor circuit for 5 seconds of a 16 second period. **Note: For additional help with this code, view the Failure Records.** **Possible Causes:** • MAP sensor signal circuit is open (an intermittent fault) • MAP sensor signal circuit is shorted to VREF (intermittent fault) • MAP sensor is damaged or has failed

DTC	Trouble Code Title, Conditions & Possible Causes
DTC: P1107 **1T CCM, MIL: YES** **1995, 1996, 1997, 1998, 1999,** **2000, 2001, 2002, 2003, 2004,** **2005** **Models:** NSX, NSX-T **Engines:** All **Transmissions:** A/T	**BARO Pressure Sensor Circuit Low Input** Key on or engine running; and the PCM detected an unexpected "low" voltage on the BARO sensor circuit during the CCM test. **Possible Causes:** • BARO sensor signal circuit is shorted to signal ground • BARO sensor signal circuit is shorted to chassis ground • BARO sensor is damaged (it may be shorted internally) • BARO sensor signal circuit to the TCM is open or grounded • TCM or the PCM has failed
DTC: P1107 **1T CCM, MIL: YES** **1995, 1996, 1997, 1998** **Models:** 2.5TL **Engines:** All **Transmissions:** A/T	**BARO Pressure Sensor Circuit Low Input** Key on or engine running; and the PCM detected an unexpected "low" voltage on the BARO sensor circuit during the CCM test. **Possible Causes:** • BARO sensor signal circuit is shorted to signal ground • BARO sensor signal circuit is shorted to chassis ground • BARO sensor is damaged (it may be shorted internally) • BARO sensor signal circuit to the TCM is open or grounded • TCM or the PCM has failed
DTC: P1107 **1T CCM, MIL: YES** **1996, 1997, 1998, 1999, 2000,** **2001, 2002, 2003, 2004, 2005,** **2006** **Models:** 2.2CL, 2.3CL, 3.0CL, 3.2CL, 3.2TL, TL, 3.5RL, RL, Integra, MDX, RSX **Engines:** All **Transmissions:** A/T	**BARO Pressure Sensor Circuit Low Input** Key on or engine running; and the PCM detected an unexpected "low" voltage on the BARO sensor circuit during the CCM test. **Possible Causes:** • BARO sensor signal circuit is shorted to signal ground • BARO sensor signal circuit is shorted to chassis ground • BARO sensor is damaged (it may be shorted internally) • BARO sensor signal circuit to the TCM is open or grounded • TCM or the PCM has failed
DTC: P1107 **1T CCM, MIL: NO** **1996, 1997, 1998, 1999** **Models:** SLX **Engines:** 3.2L VIN V, 3.5L VIN X **Transmissions:** A/T	**MAP Sensor Circuit Low Input (Intermittent)** DTC P0121, P0122, P0123 not set, engine running, engine speed less than 1000 rpm and throttle angle over 1%, or the engine speed more than 1000 rpm and throttle angle more than 2%, and the PCM detected an unexpected low value (less than 11 kPa) on the MAP sensor circuit for 5 seconds of a 16 second period. **Note: For additional help with this code, view the Failure Records.** **Possible Causes:** • MAP sensor signal circuit shorted to ground (intermittent fault) • MAP sensor VREF circuit is open or shorted to ground • MAP sensor is damaged or has failed
DTC: P1108 **1T CCM, MIL: YES** **1995, 1996, 1997, 1998, 1999,** **2000, 2001, 2002, 2003, 2004,** **2005** **Models:** NSX, NSX-T **Engines:** All **Transmissions:** A/T	**BARO Pressure Sensor Circuit High Input** Key on or engine running; and the PCM detected an unexpected "high" voltage on the BARO sensor circuit during the CCM test. **Possible Causes:** • BARO sensor signal circuit shorted to VREF • BARO sensor signal circuit is shorted to system power (B+) • BARO sensor is damaged (it may be open internally) • BARO sensor signal circuit to the TCM is shorted to power • TCM or the PCM has failed
DTC: P1108 **1T CCM, MIL: YES** **1995, 1996, 1997, 1998** **Models:** 2.5TL **Engines:** All **Transmissions:** A/T	**BARO Pressure Sensor Circuit High Input** Key on or engine running; and the PCM detected an unexpected "high" voltage on the BARO sensor circuit during the CCM test. **Possible Causes:** • BARO sensor signal circuit shorted to VREF • BARO sensor signal circuit is shorted to system power (B+) • BARO sensor is damaged (it may be open internally) • BARO sensor signal circuit to the TCM is shorted to power • TCM or the PCM has failed
DTC: P1108 **1T CCM, MIL: YES** **1996, 1997, 1998, 1999, 2000,** **2001, 2002, 2003, 2004, 2005,** **2006** **Models:** 2.2CL, 2.3CL, 3.0CL, 3.2CL, 3.2TL, TL, 3.5RL, RL, Integra, MDX, RSX **Engines:** All **Transmissions:** A/T	**BARO Pressure Sensor Circuit High Input** Key on or engine running; and the PCM detected an unexpected "high" voltage on the BARO sensor circuit during the CCM test. **Possible Causes:** • BARO sensor signal circuit shorted to VREF • BARO sensor signal circuit is shorted to system power (B+) • BARO sensor is damaged (it may be open internally) • BARO sensor signal circuit to the TCM is shorted to power • TCM or the PCM has failed

DTC	Trouble Code Title, Conditions & Possible Causes
DTC: P1111 **1T CCM, MIL: NO** **1996, 1997, 1998, 1999** **Models:** SLX **Engines:** 3.2L VIN V, 3.5L VIN X **Transmissions:** A/T	**IAT Sensor Circuit High Input (Intermittent)** Engine started, engine runtime over 4 minutes, ECT sensor more than 140°F, vehicle speed under 20 mph, MAF sensor less than 20 g/sec, and the PCM detected an unexpected high voltage signal of over 4.90v [Scan Tool reads -38°F] on the IAT sensor circuit for 2.5 seconds over a 25 second time period. **Note: For additional help with this code, view the Failure Records.** **Possible Causes:** • IAT sensor signal circuit is open (an intermittent fault) • IAT sensor is damaged or has failed (an intermittent fault) • PCM has failed
DTC: P1112 **1T CCM, MIL: NO** **1996, 1997, 1998, 1999** **Models:** SLX **Engines:** 3.2L VIN V, 3.5L VIN X **Transmissions:** A/T	**IAT Sensor Circuit Low Input (Intermittent)** Engine started, engine runtime over 4 minutes, ECT sensor more than 140°F, vehicle driven to a speed of over 20 mph, Calculated airflow less than 20 g/sec, and the PCM detected an unexpected low voltage of under 0.10v on the IAT sensor circuit (Scan Tool reads 298°F) for 2.5 seconds over a 25 second time period. **Note: For additional help with this code, view the Failure Records.** **Possible Causes:** • IAT sensor signal circuit is shorted to ground (intermittent fault) • IAT sensor is damaged or has failed (an intermittent fault) • PCM has failed
DTC: P1114 **1T CCM, MIL: NO** **1996, 1997, 1998, 1999** **Models:** SLX **Engines:** 3.2L VIN V, 3.5L VIN X **Transmissions:** A/T	**ECT Sensor Circuit Low Input (Intermittent)** Engine started, engine runtime 1 minute, and the PCM detected an unexpected low voltage of under 0.10v [Scan Tool read 302°F] on the ECT sensor circuit for 10 seconds over a 100 second period. **Note: For additional help with this code, view the Failure Records.** **Possible Causes:** • ECT sensor signal circuit shorted to ground (intermittent fault) • ECT sensor is damaged or has failed • PCM has failed
DTC: P1115 **1T CCM, MIL: NO** **1996, 1997, 1998, 1999** **Models:** SLX **Engines:** 3.2L VIN V, 3.5L VIN X **Transmissions:** A/T	**ECT Sensor Circuit High Input (Intermittent)** Engine started, engine runtime 1 minute, and the PCM detected an unexpected high voltage of over 4.90v [Scan Tool reads -38°F] on the ECT sensor circuit for 10 seconds over a 100 second period. **Note: For additional help with this code, view the Failure Records.** **Possible Causes:** • ECT sensor signal circuit is open (an intermittent fault) • ECT sensor is damaged or has failed (an intermittent fault) • PCM has failed
DTC: P1121 **1T CCM, MIL: YES** **1996, 1997, 1998, 1999, 2000, 2001, 2002, 2003, 2004, 2005, 2006** **Models:** 3.2TL, TL **Engines:** All **Transmissions:** A/T, M/T	**TP Sensor Signal Lower Than Expected** Engine started, and the PCM detected the TP sensor was lower than the expected value with the throttle wide open (i.e., the Scan Tool reads less than 14.1% under these conditions) during the CCM Rationality Test. **Possible Causes:** • TP sensor signal circuit is shorted to ground between the PCM and the TCM • TP sensor is damaged or has failed • PCM has failed
DTC: P1121 **1T CCM, MIL: YES** **1997, 1998, 1999, 2000, 2001, 2002, 2003, 2004, 2005, 2006** **Models:** 2.3CL, 2.5TL, 3.0CL, 3.2CL, 3.5RL, RL, Integra, MDX, RSX **Engines:** All **Transmissions:** A/T, M/T	**TP Sensor Signal Lower Than Expected** Engine started, and the PCM detected the TP sensor was lower than the expected value with the throttle wide open (i.e., the Scan Tool reads less than 14.1% under these conditions) during the CCM Test. **Possible Causes:** • TP sensor signal circuit is shorted to ground between the PCM and the TCM • TP sensor is damaged or has failed • PCM has failed
DTC: P1121 **1T CCM, MIL: NO** **1996, 1997, 1998, 1999** **Models:** SLX **Engines:** 3.2L VIN V, 3.5 VIN X **Transmissions:** A/T	**Throttle Position Sensor Intermittent High Input** Engine started, and the PCM detected an unexpected high signal of over 4.90v on the TP sensor circuit for 0.15 seconds over a 1.5 second period during the CCM continuous test. Note: For additional help with this code, view the Failure Records. **Possible Causes:** • TP sensor signal circuit is open between the sensor and PCM • TP sensor ground circuit is open between the sensor and PCM • TP sensor signal circuit is shorted to VREF or system power • TP sensor is damaged or has failed

DTC	Trouble Code Title, Conditions & Possible Causes
DTC: P1122 **1T CCM, MIL: YES** **1996, 1997, 1998, 1999, 2000,** **2001, 2002, 2003, 2004, 2005,** **2006** **Models:** 3.2TL, TL **Engines:** All **Transmissions:** A/T, M/T	**TP Sensor Signal Higher Than Expected** Engine started, and the PCM detected the TP sensor was more than the expected value with the throttle wide open (i.e., the Scan Tool reads more than 16.5% under these conditions) during the CCM Rationality Test. **Possible Causes:** • TP sensor signal circuit is open or shorted to VREF between the PCM and the TCM • TP sensor is damaged or has failed • PCM has failed
DTC: P1122 **1T CCM, MIL: YES** **1997, 1998, 1999, 2000, 2001,** **2002, 2003, 2004, 2005, 2006** **Models:** 2.3CL, 2.5TL, 3.0CL, 3.2CL, 3.5RL, RL, Integra, MDX, RSX **Engines:** All **Transmissions:** A/T, M/T	**TP Sensor Signal Higher Than Expected** Engine started, and the PCM detected the TP sensor was more than the expected value with the throttle wide open (i.e., the Scan Tool reads more than 16.5% under these conditions) during the CCM Rationality test. **Possible Causes:** • TP sensor signal circuit is open or shorted to VREF between the PCM and the TCM • TP sensor is damaged or has failed • PCM has failed
DTC: P1122 **1T CCM, MIL: YES** **2002, 2003, 2004, 2005, 2006** **Models:** RSX **Engines:** All **Transmissions:** A/T, M/T	**TP Sensor Signal Higher Than Expected** Engine started, and the PCM detected the TP sensor was more than the expected value with the throttle wide-open (i.e., the Scan Tool reads more than 16.5% under these conditions) in the CCM test. **Possible Causes:** • TP sensor signal circuit is open or shorted to VREF between the PCM and the TCM • TP sensor is damaged or has failed • PCM has failed • TSB 01-038 (10/29/01) contains a repair procedure for the code
DTC: P1122 **1T CCM, MIL: NO** **1996, 1997, 1998, 1999** **Models:** SLX **Engines:** 3.2L VIN V, 3.5 VIN X **Transmissions:** A/T	**Throttle Position Sensor Intermittent Low Input** Engine started, and the PCM detected an unexpected low signal of less than 0.22v on the TP sensor circuit for 0.15 seconds over a 1.5 second period during the CCM continuous test. Note: For additional help with this code, view the Failure Records. **Possible Causes:** • TP sensor signal circuit is shorted to ground • TP sensor VREF circuit shorted to ground (test other sensors) • TP sensor is damaged or has failed
DTC: P1128 **1T CCM, MIL: YES** **1997, 1998, 1999, 2000, 2001,** **2002, 2003, 2004, 2005** **Models:** NSX, NSX-T **Engines:** All **Transmissions:** A/T	**MAP Sensor Signal Less Than Expected** Engine started, engine running at cruise speed and then back to idle speed, and the PCM detected the MAP sensor signal was lower than the expected value during the CCM Rationality test. **Possible Causes:** • MAP sensor signal circuit shorted to ground (intermittent fault) • MAP sensor vacuum line bent or plugged at intake manifold • MAP sensor is damaged or it is out-of-calibration • PCM has failed
DTC: P1128 **1T CCM, MIL: YES** **1996, 1997,1998, 1999, 2000,** **2001, 2002, 2003, 2004, 2005,** **2006** **Models:** 3.2TL, TL **Engines:** All **Transmissions:** A/T, M/T	**MAP Sensor Signal Less Than Expected** Engine started, engine running at cruise speed and then back to idle speed, and the PCM detected a MAP sensor signal was less than the expected value during the CCM Rationality Test. **Possible Causes:** • MAP sensor signal circuit shorted to ground (intermittent fault) • MAP sensor vacuum line bent or plugged at intake manifold • MAP sensor is damaged or it is out-of-calibration • PCM has failed
DTC: P1128 **1T CCM, MIL: YES** **1997, 1998, 1999, 2000, 2001,** **2002, 2003, 2004, 2005, 2006** **Models:** 2.3CL, 3.0CL, 3.2CL, 3.5RL, RL, Integra, MDX, RSX **Engines:** All **Transmissions:** A/T, M/T	**MAP Sensor Signal Lower Than Expected** Engine started, engine running at cruise speed and then back to idle speed, and the PCM detected a MAP sensor signal was less than the expected value during the CCM Rationality Test. **Possible Causes:** • MAP sensor signal circuit shorted to ground (intermittent fault) • MAP sensor vacuum line bent or plugged at intake manifold • MAP sensor is damaged or it is out-of-calibration • PCM has failed

DTC	Trouble Code Title, Conditions & Possible Causes
DTC: P1129 **1T CCM, MIL: YES** **1997, 1998, 1999, 2000, 2001, 2002, 2003, 2004, 2005** **Models:** NSX, NSX-T **Engines:** All **Transmissions:** A/T	**MAP Sensor Signal Higher Than Expected** Engine started, engine running at cruise speed and then back to idle speed, and the PCM detected a MAP sensor signal was higher than the expected value during the CCM Rationality Test. **Possible Causes:** • Check for signs of a vacuum leak at the PCV valve or hose, the brake booster and the throttle body area • MAP sensor signal circuit shorted to VREF (intermittent fault) • MAP sensor is damaged or it is out-of-calibration • PCM has failed
DTC: P1129 **1T CCM, MIL: YES** **1996, 1997,1998, 1999, 2000, 2001, 2002, 2003, 2004, 2005, 2006** **Models:** 3.2TL, TL **Engines:** All **Transmissions:** A/T, M/T	**MAP Sensor Value Higher Than Expected** Engine started, engine running at cruise speed and then back to idle speed, and the PCM detected a MAP sensor signal was higher than the expected value during the CCM Rationality Test. **Possible Causes:** • Check for signs of a vacuum leak at the PCV valve or hose, the engine mount vacuum hose, brake booster and throttle body • MAP sensor signal circuit shorted to VREF (intermittent fault) • MAP sensor is damaged or it is out-of-calibration • PCM has failed
DTC: P1129 **1T CCM, MIL: YES** **1997, 1998, 1999, 2000, 2001, 2002, 2003, 2004, 2005, 2006** **Models:** 2.3CL, 3.0CL, 3.2CL, 3.5RL, RL, Integra, MDX, RSX **Engines:** All **Transmissions:** A/T, M/T	**MAP Sensor Value Higher Than Expected** Engine started, engine running at cruise speed and then back to idle speed, and the PCM detected a MAP sensor signal was higher than the expected value during the CCM Rationality Test. **Possible Causes:** • Check for signs of a vacuum leak at the PCV valve or hose, the engine mount vacuum hose, brake booster and throttle body • MAP sensor signal circuit shorted to VREF (intermittent fault) • MAP sensor is damaged or it is out-of-calibration • PCM has failed
DTC: P1133 **2T O2S2, MIL: YES** **1996, 1997, 1998, 1999** **Models:** SLX **Engines:** 3.2L VIN V, 3.5L VIN X **Transmissions:** A/T	**HO2S-11 (Bank 1 Sensor 1) Insufficient Switching** DTC P0101, P0102, P0103, P0106, P0107, P0108, P0117, P0118, P0121, P0122, P0123, P0131, P0132, P0133, P0134, P0135, P0300, P0301-P0306, P0441 and P1441 not set, engine speed from 1500-3000 rpm in closed loop for 1 minute, system voltage from 11-16v, ECT sensor more than 122°F, MAF sensor from 9-42 g/sec, Purge duty cycle over 2%, conditions met for 3 seconds, and the PCM detected less than 23 rich-to-lean or lean-to-rich switches on the HO2S-11 signal circuit. **Possible Causes:** • Air leaks after the MAF sensor, or in the EGR or PCV system • Exhaust leaks before or near where the front HO2S is mounted • Fuel control sensor is out of calibration (i.e., ECT, IAT or MAP) • Fuel delivery system supplying too much or too little fuel during cruise or idle periods (e.g., faulty fuel pump, or dirty fuel filter) • Fuel injector (one or more) dirty, leaking or sticking • Fuel pressure regulator leaking, damaged or has failed • HO2S is contaminated, deteriorated or it has failed
DTC: P1134 **2T O2S2, MIL: YES** **1996, 1997, 1998, 1999** **Models:** SLX **Engines:** 3.2L VIN V, 3.5L VIN X **Transmissions:** A/T	**HO2S-11 (Bank 1 Sensor 1) Transition Time Ratio Error** DTC P0101, P0102, P0103, P0106, P0107, P0108, P0117, P0118, P0121, P0122, P0123, P0131, P0132, P0133, P0134, P0135, P0300, P0301-P0306, P0441 and P1441 not set, engine speed from 1500-3000 rpm in closed loop for 1 minute, system voltage from 11-16v, ECT sensor more than 122-167°F, MAF sensor at 18-42 g/sec, Purge duty cycle over 2%, conditions met for 3 seconds, and the PCM detected the HO2S-11 transition ratio from lean-to-rich and rich-to-lean was less than 0.44 or more than 3.8 during the test. **Possible Causes:** • Air leaks after the MAF sensor, or in the EGR or PCV system • Exhaust leaks before or near where the front HO2S is mounted • Fuel control sensor is out of calibration (i.e., ECT, IAT or MAP) • Fuel delivery system supplying too much or too little fuel during cruise or idle periods (e.g., faulty fuel pump, or dirty fuel filter) • Fuel injector (one or more) dirty, leaking or sticking • Fuel pressure regulator leaking, damaged or has failed • HO2S is contaminated, deteriorated or it has failed

DTC	Trouble Code Title, Conditions & Possible Causes
DTC: P1153 **2T CCM, MIL: YES** **1996, 1997, 1998, 1999** **Models:** SLX **Engines:** 3.2L VIN V, 3.5L VIN X **Transmissions:** A/T	**HO2S-21 (Bank 2 Sensor 1) Insufficient Switching** DTC P0101, P0102, P0103, P0106, P0107, P0108, P0117, P0118, P0121, P0122, P0123, P0131, P0132, P0133, P0134, P0135, P0300, P0301-P0306, P0441 and P1441 not set, system voltage from 11-16v, engine speed from 1500-3000 rpm in closed loop for 1 minute, ECT sensor more than 122-167°F, MAF sensor from 9-42 g/sec, Purge duty cycle over 2%, conditions met for 3 seconds, and the PCM detected less than 27 rich-to-lean or lean-to-rich switches on the HO2S-21 signal circuit in the Oxygen Sensor Monitor test. **Possible Causes:** • Air leaks after the MAF sensor, or in the EGR or PCV system • Exhaust leaks before or near where the front HO2S is mounted • Fuel control sensor is out of calibration (i.e., ECT, IAT or MAP) • Fuel delivery system supplying too much or too little fuel during cruise or idle periods (e.g., faulty fuel pump, or dirty fuel filter) • Fuel injector (one or more) dirty, leaking or sticking • Fuel pressure regulator leaking, damaged or has failed • HO2S is contaminated, deteriorated or it has failed
DTC: P1154 **2T O2S2, MIL: YES** **1996, 1997, 1998, 1999** **Models:** SLX **Engines:** 3.2L VIN V, 3.5L VIN X **Transmissions:** A/T	**HO2S-21 (Bank 2 Sensor 1) Transition Time Ratio** DTC P0101, P0102, P0103, P0106, P0107, P0108, P0117, P0118, P0121, P0122, P0123, P0131, P0132, P0133, P0134, P0135, P0300, P0301-P0306, P0441 and P1441 not set, system voltage from 10-16v, engine speed from 1500-3000 rpm in closed loop for 1 minute, ECT sensor more than 122-167°F, MAF sensor from 9-42 g/sec, Purge duty cycle over 2%, conditions met for 3 seconds, then 90 seconds after entering closed loop control, the PCM detected the transition time ratio to switch from lean-to-rich or rich to lean from the HO2S-21 was less than 0.44 or more than 3.8 during the test. **Possible Causes:** • Air leaks at the exhaust manifold or exhaust pipes • HO2S signal circuit is open or shorted to ground (intermittent) • HO2S heater power circuit is open, or the heater has failed • HO2S contaminated with wrong fuel, has deteriorated or failed
DTC: P1162 **1T CCM, MIL: YES** **2002, 2003, 2004, 2005, 2006** **Models:** RSX **Engines:** All **Transmissions:** A/T, M/T	**Air Fuel Sensor (Bank 1 Sensor 1) Circuit Malfunction** DTC P0131, P0132, P0133, P1163 not set, vehicle driven while in closed loop at over 55 mph in D4 for 1-2 minutes, and the PCM detected an unexpected voltage condition on the A/F circuit during the HO2S Monitor test. **Possible Causes:** • A/F sensor may be contaminated or may have failed • Fuel supply system is too lean (exhaust leaks in front of HO2S) • Fuel supply system is too rich (fuel filter is clogged or dirty) • PCM has failed
DTC: P1163 **1T CCM, MIL: YES** **2002, 2003, 2004, 2005, 2006** **Models:** RSX **Engines:** All **Transmissions:** A/T, M/T	**Air Fuel Sensor (Bank 1 Sensor 1) Slow Response** Engine at idle speed, then accelerated to 55 mph for 5 seconds, then back to idle speed for 5 seconds, and the PCM detected the A/F sensor response time was too slow, or the rich-lean or lean-rich response switch rate was too slow during the HO2S Monitor test. **Possible Causes:** • A/F sensor may be contaminated or may have failed • Fuel supply system is too lean (exhaust leaks in front of LAF) • Fuel supply system is too rich (fuel filter is clogged or dirty) • PCM has failed • TSB 01-038 (10/29/01) contains a repair procedure for the code
DTC: P1164 **1T CCM, MIL: YES** **2002, 2003, 2004, 2005, 2006** **Models:** RSX **Engines:** All **Transmissions:** A/T, M/T	**Air Fuel Sensor (Bank 1 Sensor 1) Range/Performance** Engine speed over 1500 rpm in 4th gear in closed loop, then a quick acceleration to WOT, followed by a 5 second deceleration period with the throttle closed, and the PCM detected the A/F sensor response time or the R-L or L-R switch rate was too slow. **Possible Causes:** • A/F sensor may be contaminated or may have failed • Fuel supply system too lean (exhaust leaks in front of the LAF) • Fuel supply system too rich (the fuel filter may be very dirty) • PCM has failed
DTC: P1166 **1T O2S HTR1, MIL: YES** **2002, 2003, 2004, 2005, 2006** **Models:** RSX **Engines:** All **Transmissions:** A/T, M/T	**Air Fuel Sensor (Bank 1 Sensor 1) Heater Circuit Malfunction** Engine runtime more than 80 seconds; and the PCM detected an unexpected voltage condition on the A/F sensor heater circuit. **Possible Causes:** • A/F sensor heater circuit is open or shorted to ground • A/F sensor heater is damaged or has failed • PCM has failed

DTC	Trouble Code Title, Conditions & Possible Causes
DTC: P1167 **1T O2S HTR1, MIL: YES** **2002, 2003, 2004, 2005, 2006** **Models:** RSX **Engines:** All **Transmissions:** A/T, M/T	**Air Fuel Sensor (Bank 1 Sensor 1) Heater Circuit Malfunction** Engine runtime over 80 seconds, and the PCM detected an unexpected voltage condition on the A/F sensor Heater circuit. **Possible Causes:** • A/F sensor heater power supply circuit is open (check the fuse) • A/F sensor heater circuit is shorted to system power (B+) • A/F sensor heater is damaged or has failed • PCM has failed
DTC: P1171 **1T FUEL, MIL: YES** **1996, 1997, 1998, 1999** **Models:** SLX **Engines:** 3.2L VIN V, 3.5L VIN X **Transmissions:** A/T	**Fuel System Lean During Acceleration Detected** DTC P0131, P0132, P0133, P0134 and P1133 not set, ECT sensor more than 140°F, engine running in Power Enrichment mode in closed loop, and the PCM detected the HO2S-11 signal indicated less than 400 mv for 5 seconds during the Fuel System Monitor test. **Possible Causes:** • Air intake leaks in the engine, or in the PCV system (valve) • Air leaks at the EGR gasket, or at the EGR valve diaphragm • Base engine "mechanical" fault affecting one or more cylinders • Exhaust leaks before or near where the front HO2S is mounted • Fuel injectors (one or more) restricted (allowing too little fuel) • Fuel delivery system supplying too little fuel during acceleration periods (e.g., faulty fuel pump, dirty or restricted fuel filter) • Fuel control sensor out of calibration (i.e., IAT, MAF or MAP) • HO2S is contaminated, deteriorated or it has failed • Vehicle driven low on fuel or until it ran out of fuel
DTC: P1201 **2T MISFIRE, MIL: YES** **1995, 1996, 1997, 1998, 1999,** **2000, 2001, 2002, 2003, 2004,** **2005, 2006** **Models:** 3.5RL, RL, NSX, NSX-T **Engines:** All **Transmissions:** A/T, M/T	**Cylinder 1 Misfire Detected** Engine runtime 10 seconds, engine under positive load conditions, and the PCM detected a random misfire condition in one cylinder during the 200 (Catalyst) or 1000 revolution (Emission) test range. **Note: If the misfire is severe, the MIL will flash on/off on the 1st trip!** **Possible Causes:** • Air leaks (intake) or exhaust leaks affecting only one cylinder • Base engine (compression) problem affecting only one cylinder • Fuel metering (fuel injector dirty) problem affecting one cylinder • Ignition system (spark plug or wire) fault affecting one cylinder
DTC: P1201 **2T MISFIRE, MIL: YES** **1996, 1997, 1998** **Models:** 3.2TL **Engines:** All **Transmissions:** A/T, M/T	**Cylinder 1 Misfire Detected** Engine runtime 10 seconds, engine under positive load conditions, and the PCM detected a random misfire condition in one cylinder during the 200 (Catalyst) or 1000 revolution (Emission) test range. **Note: If the misfire is severe, the MIL will flash on/off on the 1st trip!** **Possible Causes:** • Air leaks (intake) or exhaust leaks affecting only Cylinder 1 • Base engine (compression) problem affecting Cylinder 1 • Fuel metering (fuel injector dirty) problem affecting Cylinder 1 • Ignition system (spark plug or plug wire) problem on Cylinder 1
DTC: P1202 **2T MISFIRE, MIL: YES** **1995, 1996, 1997, 1998, 1999,** **2000, 2001, 2002, 2003, 2004,** **2005, 2006** **Models:** 3.5RL, RL, NSX, NSX-T **Engines:** All **Transmissions:** A/T, M/T	**Cylinder 2 Misfire Detected** Engine runtime 10 seconds, engine under positive load conditions, and the PCM detected a random misfire condition in one cylinder during the 200 (Catalyst) or 1000 revolution (Emission) test range. **Note: If the misfire is severe, the MIL will flash on/off on the 1st trip!** **Possible Causes:** • Air leaks (intake) or exhaust leaks affecting only one cylinder • Base engine (compression) problem affecting only one cylinder • Fuel metering (fuel injector dirty) problem affecting one cylinder • Ignition system (spark plug or wire) fault affecting one cylinder
DTC: P1202 **2T MISFIRE, MIL: YES** **1996, 1997, 1998** **Models:** 3.2TL **Engines:** All **Transmissions:** A/T, M/T	**Cylinder 2 Misfire Detected** Engine runtime 10 seconds, engine under positive load conditions, and the PCM detected a random misfire condition in one cylinder during the 200 (Catalyst) or 1000 revolution (Emission) test range. **Note: If the misfire is severe, the MIL will flash on/off on the 1st trip!** **Possible Causes:** • Air leaks (intake) or exhaust leaks affecting only one cylinder • Base engine (compression) problem affecting only one cylinder • Fuel metering (fuel injector dirty) problem affecting one cylinder • Ignition system (spark plug or wire) fault affecting one cylinder

DTC	Trouble Code Title, Conditions & Possible Causes

DTC: P1203
2T MISFIRE, MIL: YES
1995, 1996, 1997, 1998, 1999,
2000, 2001, 2002
Models: 3.5RL, NSX, NSX-T
Engines: All
Transmissions: A/T, M/T
	Cylinder 3 Misfire Detected
	Engine runtime 10 seconds, engine under positive load conditions, and the PCM detected a random misfire condition in one cylinder during the 200 (Catalyst) or 1000 revolution (Emission) test range.
	Note: If the misfire is severe, the MIL will flash on/off on the 1st trip!
	Possible Causes:
	    • Air leaks (intake) or exhaust leaks affecting only one cylinder
	    • Base engine (compression) problem affecting only one cylinder
	    • Fuel metering (fuel injector dirty) problem affecting one cylinder
	    • Ignition system (spark plug or wire) fault affecting one cylinder

DTC: P1203
2T MISFIRE, MIL: YES
1996, 1997, 1998
Models: 3.2TL
Engines: All
Transmissions: A/T, M/T
	Cylinder 3 Misfire Detected
	Engine runtime 10 seconds, engine under positive load conditions, and the PCM detected a random misfire condition in one cylinder during the 200 (Catalyst) or 1000 revolution (Emission) test range.
	Note: If the misfire is severe, the MIL will flash on/off on the 1st trip!
	Possible Causes:
	    • Air leaks (intake) or exhaust leaks affecting only one cylinder
	    • Base engine (compression) problem affecting only one cylinder
	    • Fuel metering (fuel injector dirty) problem affecting one cylinder
	    • Ignition system (spark plug or wire) fault affecting one cylinder

DTC: P1204
2T MISFIRE, MIL: YES
1995, 1996, 1997, 1998, 1999,
2000, 2001, 2002, 2003, 2004,
2005, 2006
Models: 3.5RL, RL, NSX, NSX-T
Engines: All
Transmissions: A/T, M/T
	Cylinder 4 Misfire Detected
	Engine runtime 10 seconds, engine under positive load conditions, and the PCM detected a random misfire condition in one cylinder during the 200 (Catalyst) or 1000 revolution (Emission) test range.
	Note: If the misfire is severe, the MIL will flash on/off on the 1st trip!
	Possible Causes:
	    • Air leaks (intake) or exhaust leaks affecting only one cylinder
	    • Base engine (compression) problem affecting only one cylinder
	    • Fuel metering (fuel injector dirty) problem affecting one cylinder
	    • Ignition system (spark plug or wire) fault affecting one cylinder

DTC: P1204
2T MISFIRE, MIL: YES
1996, 1997, 1998
Models: 3.2TL
Engines: All
Transmissions: A/T, M/T
	Cylinder 4 Misfire Detected
	Engine runtime 10 seconds, engine under positive load conditions, and the PCM detected a random misfire condition in one cylinder during the 200 (Catalyst) or 1000 revolution (Emission) test range.
	Note: If the misfire is severe, the MIL will flash on/off on the 1st trip!
	Possible Causes:
	    • Air leaks (intake) or exhaust leaks affecting only one cylinder
	    • Base engine (compression) problem affecting only one cylinder
	    • Fuel metering (fuel injector dirty) problem affecting one cylinder
	    • Ignition system (spark plug or wire) fault affecting one cylinder

DTC: P1205
2T MISFIRE, MIL: YES
1995, 1996, 1997, 1998, 1999,
2000, 2001, 2002, 2003, 2004,
2005, 2006
Models: 3.5RL, RL, NSX, NSX-T
Engines: All
Transmissions: A/T, M/T
	Cylinder 5 Misfire Detected
	Engine runtime 10 seconds, engine under positive load conditions, and the PCM detected a random misfire condition in one cylinder during the 200 (Catalyst) or 1000 revolution (Emission) test range.
	Note: If the misfire is severe, the MIL will flash on/off on the 1st trip!
	Possible Causes:
	    • Air leaks (intake) or exhaust leaks affecting only one cylinder
	    • Base engine (compression) problem affecting only one cylinder
	    • Fuel metering (fuel injector dirty) problem affecting one cylinder
	    • Ignition system (spark plug or wire) fault affecting one cylinder

DTC: P1205
2T MISFIRE, MIL: YES
1996, 1997, 1998
Models: 3.2TL
Engines: All
Transmissions: A/T, M/T
	Cylinder 5 Misfire Detected
	Engine runtime 10 seconds, engine under positive load conditions, and the PCM detected a random misfire condition in one cylinder during the 200 (Catalyst) or 1000 revolution (Emission) test range.
	Note: If the misfire is severe, the MIL will flash on/off on the 1st trip!
	Possible Causes:
	    • Air leaks (intake) or exhaust leaks affecting only one cylinder
	    • Base engine (compression) problem affecting only one cylinder
	    • Fuel metering (fuel injector dirty) problem affecting one cylinder
	    • Ignition system (spark plug or wire) fault affecting one cylinder

DTC	Trouble Code Title, Conditions & Possible Causes
DTC: P1206 **2T MISFIRE, MIL: YES** **1995, 1996, 1997, 1998, 1999, 2000, 2001, 2002, 2003, 2004, 2005, 2006** **Models:** 3.5RL, RL, NSX, NSX-T **Engines:** All **Transmissions:** A/T, M/T	**Cylinder 6 Misfire Detected** Engine runtime 10 seconds, engine under positive load conditions, and the PCM detected a random misfire condition in one cylinder during the 200 (Catalyst) or 1000 revolution (Emission) test range. **Note: If the misfire is severe, the MIL will flash on/off on the 1st trip!** **Possible Causes:** • Air leaks (intake) or exhaust leaks affecting only one cylinder • Base engine (compression) problem affecting only one cylinder • Fuel metering (fuel injector dirty) problem affecting one cylinder • Ignition system (spark plug or wire) fault affecting one cylinder
DTC: P1206 **2T MISFIRE, MIL: YES** **1996, 1997, 1998** **Models:** 3.2TL **Engines:** All **Transmissions:** A/T, M/T	**Cylinder 6 Misfire Detected** Engine runtime 10 seconds, engine under positive load conditions, and the PCM detected a random misfire condition in one cylinder during the 200 (Catalyst) or 1000 revolution (Emission) test range. Note: If the misfire is severe, the MIL will flash on/off on the 1st trip! **Possible Causes:** • Air leaks (intake) or exhaust leaks affecting only one cylinder • Base engine (compression) problem affecting only one cylinder • Fuel metering (fuel injector dirty) problem affecting one cylinder • Ignition system (spark plug or wire) fault affecting one cylinder
DTC: P1241 **1T CCM, MIL: YES** **1995, 1996, 1997, 1998, 1999, 2000, 2001, 2002, 2003, 2004, 2005** **Models:** NSX, NSX-T **Engines:** All **Transmissions:** A/T, M/T	**Throttle Valve Control Motor 1 Circuit Malfunction** Engine started, and the PCM detected an unexpected voltage condition on the Throttle Valve Control Motor 1 circuit. Note: This throttle valve is part of the NSX Drive by Wire system. **Possible Causes:** • Throttle valve control motor 1 circuit open or shorted to ground • Throttle valve control motor 1 circuit is shorted to system power • Throttle valve control motor 1 is damaged or has failed • PCM has failed
DTC: P1242 **1T CCM, MIL: YES** **1995, 1996, 1997, 1998, 1999, 2000, 2001, 2002, 2003, 2004, 2005** **Models:** NSX, NSX-T **Engines:** All **Transmissions:** A/T, M/T	**Throttle Valve Control Motor 2 Circuit Malfunction** Engine started, and the PCM detected an unexpected voltage condition on the Throttle Valve Control Motor 2 circuit. Note: This throttle valve is part of the NSX Drive by Wire system. **Possible Causes:** • Throttle valve control motor 2 circuit open or shorted to ground • Throttle valve control motor 2 circuit is shorted to system power • Throttle valve control motor 2 is damaged or has failed • PCM has failed
DTC: P1243 **1T CCM, MIL: YES** **1995, 1996, 1997, 1998, 1999, 2000, 2001, 2002, 2003, 2004, 2005** **Models:** NSX, NSX-T **Engines:** All **Transmissions:** A/T, M/T	**Throttle Position Insufficient** Engine running at road load, with at least one wide-open throttle event, and the PCM detected an insufficient throttle position value. **Possible Causes:** • Throttle valve is stuck, damaged or has failed • TP sensor is damaged or has failed • Throttle valve control motor is damaged or has failed
DTC: P1244 **1T CCM, MIL: YES** **1995, 1996, 1997, 1998, 1999, 2000, 2001, 2002, 2003, 2004, 2005** **Models:** NSX, NSX-T **Engines:** All **Transmissions:** A/T, M/T	**Insufficient Closed Throttle Position Detected** Engine running at road load, with at least one wide-open throttle event, and the PCM detected insufficient closed throttle position. **Possible Causes:** • Throttle valve is stuck, damaged or has failed • TP sensor is damaged or has failed
DTC: P1246 **1T CCM, MIL: YES** **1995, 1996, 1997, 1998, 1999, 2000, 2001, 2002, 2003, 2004, 2005** **Models:** NSX, NSX-T **Engines:** All **Transmissions:** A/T, M/T	**Accelerator Position Sensor 1 Circuit Malfunction** Key on or engine running; and the PCM detected an unexpected voltage condition on the Accelerator Pedal Position Sensor 1 circuit. **Possible Causes:** • APP sensor 1 signal circuit is open • APP sensor 1 signal circuit is shorted to ground • APP sensor 1 signal circuit shorted to VREF or system power • APP sensor is damaged or has failed • PCM has failed

DTC	Trouble Code Title, Conditions & Possible Causes
DTC: P1247 **1T CCM, MIL: YES** **1995, 1996, 1997, 1998, 1999, 2000, 2001, 2002, 2003, 2004, 2005** **Models:** NSX, NSX-T **Engines:** All **Transmissions:** A/T, M/T	**Accelerator Position Sensor 2 Circuit Malfunction** Key on or engine running; and the PCM detected an unexpected voltage condition on the Accelerator Pedal Position Sensor 2 circuit. **Possible Causes:** • APP sensor 2 signal circuit is open • APP sensor 2 signal circuit is shorted to ground • APP sensor 2 signal circuit shorted to VREF or system power • APP sensor is damaged or has failed • PCM has failed
DTC: P1248 **1T CCM, MIL: YES** **1995, 1996, 1997, 1998, 1999, 2000, 2001, 2002, 2003, 2004, 2005** **Models:** NSX, NSX-T **Engines:** All **Transmissions:** A/T, M/T	**Accelerator Pedal Position Sensor Correlation** Key on or engine running; and the PCM detected the correlation between the Accelerator Pedal Position Sensor 1 and 2 signals were incorrect or not plausible. **Possible Causes:** • APP sensor is damaged or has failed • APP sensor 2 is damaged or has failed • Throttle plate or linkage is jammed or bent • PCM has failed
DTC: P1259 **1T CCM, MIL: YES** **1995, 1996, 1997, 1998, 1999, 2000, 2001, 2002, 2003, 2004, 2005** **Models:** NSX, NSX-T **Engines:** All **Transmissions:** A/T, M/T	**VTEC System Malfunction (Bank 1)** Engine running in closed loop, then accelerated in 1st gear to over 6000 rpm for 2 seconds, and the PCM detected a fault in the VTEC front solenoid or the VTEC pressure switch during the CCM test. **Possible Causes:** • VTEC front solenoid control circuit is open or shorted to ground • VTEC front solenoid is damaged or has failed • VTEC front pressure switch circuit is open or shorted to ground • VTEC front pressure switch is damaged or has failed • PCM has failed
DTC: P1259 **1T CCM, MIL: YES** **1996, 1997, 1998, 1999, 2000, 2001** **Models:** 2.3CL, 3.0CL, Integra **Engines:** 2.3L VIN YA3, 3.0L VIN YA2, 1.8L VIN DB8, 1.8L VIN DC2 **Transmissions:** A/T, M/T	**VTEC System Malfunction** Engine running under hard acceleration, than back to road load, and the PCM detected a fault in the operation of the VTEC system. **Possible Causes:** • VTEC solenoid control circuit is open or shorted to ground • VTEC solenoid is damaged or has failed • VTEC pressure switch circuit is open or shorted to ground • VTEC pressure switch is damaged or has failed • PCM has failed
DTC: P1259 **1T CCM, MIL: YES** **2002, 2003, 2004, 2005, 2006** **Models:** RSX **Engines:** All **Transmissions:** A/T, M/T	**VTEC System Malfunction** Engine running under hard acceleration, than back to road load, and the PCM detected a fault in the operation of the VTEC system. **Possible Causes:** • VTEC solenoid control circuit is open or shorted to ground • VTEC solenoid is damaged or has failed • VTEC pressure switch circuit is open or shorted to ground • VTEC pressure switch is damaged or has failed • PCM has failed
DTC: P1279 **1T CCM, MIL: YES** **1995, 1996, 1997, 1998, 1999, 2000, 2001, 2002, 2003, 2004, 2005** **Models:** NSX, NSX-T **Engines:** All **Transmissions:** A/T, M/T	**VTEC System Malfunction (Bank 2)** Engine running in closed loop, then accelerated in 1st gear to over 6000 rpm for 2 seconds, and the PCM detected a fault in the VTEC rear solenoid or the VTEC rear pressure switch during the CCM test. **Possible Causes:** • VTEC rear solenoid control circuit is open or shorted to ground • VTEC rear solenoid is damaged or has failed • VTEC rear pressure switch circuit is open or shorted to ground • VTEC rear pressure switch is damaged or has failed • PCM has failed
DTC: P1297 **1T CCM, MIL: YES** **1996, 1997, 1998, 1999, 2000, 2001** **Models:** 2.2CL, 2.3CL, 3.0CL, Integra **Engines:** All **Transmissions:** A/T	**Electrical Load Detector Circuit Low Input** Engine running at hot idle speed or at cruise speed, headlights "on", and the PCM detected the ELD signal was less than a stored value. **Possible Causes:** • ELD sensor signal circuit is open or shorted to ground • ELD sensor power circuit is open or shorted to ground • ELD sensor is damaged or has failed • PCM has failed

DTC	Trouble Code Title, Conditions & Possible Causes
DTC: P1297 **1T CCM, MIL: YES** **1995, 1996, 1997, 1998** **Models:** 2.5TL **Engines:** All **Transmissions:** A/T	**Electrical Load Detector Circuit Low Input** Engine running at hot idle speed or at cruise speed, headlights "on", and the PCM detected the ELD signal was less than a stored value. **Possible Causes:** • ELD sensor signal circuit is open or shorted to ground • ELD sensor power circuit is open or shorted to ground • ELD sensor is damaged or has failed • PCM has failed
DTC: P1297 **1T CCM, MIL: YES** **1999, 2000, 2001, 2002, 2003,** **2004, 2005, 2006** **Models:** 3.2CL, 3.2TL, TL, MDX, RSX **Engines:** All **Transmissions:** A/T	**Electrical Load Detector Circuit Low Input** Engine running at hot idle speed or at cruise speed, headlights "on", and the PCM detected the ELD signal was less than a stored value. **Possible Causes:** • ELD sensor signal circuit is open or shorted to ground • ELD sensor power circuit is open or shorted to ground • ELD sensor is damaged or has failed • PCM has failed
DTC: P1298 **1T CCM, MIL: YES** **1996, 1997, 1998, 1999, 2000,** **2001** **Models:** 2.2CL, 2.3CL, 3.0CL, Integra **Engines:** All **Transmissions:** A/T	**Electrical Load Detector Circuit High Input** Engine running at hot idle speed or at cruise speed, headlights "on", and the PCM detected the ELD signal was more than a stored value. **Possible Causes:** • ELD sensor signal circuit is shorted to VREF • ELD sensor signal circuit is shorted to system power (B+) • ELD sensor is damaged or has failed • PCM has failed
DTC: P1298 **1T CCM, MIL: YES** **1995, 1996, 1997, 1998** **Models:** 2.5TL **Engines:** All **Transmissions:** A/T	**Electrical Load Detector Circuit High Input** Engine running at hot idle speed or at cruise speed, headlights "on", and the PCM detected the ELD signal was more than a stored value. **Possible Causes:** • ELD sensor signal circuit is shorted to VREF • ELD sensor signal circuit is shorted to system power (B+) • ELD sensor is damaged or has failed • PCM has failed
DTC: P1298 **1T CCM, MIL: YES** **1999, 2000, 2001, 2002, 2003,** **2004, 2005, 2006** **Models:** 3.2CL, 3.2TL, TL, MDX, RSX **Engines:** All **Transmissions:** A/T	**Electrical Load Detector Circuit High Input** Engine running at hot idle speed or at cruise speed, headlights "on", and the PCM detected the ELD signal was more than a stored value. **Possible Causes:** • ELD sensor signal circuit is shorted to VREF • ELD sensor signal circuit is shorted to system power (B+) • ELD sensor is damaged or has failed • PCM has failed
DTC: P1300 **1T CCM, MIL: YES** **1995, 1996, 1997, 1998, 1999,** **2000, 2001, 2002, 2003, 2004,** **2005, 2006** **Models:** NSX, NSX-T, 3.5RL, RL **Engines:** All **Transmissions:** A/T, M/T	**Random Misfire Detected** DTC P0107, P0108, P0131, P0132, P0171, P0172, P1128, P0335, P0336, P0505, P1128, P1129, P1259, P1361, P1362, P1366, P1367 and P1519 not set, engine running under positive torque conditions, and the PCM detected a misfire in 2 or more cylinders. **Note: If the misfire is severe, the MIL will flash on/off on the 1st trip!** **Possible Causes:** • CKP or CMP sensor signal erratic or intermittent • Fuel system problem affecting more than one cylinder • Ignition system problem affecting more than one cylinder • Base engine mechanical fault affecting more than one cylinder
DTC: P1300 **1T CCM, MIL: YES** **1997** **Models:** 3.0CL **Engines:** All **Transmissions:** A/T, M/T	**Random Misfire Detected** DTC P0107, P0108, P0131, P0132, P0171, P0172, P1128, P0335, P0336, P0505, P1128, P1129, P1259, P1361, P1362, P1366, P1367 and P1519 not set, engine running under positive torque conditions, and the PCM detected a misfire in 2 or more cylinders. **Note: If the misfire is severe, the MIL will flash on/off on the 1st trip!** **Possible Causes:** • CKP or CMP sensor signal erratic or intermittent • Fuel system problem affecting more than one cylinder • Ignition system problem affecting more than one cylinder • Base engine mechanical fault affecting more than one cylinder

DTC	Trouble Code Title, Conditions & Possible Causes
DTC: P1300 **1T CCM, MIL: YES** **1996, 1997, 1998** **Models:** 3.2TL **Engines:** All **Transmissions:** A/T, M/T	**Random Misfire Detected** DTC P0107, P0108, P0131, P0132, P0171, P0172, P1128, P0335, P0336, P0505, P1128, P1129, P1259, P1361, P1362, P1366, P1367 and P1519 not set, engine running under positive torque conditions, and the PCM detected a misfire in 2 or more cylinders. **Note: If the misfire is severe, the MIL will flash on/off on the 1st trip!** **Possible Causes:** • CKP or CMP sensor signal erratic or intermittent • Fuel system problem affecting more than one cylinder • Ignition system problem affecting more than one cylinder • Base engine mechanical fault affecting more than one cylinder
DTC: P1301 **1T MISFIRE, MIL: YES** **1995, 1996, 1997, 1998, 1999,** **2000, 2001, 2002, 2003, 2004,** **2005, 2006** **Models:** NSX, NSX-T, 3.5RL, RL **Engines:** All **Transmissions:** A/T, M/T	**Cylinder 1 Misfire Detected** DTC P0107, P0108, P0131, P0132, P0171, P0172, P1128, P0335, P0336, P0505, P1128, P1129, P1259, P1361, P1362, P1366, P1367 and P1519 not set, engine running under positive torque conditions, and the PCM detected a misfire condition in one cylinder. **Note: If the misfire is severe, the MIL will flash on/off on the 1st trip!** **Possible Causes:** • Base engine (mechanical) problem affecting only one cylinder • Fuel system problem affecting only one cylinder • Ignition system problem affecting one cylinder
DTC: P1301 **1T MISFIRE, MIL: YES** **1996, 1997, 1998** **Models:** 3.2TL **Engines:** All **Transmissions:** A/T, M/T	**Cylinder 1 Misfire Detected** DTC P0107, P0108, P0131, P0132, P0171, P0172, P1128, P0335, P0336, P0505, P1128, P1129, P1259, P1361, P1362, P1366, P1367 and P1519 not set, engine running under positive torque conditions, and the PCM detected a misfire condition in one cylinder. **Note: If the misfire is severe, the MIL will flash on/off on the 1st trip!** **Possible Causes:** • Base engine (mechanical) problem affecting only one cylinder • Fuel system problem affecting only one cylinder • Ignition system problem affecting one cylinder
DTC: P1302 **1T MISFIRE, MIL: YES** **1995, 1996, 1997, 1998, 1999,** **2000, 2001, 2002, 2003, 2004,** **2005, 2006** **Models:** NSX, NSX-T, 3.5RL, RL **Engines:** All **Transmissions:** A/T, M/T	**Cylinder 2 Misfire Detected** DTC P0107, P0108, P0131, P0132, P0171, P0172, P1128, P0335, P0336, P0505, P1128, P1129, P1259, P1361, P1362, P1366, P1367 and P1519 not set, engine running under positive torque conditions, and the PCM detected a misfire condition in one cylinder. **Note: If the misfire is severe, the MIL will flash on/off on the 1st trip!** **Possible Causes:** • Base engine (mechanical) problem affecting only one cylinder • Fuel system problem affecting only one cylinder • Ignition system problem affecting one cylinder
DTC: P1302 **1T MISFIRE, MIL: YES** **1996, 1997, 1998** **Models:** 3.2TL **Engines:** All **Transmissions:** A/T, M/T	**Cylinder 2 Misfire Detected** DTC P0107, P0108, P0131, P0132, P0171, P0172, P1128, P0335, P0336, P0505, P1128, P1129, P1259, P1361, P1362, P1366, P1367 and P1519 not set, engine running under positive torque conditions, and the PCM detected a misfire condition in one cylinder. **Note: If the misfire is severe, the MIL will flash on/off on the 1st trip!** **Possible Causes:** • Fuel system problem affecting only Cylinder 2 • Ignition system problem affecting Cylinder 2 • Base engine (mechanical) problem affecting only Cylinder 2
DTC: P1303 **1T MISFIRE, MIL: YES** **1995, 1996, 1997, 1998, 1999,** **2000, 2001, 2002, 2003, 2004,** **2005, 2006** **Models:** NSX, NSX-T, 3.5RL, RL **Engines:** All **Transmissions:** A/T, M/T	**Cylinder 3 Misfire Detected** DTC P0107, P0108, P0131, P0132, P0171, P0172, P1128, P0335, P0336, P0505, P1128, P1129, P1259, P1361, P1362, P1366, P1367 and P1519 not set, engine running under positive torque conditions, and the PCM detected a misfire condition in one cylinder. **Note: If the misfire is severe, the MIL will flash on/off on the 1st trip!** **Possible Causes:** • Base engine (mechanical) problem affecting only one cylinder • Fuel system problem affecting only one cylinder • Ignition system problem affecting one cylinder
DTC: P1303 **1T MISFIRE, MIL: YES** **1996, 1997, 1998** **Models:** 3.2TL **Engines:** All **Transmissions:** A/T, M/T	**Cylinder 3 Misfire Detected** DTC P0107, P0108, P0131, P0132, P0171, P0172, P1128, P0335, P0336, P0505, P1128, P1129, P1259, P1361, P1362, P1366, P1367 and P1519 not set, engine running under positive torque conditions, and the PCM detected a misfire condition in one cylinder. **Note: If the misfire is severe, the MIL will flash on/off on the 1st trip!** **Possible Causes:** • Base engine (mechanical) problem affecting only one cylinder • Fuel system problem affecting only one cylinder • Ignition system problem affecting one cylinder

DTC	Trouble Code Title, Conditions & Possible Causes
DTC: P1304 **1T MISFIRE, MIL: YES** **1995, 1996, 1997, 1998, 1999,** **2000, 2001, 2002, 2003, 2004,** **2005, 2006** **Models:** NSX, NSX-T, 3.5RL, RL **Engines:** All **Transmissions:** A/T, M/T	**Cylinder 4 Misfire Detected** DTC P0107, P0108, P0131, P0132, P0171, P0172, P1128, P0335, P0336, P0505, P1128, P1129, P1259, P1361, P1362, P1366, P1367 and P1519 not set, engine running under positive torque conditions, and the PCM detected a misfire condition in one cylinder. **Note: If the misfire is severe, the MIL will flash on/off on the 1st trip!** **Possible Causes:** • Base engine (mechanical) problem affecting only one cylinder • Fuel system problem affecting only one cylinder • Ignition system problem affecting one cylinder
DTC: P1304 **1T MISFIRE, MIL: YES** **1996, 1997, 1998** **Models:** 3.2TL **Engines:** All **Transmissions:** A/T, M/T	**Cylinder 4 Misfire Detected** DTC P0107, P0108, P0131, P0132, P0171, P0172, P1128, P0335, P0336, P0505, P1128, P1129, P1259, P1361, P1362, P1366, P1367 and P1519 not set, engine running under positive torque conditions, and the PCM detected a misfire condition in one cylinder. **Note: If the misfire is severe, the MIL will flash on/off on the 1st trip!** **Possible Causes:** • Base engine (mechanical) problem affecting only one cylinder • Fuel system problem affecting only one cylinder • Ignition system problem affecting one cylinder
DTC: P1305 **1T MISFIRE, MIL: YES** **1995, 1996, 1997, 1998, 1999,** **2000, 2001, 2002, 2003, 2004,** **2005, 2006** **Models:** NSX, NSX-T, 3.5RL, RL **Engines:** All **Transmissions:** A/T, M/T	**Cylinder 5 Misfire Detected** DTC P0107, P0108, P0131, P0132, P0171, P0172, P1128, P0335, P0336, P0505, P1128, P1129, P1259, P1361, P1362, P1366, P1367 and P1519 not set, engine running under positive torque conditions, and the PCM detected a misfire condition in one cylinder. **Note: If the misfire is severe, the MIL will flash on/off on the 1st trip!** **Possible Causes:** • Base engine (mechanical) problem affecting only one cylinder • Fuel system problem affecting only one cylinder • Ignition system problem affecting one cylinder
DTC: P1305 **1T MISFIRE, MIL: YES** **1996, 1997, 1998** **Models:** 3.2TL **Engines:** All **Transmissions:** A/T, M/T	**Cylinder 5 Misfire Detected** DTC P0107, P0108, P0131, P0132, P0171, P0172, P1128, P0335, P0336, P0505, P1128, P1129, P1259, P1361, P1362, P1366, P1367 and P1519 not set, engine running under positive torque conditions, and the PCM detected a misfire condition in one cylinder. **Note: If the misfire is severe, the MIL will flash on/off on the 1st trip!** **Possible Causes:** • Base engine (mechanical) problem affecting only one cylinder • Fuel system problem affecting only one cylinder • Ignition system problem affecting one cylinder
DTC: P1306 **1T MISFIRE, MIL: YES** **1995, 1996, 1997, 1998, 1999,** **2000, 2001, 2002, 2003, 2004,** **2005, 2006** **Models:** NSX, NSX-T, 3.5RL, RL **Engines:** All **Transmissions:** A/T, M/T	**Cylinder 6 Misfire Detected** DTC P0107, P0108, P0131, P0132, P0171, P0172, P1128, P0335, P0336, P0505, P1128, P1129, P1259, P1361, P1362, P1366, P1367 and P1519 not set, engine running under positive torque conditions, and the PCM detected a misfire condition in one cylinder. **Note: If the misfire is severe, the MIL will flash on/off on the 1st trip!** **Possible Causes:** • Base engine (mechanical) problem affecting only one cylinder • Fuel system problem affecting only one cylinder • Ignition system problem affecting one cylinder
DTC: P1306 **1T MISFIRE, MIL: YES** **1996, 1997, 1998** **Models:** 3.2TL **Engines:** All **Transmissions:** A/T, M/T	**Cylinder 6 Misfire Detected** DTC P0107, P0108, P0131, P0132, P0171, P0172, P1128, P0335, P0336, P0505, P1128, P1129, P1259, P1361, P1362, P1366, P1367 and P1519 not set, engine running under positive torque conditions, and the PCM detected a misfire condition in one cylinder. **Note: If the misfire is severe, the MIL will flash on/off on the 1st trip!** **Possible Causes:** • Base engine (mechanical) problem affecting only one cylinder • Fuel system problem affecting only one cylinder • Ignition system problem affecting one cylinder
DTC: P1316 **1T CCM, MIL: YES** **1995, 1996, 1997, 1998, 1999,** **2000, 2001, 2002, 2003, 2004,** **2005, 2006** **Models:** NSX, NSX-T, 3.5RL, RL **Engines:** All **Transmissions:** A/T, M/T	**Spark Plug Detection Module Circuit Fault (Bank 2)** Engine started, and the PCM detected an unexpected condition on the Spark Plug Detection Module circuit for Cylinder Bank 2. **Possible Causes:** • Spark Plug Detection Module circuit open or shorted to ground • Spark Plug Detection Module circuit is shorted to system power • Spark Plug Detection Module is damaged or has failed • PCM has failed

DTC	Trouble Code Title, Conditions & Possible Causes
DTC: P1316 **1T CCM, MIL: YES** **1996, 1997, 1998** **Models:** 3.2TL **Engines:** All **Transmissions:** A/T, M/T	**Spark Plug Detection Module Circuit Fault (Bank 2)** Engine started, and the PCM detected an unexpected condition on the Spark Plug Detection Module circuit for Cylinder Bank 2. **Possible Causes:** • Spark Plug Detection Module circuit open or shorted to ground • Spark Plug Detection Module circuit is shorted to system power • Spark Plug Detection Module is damaged or has failed • PCM has failed
DTC: P1317 **1T CCM, MIL: YES** **1995, 1996, 1997, 1998, 1999,** **2000, 2001, 2002, 2003, 2004,** **2005, 2006** **Models:** NSX, NSX-T, 3.5RL, RL **Engines:** All **Transmissions:** A/T, M/T	**Spark Plug Detection Module Circuit Fault (Bank 1)** Engine started, and the PCM detected an unexpected condition on the Spark Plug Detection Module circuit for Cylinder Bank 1. **Possible Causes:** • Spark Plug Detection Module circuit open or shorted to ground • Spark Plug Detection Module circuit is shorted to system power • Spark Plug Detection Module is damaged or has failed • PCM has failed
DTC: P1317 **1T CCM, MIL: YES** **1996, 1997, 1998** **Models:** 3.2TL **Engines:** All **Transmissions:** A/T, M/T	**Spark Plug Detection Module Reset Circuit Fault (Bank 1)** Engine started, and the PCM detected an unexpected condition on the Spark Plug Detection Module circuit for Cylinder Bank 1. **Possible Causes:** • Spark Plug Detection Module circuit open or shorted to ground • Spark Plug Detection Module circuit is shorted to system power • Spark Plug Detection Module is damaged or has failed • PCM has failed
DTC: P1318 **1T CCM, MIL: YES** **1995, 1996, 1997, 1998, 1999,** **2000, 2001, 2002, 2003, 2004,** **2005, 2006** **Models:** NSX, NSX-T, 3.5RL, RL **Engines:** All **Transmissions:** A/T, M/T	**Spark Plug Detection Module Reset Circuit Fault (Bank 2)** Engine started, and the PCM detected an unexpected condition on the Spark Plug Detection Module Reset circuit for Cylinder Bank 2. **Possible Causes:** • Spark Plug Detection Module circuit open or shorted to ground • Spark Plug Detection Module circuit is shorted to system power • Spark Plug Detection Module is damaged or has failed • PCM has failed
DTC: P1318 **1T CCM, MIL: YES** **1996, 1997, 1998** **Models:** 3.2TL **Engines:** All **Transmissions:** A/T, M/T	**Spark Plug Detection Module Reset Circuit Fault (Bank 2)** Engine started, and the PCM detected an unexpected condition on the Spark Plug Detection Module Reset circuit for Cylinder Bank 2. **Possible Causes:** • Spark Plug Detection Module circuit open or shorted to ground • Spark Plug Detection Module is damaged or has failed • PCM has failed
DTC: P1319 **1T CCM, MIL: YES** **1995, 1996, 1997, 1998, 1999,** **2000, 2001, 2002, 2003, 2004,** **2005, 2006** **Models:** NSX, NSX-T, 3.5RL, RL **Engines:** All **Transmissions:** A/T, M/T	**Spark Plug Detection Module Reset Circuit Fault (Bank 1)** Engine started, and the PCM detected an unexpected condition on the Spark Plug Detection Module Reset circuit for Cylinder Bank 1. **Possible Causes:** • Spark Plug Detection Module circuit open or shorted to ground • Spark Plug Detection Module circuit is shorted to system power • Spark Plug Detection Module is damaged or has failed • PCM has failed
DTC: P1319 **1T CCM, MIL: YES** **1996, 1997, 1998** **Models:** 3.2TL **Engines:** All **Transmissions:** A/T, M/T	**Spark Plug Detection Module Reset Circuit Fault (Bank 1)** Engine started, and the PCM detected an unexpected condition on the Spark Plug Detection Module Reset circuit for Cylinder Bank 1. **Possible Causes:** • Spark Plug Detection Module circuit open or shorted to ground • Spark Plug Detection Module circuit is shorted to system power • Spark Plug Detection Module is damaged or has failed • PCM has failed
DTC: P1336 **1T CCM, MIL: YES** **1995, 1996, 1997, 1998** **Models:** 2.5TL **Engines:** All **Transmissions:** A/T	**Crankshaft Speed Fluctuation Sensor Circuit Malfunction** Engine started, and PCM detected an interruption of the crankshaft speed fluctuation (CSF) sensor signal during the test. **Possible Causes:** • CSF signal circuit is open • CSF signal is shorted to ground or to system power • CSF pickup assembly or its pulse rotor is damaged • PCM has failed

DTC	Trouble Code Title, Conditions & Possible Causes
DTC: P1336 **1T CCM, MIL: YES** **1996, 1997, 1998, 1999, 2000, 2001, 2002, 2003, 2004, 2005, 2006** **Models:** 3.5RL, RL **Engines:** All **Transmissions:** A/T, M/T	**Crankshaft Position Sensor 2 Circuit Malfunction** Engine started, and the PCM detected an unexpected (intermittent) interruption of the Crankshaft position (CKP) 'B' sensor signal. **Possible Causes:** • CSF2 signal circuit is open, short to ground or to system power • CSF2 is damaged or has failed • CSF2 pickup assembly or its pulse rotor is damaged • PCM has failed
DTC: P1336 **1T CCM, MIL: YES** **1996, 1997, 1998, 1999, 2000, 2001, 2002** **Models:** 2.2CL, 2.3CL, 3.0CL, 3.2CL, 3.2TL **Engines:** All **Transmissions:** A/T	**Crankshaft Speed Fluctuation Sensor Circuit Malfunction** Engine started, and PCM detected an unexpected (intermittent) interruption of the crankshaft speed fluctuation (CSF) sensor signal. **Possible Causes:** • CSF signal circuit is open or shorted to ground • CSF signal circuit is shorted to VREF or system power (B+) • CSF is damaged or has failed • CSF pickup assembly or its pulse rotor is damaged • PCM has failed
DTC: P1336 **1T CCM, MIL: YES** **1996, 1997, 1998, 1999, 2000, 2001** **Models:** Integra **Engines:** All **Transmissions:** A/T, M/T	**Crankshaft Speed Fluctuation Sensor Circuit Malfunction** Engine started, and PCM detected an unexpected (intermittent) interruption of the crankshaft speed fluctuation (CSF) sensor signal. **Possible Causes:** • CSF signal circuit is open or shorted to ground • CSF signal circuit is shorted to VREF or system power (B+) • CSF is damaged or has failed • CSF pickup assembly or its pulse rotor is damaged • PCM has failed
DTC: P1336 **1T CCM, MIL: YES** **1995, 1996, 1997, 1998, 1999, 2000, 2001, 2002, 2003, 2004, 2005** **Models:** NSX, NSX-T **Engines:** All **Transmissions:** A/T, M/T	**Crankshaft Position Sensor 'B' Circuit Malfunction** Engine started, and PCM detected an interruption of the CKP Sensor 'B' signal during the CCM test. **Possible Causes:** • CSF 'B' signal circuit is open or shorted to ground • CSF 'B' signal circuit is shorted to VREF or system power (B+) • CSF 'B' is damaged or has failed • CSF 'B' pickup assembly or its pulse rotor is damaged • PCM has failed
DTC: P1337 **1T CCM, MIL: YES** **1995, 1996, 1997, 1998** **Models:** 2.5TL **Engines:** All **Transmissions:** A/T	**Crankshaft Speed Fluctuation Sensor No Signal** Engine cranking or running; and the PCM did not detect any signals from the Crankshaft Speed Fluctuation (CSF) sensor during the test. **Possible Causes:** • CKP signal circuit is open or shorted to ground • CKP signal circuit is shorted to VREF or system power (B+) • CKP is damaged or has failed • CKP pickup assembly or its pulse rotor is damaged • PCM has failed
DTC: P1337 **1T CCM, MIL: YES** **1995, 1996, 1997, 1998, 1999, 2000, 2001, 2002, 2003, 2004, 2005** **Models:** NSX, NSX-T **Engines:** All **Transmissions:** A/T, M/T	**Crankshaft Position Sensor 'B' Circuit No Signal** Engine cranking or running; and the PCM did not detect any signals from the Crankshaft Position (CKP) 'B' sensor during the CCM test. **Possible Causes:** • CKP Sensor 'B' signal circuit is open or shorted to ground • CKP Sensor 'B' signal circuit is shorted to VREF or power • CKP Sensor 'B' is damaged or has failed • CKP Sensor 'B' pickup assembly or its pulse rotor is damaged • PCM has failed
DTC: P1337 **1T CCM, MIL: YES** **1996, 1997, 1998, 1999, 2000, 2001, 2002, 2003, 2004, 2005, 2006** **Models:** 3.5RL, RL **Engines:** All **Transmissions:** A/T, M/T	**Crankshaft Position Sensor 2 Circuit No Signal** Engine cranking or running; and the PCM did not detect any signals from the Crankshaft Position (CKP) 2 sensor during the CCM test. **Possible Causes:** • CKP Sensor 2 signal circuit is open or shorted to ground • CKP Sensor 2 signal circuit is shorted to VREF or power • CKP Sensor 2 is damaged or has failed • CKP Sensor 2 pickup assembly or its pulse rotor is damaged • PCM has failed

DTC	Trouble Code Title, Conditions & Possible Causes
DTC: P1337 **1T CCM, MIL: YES** **1996, 1997, 1998, 1999, 2000, 2001, 2002** **Models:** 2.2CL, 2.3CL, 3.0CL, 3.2CL, 3.2TL **Engines:** All **Transmissions:** A/T	**Crankshaft Speed Fluctuation Sensor No Signal** Engine cranking or running; and the PCM did not detect any signals from the Crankshaft Speed Fluctuation (CSF) sensor during the test. **Possible Causes:** • CSF signal circuit is open or shorted to ground • CSF signal circuit is shorted to VREF or system power (B+) • CSF is damaged or has failed • CSF pickup assembly or its pulse rotor is damaged • PCM has failed
DTC: P1337 **1T CCM, MIL: YES** **1996, 1997, 1998, 1999, 2000, 2001** **Models:** Integra **Engines:** All **Transmissions:** A/T, M/T	**Crankshaft Speed Fluctuation Sensor No Signal** Engine cranking or running; and the PCM did not detect any signals from the Crankshaft Speed Fluctuation (CSF) sensor during the test. **Possible Causes:** • CSF signal circuit is open or shorted to ground • CSF signal circuit is shorted to VREF or system power (B+) • CSF is damaged or has failed • CSF pickup assembly or its pulse rotor is damaged • PCM has failed
DTC: P1359 **1T CCM, MIL: YES** **1995, 1996, 1997, 1998** **Models:** 2.5TL **Engines:** All **Transmissions:** A/T, M/T	**CKP/TDC/CYP Sensor Circuit Malfunction** Engine started, and the PCM detected an unexpected voltage condition on the CKP/TDC sensor circuit during the CCM test. **Possible Causes:** • CKP/TDC signal circuit is open or shorted to ground • CKP/TDC signal circuit is shorted to VREF or system power • CKP/TDC pickup assembly or its pulse rotor is damaged • CKP/TDC is damaged or has failed • PCM has failed
DTC: P1359 **1T CCM, MIL: YES** **1997, 1998, 1999** **Models:** 2.2CL, 2.3CL **Engines:** All **Transmissions:** A/T, M/T	**CKP/TDC/CYP Sensor Circuit Malfunction** Engine started, and the PCM detected an unexpected voltage condition on the CKP/TDC/CYP sensor circuit during the CCM test. **Possible Causes:** • CKP/TDC/CYP signal circuit is open or shorted to ground • CKP/TDC/CYP signal circuit is shorted to system power (B+) • CKP/TDC/CYP pickup assembly or its pulse rotor is damaged • CKP/TDC/CYP is damaged or has failed • PCM has failed
DTC: P1359 **1T CCM, MIL: YES** **1996, 1997, 1998, 1999, 2000, 2001** **Models:** Integra **Engines:** All **Transmissions:** A/T, M/T	**CKP/TDC/CYP Sensor Circuit Malfunction** Engine started, and the PCM detected an unexpected voltage condition on the CKP/TDC/CYP sensor circuit during the CCM test. **Possible Causes:** • CKP/TDC/CYP signal circuit is open or shorted to ground • CKP/TDC/CYP signal circuit is shorted to system power (B+) • CKP/TDC/CYP pickup assembly or its pulse rotor is damaged • CKP/TDC/CYP is damaged or has failed • PCM has failed
DTC: P1361 **1T CCM, MIL: YES** **1995, 1996, 1997, 1998** **Models:** 2.5TL **Engines:** All **Transmissions:** A/T, M/T	**Top Dead Center Sensor Signal Intermittent** Engine started, and the PCM detected an unexpected (intermittent) interruption of the Top Dead Center (TDC) sensor signal in the test. **Possible Causes:** • TDC signal circuit is open or shorted to ground • TDC signal circuit is shorted to VREF or system power • TDC pickup assembly or its pulse rotor is damaged • TDC is damaged or has failed • PCM has failed
DTC: P1361 **1T CCM, MIL: YES** **1999, 2000, 2001, 2002** **Models:** 3.2TL **Engines:** All **Transmissions:** A/T, M/T	**Top Dead Center Sensor Signal Intermittent** Engine started, and the PCM detected an unexpected (intermittent) interruption of the Top Dead Center (TDC) sensor signal in the test. **Possible Causes:** • TDC signal circuit is open or shorted to ground • TDC signal circuit is shorted to VREF or system power • TDC pickup assembly or its pulse rotor is damaged • TDC is damaged or has failed • PCM has failed

DTC	Trouble Code Title, Conditions & Possible Causes
DTC: P1361 **1T CCM, MIL: YES** **1997, 1998, 1999, 2000, 2001,** **2002, 2003, 2004, 2005, 2006** **Models:** 2.2CL, 2.3CL, 3.0CL, 3.2CL, RSX **Engines:** All **Transmissions:** A/T	**Top Dead Center Sensor Circuit Intermittent Signal** Engine started, and the PCM detected an unexpected (intermittent) interruption of the Top Dead Center (TDC) sensor signal in the test. **Possible Causes:** • TDC signal circuit is open (intermittent fault) • TDC signal circuit is shorted to ground (intermittent fault) • TDC pickup assembly or its pulse rotor is damaged (intermittent fault) • TDC is damaged or has failed (intermittent fault) • PCM has failed
DTC: P1361 **1T CCM, MIL: YES** **1996, 1997, 1998, 1999, 2000,** **2001** **Models:** Integra **Engines:** All **Transmissions:** A/T, M/T	**Top Dead Center Sensor Intermittent Signal** Engine started, and the PCM detected an unexpected (intermittent) interruption of the Top Dead Center (TDC) sensor signal in the test. **Possible Causes:** • TDC signal circuit is open or shorted to ground • TDC signal circuit is shorted to VREF or system power • TDC pickup assembly or its pulse rotor is damaged • TDC is damaged or has failed • PCM has failed
DTC: P1362 **1T CCM, MIL: YES** **1997, 1998, 1999, 2000, 2001,** **2002, 2003, 2004, 2005, 2006** **Models:** 2.2CL, 2.3CL, 3.0CL, 3.2CL, RSX **Engines:** All **Transmissions:** A/T	**Top Dead Center 1 Sensor Circuit No Signal** Engine cranking or running; and the PCM did not receive any signals from the Top Dead Center 1(TDC1) sensor during the CCM test. **Note: The engine will start and run without the TDC sensor 1 signal.** **Possible Causes:** • TDC1 signal circuit is open or shorted to ground • TDC1 pickup assembly or its pulse rotor is damaged • TDC1 is damaged or has failed • PCM has failed
DTC: P1362 **1T CCM, MIL: YES** **1995, 1996, 1997, 1998** **Models:** 2.5TL **Engines:** All **Transmissions:** A/T, M/T	**Top Dead Center 1 Sensor No Signal** Engine cranking or running; and the PCM did not receive any signals from the Top Dead Center 1 (TDC1) sensor during the CCM test. **Note: The engine will start and run without the TDC sensor 1 signal.** **Possible Causes:** • TDC1 signal circuit is open or shorted to ground • TDC1 pickup assembly or its pulse rotor is damaged • TDC1 is damaged or has failed • PCM has failed
DTC: P1362 **1T CCM, MIL: YES** **1999, 2000, 2001, 2002** **Models:** 3.2TL **Engines:** All **Transmissions:** A/T, M/T	**Top Dead Center 1 Sensor No Signal** Engine cranking or running; and the PCM did not receive any signals from the Top Dead Center 1 (TDC1) sensor during the CCM test. **Note: The engine will start and run without the TDC sensor 1 signal.** **Possible Causes:** • TDC1 signal circuit is open or shorted to ground • TDC1 pickup assembly or its pulse rotor is damaged • TDC is damaged or has failed • PCM has failed
DTC: P1362 **1T CCM, MIL: YES** **1996, 1997, 1998, 1999, 2000,** **2001** **Models:** Integra **Engines:** All **Transmissions:** A/T, M/T	**Top Dead Center 1 Sensor No Signal** Engine cranking or running; and the PCM did not receive any signals from the Top Dead Center 1 (TDC1) sensor during the CCM test. **Note: The engine will start and run without the TDC sensor 1 signal.** **Possible Causes:** • TDC1 signal circuit is open or shorted to ground • TDC1 pickup assembly or its pulse rotor is damaged • TDC1 is damaged or has failed • PCM has failed

DTC	Trouble Code Title, Conditions & Possible Causes
DTC: P1366 **1T CCM, MIL: YES** **1997, 1998, 1999** **Models:** 3.0CL **Engines:** All **Transmissions:** A/T, M/T	**Top Dead Center Sensor 2 Circuit Intermittent Signal** Engine started, and the PCM detected an unexpected (intermittent) interruption of the Top Dead Center 2 (TDC2) sensor signal. **Possible Causes:** • TDC2 signal circuit is open (intermittent signal) • TDC2 signal circuit is shorted to ground (intermittent signal) • TDC2 pickup assembly or its pulse rotor is damaged (intermittent signal) • TDC2 is damaged or has failed (intermittent signal) • PCM has failed
DTC: P1366 **1T CCM, MIL: YES** **1999, 2000, 2001, 2002, 2003,** **2004, 2005, 2006** **Models:** 3.2CL, 3.2TL, TL, MDX **Engines:** All **Transmissions:** A/T	**Top Dead Center Sensor 2 Circuit Intermittent Signal** Engine started, and the PCM detected an unexpected (intermittent) interruption of the Top Dead Center 2 (TDC2) sensor signal. **Possible Causes:** • TDC2 signal circuit is open (intermittent fault) • TDC2 signal circuit is shorted to ground (intermittent fault) • TDC2 pickup assembly or its pulse rotor is damaged (intermittent fault) • TDC2 is damaged or has failed (intermittent fault) • PCM has failed
DTC: P1367 **1T CCM, MIL: YES** **1997, 1998, 1999** **Models:** 3.0CL **Engines:** All **Transmissions:** A/T, M/T	**Top Dead Center Sensor 2 No Signal** Engine cranking or running; and the PCM did not detect any signals from the Top Dead Center 2 (TDC2) sensor during the CCM test. **Note: The engine will start and run without the TDC sensor 2 signal.** **Possible Causes:** • TDC2 signal circuit is open or shorted to ground • TDC2 pickup assembly or its pulse rotor is damaged • TDC2 is damaged or has failed • PCM has failed
DTC: P1367 **1T CCM, MIL: YES** **1999, 2000, 2001, 2002, 2003,** **2004, 2005, 2006** **Models:** 3.0CL, 3.2CL, 3.2TL, TL, MDX **Engines:** All **Transmissions:** A/T	**Top Dead Center Sensor 2 No Signal** Engine cranking or running; and the PCM did not detect any signals from the Top Dead Center 2 (TDC2) sensor during the CCM test. **Note: The engine will start and run without the TDC sensor 2 signal.** **Possible Causes:** • TDC2 signal circuit is open or shorted to ground • TDC2 pickup assembly or its pulse rotor is damaged • TDC2 is damaged or has failed • PCM has failed
DTC: P1380 **1T CCM, MIL: NO** **2000, 2001, 2002** **Models:** SLX **Engines:** 3.5L VIN X **Transmissions:** A/T	**Antilock Brake System Rough Road System Malfunction** DTC P0300 or P0301-306 set as a hard fault, engine started, then after the vehicle was driven at over 5 mph at an engine speed less than 6250 rpm with engine load below 99%, and the PCM received a signal from the ABS module indicating a fault (100 test failure in 120 samples). Refer to the Failure Records for help with this code. **Possible Causes:** • Class 2 data line open or shorted to ground (check for loose connections or frayed wiring as the fault may be intermittent) • ABS module or the PCM has failed
DTC: P1381 **1T CCM, MIL: YES** **1995, 1996, 1997, 1998, 1999,** **2000, 2001, 2002, 2003, 2004,** **2005, 2006** **Models:** NSX, NSX-T, 3.5RL, RL **Engines:** All **Transmissions:** A/T, M/T	**Camshaft Position Sensor 1 No Signal** Engine cranking or running; and the PCM did not detect any signals from the Camshaft Position (CMP) sensor 1 during the CCM test. **Note: The engine will start and run without the CMP sensor 1 signal.** **Possible Causes:** • CMP signal circuit is open or shorted to ground • CMP pickup assembly or CMP sensor is damaged or has failed • PCM has failed
DTC: P1381 **1T CCM, MIL: YES** **1995, 1996, 1997, 1998** **Models:** 2.5TL **Engines:** All **Transmissions:** A/T, M/T	**Camshaft Position Sensor Circuit Malfunction** Engine started, and the PCM detected an unexpected (intermittent) interruption of the Camshaft Position (CMP) sensor signal. **Possible Causes:** • CMP signal circuit is open or shorted to ground • CMP signal circuit is shorted to VREF or system power • CMP pickup unit or the CMP sensor is damaged or has failed • PCM has failed

DTC	Trouble Code Title, Conditions & Possible Causes
DTC: P1381 **1T CCM, MIL: YES** **1996, 1997, 1998** **Models:** 3.2TL **Engines:** All **Transmissions:** A/T, M/T	**Camshaft Position Sensor 1 Circuit Malfunction** Engine running and the PCM detected an unexpected or intermittent interruption of the Camshaft Position 1 (CMP) sensor signal. **Possible Causes:** • CMP1signal circuit is open or shorted to ground • CMP1signal circuit is shorted to VREF or system power • CMP1pickup unit or the CMP sensor is damaged or has failed • PCM has failed
DTC: P1381 **1T CCM, MIL: NO** **2000, 2001, 2002** **Models:** SLX **Engines:** 3.5L VIN X **Transmissions:** A/T	**ABS Module Rough Road/Class 2 Serial Data Malfunction** DTC P0300, P0301-P0306 set as a hard code, engine started, then after the vehicle was driven to over 1 mph at an engine speed below 6250 rpm for 2.5 seconds with engine load below 99%, and the PCM did not receive any ABS signals from the ABS module (100 test failures in 120 test samples). Refer to the Failure Records for help with this diagnosis of this trouble code). **Possible Causes:** • Class 2 data line open or shorted to ground (check for loose connections or frayed wiring as the fault may be intermittent) • Check for a legitimate misfire condition in the engine that is not related to the ABS Rough Road information
DTC: P1382 **1T CCM, MIL: YES** **1995, 1996, 1997, 1998, 1999,** **2000, 2001, 2002, 2003, 2004,** **2005, 2006** **Models:** NSX, NSX-T, 3.5RL, RL **Engines:** All **Transmissions:** A/T, M/T	**Camshaft Position Sensor 1 No Signal** Engine cranking or running; and the PCM did not detect any signals from the Camshaft Position 1 (CMP) sensor during the CCM test. **Note: The engine will start and run without the CMP sensor signal.** **Possible Causes:** • CMP1 signal circuit is open or shorted to ground • CMP1 pickup unit or the CMP sensor is damaged or has failed • PCM has failed
DTC: P1382 **1T CCM, MIL: YES** **1995, 1996, 1997, 1998** **Models:** 2.5TL **Engines:** All **Transmissions:** A/T, M/T	**Camshaft Position Sensor No Signal** Engine cranking or running; and the PCM did not detect any signals from the Camshaft Position (CMP) sensor during the CCM test. **Note: The engine will start and run without the CMP sensor signal.** **Possible Causes:** • CMP signal circuit is open or shorted to ground • CMP pickup unit is damaged • CMP sensor is damaged or has failed • PCM has failed
DTC: P1382 **1T CCM, MIL: YES** **1996, 1997, 1998** **Models:** 3.2TL **Engines:** All **Transmissions:** A/T, M/T	**Camshaft Position Sensor 1 Circuit Malfunction** Engine started, and the PCM detected an unexpected or intermittent interruption of the Camshaft Position 1 (CMP) Sensor signal. **Possible Causes:** • CMP1 signal circuit is open or shorted to ground • CMP1 pickup unit is damaged • CMP1 sensor is damaged or has failed • PCM has failed
DTC: P1386 **1T CCM, MIL: YES** **1995, 1996, 1997, 1998, 1999,** **2000, 2001, 2002, 2003, 2004,** **2005, 2006** **Models:** NSX, NSX-T, 3.5RL, RL **Engines:** All **Transmissions:** A/T, M/T	**Camshaft Position Sensor 2 Circuit Malfunction** Engine running and the PCM detected an unexpected or intermittent interruption of the Camshaft Position 2 (CMP) sensor signal. **Possible Causes:** • CMP2 signal circuit is open or shorted to ground • CMP2 signal circuit is shorted to VREF or system power • CMP2 pickup unit or the CMP sensor is damaged or has failed • PCM has failed
DTC: P1387 **1T CCM, MIL: YES** **1995, 1996, 1997, 1998, 1999,** **2000, 2001, 2002, 2003, 2004,** **2005, 2006** **Models:** NSX, NSX-T, 3.5RL, RL **Engines:** All **Transmissions:** A/T, M/T	**Camshaft Position Sensor 2 No Signal** Engine cranking or running; and the PCM did not detect any signals from the Camshaft Position 2 (CMP) sensor during the CCM test. **Note: The engine will start and run without the CMP sensor signal.** **Possible Causes:** • CMP2 signal circuit is open or shorted to ground • CMP2 pickup unit is damaged or has failed • CMP2 sensor is damaged or has failed • PCM has failed

DTC	Trouble Code Title, Conditions & Possible Causes
DTC: P1404 **2T EGR2, MIL: YES** **1998, 1999, 2000, 2001, 2002** **Models:** SLX **Engines:** 3.5L VIN X **Transmissions:** A/T	**EGR Valve Stuck Closed Malfunction** Engine started, system voltage from 11-16v, ECT sensor from 176-248, IAT sensor less than 212°F, Desired EGR valve position is 0%, and the PCM detected the difference between the Actual and Desired EGR pintle position was over 30% for 5 seconds. The test must fail three times in one trip to set this code. **Possible Causes:** • EGR valve sticking or binding (check for deposits on the valve) • EGR valve is damaged or has failed (if the valve shows signs of excessive heat, check the converter and pipes for a restriction) • PCM has failed
DTC: P1441 **2T CCM, MIL: YES** **1996, 1997, 1998, 1999** **Models:** SLX **Engines:** 3.2L VIN V, 3.5L VIN X **Transmissions:** A/T	**EVAP Vacuum Switch Circuit High Input** DTC P0106, P0107, P0108, P0112, P0113, P0121, P0122, P0123, P1640 and P1650 not set, system voltage at 11-16v, IAT sensor more than 32°F, fuel level from 15-85%, and the PCM detected a continuous "open" purge condition during the EVAP Monitor test. **Possible Causes:** • Vacuum switch signal circuit is open between switch and PCM • Vacuum switch ground circuit open between switch and ground • Vacuum switch is damaged or has failed • PCM has failed
DTC: P1450 **1T CCM, MIL: YES** **2003, 2004, 2005, 2006** **Models:** MDX **Engines:** All **Transmissions:** A/T, M/T	**EVAP Two-Way Valve Bypass Valve Control Circuit Low Input** Key on or engine running; and the PCM detected an unexpected low voltage condition on the EVAP two-way bypass valve control circuit. **Possible Causes:** • 2-Way bypass valve control circuit is open • 2-Way bypass valve control power circuit is open (test the fuse) • 2-Way bypass control valve is damaged or has failed • PCM is damaged or has failed
DTC: P1451 **1T CCM, MIL: YES** **2003, 2004, 2005, 2006** **Models:** MDX **Engines:** All **Transmissions:** A/T, M/T	**EVAP Two-Way Valve Bypass Valve Control Circuit High Input** Key on or engine running; and the PCM detected an unexpected high voltage condition on the two-way bypass valve control circuit. **Possible Causes:** • 2-Way bypass valve control circuit is shorted to system power • 2-Way bypass control valve is damaged or has failed • PCM is damaged or has failed
DTC: P1454 **1T CCM, MIL: YES** **2003, 2004, 2005, 2006** **Models:** MDX **Engines:** All **Transmissions:** A/T, M/T	**Fuel Tank Pressure Sensor Circuit Range/Performance** Key on or engine running; and the PCM detected an unexpected voltage condition on the fuel tank pressure sensor signal circuit. **Possible Causes:** • FTP sensor circuit is open or shorted (intermittent fault) • FTP sensor air tube is clogged with debris or dirt • Fuel tank pressure sensor is damaged or has failed • PCM is damaged or has failed
DTC: P1456 **2T EVAP1, MIL: YES** **1997, 1998, 1999, 2000, 2001, 2002, 2003, 2004, 2005** **Models:** NSX, NSX-T **Engines:** All **Transmissions:** A/T, M/T	**EVAP System Leak Detected (Fuel Tank Area)** IAT sensor signal from 32-86°F at startup, vehicle driven at over 5 mph for over 2 minutes, ECT sensor signal over 154°F, then with the EVAP Control and Vent solenoids "on", the PCM detected the fuel tank pressure was incorrect due to a leak in the fuel tank area. **Possible Causes:** • Fuel tank cap damaged, loose or the wrong part number • Fuel tank leaks at the fuel fill pipe or at the fuel tank seals • Fuel vapor control valve is damaged or has failed • Fuel tank vapor recirculation valve or vapor tube is damaged • Fuel tank vapor control vent tube is damaged or has failed
DTC: P1456 **2T EVAP1, MIL: YES** **1997, 1998, 1999, 2000, 2001, 2002, 2003, 2004, 2005, 2006** **Models:** 2.3CL, 3.0CL, 3.2CL, 3.2TL, TL, MDX, RSX **Engines:** All **Transmissions:** A/T, M/T	**EVAP System Leak Detected (Fuel Tank Area)** IAT sensor signal from 32-86°F at startup, vehicle driven to a speed over 5 mph for 2-3 minutes, ECT sensor signal over 154°F, then with the EVAP Control and Vent solenoids "on", the PCM detected an incorrect fuel tank pressure value due to a leak in the fuel tank area. **Possible Causes:** • Fuel tank cap damaged, loose or the wrong part number • Fuel tank leaks at the fuel fill pipe or at the fuel tank seals • Fuel vapor control valve is damaged or has failed • Fuel tank vapor recirculation valve or vapor tube is damaged • Fuel tank vapor control vent tube is damaged or has failed • TSB 02-005 (2/18/02) contains a repair procedure for this code

DTC	Trouble Code Title, Conditions & Possible Causes
DTC: P1456 **2T EVAP1, MIL: YES** **1996, 1997, 1998, 1999, 2000,** **2001, 2002, 2003, 2004, 2005,** **2006** **Models:** 3.5RL, RL **Engines:** All **Transmissions:** A/T, M/T	**EVAP System Leak Detected (Fuel Tank Area)** IAT sensor signal from 32-86°F at startup, vehicle driven at over 5 mph for over 2 minutes, ECT sensor signal over 154°F, the with the EVAP Control and Vent solenoids "on", the PCM detected the fuel tank pressure was incorrect due to a leak in the fuel tank area. **Possible Causes:** • Fuel tank cap damaged, loose or the wrong part number • Fuel tank leaks at the fuel fill pipe or at the fuel tank seals • Fuel vapor control valve is damaged or has failed • Fuel tank vapor recirculation valve or vapor tube is damaged • Fuel tank vapor control vent tube is damaged or has failed
DTC: P1456 **2T EVAP1, MIL: YES** **1998, 1999, 2000, 2001** **Models:** Integra **Engines:** All **Transmissions:** A/T, M/T	**EVAP System Leak Detected (Fuel Tank Area)** IAT sensor signal from 32-86°F at startup vehicle driven at over 5 mph for over 2 minutes, ECT sensor signal over 154°F, then with the EVAP Vent and Control solenoids "on", and the PCM detected the fuel tank pressure was incorrect due to a leak in the fuel tank area. **Possible Causes:** • Fuel tank cap damaged, loose or the wrong part number • Fuel tank leaks at the fuel fill pipe or at the fuel tank seals • Fuel vapor control valve is damaged or has failed • Fuel tank vapor recirculation valve or vapor tube is damaged • Fuel tank vapor control vent tube is damaged or has failed
DTC: P1457 **2T EVAP1, MIL: YES** **1997, 1998, 1999, 2000, 2001,** **2002, 2003, 2004, 2005** **Models:** NSX, NSX-T **Engines:** All **Transmissions:** A/T, M/T	**EVAP System Leak Detected (Canister Area)** Cold startup completed (IAT sensor signal from 32-86°F at engine startup), vehicle driven at over 5 mph for over 2 minutes, then with ECT sensor signal more than 154°F and the EVAP Control and Vent solenoids enabled, the PCM detected the fuel tank pressure was incorrect due to a leak in the canister area during the Leak Test. **Possible Causes:** • EVAP canister is leaking, damaged or full of water • EVAP canister purge line is loose, damaged or blocked • EVAP two-way valve or ORVR vent shut valve is damaged • EVAP fuel tank vapor control valve is damaged or has failed • PCM has failed • TSB 03-001 (1/27/03) contains a repair procedure for this code
DTC: P1457 **2T EVAP1, MIL: YES** **1997, 1998, 1999, 2000, 2001,** **2002, 2003, 2004, 2005, 2006** **Models:** 2.3CL, 3.0CL, 3.2CL, 3.2TL, TL, MDX, RSX **Engines:** All **Transmissions:** A/T, M/T	**EVAP System Leak Detected (Canister Area)** Cold startup completed (IAT sensor signal from 32-86°F at engine startup), vehicle driven at over 5 mph for over 2 minutes, then with ECT sensor signal more than 154°F and the EVAP Control and Vent solenoids enabled, and the PCM detected an invalid fuel tank pressure value due to a leak in the canister area in the Leak Test. **Possible Causes:** • EVAP canister is leaking, damaged or full of water • EVAP canister purge line is loose, damaged or blocked • EVAP two-way valve or ORVR vent shut valve is damaged • EVAP fuel tank vapor control valve is damaged or has failed • PCM has failed • TSB 03-001 (1/27/03) contains a repair procedure for this code
DTC: P1457 **2T EVAP1, MIL: YES** **1996, 1997, 1998, 1999, 2000,** **2001, 2002, 2003, 2004, 2005,** **2006** **Models:** 3.5RL, RL **Engines:** All **Transmissions:** A/T, M/T	**EVAP System Leak Detected (Canister Area)** Cold startup completed (IAT sensor signal from 32-86°F at engine startup), vehicle driven at over 5 mph for over 2 minutes, then with ECT sensor signal more than 154°F and the EVAP Control and Vent solenoids enabled, the PCM detected the fuel tank pressure was incorrect due to a leak in the canister area during the Leak Test. **Possible Causes:** • EVAP canister is leaking, damaged or full of water • EVAP canister purge line is loose, damaged or blocked • EVAP two-way valve or ORVR vent shut valve is damaged • EVAP fuel tank vapor control valve is damaged or has failed • PCM has failed • TSB 03-001 (1/27/03) contains a repair procedure for this code
DTC: P1457 **2T EVAP1, MIL: YES** **1998, 1999, 2000, 2001** **Models:** Integra **Engines:** All **Transmissions:** A/T, M/T	**EVAP System Leak Detected (Canister Area)** Cold startup completed (IAT sensor signal from 32-86°F at engine startup), vehicle driven at over 5 mph for over 2 minutes, then with ECT sensor signal more than 154°F and the EVAP Control and Vent solenoids enabled, the PCM detected the fuel tank pressure was incorrect due to a leak in the canister area during the Leak Test. **Possible Causes:** • EVAP canister is leaking, damaged or full of water • EVAP canister purge line is loose, damaged or blocked • EVAP two-way valve or ORVR vent shut valve is damaged • EVAP fuel tank vapor control valve is damaged or has failed • PCM has failed • TSB 03-001 (1/27/03) contains a repair procedure for this code

DTC	Trouble Code Title, Conditions & Possible Causes
DTC: P1459 **1T CCM, MIL: YES** **1995, 1996** **Models:** NSX, NSX-T **Engines:** All **Transmissions:** A/T, M/T	**EVAP Purge Flow Switch Circuit Malfunction** Engine running at cruise speed and than back to idle speed, ECT sensor signal more than 154°F, and the PCM detected an unexpected voltage condition on Purge Flow switch circuit. **Possible Causes:** • Purge flow switch is open, shorted to ground or to power (B+) • Purge flow switch vacuum line is loose, bent or damaged • Purge flow switch is damaged or has failed • PCM has failed • TSB 03-001 (1/27/03) contains a repair procedure for this code
DTC: P1459 **1T CCM, MIL: YES** **1995, 1996, 1997, 1998** **Models:** 2.2CL, 2.5TL **Engines:** All **Transmissions:** A/T, M/T	**EVAP Purge Flow Switch Circuit Malfunction** Engine running at cruise speed and than back to idle speed, ECT sensor signal more than 154°F, and the PCM detected an unexpected voltage condition on Purge Flow switch circuit. **Possible Causes:** • Purge flow switch is open, shorted to ground or to power (B+) • Purge flow switch vacuum line is loose, bent or damaged • Purge flow switch is damaged or has failed • PCM has failed
DTC: P1460 **1T CCM, MIL: YES** **2003, 2004, 2005, 2006** **Models:** MDX **Engines:** All **Transmissions:** A/T, M/T	**Fuel Level Sensor Power Supply Circuit Malfunction** Key on or engine running; and the PCM detected an unexpected voltage condition on the fuel level sensor power supply circuit. **Possible Causes:** • Fuel level sensor power circuit is open or shorted to ground • Fuel level sensor circuit to fuel gauge assembly is damaged • Fuel level sensor is damaged or has failed • PCM is damaged or has failed
DTC: P1486 **2T ECT, MIL: YES** **2000, 2001, 2002, 2003, 2004, 2005** **Models:** NSX, NSX-T **Engines:** All **Transmissions:** A/T, M/T	**Cooling System Malfunction** DTC P0107, P0108, P0112, P0113, P0116, P0117, P0118, P0300, P0301, P0302, P0303, P0304, P0305, P0306, P0335, P0336, P0401, P0500, P0505, P1106, P1107, P1108, P1128, P1129, P1253, P1257, P1258, P1259, P1359, P1399, P1491, P1498 and P1519 not set, vehicle driven for over 10 minutes, and the PCM detected the ECT signal did not reach the correct closed loop value. **Possible Causes:** • ECT sensor is out of calibration • Check for low coolant level or incorrect coolant mixture • Cooling system component failure (thermostat stuck open)
DTC: P1486 **2T ECT, MIL: YES** **2000, 2001, 2002, 2003, 2004, 2005, 2006**	**Cooling System Malfunction** DTC P0107, P0108, P0112, P0113, P0116, P0117, P0118, P0300, P0301, P0302, P0303, P0304, P0305, P0306, P0335, P0336, P0401, P0500, P0505, P1106, P1107, P1108, P1128, P1129, P1253, P1257, P1258, P1259, P1359, P1399, P1491, P1498 and P1519 not set, vehicle driven for over 10 minutes, and the PCM detected the ECT signal did not reach the correct closed loop value. **Note: This trouble code can set if the engine remains under hot idle conditions with the hood open for an extended period of time.** **Possible Causes:** • ECT sensor is out of calibration • Check for low coolant level or incorrect coolant mixture • Cooling system component failure (thermostat stuck open) • TSB 01-016 (9/10/01) contains a repair procedure for this code
DTC: P1486 **2T ECT, MIL: YES** **1999, 2000, 2001, 2002, 2003, 2004, 2005, 2006** **Models:** 3.5RL, RL **Engines:** All **Transmissions:** A/T, M/T	**Cooling System Malfunction** DTC P0107, P0108, P0112, P0113, P0116, P0117, P0118, P0300, P0301, P0302, P0303, P0304, P0305, P0306, P0335, P0336, P0401, P0500, P0505, P1106, P1107, P1108, P1128, P1129, P1253, P1257, P1258, P1259, P1359, P1399, P1491, P1498 and P1519 not set, vehicle driven for over 10 minutes, and the PCM detected the ECT signal did not reach the correct closed loop value. **Possible Causes:** • ECT sensor is out of calibration • Check for low coolant level or incorrect coolant mixture • Cooling system component failure (thermostat stuck open)
DTC: P1491 **2T EGR1, MIL: YES** **1995, 1996, 1997, 1998, 1999, 2000, 2001, 2002, 2003, 2004, 2005** **Models:** NSX, NSX-T **Engines:** All **Transmissions:** A/T, M/T	**EGR Valve Lift Sensor Insufficient Flow Detected** Vehicle driven in closed loop at 1700-2500 rpm for over 10 minutes, and the PCM detected the EGR valve lift sensor (EGRV) signal indicated insufficient EGR flow during the EGR Monitor test. **Possible Causes:** • EGR valve lift sensor is stuck, damaged or has failed • EGR control solenoid circuit is open or shorted to ground • EGR control solenoid valve is damaged or has failed • PCM has failed

DTC	Trouble Code Title, Conditions & Possible Causes
DTC: P1491 **2T EGR1, MIL: YES** **1995, 1996, 1997, 1998** **Models:** 2.5TL **Engines:** All **Transmissions:** A/T, M/T	**EGR Valve Lift Sensor Insufficient Flow Detected** Vehicle driven in closed loop at 1700-2500 rpm for over 10 minutes, and the PCM detected the EGR valve lift sensor (EGRV) signal indicated insufficient EGR flow during the EGR Monitor test. **Possible Causes:** • EGR valve lift sensor is stuck, damaged or has failed • EGR control solenoid circuit is open or shorted to ground • EGR control solenoid valve is damaged or has failed • PCM has failed
DTC: P1491 **2T EGR1, MIL: YES** **1996, 1997, 1998, 1999, 2000, 2001, 2002, 2003, 2004, 2005, 2006** **Models:** 2.2CL, 2.3CL, 3.0CL, 3.2CL, 3.2TL, TL, 3.5RL, RL, MDX **Engines:** All **Transmissions:** A/T, M/T	**EGR Valve Lift Sensor Insufficient Flow Detected** Engine started, vehicle driven while in closed loop at 1700-2500 rpm for at least 10 minutes, and the PCM detected the EGR valve lift sensor (EGRV) signal indicated insufficient EGR flow during the EGR Monitor test. **Possible Causes:** • EGR valve lift sensor is stuck, damaged or has failed • EGR control solenoid circuit is open or shorted to ground • EGR control solenoid valve is damaged or has failed • PCM has failed
DTC: P1498 **1T CCM, MIL: YES** **1995, 1996, 1997, 1998, 1999, 2000, 2001, 2002, 2003, 2004, 2005** **Models:** NSX, NSX-T **Engines:** All **Transmissions:** A/T, M/T	**EGR Valve Lift Sensor High Input** Key on or engine running; and the PCM detected the EGR Valve Lift sensor signal was more than an allowable range stored in memory. **Possible Causes:** • EGR valve lift sensor circuit is open or shorted to power • EGR valve lift sensor is shorted to VREF or system power (B+) • EGR valve lift sensor is stuck, damaged or has failed • PCM has failed
DTC: P1498 **1T CCM, MIL: YES** **1995, 1996, 1997, 1998** **Models:** 2.5TL **Engines:** All **Transmissions:** A/T, M/T	**EGR Valve Lift Sensor Circuit High Input** Key on or engine running; and the PCM detected the EGR Valve Lift sensor signal was more than an allowable range stored in memory. **Possible Causes:** • EGR valve lift sensor circuit is open or shorted to power • EGR valve lift sensor is shorted to VREF or system power (B+) • EGR valve lift sensor is stuck, damaged or has failed • PCM has failed
DTC: P1498 **1T CCM, MIL: YES** **1996, 1997, 1998, 1999, 2000, 2001, 2002, 2003, 2004, 2005, 2006** **Models:** 2.2CL, 2.3CL, 3.0CL, 3.2CL, 3.2TL, TL, 3.5RL, RL, MDX **Engines:** All **Transmissions:** A/T, M/T	**EGR Valve Lift Sensor Circuit High Input** Key on or engine running; and the PCM detected the EGR Valve Lift sensor signal was more than an allowable range stored in memory. **Possible Causes:** • EGR valve lift sensor circuit is open or shorted to power • EGR valve lift sensor is shorted to VREF or system power (B+) • EGR valve lift sensor is stuck, damaged or has failed • PCM has failed
DTC: P1508 **1T CCM, MIL: YES** **1995, 1996, 1997, 1998** **Models:** 2.5TL **Engines:** All **Transmissions:** A/T, M/T	**Idle Air Control Valve Circuit Malfunction** Key on or engine running; and the PCM detected an unexpected voltage condition on the IAC valve control circuit during the test. **Possible Causes:** • IAC valve control circuit is open or shorted to ground • IAC valve control circuit is shorted to system power (B+) • IAC valve power circuit is open or shorted to ground • IAC valve is damaged or has failed • PCM has failed
DTC: P1508 **1T CCM, MIL: YES** **1996, 1997, 1998, 1999, 2000, 2001, 2002** **Models:** 2.2CL, 3.2CL, 3.5RL, Integra, MDX **Engines:** All **Transmissions:** A/T, M/T	**Idle Air Control Valve Circuit Malfunction** Key on or engine running; and the PCM detected an unexpected voltage condition on the IAC valve control circuit during the test. **Possible Causes:** • IAC valve control circuit is open or shorted to ground • IAC valve control circuit is shorted to system power (B+) • IAC valve power circuit is open or shorted to ground • IAC valve is damaged or has failed • PCM has failed • TSB 03-006 (3/17/03) contains a repair procedure for this code

DTC	Trouble Code Title, Conditions & Possible Causes
DTC: P1508 **1T CCM, MIL: YES** **1996, 1997, 1998** **Models:** 3.2TL **Engines:** All **Transmissions:** A/T, M/T	**Idle Air Control Valve Circuit Malfunction** Key on or engine running; and the PCM detected an unexpected voltage condition on the IAC valve control circuit during the test. **Possible Causes:** • IAC valve circuit is open, shorted to ground or to system power • IAC valve power circuit is open or shorted to ground • IAC valve is damaged or has failed • PCM is damaged • TSB 03-006 (3/17/03) contains a repair procedure for this code
DTC: P1508 **2T CCM, MIL: YES** **1996, 1997, 1998, 1999** **Models:** SLX **Engines:** 3.2L VIN V, 3.5L VIN X **Transmissions:** A/T	**Idle Speed Control System Low RPM** DTC P0101, P0102, P0103, P0106, P0107, P0108, P0112, P0113, P0117, P0118, P0121, P0122, P0123, P0171, P0172, P0201-P0206, P0351-P0356, P0401, P0440, engine runtime 120 seconds, system voltage from 11-16v, ECT sensor over 120ºF, BARO sensor over 75 kPa, Scan Tool tests all "off", vehicle speed under 1 mph with the throttle closed, and the PCM detected the Actual engine speed was 100-200 rpm less than the Desired idle speed for 5 seconds (Desired idle speed is based on ECT signal during the test). **Possible Causes:** • Fuel delivery system is too lean or too rich during the test • Inspect the engine mounts for damage to the mounts • Inspect the throttle linkage adjustment and tension • Inspect the throttle body bore for dirt or foreign material • Inspect for any air intake system leaks in the engine or hoses • IAC valve may be damaged or have failed • IAC motor control circuit is open or shorted to ground • Perform an IAC Reset function with the Scan Tool
DTC: P1509 **2T CCM, MIL: YES** **1996, 1997, 1998, 1999** **Models:** SLX **Engines:** 3.2L VIN V, 3.5L VIN X **Transmissions:** A/T	**Idle Speed Control System High RPM** DTC P0101, P0102, P0103, P0106, P0107, P0108, P0112, P0113, P0117, P0118, P0121, P0122, P0123, P0171, P0172, P0201-P0206, P0351-P0356, P0401, P0440, engine runtime 120 seconds, system voltage from 11-16v, ECT sensor over 120ºF, BARO sensor over 75 kPa, Scan Tool tests all "off", vehicle speed under 1 mph with the throttle closed, and the PCM detected the Actual engine speed was 100-200 rpm more than the Desired idle speed for 5 seconds (Desired idle speed is based on ECT signal during the test). **Possible Causes:** • Fuel delivery system is too lean or too rich during the test • Inspect the engine mounts for damage to the mounts • Inspect the throttle linkage adjustment and tension • Inspect the throttle body bore for dirt or foreign material • Inspect for any air intake system leaks in the engine or hoses • IAC valve may be damaged or have failed • IAC motor control circuit is open or shorted to ground • Perform an IAC Reset function with the Scan Tool
DTC: P1519 **1T CCM, MIL: YES** **1997, 1998, 1999, 2000, 2001,** **2002, 2003, 2004, 2005, 2006** **Models:** 2.3CL, 3.0CL, 3.2CL, MDX, RSX **Engines:** All **Transmissions:** A/T, M/T	**Idle Air Control Valve Circuit Malfunction** Key on or engine running; and the PCM detected an unexpected voltage condition on the Idle Air Control (IAC) valve control circuit. **Possible Causes:** • IAC valve circuit is open, shorted to ground or to system power • IAC valve power circuit is open or shorted to ground • IAC valve is damaged or has failed • PCM is damaged • TSB 03-006 (3/17/03) contains a repair procedure for this code
DTC: P1519 **1T CCM, MIL: YES** **1999, 2000, 2001, 2002, 2003,** **2004, 2005, 2006** **Models:** 3.2TL, TL **Engines:** All **Transmissions:** A/T, M/T	**Idle Air Control Valve Circuit Malfunction** Key on or engine running; and the PCM detected an unexpected voltage condition on the Idle Air Control (IAC) valve control circuit. **Possible Causes:** • IAC valve circuit is open, shorted to ground or to system power • IAC valve power circuit is open or shorted to ground • IAC valve is damaged or has failed • PCM is damaged • TSB 03-006 (3/17/03) contains a repair procedure for this code
DTC: P1607 **1T PCM, MIL: YES** **1995, 1996, 1997, 1998, 1999,** **2000, 2001, 2002, 2003, 2004,** **2005** **Models:** NSX, NSX-T **Engines:** All **Transmissions:** A/T, M/T	**PCM Internal Circuit 'A' Malfunction** Key on, and the PCM detected an Internal Fault 'A' condition. **Note: This trouble code indicates an internal failure in the PCM. The OEM repair procedure recommends replacing the original PCM with a "known good" PCM and then verify the code does not reset.** **Possible Causes:** • PCM is damaged or has failed

DTC	Trouble Code Title, Conditions & Possible Causes
DTC: P1607 **1T PCM, MIL: YES** **1996, 1997, 1998, 1999, 2000, 2001, 2002, 2003, 2004, 2005, 2006** **Models:** 2.2CL, 2.3CL, 3.0CL, 3.2CL, 3.2TL, TL, 3.5RL, RL, Integra, MDX, RSX **Engines:** All **Transmissions:** A/T, M/T	**PCM Internal Circuit Malfunction** Key on, and the PCM detected a fault in one of its internal circuits. **Note: This trouble code indicates an internal failure in the PCM. The OEM repair procedure recommends replacing the original PCM with a "known good" PCM and then verify the code does not reset.** **Possible Causes:** • PCM is damaged or has failed
DTC: P1608 **1T PCM, MIL: YES** **1995, 1996, 1997, 1998, 1999, 2000, 2001, 2002, 2003, 2004, 2005** **Models:** NSX, NSX-T **Engines:** All **Transmissions:** A/T, M/T	**PCM Internal Circuit 'B' Malfunction** Key on, and the PCM detected an Internal Fault 'B' condition. **Note: This trouble code indicates an internal failure in the PCM. The OEM repair procedure recommends replacing the original PCM with a "known good" PCM and then verify the code does not reset.** **Possible Causes:** • PCM is damaged or has failed
DTC: P1618 **1T PCM, MIL: NO** **1996, 1997, 1998, 1999** **Models:** SLX **Engines:** 3.2L VIN V, 3.5 VIN X **Transmissions:** A/T	**Serial Peripheral Interface Communication** Key on for 2 seconds, system voltage over 9v, and the PCM detected an internal program fault (the check sum of the data communications error) for 3 out of 6 seconds with no TCM resets during the test period of 2 seconds. **Note: For additional help with this code, view the Failure Records.** **Possible Causes:** • Check the PCM calibration to verify it is the latest calibration • Recalibrate the PCM as required • PCM may need to be replaced and reprogrammed
DTC: P1625 **1T PCM, MIL: NO** **1998, 1999** **Models:** SLX **Engines:** 3.5 VIN X **Transmissions:** A/T	**PCM Unexpected Reset** Key on, and the PCM detected a Clock or Computer Operating Properly (COP) reset or illegal software code interrupt occurred. **Note: For additional help with this code, view the Failure Records.** **Possible Causes:** • Perform a PCM Reset and retrieve the trouble codes
DTC: P1640 **1T CCM, MIL: NO** **1998, 1999, 2000, 2001, 2002** **Models:** SLX **Engines:** 3.2L VIN V, 3.5L VIN X **Transmissions:** A/T	**Output Driver Module 'A' Circuit Malfunction** DTC P1618 not set, engine started, engine running, system voltage over 13.2v for 4 seconds, and the PCM detected an open circuit condition and an unexpected high voltage condition on the Output Driver Module circuit (A/C Clutch Relay or Purge Solenoid) with the device "on" for 2.5 seconds during the CCM Rationality test. **Note: For additional help with this code, view the Failure Records.** **Possible Causes:** • One or more output device driver circuits connected to ODM 'A' has an open circuit condition • One or more output device driver circuits connected to ODM 'A' has a short-to-voltage condition • Check for an open power circuit to the related output devices • Disconnect the A/C clutch relay and Purge solenoid to find fault
DTC: P1650 **1T CCM, MIL: NO** **1998, 1999, 2000, 2001, 2002** **Models:** SLX **Engines:** 3.2L VIN V, 3.5L VIN X **Transmissions:** A/T	**Output Driver Module Circuit Malfunction** DTC P1618 not set, engine started, engine running, system voltage over 13.2v for 4 seconds, and the PCM detected the voltage on the Output Driver Module (ODM) circuit did not indicate less than 1.0 volt with the device commanded "on" for 0.5 seconds in the CCM test. **Note: For additional help with this code, view the Failure Records.** **Possible Causes:** • One or more output device driver circuits connected to ODM has an open circuit condition or has a short-to-voltage condition • Check for an open power circuit to the related output devices • Disconnect the A/C clutch relay and Purge solenoid to find fault
DTC: P1655 **1T CCM, MIL: YES** **1997** **Models:** 2.2CL **Engines:** All **Transmissions:** A/T	**PCM Internal Circuit Malfunction** Key on, and the PCM detected an Internal circuit problem. **Note: This trouble code indicates an internal failure in the PCM. The OEM repair procedure recommends replacing the original PCM with a "known good" PCM and then verify the code does not reset.** **Possible Causes:** • PCM is damaged or has failed

DTC	Trouble Code Title, Conditions & Possible Causes
DTC: P1656 1T CCM, MIL: YES 2001, 2002, 2003, 2004, 2005, 2006 Models: MDX Engines: All Transmissions: A/T	**PCM To VSA Unit Signal Line Circuit Malfunction** Engine started, and the PCM detected an unexpected voltage condition on the PCM to VSA circuit during the CCM test. **Possible Causes:** • SEFA or SEAF signal line is open or shorted to ground • SEFA or SEAF signal line is shorted to system power • TCM or PCM has failed
DTC: P1660 1T CCM, MIL: YES 1995, 1996, 1997, 1998 Models: 2.5TL Engines: All Transmissions: A/T	**A/T (TCM) FI Data Line Circuit Malfunction** Key on or engine running; and the PCM detected an unexpected voltage condition on the A/T FI Data Line circuit during the test. **Possible Causes:** • A/T FI data line is open or shorted to ground • A/T FI data line is shorted to VREF or system power • TCM or PCM has failed
DTC: P1660 1T CCM, MIL: YES 1996, 1997, 1998 Models: Integra Engines: All Transmissions: A/T	**A/T (TCM) FI Data Line Circuit Malfunction** Key on or engine running; and the PCM detected an unexpected voltage condition on the A/T FI Data Line circuit during the test. **Possible Causes:** • A/T FI data line is open or shorted to ground • A/T FI data line is shorted to VREF or system power • TCM or PCM has failed
DTC: P1671 1T PCM, MIL: YES 1995, 1996, 1997, 1998, 1999, 2000, 2001, 2002, 2003, 2004, 2005 Models: NSX, NSX-T Engines: All Transmissions: A/T	**A/T (TCM) FI Data Line No Signal** Key on or engine running; and the PCM did not detect any signals on the A/T FI Data Line circuit during the CCM test. **Possible Causes:** • A/T FI data line is open • A/T FI data line is shorted to sensor ground or chassis ground • A/T FI data line is shorted to VREF or system power • TCM or PCM has failed
DTC: P1672 1T PCM, MIL: YES 1995, 1996, 1997, 1998, 1999, 2000, 2001, 2002, 2003, 2004, 2005 Models: NSX, NSX-T Engines: All Transmissions: A/T	**A/T (TCM) FI Data Line Circuit Malfunction** Engine started, and the PCM detected an unexpected (intermittent) interruption of the TCM FI Data Line signal during the CCM test. **Possible Causes:** • A/T FI data line is open (intermittent fault) • A/T FI data line is shorted to ground (intermittent fault) • A/T FI data line is shorted to VREF or power (intermittent fault) • TCM or PCM has failed
DTC: P1676 1T PCM, MIL: YES 1995, 1996 Models: NSX, NSX-T Engines: All Transmissions: A/T	**A/T (TCM) FI Data Line Circuit Malfunction** Key on or engine running; and the PCM detected an unexpected voltage condition on the TCM FI Data Line circuit during the test. **Possible Causes:** • A/T FI data line is open or shorted to ground • A/T FI data line is shorted to VREF or system power • TCM or PCM has failed
DTC: P1676 1T PCM, MIL: YES 2000, 2001, 2002, 2003, 2004, 2005, 2006 Models: 3.2CL, 3.2TL, TL Engines: All Transmissions: A/T	**A/T (TCM) FPTDR Signal Line Circuit Malfunction** Key on or engine running; and the PCM detected an unexpected voltage condition on the TCM FPTDR Line circuit during the test. **Possible Causes:** • A/T FTPDR data line is open or shorted to ground • A/T FTPDR data line is shorted to VREF or system power • TCM or PCM has failed
DTC: P1677 1T PCM, MIL: YES 1995, 1996 Models: NSX, NSX-T Engines: All Transmissions: A/T	**A/T (TCM) FI Data Line Circuit Malfunction** Key on or engine running; and the PCM detected an unexpected voltage condition on the TCM FI Data Line circuit during the test. **Possible Causes:** • A/T FI data line is open or shorted to ground • A/T FI data line is shorted to VREF or system power • TCM or PCM has failed

DTC	Trouble Code Title, Conditions & Possible Causes
DTC: P1678 **1T PCM, MIL: YES** **2000, 2001, 2002, 2003, 2004, 2005, 2006** **Models:** 3.2CL, 3.2TL, TL **Engines:** All **Transmissions:** A/T	**A/T (TCM) FPTDR Signal Line Circuit Malfunction** Key on or engine running; and the PCM detected an unexpected (intermittent) interruption of the TCM FPTDR Line signal in the test. **Possible Causes:** • A/T FTPDR data line is open (intermittent fault) • A/T FTPDR data line is shorted to ground (intermittent fault) • A/T FTPDR data line is shorted to VREF or power (intermittent) • TCM or PCM has failed
DTC: P1683 **1T CCM, MIL: YES** **2003, 2004, 2005, 2006** **Models:** MDX **Engines:** All **Transmissions:** A/T, M/T	**Throttle Valve Default Position Spring Range/Performance** Key on or engine running; and the PCM detected an unexpected voltage condition on the throttle valve position during the CCM test. Do not check the throttle valve position spring action with the key on. **Possible Causes:** • Check for a binding throttle valve position spring • Check for dirt and sludge on the throttle body causing the throttle valve default position to indicate an incorrect value
DTC: P1684 **1T CCM, MIL: YES** **2003, 2004, 2005, 2006** **Models:** MDX **Engines:** All **Transmissions:** A/T, M/T	**Throttle Valve Default Position Spring Range/Performance** Key on or engine running; and the PCM detected an unexpected voltage condition on the throttle valve position during the CCM test. Do not check the throttle valve position spring action with the key on. **Possible Causes:** • Check for a binding throttle valve position spring • Check for dirt and sludge on the throttle body causing the throttle valve default position to indicate an incorrect value
DTC: P1690 **1T PCM, MIL: YES** **1996, 1997, 1998** **Models:** 3.2TL **Engines:** All **Transmissions:** A/T	**A/T (TCM) TCSTB Data Line Circuit Malfunction** Key on or engine running; and the PCM detected an unexpected (intermittent) interruption of the TCSTB Data Line signal in the test. **Note: This code applies to models with a Traction Control System.** **Possible Causes:** • A/T TCSTB data line is open (intermittent fault) • A/T TCSTB data line is shorted to ground (intermittent fault) • A/T TCSTB data line is shorted to VREF or power (intermittent) • TCM or PCM has failed
DTC: P1690 **1T PCM, MIL: YES** **1996, 1997, 1998, 1999, 2000** **Models:** 3.5RL **Engines:** All **Transmissions:** A/T	**A/T (TCM) TCSTB Data Line Circuit Malfunction** Key on or engine running; and the PCM detected an unexpected (intermittent) interruption of the TCSTB Data Line signal in the test. **Note: This code applies to models with a Traction Control System.** **Possible Causes:** • A/T TCSTB data line is open (intermittent fault) • A/T TCSTB data line is shorted to ground (intermittent fault) • A/T TCSTB data line is shorted to VREF or power (intermittent) • TCM or PCM has failed
DTC: P1696 **1T PCM, MIL: YES** **1996, 1997, 1998** **Models:** 3.2TL **Engines:** All **Transmissions:** A/T	**A/T (TCM) TCFC Data Line Signal Low Input** Key on or engine running; and the PCM detected an unexpected "low" input on the TCFC Data Line circuit during the CCM test. **Note: This code applies to models with a Traction Control System.** **Possible Causes:** • A/T TCFC data line is shorted to sensor ground • A/T TCFC data line is shorted to chassis ground • TCM or PCM has failed
DTC: P1696 **1T PCM, MIL: YES** **1996, 1997, 1998, 1999, 2000, 2001, 2002, 2003, 2004, 2005, 2006** **Models:** 3.5RL, RL **Engines:** All **Transmissions:** A/T	**TCFC Data Line Signal Low Input** Key on or engine running; and the PCM detected an unexpected "low" input on the TCFC Data Line circuit during the CCM test. **Note: This code applies to models with a Traction Control System.** **Possible Causes:** • A/T TCFC data line is shorted to sensor ground • A/T TCFC data line is shorted to chassis ground • TCM or PCM has failed
DTC: P1697 **1T PCM, MIL: YES** **1996, 1997, 1998** **Models:** 3.2TL **Engines:** All **Transmissions:** A/T	**TUFT Data Line Signal High Input** Key on or engine running; and the PCM detected an unexpected "high" input on the TUFT Data Line circuit during the CCM test. **Note: This code applies to models with a Traction Control System.** **Possible Causes:** • A/T TUFT data line is shorted to VREF (voltage reference) • A/T TUFT data line is shorted to system power (B+) • TCM or PCM has failed

DTC	Trouble Code Title, Conditions & Possible Causes
DTC: P1697 **1T PCM, MIL: YES** **1996, 1997, 1998, 1999, 2000,** **2001, 2002, 2003, 2004, 2005,** **2006** **Models:** 3.5RL, RL **Engines:** All **Transmissions:** A/T	**TUFT Data Line Signal High Input** Key on or engine running; and the PCM detected an unexpected "high" input on the TUFT Data Line circuit. Note: This code applies to models with a Traction Control System. **Possible Causes:** • A/T TUFT data line is shorted to VREF (voltage reference) • A/T TUFT data line is shorted to system power (B+) • TCM or PCM has failed
DTC: P1705 **1T CCM, MIL: YES** **1995, 1996, 1997, 1998, 1999,** **2000, 2001, 2002, 2003, 2004,** **2005, 2006** **Models:** NSX, NSX-T, 3.5RL, RL **Engines:** All **Transmissions:** A/T	**A/T Transmission Range Switch Low Input** Engine started, vehicle driven and the PCM detected two gear position switch inputs simultaneously (two input at the same time). Note: The D4 lamp on the dash will flash 5 times if this code is set. **Possible Causes:** • A/T gear position switch signal circuit is shorted to ground • A/T gear position switch signal circuit is shorted to another wire • A/T gear position switch is damaged or has failed • TCM or PCM has failed
DTC: P1705 **1T CCM, MIL: YES** **1995, 1996, 1997, 1998, 1999,** **2000, 2001, 2002, 2003, 2004,** **2005, 2006** **Models:** 2.5TL, 3.2TL, TL, RSX **Engines:** All **Transmissions:** A/T	**A/T Transmission Range Switch Low Input** Engine started, vehicle driven and the PCM detected two gear position switch inputs simultaneously (two input at the same time). Note: The D4 or D5 lamp on the dash will flash when this code sets. **Possible Causes:** • A/T gear position switch signal circuit is shorted to ground • A/T gear position switch signal circuit is shorted to another wire • A/T gear position switch is damaged or has failed • TCM or PCM has failed
DTC: P1706 **1T CCM, MIL: YES** **1996, 1997, 1998, 1999, 2000,** **2001, 2002, 2003, 2004, 2005,** **2006** **Models:** 2.2CL, 2.3CL, 3.0CL, 3.5RL, RL, Integra, MDX, RSX **Engines:** All **Transmissions:** A/T	**A/T Transmission Range Switch Circuit No Gear Inputs** Engine started, vehicle driven and the PCM detected an unexpected high voltage condition on the Gear Position Switch circuit. Note: The D4 or D5 indicator on the dash will flash 6 times if this code is set. **Possible Causes:** • A/T gear position switch signal circuit is shorted to ground • A/T gear position switch signal circuit is shorted to another wire • A/T gear position switch is damaged or has failed • TCM or PCM has failed
DTC: P1706 **1T CCM, MIL: YES** **1995, 1996, 1997, 1998, 1999,** **2000, 2001, 2002, 2003, 2004,** **2005** **Models:** NSX, NSX-T **Engines:** All **Transmissions:** A/T	**A/T Transmission Range Switch Circuit No Gear Inputs** Key on or engine running; and the PCM detected an unexpected "high" voltage condition on the Gear Position Switch circuit. **Possible Causes:** • A/T gear position switch signal circuit is open • A/T gear position switch ground circuit is open • A/T gear position switch is shorted to VREF or system power • A/T gear position switch is damaged or has failed
DTC: P1706 **1T CCM, MIL: YES** **1995, 1996, 1997, 1998, 1999,** **2000, 2001, 2002, 2003, 2004,** **2005, 2006** **Models:** 2.5TL, 3.2TL, TL **Engines:** All **Transmissions:** A/T	**A/T Transmission Range Switch Circuit No Gear Inputs** Key on or engine running; and the PCM detected an unexpected "high" voltage condition on the Gear Position Switch circuit. **Possible Causes:** • A/T gear position switch signal circuit is open • A/T gear position switch ground circuit is open • A/T gear position switch is shorted to VREF or system power • A/T gear position switch is damaged or has failed

DTC	Trouble Code Title, Conditions & Possible Causes
DTC: P1706 **1T CCM, MIL: YES** **1996, 1997, 1998, 1999, 2000, 2001, 2002, 2003, 2004, 2005, 2006** **Models:** 2.2CL, 2.3CL, 3.0CL, Integra, MDX, RSX **Engines:** All **Transmissions:** A/T	**A/T Transmission Range Switch Circuit No Gear Inputs** Key on or engine running; and the PCM detected an unexpected "high" voltage condition on the Gear Position Switch circuit. **Possible Causes:** • A/T gear position switch signal circuit is open • A/T gear position switch ground circuit is open • A/T gear position switch is shorted to VREF or system power • A/T gear position switch is damaged or has failed
DTC: P1709 **1T CCM, MIL: YES** **1999, 2000, 2001, 2002, 2003, 2004, 2005, 2006** **Models:** 3.2CL, 3.2TL, TL, RSX **Engines:** All **Transmissions:** A/T	**A/T Transmission Gear Switch Circuit Malfunction** Key on or engine running; and the PCM detected more than one signal on the Position Switch circuit at the same time during the test. **Possible Causes:** • A/T gear position switch signal circuit is shorted to another wire inside the switch or in the wiring harness (i.e., the P/N signal is present along with a Drive, 1st, 2nd, 3rd, 4th or Reverse signal) • A/T gear position switch is damaged or has failed • TCM or PCM has failed
DTC: P1710 **1T CCM, MIL: NO** **2000, 2001, 2002, 2003, 2004, 2005, 2006** **Models:** 3.2CL, 3.2TL, TL, MDX, RSX **Engines:** All **Transmissions:** A/T	**A/T First Hold Switch Circuit Malfunction** Key on or engine running; and the PCM detected an unexpected voltage condition on the First Hold switch circuit during the test. **Possible Causes:** • First hold witch signal circuit is open • First hold switch is shorted to ground • First hold switch ground circuit is open • First hold switch is shorted to VREF or system power • First hold switch is damaged or has failed • TCM or PCM has failed
DTC: P1717 **1T CCM, MIL: YES** **2002, 2003, 2004, 2005, 2006** **Models:** RSX **Engines:** All **Transmissions:** A/T	**A/T Transmission Range Switch Circuit Malfunction** P1705 and P1706 not set, engine started, vehicle driven, and the PCM detected an unexpected voltage condition on the Transmission Range Switch ATP RVS circuit during the CCM test. **Possible Causes:** • Transmission range switch circuit is open or shorted to ground • Transmission range switch circuit is shorted to VREF or power • Transmission range switch connector is disconnected • TCM or PCM has failed
DTC: P1738 **1T CCM, MIL: NO** **1998, 1999** **Models:** 2.3CL, 3.2TL **Engines:** All **Transmissions:** A/T	**A/T 2nd Pressure Switch Circuit Malfunction** Engine running in gear, and the PCM detected an unexpected voltage condition on the 2nd pressure switch circuit during the test. **Possible Causes:** • A/T 2nd pressure switch signal circuit is open • A/T 2nd pressure switch signal circuit is shorted to ground • A/T 2nd pressure switch signal circuit shorted to system power • A/T 2nd pressure switch is damaged or has failed • TCM or PCM has failed
DTC: P1739 **1T CCM, MIL: NO** **1999, 2000, 2001, 2002, 2003, 2004, 2005, 2006** **Models:** 3.2CL, 3.2TL, TL, MDX, RSX **Engines:** All **Transmissions:** A/T	**A/T 3rd Pressure Switch Circuit Malfunction** Engine running in gear, and the PCM detected an unexpected voltage condition on the 3rd pressure switch circuit during the test. **Possible Causes:** • A/T 3rd pressure switch signal circuit is open • A/T 3rd pressure switch signal circuit is shorted to ground • A/T 3rd pressure switch signal circuit shorted to system power • A/T 3rd pressure switch is damaged or has failed • TCM or PCM has failed
DTC: P1740 **1T CCM, MIL: NO** **1996, 1997, 1998, 1999, 2000, 2001, 2002, 2003, 2004, 2005, 2006** **Models:** 3.2CL, 3.2TL, TL, MDX, RSX **Engines:** All **Transmissions:** A/T	**A/T 4th Pressure Switch Circuit Malfunction** Engine running in gear, and the PCM detected an unexpected voltage condition on the 4th pressure switch circuit during the test. **Possible Causes:** • A/T 4th pressure switch signal circuit is open • A/T 4th pressure switch signal circuit is shorted to ground • A/T 4th pressure switch signal circuit shorted to system power • A/T 4th pressure switch is damaged or has failed • TCM or PCM has failed

DTC	Trouble Code Title, Conditions & Possible Causes
DTC: P1750 **1T CCM, MIL: YES** **1996, 1997, 1998, 1999, 2000,** **2001, 2002, 2003, 2004, 2005,** **2006** **Models:** 3.2CL, 3.2TL, TL, MDX, RSX **Engines:** All **Transmissions:** A/T	**A/T Hydraulic Pressure Control Malfunction (Mechanical)** Engine started, engine running in gear, and the PCM detected a mechanical fault existed in the hydraulic portion of the A/T (as it relates to Shift Solenoids 'A' and 'B') during the CCM test. **Note: The D4 lamp on the dash will flash when this code sets.** **Possible Causes:** • Shift Solenoid 'A' is damaged or has failed • Shift Solenoid 'B' is damaged or has failed • The transmission is damaged or has failed
DTC: P1751 **1T CCM, MIL: YES** **1996, 1997, 1998, 1999, 2000,** **2001, 2002, 2003, 2004, 2005,** **2006** **Models:** 3.2CL, 3.2TL, TL, MDX, RSX **Engines:** All **Transmissions:** A/T	**A/T Hydraulic Pressure Control Malfunction (Mechanical)** Engine started, engine running in gear, and the PCM detected a mechanical fault existed in the hydraulic portion of the A/T (as it relates to Clutch Pressure Control Solenoids 'A' and 'B') during the CCM test. **Note: The D4 lamp on the dash will flash once when this code is set.** **Possible Causes:** • Clutch Pressure Control Solenoid 'A' is damaged or has failed • Clutch Pressure Control Solenoid 'B' is damaged or has failed • Shift Solenoid 'B' is damaged or has failed • The transmission is damaged or has failed
DTC: P1753 **1T CCM, MIL: YES** **1995, 1996, 1997, 1998, 1999,** **2000, 2001, 2002, 2003, 2004,** **2005, 2006** **Models:** NSX, NSX-T, 3.5RL, RL **Engines:** All **Transmissions:** A/T	**A/T Lockup Solenoid Valve 'A' Circuit Malfunction** Vehicle driven in 1st, 2nd, 3rd and 4th gears, and the PCM detected an unexpected voltage condition on the Solenoid 'A' control circuit. **Note: The D4 lamp on the dash will flash once when this code is set.** **Possible Causes:** • Lockup solenoid 'A' control circuit is open or shorted to ground • Lockup solenoid 'A' circuit is shorted to system power • Lockup solenoid 'A' connector is disconnected or damaged • Lockup solenoid 'A' is damaged or has failed • TCM or PCM has failed
DTC: P1753 **1T CCM, MIL: YES** **1995, 1996, 1997, 1998** **Models:** 2.5TL **Engines:** All **Transmissions:** A/T	**A/T Lockup Solenoid Valve 'A' Circuit Malfunction** Vehicle driven in 1st, 2nd, 3rd and 4th gears, and the PCM detected an unexpected voltage condition on the Solenoid 'A' control circuit. **Note: The D4 lamp on the dash will flash once when this code is set.** **Possible Causes:** • Lockup solenoid 'A' control circuit is open or shorted to ground • Lockup solenoid 'A' circuit is shorted to system power • Lockup solenoid 'A' connector is disconnected or damaged • Lockup solenoid 'A' is damaged or has failed • TCM or PCM has failed
DTC: P1753 **1T CCM, MIL: YES** **1996, 1997, 1998, 1999, 2000,** **2001, 2002, 2003, 2004, 2005,** **2006** **Models:** 2.2CL, 2.3CL, 3.0CL, 3.2CL, 3.2TL, TL, MDX, Integra, RSX **Engines:** All **Transmissions:** A/T	**A/T Lockup Solenoid Valve 'A' Circuit Malfunction** Vehicle driven in 1st, 2nd, 3rd and 4th gears, and the PCM detected an unexpected voltage condition on the Solenoid 'A' control circuit. **Note: The D4 lamp on the dash will flash once when this code is set.** **Possible Causes:** • Lockup solenoid 'A' control circuit is open or shorted to ground • Lockup solenoid 'A' circuit is shorted to system power • Lockup solenoid 'A' connector is disconnected or damaged • Lockup solenoid 'A' is damaged or has failed • TCM or PCM has failed
DTC: P1758 **1T CCM, MIL: YES** **1995, 1996, 1997, 1998, 1999,** **2000, 2001, 2002, 2003, 2004,** **2005, 2006** **Models:** NSX, NSX-T, 3.5RL, RL **Engines:** All **Transmissions:** A/T	**A/T Lockup Solenoid Valve 'B' Circuit Malfunction** Vehicle driven in 1st, 2nd, 3rd and 4th gears, and the PCM detected an unexpected voltage condition on the Solenoid 'B' circuit during the CCM test. **Note: The D4 lamp on the dash will flash twice when this code is set.** **Possible Causes:** • Lockup solenoid 'B' control circuit is open or shorted to ground • Lockup solenoid 'B' circuit is shorted to system power • Lockup solenoid 'B' connector is disconnected or damaged • Lockup solenoid 'B' is damaged or has failed • TCM or PCM has failed

DTC	Trouble Code Title, Conditions & Possible Causes
DTC: P1758 **1T CCM, MIL: YES** **1995, 1996, 1997, 1998** **Models:** 2.5TL **Engines:** All **Transmissions:** A/T	**A/T Lockup Solenoid Valve 'B' Circuit Malfunction** Vehicle driven in 1st, 2nd, 3rd and 4th gears, and the PCM detected an unexpected voltage condition on the Solenoid 'B' circuit during the CCM test. **Note: The D4 lamp on the dash will flash twice when this code is set.** **Possible Causes:** • Lockup solenoid 'B' control circuit is open or shorted to ground • Lockup solenoid 'B' circuit is shorted to system power • Lockup solenoid 'B' is damaged or has failed • TCM or PCM has failed
DTC: P1758 **1T CCM, MIL: YES** **1996, 1997, 1998, 1999, 2000,** **2001, 2002, 2003, 2004, 2005,** **2006** **Models:** 3.2CL, 3.2TL, TL, MDX, Integra, RSX **Engines:** All **Transmissions:** A/T	**A/T Lockup Solenoid Valve 'B' Circuit Malfunction** Vehicle driven in 1st, 2nd, 3rd and 4th gears, and the PCM detected an unexpected voltage condition on the Solenoid 'B' control circuit. **Note: The D4 lamp on the dash will flash twice when this code is set.** **Possible Causes:** • Lockup solenoid 'B' control circuit is open or shorted to ground • Lockup solenoid 'B' circuit is shorted to system power • Lockup solenoid 'B' connector is disconnected or damaged • Lockup solenoid 'B' is damaged or has failed • TCM or PCM has failed
DTC: P1768 **1T CCM, MIL: YES** **1995, 1996, 1997, 1998, 1999,** **2000, 2001, 2002, 2003, 2004,** **2005, 2006** **Models:** NSX, NSX-T, 3.5RL, RL **Engines:** All **Transmissions:** A/T	**Automatic Transaxle Linear Solenoid Circuit Malfunction** Key on or engine running; and the PCM detected an unexpected voltage condition on the A/T Linear Solenoid circuit to the TCM. Note: The D4 lamp on the dash will flash 16 times if this code is set. **Possible Causes:** • A/T linear solenoid control circuit to the TCM is open • A/T linear solenoid circuit to the TCM is shorted to ground • A/T linear solenoid connector is disconnected or damaged • A/T linear solenoid circuit to the TCM is shorted to VREF • TCM or PCM has failed
DTC: P1768 **1T CCM, MIL: YES** **1995, 1996, 1997, 1998** **Models:** 2.5TL **Engines:** All **Transmissions:** A/T	**Automatic Transaxle Linear Solenoid Circuit Malfunction** Key on or engine running; and the PCM detected an unexpected voltage condition on the A/T Linear Solenoid circuit to the TCM. Note: The D4 lamp on the dash will flash 16 times if this code is set. **Possible Causes:** • A/T linear solenoid control circuit to the TCM is open • A/T linear solenoid circuit to the TCM is shorted to ground • A/T linear solenoid connector is disconnected or damaged • A/T linear solenoid circuit to the TCM is shorted to VREF • TCM or PCM has failed
DTC: P1768 **1T CCM, MIL: YES** **1996, 1997, 1998, 1999, 2000,** **2001, 2002, 2003, 2004, 2005,** **2006** **Models:** 2.2CL, 2.3CL, 3.0CL, 3.2CL, 3.2TL, TL, MDX, RSX **Engines:** All **Transmissions:** A/T	**A/T Lockup Solenoid Valve 'A' Circuit Malfunction** Vehicle driven in 1st, 2nd, 3rd and 4th gears, and the PCM detected an unexpected voltage condition on the Solenoid 'A' control circuit. **Note: The D4 lamp on the dash will flash 16 times if this code is set.** **Possible Causes:** • Lockup solenoid 'A' control circuit is open or shorted to ground • Lockup solenoid 'A' circuit is shorted to system power • Lockup solenoid 'A' is damaged or has failed • TCM or PCM has failed
DTC: P1773 **1T CCM, MIL: YES** **1997, 1998, 1999** **Models:** 3.0CL **Engines:** All **Transmissions:** A/T	**A/T Lockup Solenoid Valve 'B' Circuit Malfunction** Vehicle driven in 1st, 2nd, 3rd and 4th gears, and the PCM detected an unexpected voltage condition on the Solenoid 'B' control circuit. **Note: The D4 lamp on the dash will blink when this code is set.** **Possible Causes:** • Lockup solenoid 'B' control circuit is open or shorted to ground • Lockup solenoid 'B' circuit is shorted to system power • Lockup solenoid 'B' is damaged or has failed • TCM or PCM has failed

DTC	Trouble Code Title, Conditions & Possible Causes
DTC: P1773 **1T CCM, MIL: YES** **1999, 2000** **Models:** Integra **Engines:** All **Transmissions:** A/T	**A/T Lockup Solenoid Valve 'B' Circuit Malfunction** Vehicle driven in 1st, 2nd, 3rd and 4th gears, and the PCM detected an unexpected voltage condition on the Solenoid 'B' control circuit. **Note: The D4 lamp on the dash will blink when this code is set.** **Possible Causes:** • Lockup solenoid 'B' control circuit is open or shorted to ground • Lockup solenoid 'B' circuit is shorted to system power • Lockup solenoid 'B' is damaged or has failed • TCM or PCM has failed
DTC: P1773 **1T CCM, MIL: YES** **2000, 2001, 2002, 2003, 2004,** **2005, 2006** **Models:** 3.2CL, 3.2TL, TL **Engines:** All **Transmissions:** A/T	**A/T Lockup Solenoid Valve 'B' Circuit Malfunction** Vehicle driven in 1st, 2nd, 3rd and 4th gears, and the PCM detected an unexpected voltage condition on the Solenoid 'B' control circuit. **Note: The D4 lamp on the dash will blink when this code is set.** **Possible Causes:** • Lockup solenoid 'B' control circuit is open or shorted to ground • Lockup solenoid 'B' circuit is shorted to system power • Lockup solenoid 'B' is damaged or has failed • TCM or PCM has failed
DTC: P1778 **1T CCM, MIL: YES** **2001, 2002, 2003, 2004, 2005,** **2006** **Models:** 3.2CL, 3.2TL, TL **Engines:** All **Transmissions:** A/T	**A/T Lockup Solenoid Valve 'C' Circuit Malfunction** Vehicle driven in 1st, 2nd, 3rd and 4th gears, and the PCM detected an unexpected voltage condition on the Solenoid 'C' control circuit. **Note: The D4 lamp on the dash will blink when this code is set.** **Possible Causes:** • Lockup solenoid 'C' control circuit is open or shorted to ground • Lockup solenoid 'C' circuit is shorted to system power • Lockup solenoid 'C' is damaged or has failed • TCM or PCM has failed
DTC: P1786 **1T CCM, MIL: YES** **1995, 1996, 1997, 1998** **Models:** 2.5TL **Engines:** All **Transmissions:** A/T	**A/T FAS Signal Circuit Malfunction** Key on or engine running; and the PCM detected an unexpected voltage condition on the A/T FAS signal circuit to the TCM. **Possible Causes:** • A/T FAS signal circuit to the TCM is open • A/T FAS signal circuit to the TCM is shorted to ground • A/T FAS signal circuit to the TCM is shorted to VREF • TCM or PCM has failed
DTC: P1786 **1T CCM, MIL: YES** **1996, 1997, 1998, 1999** **Models:** Integra **Engines:** All **Transmissions:** A/T	**A/T FAS Signal Circuit Malfunction** Key on or engine running; and the PCM detected an unexpected voltage condition on the A/T FAS signal circuit to the TCM. **Possible Causes:** • A/T FAS signal circuit to the TCM is open • A/T FAS signal circuit to the TCM is shorted to ground • A/T FAS signal circuit to the TCM is shorted to VREF • TCM or PCM has failed
DTC: P1787 **1T CCM, MIL: YES** **1995, 1996, 1997, 1998** **Models:** 2.5TL **Engines:** All **Transmissions:** A/T	**A/T FFS Signal Circuit Malfunction** Key on or engine running; and the PCM detected an unexpected voltage condition on the A/T FFS signal circuit to the TCM. **Possible Causes:** • A/T FFS signal circuit to the TCM is open • A/T FFS signal circuit to the TCM is shorted to ground • A/T FFS signal circuit to the TCM is shorted to VREF • TCM or PCM has failed
DTC: P1788 **1T CCM, MIL: YES** **1995, 1996, 1997, 1998, 1999,** **2000, 2001, 2002, 2003, 2004,** **2005** **Models:** NSX, NSX-T **Engines:** All **Transmissions:** A/T	**A/T ATSDLA Signal Circuit Malfunction** Key on or engine running; and the PCM detected an unexpected voltage condition on the A/T ATSDLA signal circuit to the TCM. **Note: The D4 lamp on the dash will flash 19 times if this code is set.** **Possible Causes:** • A/T ATSDLA signal circuit to the TCM is open • A/T ATSDLA signal circuit to the TCM is shorted to ground • A/T ATSDLA signal circuit to the TCM is shorted to VREF • TCM or PCM has failed

DTC	Trouble Code Title, Conditions & Possible Causes
DTC: P1790 **1T CCM, MIL: YES** 1995, 1996, 1997, 1998, 1999, 2000, 2001, 2002, 2003, 2004, 2005 **Models:** NSX, NSX-T **Engines:** All **Transmissions:** A/T	**A/T Throttle Position Sensor Circuit Malfunction** Key on or engine running; and the PCM detected an unexpected voltage condition on the TP sensor signal circuit to the TCM. **Note: The lockup clutch does not engage when this code is set.** **Note: The D4 lamp on the dash will flash 3 times if this code is set.** **Possible Causes:** • TP sensor signal circuit to the TCM is open • TP sensor signal circuit to the TCM is shorted to ground • TP sensor signal circuit to the TCM is shorted to VREF • TP sensor connector is disconnected or damaged • TCM or PCM has failed
DTC: P1790 **1T CCM, MIL: YES** 1995, 1996, 1997, 1998 **Models:** 2.5TL **Engines:** All **Transmissions:** A/T	**A/T Throttle Position Sensor Circuit Malfunction** Key on or engine running; and the PCM detected an unexpected voltage condition on the TP sensor signal circuit to the TCM. **Note: The lockup clutch does not engage when this code is set.** **Possible Causes:** • TP sensor signal circuit to the TCM is open • TP sensor signal circuit to the TCM is shorted to ground • TP sensor signal circuit to the TCM is shorted to VREF • TCM or PCM has failed
DTC: P1790 **1T CCM, MIL: YES** 1996, 1997, 1998, 1999 **Models:** Integra **Engines:** All **Transmissions:** A/T	**A/T Throttle Position Sensor Circuit Malfunction** Key on or engine running; and the PCM detected an unexpected voltage condition on the TP sensor signal circuit to the TCM. **Note: The lockup clutch does not engage when this code is set.** **Possible Causes:** • TP sensor signal circuit to the TCM is open • TP sensor signal circuit to the TCM is shorted to ground • TP sensor signal circuit to the TCM is shorted to VREF • TCM or PCM has failed
DTC: P1790 **2T PCM, MIL: YES** 1996, 1997, 1998, 1999 **Models:** SLX **Engines:** All **Transmissions:** A/T	**PCM ROM (Transmission Side) Check Sum Error** Key on, and the PCM detected a ROM Check Sum Error occurred in the Transmission Side of the controller for 1 second. **Note: For additional help with this code, view the Failure Records.** **Possible Causes:** • Reprogram the Transmission EEPROM • Recheck for the trouble code, and if it resets, the PCM may need to be replaced and reprogrammed
DTC: P1791 **1T CCM, MIL: YES** 1995, 1996, 1997, 1998 **Models:** 2.5TL **Engines:** All **Transmissions:** A/T	**A/T Vehicle Speed Sensor Circuit Malfunction** Vehicle driven for 15 seconds, and the PCM detected an unexpected voltage condition on the vehicle speed signal circuit to the TCM. **Note: The lockup clutch does not engage when this code is set.** **Note: The D4 lamp on the dash will flash 4 times if this code is set.** **Possible Causes:** • VSS signal circuit to the TCM is open • VSS signal circuit is shorted to ground • VSS signal circuit to the TCM is shorted to VREF • VSS connector is disconnected or damaged • TCM or PCM has failed
DTC: P1791 **1T CCM, MIL: YES** 1995, 1996, 1997, 1998, 1999, 2000, 2001, 2002, 2003, 2004, 2005, 2006 **Models:** NSX, NSX-T, 3.5RL, RL **Engines:** All **Transmissions:** A/T	**A/T Vehicle Speed Sensor Circuit Malfunction** Vehicle driven for 15 seconds, and the PCM detected an unexpected voltage condition on the vehicle speed signal circuit to the TCM. Note: The lockup clutch does not engage when this code is set. **Note: The D4 lamp on the dash will flash 4 times if this code is set.** **Possible Causes:** • VSS signal circuit to the TCM is open • VSS signal circuit is shorted to ground • VSS signal circuit to the TCM is shorted to VREF • VSS connector is disconnected or damaged • TCM or PCM has failed

DTC	Trouble Code Title, Conditions & Possible Causes

DTC: P1791
1T CCM, MIL: YES
1996, 1997, 1998, 1999
Models: 3.2TL, Integra
Engines: All
Transmissions: A/T
	A/T Vehicle Speed Sensor Circuit Malfunction
	Vehicle driven for 15 seconds, and the PCM detected an unexpected voltage condition on the vehicle speed signal circuit to the TCM.
	Note: The lockup clutch does not engage when this code is set.
	Note: The D4 lamp on the dash will flash 4 times if this code is set.
	Possible Causes:
	    • VSS signal circuit to the TCM is open
	    • VSS signal circuit is shorted to ground
	    • VSS signal circuit to the TCM is shorted to VREF
	    • VSS connector is disconnected or damaged
	    • TCM or PCM has failed

DTC: P1792
1T CCM, MIL: YES
1995, 1996, 1997, 1998, 1999,
2000, 2001, 2002, 2003, 2004,
2005
Models: NSX, NSX-T
Engines: All
Transmissions: A/T
	A/T Engine Coolant Temperature Sensor Circuit Malfunction
	Key on or engine running; and the PCM detected an unexpected voltage condition on the ECT sensor signal circuit to the TCM.
	Note: The lockup clutch does not engage when this code is set.
	Note: The D4 lamp on the dash will flash 10 times if this code is set.
	Possible Causes:
	    • ECT signal circuit to the TCM is open
	    • ECT signal circuit to the TCM is shorted to ground
	    • ECT signal circuit to the TCM is shorted to VREF
	    • ECT sensor connector is disconnected or damaged
	    • TCM or PCM has failed

DTC: P1792
1T CCM, MIL: YES
1995, 1996, 1997, 1998
Models: 2.5TL
Engines: All
Transmissions: A/T
	A/T Engine Coolant Temperature Sensor Circuit Malfunction
	Key on or engine running; and the PCM detected an unexpected voltage condition on the ECT sensor signal circuit to the TCM.
	Note: The lockup clutch does not engage when this code is set.
	Note: The D4 lamp on the dash will flash 10 times if this code is set.
	Possible Causes:
	    • ECT signal circuit to the TCM is open
	    • ECT signal circuit to the TCM is shorted to ground
	    • ECT signal circuit to the TCM is shorted to VREF
	    • ECT sensor connector is disconnected or damaged
	    • TCM or PCM has failed

DTC: P1792
1T CCM, MIL: YES
1996, 1997, 1998, 1999
Models: Integra
Engines: All
Transmissions: A/T
	A/T Engine Coolant Temperature Sensor Circuit Malfunction
	Key on or engine running; and the PCM detected an unexpected voltage condition on the ECT sensor signal circuit to the TCM.
	Note: The lockup clutch does not engage when this code is set.
	Note: The D4 lamp on the dash will flash 10 times if this code is set.
	Possible Causes:
	    • ECT signal circuit to the TCM is open
	    • ECT signal circuit to the TCM is shorted to ground
	    • ECT signal circuit to the TCM is shorted to VREF
	    • ECT sensor connector is disconnected or damaged
	    • TCM or PCM has failed

DTC: P1792
2T PCM, MIL: YES
1996, 1997, 1998, 1999
Models: SLX
Engines: All
Transmissions: A/T
	VCM EEPROM (Transmission Side) Check Sum Error
	Key on, and the PCM detected an EEPROM Check Sum Error occurred in the Transmission Side of the controller for 1 second.
	Note: For additional help with this code, view the Failure Records.
	Possible Causes:
	    • Reprogram the Transmission EEPROM
	    • Recheck for the trouble code, and if it resets, the PCM may need to be replaced and reprogrammed

DTC: P1793
1T CCM, MIL: YES
1995, 1996, 1997, 1998, 1999,
2000, 2001, 2002, 2003, 2004,
2005
Models: NSX, NSX-T
Engines: All
Transmissions: A/T
	A/T Manifold Absolute Pressure Sensor Circuit Malfunction
	Key on or engine running; and the PCM detected an unexpected voltage condition on the MAP sensor signal circuit to the TCM.
	Note: The D4 lamp on the dash will flash 12 times if this code is set.
	Possible Causes:
	    • MAP signal circuit to the TCM is open
	    • MAP signal circuit to the TCM is shorted to ground
	    • MAP signal circuit to the TCM is shorted to VREF
	    • MAP sensor connector is disconnected or damaged
	    • TCM or PCM has failed

DTC	Trouble Code Title, Conditions & Possible Causes
DTC: P1794 **1T CCM, MIL: YES** 1995, 1996, 1997, 1998 **Models:** 2.5TL **Engines:** All **Transmissions:** A/T	**A/T Barometric Pressure Sensor Circuit Malfunction** Key on or engine running; and the PCM detected an unexpected voltage condition on the BARO sensor signal circuit to the TCM. The BARO sensor is built into the PCM on this vehicle application. **Note: The D4 lamp on the dash will flash 13 times if this code is set.** **Possible Causes:** • BARO signal circuit to the TCM is open or shorted to ground • BARO signal circuit to the TCM is shorted to VREF • BARO sensor connector is disconnected or damaged • TCM or PCM has failed
DTC: P1794 **1T CCM, MIL: YES** 1996, 1997, 1998, 1999 **Models:** Integra **Engines:** All **Transmissions:** A/T	**A/T Barometric Pressure Sensor Circuit Malfunction** Key on or engine running; and the PCM detected an unexpected voltage condition on the BARO sensor signal circuit to the TCM. The BARO sensor is built into the PCM on this vehicle application. **Note: The D4 lamp on the dash will flash 13 times if this code is set.** **Possible Causes:** • BARO signal circuit to the TCM is open or shorted to ground • BARO signal circuit to the TCM is shorted to VREF • BARO sensor connector is disconnected or damaged • TCM or PCM has failed
DTC: P1795 **1T CCM, MIL: YES** 1995, 1996, 1997, 1998, 1999, 2000, 2001, 2002, 2003, 2004, 2005 **Models:** NSX, NSX-T **Engines:** All **Transmissions:** A/T	**A/T Accelerator Position Sensor Circuit Malfunction** Key on or engine running; and the PCM detected an unexpected voltage condition on the APP sensor signal circuit to the TCM. **Note: The D4 lamp on the dash will flash 20 times if this code is set.** **Possible Causes:** • APP signal circuit to the TCM is open • APP signal circuit to the TCM is shorted to ground • APP signal circuit to the TCM is shorted to VREF • APP sensor connector is disconnected or damaged • TCM or PCM has failed
DTC: P1835 **1T CCM, MIL: NO** 1998, 1999 **Models:** SLX **Engines:** 3.5L VIN X **Transmissions:** A/T	**Kick Down Switch Circuit Low Input** DTC P0122 and P0123 not set, engine started, engine running, throttle position less than 70%, and the PCM detected the Kick Down Switch remained in "on" position during the CCM test. **Note: For additional help with this code, view the Failure Records.** **Possible Causes:** • Kick down switch signal circuit is shorted to ground • Kick down switch ground circuit open between switch & ground • Kick down switch is damaged, out-of-adjustment or has failed • PCM has failed
DTC: P1850 **1T CCM, MIL: NO** 1998, 1999 **Models:** SLX **Engines:** 3.5L VIN X **Transmissions:** A/T	**A/T Brake Band Apply Solenoid Circuit Malfunction** Engine started, then with A/T Brake Band Apply solenoid commanded "on", and PCM detected the solenoid control signal was 12v, or with A/T Brake Band Apply solenoid commanded "off", the solenoid control signal was 1.34 to 1.56 seconds. **Note: This PCN controls this solenoid with a PWM type of signal.** **Possible Causes:** • A/T Brake band solenoid circuit is open or shorted to ground • A/T Brake band solenoid High circuit open or shorted to ground • A/T brake band apply solenoid is damaged or has failed • PCM has failed
DTC: P1860 **2T CCM, MIL: YES** 1998, 1999 **Models:** SLX **Engines:** 3.5L VIN X **Transmissions:** A/T	**Torque Converter Clutch PWM Solenoid Circuit Malfunction** DTC P0751, P0752, P0753, P0756, P0757 and P0758 not set, engine started, engine running, then with the TCC solenoid commanded "on", the PCM detected the solenoid control circuit signal was 12v, or with the TCC solenoid commanded "off", it detected the TCC control circuit signal was 0v for 1.25 seconds. **Possible Causes:** • TCC PWM solenoid control circuit open or shorted to ground • TCC PWM solenoid control circuit is shorted to power • TCC PWM solenoid is damaged or has failed • PCM has failed

DTC	Trouble Code Title, Conditions & Possible Causes
DTC: P1870 **2T CCM, MIL: YES** **1998, 1999** **Models:** SLX **Engines:** 3.2L VIN V, 3.5L VIN X **Transmissions:** A/T	**Transmission Component Slipping Malfunction** DTC P0722, P0723, P0742, P0751, P0752, P0753, P0756, P0757, P0758, P1860 and P1870 not set, engine started, then driven to a speed of 15-58 mph, engine speed from 1000-3500 rpm, TP sensor signal from 15-99%, MAP sensor signal from 0-70 kPa, 50 < Engine Torque < 300 Nm, gear selector in D4, TFT sensor signal from 68-302°F, speed ratio at 0.6-0.95, and the PCM detected the TCC slip speed was 250-800 rpm (event occurred 3 times within 7 seconds). **Possible Causes:** • Engine speed signal circuit open or shorted (intermittent fault) • Internal transmission component problem • TCC PWM or Shift Solenoids have failed (mechanical fault) • TR switch is damaged, out-of-adjustment or has failed

OBD II Trouble Code List (P2xxx Codes)

DTC	Trouble Code Title
DTC: P2101 **1T CCM, MIL: YES** **2003, 2004, 2005, 2006** **Models:** MDX **Engines:** All **Transmissions:** A/T, M/T	**Throttle Actuator System Malfunction** Engine started, engine running (at the speed indicated in the trouble code Freeze Frame data), and the PCM detected a signal from the TAC module indicating a problem present in the TAC system. **Possible Causes:** • Check for loose connections at TAC module (intermittent fault) • Key off - check the throttle blade for free movement (be careful not to pinch your finger while performing this inspection step) Clean the throttle body as needed and recheck for the code • Update the PCM to the latest software. If the problem is still present, substitute a known good PCM and then retest. If the problem goes away, the original PCM has failed.
DTC: P2108 **1T CCM, MIL: YES** **2003, 2004, 2005, 2006** **Models:** MDX **Engines:** All **Transmissions:** A/T, M/T	**Throttle Actuator Control Module Malfunction** Key on or engine running; and the PCM detected a signal from the TAC module indicating a problem present in the TAC module. **Possible Causes:** • Check for loose connections at TAC module (intermittent fault) • Check for loose connections at the PCM (intermittent fault) • Substitute a known good throttle body assembly and retest. If the problem goes away, the original throttle body has failed.
DTC: P2118 **1T CCM, MIL: YES** **2003, 2004, 2005, 2006** **Models:** MDX **Engines:** All **Transmissions:** A/T, M/T	**Throttle Actuator Motor Current Range/Performance** Engine started, vehicle driven with several changes in the throttle position angle, and the PCM detected a signal from the TAC module indicating a TAC motor current problem had occurred. **Possible Causes:** • Check for loose connections at TAC module (intermittent fault) • Substitute a known good throttle body assembly and retest. If the problem goes away, the original throttle body has failed. • Update the PCM to the latest software. If the problem is still present, substitute a known good PCM and then retest. If the problem goes away, the original PCM has failed.
DTC: P2122 **1T CCM, MIL: YES** **2003, 2004, 2005, 2006** **Models:** MDX **Engines:** All **Transmissions:** A/T, M/T	**Accelerator Pedal Position Sensor 1 Circuit Low Input** Key on or engine running; and the PCM detected an unexpected low voltage condition (less than 0.1v) on the APP1 sensor signal circuit. If P0641 (low VREF) is set along with P2127, repair the P0641 first. **Possible Causes:** • Check for loose connections at APP1 sensor (intermittent fault) • APP1 Sensor VREF circuit is open • APP1 Sensor signal circuit is shorted to ground • APP1 Sensor is damaged or has failed. • Update the PCM to the latest software. If the problem is still present, substitute a known good PCM and then retest. If the problem goes away, the original PCM has failed.
DTC: P2123 **1T CCM, MIL: YES** **2003, 2004, 2005, 2006** **Models:** MDX **Engines:** All **Transmissions:** A/T, M/T	**Accelerator Pedal Position Sensor 1 Circuit High Input** Key on or engine running; and the PCM detected an unexpected high voltage condition (more than 4.85v) on the APP1 sensor signal circuit. **Possible Causes:** • Check for loose connections at APP1 sensor (intermittent fault) • APP1 Sensor signal circuit is shorted to the VREF circuit • APP1 Sensor signal circuit is open between sensor and PCM • APP1 Sensor is damaged or has failed • Update the PCM to the latest software. If the problem is still present, substitute a known good PCM and then retest. If the problem goes away, the original PCM has failed.
DTC: P2127 **1T CCM, MIL: YES** **2003, 2004, 2005, 2006** **Models:** MDX **Engines:** All **Transmissions:** A/T, M/T	**Accelerator Pedal Position Sensor 2 Circuit Low Input** Key on or engine running; and the PCM detected an unexpected low voltage condition (less than 0.1v) on the APP2 sensor signal circuit. If P0641 (low VREF) is set along with P2127, repair the P0641 first. **Possible Causes:** • Check for loose connections at APP2 sensor (intermittent fault) • APP2 Sensor VREF circuit is open • APP2 Sensor signal circuit is shorted to ground • APP2 Sensor is damaged or has failed. • Update the PCM to the latest software. If the problem is still present, substitute a known good PCM and then retest. If the problem goes away, the original PCM has failed.

DTC	Trouble Code Title, Conditions & Possible Causes

DTC: P2128
1T CCM, MIL: YES
2003, 2004, 2005, 2006
Models: MDX
Engines: All
Transmissions: A/T, M/T
	Accelerator Pedal Position Sensor 2 Circuit High Input
	Key on or engine running; and the PCM detected an unexpected high voltage condition (more than 4.80v) on the APP2 sensor signal circuit.
	Possible Causes:
	    • Check for loose connections at APP2 sensor (intermittent fault)
	    • APP2 Sensor signal circuit is shorted to the VREF circuit
	    • APP2 Sensor signal circuit is open between sensor and PCM
	    • APP2 Sensor is damaged or has failed
	    • Update the PCM to the latest software. If the problem is still present, substitute a known good PCM and then retest. If the problem goes away, the original PCM has failed.

DTC: P2135
1T CCM, MIL: YES
2003, 2004, 2005, 2006
Models: MDX
Engines: All
Transmissions: A/T, M/T
	Throttle Position Sensor 1/2 Incorrect Voltage Correlation
	Key on or engine running; and the PCM detected an incorrect voltage correlation on the TP1/2 sensor circuit during the CCM test.
	Possible Causes:
	    • Check for loose connections at TAC module (intermittent fault)
	    • Clear the codes, and then use the Scan Tool ETCS test to check for any related trouble codes. If any codes are present, repair these trouble codes and then retest for the original code.

DTC: P2138
1T CCM, MIL: YES
2003, 2004, 2005, 2006
Models: MDX
Engines: All
Transmissions: A/T, M/T
	Accelerator Pedal Position Sensor 1/2 Incorrect Voltage Correlation
	Key on or engine running; and the PCM detected an incorrect voltage correlation on the APP1/2 sensor circuit in the CCM test.
	Possible Causes:
	    • Check for loose connections at APP sensor (intermittent fault)
	    • Clear the codes, and then use the Scan Tool ETCS test to check for any related trouble codes. If any codes are present, repair these trouble codes and then retest for the original code.
	    • Update the PCM to the latest software. If the problem is still present, substitute a known good PCM and then retest. If the problem goes away, the original PCM has failed.

DTC: P2176
1T CCM, MIL: YES
2003, 2004, 2005, 2006
Models: MDX
Engines: All
Transmissions: A/T, M/T
	Throttle Actuator Control System Idle Position Not Learned
	Key on or engine running; and the PCM detected an incorrect voltage correlation on the APP1/2 sensor circuit in the CCM test.
	Possible Causes:
	    • Check for loose connections at TAC module (intermittent fault)
	    • Clear the codes, and then use the Scan Tool ETCS test to check the throttle valve movement with the key on, engine off and the intake air duct removed (keep your fingers clear during the inspection step). If the throttle valve does not operate to its fully closed position, substitute a known good throttle assembly and retest for the same code. If okay, clean the throttle body.
	    • Update the PCM to the latest software. If the problem is still present, substitute a known good PCM and then retest. If the problem goes away, the original PCM has failed.

DTC: P2195
1T CCM, MIL: YES
2003, 2004, 2005, 2006
Models: MDX
Engines: All
Transmissions: A/T, M/T
	Air Fuel Sensor 1 (Bank 1 Sensor 1) Signal Stuck Lean
	Engine started, vehicle driven for several minutes at various cruise speeds, and the PCM detected the AFS1 signal was stuck "lean".
	Possible Causes:
	    • Check for a loose AFS1 assembly (not tightened) at the exhaust pipe connection. Clear the codes and retest the code.
	    • If DTC P2195 is present, replace the AFS1 and then retest.
	    • If this code is an intermittent problem, check for loose connections at the AFS1 connector.

DTC: P2197
1T CCM, MIL: YES
2003, 2004, 2005, 2006
Models: MDX
Engines: All
Transmissions: A/T, M/T
	Air Fuel Sensor 2 (Bank 2 Sensor 1) Signal Stuck Lean
	Engine started, vehicle driven for several minutes at various cruise speeds, and the PCM detected the AFS2 signal was stuck "lean".
	Possible Causes:
	    • Check for a loose AFS2 assembly (not tightened) at the exhaust pipe connection. Clear the codes and retest the code.
	    • If DTC P2197 is present, replace the AFS1 and then retest.
	    • If this code is an intermittent problem, check for loose connections at the AFS2 connector.

DTC: P2199
2T CCM, MIL: YES
2003, 2004, 2005, 2006
Models: MDX
Engines: All
Transmissions: A/T, M/T
	Intake Air Temperature Sensor 1/2 Voltage Correlation
	Engine started cold (both IAT sensors less than 77°F), and the PCM detected the voltage correlation difference between the IAT Sensor 1 and IAT Sensor 2 was more than 45°F during the CCM test period.
	Possible Causes:
	    • Check for loose connections at the IAT 1/2 (intermittent fault)
	    • Let the vehicle sit for at least 8 hours. If the code resets, replace the IAT sensor that has the largest difference from the room temperature (after testing at KOEO with a cold engine).

DTC: P2227
1T CCM, MIL: YES
2003, 2004, 2005, 2006
Models: MDX
Engines: All
Transmissions: A/T, M/T
	BARO Sensor Circuit Range/Performance
	DTC P0107, P0108, P1128 and P1129 not set, ECT sensor input over 158°F, vehicle driven in Drive at a throttle angle of 12-20 degrees, and the PCM detected a problem in the BARO sensor.
	Possible Causes:
	    • Check for loose connections at the PCM (intermittent fault)
	    • Update the PCM to the latest software. If the problem is still present, substitute a known good PCM and then retest. If the problem goes away, the original PCM has failed.

DTC	Trouble Code Title, Conditions & Possible Causes
DTC: P2228 **1T CCM, MIL: YES** **2003, 2004, 2005, 2006** **Models:** MDX **Engines:** All **Transmissions:** A/T, M/T	**BARO Sensor Circuit Low Input** Key on or engine running; and the PCM detected an unexpected low voltage (less than 1.58v [53 kPa]) on the BARO sensor signal circuit. **Possible Causes:** • Check for loose connections at the PCM (intermittent fault) • Update the PCM to the latest software. If the problem is still present, substitute a known good PCM and then retest. If the problem goes away, the original PCM has failed.
DTC: P2229 **1T CCM, MIL: YES** **2003, 2004, 2005, 2006** **Models:** MDX **Engines:** All **Transmissions:** A/T, M/T	**BARO Sensor Circuit High Input** Key on or engine running; and the PCM detected an unexpected high voltage (more than 4.50v [160 kPa]) on the BARO sensor signal circuit. **Possible Causes:** • Check for loose connections at the PCM (intermittent fault) • Update the PCM to the latest software. If the problem is still present, substitute a known good PCM and then retest. If the problem goes away, the original PCM has failed.
DTC: P2237 **1T CCM, MIL: YES** **2003, 2004, 2005, 2006** **Models:** MDX **Engines:** All **Transmissions:** A/T, M/T	**Air Fuel Sensor 1 (Bank 1 Sensor 1) I/P Line High Voltage** DTC P0134 and P2237 not set, engine started and engine running, and the PCM detected an unexpected high voltage condition on the AFS1 I/P circuit during the CCM test. **Possible Causes:** • Check for loose connections at the AFS1 unit (intermittent fault) • AFS1 ground circuit is open • AFS1 assembly is damaged or has failed • Update the PCM to the latest software. If the problem is still present, substitute a known good PCM and then retest. If the problem goes away, the original PCM has failed.
DTC: P2238 **1T CCM, MIL: YES** **2003, 2004, 2005, 2006** **Models:** MDX **Engines:** All **Transmissions:** A/T, M/T	**Air Fuel Sensor 1 (Bank 1 Sensor 1) I/P Line Low Voltage** DTC P0134 and P2237 not set, engine started and engine running, and the PCM detected an unexpected low voltage condition on the AFS1 I/P circuit during the CCM test. **Possible Causes:** • Check for loose connections at the AFS1 unit (intermittent fault) • AFS1 signal circuit is shorted to ground • AFS1 assembly is damaged or has failed • Update the PCM to the latest software. If the problem is still present, substitute a known good PCM and then retest. If the problem goes away, the original PCM has failed.
DTC: P2240 **1T CCM, MIL: YES** **2003, 2004, 2005, 2006** **Models:** MDX **Engines:** All **Transmissions:** A/T, M/T	**Air Fuel Sensor 2 (Bank 2 Sensor 1) I/P Line High Voltage** DTC P0154 and P2240 not set, engine started and engine running, and the PCM detected an unexpected high voltage condition on the AFS2 I/P circuit during the CCM test. **Possible Causes:** • Check for loose connections at the AFS2 unit (intermittent fault) • AFS2 ground circuit is open • AFS2 assembly is damaged or has failed • Update the PCM to the latest software. If the problem is still present, substitute a known good PCM and then retest. If the problem goes away, the original PCM has failed.
DTC: P2241 **1T CCM, MIL: YES** **2003, 2004, 2005, 2006** **Models:** MDX **Engines:** All **Transmissions:** A/T, M/T	**Air Fuel Sensor 2 (Bank 2 Sensor 1) I/P Line Low Voltage** DTC P0154 and P2240 not set, engine started and engine running, and the PCM detected an unexpected high voltage condition on the AFS2 I/P circuit during the CCM test. **Possible Causes:** • Check for loose connections at the AFS2 unit (intermittent fault) • AFS2 ground circuit is open • AFS2 assembly is damaged or has failed • Update the PCM to the latest software. If the problem is still present, substitute a known good PCM and then retest. If the problem goes away, the original PCM has failed.
DTC: P2243 **1T CCM, MIL: YES** **2003, 2004, 2005, 2006** **Models:** MDX **Engines:** All **Transmissions:** A/T, M/T	**Air Fuel Sensor 1 (Bank 1 Sensor 1) VCENT Line High Voltage** DTC P2243 not set, engine started and engine running, and the PCM detected an unexpected high voltage condition on the AFS1 VCENT circuit during the CCM test. **Possible Causes:** • Check for loose connections at the AFS1 unit (intermittent fault) • AFS1 VCENT ground circuit is open (Circuit A54) • AFS1 assembly is damaged or has failed • Update the PCM to the latest software. If the problem is still present, substitute a known good PCM and then retest. If the problem goes away, the original PCM has failed.

DTC	Trouble Code Title, Conditions & Possible Causes
DTC: P2245 **1T CCM, MIL: YES** **2003, 2004, 2005, 2006** **Models:** MDX **Engines:** All **Transmissions:** A/T, M/T	**Air Fuel Sensor 1 (Bank 1 Sensor 1) VCENT Line Low Voltage** Engine started, and the PCM detected an unexpected low voltage condition on the AFS1 VCENT circuit during the CCM test. **Possible Causes:** • Check for loose connections at the AFS1 unit (intermittent fault) • AFS1 VCENT signal circuit is shorted to ground • AFS1 assembly is damaged or has failed • Update the PCM to the latest software. If the problem is still present, substitute a known good PCM and then retest. If the problem goes away, the original PCM has failed.
DTC: P2247 **1T CCM, MIL: YES** **2003, 2004, 2005, 2006** **Models:** MDX **Engines:** All **Transmissions:** A/T, M/T	**Air Fuel Sensor 2 (Bank 2 Sensor 1) VCENT Line High Voltage** DTC P2247 not set, engine started and engine running, and the PCM detected an unexpected high voltage condition on the AFS2 VCENT circuit during the CCM test. **Possible Causes:** • Check for loose connections at the AFS2 unit (intermittent fault) • AFS1 VCENT ground circuit is open (Circuit A50) • AFS1 assembly is damaged or has failed • Update the PCM to the latest software. If the problem is still present, substitute a known good PCM and then retest. If the problem goes away, the original PCM has failed.
DTC: P2249 **1T CCM, MIL: YES** **2003, 2004, 2005, 2006** **Models:** MDX **Engines:** All **Transmissions:** A/T, M/T	**Air Fuel Sensor 2 (Bank 2 Sensor 1) VCENT Line Low Voltage** Engine started, and the PCM detected an unexpected low voltage condition on the AFS2 VCENT circuit during the CCM test. **Possible Causes:** • Check for loose connections at the AFS2 unit (intermittent fault) • AFS2 VCENT signal circuit is shorted to ground • AFS2 assembly is damaged or has failed • Update the PCM to the latest software. If the problem is still present, substitute a known good PCM and then retest. If the problem goes away, the original PCM has failed.
DTC: P2251 **1T CCM, MIL: YES** **2003, 2004, 2005, 2006** **Models:** MDX **Engines:** All **Transmissions:** A/T, M/T	**Air Fuel Sensor 1 (Bank 1 Sensor 1) VS Line High Voltage** Engine started, and the PCM detected an unexpected high voltage condition on the AFS1 VS circuit during the CCM test. **Possible Causes:** • Check for loose connections at the AFS1 unit (intermittent fault) • AFS1 VS circuit is open (Circuit A53) • AFS1 assembly is damaged or has failed • Update the PCM to the latest software. If the problem is still present, substitute a known good PCM and then retest. If the problem goes away, the original PCM has failed.
DTC: P2252 **1T CCM, MIL: YES** **2003, 2004, 2005, 2006** **Models:** MDX **Engines:** All **Transmissions:** A/T, M/T	**Air Fuel Sensor 1 (Bank 1 Sensor 1) VS Line Low Voltage** Engine started, and the PCM detected an unexpected low voltage condition on the AFS1 VS circuit during the CCM test. **Possible Causes:** • Check for loose connections at the AFS1 unit (intermittent fault) • AFS1 VS circuit is shorted to ground • AFS1 assembly is damaged or has failed • Update the PCM to the latest software. If the problem is still present, substitute a known good PCM and then retest. If the problem goes away, the original PCM has failed.
DTC: P2254 **1T CCM, MIL: YES** **2003, 2004, 2005, 2006** **Models:** MDX **Engines:** All **Transmissions:** A/T, M/T	**Air Fuel Sensor 2 (Bank 2 Sensor 1) VS Line High Voltage** Engine started, and the PCM detected an unexpected high voltage condition on the AFS2 VS circuit during the CCM test. **Possible Causes:** • Check for loose connections at the AFS1 unit (intermittent fault) • AFS2 VS circuit is open (Circuit A49) • AFS2 assembly is damaged or has failed • Update the PCM to the latest software. If the problem is still present, substitute a known good PCM and then retest. If the problem goes away, the original PCM has failed.
DTC: P2255 **1T CCM, MIL: YES** **2003, 2004, 2005, 2006** **Models:** MDX **Engines:** All **Transmissions:** A/T, M/T	**Air Fuel Sensor 2 (Bank 2 Sensor 1) VS Line Low Voltage** Engine started, and the PCM detected an unexpected low voltage condition on the AFS2 VS circuit during the CCM test. **Possible Causes:** • Check for loose connections at the AFS2 unit (intermittent fault) • AFS2 VS circuit is shorted to ground • AFS2 assembly is damaged or has failed • Update the PCM to the latest software. If the problem is still present, substitute a known good PCM and then retest. If the problem goes away, the original PCM has failed.

DTC	Trouble Code Title, Conditions & Possible Causes
DTC: P2279 **1T CCM, MIL: YES** **2003, 2004, 2005, 2006** **Models:** MDX **Engines:** All **Transmissions:** A/T, M/T	**Intake Air System Leak** DTC P0443 not set, engine started and engine running, and the PCM detected a problem in the intake Air System (a vacuum leak). **Possible Causes:** • Check for a vacuum leak at the PCV valve or vacuum hose • Check for a vacuum leak at EVAP purge valve vacuum hose • Check for a vacuum leak the throttle body assembly • Check for a vacuum leak at the brake booster hose • Check for a vacuum leak at the engine mount control solenoid vacuum hose • Check the camshaft timing to determine if it is correct
DTC: P2413 **2T CCM, MIL: YES** **2003, 2004, 2005, 2006** **Models:** MDX **Engines:** All **Transmissions:** A/T, M/T	**EGR System Malfunction** DTC P0651 not set, engine started and engine running, and the PCM detected a problem in the intake Air System (a vacuum leak). **Possible Causes:** • Check for loose connections at EGR valve (intermittent fault) • EGR valve position sensor ground circuit is open • EGR valve position sensor VREF circuit is open or shorted • EGR valve position sensor signal circuit is open or shorted • EGR valve position sensor is damaged or has failed
DTC: P2422 **2T CCM, MIL: YES** **2003, 2004, 2005, 2006** **Models:** MDX **Engines:** All **Transmissions:** A/T, M/T	**EVAP Canister Vent Shut Valve Close Malfunction** DTC P0651 not set, engine started and engine running, and the PCM detected a problem in the intake Air System (a vacuum leak). **Possible Causes:** • Check for loose connections at EVAP canister vent shut valve (intermittent fault) • EVAP canister is clogged or restricted • EVAP canister vent shut valve is damaged or has failed
DTC: P2552 **2T CCM, MIL: YES** **2003, 2004, 2005, 2006** **Models:** MDX **Engines:** All **Transmissions:** A/T, M/T	**Throttle Actuator Control Module Relay Malfunction** Key on or engine running; and the PCM detected an unexpected voltage condition on the TAC module relay control circuit. **Possible Causes:** • Check for loose connections at the TAC module relay (intermittent fault) • TAC module relay control circuit is shorted to power • Throttle body assembly is damaged or has failed
DTC: P2610 **1T CCM, MIL: YES** **2003, 2004, 2005, 2006** **Models:** MDX **Engines:** All **Transmissions:** A/T, M/T	**24 Hour Time Malfunction** Key on or engine running; and the PCM detected an unexpected voltage condition on the TAC module relay control circuit. **Possible Causes:** • Clear codes and retest. If the code does not reset, the problem was an intermittent fault and is not present at this time. • Update the PCM to the latest software. If the problem is still present, substitute a known good PCM and then retest. If the problem goes away, the original PCM has failed.
DTC: P2627 **1T CCM, MIL: YES** **2003, 2004, 2005, 2006** **Models:** MDX **Engines:** All **Transmissions:** A/T, M/T	**Air Fuel Sensor 1 (Bank 1 Sensor 1) Label Circuit Low Input** Engine started, vehicle driven at cruise speed for 2-3 minutes in closed loop, and the PCM detected an unexpected low voltage condition on the AFS1 Label circuit during the CCM test. **Possible Causes:** • Clear the codes and recheck for DTC P2627. If it does, check for a short to ground condition on Circuit A56. • Check for loose connections at the AFS1 (intermittent fault) • If DTC P0107, P0112, P0117, P0452, P0700, P2122, P2128 or P02630 are set, repair these trouble codes first and then retest to determine if DTC P2627 is still present.
DTC: P2628 **1T CCM, MIL: YES** **2003, 2004, 2005, 2006** **Models:** MDX **Engines:** All **Transmissions:** A/T, M/T	**Air Fuel Sensor 1 (Bank 1 Sensor 1) Label Circuit High Input** Engine started, vehicle driven at cruise speed for 2-3 minutes in closed loop, and the PCM detected an unexpected high voltage condition on the AFS1 Label circuit during the CCM test. **Possible Causes:** • Check for loose connections at the AFS1 (intermittent fault) • Check for an open circuit condition on Circuit A56 • AFS2 assembly is damaged or has failed
DTC: P2630 **1T CCM, MIL: YES** **2003, 2004, 2005, 2006** **Models:** MDX **Engines:** All **Transmissions:** A/T, M/T	**Air Fuel Sensor 2 (Bank 2 Sensor 1) Label Circuit Low Input** Engine started, vehicle driven at cruise speed for 2-3 minutes in closed loop, and the PCM detected an unexpected low voltage condition on the AFS2 Label circuit during the CCM test. **Possible Causes:** • Clear the codes and recheck for DTC P2630. If it does, check for a short to ground condition on Circuit A52. • Check for loose connections at the AFS1 (intermittent fault) • If DTC P0107, P0112, P0117, P0452, P0700, P2122, P2127 or P02628 are set, repair these trouble codes first and then retest to determine if DTC P2630 is still present.

DTC	Trouble Code Title, Conditions & Possible Causes
DTC: P2631 **1T CCM, MIL: YES** **2003, 2004, 2005, 2006** **Models:** MDX **Engines:** All **Transmissions:** A/T, M/T	**Air Fuel Sensor 2 (Bank 2 Sensor 1) Label Circuit High Input** Engine started, vehicle driven at cruise speed for 2-3 minutes in closed loop, and the PCM detected an unexpected high voltage condition on the AFS2 Label circuit during the CCM test. **Possible Causes:** • Check for loose connections at the AFS2 (intermittent fault) • Check for an open circuit condition on Circuit A52 • AFS2 assembly is damaged or has failed
DTC: P2646 **1T CCM, MIL: YES** **2003, 2004, 2005, 2006** **Models:** MDX **Engines:** All **Transmissions:** A/T, M/T	**VTEC Oil Pressure Switch Circuit Low Input** Engine started, and the PCM detected an unexpected low voltage condition on the VTEC oil pressure switch circuit in the CCM test. **Possible Causes:** • Check for loose connections at VTEC switch (intermittent fault) • VTEC oil pressure switch circuit is shorted to ground • VTEC oil pressure switch is damaged or has failed • PCM has failed (substitute known good PCM and then retest)
DTC: P2647 **1T CCM, MIL: YES** **2003, 2004, 2005, 2006** **Models:** MDX **Engines:** All **Transmissions:** A/T, M/T	**VTEC Oil Pressure Switch Circuit High Input** Engine started, and the PCM detected an unexpected high voltage condition on the VTEC oil pressure switch circuit in the CCM test. **Possible Causes:** • Check for loose connections at VTEC switch (intermittent fault) • VTEC oil pressure switch circuit is shorted to ground • VTEC oil pressure switch is damaged or has failed • PCM has failed (substitute known good PCM and then retest)
DTC: P2648 **1T CCM, MIL: YES** **2003, 2004, 2005, 2006** **Models:** MDX **Engines:** All **Transmissions:** A/T, M/T	**VTEC Solenoid Valve Circuit Low Input** Key on or engine running; and the PCM detected an unexpected low voltage condition on the VTEC solenoid valve circuit during the test. The VTEC solenoid resistance at 68°F is 14-30 ohms. **Possible Causes:** • Check for loose connections at VTEC solenoid valve (intermittent fault) • VTEC solenoid valve control circuit is shorted to ground • VTEC solenoid valve is damaged or has failed • PCM has failed (substitute known good PCM and then retest)
DTC: P2649 **1T CCM, MIL: YES** **2003, 2004, 2005, 2006** **Models:** MDX **Engines:** All **Transmissions:** A/T, M/T	**VTEC Solenoid Valve Circuit High Input** Key on or engine running; and the PCM detected an unexpected high voltage condition on the VTEC solenoid valve circuit in the test. The VTEC solenoid resistance at 68°F is 14-30 ohms. **Possible Causes:** • Check for loose connections at VTEC solenoid valve (intermittent fault) • VTEC solenoid valve control circuit is open • VTEC solenoid valve is damaged or has failed • PCM has failed (substitute known good PCM and then retest)
DTC: P2A00 **1T CCM, MIL: YES** **2003, 2004, 2005, 2006** **Models:** MDX **Engines:** All **Transmissions:** A/T, M/T	**Air Fuel Sensor 1 (Bank 1 Sensor 1) Range/Performance** Engine started, vehicle driven at over 45 mph for 2-3 minutes followed by a deceleration period of 5 seconds with the throttle closed, and the PCM detected a problem in the AFS1 signal (it was not operating with a range of 0.8-1.2v during the test period). **Possible Causes:** • Check for loose connections at the AFS1 unit (intermittent fault) • AFS1 assembly is contaminated or has failed
DTC: P2A02 **1T CCM, MIL: YES** **2003, 2004, 2005, 2006** **Models:** MDX **Engines:** All **Transmissions:** A/T, M/T	**Air Fuel Sensor 2 (Bank 2 Sensor 1) Range/Performance** Engine started, vehicle driven at over 45 mph for 2-3 minutes followed by a deceleration period of 5 seconds with the throttle closed, and the PCM detected a problem in the AFS2 signal (it was not operating with a range of 0.8-1.2v during the test period). **Possible Causes:** • Check for loose connections at the AFS1 unit (intermittent fault) • AFS2 assembly is contaminated or has failed

OBD II Trouble Code List (U0xxx Codes)

DTC	Trouble Code Title
DTC: U0073 **1T PCM, MIL: YES** **2003, 2004, 2005, 2006** **Models:** MDX, RSX **Engines:** All **Transmissions:** A/T, M/T	**FCAN Malfunction (BUS Off)** DTC U0122 and U0155 not set; and a message from a learned ID number was not detected for the five seconds from the Vehicle Navigation Module (VTM) or the VSA control module. **Possible Causes:** • Check for loose connections at the VSA control unit (intermittent fault) • Communication is open, shorted to ground or to B+ • VSA has failed (substitute known good VSA and then retest)
DTC: U0107 **1T PCM, MIL: YES** **2003, 2004, 2005, 2006** **Models:** MDX, RSX **Engines:** All **Transmissions:** A/T, M/T	**Lost Communication With TAC Module** DTC U0122 and U0155 not set; and the PCM detected a problem in the FCAN communication circuit to the Vehicle Navigation Module (VTM) or the VSA control module. **Possible Causes** • Check for loose connections at the TAC control unit (intermittent fault) • FCAN communication circuit to the TAC is open, shorted to ground or B+ • Throttle body assembly is damaged or has failed
DTC: U0114 **1T PCM, MIL: YES** **2003, 2004, 2005, 2006** **Models:** MDX, RSX **Engines:** All **Transmissions:** A/T, M/T	**FCAN Malfunction (VTM-4 Control Module To PCM)** DTC U0073 not set; and the PCM detected a problem in the FCAN communication circuit to the Vehicle Navigation Module (VTM) or the VSA control module. **Possible Causes:** • Check for loose connections at VTM-4 unit (intermittent fault) • Check for any Body Control Module trouble codes • FCAN communication circuit to the VTM-4 is open, shorted to ground or B+
DTC: U0122 **1T PCM, MIL: YES** **2003, 2004, 2005, 2006** **Models:** MDX, RSX **Engines:** All **Transmissions:** A/T, M/T	**FCAN Malfunction (VAS To PCM)** DTC U0155 not set; and the PCM detected a problem in the FCAN communication circuit to the Fuel Control Module (FCM). **Possible Causes:** • Check for loose connections at the VSA control unit (intermittent fault) • FCAN communication circuit to the VSA is open, shorted to ground or B+ • VSA has failed (substitute known good VSA and then retest)

ACURA
COMPONENT TESTING

TABLE OF CONTENTS

COMPONENT TESTING

General Testing

INTERMITTENT TESTS

Intermittent problems can be difficult to diagnose because they may or may not be present while the vehicle is available to the repair technician. An intermittent fault can appear and leave so quickly that the vehicle computer does not detect a fault and store a trouble code in memory.

A computer will have to see the fault for a specific period of time before a fault is detected and a code set. While intermittent problems may appear to be occasional by nature, they usually occur under specific conditions. Therefore, the technician must identify and duplicate these conditions. Since an intermittent fault is difficult to duplicate, it is important to follow a logical, systematic routine (or checklist) when attempting to find the faulty system, circuit or component.

Some intermittent faults can be due to a loose connection, a wiring problem or a warped circuit board. Some of these problems are caused by poor test procedures that can damage the male or female ends of a connector. To test for loose or damaged connection, take the male end of a used connector from a wiring harness and carefully push it into the female terminal to verify the opening is tight. There should be some resistance felt when inserting the male into the female part of the terminal.

The Wiggle Test

A wiggle test can be used to locate some intermittent faults. To connect to the circuits, backprobe the actuator, sensor or switch and the PCM.

To perform this test, wiggle the suspect wiring, connector, and component while watching for a change on the DVOM. If it is available, use the Min/Max record mode on the DVOM to capture faults.

This is an excellent method to use to attempt to identify a condition that could cause an intermittent fault to occur.

Intermittents

This test procedure contains instructions for isolating an intermittent fault while using a Scan Tool, Breakout Box, fuel pressure gauge, vacuum gauge and a DVOM. Actual values from the vehicle under test can be compared to a typical set of values taken from the Reference Values.

PRELIMINARY CHECKS

All preliminary checks listed below are required prior to performing this test for intermittent faults. Preliminary checks include:

- All related electrical connections
- For vacuum leaks in related components or mounting hardware
- Fuel level - both quantity and quality
- Ignition wiring connections
- Air intake system filters, tubes and gaskets
- Basic engine components (compression, valve and ignition timing, etc.)
- Any aftermarket add-on devices

Locating and Repairing Intermittent Faults

Intermittent faults are generally associated with circuit problems. In order to pinpoint the fault area, a particular component or its wiring and connectors must be thoroughly inspected and tested. Prior to starting your test sequence, turn off all accessories and vehicle lighting. Also, verify that the battery and vehicle charging system are free of problems as these areas can disguise or mask a problem. Several of these tests are discussed next.

Change Input and Verify Output Response

The purpose of this type of test is to monitor how the PCM and its output devices respond to changes in sensor or switch inputs.

- Connect the Acura Test Harness or BOB.
- Record any pin voltages that related to an intermittent code or symptom
- Create a condition to cause the selected input condition to change
- Monitor the change in the pin voltage for a particular actuator signal on the DVOM (i.e., increase the throttle angle under engine load, and watch the IAC and TP Sensor pin voltage changes)

Actuator "Click" Testing

The purpose of this type of test is to monitor a particular PCM controlled relay or solenoid while watching and listening for a change of state

- Turn the key on or start the engine to actuate the device or switch
- If necessary, remove the device and test or actuate it on the bench
- Listen to verify that certain relays (PGM-FI, A/C) actually click on and off. If a Breakout Box is connected to the PCM, measure the control circuit while turning the outputs on and then off. A voltage change to close to battery voltage should occur during the on and off transition.

Test for Open Conditions in a Harness

The purpose of this type of test is to check a suspect wiring harness for an open circuit condition.

- Turn the key off and install the Acura Test Harness or BOB
- Remove the "suspect" component from the vehicle wiring harness
- Use a DVOM (ohmmeter function on the low range) to measure from one end of the suspect circuit at the test harness or BOB to the other end of the circuit at the particular component connector pin.
- The continuity test result should read under 5 ohms (If the test leads are okay and the ohmmeter is zeroed, the actual test results will read near 0.1 ohms).

Test for Shorted Conditions in a Harness

The purpose of this type of test is to check a suspect wiring harness for a short-to-power or short-to-ground condition.

- Turn the key off and disconnect the wiring harness from the PCM
- Remove the "suspect" component from its wiring harness connector.
- Use a DVOM (ohmmeter function on the high range) at the sensor harness connector to measure between the "suspect" circuit and signal return (ground) and to vehicle power or the voltage reference circuit.
- The continuity test result read more than 10,000 ohms (Infinity, >> or OL on some meters). If the actual reading is less than this amount, the two circuits are making contact somewhere in the wiring harness.

REFERENCE CHARTS

➡ **These example charts were designed for use with the Acura Test Harness or a Breakout Box and a DVOM. Readings were collected from a "known good" vehicle at Key On, Engine Off unless otherwise noted.**

Atmospheric Pressure Sensor Chart

BARO Pressure (inches Hg)	Voltage
0" Hg	2.8-3.0v
5" Hg	2.3-2.5v
10" Hg	1.8-2.0v
15" Hg	1.3-1.5v
20"Hg	0.8-1.0v
25" Hg	0.3-0.5v

Fuel Tank Pressure Chart

Fuel Tank Pressure (in. Hg)	Voltage
0" Hg	2.8-3.0v
5" Hg	2.3-2.5v
10" Hg	1.8-2.0v
15" Hg	1.3-1.5v
20"Hg	0.8-1.0v
25" Hg	0.3-0.5v

Manifold Absolute Pressure (MAP) Sensor Chart

MAP Pressure (inches Hg)	Voltage
0" Hg	2.8-3.0v
5" Hg	2.3-2.5v
10" Hg	1.8-2.0v
15" Hg	1.3-1.5v
20"Hg	0.8-1.0v
25" Hg	0.3-0.5v

EGR Lift Sensor Chart

EGR Valve Position	Voltage
Valve Fully Closed (idle)	1.0-1.2v
Valve Fully Open (off-idle)	4.5v

Engine Coolant Temperature (ECT) Sensor Conversion Chart

Degrees F	Resistance	Voltage
0°F	14,000-20,000 ohms	4.0v
100°F	900-1200 ohms	2-3v
180°F	250-350 ohms	0.5-0.6v

Intake Air Temperature (IAT) Sensor Conversion Chart

Degrees F	Resistance	Voltage
0°F	14,000-20,000 ohms	4.0v
100°F	900-1200 ohms	2-3v
192°F	150-350 ohms	0.49-0.55v

Component Locations

2.2CL

4 Cylinder Engines

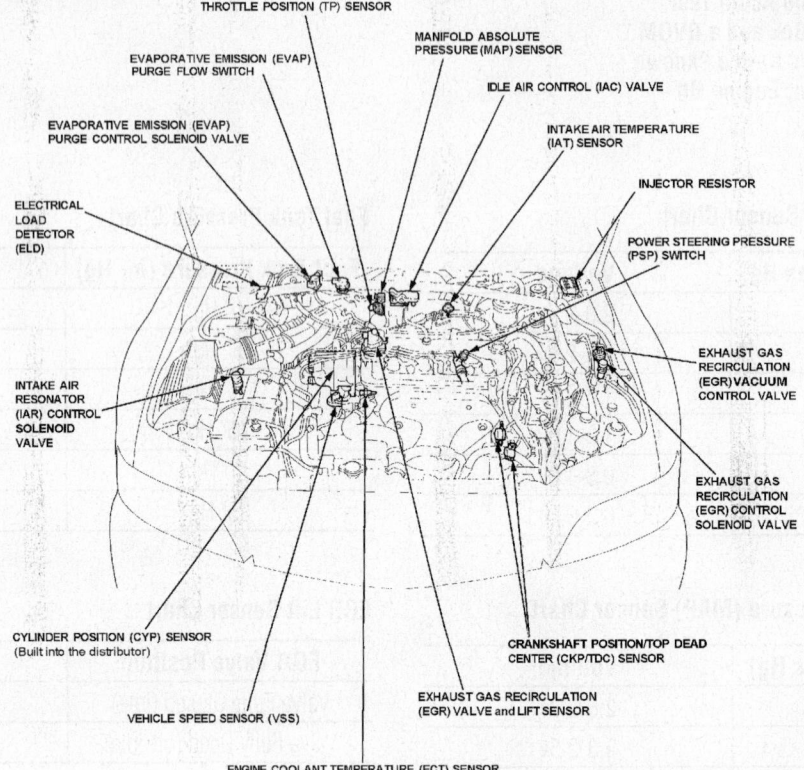

THROTTLE POSITION (TP) SENSOR

MANIFOLD ABSOLUTE PRESSURE (MAP) SENSOR

EVAPORATIVE EMISSION (EVAP) PURGE FLOW SWITCH

IDLE AIR CONTROL (IAC) VALVE

EVAPORATIVE EMISSION (EVAP) PURGE CONTROL SOLENOID VALVE

INTAKE AIR TEMPERATURE (IAT) SENSOR

INJECTOR RESISTOR

ELECTRICAL LOAD DETECTOR (ELD)

POWER STEERING PRESSURE (PSP) SWITCH

EXHAUST GAS RECIRCULATION (EGR) VACUUM CONTROL VALVE

INTAKE AIR RESONATOR (IAR) CONTROL SOLENOID VALVE

EXHAUST GAS RECIRCULATION (EGR) CONTROL SOLENOID VALVE

CYLINDER POSITION (CYP) SENSOR (Built into the distributor)

CRANKSHAFT POSITION/TOP DEAD CENTER (CKP/TDC) SENSOR

EXHAUST GAS RECIRCULATION (EGR) VALVE and LIFT SENSOR

VEHICLE SPEED SENSOR (VSS)

ENGINE COOLANT TEMPERATURE (ECT) SENSOR

29130_ACUR_G0001

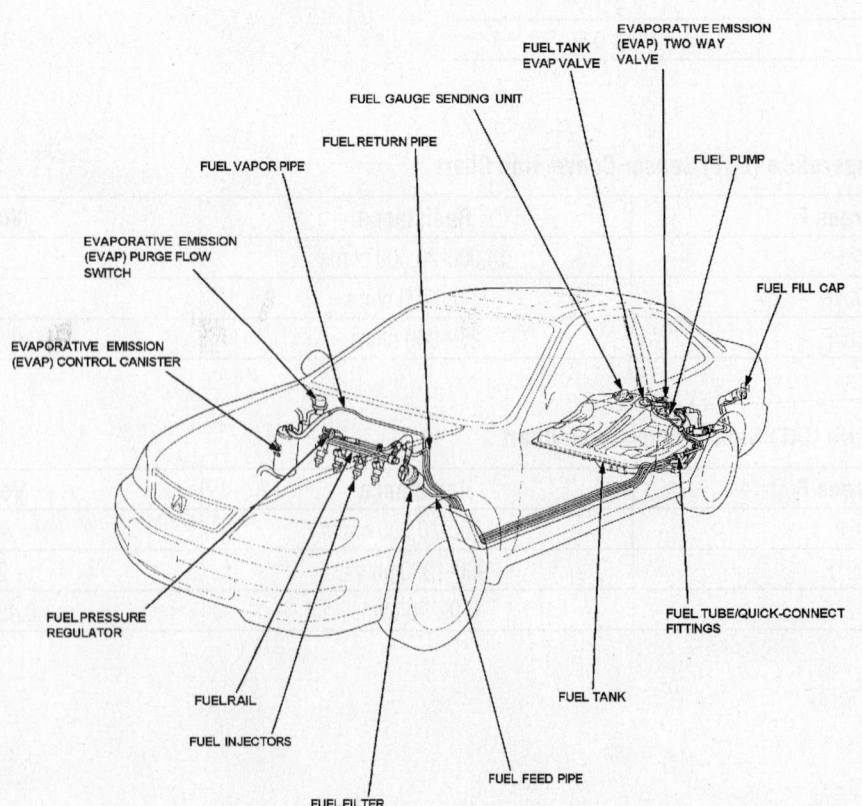

FUEL TANK EVAP VALVE

EVAPORATIVE EMISSION (EVAP) TWO WAY VALVE

FUEL GAUGE SENDING UNIT

FUEL RETURN PIPE

FUEL PUMP

FUEL VAPOR PIPE

EVAPORATIVE EMISSION (EVAP) PURGE FLOW SWITCH

FUEL FILL CAP

EVAPORATIVE EMISSION (EVAP) CONTROL CANISTER

FUEL TUBE/QUICK-CONNECT FITTINGS

FUEL PRESSURE REGULATOR

FUEL TANK

FUEL RAIL

FUEL INJECTORS

FUEL FEED PIPE

FUEL FILTER

29130_ACUR_G0003

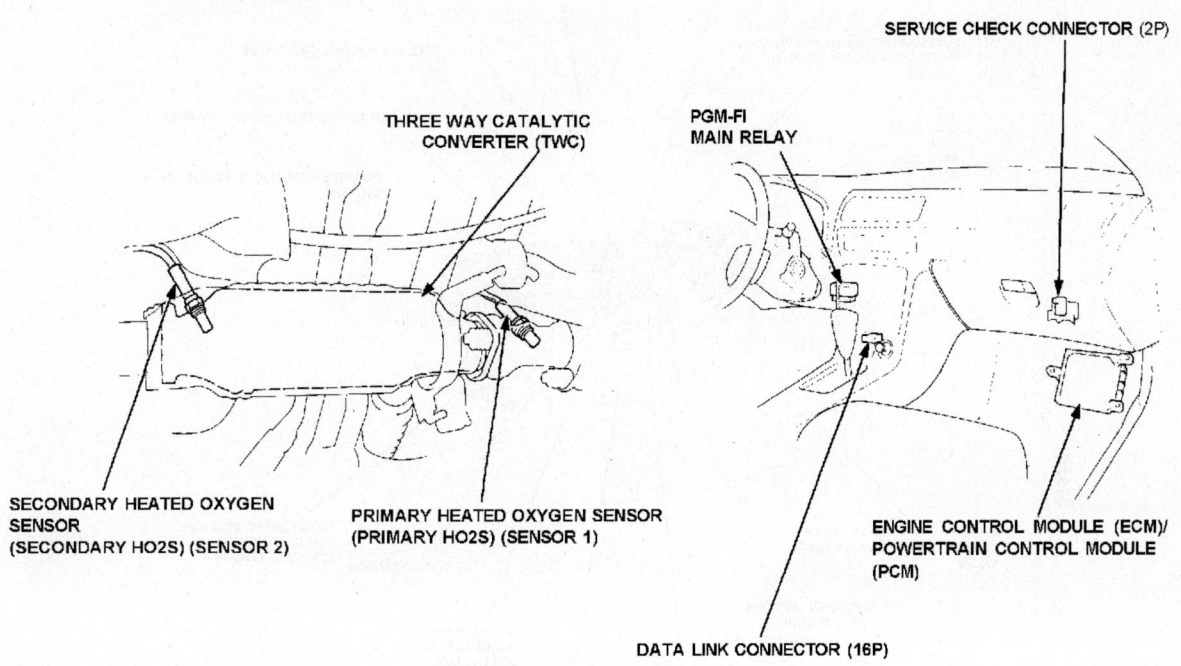

THREE WAY CATALYTIC
CONVERTER (TWC)

PGM-FI
MAIN RELAY

SERVICE CHECK CONNECTOR (2P)

SECONDARY HEATED OXYGEN
SENSOR
(SECONDARY HO2S) (SENSOR 2)

PRIMARY HEATED OXYGEN SENSOR
(PRIMARY HO2S) (SENSOR 1)

ENGINE CONTROL MODULE (ECM)/
POWERTRAIN CONTROL MODULE
(PCM)

DATA LINK CONNECTOR (16P)

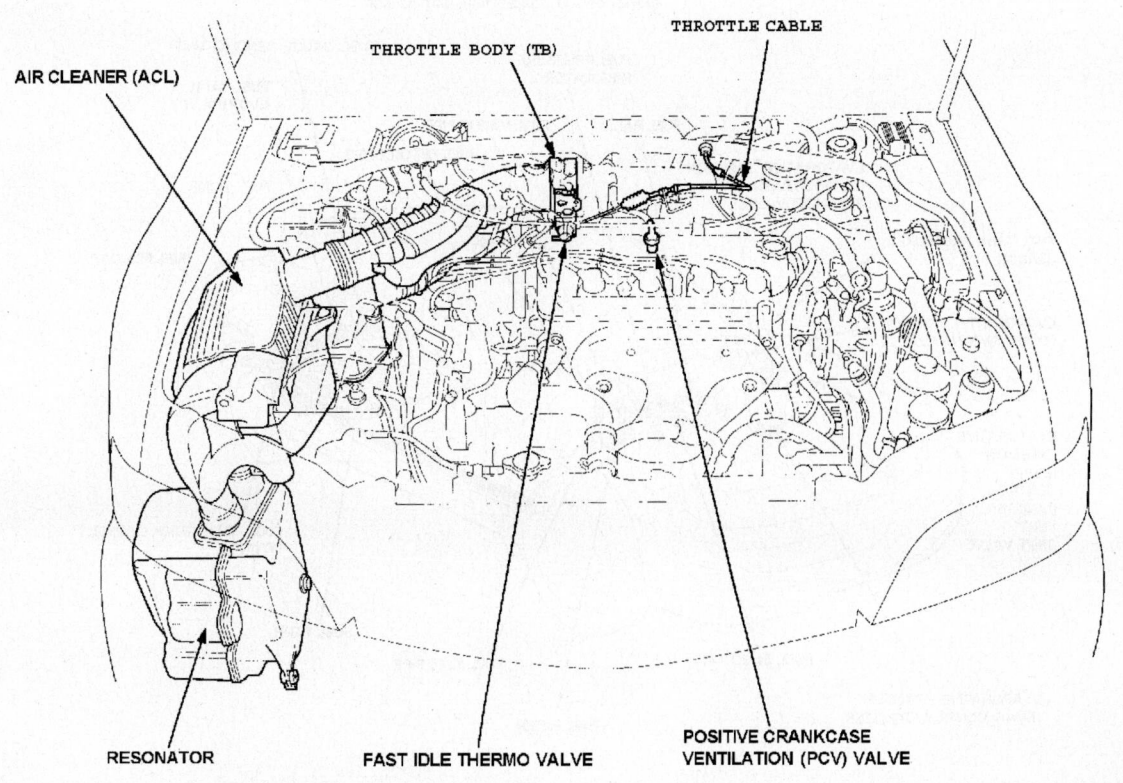

THROTTLE CABLE

AIR CLEANER (ACL)

THROTTLE BODY (TB)

RESONATOR

FAST IDLE THERMO VALVE

POSITIVE CRANKCASE
VENTILATION (PCV) VALVE

2.3CL

4 Cylinder Engines

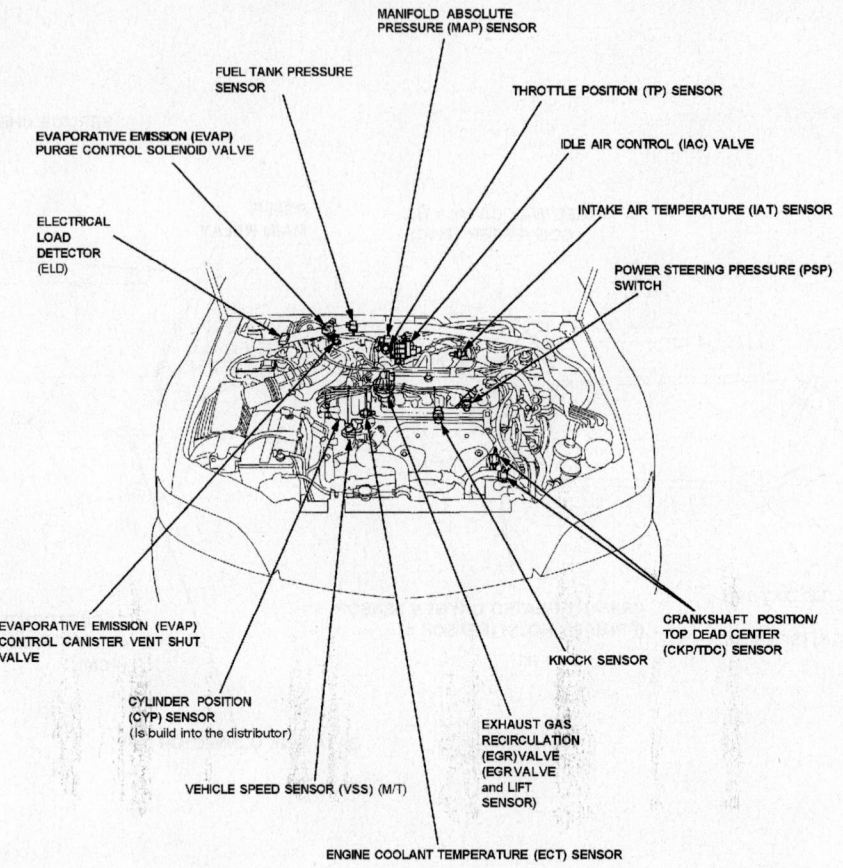

MANIFOLD ABSOLUTE
PRESSURE (MAP) SENSOR

FUEL TANK PRESSURE
SENSOR

THROTTLE POSITION (TP) SENSOR

EVAPORATIVE EMISSION (EVAP)
PURGE CONTROL SOLENOID VALVE

IDLE AIR CONTROL (IAC) VALVE

ELECTRICAL
LOAD
DETECTOR
(ELD)

INTAKE AIR TEMPERATURE (IAT) SENSOR

POWER STEERING PRESSURE (PSP)
SWITCH

EVAPORATIVE EMISSION (EVAP)
CONTROL CANISTER VENT SHUT
VALVE

CRANKSHAFT POSITION/
TOP DEAD CENTER
(CKP/TDC) SENSOR

CYLINDER POSITION
(CYP) SENSOR
(Is build into the distributor)

KNOCK SENSOR

EXHAUST GAS
RECIRCULATION
(EGR) VALVE
(EGR VALVE
and LIFT
SENSOR)

VEHICLE SPEED SENSOR (VSS) (M/T)

ENGINE COOLANT TEMPERATURE (ECT) SENSOR

29130_ACUR_G0004

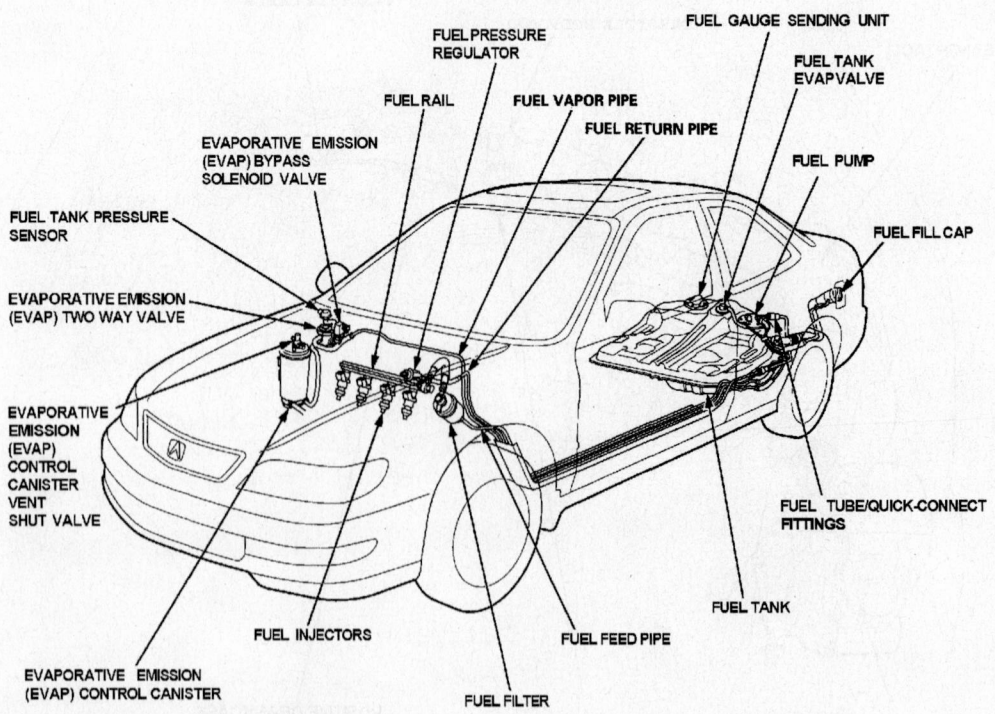

FUEL PRESSURE
REGULATOR

FUEL GAUGE SENDING UNIT

FUEL TANK
EVAP VALVE

FUEL RAIL

FUEL VAPOR PIPE

FUEL RETURN PIPE

EVAPORATIVE EMISSION
(EVAP) BYPASS
SOLENOID VALVE

FUEL PUMP

FUEL TANK PRESSURE
SENSOR

FUEL FILL CAP

EVAPORATIVE EMISSION
(EVAP) TWO WAY VALVE

EVAPORATIVE
EMISSION
(EVAP)
CONTROL
CANISTER
VENT
SHUT VALVE

FUEL TUBE/QUICK-CONNECT
FITTINGS

FUEL TANK

FUEL INJECTORS

FUEL FEED PIPE

EVAPORATIVE EMISSION
(EVAP) CONTROL CANISTER

FUEL FILTER

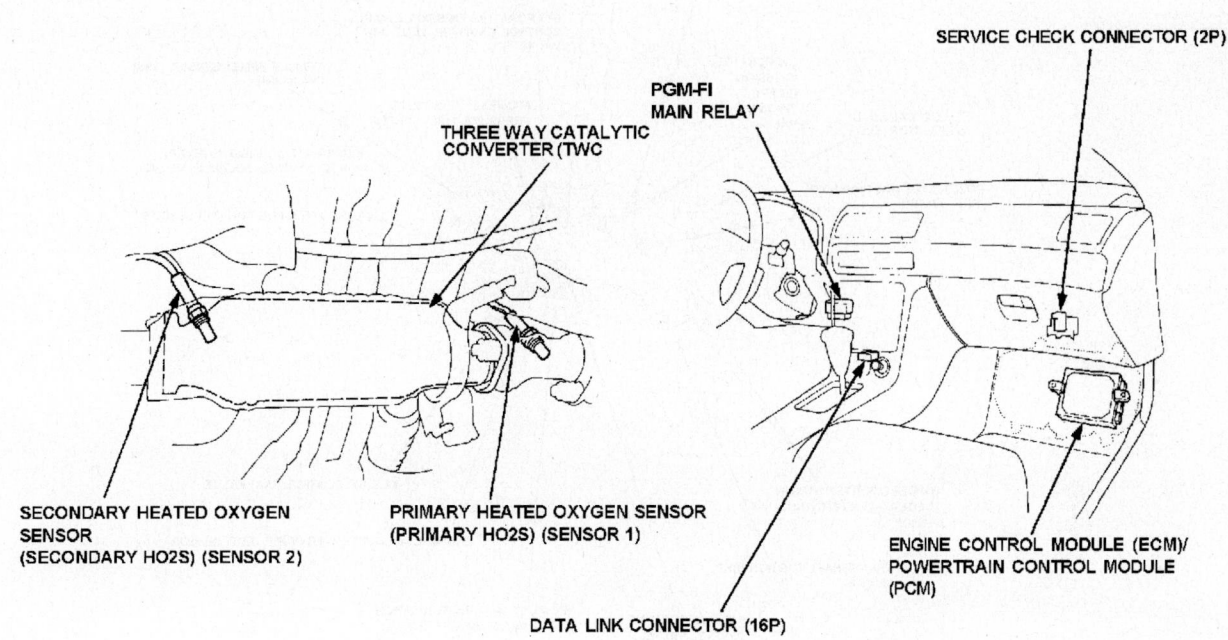

SERVICE CHECK CONNECTOR (2P)

PGM-FI
MAIN RELAY

THREE WAY CATALYTIC
CONVERTER (TWC

SECONDARY HEATED OXYGEN
SENSOR
(SECONDARY HO2S) (SENSOR 2)

PRIMARY HEATED OXYGEN SENSOR
(PRIMARY HO2S) (SENSOR 1)

ENGINE CONTROL MODULE (ECM)/
POWERTRAIN CONTROL MODULE
(PCM)

DATA LINK CONNECTOR (16P)

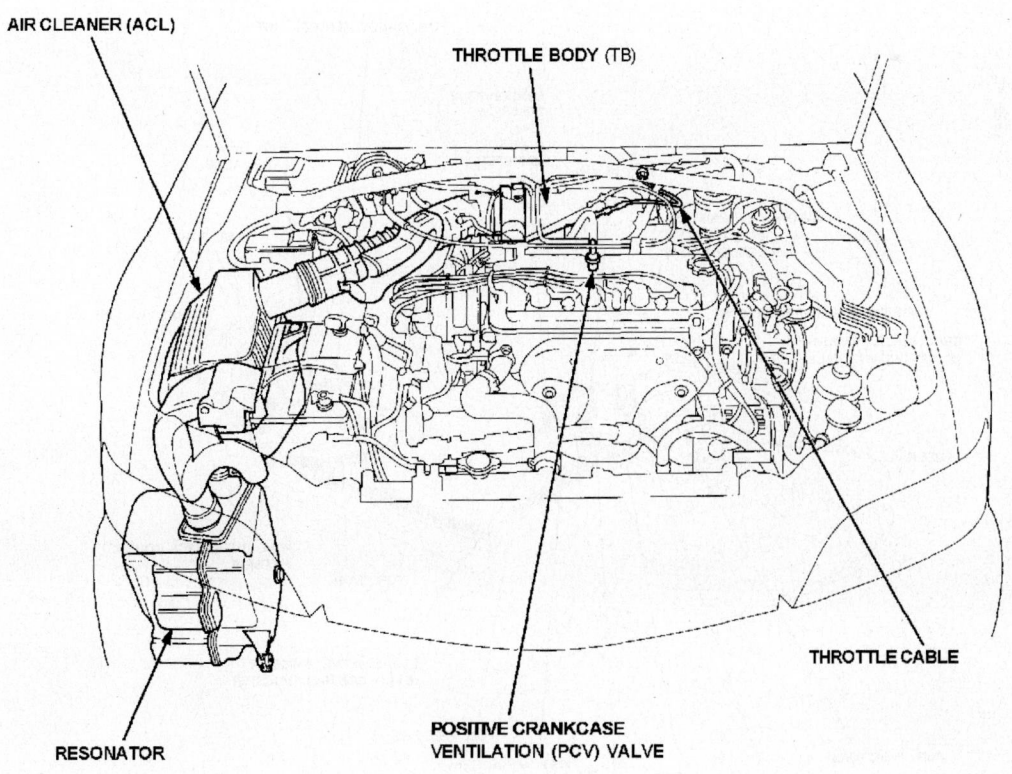

AIR CLEANER (ACL)

THROTTLE BODY (TB)

THROTTLE CABLE

RESONATOR

POSITIVE CRANKCASE
VENTILATION (PCV) VALVE

3.0CL

6 Cylinder Engines

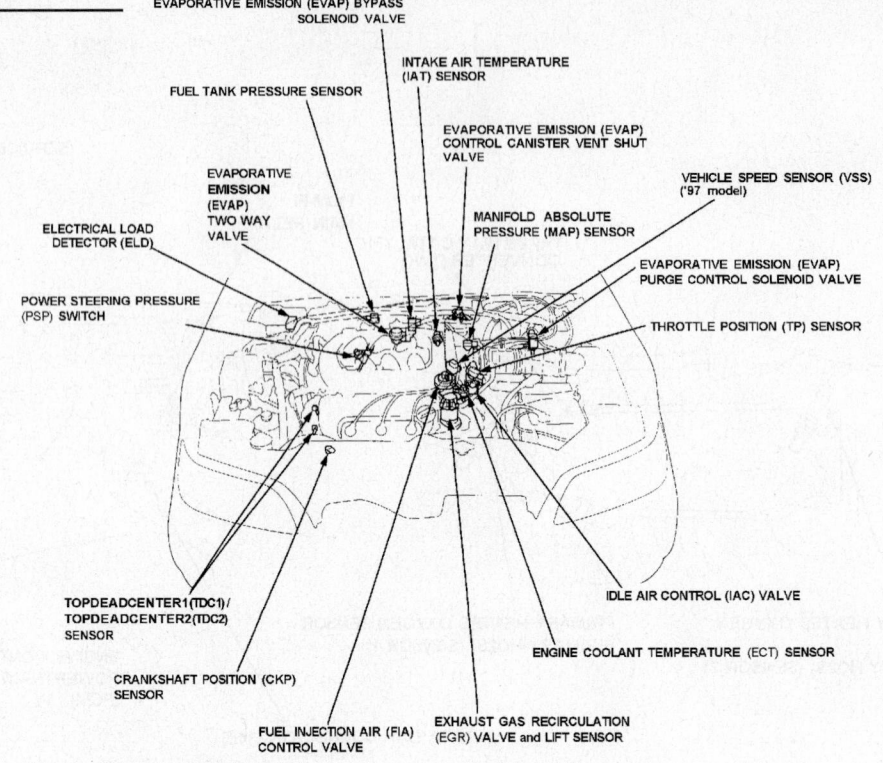

EVAPORATIVE EMISSION (EVAP) BYPASS
SOLENOID VALVE

INTAKE AIR TEMPERATURE
(IAT) SENSOR

EVAPORATIVE EMISSION (EVAP)
CONTROL CANISTER VENT SHUT
VALVE

FUEL TANK PRESSURE SENSOR

VEHICLE SPEED SENSOR (VSS)
('97 model)

EVAPORATIVE
EMISSION
(EVAP)
TWO WAY
VALVE

MANIFOLD ABSOLUTE
PRESSURE (MAP) SENSOR

ELECTRICAL LOAD
DETECTOR (ELD)

EVAPORATIVE EMISSION (EVAP)
PURGE CONTROL SOLENOID VALVE

POWER STEERING PRESSURE
(PSP) SWITCH

THROTTLE POSITION (TP) SENSOR

TOPDEADCENTER1(TDC1)/
TOPDEADCENTER2(TDC2)
SENSOR

IDLE AIR CONTROL (IAC) VALVE

ENGINE COOLANT TEMPERATURE (ECT) SENSOR

CRANKSHAFT POSITION (CKP)
SENSOR

FUEL INJECTION AIR (FIA)
CONTROL VALVE

EXHAUST GAS RECIRCULATION
(EGR) VALVE and LIFT SENSOR

29130_ACUR_G0008

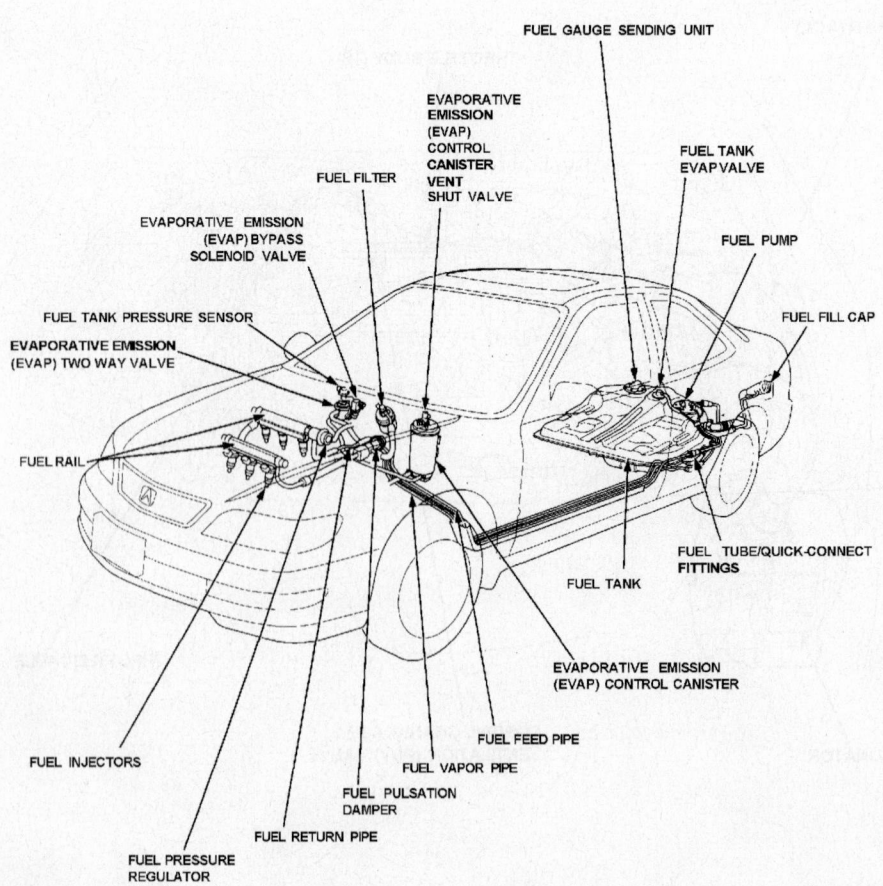

FUEL GAUGE SENDING UNIT

EVAPORATIVE
EMISSION
(EVAP)
CONTROL
CANISTER
VENT
SHUT VALVE

FUEL FILTER

FUEL TANK
EVAP VALVE

EVAPORATIVE EMISSION
(EVAP) BYPASS
SOLENOID VALVE

FUEL PUMP

FUEL TANK PRESSURE SENSOR

FUEL FILL CAP

EVAPORATIVE EMISSION
(EVAP) TWO WAY VALVE

FUEL RAIL

FUEL TUBE/QUICK-CONNECT
FITTINGS

FUEL TANK

EVAPORATIVE EMISSION
(EVAP) CONTROL CANISTER

FUEL INJECTORS

FUEL FEED PIPE

FUEL VAPOR PIPE

FUEL PULSATION
DAMPER

FUEL RETURN PIPE

FUEL PRESSURE
REGULATOR

29130_ACUR_G0009

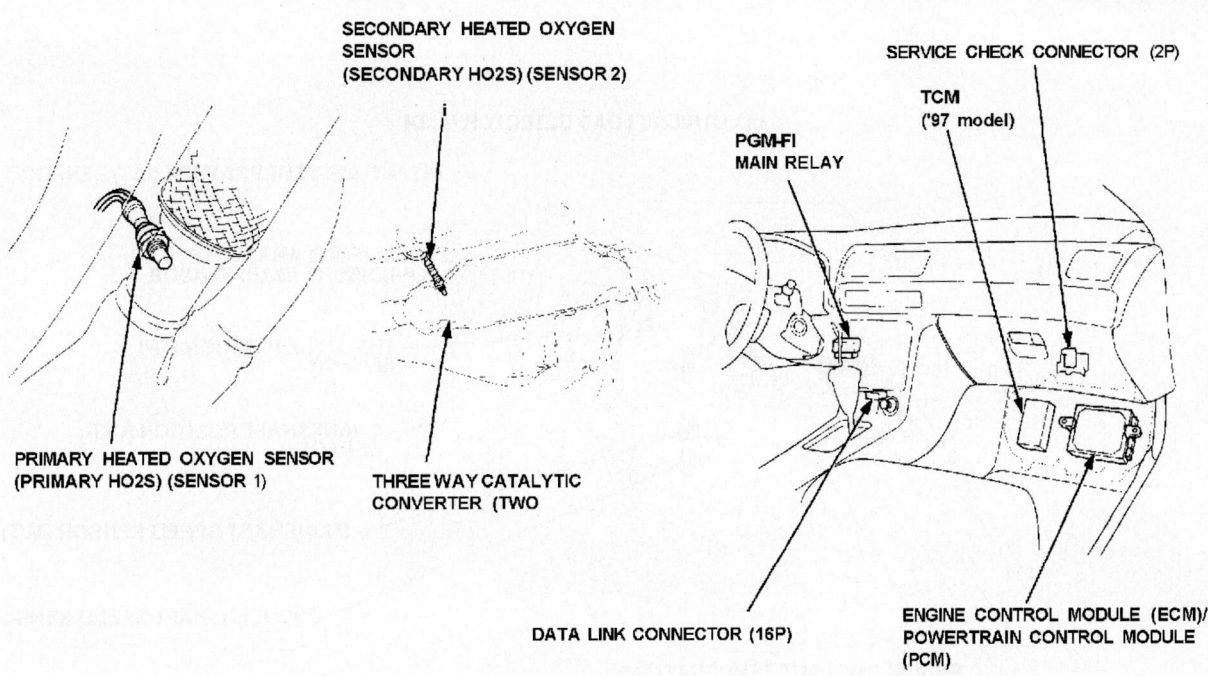

SECONDARY HEATED OXYGEN
SENSOR
(SECONDARY HO2S) (SENSOR 2)

SERVICE CHECK CONNECTOR (2P)

TCM
('97 model)

PGM-FI
MAIN RELAY

PRIMARY HEATED OXYGEN SENSOR
(PRIMARY HO2S) (SENSOR 1)

THREE WAY CATALYTIC
CONVERTER (TWO

DATA LINK CONNECTOR (16P)

ENGINE CONTROL MODULE (ECM)/
POWERTRAIN CONTROL MODULE
(PCM)

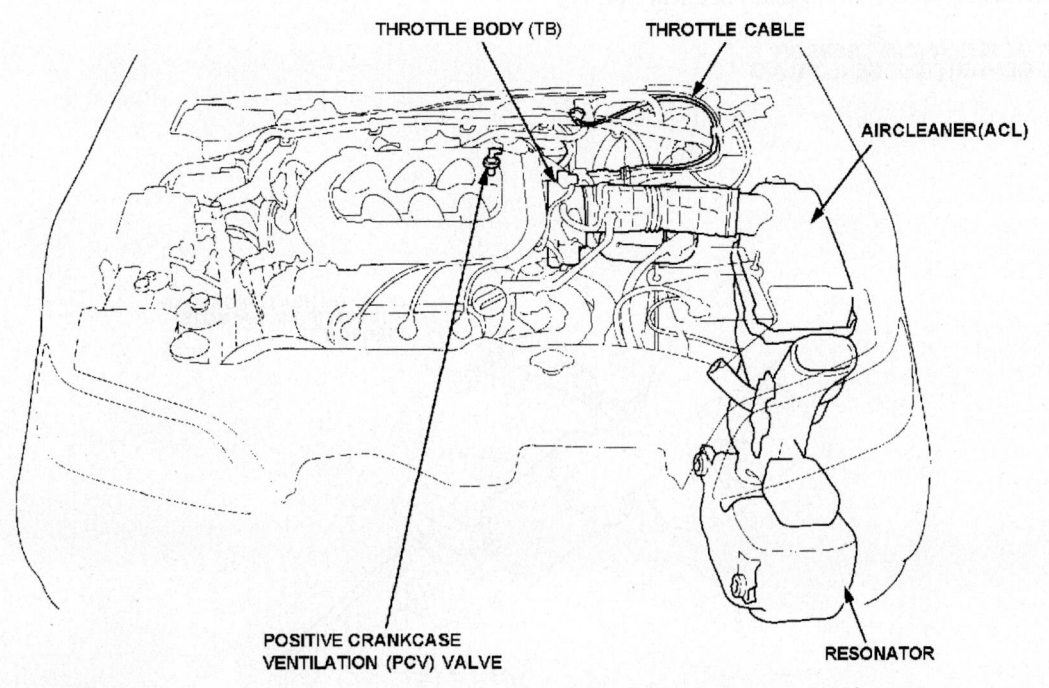

THROTTLE BODY (TB)

THROTTLE CABLE

AIRCLEANER(ACL)

POSITIVE CRANKCASE
VENTILATION (PCV) VALVE

RESONATOR

29130_ACUR_G0008

3.2CL

6 Cylinder Engines

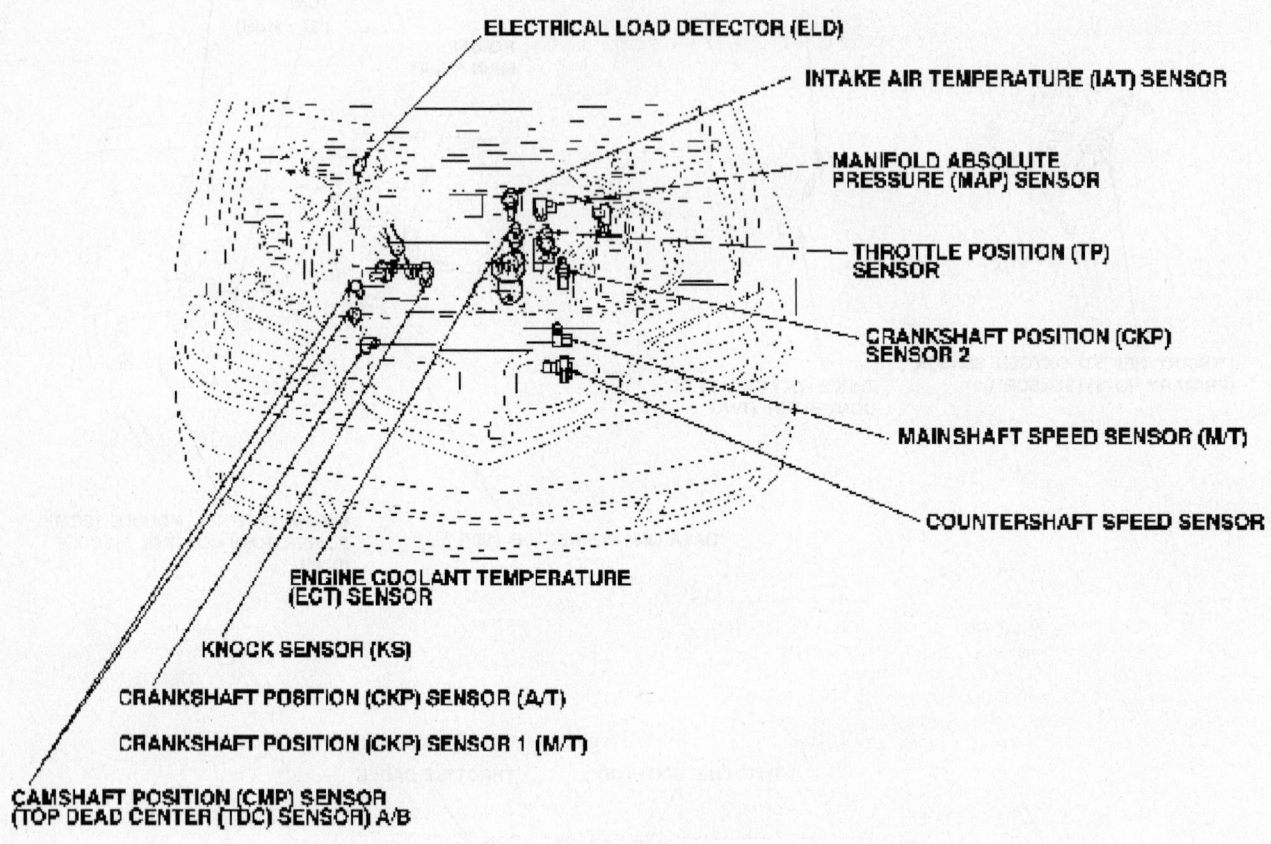

ELECTRICAL LOAD DETECTOR (ELD)

INTAKE AIR TEMPERATURE (IAT) SENSOR

MANIFOLD ABSOLUTE PRESSURE (MAP) SENSOR

THROTTLE POSITION (TP) SENSOR

CRANKSHAFT POSITION (CKP) SENSOR 2

MAINSHAFT SPEED SENSOR (M/T)

COUNTERSHAFT SPEED SENSOR (

ENGINE COOLANT TEMPERATURE (ECT) SENSOR

KNOCK SENSOR (KS)

CRANKSHAFT POSITION (CKP) SENSOR (A/T)

CRANKSHAFT POSITION (CKP) SENSOR 1 (M/T)

CAMSHAFT POSITION (CMP) SENSOR (TOP DEAD CENTER (TDC) SENSOR) A/B

29130_ACUR_G0010

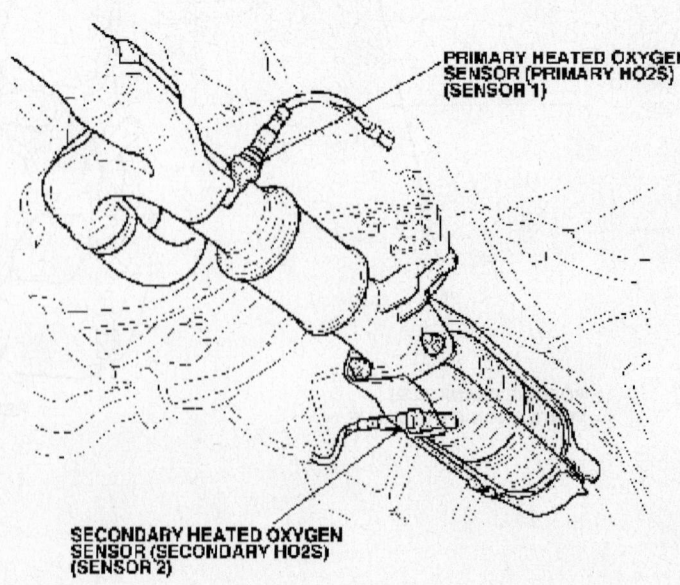

PRIMARY HEATED OXYGEN SENSOR (PRIMARY HO2S) (SENSOR 1)

SECONDARY HEATED OXYGEN SENSOR (SECONDARY HO2S) (SENSOR 2)

29130_ACUR_G0011

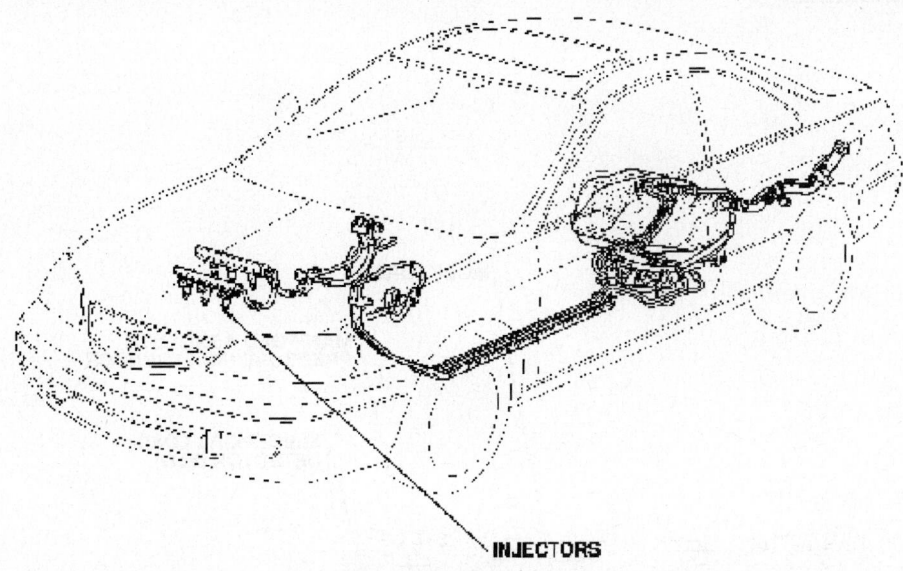

INJECTORS

29130_ACUR_G0012

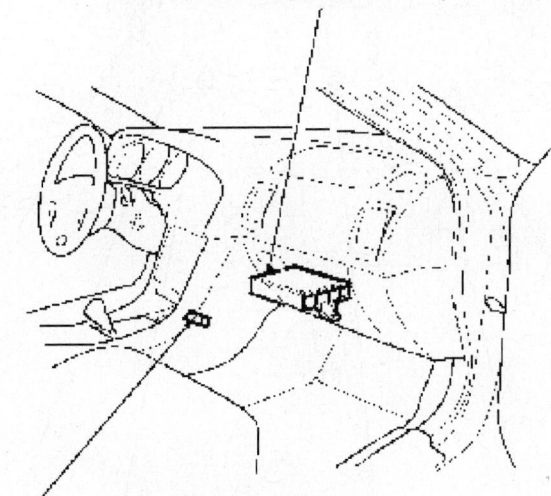

ENGINE CONTROL MODULE/
POWERTRAIN CONTROL MODULE (ECM/PCM)

DATA LINK CONNECTOR (DLC) (16P)

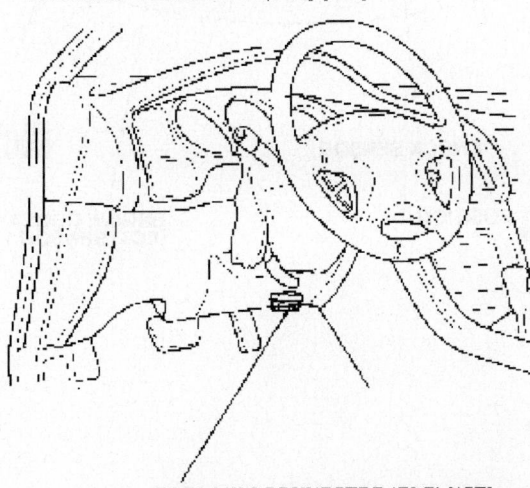

DATA LINK CONNECTOR (DLC) (16P)

29130_ACUR_G0013

RL

6 Cylinder Engines

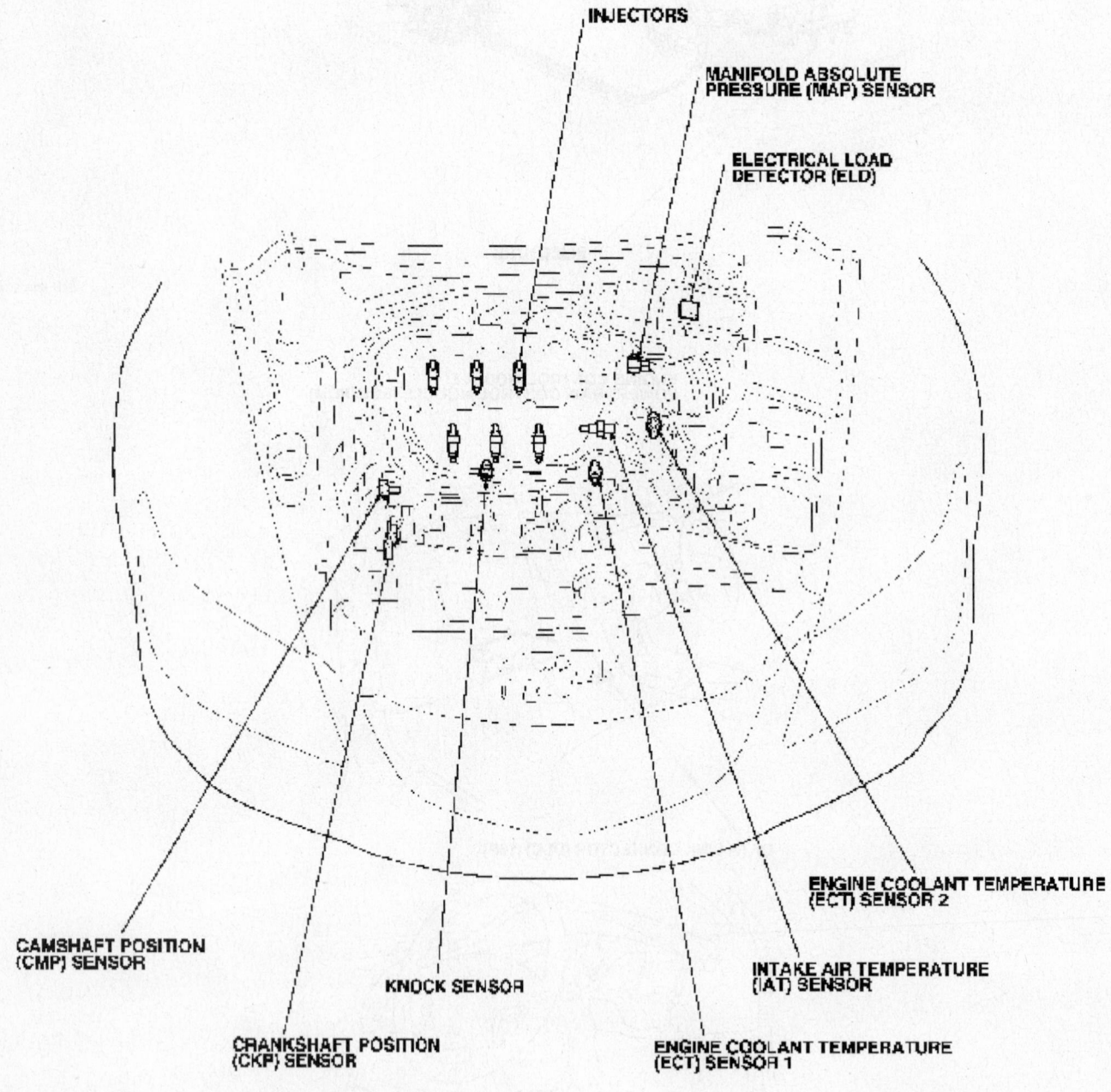

INJECTORS

MANIFOLD ABSOLUTE
PRESSURE (MAP) SENSOR

ELECTRICAL LOAD
DETECTOR (ELD)

ENGINE COOLANT TEMPERATURE
(ECT) SENSOR 2

CAMSHAFT POSITION
(CMP) SENSOR

INTAKE AIR TEMPERATURE
(IAT) SENSOR

KNOCK SENSOR

CRANKSHAFT POSITION
(CKP) SENSOR

ENGINE COOLANT TEMPERATURE
(ECT) SENSOR 1

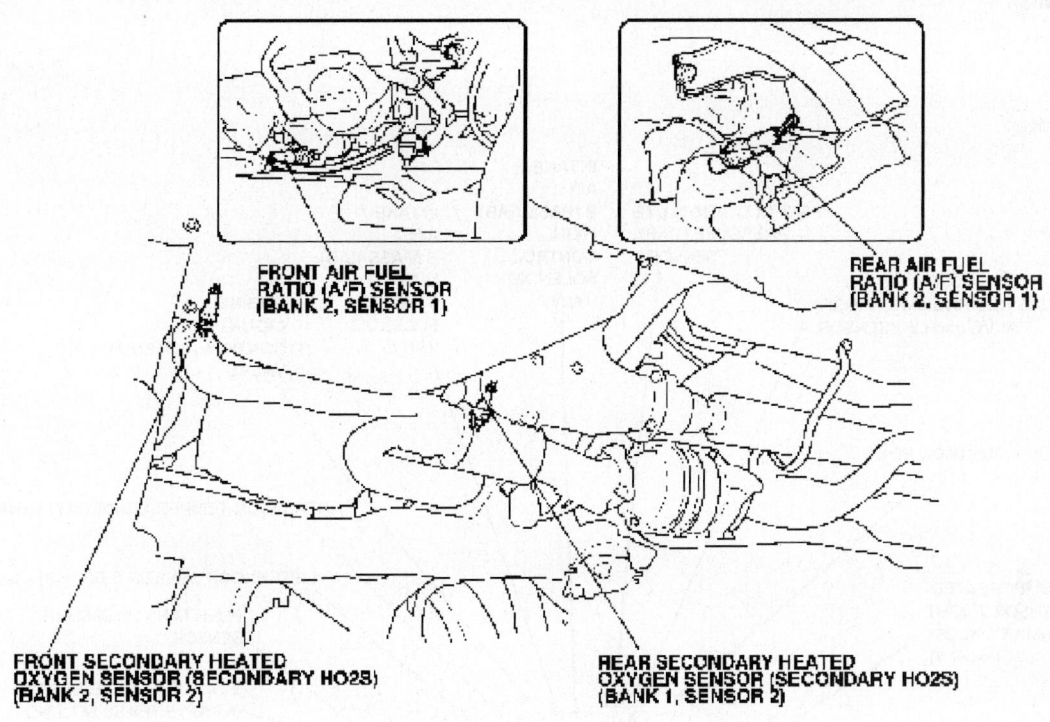

FRONT AIR FUEL
RATIO (A/F) SENSOR
(BANK 2, SENSOR 1)

REAR AIR FUEL
RATIO (A/F) SENSOR
(BANK 2, SENSOR 1)

FRONT SECONDARY HEATED
OXYGEN SENSOR (SECONDARY HO2S)
(BANK 2, SENSOR 2)

REAR SECONDARY HEATED
OXYGEN SENSOR (SECONDARY HO2S)
(BANK 1, SENSOR 2)

29130_ACUR_G0026

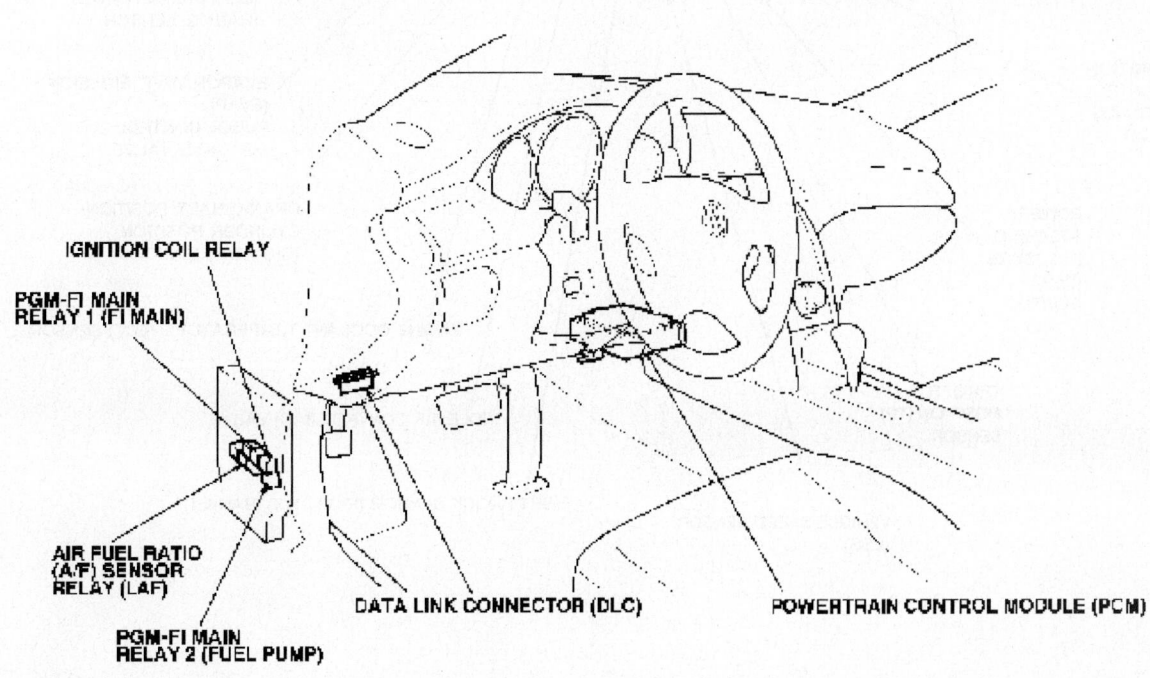

IGNITION COIL RELAY

PGM-FI MAIN
RELAY 1 (FI MAIN)

AIR FUEL RATIO
(A/F) SENSOR
RELAY (LAF)

PGM-FI MAIN
RELAY 2 (FUEL PUMP)

DATA LINK CONNECTOR (DLC)

POWERTRAIN CONTROL MODULE (PCM)

29130_ACUR_G0027

3.5RL

6 Cylinder Engines

'96 - 99 models:

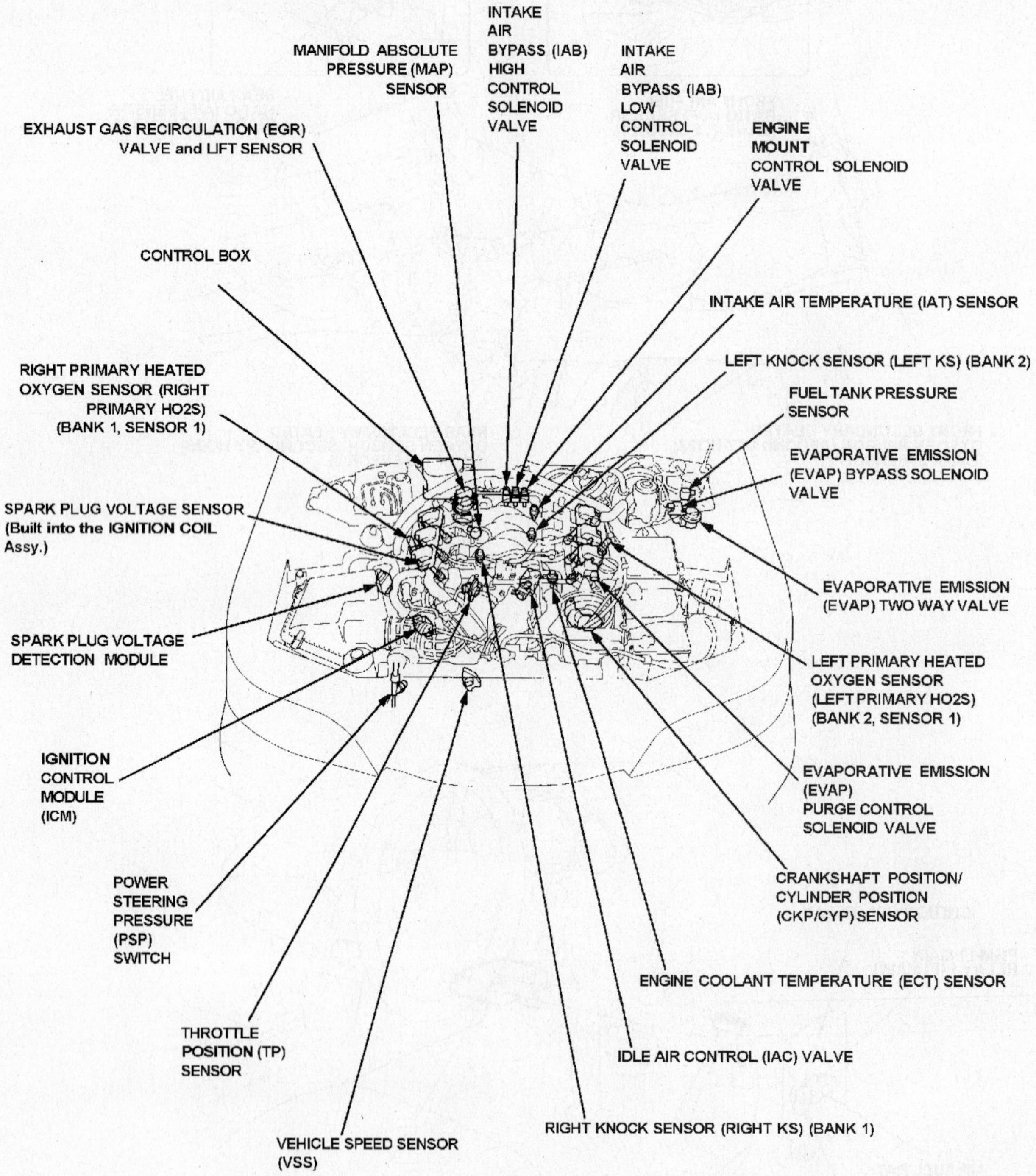

MANIFOLD ABSOLUTE
PRESSURE (MAP)
SENSOR

INTAKE
AIR
BYPASS (IAB)
HIGH
CONTROL
SOLENOID
VALVE

INTAKE
AIR
BYPASS (IAB)
LOW
CONTROL
SOLENOID
VALVE

ENGINE
MOUNT
CONTROL SOLENOID
VALVE

EXHAUST GAS RECIRCULATION (EGR)
VALVE and LIFT SENSOR

CONTROL BOX

INTAKE AIR TEMPERATURE (IAT) SENSOR

RIGHT PRIMARY HEATED
OXYGEN SENSOR (RIGHT
PRIMARY HO2S)
(BANK 1, SENSOR 1)

LEFT KNOCK SENSOR (LEFT KS) (BANK 2)

FUEL TANK PRESSURE
SENSOR

EVAPORATIVE EMISSION
(EVAP) BYPASS SOLENOID
VALVE

SPARK PLUG VOLTAGE SENSOR
(Built into the IGNITION COIL
Assy.)

EVAPORATIVE EMISSION
(EVAP) TWO WAY VALVE

SPARK PLUG VOLTAGE
DETECTION MODULE

LEFT PRIMARY HEATED
OXYGEN SENSOR
(LEFT PRIMARY HO2S)
(BANK 2, SENSOR 1)

IGNITION
CONTROL
MODULE
(ICM)

EVAPORATIVE EMISSION
(EVAP)
PURGE CONTROL
SOLENOID VALVE

POWER
STEERING
PRESSURE
(PSP)
SWITCH

CRANKSHAFT POSITION/
CYLINDER POSITION
(CKP/CYP) SENSOR

ENGINE COOLANT TEMPERATURE (ECT) SENSOR

THROTTLE
POSITION (TP)
SENSOR

IDLE AIR CONTROL (IAC) VALVE

VEHICLE SPEED SENSOR
(VSS)

RIGHT KNOCK SENSOR (RIGHT KS) (BANK 1)

'00 - 04 models:

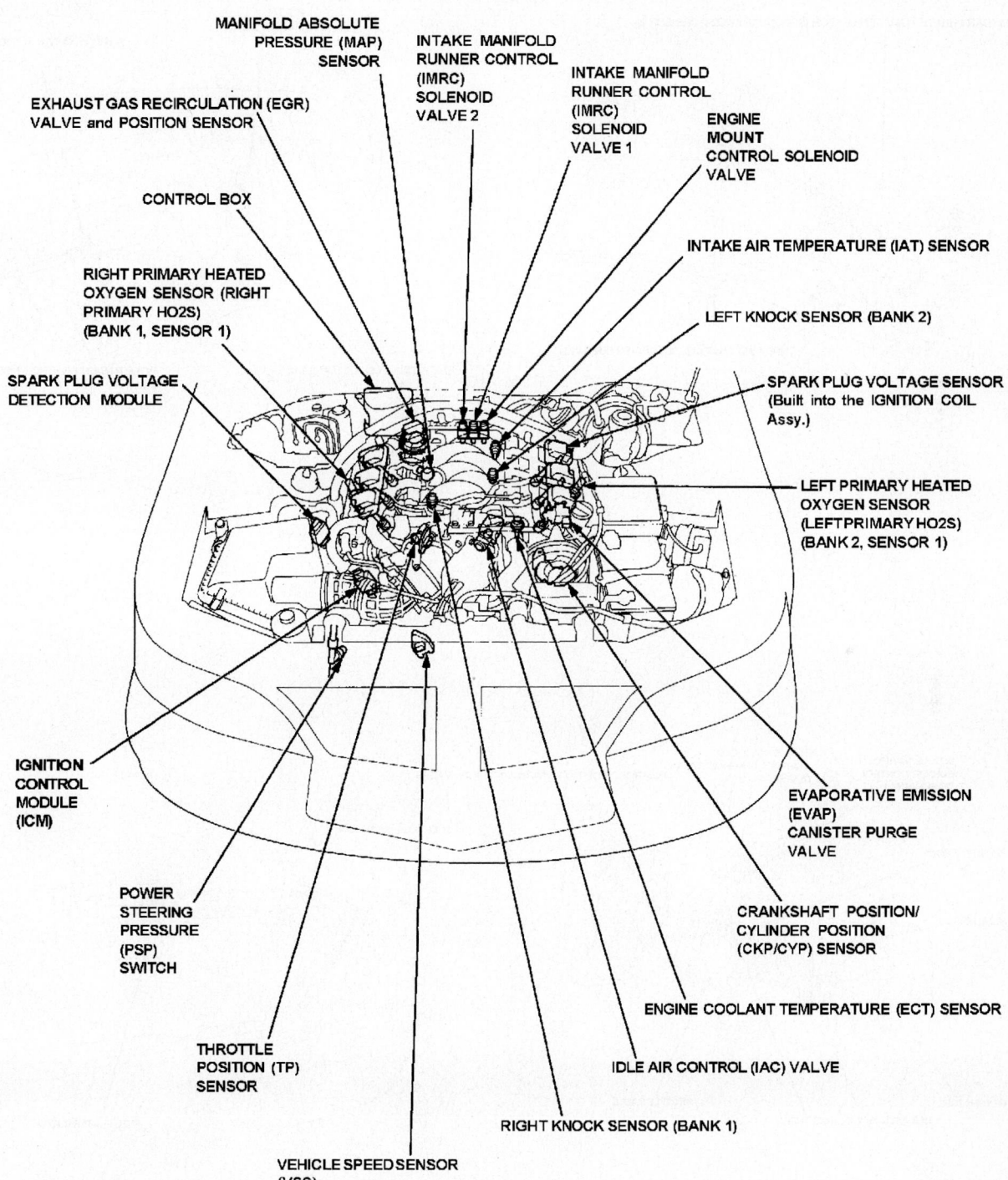

MANIFOLD ABSOLUTE
PRESSURE (MAP)
SENSOR

INTAKE MANIFOLD
RUNNER CONTROL
(IMRC)
SOLENOID
VALVE 2

INTAKE MANIFOLD
RUNNER CONTROL
(IMRC)
SOLENOID
VALVE 1

ENGINE
MOUNT
CONTROL SOLENOID
VALVE

EXHAUST GAS RECIRCULATION (EGR)
VALVE and POSITION SENSOR

CONTROL BOX

INTAKE AIR TEMPERATURE (IAT) SENSOR

RIGHT PRIMARY HEATED
OXYGEN SENSOR (RIGHT
PRIMARY HO2S)
(BANK 1, SENSOR 1)

LEFT KNOCK SENSOR (BANK 2)

SPARK PLUG VOLTAGE
DETECTION MODULE

SPARK PLUG VOLTAGE SENSOR
(Built into the IGNITION COIL
Assy.)

LEFT PRIMARY HEATED
OXYGEN SENSOR
(LEFT PRIMARY HO2S)
(BANK 2, SENSOR 1)

IGNITION
CONTROL
MODULE
(ICM)

EVAPORATIVE EMISSION
(EVAP)
CANISTER PURGE
VALVE

POWER
STEERING
PRESSURE
(PSP)
SWITCH

CRANKSHAFT POSITION/
CYLINDER POSITION
(CKP/CYP) SENSOR

ENGINE COOLANT TEMPERATURE (ECT) SENSOR

THROTTLE
POSITION (TP)
SENSOR

IDLE AIR CONTROL (IAC) VALVE

RIGHT KNOCK SENSOR (BANK 1)

VEHICLE SPEED SENSOR
(VSS)

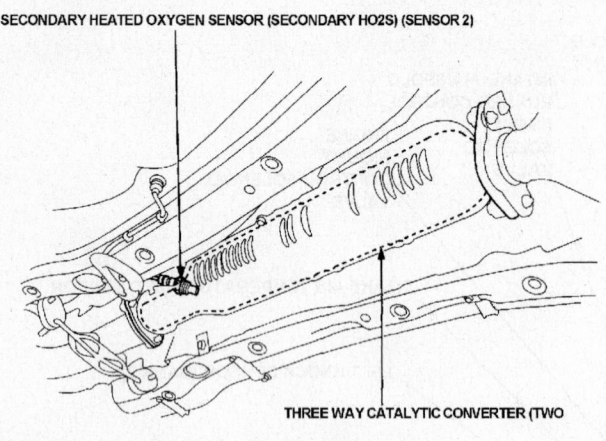

SECONDARY HEATED OXYGEN SENSOR (SECONDARY HO2S) (SENSOR 2)

THREE WAY CATALYTIC CONVERTER (TWO

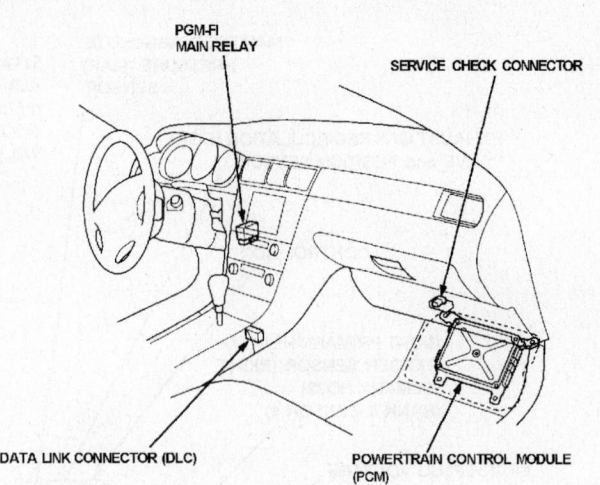

PGM-FI
MAIN RELAY

SERVICE CHECK CONNECTOR

DATA LINK CONNECTOR (DLC)

POWERTRAIN CONTROL MODULE
(PCM)

29130_ACUR_G0030

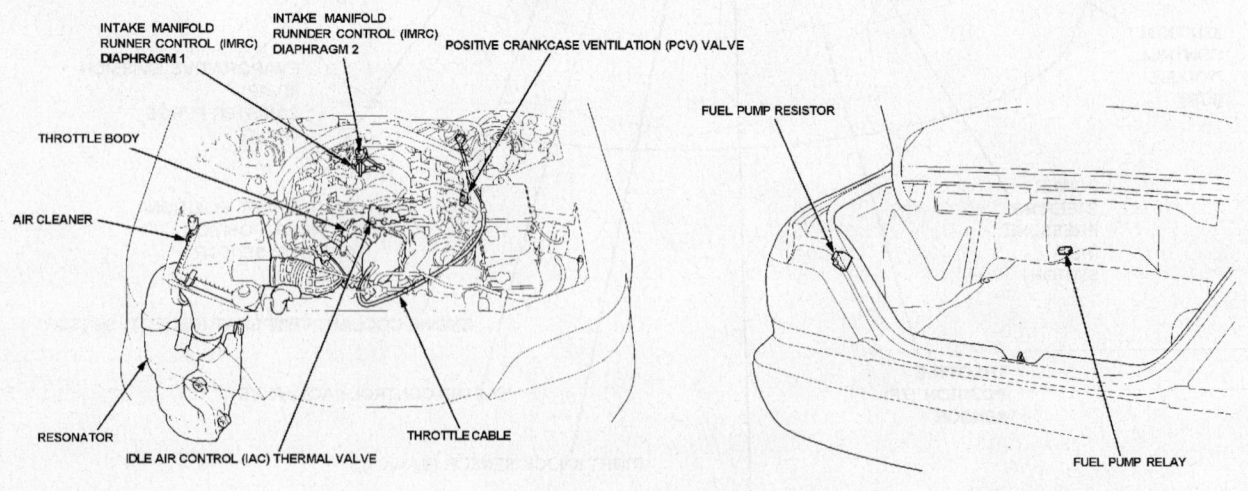

INTAKE MANIFOLD
RUNNER CONTROL (IMRC)
DIAPHRAGM 1

INTAKE MANIFOLD
RUNNDER CONTROL (IMRC)
DIAPHRAGM 2

POSITIVE CRANKCASE VENTILATION (PCV) VALVE

THROTTLE BODY

AIR CLEANER

RESONATOR

IDLE AIR CONTROL (IAC) THERMAL VALVE

THROTTLE CABLE

FUEL PUMP RESISTOR

FUEL PUMP RELAY

29130_ACUR_G0031

'96-99 models:

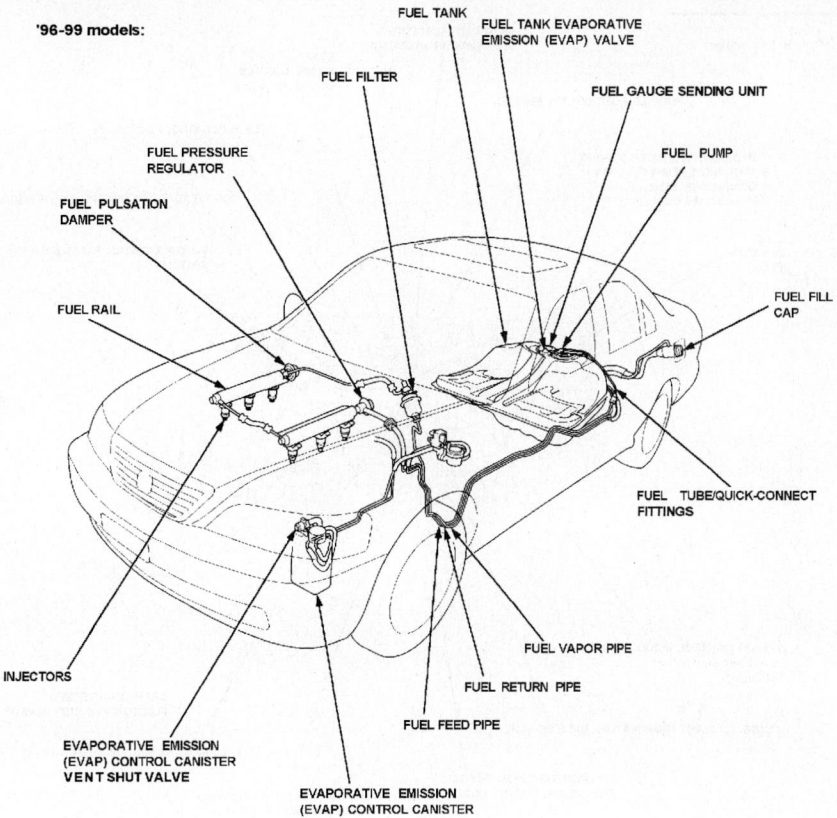

FUEL TANK

FUEL TANK EVAPORATIVE
EMISSION (EVAP) VALVE

FUEL FILTER

FUEL GAUGE SENDING UNIT

FUEL PRESSURE
REGULATOR

FUEL PUMP

FUEL PULSATION
DAMPER

FUEL FILL
CAP

FUEL RAIL

FUEL TUBE/QUICK-CONNECT
FITTINGS

FUEL VAPOR PIPE

FUEL RETURN PIPE

INJECTORS

FUEL FEED PIPE

EVAPORATIVE EMISSION
(EVAP) CONTROL CANISTER
VENT SHUT VALVE

EVAPORATIVE EMISSION
(EVAP) CONTROL CANISTER

29130_ACUR_G0032

'00 - 04 models:

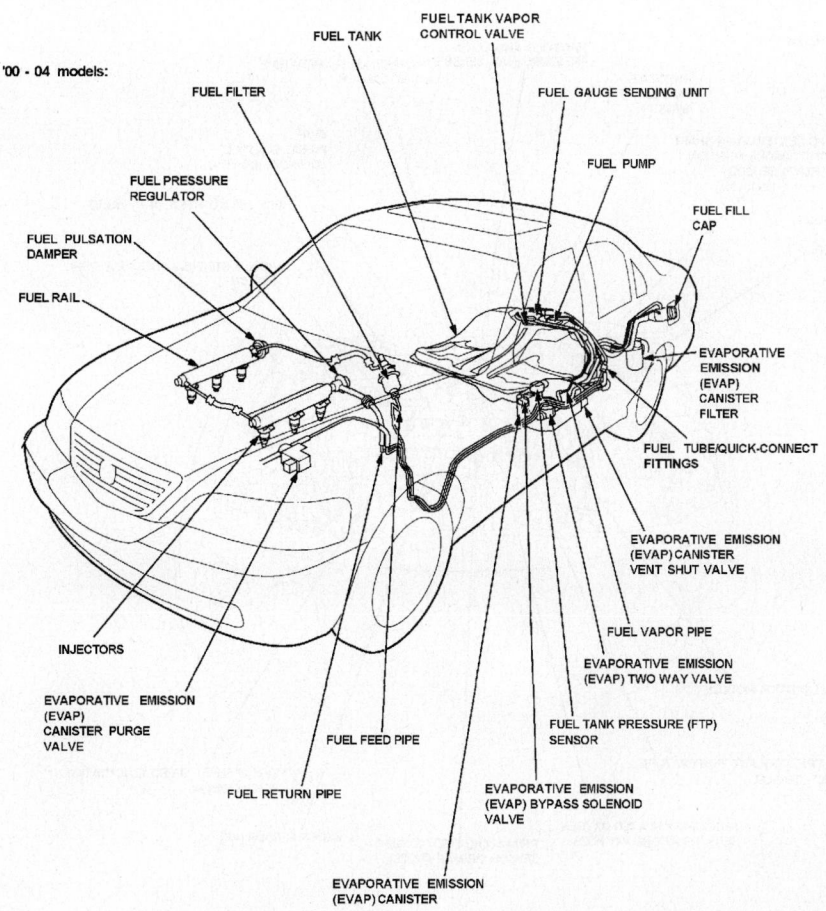

FUEL TANK VAPOR
CONTROL VALVE

FUEL TANK

FUEL FILTER

FUEL GAUGE SENDING UNIT

FUEL PRESSURE
REGULATOR

FUEL PUMP

FUEL PULSATION
DAMPER

FUEL FILL
CAP

FUEL RAIL

EVAPORATIVE
EMISSION
(EVAP)
CANISTER
FILTER

FUEL TUBE/QUICK-CONNECT
FITTINGS

EVAPORATIVE EMISSION
(EVAP) CANISTER
VENT SHUT VALVE

INJECTORS

FUEL VAPOR PIPE

EVAPORATIVE EMISSION
(EVAP) TWO WAY VALVE

EVAPORATIVE EMISSION
(EVAP)
CANISTER PURGE
VALVE

FUEL FEED PIPE

FUEL TANK PRESSURE (FTP)
SENSOR

FUEL RETURN PIPE

EVAPORATIVE EMISSION
(EVAP) BYPASS SOLENOID
VALVE

EVAPORATIVE EMISSION
(EVAP) CANISTER

29130_ACUR_G0033

INTEGRA

4 Cylinder Engines

B18B1 engine:

MANIFOLD ABSOLUTE
PRESSURE (MAP) SENSOR

EVAP
PURGE CONTROL
SOLENOID VALVE

THROTTLE POSITION (TP) SENSOR

IDLE AIR CONTROL (IAC) VALVE

TOP DEAD CENTER/CRANKSHAFT
POSITION/CYLINDER POSITION
(TDC/CKP/CYP) SENSOR
(Is built into the distributor)

INTAKE AIR TEMPERATURE (IAT) SENSOR

ELECTRICAL
LOAD
DETECTOR
(ELD)

POWER STEERING PRESSURE (PSP)
SWITCH

IGNITION CONTROL MODULE (ICM)
(Is built into the
distributor)

CRANKSHAFT SPEED
FLUCTUATION (CKF) SENSOR

ENGINE COOLANT TEMPERATURE (ECT) SENSOR

FAST IDLE THERMO VALVE (A/T model)

SECONDARY HEATED OXYGEN
SENSOR (SECONDARY HO2S)

PRIMARY HEATED OXYGEN
SENSOR (PRIMARY HO2S)

29130_ACUR_G0034

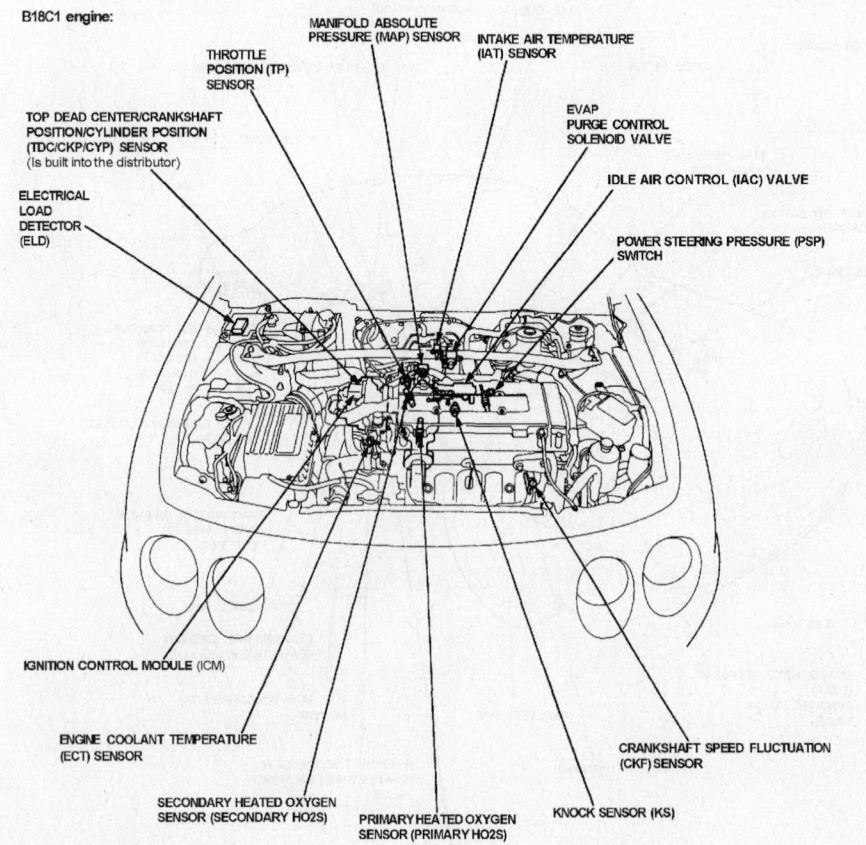

B18C1 engine:

MANIFOLD ABSOLUTE
PRESSURE (MAP) SENSOR

INTAKE AIR TEMPERATURE
(IAT) SENSOR

THROTTLE
POSITION (TP)
SENSOR

EVAP
PURGE CONTROL
SOLENOID VALVE

TOP DEAD CENTER/CRANKSHAFT
POSITION/CYLINDER POSITION
(TDC/CKP/CYP) SENSOR
(Is built into the distributor)

IDLE AIR CONTROL (IAC) VALVE

ELECTRICAL
LOAD
DETECTOR
(ELD)

POWER STEERING PRESSURE (PSP)
SWITCH

IGNITION CONTROL MODULE (ICM)

ENGINE COOLANT TEMPERATURE
(ECT) SENSOR

CRANKSHAFT SPEED FLUCTUATION
(CKF) SENSOR

SECONDARY HEATED OXYGEN
SENSOR (SECONDARY HO2S)

PRIMARY HEATED OXYGEN
SENSOR (PRIMARY HO2S)

KNOCK SENSOR (KS)

29130_ACUR_G0035

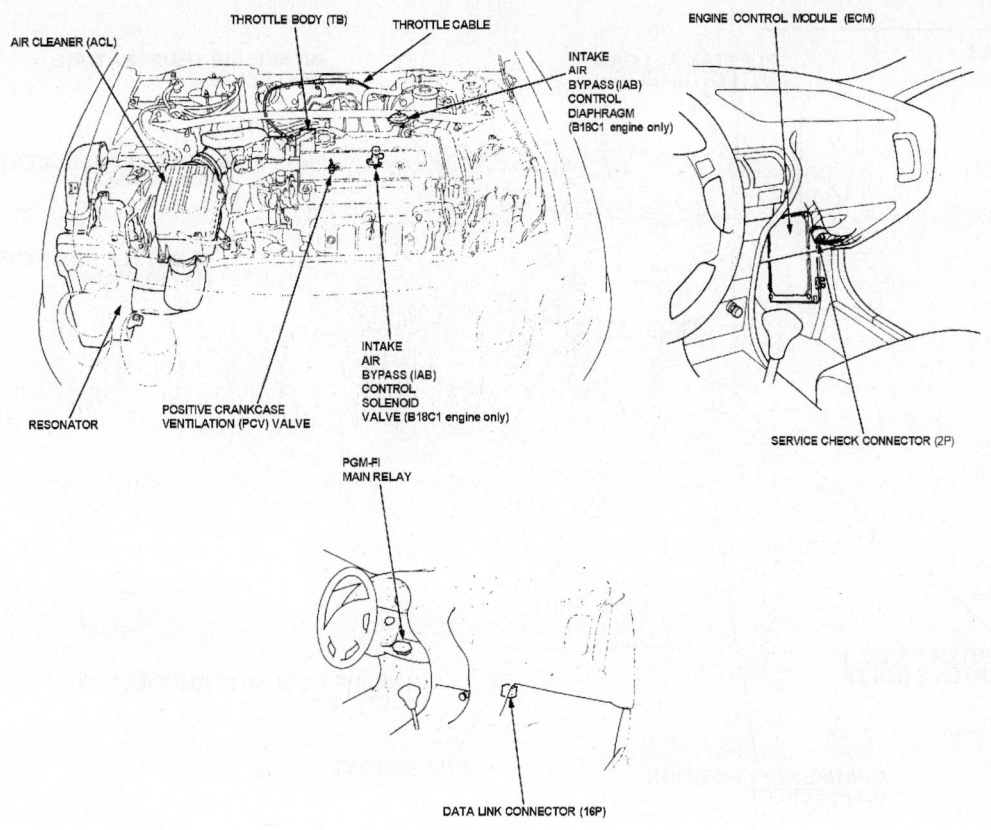

AIR CLEANER (ACL)

THROTTLE BODY (TB)

THROTTLE CABLE

ENGINE CONTROL MODULE (ECM)

INTAKE
AIR
BYPASS (IAB)
CONTROL
DIAPHRAGM
(B18C1 engine only)

RESONATOR

POSITIVE CRANKCASE
VENTILATION (PCV) VALVE

INTAKE
AIR
BYPASS (IAB)
CONTROL
SOLENOID
VALVE (B18C1 engine only)

SERVICE CHECK CONNECTOR (2P)

PGM-FI
MAIN RELAY

DATA LINK CONNECTOR (16P)

29130_ACUR_G0036

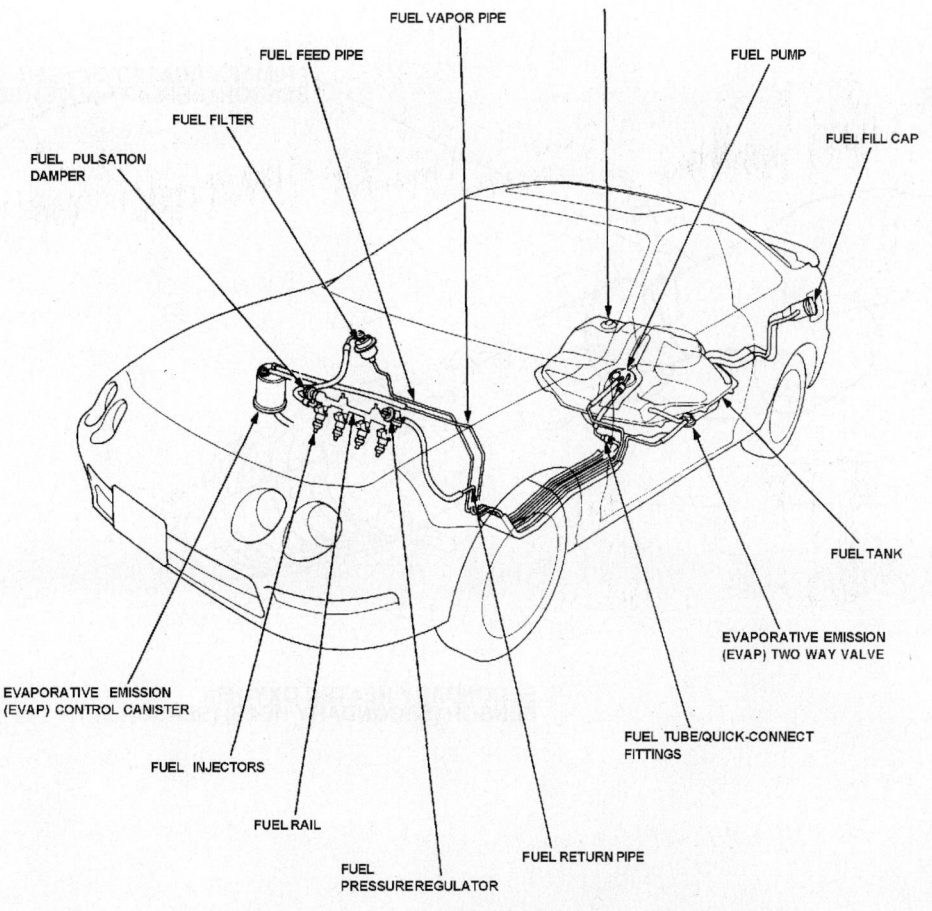

FUEL VAPOR PIPE

FUEL FEED PIPE

FUEL PUMP

FUEL FILTER

FUEL FILL CAP

FUEL PULSATION
DAMPER

FUEL TANK

EVAPORATIVE EMISSION
(EVAP) TWO WAY VALVE

EVAPORATIVE EMISSION
(EVAP) CONTROL CANISTER

FUEL TUBE/QUICK-CONNECT
FITTINGS

FUEL INJECTORS

FUEL RAIL

FUEL
PRESSURE REGULATOR

FUEL RETURN PIPE

29130_ACUR_G0037

MDX

6 Cylinder Engines

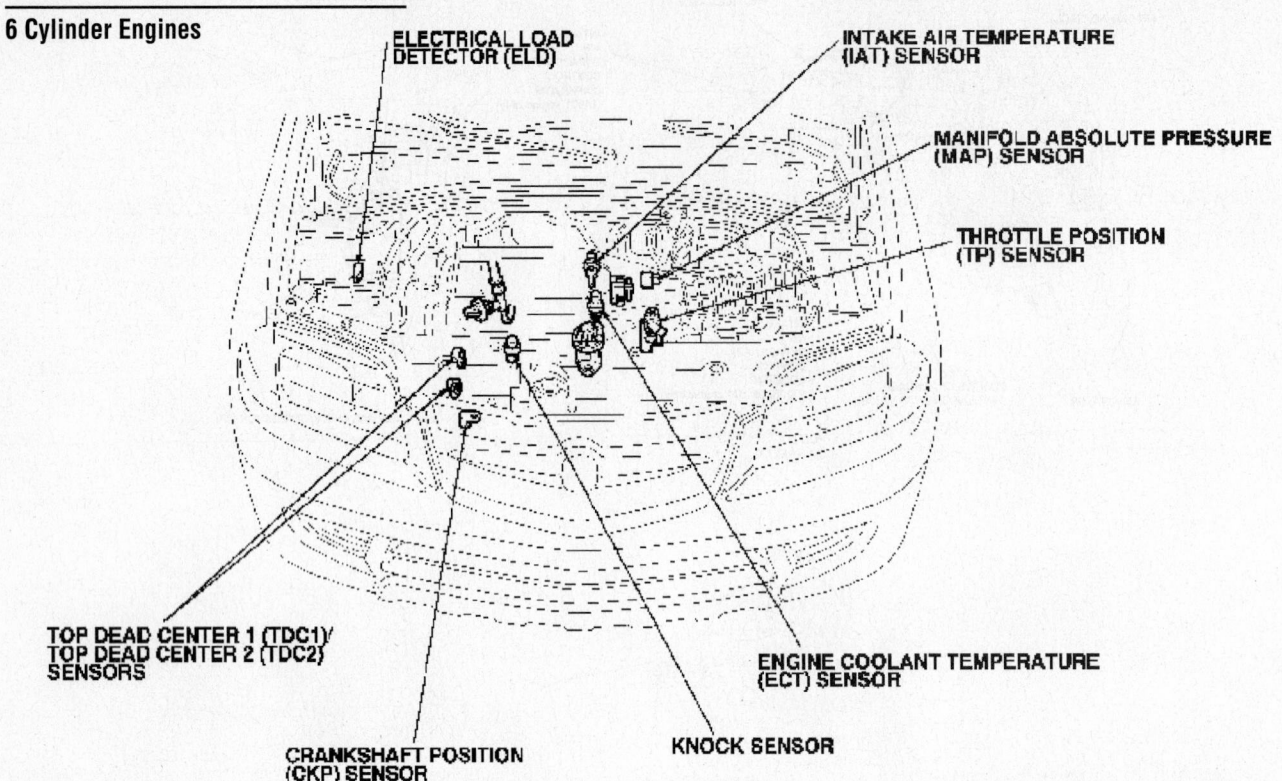

ELECTRICAL LOAD DETECTOR (ELD)

INTAKE AIR TEMPERATURE (IAT) SENSOR

MANIFOLD ABSOLUTE PRESSURE (MAP) SENSOR

THROTTLE POSITION (TP) SENSOR

TOP DEAD CENTER 1 (TDC1)/ TOP DEAD CENTER 2 (TDC2) SENSORS

CRANKSHAFT POSITION (CKP) SENSOR

KNOCK SENSOR

ENGINE COOLANT TEMPERATURE (ECT) SENSOR

29130_ACUR_G0038

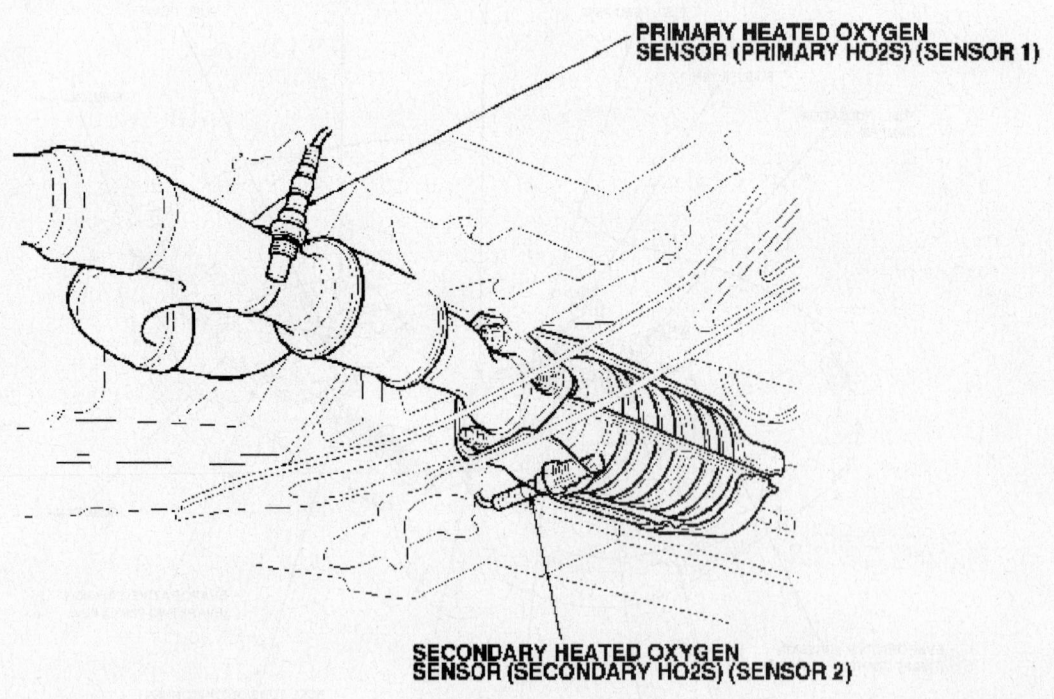

PRIMARY HEATED OXYGEN SENSOR (PRIMARY HO2S) (SENSOR 1)

SECONDARY HEATED OXYGEN SENSOR (SECONDARY HO2S) (SENSOR 2)

29130_ACUR_G0039

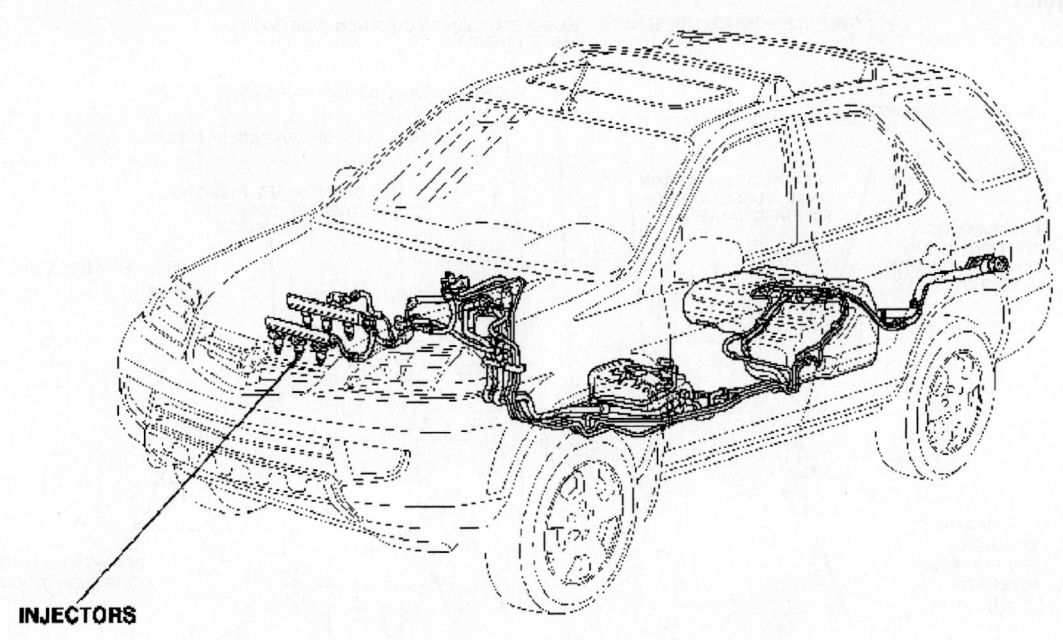

INJECTORS

29130_ACUR_G0040

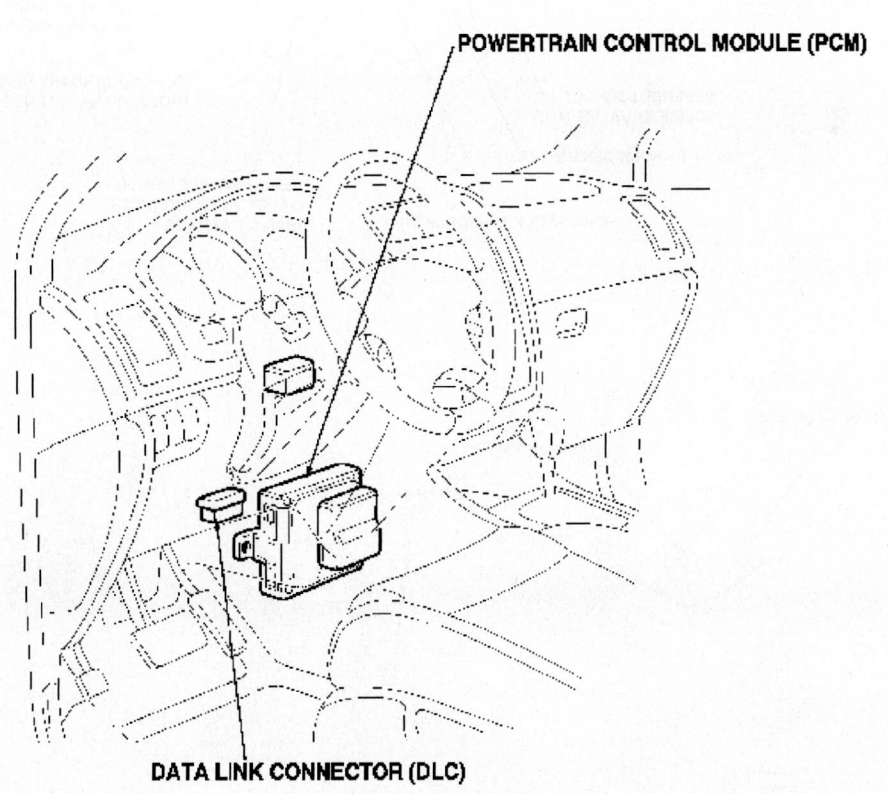

POWERTRAIN CONTROL MODULE (PCM)

DATA LINK CONNECTOR (DLC)

29130_ACUR_G0041

NSX

6 Cylinder Engines

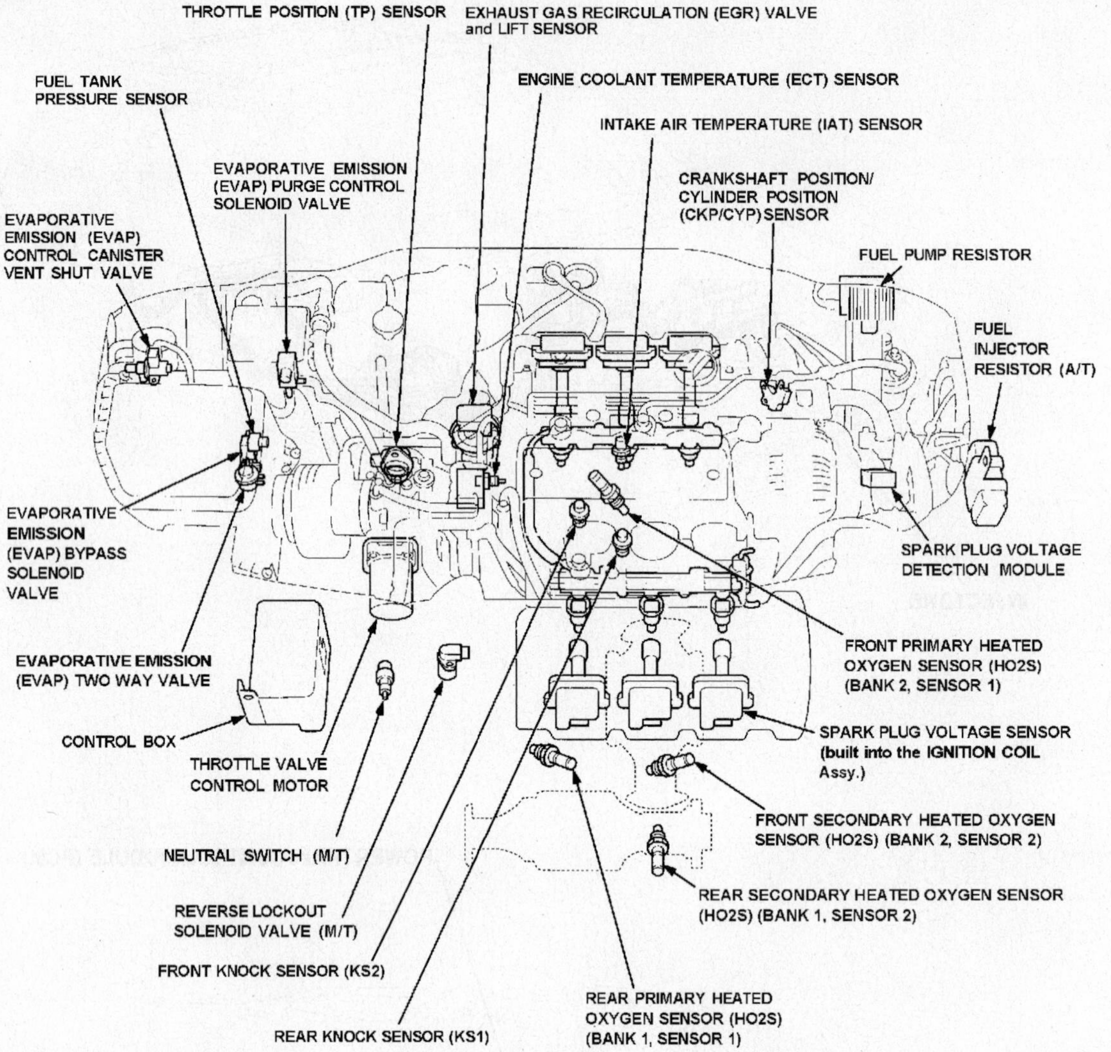

THROTTLE POSITION (TP) SENSOR

EXHAUST GAS RECIRCULATION (EGR) VALVE and LIFT SENSOR

ENGINE COOLANT TEMPERATURE (ECT) SENSOR

INTAKE AIR TEMPERATURE (IAT) SENSOR

FUEL TANK PRESSURE SENSOR

EVAPORATIVE EMISSION (EVAP) PURGE CONTROL SOLENOID VALVE

CRANKSHAFT POSITION/ CYLINDER POSITION (CKP/CYP) SENSOR

EVAPORATIVE EMISSION (EVAP) CONTROL CANISTER VENT SHUT VALVE

FUEL PUMP RESISTOR

FUEL INJECTOR RESISTOR (A/T)

EVAPORATIVE EMISSION (EVAP) BYPASS SOLENOID VALVE

SPARK PLUG VOLTAGE DETECTION MODULE

EVAPORATIVE EMISSION (EVAP) TWO WAY VALVE

FRONT PRIMARY HEATED OXYGEN SENSOR (HO2S) (BANK 2, SENSOR 1)

CONTROL BOX

SPARK PLUG VOLTAGE SENSOR (built into the IGNITION COIL Assy.)

THROTTLE VALVE CONTROL MOTOR

FRONT SECONDARY HEATED OXYGEN SENSOR (HO2S) (BANK 2, SENSOR 2)

NEUTRAL SWITCH (M/T)

REVERSE LOCKOUT SOLENOID VALVE (M/T)

REAR SECONDARY HEATED OXYGEN SENSOR (HO2S) (BANK 1, SENSOR 2)

FRONT KNOCK SENSOR (KS2)

REAR PRIMARY HEATED OXYGEN SENSOR (HO2S) (BANK 1, SENSOR 1)

REAR KNOCK SENSOR (KS1)

29130_ACUR_G0042

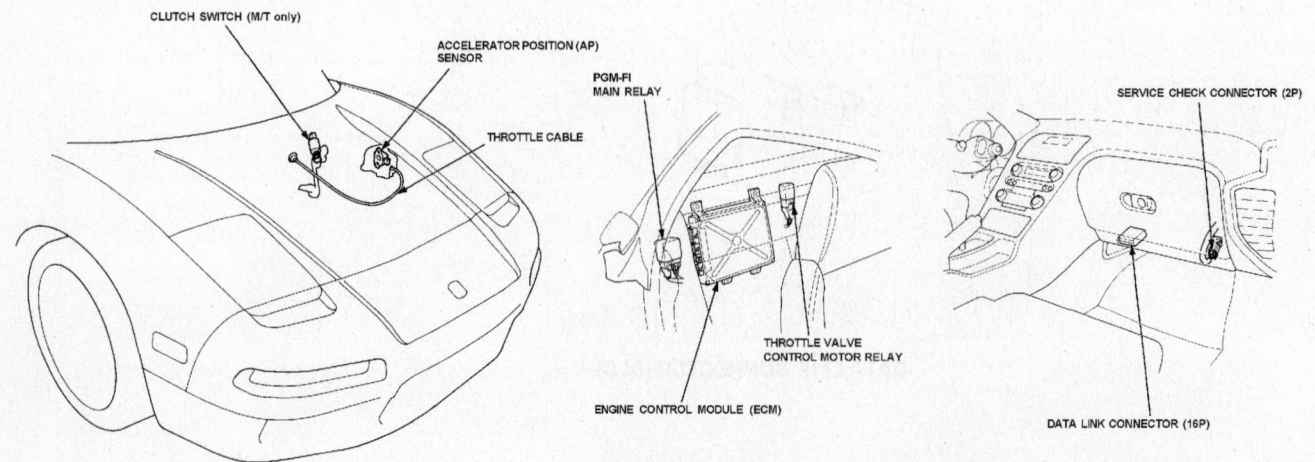

CLUTCH SWITCH (M/T only)

ACCELERATOR POSITION (AP) SENSOR

PGM-FI MAIN RELAY

SERVICE CHECK CONNECTOR (2P)

THROTTLE CABLE

THROTTLE VALVE CONTROL MOTOR RELAY

ENGINE CONTROL MODULE (ECM)

DATA LINK CONNECTOR (16P)

29130_ACUR_G0043

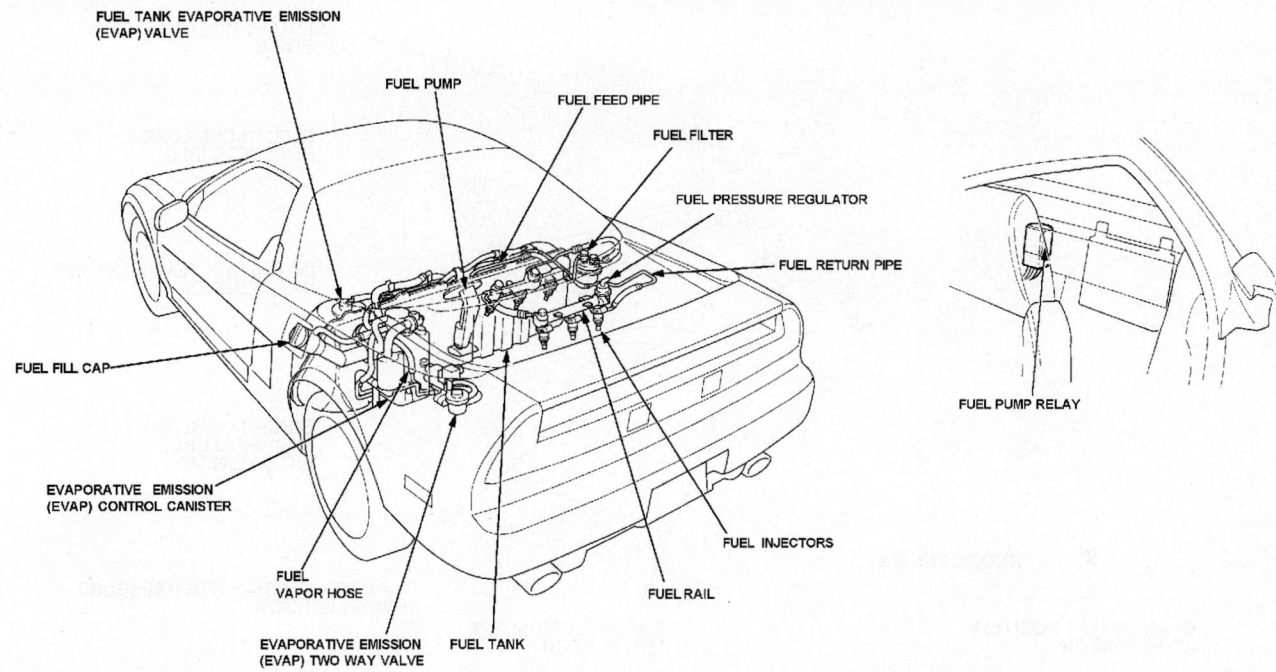

FUEL TANK EVAPORATIVE EMISSION (EVAP) VALVE

FUEL PUMP

FUEL FEED PIPE

FUEL FILTER

FUEL PRESSURE REGULATOR

FUEL RETURN PIPE

FUEL FILL CAP

EVAPORATIVE EMISSION (EVAP) CONTROL CANISTER

FUEL VAPOR HOSE

EVAPORATIVE EMISSION (EVAP) TWO WAY VALVE

FUEL TANK

FUEL RAIL

FUEL INJECTORS

FUEL PUMP RELAY

29130_ACUR_G0044

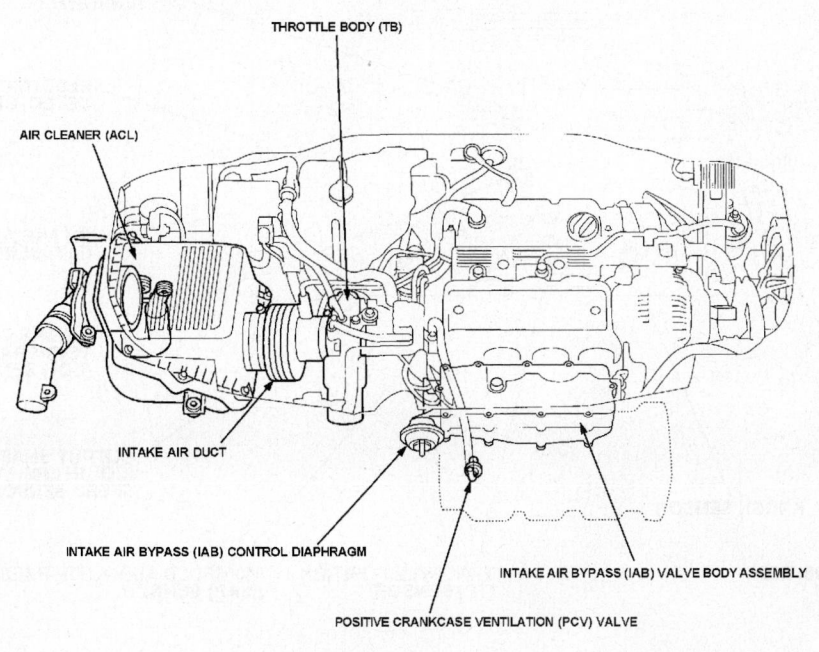

THROTTLE BODY (TB)

AIR CLEANER (ACL)

INTAKE AIR DUCT

INTAKE AIR BYPASS (IAB) CONTROL DIAPHRAGM

POSITIVE CRANKCASE VENTILATION (PCV) VALVE

INTAKE AIR BYPASS (IAB) VALVE BODY ASSEMBLY

29130_ACUR_G0045

RSX

4 Cylinder Engines

K20A2 engine

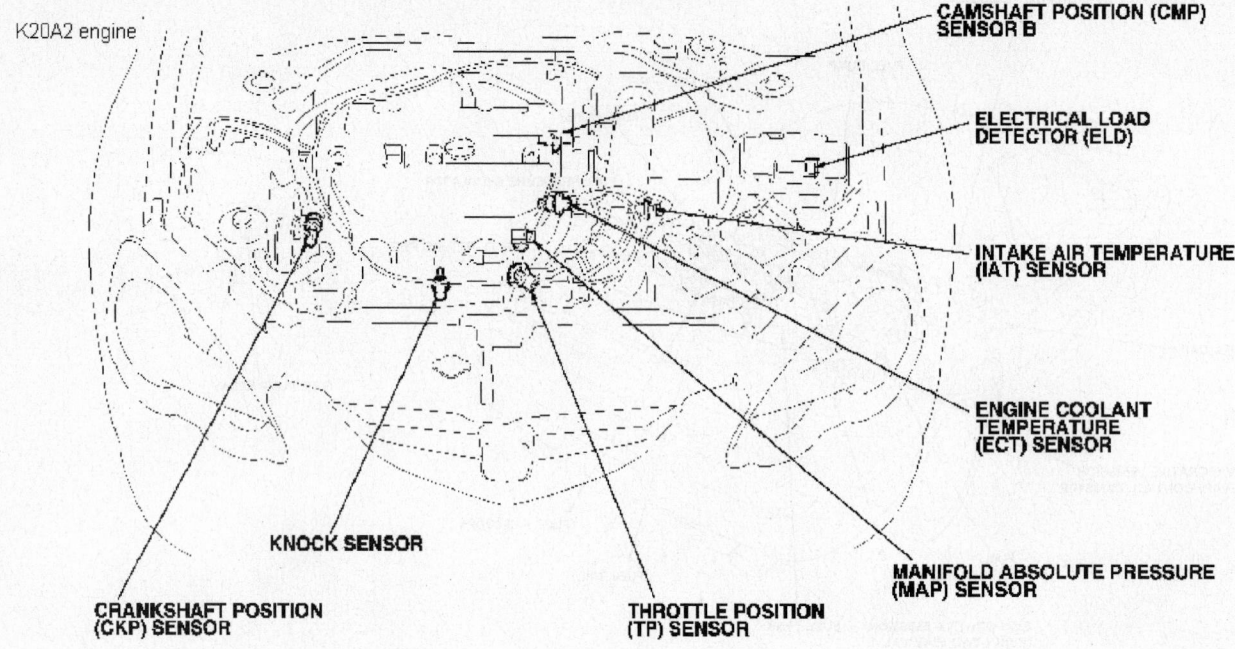

CAMSHAFT POSITION (CMP) SENSOR B

ELECTRICAL LOAD DETECTOR (ELD)

INTAKE AIR TEMPERATURE (IAT) SENSOR

ENGINE COOLANT TEMPERATURE (ECT) SENSOR

MANIFOLD ABSOLUTE PRESSURE (MAP) SENSOR

KNOCK SENSOR

THROTTLE POSITION (TP) SENSOR

CRANKSHAFT POSITION (CKP) SENSOR

29130_ACUR_G0046

K20Z1 engine

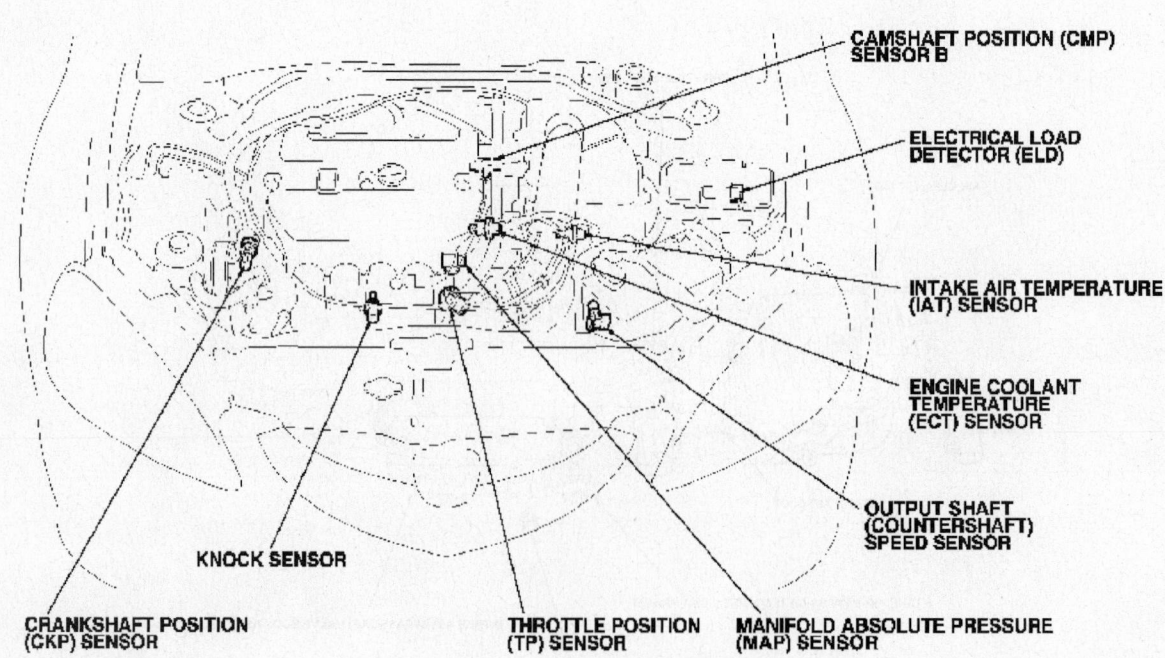

CAMSHAFT POSITION (CMP) SENSOR B

ELECTRICAL LOAD DETECTOR (ELD)

INTAKE AIR TEMPERATURE (IAT) SENSOR

ENGINE COOLANT TEMPERATURE (ECT) SENSOR

OUTPUT SHAFT (COUNTERSHAFT) SPEED SENSOR

KNOCK SENSOR

THROTTLE POSITION (TP) SENSOR

MANIFOLD ABSOLUTE PRESSURE (MAP) SENSOR

CRANKSHAFT POSITION (CKP) SENSOR

29130_ACUR_G0047

K20A3 engine-2002-2004 models

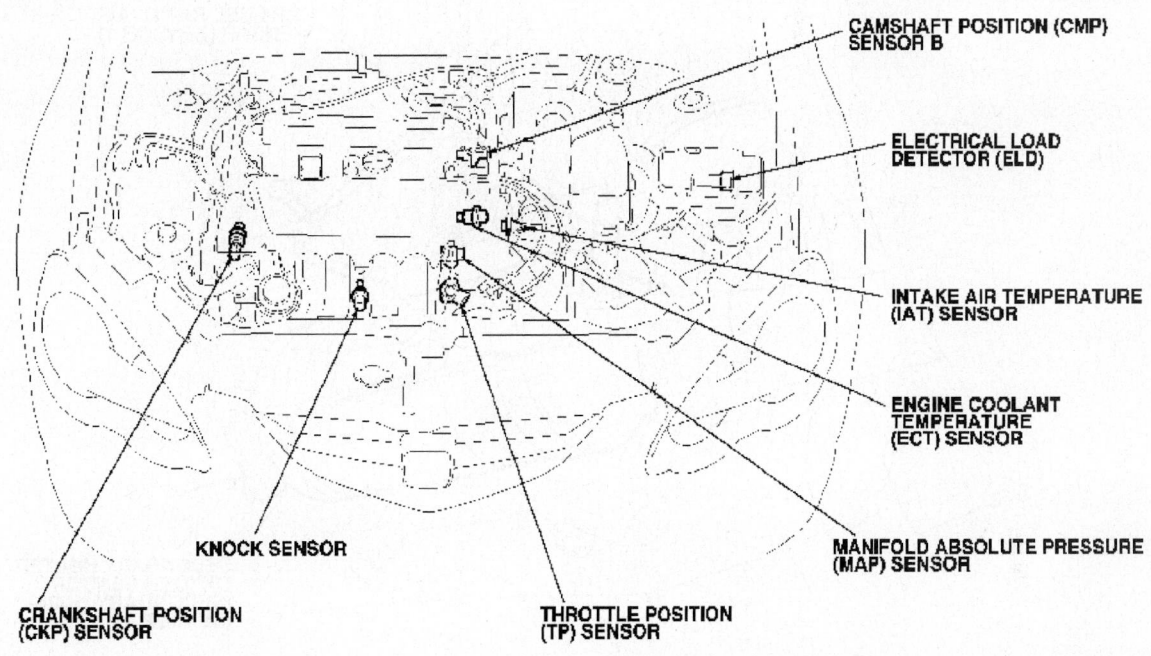

CAMSHAFT POSITION (CMP)
SENSOR B

ELECTRICAL LOAD
DETECTOR (ELD)

INTAKE AIR TEMPERATURE
(IAT) SENSOR

ENGINE COOLANT
TEMPERATURE
(ECT) SENSOR

MANIFOLD ABSOLUTE PRESSURE
(MAP) SENSOR

KNOCK SENSOR

CRANKSHAFT POSITION
(CKP) SENSOR

THROTTLE POSITION
(TP) SENSOR

29130_ACUR_G0048

K20A3 engine-2005-2006 models

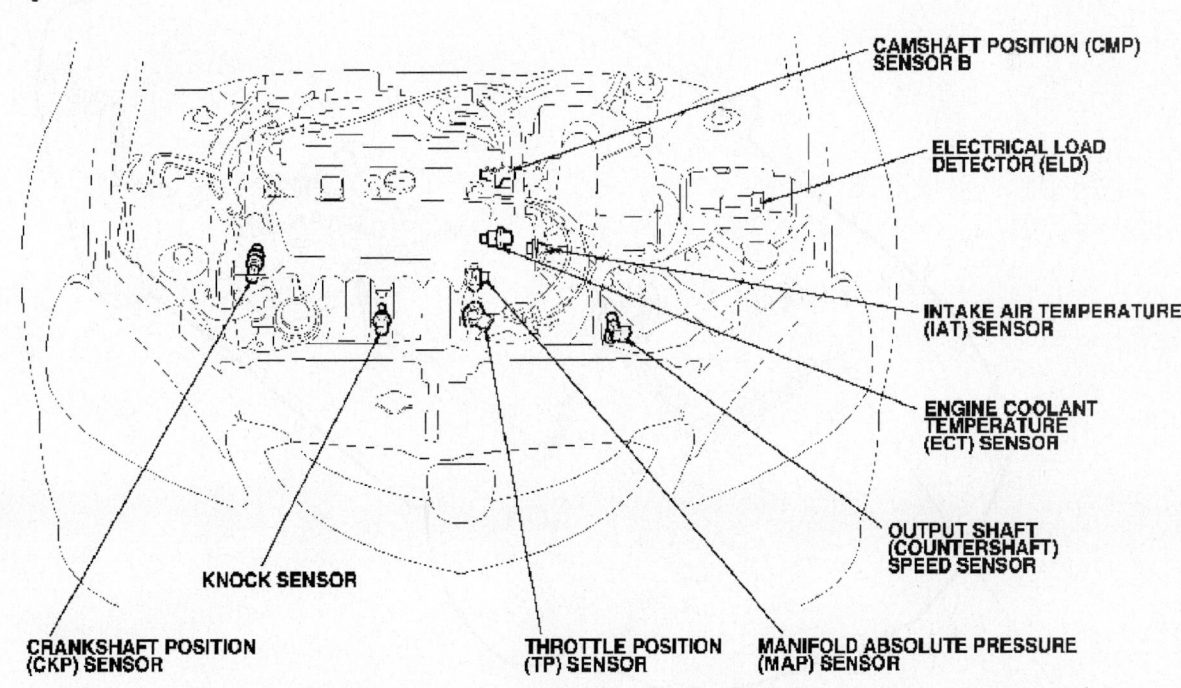

CAMSHAFT POSITION (CMP)
SENSOR B

ELECTRICAL LOAD
DETECTOR (ELD)

INTAKE AIR TEMPERATURE
(IAT) SENSOR

ENGINE COOLANT
TEMPERATURE
(ECT) SENSOR

OUTPUT SHAFT
(COUNTERSHAFT)
SPEED SENSOR

KNOCK SENSOR

CRANKSHAFT POSITION
(CKP) SENSOR

THROTTLE POSITION
(TP) SENSOR

MANIFOLD ABSOLUTE PRESSURE
(MAP) SENSOR

29130_ACUR_G0049

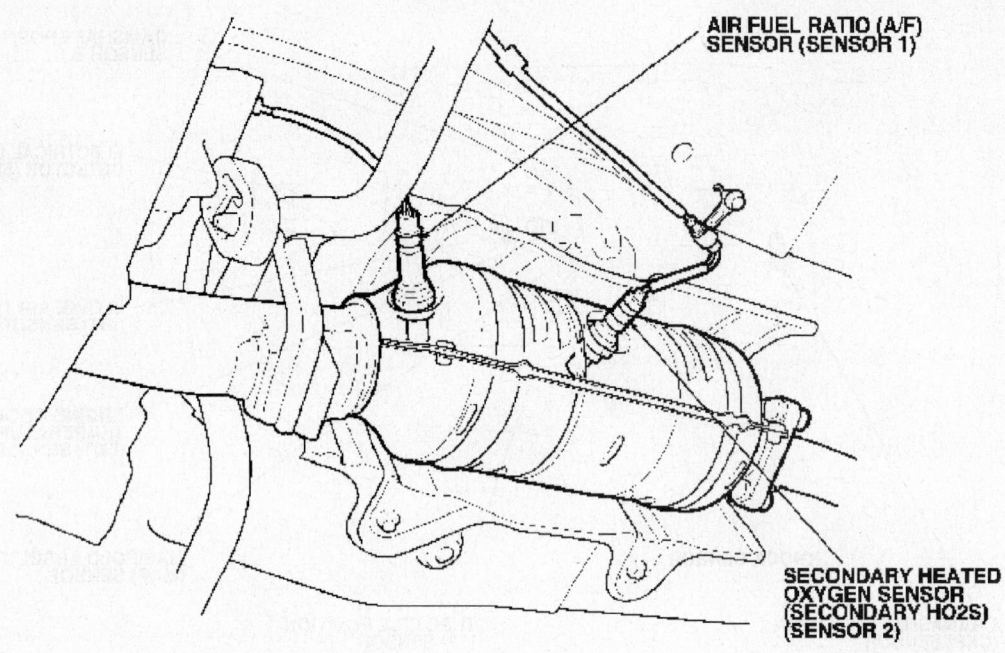

AIR FUEL RATIO (A/F)
SENSOR (SENSOR 1)

SECONDARY HEATED
OXYGEN SENSOR
(SECONDARY HO2S)
(SENSOR 2)

29130_ACUR_G0050

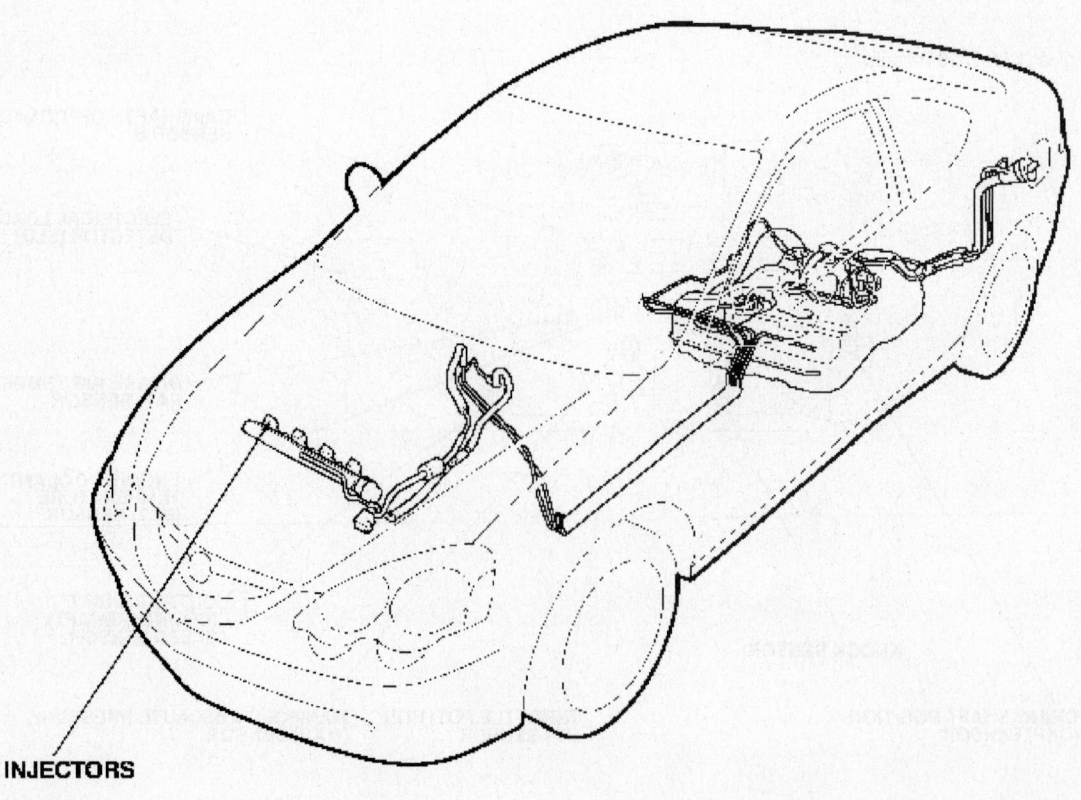

INJECTORS

29130_ACUR_G0051

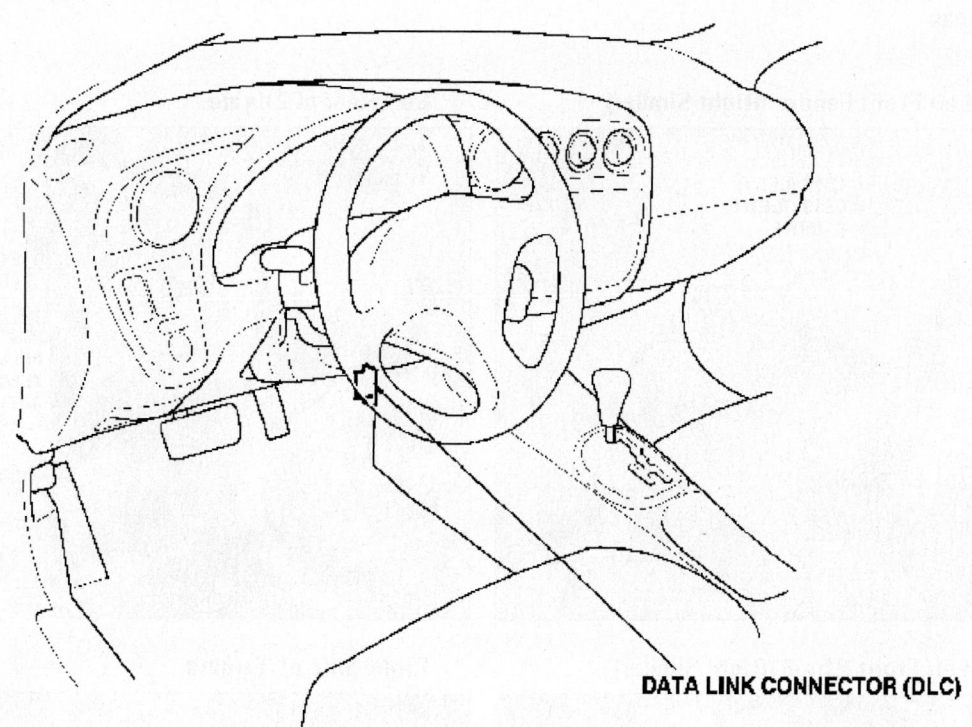

DATA LINK CONNECTOR (DLC)

29130_ACUR_G0052

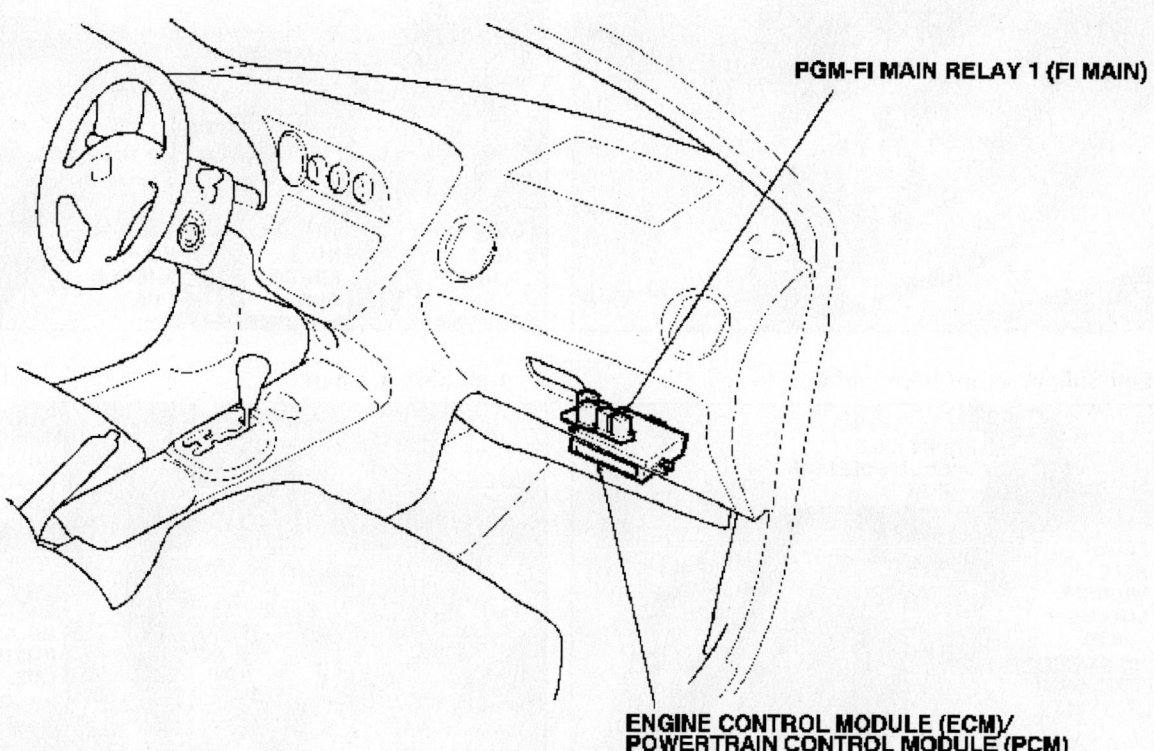

PGM-FI MAIN RELAY 1 (FI MAIN)

ENGINE CONTROL MODULE (ECM)/
POWERTRAIN CONTROL MODULE (PCM)

SLX

6 Cylinder Engines

Inside of Left Front Fender (Right Similar)

C313 (LEFT)
C219 (RIGHT)
(2-BRN)

Left Rear of Engine

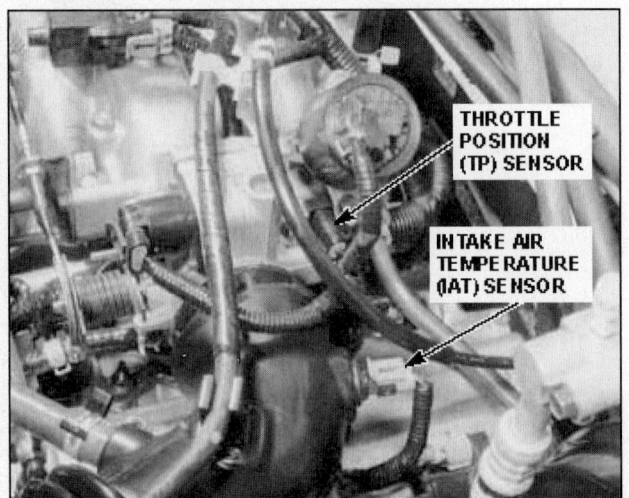

THROTTLE
POSITION
(TP) SENSOR

INTAKE AIR
TEMPERATURE
(IAT) SENSOR

Inside of Left Front Wheel (Right Similar)

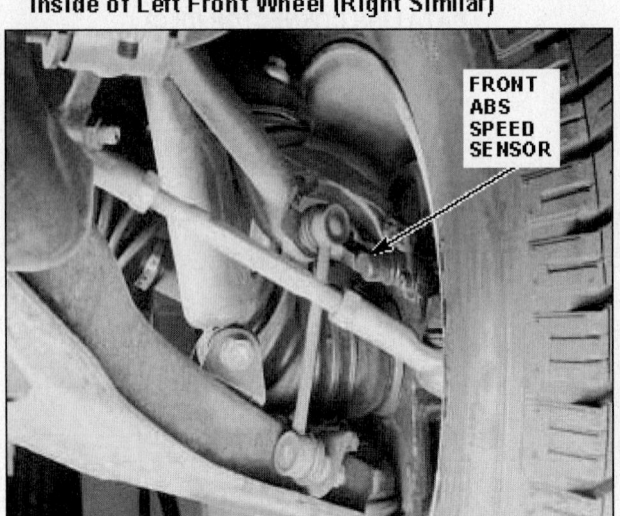

FRONT
ABS
SPEED
SENSOR

Right Side of Engine

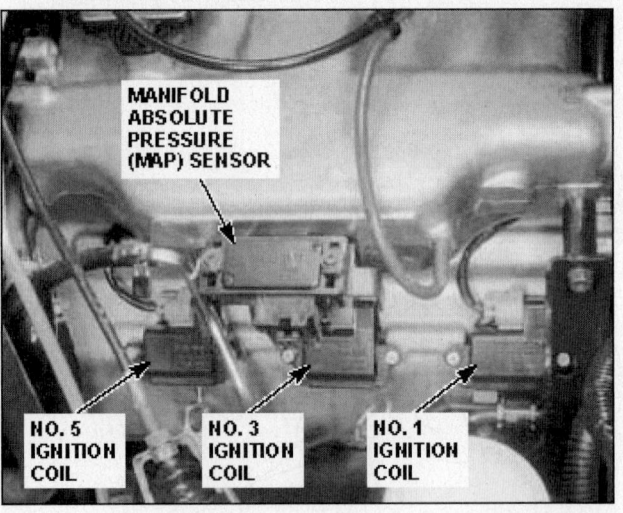

MANIFOLD
ABSOLUTE
PRESSURE
(MAP) SENSOR

NO. 5
IGNITION
COIL

NO. 3
IGNITION
COIL

NO. 1
IGNITION
COIL

Left Side of Front Differential

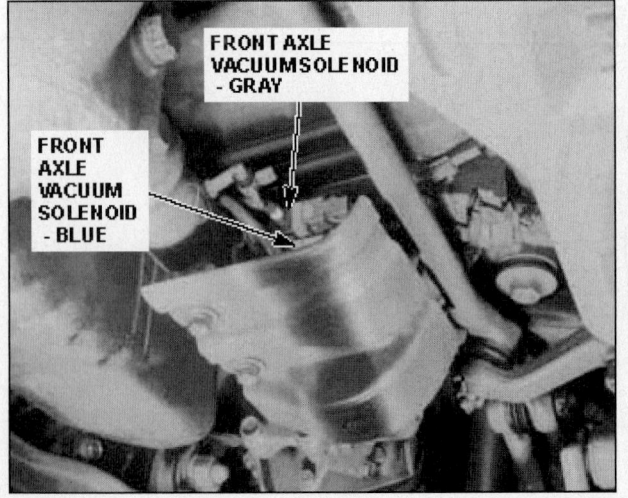

FRONT AXLE
VACUUM SOLENOID
- GRAY

FRONT
AXLE
VACUUM
SOLENOID
- BLUE

Left Side of Engine

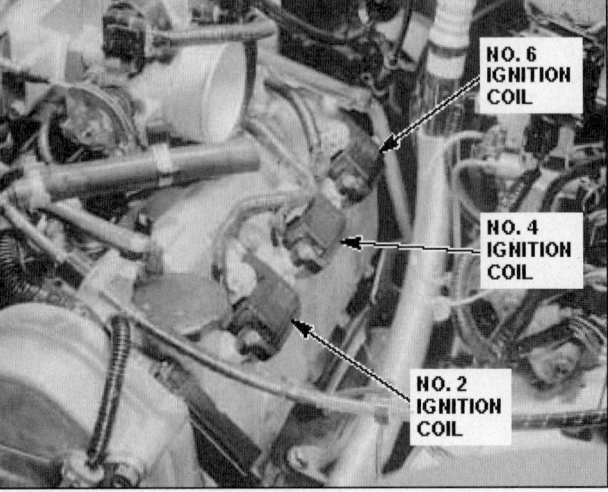

NO. 6
IGNITION
COIL

NO. 4
IGNITION
COIL

NO. 2
IGNITION
COIL

29130_ACUR_P0001

Top of Left Tailgate Door

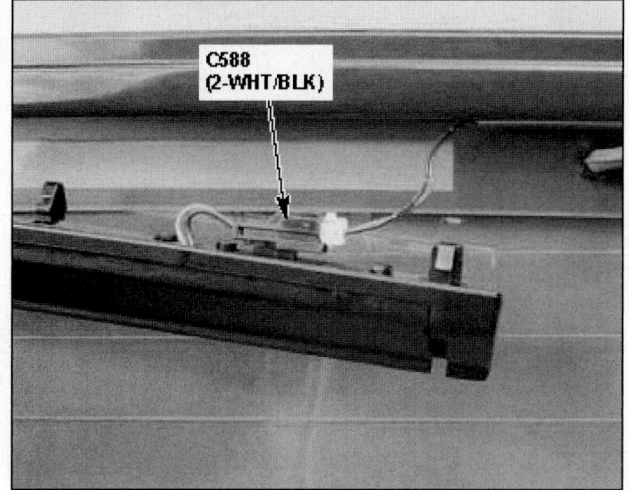

C588
(2-WHT/BLK)

Right Front Corner of Fuel Tank

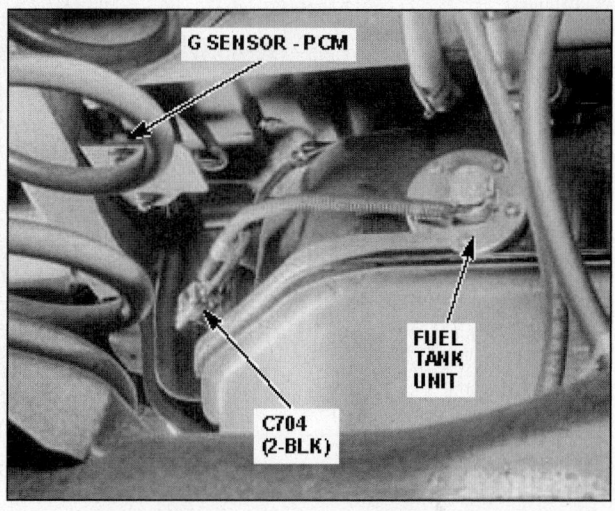

G SENSOR - PCM

FUEL
TANK
UNIT

C704
(2-BLK)

Right Rear Underside of Vehicle

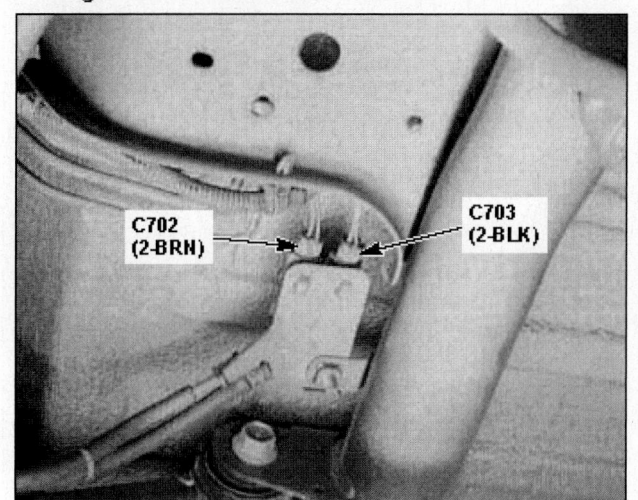

C702
(2-BRN)

C703
(2-BLK)

Inside of Left Rear Wheel (Right Similar)

REAR
ABS
SPEED
SENSOR

Center Front of Roof

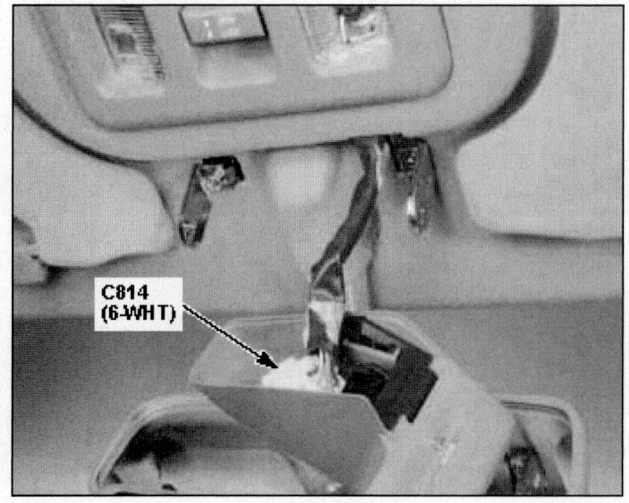

C814
(6-WHT)

Left Rear Underside of Vehicle

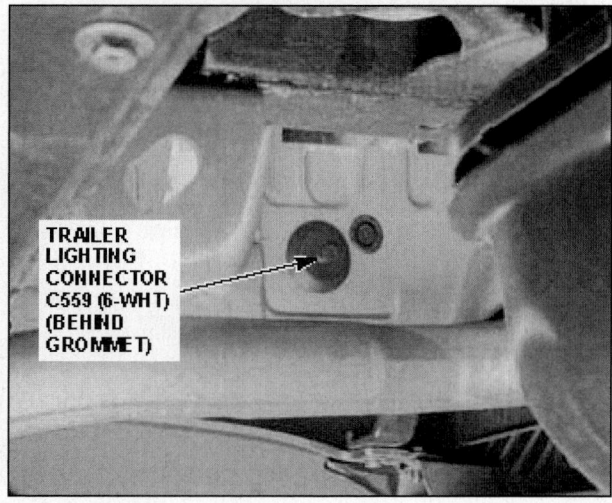

TRAILER
LIGHTING
CONNECTOR
C559 (6-WHT)
(BEHIND
GROMMET)

Underside of Driver's Seat

C724
(3-GRY)

C725
(2-GRY)

C729
(2-BLK)

Underside of Right Front Seat

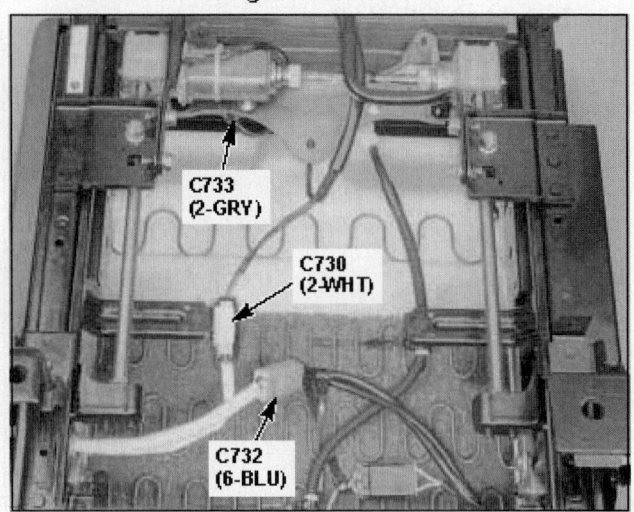

C733
(2-GRY)

C730
(2-WHT)

C732
(6-BLU)

Below Right Front Seat

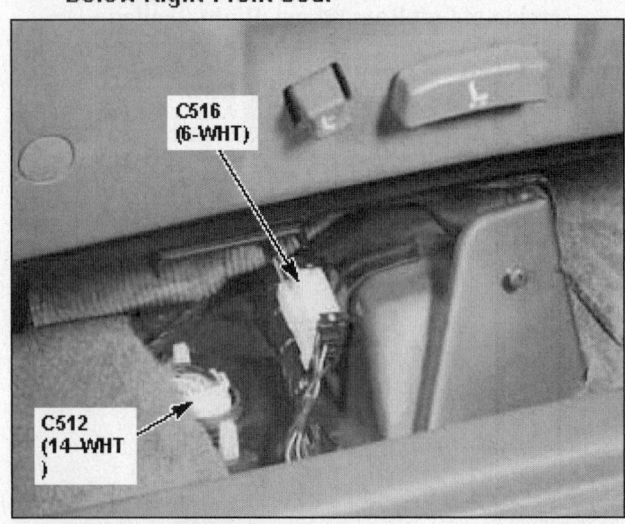

C516
(6-WHT)

C512
(14-WHT
)

Driver's Seat Back (Right Front Similar)

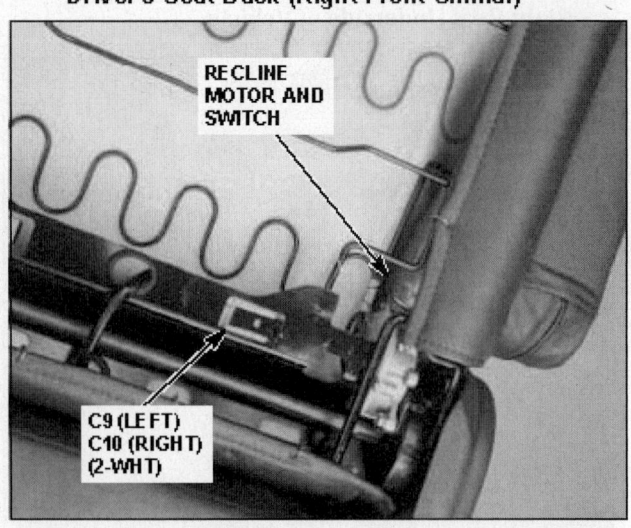

RECLINE
MOTOR AND
SWITCH

C9 (LEFT)
C10 (RIGHT)
(2-WHT)

Underside of Right Front Seat

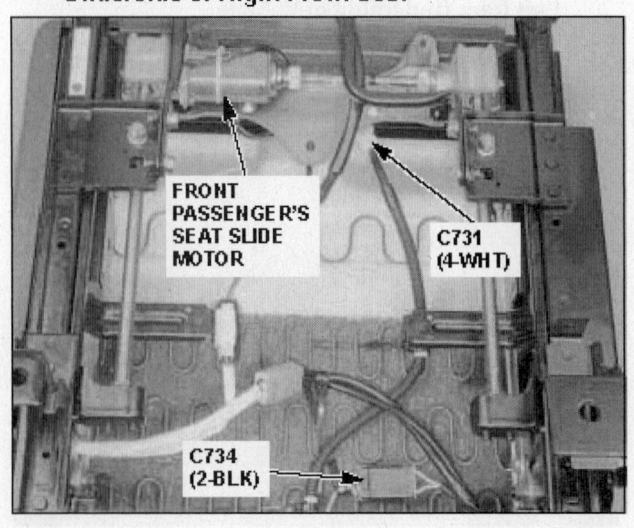

FRONT
PASSENGER'S
SEAT SLIDE
MOTOR

C731
(4-WHT)

C734
(2-BLK)

Top Rear of Engine (On Stand)

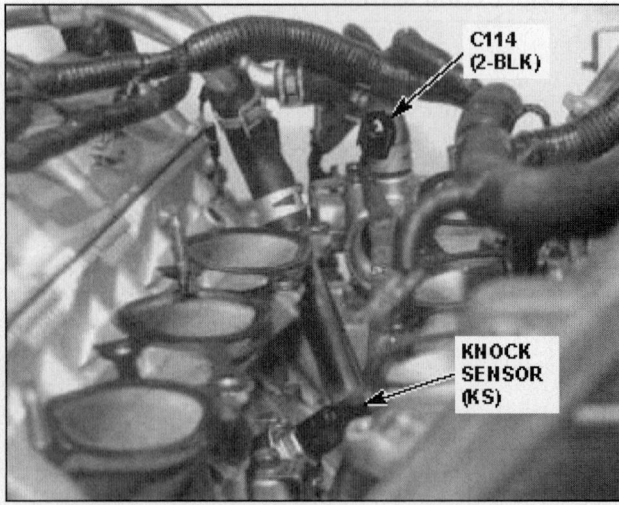

C114
(2-BLK)

KNOCK
SENSOR
(KS)

Left Rear of Transmission (M/T)

VEHICLE
SPEED
SENSOR
(VSS)

Left Side of Manual Transmission

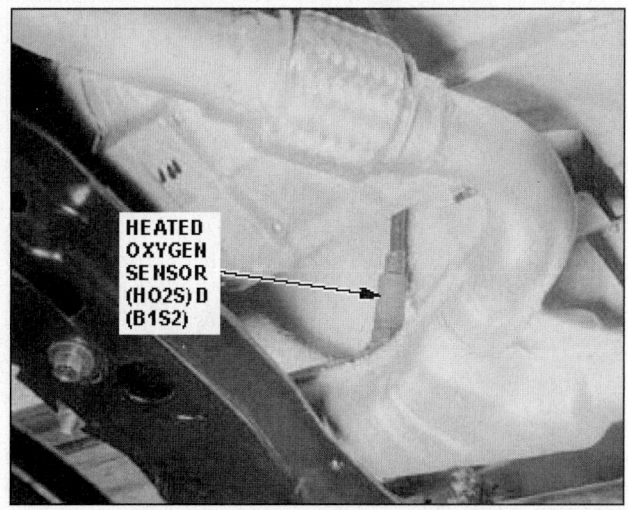

HEATED
OXYGEN
SENSOR
(HO2S) D
(B1S2)

Left Side of Transmission (M/T)

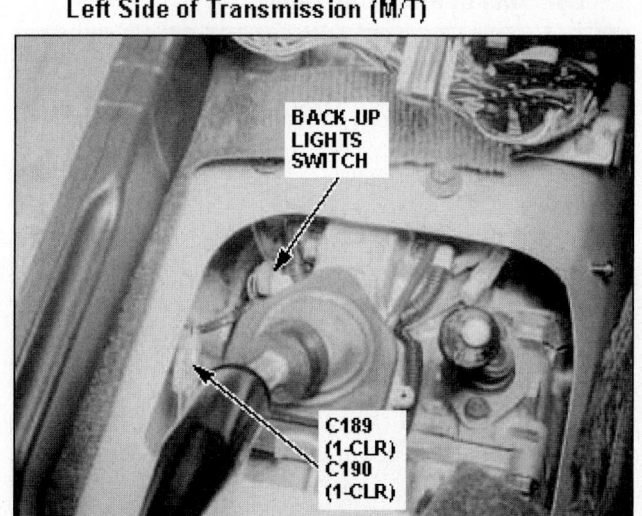

BACK-UP
LIGHTS
SWITCH

C189
(1-CLR)
C190
(1-CLR)

Right Front Corner of Engine Compartment

C329
(16-BLK)

Left Rear of Transmission (M/T)

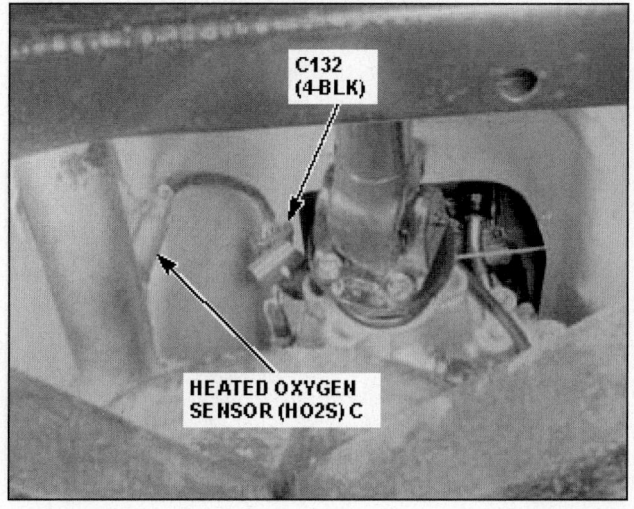

C132
(4-BLK)

HEATED OXYGEN
SENSOR (HO2S) C

Behind Front Console

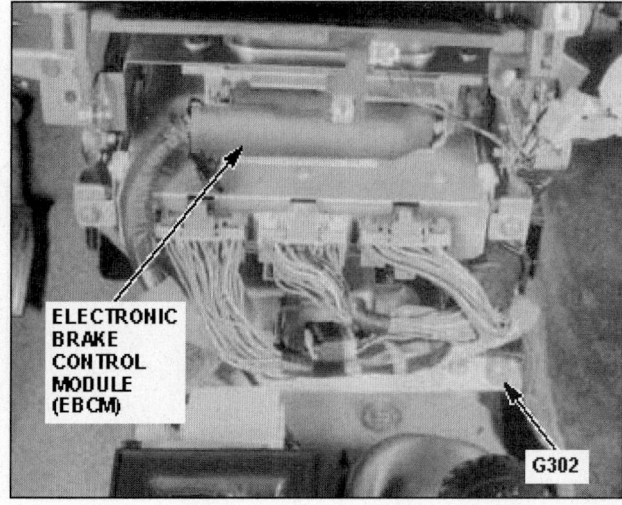

ELECTRONIC
BRAKE
CONTROL
MODULE
(EBCM)

G302

Front Left Side of Engine Compartment

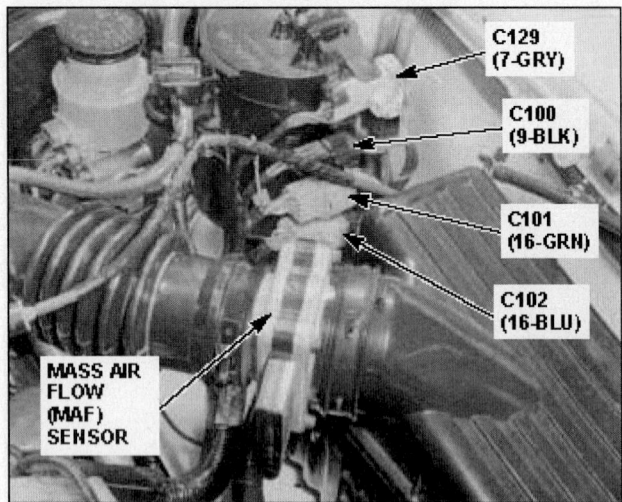

C129
(7-GRY)

C100
(9-BLK)

C101
(16-GRN)

C102
(16-BLU)

MASS AIR
FLOW
(MAF)
SENSOR

Top Center of Engine

EXHAUST GAS
RECIRCULATION
(EGR) VALVE

IDLE AIR
CONTROL
(IAC) VALVE

Top Center of Engine

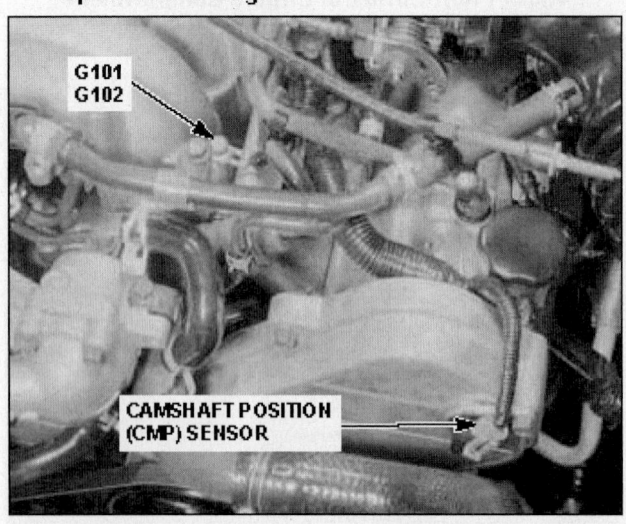

G101
G102

CAMSHAFT POSITION
(CMP) SENSOR

Left Side of Front Differential

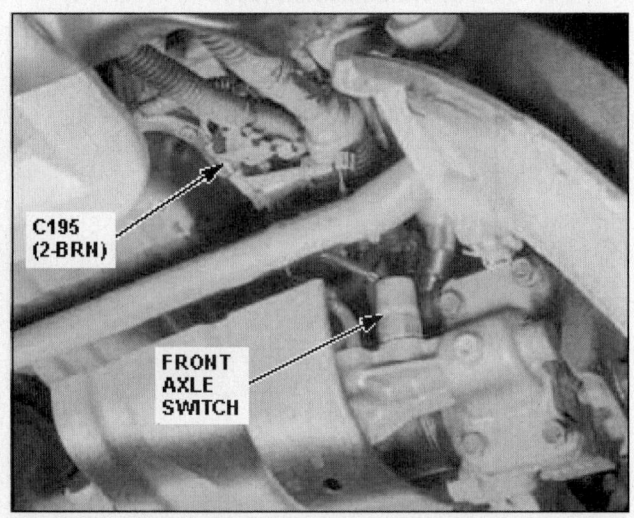

C195
(2-BRN)

FRONT
AXLE
SWITCH

Top Center of Engine

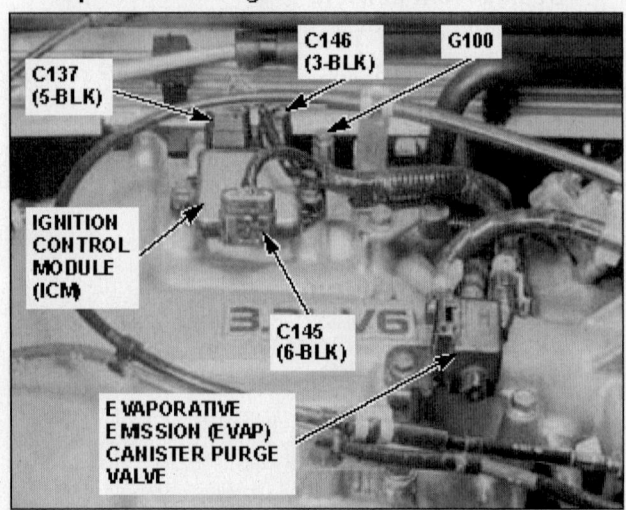

C137
(5-BLK)

C146
(3-BLK)

G100

IGNITION
CONTROL
MODULE
(ICM)

C145
(6-BLK)

EVAPORATIVE
EMISSION (EVAP)
CANISTER PURGE
VALVE

Lower Right Side of Engine

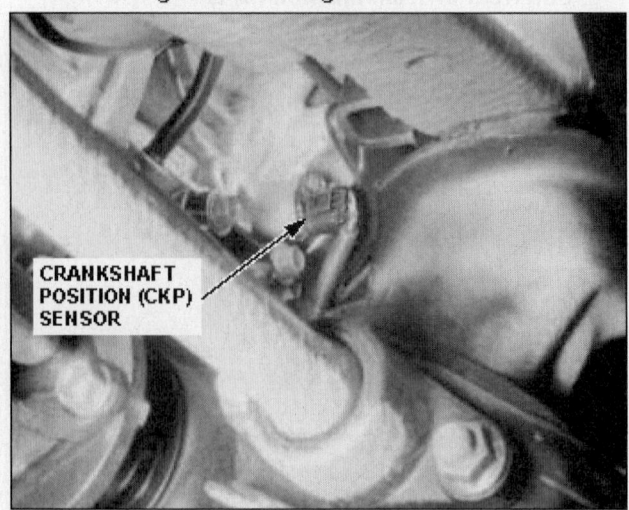

CRANKSHAFT
POSITION (CKP)
SENSOR

Left Exhaust Down Pipe

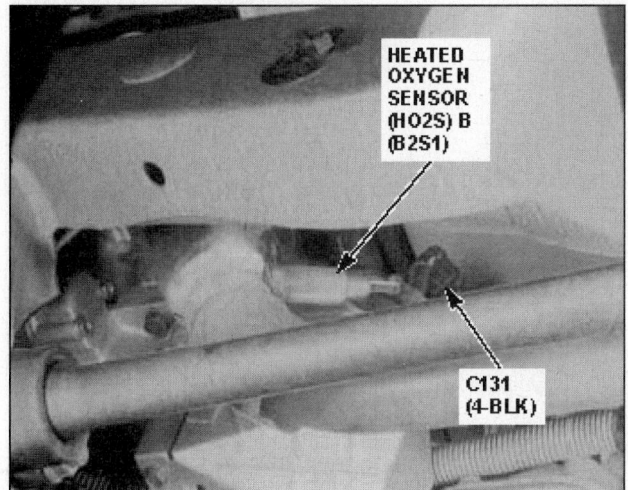

HEATED OXYGEN SENSOR (HO2S) B (B2S1)

C131 (4-BLK)

Left Rear Corner of Engine Compartment

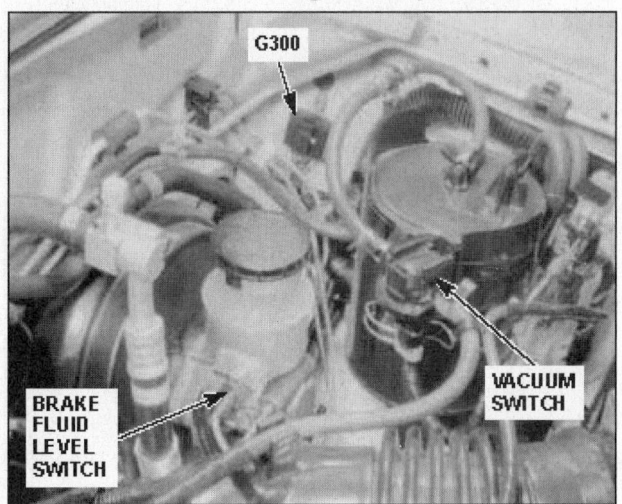

G300

BRAKE FLUID LEVEL SWITCH

VACUUM SWITCH

Rear of Engine

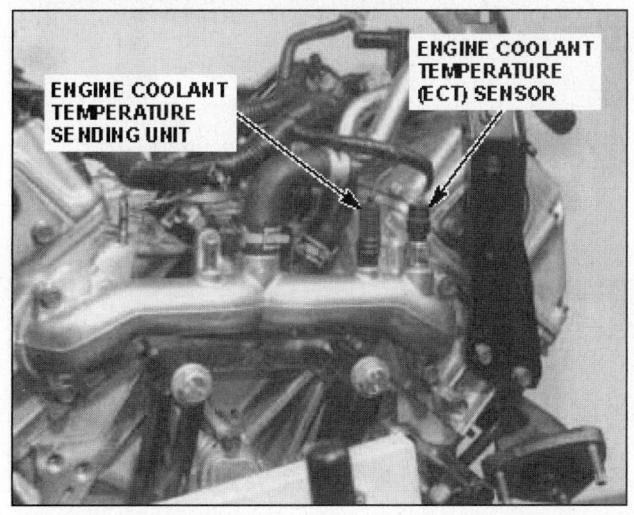

ENGINE COOLANT TEMPERATURE SENDING UNIT

ENGINE COOLANT TEMPERATURE (ECT) SENSOR

Rear of Right Catalytic Converter (A/T)

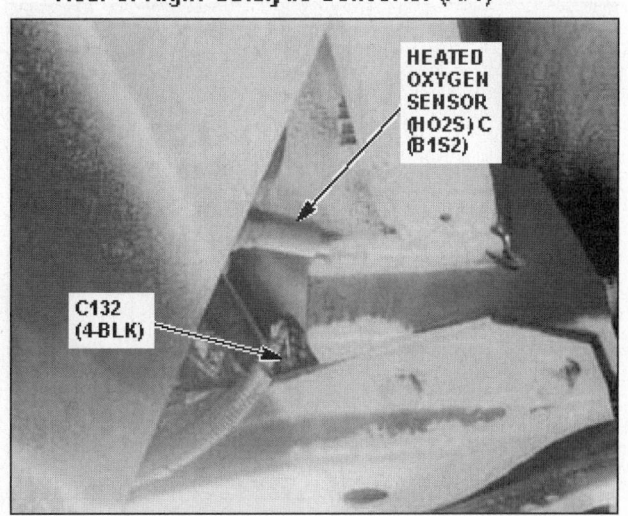

HEATED OXYGEN SENSOR (HO2S) C (B1S2)

C132 (4-BLK)

Left Front Corner of Engine Compartment

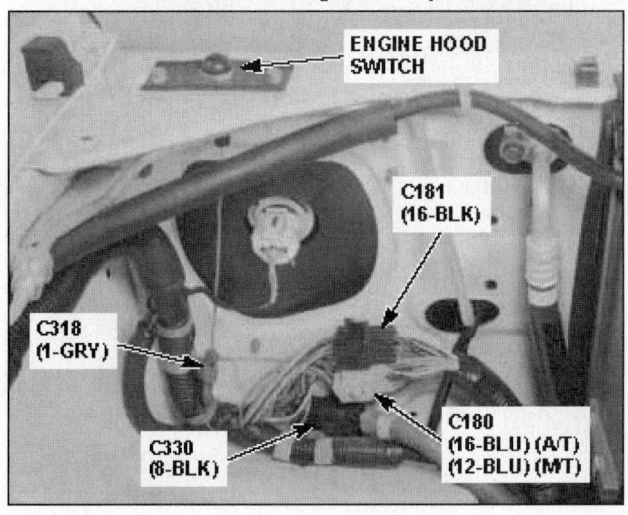

ENGINE HOOD SWITCH

C181 (16-BLK)

C318 (1-GRY)

C330 (8-BLK)

C180 (16-BLU) (A/T) (12-BLU) (MT)

Rear of Left Catalytic Converter (A/T)

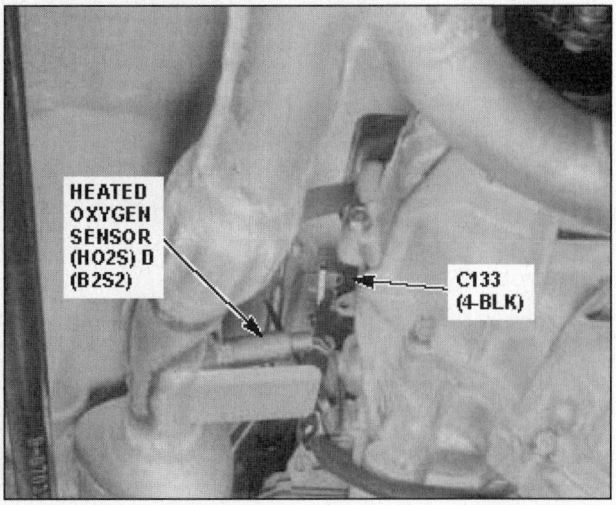

HEATED OXYGEN SENSOR (HO2S) D (B2S2)

C133 (4-BLK)

Right Side of Engine Compartment

T100 G4

Left Side of Engine Compartment

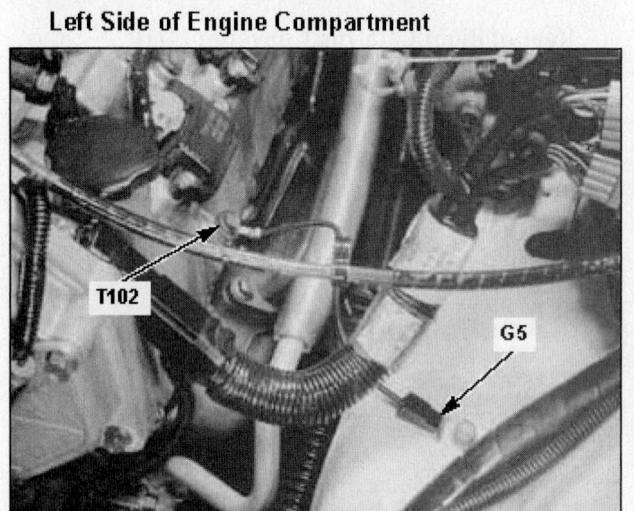

T102 G5

Right Exhaust Down Pipe

HEATED OXYGEN SENSOR (HO2S) A (B1S1)

Lower Right Front of Engine Compartment

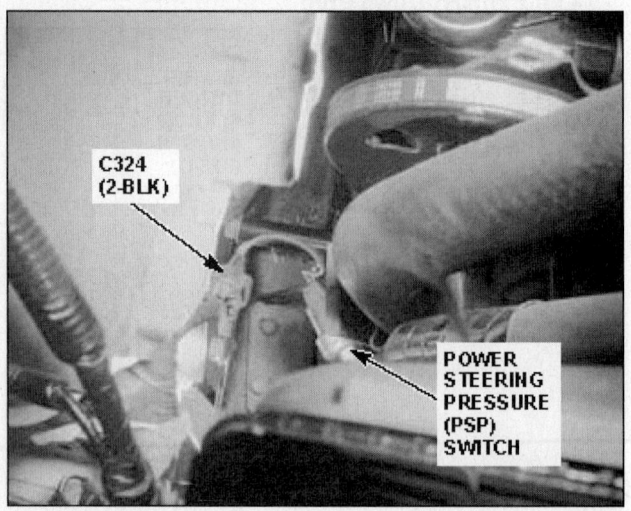

C324 (2-BLK)

POWER STEERING PRESSURE (PSP) SWITCH

Lower Front Center of Engine

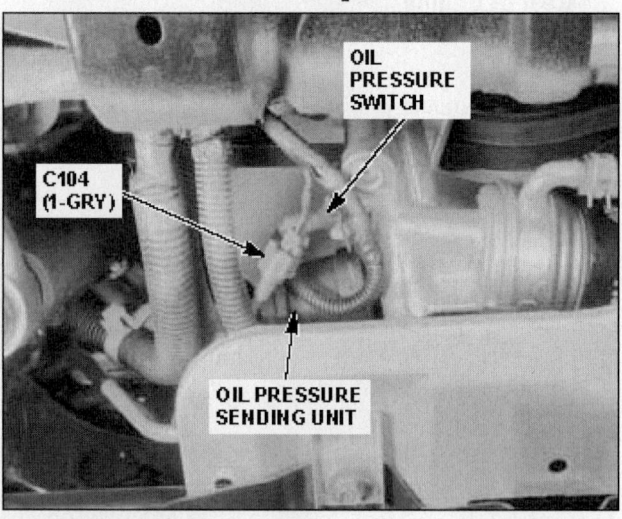

OIL PRESSURE SWITCH

C104 (1-GRY)

OIL PRESSURE SENDING UNIT

Left Underside of Engine

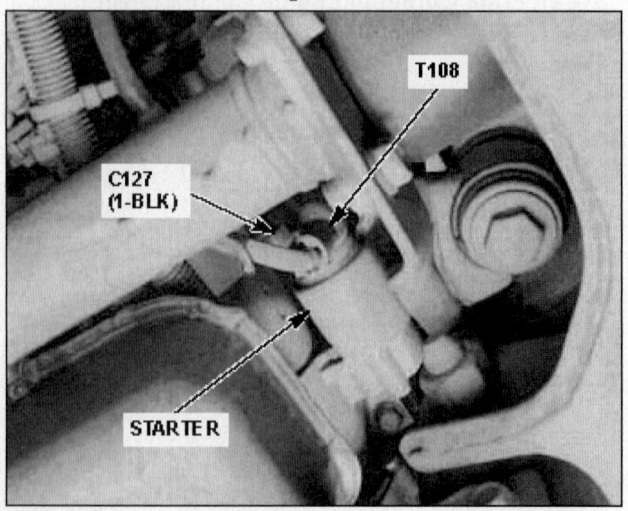

T108

C127 (1-BLK)

STARTER

TL

6 Cylinder Engines

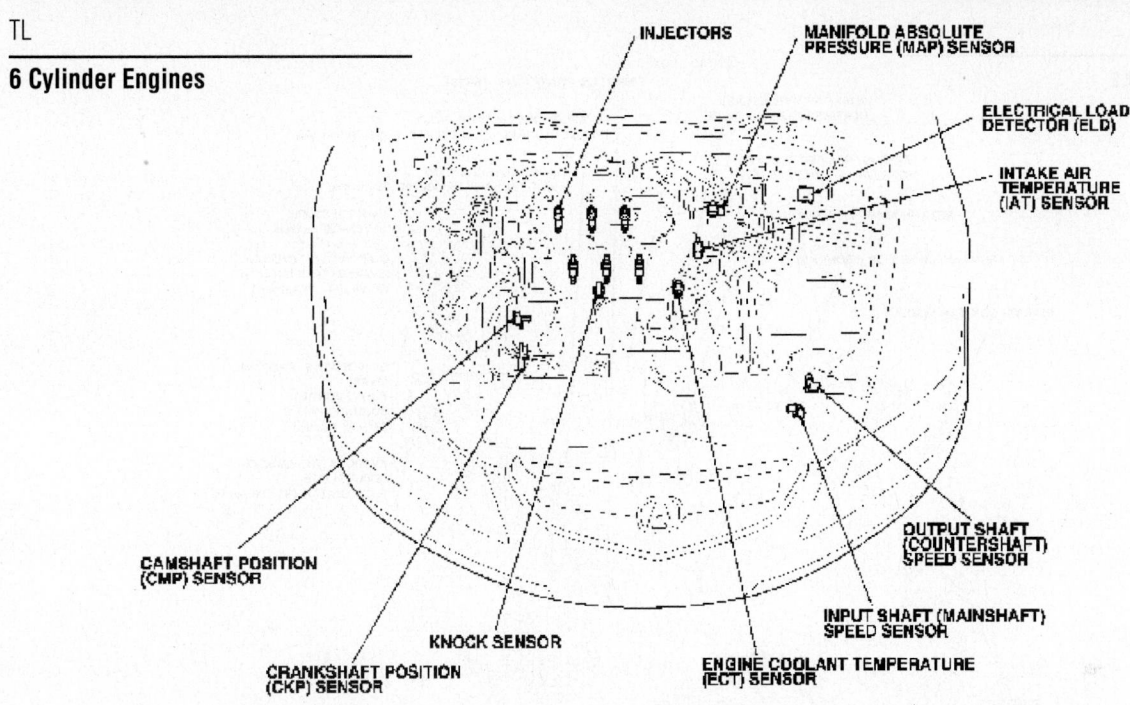

INJECTORS

MANIFOLD ABSOLUTE
PRESSURE (MAP) SENSOR

ELECTRICAL LOAD
DETECTOR (ELD)

INTAKE AIR
TEMPERATURE
(IAT) SENSOR

CAMSHAFT POSITION
(CMP) SENSOR

OUTPUT SHAFT
(COUNTERSHAFT)
SPEED SENSOR

KNOCK SENSOR

INPUT SHAFT (MAINSHAFT)
SPEED SENSOR

CRANKSHAFT POSITION
(CKP) SENSOR

ENGINE COOLANT TEMPERATURE
(ECT) SENSOR

29130_ACUR_G0014

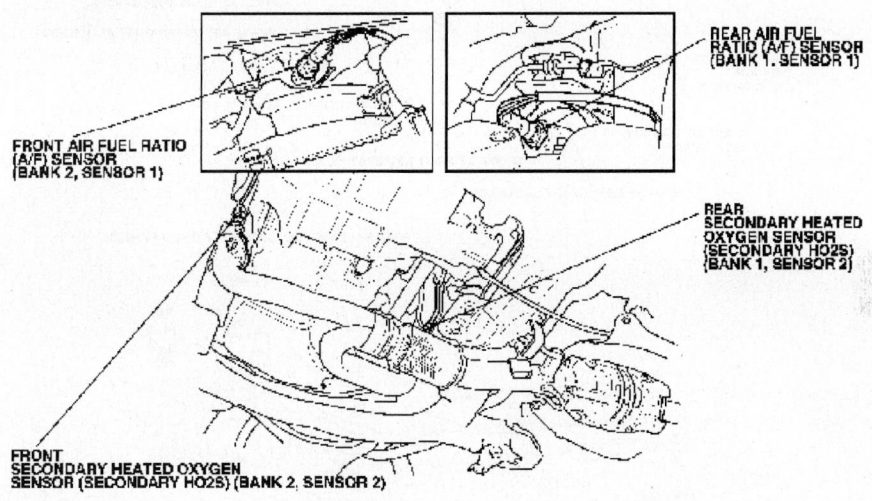

REAR AIR FUEL
RATIO (A/F) SENSOR
(BANK 1, SENSOR 1)

FRONT AIR FUEL RATIO
(A/F) SENSOR
(BANK 2, SENSOR 1)

REAR
SECONDARY HEATED
OXYGEN SENSOR
(SECONDARY HO2S)
(BANK 1, SENSOR 2)

FRONT
SECONDARY HEATED OXYGEN
SENSOR (SECONDARY HO2S) (BANK 2, SENSOR 2)

29130_ACUR_G0015

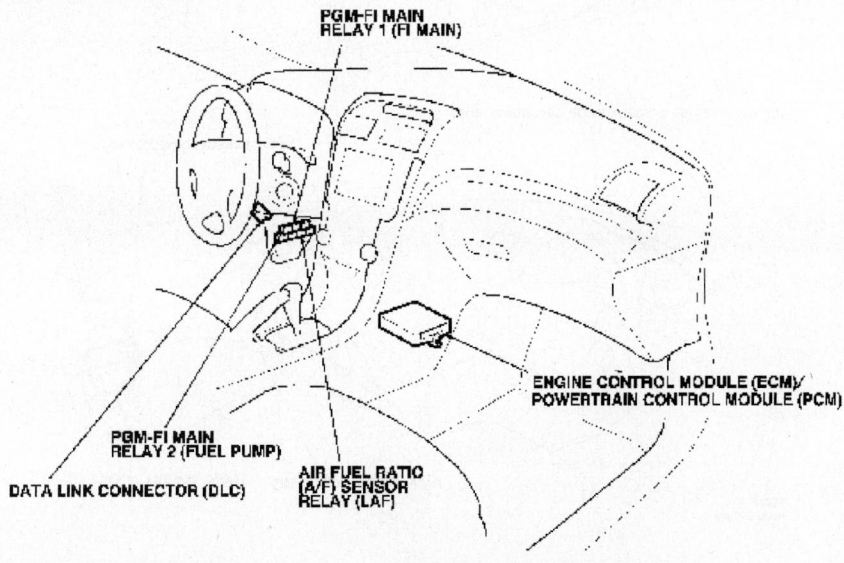

PGM-FI MAIN
RELAY 1 (FI MAIN)

ENGINE CONTROL MODULE (ECM)/
POWERTRAIN CONTROL MODULE (PCM)

PGM-FI MAIN
RELAY 2 (FUEL PUMP)

DATA LINK CONNECTOR (DLC)

AIR FUEL RATIO
(A/F) SENSOR
RELAY (LAF)

29130_ACUR_G0016

2.5TL

5 Cylinder Engines

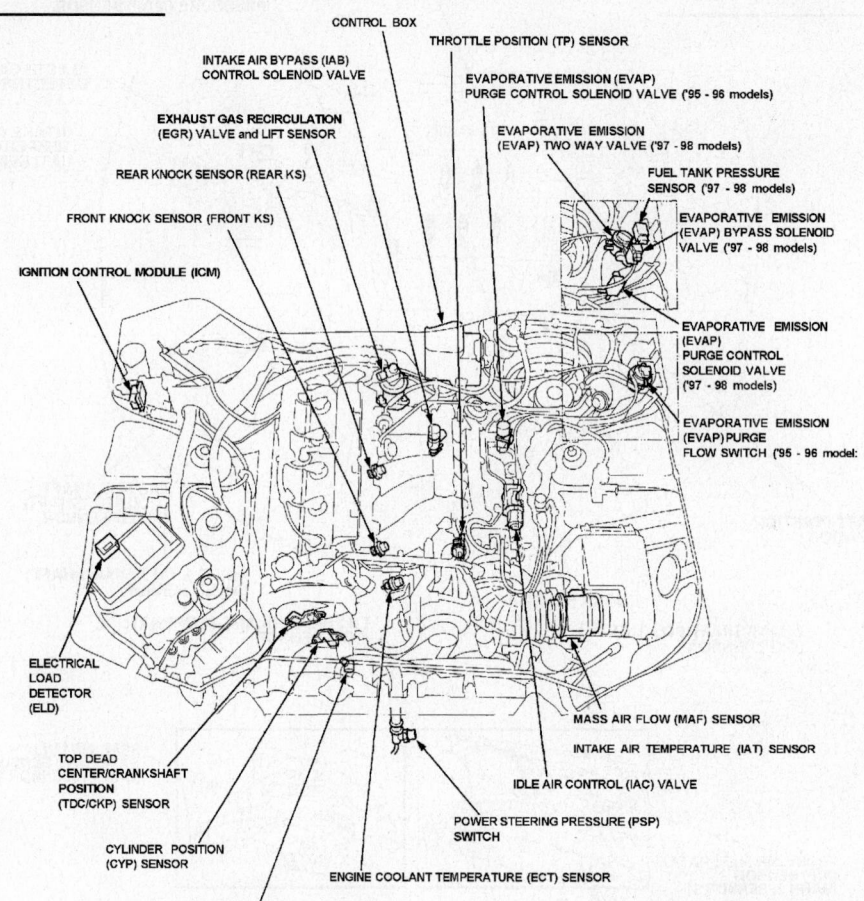

CONTROL BOX

INTAKE AIR BYPASS (IAB)
CONTROL SOLENOID VALVE

THROTTLE POSITION (TP) SENSOR

EVAPORATIVE EMISSION (EVAP)
PURGE CONTROL SOLENOID VALVE ('95 - 96 models)

EXHAUST GAS RECIRCULATION
(EGR) VALVE and LIFT SENSOR

EVAPORATIVE EMISSION
(EVAP) TWO WAY VALVE ('97 - 98 models)

REAR KNOCK SENSOR (REAR KS)

FUEL TANK PRESSURE
SENSOR ('97 - 98 models)

FRONT KNOCK SENSOR (FRONT KS)

EVAPORATIVE EMISSION
(EVAP) BYPASS SOLENOID
VALVE ('97 - 98 models)

IGNITION CONTROL MODULE (ICM)

EVAPORATIVE EMISSION
(EVAP)
PURGE CONTROL
SOLENOID VALVE
('97 - 98 models)

EVAPORATIVE EMISSION
(EVAP) PURGE
FLOW SWITCH ('95 - 96 model:

ELECTRICAL
LOAD
DETECTOR
(ELD)

MASS AIR FLOW (MAF) SENSOR

INTAKE AIR TEMPERATURE (IAT) SENSOR

TOP DEAD
CENTER/CRANKSHAFT
POSITION
(TDC/CKP) SENSOR

IDLE AIR CONTROL (IAC) VALVE

POWER STEERING PRESSURE (PSP)
SWITCH

CYLINDER POSITION
(CYP) SENSOR

ENGINE COOLANT TEMPERATURE (ECT) SENSOR

CRANKSHAFT SPEED FLUCTUATION

29130_ACUR_G0017

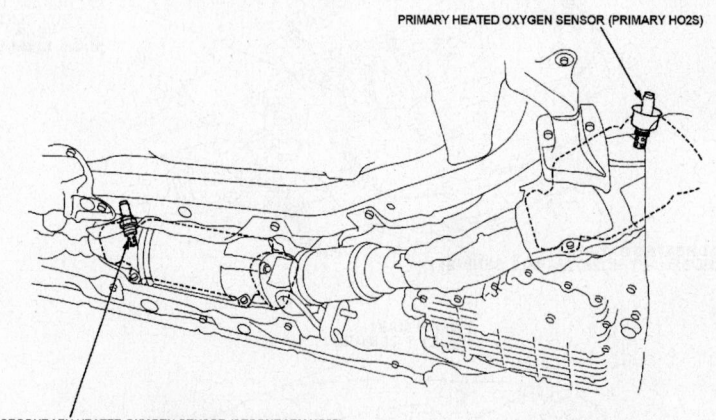

PRIMARY HEATED OXYGEN SENSOR (PRIMARY HO2S)

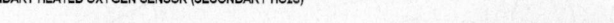

SECONDARY HEATED OXYGEN SENSOR (SECONDARY HO2S)

SERVICE CHECK CONNECTOR (2P)

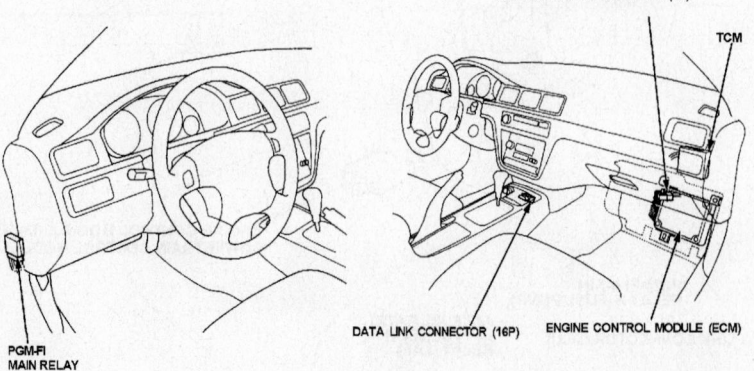

TCM

PGM-FI
MAIN RELAY

DATA LINK CONNECTOR (16P)

ENGINE CONTROL MODULE (ECM)

29130_ACUR_G0018

'95 - 96 models:

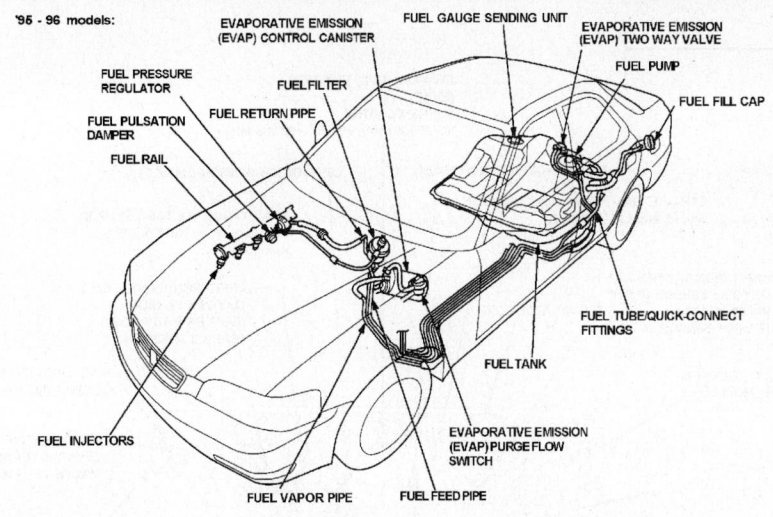

FUEL GAUGE SENDING UNIT

EVAPORATIVE EMISSION
(EVAP) CONTROL CANISTER

EVAPORATIVE EMISSION
(EVAP) TWO WAY VALVE

FUEL PUMP

FUEL PRESSURE
REGULATOR

FUEL FILTER

FUEL FILL CAP

FUEL RETURN PIPE

FUEL PULSATION
DAMPER

FUEL RAIL

FUEL TUBE/QUICK-CONNECT
FITTINGS

FUEL TANK

FUEL INJECTORS

EVAPORATIVE EMISSION
(EVAP) PURGE FLOW
SWITCH

FUEL VAPOR PIPE

FUEL FEED PIPE

'97-98 models:

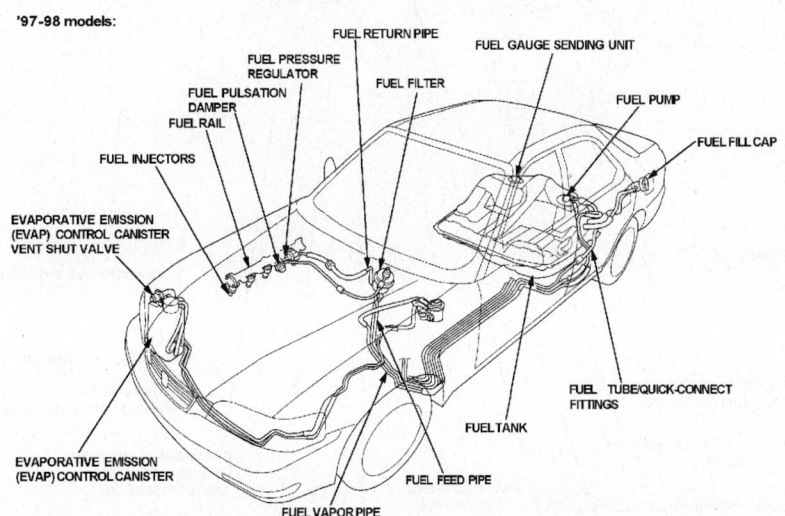

FUEL RETURN PIPE

FUEL GAUGE SENDING UNIT

FUEL PRESSURE
REGULATOR

FUEL FILTER

FUEL PUMP

FUEL PULSATION
DAMPER

FUEL RAIL

FUEL FILL CAP

FUEL INJECTORS

EVAPORATIVE EMISSION
(EVAP) CONTROL CANISTER
VENT SHUT VALVE

FUEL TUBE/QUICK-CONNECT
FITTINGS

FUEL TANK

EVAPORATIVE EMISSION
(EVAP) CONTROL CANISTER

FUEL FEED PIPE

FUEL VAPOR PIPE

29130_ACUR_G0019

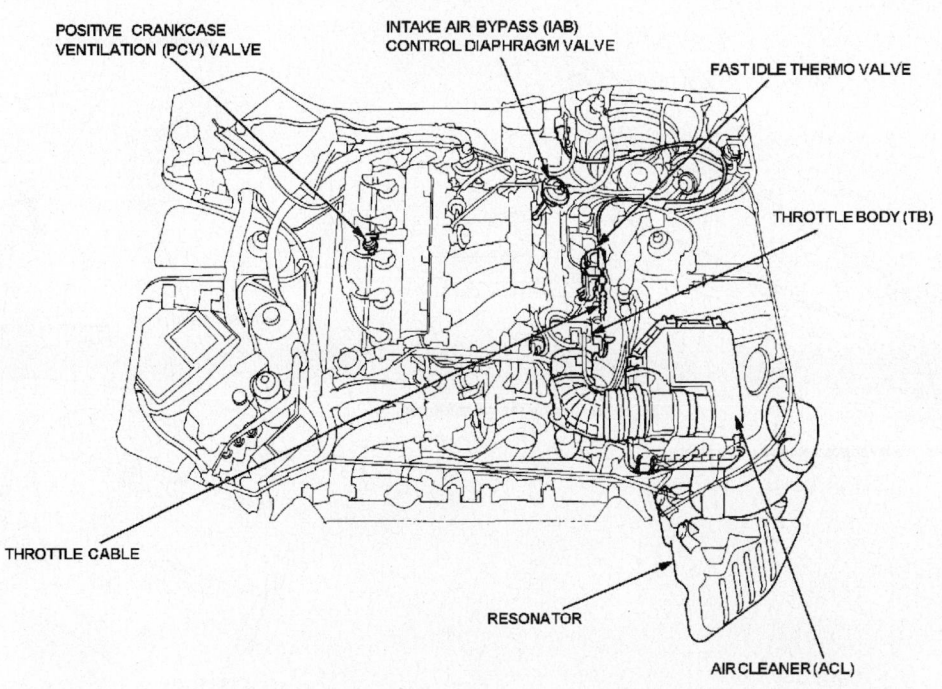

POSITIVE CRANKCASE
VENTILATION (PCV) VALVE

INTAKE AIR BYPASS (IAB)
CONTROL DIAPHRAGM VALVE

FAST IDLE THERMO VALVE

THROTTLE BODY (TB)

THROTTLE CABLE

RESONATOR

AIR CLEANER (ACL)

29130_ACUR_G0020

3.2TL

6 Cylinder Engines

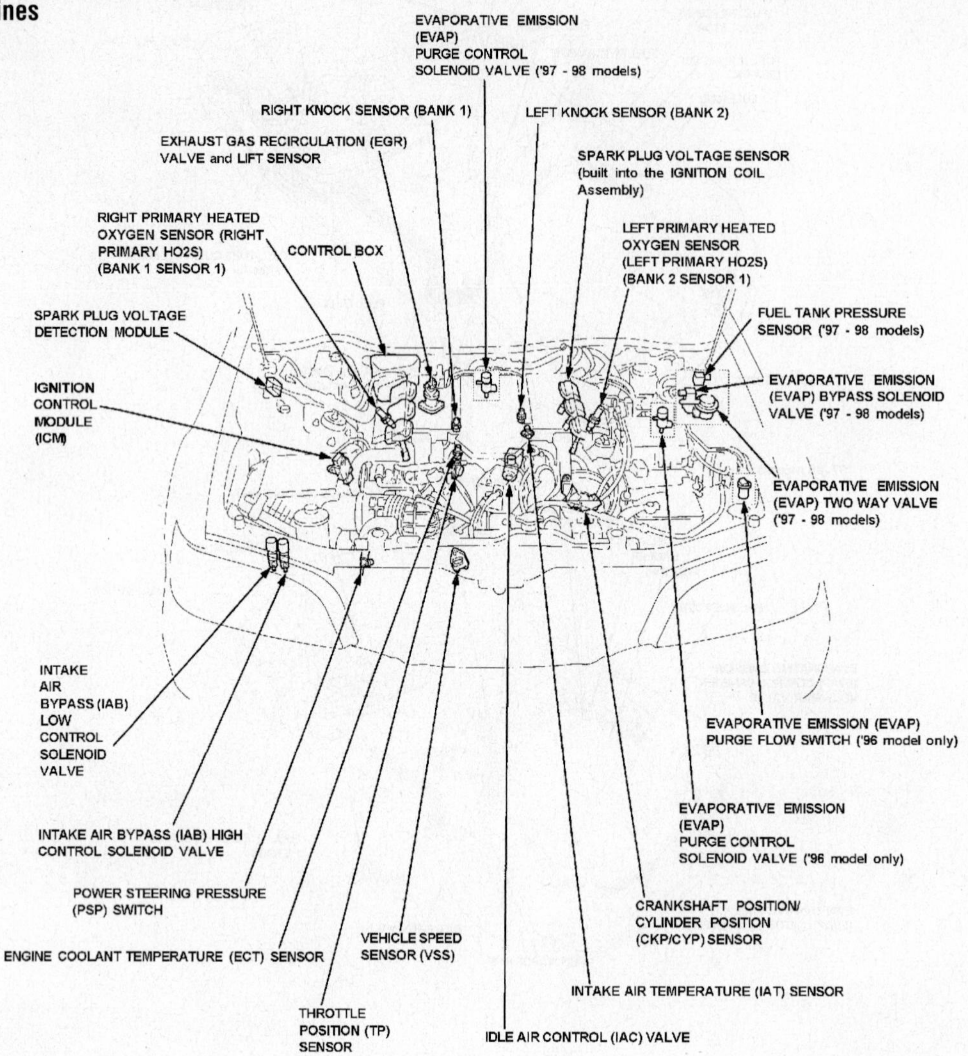

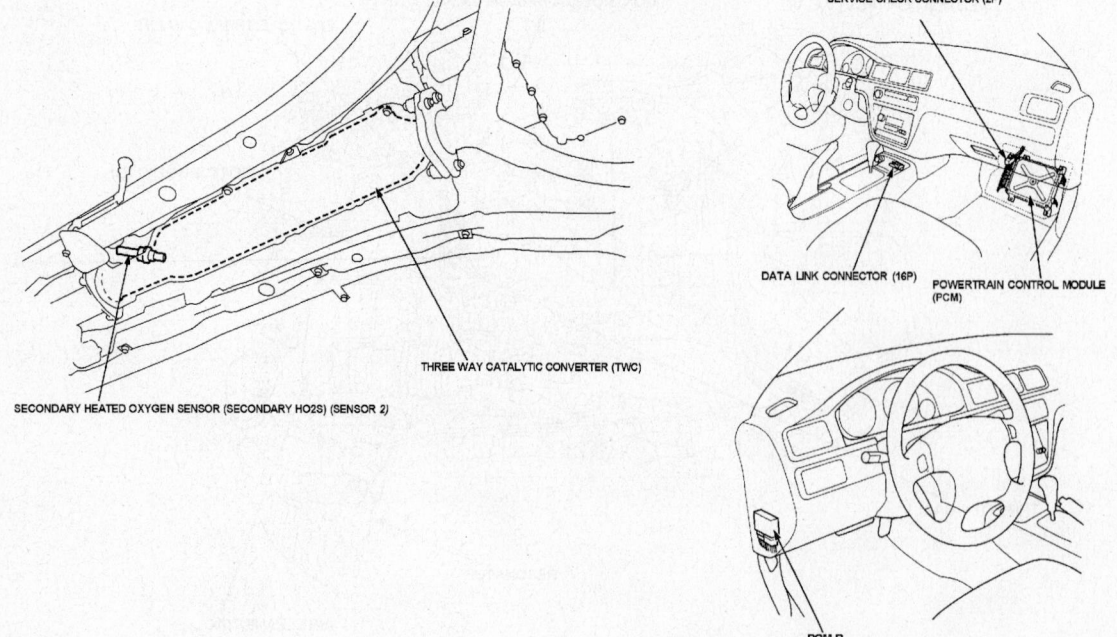

EVAPORATIVE EMISSION (EVAP) PURGE CONTROL SOLENOID VALVE ('97 - 98 models)

RIGHT KNOCK SENSOR (BANK 1)

LEFT KNOCK SENSOR (BANK 2)

EXHAUST GAS RECIRCULATION (EGR) VALVE and LIFT SENSOR

SPARK PLUG VOLTAGE SENSOR (built into the IGNITION COIL Assembly)

RIGHT PRIMARY HEATED OXYGEN SENSOR (RIGHT PRIMARY HO2S) (BANK 1 SENSOR 1)

CONTROL BOX

LEFT PRIMARY HEATED OXYGEN SENSOR (LEFT PRIMARY HO2S) (BANK 2 SENSOR 1)

SPARK PLUG VOLTAGE DETECTION MODULE

FUEL TANK PRESSURE SENSOR ('97 - 98 models)

IGNITION CONTROL MODULE (ICM)

EVAPORATIVE EMISSION (EVAP) BYPASS SOLENOID VALVE ('97 - 98 models)

EVAPORATIVE EMISSION (EVAP) TWO WAY VALVE ('97 - 98 models)

INTAKE AIR BYPASS (IAB) LOW CONTROL SOLENOID VALVE

EVAPORATIVE EMISSION (EVAP) PURGE FLOW SWITCH ('96 model only)

INTAKE AIR BYPASS (IAB) HIGH CONTROL SOLENOID VALVE

EVAPORATIVE EMISSION (EVAP) PURGE CONTROL SOLENOID VALVE ('96 model only)

POWER STEERING PRESSURE (PSP) SWITCH

CRANKSHAFT POSITION/ CYLINDER POSITION (CKP/CYP) SENSOR

ENGINE COOLANT TEMPERATURE (ECT) SENSOR

VEHICLE SPEED SENSOR (VSS)

INTAKE AIR TEMPERATURE (IAT) SENSOR

THROTTLE POSITION (TP) SENSOR

IDLE AIR CONTROL (IAC) VALVE

29130_ACUR_G0021

SERVICE CHECK CONNECTOR (2P)

THREE WAY CATALYTIC CONVERTER (TWC)

DATA LINK CONNECTOR (16P)

POWERTRAIN CONTROL MODULE (PCM)

SECONDARY HEATED OXYGEN SENSOR (SECONDARY HO2S) (SENSOR 2)

PGM-R MAIN RELAY

29130_ACUR_G0022

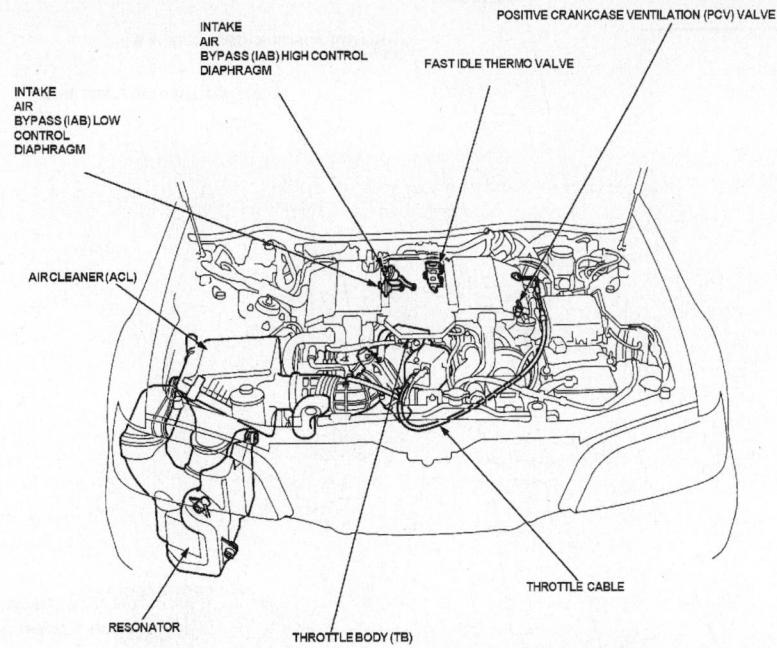

INTAKE
AIR
BYPASS (IAB) HIGH CONTROL
DIAPHRAGM

POSITIVE CRANKCASE VENTILATION (PCV) VALVE

FAST IDLE THERMO VALVE

INTAKE
AIR
BYPASS (IAB) LOW
CONTROL
DIAPHRAGM

AIR CLEANER (ACL)

THROTTLE CABLE

RESONATOR

THROTTLE BODY (TB)

29130_ACUR_G0023

'96 model:

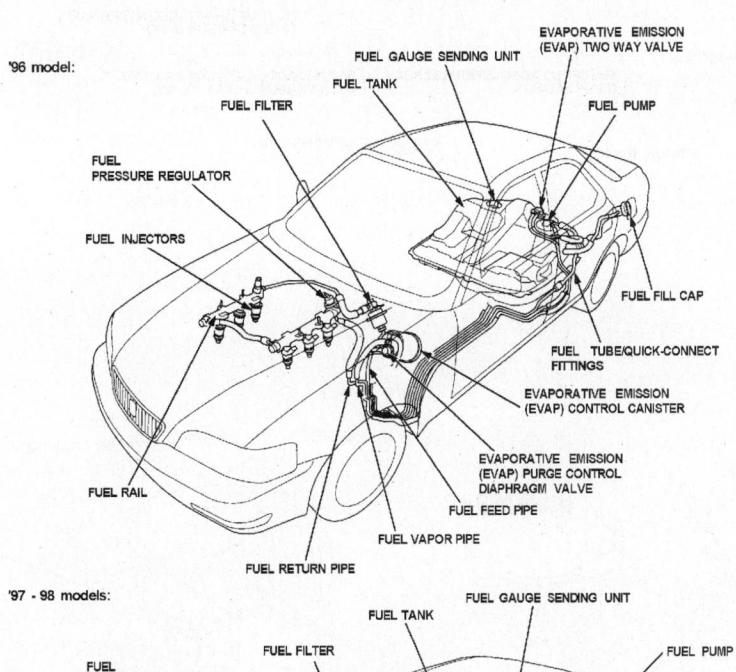

EVAPORATIVE EMISSION
(EVAP) TWO WAY VALVE

FUEL GAUGE SENDING UNIT

FUEL TANK

FUEL PUMP

FUEL FILTER

FUEL
PRESSURE REGULATOR

FUEL INJECTORS

FUEL FILL CAP

FUEL TUBE/QUICK-CONNECT
FITTINGS

EVAPORATIVE EMISSION
(EVAP) CONTROL CANISTER

EVAPORATIVE EMISSION
(EVAP) PURGE CONTROL
DIAPHRAGM VALVE

FUEL RAIL

FUEL FEED PIPE

FUEL VAPOR PIPE

FUEL RETURN PIPE

'97 - 98 models:

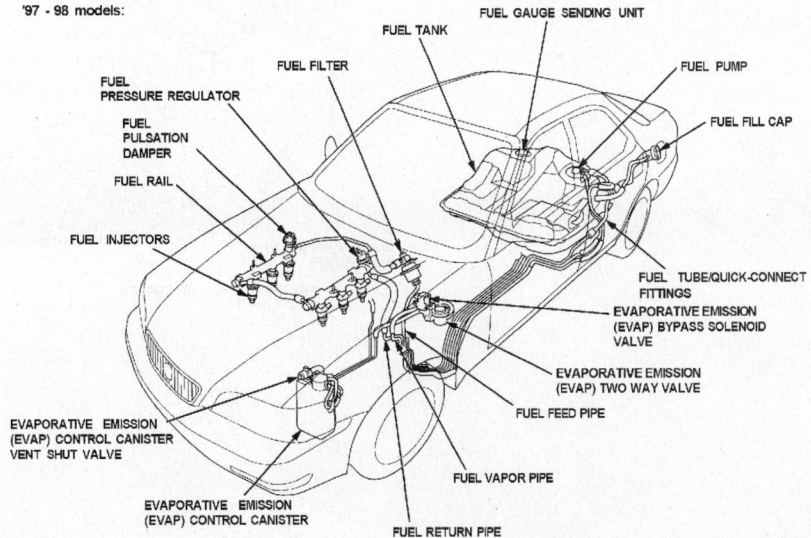

FUEL GAUGE SENDING UNIT

FUEL TANK

FUEL FILTER

FUEL PUMP

FUEL
PRESSURE REGULATOR

FUEL FILL CAP

FUEL
PULSATION
DAMPER

FUEL RAIL

FUEL INJECTORS

FUEL TUBE/QUICK-CONNECT
FITTINGS

EVAPORATIVE EMISSION
(EVAP) BYPASS SOLENOID
VALVE

EVAPORATIVE EMISSION
(EVAP) TWO WAY VALVE

FUEL FEED PIPE

EVAPORATIVE EMISSION
(EVAP) CONTROL CANISTER
VENT SHUT VALVE

EVAPORATIVE EMISSION
(EVAP) CONTROL CANISTER

FUEL VAPOR PIPE

FUEL RETURN PIPE

29130_ACUR_G0024

TSX

4 Cylinder Engines

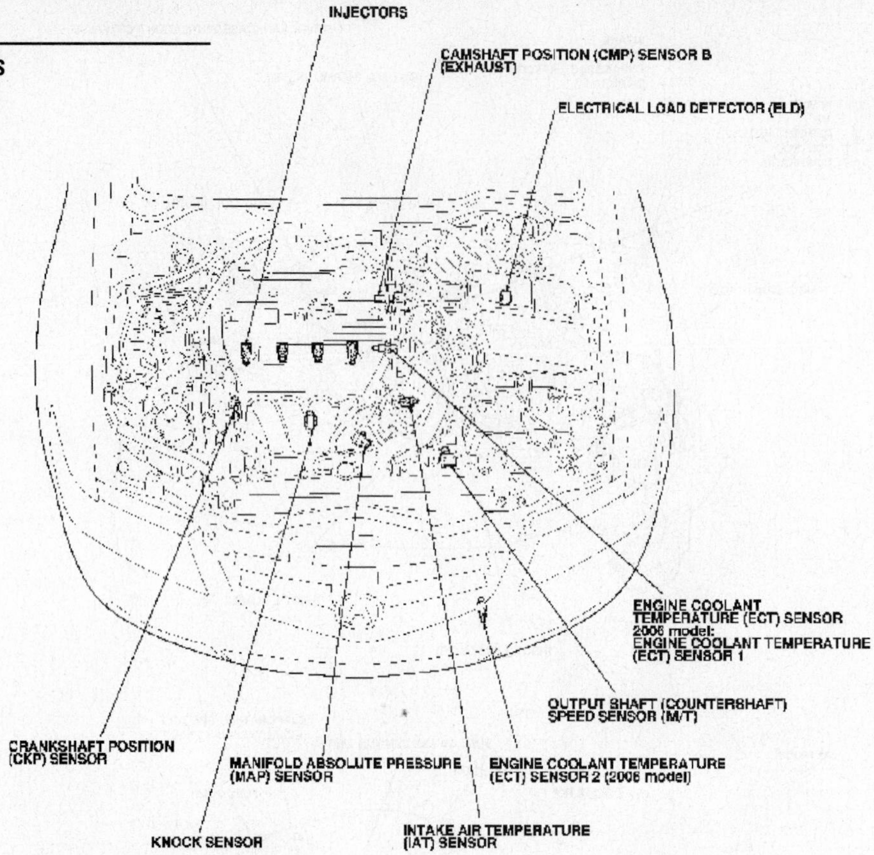

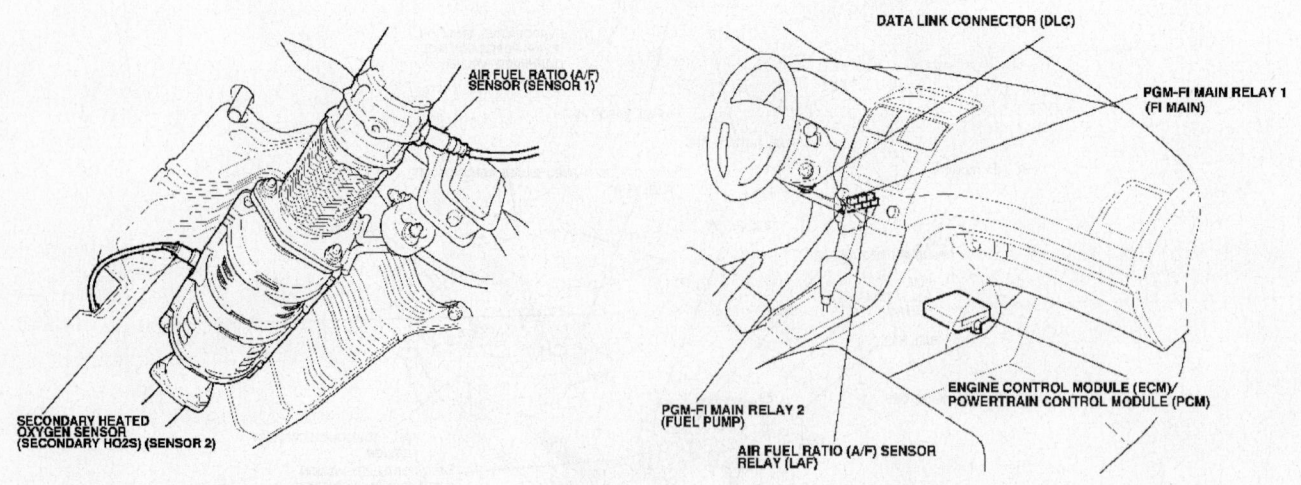

INJECTORS
CAMSHAFT POSITION (CMP) SENSOR B (EXHAUST)
ELECTRICAL LOAD DETECTOR (ELD)
ENGINE COOLANT TEMPERATURE (ECT) SENSOR
2006 model:
ENGINE COOLANT TEMPERATURE (ECT) SENSOR 1
OUTPUT SHAFT (COUNTERSHAFT) SPEED SENSOR (M/T)
CRANKSHAFT POSITION (CKP) SENSOR
MANIFOLD ABSOLUTE PRESSURE (MAP) SENSOR
ENGINE COOLANT TEMPERATURE (ECT) SENSOR 2 (2006 model)
KNOCK SENSOR
INTAKE AIR TEMPERATURE (IAT) SENSOR

29130_ACUR_G0054

AIR FUEL RATIO (A/F) SENSOR (SENSOR 1)
SECONDARY HEATED OXYGEN SENSOR (SECONDARY HO2S) (SENSOR 2)

29130_ACUR_G0055

DATA LINK CONNECTOR (DLC)
PGM-FI MAIN RELAY 1 (FI MAIN)
ENGINE CONTROL MODULE (ECM)/ POWERTRAIN CONTROL MODULE (PCM)
PGM-FI MAIN RELAY 2 (FUEL PUMP)
AIR FUEL RATIO (A/F) SENSOR RELAY (LAF)

29130_ACUR_G0056

ACURA
PIN CHARTS

TABLE OF CONTENTS

PIN CHARTS

Introduction

A Pin Voltage Table is a term used to describe a table that identifies PCM pins, wire colors of the PCM circuits, circuit descriptions and "known good" values for devices that connect to the PCM. These tables include the following information:

- Signals from various sensors (ECT, IAT, MAP, TPS, etc.)
- Signals from various switches (PNP, PSP, WOT, etc.)
- Signals from oxygen sensors (O2S, HO2S)
- Signals from output devices (IAC, INJ, TCC, etc.)
- Power & ground signals

Pin Voltage Tables

Information contained within the Pin Voltage Tables can be used to:

- Test circuits for open, short to power or short to ground faults
- Check the operation of a component before or after a repair
- Check the operation of a component or system by viewing signals on PCM input/output circuits with a DVOM or Lab Scope

Using a Breakout Box

There are several Breakout Box (BOB) designs available for use to test the PCM and its input and output circuits. However, all of them require removal of the wire harness to the PCM so that the BOB can be installed between the PCM and wire harness connector. Several breakout boxes require the use of overlays in order to allow the tool to be used on more than one year or engine type. Always verify that the correct adapter and overlays are used to prevent connection to the wrong circuits and a misdiagnosis.

Power and Ground Circuit Checks

Measurements made at the BOB are accomplished via test leads and probes from the DVOM or a Lab Scope. If any of the terminals on the PCM or BOB are damaged or loose, test measurements made at the Breakout Box will be inaccurate. To verify the PCM battery power and ground circuits are normal (correct) at the BOB, test the condition of the circuit between the battery negative (-) post and these circuits prior to starting a test sequence.

Diagnosis with Pin Voltage Tables

See figure 1.

Once an actual PCM pin voltage reading is recorded, it can be compared to an example from a vehicle with "known good" values. In the example shown the Value at Hot Idle for the EVP sensor signal (0.4v) is the "known good" value.

Wire Color Changes

Every effort has been made to obtain and list the correct circuit wire colors for all vehicles. However, running changes from the vehicle manufacturer can cause the wrong colors to be listed.

PCM Pin #	W/Color	Circuit Description (60-Pin)	Value at Hot Idle
27	BN/LG	EVP Sensor Signal	0.4v

Fig. 1 Example

INTEGRA PIN CHARTS

1990-91 Hatchback 1.8L MFI VIN DA9 [All] 18P 'A' Connector

PCM Pin #	W/Color	Circuit Description (18-Pin)	Value at Hot Idle
A1	BRN	Injector 1	2.0-3.3 ms
A2, 4	BLK	Power Ground	<0.1v
A3	RED	Injector 2	2.0-3.3 ms
A5	LT BLU	Injector 3	2.0-3.3 ms
A6	GRN	EVAP Purge Solenoid	Solenoid Off: 12v, On: 1v
A7	YEL	Injector 4	2.0-3.3 ms
A8	RED	EGR Solenoid Control	Solenoid Off: 12v, On: 1v
A9	---	Not Used	---
A10	GRN/YEL	Fuel Pressure Regulator	Solenoid Off: 12v, On: 1v
A11	BLU/YEL	Electronic Air Control Valve	Pulse Signals
A12	GRN/BLK	Fuel Pump Relay Control	Relay Off: 12v, On: 1v
A13	YEL/BLK	Main Relay Power (B+)	12-14v
A14	GRN/BLK	Fuel Pump Relay Control	Relay Off: 12v, On: 1v
A15	YEL/BLK	Main Relay Power (B+)	12-14v
A16	BRN/BLK	Power Ground	<0.1v
A17	---	Not Used	---
A18	BLK/RED	Power Ground	<0.1v

1990-91 Hatchback 1.8L MFI VIN DA9 [All] 20P 'B' Connector

PCM Pin #	W/Color	Circuit Description (20-Pin)	Value at Hot Idle
B1	WHT/GRN	Keep Alive Power (VBU)	12-14v
B2	BLU	A/T: Fast Idle Solenoid	Solenoid Off: 12v, On: 1v
B3	YEL	A/C Clutch Relay Control	Relay Off: 12v, On: 1v
B4	BLU/WHT	A/T: Control Unit Signal	---
B5	---	Not Used	---
B6	GRN/ORN	Malfunction Indicator Lamp	MIL Off: 12v, On: 1v
B7	GRN	A/T: Shift Position Switch	In P/N: 0v, Others: 12v
B8	BLU/RED	A/C Switch Signal	Relay Off: 12v, On: 1v
B9	---	Not Used	---
B10	ORN	CKP Sensor Signal	AC pulse signals
B11	---	Not Used	---
B12	WHT	CKP Sensor Ground	<0.050v
B13	BLU/WHT	Starter Switch Signal	Cranking: 9-11v
B14	BLU	Alternator Charging	Headlights off: 0v, on: 12v
B15	WHT	Igniter Signal	Digital Signals: 0-12-0v
B16	YEL/RED	Vehicle Speed Sensor	Moving: pulse signals
B17	WHT	Igniter Signal	Digital Signals: 0-12-0v
B18	---	Not Used	---
B19	RED	PSP Switch Signal	Straight: 0v, Turning: 12v
B20	BRN	Ignition Timing Adjustment	0.4-4.5v

1990-91 Hatchback 1.8L MFI VIN DA9 [All] 16P 'C' Connector

PCM Pin #	W/Color	Circuit Description (16-Pin)	Value at Hot Idle
C1	BLU/GRN	CYP Sensor Signal	AC pulse signals
C2	BLU/YEL	CYP Sensor Ground	<0.050v
C3	ORN/BLU	TDC Sensor Signal	AC pulse signals
C4	WHT/BLU	TDC Sensor Ground	<0.050v
C5	RED/YEL	IAT Sensor Signal	At 100°F: 2-3v
C6	RED/WHT	ECT Sensor Signal	At 180°F: 0.5-0.6v
C7	RED/BLU	Throttle Angle Sensor	0.5-0.6v
C8	GRN/RED	EGR Valve Lift Sensor	1.1-1.2v
C9	RED/WHT	Atmospheric Press. Sensor	Varies w/alt: 0.5-3.0v
C10	GRN/WHT	Brake Switch Signal	Brake Off: 12v, On: 0v
C11	WHT	MAP Sensor Signal	0.8-0.9v
C12	GRN	Sensor Ground	<0.050v
C13	YEL/WHT	Sensor VREF	4.9-5.1v
C14	GRN/WHT	MAP Sensor Ground	<0.050v
C15	YEL/RED	MAP Sensor VREF	4.9-5.1v
C16	WHT	O2S-11 Signal	0.1-1.1v

Standard Colors and Abbreviations

Abbreviation	Color	Abbreviation	Color	Abbreviation	Color
BLK	Black	LT BLU	Lt. Blue	TAN	Tan
BLU	Blue	LT GRN	Lt. Green	VIO	Violet
BRN	Brown	ORN	Orange	WHT	White
GRY	Gray	PNK	Pink	YEL	Yellow
GRN	Green	PPL	Purple		

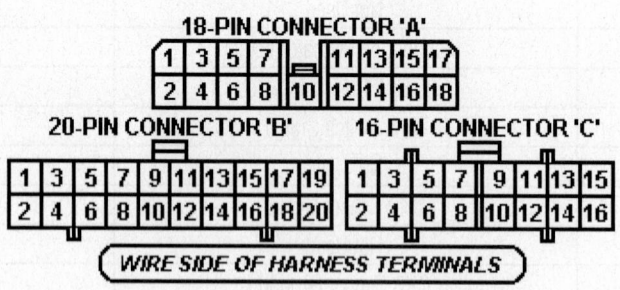

Pin Connector Graphic

05533_ADIA_G579

1992-93 Hatchback 1.7L VTEC VIN DB2 [M/T] 26P 'A' Connector

PCM Pin #	W/Color	Circuit Description (26-Pin)	Value at Hot Idle
A1	BRN	Injector 1	2.0-3.3 ms
A2	YEL	Injector 4	2.0-3.3 ms
A3	RED	Injector 2	2.0-3.3 ms
A4	GRN/YEL	VTEC Solenoid Control	0v, Hi-Speed: 12v
A5	LT BLU	Injector 3	2.0-3.3 ms
A6	ORN/BLK	HO2S-11 (B1 S1) Heater	Heater Off: 12v, On: 1v
A7	GRN/BLK	Fuel Pump Relay Control	Relay Off: 12v, On: 1v
A8	GRN/BLK	Fuel Pump Relay Control	Relay Off: 12v, On: 1v
A9	BLU/YEL	Idle Air Control Valve	Pulse Signals
A10-12	---	Not Used	---
A13	GRN/ORN	Malfunction Indicator Lamp	MIL Off: 12v, On: 1v
A14	---	Not Used	---
A15	YEL	A/C Clutch Relay Control	Relay Off: 12v, On: 1v
A16-19	---	Not Used	---
A20	GRN	EVAP Purge Solenoid	Solenoid Off: 12v, On: 1v
A21	YEL/GRN	Igniter Signal	Digital Signals: 0-12-0v
A22	YEL/GRN	Igniter Signal	Digital Signals: 0-12-0v
A23	BLK	Power Ground	<0.1v
A24	BLK	Power Ground	<0.1v
A25	YEL/BLK	Main Relay Power (B+)	12-14v
A26	BLK/RED	Chassis Ground	0.1v

1992-93 Hatchback 1.7L VTEC VIN DB2 [M/T] 16P 'B' Connector

PCM Pin #	W/Color	Circuit Description (16-Pin)	Value at Hot Idle
B1	YEL/BLK	Main Relay Power (B+)	12-14v
B2	BRN/BLK	Chassis Ground	<0.050v
B3-4	---	Not Used	---
B5	BLU/RED	A/C Pressure Switch	Relay Off: 12v, On: 1v
B6-7	---	Not Used	---
B8	RED	PSP Switch Signal	Straight: 0v, Turning: 12v
B9	BLU/WHT	Starter Switch Signal	Cranking: 9-11v
B10	YEL/RED	Vehicle Speed Sensor	Moving: pulse signals
B11	ORN	CYP Sensor Signal	AC pulse signals
B12	WHT	CYP Sensor Ground	<0.050v
B13	ORN/BLU	TDC Sensor Signal	AC pulse signals
B14	WHT/BLU	TDC Sensor Ground	<0.050v
B15	BLU/GRN	CKP Sensor Signal	AC pulse signals
B16	BLU/YEL	CKP Sensor Ground	<0.050v

1992-93 Hatchback 1.7L VTEC VIN DB2 [M/T] 22P 'D' Connector

PCM Pin #	W/Color	Circuit Description (22-Pin)	Value at Hot Idle
D1	WHT/YEL	Keep Alive Power (VBU)	12-14v
D2	GRN/WHT	Brake Switch Signal	Brake Off: 12v, On: 0v
D3	RED/BLU	Knock Sensor Signal	No Detonation: 18mv AC
D4	BRN	Service Check Connector	SCS Open: 4.80v
D5	---	Not Used	---
D6	BLU/BLK	VTEC Pressure Switch	0v, Hi-Speed: 12v
D7	LT BLU	Data Link Connector	No Scan Tool: 5v
D8	---	Not Used	---
D9	BLU	Alternator 'FR' Signal	Digital Signals: 0-5-0v
D10	---	Not Used	---
D11	RED/BLU	TP Sensor Signal	0.5-0.6v
D12	---	Not Used	---
D13	RED/WHT	ECT Sensor Signal	At 180°F: 0.5-0.6v
D14	WHT	HO2S-11 (B1 S1) Signal	0.1-1.1v
D15	RED/YEL	IAT Sensor Signal	At 100°F: 2-3v
D18	---	Not Used	---
D17	WHT	MAP Sensor Signal	0.8-0.9v
D18	---	Not Used	---
D19	YEL/RED	MAP Sensor VREF	4.9-5.1v
D20	YEL/WHT	Sensor VREF	4.9-5.1v
D21	GRN/WHT	MAP Sensor Ground	<0.050v
D22	GRN/WHT	Sensor Ground	<0.050v

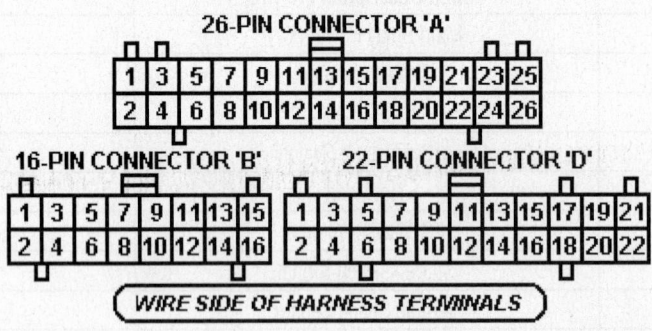

05533_ADIA_G580

Pin Connector Graphic

1992-93 Hatchback 1.8L MFI VIN DA9 [All] 26P 'A' Connector

PCM Pin #	W/Color	Circuit Description (26-Pin)	Value at Hot Idle
A1	BRN	Injector 1	2.0-3.3 ms
A2	YEL	Injector 4	2.0-3.3 ms
A3	RED	Injector 2	2.0-3.3 ms
A4	---	Not Used	---
A5	LT BLU	Injector 3	2.0-3.3 ms
A6	ORN/BLK	HO2S-11 (B1 S1) Heater	Heater Off: 12v, On: 1v
A7	GRN/BLK	Fuel Pump Relay Control	Relay Off: 12v, On: 1v
A8	GRN/BLK	Fuel Pump Relay Control	Relay Off: 12v, On: 1v
A9	BLU/YEL	Idle Air Control Valve	Pulse Signals
A10	GRN/YEL	Pressure Regulator Solenoid	Solenoid Off: 12v, On: 1v
A11	RED	EGR Valve Lift Sensor	1.1-1.2v
A12	---	Not Used	---
A13	GRN/ORN	Malfunction Indicator Lamp	MIL Off: 12v, On: 1v
A14	---	---	---
A15	YEL	A/C Clutch Relay Control	Relay Off: 12v, On: 1v
A16-19	---	Not Used	---
A20	GRN	EVAP Purge Solenoid	Solenoid Off: 12v, On: 1v
A21	YEL/GRN	Igniter Signal	Digital Signals: 0-12-0v
A22	YEL/GRN	Igniter Signal	Digital Signals: 0-12-0v
A23	BLK	Power Ground	<0.1v
A24	BLK	Power Ground	<0.1v
A25	YEL/BLK	Main Relay Power (B+)	12-14v
A26	BRN/BLK	Chassis Ground	0.1v

1992-93 Hatchback 1.8L MFI VIN DA9 [All] 16P 'B' Connector

PCM Pin #	W/Color	Circuit Description (16-Pin)	Value at Hot Idle
B1	YEL/BLK	Main Relay Power (B+)	12-14v
B2	BRN/BLK	Chassis Ground	<0.050v
B3-4	---	Not Used	---
B5	BLU/RED	A/C Pressure Switch	Relay Off: 12v, On: 1v
B6	---	Not Used	---
B7	GRN	A/T: Shift Position Switch	In P/N: 0v, Others: 12v
B8	RED	PSP Switch Signal	Straight: 0v, Turning: 12v
B9	BLU/WHT	Starter Switch Signal	Cranking: 9-11v
B10	YEL/RED	Vehicle Speed Sensor	Moving: pulse signals
B11	ORN	CYP Sensor Signal	AC pulse signals
B12	WHT	CYP Sensor Ground	<0.050v
B13	ORN/BLU	TDC Sensor Signal	AC pulse signals
B14	WHT/BLU	TDC Sensor Ground	<0.050v
B15	BLU/GRN	CKP Sensor Signal	AC pulse signals
B16	BLU/YEL	CKP Sensor Ground	<0.050v

1992-93 Hatchback 1.8L MFI VIN DA9 [All] 22P 'D' Connector

PCM Pin #	W/Color	Circuit Description (22-Pin)	Value at Hot Idle
D1	WHT/YEL	Keep Alive Power (VBU)	12-14v
D2	GRN/WHT	Brake Switch Signal	Brake Off: 12v, On: 0v
D3	---	Not Used	---
D4	BRN	Service Check Connector	SCS Open: 4.80v
D5-6	---	Not Used	---
D7	LT BLU	Data Link Connector	No Scan Tool: 5v
D8	RED/WHT	A/T: Control Unit	---
D9	BLU	Alternator 'FR' Signal	Digital Signals: 0-5-0v
D10	---	Not Used	---
D11	RED/BLU	TP Sensor Signal	0.5-0.6v
D12	YEL	EGR Valve Lift Sensor	1.1-1.2v
D13	RED/WHT	ECT Sensor Signal	At 180°F: 0.5-0.6v
D14	WHT	HO2S-11 (B1 S1) Signal	0.1-1.1v
D15	RED/YEL	IAT Sensor Signal	At 100°F: 2-3v
D16	BLU/WHT	A/T: Control Unit	---
D17	WHT	MAP Sensor Signal	0.8-0.9v
D18	---	Not Used	---
D19	YEL/RED	MAP Sensor VREF	4.9-5.1v
D20	YEL/WHT	Sensor VREF	4.9-5.1v
D21	GRN/WHT	MAP Sensor Ground	<0.050v
D22	GRN/WHT	Sensor Ground	<0.050v

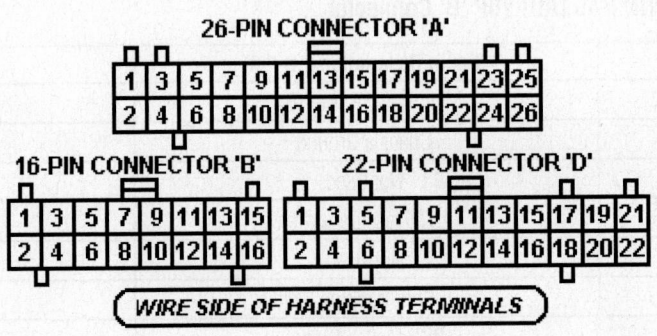

05533_ADIA_G580

Pin Connector Graphic

1990-91 GS, LS Sedan 1.8L I4 MFI VIN DB1 [All] 18P A Connector

PCM Pin #	W/Color	Circuit Description (18-Pin)	Value at Hot Idle
A1	BRN	Injector 1	2.0-3.3 ms
A2	BLK	Power Ground	<0.1v
A3	RED	Injector 2	2.0-3.3 ms
A4	BLK	Power Ground	<0.1v
A5	LT BLU	Injector 3	2.0-3.3 ms
A6	GRN	EVAP Purge Solenoid	Solenoid Off: 12v, On: 1v
A7	YEL	Injector 4	2.0-3.3 ms
A8	RED	EGR Solenoid Control	Solenoid Off: 12v, On: 1v
A9	---	Not Used	---
A10	GRN/YEL	Fuel Pressure Regulator	Solenoid Off: 12v, On: 1v
A11	BLU/YEL	Electronic Air Control Valve	Pulse Signals
A12	GRN/BLK	Fuel Pump Relay Control	Relay Off: 12v, On: 1v
A13	YEL/BLK	Main Relay Power (B+)	12-14v
A14	GRN/BLK	Fuel Pump Relay Control	Relay Off: 12v, On: 1v
A15	YEL/BLK	Main Relay Power (B+)	12-14v
A16	BRN/BLK	Power Ground	<0.1v
A17	---	Not Used	---
A18	BLK/RED	Power Ground	<0.1v

1990-91 GS, LS Sedan 1.8L I4 MFI VIN DB1 [All] 20P B Connector

PCM Pin #	W/Color	Circuit Description (20-Pin)	Value at Hot Idle
B1	WHT/GRN	Keep Alive Power (VBU)	12-14v
B2	BLU	A/T: Fast Idle Solenoid	Solenoid Off: 12v, On: 1v
B3	YEL	A/C Clutch Relay Control	Relay Off: 12v, On: 1v
B4	BLU/WHT	A/T: Control Unit Signal	---
B5	---	Not Used	---
B6	GRN/ORN	Malfunction Indicator Lamp	MIL Off: 12v, On: 1v
B7	GRN	A/T: Shift Position Switch	In P/N: 0v, Others: 12v
B8	BLU/RED	A/C Switch Signal	Relay Off: 12v, On: 1v
B9	---	Not Used	---
B10	ORN	CKP Sensor Signal	AC pulse signals
B11	---	Not Used	---
B12	WHT	CKP Sensor Ground	<0.050v
B13	BLU/WHT	Starter Switch Signal	Cranking: 9-11v
B14	BLU	Alternator Charging	Headlights off: 0v, on: 12v
B15	WHT	Igniter Signal	Digital Signals: 0-12-0v
B16	YEL/RED	Vehicle Speed Sensor	Moving: pulse signals
B17	WHT	Igniter Signal	Digital Signals: 0-12-0v
B18	---	Not Used	---
B19	RED	PSP Switch Signal	Straight: 0v, Turning: 12v
B20	BRN	Ignition Timing Adjustment	0.4-4.5v

1990-91 GS, LS Sedan 1.8L I4 MFI VIN DB1 [All] 16P C Connector

PCM Pin #	W/Color	Circuit Description (16-Pin)	Value at Hot Idle
C1	BLU/GRN	CYP Sensor Signal	AC pulse signals
C2	BLU/YEL	CYP Sensor Ground	<0.050v
C3	ORN/BLU	TDC Sensor Signal	AC pulse signals
C4	WHT/BLU	TDC Sensor Ground	<0.050v
C5	RED/YEL	IAT Sensor Signal	At 100°F: 2-3v
C6	RED/WHT	ECT Sensor Signal	At 180°F: 0.5-0.6v
C7	RED/BLU	Throttle Angle Sensor	0.5-0.6v
C8	GRN/RED	EGR Valve Lift Sensor	1.1-1.2v
C9	RED/WHT	Atmospheric Press. Sensor	Varies w/alt: 0.5-3.0v
C10	GRN/WHT	Brake Switch Signal	Brake Off: 12v, On: 0v
C11	WHT	MAP Sensor Signal	0.8-0.9v
C12	GRN	Sensor Ground	<0.050v
C13	YEL/WHT	Sensor VREF	4.9-5.1v
C14	GRN/WHT	MAP Sensor Ground	<0.050v
C15	YEL/RED	MAP Sensor VREF	4.9-5.1v
C16	WHT	O2S-11 Signal	0.1-1.1v

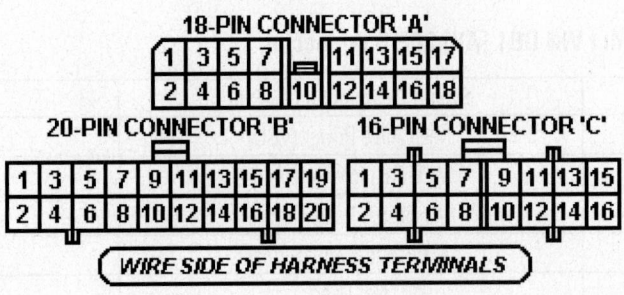

05533_ADIA_G580

Pin Connector Graphic

1992-93 GS, RS Sedan 1.8L I4 MFI VIN DB1 [All] 26P A Connector

PCM Pin #	W/Color	Circuit Description (26-Pin)	Value at Hot Idle
A1	BRN	Injector 1	2.0-3.3 ms
A2	YEL	Injector 4	2.0-3.3 ms
A3	RED	Injector 2	2.0-3.3 ms
A4	---	Not Used	---
A5	LT BLU	Injector 3	2.0-3.3 ms
A6	ORN/BLK	HO2S-11 (B1 S1) Heater	Heater Off: 12v, On: 1v
A7	GRN/BLK	Fuel Pump Relay Control	Relay Off: 12v, On: 1v
A8	GRN/BLK	Fuel Pump Relay Control	Relay Off: 12v, On: 1v
A9	BLU/YEL	Idle Air Control Valve	Pulse Signals
A10	GRN/YEL	Pressure Regulator Solenoid	Solenoid Off: 12v, On: 1v
A11	RED	EGR Valve Lift Sensor	1.1-1.2v
A12	---	Not Used	---
A13	GRN/ORN	Malfunction Indicator Lamp	MIL Off: 12v, On: 1v
A14	---	---	---
A15	YEL	A/C Clutch Relay Control	Relay Off: 12v, On: 1v
A16-19	---	Not Used	---
A20	GRN	EVAP Purge Solenoid	Solenoid Off: 12v, On: 1v
A21	YEL/GRN	Igniter Signal	Digital Signals: 0-12-0v
A22	YEL/GRN	Igniter Signal	Digital Signals: 0-12-0v
A23	BLK	Power Ground	<0.1v
A24	BLK	Power Ground	<0.1v
A25	YEL/BLK	Main Relay Power (B+)	12-14v
A26	BRN/BLK	Chassis Ground	0.1v

1992-93 GS, RS Sedan 1.8L I4 MFI VIN DB1 [All] 16P B Connector

PCM Pin #	W/Color	Circuit Description (16-Pin)	Value at Hot Idle
B1	YEL/BLK	Main Relay Power (B+)	12-14v
B2	BRN/BLK	Chassis Ground	<0.050v
B3-4	---	Not Used	---
B5	BLU/RED	A/C Pressure Switch	Relay Off: 12v, On: 1v
B6	---	Not Used	---
B7	GRN	A/T: Shift Position Switch	In P/N: 0v, Others: 12v
B8	RED	PSP Switch Signal	Straight: 0v, Turning: 12v
B9	BLU/WHT	Starter Switch Signal	Cranking: 9-11v
B10	YEL/RED	Vehicle Speed Sensor	Moving: pulse signals
B11	ORN	CYP Sensor Signal	AC pulse signals
B12	WHT	CYP Sensor Ground	<0.050v
B13	ORN/BLU	TDC Sensor Signal	AC pulse signals
B14	WHT/BLU	TDC Sensor Ground	<0.050v
B15	BLU/GRN	CKP Sensor Signal	AC pulse signals
B16	BLU/YEL	CKP Sensor Ground	<0.050v

1992-93 GS, RS Sedan 1.8L I4 MFI VIN DB1 [All] 22P D Connector

PCM Pin #	W/Color	Circuit Description (22-Pin)	Value at Hot Idle
D1	WHT/YEL	Keep Alive Power (VBU)	12-14v
D2	GRN/WHT	Brake Switch Signal	Brake Off: 12v, On: 0v
D3	---	Not Used	---
D4	BRN	Service Check Connector	SCS Open: 4.80v
D5-6	---	Not Used	---
D7	LT BLU	Data Link Connector	No Scan Tool: 5v
D8	RED/WHT	A/T: Control Unit	---
D9	BLU	Alternator 'FR' Signal	Digital Signals: 0-5-0v
D10	---	Not Used	---
D11	RED/BLU	TP Sensor Signal	0.5-0.6v
D12	YEL	EGR Valve Lift Sensor	1.1-1.2v
D13	RED/WHT	ECT Sensor Signal	At 180°F: 0.5-0.6v
D14	WHT	HO2S-11 (B1 S1) Signal	0.1-1.1v
D15	RED/YEL	IAT Sensor Signal	At 100°F: 2-3v
D16	BLU/WHT	A/T: Control Unit	N/A
D17	WHT	MAP Sensor Signal	0.8-0.9v
D18	---	Not Used	---
D19	YEL/RED	MAP Sensor VREF	4.9-5.1v
D20	YEL/WHT	Sensor VREF	4.9-5.1v
D21	GRN/WHT	MAP Sensor Ground	<0.050v
D22	GRN/WHT	Sensor Ground	<0.050v

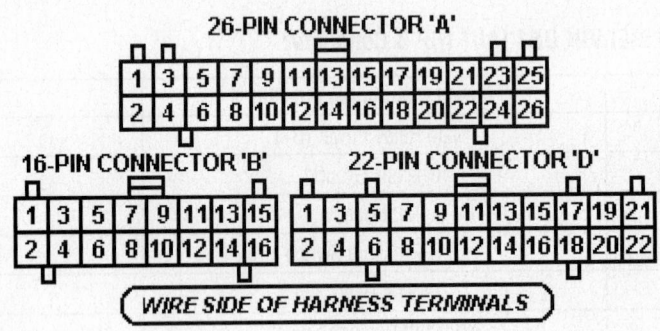

05533_ADIA_G580

Pin Connector Graphic

1994 LS, RS Sedan 1.8L I4 MFI VIN DB7 [All] 26P 'A' Connector

PCM Pin #	W/Color	Circuit Description (26-Pin)	Value at Hot Idle
A1	BRN	Injector 1	2.0-3.3 ms
A2	YEL	Injector 4	2.0-3.3 ms
A3	RED	Injector 2	2.0-3.3 ms
A4	---	Not Used	---
A5	LT BLU	Injector 3	2.0-3.3 ms
A6	ORN/BLK	HO2S-11 (B1 S1) Heater	Heater Off: 12v, On: 1v
A7	GRN/BLK	Fuel Pump Relay Control	Relay Off: 12v, On: 1v
A8	---	Not Used	---
A9	BLK/BLU	Idle Air Control Valve	Pulse Signals
A10-11	---	Not Used	---
A12	GRN	Radiator Fan Relay Control	Fan on: 1v, off: 12v
A13	GRN/ORN	Malfunction Indicator Lamp	MIL Off: 12v, On: 1v
A14	---	Not Used	---
A15	BLK/RED	A/C Clutch Relay Control	Relay Off: 12v, On: 1v
A16	WHT/GRN	Alternator Charging Signal	Headlights off: 0v, on: 12v
A17-19	---	Not Used	---
A20	GRN	EVAP Purge Solenoid	Solenoid Off: 12v, On: 1v
A21	YEL/GRN	Igniter Signal	Digital Signals: 0-12-0v
A22	---	Not Used	---
A23	BLK	Power Ground	<0.1v
A24	BLK	Power Ground	<0.1v
A25	YEL/BLK	Main Relay Power (B+)	12-14v
A26	BRN/BLK	Chassis Ground	0.1v

1994 LS, RS Sedan1.8L I4 MFI VIN DB7 [All] 16P 'B' Connector

PCM Pin #	W/Color	Circuit Description (16-Pin)	Value at Hot Idle
B1	YEL/BLK	Main Relay Power (B+)	12-14v
B2	BRN/BLK	Chassis Ground	<0.050v
B3	GRN/BLU	A/T: TCM Signal	N/A
B4	GRY	A/T: TCM Signal	N/A
B5	BLU/RED	A/C Switch Signal	Relay Off: 12v, On: 1v
B6	---	Not Used	---
B7	GRN/BLK	Park/Neutral Signal	In P/N: 0v, Others: 12v
B8	LT GRN	PSP Switch Signal	Straight: 0v, Turning: 12v
B9	BLU/WHT	Starter Switch Signal	Cranking: 9-11v
B10	ORN	Vehicle Speed Sensor	Moving: pulse signals
B11	ORN	CYP Sensor Signal	AC pulse signals
B12	WHT	CYP Sensor Ground	<0.050v
B13	ORN/BLU	TDC Sensor Signal	AC pulse signals
B14	WHT/BLU	TDC Sensor Ground	<0.050v
B15	BLU/GRN	CKP Sensor Signal	AC pulse signals
B16	BLU/YEL	CKP Sensor Ground	<0.050v

1994 LS, RS Sedan 1.8L I4 MFI VIN DB7 [All] 22P 'D' Connector

PCM Pin #	W/Color	Circuit Description (22-Pin)	Value at Hot Idle
D1	WHT/BLU	Keep Alive Power (VBU)	12-14v
D2	GRN/WHT	Brake Switch Signal	Brake Off: 12v, On: 0v
D3	---	Not Used	---
D4	BRN/WHT	Service Check Connector	SCS Open: 4.80v
D5-6	---	Not Used	---
D7	LT BLU	Data Link Connector	No Scan Tool: 5v
D8	LT GRN	A/T: TCM Signal	---
D9	WHT/RED	Alternator 'FR' Signal	Digital Signals: 0-5-0v
D10	GRN/RED	Electronic Load Detector	Varies w/Load: 2.5-3.5v
D11	RED/BLU	TP Sensor Signal	0.5-0.6v
D12	---	Not Used	---
D13	RED/WHT	ECT Sensor Signal	At 180°F: 0.5-0.6v
D14	WHT/RED	HO2S-11 (B1 S1) Signal	0.1-1.1v
D15	RED/YEL	IAT Sensor Signal	At 100°F: 2-3v
D16	WHT/BLK	A/T: TCM Signal	---
D17	WHT/YEL	MAP Sensor Signal	0.8-0.9v
D18	---	Not Used	---
D19	YEL/RED	MAP Sensor VREF	4.9-5.1v
D20	YEL/BLU	Sensor VREF	4.9-5.1v
D21	GRN/WHT	MAP Sensor Ground	<0.050v
D22	GRN/BLU	Sensor Ground	<0.050v

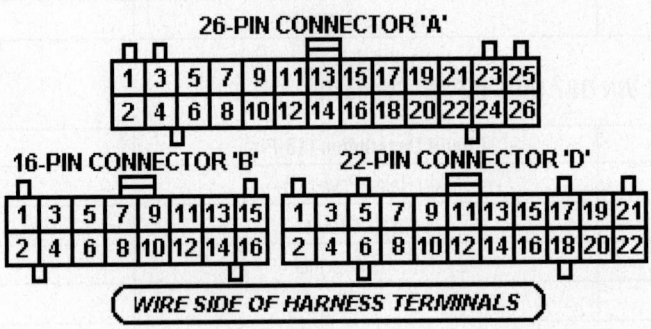

05533_ADIA_G580

Pin Connector Graphic

1994 Sedan 1.8L VTEC I4 MFI VIN DB8 [M/T] 26P 'A' Connector

PCM Pin #	W/Color	Circuit Description (26-Pin)	Value at Hot Idle
A1	BRN	Injector 1	2.0-3.3 ms
A2	YEL	Injector 4	2.0-3.3 ms
A3	RED	Injector 2	2.0-3.3 ms
A4	GRN/YEL	VTEC Solenoid Control	0v, Hi-Speed: 12v
A5	LT BLU	Injector 3	2.0-3.3 ms
A6	ORN/BLK	HO2S-11 (B1 S1) Heater	Heater Off: 12v, On: 1v
A7	GRN/BLK	Fuel Pump Relay Control	Relay Off: 12v, On: 1v
A8	---	Not Used	---
A9	BLK/BLU	Idle Air Control Valve	Pulse Signals
A10-11	---	Not Used	---
A12	GRN	Radiator Fan Relay Control	Fan on: 1v, off: 12v
A13	GRN/ORN	Malfunction Indicator Lamp	MIL Off: 12v, On: 1v
A14	---	Not Used	---
A15	BLK/RED	A/C Clutch Relay Control	Relay Off: 12v, On: 1v
A16	WHT/GRN	Alternator Charging Signal	Headlights off: 0v, on: 12v
A17-19	---	Not Used	---
A20	GRN	EVAP Purge Solenoid	Solenoid Off: 12v, On: 1v
A21	YEL/GRN	Igniter Signal	Digital Signals: 0-12-0v
A22	---	Not Used	---
A23	BLK	Power Ground	<0.1v
A24	BLK	Power Ground	<0.1v
A25	YEL/BLK	Main Relay Power (B+)	12-14v
A26	BRN/BLK	Chassis Ground	0.1v

1994 Sedan 1.8L VTEC I4 MFI VIN DB8 [M/T] 16P 'B' Connector

PCM Pin #	W/Color	Circuit Description (16-Pin)	Value at Hot Idle
B1	YEL/BLK	Main Relay Power (B+)	12-14v
B2	BRN/BLK	Chassis Ground	<0.050v
B3-4	---	Not Used	---
B5	BLU/RED	A/C Switch Signal	Relay Off: 12v, On: 1v
B6	---	Not Used	---
B7	---	Not Used	---
B8	LT GRN	PSP Switch Signal	Straight: 0v, Turning: 12v
B9	BLU/WHT	Starter Switch Signal	Cranking: 9-11v
B10	ORN	Vehicle Speed Sensor	Moving: pulse signals
B11	ORN	CYP Sensor Signal	AC pulse signals
B12	WHT	CYP Sensor Ground	<0.050v
B13	ORN/BLU	TDC Sensor Signal	AC pulse signals
B14	WHT/BLU	TDC Sensor Ground	<0.050v
B15	BLU/GRN	CKP Sensor Signal	AC pulse signals
B16	BLU/YEL	CKP Sensor Ground	<0.050v

1994 Sedan 1.8L VTEC I4 MFI VIN DB8 [M/T] 22P 'D' Connector

PCM Pin #	W/Color	Circuit Description (22-Pin)	Value at Hot Idle
D1	WHT/BLU	Keep Alive Power (VBU)	12-14v
D2	GRN/WHT	Brake Switch Signal	Brake Off: 12v, On: 0v
D3	---	Not Used	---
D4	BRN/WHT	Service Check Connector	SCS Open: 4.80v
D5	---	Not Used	---
D6	BLU/BLK	VTEC Pressure Switch	0v, Hi-Speed: 12v
D7	LT BLU	Data Link Connector	No Scan Tool: 5v
D8	---	Not Used	---
D9	WHT/RED	Alternator 'FR' Signal	Digital Signals: 0-5-0v
D10	GRN/RED	Electronic Load Detector	Varies w/Load: 2.5-3.5v
D11	RED/BLU	TP Sensor Signal	0.5-0.6v
D12	---	Not Used	---
D13	RED/WHT	ECT Sensor Signal	At 180°F: 0.5-0.6v
D14	WHT/RED	HO2S-11 (B1 S1) Signal	0.1-1.1v
D15	RED/YEL	IAT Sensor Signal	At 100°F: 2-3v
D16	---	Not Used	---
D17	WHT/YEL	MAP Sensor Signal	0.8-0.9v
D18	---	Not Used	---
D19	YEL/RED	MAP Sensor VREF	4.9-5.1v
D20	YEL/BLU	Sensor VREF	4.9-5.1v
D21	GRN/WHT	MAP Sensor Ground	<0.050v
D22	GRN/BLU	Sensor Ground	<0.050v

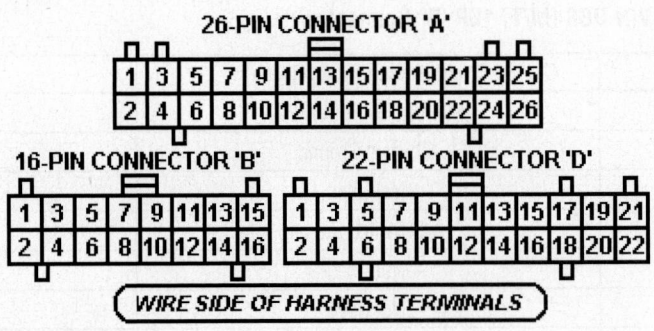

Pin Connector Graphic

1995 RS, SE Sedan 1.8L I4 MFI VIN DB7 [All] 26P 'A' Connector

PCM Pin #	W/Color	Circuit Description (26-Pin)	Value at Hot Idle
A1	BRN	Injector 1	2.0-3.3 ms
A2	YEL	Injector 4	2.0-3.3 ms
A3	RED	Injector 2	2.0-3.3 ms
A4	---	Not Used	---
A5	LT BLU	Injector 3	2.0-3.3 ms
A6	ORN/BLK	HO2S-11 (B1 S1) Heater	Heater Off: 12v, On: 1v
A7	GRN/BLK	Fuel Pump Relay Control	Relay Off: 12v, On: 1v
A8	---	Not Used	---
A9	BLK/BLU	Idle Air Control Valve	Pulse Signals
A10-11	---	Not Used	---
A12	GRN	Radiator Fan Relay Control	Fan on: 1v, off: 12v
A13	GRN/ORN	Malfunction Indicator Lamp	MIL Off: 12v, On: 1v
A14	---	Not Used	---
A15	BLK/RED	A/C Clutch Relay Control	Relay Off: 12v, On: 1v
A16	WHT/GRN	Alternator Charging Signal	Lights off: 12v, off: 0v
A17-18	---	Not Used	---
A19	YEL/RED	A/T: TCM Signal	N/A
A20	GRN	EVAP Purge Solenoid	Solenoid Off: 12v, On: 1v
A21	YEL/GRN	Igniter Signal	Digital Signals: 0-12-0v
A22	---	Not Used	---
A23	BLK	Power Ground	<0.1v
A24	BLK	Power Ground	<0.1v
A25	YEL/BLK	Main Relay Power (B+)	12-14v
A26	BRN/BLK	Chassis Ground	0.1v

1995 RS, SE Sedan 1.8L I4 MFI VIN DB7 [All] 16P 'B' Connector

PCM Pin #	W/Color	Circuit Description (16-Pin)	Value at Hot Idle
B1	YEL/BLK	Main Relay Power (B+)	12-14v
B2	BRN/BLK	Chassis Ground	<0.050v
B3	GRN/BLU	A/T: TCM Signal	N/A
B4	GRY	A/T: TCM Signal	N/A
B5	BLU/RED	A/C Switch Signal	Relay Off: 12v, On: 1v
B6	---	Not Used	---
B7	GRN/BLK	Park/Neutral Switch	In P/N: 1v, others: 12v
B8	LT GRN	PSP Switch Signal	Straight: 0v, Turning: 12v
B9	BLU/WHT	Starter Switch Signal	Cranking: 9-11v
B10	ORN	Vehicle Speed Sensor	Moving: pulse signals
B11	ORN	CYP Sensor Signal	AC pulse signals
B12	WHT	CYP Sensor Ground	<0.050v
B13	ORN/BLU	TDC Sensor Signal	AC pulse signals
B14	WHT/BLU	TDC Sensor Ground	<0.050v
B15	BLU/GRN	CKP Sensor Signal	AC pulse signals
B16	BLU/YEL	CKP Sensor Ground	<0.050v

1995 RS, SE Sedan 1.8L I4 MFI VIN DB7 [All] 22P 'D' Connector

PCM Pin #	W/Color	Circuit Description (22-Pin)	Value at Hot Idle
D1	WHT/BLU	Keep Alive Power (VBU)	12-14v
D2	GRN/WHT	Brake Switch Signal	Brake Off: 12v, On: 0v
D3	---	Not Used	---
D4	BRN/WHT	Service Check Connector	SCS Open: 4.80v
D5-6	---	Not Used	---
D7	LT BLU	Data Link Connector	Jumped: 2.50v
D8	LT GRN	A/T: TCM Signal	---
D9	WHT/RED	Alternator 'FR' Signal	Digital Signals: 0-5-0v
D10	GRN/RED	Electronic Load Detector	Varies w/Load: 2.5-3.5v
D11	RED/BLU	TP Sensor Signal	0.5-0.6v
D12	---	Not Used	---
D13	RED/WHT	ECT Sensor Signal	At 180°F: 0.5-0.6v
D14	WHT/RED	HO2S-11 (B1 S1) Signal	0.1-1.1v
D15	RED/YEL	IAT Sensor Signal	At 100°F: 2-3v
D16	WHT/BLK	A/T: TCM Signal	---
D17	WHT/YEL	MAP Sensor Signal	0.8-0.9v
D18	---	Not Used	---
D19	YEL/RED	MAP Sensor VREF	4.9-5.1v
D20	YEL/BLU	Sensor VREF	4.9-5.1v
D21	GRN/WHT	MAP Sensor Ground	<0.050v
D22	GRN/BLU	Sensor Ground	<0.050v

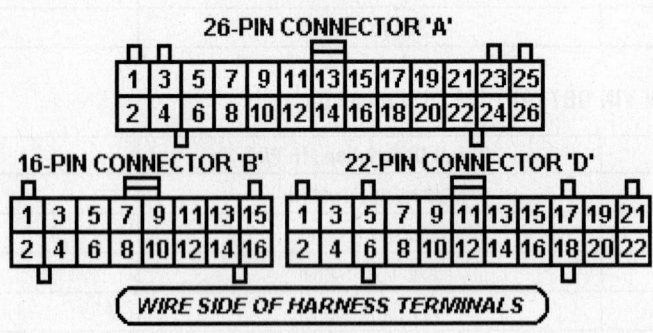

05533_ADIA_G580

Pin Connector Graphic

1995 Sedan 1.8L VTEC I4 MFI VIN DB8 [M/T] 26P 'A' Connector

PCM Pin #	W/Color	Circuit Description (26-Pin)	Value at Hot Idle
A1	BRN	Injector 1	2.0-3.3 ms
A2	YEL	Injector 4	2.0-3.3 ms
A3	RED	Injector 2	2.0-3.3 ms
A4	GRN/YEL	VTEC Solenoid Control	0v, Hi-Speed: 12v
A5	LT BLU	Injector 3	2.0-3.3 ms
A6	ORN/BLK	HO2S-11 (B1 S1) Heater	Heater Off: 12v, On: 1v
A7	GRN/BLK	Fuel Pump Relay Control	Relay Off: 12v, On: 1v
A8	---	Not Used	---
A9	BLK/BLU	Idle Air Control Valve	Pulse Signals
A10-11	---	Not Used	---
A12	GRN	Radiator Fan Relay Control	Fan on: 1v, off: 12v
A13	GRN/ORN	Malfunction Indicator Lamp	MIL Off: 12v, On: 1v
A14	---	Not Used	---
A15	BLK/RED	A/C Clutch Relay Control	Relay Off: 12v, On: 1v
A16	WHT/GRN	Alternator Charging Signal	Lights Off: 12v, On: 1v
A17-19	---	Not Used	---
A20	GRN	EVAP Purge Solenoid	Solenoid Off: 12v, On: 1v
A21	YEL/GRN	Igniter Signal	Digital Signals: 0-12-0v
A22	---	Not Used	---
A23	BLK	Power Ground	<0.1v
A24	BLK	Power Ground	<0.1v
A25	YEL/BLK	Main Relay Power (B+)	12-14v
A26	BRN/BLK	Chassis Ground	<0.1v

1995 Sedan 1.8L VTEC I4 MFI VIN DB8 [M/T] 16P 'B' Connector

PCM Pin #	W/Color	Circuit Description (16-Pin)	Value at Hot Idle
B1	YEL/BLK	Main Relay Power (B+)	12-14v
B2	BRN/BLK	Power Ground	<0.1v
B3-4	---	Not Used	---
B5	BLU/RED	A/C Switch Signal	A/C on: 12v
B6-7	---	Not Used	---
B8	LT GRN	PSP Switch Signal	Straight: 0v, Turning: 12v
B9	BLU/WHT	Starter Switch Signal	Cranking: 9-11v
B10	ORN	Vehicle Speed Sensor	Moving: pulse signals
B11	ORN	CYP Sensor Signal	AC pulse signals
B12	WHT	CYP Sensor Ground	<0.050v
B13	ORN/BLU	TDC Sensor Signal	AC pulse signals
B14	WHT/BLU	TDC Sensor Ground	<0.050v
B15	BLU/GRN	CKP Sensor Signal	AC pulse signals
B16	BLU/YEL	CKP Sensor Ground	<0.050v

1995 Sedan 1.8L VTEC I4 MFI VIN DB8 [M/T] 22P 'D' Connector

PCM Pin #	W/Color	Circuit Description (22-Pin)	Value at Hot Idle
D1	WHT/BLU	Keep Alive Power (VBU)	12-14v
D2	GRN/WHT	Brake Switch Signal	Brake Off: 12v, On: 0v
D3	---	Not Used	---
D4	BRN/WHT	Service Check Connector	SCS Open: 4.80v
D5	---	Not Used	---
D6	BLU/BLK	VTEC Pressure Switch	0v, Hi-Speed: 12v
D7	LT BLU	Data Link Connector	Jumped: 2.50v
D8	---	Not Used	---
D9	WHT/RED	Alternator 'FR' Signal	Digital Signals: 0-5-0v
D10	GRN/RED	Electronic Load Detector	Varies w/Load: 2.5-3.5v
D11	RED/BLU	TP Sensor Signal	0.5-0.6v
D12	---	Not Used	---
D13	RED/WHT	ECT Sensor Signal	At 180°F: 0.5-0.6v
D14	WHT/RED	HO2S-11 (B1 S1) Signal	0.1-1.1v
D15	RED/YEL	IAT Sensor Signal	At 100°F: 2-3v
D16	---	Not Used	---
D17	WHT/YEL	MAP Sensor Signal	0.8-0.9v
D18	---	Not Used	---
D19	YEL/RED	MAP Sensor VREF	4.9-5.1v
D20	YEL/BLU	Sensor VREF	4.9-5.1v
D21	GRN/WHT	MAP Sensor Ground	<0.050v
D22	GRN/BLU	Sensor Ground	<0.050v

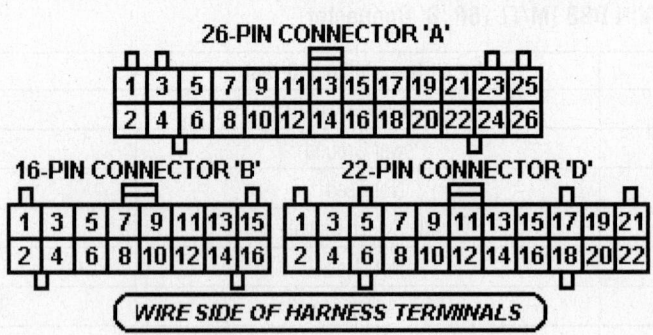

05533_ADIA_G580

Pin Connector Graphic

Standard Colors and Abbreviations

Abbreviation	Color	Abbreviation	Color	Abbreviation	Color
BLK	Black	LT BLU	Lt. Blue	TAN	Tan
BLU	Blue	LT GRN	Lt. Green	VIO	Violet
BRN	Brown	ORN	Orange	WHT	White
GRY	Gray	PNK	Pink	YEL	Yellow
GRN	Green	PPL	Purple		

1996-01 LS, RS Sedan 1.8L I4 MFI VIN DB7 [All] 32P 'A' Connector

PCM Pin #	W/Color	Circuit Description (32-Pin)	Value at Hot Idle
A1	YEL	Injector 4	2.0-3.3 ms
A2	BLU	Injector 3	2.0-3.3 ms
A3	RED	Injector 2	2.0-3.3 ms
A4	BRN	Injector 1	2.0-3.3 ms
A5	GRN/RED	HO2S-12 (B1 S2) Heater	Digital Signals: 0-12-0v
A6	ORN/BLK	HO2S-11 (B1 S1) Heater	Digital Signals: 0-12-0v
A7-8	---	Not Used	---
A9	BRN/BLK	Logic Ground	0.1v
A10	BLK	Power Ground	<0.1v
A11	YEL/BLK	Main Relay Power (B+)	12-14v
A12	BLK/BLU	Idle Air Control Valve	Pulse Signals
A13-14	---	Not Used	---
A15	RED	EVAP Purge Solenoid	Pulse Signals
A16	GRN/BLK	Fuel Pump Relay Control	Relay Off: 12v, On: 1v
A17	BLK/RED	A/C Clutch Relay Control	Relay Off: 12v, On: 1v
A18	GRN/ORN	Malfunction Indicator Lamp	MIL Off: 12v, On: 1v
A19	WHT/GRN	Alternator Charging Signal	Headlights off: 0v, on: 12v
A20	YEL/GRN	Igniter Signal	Digital Signals: 0-12-0v
A21	---	Not Used	---
A22	BRN/BLK	Logic Ground	0.1v
A23	BLK	Power Ground	<0.1v
A24	YEL/BLK	Main Relay Power (B+)	12-14v
A25	WHT/BLK	TCM VREF	4.9-5.1v
A26	---	Not Used	---
A27	GRN	Radiator Fan Relay Control	Fan on: 1v, off: 12v
A28-32	---	Not Used	---

32-PIN CONNECTOR 'A'

1	2	3	4		5	6	7		8	9	10		11
12	13	14	15	16	17	18	19	20	21	22	23		24
	25	26	27		28	29	30	31		32			

31-PIN CONNECTOR 'C'   **16-PIN CONNECTOR 'D'**

1	2	3	4		5	6	7		8	9	10		1		2	3	4		5
11	12	13	14	15	16	17	18	19	20	21	22		6	7	8	9	10	11	12
	23	24	25		26	27	28		29	30	31		13	14	15		16		

WIRE SIDE OF HARNESS TERMINALS

05533_ADIA_G581

Pin Connector Graphic

1996-01 LS, RS Sedan 1.8L I4 MFI VIN DB7 [All] 31P 'C' Connector

PCM Pin #	W/Color	Circuit Description (31-Pin)	Value at Hot Idle
C1	BLU/RED	CKF Sensor	AC pulse signals
C2	BLU/GRN	CKP Sensor Signal	AC pulse signals
C3	ORN/BLU	TDC Sensor Signal	AC pulse signals
C4	ORN	CYP Sensor Signal	AC pulse signals
C5	BLU/RED	A/C Switch Signal	A/C Off: 5v, On: 0v
C6	BLU/WHT	Starter Switch Signal	Cranking: 9-11v
C7	BRN/WHT	Service Check Connector	SCS Open: 4.80v
C8	GRN/WHT	K-Line Signal	12v
C9	YEL	FAS TCM Signal	At idle: 4.5-5.5v
C10	WHT/BLU	Keep Alive Power (VBU)	12-14v
C11	WHT/RED	CKF Sensor Ground	<0.050v
C12	BLU/YEL	CKP Sensor Ground	<0.050v
C13	WHT/BLU	TDC Sensor Ground	<0.050v
C14	WHT	CYP Sensor Ground	<0.050v
C15	---	Not Used	---
C16	GRN	PSP Switch Signal	Straight: 0v, Turning: 12v
C17	WHT/RED	Alternator 'FR' Signal	Digital Signals: 0-5-0v
C18	ORN	Vehicle Speed Sensor	Moving: pulse signals
C19	BLU	ATCHK TCM Signal	Pulse Signals
C20	---	Not Used	---
C21	LT GRN	TCM BARO Signal	Varies: 0.5-4.9v
C22-26	---	Not Used	---
C27	GRY	AFSB TCM Signal	At idle: 4.5-5.5v
C28	GRN/BLU	AFSA TCM Signal	At idle: 4.5-5.5v
C29	GRN/BLK	A/T: Gear Position Switch	In P/N: 0v, Others: 12v
C30-31	---	Not Used	---

1996-01 LS, RS Sedan 1.8L I4 MFI VIN DB7 [All] 16P 'D' Connector

PCM Pin #	W/Color	Circuit Description (16-Pin)	Value at Hot Idle
D1	RED/BLK	TP Sensor Signal	0.5-0.6v
D2	RED/WHT	ECT Sensor Signal	At 180°F: 0.5-0.6v
D3	WHT/YEL	MAP Sensor Signal	0.8-0.9v
D4	YEL/WHT	MAP Sensor VREF	4.9-5.1v
D5	GRN/WHT	Brake Switch Signal	Brake Off: 12v, On: 0v
D6	---	Not Used	---
D7	WHT/RED	HO2S-11 (B1 S1) Signal	0.1-1.1v
D8	RED/YEL	IAT Sensor Signal	At 100°F: 2-3v
D9	---	Not Used	---
D10	YEL/BLU	Sensor VREF	4.9-5.1v
D11	GRN/BLU	Sensor Ground	<0.050v
D12	GRN/WHT	MAP Sensor Ground	<0.050v
D13	ORN/BLU	HO2S-12 Ground	<0.050v
D14	BLU/GRN	HO2S-12 (B1 S2) Signal	0.1-1.1v
D15	---	Not Used	---
D16	GRN/RED	Electric Load Detector	Varies: 2.5-3.5v

1996-01 Sedan 1.8L VTEC I4 MFI VIN DB8 [M/T] 32P 'A' Connector

PCM Pin #	W/Color	Circuit Description (32-Pin)	Value at Hot Idle
A1	YEL	Injector 4	2.0-3.3 ms
A2	BLU	Injector 3	2.0-3.3 ms
A3	RED	Injector 2	2.0-3.3 ms
A4	BRN	Injector 1	2.0-3.3 ms
A5	GRN/RED	HO2S-12 (B1 S2) Heater	Digital Signals: 0-12-0v
A6	ORN/BLK	HO2S-11 (B1 S1) Heater	Digital Signals: 0-12-0v
A7	---	Not Used	---
A8	GRN/YEL	VTEC Solenoid Control	0v, Hi-Speed: 12v
A9	BRN/BLK	Logic Ground	0.1v
A10	BLK	Power Ground	<0.1v
A11	YEL/BLK	Main Relay Power (B+)	12-14v
A12	BLK/BLU	Idle Air Control Valve	Pulse Signals
A13-14	---	Not Used	---
A15	RED	EVAP Purge Solenoid	Pulse Signals
A16	GRN/BLK	Fuel Pump Relay Control	Relay Off: 12v, On: 1v
A17	BLK/RED	A/C Clutch Relay Control	Relay Off: 12v, On: 1v
A18	GRN/ORN	Malfunction Indicator Lamp	MIL Off: 12v, On: 1v
A19	WHT/GRN	Alternator Charging Signal	Headlights off: 0v, on: 12v
A20	YEL/GRN	Igniter Signal	Digital Signals: 0-12-0v
A21	---	Not Used	---
A22	BRN/BLK	Logic Ground	0.1v
A23	BLK	Power Ground	<0.1v
A24	YEL/BLK	Main Relay Power (B+)	12-14v
A25	---	Not Used	---
A26	PNK/BLU	Intake Air Bypass Solenoid	Solenoid Off: 12v, On: 1v
A27	GRN	Radiator Fan Relay Control	Fan on: 1v, off: 12v
A28-32	---	Not Used	---

05533_ADIA_G581

Pin Connector Graphic

1996-01 Sedan 1.8L VTEC I4 MFI VIN DB8 [M/T] 31P 'C' Connector

PCM Pin #	W/Color	Circuit Description (31-Pin)	Value at Hot Idle
C1	BLU/RED	CKF Sensor	AC pulse signals
C2	BLU/GRN	CKP Sensor Signal	AC pulse signals
C3	ORN/BLU	TDC Sensor Signal	AC pulse signals
C4	ORN	CYP Sensor Signal	AC pulse signals
C5	BLU/RED	A/C Switch Signal	A/C Off: 5v, On: 0v
C6	BLU/WHT	Starter Switch Signal	Cranking: 9-11v
C7	BRN/WHT	Service Check Connector	SCS Open: 4.80v
C8	GRN/WHT	K-Line Signal	12v
C9	---	Not Used	---
C10	WHT/BLU	Keep Alive Power (VBU)	12-14v
C11	WHT/RED	CKF Sensor Ground	<0.050v
C12	BLU/YEL	CKP Sensor Ground	<0.050v
C13	WHT/BLU	TDC Sensor Ground	<0.050v
C14	WHT	CYP Sensor Ground	<0.050v
C15	BLU/BLK	VTEC Pressure Switch	0v, Hi-Speed: 12v
C16	GRN	PSP Switch Signal	Straight: 0v, Turning: 12v
C17	WHT/RED	Alternator 'FR' Signal	Digital Signals: 0-5-0v
C18	ORN	Vehicle Speed Sensor	Moving: pulse signals
C19-31	---	Not Used	---

1996-01 Sedan 1.8L VTEC I4 MFI VIN DB8 [M/T] 16P 'D' Connector

PCM Pin #	W/Color	Circuit Description (16-Pin)	Value at Hot Idle
D1	RED/BLK	TP Sensor Signal	0.5-0.6v
D2	RED/WHT	ECT Sensor Signal	At 180°F: 0.5-0.6v
D3	WHT/YEL	MAP Sensor Signal	0.8-0.9v
D4	YEL/WHT	MAP Sensor VREF	4.9-5.1v
D5	GRN/WHT	Brake Switch Signal	Brake Off: 12v, On: 0v
D6	RED/BLU	Knock Sensor Signal	No Detonation: 18mv AC
D7	WHT/RED	HO2S-11 (B1 S1) Signal	0.1-1.1v
D8	RED/YEL	IAT Sensor Signal	At 100°F: 2-3v
D9	---	Not Used	---
D10	YEL/BLU	Sensor VREF	4.9-5.1v
D11	GRN/BLU	Sensor Ground	<0.050v
D12	GRN/WHT	MAP Sensor Ground	<0.050v
D13	ORN/BLU	HO2S-12 Ground	<0.050v
D14	BLU/GRN	HO2S-12 (B1 S2) Signal	0.1-1.1v
D16	GRN/RED	Electric Load Detector	Varies w/Load: 2.5-3.5v

1994-95 RS Coupe 1.8L I4 MFI VIN DC4 [A/T] 26P 'A' Connector

PCM Pin #	W/Color	Circuit Description (26-Pin)	Value at Hot Idle
A1	BRN	Injector 1	2.0-3.3 ms
A2	YEL	Injector 4	2.0-3.3 ms
A3	RED	Injector 2	2.0-3.3 ms
A4	---	Not Used	---
A5	LT BLU	Injector 3	2.0-3.3 ms
A6	ORN/BLK	HO2S-11 (B1 S1) Heater	Heater Off: 12v, On: 1v
A7	GRN/BLK	Fuel Pump Relay Control	Relay Off: 12v, On: 1v
A8	---	Not Used	---
A9	BLK/BLU	Idle Air Control Valve	Pulse Signals
A10-11	---	Not Used	---
A12	GRN	Radiator Fan Relay Control	Fan on: 1v, off: 12v
A13	GRN/ORN	Malfunction Indicator Lamp	MIL Off: 12v, On: 1v
A14	---	Not Used	---
A15	BLK/RED	A/C Clutch Relay Control	Relay Off: 12v, On: 1v
A16	WHT/GRN	Alternator Charging Signal	Headlights off: 0v, on: 12v
A17-18	---	Not Used	---
A19	YEL/RED	A/T: TCM Signal	N/A
A20	GRN	EVAP Purge Solenoid	Solenoid Off: 12v, On: 1v
A21	YEL/GRN	Igniter Signal	Digital Signals: 0-12-0v
A22	---	Not Used	---
A23	BLK	Power Ground	<0.1v
A24	BLK	Power Ground	<0.1v
A25	YEL/BLK	Main Relay Power (B+)	12-14v
A26	BRN/BLK	Chassis Ground	0.1v

1994-95 RS Coupe 1.8L I4 MFI VIN DC4 [A/T] 16P 'B' Connector

PCM Pin #	W/Color	Circuit Description (16-Pin)	Value at Hot Idle
B1	YEL/BLK	Main Relay Power (B+)	12-14v
B2	BRN/BLK	Chassis Ground	<0.050v
B3	GRN/BLU	A/T: TCM Signal	N/A
B4	GRY	A/T: TCM Signal	N/A
B5	BLU/RED	A/C Switch Signal	Relay Off: 12v, On: 1v
B6	---	Not Used	---
B7	GRN/BLK	Park/Neutral Signal	In P/N: 0v, Others: 12v
B8	LT GRN	PSP Switch Signal	Straight: 0v, Turning: 12v
B9	BLU/WHT	Starter Switch Signal	Cranking: 9-11v
B10	ORN	Vehicle Speed Sensor	Moving: pulse signals
B11	ORN	CYP Sensor Signal	AC pulse signals
B12	WHT	CYP Sensor Ground	<0.050v
B13	ORN/BLU	TDC Sensor Signal	AC pulse signals
B14	WHT/BLU	TDC Sensor Ground	<0.050v
B15	BLU/GRN	CKP Sensor Signal	AC pulse signals
B16	BLU/YEL	CKP Sensor Ground	<0.050v

1994-95 RS Coupe 1.8L I4 MFI VIN DC4 [A/T] 22P 'D' Connector

PCM Pin #	W/Color	Circuit Description (22-Pin)	Value at Hot Idle
D1	WHT/BLU	Keep Alive Power (VBU)	12-14v
D2	GRN/WHT	Brake Switch Signal	Brake Off: 12v, On: 0v
D3	---	Not Used	---
D4	BRN/WHT	Service Check Connector	SCS Open: 4.80v
D5-6	---	Not Used	---
D7	LT BLU	Data Link Connector	No Scan Tool: 5v
D8	LT GRN	A/T: TCM Signal	N/A
D9	WHT/RED	Alternator 'FR' Signal	Digital Signals: 0-5-0v
D10	GRN/RED	Electronic Load Detector	Varies w/Load: 2.5-3.5v
D11	RED/BLU	TP Sensor Signal	0.5-0.6v
D12	---	Not Used	---
D13	RED/WHT	ECT Sensor Signal	At 180°F: 0.5-0.6v
D14	WHT/RED	HO2S-11 (B1 S1) Signal	0.1-1.1v
D15	RED/YEL	IAT Sensor Signal	At 100°F: 2-3v
D16	WHT/BLK	A/T: TCM Signal	---
D17	WHT/YEL	MAP Sensor Signal	0.8-0.9v
D18	---	Not Used	---
D19	YEL/RED	MAP Sensor VREF	4.9-5.1v
D20	YEL/BLU	Sensor VREF	4.9-5.1v
D21	GRN/WHT	MAP Sensor Ground	<0.050v
D22	GRN/BLU	Sensor Ground	<0.050v

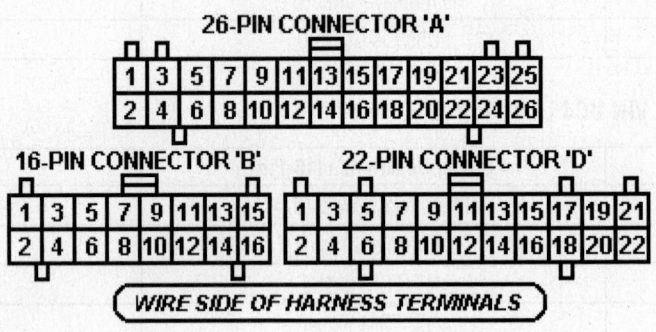

Pin Connector Graphic

05533_ADIA_G580

1994-95 RS Coupe 1.8L I4 MFI VIN DC4 [M/T] 26P 'A' Connector

PCM Pin #	W/Color	Circuit Description (26-Pin)	Value at Hot Idle
A1	BRN	Injector 1	2.0-3.3 ms
A2	YEL	Injector 4	2.0-3.3 ms
A3	RED	Injector 2	2.0-3.3 ms
A4	---	Not Used	---
A5	LT BLU	Injector 3	2.0-3.3 ms
A6	ORN/BLK	HO2S-11 (B1 S1) Heater	Heater Off: 12v, On: 1v
A7	GRN/BLK	Fuel Pump Relay Control	Relay Off: 12v, On: 1v
A8	---	Not Used	---
A9	BLK/BLU	Idle Air Control Valve	Pulse Signals
A10-11	---	Not Used	---
A12	GRN	Radiator Fan Relay Control	Fan on: 1v, off: 12v
A13	GRN/ORN	Malfunction Indicator Lamp	MIL Off: 12v, On: 1v
A14	---	Not Used	---
A15	BLK/RED	A/C Clutch Relay Control	Relay Off: 12v, On: 1v
A16	WHT/GRN	Alternator Charging Signal	Headlights off: 0v, on: 12v
A17-19	---	Not Used	---
A20	GRN	EVAP Purge Solenoid	Solenoid Off: 12v, On: 1v
A21	YEL/GRN	Igniter Signal	Digital Signals: 0-12-0v
A22	---	Not Used	---
A23	BLK	Power Ground	<0.1v
A24	BLK	Power Ground	<0.1v
A25	YEL/BLK	Main Relay Power (B+)	12-14v
A26	BRN/BLK	Chassis Ground	0.1v

1994-95 RS Coupe 1.8L I4 MFI VIN DC4 [M/T] 16P 'B' Connector

PCM Pin #	W/Color	Circuit Description (16-Pin)	Value at Hot Idle
B1	YEL/BLK	Main Relay Power (B+)	12-14v
B2	BRN/BLK	Chassis Ground	<0.050v
B3-4	---	Not Used	---
B5	BLU/RED	A/C Switch Signal	Relay Off: 12v, On: 1v
B6	---	Not Used	---
B7	---	Not Used	---
B8	LT GRN	PSP Switch Signal	Straight: 0v, Turning: 12v
B9	BLU/WHT	Starter Switch Signal	Cranking: 9-11v
B10	ORN	Vehicle Speed Sensor	Moving: pulse signals
B11	ORN	CYP Sensor Signal	AC pulse signals
B12	WHT	CYP Sensor Ground	<0.050v
B13	ORN/BLU	TDC Sensor Signal	AC pulse signals
B14	WHT/BLU	TDC Sensor Ground	<0.050v
B15	BLU/GRN	CKP Sensor Signal	AC pulse signals
B16	BLU/YEL	CKP Sensor Ground	<0.050v

1994-95 RS Coupe 1.8L I4 MFI VIN DC4 [M/T] 22P 'D' Connector

PCM Pin #	W/Color	Circuit Description (22-Pin)	Value at Hot Idle
D1	WHT/BLU	Keep Alive Power (VBU)	12-14v
D2	GRN/WHT	Brake Switch Signal	Brake Off: 12v, On: 0v
D3	---	Not Used	---
D4	BRN/WHT	Service Check Connector	SCS open: 4.80v
D5-6	---	Not Used	---
D7	LT BLU	Data Link Connector	No Scan Tool: 5v
D8	---	Not Used	---
D9	WHT/RED	Alternator 'FR' Signal	Digital Signals: 0-5-0v
D10	GRN/RED	Electronic Load Detector	Varies w/Load: 2.5-3.5v
D11	RED/BLU	TP Sensor Signal	0.5-0.6v
D12	---	Not Used	---
D13	RED/WHT	ECT Sensor Signal	At 180°F: 0.5-0.6v
D14	WHT/RED	HO2S-11 (B1 S1) Signal	0.1-1.1v
D15	RED/YEL	IAT Sensor Signal	At 100°F: 2-3v
D16	---	Not Used	---
D17	WHT/YEL	MAP Sensor Signal	0.8-0.9v
D18	---	Not Used	---
D19	YEL/RED	MAP Sensor VREF	4.9-5.1v
D20	YEL/BLU	Sensor VREF	4.9-5.1v
D21	GRN/WHT	MAP Sensor Ground	<0.050v
D22	GRN/BLU	Sensor Ground	<0.050v

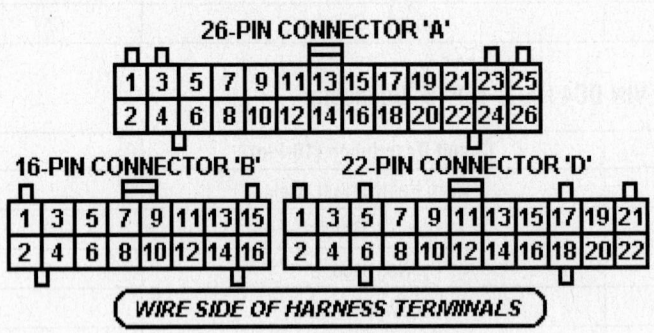

Pin Connector Graphic

05533_ADIA_G580

1994-95 GS-R Coupe 1.8L MFI VIN DC2 [M/T] 26P 'A' Connector

PCM Pin #	W/Color	Circuit Description (26-Pin)	Value at Hot Idle
A1	BRN	Injector 1	2.0-3.3 ms
A2	YEL	Injector 4	2.0-3.3 ms
A3	RED	Injector 2	2.0-3.3 ms
A4	GRN/YEL	VTEC Solenoid Control	0v, Hi-Speed: 12v
A5	LT BLU	Injector 3	2.0-3.3 ms
A6	ORN/BLK	HO2S-11 (B1 S1) Heater	Heater Off: 12v, On: 1v
A7	GRN/BLK	Fuel Pump Relay Control	Relay Off: 12v, On: 1v
A8	---	Not Used	---
A9	BLK/BLU	Idle Air Control Valve	Pulse Signals
A10-11	---	Not Used	---
A12	GRN	Radiator Fan Relay Control	Fan on: 1v, off: 12v
A13	GRN/ORN	Malfunction Indicator Lamp	MIL Off: 12v, On: 1v
A14	---	Not Used	---
A15	BLK/RED	A/C Clutch Relay Control	Relay Off: 12v, On: 1v
A16	WHT/GRN	Alternator Charging Signal	Headlights off: 0v, on: 12v
A17-19	---	Not Used	---
A20	GRN	EVAP Purge Solenoid	Solenoid Off: 12v, On: 1v
A21	YEL/GRN	Igniter Signal	Digital Signals: 0-12-0v
A22	---	Not Used	---
A23	BLK	Power Ground	<0.1v
A24	BLK	Power Ground	<0.1v
A25	YEL/BLK	Main Relay Power (B+)	12-14v
A26	BRN/BLK	Chassis Ground	0.1v

1994-95 GS-R Coupe 1.8L MFI VIN DC2 [M/T] 16P 'B' Connector

PCM Pin #	W/Color	Circuit Description (16-Pin)	Value at Hot Idle
B1	YEL/BLK	Main Relay Power (B+)	12-14v
B2	BRN/BLK	Chassis Ground	<0.050v
B3-4	---	Not Used	---
B5	BLU/RED	A/C Switch Signal	Relay Off: 12v, On: 1v
B6	---	Not Used	---
B7	---	Not Used	---
B8	LT GRN	PSP Switch Signal	Straight: 0v, Turning: 12v
B9	BLU/WHT	Starter Switch Signal	Cranking: 9-11v
B10	ORN	Vehicle Speed Sensor	Moving: pulse signals
B11	ORN	CYP Sensor Signal	AC pulse signals
B12	WHT	CYP Sensor Ground	<0.050v
B13	ORN/BLU	TDC Sensor Signal	AC pulse signals
B14	WHT/BLU	TDC Sensor Ground	<0.050v
B15	BLU/GRN	CKP Sensor Signal	AC pulse signals
B16	BLU/YEL	CKP Sensor Ground	<0.050v

1994-95 GS-R Coupe 1.8L MFI VIN DC2 [M/T] 22P 'D' Connector

PCM Pin #	W/Color	Circuit Description (22-Pin)	Value at Hot Idle
D1	WHT/BLU	Keep Alive Power (VBU)	12-14v
D2	GRN/WHT	Brake Switch Signal	Brake Off: 12v, On: 0v
D3	---	Not Used	---
D4	BRN/WHT	Service Check Connector	SCS Open: 4.80v
D5	---	Not Used	---
D6	BLU/BLK	VTEC Pressure Switch	0v, Hi-Speed: 12v
D7	LT BLU	Data Link Connector	No Scan Tool: 5v
D8	---	Not Used	---
D9	WHT/RED	Alternator 'FR' Signal	Digital Signals: 0-5-0v
D10	GRN/RED	Electronic Load Detector	Varies w/Load: 2.5-3.5v
D11	RED/BLU	TP Sensor Signal	0.5-0.6v
D12	---	Not Used	---
D13	RED/WHT	ECT Sensor Signal	At 180°F: 0.5-0.6v
D14	WHT/RED	HO2S-11 (B1 S1) Signal	0.1-1.1v
D15	RED/YEL	IAT Sensor Signal	At 100°F: 2-3v
D16	---	Not Used	---
D17	WHT/YEL	MAP Sensor Signal	0.8-0.9v
D18	---	Not Used	---
D19	YEL/RED	MAP Sensor VREF	4.9-5.1v
D20	YEL/BLU	Sensor VREF	4.9-5.1v
D21	GRN/WHT	MAP Sensor Ground	<0.050v
D22	GRN/BLU	Sensor Ground	<0.050v

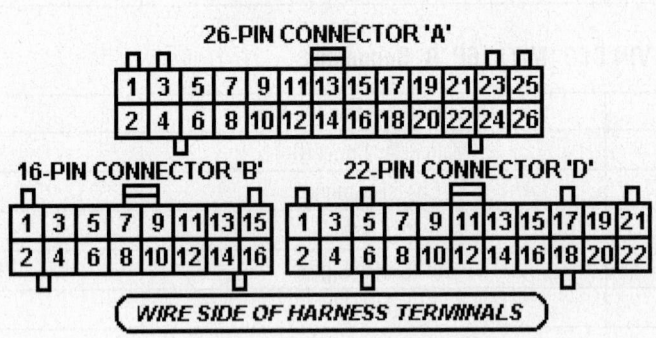

Pin Connector Graphic

05533_ADIA_G580

Standard Colors and Abbreviations

Abbreviation	Color	Abbreviation	Color	Abbreviation	Color
BLK	Black	LT BLU	Lt. Blue	TAN	Tan
BLU	Blue	LT GRN	Lt. Green	VIO	Violet
BRN	Brown	ORN	Orange	WHT	White
GRY	Gray	PNK	Pink	YEL	Yellow
GRN	Green	PPL	Purple		

1996-03 LS-RS-SE Coupe 1.8L MFI VIN DC4 A/T 32P 'A' Connector

PCM Pin #	W/Color	Circuit Description (32-Pin)	Value at Hot Idle
A1	YEL	Injector 4	2.0-3.3 ms
A2	BLU	Injector 3	2.0-3.3 ms
A3	RED	Injector 2	2.0-3.3 ms
A4	BRN	Injector 1	2.0-3.3 ms
A5	GRN/RED	HO2S-12 (B1 S2) Heater	Digital Signals: 0-12-0v
A6	ORN/BLK	HO2S-11 (B1 S1) Heater	Digital Signals: 0-12-0v
A7-8	---	Not Used	---
A9	BRN/BLK	Logic Ground	0.1v
A10	BLK	Power Ground	<0.1v
A11	YEL/BLK	Main Relay Power (B+)	12-14v
A12	BLK/BLU	Idle Air Control Valve	Pulse Signals
A13-14	---	Not Used	---
A15	RED	EVAP Purge Solenoid	Pulse Signals
A16	GRN/BLK	Fuel Pump Relay Control	Relay Off: 12v, On: 1v
A17	BLK/RED	A/C Clutch Relay Control	Relay Off: 12v, On: 1v
A18	GRN/ORN	Malfunction Indicator Lamp	MIL Off: 12v, On: 1v
A19	WHT/GRN	Alternator Charging Signal	Headlights off: 0v, on: 12v
A20	YEL/GRN	Igniter Signal	Digital Signals: 0-12-0v
A21	---	Not Used	---
A22	BRN/BLK	Logic Ground	0.1v
A23	BLK	Power Ground	<0.1v
A24	YEL/BLK	Main Relay Power (B+)	12-14v
A25	WHT/BLK	TCM VREF	4.9-5.1v
A26	---	Not Used	---
A27	GRN	Radiator Fan Relay Control	Fan on: 1v, off: 12v
A28-32	---	Not Used	---

Pin Connector Graphic

1996-03 LS-RS-SE Coupe 1.8L MFI VIN DC4 A/T 31P 'C' Connector

PCM Pin #	W/Color	Circuit Description (31-Pin)	Value at Hot Idle
C1	BLU/RED	CKF Sensor	AC pulse signals
C2	BLU/GRN	CKP Sensor Signal	AC pulse signals
C3	ORN/BLU	TDC Sensor Signal	AC pulse signals
C4	ORN	CYP Sensor Signal	AC pulse signals
C5	BLU/RED	A/C Switch Signal	A/C Off: 5v, On: 0v
C6	BLU/WHT	Starter Switch Signal	Cranking: 9-11v
C7	BRN/WHT	Service Check Connector	SCS Open: 4.80v
C8	GRN/WHT	K-Line Signal	12v
C9	YEL	FAS TCM Signal	At idle: 4.5-5.5v
C10	WHT/BLU	Keep Alive Power (VBU)	12-14v
C11	WHT/RED	CKF Sensor Ground	<0.050v
C12	BLU/YEL	CKP Sensor Ground	<0.050v
C13	WHT/BLU	TDC Sensor Ground	<0.050v
C14	WHT	CYP Sensor Ground	<0.050v
C15	---	Not Used	---
C16	GRN	PSP Switch Signal	Straight: 0v, Turning: 12v
C17	WHT/RED	Alternator 'FR' Signal	Digital Signals: 0-5-0v
C18	ORN	Vehicle Speed Sensor	Moving: pulse signals
C19	BLU	ATCHK TCM Signal	KOEO on: pulses
C20	---	Not Used	---
C21	LT GRN	TCM BARO Signal	KOEO on: 2.5-3.5v
C22-26	---	Not Used	---
C27	GRY	AFSB TCM Signal	At idle: 4.5-5.5v
C28	GRN/BLU	AFSA TCM Signal	At idle: 4.5-5.5v
C29	GRN/BLK	A/T: Gear Position Switch	In P/N: 0v, Others: 12v
C30-31	---	Not Used	---

1996-03 LS-RS-SE Coupe 1.8L MFI VIN DC4 A/T 16P 'D' Connector

PCM Pin #	W/Color	Circuit Description (16-Pin)	Value at Hot Idle
D1	RED/BLK	TP Sensor Signal	0.5-0.6v
D2	RED/WHT	ECT Sensor Signal	At 180°F: 0.5-0.6v
D3	WHT/YEL	MAP Sensor Signal	0.8-0.9v
D4	YEL/WHT	MAP Sensor VREF	4.9-5.1v
D5	GRN/WHT	Brake Switch Signal	Brake Off: 12v, On: 0v
D6	---	Not Used	---
D7	WHT/RED	HO2S-11 (B1 S1) Signal	0.1-1.1v
D8	RED/YEL	IAT Sensor Signal	At 100°F: 2-3v
D9	---	Not Used	---
D10	YEL/BLU	Sensor VREF	4.9-5.1v
D11	GRN/BLU	Sensor Ground	<0.050v
D12	GRN/WHT	MAP Sensor Ground	<0.050v
D13	ORN/BLU	HO2S-12 Ground	<0.050v
D14	BLU/GRN	HO2S-12 (B1 S2) Signal	0.1-1.1v
D15	---	Not Used	---
D16	GRN/RED	Electric Load Detector	Varies: 2.5-3.5v

1996-03 LS-RS-SE Coupe 1.8L MFI VIN DC4 [M/T] 32P A Connector

PCM Pin #	W/Color	Circuit Description (32-Pin)	Value at Hot Idle
A1	YEL	Injector 4	2.0-3.3 ms
A2	BLU	Injector 3	2.0-3.3 ms
A3	RED	Injector 2	2.0-3.3 ms
A4	BRN	Injector 1	2.0-3.3 ms
A5	GRN/RED	HO2S-12 (B1 S2) Heater	Digital Signals: 0-12-0v
A6	ORN/BLK	HO2S-11 (B1 S1) Heater	Digital Signals: 0-12-0v
A7-8	---	Not Used	---
A9	BRN/BLK	Logic Ground	0.1v
A10	BLK	Power Ground	<0.1v
A11	YEL/BLK	Main Relay Power (B+)	12-14v
A12	BLK/BLU	Idle Air Control Valve	Pulse Signals
A13-14	---	Not Used	---
A15	RED	EVAP Purge Solenoid	Pulse Signals
A16	GRN/BLK	Fuel Pump Relay Control	Relay Off: 12v, On: 1v
A17	BLK/RED	A/C Clutch Relay Control	Relay Off: 12v, On: 1v
A18	GRN/ORN	Malfunction Indicator Lamp	MIL Off: 12v, On: 1v
A19	WHT/GRN	Alternator Charging Signal	Headlights off: 0v, on: 12v
A20	YEL/GRN	Igniter Signal	Digital Signals: 0-12-0v
A21	---	Not Used	---
A22	BRN/BLK	Logic Ground	0.1v
A23	BLK	Power Ground	<0.1v
A24	YEL/BLK	Main Relay Power (B+)	12-14v
A25	---	Not Used	---
A26	---	Not Used	---
A27	GRN	Radiator Fan Relay Control	Fan on: 1v, off: 12v
A28-32	---	Not Used	---

32-PIN CONNECTOR 'A'

1	2	3	4		5	6	7		8	9	10		11
12	13	14	15	16	17	18	19	20	21	22	23		24
	25	26	27		28	29	30	31		32			

31-PIN CONNECTOR 'C'   16-PIN CONNECTOR 'D'

WIRE SIDE OF HARNESS TERMINALS

05533_ADIA_G581

Pin Connector Graphic

1996-03 LS-RS-SE Coupe 1.8L MFI VIN DC4 [M/T] 31P C Connector

PCM Pin #	W/Color	Circuit Description (31-Pin)	Value at Hot Idle
C1	BLU/RED	CKF Sensor	AC pulse signals
C2	BLU/GRN	CKP Sensor Signal	AC pulse signals
C3	ORN/BLU	TDC Sensor Signal	AC pulse signals
C4	ORN	CYP Sensor Signal	AC pulse signals
C5	BLU/RED	A/C Switch Signal	A/C Off: 5v, On: 0v
C6	BLU/WHT	Starter Switch Signal	Cranking: 9-11v
C7	BRN/WHT	Service Check Connector	SCS Open: 4.80v
C8	GRN/WHT	K-Line Signal	12v
C9	---	Not Used	---
C10	WHT/BLU	Keep Alive Power (VBU)	12-14v
C11	WHT/RED	CKF Sensor Ground	<0.050v
C12	BLU/YEL	CKP Sensor Ground	<0.050v
C13	WHT/BLU	TDC Sensor Ground	<0.050v
C14	WHT	CYP Sensor Ground	<0.050v
C15	---	Not Used	---
C16	GRN	PSP Switch Signal	Straight: 0v, Turning: 12v
C17	WHT/RED	Alternator 'FR' Signal	Digital Signals: 0-5-0v
C18	ORN	Vehicle Speed Sensor	Moving: pulse signals
C19-31	---	Not Used	---

1996-03 LS-RS-SE Coupe 1.8L MFI VIN DC4 [M/T] 16P D Connector

PCM Pin #	W/Color	Circuit Description (16-Pin)	Value at Hot Idle
D1	RED/BLK	TP Sensor Signal	0.5-0.6v
D2	RED/WHT	ECT Sensor Signal	At 180°F: 0.5-0.6v
D3	WHT/YEL	MAP Sensor Signal	0.8-0.9v
D4	YEL/WHT	MAP Sensor VREF	4.9-5.1v
D5	GRN/WHT	Brake Switch Signal	Brake Off: 12v, On: 0v
D6	---	Not Used	---
D7	WHT/RED	HO2S-11 (B1 S1) Signal	0.1-1.1v
D8	RED/YEL	IAT Sensor Signal	At 100°F: 2-3v
D9	---	Not Used	---
D10	YEL/BLU	Sensor VREF	4.9-5.1v
D11	GRN/BLU	Sensor Ground	<0.050v
D12	GRN/WHT	MAP Sensor Ground	<0.050v
D13	ORN/BLU	HO2S-12 Ground	<0.050v
D14	BLU/GRN	HO2S-12 (B1 S2) Signal	0.1-1.1v
D15	---	Not Used	---
D16	GRN/RED	Electric Load Detector	Varies: 2.5-3.5v

1996-03 GS-R Coupe 1.8L I4 MFI VIN DC2 [M/T] 32P 'A' Connector

PCM Pin #	W/Color	Circuit Description (32-Pin)	Value at Hot Idle
A1	YEL	Injector 4	2.0-3.3 ms
A2	BLU	Injector 3	2.0-3.3 ms
A3	RED	Injector 2	2.0-3.3 ms
A4	BRN	Injector 1	2.0-3.3 ms
A5	GRN/RED	HO2S-12 (B1 S2) Heater	Digital Signals: 0-12-0v
A6	ORN/BLK	HO2S-11 (B1 S1) Heater	Digital Signals: 0-12-0v
A7	---	Not Used	---
A8	GRN/YEL	VTEC Solenoid Control	0v, Hi-Speed: 12v
A9	BRN/BLK	Logic Ground	0.1v
A10	BLK	Power Ground	<0.1v
A11	YEL/BLK	Main Relay Power (B+)	12-14v
A12	BLK/BLU	Idle Air Control Valve	Pulse Signals
A13-14	---	Not Used	---
A15	RED	EVAP Purge Solenoid	Pulse Signals
A16	GRN/BLK	Fuel Pump Relay Control	Relay Off: 12v, On: 1v
A17	BLK/RED	A/C Clutch Relay Control	Relay Off: 12v, On: 1v
A18	GRN/ORN	Malfunction Indicator Lamp	MIL Off: 12v, On: 1v
A19	WHT/GRN	Alternator Charging Signal	Headlights off: 0v, on: 12v
A20	YEL/GRN	Igniter Signal	Digital Signals: 0-12-0v
A21	---	Not Used	---
A22	BRN/BLK	Logic Ground	<0.050v
A23	BLK	Power Ground	<0.1v
A24	YEL/BLK	Main Relay Power (B+)	12-14v
A25	---	Not Used	---
A26	PNK/BLU	Intake Air Bypass Solenoid	Solenoid Off: 12v, On: 1v
A27	GRN	Radiator Fan Relay Control	Fan on: 1v, off: 12v
A28-32	---	Not Used	---

05533_ADIA_G582

Pin Connector Graphic

1996-03 GS-R Coupe 1.8L I4 MFI VIN DC2 [M/T] 31P 'C' Connector

PCM Pin #	W/Color	Circuit Description (31-Pin)	Value at Hot Idle
C1	BLU/RED	CKF Sensor	AC pulse signals
C2	BLU/GRN	CKP Sensor Signal	AC pulse signals
C3	ORN/BLU	TDC Sensor Signal	AC pulse signals
C4	ORN	CYP Sensor Signal	AC pulse signals
C5	BLU/RED	A/C Switch Signal	A/C Off: 5v, On: 0v
C6	BLU/WHT	Starter Switch Signal	Cranking: 9-11v
C7	BRN/WHT	Service Check Connector	SCS Open: 4.80v
C8	GRN/WHT	K-Line Signal	12v
C9	---	Not Used	---
C10	WHT/BLU	Keep Alive Power (VBU)	12-14v
C11	WHT/RED	CKF Sensor Ground	<0.050v
C12	BLU/YEL	CKP Sensor Ground	<0.050v
C13	WHT/BLU	TDC Sensor Ground	<0.050v
C14	WHT	CYP Sensor Ground	<0.050v
C15	BLU/BLK	VTEC Pressure Switch	0v, Hi-Speed: 12v
C16	GRN	PSP Switch Signal	Straight: 0v, Turning: 12v
C17	WHT/RED	Alternator 'FR' Signal	Digital Signals: 0-5-0v
C18	ORN	Vehicle Speed Sensor	Moving: pulse signals
C19-31	---	Not Used	---

1996-03 GS-R Coupe 1.8L I4 MFI VIN DC2 [M/T] 16P 'D' Connector

PCM Pin #	W/Color	Circuit Description (16-Pin)	Value at Hot Idle
D1	RED/BLK	TP Sensor Signal	0.5-0.6v
D2	RED/WHT	ECT Sensor Signal	At 180°F: 0.5-0.6v
D3	WHT/YEL	MAP Sensor Signal	0.8-0.9v
D4	YEL/WHT	MAP Sensor VREF	4.9-5.1v
D5	GRN/WHT	Brake Switch Signal	Brake Off: 12v, On: 0v
D6	RED/BLU	Knock Sensor Signal	No Detonation: 18mv AC
D7	WHT/RED	HO2S-11 (B1 S1) Signal	0.1-1.1v
D8	RED/YEL	IAT Sensor Signal	At 100°F: 2-3v
D10	YEL/BLU	Sensor VREF	4.9-5.1v
D11	GRN/BLU	Sensor Ground	<0.050v
D12	GRN/WHT	MAP Sensor Ground	<0.050v
D13	ORN/BLU	HO2S-12 Ground	<0.050v
D14	BLU/GRN	HO2S-12 (B1 S2) Signal	0.1-1.1v
D16	GRN/RED	Electric Load Detector	Varies w/Load: 2.5-3.5v

1997 GS & LS 1.8L I4 MFI VIN DB7, DB8 [All] 32P 'A' Connector

PCM Pin #	W/Color	Circuit Description (32-Pin)	Value at Hot Idle
A1	YEL	Injector 4	2.0-3.3 ms
A2	BLU	Injector 3	2.0-3.3 ms
A3	RED	Injector 2	2.0-3.3 ms
A4	BRN	Injector 1	2.0-3.3 ms
A5	GRN/RED	HO2S-12 (B1 S2) Heater	Digital Signals: 0-12-0v
A6	ORN/BLK	HO2S-11 (B1 S1) Heater	Digital Signals: 0-12-0v
A7-8	---	Not Used	---
A9	BRN/BLK	Logic Ground (LG1)	<0.050v
A10	BLK	Power Ground	<0.1v
A11	YEL/BLK	Main Relay Power (B+)	12-14v
A12	BLK/BLU	Idle Air Control Valve	Pulse Signals
A13-14	---	Not Used	---
A15	RED	EVAP Purge Solenoid	Pulse Signals
A16	GRN/BLU	Fuel Pump Relay Control	Relay Off: 12v, On: 1v
A17	BLK/RED	A/C Clutch Relay Control	Relay Off: 12v, On: 1v
A18	GRN/ORN	Malfunction Indicator Lamp	MIL Off: 12v, On: 1v
A19	WHT/GRN	Alternator Charging Signal	Lights on: 0v: off: 12v
A20	YEL/GRN	Igniter Control Signal	Digital Signals: 0-12-0v
A21	---	Not Used	---
A22	BRN/BLK	Logic Ground (LG2)	<0.050v
A23	BLK	Power Ground	<0.1v
A24	YEL/BLK	Main Relay Power (B+)	12-14v
A25	WHT/BLK	A/T: TCM VREF	4.9-5.1v
A26	---	Not Used	---
A27	GRN	Radiator Fan Relay Control	Fan on: 1v, off: 12v
A28-32	---	---	---

05533_ADIA_G582

Pin Connector Graphic

1997 GS & LS 1.8L I4 MFI VIN DB7, DB8 [All] 31P 'C' Connector

PCM Pin #	W/Color	Circuit Description (31-Pin)	Value at Hot Idle
C1	BLU/RED	CKF Sensor Signal	AC pulse signals
C2	BLU	CKP Sensor Signal	AC pulse signals
C3	ORN/BLU	TDC Sensor Signal	AC pulse signals
C4	YEL	CYP Sensor Signal	AC pulse signals
C5	BLU/RED	A/C Switch Signal	A/C Off: 5v, On: 0v
C6	BLU/WHT	Starter Switch Signal	Cranking: 9-11v
C7	BRN/WHT	Service Check Connector	SCS Open: 4.80v
C8	GRN/WHT	K-Line Signal	12v
C9	YEL	FAS TCM Signal	At idle: 4.5-5.5v
C10	WHT/BLU	Keep Alive Power (VBU)	12-14v
C11	WHT/RED	CKF Sensor Ground	<0.050v
C12	BLU/YEL	CKP Sensor Ground	<0.050v
C13	WHT/BLU	TDC Sensor Ground	<0.050v
C14	WHT	CYP Sensor Ground	<0.050v
C15	---	Not Used	---
C16	GRN	PSP Switch Signal	Straight: 0v, Turning: 12v
C17	WHT/RED	Alternator 'FR' Signal	Digital Signals: 0-5-0v
C18	ORN	Vehicle Speed Sensor	Moving: pulse signals
C19	BLU	A/T: BARO to TCM Signal	Varies w/alt: 0-5-4.5v
C20	---	Not Used	---
C21	LT GRN	BARO Signal to TCM	Varies: 0.5-4.9v
C22-26	---	Not Used	---
C27	GRY	A/T: TCM Spark Retard 'A'	At idle: 5v, Shifting: 0v
C28	GRN/BLU	A/T: TCM Spark Retard 'B'	At idle: 5v, Shifting: 0v
C29	GRN/BLK	A/T: Gear Position Switch	In P/N: 0v, Others: 12v
C30-31	---	Not Used	---

1997 GS & LS 1.8L I4 MFI VIN DB7, DB8 [All] 16P 'D' Connector

PCM Pin #	W/Color	Circuit Description (16-Pin)	Value at Hot Idle
D1	RED/BLK	TP Sensor Signal	0.5-0.6v
D2	RED/WHT	ECT Sensor Signal	At 180°F: 0.5-0.6v
D3	WHT/YEL	MAP Sensor Signal	0.8-0.9v
D4	YEL/WHT	MAP Sensor VREF	4.9-5.1v
D5	GRN/WHT	Brake Switch Signal	Brake Off: 12v, On: 0v
D6	---	Not Used	---
D7	WHT	HO2S-11 (B1 S1) Signal	0.1-1.1v
D8	RED/YEL	IAT Sensor Signal	At 100°F: 2-3v
D9	---	Not Used	---
D10	YEL/BLU	Sensor VREF	4.9-5.1v
D11	GRN/BLU	Sensor Ground	<0.050v
D12	GRN/WHT	MAP Sensor Ground	<0.050v
D13	ORN/BLU	HO2S-12 Ground	<0.050v
D14	BLU/RED	HO2S-12 (B1 S2) Signal	0.1-1.1v
D15	---	Not Used	---
D16	GRN/RED	Electric Load Detector	Varies: 2.5-3.5v

1998-99 GS/LS 1.8L I4 MFI VIN DB7, DB8 [All] 32P 'A' Connector

PCM Pin #	W/Color	Circuit Description (32-Pin)	Value at Hot Idle
A1	YEL	Injector 4	2.0-3.3 ms
A2	BLU	Injector 3	2.0-3.3 ms
A3	RED	Injector 2	2.0-3.3 ms
A4	BRN	Injector 1	2.0-3.3 ms
A5	GRN/RED	HO2S-12 (B1 S2) Heater	Digital Signals: 0-12-0v
A6	ORN/BLK	HO2S-11 (B1 S1) Heater	Digital Signals: 0-12-0v
A7-8	---	Not Used	---
A9	BRN/BLK	Logic Ground (LG1)	<0.050v
A10	BLK	Power Ground	<0.1v
A11	YEL/BLK	Main Relay Power (B+)	12-14v
A12	BLK/BLU	Idle Air Control Valve	Pulse Signals
A13-14	---	Not Used	---
A15	RED	EVAP Purge Solenoid	Pulse Signals
A16	GRN/BLU	Fuel Pump Relay Control	Relay Off: 12v, On: 1v
A17	BLK/RED	A/C Clutch Relay Control	Relay Off: 12v, On: 1v
A18	GRN/ORN	Malfunction Indicator Lamp	MIL Off: 12v, On: 1v
A19	WHT/GRN	Alternator Charging Signal	Lights on: 0v: off: 12v
A20	YEL/GRN	Igniter Control Signal	Digital Signals: 0-12-0v
A21	---	Not Used	---
A22	BRN/BLK	Logic Ground (LG2)	<0.050v
A23	BLK	Power Ground	<0.1v
A24	YEL/BLK	Main Relay Power (B+)	12-14v
A25	WHT/BLK	A/T: TCM VREF	4.9-5.1v
A26	---	Not Used	---
A27	GRN	Radiator Fan Relay Control	Fan on: 1v, off: 12v
A28	BLU	EVAP Bypass Solenoid	Solenoid Off: 12v, On: 1v
A29	LT GRN	EVAP Vent Shut Control	Solenoid Off: 12v, On: 1v
A30-32	---	---	---

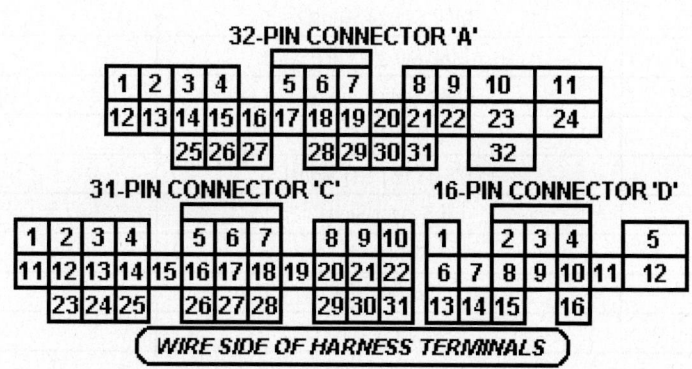

Pin Connector Graphic

05533_ADIA_G582

1998-99 GS/LS 1.8L I4 MFI VIN DB7, DB8 [All] 31P 'C' Connector

PCM Pin #	W/Color	Circuit Description (31-Pin)	Value at Hot Idle
C1	BLU/RED	CKF Sensor Signal	AC pulse signals
C2	BLU	CKP Sensor Signal	AC pulse signals
C3	GRN	TDC Sensor Signal	AC pulse signals
C4	YEL	CYP Sensor Signal	AC pulse signals
C5	BLU/RED	A/C Switch Signal	A/C Off: 5v, On: 0v
C6	BLU/WHT	Starter Switch Signal	Cranking: 9-11v
C7	BRN/WHT	Service Check Connector	SCS Open: 4.80v
C8	GRN/WHT	K-Line Signal	12v
C9	YEL	FAS TCM Signal	At idle: 4.5-5.5v
C10	WHT/BLU	Keep Alive Power (VBU)	12-14v
C11	WHT/RED	CKF Sensor Ground	<0.050v
C12	BLU/YEL	CKP Sensor Ground	<0.050v
C13	WHT/BLU	TDC Sensor Ground	<0.050v
C14	WHT	CYP Sensor Ground	<0.050v
C15	---	Not Used	---
C16	GRN	PSP Switch Signal	Straight: 0v, Turning: 12v
C17	WHT/RED	Alternator 'FR' Signal	Digital Signals: 0-5-0v
C18	ORN	Vehicle Speed Sensor	Moving: pulse signals
C19	BLU	A/T: BARO to TCM Signal	Varies w/alt: 0-5-4.5v
C20	---	Not Used	---
C21	LT GRN	BARO Signal to TCM	Varies: 0.5-4.9v
C22-26	---	Not Used	---
C27	GRY	A/T: TCM Spark Retard 'A'	At idle: 5v, Shifting: 0v
C28	GRN/BLU	A/T: TCM Spark Retard 'B'	At idle: 5v, Shifting: 0v
C29	GRN/BLK	A/T: Gear Position Switch	In P/N: 0v, Others: 12v
C30-31	---	Not Used	---

1998-99 GS/LS 1.8L I4 MFI VIN DB7, DB8 [All] 16P 'D' Connector

PCM Pin #	W/Color	Circuit Description (16-Pin)	Value at Hot Idle
D1	RED/BLK	TP Sensor Signal	0.5-0.6v
D2	RED/WHT	ECT Sensor Signal	At 180°F: 0.5-0.6v
D3	WHT/YEL	MAP Sensor Signal	0.8-0.9v
D4	YEL/WHT	MAP Sensor VREF	4.9-5.1v
D5	GRN/WHT	Brake Switch Signal	Brake Off: 12v, On: 0v
D6	---	Not Used	---
D7	WHT	HO2S-11 (B1 S1) Signal	0.1-1.1v
D8	RED/YEL	IAT Sensor Signal	At 100°F: 2-3v
D9	---	Not Used	---
D10	YEL/BLU	Sensor VREF	4.9-5.1v
D11	GRN/BLU	Sensor Ground	<0.050v
D12	GRN/WHT	MAP Sensor Ground	<0.050v
D13	ORN/BLU	HO2S-12 Ground	<0.050v
D14	BLU/RED	HO2S-12 (B1 S2) Signal	0.1-1.1v
D15	LT GRN	Fuel Tank Pressure Sensor	Fuel Cap off: 2.5v
D16	GRN/RED	Electric Load Detector	Varies: 2.5-3.5v

2000-01 GS 1.8L MFI VIN DB7, DB8 [All] 32P 'A' Connector

PCM Pin #	W/Color	Circuit Description (32-Pin)	Value at Hot Idle
A1-2	---	Not Used	---
A3	BLU	EVAP Bypass Solenoid	Solenoid Off: 12v, On: 1v
A4	LT GRN	EVAP Vent Shut Control	Solenoid Off: 12v, On: 1v
A5	---	Not Used	---
A6	RED	EVAP Purge Solenoid	Pulse Signals
A7	---	Not Used	---
A8	GRN/RED	HO2S-12 (B1 S2) Heater	Digital Signals: 0-12-0v
A9	---	Not Used	---
A10	BRN/WHT	Service Check Connector	SCS Open: 4.80v
A11	---	Not Used	---
A12	PNK	Immobilizer Indicator Control	Lamp On: 1v, Off: 12v
A13	BLU	IMOEN Immobilizer Signal	Digital Signals
A14	GRN/BLK	A/T: D4 Indicator Light	In D4: 0v, Others: 12v
A15	GRN/YEL	Fuel Pump Relay Control	Relay Off: 12v, On: 1v
A16	---	Not Used	---
A17	BLK/RED	A/C Clutch Relay Control	Relay Off: 12v, On: 1v
A18	GRN/ORN	Malfunction Indicator Light	MIL Off: 12v, On: 1v
A19	BLU	Engine Speed (NEP) Signal	Digital Signals
A20	GRN	Radiator Fan Relay Control	Fan on: 1v, off: 12v
A21	GRN/WHT	K-Line Signal	12v
A22	---	Not Used	---
A23	BLU/RED	HO2S-12 (B1 S2) Signal	0.1-1.1v
A24	BLU/WHT	Starter Switch Signal	Cranking: 9-11v
A25	RED	INOCD Immobilizer Code	Digital Signals
A26	GRN	PSP Switch Signal	Straight: 0v, Turning: 12v
A27	BLU/RED	A/C Switch Signal	A/C Off: 5v, On: 0v
A28	WHT/RED	A/T: Interlock Control Unit	With Brake on: 12v
A29	LT GRN	Fuel Tank Pressure Sensor	Fuel Cap off: 2.5v
A30	GRN/RED	Electric Load Detector	Varies w/Load: 2.5-3.5v
A31	---	Not Used	---
A32	GRN/WHT	Brake Switch Signal	Brake Off: 12v, On: 0v

2000-01 GS 1.8L MFI VIN DB7, DB8 [All] 25P 'B' Connector

PCM Pin #	W/Color	Circuit Description (25-Pin)	Value at Hot Idle
B1	YEL/BLK	Main Relay Power (B+)	12-14v
B2	BLK	Power Ground	<0.1v
B3	RED	Injector 2	2.0-3.3 ms
B4	BLU	Injector 3	2.0-3.3 ms
B5	YEL	Injector 4	2.0-3.3 ms
B6-7	---	Not Used	---
B8	WHT/GRN	A/T: Clutch Solenoid LS-	AC Pulse Signals
B9	YEL/BLK	Main Relay Power (B+)	12-14v
B10	BLK	Power Ground	<0.1v
B11	BRN	Injector 1	2.0-3.3 ms
B12	---	Not Used	---
B13	YEL/GRN	Ignition Control Module Signal	Digital Signals: 0-12-0v
B14-15	---	Not Used	---
B16	PNK/BLU	Intake Air Bypass Solenoid	Solenoid Off: 12v, On: 1v
B17	RED/BLU	A/T: Clutch Solenoid LS+	AC Pulse Signals
B18-19	---	Not Used	---
B20	BRN/BLK	Logic Ground (LG1)	<0.050v
B21	WHT/BLU	Keep Alive Power (VBU)	12-14v
B22	BRN/BLK	Logic Ground (LG2)	<0.050v
B23	BLK/BLU	Idle Air Control Valve	DC Pulse Signals

2000-01 GS 1.8L MFI VIN DB7, DB8 [All] 31P 'C' Connector

PCM Pin #	W/Color	Circuit Description (31-Pin)	Value at Hot Idle
C1	BRN/BLK	HO2S-11 (B1 S1) Heater	Digital Signals: 0-12-0v
C2	WHT/GRN	Alternator Charging Signal	Headlights off: 0v, on: 12v
C3	RED/BLU	Knock Sensor Signal	No Detonation: 18mv AC
C4	---	Not Used	---
C5	WHT/RED	Alternator 'FR' Signal	Digital Signals: 0-5-0v
C6	---	Not Used	---
C7	GRN/WHT	MAP Sensor Ground	<0.050v
C8	BLU	CKP Sensor Signal	AC Pulse Signals
C9	BLU/YEL	CKP Sensor Ground	<0.050v
C10-15	---	Not Used	---
C16	WHT	HO2S-11 (B1 S1) Signal	0.1-1.1v
C17	WHT/YEL	MAP Sensor Signal	0.8-0.9v
C18	GRN/BLU	Sensor Ground (SG2)	<0.050v
C19	YEL/RED	MAP Sensor VREF (VCC1)	4.9-5.1v
C20	GRN	TDC Sensor 'P' Signal	A/C Pulse Signals
C21	WHT/BLU	TDC Sensor 'N' Signal	<0.050v
C22	BLU/RED	CKF Sensor 'P' Signal	A/C Pulse Signals
C23	ORN	Vehicle Speed Sensor	Digital Signals: 0-5-0v
C24	---	Not Used	---
C25	RED/YEL	IAT Sensor Signal	At 100°F: 2-3v
C26	RED/WHT	ECT Sensor Signal	At 180°F: 0.5-0.6v
C27	RED/BLK	TP Sensor Signal	0.5-0.6v
C28	YEL/BLU	Sensor VREF (VCC2)	4.9-5.1v
C29	YEL	CYP Sensor 'P' Signal	AC Pulse Signals
C30	WHT	CYP Sensor Ground	<0.050v
C31	WHT/RED	CKF Sensor Ground	<0.050v

2000-01 GS 1.8L MFI VIN DB7, DB8 [All] 16P 'D' Connector

PCM Pin #	W/Color	Circuit Description (16-Pin)	Value at Hot Idle
D1	YEL	A/T: Lockup Solenoid 'A'	LSA Off: 0v, On: 12v
D2	GRN/WHT	A/T: Shift Solenoid 'B'	SSB Off: 0v, On: 12v
D3	GRN/BLK	A/T: Lockup Solenoid 'B'	LSB Off: 0v, On: 12v
D4	---	Not Used	---
D5	BLK/YEL	A/T: Solenoid Feed (B+)	Full: 12v, Partial: Pulses
D6	GRN/RED	A/T: Gear Position Switch	In 'R': 0v, Others: 12v
D7	BLU/YEL	A/T: Shift Solenoid 'A'	SSA Off: 0v, On: 12v
D8	GRN/BLU	A/T: Gear Position Switch	In D3: 0v, Others: 12v
D9	YEL	A/T: Gear Position Switch	In D4: 0v, Others: 12v
D10	BLU	Countershaft Speed Sensor P	Moving: AC Pulse Signals
D11	RED	Mainshaft Speed Sensor 'P'	AC Pulse Signals
D12	WHT	Mainshaft Speed Sensor 'N'	AC Pulse Signals
D13	GRN/BLK	A/T: Gear Position Switch	In P/N: 0v, Others: 12v
D14	GRN/YEL	A/T: Gear Position Switch	In D2: 0v, Others: 12v
D15	GRN/WHT	A/T: Gear Position Switch	In D1: 0v, Others: 12v
D16	GRN	Countershaft Speed Sensor N	Moving: AC Pulse Signals

Standard Colors and Abbreviations

Abbreviation	Color	Abbreviation	Color	Abbreviation	Color
BLK	Black	LT BLU	Lt. Blue	TAN	Tan
BLU	Blue	LT GRN	Lt. Green	VIO	Violet
BRN	Brown	ORN	Orange	WHT	White
GRY	Gray	PNK	Pink	YEL	Yellow
GRN	Green	PPL	Purple		

1997-99 GS-R, Type R 1.8L VTEC VIN DC2 [All] 32P 'A' Connector

PCM Pin #	W/Color	Circuit Description (32-Pin)	Value at Hot Idle
A1	YEL	Injector 4	2.0-3.3 ms
A2	BLU	Injector 3	2.0-3.3 ms
A3	RED	Injector 2	2.0-3.3 ms
A4	BRN	Injector 1	2.0-3.3 ms
A5	GRN/RED	HO2S-12 (B1 S2) Heater	Digital Signals: 0-12-0v
A6	ORN/BLK	HO2S-11 (B1 S1) Heater	Digital Signals: 0-12-0v
A7	---	Not Used	---
A8	GRN/YEL	VTEC Solenoid Control	0v, Hi-Speed: 12v
A9	BRN/BLK	Logic Ground	0.1v
A10	BLK	Power Ground (PG1)	<0.1v
A11	YEL/BLK	Main Relay Power (B+)	12-14v
A12	BLK/BLU	Idle Air Control Valve	Pulse Signals
A13-14	---	Not Used	---
A15	RED	EVAP Purge Solenoid	Pulse Signals
A16	GRN/BLU	Fuel Pump Relay Control	Relay Off: 12v, On: 1v
A17	BLK/RED	A/C Clutch Relay Control	Relay Off: 12v, On: 1v
A18	GRN/ORN	Malfunction Indicator Light	MIL Off: 12v, On: 1v

1997-99 GS-R, Type R 1.8L VTEC VIN DC2 [All] 32P 'A' Connector, , _continued_

A19	WHT/GRN	Alternator Charging Signal	Lights on: 0v, off: 12v
A20	YEL/GRN	Igniter Control Signal	Digital Signals: 0-12-0v
A21	---	Not Used	---
A22	BRN/BLK	Logic Ground (LG2)	0.1v
A23	BLK	Power Ground (PG2)	<0.1v
A24	YEL/BLK	Main Relay Power (B+)	12-14v
A25	WHT/BLK	Reference Voltage	4.9-5.1v
A26	PNK/BLU	Intake Air Bypass Solenoid	Solenoid off: 12v, on: 1v
A27	GRN	Radiator Fan Relay Control	Fan on: 1v, off: 12v
A28 ('98-'99)	BLU	EVAP Bypass Solenoid	Solenoid off: 12v, on: 1v
A29 ('98-'99)	LT GRN	EVAP Vent Shut Control	Solenoid off: 12v, on: 1v
A30-32	---	Not Used	---

1997-99 GS-R, Type R 1.8L VTEC VIN DC2 [All] 31P 'C' Connector

PCM Pin #	W/Color	Circuit Description (31-Pin)	Value at Hot Idle
C1	BLU/RED	CKF Sensor Signal	AC pulse signals
C2 ('97)	BLU/GRN	CKP Sensor Signal	AC pulse signals
C2 ('98-'99)	BLU	CKP Sensor Signal	AC pulse signals
C3 ('97)	ORN/BLU	TDC Sensor Signal	AC pulse signals
C3 ('98-'99)	GRN	TDC Sensor Signal	AC pulse signals
C4 ('97)	ORN	CYP Sensor Signal	AC pulse signals
C4 ('98-'99)	YEL	CYP Sensor Signal	AC pulse signals
C5	BLU/RED	A/C Switch Signal	A/C Off: 5v, On: 0v
C6	BLU/WHT	Starter Switch Signal	Cranking: 9-11v
C7	BRN/WHT	Service Check Connector	SCS Open: 4.80v
C8	GRN/WHT	K-Line Signal	12v
C9	YEL	A/T: Feedback Signal	At idle: 5v, Shifting: 0v
C10	WHT/BLU	Keep Alive Power (VBU)	12-14v
C11	WHT/RED	CKF Sensor Ground	<0.050v
C12	BLU/YEL	CKP Sensor Ground	<0.050v
C13	WHT/BLU	TDC Sensor Ground	<0.050v
C14	WHT	CYP Sensor Ground	<0.050v
C15	BLU/BLK	VTEC Pressure Switch	0v, Hi-Speed: 12v
C16	GRN	PSP Switch Signal	Straight: 0v, Turned: 12v
C17	WHT/RED	Alternator 'FR' Signal	Digital Signals: 0-5-0v
C18	ORN	Vehicle Speed Sensor	Moving: pulse signals
C19	BLU	A/T: FI Data Line (CHK)	Serial Data
C20	---	Not Used	---
C21	LT GRN	A/T: BARO Signal to TCM	Varies w/alt: 0.5-4.9v
C22-26	---	Not Used	---
C27	GRY	A/T: FI Signal 'B'	At idle: 5v, Shifting: 0v
C28	GRN/BLU	A/T: FI Signal 'A'	At idle: 5v, Shifting: 0v
C29	GRN/BLK	A/T: Gear Position Switch	In P/N: 0v, Others: 12v
C30-31	---	Not Used	---

1997-99 GS-R, Type R 1.8L VTEC VIN DC2 [All] 16P 'D' Connector

PCM Pin #	W/Color	Circuit Description (16-Pin)	Value at Hot Idle
D1	RED/BLK	TP Sensor Signal	0.5-0.6v
D2	RED/WHT	ECT Sensor Signal	At 180°F: 0.5-0.6v
D3	WHT/YEL	MAP Sensor Signal	0.8-0.9v
D4	YEL/WHT	MAP Sensor VREF (VCC1)	4.9-5.1v
D5	GRN/WHT	Brake Switch Signal	Brake Off: 12v, On: 0v
D6	RED/BLU	Knock Sensor Signal	No detonation: 18mv AC
D7	WHT/RED	HO2S-11 (B1 S1) Signal	0.1-1.1v
D8	RED/YEL	IAT Sensor Signal	At 100°F: 2-3v
D9	---	Not Used	---
D10	YEL/BLU	Sensor VREF (VCC2)	4.9-5.1v
D11	GRN/BLU	Sensor Ground (SG2)	<0.050v
D12	GRN/WHT	MAP Sensor Ground (SG1)	<0.050v
D13	ORN/BLU	HO2S-12 Ground	<0.050v
D14	BLU/GRN	HO2S-12 (B1 S2) Signal	0.1-1.1v
D15 ('98-'99)	LT GRN	Fuel Tank Pressure Sensor	Fuel Cap off: 2.5v
D16	GRN/RED	Electric Load Detector	Varies w/Load: 2.5-3.5v

2000-03 GS-R 1.8L VTEC VIN DC2, DC4 [All] 32P 'A' Connector

PCM Pin #	W/Color	Circuit Description (32-Pin)	Value at Hot Idle
A1-2	---	Not Used	---
A3	BLU	EVAP Bypass Solenoid	Solenoid Off: 12v, On: 1v
A4	LT GRN	EVAP Vent Shut Control	Solenoid Off: 12v, On: 1v
A5	---	Not Used	---
A6	RED	EVAP Purge Solenoid	Pulse Signals
A7	---	Not Used	---
A8	GRN/RED	HO2S-12 (B1 S2) Heater	Digital Signals: 0-12-0v
A9	---	Not Used	---
A10	BRN/WHT	Service Check Connector	SCS Open: 4.80v
A11	---	Not Used	---
A12	PNK	Immobilizer Indicator Control	Lamp On: 1v, Off: 12v
A13	BLU	IMOEN Immobilizer Signal	Digital Signals
A14	GRN/BLK	A/T: D4 Indicator Light	In D4: 0v, Others: 12v
A15	GRN/YEL	Fuel Pump Relay Control	Relay Off: 12v, On: 1v
A16	---	Not Used	---
A17	BLK/RED	A/C Clutch Relay Control	Relay Off: 12v, On: 1v
A18	GRN/ORN	Malfunction Indicator Light	MIL Off: 12v, On: 1v
A19	BLU	Engine Speed (NEP) Signal	Digital Signals
A20	GRN	Radiator Fan Relay Control	Fan on: 1v, off: 12v
A21	GRN/WHT	K-Line Signal	12v
A22	---	Not Used	---
A23	BLU/RED	HO2S-12 (B1 S2) Signal	0.1-1.1v

2000-03 GS-R 1.8L VTEC VIN DC2, DC4 [All] 32P 'A' Connector, *continued*

A24	BLU/WHT	Starter Switch Signal	Cranking: 9-11v
A25	RED	INOCD Immobilizer Code	Digital Signals
A26	GRN	PSP Switch Signal	Straight: 0v, Turning: 12v
A27	BLU/RED	A/C Switch Signal	A/C Off: 5v, On: 0v
A28	WHT/RED	A/T: Interlock Control Unit	With Brake on: 12v
A29	LT GRN	Fuel Tank Pressure Sensor	Fuel Cap off: 2.5v
A30	GRN/RED	Electric Load Detector	Varies w/Load: 2.5-3.5v
A31	---	Not Used	---
A32	GRN/WHT	Brake Switch Signal	Brake Off: 12v, On: 0v

2000-03 GS-R 1.8L VTEC VIN DC2, DC4 [All] 25P 'B' Connector

PCM Pin #	W/Color	Circuit Description (25-Pin)	Value at Hot Idle
B1	YEL/BLK	Main Relay Power (B+)	12-14v
B2	BLK	Power Ground	<0.1v
B3	RED	Injector 2	2.0-3.3 ms
B4	BLU	Injector 3	2.0-3.3 ms
B5	YEL	Injector 4	2.0-3.3 ms
B6-7	---	Not Used	---
B8	WHT/GRN	A/T: Clutch Solenoid LS-	Pulse Signals
B9	YEL/BLK	Main Relay Power (B+)	12-14v
B10	BLK	Power Ground	<0.1v
B11	BRN	Injector 1	2.0-3.3 ms
B12	GRN/YEL	VTEC Solenoid Control	0v, Hi-Speed: 12v
B13	YEL/GRN	Ignition Control Module Signal	Digital Signals: 0-12-0v
B14-15	---	Not Used	---
B16	PNK/BLU	Intake Air Bypass Solenoid	Solenoid Off: 12v, On: 1v
B17	RED/BLU	A/T: Clutch Solenoid LS+	Pulse Signals
B18-19	---	Not Used	---
B20	BRN/BLK	Logic Ground (LG1)	<0.050v
B21	WHT/BLK	Keep Alive Power (VBU)	12-14v
B22	BRN/BLK	Logic Ground (LG2)	<0.050v
B23	BLK/BLU	Idle Air Control Valve	Pulse Signals

2000-03 GS-R 1.8L VTEC VIN DC2, DC4 [All] 31P 'C' Connector

PCM Pin #	W/Color	Circuit Description (31-Pin)	Value at Hot Idle
C1	BRN/BLK	HO2S-11 (B1 S1) Heater	Digital Signals: 0-12-0v
C2	WHT/GRN	Alternator Charging Signal	Headlights off: 0v, on: 12v
C3	RED/BLU	Knock Sensor Signal	No Detonation: 18mv AC
C4	---	Not Used	---
C5	WHT/RED	Alternator 'FR' Signal	Digital Signals: 0-5-0v
C6	---	Not Used	---
C7	GRN/WHT	MAP Sensor Ground	<0.050v
C8	BLU	CKP Sensor Signal	AC Pulse Signals
C9	BLU/YEL	CKP Sensor Ground	<0.050v
C10	BLU/BLK	VTEC Pressure Switch	0v, Hi-Speed: 12v
C11-15	---	Not Used	---
C16	WHT	HO2S-11 (B1 S1) Signal	0.1-1.1v
C17	WHT/YEL	MAP Sensor Signal	0.8-0.9v
C18	GRN/BLU	Sensor Ground (SG2)	<0.050v
C19	YEL/RED	MAP Sensor VREF (VCC1)	4.9-5.1v
C20	GRN	TDC Sensor Signal	AC Pulse Signals
C21	WHT/BLU	TDC Sensor Ground	<0.050v
C22	BLU/RED	CKF Sensor Signal	AC Pulse Signals
C23	ORN	Vehicle Speed Sensor	Moving: pulse signals
C24	---	Not Used	---
C25	RED/YEL	IAT Sensor Signal	At 100°F: 2-3v
C26	RED/WHT	ECT Sensor Signal	At 180°F: 0.5-0.6v
C27	RED/BLK	TP Sensor Signal	0.5-0.6v
C28	YEL/BLU	Sensor VREF (VCC2)	4.9-5.1v
C29	YEL	CYP Sensor Signal	AC Pulse Signals
C30	WHT	CYP Sensor Ground	<0.050v
C31	WHT/RED	CKF Sensor Ground	<0.050v

2000-03 GS-R 1.8L VTEC VIN DC2, DC4 [All] 16P 'D' Connector

PCM Pin #	W/Color	Circuit Description (16-Pin)	Value at Hot Idle
D1	YEL	A/T: Lockup Solenoid 'A'	LSA Off: 0v, On: 12v
D2	GRN/WHT	A/T: Shift Solenoid 'B'	SSB Off: 0v, On: 12v
D3	GRN/BLK	A/T: Lockup Solenoid 'B'	LSB Off: 0v, On: 12v
D4	---	Not Used	---
D5	BLK/YEL	A/T: Solenoid Feed (B+)	Full: 12v, Partial: Pulses
D6	GRN/RED	A/T: Gear Position Switch	In 'R': 0v, Others: 12v
D7	BLU/YEL	A/T: Shift Solenoid 'A'	SSA Off: 0v, On: 12v
D8	GRN/BLU	A/T: Gear Position Switch	In D3: 0v, Others: 12v
D9	YEL	A/T: Gear Position Switch	In D4: 0v, Others: 12v
D10	BLU	Countershaft Speed Sensor P	Moving: AC Pulse Signals
D11	RED	Mainshaft Speed Sensor 'P'	AC Pulse Signals
D12	WHT	Mainshaft Speed Sensor 'N'	AC Pulse Signals
D13	GRN/BLK	A/T: Gear Position Switch	In P/N: 0v, Others: 12v
D14	GRN/YEL	A/T: Gear Position Switch	In D2: 0v, Others: 12v
D15	GRN/WHT	A/T: Gear Position Switch	In D1: 0v, Others: 12v
D16	GRN	Countershaft Speed Sensor N	Moving: AC Pulse Signals

Standard Colors and Abbreviations

Abbreviation	Color	Abbreviation	Color	Abbreviation	Color
BLK	Black	LT BLU	Lt. Blue	TAN	Tan
BLU	Blue	LT GRN	Lt. Green	VIO	Violet
BRN	Brown	ORN	Orange	WHT	White
GRY	Gray	PNK	Pink	YEL	Yellow
GRN	Green	PPL	Purple		

1990 LS Coupe 2.7L V6 MFI VIN KA3 [All] 18P 'A' Connector

PCM Pin #	W/Color	Circuit Description (18-Pin)	Value at Hot Idle
A1	BRN	Injector 1	2.0-3.3 ms
A2	BLK, BLK	Power Ground	<0.1v
A3	RED	Injector 3	2.0-3.3 ms
A4	BLK, BLK	Power Ground	<0.1v
A5	ORN	Injector 3	2.0-3.3 ms
A6	YEL	Injector 6	2.0-3.3 ms
A7	WHT/BLU	Injector 4	2.0-3.3 ms
A8	BLK/RED	Injector 5	2.0-3.3 ms
A9	---	Not Used	---
A10	WHT	EGR Solenoid Control	Solenoid Off: 12v, On: 1v
A11	BLU/RED	Electronic Air Control Valve	Pulse Signals
A12	GRN/BLK	Fuel Pump Relay Control	Relay Off: 12v, On: 1v
A13	LT GRN	Pressure Regulator Solenoid	Solenoid Off: 12v, On: 1v
A14	RED/BLU	Bypass & Resonator Control	Solenoid Off: 12v, On: 1v
A15	YEL/BLK	Main Relay Power (B+)	12-14v
A16	BRN/WHT	Chassis Ground	<0.050v
A17	YEL/BLU	Keep Alive Power (VBU)	12-14v
A18	BRN/BLK	Chassis Ground	<0.050v

1990 LS Coupe 2.7L V6 MFI VIN KA3 [All] 20P 'B' Connector

PCM Pin #	W/Color	Circuit Description (20-Pin)	Value at Hot Idle
B1	PNK	Air Suction Solenoid	Solenoid Off: 12v, On: 1v
B2	RED/BLU	A/C Pressure Switch	A/C Off: 12v, On: 1v
B3	GRN	Bypass Control Solenoid 'B'	Solenoid Off: 12v, On: 1v
B4	BLU	Cooling Fan Control Unit	N/A
B5	PNK	Information Center Control	N/A
B6	BLU	Malfunction Indicator Lamp	MIL Off: 12v, On: 1v
B7	BLU	M/T: Neutral Switch	In 'N': 0v, Others: 12v
B7	LT GRN	A/T: Shift Position Indicator	In P/N: 0v, Others: 12v
B8	RED/BLU	A/C Clutch Relay Control	Relay Off: 12v, On: 1v
B9	PNK	A/T: Control Unit Signal	---
B9	PNK	M/T: Clutch Switch Signal	Clutch in: 0v, out: 12v
B10	BLU/BLK	Cooling Fan Control Unit	---
B11	YEL/RED	Vehicle Speed Sensor	Moving: pulse signals
B12	RED	PSP Switch Signal	Straight: 0v, Turning: 12v
B13	BLK/WHT	Starter Switch Signal	Cranking: 9-11v
B14	WHT/RED	Alternator Control Signal	Headlights off: 0v, on: 12v
B15	RED/BLU	Igniter Signal	Digital Signals: 0-12-0v
B16	RED/BLU	Igniter Signal	Digital Signals: 0-12-0v
B17	ORN	A/T: Control Unit Signal	---
B18	BRN	Ignition Timing Adjustment	0.4-4.5v
B19	BLU/GRN	CKP Sensor Signal	AC pulse signals
B20	BLU/YEL	CKP Sensor Ground	<0.050v

1990 LS Coupe 2.7L V6 MFI VIN KA3 [All] 16P 'C' Connector

PCM Pin #	W/Color	Circuit Description (16-Pin)	Value at Hot Idle
C1	ORN	CYP Sensor Signal	AC pulse signals
C2	WHT	CYP Sensor Ground	<0.050v
C3	ORN/BLU	TDC Sensor Signal	AC pulse signals
C4	WHT/BLU	TDC Sensor Ground	<0.050v
C5	RED/BLK	IAT Sensor Signal	At 100°F: 2-3v
C6	RED/WHT	ECT Sensor Signal	At 180°: 0.5-0.6v
C7	RED/YEL	Throttle Angle Sensor	0.5-0.6v
C8	WHT/GRN	EGR Valve Lift Sensor	1.1-1.2v
C9	RED	Atmospheric Press. Sensor	2.76-2.96v sea level
C10	RED/BLU	O2S-21 Signal	0.1-1.1v
C11	WHT	MAP Sensor Signal	0.8-0.9v
C12	WHT	O2S-11 Signal	0.1-1.1v
C13	YEL/WHT	Sensor VREF	4.9-5.1v
C14	GRN/WHT	Sensor Ground	<0.050v
C15	YEL/WHT	MAP Sensor VREF	4.9-5.1v
C16	GRN/WHT	MAP Sensor Ground	<0.050v

1990 LS Coupe 2.7L V6 MFI VIN KA3 [All] 5P 'D' Connector

PCM Pin #	W/Color	Circuit Description (5-Pin)	Value at Hot Idle
D1	---	Not Used	---
D2	---	Not Used	---
D3	BLU/GRN	A/T: Control Unit	N/A
D4	BLU/RED	A/T: Control Unit	N/A
D5	BLU/WHT	A/T: Control Unit	N/A

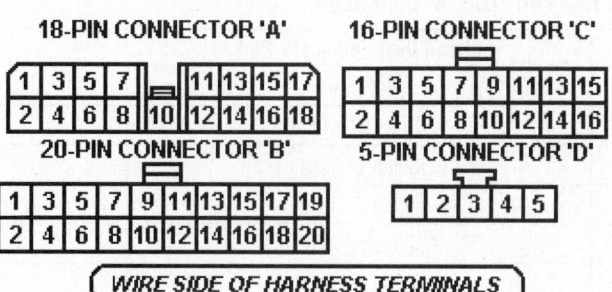

Pin Connector Graphic

Pin Connector Graphic

05533_ADIA_G585

1991-93 L Coupe 3.2L V6 MFI VIN KA8 [All] 26P 'A' Connector

PCM Pin #	W/Color	Circuit Description (26-Pin)	Value at Hot Idle
A1	BRN	Injector 1	2.0-3.3 ms
A3	RED	Injector 3	2.0-3.3 ms
A2	WHT/BLU	Injector 4	2.0-3.3 ms
A4	BLK/RED	Injector 5	2.0-3.3 ms
A5	ORN	Injector 3	2.0-3.3 ms
A6	YEL	Injector 6	2.0-3.3 ms
A7	GRN/BLK	Fuel Pump Relay Control	Relay Off: 12v, On: 1v
A8	GRN	Cooling Fan Control Unit	N/A
A9	BLU/RED	Electronic Air Control Valve	Pulse Signals
A10	GRN/BLU	HO2S-21 (B2 S1) Heater	Heater Off: 12v, On: 1v
A11	WHT	EGR Solenoid Control	Solenoid Off: 12v, On: 1v
A12	GRN/RED	HO2S-11 (B1 S1) Heater	Heater Off: 12v, On: 1v
A13	BLU	Malfunction Indicator Lamp	MIL Off: 12v, On: 1v
A14	RED/BLU	Bypass Low Solenoid	Solenoid Off: 12v, On: 1v
A15	RED/BLU	A/C Clutch Relay Control	Relay Off: 12v, On: 1v
A17	GRY	Air Suction Solenoid	Solenoid Off: 12v, On: 1v
A18	YEL	Bypass High Solenoid	Solenoid Off: 12v, On: 1v
A19	LT GRN	Press. Regulator Solenoid	Solenoid Off: 12v, On: 1v
A20	LT GRN	EVAP Purge Solenoid	Solenoid Off: 12v, On: 1v
A21	PNK	Igniter 1 Signal	Digital Signals: 0-12-0v
A22	BRN	Igniter 2 Signal	Digital Signals: 0-12-0v
A23	BLK	Power Ground	<0.1v
A24	BLK	Power Ground	<0.1v
A25	YEL/BLK	Main Relay Power (B+)	12-14v
A26	BRN/BLK	Chassis Ground	<0.050v

1991-93 L Coupe 3.2L V6 MFI VIN KA8 [All] 16P 'B' Connector

PCM Pin #	W/Color	Circuit Description (16-Pin)	Value at Hot Idle
B1	---	Not Used	---
B2	BRN/BLK	Chassis Ground	<0.050v
B3	RED	Igniter 6 Signal	Digital Signals: 0-12-0v
B4	GRY	Igniter 5 Signal	Digital Signals: 0-12-0v
B5	GRN	PSP Switch Signal	Straight: 0v, Turning: 12v
B6	GRN	Igniter 4 Signal	Digital Signals: 0-12-0v
B7	WHT	M/T: Neutral Switch	In 'N': 0v, Others: 12v
B7	LT GRN	A/T: Shift Position Indicator	In P/N: 0v, Others: 12v
B8	BLU	Igniter 3 Signal	Digital Signals: 0-12-0v
B9	BLU/GRN	CYP Sensor 2 Signal 'P'	AC pulse signals
B10	BLU/YEL	CYP Sensor Ground 2 'N'	<0.050v
B11	ORN/BLU	CYP Sensor 1 Signal 'P'	AC pulse signals
B12	WHT/BLU	CYP Sensor Ground 1 'N'	<0.050v
B13	ORN	CKP Sensor 2 Signal 'P'	AC pulse signals
B14	WHT	CKP Sensor 2 Ground 'N'	<0.050v
B15	ORN/BLU	CKP Sensor 1 Signal 'P'	AC pulse signals
B16	WHT/BLU	CKP Sensor 1 Ground 'N'	<0.050v

1991-93 L Coupe 3.2L V6 MFI VIN KA8 [All] 12P 'C' Connector

PCM Pin #	W/Color	Circuit Description (12-Pin)	Value at Hot Idle
C1	YEL/BLK	Main Relay Power (B+)	12-14v
C2	YEL/RED	Vehicle Speed Sensor	Moving: pulse signals
C3	BLU/BLK	Cooling Fan Control Unit	N/A
C4	BLU	Tachometer Signal	Pulse Signals
C5	RED/BLU	A/C Pressure Switch	A/C Off: 12v, On: 1v
C6	---	Not Used	---
C7	PNK	M/T: Clutch Switch	In 'N': 0v, Others: 12v
C8	---	Not Used	---
C9	PNK	Service Check Connector	SCS open: 4.80v
C10	---	Not Used	---
C11	BLK/WHT	Starter Switch Signal	Cranking: 9-11v

1991-93 L Coupe 3.2L V6 MFI VIN KA8 [All] 22P 'D' Connector

PCM Pin #	W/Color	Circuit Description (22-Pin)	Value at Hot Idle
D1	YEL/BLK	Keep Alive Power (VBU)	12-14v
D2	GRN/WHT	Brake Switch Signal	Brake Off: 12v, On: 0v
D3	WHT	Knock Sensor Signal (Bank 1)	No Detonation: 18mv AC
D4	RED/BLU	Knock Sensor Signal (Bank 2)	No Detonation: 18mv AC
D5-7	---	Not Used	---
D8	BLU/YEL	Ignition Timing Adjustment	0.4-4.5v
D9	WHT/RED	Alternator Control Signal	Headlights off: 0v, on: 12v
D10	---	Not Used	---
D11	RED/BLU	TP Sensor Signal	0.5-0.6v
D12	BLK/WHT	EGR Valve Lift Sensor	1.1-1.2v
D13	RED/WHT	ECT Sensor Signal	At 180°: 0.5-0.6v
D14	WHT	HO21-21 (B2 S1) Signal	0.1-1.1v
D15	RED/YEL	IAT Sensor Signal	At 100°F: 2-3v
D16	RED/BLU	HO2S-11 (B1 S1) Signal	0.1-1.1v
D17	RED	MAP Sensor Signal	0.8-0.9v
D18	---	Not Used	---
D19	YEL/WHT	MAP Sensor VREF	4.9-5.1v
D20	YEL/WHT	Sensor VREF	4.9-5.1v
D21	GRN/WHT	MAP Sensor Ground	<0.050v
D22	GRN/YEL	Sensor Ground	<0.050v

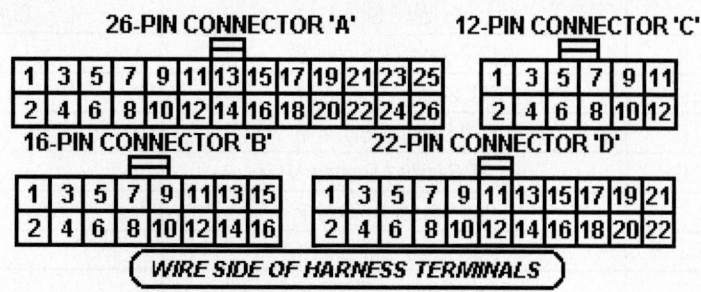

Pin Connector Graphic

1994-95 L, LS Coupe 3.2L MFI VIN KA8 [A/T] 26P 'A' Connector

PCM Pin #	W/Color	Circuit Description (26-Pin)	Value at Hot Idle
A1	BRN	Injector 1	2.0-3.3 ms
A2	WHT/BLU	Injector 4	2.0-3.3 ms
A3	RED	Injector 2	2.0-3.3 ms
A4	BLK/RED	Injector 5	2.0-3.3 ms
A5	ORN	Injector 3	2.0-3.3 ms
A6	YEL	Injector 6	2.0-3.3 ms
A7	GRN/BLK	Fuel Pump Relay Control	Relay Off: 12v, On: 1v
A8	---	Not Used	---
A9	BLU/RED	Idle Air Control Valve	Pulse Signals
A10	GRN/BLU	HO2S-21 (B2 S1) Heater	Heater Off: 12v, On: 1v
A11	WHT	EGR Solenoid Control	Solenoid Off: 12v, On: 1v
A12	GRN/RED	HO2S-11 (B1 S1) Heater	Heater Off: 12v, On: 1v
A13	BLU	Malfunction Indicator Lamp	MIL Off: 12v, On: 1v
A14	RED/BLU	Bypass Low Solenoid	Solenoid Off: 12v, On: 1v
A15	RED/BLU	A/C Clutch Relay Control	Relay Off: 12v, On: 1v
A17	GRY	Air Suction Solenoid	Solenoid Off: 12v, On: 1v
A18	YEL	Bypass High Solenoid	Solenoid Off: 12v, On: 1v
A19	LT GRN	Press. Regulator Solenoid	Solenoid Off: 12v, On: 1v
A20	LT GRN	EVAP Purge Solenoid	Solenoid Off: 12v, On: 1v
A21	PNK	Igniter 1 Signal	Digital Signals: 0-12-0v
A22	BRN	Igniter 2 Signal	Digital Signals: 0-12-0v
A23	BLK	Power Ground	<0.1v
A24	BLK	Power Ground	<0.1v
A25	YEL/BLK	Main Relay Power (B+)	12-14v
A26	BRN/BLK	Chassis Ground	<0.050v

1994-95 L, LS Coupe 3.2L MFI VIN KA8 [A/T] 16P 'B' Connector

PCM Pin #	W/Color	Circuit Description (16-Pin)	Value at Hot Idle
B1	---	Not Used	---
B2	BRN/BLK	Chassis Ground	<0.050v
B3	RED	Igniter 6 Signal	Digital Signals: 0-12-0v
B4	GRY	Igniter 5 Signal	Digital Signals: 0-12-0v
B5	GRN	PSP Switch Signal	Straight: 0v, Turning: 12v
B6	GRN	Igniter 4 Signal	Digital Signals: 0-12-0v
B7	LT GRN	Park Neutral Switch Signal	In P/N: 0v, Others: 12v
B8	BLU	Igniter 3 Signal	Digital Signals: 0-12-0v
B9	BLU/GRN	CYP Sensor 2 Signal 'P'	AC pulse signals
B10	BLU/YEL	CYP Sensor 2 Ground 'N'	<0.050v
B11	ORN/BLU	CYP Sensor 1 Signal 'P'	AC pulse signals
B12	WHT/BLU	CYP Sensor 1 Ground 'N'	<0.050v
B13	ORN	CKP Sensor 2 Signal 'P'	AC pulse signals
B14	WHT	CKP Sensor 2 Ground 'N'	<0.050v
B15	ORN/BLU	CKP Sensor 1 Signal 'P'	AC pulse signals
B16	WHT/BLU	CKP Sensor 1 Ground 'N'	<0.050v

1994-95 L, LS Coupe 3.2L MFI VIN KA8 [A/T] 12P 'C' Connector

PCM Pin #	W/Color	Circuit Description (12-Pin)	Value at Hot Idle
C1	YEL/BLK	Main Relay Power (B+)	12-14v
C2	YEL/RED	Vehicle Speed Sensor	Moving: pulse signals
C3	BLU/BLK	Cooling Fan Switch	A/C Off: 12v, On: 1v
C4	BLU	Tachometer Signal	Pulse Signals
C5	RED/BLU	A/C Pressure Switch	Relay Off: 12v, On: 1v
C6	GRN/BLK	TCM Signal	N/A
C8	GRY/BLU	TCM Signal	N/A
C9	WHT	Service Check Connector	SCS Open: 4.80v
C11	BLK/WHT	Starter Switch Signal	Cranking: 9-11v
C12	GRY/YEL	TCM Signal	N/A

1994-95 L, LS Coupe 3.2L MFI VIN KA8 [A/T] 22P 'D' Connector

PCM Pin #	W/Color	Circuit Description (22-Pin)	Value at Hot Idle
D1	YEL/BLU	Keep Alive Power (VBU)	12-14v
D2	GRN/WHT	Brake Switch Signal	Brake Off: 12v, On: 0v
D3	WHT	Knock Sensor Signal (Bank 1)	No Detonation: 18mv AC
D4	RED/BLU	Knock Sensor Signal (Bank 2)	No Detonation: 18mv AC
D5	ORN/RED	TCM Signal	N/A
D6-7	---	Not Used	---
D8	BLU/YEL	Ignition Timing Adjustment	0.4-4.5v
D9	WHT/RED	Alternator Control Signal	Headlights off: 0v, on: 12v
D10	---	Not Used	---
D11	RED/BLU	TP Sensor Signal	0.5-0.6v
D12	BLK/WHT	EGR Valve Lift Sensor	1.1-1.2v
D13	RED/WHT	ECT Sensor Signal	At 180°: 0.5-0.6v
D14	WHT	HO2S-11 (B1 S1) Signal	0.1-1.1v
D15	RED/YEL	IAT Sensor Signal	At 100°F: 2-3v
D16	RED/BLU	HO21-21 (B2 S1) Signal	0.1-1.1v
D17	RED	MAP Sensor Signal	0.8-0.9v
D18	BLU	TCM Signal	N/A
D19	YEL/WHT	MAP Sensor VREF	4.9-5.1v
D20	YEL/WHT	Sensor VREF	4.9-5.1v
D21	GRN/WHT	MAP Sensor Ground	<0.050v
D22	GRN/WHT	Sensor Ground	<0.050v

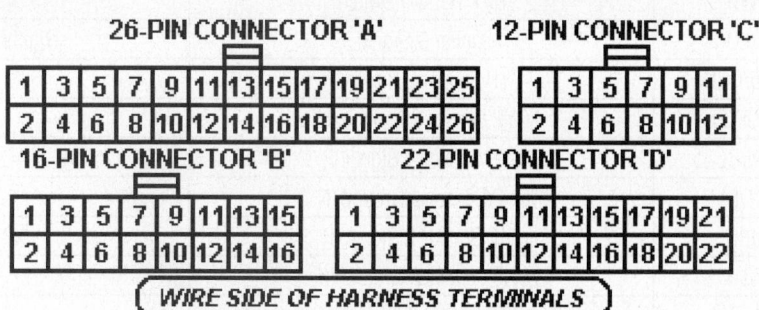

Pin Connector Graphic

05533_ADIA_586

1994-95 L, LS Coupe 3.2L MFI VIN KA8 [M/T] 26P 'A' Connector

PCM Pin #	W/Color	Circuit Description (26-Pin)	Value at Hot Idle
A1	BRN	Injector 1	2.0-3.3 ms
A3	RED	Injector 2	2.0-3.3 ms
A2	WHT/BLU	Injector 4	2.0-3.3 ms
A4	BLK/RED	Injector 5	2.0-3.3 ms
A5	ORN	Injector 3	2.0-3.3 ms
A6	YEL	Injector 6	2.0-3.3 ms
A7	GRN/BLK	Fuel Pump Relay Control	Relay Off: 12v, On: 1v
A8	RED/WHT	M/T: Reverse Lockout Relay	Relay Off: 12v, On: 1v
A9	BLU/RED	Idle Air Control Valve	Pulse Signals
A10	GRN/BLU	HO2S-21 (B2 S1) Heater	Heater Off: 12v, On: 1v
A11	WHT	EGR Solenoid Control	Solenoid Off: 12v, On: 1v
A12	GRN/RED	HO2S-11 (B1 S1) Heater	Heater Off: 12v, On: 1v
A13	BLU	Malfunction Indicator Lamp	MIL Off: 12v, On: 1v
A14	RED/BLU	Bypass Low Solenoid	Solenoid Off: 12v, On: 1v
A15	RED/BLU	A/C Clutch Relay Control	Relay Off: 12v, On: 1v
A16	---	Not Used	---
A17	GRY	Air Suction Solenoid	Solenoid Off: 12v, On: 1v
A18	YEL	Bypass High Solenoid	Solenoid Off: 12v, On: 1v
A19	LT GRN	Press. Regulator Solenoid	Solenoid Off: 12v, On: 1v
A20	LT GRN	EVAP Purge Solenoid	Solenoid Off: 12v, On: 1v
A21	PNK	Igniter 1 Signal	Digital Signals: 0-12-0v
A22	BRN	Igniter 2 Signal	Digital Signals: 0-12-0v
A23	BLK	Power Ground	<0.1v
A24	BLK	Power Ground	<0.1v
A25	YEL/BLK	Main Relay Power (B+)	12-14v
A26	BRN/BLK	Chassis Ground	<0.050v

1994-95 L, LS Coupe 3.2L MFI VIN KA8 [M/T] 16P 'B' Connector

PCM Pin #	W/Color	Circuit Description (16-Pin)	Value at Hot Idle
B1	---	Not Used	---
B2	BRN/BLK	Chassis Ground	<0.050v
B3	RED	Igniter 6 Signal	Digital Signals: 0-12-0v
B4	GRY	Igniter 5 Signal	Digital Signals: 0-12-0v
B5	GRN	PSP Switch Signal	Straight: 0v, Turning: 12v
B6	GRN	Igniter 4 Signal	Digital Signals: 0-12-0v
B7	WHT	M/T: Neutral Switch	In 'N': 0v, Others: 12v
B8	BLU	Igniter 3 Signal	Digital Signals: 0-12-0v
B9	BLU/GRN	CYP Sensor 2 Signal 'P'	AC pulse signals
B10	BLU/YEL	CYP Sensor 2 Ground 'N'	<0.050v
B11	ORN/BLU	CYP Sensor 1 Signal 'P'	AC pulse signals
B12	WHT/BLU	CYP Sensor 1 Ground 'N'	<0.050v
B13	ORN	CKP Sensor 2 Signal 'P'	AC pulse signals
B14	WHT	CKP Sensor 2 Ground 'N'	<0.050v
B15	ORN/BLU	CKP Sensor 1 Signal 'P'	AC pulse signals
B16	WHT/BLU	CKP Sensor 1 Ground 'N'	<0.050v

1994-95 L, LS Coupe 3.2L MFI VIN KA8 [M/T] 12P 'C' Connector

PCM Pin #	W/Color	Circuit Description (12-Pin)	Value at Hot Idle
C1	YEL/BLK	Main Relay Power (B+)	12-14v
C2	YEL/RED	Vehicle Speed Sensor	Moving: pulse signals
C3	BLU/BLK	Cooling Fan Switch	A/C Off: 12v, On: 1v
C4	BLU	Tachometer Signal	Pulse Signals
C5	RED/BLU	A/C Pressure Switch	Relay Off: 12v, On: 1v
C6	---	Not Used	---
C7	PNK	M/T: Clutch Switch	In 'N': 0v, Others: 12v
C8	---	Not Used	---
C9	WHT	Service Check Connector	SCS Open: 4.80v
C10	---	Not Used	---
C11	BLK/WHT	Starter Switch Signal	Cranking: 9-11v
C12	---	Not Used	---

1994-95 L, LS Coupe 3.2L MFI VIN KA8 [M/T] 22P 'D' Connector

PCM Pin #	W/Color	Circuit Description (22-Pin)	Value at Hot Idle
D1	YEL/BLU	Keep Alive Power (VBU)	12-14v
D2	GRN/WHT	Brake Switch Signal	Brake Off: 12v, On: 0v
D3	WHT	Knock Sensor Signal (Bank 1)	No Detonation: 18mv AC
D4	RED/BLU	Knock Sensor Signal (Bank 2)	No Detonation: 18mv AC
D5-7	---	Not Used	---
D8	BLU/YEL	Ignition Timing Adjustment	0.4-4.5v
D9	WHT/RED	Alternator Control Signal	Headlights off: 0v, on: 12v
D10	RED/BLU	Knock Sensor Signal (Bank 2)	No Detonation: 18mv AC .
D11	RED/BLU	TP Sensor Signal	0.5-0.6v
D12	BLK/WHT	EGR Valve Lift Sensor	1.1-1.2v
D13	RED/WHT	ECT Sensor Signal	At 180°: 0.5-0.6v
D14	WHT	HO2S-11 (B1 S1) Signal	0.1-1.1v
D15	RED/YEL	IAT Sensor Signal	At 100°F: 2-3v
D16	RED/BLU	HO21-21 (B2 S1) Signal	0.1-1.1v
D17	RED	MAP Sensor Signal	0.8-0.9v
D18	---	Not Used	---
D19	YEL/WHT	MAP Sensor VREF	4.9-5.1v
D20	YEL/WHT	Sensor VREF	4.9-5.1v
D21	GRN/WHT	MAP Sensor Ground	<0.050v
D22	GRN/WHT	Sensor Ground	<0.050v

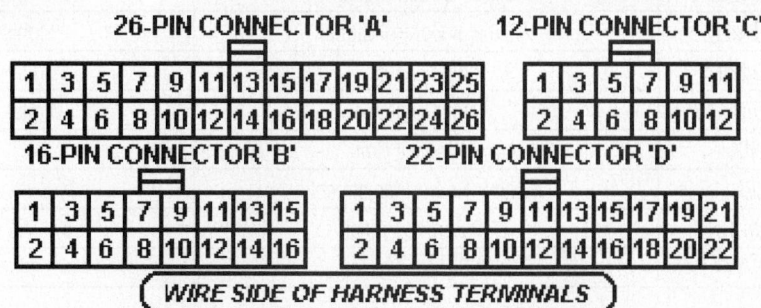

Pin Connector Graphic

05533_ADIA_G586

1990 L, LS Sedan 2.7L V6 MFI VIN KA4 [All] 18P 'A' Connector

PCM Pin #	W/Color	Circuit Description (18-Pin)	Value at Hot Idle
A1	BRN	Injector 1	2.0-3.3 ms
A2	BLK	Power Ground	<0.1v
A3	RED	Injector 2	2.0-3.3 ms
A4	BLK	Power Ground	<0.1v
A5	ORN	Injector 3	2.0-3.3 ms
A6	YEL	Injector 6	2.0-3.3 ms
A7	WHT/BLU	Injector 4	2.0-3.3 ms
A8	BLK/RED	Injector 5	2.0-3.3 ms
A10	WHT	EGR Solenoid Control	Solenoid Off: 12v, On: 1v
A11	BLU/RED	Electronic Air Control Valve	Pulse Signals
A12	GRN/BLK	Fuel Pump Relay Control	Relay Off: 12v, On: 1v
A13	LT GRN	Pressure Regulator Solenoid	Solenoid Off: 12v, On: 1v
A14	RED/BLU	Bypass & Resonator Control	Solenoid Off: 12v, On: 1v
A15	YEL/BLK	Main Relay Power (B+)	12-14v
A16	BRN/WHT	Chassis Ground	<0.050v
A17	YEL/BLU	Keep Alive Power (VBU)	12-14v
A18	BRN/WHT	Chassis Ground	<0.050v

1990 L, LS Sedan 2.7L V6 MFI VIN KA4 [All] 20P 'B' Connector

PCM Pin #	W/Color	Circuit Description (20-Pin)	Value at Hot Idle
B1	PNK	Air Suction Solenoid	Solenoid Off: 12v, On: 1v
B2	RED/BLU	A/C Pressure Switch	A/C Off: 12v, On: 1v
B3	GRN	Bypass Control Solenoid 'B'	Solenoid Off: 12v, On: 1v
B4	BLU	Cooling Fan Control Unit	N/A
B5	PNK	Information Center Control	N/A
B6	BLU	Malfunction Indicator Lamp	MIL Off: 12v, On: 1v
B7	BLU	M/T: Neutral Switch	In 'N': 0v, Others: 12v
B7	LT GRN	A/T: Shift Position Indicator	In P/N: 0v, Others: 12v
B8	RED/BLU	A/C Clutch Relay Control	Relay Off: 12v, On: 1v
B9	PNK	M/T: Clutch Switch Signal	Clutch in: 0v, out: 12v
B9	PNK	A/T: Control Unit Signal	N/A
B10	BLU/BLK	Cooling Fan Control Unit	N/A
B11	YEL/RED	Vehicle Speed Sensor	Moving: pulse signals
B12	RED	PSP Switch Signal	Straight: 0v, Turning: 12v
B13	BLK/WHT	Starter Switch Signal	Cranking: 9-11v
B14	WHT/RED	Alternator Control Signal	Lights Off: 12v, On: 1v
B15	RED/BLU	Igniter Signal	Digital Signals: 0-12-0v
B16	RED/BLU	Igniter Signal	Digital Signals: 0-12-0v
B17	ORN	A/T: Control Unit	---
B18	BRN	Ignition Timing Adjustment	0.4-4.5v
B19	BLU/GRN	CKP Sensor Signal	AC pulse signals
B20	BLU/YEL	CKP Sensor Ground	<0.050v

1990 L, LS Sedan 2.7L V6 MFI VIN KA4 [All] 16P 'C' Connector

PCM Pin #	W/Color	Circuit Description (16-Pin)	Value at Hot Idle
C1	ORN	CYP Sensor Signal	AC pulse signals
C2	WHT	CYP Sensor Ground	<0.050v
C3	ORN/BLU	TDC Sensor Signal	AC pulse signals
C4	WHT/BLU	TDC Sensor Ground	<0.050v
C5	RED/BLK	IAT Sensor Signal	At 100°F: 2-3v
C6	RED/WHT	ECT Sensor Signal	At 180°: 0.5-0.6v
C7	RED/YEL	Throttle Angle Sensor	0.5-0.6v
C8	WHT/GRN	EGR Valve Lift Sensor	1.1-1.2v
C9	RED	Atmospheric Press. Sensor	2.76-2.96v sea level
C10	RED/BLU	O2S-21 Signal	0.1-1.1v
C11	WHT	MAP Sensor Signal	0.8-0.9v
C12	WHT	O2S-11 Signal	0.1-1.1v
C13	YEL/WHT	Sensor VREF	4.9-5.1v
C14	GRN/WHT	Sensor Ground	<0.050v
C15	YEL/WHT	MAP Sensor VREF	4.9-5.1v
C16	GRN/WHT	MAP Sensor Ground	<0.050v

1990 L, LS Sedan 2.7L V6 MFI VIN KA4 [All] 5P 'D' Connector

PCM Pin #	W/Color	Circuit Description (5-Pin)	Value at Hot Idle
D1	---	Not Used	---
D2	---	Not Used	---
D3	BLU/GRN	A/T: Control Unit	N/A
D4	BLU/RED	A/T: Control Unit	N/A
D5	BLU/WHT	A/T: Control Unit	N/A

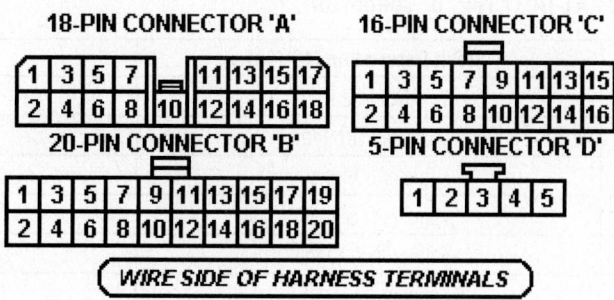

Pin Connector Graphic

05533_ADIA_G585

1991-93 L, LS Sedan 3.2L MFI VIN KA7 [A/T] 26P 'A' Connector

PCM Pin #	W/Color	Circuit Description (26-Pin)	Value at Hot Idle
A1	BRN	Injector 1	2.0-3.3 ms
A2	WHT/BLU	Injector 4	2.0-3.3 ms
A3	RED	Injector 2	2.0-3.3 ms
A4	BLK/RED	Injector 5	2.0-3.3 ms
A5	ORN	Injector 3	2.0-3.3 ms
A6	YEL	Injector 6	2.0-3.3 ms
A7	GRN/BLK	Fuel Pump Relay Control	Relay Off: 12v, On: 1v
A8	GRN	Cooling Fan Control Unit	N/A
A9	BLU/RED	Idle Air Control Valve	Pulse Signals
A10	GRN/BLU	HO2S-21 (B2 S1) Heater	Heater Off: 12v, On: 1v
A11	WHT	EGR Solenoid Control	Solenoid Off: 12v, On: 1v
A12	GRN/RED	HO2S-11 (B1 S1) Heater	Heater Off: 12v, On: 1v
A13	BLU	Malfunction Indicator Lamp	MIL Off: 12v, On: 1v
A14	RED/BLU	Bypass Low Solenoid	Solenoid Off: 12v, On: 1v
A15	RED/BLU	A/C Clutch Relay Control	Relay Off: 12v, On: 1v
A17	GRY	Air Suction Solenoid	Solenoid Off: 12v, On: 1v
A18	YEL	Bypass High Solenoid	Solenoid Off: 12v, On: 1v
A19	LT GRN	Press. Regulator Solenoid	Solenoid Off: 12v, On: 1v
A20	LT GRN	EVAP Purge Solenoid	Solenoid Off: 12v, On: 1v
A21	PNK	Igniter 1 Signal	Digital Signals: 0-12-0v
A22	BRN	Igniter 2 Signal	Digital Signals: 0-12-0v
A23	BLK	Power Ground	<0.1v
A24	BLK	Power Ground	<0.1v
A25	YEL/BLK	Main Relay Power (B+)	12-14v
A26	BRN/BLK	Chassis Ground	<0.050v

1991-93 L, LS Sedan 3.2L MFI VIN KA7 [A/T] 16P 'B' Connector

PCM Pin #	W/Color	Circuit Description (16-Pin)	Value at Hot Idle
B1	---	Not Used	---
B2	BRN/BLK	Chassis Ground	<0.050v
B3	RED	Igniter 6 Signal	Digital Signals: 0-12-0v
B4	GRY	Igniter 5 Signal	Digital Signals: 0-12-0v
B5	GRN	PSP Switch Signal	Straight: 0v, Turning: 12v
B6	GRN	Igniter 4 Signal	Digital Signals: 0-12-0v
B8	BLU	Igniter 3 Signal	Digital Signals: 0-12-0v
B7	LT GRN	A/T: Park Neutral Indicator	In P/N: 0v, Others: 12v
B9	BLU/GRN	CYP Sensor 2 Signal 'P'	AC pulse signals
B10	BLU/YEL	CYP Sensor 2 Ground 'N'	<0.050v
B11	ORN/BLU	CYP Sensor 1 Signal 'P'	AC pulse signals
B12	WHT/BLU	CYP Sensor 1 Ground 'N'	<0.050v
B13	ORN	CKP Sensor 2 Signal 'P'	AC pulse signals
B14	WHT	CKP Sensor 2 Ground 'N'	<0.050v
B15	ORN/BLU	CKP Sensor 1 Signal 'P'	AC pulse signals
B16	WHT/BLU	CKP Sensor 1 Ground 'N'	<0.050v

1991-93 L, LS Sedan 3.2L MFI VIN KA7 [A/T] 12P 'C' Connector

PCM Pin #	W/Color	Circuit Description (12-Pin)	Value at Hot Idle
C1	YEL/BLK	Main Relay Power (B+)	12-14v
C2	YEL/RED	Vehicle Speed Sensor	Moving: pulse signals
C3	BLU/BLK	Cooling Fan Control Unit	---
C4	BLU	Tachometer Signal	Pulse Signals
C5	RED/BLU	A/C Pressure Switch	A/C Off: 12v, On: 1v
C6-8	---	Not Used	---
C9	WHT	Service Check Connector	SCS Open: 4.80v
C10	---	Not Used	---
C11	BLK/WHT	Starter Switch Signal	Cranking: 9-11v
C12	---	Not Used	---

1991-93 L, LS Sedan 3.2L MFI VIN KA7 [A/T] 22P 'D' Connector

PCM Pin #	W/Color	Circuit Description (22-Pin)	Value at Hot Idle
D1	YEL/BLK	Keep Alive Power (VBU)	12-14v
D2	GRN/WHT	Brake Switch Signal	Brake Off: 12v, On: 0v
D3	WHT	Knock Sensor Signal (Bank 1)	No Detonation: 18mv AC
D4	RED/BLU	Knock Sensor Signal (Bank 2)	No Detonation: 18mv AC
D5-7	---	Not Used	---
D8	BLU/YEL	Ignition Timing Adjustment	0.4-4.5v
D9	WHT/RED	Alternator Control Signal	Headlights off: 0v, on: 12v
D10	---	Not Used	---
D11	RED/BLU	TP Sensor Signal	0.5-0.6v
D12	BLK/WHT	EGR Valve Lift Sensor	1.1-1.2v
D13	RED/WHT	ECT Sensor Signal	At 180°: 0.5-0.6v
D14	WHT	HO2S-11 (B1 S1) Signal	0.1-1.1v
D15	RED/YEL	IAT Sensor Signal	At 100°F: 2-3v
D16	RED/BLU	HO21-21 (B2 S1) Signal	0.1-1.1v
D17	RED	MAP Sensor Signal	0.8-0.9v
D18	---	Not Used	---
D19	YEL/WHT	MAP Sensor VREF	4.9-5.1v
D20	YEL/WHT	Sensor VREF	4.9-5.1v
D21	GRN/WHT	MAP Sensor Ground	<0.050v
D22	GRN/YEL	Sensor Ground	<0.050v

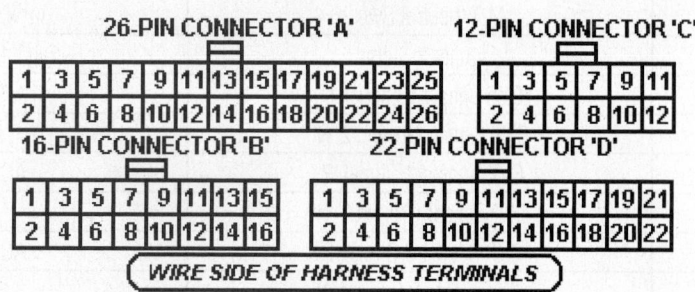

Pin Connector Graphic

1991-93 L, LS Sedan 3.2L MFI VIN KA7 [M/T] 26P 'A' Connector

PCM Pin #	W/Color	Circuit Description (26-Pin)	Value at Hot Idle
A1	BRN	Injector 1	2.0-3.3 ms
A2	WHT/BLU	Injector 4	2.0-3.3 ms
A3	RED	Injector 2	2.0-3.3 ms
A4	BLK/RED	Injector 5	2.0-3.3 ms
A5	ORN	Injector 3	2.0-3.3 ms
A6	YEL	Injector 6	2.0-3.3 ms
A7	GRN/BLK	Fuel Pump Relay Control	Relay Off: 12v, On: 1v
A8	GRN	Cooling Fan Control Unit	---
A9	BLU/RED	Electronic Air Control Valve	Pulse Signals
A10	GRN/BLU	HO2S-21 (B2 S1) Heater	Heater Off: 12v, On: 1v
A11	WHT	EGR Solenoid Control	Solenoid Off: 12v, On: 1v
A12	GRN/RED	HO2S-11 (B1 S1) Heater	Heater Off: 12v, On: 1v
A13	BLU	Malfunction Indicator Lamp	MIL Off: 12v, On: 1v
A14	RED/BLU	Bypass Low Solenoid	Solenoid Off: 12v, On: 1v
A15	RED/BLU	A/C Clutch Relay Control	Relay Off: 12v, On: 1v
A17	GRY	Air Suction Solenoid	Solenoid Off: 12v, On: 1v
A18	YEL	Bypass High Solenoid	Solenoid Off: 12v, On: 1v
A19	LT GRN	Pressure Regulator Solenoid	Solenoid Off: 12v, On: 1v
A20	LT GRN	EVAP Purge Solenoid	Solenoid Off: 12v, On: 1v
A21	PNK	Igniter 1 Signal	Digital Signals: 0-12-0v
A22	BLK	Igniter 2 Signal	Digital Signals: 0-12-0v
A23	BLK	Power Ground	<0.1v
A24	BLK	Power Ground	<0.1v
A25	YEL/BLK	Main Relay Power (B+)	12-14v
A26	BRN/BLK	Chassis Ground	<0.050v

1991-93 L, LS Sedan 3.2L MFI VIN KA7 [M/T] 16P 'B' Connector

PCM Pin #	W/Color	Circuit Description (16-Pin)	Value at Hot Idle
B1	---	Not Used	---
B2	BRN/BLK	Chassis Ground	<0.050v
B3	RED	Igniter 6 Signal	Digital Signals: 0-12-0v
B4	GRY	Igniter 5 Signal	Digital Signals: 0-12-0v
B5	GRN	PSP Switch Signal	Straight: 0v, Turning: 12v
B6	GRN	Igniter 4 Signal	Digital Signals: 0-12-0v
B7	WHT	M/T: Neutral Switch	In 'N': 0v, Others: 12v
B8	BLU	Igniter 3 Signal	Digital Signals: 0-12-0v
B9	BLU/GRN	CYP Sensor 2 Signal 'P'	AC pulse signals
B10	BLU/YEL	CYP Sensor Ground 2 'N'	<0.050v
B11	ORN/BLU	CYP Sensor 1 Signal 'P'	AC pulse signals
B12	WHT/BLU	CYP Sensor Ground 1 'N'	<0.050v
B13	ORN	CKP Sensor 2 Signal 'P'	AC pulse signals
B14	WHT	CKP Sensor 2 Ground 'N'	<0.050v
B15	ORN/BLU	CKP Sensor 1 Signal 'P'	AC pulse signals
B16	WHT/BLU	CKP Sensor 1 Ground 'N'	<0.050v

1991-93 L, LS Sedan 3.2L MFI VIN KA7 [M/T] 12P 'C' Connector

PCM Pin #	W/Color	Circuit Description (12-Pin)	Value at Hot Idle
C1	YEL/BLK	Main Relay Power (B+)	12-14v
C2	YEL/RED	Vehicle Speed Sensor	Moving: pulse signals
C3	BLU/BLK	Cooling Fan Control Unit	N/A
C4	BLU	Tachometer Signal	Pulse Signals
C5	RED/BLU	A/C Pressure Switch	A/C Off: 12v, On: 1v
C6	---	Not Used	---
C7	PNK	M/T: Clutch Switch	In 'N': 0v, Others: 12v
C8	---	Not Used	---
C9	PNK	Service Check Connector	SCS Open: 4.80v
C10	---	Not Used	---
C11	BLK/WHT	Starter Switch Signal	Cranking: 9-11v
C12	---	Not Used	---

1991-93 L, LS Sedan 3.2L MFI VIN KA7 [M/T] 22P 'D' Connector

PCM Pin #	W/Color	Circuit Description (22-Pin)	Value at Hot Idle
D1	YEL/BLK	Keep Alive Power (VBU)	12-14v
D2	GRN/WHT	Brake Switch Signal	Brake Off: 12v, On: 0v
D3	WHT	Knock Sensor Signal (Bank 1)	No Detonation: 18mv AC
D4	RED/BLU	Knock Sensor Signal (Bank 2)	No Detonation: 18mv AC
D5-7	---	Not Used	---
D8	BLU/YEL	Ignition Timing Adjustment	0.4-4.5v
D9	WHT/RED	Alternator Control Signal	Headlights off: 0v, on: 12v
D10	---	Not Used	---
D11	RED/BLU	TP Sensor Signal	0.5-0.6v
D12	BLK/WHT	EGR Valve Lift Sensor	1.1-1.2v
D13	RED/WHT	ECT Sensor Signal	At 180°: 0.5-0.6v
D14	WHT	HO2S-11 (B1 S1) Signal	0.1-1.1v
D15	RED/YEL	IAT Sensor Signal	At 100°F: 2-3v
D16	RED/BLU	HO21-21 (B2 S1) Signal	0.1-1.1v
D17	RED	MAP Sensor Signal	0.8-0.9v
D18	---	Not Used	---
D19	YEL/WHT	MAP Sensor VREF	4.9-5.1v
D20	YEL/WHT	Sensor VREF	4.9-5.1v
D21	GRN/WHT	MAP Sensor Ground	<0.050v
D22	GRN/YEL	Sensor Ground	<0.050v

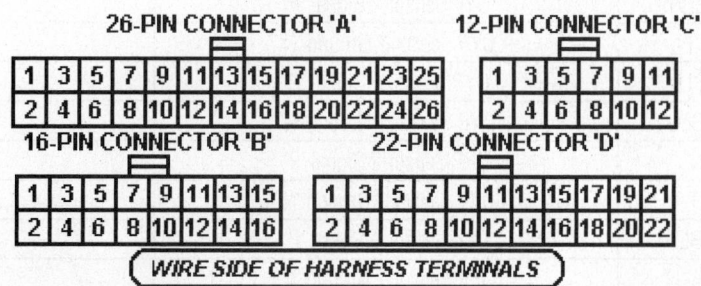

Pin Connector Graphic

05533_ADIA_G586

1994-95 L, LS Sedan 3.2L MFI VIN KA7 [A/T] 26P 'A' Connector

PCM Pin #	W/Color	Circuit Description (26-Pin)	Value at Hot Idle
A1	BRN	Injector 1	2.0-3.3 ms
A2	WHT/BLU	Injector 4	2.0-3.3 ms
A3	RED	Injector 2	2.0-3.3 ms
A4	BLK/RED	Injector 5	2.0-3.3 ms
A5	ORN	Injector 3	2.0-3.3 ms
A6	YEL	Injector 6	2.0-3.3 ms
A7	GRN/BLK	Fuel Pump Relay Control	Relay Off: 12v, On: 1v
A9	BLU/RED	Idle Air Control Valve	Pulse Signals
A10	GRN/BLU	HO2S-21 (B2 S1) Heater	Heater Off: 12v, On: 1v
A11	WHT	EGR Solenoid Control	Solenoid Off: 12v, On: 1v
A12	GRN/RED	HO2S-11 (B1 S1) Heater	Heater Off: 12v, On: 1v
A13	BLU	Malfunction Indicator Lamp	MIL Off: 12v, On: 1v
A14	RED/BLU	Bypass Low Solenoid	Solenoid Off: 12v, On: 1v
A15	RED/BLU	A/C Clutch Relay Control	Relay Off: 12v, On: 1v
A17	GRY	Air Suction Solenoid	Solenoid Off: 12v, On: 1v
A18	YEL	Bypass High Solenoid	Solenoid Off: 12v, On: 1v
A19	LT GRN	Pressure Regulator Solenoid	Solenoid Off: 12v, On: 1v
A20	LT GRN	EVAP Purge Solenoid	Solenoid Off: 12v, On: 1v
A21	PNK	Igniter 1 Signal	Digital Signals: 0-12-0v
A22	BRN	Igniter 2 Signal	Digital Signals: 0-12-0v
A23	BLK	Power Ground	<0.1v
A24	BLK	Power Ground	<0.1v
A25	YEL/BLK	Main Relay Power (B+)	12-14v
A26	BRN/BLK	Chassis Ground	<0.050v

1994-95 L, LS Sedan 3.2L MFI VIN KA7 [A/T] 16P 'B' Connector

PCM Pin #	W/Color	Circuit Description (16-Pin)	Value at Hot Idle
B1	---	Not Used	---
B2	BRN/BLK	Chassis Ground	<0.050v
B3	RED	Igniter 6 Signal	Digital Signals: 0-12-0v
B4	GRY	Igniter 5 Signal	Digital Signals: 0-12-0v
B5	GRN	PSP Switch Signal	Straight: 0v, Turning: 12v
B6	BLU	Igniter 4 Signal	Digital Signals: 0-12-0v
B7	LT GRN	A/T: Neutral Switch	In P/N: 0v, Others: 12v
B8	BLU	Igniter 3 Signal	Digital Signals: 0-12-0v
B9	BLU/GRN	CYP Sensor 2 Signal 'P'	AC pulse signals
B10	BLU/YEL	CYP Sensor 2 Ground 'N'	<0.050v
B11	ORN/BLU	CYP Sensor 1 Signal 'P'	AC pulse signals
B12	WHT/BLU	CYP Sensor 1 Ground 'N'	<0.050v
B13	ORN	CKP Sensor 2 Signal 'P'	AC pulse signals
B14	WHT	CKP Sensor 2 Ground 'N'	<0.050v
B15	ORN/BLU	CKP Sensor 1 Signal 'P'	AC pulse signals
B16	WHT/BLU	CKP Sensor 1 Ground 'N'	<0.050v

1994-95 L, LS Sedan 3.2L MFI VIN KA7 [A/T] 12P 'C' Connector

PCM Pin #	W/Color	Circuit Description (12-Pin)	Value at Hot Idle
C1	YEL/BLK	Main Relay Power (B+)	12-14v
C2	YEL/RED	Vehicle Speed Sensor	Moving: pulse signals
C3	BLU/BLK	Cooling Fan Switch	A/C Off: 12v, On: 1v
C4	BLU	Tachometer Signal	Pulse Signals
C5	RED/BLU	A/C Pressure Switch	Relay Off: 12v, On: 1v
C6	GRN/BLK	TCM Signal	---
C7	---	Not Used	---
C8	GRY/BLU	TCM Signal	---
C9	WHT	Service Check Connector	SCS Open: 4.80v
C10	---	Not Used	---
C11	BLK/WHT	Starter Switch Signal	Cranking: 9-11v
C12	GRY/YEL	TCM Signal	---

1994-95 L, LS Sedan 3.2L MFI VIN KA7 [A/T] 22P 'D' Connector

PCM Pin #	W/Color	Circuit Description (22-Pin)	Value at Hot Idle
D1	YEL/BLU	Keep Alive Power (VBU)	12-14v
D2	GRN/WHT	Brake Switch Signal	Brake Off: 12v, On: 0v
D3	WHT	Knock Sensor Signal (Bank 1)	No Detonation: 18mv AC
D4	RED/BLU	Knock Sensor Signal (Bank 2)	No Detonation: 18mv AC
D5	ORN/RED	TCM Signal	N/A
D6-7	---	Not Used	---
D8	BLU/YEL	Ignition Timing Adjustment	0.4-4.5v
D9	WHT/RED	Alternator Control Signal	Headlights off: 0v, on: 12v
D10	---	Not Used	---
D11	RED/BLU	TP Sensor Signal	0.5-0.6v
D12	BLK/WHT	EGR Valve Lift Sensor	1.1-1.2v
D13	RED/WHT	ECT Sensor Signal	At 180°: 0.5-0.6v
D14	WHT	HO2S-11 (B1 S1) Signal	0.1-1.1v
D15	RED/YEL	IAT Sensor Signal	At 100°F: 2-3v
D16	RED/BLU	HO21-21 (B2 S1) Signal	0.1-1.1v
D17	RED	MAP Sensor Signal	0.8-0.9v
D18	BLU	TCM Signal	N/A
D19	YEL/WHT	MAP Sensor VREF	4.9-5.1v
D20	YEL/WHT	Sensor VREF	4.9-5.1v
D21	GRN/WHT	MAP Sensor Ground	<0.050v
D22	GRN/WHT	Sensor Ground	<0.050v

26-PIN CONNECTOR 'A'

1	3	5	7	9	11	13	15	17	19	21	23	25
2	4	6	8	10	12	14	16	18	20	22	24	26

12-PIN CONNECTOR 'C'

1	3	5	7	9	11
2	4	6	8	10	12

16-PIN CONNECTOR 'B'

1	3	5	7	9	11	13	15
2	4	6	8	10	12	14	16

22-PIN CONNECTOR 'D'

1	3	5	7	9	11	13	15	17	19	21
2	4	6	8	10	12	14	16	18	20	22

WIRE SIDE OF HARNESS TERMINALS

Pin Connector Graphic

05533_ADIA_G586

1994-95 L, LS Sedan 3.2L MFI VIN KA7 [M/T] 26P 'A' Connector

PCM Pin #	W/Color	Circuit Description (26-Pin)	Value at Hot Idle
A1	BRN	Injector 1	2.0-3.3 ms
A2	WHT/BLU	Injector 4	2.0-3.3 ms
A3	RED	Injector 2	2.0-3.3 ms
A4	BLK/RED	Injector 5	2.0-3.3 ms
A5	ORN	Injector 3	2.0-3.3 ms
A6	YEL	Injector 6	2.0-3.3 ms
A7	GRN/BLK	Fuel Pump Relay Control	Relay Off: 12v, On: 1v
A8	RED/WHT	M/T: Reverse Lockout Relay	R/L on: 1v, off: 12v
A9	BLU/RED	Idle Air Control Valve	Pulse Signals
A10	GRN/BLU	HO2S-21 (B2 S1) Heater	Heater Off: 12v, On: 1v
A11	WHT	EGR Solenoid Control	Solenoid Off: 12v, On: 1v
A12	GRN/RED	HO2S-11 (B1 S1) Heater	Heater Off: 12v, On: 1v
A13	BLU	Malfunction Indicator Lamp	MIL Off: 12v, On: 1v
A14	RED/BLU	Bypass Low Solenoid	Solenoid Off: 12v, On: 1v
A15	RED/BLU	A/C Clutch Relay Control	Relay Off: 12v, On: 1v
A17	GRY	Air Suction Solenoid	Solenoid Off: 12v, On: 1v
A18	YEL	Bypass High Solenoid	Solenoid Off: 12v, On: 1v
A19	LT GRN	Pressure Regulator Solenoid	Solenoid Off: 12v, On: 1v
A20	LT GRN	EVAP Purge Solenoid	Solenoid Off: 12v, On: 1v
A21	PNK	Igniter 1 Signal	Digital Signals: 0-12-0v
A22	PNK	Igniter 2 Signal	Digital Signals: 0-12-0v
A23	BLK	Power Ground	<0.1v
A24	BLK	Power Ground	<0.1v
A25	YEL/BLK	Main Relay Power (B+)	12-14v
A26	BRN/BLK	Chassis Ground	<0.050v

1994-95 L, LS Sedan 3.2L MFI VIN KA7 [M/T] 16P 'B' Connector

PCM Pin #	W/Color	Circuit Description (16-Pin)	Value at Hot Idle
B1	---	Not Used	---
B2	BRN/BLK	Chassis Ground	<0.050v
B3	RED	Igniter 6 Signal	Digital Signals: 0-12-0v
B4	GRY	Igniter 5 Signal	Digital Signals: 0-12-0v
B5	GRN	PSP Switch Signal	Straight: 0v, Turning: 12v
B6	GRN	Igniter 4 Signal	Digital Signals: 0-12-0v
B8	BLU	Igniter 3 Signal	Digital Signals: 0-12-0v
B7	LT GRN	M/T Neutral Switch	In 'N': 0v, Others: 12v
B9	BLU/GRN	CYP Sensor 2 Signal 'P'	AC pulse signals
B10	BLU/YEL	CYP Sensor 2 Ground 'N'	<0.050v
B11	ORN/BLU	CYP Sensor 1 Signal 'P'	AC pulse signals
B12	WHT/BLU	CYP Sensor 1 Ground 'N'	<0.050v
B13	ORN	CKP Sensor 2 Signal 'P'	AC pulse signals
B14	WHT	CKP Sensor 2 Ground 'N'	<0.050v
B15	ORN/BLU	CKP Sensor 1 Signal 'P'	AC pulse signals
B16	WHT/BLU	CKP Sensor 1 Ground 'N'	<0.050v

1994-95 L, LS Sedan 3.2L MFI VIN KA7 [M/T] 12P 'C' Connector

PCM Pin #	W/Color	Circuit Description (12-Pin)	Value at Hot Idle
C1	YEL/BLK	Main Relay Power (B+)	12-14v
C2	YEL/RED	Vehicle Speed Sensor	Moving: pulse signals
C3	BLU/BLK	Cooling Fan Switch	A/C Off: 12v, On: 1v
C4	BLU	Tachometer Signal	Pulse Signals
C5	RED/BLU	A/C Pressure Switch	Relay Off: 12v, On: 1v
C6	---	Not Used	---
C7	PNK	M/T Clutch Switch	In 'N': 0v, Others: 12v
C8	---	Not Used	---
C9	WHT	Service Check Connector	SCS Open: 4.80v
C10	---	Not Used	---
C11	BLK/WHT	Starter Switch Signal	Cranking: 9-11v
C12	---	Not Used	---

1994-95 L, LS Sedan 3.2L MFI VIN KA7 [M/T] 22P 'D' Connector

PCM Pin #	W/Color	Circuit Description (22-Pin)	Value at Hot Idle
D1	YEL/BLU	Keep Alive Power (VBU)	12-14v
D2	GRN/WHT	Brake Switch Signal	Brake Off: 12v, On: 0v
D3	WHT	Knock Sensor Signal (Bank 1)	No Detonation: 18mv AC
D4	RED/BLU	Knock Sensor Signal (Bank 2)	No Detonation: 18mv AC
D5-7	---	Not Used	---
D8	BLU/YEL	Ignition Timing Adjustment	0.4-4.5v
D9	WHT/RED	Alternator Control Signal	Headlights off: 0v, on: 12v
D10	---	Not Used	---
D11	RED/BLU	TP Sensor Signal	0.5-0.6v
D12	BLK/WHT	EGR Valve Lift Sensor	1.1-1.2v
D13	RED/WHT	ECT Sensor Signal	At 180°: 0.5-0.6v
D14	WHT	HO2S-11 (B1 S1) Signal	0.1-1.1v
D15	RED/YEL	IAT Sensor Signal	At 100°F: 2-3v
D16	RED/BLU	HO21-21 (B2 S1) Signal	0.1-1.1v
D17	RED	MAP Sensor Signal	0.8-0.9v
D18	---	Not Used	---
D19	YEL/WHT	MAP Sensor VREF	4.9-5.1v
D20	YEL/WHT	Sensor VREF	4.9-5.1v
D21	GRN/WHT	MAP Sensor Ground	<0.050v
D22	GRN/WHT	Sensor Ground	<0.050v

26-PIN CONNECTOR 'A'

1	3	5	7	9	11	13	15	17	19	21	23	25
2	4	6	8	10	12	14	16	18	20	22	24	26

12-PIN CONNECTOR 'C'

1	3	5	7	9	11
2	4	6	8	10	12

16-PIN CONNECTOR 'B'

1	3	5	7	9	11	13	15
2	4	6	8	10	12	14	16

22-PIN CONNECTOR 'D'

1	3	5	7	9	11	13	15	17	19	21
2	4	6	8	10	12	14	16	18	20	22

WIRE SIDE OF HARNESS TERMINALS

Pin Connector Graphic

05533_ADIA_G586

1991-94 Coupe 3.0L V6 MFI VIN NA1 [A/T] 26P 'A' Connector

PCM Pin #	W/Color	Circuit Description (26-Pin)	Value at Hot Idle
A1, A3	BRN, RED	Injector 1, Injector 3 Control	2.0-3.3 ms
A2	WHT/BLU	Injector 4 Control	2.0-3.3 ms
A4	BLK/RED	Injector 5 Control	2.0-3.3 ms
A5, A6	ORN, YEL	Injector 3, Injector 6 Control	2.0-3.3 ms
A7, A8	GRN/BLK	Fuel Pump Relay Control	Relay Off: 12v, On: 1v
A9	BLU/RED	Electronic Air Control Valve	Pulse Signals
A10	BLK	HO2S-21 (B2 S1) Heater	Heater Off: 12v, On: 1v
A11	RED	EGR Solenoid Control	EGRS On: <1v
A12	BLK/BLU	HO2S-11 (B1 S1) Heater	Heater Off: 12v, On: 1v
A13	BLU	Malfunction Indicator Lamp	MIL Off: 12v, On: 1v
A14	RED/BLU	Chamber Volume Solenoid	CVS On: <1v
A15	RED/BLU	A/C Clutch Relay Control	Relay Off: 12v, On: 1v
A16	PNK	A/T: Control Unit Signal	Digital Signals
A18	BLU/RED	Radiator Fan High Relay	Fan on: 1v, off: 12v
A20	GRN	EVAP Purge Solenoid	Solenoid On: 1v, off: 12v
A21, A22	WHT	Igniter 1 Signal	At idle: 5 dwell
A23, A24	BLK	Power Ground	<0.1v
A25	YEL/BLK	Main Relay Power (B+)	12-14v
A26	BRN/BLK	Chassis Ground	<0.050v

1991-94 Coupe 3.0L V6 MFI VIN NA1 [A/T] 16P 'B' Connector

PCM Pin #	W/Color	Circuit Description (16-Pin)	Value at Hot Idle
B1	---	Not Used	---
B2	BRN/BLK	Chassis Ground	<0.050v
B3	WHT/RED	Igniter 6 Signal	Digital Signals: 0-12-0v
B4	WHT/YEL	Igniter 5 Signal	Digital Signals: 0-12-0v
B5	GRN	PSP Switch Signal	Straight: 0v, Turning: 12v
B6	WHT/BLU	Igniter 4 Signal	Digital Signals: 0-12-0v
B7	LT GRN	Park Neutral Switch Signal	In P/N: 0v, Others: 12v
B8	WHT/BLK	Igniter 3 Signal	Digital Signals: 0-12-0v
B9	ORN	CYP Sensor 2 Signal	AC pulse signals
B10	WHT	CYP Sensor 2 Ground	<0.050v
B11	BLU/GRN	CYP Sensor 1 Signal	AC pulse signals
B12	BLU/YEL	CYP Sensor 1 Ground	<0.050v
B13	ORN/BLU	CKP Sensor 2 Signal	AC pulse signals
B14	WHT/BLU	CKP Sensor 2 Ground	<0.050v
B15	ORN/BLU	CKP Sensor 1 Signal	AC pulse signals
B16	WHT/BLU	CKP Sensor 1 Ground	<0.050v

26-PIN CONNECTOR 'A' **16-PIN CONNECTOR 'B'** **12-PIN CONNECTOR 'C'**

| 1 | 2 | 3 | 4 | 5 | 6 | 7 | 8 | 9 | 10 | 11 | 12 | 13 |
| 14 | 15 | 16 | 17 | 18 | 19 | 20 | 21 | 22 | 23 | 24 | 25 | 26 |

| 1 | 2 | 3 | 4 | 5 | 6 | 7 | 8 |
| 9 | 10 | 11 | 12 | 13 | 14 | 15 | 16 |

| 1 | 2 | 3 | 4 | 5 | 6 |
| 7 | 8 | 9 | 10 | 11 | 12 |

22-PIN CONNECTOR 'D' **8-PIN CONNECTOR 'F'**

| 1 | 2 | 3 | 4 | 5 | 6 | 7 | 8 | 9 | 10 | 11 |
| 12 | 13 | 14 | 15 | 16 | 17 | 18 | 19 | 20 | 21 | 22 |

| 1 | 2 | 3 | 4 |
| 5 | 6 | 7 | 8 |

WIRE SIDE OF HARNESS TERMINALS

05533_ADIA_G587

Pin Connector Graphic

1991-94 Coupe 3.0L V6 MFI VIN NA1 [A/T] 12P 'C' Connector

PCM Pin #	W/Color	Circuit Description (12-Pin)	Value at Hot Idle
C1	YEL/BLK	Main Relay Power (B+)	12-14v
C2	YEL/RED	Vehicle Speed Sensor	Moving: pulse signals
C3	BLU/BLK	Cooling Fan Control Unit	N/A
C4	BLU	Tachometer Signal	Pulse Signals
C5	RED/BLU	A/C Pressure Switch	A/C Off: 12v, On: 1v
C6	WHT/BLK	TCS Signal	Digital Signals
C7	---	Not Used	---
C8	GRN/YEL	TCS Signal	Digital Signals
C9	BLU	Service Check Connector	SCS Open: 4.80v
C10	WHT/RED	TCS Signal	Digital Signals
C11	BLK/WHT	Starter Switch Signal	Cranking: 9-11v
C12	GRY	TCS Signal	Digital Signals

1991-94 Coupe 3.0L V6 MFI VIN NA1 [A/T] 22P 'D' Connector

PCM Pin #	W/Color	Circuit Description (22-Pin)	Value at Hot Idle
D1	WHT/YEL	Keep Alive Power (VBU)	12-14v
D2	---	Not Used	---
D3	ORN	Rear Knock Sensor Signal	No Detonation: 18mv AC
D4	WHT	Front Knock Sensor Signal	No Detonation: 18mv AC
D5	ORN/RED	TCS Signal	Digital Signals
D6	BLK/RED	Fuel Pump Resistor	N/A
D8	BRN	Ignition Timing Adjustment	0.4-4.5v
D9	WHT/RED	Alternator Control Signal	Headlights off: 0v, on: 12v
D10	---	Not Used	---
D11	RED/YEL	TP Sensor Signal	0.5-0.6v
D12	WHT/GRN	EGR Valve Lift Sensor	1.1-1.2v
D13	RED/WHT	ECT Sensor Signal	At 180°: 0.5-0.6v
D14	WHT	HO2S-11 (B1 S1) Signal	0.1-1.1v
D15	RED/BLK	IAT Sensor Signal	At 100°F: 2-3v
D16	RED/BLU	HO21-21 (B2 S1) Signal	0.1-1.1v
D17	WHT	MAP Sensor Signal	0.8-0.9v
D18	BLU	A/T: Control Unit VREF	4.9-5.1v
D19	YEL/WHT	MAP Sensor VREF	4.9-5.1v
D20	YEL/WHT	Sensor VREF	4.9-5.1v
D21	GRN/WHT	MAP Sensor Ground	<0.050v
D22	GRN/WHT	Sensor Ground	<0.050v

1991-94 Coupe 3.0L V6 MFI VIN NA1 [A/T] 8P 'F' Connector

PCM Pin #	W/Color	Circuit Description (8-Pin)	Value at Hot Idle
F1	WHT/GRN	A/T: Control Unit Signal	Digital Signals
F2	BLU	Front Valve Oil Pressure Sw.	Open: 12v, Closed: 1v
F3	WHT/RED	A/T: Control Unit Signal	Digital Signals
F4	BLU/BLK	Rear Valve Oil Pressure Sw.	Open: 12v, Closed: 1v
F5	YEL	Front Spool Solenoid Valve	Solenoid On: 1v, off: 12v
F6	BLU/RED	Acceleration Pedal Sensor 1	ACP closed: 0.5v
F7	BLU/YEL	Rear Spool Solenoid Valve	Solenoid On: 1v, off: 12v
F8	---	Not Used	---

1991-94 Coupe 3.0L V6 MFI VIN NA1 [M/T] 26P 'A' Connector

PCM Pin #	W/Color	Circuit Description (26-Pin)	Value at Hot Idle
A1	BRN	Injector 1	2.0-3.3 ms
A2	WHT/BLU	Injector 4	2.0-3.3 ms
A3	RED	Injector 3	2.0-3.3 ms
A4	BLK/RED	Injector 5	2.0-3.3 ms
A5	ORN	Injector 3	2.0-3.3 ms
A6	YEL	Injector 6	2.0-3.3 ms
A7	GRN/BLK	Fuel Pump Relay Control	Relay Off: 12v, On: 1v
A8	RED	Fuel Pump Relay Control	Relay Off: 12v, On: 1v
A9	BLU/RED	Electronic Air Control Valve	Pulse Signals
A10	BLK	HO2S-21 (B2 S1) Heater	Heater Off: 12v, On: 1v
A11	RED	EGR Solenoid Control	EGRS On: <1v
A12	RED/BLU	HO2S-11 (B1 S1) Heater	Heater Off: 12v, On: 1v
A13	BLU	Malfunction Indicator Lamp	MIL Off: 12v, On: 1v
A14	RED/BLU	Chamber Volume Solenoid	CVS On: <1v
A15	RED/BLU	A/C Clutch Relay Control	Relay Off: 12v, On: 1v
A17	---	Not Used	---
A18	BLU/RED	Radiator Fan High Relay	Relay On: <1v
A19	---	Not Used	---
A20	GRN	EVAP Purge Solenoid	Solenoid On: 1v, off: 12v
A21	WHT	Igniter 1 Signal	At idle: 5 dwell
A22	WHT/GRN	Igniter 2 Signal	At idle: 5 dwell
A23	BLK	Power Ground	<0.1v
A24	BLK	Power Ground	<0.1v
A25	YEL/BLK	Main Relay Power (B+)	12-14v
A26	BRN/BLK	Chassis Ground	<0.050v

1991-94 Coupe 3.0L V6 MFI VIN NA1 [M/T] 16P 'B' Connector

PCM Pin #	W/Color	Circuit Description (16-Pin)	Value at Hot Idle
B1	---	Not Used	---
B2	BRN/WHT	Chassis Ground	<0.050v
B3	WHT/RED	Igniter 6 Signal	Digital Signals: 0-12-0v
B4	WHT/YEL	Igniter 5 Signal	Digital Signals: 0-12-0v
B5	GRN	PSP Switch Signal	Straight: 0v, Turning: 12v
B6	WHT/BLU	Igniter 4 Signal	Digital Signals: 0-12-0v
B7	LT GRN	M/T: Neutral Position Switch	In N: 0v, others: 12v
B8	WHT/BLK	Igniter 3 Signal	Digital Signals: 0-12-0v
B9	ORN	CYP 2 Signal	AC pulse signals
B10	WHT	CYP 2 Signal	AC pulse signals
B11	BLU/GRN	CYP Sensor 1 Signal	AC pulse signals
B12	BLU/YEL	CYP Sensor 1 Ground	<0.050v
B13	ORN/BLU	CKP Sensor 2 Signal	AC pulse signals
B14	WHT/BLU	CKP Sensor 2 Ground	<0.050v
B15	ORN/BLU	CKP Sensor 1 Signal	AC pulse signals
B16	WHT/BLU	CKP Sensor 1 Ground	<0.050v

1991-94 Coupe 3.0L V6 MFI VIN NA1 [M/T] 12P 'C' Connector

PCM Pin #	W/Color	Circuit Description (12-Pin)	Value at Hot Idle
C1	YEL/BLK	Main Relay Power (B+)	12-14v
C2	YEL/RED	Vehicle Speed Sensor	Moving: pulse signals
C3	BLU/BLK	Cooling Fan Control Unit	N/A
C4	BLU	Tachometer Signal	Pulse Signals
C5	RED/BLU	A/C Pressure Switch	A/C Off: 12v, On: 1v
C6	---	Not Used	---
C7	PNK	Clutch Switch	With Clutch In: 0v
C8-9	---	Not Used	---
C10	BLU	Service Check Connector	SCS Open: 4.80v
C11	BLK/WHT	Starter Switch Signal	Cranking: 9-11v
C12	---	Not Used	---

1991-94 Coupe 3.0L V6 MFI VIN NA1 [M/T] 22P 'D' Connector

PCM Pin #	W/Color	Circuit Description (22-Pin)	Value at Hot Idle
D1	WHT/YEL	Keep Alive Power (VBU)	12-14v
D2	---	Not Used	---
D3	ORN	Rear Knock Sensor Signal	No Detonation: 18mv AC
D4	WHT	Front Knock Sensor Signal	No Detonation: 18mv AC
D5	ORN/RED	TCS Signal	Digital Signals
D6	BLK/RED	Fuel Pump Resistor	N/A
D7	---	Not Used	---
D8	BRN	Ignition Timing Adjustment	0.4-4.5v
D9	WHT/RED	Alternator Control Signal	Headlights off: 0v, on: 12v
D10	---	Not Used	---
D11	RED/YEL	TP Sensor Signal	0.5-0.6v
D12	WHT/GRN	EGR Valve Lift Sensor	1.1-1.2v
D13	RED/WHT	ECT Sensor Signal	At 180°: 0.5-0.6v
D14	WHT	HO2S-11 (B1 S1) Signal	0.1-1.1v
D15	RED/BLK	IAT Sensor Signal	At 100°F: 2-3v
D16	RED/BLU	HO21-21 (B2 S1) Signal	0.1-1.1v
D17	WHT	MAP Sensor Signal	0.8-0.9v
D18	---	Not Used	---
D19	YEL/WHT	MAP Sensor VREF	4.9-5.1v
D20	YEL/WHT	Sensor VREF	4.9-5.1v
D21	GRN/WHT	MAP Sensor Ground	<0.050v
D22	GRN/WHT	Sensor Ground	<0.050v

26-PIN CONNECTOR 'A'

1	3	5	7	9	11	13	15	17	19	21	23	25
2	4	6	8	10	12	14	16	18	20	22	24	26

12-PIN CONNECTOR 'C'

1	3	5	7	9	11
2	4	6	8	10	12

16-PIN CONNECTOR 'B'

1	3	5	7	9	11	13	15
2	4	6	8	10	12	14	16

22-PIN CONNECTOR 'D'

1	3	5	7	9	11	13	15	17	19	21
2	4	6	8	10	12	14	16	18	20	22

WIRE SIDE OF HARNESS TERMINALS

05533_ADIA_G588

Pin Connector Graphic

1995-99 Coupe 3.0L V6 VTEC VIN NA1 [A/T] 26P 'A' Connector

PCM Pin #	W/Color	Circuit Description (26-Pin)	Value at Hot Idle
A1	BRN	Injector 1	2.0-3.3 ms
A2	RED	Injector 2	2.0-3.3 ms
A3	BLU	Injector 3	2.0-3.3 ms
A4 ('97)	---	Not Used	---
A4 ('98-'99)	BLK/YEL	EVAP Canister Vent Valve	Solenoid on: 1v, off: 12v
A5	GRN/BLK	Fuel Pump Relay Control 1	Relay Off: 12v, On: 1v
A6	GRN	EGR Solenoid Control	Solenoid Off: 12v, On: 1v
A7	BLU	Malfunction Indicator Lamp	MIL Off: 12v, On: 1v
A8	RED/BLU	A/C Clutch Relay Control	Relay Off: 12v, On: 1v
A9	GRY	HO2S-11 (B1 S1) Heater	Digital Signals: 0-12-0v
A10	LT GRN	HO2S-21 (B2 S1) Heater	Digital Signals: 0-12-0v
A11	PNK	Ignition Coil 1 Control	Digital Signals: 0-12-0v
A12	BLK	Power Ground	<0.1v
A13	YEL/BLK	Main Relay Power (B+)	12-14v
A14	YEL	Injector 4	2.0-3.3 ms
A15	BLK/RED	Injector 5	2.0-3.3 ms
A16	WHT/BLU	Injector 6	2.0-3.3 ms
A17 ('97)	---	Not Used	---
A17 ('98-'99)	BLK/ORN	EVAP Bypass Solenoid Valve	Solenoid on: 1v, off: 12v
A18	BLU/YEL	Front VTEC Solenoid	Idle: 0v, Hi-Speed: 12v
A19	GRN/YEL	Rear VTEC Solenoid	Idle: 0v, Hi-Speed: 12v
A20	PNK/BLU	IAB Solenoid Control	Solenoid off: 12v, on: 1v
A21	RED	EVAP Purge Solenoid	Pulse Signals
A22	GRN/RED	HO2S-12 (B1 S2) Heater	Digital Signals: 0-12-0v
A23	ORN/BLK	HO2S-22 Heater Control	Digital Signals: 0-12-0v
A24	BRN	Ignition Coil 2 Control	Digital Signals: 0-12-0v
A25	BLK	Power Ground	<0.1v
A26	BRN/BLK	Logic Ground	<0.050v

1995-99 Coupe 3.0L V6 VTEC VIN NA1 [A/T] 16P 'B' Connector

PCM Pin #	W/Color	Circuit Description (16-Pin)	Value at Hot Idle
B1	BLU	Front VTEC Pressure Switch	Switch on: 0v, off: 12v
B2	RED	Ignition Coil 6 Control	Digital Signals: 0-12-0v
B3	BLU/BLK	Rear VTEC Pressure Switch	At Low Speed: 0.1v
B4	RED	A/T: Gear Position Switch	P/N: 0v, gear: <5v
B5	ORN	CYP Sensor 2 Signal	AC pulse signals
B6	WHT	CYP Sensor 1 Signal	AC pulse signals
B7	ORN/BLU	CKP Sensor 2 Signal	AC pulse signals
B8	BLU/GRN	CKP Sensor 1 Signal	AC pulse signals
B9	BRN/BLK	Logic Ground	<0.050v
B10	GRY	Ignition Coil 5 Control	Digital Signals: 0-12-0v
B11	GRN	Ignition Coil 4 Control	Digital Signals: 0-12-0v
B12	BLU	Ignition Coil 3 Control	Digital Signals: 0-12-0v
B13	ORN/BLU	CYP Sensor 2 Ground	<0.050v
B14	WHT/BLU	CYP Sensor 1 Ground	<0.050v
B15	WHT/BLU	CKP Sensor 2 Ground	<0.050v
B16	BLU/YEL	CKP Sensor 1 Ground	<0.050v

1995-99 Coupe 3.0L V6 VTEC VIN NA1 [A/T] 12P 'C' Connector

PCM Pin #	W/Color	Circuit Description (12-Pin)	Value at Hot Idle
C1	YEL/BLK	Main Relay Power (B+)	12-14v
C2	BLU/BLK	A/C Switch Signal	A/C Off: 12v, On: 1v
C3	RED/GRN	A/C Pressure Switch Signal	A/C Off: 4-5v, On: 0v
C5	BLU	Service Check Connector	SCS Open: 4.80v
C6	BLK/WHT	Starter Switch Signal	Cranking: 9-11v
C7	ORN	Vehicle Speed Sensor	Moving: pulse signals
C8	GRN	Engine Speed Pulse Signal	Pulses
C9	GRN/YEL	Front Peak Hold Reset	Pulses
C10	GRN/BLK	Rear Peak Hold Reset	Pulses
C11	YEL/RED	Front Misfire Pulse	Pulses
C12	YEL	Rear Misfire Pulse	Pulses

1995-99 Coupe 3.0L V6 VTEC VIN NA1 [A/T] 22P 'D' Connector

PCM Pin #	W/Color	Circuit Description (22-Pin)	Value at Hot Idle
D1	WHT/YEL	Keep Alive Power (VBU)	12-14v
D2	RED/BLU	Knock Sensor 1 Signal	No detonation: 18mv AC
D3 ('97)	BRN	EVAP Purge Flow Switch	Switch on: 0v, off: 12v
D4	YEL/GRN	K-Line Signal	12v
D5	WHT/RED	Alternator 'FR' Signal	Digital Signals: 0-5-0v
D6 ('97)	---	Not Used	---
D6 ('98-'99)	BLU	Fuel Tank Pressure Sensor	Fuel Cap off: 2.5v
D7	RED/WHT	ECT Sensor Signal	At 180°F: 0.5-0.6v
D8	RED/YEL	IAT Sensor Signal	At 100°F: 2-3v
D9	WHT/YEL	MAP Sensor Signal	0.8-0.9v
D10	YEL/BLU	Sensor VREF	4.9-5.1v

1995-99 Coupe 3.0L V6 VTEC VIN NA1 [A/T] 22P 'D' Connector, *continued*

D11	GRN/BLU	Sensor Ground	<0.050v
D12	RED/BLK	TP Sensor Signal	0.5-0.6v
D13	WHT	Knock Sensor 2 Signal	No detonation: 18mv AC
D14-15	---	Not Used	---
D16	GRN	HO2S-21 (B2 S1) Signal	0.1-1.1v
D17	WHT/RED	HO2S-22 (B2 S2) Signal	0.1-1.1v
D18	BLU/RED	HO2S-11 (B1 S1) Signal	0.1-1.1v
D19	WHT	HO2S-12 (B1 S2) Signal	0.1-1.1v
D20	WHT/BLK	EGR Valve Lift Sensor	1.1-1.2v
D21	BLU/GRN	HO2S-21 (B2 S1) Ground	<0.050v
D22	WHT	HO2S-22 (B2 S2) Ground	<0.050v

Standard Colors and Abbreviations

Abbreviation	Color	Abbreviation	Color	Abbreviation	Color
BLK	Black	LT BLU	LT Blue	TAN	Tan
BLU	Blue	LT GRN	LT Green	VIO	Violet
BRN	Brown	ORN	Orange	WHT	White
GRY	Gray	PNK	Pink	YEL	Yellow
GRN	Green	PPL	Purple		

1995-99 Coupe 3.0L V6 VTEC VIN NA1 [A/T] 26P 'F' Connector

PCM Pin #	W/Color	Circuit Description (26-Pin)	Value at Hot Idle
F1	YEL/WHT	Sensor VREF	4.9-5.1v
F2	GRN/WHT	Sensor Ground	<0.050v
F3 ('97)	---	Not Used	---
F3 ('98-'99)	BRN	INOCD Immobilizer Code	Digital Signals
F4	RED	Fuel Pump Relay Control	Relay Off: 12v, On: 1v
F5, 17, 19-20	---	Not Used	---
F6	GRY	Brake Switch Signal 2	B/P depressed: 0v
F7	GRN/BLK	Cruise Resume Switch	CR/S released: 0v
F8	GRN/RED	Cruise Control Set Switch	CS/S released: 0v
F9	LT GRN	Cruise Control Main Switch	M/S on: 12v, off: 0v
F10	BLU/ORN	C/C A/T: Gear Position Switch	In 'D', D3, D2: 0v
F11	BLU/BLK	Cruise Indicator Light	Light on: 0v, off: 12v
F12	YEL/RED	Accelerator Position Sensor 2	ACP closed: 0.5v
F13	BLU/RED	Accelerator Position Sensor 1	ACP closed: 0.5v
F14	YEL/WHT	Sensor VREF	4.9-5.1v
F15	GRN/WHT	Sensor Ground	<0.050v
F16	GRN/WHT	Brake Switch Signal 1	Brake Off: 12v, On: 0v
F18	BLU	TCM VREF	4.9-5.1v
F21	GRN/YEL	A/T: FI Data Line 'B'	Pulses
F22	WHT/YEL	A/T: FI Data Line 'A'	Pulses
F24	RED/BLU	TCS Data Line	Pulses

1995-99 Coupe 3.0L V6 VTEC VIN NA1 [A/T] 12P 'G' Connector

PCM Pin #	W/Color	Circuit Description (12-Pin)	Value at Hot Idle
G2, G3	GRN/RED	IGM1, IGM2 Power Source	Key on: 12v, off: 0v
G4	BRN	Motor Phase Out 1	Key on: 0v or pulse
G5	WHT/GRN	Motor Phase Out 2	Key on: 0v or pulse
G7, G8	BLK	Power Ground	<0.1v
G9	ORN	PCM PWR to Motor 1, 3	KOEO: pulse
G10	GRN	PCM PWR to Motor 2, 4	KOEO: pulse
G11	YEL	Motor Phase Out 3	Key on: 0v or pulse
G12	RED	Motor Phase Out 4	Key on: 0v or pulse

Pin Connector Graphic

05533_ADIA_G589

1995-99 Coupe 3.0L V6 VTEC VIN NA1 [M/T] 26P 'A' Connector

PCM Pin #	W/Color	Circuit Description (26-Pin)	Value at Hot Idle
A1	BRN	Injector 1	2.0-3.3 ms
A2	RED	Injector 2	2.0-3.3 ms
A3	BLU	Injector 3	2.0-3.3 ms
A4 ('97)	---	Not Used	---
A4 ('98-'99)	BLK/YEL	EVAP Canister Vent Valve	Solenoid on: 1v, off: 12v
A5	GRN/BLK	Fuel Pump Relay Control 1	Relay Off: 12v, On: 1v
A6	GRN	EGR Solenoid Control	Solenoid Off: 12v, On: 1v
A7	BLU	Malfunction Indicator Lamp	MIL Off: 12v, On: 1v
A8	RED/BLU	A/C Clutch Relay Control	Relay Off: 12v, On: 1v
A9	GRY	HO2S-11 (B1 S1) Heater	Digital Signals: 0-12-0v
A10	LT GRN	HO2S-12 (B1 S2) Heater	Digital Signals: 0-12-0v
A11	PNK	Ignition Coil 1	Digital Signals: 0-12-0v
A12	BLK	Power Ground	<0.1v
A13	YEL/BLK	Main Relay Power (B+)	12-14v
A14	YEL	Injector 4	2.0-3.3 ms
A15	BLK/RED	Injector 5	2.0-3.3 ms
A16	WHT/BLU	Injector 6	2.0-3.3 ms
A17 ('97)	---	Not Used	---
A17 ('98-'99)	BLK/ORN	EVAP Bypass Solenoid Valve	Solenoid on: 1v, off: 12v
A18	BLU/YEL	Front VTEC Solenoid	Idle: 0v, Hi-Speed: 12v

1995-99 Coupe 3.0L V6 VTEC VIN NA1 [M/T] 26P 'A' Connector, *continued*

A19	GRN/YEL	Rear VTEC Solenoid	Idle: 0v, Hi-Speed: 12v
A20	PNK/BLU	IAB Solenoid Control	Solenoid Off: 12v, On: 1v
A21	RED/YEL	EVAP Purge Solenoid	Pulse Signals
A22	GRN/RED	HO2S-11 (B1 S1) Heater	Digital Signals: 0-12-0v
A23	ORN/BLK	HO2S-12 (B1 S2) Heater	Digital Signals: 0-12-0v
A24	BRN	Ignition Coil 2	Digital Signals: 0-12-0v
A25	BLK	Power Ground	<0.1v
A26	BRN/BLK	Logic Ground	<0.050v

1995-99 Coupe 3.0L V6 VTEC VIN NA1 [M/T] 16P 'B' Connector

PCM Pin #	W/Color	Circuit Description (16-Pin)	Value at Hot Idle
B1	BLU	Front VTEC Pressure Switch	Switch on: 0v, off: 12v
B2	RED	Ignition Coil 6 Control	Digital Signals: 0-12-0v
B3	BLU/BLK	Rear VTEC Pressure Switch	At Low Speed: 0.1v
B4	LT GRN	Neutral Switch Signal	In 'N': 0v, Others: 12v
B5	ORN	CYP Sensor Signal 2	AC pulse signals
B6	WHT	CYP Sensor Signal 1	AC pulse signals
B7	ORN/BLU	CKP Sensor 2 Signal	AC pulse signals
B8	BLU/GRN	CKP Sensor 1 Signal	AC pulse signals
B9	BRN/BLK	Logic Ground	<0.050v
B10	GRY	Ignition Coil 5	Digital Signals: 0-12-0v
B11	GRN	Ignition Coil 4	Digital Signals: 0-12-0v
B12	BLU	Ignition Coil 3	Digital Signals: 0-12-0v
B13	ORN/BLU	CYP Sensor 2 Ground	<0.050v
B14	WHT/BLU	CYP Sensor 1 Ground	<0.050v
B15	WHT/BLU	CKP Sensor 2 Ground	<0.050v
B16	BLU/YEL	CKP Sensor 1 Ground	<0.050v

1995-99 Coupe 3.0L V6 VTEC VIN NA1 [M/T] 12P 'C' Connector

PCM Pin #	W/Color	Circuit Description (12-Pin)	Value at Hot Idle
C1	YEL/BLK	Main Relay Power (B+)	12-14v
C2	BLU/BLK	A/C Switch Signal	A/C Off: 12v, On: 1v
C3	RED/GRN	A/C Pressure Switch Signal	A/C Off: 4-5v, On: 0v
C4	PNK	Clutch Switch Signal	With Clutch In: 0v
C5	BLU	Service Check Connector	SCS Open: 4.80v
C6	BLK/WHT	Starter Switch Signal	Cranking: 9-11v
C7	ORN	Vehicle Speed Sensor	Moving: pulse signals
C8	GRN	Engine Speed Pulse Signal	Pulses
C9	GRN/YEL	Front Peak Hold Reset	Pulses
C10	GRN/BLK	Rear Peak Hold Reset	Pulses
C11	YEL/RED	Front Misfire Pulse	Pulses
C12	YEL	Rear Misfire Pulse	Pulses

1995-99 Coupe 3.0L V6 VTEC VIN NA1 [M/T] 22P 'D' Connector

PCM Pin #	W/Color	Circuit Description (22-Pin)	Value at Hot Idle
D1	WHT/YEL	Keep Alive Power (VBU)	12-14v
D2	RED/BLU	Knock Sensor 1 Signal	No detonation: 18mv AC
D3 ('97)	BRN	EVAP Purge Flow Switch	Switch on: 0v, off: 12v
D4	YEL/GRN	K-Line Signal	12v
D5	WHT/RED	Alternator 'FR' Signal	Digital Signals: 0-5-0v
D6 ('97)	---	Not Used	---
D6 ('98-'99)	BLU	Fuel Tank Pressure Sensor	Fuel Cap off: 2.5v
D7	RED/WHT	ECT Sensor Signal	At 180°F: 0.5-0.6v
D8	RED/YEL	IAT Sensor Signal	At 100°F: 2-3v
D9	WHT/YEL	MAP Sensor Signal	0.8-0.9v
D10	YEL/BLU	Sensor VREF	4.9-5.1v
D11	GRN/BLU	Sensor Ground	<0.050v
D12	RED/BLK	TP Sensor Signal	0.5-0.6v
D13	WHT	Knock Sensor 2 Signal	No detonation: 18mv AC
D14	---	Not Used	---
D15	---	Not Used	---
D16	GRN	HO2S-21 (B2 S1) Signal	0.1-1.1v
D17	WHT/RED	HO2S-22 (B2 S2) Signal	0.1-1.1v
D18	BLU/RED	HO2S-11 (B1 S1) Signal	0.1-1.1v
D19	WHT	Rear HO2S-11 (B1 S1) Signal	0.1-1.1v
D20	WHT/BLK	EGR Valve Lift Sensor	1.1-1.2v
D21	BLU/GRN	HO2S-21 (B2 S1) Ground	<0.050v
D22	WHT	HO2S-22 (B2 S2) Ground	<0.050v

1995-99 Coupe 3.0L V6 VTEC VIN NA1 [M/T] 26P 'F' Connector

PCM Pin #	W/Color	Circuit Description (26-Pin)	Value at Hot Idle
F1	YEL/WHT	Sensor VREF (VCC1)	4.9-5.1v
F2	GRN/WHT	Sensor Ground	<0.050v
F3 ('97)	---	Not Used	---
F3 ('98-'99)	BRN	INOCD Immobilizer Code	Digital Signals
F4	RED	Fuel Pump Relay Control	Relay Off: 12v, On: 1v
F5	---	Not Used	---
F6	GRY	Brake Switch Signal 2	B/P depressed: 0v
F7	GRN/BLK	Cruise Resume Switch	CR/S released: 0v
F8	GRN/RED	Cruise Control Set Switch	CS/S released: 0v
F9	LT GRN	Cruise Control Main Switch	M/S on: 12v, off: 0v
F10	BLU/ORN	Cruise Control Clutch Switch	Clutch in: 0v, out: 8v
F11	BLU/BLK	Cruise Indicator Light	Light on: 0v, off: 12v
F12	YEL/RED	ACP Sensor 2 Signal (AP2)	ACP closed: 0.5v
F13	BLU/RED	ACP Sensor 1 Signal (AP1)	ACP closed: 0.5v
F14	YEL/WHT	Sensor VREF (VCC3)	4.9-5.1v
F15	GRN/WHT	Sensor Ground (SG3)	<0.050v
F16	GRN/WHT	Brake Switch Signal 1	Brake Off: 12v, On: 0v
F17-26	---	Not Used	---

1995-99 Coupe 3.0L V6 VTEC VIN NA1 [M/T] 12P 'G' Connector

PCM Pin #	W/Color	Circuit Description (12-Pin)	Value at Hot Idle
G1	---	Not Used	---
G2	GRN/RED	IGM1 Power Source	12-14v
G3	GRN/RED	IGM2 Power Source	12-14v
G4	BRN	Motor Phase Out 1	Pulse Signals
G5	WHT/GRN	Motor Phase Out 2	Pulse Signals
G7	BLK	Power Ground	<0.1v
G8	BLK	Power Ground	<0.1v
G9	ORN	PCM PWR to Motor 1, 3	Pulse Signals
G10	GRN	PCM PWR to Motor 2, 4	Pulse Signals
G11	YEL	Motor Phase Out 3	Pulse Signals
G12	RED	Motor Phase Out 4	Pulse Signals

Pin Connector Graphic

2000-03 Coupe 3.0L V6 VTEC VIN NA1 [All] 26P 'A' Connector

PCM Pin #	W/Color	Circuit Description (26-Pin)	Value at Hot Idle
A1	BRN	Injector 1 Control	2.0-3.3 ms
A2	RED	Injector 2 Control	2.0-3.3 ms
A3	BLU	Injector 3 Control	2.0-3.3 ms
A4	BLK/YEL	EVAP Canister Vent Valve	Solenoid on: 1v, off: 12v
A5	GRN/BLK	Fuel Pump Relay Control 1	Relay Off: 12v, On: 1v
A6	GRN	EGR Solenoid Control	Solenoid Off: 12v, On: 1v
A7	BLU	Malfunction Indicator Lamp	MIL Off: 12v, On: 1v
A8	RED/BLU	A/C Clutch Relay Control	Relay Off: 12v, On: 1v
A9	GRY	HO2S-11 (B1 S1) Heater	Digital Signals: 0-12-0v
A10	LT GRN	HO2S-21 (B2 S1) Heater	Digital Signals: 0-12-0v
A11	PNK	Ignition Coil 1 Control	Digital Signals: 0-12-0v
A12	BLK	Power Ground (PG1)	<0.1v
A13	YEL/BLK	Main Relay Power (B+)	12-14v
A14	YEL	Injector 4 Control	2.0-3.3 ms
A15	BLK/RED	Injector 5 Control	2.0-3.3 ms
A16	WHT/BLU	Injector 6 Control	2.0-3.3 ms
A17	ORN	EVAP Bypass Solenoid Valve	Solenoid on: 1v, off: 12v
A18	YEL/GRN	Front VTEC Solenoid	Idle: 0v, Hi-Speed: 12v
A19	---	Not Used	---
A20	PNK/BLU	IAB Solenoid Control	Solenoid Off: 12v, On: 1v
A21	RED/YEL	EVAP Purge Solenoid	Pulse Signals
A22	GRN/RED	HO2S-12 (B1 S2) Heater	Digital Signals: 0-12-0v
A23	ORN/BLK	HO2S-22 Heater Control	Digital Signals: 0-12-0v
A24	BRN	Ignition Coil 2 Control	Digital Signals: 0-12-0v
A25	BLK	Power Ground (PG2)	<0.1v
A26	BRN/BLK	Logic Ground (LG1)	<0.050v

2000-03 Coupe 3.0L V6 VTEC VIN NA1 [All] 16P 'B' Connector

PCM Pin #	W/Color	Circuit Description (16-Pin)	Value at Hot Idle
B1	BLU	Front VTEC Pressure Switch	Switch on: 0v, off: 12v
B2	RED	Ignition Coil 6 Control	Digital Signals: 0-12-0v
B3	BLU/BLK	Rear VTEC Pressure Switch	At Low Speed: 0.1v
B4	RED	A/T: Gear Position Switch	In P/N: 0v, others: 5v
B4	LT GRN	M/T: Neutral Position Switch	In 'N': 0v, others: 5v
B5	ORN	CYP Sensor 2 Signal	AC pulse signals
B6	WHT	CYP Sensor 1 Signal	AC pulse signals
B7	ORN/BLU	CKP Sensor 2 Signal	AC pulse signals
B8	BLU/GRN	CKP Sensor 1 Signal	AC pulse signals
B9	BRN/BLK	Logic Ground	<0.050v
B10	GRY	Ignition Coil 5 Control	Digital Signals: 0-12-0v
B11	GRN	Ignition Coil 4 Control	Digital Signals: 0-12-0v
B12	BLU	Ignition Coil 3 Control	Digital Signals: 0-12-0v

2000-03 Coupe 3.0L V6 VTEC VIN NA1 [All] 16P 'B' Connector, *continued*

B13	ORN/BLU	CYP Sensor 2 Ground	<0.050v
B14	WHT/BLU	CYP Sensor 1 Ground	<0.050v
B15	WHT/BLU	CKP Sensor 2 Ground	<0.050v
B16	BLU/YEL	CKP Sensor 1 Ground	<0.050v

2000-03 Coupe 3.0L V6 VTEC VIN NA1 [All] 12P 'C' Connector

PCM Pin #	W/Color	Circuit Description (12-Pin)	Value at Hot Idle
C1	YEL/BLK	Main Relay Power (B+)	12-14v
C2	BLU/BLK	A/C Switch Signal	A/C Off: 12v, On: 1v
C3	RED/GRN	A/C Pressure Switch Signal	A/C Off: 4-5v, On: 0v
C4	PNK	M/T: Clutch Engage Switch	Clutch in: 5v, out: 0v
C5	BLU	Service Check Connector	SCS Open: 4.80v
C6	BLK/WHT	Starter Switch Signal	Cranking: 9-11v
C7	ORN	Vehicle Speed Sensor	Moving: pulse signals
C8	GRN	Engine Speed Pulse Signal	Pulses
C9	GRN/YEL	Front Peak Hold Reset	Pulses
C10	GRN/BLK	Rear Peak Hold Reset	Pulses
C11	YEL/RED	Front Misfire Pulse	Pulses
C12	YEL	Rear Misfire Pulse	Pulses

2000-03 Coupe 3.0L V6 VTEC VIN NA1 [All] 22P 'D' Connector

PCM Pin #	W/Color	Circuit Description (22-Pin)	Value at Hot Idle
D1	WHT/YEL	Keep Alive Power (VBU)	12-14v
D2	RED/BLU	Knock Sensor 1 Signal	No detonation: 18mv AC
D3	WHT/BLK	Air Pump Current Sensor	Pump on: Pulse Signals
D4	YEL/GRN	K-Line Signal	12v
D5	WHT/RED	Alternator 'FR' Signal	Digital Signals: 0-5-0v
D6	BLU	Fuel Tank Pressure Sensor	Fuel Cap off: 2.5v
D7	RED/WHT	ECT Sensor Signal	At 180°F: 0.5-0.6v
D8	RED/YEL	IAT Sensor Signal	At 100°F: 2-3v
D9	WHT/YEL	MAP Sensor Signal	0.8-0.9v
D10	YEL/BLU	Sensor VREF (VCC2)	4.9-5.1v
D11	GRN/BLU	Sensor Ground	<0.050v
D12	RED/BLK	TP Sensor Signal	0.5-0.6v
D13	WHT	Knock Sensor 2 Signal	No detonation: 18mv AC
D14-15	---	Not Used	---
D16	GRN	HO2S-21 (B2 S1) Signal	0.1-1.1v
D17	WHT/RED	HO2S-22 (B2 S2) Signal	0.1-1.1v
D18	BLU/RED	HO2S-11 (B1 S1) Signal	0.1-1.1v
D19	WHT	HO2S-12 (B1 S2) Signal	0.1-1.1v
D20	WHT/BLK	EGR Valve Lift Sensor	1.1-1.2v
D21	BLU/GRN	HO2S-21 (B2 S1) Ground	<0.050v
D22	WHT	HO2S-22 (B2 S2) Ground	<0.050v

Standard Colors and Abbreviations

Abbreviation	Color	Abbreviation	Color	Abbreviation	Color
BLK	Black	LT BLU	LT Blue	TAN	Tan
BLU	Blue	LT GRN	LT Green	VIO	Violet
BRN	Brown	ORN	Orange	WHT	White
GRY	Gray	PNK	Pink	YEL	Yellow
GRN	Green	PPL	Purple		

2000-03 Coupe 3.0L V6 VTEC VIN NA1 [All] 26P 'F' Connector

PCM Pin #	W/Color	Circuit Description (26-Pin)	Value at Hot Idle
F1	YEL/WHT	Sensor VREF (VCC1)	4.9-5.1v
F2, F15	GRN/WHT	Sensor Ground SG1, SG3	<0.050v
F3	BRN	INOCD Immobilizer Code	Digital Signals
F4	RED	Fuel Pump Relay Control	Relay Off: 12v, On: 1v
F6	GRY	Brake Switch Signal 2	B/P depressed: 0v
F7	GRN/BLK	Cruise Resume Switch	CR/S released: 0v
F8	GRN/RED	Cruise Control Set Switch	CS/S released: 0v
F9	LT GRN	Cruise Control Main Switch	M/S on: 12v, off: 0v
F10	BLU/ORN	C/C A/T: Gear Position Switch	In 'D', D3, D2: 0v
F10	BLU/ORN	C/C M/T: Clutch Switch	Clutch in: 0v, out: 8v
F11	BLU/BLK	Cruise Indicator Light	Light on: 0v, off: 12v
F12	YEL/RED	Accelerator Position Sensor 2	ACP closed: 0.5v
F13	BLU/RED	Accelerator Position Sensor 1	ACP closed: 0.5v
F14	YEL/WHT	Sensor VREF (VCC1)	4.9-5.1v
F16	GRN/WHT	Brake Switch Signal 1	Brake Off: 12v, On: 0v
F17	RED/WHT	Reverse Lockout Relay Out	VSS over 30 mph: 0v
F18	BLU	TCM VREF	4.9-5.1v
F19	BLU/YEL	Front VTEC Solenoid Valve	0v, Hi-Speed: 12v
F20	GRY/YEL	Rear VTEC Solenoid Valve	0v, Hi-Speed: 12v
F21	GRN/YEL	A/T: FI Data Line 'B'	Pulse Signals
F22	WHT/YEL	A/T: FI Data Line 'A'	Pulse Signals
F23	BLU/WHT	Air Pump Relay Control	Relay Off: 12v, On: 1v
F24	RED/BLU	TCS Data Line	Pulse Signals
F25	RED	Air Control Solenoid Valve	Pump on: 1v, off: 12v

2000-03 Coupe 3.0L V6 VTEC VIN NA1 [All] 12P 'G' Connector

PCM Pin #	W/Color	Circuit Description (12-Pin)	Value at Hot Idle
G1, G9	---	Not Used	---
G2, G3	GRN/RED	IGM1, IGM2 Power Source	Key on: 12v, off: 0v
G4	BRN	Motor Phase Out 1	Key on: 0v or pulse
G5	WHT/GRN	Motor Phase Out 2	Key on: 0v or pulse
G7, G8	BLK	Power Ground	<0.1v
G9	ORN	PCM PWR to Motor 1, 3	KOEO: pulse
G10	GRN	PCM PWR to Motor 2, 4	KOEO: pulse
G11	YEL	Motor Phase Out 3	Key on: 0v or pulse
G12	RED	Motor Phase Out 4	Key on: 0v or pulse

05533_ADIA_G589

Pin Connector Graphic

1997-99 Coupe 3.2L V6 VTEC VIN NA2 [A/T] 26P 'A' Connector

PCM Pin #	W/Color	Circuit Description (26-Pin)	Value at Hot Idle
A1	BRN	Injector 1	2.0-3.3 ms
A2	RED	Injector 2	2.0-3.3 ms
A3	BLU	Injector 3	2.0-3.3 ms
A4 ('97)	---	Not Used	---
A4 ('98-'99)	BLK/YEL	EVAP Canister Vent Valve	Solenoid on: 1v, off: 12v
A5	GRN/BLK	Fuel Pump Relay Control 1	Relay Off: 12v, On: 1v
A6	GRN	EGR Solenoid Control	Solenoid Off: 12v, On: 1v
A7	BLU	Malfunction Indicator Lamp	MIL Off: 12v, On: 1v
A8	RED/BLU	A/C Clutch Relay Control	Relay Off: 12v, On: 1v
A9	GRY	HO2S-11 (B1 S1) Heater	Digital Signals: 0-12-0v
A10	LT GRN	HO2S-21 (B2 S1) Heater	Digital Signals: 0-12-0v
A11	PNK	Ignition Coil 1 Control	Digital Signals: 0-12-0v
A12	BLK	Power Ground	<0.1v
A13	YEL/BLK	Main Relay Power (B+)	12-14v
A14	YEL	Injector 4	2.0-3.3 ms
A15	BLK/RED	Injector 5	2.0-3.3 ms
A16	WHT/BLU	Injector 6	2.0-3.3 ms
A17 ('97)	---	Not Used	---
A17 ('98-'99)	BLK/ORN	EVAP Bypass Solenoid Valve	Solenoid on: 1v, off: 12v
A18	BLU/YEL	Front VTEC Solenoid	Idle: 0v, Hi-Speed: 12v
A19	GRN/YEL	Rear VTEC Solenoid	Idle: 0v, Hi-Speed: 12v
A20	PNK/BLU	IAB Solenoid Control	Solenoid Off: 12v, On: 1v
A21	RED	EVAP Purge Solenoid	Pulse Signals
A22	GRN/RED	HO2S-12 (B1 S2) Heater	Digital Signals: 0-12-0v
A23	ORN/BLK	HO2S-22 Heater Control	Digital Signals: 0-12-0v
A24	BRN	Ignition Coil 2 Control	Digital Signals: 0-12-0v
A25	BLK	Power Ground	<0.1v
A26	BRN/BLK	Logic Ground	<0.050v

1997-99 Coupe 3.2L V6 VTEC VIN NA2 [A/T] 16P 'B' Connector

PCM Pin #	W/Color	Circuit Description (16-Pin)	Value at Hot Idle
B1	BLU	Front VTEC Pressure Switch	Switch on: 0v, off: 12v
B2	RED	Ignition Coil 6 Control	Digital Signals: 0-12-0v
B3	BLU/BLK	Rear VTEC Pressure Switch	At Low Speed: 0.1v
B4	RED	A/T: Gear Position Switch	P/N: 0v, gear: <5v
B5	ORN	CYP Sensor 2 Signal	AC pulse signals
B6	WHT	CYP Sensor 1 Signal	AC pulse signals
B7	ORN/BLU	CKP Sensor 2 Signal	AC pulse signals
B8	BLU/GRN	CKP Sensor 1 Signal	AC pulse signals
B9	BRN/BLK	Logic Ground	<0.050v
B10	GRY	Ignition Coil 5 Control	Digital Signals: 0-12-0v
B11	GRN	Ignition Coil 4 Control	Digital Signals: 0-12-0v
B12	BLU	Ignition Coil 3 Control	Digital Signals: 0-12-0v
B13	ORN/BLU	CYP Sensor 2 Ground	<0.050v
B14	WHT/BLU	CYP Sensor 1 Ground	<0.050v
B15	WHT/BLU	CKP Sensor 2 Ground	<0.050v
B16	BLU/YEL	CKP Sensor 1 Ground	<0.050v

1997-99 Coupe 3.2L V6 VTEC VIN NA2 [A/T] 12P 'C' Connector

PCM Pin #	W/Color	Circuit Description (12-Pin)	Value at Hot Idle
C1	YEL/BLK	Main Relay Power (B+)	12-14v
C2	BLU/BLK	A/C Switch Signal	A/C Off: 12v, On: 1v
C3	RED/GRN	A/C Pressure Switch Signal	A/C Off: 4-5v, On: 0v
C5	BLU	Service Check Connector	SCS Open: 4.80v
C6	BLK/WHT	Starter Switch Signal	Cranking: 9-11v
C7	ORN	Vehicle Speed Sensor	Moving: pulse signals
C8	GRN	Engine Speed Pulse Signal	Pulse Signals
C9	GRN/YEL	Front Peak Hold Reset	Pulse Signals
C10	GRN/BLK	Rear Peak Hold Reset	Pulse Signals
C11	YEL/RED	Front Misfire Pulse	Pulse Signals
C12	YEL	Rear Misfire Pulse	Pulse Signals

1997-99 Coupe 3.2L V6 VTEC VIN NA2 [A/T] 22P 'D' Connector

PCM Pin #	W/Color	Circuit Description (22-Pin)	Value at Hot Idle
D1	WHT/YEL	Keep Alive Power (VBU)	12-14v
D2	RED/BLU	Knock Sensor 1 Signal	No detonation: 18mv AC
D3 ('97)	BRN	EVAP Purge Flow Switch	Switch on: 0v, off: 12v
D4	YEL/GRN	K-Line Signal	12v
D5	WHT/RED	Alternator 'FR' Signal	Digital Signals: 0-5-0v
D6 ('97)	---	Not Used	---
D6 ('98-'99)	BLU	Fuel Tank Pressure Sensor	Fuel Cap off: 2.5v
D7	RED/WHT	ECT Sensor Signal	At 180°F: 0.5-0.6v
D8	RED/YEL	IAT Sensor Signal	At 100°F: 2-3v
D9	WHT/YEL	MAP Sensor Signal	0.8-0.9v
D10	YEL/BLU	Sensor VREF	4.9-5.1v
D11	GRN/BLU	Sensor Ground	<0.050v
D12	RED/BLK	TP Sensor Signal	0.5-0.6v
D13	WHT	Knock Sensor 2 Signal	No detonation: 18mv AC
D14-15	---	Not Used	---
D16	GRN	HO2S-21 (B2 S1) Signal	0.1-1.1v
D17	WHT/RED	HO2S-22 (B2 S2) Signal	0.1-1.1v
D18	BLU/RED	HO2S-11 (B1 S1) Signal	0.1-1.1v
D19	WHT	HO2S-12 (B1 S2) Signal	0.1-1.1v
D20	WHT/BLK	EGR Valve Lift Sensor	1.1-1.2v
D21	BLU/GRN	HO2S-21 (B2 S1) Ground	<0.050v
D22	WHT	HO2S-22 (B2 S2) Ground	<0.050v

Standard Colors and Abbreviations

Abbreviation	Color	Abbreviation	Color	Abbreviation	Color
BLK	Black	LT BLU	LT Blue	TAN	Tan
BLU	Blue	LT GRN	LT Green	VIO	Violet
BRN	Brown	ORN	Orange	WHT	White
GRY	Gray	PNK	Pink	YEL	Yellow
GRN	Green	PPL	Purple		

1997-99 Coupe 3.2L V6 VTEC VIN NA2 [A/T] 26P 'F' Connector

PCM Pin #	W/Color	Circuit Description (26-Pin)	Value at Hot Idle
F1	YEL/WHT	Sensor VREF	4.9-5.1v
F2	GRN/WHT	Sensor Ground	<0.050v
F3 ('97)	---	Not Used	---
F3 ('98-'99)	BRN	INOCD Immobilizer Code	Digital Signals
F4	RED	Fuel Pump Relay Control	Relay Off: 12v, On: 1v
F5, 17, 19-20	---	Not Used	---
F6	GRY	Brake Switch Signal 2	B/P depressed: 0v
F7	GRN/BLK	Cruise Resume Switch	CR/S released: 0v
F8	GRN/RED	Cruise Control Set Switch	CS/S released: 0v
F9	LT GRN	Cruise Control Main Switch	M/S on: 12v, off: 0v
F10	BLU/ORN	C/C A/T: Gear Position Switch	In 'D', D3, D2: 0v
F11	BLU/BLK	Cruise Indicator Light	Light on: 0v, off: 12v
F12	YEL/RED	Accelerator Position Sensor 2	ACP closed: 0.5v
F13	BLU/RED	Accelerator Position Sensor 1	ACP closed: 0.5v
F14	YEL/WHT	Sensor VREF	4.9-5.1v
F15	GRN/WHT	Sensor Ground	<0.050v
F16	GRN/WHT	Brake Switch Signal 1	Brake Off: 12v, On: 0v
F18	BLU	TCM VREF	4.9-5.1v
F21	GRN/YEL	A/T: FI Data Line 'B'	Pulses
F22	WHT/YEL	A/T: FI Data Line 'A'	Pulses
F24	RED/BLU	TCS Data Line	Pulses

1997-99 Coupe 3.2L V6 VTEC VIN NA2 [A/T] 12P 'G' Connector

PCM Pin #	W/Color	Circuit Description (12-Pin)	Value at Hot Idle
G2, G3	GRN/RED	IGM1, IGM2 Power Source	Key on: 12v, off: 0v
G4	BRN	Motor Phase Out 1	Key on: 0v or pulse
G5	WHT/GRN	Motor Phase Out 2	Key on: 0v or pulse
G7, G8	BLK	Power Ground	<0.1v
G9	ORN	PCM PWR to Motor 1, 3	KOEO: pulse
G10	GRN	PCM PWR to Motor 2, 4	KOEO: pulse
G11	YEL	Motor Phase Out 3	Key on: 0v or pulse
G12	RED	Motor Phase Out 4	Key on: 0v or pulse

WIRE SIDE OF HARNESS TERMINALS

05533_ADIA_G589

Pin Connector Graphic

1997-99 Coupe 3.2L V6 VTEC VIN NA2 [M/T] 26P 'A' Connector

PCM Pin #	W/Color	Circuit Description (26-Pin)	Value at Hot Idle
A1	BRN	Injector 1	2.0-3.3 ms
A2	RED	Injector 2	2.0-3.3 ms
A3	BLU	Injector 3	2.0-3.3 ms
A4 ('97)	---	Not Used	---
A4 ('98-'99)	BLK/YEL	EVAP Canister Vent Valve	Solenoid on: 1v, off: 12v
A5	GRN/BLK	Fuel Pump Relay Control 1	Relay Off: 12v, On: 1v
A6	GRN	EGR Solenoid Control	Solenoid Off: 12v, On: 1v
A7	BLU	Malfunction Indicator Lamp	MIL Off: 12v, On: 1v
A8	RED/BLU	A/C Clutch Relay Control	Relay Off: 12v, On: 1v
A9	GRY	HO2S-11 (B1 S1) Heater	Digital Signals: 0-12-0v
A10	LT GRN	HO2S-12 (B1 S2) Heater	Digital Signals: 0-12-0v
A11	PNK	Ignition Coil 1	Digital Signals: 0-12-0v
A12	BLK	Power Ground	<0.1v
A13	YEL/BLK	Main Relay Power (B+)	12-14v
A14	YEL	Injector 4	2.0-3.3 ms
A15	BLK/RED	Injector 5	2.0-3.3 ms
A16	WHT/BLU	Injector 6	2.0-3.3 ms
A17 ('97)	---	Not Used	---
A17 ('98-'99)	BLK/ORN	EVAP Bypass Solenoid Valve	Solenoid on: 1v, off: 12v
A18	BLU/YEL	Front VTEC Solenoid	Idle: 0v, Hi-Speed: 12v
A19	GRN/YEL	Rear VTEC Solenoid	Idle: 0v, Hi-Speed: 12v
A20	PNK/BLU	IAB Solenoid Control	Solenoid Off: 12v, On: 1v
A21	RED/YEL	EVAP Purge Solenoid	Pulse Signals
A22	GRN/RED	HO2S-11 (B1 S1) Heater	Digital Signals: 0-12-0v
A23	ORN/BLK	HO2S-12 (B1 S2) Heater	Digital Signals: 0-12-0v
A24	BRN	Ignition Coil 2	Digital Signals: 0-12-0v
A25	BLK	Power Ground	<0.1v
A26	BRN/BLK	Logic Ground	<0.050v

1997-99 Coupe 3.2L V6 VTEC VIN NA2 [M/T] 16P 'B' Connector

PCM Pin #	W/Color	Circuit Description (16-Pin)	Value at Hot Idle
B1	BLU	Front VTEC Pressure Switch	Switch on: 0v, off: 12v
B2	RED	Ignition Coil 6 Control	Digital Signals: 0-12-0v
B3	BLU/BLK	Rear VTEC Pressure Switch	At Low Speed: 0.1v
B4	LT GRN	Neutral Switch Signal	In 'N': 0v, Others: 12v
B5	ORN	CYP Sensor Signal 2	AC pulse signals
B6	WHT	CYP Sensor Signal 1	AC pulse signals
B7	ORN/BLU	CKP Sensor 2 Signal	AC pulse signals
B8	BLU/GRN	CKP Sensor 1 Signal	AC pulse signals
B9	BRN/BLK	Logic Ground	<0.050v
B10	GRY	Ignition Coil 5	Digital Signals: 0-12-0v
B11	GRN	Ignition Coil 4	Digital Signals: 0-12-0v
B12	BLU	Ignition Coil 3	Digital Signals: 0-12-0v
B13	ORN/BLU	CYP Sensor 2 Ground	<0.050v
B14	WHT/BLU	CYP Sensor 1 Ground	<0.050v
B15	WHT/BLU	CKP Sensor 2 Ground	<0.050v
B16	BLU/YEL	CKP Sensor 1 Ground	<0.050v

1997-99 Coupe 3.2L V6 VTEC VIN NA2 [M/T] 12P 'C' Connector

PCM Pin #	W/Color	Circuit Description (12-Pin)	Value at Hot Idle
C1	YEL/BLK	Main Relay Power (B+)	12-14v
C2	BLU/BLK	A/C Switch Signal	A/C Off: 12v, On: 1v
C3	RED/GRN	A/C Pressure Switch Signal	A/C Off: 4-5v, On: 0v
C4	PNK	Clutch Switch Signal	With Clutch In: 0v
C5	BLU	Service Check Connector	SCS Open: 4.80v
C6	BLK/WHT	Starter Switch Signal	Cranking: 9-11v
C7	ORN	Vehicle Speed Sensor	Moving: pulse signals
C8	GRN	Engine Speed Pulse Signal	Pulses
C9	GRN/YEL	Front Peak Hold Reset	Pulses
C10	GRN/BLK	Rear Peak Hold Reset	Pulses
C11	YEL/RED	Front Misfire Pulse	Pulses
C12	YEL	Rear Misfire Pulse	Pulses

1997-99 Coupe 3.2L V6 VTEC VIN NA2 [M/T] 22P 'D' Connector

PCM Pin #	W/Color	Circuit Description (22-Pin)	Value at Hot Idle
D1	WHT/YEL	Keep Alive Power (VBU)	12-14v
D2	RED/BLU	Knock Sensor 1 Signal	No detonation: 18mv AC
D3 ('97)	BRN	EVAP Purge Flow Switch	Switch on: 0v, off: 12v
D4	YEL/GRN	K-Line Signal	12v
D5	WHT/RED	Alternator 'FR' Signal	Digital Signals: 0-5-0v
D6 ('97)	---	Not Used	---
D6 ('98-'99)	BLU	Fuel Tank Pressure Sensor	Fuel Cap off: 2.5v
D7	RED/WHT	ECT Sensor Signal	At 180°F: 0.5-0.6v
D8	RED/YEL	IAT Sensor Signal	At 100°F: 2-3v
D9	WHT/YEL	MAP Sensor Signal	0.8-0.9v
D10	YEL/BLU	Sensor VREF	4.9-5.1v
D11	GRN/BLU	Sensor Ground	<0.050v
D12	RED/BLK	TP Sensor Signal	0.5-0.6v
D13	WHT	Knock Sensor 2 Signal	No detonation: 18mv AC
D14	---	Not Used	---
D15	---	Not Used	---
D16	GRN	HO2S-21 (B2 S1) Signal	0.1-1.1v
D17	WHT/RED	HO2S-22 (B2 S2) Signal	0.1-1.1v
D18	BLU/RED	HO2S-11 (B1 S1) Signal	0.1-1.1v
D19	WHT	Rear HO2S-11 (B1 S1) Signal	0.1-1.1v
D20	WHT/BLK	EGR Valve Lift Sensor	1.1-1.2v
D21	BLU/GRN	HO2S-21 (B2 S1) Ground	<0.050v
D22	WHT	HO2S-22 (B2 S2) Ground	<0.050v

1997-99 Coupe 3.2L V6 VTEC VIN NA2 [M/T] 26P 'F' Connector

PCM Pin #	W/Color	Circuit Description (26-Pin)	Value at Hot Idle
F1	YEL/WHT	Sensor VREF (VCC1)	4.9-5.1v
F2	GRN/WHT	Sensor Ground	<0.050v
F3 ('97)	---	Not Used	---
F3 ('98-'99)	BRN	INOCD Immobilizer Code	Digital Signals
F4	RED	Fuel Pump Relay Control	Relay Off: 12v, On: 1v
F5	---	Not Used	---
F6	GRY	Brake Switch Signal 2	B/P depressed: 0v
F7	GRN/BLK	Cruise Resume Switch	CR/S released: 0v
F8	GRN/RED	Cruise Control Set Switch	CS/S released: 0v
F9	LT GRN	Cruise Control Main Switch	M/S on: 12v, off: 0v
F10	BLU/ORN	Cruise Control Clutch Switch	Clutch in: 0v, out: 8v
F11	BLU/BLK	Cruise Indicator Light	Light on: 0v, off: 12v
F12	YEL/RED	ACP Sensor 2 Signal (AP2)	ACP closed: 0.5v
F13	BLU/RED	ACP Sensor 1 Signal (AP1)	ACP closed: 0.5v
F14	YEL/WHT	Sensor VREF (VCC3)	4.9-5.1v
F15	GRN/WHT	Sensor Ground (SG3)	<0.050v
F16	GRN/WHT	Brake Switch Signal 1	Brake Off: 12v, On: 0v
F17-26	---	Not Used	---

1997-99 Coupe 3.2L V6 VTEC VIN NA2 [M/T] 12P 'G' Connector

PCM Pin #	W/Color	Circuit Description (12-Pin)	Value at Hot Idle
G1	---	Not Used	---
G2	GRN/RED	IGM1 Power Source	12-14v
G3	GRN/RED	IGM2 Power Source	12-14v
G4	BRN	Motor Phase Out 1	Pulse Signals
G5	WHT/GRN	Motor Phase Out 2	Pulse Signals
G7	BLK	Power Ground	<0.1v
G8	BLK	Power Ground	<0.1v
G9	ORN	PCM PWR to Motor 1, 3	Pulse Signals
G10	GRN	PCM PWR to Motor 2, 4	Pulse Signals
G11	YEL	Motor Phase Out 3	Pulse Signals
G12	RED	Motor Phase Out 4	Pulse Signals

05533_ADIA_G589

Pin Connector Graphic

2000-05 Coupe 3.2L V6 VTEC VIN NA2 [All] 26P 'A' Connector

PCM Pin #	W/Color	Circuit Description (26-Pin)	Value at Hot Idle
A1	BRN	Injector 1 Control	2.0-3.3 ms
A2	RED	Injector 2 Control	2.0-3.3 ms
A3	BLU	Injector 3 Control	2.0-3.3 ms
A4	BLK/YEL	EVAP Canister Vent Valve	Solenoid on: 1v, off: 12v
A5	GRN/BLK	Fuel Pump Relay Control 1	Relay Off: 12v, On: 1v
A6	GRN	EGR Solenoid Control	Solenoid Off: 12v, On: 1v
A7	BLU	Malfunction Indicator Lamp	MIL Off: 12v, On: 1v
A8	RED/BLU	A/C Clutch Relay Control	Relay Off: 12v, On: 1v
A9	GRY	HO2S-11 (B1 S1) Heater	Digital Signals: 0-12-0v
A10	LT GRN	HO2S-21 (B2 S1) Heater	Digital Signals: 0-12-0v
A11	PNK	Ignition Coil 1 Control	Digital Signals: 0-12-0v
A12	BLK	Power Ground (PG1)	<0.1v
A13	YEL/BLK	Main Relay Power (B+)	12-14v
A14	YEL	Injector 4 Control	2.0-3.3 ms
A15	BLK/RED	Injector 5 Control	2.0-3.3 ms
A16	WHT/BLU	Injector 6 Control	2.0-3.3 ms
A17	ORN	EVAP Bypass Solenoid Valve	Solenoid on: 1v, off: 12v
A18	YEL/GRN	Front VTEC Solenoid	Idle: 0v, Hi-Speed: 12v
A19	---	Not Used	---
A20	PNK/BLU	IAB Solenoid Control	Solenoid Off: 12v, On: 1v
A21	RED/YEL	EVAP Purge Solenoid	Pulse Signals
A22	GRN/RED	HO2S-12 (B1 S2) Heater	Digital Signals: 0-12-0v
A23	ORN/BLK	HO2S-22 Heater Control	Digital Signals: 0-12-0v
A24	BRN	Ignition Coil 2 Control	Digital Signals: 0-12-0v
A25	BLK	Power Ground (PG2)	<0.1v
A26	BRN/BLK	Logic Ground (LG1)	<0.050v

2000-05 Coupe 3.2L V6 VTEC VIN NA2 [All] 16P 'B' Connector

PCM Pin #	W/Color	Circuit Description (16-Pin)	Value at Hot Idle
B1	BLU	Front VTEC Pressure Switch	Switch on: 0v, off: 12v
B2	RED	Ignition Coil 6 Control	Digital Signals: 0-12-0v
B3	BLU/BLK	Rear VTEC Pressure Switch	At Low Speed: 0.1v
B4	RED	A/T: Gear Position Switch	In P/N: 0v, others: 5v
B4	LT GRN	M/T: Neutral Position Switch	In 'N': 0v, others: 5v
B5	ORN	CYP Sensor 2 Signal	AC pulse signals
B6	WHT	CYP Sensor 1 Signal	AC pulse signals
B7	ORN/BLU	CKP Sensor 2 Signal	AC pulse signals
B8	BLU/GRN	CKP Sensor 1 Signal	AC pulse signals
B9	BRN/BLK	Logic Ground	<0.050v
B10	GRY	Ignition Coil 5 Control	Digital Signals: 0-12-0v
B11	GRN	Ignition Coil 4 Control	Digital Signals: 0-12-0v
B12	BLU	Ignition Coil 3 Control	Digital Signals: 0-12-0v

2000-05 Coupe 3.2L V6 VTEC VIN NA2 [All] 16P 'B' Connector, *continued*

B13	ORN/BLU	CYP Sensor 2 Ground	<0.050v
B14	WHT/BLU	CYP Sensor 1 Ground	<0.050v
B15	WHT/BLU	CKP Sensor 2 Ground	<0.050v
B16	BLU/YEL	CKP Sensor 1 Ground	<0.050v

2000-05 Coupe 3.2L V6 VTEC VIN NA2 [All] 12P 'C' Connector

PCM Pin #	W/Color	Circuit Description (12-Pin)	Value at Hot Idle
C1	YEL/BLK	Main Relay Power (B+)	12-14v
C2	BLU/BLK	A/C Switch Signal	A/C Off: 12v, On: 1v
C3	RED/GRN	A/C Pressure Switch Signal	A/C Off: 4-5v, On: 0v
C4	PNK	M/T: Clutch Engage Switch	Clutch in: 5v, out: 0v
C5	BLU	Service Check Connector	SCS Open: 4.80v
C6	BLK/WHT	Starter Switch Signal	Cranking: 9-11v
C7	ORN	Vehicle Speed Sensor	Moving: pulse signals
C8	GRN	Engine Speed Pulse Signal	Pulses
C9	GRN/YEL	Front Peak Hold Reset	Pulses
C10	GRN/BLK	Rear Peak Hold Reset	Pulses
C11	YEL/RED	Front Misfire Pulse	Pulses
C12	YEL	Rear Misfire Pulse	Pulses

2000-05 Coupe 3.2L V6 VTEC VIN NA2 [All] 22P 'D' Connector

PCM Pin #	W/Color	Circuit Description (22-Pin)	Value at Hot Idle
D1	WHT/YEL	Keep Alive Power (VBU)	12-14v
D2	RED/BLU	Knock Sensor 1 Signal	No detonation: 18mv AC
D3	WHT/BLK	Air Pump Current Sensor	Pump on: Pulse Signals
D4	YEL/GRN	K-Line Signal	12v
D5	WHT/RED	Alternator 'FR' Signal	Digital Signals: 0-5-0v
D6	BLU	Fuel Tank Pressure Sensor	Fuel Cap off: 2.5v
D7	RED/WHT	ECT Sensor Signal	At 180°F: 0.5-0.6v
D8	RED/YEL	IAT Sensor Signal	At 100°F: 2-3v
D9	WHT/YEL	MAP Sensor Signal	0.8-0.9v
D10	YEL/BLU	Sensor VREF (VCC2)	4.9-5.1v
D11	GRN/BLU	Sensor Ground	<0.050v
D12	RED/BLK	TP Sensor Signal	0.5-0.6v
D13	WHT	Knock Sensor 2 Signal	No detonation: 18mv AC
D14-15	---	Not Used	---
D16	GRN	HO2S-21 (B2 S1) Signal	0.1-1.1v
D17	WHT/RED	HO2S-22 (B2 S2) Signal	0.1-1.1v
D18	BLU/RED	HO2S-11 (B1 S1) Signal	0.1-1.1v
D19	WHT	HO2S-12 (B1 S2) Signal	0.1-1.1v
D20	WHT/BLK	EGR Valve Lift Sensor	1.1-1.2v
D21	BLU/GRN	HO2S-21 (B2 S1) Ground	<0.050v
D22	WHT	HO2S-22 (B2 S2) Ground	<0.050v

Standard Colors and Abbreviations

Abbreviation	Color	Abbreviation	Color	Abbreviation	Color
BLK	Black	LT BLU	LT Blue	TAN	Tan
BLU	Blue	LT GRN	LT Green	VIO	Violet
BRN	Brown	ORN	Orange	WHT	White
GRY	Gray	PNK	Pink	YEL	Yellow
GRN	Green	PPL	Purple		

2000-05 Coupe 3.2L V6 VTEC VIN NA2 [All] 26P 'F' Connector

PCM Pin #	W/Color	Circuit Description (26-Pin)	Value at Hot Idle
F1	YEL/WHT	Sensor VREF (VCC1)	4.9-5.1v
F2, F15	GRN/WHT	Sensor Ground SG1, SG3	<0.050v
F3	BRN	INOCD Immobilizer Code	Digital Signals
F4	RED	Fuel Pump Relay Control	Relay Off: 12v, On: 1v
F6	GRY	Brake Switch Signal 2	B/P depressed: 0v
F7	GRN/BLK	Cruise Resume Switch	CR/S released: 0v
F8	GRN/RED	Cruise Control Set Switch	CS/S released: 0v
F9	LT GRN	Cruise Control Main Switch	M/S on: 12v, off: 0v
F10	BLU/ORN	C/C A/T: Gear Position Switch	In 'D', D3, D2: 0v
F10	BLU/ORN	C/C M/T: Clutch Switch	Clutch in: 0v, out: 8v
F11	BLU/BLK	Cruise Indicator Light	Light on: 0v, off: 12v
F12	YEL/RED	Accelerator Position Sensor 2	ACP closed: 0.5v
F13	BLU/RED	Accelerator Position Sensor 1	ACP closed: 0.5v
F14	YEL/WHT	Sensor VREF (VCC1)	4.9-5.1v
F16	GRN/WHT	Brake Switch Signal 1	Brake Off: 12v, On: 0v
F17	RED/WHT	Reverse Lockout Relay Out	VSS over 30 mph: 0v
F18	BLU	TCM VREF	4.9-5.1v
F19	BLU/YEL	Front VTEC Solenoid Valve	0v, Hi-Speed: 12v
F20	GRY/YEL	Rear VTEC Solenoid Valve	0v, Hi-Speed: 12v
F21	GRN/YEL	A/T: FI Data Line 'B'	Pulse Signals
F22	WHT/YEL	A/T: FI Data Line 'A'	Pulse Signals
F23	BLU/WHT	Air Pump Relay Control	Relay Off: 12v, On: 1v
F24	RED/BLU	TCS Data Line	Pulse Signals
F25	RED	Air Control Solenoid Valve	Pump on: 1v, off: 12v

2000-05 Coupe 3.2L V6 VTEC VIN NA2 [All] 12P 'G' Connector

PCM Pin #	W/Color	Circuit Description (12-Pin)	Value at Hot Idle
G1, G9	---	Not Used	---
G2, G3	GRN/RED	IGM1, IGM2 Power Source	Key on: 12v, off: 0v
G4	BRN	Motor Phase Out 1	Key on: 0v or pulse
G5	WHT/GRN	Motor Phase Out 2	Key on: 0v or pulse
G7, G8	BLK	Power Ground	<0.1v
G9	ORN	PCM PWR to Motor 1, 3	KOEO: pulse
G10	GRN	PCM PWR to Motor 2, 4	KOEO: pulse
G11	YEL	Motor Phase Out 3	Key on: 0v or pulse
G12	RED	Motor Phase Out 4	Key on: 0v or pulse

05533_ADIA_G589

Pin Connector Graphic

1992 Sedan 2.5L I5 MFI SOHC VIN CC2 [A/T] 26P 'A' Connector

PCM Pin #	W/Color	Circuit Description (26-Pin)	Value at Hot Idle
A1	BRN	Injector 1	2.0-3.3 ms
A2	YEL	Injector 4	2.0-3.3 ms
A3	RED	Injector 2	2.0-3.3 ms
A4	GRN	Injector 5	2.0-3.3 ms
A5	LT BLU	Injector 3	2.0-3.3 ms
A6	PNK/WHT	HO2S-11 (B1 S1) Heater	Heater Off: 12v, On: 1v
A7	GRN/BLK	Fuel Pump Relay Control	Relay Off: 12v, On: 1v
A8	GRN/BLK	Fuel Pump Relay Control	Relay Off: 12v, On: 1v
A9	BLK/BLU	Electronic Air Control Valve	Pulse Signals
A10	ORN	EVAP Purge Solenoid	Solenoid Off: 12v, On: 1v
A11	RED	EGR Solenoid Control	Solenoid Off: 12v, On: 1v
A12	GRN/YEL	Fan Timing Unit Signal	Relay Off: 12v, On: 1v
A13	GRN/RED	Malfunction Indicator Lamp	MIL Off: 12v, On: 1v
A15	RED/BLU	A/C Clutch Relay Control	Relay Off: 12v, On: 1v
A16	WHT/GRN	Alternator Control Signal	Lights Off: 12v, On: 1v
A17	BLK/RED	Bypass Solenoid Signal	Solenoid Off: 12v, On: 1v
A18	PNK	A/T: Shift Acknowledge Signal	Digital Signals
A21	YEL/GRN	Igniter 1 Signal	Digital Signals: 0-12-0v
A22	YEL/GRN	Igniter 2 Signal	Digital Signals: 0-12-0v
A23	BLK	Power Ground	<0.1v
A24	BLK	Power Ground	<0.1v
A25	YEL/BLK	Main Relay Power (B+)	12-14v
A26	BLK/RED	Logic Ground	<0.050v

1992 Sedan 2.5L I5 MFI SOHC VIN CC2 [A/T] 16P 'B' Connector

PCM Pin #	W/Color	Circuit Description (16-Pin)	Value at Hot Idle
B1	YEL/BLK	Main Relay Power (B+)	12-14v
B2	BRN/BLK	Logic Ground	<0.050v
B3	WHT/GRN	A/T: Upshift/Downshift Signal	Digital Signals
B4	WHT/RED	A/T: Upshift/Downshift Signal	Digital Signals
B5	BLU/BLK	A/C Pressure Switch	A/C Off: 12v, On: 1v
B6	---	Not Used	---
B7	YEL/GRN	A/T: Neutral Indicator	In 'N': 0v, Others: 12v
B8	RED	PSP Switch Signal	Straight: 0v, Turning: 12v
B9	BLU/RED	Start (Cranking) Signal	Cranking: 9-11v
B10	ORN	Vehicle Speed Sensor	Moving: pulse signals
B11	ORN	CYP Sensor Signal	AC pulse signals
B12	WHT	CYP Sensor Ground	<0.050v
B13	ORN/BLU	TDC Sensor Signal	AC pulse signals
B14	WHT/BLU	TDC Sensor Ground	<0.050v
B15	YEL/GRN	CKP Sensor Signal	AC pulse signals
B16	BLU/YEL	CKP Sensor Ground	<0.050v

1992 Sedan 2.5L I5 MFI SOHC VIN CC2 [A/T] 12P 'C' Connector

PCM Pin #	W/Color	Circuit Description (12-Pin)	Value at Hot Idle
C1	ORN/WHT	Rear Knock Sensor Signal	No Detonation: 18mv AC
C2	---	Not Used	---
C3	RED/BLU	Front Knock Sensor Signal	No Detonation: 18mv AC
C4-12	---	Not Used	---

1992 Sedan 2.5L I5 MFI SOHC VIN CC2 [A/T] 22P 'D' Connector

PCM Pin #	W/Color	Circuit Description (22-Pin)	Value at Hot Idle
D1	WHT/GRN	Keep Alive Power (VBU)	12-14v
D2	GRN/WHT	Brake Switch Signal	Brake Off: 12v, On: 0v
D3	---	Not Used	---
D4	BRN	Service Check Connector	SCS Open: 4.80v
D5-7	---	Not Used	---
D8	BRN	Ignition Timing Adjustment	0.4-4.5v
D9	WHT/RED	Alternator 'FR' Signal	Digital Signals: 0-5-0v
D10	GRN/RED	Electronic Load Detector	Varies w/Load: 2.5-3.5v
D11	RED/YEL	Throttle Angle Sensor	0.5-0.6v
D12	WHT/BLK	EGR Valve Lift Sensor	1.1-1.2v
D13	YEL/GRN	ECT Sensor Signal	At 180°F: 0.5-0.6v
D14	WHT	HO2S Sensor	0.1-1.1v
D15	WHT/YEL	IAT Sensor Signal	At 100°F: 2-3v
D16	---	Not Used	---
D17	WHT/BLU	MAP Sensor Signal	0.8-0.9v
D18	BLU/WHT	A/T: Control Unit VREF	4.9-5.1v
D19	YEL/WHT	MAP Sensor VREF	4.9-5.1v
D20	YEL/WHT	Sensor VREF	4.9-5.1v
D21	GRN/WHT	MAP Sensor Ground	<0.050v
D22	GRN/WHT	Sensor Ground	<0.050v

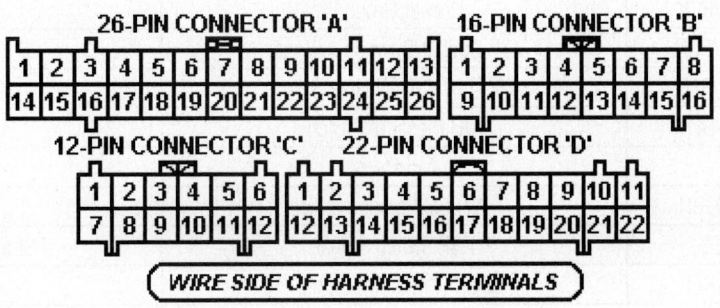

Pin Connector Graphic

05533_ADIA_G590

1993 Sedan 2.5L I5 MFI SOHC VIN CC2 [A/T] 26P 'A' Connector

PCM Pin #	W/Color	Circuit Description (26-Pin)	Value at Hot Idle
A1	BRN	Injector 1	2.0-3.3 ms
A2	YEL	Injector 4	2.0-3.3 ms
A3	RED	Injector 2	2.0-3.3 ms
A4	GRN	Injector 5	2.0-3.3 ms
A5	BLU	Injector 3	2.0-3.3 ms
A6	PNK/WHT	HO2S-11 (B1 S1) Heater	Heater Off: 12v, On: 1v
A7	GRN/BLK	Fuel Pump Relay Control	Relay Off: 12v, On: 1v
A8	GRN/BLK	Fuel Pump Relay Control	Relay Off: 12v, On: 1v
A9	BLK/BLU	Idle Air Control Valve	Pulse Signals
A10	ORN	EVAP Purge Solenoid	Solenoid Off: 12v, On: 1v
A11	RED	EGR Solenoid Control	Solenoid Off: 12v, On: 1v
A12	GRN/YEL	Fan Timing Unit Signal	Relay Off: 12v, On: 1v
A13	GRN/RED	Malfunction Indicator Lamp	MIL Off: 12v, On: 1v
A15	RED/BLU	A/C Clutch Relay Control	Relay Off: 12v, On: 1v
A16	WHT/GRN	Alternator Control Signal	Lights Off: 12v, On: 1v
A17	BLU/RED	Bypass Solenoid Signal	Solenoid Off: 12v, On: 1v
A18	PNK	A/T: Shift Acknowledge Signal	Digital Signals
A19-20	BLK/RED	Power Ground	<0.1v
A21	YEL/GRN	Igniter 1 Signal	Digital Signals: 0-12-0v
A22	YEL/GRN	Igniter 2 Signal	Digital Signals: 0-12-0v
A23	BLK	Power Ground	<0.1v
A24	BLK	Power Ground	<0.1v
A25	YEL/BLK	Main Relay Power (B+)	12-14v
A26	BLK/RED	Chassis Ground	<0.050v

1993 Sedan 2.5L I5 MFI SOHC VIN CC2 [A/T] 16P 'B' Connector

PCM Pin #	W/Color	Circuit Description (16-Pin)	Value at Hot Idle
B1	YEL/BLK	Main Relay Power (B+)	12-14v
B2	BRN/BLK	Power Ground	<0.1v
B3	WHT/GRN	A/T: Upshift/Downshift Signal	Digital Signals
B4	WHT/RED	A/T: Upshift/Downshift Signal	Digital Signals
B5	BLU/BLK	A/C Pressure Switch	A/C Off: 12v, On: 1v
B6	---	Not Used	---
B7	YEL/GRN	A/T: Neutral Position Indicator	In P/N: 0v, Others: 12v
B8	RED	PSP Switch Signal	Straight: 0v, Turning: 12v
B9	BLU/RED	Start (Cranking) Signal	Cranking: 9-11v
B10	ORN	Vehicle Speed Sensor	Moving: pulse signals
B11	ORN	CYP Sensor Signal	AC pulse signals
B12	WHT	CYP Sensor Ground	<0.050v
B13	ORN/BLU	TDC Sensor Signal	AC pulse signals
B14	WHT/BLU	TDC Sensor Ground	<0.050v
B15	YEL/GRN	CKP Sensor Signal	AC pulse signals
B16	BLU/YEL	CKP Sensor Ground	<0.050v

1993 Sedan 2.5L I5 MFI SOHC VIN CC2 [A/T] 12P 'C' Connector

PCM Pin #	W/Color	Circuit Description (12-Pin)	Value at Hot Idle
C1	ORN/WHT	Rear Knock Sensor Signal	No Detonation: 18mv AC
C2	---	Not Used	---
C3	RED/BLU	Front Knock Sensor Signal	No Detonation: 18mv AC
C4-12	---	Not Used	---

1993 Sedan 2.5L I5 MFI SOHC VIN CC2 [A/T] 22P 'D' Connector

PCM Pin #	W/Color	Circuit Description (22-Pin)	Value at Hot Idle
D1	WHT/GRN	Keep Alive Power (VBU)	12-14v
D2	GRN/WHT	Brake Switch Signal	Brake Off: 12v, On: 0v
D3	---	Not Used	---
D4	BRN	Service Check Connector	SCS Open: 4.80v
D5-7	---	Not Used	---
D8	BRN	Ignition Timing Adjustment	0.4-4.5v
D9	WHT/RED	Alternator Control Signal	Lights Off: 12v, On: 1v
D10	GRN/RED	Electronic Load Detector	Varies: 2.5-3.5v
D11	RED/YEL	TP Sensor Signal	0.5-0.6v
D12	WHT/BLK	EGR Valve Lift Sensor	1.1-1.2v
D13	YEL/GRN	ECT Sensor Signal	At 180°F: 0.5-0.6v
D14	WHT	HO2S-11 (B1 S1) Signal	0.1-1.1v
D15	WHT/YEL	IAT Sensor Signal	At 100°F: 2-3v
D16	---	Not Used	---
D17	WHT/BLU	MAP Sensor Signal	0.8-0.9v
D18	BLU/WHT	A/T: Control Unit VREF	4.9-5.1v
D19	YEL/WHT	MAP Sensor VREF	4.9-5.1v
D20	YEL/WHT	Sensor VREF	4.9-5.1v
D21	GRN/WHT	MAP Sensor Ground	<0.050v
D22	GRN/WHT	Sensor Ground	<0.050v

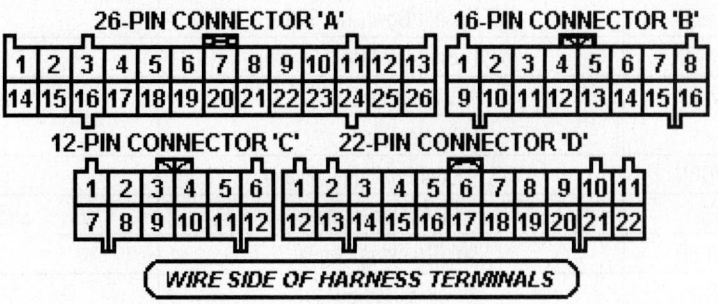

05533_ADIA_G590

Pin Connector Graphic

1994 Sedan 2.5L I5 MFI SOHC VIN CC2 [A/T] 26P 'A' Connector

PCM Pin #	W/Color	Circuit Description (26-Pin)	Value at Hot Idle
A1	BRN	Injector 1	2.0-3.3 ms
A2	YEL	Injector 4	2.0-3.3 ms
A3	RED	Injector 2	2.0-3.3 ms
A4	GRN	Injector 5	2.0-3.3 ms
A5	BLU	Injector 3	2.0-3.3 ms
A6	PNK/WHT	HO2S-11 (B1 S1) Heater	Heater Off: 12v, On: 1v
A7	GRN/BLK	Fuel Pump Relay Control	Relay Off: 12v, On: 1v
A8	GRN/BLK	Fuel Pump Relay Control	Relay Off: 12v, On: 1v
A9	BLK/BLU	Electronic Air Control Valve	Pulse Signals
A10	ORN	EVAP Purge Solenoid	Solenoid Off: 12v, On: 1v
A11	RED	EGR Solenoid Control	Solenoid Off: 12v, On: 1v
A12	GRN/YEL	Fan Timing Unit Signal	Relay Off: 12v, On: 1v
A13	GRN/RED	Malfunction Indicator Lamp	MIL Off: 12v, On: 1v
A15	RED/BLU	A/C Clutch Relay Control	Relay Off: 12v, On: 1v
A16	WHT/GRN	Alternator Control Signal	Lights Off: 12v, On: 1v
A17	BLU/RED	Bypass Solenoid Signal	Solenoid Off: 12v, On: 1v
A18	PNK	A/T: Shift Acknowledge Signal	Digital Signals
A20	GRN	EVAP Purge Solenoid	Solenoid Off: 12v, On: 1v
A21	YEL/GRN	Igniter 1 Signal	Digital Signals: 0-12-0v
A22	YEL/GRN	Igniter 2 Signal	Digital Signals: 0-12-0v
A23	BLK	Power Ground	<0.1v
A24	BLK	Power Ground	<0.1v
A25	YEL/BLK	Main Relay Power (B+)	12-14v
A26	BLK/RED	Chassis Ground	<0.050v

1994 Sedan 2.5L I5 MFI SOHC VIN CC2 [A/T] 16P 'B' Connector

PCM Pin #	W/Color	Circuit Description (16-Pin)	Value at Hot Idle
B1	YEL/BLK	Main Relay Power (B+)	12-14v
B2	BRN/BLK	Power Ground	<0.1v
B3	WHT/GRN	A/T: Upshift/Downshift Signal	Digital Signals
B4	WHT/GRN	A/T: Upshift/Downshift Signal	Digital Signals
B5	BLU/BLK	A/C Pressure Switch	A/C Off: 12v, On: 1v
B6	---	Not Used	---
B7	YEL/GRN	A/T: Neutral Position Indicator	In 'N': 0v, Others: 12v
B8	RED	PSP Switch Signal	Straight: 0v, Turning: 12v
B9	BLU/RED	Start (Cranking) Signal	Cranking: 9-11v
B10	ORN	Vehicle Speed Sensor	Moving: pulse signals
B11	ORN	CYP Sensor Signal	AC pulse signals
B12	WHT	CYP Sensor Ground	<0.050v
B13	ORN/BLU	TDC Sensor Signal	AC pulse signals
B14	WHT/BLU	TDC Sensor Ground	<0.050v
B15	BLU/GRN	CKP Sensor Signal	AC pulse signals
B16	BLU/YEL	CKP Sensor Ground	<0.050v

1994 Sedan 2.5L I5 MFI SOHC VIN CC2 [A/T] 12P 'C' Connector

PCM Pin #	W/Color	Circuit Description (12-Pin)	Value at Hot Idle
C1	ORN/WHT	Rear Knock Sensor Signal	No Detonation: 18mv AC
C2	---	Not Used	---
C3	RED/BLU	Front Knock Sensor Signal	No Detonation: 18mv AC
C4-12	---	Not Used	---

1994 Sedan 2.5L I5 MFI SOHC VIN CC2 [A/T] 22P 'D' Connector

PCM Pin #	W/Color	Circuit Description (22-Pin)	Value at Hot Idle
D1	WHT/GRN	Keep Alive Power (VBU)	12-14v
D2	GRN/WHT	Brake Switch Signal	Brake Off: 12v, On: 0v
D3	---	Not Used	---
D4	BRN	Service Check Connector	SCS Open: 4.80v
D5-7	---	Not Used	---
D8	BRN	Ignition Timing Adjustment	0.4-4.5v
D9	WHT/RED	Alternator Control Signal	Lights Off: 12v, On: 1v
D10	GRN/RED	Electronic Load Detector	Varies: 2.5-3.5v
D11	RED/YEL	TP Sensor Signal	0.5-0.6v
D12	WHT/BLK	EGR Valve Lift Sensor	1.1-1.2v
D13	YEL/GRN	ECT Sensor Signal	At 180°F: 0.5-0.6v
D14	WHT	HO2S-11 (B1 S1) Signal	0.1-1.1v
D15	WHT/YEL	IAT Sensor Signal	At 100°F: 2-3v
D16	---	Not Used	---
D17	WHT/BLU	MAP Sensor Signal	0.8-0.9v
D18	---	A/T: Control Unit VREF	4.9-5.1v
D19	YEL/WHT	MAP Sensor VREF	4.9-5.1v
D20	YEL/WHT	Sensor VREF	4.9-5.1v
D21	GRN/WHT	MAP Sensor Ground	<0.050v
D22	GRN/WHT	Sensor Ground	<0.050v

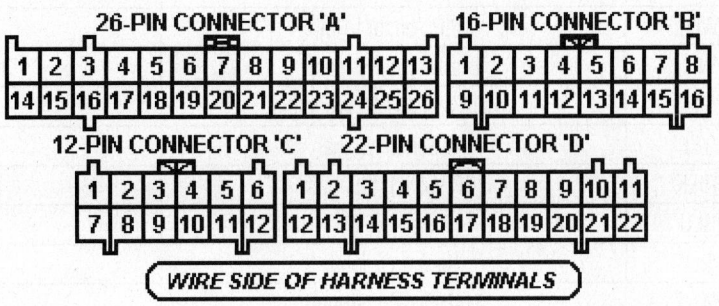

05533_ADIA_G590

Pin Connector Graphic

1997 Coupe 2.2L I4 VTEC VIN YA1 [A/T] 32P 'A' Connector

PCM Pin #	W/Color	Circuit Description (32-Pin)	Value at Hot Idle
A1	YEL	Injector 4	2.0-3.3 ms
A3	BLU	Injector 3	2.0-3.3 ms
A2	RED	Injector 2	2.0-3.3 ms
A1	BRN	Injector 1	2.0-3.3 ms
A5	ORN/BLU	HO2S-12 (B1 S2) Heater	Digital Signals: 0-12-0v
A6	ORN/BLK	HO2S-11 (B1 S1) Heater	Digital Signals: 0-12-0v
A7	RED	EGR Solenoid Control	Solenoid Off: 12v, On: 1v
A8	GRN/YEL	VTEC Solenoid Control	Idle; 0v, hi-rpm: 12v
A9, A22	BRN/BLK	Logic Ground	<0.050v
A10, A23	BLK	Power Ground	<0.1v
A11	YEL/BLK	Main Relay Power (B+)	12-14v
A12	BLK/BLU	Idle Air Control Valve	Pulse Signals
A13	GRN/WHT	Engine Mount Solenoid	Solenoid Off: 12v, On: 1v
A15	RED/YEL	EVAP Purge Solenoid	Pulse Signals
A16	GRN/BLK	Fuel Pump Relay Control	Relay Off: 12v, On: 1v
A17	RED/BLU	A/C Clutch Relay Control	Relay Off: 12v, On: 1v
A18	GRN/RED	Malfunction Indicator Lamp	MIL Off: 12v, On: 1v
A19	WHT/GRN	Alternator Charging Signal	Lights Off: 12v, On: 1v
A20	YEL/GRN	Igniter Signal	Digital Signals: 0-12-0v
A24	YEL/BLK	Main Relay Power (B+)	12-14v
A25	ORN/GRN	IAR Solenoid	Solenoid Off: 12v, On: 1v
A27	GRN	Radiator Fan Relay Control	Fan on: 1v, off: 12v

1997 Coupe 2.2L I4 VTEC VIN YA1 [A/T] 16P 'D' Connector

PCM Pin #	W/Color	Circuit Description (16-Pin)	Value at Hot Idle
D1	RED/BLK	TP Sensor Signal	0.5-0.6v
D2	RED/WHT	ECT Sensor Signal	At 180°F: 0.5-0.6v
D3	WHT/YEL	MAP Sensor Signal	0.8-0.9v
D4	YEL/WHT	MAP Sensor VREF	4.9-5.1v
D5	GRN/WHT	Brake Switch Signal	Brake on: 0v, off: 12v
D7	WHT/RED	HO2S-11 (B1 S1) Signal	0.1-1.1v
D8	RED/YEL	IAT Sensor Signal	At 100°F: 2-3v
D9	WHT/BLK	EGR Valve Lift Sensor	1.1-1.2v
D10	YEL/BLU	Sensor VREF	4.9-5.1v
D11	GRN/BLK	Sensor Ground	<0.050v
D12	GRN/WHT	MAP Sensor Ground	<0.050v
D13	RED/WHT	HO2S-22 (B2 S2) Ground	<0.050v
D14	WHT/RED	HO2S-12 (B1 S2) Signal	0.1-1.1v
D16	GRN/RED	Electronic Load Detector	Varies: 2.5-3.5v

WIRE SIDE OF HARNESS TERMINALS

05533_ADIA_G591

Pin Connector Graphic

1997 Coupe 2.2L I4 VTEC VIN YA1 [A/T] 25P 'B' Connector

PCM Pin #	W/Color	Circuit Description (25-Pin)	Value at Hot Idle
B1-2	---	Not Used	---
B3	BLU/YEL	A/T: Shift Solenoid 'A'	SSA Off: 12v, On: 1v
B4	GRN/BLK	A/T: Lockup Solenoid 'B'	LSB on: 1v, off: 12v
B5	YEL	A/T: Lockup Solenoid 'A'	LSA on: 1v, off: 12v
B6-7	---	Not Used	---
B8	GRN/BLU	A/T: Gear Position Switch	In D3: 0v, other: 12v
B9-10	---	Not Used	---
B11	GRN/WHT	A/T: Shift Solenoid 'B'	SSB Off: 12v, On: 1v
B12	WHT/GRN	Interlock Control Unit Signal	Key & Brake on: 12v
B13	BLU/RED	A/T: D4 Indicator Light	Light on: 12v, off: 0v
B14	WHT/BLU	Mainshaft Speed Sensor 'N'	<0.050v
B15	ORN/BLU	Mainshaft Speed Sensor 'P'	Pulses
B16	GRN/RED	A/T: Gear Position Switch	In 'D': 0v, others: 12v
B17	GRN/YEL	A/T: Gear Position Switch	In D2: 0v, other: 12v
B18	GRN/WHT	A/T: Gear Position Switch	In D1: 0v, other: 12v
B19-21	---	Not Used	---
B22	BLU/YEL	Countershaft Speed Sensor N	<0.050v
B23	BLU/GRN	Countershaft Speed Sensor P	Wheels turn: pulses
B24	GRN/BLK	A/T: Gear Position Switch	In D4: 0v, other: 12v
B25	LT GRN	A/T: Gear Position Switch	In P/N: 0v, Others: 12v

1997 Coupe 2.2L I4 VTEC VIN YA1 [A/T] 31P 'C' Connector

PCM Pin #	W/Color	Circuit Description (31-Pin)	Value at Hot Idle
C1	---	Not Used	---
C2	BLU	CKP Sensor Signal	AC pulse signals
C3	GRN	TDC Sensor Signal	AC pulse signals
C4	YEL	CYP Sensor Signal	AC pulse signals
C5	RED/WHT	A/C Switch Signal	A/C Off: 12v, On: 1v
C6	BLK/GRN	Starter Switch Signal	Cranking: 9-11v
C7	RED	Service Check Connector	SCS Open: 4.80v
C8	LT GRN	K-Line Signal	12v
C9	---	Not Used	---
C10	WHT/YEL	Keep Alive Power (VBU)	12-14v
C11	---	Not Used	---
C12	WHT	CKP Sensor Ground	<0.050v
C13	RED	TDC Sensor Ground	<0.050v
C14	BLK	CYP Sensor Ground	<0.050v
C15	BLU/BLK	VTEC Pressure Switch	Switch on: 0v, off: 12v
C16	GRN	PSP Switch Signal	Straight: 0v, Turning: 12v
C17	WHT/RED	Alternator 'FR' Signal	Idle: 3-5v, Acc: 2-4v
C18	ORN	Vehicle Speed Sensor	Moving: pulse signals
C19	---	Not Used	---
C20	BRN	EVAP Purge Flow Switch	Switch on: 0v, off: 12v
C21-31	---	Not Used	---

1998-99 Coupe 2.3L I4 VTEC VIN YA3 [A/T] 32P 'A' Connector

PCM Pin #	W/Color	Circuit Description (32-Pin)	Value at Hot Idle
A1, A11	---	Not Used	---
A2	GRN/WHT	Engine Mount Solenoid	0v, Off-Idle: 12v
A3	BLU/BLK	EVAP Bypass Solenoid	Solenoid Off: 12v, On: 1v
A4	ORN/GRN	EVAP Vent Solenoid	Solenoid Off: 12v, On: 1v
A5	BLU/WHT	Cruise Control Signal	Cruise On: pulse signals
A6	RED/YEL	EVAP Purge Solenoid	Pulse Signals
A8	ORN/BLU	HO2S-12 (B1 S2) Heater	Digital Signals: 0-12-0v
A9	ORN	Vehicle Speed Sensor	Moving: pulse signals
A10	RED	Service Check Connector	SCS Open: 4.80v
A12	PNK	Immobilizer Indicator Control	Lamp On: 1v, Off: 12v
A13	ORN/BLU	Immobilizer Enable Signal	Digital Signals
A14	BLU/RED	A/T: D4 Indicator Light	In D4: 0v, Off: 12v
A15	YEL/BLK	Immobilizer Fuel Pump Relay	Relay Off: 12v, On: 1v
A17	RED/BLU	A/C Clutch Relay Control	Relay Off: 12v, On: 1v
A18	GRN/RED	Malfunction Indicator Lamp	MIL Off: 12v, On: 1v
A19	BLU	Engine Speed Pulse	Pulses
A20	GRN	Radiator Fan Relay Control	Relay Off: 12v, On: 1v
A21	LT BLU	K-Line Signal	12v
A16, A22	---	Not Used	---
A23	WHT/RED	HO2S-12 (B1 S2) Signal	0.1-1.1v
A24	BLU/RED	Starter Switch Signal	Cranking: 9-11v
A25	BLU/GRN	INOCD Immobilizer Code	Digital Signals
A26	GRN/YEL	PSP Switch Signal	Straight: 0v, Turning: 12v
A27	RED/WHT	A/C Switch Signal	A/C Off: 12v, On: 1v
A28	WHT/GRN	Interlock Control Unit Signal	Key & Brake on: 12v
A29	LT GRN	Fuel Tank Pressure Sensor	Fuel Cap off: 2.5v
A30	GRN/RED	Electronic Load Detector	Varies: 2.5-3.5v
A31	---	Not Used	---
A32	GRN/WHT	Brake Switch Signal	Brake on: 0v, off: 12v

1998-99 Coupe 2.3L I4 VTEC VIN YA3 [A/T] 16P 'D' Connector

PCM Pin #	W/Color	Circuit Description (16-Pin)	Value at Hot Idle
D1	YEL	A/T: Lockup Solenoid	LUS Off: 12v, On: 1v
D2	GRN/WHT	A/T: Shift Solenoid 'B'	SSB Off: 12v, On: 1v
D3	GRN	A/T Shift Solenoid 'C'	SSA Off: 12v, On: 1v
D4	---	Not Used	---
D5	BLK/YEL	Solenoid Feed (B+)	Solenoid on: 12-14v
D6	GRN/RED	A/T: Gear Position Switch	In 'D': 0v, others: 12v
D7	BLU/YEL	A/T: Shift Solenoid 'A'	SSA Off: 12v, On: 1v
D8	GRN/BLU	A/T: Gear Position Switch	In D3: 0v, Others: 12v
D9	GRN/BLK	A/T: Gear Position Switch	In D4: 0v, Others: 12v
D10	BLU	Countershaft Speed Sensor P	Wheels turning: 12v
D11	RED/BLU	Mainshaft Speed Sensor 'P'	Moving: pulse signals
D12	WHT	Mainshaft Speed, Ground 'N'	<0.050v
D13	LT GRN	A/T: Gear Position Switch	In P/N: 0v, Others: 12v
D14	GRN/YEL	A/T: Gear Position Switch	In D2: 0v, Others: 12v
D15	GRN/WHT	A/T: Gear Position Switch	In D1: 0v, Others: 12v
D16	GRN	Countershaft Speed Sensor N	<0.050v

1998-99 Coupe 2.3L I4 VTEC VIN YA3 [All] 25P 'B' Connector

PCM Pin #	W/Color	Circuit Description (25-Pin)	Value at Hot Idle
B1, B9	YEL/BLK	Main Relay Power (B+)	12-14v
B2, B10	BLK	Power Ground	<0.1v
B3, B4	RED, BLU	Injector 2, Injector 3	2.0-3.3 ms
B5, B11	YEL, BRN	Injector 4, Injector 1	2.0-3.3 ms
B7	PNK	E-EGR Solenoid Control	Digital Signals: 0-12-0v
B8	WHT	A/T: Clutch Solenoid 'A-'	Pulse Signals
B12	GRN/YEL	VTEC Solenoid Control	Idle: 0v, Hi-Speed: 12v
B13	YEL/GRN	Igniter Control Signals	Digital Signals: 0-12-0v
B14	BLU/BLK	A/T: 2nd Clutch Pressure Sw.	Open: 12v, Closed: 1v
B17	RED	A/T: Clutch Solenoid 'A+'	Pulse Signals
B18	GRN	A/T: Clutch Solenoid 'B-'	Pulse Signals
B20, 22	BRN/BLK	Logic Ground (LG1, LG2)	<0.050v
B21	WHT/YEL	Keep Alive Power (VBU)	12-14v
B23	BLK/BLU	Idle Air Control Valve	Pulse Signals
B24	BLU/WHT	3rd Clutch Press. Switch	Open: 12v, Closed: 1v
B25	ORN	A/T" Clutch Solenoid 'B+'	Pulse Signals

1998-99 Coupe 2.3L I4 VTEC VIN YA3 [All] 31P 'C' Connector

PCM Pin #	W/Color	Circuit Description (31-Pin)	Value at Hot Idle
C1	BLK/WHT	HO2S-11 (B1 S1) Heater	Digital Signals: 0-12-0v
C2	WHT/GRN	Alternator Charging Signal	Lights Off: 12v, On: 1v
C3	RED/BLU	Knock Sensor Signal	No Detonation: 18mv AC
C5	WHT/RED	Alternator 'FR' Signal	Digital Signals: 0-5-0v
C6	WHT/BLK	EGR Valve Lift Sensor	1.1-1.2v
C7, C18	GRN/WHT	MAP Sensor, Sensor Ground	<0.050v
C8	BLU	CKP Sensor Signal	AC pulse signals
C9	WHT	CKP Sensor Ground	<0.050v
C10	BLU/BLK	VTEC Pressure Switch	0v, Off-Idle: 12v
C16	WHT	HO2S-11 (B1 S1) Signal	0.1-1.1v
C17	RED/GRN	MAP Sensor Signal	0.8-0.9v
C19	YEL/RED	MAP Sensor VREF	4.9-5.1v
C20	GRN	TDC Sensor Signal	AC pulse signals
C21	RED	TDC Sensor Ground	<0.050v
C23	BLU/WHT	Vehicle Speed Sensor	Moving: pulse signals
C25	RED/YEL	IAT Sensor Signal	At 100ºF: 2-3v
C26	RED/WHT	ECT Sensor Signal	At 180ºF: 0.5-0.6v
C27	RED/BLK	TP Sensor Signal	0.5-0.6v
C28	YEL/BLU	Sensor VREF (VCC2)	4.9-5.1v
C29	YEL	CYP Sensor Signal	AC pulse signals
C30	BLK	CYP Sensor Ground	<0.050v

05533_ADIA_G591

Pin Connector Graphic

1997 Coupe 3.0L V6 VTEC VIN YA2 [A/T] 32P 'A' Connector

PCM Pin #	W/Color	Circuit Description (32-Pin)	Value at Hot Idle
A1	YEL	Injector 4 Control	2.0-3.3 ms
A2	BLU	Injector 3 Control	2.0-3.3 ms
A3	RED	Injector 2 Control	2.0-3.3 ms
A4	BRN	Injector 1 Control	2.0-3.3 ms
A5	BLK/WHT	HO2S-12 (B1 S2) Heater	Digital Signals: 0-12-0v
A6	BLK/WHT	HO2S-11 (B1 S1) Heater	Digital Signals: 0-12-0v
A7	PNK	EGR Solenoid Control	Solenoid Off: 12v, On: 1v
A8	GRN/YEL	VTEC Solenoid Control	Solenoid Off: 12v, On: 1v
A9	BRN/BLK	Logic Ground	<0.050v
A10	BLK	Power Ground	<0.1v
A11	YEL/BLK	Main Relay Power (B+)	12-14v
A12	BLK/BLU	Idle Air Control Valve	Pulse Signals
A13	GRN/WHT	Engine Mount Solenoid	Solenoid Off: 12v, On: 1v
A14	BLK/RED	Injector 5	2.0-3.3 ms
A15	RED/YEL	EVAP Purge Solenoid	Pulse Signals
A16	GRN/BLK	Fuel Pump Relay Control	Relay Off: 12v, On: 1v
A17	RED/BLU	A/C Clutch Relay Control	Relay Off: 12v, On: 1v
A18	GRN/RED	Malfunction Indicator Lamp	MIL Off: 12v, On: 1v
A19	WHT/GRN	Alternator Charging Signal	Lights Off: 12v, On: 1v
A20	YEL/GRN	Igniter Signal	Digital Signals: 0-12-0v
A21	---	Not Used	---
A22	BRN/BLK	Logic Ground	<0.050v
A23	BLK	Power Ground	<0.1v
A24	YEL/BLK	Main Relay Power (B+)	12-14v
A25	WHT/BLU	Injector 6	2.0-3.3 ms
A26	---	Not Used	---
A27	GRN	Radiator Fan Relay Control	Fan on: 1v, off: 12v
A28	BLU	EVAP Bypass Solenoid	Solenoid Off: 12v, On: 1v
A29	GRN/YEL	EVAP Vent Solenoid	Solenoid Off: 12v, On: 1v
A30-32	---	Not Used	---

Pin Connector Graphic

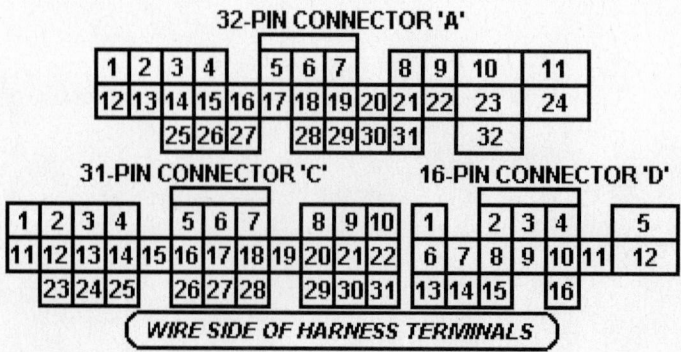

05533_ADIA_G592

1997 Coupe 3.0L V6 VTEC VIN YA2 [A/T] 31P 'C' Connector

PCM Pin #	W/Color	Circuit Description (31-Pin)	Value at Hot Idle
C1	GRN/BLK	TCM VREF	4.9-5.1v
C2	BLU	CKP Sensor Signal	AC pulse signals
C3	GRN	TDC1 Sensor Signal	AC pulse signals
C4	YEL	TDC2 Sensor Signal	AC pulse signals
C5	RED/WHT	A/C Switch Signal	A/C Off: 12v, On: 1v
C6	BLK/GRN	Starter Switch Signal	Cranking: 9-11v
C7	RED	Service Check Connector	SCS Open: 4.80v
C8	LT GRN	K-Line Signal	12v
C9	---	Not Used	---
C10	WHT/YEL	Keep Alive Power (VBU)	12-14v
C11	---	Not Used	---
C12	WHT	CKP Sensor Ground	<0.050v
C13	RED	TDC1 Sensor Ground	<0.050v
C14	BLK	TDC2 Sensor Ground	<0.050v
C15	BLU/BLK	VTEC Pressure Switch	Switch on: 0v, off: 12v
C16	GRN	PSP Switch Signal	Straight: 0v, Turning: 12v
C17	WHT/RED	Alternator 'FR' Signal	Digital Signals: 0-5-0v
C18	BLU/WHT	Vehicle Speed Sensor	Moving: pulse signals
C19-25	---	Not Used	---
C26	BLU	Engine Speed Pulse Signal	Pulses
C27-28	---	Not Used	---
C29	LT GRN	A/T: Gear Position Switch	In P/N: 0v, Others: 12v
C30	BLU/RED	SEAF ECM Input To TCM	Pulse Signals
C31	RED/BLU	SEAF ECM Output To TCM	Pulse Signals

1997 Coupe 3.0L V6 VTEC VIN YA2 [A/T] 16P 'D' Connector

PCM Pin #	W/Color	Circuit Description (16-Pin)	Value at Hot Idle
D1	RED/BLK	TP Sensor Signal	0.5-0.6v
D2	RED/WHT	ECT Sensor Signal	At 180°F: 0.5-0.6v
D3	RED/GRN	MAP Sensor Signal	0.8-0.9v
D4	YEL/RED	MAP Sensor VREF	4.9-5.1v
D5	GRN/WHT	Brake Switch Signal	Brake on: 0v, off: 12v
D6	---	Not Used	---
D7	WHT	HO2S-11 (B1 S1) Signal	0.1-1.1v
D8	RED/YEL	IAT Sensor Signal	At 100°F: 2-3v
D9	WHT/BLK	EGR Valve Lift Sensor	1.1-1.2v
D10	YEL/BLU	Sensor VREF	4.9-5.1v
D11	GRN/BLK	Sensor Ground	<0.050v
D12	GRN/WHT	MAP Sensor Ground	<0.050v
D14	WHT/RED	HO2S-12 (B1 S2) Signal	0.1-1.1v
D15	LT GRN	Fuel Tank Pressure Sensor	Fuel Cap off: 2.5v
D16	GRN/RED	Electronic Load Detector	Varies: 2.5-3.5v

1998-99 Coupe 3.0L V6 VTEC VIN YA2 [A/T] 32P 'A' Connector

PCM Pin #	W/Color	Circuit Description (32-Pin)	Value at Hot Idle
A1	---	Not Used	---
A2	GRN/WHT	Engine Mount Solenoid	Solenoid Off: 12v, On: 1v
A3	BLU/BLK	EVAP Bypass Solenoid	Solenoid Off: 12v, On: 1v
A4	ORN/GRN	EVAP Vent Solenoid	Solenoid Off: 12v, On: 1v
A5	BLU/WHT	Cruise Control Signal	Cruise On: pulse signals
A6	RED/YEL	EVAP Purge Solenoid	Pulse Signals
A8	BLK/WHT	HO2S-12 (B1 S2) Heater	Digital Signals: 0-12-0v
A9	ORN	Vehicle Speed Sensor	Moving: pulse signals
A10	RED	Service Check Connector	SCS Open: 4.80v
A11	---	Not Used	---
A12	PNK	Immobilizer Indicator Control	Lamp On: 1v, Off: 12v
A13	ORN/BLU	Immobilizer Enable Signal	Digital Signals
A14	BLU/RED	A/T: D4 Light Switch	In D4: 0v, Off: 12v
A15	GRN/BLK	IMO Fuel Pump Relay Control	Relay Off: 12v, On: 1v
A17	RED/BLU	A/C Clutch Relay Control	Relay Off: 12v, On: 1v
A18	GRN/RED	Malfunction Indicator Lamp	MIL Off: 12v, On: 1v
A19	BLU	Engine Speed Pulse	Pulses
A20	GRN	Radiator Fan Relay Control	Relay Off: 12v, On: 1v
A21	LT BLU	K-Line Signal	12v
A23	WHT/RED	HO2S-12 (B1 S2) Signal	0.1-1.1v
A24	BLK/GRN	Starter Switch Signal	Cranking: 9-11v
A25	BLU/GRN	INOCD Immobilizer Code	Digital Signals
A26	GRN/YEL	PSP Switch Signal	Straight: 0v, Turning: 12v
A27	RED/WHT	A/C Switch Signal	A/C Off: 12v, On: 1v
A28	WHT/GRN	Interlock Control Unit Signal	Key & Brake on: 12v
A29	LT GRN	Fuel Tank Pressure Sensor	Fuel Cap off: 2.5v
A30	GRN/RED	Electronic Load Detector	Varies: 2.5-3.5v
A31	---	Not Used	---
A32	GRN/WHT	Brake Switch Signal	Brake on: 0v, off: 12v

1998-99 Coupe 3.0L V6 VTEC VIN YA2 [A/T] 16P 'D' Connector

PCM Pin #	W/Color	Circuit Description (16-Pin)	Value at Hot Idle
D1	YEL	A/T: Lockup Solenoid	Solenoid Off: 12v, On: 1v
D2	GRN/WHT	A/T: Shift Solenoid 'B'	Solenoid Off: 12v, On: 1v
D3	GRN	A/T Shift Solenoid 'C'	Solenoid Off: 12v, On: 1v
D5	BLK/YEL	Solenoid Feed (B+)	12-14v
D6	GRN/RED	A/T: Gear Position Switch	In 'D': 0v, others: 12v
D7	BLU/YEL	A/T: Shift Solenoid 'A'	On: <1v, off: 12v
D8	GRN/BLU	A/T: Gear Position Switch	In D3: 0v, other: 12v
D9	GRN/BLK	A/T: Gear Position Switch	In D4: 0v, other: 12v
D10	BLU	Countershaft Speed Sensor P	Moving: AC Pulses
D11	RED/BLU	Mainshaft Speed Sensor 'P'	Pulses
D12	WHT	Mainshaft Speed Sensor 'N'	<0.050v
D13	LT GRN	A/T: Gear Position Switch	In P/N: 0v, Others: 12v
D14	GRN/YEL	A/T: Gear Position Switch	In D2: 0v, other: 12v
D15	GRN/WHT	A/T: Gear Position Switch	In D1: 0v, other: 12v
D16	GRN	Countershaft Speed Sensor N	<0.050v

1998-99 Coupe 3.0L V6 VTEC VIN YA2 [A/T] 25P 'B' Connector

PCM Pin #	W/Color	Circuit Description (25-Pin)	Value at Hot Idle
B1, B9	YEL/BLK	Main Relay Power (B+)	12-14v
B2, B10	BLK	Power Ground	<0.1v
B3, B4	BLK	Injector 5, Injector 4	2.0-3.3 ms
B5, B6	RED, WHT	Injector 2, Injector 6	2.0-3.3 ms
B7	PNK	EGR Solenoid Control	Solenoid Off: 12v, On: 1v
B8	WHT	A/T: Clutch Solenoid 'A-'	Pulse Signals
B11, 15	BRN, BLU	Injector 1, Injector 3	2.0-3.3 ms
B12	GRN/YEL	VTEC Solenoid Control	Idle: 0v, Hi-Speed: 12v
B13	YEL/GRN	Igniter Signal	Digital Signals: 0-12-0v
B14	BLU/BLK	A/T: 2nd Clutch Pressure Sw.	Open: 12v, Closed: 1v
B17	RED	A/T: Clutch Solenoid 'A+'	Pulse Signals
B18	ORN	A/T: Clutch Solenoid 'B-'	Pulse Signals
B20, 22	BRN/BLK	Logic Ground	<0.050v
B21	WHT/YEL	Keep Alive Power (VBU)	12-14v
B23	BLK/BLU	Idle Air Control Valve	Pulse Signals
B24	BLU/WHT	3rd Clutch Press. Switch	Open: 12v, Closed: 1v
B25	ORN	A/T: Clutch Solenoid 'B+'	Pulse Signals

1998-99 Coupe 3.0L V6 VTEC VIN YA2 [A/T] 31P 'C' Connector

PCM Pin #	W/Color	Circuit Description (31-Pin)	Value at Hot Idle
C1	BLK/WHT	HO2S-11 (B1 S1) Heater	Digital Signals: 0-12-0v
C2	WHT/GRN	Alternator Charging Signal	Light on: 0v, off: 12v
C5	WHT/RED	Alternator 'FR' Signal	Digital Signals: 0-5-0v
C6	WHT/BLK	EGR Valve Lift Sensor	1.1-1.2v
C7	GRN/WHT	MAP Sensor Ground	<0.050v
C8	BLU	CKP Sensor Signal	AC pulse signals
C9	WHT	CKP Sensor Ground	<0.050v
C10	BLU/BLK	VTEC Pressure Switch	Switch on: 0v, off: 12v
C16	WHT	HO2S-11 (B1 S1) Signal	0.1-1.1v
C17	RED/GRN	MAP Sensor Signal	0.8-0.9v
C18	GRN/BLK	Sensor Ground	<0.050v
C19	YEL/RED	MAP Sensor VREF	4.9-5.1v
C20	GRN	TDC1 Sensor Signal	AC pulse signals
C21	RED	TDC1 Sensor Ground	<0.050v
C25	RED/YEL	IAT Sensor Signal	At 100°F: 2-3v
C26	RED/WHT	ECT Sensor Signal	At 180°F: 0.5-0.6v
C27	RED/BLK	TP Sensor Signal	0.5-0.6v
C28	YEL/BLU	Sensor VREF	4.9-5.1v
C29	YEL	TDC2 Sensor Signal	AC pulse signals
C30	BLK	TDC2 Sensor Ground	<0.050v

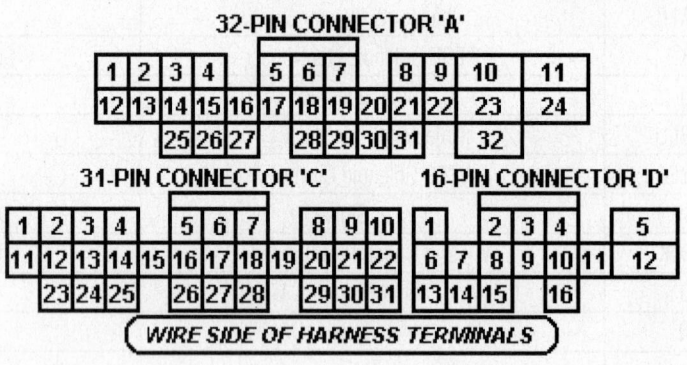

Pin Connector Graphic

05533_ADIA_G591

2001-03 Sedan 3.2L V6 VTEC VIN YA4 [A/T] 32P 'A' Connector

PCM Pin #	W/Color	Circuit Description (32-Pin)	Value at Hot Idle
A1	WHT/RED	HO2S-22 (B2 S2) Signal	0.1-1.1v
A2	GRN/WHT	Engine Mount Solenoid	0v, Off-Idle: 12v
A3	BLU	EVAP Bypass Solenoid	Solenoid Off: 12v, On: 1v
A4	GRN/WHT	EVAP Vent Solenoid	Solenoid Off: 12v, On: 1v
A5	BLU/GRN	Cruise Control Signal	Cruise On: pulse signals
A6	RED/YEL	EVAP Purge Solenoid	Pulse Signals
A7	ORN/WHT	Reference Voltage	4.9-5.1v
A8	BLK/WHT	HO2S-12 (B1 S2) Heater	Digital Signals: 0-12-0v
A9	BLU/WHT	Vehicle Speed Sensor	Moving: pulse signals
A10	BRN	Service Check Connector	SCS Open: 4.80v
A11-12	---	Not Used	---
A13	LT BLU	Intake Manifold Runner J32A2	Solenoid Off: 12v, On: 1v
A14	GRN/BLK	D5 Indicator Light	In D5: 0.1v, off: 12v
A15	GRN/YEL	Fuel Pump Relay Control	Relay Off: 12v, On: 1v
A16	---	Not Used	---
A17	RED	A/C Clutch Relay Control	Relay Off: 12v, On: 1v
A18	GRN/ORN	Malfunction Indicator Light	MIL Off: 12v, On: 1v
A19	BLU	Engine Speed Signal	Pulses
A20	GRN	Radiator Fan Relay Control	Relay Off: 12v, On: 1v
A21	GRY	K-Line Signal	12v
A22	RED/GRN	SEAF Serial Data Line J32A2	Digital Signals
A23	---	Not Used	---
A24	BLU/ORN	Starter Switch Signal	Cranking: 9-11v
A25	---	Not Used	---
A26	GRN	PSP Switch Signal	Straight: 0v, Turning: 12v
A27	BLU/RED	A/C Switch Signal	A/C Off: 12v, On: 1v
A28	WHT/RED	Interlock Control Unit Signal	Key & Brake on: 12v
A29	LT GRN	Fuel Tank Pressure Sensor	Fuel Cap off: 2.5v
A30	GRN/RED	Electronic Load Detector	Varies: 2.5-3.5v
A31	YEL/GRN	ECT Sensor Signal to TCM	Digital Signals
A32	WHT/BLK	Brake Switch Signal	Brake on: 0v, off: 12v

Pin Connector Graphic

2001-03 Sedan 3.2L V6 VTEC VIN YA4 [A/T] 25P 'B' Connector

PCM Pin #	W/Color	Circuit Description (25-Pin)	Value at Hot Idle
B1, B9	YEL/BLK	Main Relay Power (B+)	12-14v
B2, B10	BLK	Power Ground PG1, PG2	<0.1v
B3	BLK/RED	Injector 5 Control	2.0-3.3 ms
B4	YEL	Injector 4 Control	2.0-3.3 ms
B5	RED	Injector 2 Control	2.0-3.3 ms
B6	WHT/BLU	Injector 6 Control	2.0-3.3 ms
B7	BLU/RED	E-EGR Control Solenoid	Digital Signals: 0-12-0v
B8	WHT	A/T: Clutch Solenoid 'A-'	Pulse Signals
B11	BRN	Injector 1 Control	2.0-3.3 ms
B12	GRN/YEL	VTEC Solenoid Control	Idle: 0v, Hi-Speed: 12v
B13	GRN/RED	A/T: Clutch Solenoid 'C+'	Pulse Signals
B14	BLU/WHT	A/T: Gear Position Switch	In P/N: 0v, Others: 12v
B15	BLU	Injector 3 Control	2.0-3.3 ms
B16	---	Not Used	---
B17	RED	A/T: Clutch Solenoid 'A+'	Pulse Signals
B18	GRN	A/T: Clutch Solenoid 'B-'	Pulse Signals
B19	BLU/YEL	A/T: 4th Oil Pressure Switch	12-14v
B20	BRN/YEL	Logic Ground (LG1)	<0.050v
B21	WHT/RED	Keep Alive Power (VBU)	12-14v
B22	BRN/YEL	Logic Ground (LG2)	<0.050v
B23	BLK/RED	Idle Air Control Valve	Pulse Signals
B24	RED/BLU	A/T: Clutch Solenoid 'A-'	Pulse Signals
B25	BRN/BLU	A/T: Clutch Solenoid 'B+'	Pulse Signals

2001-03 Sedan 3.2L V6 VTEC VIN YA4 [A/T] 16P 'D' Connector

PCM Pin #	W/Color	Circuit Description (16-Pin)	Value at Hot Idle
D1	YEL	A/T: Lockup Solenoid	LUS Off: 12v, On: 1v
D2	GRN/WHT	A/T: Shift Solenoid 'B'	SSB Off: 12v, On: 1v
D3	GRN	A/T" Shift Solenoid 'C'	SSA Off: 12v, On: 1v
D4	RED/BLK	A/T: 'N' Gear Position Switch	In 'N': 0v, Others: 12v
D5	BLK/YEL	A/T: Solenoid Feed (B+)	12-14v
D6	WHT	A/T: 'R' Gear Position Switch	In 'R': 0v, Others: 12v
D7	BLU/YEL	A/T: Shift Solenoid 'A'	SSA Off: 12v, On: 1v
D8	YEL	A/T: D4 Gear Position Switch	In D4: 0v, others: 12v
D9	YEL/GRN	A/T: D5 Gear Position Switch	In D5: 0v, Others: 12v
D10	BLU	Countershaft Speed Sensor P	Wheels turning: 12v
D11	RED	Mainshaft Speed Sensor 'P'	Moving: pulse signals
D12	WHT	Mainshaft Speed, Ground 'N'	<0.050v
D13	BLU/WHT	A/T: 3rd Clutch Pressure Sw.	Open: 12v, Closed: 1v
D14	RED	A/T: D3 Gear Position Switch	In D3: 0v, Others: 12v
D15	BLU	A/T: D2 Gear Position Switch	In D2: 0v, Others: 12v
D16	GRN	Countershaft Speed Sensor N	<0.050v

32P 'A'   25P 'B'   31P 'C'   16P 'D'   20P 'E' Connector

WIRE SIDE OF HARNESS TERMINALS

05533_ADIA_G593

Pin Connector Graphic

2001-03 Sedan 3.2L V6 VTEC VIN YA4 [A/T] 31P 'C' Connector

PCM Pin #	W/Color	Circuit Description (31-Pin)	Value at Hot Idle
C1	BLK/WHT	HO2S-11 (B1 S1) Heater	Digital Signals: 0-12-0v
C2	WHT/GRN	Alternator Charging Signal	Lights on: 0v, off: 12v
C3	WHT/BLU	Ignition Coil 3 Control	Digital Signals: 0-12-0v
C4	YEL/GRN	Ignition Coil 1 Control	Digital Signals: 0-12-0v
C5	WHT/RED	Alternator 'FR' Signal	Digital Signal: 0-5-0-5v
C6	WHT/BLK	EGR Valve Lift Sensor	1.2v
C7	GRN/WHT	MAP Sensor Ground (SG1)	<0.050v
C8	BLU	CKP Sensor 'P' Signal	AC pulse signals
C9	WHT	CKP Sensor 'N' Signal	<0.050v
C10	BLU/BLK	VTEC Pressure Switch	Idle: 0v, Cruise: 12v
C11	---	Not Used	---
C12	BLK/RED	Ignition Coil 5 Control	Digital Signals: 0-12-0v
C13	BRN	Ignition Coil 4 Control	Digital Signals: 0-12-0v
C14	BLU/RED	Ignition Coil 2 Control	Digital Signals: 0-12-0v
C15	---	Not Used	---
C16	WHT	HO2S-11 (B1 S1) Signal	0.1-1.1v
C17	GRN/RED	MAP Sensor Signal	0.8-0.9v
C18	GRN/YEL	Sensor Ground (SG2)	<0.050v
C19	YEL/RED	MAP Sensor VREF (VCC1)	4.9-5.1v
C20	GRN	TDC1 Sensor 'P' Signal	AC pulse signals
C21	RED	TDC1 Sensor 'N' Signal	<0.050v
C22	RED/BLU	Knock Sensor Signal	No Detonation: 18mv AC
C23	BRN/WHT	Ignition Coil 6 Control	Digital Signals: 0-12-0v
C24	BLU/YEL	A/T: TFT Sensor Signal	0.1-4.2v
C25	RED/YEL	IAT Sensor Signal	Varies w/temp: 0.5-4.9v
C26	RED/WHT	ECT Sensor Signal	At 180°F: 0.5-0.6v
C27	RED/BLK	TP Sensor Signal	0.5-0.6v
C28	YEL/BLU	Sensor VREF (VCC2)	4.9-5.1v
C29	YEL	TDC2 Sensor 'P' Signal	AC pulse signals
C30	BLK	TDC2 Sensor 'N' Signal	<0.050v

2001-03 Sedan 3.2L V6 VTEC VIN YA4 [A/T] 20P 'E' Connector

PCM Pin #	W/Color	Circuit Description (20-Pin)	Value at Hot Idle
E1	PNK/WHT	Immobilizer Indicator Lamp	Lamp Off: 12v, On: 1v
E2	RED	IMOCD Immobilizer Code	Digital Signals
E3-4	---	Not Used	---
E5	YEL/GRN	TP Sensor Signal to TCM	0.5-0.6v
E6	ORN/GRN	Voltage Reference (ABS)	4.9-5.1v
E7	BLU	PCM to Frame Inhibition	2.5v (normal signal)
E8	BLU/BLK	A/T: Drive Shift Indicator 'C'	2nd, 4th: 0v, 1st, 3rd: 12v
E9	BLU/GRN	A/T: Drive Shift Indicator 'B'	In 4th: 12v, 1st & 3rd: 0v
E10	BLU/RED	A/T: Drive Shift Indicator 'A'	In 4th: 12v, 1st - 3rd: 0v
E11	---	Not Used	---
E12	BLU/WHT	IMOEN Immobilizer Enable	Digital Signals
E13, E15	---	Not Used	---
E14	BRN	A/T: First Hold Down Switch	LH Switch on: 0v, off: 12v
E16	PNK	Frame to PCM Torque Down	TCS On: 5v, Off: 2.5v
E17	LT GRN	A/T: 'P' Gear Position Switch	In 'P': 4v, Others: 0v
E18	ORN	Torque Mgmt. (-) Signal	In Neutral: 12v
E19	YEL	Torque Mgmt. (+) Signal	In Sports Shift: 0v
E20	RED	Torque Mgmt. Mode Signal	In Sports Shift: 0v

RL PIN CHARTS

1996-99 Sedan 3.5L V6 MFI VIN KA9 [A/T] 26P 'A' Connector

PCM Pin #	W/Color	Circuit Description (26-Pin)	Value at Hot Idle
A1	BRN	Injector 1	2.0-3.3 ms
A2	RED	Injector 2	2.0-3.3 ms
A3	BLU	Injector 3	2.0-3.3 ms
A4	YEL/GRN	Fuel Pump Relay 1 Control	Relay Off: 12v, On: 1v
A5	BLK/BLU	Idle Air Control Valve	Pulse Signals
A6	RED	EGR Solenoid Control	At off-idle (hot): 0-12-0v
A7	GRN/RED	Malfunction Indicator Lamp	MIL Off: 12v, On: 1v
A8	RED/WHT	A/C Clutch Relay Control	Relay Off: 12v, On: 1v
A9	---	Not Used	---
A10	RED/GRN	Fuel Pressure Regulator	Solenoid Off: 12v, On: 1v
A11	PNK	Ignition Coil 1 Control	Digital Signals: 0-12-0v
A12	BLK	Power Ground	<0.1v
A13	YEL/BLK	Main Relay Power (B+)	12-14v
A14	YEL	Injector 4	2.0-3.3 ms
A15	BLK/RED	Injector 5	2.0-3.3 ms
A16	WHT/BLU	Injector 6	2.0-3.3 ms
A17	LT GRN	Engine Mount Solenoid	Solenoid Off: 12v, On: 1v
A18	GRN/RED	HO2S-11 (B1 S1) Heater	Digital Signals: 0-12-0v
A19	BLK/WHT	HO2S-21 (B2 S1) Heater	Digital Signals: 0-12-0v
A20	RED/BLU	IAB Solenoid Low Control	Under 3200 rpm: 0v
A21	BLK/WHT	HO2S-12 (B1 S2) Heater	Digital Signals: 0-12-0v
A22	YEL	IAB Solenoid High Control	Under 3200 rpm: 0v
A23	RED/YEL	EVAP Purge Solenoid	<90°F: 12v, >90°F: 0v
A24	BRN	Ignition Coil 2 Control	Digital Signals: 0-12-0v
A25	BLK	Power Ground (PG1)	<0.1v
A26	BRN/BLK	Logic Ground (LG1)	<0.050v

1996-99 Sedan 3.5L V6 MFI VIN KA9 [A/T] 16P 'B' Connector

PCM Pin #	W/Color	Circuit Description (16-Pin)	Value at Hot Idle
B1	YEL/BLK	Main Relay Power (B+)	12-14v
B2	RED	Ignition Coil 6 Control	Digital Signals: 0-12-0v
B3	GRN	PSP Switch Signal	Straight: 0v, Turning: 12v
B4	---	Not Used	---
B5	YEL/RED	CYP Sensor 2 Signal	AC pulse signals
B6	YEL	CYP Sensor 1 Signal	AC pulse signals
B7	BLU/RED	CKP Sensor 2 Signal	AC pulse signals
B8	BLU	CKP Sensor 1 Signal	AC pulse signals
B9	BRN/BLK	Logic Ground	<0.050v
B10	GRY	Ignition Coil 5 Control	Digital Signals: 0-12-0v
B11	GRN	Ignition Coil 4 Control	Digital Signals: 0-12-0v
B12	BLU	Ignition Coil 3 Control	Digital Signals: 0-12-0v
B13	BLU/WHT	Vehicle Speed Sensor	Moving: pulse signals
B14	---	Not Used	---
B15	WHT/RED	CKP Sensor 2 Ground	<0.050v
B16	WHT	CKP Sensor 1 Ground	<0.050v

1996-99 Sedan 3.5L V6 MFI VIN KA9 [A/T] 12P 'C' Connector

PCM Pin #	W/Color	Circuit Description (12-Pin)	Value at Hot Idle
C1	---	Not Used	---
C2	PNK	A/C Switch Signal	A/C Off: 12v, On: 1v
C3	GRN/RED	A/C Pressure Switch Signal	A/C Off: 4-5v, On: 0v
C4	GRN	Fuel Pump Relay 2 Control	Relay Off: 12v, On: 1v
C5	RED	Service Check Connector	SCS Open: 4.80v
C6	BLU/RED	Starter Switch Signal	Cranking: 9-11v
C7	---	Not Used	---
C8	BLU	Engine Speed Pulse Signal	Pulses
C9	GRN/RED	Traction Control Inhibit Signal	Pulses
C10	BLK/ORN	T/C Fuel Cut Signal	Pulses
C11	WHT	INOCD Immobilizer Code	Digital Signals
C12	PNK/BLK	T/C Standby Signal	Pulses

1996-99 Sedan 3.5L V6 MFI VIN KA9 [A/T] 22P 'D' Connector

PCM Pin #	W/Color	Circuit Description (22-Pin)	Value at Hot Idle
D1	WHT/YEL	Keep Alive Power (VBU)	12-14v
D2	RED/BLU	Knock Sensor Signal (Bank 1)	No Detonation: 18mv AC
D3	GRN/ORN	Barometric Output Signal	3v (at sea level)
D4	LT GRN	K-Line Signal	12v
D5	WHT/RED	Alternator 'FR' Signal	Digital Signals: 0-5-0v
D6	RED/BLK	TP Sensor Signal	0.5-0.6v
D7	RED/WHT	ECT Sensor Signal	At 180°F: 0.5-0.6v

1996-99 Sedan 3.5L V6 MFI VIN KA9 [A/T] 22P 'D' Connector, *continued*

D8	RED/YEL	IAT Sensor Signal	At 100°F: 2-3v
D9	RED/GRN	MAP Sensor Signal	0.8-0.9v
D10	YEL/RED	MAP Sensor VREF	4.9-5.1v
D11	GRN/WHT	MAP Sensor Ground	<0.050v
D12	BLK	CYP Sensor 1 Ground	<0.050v
D13	YEL	Knock Sensor Signal (Bank 2)	No Detonation: 18mv AC
D14	BLK/RED	CYP Sensor 2 Ground	<0.050v
D15	GRN/YEL	Fuel Tank Pressure Sensor	Fuel Cap off: 2.5v
D16	---	Not Used	---
D17	WHT/BLK	EGR Valve Lift Sensor	1.1-1.2v
D18	BLU/RED	HO21-21 (B2 S1) Signal	0.1-1.1v
D19	WHT	HO2S-11 (B1 S1) Signal	0.1-1.1v
D20	BLU/WHT	TCS VREF	4.9-5.1v
D21	YEL/BLU	Sensor VREF (VCC2)	4.9-5.1v
D22	GRN/BLK	Sensor Ground (SG2)	<0.050v

1996-99 Sedan 3.5L V6 MFI VIN KA9 [A/T] 26P 'E' Connector

PCM Pin #	W/Color	Circuit Description (26-Pin)	Value at Hot Idle
E1	BLK/YEL	Solenoid Feed (B+) (VB SOL)	12-14v
E2	GRN/WHT	Brake Switch Signal	Brake Off: 12v, On: 0v
E3	RED	Linear Solenoid Valve 'P'	Pulses
E4	RED	Mainshaft Speed Sensor	Pulses
E5	BLU	Countershaft Speed Sensor	Wheels turning: 12v
E6	PNK/BLU	TCS Shift Control Signal	Pulses
E7	WHT	A/T: Gear Position Switch	In 'D': 0v, others: 12v
E8	YEL	A/T: Gear Position Switch	In D4: 0v, other: 12v
E9	GRN	A/T: Gear Position Switch	In D3: 0v, other: 12v
E10	BLU	A/T: Gear Position Switch	In D2: 0v, other: 12v
E11	BRN	A/T: Gear Position Switch	In D1: 0v, other: 12v
E12	BLU/YEL	Shift Control Solenoid 'A'	KOER in D2/D3: 12v
E13	YEL	Lockup Control Solenoid 'A'	L/Up on: 12v, off: 0v
E16	WHT	Linear Solenoid Valve 'N'	<0.050v
E17	WHT	Mainshaft Speed Sensor 'N'	<0.050v
E18	GRN	Countershaft Speed Sensor N	<0.050v
E19	WHT/RED	Serial Data Line 'B'	N/A
E20	WHT/GRN	Serial Data Line 'A'	Digital Signals
E22	WHT/GRN	Interlock Control Unit Signal	Key & Brake on: 12v
E23	BRN/YEL	A/T: D4 Indicator Light	D4 lamp on: 0v, off: 12v
E24	LT GRN	A/T: Gear Position Switch	In P/N: 0v, Others: 12v
E25	GRN/WHT	Shift Control Solenoid 'B'	SSB Off: 0v, On: 12v
E26	GRN/BLK	Lockup Control Solenoid 'B'	LSB Off: 0v, On: 12v

1996-99 Sedan 3.5L V6 MFI VIN KA9 [A/T] 8P 'F' Connector

PCM Pin #	W/Color	Circuit Description (8-Pin)	Value at Hot Idle
F1	PNK/BLK	EVAP Bypass Solenoid	Solenoid Off: 12v, On: 1v
F2	WHT/RED	HO2S-12 (B1 S2) Signal	0.1-1.1v
F3	BLU	Right Peak Hold Reset	Digital Signals
F4	YEL	Right Misfire Pulse	Digital Signals
F5	RED/WHT	EVAP Vent Shut Solenoid	Solenoid Off: 12v, On: 1v
F6	GRN/WHT	HO2S Ground	<0.050v
F7	GRN/YEL	Left Peak Hold Reset	Digital Signals
F8	RED	Left Misfire Pulse	Digital Signals

Pin Connector Graphic

2000-04 Sedan 3.5L V6 MFI VIN KA9 [A/T] 26P 'A' Connector

PCM Pin #	W/Color	Circuit Description (26-Pin)	Value at Hot Idle
A1	BRN	Injector 1 Control	2.0-3.3 ms
A2	RED	Injector 2 Control	2.0-3.3 ms
A3	BLU	Injector 3 Control	2.0-3.3 ms
A4	YEL/GRN	Fuel Pump Relay 1 Control	Relay Off: 12v, On: 1v
A5	BLK/BLU	Idle Air Control Valve	Pulse Signals
A6	RED	Electronic EGR Solenoid Control	At off-idle (hot): 0-12-0v
A7	GRN/RED	Malfunction Indicator Lamp	MIL Off: 12v, On: 1v
A8	RED/WHT	A/C Clutch Relay Control	Relay Off: 12v, On: 1v
A9	---	Not Used	---
A10	RED/GRN	Fuel Pressure Regulator	Solenoid Off: 12v, On: 1v
A11	PNK	Ignition Coil 1 Control	Digital Signals: 0-12-0v
A12	BLK	Power Ground	<0.1v
A13	YEL/BLK	Main Relay Power (B+)	12-14v
A14	YEL	Injector 4	2.0-3.3 ms
A15	BLK/RED	Injector 5	2.0-3.3 ms
A16	WHT/BLU	Injector 6	2.0-3.3 ms
A17	LT GRN	Engine Mount Solenoid	Solenoid Off: 12v, On: 1v
A18	GRN/RED	HO2S-11 (B1 S1) Heater	Digital Signals: 0-12-0v
A19	BLK/WHT	HO2S-21 (B2 S1) Heater	Digital Signals: 0-12-0v
A20	RED/BLU	IAB Solenoid Low Control	Under 3200 rpm: 0v

2000-04 Sedan 3.5L V6 MFI VIN KA9 [A/T] 26P 'A' Connector, *continued*

A21	BLK/WHT	HO2S-12 (B1 S2) Heater	Digital Signals: 0-12-0v
A22	YEL	IAB Solenoid High Control	Under 3200 rpm: 0v
A23	RED/YEL	EVAP Purge Solenoid	<140°F: 12v, >140°F: 0v
A24	BRN	Ignition Coil 2 Control	Digital Signals: 0-12-0v
A25	BLK	Power Ground (PG2)	<0.1v
A26	BRN/BLK	Logic Ground (LG1)	<0.050v

2000-04 Sedan 3.5L V6 MFI VIN KA9 [A/T] 16P 'B' Connector

PCM Pin #	W/Color	Circuit Description (16-Pin)	Value at Hot Idle
B1	YEL/BLK	Main Relay Power (B+)	12-14v
B2	RED	Ignition Coil 6 Control	Digital Signals: 0-12-0v
B3	GRN	PSP Switch Signal	Straight: 0v, Turning: 12v
B4	YEL/GRN	MTRTW	With ignition switch ON; duty cycle
B5	YEL/RED	CYP Sensor 2 'P' Signal	AC pulse signals
B6	YEL	CYP Sensor 1 'P' Signal	AC pulse signals
B7	BLU/RED	CKP Sensor 2 'P' Signal	AC pulse signals
B8	BLU	CKP Sensor 1 'P' Signal	AC pulse signals
B9	BRN/BLK	Logic Ground (LG2)	<0.050v
B10	GRY	Ignition Coil 5 Control	Digital Signals: 0-12-0v
B11	GRN	Ignition Coil 4 Control	Digital Signals: 0-12-0v
B12	BLU	Ignition Coil 3 Control	Digital Signals: 0-12-0v
B13	BLU/WHT	Vehicle Speed Sensor	Moving: pulse signals
B14	---	Not Used	---
B15	WHT/RED	Ground for CKP Sensor 2	<0.050v
B16	WHT	Ground for CKP Sensor 1	<0.050v

2000-04 Sedan 3.5L V6 MFI VIN KA9 [A/T] 12P 'C' Connector

PCM Pin #	W/Color	Circuit Description (12-Pin)	Value at Hot Idle
C1	---	Not Used	---
C2	PNK	A/C Switch Signal	A/C Off: 12v, On: 1v
C3	GRN/RED	A/C Pressure Switch Signal	A/C Off: 4-5v, On: 0v
C4	GRN	Fuel Pump Relay 2 Control	Relay Off: 12v, On: 1v
C5	RED	Service Check Connector	SCS Open: 4.80v
C6	BLU/RED	Starter Switch Signal	Cranking: 9-11v
C7	---	Not Used	---
C8	BLU	Engine Speed Pulse Signal	Pulse Signals
C9	GRN/RED	Traction Control Inhibit Signal	Pulse Signals
C10	BLK/ORN	T/C Fuel Cut Signal	Pulse Signals
C11	WHT	INOCD Immobilizer Code	Digital Signals
C12	PNK/BLK	T/C Standby Signal	Pulse Signals

2000-04 Sedan 3.5L V6 MFI VIN KA9 [A/T] 22P 'D' Connector

PCM Pin #	W/Color	Circuit Description (22-Pin)	Value at Hot Idle
D1	WHT/YEL	Keep Alive Power (VBU)	12-14v
D2	RED/BLU	Knock Sensor Signal (Bank 1)	No Detonation: 18mv AC
D3	GRN/ORN	Barometric Output Signal	3v (at sea level)
D4	LT GRN	K-Line Signal	12v
D5	WHT/RED	Alternator 'FR' Signal	Digital Signals: 0-5-0v
D6	RED/BLK	TP Sensor Signal	0.5-0.6v
D7	RED/WHT	ECT Sensor Signal	At 180°F: 0.5-0.6v
D8	RED/YEL	IAT Sensor Signal	At 100°F: 2-3v
D9	RED/GRN	MAP Sensor Signal	0.8-0.9v
D10	YEL/RED	MAP Sensor VREF	4.9-5.1v
D11	GRN/WHT	MAP Sensor Ground	<0.050v
D12	BLK	CYP Sensor 1 Ground	<0.050v
D13	YEL	Knock Sensor Signal (Bank 2)	No Detonation: 18mv AC
D14	BLK/RED	CYP Sensor 2 Ground	<0.050v
D15	GRN/YEL	Fuel Tank Pressure Sensor	Fuel Cap off: 2.5v
D16	---	Not Used	---
D17	WHT/BLK	EGR Valve Lift Sensor	1.1-1.2v
D18	BLU/RED	HO21-21 (B2 S1) Signal	0.1-1.1v
D19	WHT	HO2S-11 (B1 S1) Signal	0.1-1.1v
D20	BLU/WHT	TCS VREF	4.9-5.1v
D21	YEL/BLU	Sensor VREF (VCC2)	4.9-5.1v
D22	GRN/BLK	Sensor Ground (SG2)	<0.050v

Pin Connector Graphic

2000-04 Sedan 3.5L V6 MFI VIN KA9 [A/T] 26P 'E' Connector

PCM Pin #	W/Color	Circuit Description (26-Pin)	Value at Hot Idle
E1	BLK/YEL	Solenoid Feed (B+) (VB SOL)	12-14v
E2	GRN/WHT	Brake Switch Signal	Brake Off: 12v, On: 0v
E3	RED	Linear Solenoid Valve 'P'	Pulses
E4	RED	Mainshaft Speed Sensor	Pulses
E5	BLU	Countershaft Speed Sensor	Wheels turning: 12v
E6	PNK/BLU	TCS Shift Control Signal	Pulses
E7	WHT	A/T: Gear Position Switch	In 'D': 0v, others: 12v
E8	YEL	A/T: Gear Position Switch	In D4: 0v, other: 12v
E9	GRN	A/T: Gear Position Switch	In D3: 0v, other: 12v
E10	BLU	A/T: Gear Position Switch	In D2: 0v, other: 12v
E11	BRN	A/T: Gear Position Switch	In D1: 0v, other: 12v
E12	BLU/YEL	Shift Control Solenoid 'A'	KOER in D2/D3: 12v
E13	YEL	Lockup Control Solenoid 'A'	L/Up on: 12v, off: 0v
E16	WHT	Linear Solenoid Valve 'N'	<0.050v
E17	WHT	Mainshaft Speed Sensor 'N'	<0.050v
E18	GRN	Countershaft Speed Sensor N	<0.050v
E19	WHT/RED	Serial Data Line 'B'	Digital Signals
E20	WHT/GRN	Serial Data Line 'A'	Digital Signals
E22	WHT/GRN	Interlock Control Unit Signal	Key & Brake on: 12v
E23	BRN/YEL	A/T: D4 Indicator Light	D4 lamp on: 0v, off: 12v
E24	LT GRN	A/T: Gear Position Switch	In P/N: 0v, Others: 12v
E25	GRN/WHT	Shift Control Solenoid 'B'	SSB Off: 0v, On: 12v
E26	GRN/BLK	Lockup Control Solenoid 'B'	LSB Off: 0v, On: 12v

2000-04 Sedan 3.5L V6 MFI VIN KA9 [A/T] 8P 'F' Connector

PCM Pin #	W/Color	Circuit Description (8-Pin)	Value at Hot Idle
F1	PNK/BLK	EVAP Bypass Solenoid	Solenoid Off: 12v, On: 1v
F2	WHT/RED	HO2S-12 (B1 S2) Signal	0.1-1.1v
F3	BLU	Right Peak Hold Reset	Digital Signals
F4	YEL	Right Misfire Pulse	Digital Signals
F5	RED/WHT	EVAP Vent Shut Solenoid	Solenoid Off: 12v, On: 1v
F6	GRN/WHT	HO2S-12 Ground	<0.050v
F7	GRN/YEL	Left Peak Hold Reset	Digital Signals
F8	RED	Left Misfire Pulse	Digital Signals

Pin Connector Graphic

26-PIN CONNECTOR 'A'

16-PIN CONNECTOR 'B'   12-PIN CONNECTOR 'C'

22-PIN CONNECTOR 'D'

26-PIN CONNECTOR 'E'

8-PIN 'F'

WIRE SIDE OF HARNESS TERMINALS

05533_ADIA_G594

2005-06 Sedan 3.5L V6 'A' Connector (Gray, on PCM Wire Harness)

PCM Pin #	W/Color	Circuit Description (31-Pin)	Value at Hot Idle
A1	BLK	Ground for PCM	Less than 1.0v at all times
A2	BLK	Ground for PCM	Less than 1.0v at all times
A3	BRN/YEL	Ground for PCM	Less than 1.0v at all times
A4	YEL/BLK	PCM Power	B+
A5	RED/YEL	IAT Sensor Signal	0.1v to 4.8v
A6	GRN/WHT	Front O2 Sensor Heater Power	0v
A7	BLK/WHT	Rear O2 Sensor Heater Power	0v
A8	GRN/RED	Front Secondary HO2S Heater Power	Duty cycle
A9	BLK/WHT	Rear Secondary HO2S Heater Power	Duty cycle
A10	BLK/YEL	Solenoid Power Source	B+
A11	WHT/BLU	Detects Alternator 'L' Signal	0v
A12	WHT/RED	Detects Alternator 'FR' Signal	0v to 5v
A13	LT BLU	Drives Radiator Fan Relay	With radiator fan ON: About 0v
A14	RED/WHT	Detects IAT Sensor Signal	0.1v to 4.8v
A15	BRN/WHT	Drives No. 6 Ignition Coil	Pulses
A16	BLK/RED	Drives No. 5 Ignition Coil	Pulses
A17	BRN	Drives No. 4 Ignition Coil	Pulses
A18	WHT/BLU	Drives No. 3 Ignition Coil	Pulses
A19	BLU/WHT	Drives No. 2 Ignition Coil	Pulses
A20	YEL/GRN	Drives No. 1 Ignition Coil	Pulses
A21	BRN	Drives No. 1 Injector	Duty cycle
A22	GRN/YEL	Drives Rocker Arm VTEC Solenoid	About 0v
A23	BLU/BLK	Detects Rocker Arm VTEC Oil Pressure Signal	About 0v
A24	WHT/GRN	Sends Alternator Control Signal	B+ - varies with electrical load
A25	GRN	Sends Throttle Actuator Signal	Serial data
A26	BLU	Detects Throttle Actuator Signal	Serial data
A27	WHT/BLU	Drives No. 6 Injector	Duty cycle
A28	BLK/RED	Drives No. 5 Injector	Duty cycle
A29	YEL	Drives No. 4 Injector	Duty cycle
A30	BLU	Drives No. 3 Injector	Duty cycle
A31	RED	Drives No. 2 Injector	Duty cycle

2005-06 Sedan 3.5L V6 'B' Connector (White, on PCM Wire Harness)

PCM Pin #	W/Color	Circuit Description (24-Pin)	Value at Hot Idle
B1	BLU/RED	Drives CKP Sensor 'B' Signal	Pulses
B2	RED/YEL	Drives EVAP Canister Purge Valve	With engine running, engine coolant below 140 °F (60 °C): B+ With engine running, engine coolant above 140 °F(60 °C): duty controlled
B3	WHT/RED	Ground for Intake Manifold Tuning (IMT) Actuator	B-
B4	WHT/BLU	Drives IMT Actuator	B+
B5	BLU/RED	Drives EGR Valve	0v
B6	RED/BLU	Detects Knock Sensor Signal	With engine knocking: Pulses
B7	BLU/YEL	Drives Engine Mount Control Solenoid	About 0v
B8	YEL	Detects CMP Sensor Signal	Pulses
B9	GRN	Detects Rear Secondary HO2S (Bank 1, Sensor 2)	n/a
B10	N/A		
B11	WHT/BLK	Detects IMT Actuator Position	About 5.0v
B12	BLU	Detects Rear A/F Sensor (Bank 1, Sensor 1) VS CEL	3.4v to 4.8v
B13	RED	Reference Voltage Supply for Rear A/F Sensor (Bank 1, Sensor 1)	3.4v to 4.8v
B14	GRN	Detects Rear A/F Sensor (Bank 1, Sensor 1) Pump Cell	2.0v to 5.6v
B15	BRN/YEL	Ground Circuit for PCM	Less than 1.0v
B16	N/A		
B17	BLU	Detects CKP Sensor A Signal	Pulses
B18	WHT	Detects Front HO2S (Bank 2, Sensor 2) Signal	n/a
B19	BLU	Detects Transaxle Countershaft Speed	Pulses (with wheels rotating)
B20	RED/BLU	Detects Front A/F Sensor (Bank 2, sensor 1) VS CELL signal	3.4v to 4.8v
B21	RED/WHT	Reference Voltage Supply for Front A/F Sensor (Bank 2, Sensor 1)	0.1v to 4.8v
B22	GRN/RED	Detects Front A/F Sensor (Bank 2, Sensor 1) Pump Cell	2.0v to 5.6v
B23	RED/BLK	Detects Transmission Range Neutral Sensor	In neutral: About 0v
B24	GRN/RED	Detects ECT Sensor 2 Signal	0.1v to 4.8v

2005-06 Sedan 3.5L V6 'C' Connector (Black, On PCM Wire Harness)

PCM Pin #	W/Color	Circuit Description (22-Pin)	Value at Hot Idle
C1	BLU/YEL	Drives Shift Solenoid A	In D, D3, M and R: B+
C2	GRN/RED	Drives Shift Solenoid D	In P, N, D or D3: 0v or B+
C3	BLU/YEL	Detects ATF Temperature Signal	About 0.2v to 4.0v (1.8v at operating temp)
C4	GRN/WHT	Sensor Ground	Less than 1.0v
C5	GRN/RED	Detects MAP Sensor Signal	About 1.0v, depending on engine speed
C6	YEL/RED	Provides Sensor Pullup Voltage	5.0v
C7	RED	Drives A/T Clutch Pressure Solenoid A	Pulses
C8	GRN/WHT	Drives Shift Solenoid Valve B	0v or B+
C9	BLU/WHT	Detects 3rd Clutch Transmission Fluid Pressure Switch Input	5.0v (without 3rd clutch pressure)
C10	YEL/GRN	Detects Transmission Range Switch D Input	In D: About 0v
C11	BLU/BLK	Detects 2nd Clutch Transmission Fluid Pressure Switch Input	About 5.0v (without 2nd clutch pressure)
C12	GRN/YEL	Sensor Ground	Less than 1.0v
C13	WHT/BLK	Detects EGR Valve Position Sensor Signal	1.2v to 3.0v
C14	YEL/BLU	Provides Sensor Pullup Voltage	About 5v
C15	BRN/WHT	Drives A/T Clutch Pressure Control Solenoid B	Pulses
C16	GRN	Drives Shift Solenoid Valve C	0v or B+
C17	N/A		
C18	RED	Detects Transmission Range D3 Signal	In D3: About 0v; else about 5v
C19	BLU/YEL	Detects Transmission Range D/D3 Signal	In D, D3: About 0v, else B+
C20	BLU/YEL	Detects 4th Clutch Transmission Fluid Pressure Switch Input	About 0v without 4th clutch pressure
C21	RED/WHT	Detects Transmission Range P, R, N Sensor Input	In P, R, N: About 0v, else B+
C22	GRN/RED	Drives A/T Clutch Pressure Control Solenoid C	Pulses

TL PIN CHARTS

1995-96 Sedan 2.5L I5 MFI VIN UA2 [A/T] 26P 'A' Connector

PCM Pin #	W/Color	Circuit Description (26-Pin)	Value at Hot Idle
A1	BRN	Injector 1	2.0-3.3 ms
A2	RED	Injector 2	2.0-3.3 ms
A3	BLU	Injector 3	2.0-3.3 ms
A4	GRN/BLK	Fuel Pump Relay Control	Relay Off: 12v, On: 1v
A5	BLK/BLU	Idle Air Control Valve	Pulse Signals
A6	RED	EGR Solenoid Control	Solenoid Off: 12v, On: 1v
A7	GRN/RED	Malfunction Indicator Lamp	MIL Off: 12v, On: 1v
A8	RED/BLU	A/C Clutch Relay Control	Relay Off: 12v, On: 1v
A9	PNK/BLU	Intake Air Bypass Solenoid	Solenoid Off: 12v, On: 1v
A10	RED/GRN	Fuel Cutoff Feedback Signal	Pulses
A11	YEL/GRN	Igniter Signal	Digital Signals: 0-12-0v
A12	BLK	Power Ground	<0.1v
A13	YEL/BLK	Main Relay Power (B+)	12-14v
A14	YEL	Injector 4	2.0-3.3 ms
A15	BLK/RED	Injector 5	2.0-3.3 ms
A16	ORN/BLK	HO2S-11 (B1 S1) Heater	Digital Signals: 0-12-0v
A17	GRN/RED	HO2S-12 (B1 S2) Heater	Digital Signals: 0-12-0v
A18	RED	EVAP Purge Solenoid	Pulse Signals
A19	GRN/YEL	Radiator Fan Relay Control	Fan on: 1v: off: 12v
A20	GRN/WHT	Engine Mount Solenoid	Solenoid Off: 12v, On: 1v
A21	WHT/GRN	Alternator Charging Signal	Lights Off: 12v, On: 1v
A22	PNK	TCM A/T Feedback Signal	5v
A25	BLK	Power Ground	<0.1v
A26	BRN/BLK	Logic Ground	<0.050v

1995-96 Sedan 2.5L I5 MFI VIN UA2 [A/T] 12P 'C' Connector

PCM Pin #	W/Color	Circuit Description (12-Pin)	Value at Hot Idle
C1	WHT/GRN	Rear Knock Sensor Signal	No Detonation: 18mv AC
C2	RED/BLU	Front Knock Sensor Signal	No Detonation: 18mv AC
C3	---	Not Used	---
C4	WHT/RED	CKF Sensor	AC pulse signals
C5	BLU/RED	CKF Sensor Ground	<0.050v
C6-12	---	Not Used	---

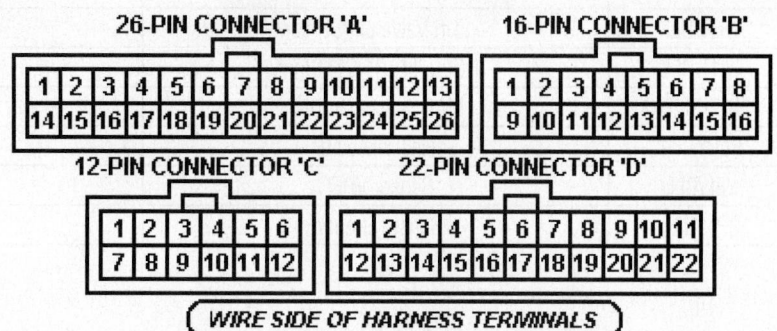

05533_ADIA_G595

Pin Connector Graphic

1995-96 Sedan 2.5L I5 MFI VIN UA2 [A/T] 16P 'B' Connector

PCM Pin #	W/Color	Circuit Description (16-Pin)	Value at Hot Idle
B1	YEL/BLK	Main Relay Power (B+)	12-14v
B2	WHT/GRN	TCM A/T FI 'A' Signal	5v
B3	BLU/BLK	A/C Switch Signal	A/C Off: 12v, On: 1v
B4	YEL/GRN	A/T: Gear Position Switch	In P/N: 0v, Others: 12v
B5	BLU/RED	Starter Switch Signal	Cranking: 9-11v
B6	ORN	CYP Sensor Ground	<0.050v
B7	ORN/BLU	TDC Sensor Ground	<0.050v
B8	BLU/GRN	CKP Sensor Ground	<0.050v
B9	BRN/BLK	Logic Ground	<0.050v
B10	WHT/RED	TCM A/T FI ' B' Signal	5v
B11	BRN	EVAP Purge Flow Switch	Switch on: 12v, off: 0.1v
B12	GRN	PSP Switch Signal	Straight: 0v, Turning: 12v
B13	ORN	Vehicle Speed Sensor	Moving: pulse signals
B14	WHT	CYP Sensor Signal	AC pulse signals
B15	WHT/BLU	TDC Sensor Signal	AC pulse signals
B16	BLU/YEL	CKP Sensor Signal	AC pulse signals

1995-96 Sedan 2.5L I5 MFI VIN UA2 [A/T] 22P 'D' Connector

PCM Pin #	W/Color	Circuit Description (22-Pin)	Value at Hot Idle
D1	WHT/GRN	Keep Alive Power (VBU)	12-14v
D2	BLU	TCM A/T FI Data Signal	5v
D3	RED/BLU	Barometric Output Signal	3v (at sea level)
D4	GRN/YEL	K-Line Signal	12v
D5	WHT/RED	Alternator 'FR' Signal	Digital Signals: 0-5-0v
D6	RED/BLK	TP Sensor Signal	0.5-0.6v
D7	RED/WHT	ECT Sensor Signal	At 180°F: 0.5-0.6v
D8	RED/YEL	IAT Sensor Signal	At 100°F: 2-3v
D9	WHT/BLU	MAF Sensor Signal	In P/N: 1.1-1.6v
D10	YEL/WHT	Sensor VREF	4.9-5.1v
D11	GRN/WHT	Sensor Ground	<0.050v
D12	GRN/WHT	Brake Switch Signal	Brake Off: 12v, On: 0v
D13	ORN	Service Check Connector	SCS Open: 4.80v
D14	BLU/RED	HO2S-12 (B1 S2) Signal	0.1-1.1v
D15	---	Not Used	---
D16	GRN/RED	Electric Load Detector	Varies w/Load: 2.5-3.5v
D17	WHT/BLK	EGR Valve Lift Sensor	1.1-1.2v
D18	WHT/RED	HO2S-11 (B1 S1) Signal	0.1-1.1v
D19	BLU/RED	MAF Sensor Ground	<0.050v
D20	BLU/WHT	Solenoid Feed (B+)	4.9-5.1v
D21	YEL/BLU	Sensor VREF	4.9-5.1v
D22	GRN/BLU	Sensor Ground	<0.050v

1997-98 Sedan 2.5L I5 MFI VIN UA2 [A/T] 26P 'A' Connector

PCM Pin #	W/Color	Circuit Description (26-Pin)	Value at Hot Idle
A1	BRN	Injector 1	2.0-3.3 ms
A2	RED	Injector 2	2.0-3.3 ms
A3	BLU	Injector 3	2.0-3.3 ms
A4	GRN/BLK	Fuel Pump Relay Control	Relay Off: 12v, On: 1v
A5	BLK/BLU	Idle Air Control Valve	Pulse Signals
A6	RED	EGR Solenoid Control	Solenoid Off: 12v, On: 1v
A7	GRN/RED	Malfunction Indicator Lamp	MIL Off: 12v, On: 1v
A8	RED/BLU	A/C Clutch Relay Control	Relay Off: 12v, On: 1v
A9	PNK/BLU	Intake Air Bypass Solenoid	Solenoid Off: 12v, On: 1v
A10	RED/GRN	Fuel Cutoff Feedback Signal	Pulses
A11	YEL/GRN	Igniter Signal	Digital Signals: 0-12-0v
A12	BLK	Power Ground	<0.1v
A13	YEL/BLK	Main Relay Power (B+)	12-14v
A14	YEL	Injector 4	2.0-3.3 ms
A15	BLK/RED	Injector 5	2.0-3.3 ms
A16	ORN/BLK	HO2S-11 (B1 S1) Heater	Digital Signals: 0-12-0v
A17	GRN/RED	HO2S-12 (B1 S2) Heater	Digital Signals: 0-12-0v
A18	RED	EVAP Purge Solenoid	Pulse Signals
A19	GRN/YEL	Radiator Fan Relay Control	Fan on: 1v: off: 12v
A20	GRN/WHT	Engine Mount Solenoid	Solenoid Off: 12v, On: 1v
A21	WHT/GRN	Alternator Charging Signal	Lights Off: 12v, On: 1v
A22	PNK	TCM A/T Feedback Signal	5v
A23	YEL/BLU	EVAP Vent Solenoid	Solenoid Off: 12v, On: 1v
A24	WHT/BLU	EVAP Bypass Solenoid	Solenoid Off: 12v, On: 1v
A25	BLK	Power Ground	<0.1v
A26	BRN/BLK	Logic Ground	<0.050v

1997-98 Sedan 2.5L I5 MFI VIN UA2 [A/T] 12P 'C' Connector

PCM Pin #	W/Color	Circuit Description (12-Pin)	Value at Hot Idle
C1	WHT/GRN	Rear Knock Sensor Signal	No Detonation: 18mv AC
C2	RED/BLU	Front Knock Sensor Signal	No Detonation: 18mv AC
C3	---	Not Used	---
C4	WHT/RED	CKF Sensor	AC pulse signals
C5	BLU/RED	CKF Sensor Ground	<0.050v

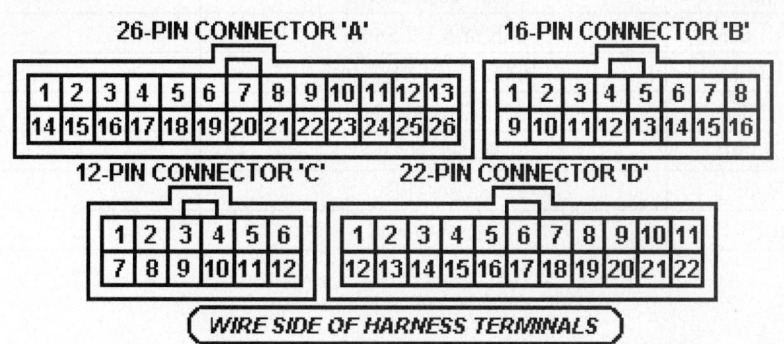

Pin Connector Graphic

05533_ADIA_G595

1997-98 Sedan 2.5L I5 MFI VIN UA2 [A/T] 16P 'B' Connector

PCM Pin #	W/Color	Circuit Description (16-Pin)	Value at Hot Idle
B1	YEL/BLK	Main Relay Power (B+)	12-14v
B2	WHT/GRN	TCM A/T FI 'A' Signal	5v
B3	BLU/BLK	A/C Switch Signal	A/C Off: 12v, On: 1v
B4	YEL/GRN	A/T: Gear Position Switch	In P/N: 0v, Others: 12v
B5	BLU/RED	Starter Switch Signal	Cranking: 9-11v
B6	ORN	CYP Sensor Ground	<0.050v
B7	ORN/BLU	TDC Sensor Ground	<0.050v
B8	BLU/GRN	CKP Sensor Ground	<0.050v
B9	BRN/BLK	Logic Ground	<0.050v
B10	WHT/RED	TCM A/T FI ' B' Signal	5v
B11	---	Not Used	---
B12	GRN	PSP Switch Signal	Straight: 0v, Turning: 12v
B13	ORN	Vehicle Speed Sensor	Moving: pulse signals
B14	WHT	CYP Sensor Signal	AC pulse signals
B15	WHT/BLU	TDC Sensor Signal	AC pulse signals
B16	BLU/YEL	CKP Sensor Signal	AC pulse signals

1997-98 Sedan 2.5L I5 MFI VIN UA2 [A/T] 22P 'D' Connector

PCM Pin #	W/Color	Circuit Description (22-Pin)	Value at Hot Idle
D1	WHT/GRN	Keep Alive Power (VBU)	12-14v
D2	BLU	TCM A/T FI Data Signal	5v
D3	RED/BLU	Barometric Output Signal	3v (at sea level)
D4	GRN/YEL	K-Line Signal	12v
D5	WHT/RED	Alternator 'FR' Signal	Digital Signals: 0-5-0v
D6	RED/BLK	TP Sensor Signal	0.5-0.6v
D7	RED/WHT	ECT Sensor Signal	At 180°F: 0.5-0.6v
D8	RED/YEL	IAT Sensor Signal	At 100°F: 2-3v
D9	WHT/BLU	MAF Sensor Signal	In P/N: 1.1-1.6v
D10	YEL/WHT	Sensor VREF	4.9-5.1v
D11	GRN/WHT	Sensor Ground	<0.050v
D12	GRN/WHT	Brake Switch Signal	Brake Off: 12v, On: 0v
D13	ORN	Service Check Connector	SCS Open: 4.80v
D14	BLU/RED	HO2S-12 (B1 S2) Signal	0.1-1.1v
D15	PNK	Fuel Tank Pressure Sensor	Fuel Cap off: 2.5v
D16	GRN/RED	Electric Load Detector	Varies w/Load: 2.5-3.5v
D17	WHT/BLK	EGR Valve Lift Sensor	1.1-1.2v
D18	WHT/RED	HO2S-11 (B1 S1) Signal	0.1-1.1v
D19	BLU/RED	MAF Sensor Ground	<0.050v
D20	BLU/WHT	Solenoid Feed (B+)	4.9-5.1v
D21	YEL/BLU	Sensor VREF	4.9-5.1v
D22	GRN/BLU	Sensor Ground	<0.050v

1996 Sedan 3.2L V6 MFI VIN UA3 [A/T] 26P 'A' Connector

PCM Pin #	W/Color	Circuit Description (26-Pin)	Value at Hot Idle
A1	BRN	Injector 1	2.0-3.3 ms
A2	RED	Injector 2	2.0-3.3 ms
A3	BLU	Injector 3	2.0-3.3 ms
A4	GRN/BLK	Fuel Pump Relay Control	Relay Off: 12v, On: 1v
A5	BLK/BLU	Idle Air Control Valve	Pulse Signals
A6	RED	EGR Solenoid Control	Solenoid Off: 12v, On: 1v
A7	GRN/RED	Malfunction Indicator Lamp	MIL Off: 12v, On: 1v
A8	RED/BLU	A/C Clutch Relay Control	Relay Off: 12v, On: 1v
A9	---	Not Used	---
A10	GRN	Pressure Regulator Solenoid	Solenoid Off: 12v, On: 1v
A11	PNK	Ignition Coil 1 Control	Digital Signals: 0-12-0v
A12	BLK	Power Ground	<0.1v
A13	YEL/BLK	Main Relay Power (B+)	12-14v
A14	YEL	Injector 4	2.0-3.3 ms
A15	BLK/RED	Injector 5	2.0-3.3 ms
A16	WHT/BLU	Injector 6	2.0-3.3 ms
A17	---	Not Used	---
A18	GRN/RED	HO2S-21 (B2 S1) Heater	Digital Signals: 0-12-0v
A19	ORN/BLK	HO2S-12 (B1 S2) Heater	Digital Signals: 0-12-0v
A20	RED/BLU	IAB Solenoid 1 Control	Solenoid Off: 12v, On: 1v
A21	ORN/BLK	HO2S-22 (B2 S2) Heater	Digital Signals: 0-12-0v
A22	YEL	IAB Solenoid 2 Control	Solenoid Off: 12v, On: 1v
A23	RED	EVAP Purge Solenoid	Pulse Signals
A24	BRN	Ignition Coil 2 Control	Digital Signals: 0-12-0v
A25	BLK	Power Ground	<0.1v
A26	BRN/BLK	Logic Ground	<0.050v

1996 Sedan 3.2L V6 MFI VIN UA3 [A/T] 16P 'B' Connector

PCM Pin #	W/Color	Circuit Description (16-Pin)	Value at Hot Idle
B1	YEL/BLK	Main Relay Power (B+)	12-14v
B2	RED	Ignition Coil 6 Control	Digital Signals: 0-12-0v
B3	GRN	PSP Switch Signal	Straight: 0v, Turning: 12v
B4	---	Not Used	---
B5	ORN	CYP Sensor 2 Signal	AC pulse signals
B6	WHT	CYP Sensor 1 Signal	AC pulse signals
B7	ORN/BLU	CKP Sensor 2 Signal	AC pulse signals
B8	BLU/GRN	CKP Sensor 1 Signal	AC pulse signals
B9	BRN/BLK	Logic Ground	<0.050v
B10	GRY	Ignition Coil 5 Control	Digital Signals: 0-12-0v
B11	GRN	Ignition Coil 4 Control	Digital Signals: 0-12-0v
B12	BLU	Ignition Coil 3 Control	Digital Signals: 0-12-0v
B13	ORN	Vehicle Speed Sensor	Moving: pulse signals
B14	---	Not Used	---
B15	WHT/BLU	CKP Sensor 2 Ground	<0.050v
B16	BLU/YEL	CKP Sensor 1 Ground	<0.050v

1996 Sedan 3.2L V6 MFI VIN UA3 [A/T] 12P 'C' Connector

PCM Pin #	W/Color	Circuit Description (12-Pin)	Value at Hot Idle
C1	---	Not Used	---
C2	BLU/BLK	A/C Switch Signal	A/C Off: 12v, On: 1v
C3	RED/BLU	A/C Switch Signal 'B'	A/C Off: 4-5v, On: 0v
C4	---	Not Used	---
C5	ORN	Service Check Connector	SCS Open: 4.80v
C6	BLU/RED	Starter Switch Signal	Cranking: 9-11v
C7	---	Not Used	---
C8	BLU	Engine Speed Pulse Sensor	Pulses
C9	GRN/BLK	Traction Control Inhibit Signal	Pulses
C10	GRY/GRN	T/C Fuel Cut Signal	Pulses
C11	---	Not Used	---
C12	LT BLU	T/C Standby Signal	Pulses

1996 Sedan 3.2L V6 MFI VIN UA3 [A/T] 22P 'D' Connector

PCM Pin #	W/Color	Circuit Description (22-Pin)	Value at Hot Idle
D1	WHT/GRN	Keep Alive Power (VBU)	12-14v
D2	RED/BLU	Knock Sensor Signal (Bank 1)	No Detonation: 18mv AC
D3	ORN/RED	Barometric Output Signal	3v (at sea level)
D4	GRN/YEL	K-Line Signal	12v
D5	WHT/RED	Alternator 'FR' Signal	Digital Signals: 0-5-0v
D6	RED/BLK	TP Sensor Signal	0.5-0.6v
D7	RED/WHT	ECT Sensor Signal	At 180°F: 0.5-0.6v
D8	RED/YEL	IAT Sensor Signal	At 100°F: 2-3v
D9	WHT/YEL	MAP Sensor Signal	0.8-0.9v
D10	YEL/WHT	MAP Sensor VREF	4.9-5.1v
D11	GRN/WHT	MAP Sensor Ground	<0.050v
D12	WHT/BLU	CYP Sensor 1 Ground	<0.050v
D13	WHT/GRN	Knock Sensor Signal (Bank 2)	No Detonation: 18mv AC
D14	ORN/BLU	CYP Sensor 2 Ground	<0.050v
D15	BRN	EVAP Purge Flow Switch	Switch on: 0v, off: 12v
D16	---	Not Used	---
D17	WHT/BLK	EGR Valve Lift Sensor	1.1-1.2v
D18	WHT/RED	HO2S-11 (B1 S1) Signal	0.1-1.1v
D19	BLU/RED	HO21-21 (B2 S1) Signal	0.1-1.1v
D20	BLU	TCS VREF	4.9-5.1v
D21	YEL/BLU	Sensor VREF	4.9-5.1v
D22	GRN/BLU	Sensor Ground	<0.050v

1996 Sedan 3.2L V6 MFI VIN UA3 [A/T] 26P 'E' Connector

PCM Pin #	W/Color	Circuit Description (26-Pin)	Value at Hot Idle
E1	BLK/YEL	Solenoid Feed (B+)	12-14v
E2	GRN/WHT	Brake Switch Signal	Brake Off: 12v, On: 0v
E3	RED	Linear Solenoid Valve 'P'	AC Pulse Signals
E4	ORN/BLU	Mainshaft Speed Sensor 'P'	AC Pulse Signals
E5	BLU/GRN	Countershaft Speed Sensor	Wheels turning: 12v
E6	BLU/RED	TCS Shift Control Signal	Pulse Signals
E7	WHT	A/T: Gear Position Switch	In 'D': 0v, others: 12v
E8	YEL	A/T: Gear Position Switch	In D4: 0v, other: 12v
E9	GRN	A/T: Gear Position Switch	In D3: 0v, other: 12v
E10	BLU	A/T: Gear Position Switch	In D2: 0v, other: 12v
E11	BRN	A/T: Gear Position Switch	In D1: 0v, other: 12v
E12	BLU/YEL	Shift Control Solenoid 'A'	In D2/D3: 12v
E13	YEL	Lockup Control Solenoid 'A'	LSA Off: 0v, On: 12v
E14-15	---	Not Used	---
E16	WHT	Linear Solenoid Valve 'N'	Pulses
E17	WHT/BLU	Mainshaft Speed Sensor 'N'	<0.050v
E18	BLU/YEL	Countershaft Speed Sensor N	<0.050v
E19-21	---	Not Used	---
E22	WHT/GRN	Interlock Control Unit Signal	Key & Brake on: 12v
E23	GRN/BLK	A/T: D4 Indicator Light	D4 Light On: 0v, Off: 12v
E24	YEL/GRN	A/T: Gear Position Switch	In P/N: 0v, Others: 12v
E25	GRN/WHT	Shift Control Solenoid 'B'	In D1/D2: 12v
E26	GRN/BLK	Lockup Control Solenoid 'B'	LSB Off: 0v, On: 12v

1996 Sedan 3.2L V6 MFI VIN UA3 [A/T] 8P 'F' Connector

PCM Pin #	W/Color	Circuit Description (8-Pin)	Value at Hot Idle
F1	---	Not Used	---
F2	WHT/RED	HO2S-12 (B1 S2) Signal	0.1-1.1v
F3	GRN/BLK	Right Peak Hold Reset	Digital Signals
F4	YEL	Right Misfire Pulse	Digital Signals
F6	GRN/WHT	HO2S-12 Ground	<0.050v
F7	GRN/YEL	Left Peak Hold Reset	Pulses
F8	YEL/RED	Left Misfire Pulse	Pulses

WIRE SIDE OF HARNESS TERMINALS

05533_ADIA_G594

Pin Connector Graphic

1997-98 Sedan 3.2L V6 MFI VIN UA3 [A/T] 26P 'A' Connector

PCM Pin #	W/Color	Circuit Description (26-Pin)	Value at Hot Idle
A1	BRN	Injector 1	2.0-3.3 ms
A2	RED	Injector 2	2.0-3.3 ms
A3	BLU	Injector 3	2.0-3.3 ms
A4	GRN/BLK	Fuel Pump Relay Control	Relay Off: 12v, On: 1v
A5	BLK/BLU	Idle Air Control Valve	Pulse Signals
A6	RED	EGR Solenoid Control	Solenoid Off: 12v, On: 1v
A7	GRN/RED	Malfunction Indicator Lamp	MIL Off: 12v, On: 1v
A8	RED/BLU	A/C Clutch Relay Control	Relay Off: 12v, On: 1v
A9	---	Not Used	---
A10	GRN	Fuel Pressure Regulator	Solenoid Off: 12v, On: 1v
A11	PNK	Ignition Coil 1 Control	Digital Signals: 0-12-0v
A12	BLK	Power Ground	<0.1v
A13	YEL/BLK	Main Relay Power (B+)	12-14v
A14	YEL	Injector 4	2.0-3.3 ms
A15	BLK/RED	Injector 5	2.0-3.3 ms
A16	WHT/BLU	Injector 6	2.0-3.3 ms
A17	---	Not Used	---
A18	GRN/RED	HO2S-21 (B2 S1) Heater	Digital Signals: 0-12-0v
A19	ORN/BLK	HO2S-12 (B1 S2) Heater	Digital Signals: 0-12-0v
A20	RED/BLU	IAB Solenoid 1 Control	Solenoid Off: 12v, On: 1v
A21	ORN/BLK	HO2S-22 (B2 S2) Heater	Digital Signals: 0-12-0v
A22	YEL	IAB Solenoid 2 Control	Solenoid Off: 12v, On: 1v
A23	RED/YEL	EVAP Purge Solenoid	Pulse Signals
A24	BRN	Ignition Coil 2 Control	Digital Signals: 0-12-0v
A25	BLK	Power Ground	<0.1v
A26	BRN/BLK	Logic Ground	<0.050v

1997-98 Sedan 3.2L V6 MFI VIN UA3 [A/T] 16P 'B' Connector

PCM Pin #	W/Color	Circuit Description (16-Pin)	Value at Hot Idle
B1	YEL/BLK	Main Relay Power (B+)	12-14v
B2	RED	Ignition Coil 6 Control	Digital Signals: 0-12-0v
B3	GRN	PSP Switch Signal	Straight: 0v, Turning: 12v
B4	---	Not Used	---
B5	ORN	CYP Sensor 2 Signal	AC pulse signals
B6	WHT	CYP Sensor 1 Signal	AC pulse signals
B7	ORN/BLU	CKP Sensor 2 Signal	AC pulse signals
B8	BLU/GRN	CKP Sensor 1 Signal	AC pulse signals
B9	BRN/BLK	Logic Ground	<0.050v
B10	GRY	Ignition Coil 5 Control	Digital Signals: 0-12-0v
B11	GRN	Ignition Coil 4 Control	Digital Signals: 0-12-0v
B12	BLU	Ignition Coil 3 Control	Digital Signals: 0-12-0v
B13	ORN	Vehicle Speed Sensor	Moving: pulse signals
B14	---	Not Used	---
B15	WHT/BLU	CKP Sensor 2 Ground	<0.050v
B16	BLU/YEL	CKP Sensor 1 Ground	<0.050v

1997-98 Sedan 3.2L V6 MFI VIN UA3 [A/T] 16P 'C' Connector

PCM Pin #	W/Color	Circuit Description (16-Pin)	Value at Hot Idle
C1	---	Not Used	---
C2	BLU/BLK	A/C Switch Signal	A/C Off: 12v, On: 1v
C3	RED/BLU	A/C Switch Signal 'B'	A/C Off: 4-5v, On: 0v
C4	---	Not Used	---
C5	ORN	Service Check Connector	SCS Open: 4.80v
C6	BLU/RED	Starter (Cranking) Signal	Cranking: 9-11v
C7	---	Not Used	---
C8	BLU	Engine Speed Pulse Sensor	Pulses
C9	GRN/BLK	Traction Control Inhibit Signal	Pulses
C10	GRY/GRN	T/C Fuel Cut Signal	Pulses
C11	---	Not Used	---
C12	LT BLU	T/C Standby Signal	Pulses
C13-16	---	Not Used	---

1997-98 Sedan 3.2L V6 MFI VIN UA3 [A/T] 22P 'D' Connector

PCM Pin #	W/Color	Circuit Description (22-Pin)	Value at Hot Idle
D1	WHT/GRN	Keep Alive Power (VBU)	12-14v
D2	RED/BLU	Knock Sensor Signal (Bank 1)	No Detonation: 18mv AC
D3	ORN/RED	Barometric Output Signal	3v (at sea level)
D4	GRN/YEL	K-Line Signal	12v
D5	WHT/RED	Alternator 'FR' Signal	Digital Signals: 0-5-0v
D6	RED/BLK	TP Sensor Signal	0.5-0.6v
D7	RED/WHT	ECT Sensor Signal	At 180°F: 0.5-0.6v
D8	RED/YEL	IAT Sensor Signal	At 100°F: 2-3v
D9	WHT/YEL	MAP Sensor Signal	0.8-0.9v
D10	YEL/WHT	MAP Sensor VREF	4.9-5.1v
D11	GRN/WHT	MAP Sensor Ground	<0.050v
D12	WHT/BLU	CYP Sensor 1 Ground	<0.050v
D13	WHT/GRN	Knock Sensor Signal (Bank 2)	No Detonation: 18mv AC
D14	ORN/BLU	CYP Sensor 2 Ground	<0.050v
D15	PNK	Fuel Tank Pressure Sensor	Fuel Cap off: 2.5v
D17	WHT/BLK	EGR Valve Lift Sensor	1.1-1.2v
D18	WHT/RED	HO2S-11 (B1 S1) Signal	0.1-1.1v
D19	BLU/RED	HO21-21 (B2 S1) Signal	0.1-1.1v
D20	BLU	TCS VREF	4.9-5.1v
D21	YEL/BLU	Sensor VREF	4.9-5.1v
D22	GRN/BLU	Sensor Ground	<0.050v

Standard Colors and Abbreviations

Abbreviation	Color	Abbreviation	Color	Abbreviation	Color
BLK	Black	LT BLU	Lt. Blue	TAN	Tan
BLU	Blue	LT GRN	Lt. Green	VIO	Violet
BRN	Brown	ORN	Orange	WHT	White
GRY	Gray	PNK	Pink	YEL	Yellow
GRN	Green	PPL	Purple		

1997-98 Sedan 3.2L V6 MFI VIN UA3 [A/T] 26P 'E' Connector

PCM Pin #	W/Color	Circuit Description (26-Pin)	Value at Hot Idle
E1	BLK/YEL	Solenoid Feed (B+)	Sol. On: 12v, off: 0v
E2	GRN/WHT	Brake Switch Signal	Brake Off: 12v, On: 0v
E3	RED	Linear Solenoid Valve 'P'	Pulses
E4	WHT/GRN	Mainshaft Speed Sensor 'P'	Pulses
E5	RED/BLU	Countershaft Speed Sensor P	Moving: AC Pulses
E6	BLU/RED	TCS Shift Control Signal	Pulses
E7	WHT	A/T: Gear Position Switch	In 'D': 0v, others: 12v
E8	YEL	A/T: Gear Position Switch	In D4: 0v, other: 12v
E9	GRN	A/T: Gear Position Switch	In D3: 0v, other: 12v
E10	BLU	A/T: Gear Position Switch	In D2: 0v, other: 12v
E11	BRN	A/T: Gear Position Switch	In D1: 0v, other: 12v
E12	BLU/YEL	Shift Control Solenoid 'A'	SSA Off: 0v, On: 12v
E13	YEL	Lockup Control Solenoid 'A'	L/U on: 12v, off: 0v
E14-15	---	Not Used	---
E16	WHT	Linear Solenoid Valve 'N'	Pulse Signals
E17	WHT/BLU	Mainshaft Speed Sensor 'N'	<0.050v
E18	BLU/YEL	Countershaft Speed Sensor N	<0.050v
E19-21	---	Not Used	---
E22	WHT/GRN	Interlock Control Unit Signal	Key & Brake on: 12v
E23	GRN/BLK	A/T: D4 Indicator Light	D4 Light On: 0v, Off: 12v
E24	YEL/GRN	A/T: Gear Position Switch	In P/N: 0v, Others: 12v
E25	GRN/WHT	Shift Control Solenoid 'B'	SSB Off: 0v, On: 12v
E26	GRN/BLK	Lockup Control Solenoid 'B'	LUS on: 12v, off: 0v

1997-98 Sedan 3.2L V6 MFI VIN UA3 [A/T] 8P 'F' Connector

PCM Pin #	W/Color	Circuit Description (8-Pin)	Value at Hot Idle
F1	WHT/BLU	EVAP Bypass Solenoid	Solenoid Off: 12v, On: 1v
F2	WHT/RED	Rear HO2S-2 Signal	0.1-1.1v
F3	GRN/BLK	Right Peak Hold Reset	Digital Signals
F4	YEL	Right Misfire Pulse	Digital Signals
F5	YEL/BLU	EVAP Vent Shut Solenoid	Solenoid Off: 12v, On: 1v
F6	GRN/WHT	HO2S-12 Ground	<0.050v
F7	GRN/YEL	Left Peak Hold Reset	Digital Signals
F8	YEL/RED	Left Misfire Pulse	Digital Signals

Pin Connector Graphic

05533_ADIA_G594

1999 Sedan 3.2L V6 MFI VIN UA3 [A/T] 32P 'A' Connector

PCM Pin #	W/Color	Circuit Description (32-Pin)	Value at Hot Idle
A1	---	Not Used	---
A2	GRN/WHT	Engine Mount Solenoid	0v, Off-Idle: 12v
A3	BLU	EVAP Bypass Solenoid	Solenoid Off: 12v, On: 1v
A4	GRN/WHT	EVAP Vent Solenoid	Solenoid Off: 12v, On: 1v
A5	BLU/GRN	Cruise Control Signal	Cruise On: pulse signals
A6	RED/YEL	EVAP Purge Solenoid	Pulse Signals
A7	ORN/WHT	Reference Voltage	4.9-5.1v
A8	BLK/WHT	HO2S-12 (B1 S2) Heater	Digital Signals: 0-12-0v
A9	BLU/WHT	Vehicle Speed Sensor	Moving: pulse signals
A10	BRN	Service Check Connector	SCS Open: 4.80v
A11	LT GRN	Gear Position Signal Out	In 'P': 4v, Others: 0v
A12	PNK	Immobilizer Indicator Control	Lamp On: 1v, Off: 12v
A13	BLU	Immobilizer Enable Signal	Digital Signals
A14	GRN/BLK	D4 Indicator Light	In D4: 0v, Off: 12v
A15	GRN/YEL	Immobilizer Fuel Pump Relay	Relay Off: 12v, On: 1v
A17	RED	A/C Clutch Relay Control	Relay Off: 12v, On: 1v
A18	GRN/ORN	Malfunction Indicator Light	MIL Off: 12v, On: 1v
A19	BLU	Engine Speed Signal	Pulses
A20	GRN	Radiator Fan Relay Control	Relay Off: 12v, On: 1v
A21	GRY	K-Line Signal	12v
A22	---	Not Used	---
A23	WHT/RED	HO2S-12 (B1 S2) Signal	0.1-1.1v
A24	BLU/ORN	Starter Switch Signal	Cranking: 9-11v
A25	BLU/GRN	INOCD Immobilizer Code	Digital Signals
A26	GRN	PSP Switch Signal	Straight: 0v, Turning: 12v
A27	BLU/RED	A/C Switch Signal	A/C Off: 12v, On: 1v
A28	WHT/RED	Interlock Control Unit Signal	Key & Brake on: 12v
A29	LT GRN	Fuel Tank Pressure Sensor	Fuel Cap off: 2.5v
A30	GRN/RED	Electronic Load Detector	Varies: 2.5-3.5v
A31	RED/BLU	Throttle Position Sensor Out	0.5v
A32	GRN/WHT	Brake Switch Signal	Brake on: 0v, off: 12v

Pin Connector Graphic

05533_ADIA_G591

1999 Sedan 3.2L V6 MFI VIN UA3 [A/T] 25P 'B' Connector

PCM Pin #	W/Color	Circuit Description (25-Pin)	Value at Hot Idle
B1	YEL/BLK	Main Relay Power (B+)	12-14v
B2	BLK	Power Ground (PG1)	<0.1v
B3	BLK/RED	Injector 5 Control	2.0-3.3 ms
B4	YEL	Injector 4 Control	2.0-3.3 ms
B5	RED	Injector 2 Control	2.0-3.3 ms
B6	WHT/BLU	Injector 6 Control	2.0-3.3 ms
B7	PNK	E-EGR Control Solenoid	Digital Signals: 0-12-0v
B8	WHT	A/T: Clutch Solenoid 'A-'	Pulse Signals
B9	YEL/BLK	Main Relay Power (B+)	12-14v
B10	BLK	Power Ground (PG2)	<0.1v
B11	BRN	Injector 1 Control	2.0-3.3 ms
B12	GRN/YEL	VTEC Solenoid Control	Idle: 0v, Hi-Speed: 12v
B13	---	Not Used	---
B14	BLU/WHT	A/T: Gear Position Switch	In P/N: 0v, Others: 12v
B15	BLU	Injector 3 Control	2.0-3.3 ms
B16	---	Not Used	---
B17	RED	A/T: Clutch Solenoid 'A+'	Pulse Signals
B18	GRN	A/T: Clutch Solenoid 'B-'	Pulse Signals
B19	---	Not Used	---
B20	BRN/BLK	Logic Ground (LG1)	<0.050v
B21	WHT/YEL	Keep Alive Power (VBU)	12-14v
B22	BRN/BLK	Logic Ground (LG2)	<0.050v
B23	BLK/BLU	Idle Air Control Valve	Pulse Signals
B24	WHT/RED	A/T: 3rd Clutch Pressure Sw.	Open: 12v, Closed: 1v
B25	GRN	A/T: Clutch Solenoid 'B+'	Pulse Signals

05533_ADIA_G591

Pin Connector Graphic

1999 Sedan 3.2L V6 MFI VIN UA3 [A/T] 31P 'C' Connector

PCM Pin #	W/Color	Circuit Description (31-Pin)	Value at Hot Idle
C1	BLK/WHT	HO2S-11 (B1 S1) Heater	Digital Signals: 0-12-0v
C2	WHT/GRN	Alternator Charging Signal	Lights on: 0v, off: 12v
C3	BLU	Ignition Coil 3 Control	Digital Signals: 0-12-0v
C4	YEL/GRN	Ignition Coil 1 Control	Digital Signals: 0-12-0v
C5	WHT/RED	Alternator 'FR' Signal	Digital Signal: 0-5-0-5v
C6	WHT/BLK	EGR Valve Lift Sensor	1.2v
C7	GRN/WHT	MAP Sensor Ground (SG1)	<0.050v
C8	BLU	CKP Sensor Signal	AC pulse signals
C9	WHT	CKP Sensor Ground	<0.050v
C10	BLU/BLK	VTEC Pressure Switch	Idle: 0v, Cruise: 12v
C11	---	Not Used	---
C12	BLK/RED	Ignition Coil 5 Control	Digital Signals: 0-12-0v
C13	YEL	Ignition Coil 4 Control	Digital Signals: 0-12-0v
C14	RED	Ignition Coil 2 Control	Digital Signals: 0-12-0v
C15, C24	---	Not Used	---
C16	WHT	HO2S-11 (B1 S1) Signal	0.1-1.1v
C17	RED/GRN	MAP Sensor Signal	0.8-0.9v
C18	GRN/BLK	Sensor Ground (SG2)	<0.050v
C19	YEL/RED	MAP Sensor VREF (VCC1)	4.9-5.1v
C20	GRN	TDC1 Sensor Signal	AC pulse signals
C21	RED	TDC1 Sensor Ground	<0.050v
C22	RED/BLU	Knock Sensor Signal	No Detonation: 18mv AC
C23	WHT/BLU	Ignition Coil 6 Control	Digital Signals: 0-12-0v
C25	RED/YEL	IAT Sensor Signal	Varies w/temp: 0.5-4.9v
C26	RED/WHT	ECT Sensor Signal	At 180°F: 0.5-0.6v
C27	RED/BLK	TP Sensor	0.5-0.6v
C28	YEL/BLU	Sensor VREF (VCC2)	4.9-5.1v
C29	YEL	TDC2 Sensor Signal	AC pulse signals
C30	BLK	TDC2 Sensor Ground	<0.050v

1999 Sedan 3.2L V6 MFI VIN UA3 [A/T] 16P 'D' Connector

PCM Pin #	W/Color	Circuit Description (16-Pin)	Value at Hot Idle
D1	YEL	A/T: Lockup Solenoid	LUS Off: 12v, On: 1v
D2	GRN/WHT	A/T: Shift Solenoid 'B'	SSB Off: 12v, On: 1v
D3	GRN	A/T" Shift Solenoid 'C'	SSA Off: 12v, On: 1v
D4	RED/BLK	A/T: Gear Position Switch	In 'N': 0v, Others: 12v
D5	BLK/YEL	Solenoid Feed (B+)	12-14v
D6	WHT	A/T: Gear Position Switch	In 'R': 0v, Others: 12v
D7	BLU/YEL	A/T: Shift Solenoid 'A'	SSA Off: 12v, On: 1v
D8	PNK	A/T: Gear Position Switch	In D3: 0v, Others: 12v
D9	YEL	A/T: Gear Position Switch	In D4: 0v, Others: 12v
D10	BLU	Countershaft Speed Sensor P	Wheels turning: 12v
D11	RED	Mainshaft Speed Sensor 'P'	Moving: pulse signals
D12	WHT	Mainshaft Speed, Ground 'N'	<0.050v
D13	BLU/BLK	A/T: 2nd Clutch Pressure Sw.	Open: 12v, Closed: 1v
D14	BLU	A/T: Gear Position Switch	In D2: 0v, Others: 12v
D15	BRN	A/T: Gear Position Switch	In D1: 0v, Others: 12v
D16	GRN	Countershaft Speed Sensor N	<0.050v

2000-03 Sedan 3.2L V6 MFI VIN UA5 [A/T] 32P 'A' Connector

PCM Pin #	W/Color	Circuit Description (32-Pin)	Value at Hot Idle
A1	WHT/RED	HO2S-22 (B2 S2) Signal	0.1-1.1v
A2	GRN/WHT	Engine Mount Solenoid	0v, Off-Idle: 12v
A3	BLU	EVAP Bypass Solenoid	Solenoid Off: 12v, On: 1v
A4	GRN/WHT	EVAP Vent Solenoid	Solenoid Off: 12v, On: 1v
A5	BLU/GRN	Cruise Control Signal	Cruise On: pulse signals
A6	RED/YEL	EVAP Purge Solenoid	Pulse Signals
A7	ORN/WHT	Reference Voltage	4.9-5.1v
A8	BLK/WHT	HO2S-12 (B1 S2) Heater	Digital Signals: 0-12-0v
A9	BLU/WHT	Vehicle Speed Sensor	Moving: pulse signals
A10	BRN	Service Check Connector	SCS Open: 4.80v
A11-13	---	Not Used	---
A14	GRN/BLK	D4 Indicator Light	In D4: 0v, Off: 12v
A15	GRN/YEL	Fuel Pump Relay Control	Relay Off: 12v, On: 1v
A16	---	Not Used	---
A17	RED	A/C Clutch Relay Control	Relay Off: 12v, On: 1v
A18	GRN/ORN	Malfunction Indicator Light	MIL Off: 12v, On: 1v
A19	BLU	Engine Speed Signal	Pulse Signals
A20	GRN	Radiator Fan Relay Control	Relay Off: 12v, On: 1v
A21	GRY	K-Line Signal	12v
A22-23	---	Not Used	---

2000-03 Sedan 3.2L V6 MFI VIN UA5 [A/T] 32P 'A' Connector, cont'd

A24	BLU/ORN	Starter Switch Signal	Cranking: 9-11v
A25	---	Not Used	---
A26	GRN	PSP Switch Signal	Straight: 0v, Turning: 12v
A27	BLU/RED	A/C Switch Signal	A/C Off: 12v, On: 1v
A28	WHT/RED	Interlock Control Unit Signal	Key & Brake on: 12v
A29	LT GRN	Fuel Tank Pressure Sensor	Fuel Cap off: 2.5v
A30	GRN/RED	Electronic Load Detector	Varies: 2.5-3.5v
A31	YEL/GRN	ECT Sensor Signal to TCM	Digital Signals
A32	WHT/BLK	Brake Switch Signal	Brake on: 0v, off: 12v

05533_ADIA_G593

Pin Connector Graphic

2000-03 Sedan 3.2L V6 MFI VIN UA5 [A/T] 25P 'B' Connector

PCM Pin #	W/Color	Circuit Description (25-Pin)	Value at Hot Idle
B1, B9	YEL/BLK	Main Relay Power (B+)	12-14v
B2, B10	BLK	Power Ground (PG1, PG2)	<0.1v
B3	BLK/RED	Injector 5 Control	2.0-3.3 ms
B4	YEL	Injector 4 Control	2.0-3.3 ms
B5	RED	Injector 2 Control	2.0-3.3 ms
B6	WHT/BLU	Injector 6 Control	2.0-3.3 ms
B7	BLU/RED	E-EGR Control Solenoid	Digital Signals: 0-12-0v
B8	WHT	A/T: Clutch Solenoid 'A-'	Pulse Signals
B11	BRN	Injector 1 Control	2.0-3.3 ms
B12	GRN/YEL	VTEC Solenoid Control	Idle: 0v, Hi-Speed: 12v
B13	GRN/RED	A/T: Clutch Solenoid 'C+'	Pulse Signals
B14	BLU/WHT	A/T: Gear Position Switch	In P/N: 0v, Others: 12v
B15	BLU	Injector 3 Control	2.0-3.3 ms
B16	---	Not Used	---
B17	RED	A/T: Clutch Solenoid 'A+'	Pulse Signals
B18	GRN	A/T: Clutch Solenoid 'B-'	Pulse Signals
B19	BLU/YEL	A/T: 4th Oil Pressure Switch	12-14v
B20	BRN/YEL	Logic Ground (LG1)	<0.050v
B21	WHT/RED	Keep Alive Power (VBU)	12-14v
B22	BRN/BLK	Logic Ground (LG2)	<0.050v
B23	BLK/RED	Idle Air Control Valve	Pulse Signals
B24	RED/BLU	A/T: Clutch Solenoid 'A-'	Pulse Signals
B25	BRN/BLU	A/T: Clutch Solenoid 'B+'	Pulse Signals

2000-03 Sedan 3.2L V6 MFI VIN UA5 [A/T] 16P 'D' Connector

PCM Pin #	W/Color	Circuit Description (16-Pin)	Value at Hot Idle
D1	YEL	A/T: Lockup Solenoid	LUS Off: 12v, On: 1v
D2	GRN/WHT	A/T: Shift Solenoid 'B'	SSB Off: 12v, On: 1v
D3	GRN	A/T" Shift Solenoid 'C'	SSA Off: 12v, On: 1v
D4	RED/BLK	A/T: 'N' Gear Position Switch	In 'N': 0v, Others: 12v
D5	BLK/YEL	Solenoid Feed (B+)	12-14v
D6	WHT	A/T: 'R' Gear Position Switch	In 'R': 0v, Others: 12v
D7	BLU/YEL	A/T: Shift Solenoid 'A'	SSA Off: 12v, On: 1v
D8	YEL	A/T: D4 Gear Position Switch	In D4: 0v, others: 12v
D9	YEL/GRN	A/T: D5 Gear Position Switch	In D5: 0v, Others: 12v
D10	BLU	Countershaft Speed Sensor P	Wheels turning: 12v
D11	RED	Mainshaft Speed Sensor 'P'	Moving: pulse signals
D12	WHT	Mainshaft Speed, Ground 'N'	<0.050v
D13	BLU/WHT	A/T: 3rd Clutch Pressure Sw.	Open: 12v, Closed: 1v
D14	RED	A/T: D3 Gear Position Switch	In D3: 0v, Others: 12v
D15	BLU	A/T: D2 Gear Position Switch	In D2: 0v, Others: 12v
D16	GRN	Countershaft Speed Sensor N	<0.050v

05533_ADIA_G593

Pin Connector Graphic

2000-03 Sedan 3.2L V6 MFI VIN UA5 [A/T] 31P 'C' Connector

PCM Pin #	W/Color	Circuit Description (31-Pin)	Value at Hot Idle
C1	BLK/WHT	HO2S-11 (B1 S1) Heater	Digital Signals: 0-12-0v
C2	WHT/GRN	Alternator Charging Signal	Lights on: 0v, off: 12v
C3	WHT/BLU	Ignition Coil 3 Control	Digital Signals: 0-12-0v
C4	YEL/GRN	Ignition Coil 1 Control	Digital Signals: 0-12-0v
C5	WHT/RED	Alternator 'FR' Signal	Digital Signal: 0-5-0-5v
C6	WHT/BLK	EGR Valve Lift Sensor	1.2v
C7	GRN/WHT	MAP Sensor Ground (SG1)	<0.050v
C8	BLU	CKP Sensor 'P' Signal	AC pulse signals
C9	WHT	CKP Sensor 'N' Signal	<0.050v
C10	BLU/BLK	VTEC Pressure Switch	Idle: 0v, Cruise: 12v
C11	---	Not Used	---
C12	BLK/RED	Ignition Coil 5 Control	Digital Signals: 0-12-0v
C13	YEL	Ignition Coil 4 Control	Digital Signals: 0-12-0v
C14	RED	Ignition Coil 2 Control	Digital Signals: 0-12-0v
C15	---	Not Used	---

2000-03 Sedan 3.2L V6 MFI VIN UA5 [A/T] 31P 'C' Connector, continued

C16	WHT	HO2S-11 (B1 S1) Signal	0.1-1.1v
C17	GRN/RED	MAP Sensor Signal	0.8-0.9v
C18	GRN/YEL	Sensor Ground (SG2)	<0.050v
C19	YEL/RED	MAP Sensor VREF (VCC1)	4.9-5.1v
C20	GRN	TDC1 Sensor 'P' Signal	AC pulse signals
C21	RED	TDC1 Sensor 'N' Signal	<0.050v
C22	RED/BLU	Knock Sensor Signal	No Detonation: 18mv AC
C23	BLU/WHT	Ignition Coil 6 Control	Digital Signals: 0-12-0v
C24	BLU/YEL	A/T: TFT Sensor Signal	0.1-4.2v
C25	RED/YEL	IAT Sensor Signal	Varies w/temp: 0.5-4.9v
C26	RED/WHT	ECT Sensor Signal	At 180°F: 0.5-0.6v
C27	RED/BLK	TP Sensor	0.5-0.6v
C28	YEL/BLU	Sensor VREF (VCC2)	4.9-5.1v
C29	YEL	TDC2 Sensor 'P' Signal	AC pulse signals
C30	BLK	TDC2 Sensor 'N' Signal	<0.050v

2000-03 Sedan 3.2L V6 MFI VIN UA5 [A/T] 20P 'E' Connector

PCM Pin #	W/Color	Circuit Description (20-Pin)	Value at Hot Idle
E1	PNK	Immobilizer Indicator Lamp	Lamp Off: 12v, On: 1v
E2	RED	IMOCD Immobilizer Code	Digital Signals
E3-4	---	Not Used	---
E5	RED/BLU	TP Sensor Signal to TCM	0.5-0.6v
E6	ORN/WHT	Voltage Reference (ABS)	4.9-5.1v
E7	BLU	PCM to Frame Inhibition	2.5v (normal signal)
E8	BLU/BLK	A/T: Drive Shift Indicator 'C'	2nd, 4th: 0v, 1st, 3rd: 12v
E9	BLU/GRN	A/T: Drive Shift Indicator 'B'	In 4th: 12v, 1st & 3rd: 0v
E10	BLU/RED	A/T: Drive Shift Indicator 'A'	In 4th: 12v, 1st - 3rd: 0v
E11	---	Not Used	---
E12	BLU	IMOEN Immobilizer Enable	Digital Signals
E13, E15	---	Not Used	---
E14	BRN	A/T: First Hold Down Switch	LH Switch on: 0v, off: 12v
E16	PNK	Frame to PCM Torque Down	TCS On: 5v, Off: 2.5v
E17	LT GRN	A/T: 'P' Gear Position Switch	In 'P': 4v, Others: 0v
E18	ORN	Torque Mgmt. (-) Signal	In Neutral: 12v
E19	YEL	Torque Mgmt. (+) Signal	In Sports Shift: 0v
E20	RED	Torque Mgmt. Mode Signal	In Sports Shift: 0v

2004-05 Sedan 3.2L V6 MFI VIN UA6 [A/T] 31P 'A' Connector (Gray, On Dashboard Wire Harness Right Branch)

PCM Pin #	W/Color	Circuit Description (31-Pin)	Value at Hot Idle
A1	GRN/BLK	Sensor Ground	Less than 1.0v
A2	GRN/WHT	Drives PGM-FI Main Relay 2	B+
A3	WHT/RED	Drives Throttle Actuator Control Module Relay	0v
A4	ORN	Drives A/F Sensor Relay	0v
A5	RED/YEL	Drives PGM-FI Main Relay 1	0v
A6	GRN/YEL	Drives Condenser Fan Relay	B+
A7	GRN	Drives Fan Control Relay (Radiator)	B+
A8	WHT/BLK	Detects Brake Pedal Position Switch	With brake pedal pressed: B+
A9	BRN/YEL	Detects Brake Pedal Position Switch	With brake pedal released: B+
A10	LT GRN/RED	Drives EVAP Canister Vent Shut Valve	B+
A11	BLU/WHT	Sends VSS Signal	With wheels turning: Pulses
A12	YEL/WHT	Provides Sensor Pullup Voltage	About 5v
A13	WHT/BLU	Detects Downshift Switch Signal	In M position and shift lever pushed toward downshift position: 0v In M position and shift lever in neutral position: about 5v
A14	LT BLU	Sends and Receives Scan Tool Signal	Pulses or B+
A15	WHT	Sends CAN Communication Signal	Pulses
A16	BRN	Detects Service Check Signal	With the service check signal shorted with the HDS: about 0v With the service check signal opened: about 5v
A17	BLU/YEL	Detects PSP Switch Signal	At idle with steering wheel in straight ahead position: about 0v At idle with steering wheel at full lock: B+
A18	BLU/BLK	Detects ELD Signal	0.1v to 4.8v
A19	BRN/WHT	Detects Upshift Switch Signal	In M position and shift lever pushed toward upshift position: 0v In M position and shift lever in neutral position: About 5v
A20	NC		
A21	NC		
A22	YEL/BLK	Drives Shift Lock Solenoid	With ignition switch ON (II), in P, brake pedal pressed, and accelerator released: B+
A23	NC		
A24	RED/WHT	Detects Write Enable Signal	About 0v
A25	RED/BLU	Detects Immobilizer Signal	n/a
A26	RED	Sends Communication Signal	Pulses
A27	RED	Drives A/C Clutch Relay	With compressor ON: About 0v With compressor OFF: B+
A28	NC		
A29	LT GRN	Detects FTP Sensor Signal	Fuel fill cap open: About 2.5v
A30	NC		
A31	LT GRN/RED	Sends Inhibit Signal	B+

2004-05 Sedan 3.2L V6 MFI VIN UA6 [A/T] 24P 'B' Connector

PCM Pin #	W/Color	Circuit Description (24-Pin)	Value at Hot Idle
B1	NC		
B2	RED/YEL	Drives EVAP Canister Purge Valve	Engine coolant below 149°F (65°C): B+ Engine coolant above 149°F 65°C): Duty controlled
B3	WHT/RED	Ground for IMT Actuator	0v
B4	WHT/BLU	Drives IMT Actuator	0v
B5	BLU/RED	Drives EGR Valve	About 0v
B6	RED/BLU	Detects Knock Sensor Signal	With engine knocking: Pulses
B7	BLU/YEL	Drives Engine Mount Control Solenoid Valve	About 0v
B8	NC		
B9	GRN	Detects Rear Secondary HO2S (Bank 1, Sensor 2) Signal	
B10	WHT	Detects Rear A/F Sensor (Bank 1, Sensor 1) LABEL Signal	0.4v to 4.6v
B11	WHT/BLK	Detects IMT Valve Position	About 5v
B12	BLU	Detects Rear A/F Sensor (Bank 1, Sensor 1) VS CELL Signal	3.4v to 4.8v
B13	RED	Reference Voltage for Rear A/F Sensor (Bank 1, Sensor 1)	3.4v to 3.8v
B14	GRN	Detects Rear A/F Sensor (Bank 1, Sensor 1) Pump Cell	2.0v to 5.6v
B15	BRN/YEL	Ground Circuit for the PCM	Less than 1.0v
B16	NC		
B17	NC		
B18	WHT	Detects Front Secondary HO2S (Bank 2, Sensor 2) Signal	
B19	WHT/RED	Detects Front A/F (Bank 2, Sensor 1) LABEL Signal	0.4v to 4.6v
B20	RED/BLU	Detects Front A/F Sensor (Bank 2, Sensor 1) VS CELL Signal	3.4v to 4.8v
B21	RED/WHT	Reference Voltage for Front A/F Sensor (Bank 2, Sensor 1)	3.4v to 4.8v
B22	GRN/RED	Detects Front A/F Sensor (Bank 2, Sensor 1) Pump Cell	2.0v to 5.6v
B23	RED/BLK	Detects Transmission Range Switch N Signal	In Neutral: 0v, else B+
B24	NC		

2004-05 Sedan 3.2L V6 MFI VIN UA6 [A/T] 22P 'C' Connector (White, On PCM Harness, Dashboard Branch)

PCM Pin #	W/Color	Circuit Description (22-Pin)	Value at Hot Idle
C1	BLU/YEL	Drives Shift Solenoid A	In 1, 2, L or D (gears 1, 2, 5): B+; else 0v
C2	NC	--	
C3	BLU/YEL	Detects ATF Temperature Signal	About 0.2v to 4.0v
C4	NC	--	
C5	BLU/WHT	Detects 3rd Clutch Transmission Fluid Pressure Switch Input	Without 3rd clutch pressure: B+, else about 0v
C6	YEL	Drives TCC Solenoid Valve	0v
C7	RED	Drives A/T Clutch Pressure Control Solenoid A	Duty cycle
C8	GRN/WHT	Drives Shift Solenoid B	B+
C9	NC	--	
C10	YEL/GRN	Detects Transmission Range Switch D Signal Input	In D: 0v
C11	WHT	Detects Transmission Range Switch R Signal Input	In R: 0v
C12	NC	--	
C13	BLU/YEL	Detects 4th Clutch Transmission Fluid Pressure Switch Input	About 0v
C14	NC	--	
C15	BRN/WHT	Drives A/T Clutch Pressure Control Solenoid Valve B	Duty cycle
C16	GRN	Drives Shift Solenoid Valve C	With shift selector in 1, D, L: B+, else 0v
C17	BLU/BLK	Detects Transmission Range Switch Park Signal	In P: 0v, else B+
C18	RED	Detects Transmission Range L Signal	In L: 0v, else B+
C19	BLU/YEL	Detects Transmission Range D/L Signal	In D or L: 0v, else B+
C20	ORN	Detects Sequential Sportshift Switch Signal	In M: 0v, else B+
C21	NC	--	
C22	GRN/RED	Drives A/T Clutch Pressure Control Solenoid Valve C	Duty cycle

2004-05 Sedan 3.2L V6 MFI VIN UA6 [A/T] 17P 'D' Connector (White, On PCM Harness, Dashboard Branch)

PCM Pin #	W/Color	Circuit Description (17-Pin)	Value at Hot Idle
D1	BLU/RED	Detects CKP Sensor B Signal	Pulses
D2	YEL/GRN	Provides Sensor Pullup Voltage	About 5v
D3	RED/BLU	Detects APP Sensor Signal A	About 0.5v
D4	GRN/WHT	Sensor Ground	Less than 1.0v
D5	GRN/RED	Detects MAP Sensor Signal	About 1.0v (varies with engine RPM)
D6	YEL/RED	Provides Sensor Pullup Voltage	About 5v
D7	YEL	Detects CMP Sensor Signal	Pulses
D8	RED	Detects Input Shaft Sensor Signal	With transmission in N: About 2.5v
D9	RED/YEL	Detects APP Sensor B Signal	About 0.2v
D10	GRN/YEL	Sensor Ground	Less than 1.0v
D11	WHT/BLK	Detects EGR Valve Position Sensor Signal	1.2v to 3.0v
D12	YEL/BLU	Provides Sensor Pullup Voltage	About 5v
D13	BLU	Detects CKP Sensor A Signal	Pulses
D14	BLU	Detects Output Shaft Speed Sensor Signal	With wheels rotating: Pulses
D15	GRN	Sensor Ground	Less than 1.0v
D16	RED/YEL	Detects IAT Sensor Signal	About 0.1v to 4.8v
D17	RED/WHT	Detects ECT Sensor Signal	About 0.1v to 4.8v

2004-05 Sedan 3.2L V6 MFI VIN UA6 [A/T] 31P 'E' Connector (White, On PCM Wire Harness Dashboard Branch)

PCM Pin #	W/Color	Circuit Description (31-Pin)	Value at Hot Idle
E1	BLK	Ground Circuit for PCM	Less than 1.0v
E2	BLK	Ground Circuit for PCM	Less than 1.0v
E3	BRN/YEL	Ground Circuit for PCM	Less than 1.0v
E4	YEL/BLK	Power for PCM	B+
E5	BLK/GRN	Detects Ignition Signal	B+
E6	GRN/WHT	Drives Front A/F Sensor Heater (Bank 2, Sensor 1)	About 0v or pulses
E7	BLK/WHT	Drives Rear A/F Sensor Heat (Bank 1, Sensor 1)	About 0v or pulses
E8	GRN/RED	Drives Front Secondary HO2S Heater (Bank 2, Sensor 2)	Duty cycle
E9	BLK/WHT	Drives Rear Secondary HO2S Heater (Bank 2, Sensor 2)	Duty cycle
E10	BLK/YEL	Power Source for Solenoid Valve	B+
E11	WHT/BLU	Detects Alternator L Signal	About 0v
E12	WHT/RED	Detects Alternator FR Signal	0v to 5v, depending on electrical load
E13	NC	--	
E14	RED/GRN	Detects A/C Pressure Switch Signal	With A/C pressure switch on: About 0v, else B+
E15	BRN/WHT	Drives No. 6 Ignition Coil	Pulses
E16	BLK/RED	Drives No. 5 Ignition Coil	Pulses
E17	BRN	Drives No. 4 Ignition Coil	Pulses
E18	WHT/BLU	Drives No. 3 Ignition Coil	Pulses
E19	BLU/RED	Drives No. 2 Ignition Coil	Pulses
E20	YEL/GRN	Drives No. 1 Ignition Coil	Pulses
E21	BRN	Drives No. 1 Fuel Injector	Duty cycle
E22	GRN/YEL	Drives VTEC Solenoid Valve	About 0v
E23	BLU/BLK	Detects VTEC Oil Pressure Switch Signal	About 0v
E24	WHT/GRN	Sends Alternator Control Signal	B+ (depending on electrical load)
E25	GRN	Sends Throttle Actuator Control	Serial data
E26	BLU	Detects Throttle Actuator Control	Serial data
E27	WHT/BLU	Drives No. 6 Fuel Injector	Duty cycle
E28	BLK/RED	Drives No. 5 Fuel Injector	Duty cycle
E29	YEL	Drives No. 4 Fuel Injector	Duty cycle
E30	BLU	Drives No. 3 Fuel Injector	Duty cycle
E31	RED	Drives No. 2 Fuel Injector	Duty cycle

2004-05 Sedan 3.2L V6 MFI VIN UA6 [M/T] 31P 'A' Connector (Gray, On Dashboard Wire Harness Right Branch)

PCM Pin #	W/Color	Circuit Description (31-Pin)	Value at Hot Idle
A1	GRN/BLK	Sensor Ground	Less than 1.0v
A2	GRN/WHT	Drives PGM-FI Main Relay 2	B+
A3	WHT/RED	Drives Throttle Actuator Control Module Relay	0v
A4	ORN	Drives A/F Sensor Relay	0v
A5	RED/YEL	Drives PGM-FI Main Relay 1	0v
A6	GRN/YEL	Drives Condenser Fan Relay	B+
A7	GRN	Drives Fan Control Relay (Radiator)	B+
A8	WHT/BLK	Detects Brake Pedal Position Switch	With brake pedal pressed: B+
A9	BRN/YEL	Detects Brake Pedal Position Switch	With brake pedal released: B+
A10	LT GRN/RED	Drives EVAP Canister Vent Shut Valve	B+
A11	BLU/WHT	Sends VSS Signal	With wheels turning: Pulses
A12	YEL/WHT	Provides Sensor Pullup Voltage	About 5v
A13	NC	--	--
A14	LT BLU	Sends and Receives Scan Tool Signal	Pulses or B+
A15	WHT	Sends CAN Communication Signal	Pulses
A16	BRN	Detects Service Check Signal	With the service check signal shorted with the HDS: about 0v With the service check signal opened: about 5v
A17	BLU/YEL	Detects PSP Switch Signal	At idle with steering wheel in straight ahead position: about 0v At idle with steering wheel at full lock: B+
A18	BLU/BLK	Detects ELD Signal	0.1v to 4.8v
A19	NC	--	--
A20	NC	--	--
A21	NC	--	--
A22	WHT	Drives Reverse Lock Solenoid	B+
A23	NC	--	--
A24	RED/WHT	Detects Write Enable Signal	About 0v
A25	RED/BLU	Detects Immobilizer Signal	n/a
A26	RED	Sends Communication Signal	Pulses
A27	RED	Drives A/C Clutch Relay	With compressor ON: About 0v With compressor OFF: B+
A28	NC	--	--
A29	LT GRN	Detects FTP Sensor Signal	Fuel fill cap open: About 2.5v
A30	LT BLU	Detects Clutch Pedal Position Switch	Clutch pedal released: About 0v, else B+
A31	NC	--	--

2004-05 Sedan 3.2L V6 MFI VIN UA6 [M/T] 24P 'B' Connector

PCM Pin #	W/Color	Circuit Description (24-Pin)	Value at Hot Idle
B1	NC	--	--
B2	RED/YEL	Drives EVAP Canister Purge Valve	Engine coolant below 149°F (65°C): B+ Engine coolant above 149°F 65°C): Duty controlled
B3	WHT/RED	Ground for IMT Actuator	0v
B4	WHT/BLU	Drives IMT Actuator	0v
B5	BLU/RED	Drives EGR Valve	About 0v
B6	RED/BLU	Detects Knock Sensor Signal	With engine knocking: Pulses
B7	BLU/YEL	Drives Engine Mount Control Solenoid Valve	About 0v
B8	NC	--	--
B9	GRN	Detects Rear Secondary HO2S (Bank 1, Sensor 2) Signal	--
B10	WHT	Detects Rear A/F Sensor (Bank 1, Sensor 1) LABEL Signal	0.4v to 4.6v
B11	WHT/BLK	Detects IMT Valve Position	About 5v
B12	BLU	Detects Rear A/F Sensor (Bank 1, Sensor 1) VS CELL Signal	3.4v to 4.8v
B13	RED	Reference Voltage for Rear A/F Sensor (Bank 1, Sensor 1)	3.4v to 3.8v
B14	GRN	Detects Rear A/F Sensor (Bank 1, Sensor 1) Pump Cell	2.0v to 5.6v
B15	BRN/YEL	Ground Circuit for the PCM	Less than 1.0v
B16	NC	--	--
B17	NC	--	--
B18	WHT	Detects Front Secondary HO2S (Bank 2, Sensor 2) Signal	--
B19	WHT/RED	Detects Front A/F (Bank 2, Sensor 1) LABEL Signal	0.4v to 4.6v
B20	RED/BLU	Detects Front A/F Sensor (Bank 2, Sensor 1) VS CELL Signal	3.4v to 4.8v
B21	RED/WHT	Reference Voltage for Front A/F Sensor (Bank 2, Sensor 1)	3.4v to 4.8v
B22	GRN/RED	Detects Front A/F Sensor (Bank 2, Sensor 1) Pump Cell	2.0v to 5.6v
B23	NC	--	--
B24	NC	--	--

2004-05 Sedan 3.2L V6 MFI VIN UA6 [M/T] 17P 'D' Connector (White, On PCM Harness, Dashboard Branch)

PCM Pin #	W/Color	Circuit Description (17-Pin)	Value at Hot Idle
D1	BLU/RED	Detects CKP Sensor B Signal	Pulses
D2	YEL/GRN	Provides Sensor Pullup Voltage	About 5v
D3	RED/BLU	Detects APP Sensor Signal A	About 0.5v
D4	GRN/WHT	Sensor Ground	Less than 1.0v
D5	GRN/RED	Detects MAP Sensor Signal	About 1.0v (varies with engine RPM)
D6	YEL/RED	Provides Sensor Pullup Voltage	About 5v
D7	YEL	Detects CMP Sensor Signal	Pulses
D8	NC	--	--
D9	RED/YEL	Detects APP Sensor B Signal	About 0.2v
D10	GRN/YEL	Sensor Ground	Less than 1.0v
D11	WHT/BLK	Detects EGR Valve Position Sensor Signal	1.2v to 3.0v
D12	YEL/BLU	Provides Sensor Pullup Voltage	About 5v
D13	BLU	Detects CKP Sensor A Signal	Pulses
D14	BLU	Detects Output Shaft Speed Sensor Signal	With wheels rotating: Pulses
D15	GRN	Sensor Ground	Less than 1.0v
D16	RED/YEL	Detects IAT Sensor Signal	About 0.1v to 4.8v
D17	RED/WHT	Detects ECT Sensor Signal	About 0.1v to 4.8v

2004-05 Sedan 3.2L V6 MFI VIN UA6 [M/T] 31P 'E' Connector (White, On PCM Wire Harness Dashboard Branch)

PCM Pin #	W/Color	Circuit Description (31-Pin)	Value at Hot Idle
E1	BLK	Ground Circuit for PCM	Less than 1.0v
E2	BLK	Ground Circuit for PCM	Less than 1.0v
E3	BRN/YEL	Ground Circuit for PCM	Less than 1.0v
E4	YEL/BLK	Power for PCM	B+
E5	BLK/GRN	Detects Ignition Signal	B+
E6	GRN/WHT	Drives Front A/F Sensor Heater (Bank 2, Sensor 1)	About 0v or pulses
E7	BLK/WHT	Drives Rear A/F Sensor Heat (Bank 1, Sensor 1)	About 0v or pulses
E8	GRN/RED	Drives Front Secondary HO2S Heater (Bank 2, Sensor 2)	Duty cycle
E9	BLK/WHT	Drives Rear Secondary HO2S Heater (Bank 2, Sensor 2)	Duty cycle
E10	BLK/YEL	Power Source for Solenoid Valve	B+
E11	WHT/BLU	Detects Alternator L Signal	About 0v
E12	WHT/RED	Detects Alternator FR Signal	0v to 5v, depending on electrical load
E13	NC	--	
E14	RED/GRN	Detects A/C Pressure Switch Signal	With A/C pressure switch on: About 0v, else B+
E15	BRN/WHT	Drives No. 6 Ignition Coil	Pulses
E16	BLK/RED	Drives No. 5 Ignition Coil	Pulses
E17	BRN	Drives No. 4 Ignition Coil	Pulses
E18	WHT/BLU	Drives No. 3 Ignition Coil	Pulses
E19	BLU/RED	Drives No. 2 Ignition Coil	Pulses
E20	YEL/GRN	Drives No. 1 Ignition Coil	Pulses
E21	BRN	Drives No. 1 Fuel Injector	Duty cycle
E22	GRN/YEL	Drives VTEC Solenoid Valve	About 0v
E23	BLU/BLK	Detects VTEC Oil Pressure Switch Signal	About 0v
E24	WHT/GRN	Sends Alternator Control Signal	B+ (depending on electrical load)
E25	GRN	Sends Throttle Actuator Control	Serial data
E26	BLU	Detects Throttle Actuator Control	Serial data
E27	WHT/BLU	Drives No. 6 Fuel Injector	Duty cycle
E28	BLK/RED	Drives No. 5 Fuel Injector	Duty cycle
E29	YEL	Drives No. 4 Fuel Injector	Duty cycle
E30	BLU	Drives No. 3 Fuel Injector	Duty cycle
E31	RED	Drives No. 2 Fuel Injector	Duty cycle

MDX PIN CHARTS

2001-03 Utility 3.5L V6 VTEC VIN YD1 [A/T] 32-Pin 'A' Connector

PCM Pin #	W/Color	Circuit Description (32-Pin)	Value at Hot Idle
A1	WHT/RED	HO2S-22 (B2 S2) Signal	0.1-1.1v
A2	RED/GRN	ATF Indicator Light	Lamp Off: 12v, On: 1v
A3	ORN/WHT	EVAP Bypass Solenoid	Solenoid Off: 12v, on: 1v
A4	GRN/WHT	EVAP Vent Solenoid	Solenoid Off: 12v, on: 1v
A5	BLU/GRN	Cruise Control Signal	Cruise On: pulse signals
A6	RED/YEL	EVAP Purge Solenoid	Pulse Signals
A7	BLK	Serial Data Line (VTM-4 Unit)	Digital Signals
A8	BLK/WHT	HO2S-12 (B1 S2) Heater	Digital Signals: 0-12-0v
A9	BLU/WHT	Vehicle Speed Sensor	Digital Signals: 0-5-0v
A10	BRN	Service Check Connector	SCS Open: 4.80v
A11	PNK	FUP Communication Signals	Digital Signals
A12	---	Not Used	---
A13	GRN/WHT	Intake Manifold Runner	Less than 3200 rpm: 5v
A14	GRN/BLK	A/T: D5 Indicator Light	In D5: 0v, Others: 12v
A15	GRN/YEL	Immobilizer Fuel Pump Relay	Relay Off: 12v, On: 1v
A16	---	Not Used	---
A17	RED	A/C Clutch Relay Control	Relay Off: 12v, On: 1v
A18	GRN/ORN	Malfunction Indicator Light	MIL Off: 12v, On: 1v
A19	GRN/BLU	Engine Speed Signal	Pulse Signals
A20	GRN	Radiator Fan Relay Control	Relay Off: 12v, On: 1v
A21	GRY	K-Line Signal	12v
A22	WHT	Serial Data Line (VTM-4 Unit)	Digital Signals
A23	---	Not Used	---
A24	BLU/ORN	Starter Switch Signal	Cranking: 9-11v
A25	---	Not Used	---
A26	LT BLU	PSP Switch Signal	Straight: 0v, Turning: 12v
A27	BLU/RED	A/C Switch Signal	A/C Off: 12v, On: 0v
A28	WHT/GRN	Interlock Control Unit Signal	Key and Brake On: 12v
A29	LT GRN	Fuel Tank Pressure Sensor	With Fuel Cap Off: 2.5v
A30	GRN/RED	Electronic Load Detector	Varies 2.5-3.5v
A31	YEL/GRN	ECT Signal to ECT Gauge	Digital Signals
A32	WHT/BLK	Brake Switch Signal	Brake Off: 12v, On: 0v

Pin Connector Graphic

05533_ADIA_G638

2001-03 Utility 3.5L V6 VTEC VIN YD1 [A/T] 25-Pin 'B' Connector

PCM Pin #	W/Color	Circuit Description (25-Pin)	Value at Hot Idle
B1, B9	YEL/BLK	B+ Main Relay IGP1, IGP2	12-14v
B2, B10	BLK	Power Ground PG1, PG2	<0.1v
B3	BLK/RED	Injector 5 Control	2.0-3.3 ms
B4	YEL	Injector 4 Control	2.0-3.3 ms
B5	RED	Injector 2 Control	2.0-3.3 ms
B6	WHT/BLU	Injector 6 Control	2.0-3.3 ms
B7	BLU/RED	E-EGR Control Solenoid	Digital Signals: 0-12-0v
B8	WHT	A/T: Clutch Solenoid 'A-'	Pulse Signals
B11	BRN	Injector 1 Control	2.0-3.3 ms
B12	GRN/YEL	VTEC Solenoid Control	Idle: 0v, Hi-Speed: 12v
B13	GRN/RED	A/T: Clutch Solenoid 'C+'	Pulse Signals
B14	BLU/WHT	A/T: Gear Position Switch	In P/N: 0v, Others: 5v
B15	BLU	Injector 3 Control	2.0-3.3 ms
B16	---	Not Used	---
B17	RED	A/T: Clutch Solenoid 'A+'	Pulse Signals
B18	GRN	A/T: Clutch Solenoid 'B-'	Pulse Signals
B19	BLU/YEL	A/T: 4th Oil Pressure Switch	12-14v
B20, B22	BRN/YEL	Logic Ground (LG1), (LG2)	<0.050v
B21	WHT/RED	Keep Alive Power (VBU)	12-14v
B23	BLK/RED	Idle Air Control Valve	Pulse Signals
B24	RED/BLU	A/T: Clutch Solenoid 'A-'	Pulse Signals
B25	BRN/WHT	A/T: Clutch Solenoid 'B+'	Pulse Signals

2001-03 Utility 3.5L V6 VTEC VIN YD1 [A/T] 20-Pin 'E' Connector

PCM Pin #	W/Color	Circuit Description (20-Pin)	Value at Hot Idle
E1	PNK	Immobilizer Indicator Lamp	Lamp Off: 12v, On: 1v
E2	RED	IMOCD Immobilizer Code	Digital Signals
E3-12	---	Not Used	---
E12	BLU	Immobilizer Enable Signal	Digital Signals
E14	BRN	A/T: First Hold Down Switch	LH Switch on: 0v, off: 12v
E16-20	---	Not Used	---

Pin Connector Graphic

05533_ADIA_G638

2001-03 Utility 3.5L V6 MFI VIN YD1 [A/T] 31-Pin 'C' Connector

PCM Pin #	W/Color	Circuit Description (31-Pin)	Value at Hot Idle
C1	BLK/WHT	HO2S-11 (B1 S1) Heater	Digital Signals: 0-12-0v
C2	WHT/GRN	Alternator Charging Signal	Lights on: 0v, off: 12v
C3	WHT/BLU	Ignition Coil 3 Control	Digital Signals: 0-12-0v
C4	YEL/GRN	Ignition Coil 1 Control	Digital Signals: 0-12-0v
C5	WHT/RED	Alternator 'FR' Signal	Digital Signal: 0-5-0-5v
C6	WHT/BLK	EGR Valve Lift Sensor	1.2-2.0v
C7	GRN/WHT	Sensor Ground (SG1)	<0.050v
C8	BLU	CKP Sensor 'P' Signal	0.900v AC
C9	WHT	CKP Sensor 'N' Signal	<0.050v
C10	BLU/BLK	VTEC Pressure Switch	Idle: 0v, Cruise: 12v
C11, 15	---	Not Used	---
C12	BLK/RED	Ignition Coil 5 Control	Digital Signals: 0-12-0v
C13	BRN	Ignition Coil 4 Control	Digital Signals: 0-12-0v
C14	BLU/RED	Ignition Coil 2 Control	Digital Signals: 0-12-0v
C16	WHT	HO2S-11 (B1 S1) Signal	Varies: 0.1-1v
C17	GRN/RED	MAP Sensor Signal	0.8-0.9v
C18	GRN/YEL	Sensor Ground (SG2)	<0.050v
C19	YEL/RED	Sensor VREF (VCC1)	4.9-5.1v
C20	GRN	TDC1 Sensor 'P' Signal	1.00v AC
C21	RED	TDC1 Sensor 'N' Signal	<0.050v
C22	RED/BLU	Knock Sensor Signal	No Detonation: 18mv AC
C23	BRN/WHT	Ignition Coil 6 Control	Digital Signals: 0-12-0v
C24	BLU/YEL	A/T: TFT Sensor Signal	0.1-4.2v
C25	RED/YEL	IAT Sensor Signal	Varies w/temp: 0.5-4.8v
C26	RED/WHT	ECT Sensor Signal	At 180F: 0.5-0.6v
C27	RED/BLK	TP Sensor	0.5-0.6v
C28	YEL/BLU	Sensor VREF (VCC2)	4.9-5.1v
C29	YEL	TDC2 Sensor 'P' Signal	0.250v AC
C30	BLK	TDC2 Sensor 'N' Signal	<0.050v

2001-03 Utility 3.5L V6 VTEC VIN YD1 [A/T] 16-Pin 'D' Connector

PCM Pin #	W/Color	Circuit Description (16-Pin)	Value at Hot Idle
D1	YEL	A/T: Lockup Solenoid	LUS Off: 12v, On: 1v
D2	GRN/WHT	A/T: Shift Solenoid 'B'	SSB Off: 12v, On: 1v
D3	GRN	A/T" Shift Solenoid C	SSA Off: 12v, On: 1v
D4	RED/BLK	A/T: 'N' Gear Position Switch	In 'N': 0v, Others: 5v
D5	BLK/YEL	Solenoid Feed (B+)	12-14v
D6	WHT	A/T: 'R' Gear Position Switch	In 'R': 0v, Others: 5v
D7	BLU/YEL	A/T: Shift Solenoid 'A'	SSA Off: 12v, On: 1v
D8	YEL	A/T: D4 Gear Position Switch	In D4: 0v, Others: 5v
D9	YEL/GRN	A/T: D5 Gear Position Switch	In D5: 0v, Others: 5v
D10	BLU	Countershaft Speed Sensor P	Wheels turning: pulses
D11	RED	Mainshaft Speed Sensor 'P'	Pulse Signals
D12	WHT	Mainshaft Speed Sensor 'N'	<0.050v
D13	BLU/WHT	A/T: 3rd Oil Pressure Switch	12-14v
D14	RED	A/T: D3 Gear Position Switch	In D3: 0v, Others: 5v
D15	BLU	A/T: D2 Gear Position Switch	In D2: 0v, Others: 5v
D16	GRN	Countershaft Speed Sensor N	<0.050v

SLX PIN CHARTS

1996-97 4-Door 3.2L V6 4WD VIN V [A/T] 32P Red 'A' Connector

PCM Pin #	W/Color	Circuit Description (16-Pin)	Value at Hot Idle
A1	RED	Reference Voltage 'A'	4.9-5.1v
A2	YEL	Knock Sensor Signal	No detonation: 18 mv AC
A3	---	Not Used	---
A4	RED/WHT	Keep Alive Power	12-14v
A5	BLU	Idle Air Control 'A' High	Pulse Signals
A6	BLU/WHT	Idle Air Control 'A' Low	Pulse Signals
A7	BLU/BLK	Idle Air Control 'B' Low	Pulse Signals
A8	BLU/RED	Idle Air Control 'B' High	Pulse Signals
A9	PNK	Transmission Fluid Lamp	Lamp Off: 12v, On: 1v
A10	PNK/GRY	Winter Lamp	Lamp On: 1v. off: 12v
A11	GRN/WHT	Power Lamp	Lamp On: 1v. off: 12v
A12	PPL/WHT	Antilock Brake Light	Lamp Off: 12v, On: 1v
A13	PPL	Malfunction Indicator Lamp	MIL Off: 12v, On: 1v
A14	PNK	Check Transmission Light	Lamp on: 1v, off: 12v
A15	RED/BLU	EVAP Purge Solenoid	Digital Signals: 0-12-0v
A16	BRN/YEL	A/T: Shift Low Band Apply	Sol. on: 1v, off: 12v

1996-97 4-Door 3.2L V6 4WD VIN V [A/T] 32P Red 'B' Connector

PCM Pin #	W/Color	Circuit Description (16-Pin)	Value at Hot Idle
B1	WHT	Reference Voltage 'B'	4.9-5.1v
B2	RED/WHT	Igniter 4 Control	Digital Signals: 0-12-0v
B3	RED/BLU	Igniter 2 Control	Digital Signals: 0-12-0v
B4	RED/GRN	Igniter 6 Control	Digital Signals: 0-12-0v
B5	---	Not Used	---
B6	---	Not Used	---
B7	YEL/RED	Linear EGR Sensor Signal	0.6-0.7v
B8	YEL/GRN	IAT Sensor Signal	Varies w/temp: 0.5-4.8v
B9	---	Not Used	---
B10	GRN/RED	Rough Road Sensor	2.5v
B11	GRN/YEL	PSP Switch Signal	Straight: 12v, Turning: 0v
B12	GRN	Illuminated Switch Signal	Switch on: 0.1v, off: 12v
B13	ORN/GRN	Class 2 Serial Data Link	0v
B14	GRN/RED	A/C Clutch Relay	Relay Off: 12v, On: 1v
B15	---	Not Used	---
B16	---	Not Used	---

Pin Connector Graphic

1996-97 4-Door 3.2L 4WD VIN V [A/T] 32P White 'C' Connector

PCM Pin #	W/Color	Circuit Description (16-Pin)	Value at Hot Idle
C1	GRN/RED	Injector 4 Control	2.0-3.3 ms
C2	BRN/BLK	A/T: Shift Solenoid 'B'	SSB Off: 12v, On: 1v
C3	GRN/YEL	Injector 6 Control	2.0-3.3 ms
C4	RED	Igniter 1 Control	Digital Signals: 0-12-0v
C5	YEL	CKP Sensor Signal	Digital Signals: 0-5-0v
C7	BLK/BLU	Power Ground	<0.1v
C8	BLK/PNK	Power Ground	<0.1v
C9	BLK/BLU	Power Ground	<0.1v
C10	BLK/RED	Tachometer Signal	Pulse Signals
C11	---	Not Used	---
C12	WHT/BLU	Alternator Charge Signal	12-14v
C13	YEL/RED	Fuel Gauge PWM Signal	Duty Cycle: 0-100%
C14	PNK	HO2S-21 (B2 S1) Signal	0.1-1.1v
C15	BLU	HO2S-21 (B2 S1) Ground	<0.050v
C16	GRN	HO2S-22 (B2 S2) Signal	0.1-1.1v

1996-97 4-Door 3.2L 4WD VIN V [A/T] 32P White 'D' Connector

PCM Pin #	W/Color	Circuit Description (16-Pin)	Value at Hot Idle
D1	GRN/ORN	Injector 2 Control	2.0-3.3 ms
D2	BRN/BLU	Torque Converter Clutch	TCC on: 1v, off: 12v
D3	GRN/WHT	Injector 1 Control	2.0-3.3 ms
D4	ORN	Serial Data (8192 Baud)	Digital Signals
D5	RED/YEL	Igniter 5 Control	Digital Signals: 0-12-0v
D6	RED/BLU	Igniter 3 Control	Digital Signals: 0-12-0v
D7	WHT/BLK	VSS (4096 pulses per mile)	Moving: 0-5-0-5v
D8	GRN	Sensor 'A' Ground	<0.050v
D9	BLU/YEL	Sensor 'B' Ground	<0.050v
D10	YEL/BLU	MAF Sensor Signal	7-9v
D11	BLU	CMP Sensor Signal	Digital Signals: 0-5-0v
D12	RED	HO2S-12 (B1 S2) Ground	<0.050v
D13	WHT	HO2S-12 (B1 S2) Signal	0.1-1.1v
D14	GRN	HO2S-11 (B1 S1) Ground	<0.050v
D15	RED	HO2S-11 (B1 S1) Signal	0.1-1.1v
D16	BLU	HO2S-22 (B2 S2) Ground	<0.050v

Pin Connector Graphic

1996-97 4-Door 3.2L V6 4WD VIN V [A/T] 32P Blue 'E' Connector

PCM Pin #	W/Color	Circuit Description (16-Pin)	Value at Hot Idle
E1	RED	Output Speed Sensor Signal	Moving: AC Pulses
E2	WHT	Output Speed Sensor Ground	<0.050
E3	PPL/RED	A/T: Pressure Control Low	Pulse Signals
E4	PPL/WHT	A/T: Pressure Control High	Pulse Signals
E5	BLK/YEL	Linear EGR Solenoid "High"	Pulse Signals
E6	YEL/PNK	Linear EGR Solenoid "Low"	Pulse Signals
E7	PNK	A/T: Range Signal 'B'	In P/N: 1v, others: 12v
E8	BLU	TP Sensor Signal	0.5-0.6v
E9	BLU/RED	ECT Sensor Signal	At 180F: 0.54v
E10	---	Not Used	---
E11	YEL/RED	CKP Sensor VREF	4.9-5.1v
E12	PNK/BLU	A/T: Range Signal 'A'	In P/N: 12-14v
E13	RED/WHT	Fuel Pump Relay Control	Relay on: 12v, off: 0v
E14	BRN/WHT	A/T: Shift Solenoid Feed (B+)	12-14v
E15	GRN/ORN	A/C Switch Signal	Switch on: 12v, off: 0v
E16	RED/BLU	B+ from Main Relay	12-14v

1996-97 4-Door 3.2L V6 4WD VIN V [A/T] 32P Blue 'F' Connector

PCM Pin #	W/Color	Circuit Description (16-Pin)	Value at Hot Idle
F1	---	Not Used	---
F2	PNK/YEL	A/T: Range Signal 'C'	In P/N: 1v, others: 12v
F3	PNK/BLK	A/T: Range Signal 'P'	In P/N: 1v, others: 12v
F4	GRN/YEL	Brake Switch Signal	Brake on: 12v, off: 0v
F5	PPL/GRN	Power Switch Signal	Switch on: 0.1v, off: 12v
F6	PPL/GRN	Winter Switch Signal	Switch on: 0.1v, off: 12v
F7	GRN/YEL	A/T: TOT Sensor Signal	Varies w/temp: 0.5-4.8v
F8	RED	MAP Sensor Signal	0.6-1.3v
F9	YEL/BRN	Vacuum Switch Signal	Switch On: 0.1v, off: 12v
F10	GRY/BLU	Cruise Control	Switch On: 0.1v, off: 12v
F11	BLK	A/T: Kickdown Switch	Switch On: 0.1v, off: 12v
F12	---	Not Used	---
F13	GRN	Injector 3 Control	2.0-3.3 ms
F14	BRN/RED	A/T: Shift Solenoid 'A'	SSA Off: 12v, On: 1v
F15	GRN/BLK	Injector 5 Control	2.0-3.3 ms
F16	RED/BLU	B+ from Main Relay	12-14v

32-PIN BLUE-CONNECTOR ROW 'E' & 'F'

E | 16 | 15 | 14 | 13 | 12 | 11 | / | 9 | 8 | 7 | 6 | 5 | 4 | 3 | 2 | 1

F | 16 | 15 | 14 | 13 | / | 11 | 10 | 9 | 8 | 7 | 6 | 5 | 4 | 3 | 2 | /

WIRE SIDE OF HARNESS TERMINALS

05533_ADIA_G641

Pin Connector Graphic

1996-97 4-Door 3.2L V6 4WD VIN V [M/T] 32P Red 'A' Connector

PCM Pin #	W/Color	Circuit Description (16-Pin)	Value at Hot Idle
A1	RED	Reference Voltage 'A'	4.9-5.1v
A2	YEL	Knock Sensor Signal	No detonation: 18 mv AC
A3	---	Not Used	---
A4	RED/WHT	Keep Alive Power	12-14v
A5	BLU	Idle Air Control 'A' High	Pulse Signals
A6	BLU/WHT	Idle Air Control 'A' Low	Pulse Signals
A7	BLU/BLK	Idle Air Control 'B' Low	Pulse Signals
A8	BLU/RED	Idle Air Control 'B' High	Pulse Signals
A9-12	---	Not Used	---
A13	PPL	Malfunction Indicator Lamp	MIL Off: 12v, On: 1v
A14	---	Not Used	---
A15	RED/BLU	EVAP Purge Solenoid	Digital Signals: 0-12-0v
A6	---	Not Used	---

1996-97 4-Door 3.2L V6 4WD VIN V [M/T] 32P Red 'B' Connector

PCM Pin #	W/Color	Circuit Description (16-Pin)	Value at Hot Idle
B1	WHT	Reference Voltage 'B'	4.9-5.1v
B2	RED/WHT	Igniter 4 Control	Digital Signals: 0-12-0v
B3	RED/BLU	Igniter 2 Control	Digital Signals: 0-12-0v
B4	RED/GRN	Igniter 6 Control	Digital Signals: 0-12-0v
B5	---	Not Used	---
B6	---	Not Used	---
B7	YEL/RED	Linear EGR Sensor Signal	0.5-0.6v
B8	YEL/GRN	IAT Sensor Signal	Varies w/temp: 0.5-4.8v
B9	---	Not Used	---
B10	GRN/RED	Rough Road Sensor	2.5v
B11	GRN/YEL	PSP Switch Signal	Straight: 12v, Turning: 0v
B12	GRN	Illuminated Switch Signal	Switch on: 0.1v, off: 12v
B13	ORN/GRN	Class 2 Serial Data Link	0v
B14	GRN/RED	A/C Clutch Relay Control	Relay on: 12v, off: 0v
B15	---	Not Used	---
B16	---	Not Used	---

05533_ADIA_G642

Pin Connector Graphic

Standard Colors and Abbreviations

Abbreviation	Color	Abbreviation	Color	Abbreviation	Color
BLK	Black	LT BLU	Lt. Blue	TAN	Tan
BLU	Blue	LT GRN	Lt. Green	VIO	Violet
BRN	Brown	ORN	Orange	WHT	White
GRY	Gray	PNK	Pink	YEL	Yellow
GRN	Green	PPL	Purple		

1996-97 4-Door 3.2L 4WD VIN V [M/T] 32P White 'C' Connector

PCM Pin #	W/Color	Circuit Description (16-Pin)	Value at Hot Idle
C1	GRN/RED	Injector 4 Control	2.0-3.3 ms
C2	---	Not Used	---
C3	GRN/YEL	Injector 6 Control	2.0-3.3 ms
C4	RED	Igniter 1 Control	Digital Signals: 0-12-0v
C5	YEL	CKP Sensor Signal	Digital Signals: 0-5-0v
C6	---	Not Used	---
C7	BLK/BLU	Power Ground	<0.1v
C8	BLK/PNK	Power Ground	<0.1v
C9	BLK/BLU	Power Ground	<0.1v
C10	BLK/RED	Tachometer Signal	Pulse Signals
C11	ORN/BLU	Upshift Lamp Control	Lamp on: 1v, off: 12v
C12	WHT/BLU	Alternator Charge Signal	12-14v
C13	YEL/RED	Fuel Gauge PWM Signal	Duty Cycle: 0-100%
C14	PNK	HO2S-21 (B2 S1) Signal	0.1-1.1v
C15	BLU	HO2S-21 (B2 S1) Ground	<0.050v
C16	GRN	HO2S-22 (B2 S2) Signal	0.1-1.1v

1996-97 4-Door 3.2L 4WD VIN V [M/T] 32P White 'D' Connector

PCM Pin #	W/Color	Circuit Description (16-Pin)	Value at Hot Idle
D1	GRN/ORN	Injector 2 Control	2.0-3.3 ms
D2	---	Not Used	---
D3	GRN/WHT	Injector 1 Control	2.0-3.3 ms
D4	ORN	Serial Data (8192 Baud)	Digital Signals
D5	RED/YEL	Igniter 5 Control	Digital Signals: 0-12-0v
D6	RED/BLU	Igniter 3 Control	Digital Signals: 0-12-0v
D7	WHT/BLK	VSS (4096 pulses per mile)	Moving: 0-5-0-5v
D8	GRN	Sensor 'A' Ground	<0.050v
D9	BLU/YEL	Sensor 'B' Ground	<0.050v
D10	YEL/BLU	MAF Sensor Signal	7-9v
D11	BLU	CMP Sensor Signal	Digital Signals: 0-5-0v
D12	GRN	HO2S-13 (B1 S3) Ground	<0.050v
D13	BLU	HO2S-13 (B1 S3) Signal	0.1-1.1v
D14	GRN	HO2S-11 (B1 S1) Ground	<0.050v
D15	RED	HO2S-11 (B1 S1) Signal	0.1-1.1v
D16	BLU	HO2S-22 (B2 S2) Ground	<0.050v

32-PIN WHITE-CONNECTOR ROW 'C' & 'D'

C	16	15	14	13	12	11	10	9	8	7	/	5	4	3	/	1
D	16	15	14	13	12	11	10	9	8	7	6	5	4	3	2	1

WIRE SIDE OF HARNESS TERMINALS

05533_ADIA_G643

Pin Connector Graphic
1996-97 4-Door 3.2L V6 4WD VIN V [M/T] 32P Blue 'E' Connector

PCM Pin #	W/Color	Circuit Description (16-Pin)	Value at Hot Idle
E1-4	---	Not Used	---
E5	RED/GRN	Linear EGR Solenoid "High"	Pulse Signals
E6	YEL	Linear EGR Solenoid "Low"	Pulse Signals
E7	---	Not Used	---
E8	BLU	TP Sensor Signal	0.5-0.6v
E9	BLU/RED	ECT Sensor Signal	At 180F: 0.54v
E10	---	Not Used	---
E11	YEL/RED	CKP Sensor VREF	4.9-5.1v
E12	---	Not Used	---
E13	PNK/WHT	Fuel Pump Relay Control	Relay on: 12v, off: 0v
E14	---	Not Used	---
E15	GRN/BLK	A/C Switch Signal	Switch on: 12v, off: 0v
E16	RED/BLU	B+ from Main Relay	12-14v

1996-97 4-Door 3.2L V6 4WD VIN V [M/T] 32P Blue 'F' Connector

PCM Pin #	W/Color	Circuit Description (16-Pin)	Value at Hot Idle
F1-7	---	Not Used	---
F8	RED	MAP Sensor Signal	0.6-1.3v
F9	---	Not Used	---
F10	GRY/BLU	Cruise Control	Switch On: 0.1v, off: 12v
F11	---	Not Used	---
F12	---	Not Used	---
F13	WHT/GRN	Injector 3 Control	2.0-3.3 ms
F14	---	Not Used	---
F15	GRN/BLK	Injector 5 Control	2.0-3.3 ms
F16	RED/BLU	B+ from Main Relay	12-14v

Standard Colors and Abbreviations

32-PIN BLUE-CONNECTOR ROW 'E' & 'F'

E	16	15	/	13	/	11	/	9	8	/	6	5	/	/	/	/
F	16	15	/	13	/	/	10	/	8	/	/	/	/	/	/	/

WIRE SIDE OF HARNESS TERMINALS

05533_ADIA_G644

Pin Connector Graphic

Abbreviation	Color	Abbreviation	Color	Abbreviation	Color
BLK	Black	LT BLU	Lt. Blue	TAN	Tan
BLU	Blue	LT GRN	Lt. Green	VIO	Violet
BRN	Brown	ORN	Orange	WHT	White
GRY	Gray	PNK	Pink	YEL	Yellow
GRN	Green	PPL	Purple		

1998-99 4-Door 3.5L V6 MFI VIN X [All] 32P Red 'A' Connector

PCM Pin #	W/Color	Circuit Description (16-Pin)	Value at Hot Idle
A1	RED	Reference Voltage 'A'	4.9-5.1v
A2	YEL	Knock Sensor Signal	No detonation: 18 mv AC
A3	---	Not Used	---
A4	WHT	Keep Alive Power	12-14v
A5	BLU	Idle Air Control 'A' High	Pulse Signals
A6	BLU/WHT	Idle Air Control 'A' Low	Pulse Signals
A7	BLU/BLK	Idle Air Control 'B' Low	Pulse Signals
A8	BLU/RED	Idle Air Control 'B' High	Pulse Signals
A9	ORN/BLU	A/T: Transmission Fluid Lamp	Lamp on: 1v, off: 12
A10	PNK/GRN	A/T: Winter Lamp	Lamp on: 1v, off: 12
A11	GRY/WHT	A/T: Power Lamp	Lamp on: 1v, off: 12
A12	PPL/WHT	Antilock Brake Light	Lamp Off: 12v, On: 1v
A13	PPL	Malfunction Indicator Lamp	MIL Off: 12v, On: 1v
A14	ORN/BLK	A/T: Check Transmission Light	Lamp on: 1v, off: 12v
A14	ORN/BLU	M/T: Upshift Lamp Relay	Relay on: 1v, off: 12
A15	RED/BLU	EVAP Purge Solenoid	Digital Signals: 0-12-0v
A16	BRN/YEL	A/T: Shift Solenoid Feed (B+)	12-14v
A16	BRN/YEL	M/T: Air Pump Feed (B+)	12-14v

1998-99 4-Door 3.5L V6 MFI VIN X [All] 32P Red 'B' Connector

PCM Pin #	W/Color	Circuit Description (16-Pin)	Value at Hot Idle
B1	RED/YEL	Reference Voltage 'B'	4.9-5.1v
B2	RED/WHT	Igniter 4 Control	Digital Signals: 0-12-0v
B3	RED/BLK	Igniter 2 Control	Digital Signals: 0-12-0v
B4	RED/GRN	Igniter 6 Control	Digital Signals: 0-12-0v
B5	YEL/PPL	Fuel Level Sensor	Tank empty: 1.8v
B6	GRY/RED	Fuel Tank Pressure Sensor	With Fuel Cap Off: 2.5v
B7	YEL/RED	Linear EGR Sensor Signal	0.5-0.6v
B8	YEL/GRN	IAT Sensor Signal	Varies w/temp: 0.5-4.8v
B9	---	Not Used	---
B10	---	Not Used	---
B11	GRN/YEL	PSP Switch Signal	Straight: 12v, Turning: 0v
B12	GRN	Illuminated Switch Signal	Switch on: 0v, off: 12v
B13	ORN/GRN	Class 2 Serial Data Link	0v
B14	GRN/BLK	A/C Clutch Relay Control	Relay Off: 12v, On: 1v
B15	YEL/BRN	Low Fuel Lamp	Lamp on: 1v, off: 12v
B16	YEL/RED	EVAP Vent Solenoid	Tank Empty: 5.7v

32-PIN RED-CONNECTOR ROW 'A' & 'B'

| A | 16 | 15 | 14 | 13 | 12 | 11 | 10 | 9 | 8 | 7 | 6 | 5 | 4 | / | 2 | 1 |
| B | 16 | 15 | 14 | 13 | 12 | 11 | / | / | 8 | 7 | 6 | 5 | 4 | 3 | 2 | 1 |

WIRE SIDE OF HARNESS TERMINALS

05533_ADIA_G645

Pin Connector Graphic
1998-99 4-Door 3.5L V6 MFI VIN X [All] 32P White 'C' Connector

PCM Pin #	W/Color	Circuit Description (16-Pin)	Value at Hot Idle
C1	GRN/RED	Injector 4 Control	2.0-3.3 ms
C2	BRN/BLK	A/T: Shift Solenoid 'B'	SSB Off: 12v, On: 1v
C3	GRN/YEL	Injector 6 Control	2.0-3.3 ms
C4	RED	Igniter 1 Control	Digital Signals: 0-12-0v
C5	YEL	CKP Sensor Signal	Digital Signals: 0-5-0v
C7	BLK/BLU	Power Ground	<0.1v
C8	BLK/PNK	Power Ground	<0.1v
C9	BLK/BLU	Power Ground	<0.1v
C10	BLK/RED	Tachometer Signal	Pulse Signals
C11	YEL/BLK	Variable Intake Manifold Sol.	12-14v (at 3600 rpm)
C12	WHT/BLU	Alternator Charge Signal	12-14v
C13	BLU/PNK	Fuel Gauge PWM Signal	Duty cycle: 0-100%
C14	PNK	HO2S-21 (B2 S1) Signal	0.1-1.1v
C15	BLU	HO2S-21 (B2 S1) Ground	<0.050v
C16	GRN	HO2S-22 (B2 S2) Signal	0.1-1.1v

1998-99 4-Door 3.5L V6 MFI VIN X [All] 32P White 'D' Connector

PCM Pin #	W/Color	Circuit Description (16-Pin)	Value at Hot Idle
D1	GRN/ORN	Injector 2 Control	2.0-3.3 ms
D2	BRN/BLU	A/T: Torque Converter Clutch	TCC on: 1v, off: 12v
D3	GRN/WHT	Injector 1 Control	2.0-3.3 ms
D4	ORN	Serial Data (8192 Baud)	Digital Signals
D5	RED/YEL	Igniter 5 Control	Digital Signals: 0-12-0v
D6	RED/BLU	Igniter 3 Control	Digital Signals: 0-12-0v
D7	GRN/WHT	VSS (4096 pulses per mile)	Moving: 0-5-0-5v
D8	GRN	VREF Sensor 'A' Ground	<0.050v
D9	GRY	VREF Sensor 'B' Ground	<0.050v
D10	YEL	MAF Sensor Signal	7-9v
D11	WHT	CMP Sensor Signal	Digital Signals: 0-5-0v
D12 [A/T]	RED	HO2S-12 (B1 S2) Ground	<0.050v
D12 [M/T]	RED	HO2S-13 (B1 S3) Ground	<0.050v
D13 [A/T]	WHT	HO2S-12 (B1 S2) Signal	0.1-1.1v
D13 [M/T]	WHT	HO2S-13 (B1 S3) Signal	0.1-1.1v
D14	GRN	HO2S-11 (B1 S1) Ground	<0.050v
D15	RED	HO2S-11 (B1 S1) Signal	0.1-1.1v
D16 [A/T]	BLU	HO2S-22 (B2 S2) Ground	<0.050v
D16 [M/T]	BLU	HO2S-12 (B1 S2) Ground	<0.050v

32-PIN WHITE-CONNECTOR ROW 'C' & 'D'

C	16	15	14	13	12	/	10	9	8	7	/	5	4	3	2	1
D	16	15	14	13	12	11	10	9	8	7	6	5	4	3	2	1

WIRE SIDE OF HARNESS TERMINALS

05533_ADIA_G640

Pin Connector Graphic
1998-99 4-Door 3.5L V6 MFI VIN X [All] 32P Blue 'E' Connector

PCM Pin #	W/Color	Circuit Description (16-Pin)	Value at Hot Idle
E1	RED	Output Shaft Sensor Signal	AC Pulses
E2	WHT	Output Shaft Sensor Ground	<0.050v
E3	PPL/RED	A/T: Pressure Control Low	Pulse Signals
E4	PPL/WHT	A/T: Pressure Control High	Pulse Signals
E5	BLK/YEL	Linear EGR Control High	12-14v
E6	YEL	Linear EGR Control Low	Pulse Signals
E7	BLU/YEL	A/T: Range Switch 'B' Signal	In P/N: 0v, others: 12v
E8	BLU	TP Sensor Signal	0.5-0.8v
E9	BLU/RED	ECT Sensor Signal	At 180F: 0.54v
E10	---	Not Used	---
E11	YEL/RED	CKP Sensor VREF	4.9-5.1v
E12	BLU/WHT	A/T: Range Switch 'A' Signal	In P/N: 0v, others: 12v
E13	RED/WHT	Fuel Pump Relay Control	Relay on: 12v, off: 0v
E14	BRN/WHT	A/T: Shift Solenoid Feed (B+)	12-14v
E15	GRN/ORN	A/C Switch Signal	Switch on: 12, off: 0v
E16	RED/BLU	B+ from Main Relay	12-14v

1998-99 4-Door 3.5L V6 MFI VIN X [All] 32P Blue 'F' Connector

PCM Pin #	W/Color	Circuit Description (16-Pin)	Value at Hot Idle
F1	---	Not Used	---
F2	BLU/BLK	A/T: Range Signal 'C'	In P/N: 0v, others: 12v
F3	YEL/GRN	A/T: Range Signal 'P'	In P/N: 0v, others: 12v
F4	GRN/YEL	Brake Switch Signal	Switch on: 12v, Off: 0v
F5	PPL/RED	Power Switch Signal	Switch on: 0.1v, off: 12v
F6	PPL/GRN	Winter Switch Signal	Switch on: 0.1v, off: 12v
F7	RED/BLK	A/T: TOT Sensor Signal	Varies w/temp: 0.5-4.8v
F8	GRY/BLK	MAP Sensor Signal	0.6-1.3v
F9	---	Not Used	---
F10	GRY/BLU	Cruise Control	C/C on: 0.1v, off: 12v
F11	LT BLU	A/T: Kickdown Switch	Switch on: 0.1v, off: 12v
F12	ORN/BLU	PCM Diagnostic Enable	Digital Signals
F13	GRN	Injector 3 Control	2.0-3.3 ms
F14	BRN/RED	A/T: Shift Solenoid 'A'	SSA Off: 12v, On: 1v
F15	GRN/BLK	Injector 5 Control	2.0-3.3 ms
F16	RED/BLU	B+ from Main Relay	12-14v

32-PIN BLUE-CONNECTOR ROW 'E' & 'F'

E	16	15	14	13	12	11	/	9	8	7	6	5	4	3	2	1
F	16	15	14	13	12	11	10	/	8	7	6	5	4	3	2	/

WIRE SIDE OF HARNESS TERMINALS

Pin Connector Graphic

05533_ADIA_G646

HONDA
DIAGNOSTIC TROUBLE CODES

8

TABLE OF CONTENTS

OBD II VEHICLE APPLICATIONS

HONDA

Accord
1996–2006

2.2L I4 SOHC MFI VTEC (F22B1, F22B2) ....VIN CD5
2.2L I4 SOHC MFI VTEC (F22B1, F22B2) ....VIN CD7
2.7L V6 SOHC MFI (C27A4)............................VIN CE6
2.2L I4 SOHC SMFI (F22B1 VTEC, F22B2) ....VIN CE1
2.3L I4 SOHC MFI VTEC (F23A1) ....VIN CG5
2.3L I4 SOHC MFI VTEC (F23A4) ....VIN CG6
2.3L I4 SOHC MFI (F23A5) ....VIN CF8
3.0L V6 SOHC MFI VTEC (J30A1) ....VIN CG1

Civic
1996–2006

1.6L I4 SOHC SMFI 16v (D16Y7) ....VIN EJ6
1.6L I4 SOHC SMFI 16v VTEC-E (D16Y5) ....VIN EJ7
1.6L I4 SOHC SMFI 16v VTEC (D16Y8) ....VIN EJ8
1.7L I4 SOHC SMFI 16v (D17A1) ....VIN ES1
1.7L I4 SOHC SMFI 16v VTEC (D17A2) ....VIN ES2
1.7L I4 SOHC SMFI 16v (D17A7) ....VIN EN2

Civic del Sol
1996–1997

1.5L I4 DOHC SMFI 16v VTEC (B16A2) ....VIN EG2
1.6L I4 SOHC SMFI 16v (D16Y7) ....VIN EH6
1.6L I4 SOHC SMFI 16v VTEC (D16Y8) ....VIN EH6

CR-V
1997–2006

2.0L I4 DOHC SMFI (B20B4) ....VIN RD1

Element
2003–2006

FRONT WHEEL DRIVE

2.4L I4 DOHC MFI (K24A4) ....VIN YH1

FOUR WHEEL DRIVE

2.4L I4 DOHC MFI (K24A4) ....VIN YH2

Insight
2000–2006

1.0L I3 SOHC MFI (ECA1) ....VIN ZE1

Odyssey
1996–2006

2.2L I4 SOHC SMFI (F22B6) ....VIN RA1
2.3L I4 SOHC SMFI (F23A7) ....VIN RA3
3.5L V6 SOHC SMFI (J35A1) ....VIN RL1
3.5L V6 SOHC SMFI (J35A6, J35A7) ....VIN RL3
3.5L V6 SOHC SMFI (J35A6) ....VIN RL4

Passport
1996–2002

2.6L I4 MFI (4ZE1) ....VIN E
3.2L V6 MFI (6VD1) ....VIN V
3.2L V6 DOHC SMFI (6VD1) ....VIN W

Pilot
2003–2006

3.5L V6 SOHC MFI (J35A4) ....VIN YF1

Prelude
1996–2001

2.2L I4 SOHC SMFI (F22A1) ....VIN BA8
2.2L I4 DOHC SMFI VTEC (H22A1) ....VIN BB1
2.3L I4 DOHC SMFI (H23A1) ....VIN BB2
2.2L I4 DOHC SMFI VTEC (H22A4) ....VIN BB6

Ridgeline
2006

3.5L V6 SOHC MFI (J35A9) ....VIN YK1

S2000
2000–2006

2.0L I4 DOHC MFI (F20C1) ....VIN AP1

DIAGNOSTIC TROUBLE CODES

OBD II Trouble Code List (P0xxx Codes)

DTC	Trouble Code Title, Conditions & Possible Causes
DTC: P0101 **2T CCM, MIL: Yes** **1996, 1997, 1998, 1999, 2000, 2001, 2002** **Models:** Passport **Engines:** 3.2L VIN V, 3.2L VIN W **Transmissions:** A/T, M/T	**Mass Airflow Sensor Range/Performance** DTC P0106, P0107, P0108, P0121, P0122 and P0123 not set, system voltage at 11-16v, engine running at a stable idle speed, throttle angle stable (±1%), Calculated airflow from 25-40 g/sec, conditions met for 1 second, and the PCM detected a MAF sensor frequency that was significantly higher or lower than a "predicted" MAF airflow based on throttle position and engine speed for 12.5 seconds over a 25 second period during the CCM Rationality test. **Possible Causes:** • Air leaks after the MAF sensor, or in the EGR or PCV system • Engine oil cap missing, engine oil dipstick not fully seated • MAF sensor is contaminated, dirty or out-of-calibration • MAF sensor ground circuit has high resistance • MAF minimum airflow rate to low at idle or during deceleration • MAP or TP sensor signal skewed, stuck or out of calibration • High signal interference (i.e., electrical noise from the ignition) • PCM has failed
DTC: P0102 **1T CCM, MIL: Yes** **1996, 1997, 1998, 1999, 2000, 2001, 2002** **Models:** Passport **Engines:** 3.2L VIN V, 3.2L VIN W **Transmissions:** A/T, M/T	**MAF Sensor Circuit Low Frequency** Engine started, engine speed over 500 rpm for 10 seconds, system voltage over 11.5v, the PCM detected the MAF sensor frequency was less than 1000 Hz for a total of 50% of the last 100 samples in the CCM Rationality test (a sample is taken every cylinder event). **Possible Causes:** • MAF sensor signal is shorted to ground • MAF sensor power circuit is open • MAF sensor is contaminated, dirty or is damaged • PCM has failed
DTC: P0103 **1T CCM, MIL: Yes** **1996, 1997, 1998, 1999, 2000, 2001, 2002** **Models:** Passport **Engines:** 3.2L VIN V, 3.2L VIN W **Transmissions:** A/T, M/T	**MAF Sensor Circuit High Frequency** Engine started, engine speed over 500 rpm for 10 seconds, system voltage over 11.5v, the PCM detected the MAF sensor frequency was more than 10,000 Hz for a total of 50% of the last 200 samples in the CCM Rationality test (a sample is taken every cylinder event). **Possible Causes:** • RFI or EMI interference from the Generator or Ignition system • RFI or EMI interference from the an Ignition system component • MAF sensor is contaminated, dirty or is damaged • PCM has failed
DTC: P0106 **2T CCM, MIL: Yes** **1995** **Models:** Accord **Engines:** 2.7L VIN CE6 **Transmissions:** A/T, M/T	**Manifold Air Pressure Sensor Range/Performance** Engine started, engine runtime over 1 second, and the PCM detected the MAP sensor was more 11.8" Hg during the test period. **Possible Causes:** • MAP sensor source vacuum line is leaking or disconnected • MAP sensor source vacuum line is plugged at intake manifold • MAP sensor is damaged, out-of-calibration or has failed • PCM has failed
DTC: P0106 **2T CCM, MIL: Yes** **1996, 1997, 1998, 1999, 2000, 2001, 2002, 2003, 2004, 2005, 2006** **Models:** Accord, Civic, Civic del Sol, Insight, Odyssey, Pilot, Prelude, Ridgeline, S2000 **Engines:** All **Transmissions:** A/T, M/T	**Manifold Air Pressure Sensor Range/Performance** Engine started, engine runtime over 1 second, and the PCM detected the MAP sensor was more 11.8" Hg during the test period. **Possible Causes:** • MAP sensor source vacuum line is leaking or disconnected • MAP sensor source vacuum line is plugged at intake manifold • MAP sensor is damaged, out-of-calibration or has failed • PCM has failed

DTC	Trouble Code Title, Conditions & Possible Causes
DTC: P0106 **2T CCM, MIL: Yes** **1996** **Models:** Passport **Engines:** 2.6L VIN E **Transmissions:** M/T	**MAP Sensor Signal Range/Performance** DTC P0121, P0122 and P0123 not set, engine running with the engine speed stable ±100 rpm, throttle angle stable with any change less than 1%, IAC counts stable with change less than 10 counts, EGR flow stable with any change less than 4%, no change in the A/C clutch, Brake switch, Power Steering switch and TCC status for 1 second, and the PCM detected the MAP sensor signal varied by more than 10 kPa for 10 seconds over a 20 second time period. **Possible Causes:** • MAP sensor circuit open or shorted to ground (intermittent) • MAP sensor source vacuum line is leaking or restricted • MAP sensor source vacuum line is plugged at intake manifold • MAP sensor is damaged, out-of-calibration or has failed • PCM has failed
DTC: P0106 **2T CCM, MIL: Yes** **1996, 1997, 1998, 1999, 2000, 2001, 2002** **Models:** Passport **Engines:** 3.2L VIN V, 3.2L VIN W **Transmissions:** A/T, M/T	**MAP Sensor Signal Range/Performance** DTC P0121, P0122 and P0123 not set, engine speed stable (±100 rpm), throttle angle stable (±1%), IAC counts steady (±10 counts), EGR flow stable (±4%), no change in the A/C clutch, PSPS, Brake switch or TCC status, conditions met for 1 second, and the PCM detected the Actual MAP sensor value varied more than 10 kPa from the Expected MAP value for 10 seconds over a 20 second period. **Possible Causes:** • MAP sensor circuit open or shorted to ground (intermittent) • MAP sensor source vacuum line is leaking or restricted • MAP sensor source vacuum line is plugged at intake manifold • MAP sensor is damaged, out-of-calibration or has failed • PCM has failed
DTC: P0107 **1T CCM, MIL: Yes** **1995** **Models:** Accord **Engines:** 2.7L VIN CE6 **Transmissions:** A/T, M/T	**Manifold Air Pressure Sensor Circuit Low Input** Engine started, engine running in closed loop, and the PCM detected the MAP sensor was near 0.0" Hg during the CCM test. **Note: The key on, engine off MAP sensor input should be near 2.9v.** **Possible Causes:** • MAP sensor 5-volt power circuit open or shorted to ground • MAP Sensor signal circuit is shorted to ground • MAP Sensor is damaged or has failed • PCM has failed
DTC: P0107 **1T CCM, MIL: Yes** **1996, 1997, 1998, 1999, 2000, 2001, 2002, 2003, 2004, 2005, 2006** **Models:** Accord, Civic, Civic del Sol, Insight, Odyssey, Pilot, Prelude, Ridgeline, S2000 **Engines:** All **Transmissions:** A/T, M/T	**Manifold Air Pressure Sensor Circuit Low Input** Engine started, engine running in closed loop, and the PCM detected the MAP sensor was near 0.0" Hg during the CCM test. **Note: The key on, engine off MAP sensor input should be near 2.9v.** **Possible Causes:** • MAP sensor 5-volt power circuit open or shorted to ground • MAP Sensor signal circuit is shorted to ground • MAP Sensor is damaged or has failed • PCM has failed
DTC: P0107 **1T CCM, MIL: Yes** **2003, 2004, 2005, 2006** **Models:** Element **Engines:** All **Transmissions:** All	**Manifold Air Pressure Sensor Circuit Low Input** Engine started, engine running in closed loop, and the PCM detected the MAP sensor was near 0.0" Hg during the CCM test. **Note: The key on, engine off MAP sensor input should be near 2.9v.** **Possible Causes:** • MAP sensor 5-volt power circuit open or shorted to ground • MAP Sensor signal circuit is shorted to ground • MAP Sensor is damaged or has failed • PCM has failed
DTC: P0107 **2T CCM, MIL: Yes** **1996** **Models:** Passport **Engines:** 2.6L VIN E **Transmissions:** M/T	**MAP Sensor Circuit Low Input** DTC P0121, P0122 and P0123 not set, engine started, then with the engine speed below 1000 rpm and throttle angle over 1%, or with the engine speed above 1000 rpm and throttle angle over 2%, the PCM detected MAP sensor was under 0.04v (11 kPa) for 10 seconds over a 16 second time period during the CCM test. **Possible Causes:** • MAP sensor circuit shorted to ground between sensor and PCM • MAP sensor power circuit is open or shorted to ground • MAP sensor is damaged or has failed • PCM has failed

DTC	Trouble Code Title, Conditions & Possible Causes
DTC: P0107 **1T CCM, MIL: Yes** **1996, 1997, 1998, 1999, 2000, 2001, 2002** **Models:** Passport **Engines:** 3.2L VIN V, 3.2L VIN W **Transmissions:** A/T, M/T	**MAP Sensor Circuit Low Input** DTC P0121, P0122 and P0123 not set, engine started, system voltage from 11-16v, then with the engine speed below 1000 rpm and throttle angle over 1%, or with the engine speed over 1000 rpm and throttle angle over 2%, the PCM detected the MAP sensor was less than 0.04v (11 kPa) for 10 seconds over a 20 second period. **Possible Causes:** • MAP sensor circuit shorted to ground between sensor and PCM • MAP sensor power circuit is open or shorted to ground • MAP sensor is damaged or has failed • PCM has failed
DTC: P0108 **1T CCM, MIL: Yes** **1995** **Models:** Accord **Engines:** 2.7L VIN CE6 **Transmissions:** A/T, M/T	**Manifold Air Pressure Sensor Circuit High Input** Engine started, engine running in closed loop, and the PCM detected the MAP sensor was near 29.9" Hg during the test period. **Note: The key on, engine off MAP sensor input should be near 2.9v.** **Possible Causes:** • MAP sensor signal circuit is open, or the ground circuit is open • MAP sensor signal circuit shorted to 5v VREF or system power • MAP sensor is damaged (due to an open circuit) or has failed • PCM has failed
DTC: P0108 **1T CCM, MIL: Yes** **1996, 1997, 1998, 1999, 2000, 2001, 2002, 2003, 2004, 2005, 2006** **Models:** Accord, Civic, Civic del Sol, CR-V, Insight, Odyssey, Pilot, Prelude, Ridgeline, S2000 **Engines:** All **Transmissions:** A/T, M/T	**Manifold Air Pressure Sensor Circuit High Input** Engine started, engine running in closed loop, and the PCM detected the MAP sensor was near 29.9" Hg during the test period. **Note: The key on, engine off MAP sensor input should be near 2.9v.** **Possible Causes:** • MAP sensor signal circuit is open, or the ground circuit is open • MAP sensor signal circuit shorted to 5v VREF or system power • MAP sensor is damaged (due to an open circuit) or has failed • PCM has failed
DTC: P0108 **1T CCM, MIL: Yes** **2003, 2004, 2005, 2006** **Models:** Element **Engines:** All **Transmissions:** All	**Manifold Air Pressure Sensor Circuit High Input** Engine started, engine running in closed loop, and the PCM detected the MAP sensor was near 29.9" Hg during the test period. **Note: The key on, engine off MAP sensor input should be near 2.9v.** **Possible Causes:** • MAP sensor signal circuit is open, or the ground circuit is open • MAP sensor signal circuit shorted to 5v VREF or system power • MAP sensor is damaged (due to an open circuit) or has failed • PCM has failed
DTC: P0108 **2T CCM, MIL: Yes** **1996** **Models:** Passport **Engines:** 2.6L VIN E **Transmissions:** M/T	**MAP Sensor Circuit High Input** DTC P0121, P0122 and P0123 not set, engine started, engine speed less than 1000 rpm and the throttle angle less than 3%, or with engine speed more than 1000 rpm and the throttle angle less than 10%, the PCM detected the MAP sensor signal was more than 4.40v (90 kPa) for 10 seconds over a 16 second time period. **Possible Causes:** • MAP sensor circuit is open between the sensor and the PCM • MAP sensor signal circuit is shorted to VREF or system power • MAP sensor ground circuit is open between sensor and PCM • MAP sensor is damaged or has failed • PCM has failed
DTC: P0108 **1T CCM, MIL: Yes** **1996, 1997, 1998, 1999, 2000, 2001, 2002** **Models:** Passport **Engines:** 3.2L VIN V, 3.2L VIN W **Transmissions:** A/T, M/T	**MAP Sensor Circuit High Input** DTC P0121, P0122 and P0123 not set, engine started, engine speed less than 1000 rpm and the throttle angle less than 3%, or with engine speed more than 1000 rpm and the throttle angle less than 10%, the PCM detected the MAP sensor was more than 4.40v (90 kPa) for 10 seconds over a 16 second period during the test. **Possible Causes:** • MAP sensor circuit is open between the sensor and the PCM • MAP sensor signal circuit is shorted to VREF or system power • MAP sensor ground circuit is open between sensor and PCM • MAP sensor is damaged or has failed • PCM has failed

DTC	Trouble Code Title, Conditions & Possible Causes
DTC: P0111 **2T CCM, MIL: Yes** **1996, 1997, 1998, 1999, 2000,** **2001, 2002, 2003, 2004, 2005,** **2006** **Models:** Accord, Civic, Civic del Sol, CR-V, Odyssey, Pilot, Prelude, Ridgeline, S2000 **Engines:** All **Transmissions:** A/T, M/T	**Intake Air Temperature Sensor Range/Performance** Engine started, engine runtime over 10 minutes, and the PCM detected the Intake Air Temperature (IAT) sensor signal changed too much in too short a period of time during the CCM test. **Possible Causes:** • IAT sensor signal value less than 1.0v • IAT sensor ground circuit has high resistance • IAT sensor signal circuit has high resistance • IAT sensor is damaged or has failed • PCM has failed
DTC: P0112 **1T CCM, MIL: Yes** **1995** **Models:** Accord **Engines:** 2.7L VIN CE6 **Transmissions:** A/T, M/T	**Intake Air Temperature Sensor Circuit Low Input** Key on or engine running, and the PCM detected the IAT sensor signal indicated less than 0.1v (Scan Tool reads over 302°F). **Possible Causes:** • IAT sensor signal shorted to chassis ground • IAT sensor signal shorted to sensor ground circuit • IAT sensor has an internal failure (it is shorted) or has failed • PCM has failed
DTC: P0112 **1T CCM, MIL: Yes** **1996, 1997, 1998, 1999, 2000,** **2001, 2002, 2003, 2004, 2005,** **2006** **Models:** Accord, Civic, Civic del Sol, CR-V, Insight, Odyssey, Pilot, Prelude, Ridgeline, S2000 **Engines:** All **Transmissions:** A/T, M/T	**Intake Air Temperature Sensor Circuit Low Input** Key on or engine running, and the PCM detected the IAT sensor signal indicated less than 0.1v (Scan Tool reads over 302°F). **Possible Causes:** • IAT sensor signal shorted to chassis ground • IAT sensor signal shorted to sensor ground circuit • IAT sensor has an internal failure (it is shorted) or has failed • PCM has failed
DTC: P0112 **1T CCM, MIL: Yes** **2003, 2004, 2005, 2006** **Models:** Element **Engines:** All **Transmissions:** All	**Intake Air Temperature Sensor Circuit Low Input** Key on or engine running, and the PCM detected the IAT sensor signal indicated less than 0.1v (Scan Tool reads over 302°F). **Possible Causes:** • IAT sensor signal shorted to chassis ground • IAT sensor signal shorted to sensor ground circuit • IAT sensor has an internal failure (it is shorted) or has failed • PCM has failed
DTC: P0112 **1T CCM, MIL: Yes** **1996** **Models:** Passport **Engines:** 2.6L VIN E **Transmissions:** M/T	**Intake Air Temperature Sensor Circuit Low Input** DTC P0502 not set, engine started, engine runtime over 2 minutes, vehicle speed over 30 mph, and the PCM detected the IAT sensor indicated less than 0.10v (Scan Tool reads 298°F) for 12.5 seconds over a 20 second period during the CCM test. **Possible Causes:** • IAT sensor circuit shorted to ground between sensor and PCM • IAT sensor is damaged, out-of-calibration or has failed • PCM has failed
DTC: P0112 **1T CCM, MIL: Yes** **1996, 1997, 1998, 1999, 2000,** **2001, 2002** **Models:** Passport **Engines:** 3.2L VIN V, 3.2L VIN W **Transmissions:** A/T, M/T	**Intake Air Temperature Sensor Circuit Low Input** DTC P0502 not set, engine started, engine runtime over 2 minutes, vehicle speed over 30 mph, and the PCM detected the IAT sensor indicated less than 0.10v (Scan Tool reads 298°F) for 12.5 seconds over a 20 second period during the CCM test. **Possible Causes:** • IAT sensor circuit shorted to ground between sensor and PCM • IAT sensor is damaged, out-of-calibration or has failed • PCM has failed
DTC: P0113 **1T CCM, MIL: Yes** **1995** **Models:** Accord **Engines:** 2.7L VIN CE6 **Transmissions:** A/T, M/T	**Intake Air Temperature Sensor Circuit High Input** Key on or engine running, and the PCM detected the IAT sensor signal indicated more than 4.90 (Scan Tool reads less than -4°F). **Possible Causes:** • IAT sensor signal shorted to VREF or system power • IAT sensor signal circuit is open • IAT sensor ground circuit is open • Sensor has an internal failure (it is open) • PCM has failed

DTC	Trouble Code Title, Conditions & Possible Causes

DTC: P0113
1T CCM, MIL: Yes
1996, 1997, 1998, 1999, 2000,
2001, 2002, 2003, 2004, 2005,
2006
Models: Accord, Civic, Civic del Sol, CR-V, Insight, Odyssey, Pilot, Prelude, Ridgeline, S2000
Engines: All
Transmissions: A/T, M/T
	Intake Air Temperature Sensor Circuit High Input
	Key on or engine running, and the PCM detected the IAT sensor signal indicated more than 4.90 (Scan Tool reads less than -4°F).
	Possible Causes:
	  • IAT sensor signal shorted to VREF or system power
	  • IAT sensor signal circuit is open
	  • IAT sensor ground circuit is open
	  • Sensor has an internal failure (it is open)
	  • PCM has failed

DTC: P0113
1T CCM, MIL: Yes
2003, 2004, 2005, 2006
Models: Element
Engines: All
Transmissions: All
	Intake Air Temperature Sensor Circuit High Input
	Key on or engine running, and the PCM detected the IAT sensor signal indicated more than 4.90 (Scan Tool reads less than -4°F).
	Possible Causes:
	  • IAT sensor signal shorted to VREF or system power
	  • IAT sensor signal circuit is open
	  • IAT sensor ground circuit is open
	  • Sensor has an internal failure (it is open)
	  • PCM has failed

DTC: P0113
1T CCM, MIL: Yes
1996
Models: Passport
Engines: 2.6L VIN E
Transmissions: M/T
	Intake Air Temperature Sensor Circuit High Input
	DTC P0502 not set, engine started, engine runtime over 4 minutes, vehicle speed over 20 mph, ECT sensor more than 140°F, MAF sensor less than 20 g/sec, and the PCM detected the IAT sensor indicated more than 4.90v (Scan Tool reads -38°F) for 12.5 second over a 25 second period during the CCM test.
	Possible Causes:
	  • IAT sensor signal circuit is open between the sensor and PCM
	  • IAT sensor ground circuit is open between the sensor and PCM
	  • IAT sensor is damaged, out-of-calibration or has failed
	  • PCM has failed

DTC: P0113
1T CCM, MIL: Yes
1996, 1997, 1998, 1999, 2000,
2001, 2002
Models: Passport
Engines: 3.2L VIN V, 3.2L VIN W
Transmissions: A/T, M/T
	Intake Air Temperature Sensor Circuit High Input
	DTC P0502 not set, engine started, engine runtime over 4 minutes, vehicle speed over 20 mph, ECT sensor more than 140°F, MAF sensor less than 20 g/sec, and the PCM detected the IAT sensor indicate more than 4.90v (Scan Tool reads -38°F) for 12.5 second over a 25 second period during the CCM test.
	Possible Causes:
	  • IAT sensor signal circuit is open between the sensor and PCM
	  • IAT sensor ground circuit is open between the sensor and PCM
	  • IAT sensor is damaged, out-of-calibration or has failed
	  • PCM has failed

DTC: P0116
1T CCM, MIL: Yes
1995
Models: Accord
Engines: 2.7L VIN CE6
Transmissions: A/T, M/T
	Engine Coolant Temperature Sensor Range/Performance
	Key on or engine running, and the PCM detected the Engine Coolant Temperature (ECT) sensor signal changed too much too quickly.
	Note: The ECT sensor should read 0.47v-0.78v at hot idle speed.
	Possible Causes:
	  • ECT sensor ground circuit has high resistance
	  • ECT sensor signal circuit has high resistance
	  • ECT sensor is damaged or has failed
	  • PCM has failed

DTC: P0116
1T CCM, MIL: Yes
1996, 1997, 1998, 1999, 2000,
2001, 2002, 2003, 2004, 2005,
2006
Models: Accord, Civic, Civic del Sol, CR-V, Insight, Odyssey, Pilot, Prelude, Ridgeline, S2000
Engines: All
Transmissions: A/T, M/T
	Engine Coolant Temperature Sensor Range/Performance
	Engine running, and the PCM detected the Engine Coolant Temperature (ECT) sensor signal changed too much too quickly.
	Note: The ECT sensor should read 0.47v-0.78v at hot idle speed.
	Possible Causes:
	  • ECT sensor signal circuit is open (an intermittent fault)
	  • ECT sensor signal circuit is shorted to ground (intermittent)
	  • ECT sensor is damaged or has failed
	  • PCM has failed

DTC: P0116
1T CCM, MIL: Yes
2003, 2004, 2005, 2006
Models: Element
Engines: All
Transmissions: All
	Engine Coolant Temperature Sensor Range/Performance
	Engine running, and the PCM detected the Engine Coolant Temperature (ECT) sensor signal changed too much too quickly.
	Note: The ECT sensor should read 0.47v-0.78v at hot idle speed.
	Possible Causes:
	  • ECT sensor signal circuit is open (an intermittent fault)
	  • ECT sensor signal circuit is shorted to ground (intermittent)
	  • ECT sensor is damaged or has failed
	  • PCM has failed

DTC	Trouble Code Title, Conditions & Possible Causes
DTC: P0117 **1T CCM, MIL: Yes** **1995** **Models:** Accord **Engines:** 2.7L VIN CE6 **Transmissions:** A/T, M/T	**Engine Coolant Temperature Sensor Circuit Low Input** Key on or engine running, and the PCM detected the Engine Coolant Temperature (ECT) sensor signal indicated more than 302°F (0.1v). **Note: The normal range of the ECT sensor is from 0.47v to 0.78v.** **Possible Causes:** • ECT sensor signal shorted to chassis ground • ECT sensor signal shorted to sensor ground circuit • ECT sensor has an internal failure (it is shorted) or has failed • PCM has failed
DTC: P0117 **1T CCM, MIL: Yes** **1996, 1997, 1998, 1999, 2000,** **2001, 2002, 2003, 2004, 2005,** **2006** **Models:** Accord, Civic, Civic del Sol, CR-V, Insight, Odyssey, Pilot, Prelude, Ridgeline, S2000 **Engines:** All **Transmissions:** A/T, M/T	**Engine Coolant Temperature Sensor Circuit Low Input** Key on or engine running, and the PCM detected the Engine Coolant Temperature (ECT) sensor signal indicated more than 302°F (0.1v). **Note: The normal range of the ECT sensor is from 0.47v to 0.78v.** **Possible Causes:** • ECT sensor signal shorted to chassis ground • ECT sensor signal shorted to sensor ground circuit • ECT sensor has an internal failure (it is shorted) or has failed • PCM has failed
DTC: P0117 **1T CCM, MIL: Yes** **1996** **Models:** Passport **Engines:** 2.6L VIN E **Transmissions:** M/T	**Engine Coolant Temperature Sensor Circuit Low Input** Engine started, engine runtime over 1 minute, and the PCM detected the ECT sensor indicated less than 0.10v (Scan Tool reads 302°F) for 6.25 seconds for 50 seconds over a 100 second period during the CCM test. **Possible Causes:** • ECT sensor circuit shorted to ground between sensor and PCM • ECT sensor is damaged, out-of-calibration or has failed • PCM has failed
DTC: P0117 **1T CCM, MIL: Yes** **1996, 1997, 1998, 1999, 2000,** **2001, 2002** **Models:** Passport **Engines:** 3.2L VIN V, 3.2L VIN W **Transmissions:** A/T, M/T	**Engine Coolant Temperature Sensor Circuit Low Input** Engine started, engine runtime over 1 minute, and the PCM detected the ECT sensor indicated less than 0.10v (Scan Tool reads 302°F) for 6.25 seconds for 50 seconds over a 100 second period during the CCM test. **Possible Causes:** • ECT sensor circuit shorted to ground between sensor and PCM • ECT sensor is damaged, out-of-calibration or has failed • PCM has failed
DTC: P0118 **1T CCM, MIL: Yes** **1995** **Models:** Accord **Engines:** 2.7L VIN CE6 **Transmissions:** A/T, M/T	**Engine Coolant Temperature Sensor Circuit High Input** Key on or engine running, and the PCM detected the Engine Coolant Temperature (ECT) sensor signal indicated less than -4°F (4.9v). **Note: The normal range of the ECT sensor is from 0.47v to 0.78v.** **Possible Causes:** • IAT sensor signal shorted to VREF or system power • IAT sensor signal circuit is open or the ground circuit is open • Sensor has an internal failure (it is open) • PCM has failed
DTC: P0118 **1T CCM, MIL: Yes** **2003, 2004, 2005, 2006** **Models:** Element **Engines:** All **Transmissions:** All	**Engine Coolant Temperature Sensor Circuit High Input** Key on or engine running, and the PCM detected the Engine Coolant Temperature (ECT) sensor signal indicated less than -4°F (4.9v). **Note: The normal range of the ECT sensor is from 0.47v to 0.78v.** **Possible Causes:** • IAT sensor signal shorted to VREF or system power • IAT sensor signal circuit is open • IAT sensor ground circuit is open • Sensor has an internal failure (it is open) • PCM has failed
DTC: P0118 **1T CCM, MIL: Yes** **1996** **Models:** Passport **Engines:** 2.6L VIN E **Transmissions:** M/T	**Engine Coolant Temperature Sensor Circuit High Input** Engine started, engine runtime over 90 seconds, and the PCM detected the ECT sensor was more than 4.90v (Scan Tool reads -38°F) for 50 seconds over a 100 second period during the CCM test. **Possible Causes:** • ECT sensor signal circuit is open between the sensor and PCM • ECT sensor ground circuit open between the sensor and PCM • ECT sensor is damaged, out-of-calibration or has failed • PCM has failed

DTC	Trouble Code Title, Conditions & Possible Causes
DTC: P0118 **1T CCM, MIL: Yes** **1996, 1997, 1998, 1999, 2000, 2001, 2002** **Models:** Passport **Engines:** 3.2L VIN V, 3.2L VIN W **Transmissions:** A/T, M/T	**Engine Coolant Temperature Sensor Circuit High Input** Engine started, engine runtime over 90 seconds, and the PCM detected the ECT sensor was more than 4.90v (Scan Tool reads -38°F) for 50 seconds over a 100 second period during the CCM test. **Possible Causes:** • ECT sensor signal circuit is open between the sensor and PCM • ECT sensor ground circuit open between the sensor and PCM • ECT sensor is damaged, out-of-calibration or has failed • PCM has failed
DTC: P0121 **2T CCM, MIL: Yes** **1996** **Models:** Passport **Engines:** 2.6L VIN E **Transmissions:** M/T	**Throttle Position Sensor Circuit Range/Performance** DTC P0106, P0107, P0108, P0122 and P0123 not set, engine started, MAP sensor signal less than 55 kPa, throttle angle stable with any change less than 1%, and the PCM detected the Actual throttle angle was not close to the Predicted throttle angle for 12.5 seconds over a 25 second period during the CCM Rationality test. **Possible Causes:** • TP sensor signal circuit is open to the PCM (intermittent fault) • TP sensor ground circuit is open (an intermittent fault) • MAP sensor damaged or out-of-calibration • Throttle body is damaged or throttle linkage is bent or binding • TP sensor is damaged or has failed
DTC: P0121 **2T CCM, MIL: Yes** **1996, 1997, 1998, 1999, 2000, 2001, 2002** **Models:** Passport **Engines:** 3.2L VIN V, 3.2L VIN W **Transmissions:** A/T, M/T	**Throttle Position Sensor Circuit Range/Performance** DTC P0121, P0122, P0123 and P1122 not set, engine started, MAP sensor less than 55 kPa, throttle angle stable (±1%), and the PCM detected the Actual throttle angle was not close to the Predicted throttle angle for 12.5 seconds over a 25 second period in the test. **Possible Causes:** • TP sensor signal circuit is open to the PCM (intermittent fault) • TP sensor ground circuit is open (an intermittent fault) • MAP sensor damaged or out-of-calibration • Throttle body is damaged or throttle linkage is bent or binding • TP sensor is damaged or has failed
DTC: P0122 **1T CCM, MIL: Yes** **1995** **Models:** Accord **Engines:** 2.7L VIN CE6 **Transmissions:** A/T, M/T	**Throttle Position Sensor Circuit Low Input** Engine running in closed loop conditions, and the PCM detected the closed throttle TP signal was less than 0.16v (less than 10% open). **Possible Causes:** • TP sensor signal circuit is shorted to ground • TP sensor VREF circuit is open • TP sensor VREF circuit is shorted to ground • TP sensor is damaged (it may be shorted internally) • PCM has failed
DTC: P0122 **1T CCM, MIL: Yes** **1996, 1997, 1998, 1999, 2000, 2001, 2002, 2003, 2004, 2005, 2006** **Models:** Accord, Civic, Civic del Sol, CR-V, Insight, Odyssey, Pilot, Prelude, Ridgeline, S2000 **Engines:** All **Transmissions:** A/T, M/T	**Throttle Position Sensor Circuit Low Input** Engine running in closed loop conditions, and the PCM detected the closed throttle TP signal was less than 0.16v (less than 10% open). **Possible Causes:** • TP sensor signal circuit is shorted to ground • TP sensor VREF circuit is open • TP sensor VREF circuit is shorted to ground • TP sensor is damaged (it may be shorted internally) • PCM has failed
DTC: P0122 **1T CCM, MIL: Yes** **2003, 2004, 2005, 2006** **Models:** Element **Engines:** All **Transmissions:** All	**Throttle Position Sensor Circuit Low Input** Engine running in closed loop conditions, and the PCM detected the closed throttle TP signal was less than 0.16v (less than 10% open). **Possible Causes:** • TP sensor signal circuit is shorted to ground • TP sensor VREF circuit is open • TP sensor VREF circuit is shorted to ground • TP sensor is damaged (it may be shorted internally) • PCM has failed
DTC: P0122 **1T CCM, MIL: Yes** **1996** **Models:** Passport **Engines:** 2.6L VIN E **Transmissions:** M/T	**Throttle Position Sensor Circuit Low Input** Key on or engine running, and the PCM detected the TP sensor indicated less than 0.22v for 0.78 seconds over a 1.5 second period. **Possible Causes:** • TP sensor signal circuit is grounded between sensor and PCM • Throttle body is damaged • Throttle linkage is bent or binding (sticking) • TP sensor is damaged or has failed

DTC	Trouble Code Title, Conditions & Possible Causes

DTC: P0122
1T CCM, MIL: Yes
1996, 1997, 1998, 1999, 2000, 2001, 2002
Models: Passport
Engines: 3.2L VIN V, 3.2L VIN W
Transmissions: A/T, M/T
	Throttle Position Sensor Circuit Low Input
	Key on or engine running, and the PCM detected the TP sensor indicated less than 0.22v for 0.78 seconds over a 1.5 second period.
	Possible Causes:
	    • TP sensor signal circuit is grounded between sensor and PCM
	    • Throttle body is damaged
	    • Throttle linkage is bent or binding (sticking)
	    • TP sensor is damaged or has failed

DTC: P0123
1T CCM, MIL: Yes
1995
Models: Accord
Engines: 2.7L VIN CE6
Transmissions: A/T, M/T
	Throttle Position Sensor Circuit High Input
	Engine running in closed loop conditions, and the PCM detected the wide-open-throttle TP signal was more than 4.60v (more than 90%).
	Possible Causes:
	    • TP sensor signal circuit is shorted to VREF or to system power
	    • TP sensor ground circuit is open between sensor and the PCM
	    • TP sensor is damaged or has failed
	    • PCM has failed

DTC: P0123
1T CCM, MIL: Yes
1996, 1997, 1998, 1999, 2000, 2001, 2002, 2003, 2004, 2005, 2006
Models: Accord, Civic, Civic del Sol, CR-V, Insight, Odyssey, Pilot, Prelude, Ridgeline, S2000
Engines: All
Transmissions: A/T, M/T
	Throttle Position Sensor Circuit High Input
	Engine running in closed loop conditions, and the PCM detected the wide-open-throttle TP signal was more than 4.60v (more than 90%).
	Possible Causes:
	    • TP sensor signal circuit is shorted to VREF
	    • TP sensor signal circuit is shorted to system power
	    • TP sensor ground circuit is open between sensor and the PCM
	    • TP sensor is damaged or has failed
	    • PCM has failed

DTC: P0123
1T CCM, MIL: Yes
2003, 2004, 2005, 2006
Models: Element
Engines: All
Transmissions: All
	Throttle Position Sensor Circuit High Input
	Engine running in closed loop conditions, and the PCM detected the wide-open-throttle TP signal was more than 4.60v (more than 90%).
	Possible Causes:
	    • TP sensor signal circuit is shorted to VREF
	    • TP sensor signal circuit is shorted to system power
	    • TP sensor ground circuit is open between sensor and the PCM
	    • TP sensor is damaged or has failed
	    • PCM has failed

DTC: P0123
1T CCM, MIL: Yes
1996
Models: Passport
Engines: 2.6L VIN E
Transmissions: M/T
	Throttle Position Sensor Circuit High Input
	Key on or engine running, and the PCM detected the TP sensor signal was more than 4.78v for 0.78 seconds over a 1.5 second period during the CCM test.
	Possible Causes:
	    • TP sensor signal circuit is open
	    • TP sensor ground circuit is open
	    • TP sensor signal circuit is shorted to VREF or system power
	    • TP sensor is damaged or has failed
	    • PCM has failed

DTC: P0123
2T CCM, MIL: Yes
1996, 1997, 1998, 1999, 2000, 2001, 2002
Models: Passport
Engines: 3.2L VIN V, 3.2L VIN W
Transmissions: A/T, M/T
	Throttle Position Sensor Circuit High Input
	Key on or engine running, and the PCM detected the Throttle Position (TP) sensor was more than 4.88v for 0.78 seconds out of a 1.5 second period during the CCM test.
	Possible Causes:
	    • TP sensor signal circuit or ground circuit is open
	    • TP sensor signal circuit is shorted to VREF or system power
	    • TP sensor is damaged or has failed
	    • PCM has failed

DTC	Trouble Code Title, Conditions & Possible Causes

DTC: P0125
2T CCM, MIL: Yes
1996
Models: Passport
Engines: 2.6L VIN E
Transmissions: M/T
	Insufficient Coolant Temperature For Closed Loop
	DTC P0112, P0113, P0117, P0118, P1111, P1112, P1114 and P1115 not set, engine started, ECT sensor less than -20°F and the IAT sensor less than 50°F at startup, the PCM detected the ECT sensor signal was less than 84°F after 5 minutes; or with the ECT and IAT sensor both more than 50°F at startup, the PCM detected the ECT sensor signal was less than 84°F after 2 minutes, condition met 20 times during the CCM Rationality test.
	Possible Causes:
	   • Inspect for low coolant level or an incorrect coolant mixture
	   • Check the operation of the thermostat (it may be stuck open)
	   • ECT sensor is damaged or out-of-calibration (it is "skewed")
	   • ECT sensor signal circuit has high resistance
	   • ECT sensor has failed
	   • PCM has failed

DTC: P0125
2T CCM, MIL: Yes
2003, 2004, 2005, 2006
Models: Element
Engines: All
Transmissions: All
	Engine Coolant Temperature Sensor Malfunction/Slow Response
	DTC P0112, P0113, P0117, P0118, P1111, P1112, P1114 and P1115 not set, engine started, ECT sensor less than -20°F and the IAT sensor less than 50°F at startup, the PCM detected the ECT sensor signal was less than 86°F after 5 minutes; or with the ECT and IAT sensor both more than 50°F at startup, the PCM detected the ECT sensor signal was less than 86°F after 2 minutes, condition met 20 times during the CCM Rationality test.
	Possible Causes:
	   • Inspect for low coolant level or an incorrect coolant mixture
	   • Check the operation of the thermostat (it may be stuck open)
	   • ECT sensor is damaged or out-of-calibration (it is "skewed")
	   • ECT sensor signal circuit has high resistance
	   • ECT sensor has failed
	   • PCM has failed

DTC: P0125
2T ECT, MIL: Yes
1996, 1997, 1998, 1999, 2000,
2001, 2002
Models: Passport
Engines: 3.2L VIN V, 3.2L VIN W
Transmissions: A/T, M/T
	Insufficient Coolant Temperature For Closed Loop
	DTC P0112, P0113, P0117, P0118, P1111, P1112, P1114 and P1115 not set, engine started, ECT and IAT sensors from 14-82°F at startup, then with the IAT sensor from 17-50°F at startup, the PCM detected the ECT signal did not reach 84°F after 20 minutes; or with the IAT sensor more than 50°F at startup, the PCM detected the ECT sensor signal was less than 84°F after 2 minutes, condition met at least 20 times during the CCM Rationality test.
	Possible Causes:
	   • Inspect for low coolant level or an incorrect coolant mixture
	   • Check the operation of the thermostat (it may be stuck open)
	   • ECT sensor is damaged or out-of-calibration (it is "skewed")
	   • ECT sensor signal circuit has high resistance
	   • ECT sensor has failed
	   • PCM has failed

DTC: P0128
2T ECT, MIL: Yes
2000, 2001, 2002, 2003, 2004,
2005, 2006
Models: Accord
Engines: All
Transmissions: A/T, M/T
	Thermostat Range/Performance
	DTC P0107, P0108, P0112, P0113, P0116-118, P0335, P0336, P0300-P0306, P0401, P0505, P1106-P1108, P1259 and P1519, engine running at road load for 10 minutes, and the PCM detected the ECT sensor input did not reach the correct closed loop value.
	Note: It is possible for this code to set if the engine is left running while the hood is open for an extended period in a warm climate.
	Possible Causes:
	   • Inspect for low coolant level or for an incorrect coolant mixture
	   • Check the operation of the thermostat (it may be stuck open)
	   • TSB 01-064 (9/11/01) contains a repair procedure for this code

DTC: P0128
2T ECT, MIL: Yes
2000, 2001, 2002, 2003, 2004,
2005, 2006
Models: Civic, Civic del Sol, CR-V,
Odyssey, Pilot, Prelude, Ridgeline,
S2000
Engines: All
Transmissions: A/T, M/T
	Thermostat Range/Performance
	DTC P0107, P0108, P0112, P0113, P0116-118, P0335, P0336, P0300-P0306, P0401, P0505, P1106-P1108, P1259 and P1519 not set, engine running at road load for 10 minutes, and the PCM detected the ECT sensor input did not reach the correct closed loop value.
	Note: It is possible for this code to set if the engine is left running while the hood is open for an extended period in a warm climate.
	Possible Causes:
	   • Inspect for low coolant level or for an incorrect coolant mixture
	   • Check the operation of the thermostat (it may be stuck open)
	   • TSB 01-064 (9/11/01) contains a repair procedure for this code

DTC	Trouble Code Title, Conditions & Possible Causes
DTC: P0128 **2T ECT, MIL: Yes** **2003, 2004, 2005, 2006** **Models:** Element **Engines:** All **Transmissions:** All	**Cooling System Malfunction** DTC P0112, P0113, P0116, P0117, P0118, P0125, P0300, P0301, P0302, P0303, P0304, P0335, P0339, P2227, P2228, P2229, P2646, P2647, P2648, P2649, P0506, P0507, P0511 not set, engine running at road load for 10 minutes, and the PCM detected the ECT sensor input did not reach the correct closed loop value. **Note: It is possible for this code to set if the engine is left running while the hood is open for an extended period in a warm climate.** **Possible Causes:** • Inspect for low coolant level or for an incorrect coolant mixture • Check the operation of the thermostat (it may be stuck open)
DTC: P0128 **2T ECT, MIL: Yes** **2000, 2001, 2002** **Models:** Passport **Engines:** All **Transmissions:** A/T, M/T	**Thermostat Malfunction** DTC P0101, P0102, P0103, P0112, P0113, P0117, P0118 and P0502 not set, and the PCM detected under one of these cases: Cold Case Startup Conditions IAT sensor from 20-50ºF, and the PCM detected it took over 263 seconds to reach a stabilized thermostat regulated temperature. Warm Case Startup Conditions IAT sensor from 50-128ºF, and the PCM detected it took over 239 seconds to reach a stabilized thermostat regulated temperature. **Possible Causes:** • Check the operation of the thermostat (it may be stuck open) • ECT sensor is damaged or out-of-calibration (it is "skewed") • PCM has failed
DTC: P0131 **1T CCM, MIL: Yes** **1995** **Models:** Accord **Engines:** 2.7L VIN CE6 **Transmissions:** A/T	**HO2S-11 (Bank 1 Sensor 1) Circuit Low Input** Engine running in closed loop in D4 position at cruise speed, and the PCM detected the HO2S signal was fixed at less than 0.50v. **Note: The actual value to set the code is stored in the PCM memory.** **Possible Causes:** • HO2S signal circuit is open or it is shorted to ground • HO2S may be contaminated or may have failed • Fuel supply system is too lean (fuel filter is clogged or dirty) • PCM has failed
DTC: P0131 **1T CCM, MIL: Yes** **1995** **Models:** Accord **Engines:** 2.7L VIN CE6 **Transmissions:** M/T	**HO2S-11 (Bank 1 Sensor 1) Circuit Low Input** Engine running in closed loop in 4th at cruise speed, and the PCM detected the HO2S signal was fixed at less than 0.50v. **Note: The actual value to set the code is stored in the PCM memory.** **Possible Causes:** • HO2S signal circuit is open or it is shorted to ground • HO2S may be contaminated or may have failed • Fuel supply system is too lean (fuel filter is clogged or dirty) • PCM has failed
DTC: P0131 **1T CCM, MIL: Yes** **1996, 1997, 1998, 1999, 2000, 2001, 2002, 2003, 2004, 2005, 2006** **Models:** Accord, Civic, Civic del Sol, CR-V, Odyssey, Pilot, Prelude, Ridgeline, S2000 **Engines:** All **Transmissions:** A/T	**HO2S-11 (Bank 1 Sensor 1) Circuit Low Input** Engine running in closed loop in D4 position at cruise speed, and the PCM detected the HO2S signal was fixed at less than 0.50v. **Note: The actual value to set the code is stored in the PCM memory.** **Possible Causes:** • HO2S signal circuit is open or it is shorted to ground • HO2S may be contaminated or may have failed • Fuel supply system is too lean (fuel filter is clogged or dirty) • PCM has failed
DTC: P0131 **1T CCM, MIL: Yes** **1996, 1997, 1998, 1999, 2000, 2001, 2002, 2003, 2004, 2005, 2006** **Models:** Accord, Civic, Civic del Sol, CR-V, Odyssey, Pilot, Prelude, Ridgeline, S2000 **Engines:** All **Transmissions:** M/T	**HO2S-11 (Bank 1 Sensor 1) Circuit Low Input** Engine running in closed loop in 4th or 6th gear at cruise speed, and the PCM detected the HO2S signal was fixed at less than 0.50v. **Note: The actual value to set the code is stored in the PCM memory.** **Possible Causes:** • HO2S signal circuit is open or it is shorted to ground • HO2S may be contaminated or may have failed • Fuel supply system is too lean (fuel filter is clogged or dirty) • PCM has failed

DTC	Trouble Code Title, Conditions & Possible Causes
DTC: P0131 **1T CCM, MIL: Yes** **1996** **Models:** Passport **Engines:** 2.6L VIN E **Transmissions:** M/T	**HO2S-11 (Bank 1 Sensor 1) Circuit Low Input** DTC P0106, P0107, P0108, P0112, P0113, P0117, P0118, P0121, P0122, P0123, P0171, P0172 and P0300 not set, no Intrusive or Device Control tests active, engine started, engine running in closed loop with the A/F ratio command at 14.5-14.8:1, throttle angle from 3-19% for 5 seconds, the PCM detected the HO2S-11 signal was less than 26 mv for 77 seconds over a 90 second period in the test. **Possible Causes:** • HO2S signal circuit is open or shorted to ground • HO2S contaminated by water or wrong fuel, or it has failed • PCM has failed
DTC: P0131 **1T CCM, MIL: Yes** **1996, 1997, 1998, 1999, 2000,** **2001, 2002** **Models:** Passport **Engines:** 3.2L VIN V, 3.2L VIN W **Transmissions:** A/T, M/T	**HO2S-11 (Bank 1 Sensor 1) Circuit Low Input** DTC P0106, P0107, P0108, P0112, P0113, P0117, P0118, P0121, P0122, P0123, P0171, P0172 and P0300, P0301-P0306 not set, engine running in closed loop with the A/F ratio from 14.5-14.8:1, ECT sensor more than 140°F, throttle angle from 3-19% for 5 seconds, the PCM detected the HO2S-11 signal was less than 26 mv for 77 seconds over a 90 second period during the test. **Possible Causes:** • HO2S signal circuit is open or shorted to ground • HO2S contaminated by water or wrong fuel, or it has failed • PCM has failed
DTC: P0132 **1T CCM, MIL: Yes** **1995** **Models:** Accord **Engines:** 2.7L VIN CE6 **Transmissions:** A/T	**HO2S-11 (Bank 1 Sensor 1) Circuit High Input** Engine running in closed loop in D4 position at cruise speed, and the PCM detected the HO2S signal was fixed at more than 0.90v **Note: The actual value to set the code is stored in the PCM memory.** **Possible Causes:** • HO2S signal tracking (wet/oily) in connector causing a short between the signal circuit and heater power circuit • HO2S signal circuit is open, or the ground circuit is open • HO2S heater supply circuit is open • PCM has failed
DTC: P0132 **1T CCM, MIL: Yes** **1995** **Models:** Accord **Engines:** 2.7L VIN CE6 **Transmissions:** M/T	**HO2S-11 (Bank 1 Sensor 1) Circuit High Input** Engine running in closed loop in 4th gear at cruise speed, and the PCM detected the HO2S signal was fixed at more than 0.90v. **Note: The actual value to set the code is stored in the PCM memory.** **Possible Causes:** • HO2S signal tracking (wet/oily) in connector causing a short between the signal circuit and heater power circuit • HO2S signal circuit is open, or the ground circuit is open • HO2S heater supply circuit is open • PCM has failed
DTC: P0132 **1T CCM, MIL: Yes** **1996, 1997, 1998, 1999, 2000,** **2001, 2002, 2003, 2004, 2005,** **2006** **Models:** Accord, Civic, Civic del Sol, CR-V, Odyssey, Prelude, S2000 **Engines:** All **Transmissions:** A/T	**HO2S-11 (Bank 1 Sensor 1) Circuit High Input** Engine running in closed loop in D4 position at cruise speed, and the PCM detected the HO2S signal was fixed at more than 0.90v **Note: The actual value to set the code is stored in the PCM memory.** **Possible Causes:** • HO2S signal tracking (wet/oily) in connector causing a short between the signal circuit and heater power circuit • HO2S signal circuit is open, or the ground circuit is open • HO2S heater supply circuit is open • PCM has failed
DTC: P0132 **1T CCM, MIL: Yes** **1996, 1997, 1998, 1999, 2000,** **2001, 2002, 2003, 2004, 2005,** **2006** **Models:** Accord, Civic, Civic del Sol, CR-V, Odyssey, Pilot, Prelude, Ridgeline, S2000 **Engines:** All **Transmissions:** M/T	**HO2S-11 (Bank 1 Sensor 1) Circuit High Input** Engine running in closed loop in 4th or 6th gear at cruise speed, and the PCM detected the HO2S signal was fixed at more than 0.90v. **Note: The actual value to set the code is stored in the PCM memory.** **Possible Causes:** • HO2S signal tracking (wet/oily) in connector causing a short between the signal circuit and heater power circuit • HO2S signal circuit is open, or the ground circuit is open • HO2S heater supply circuit is open • PCM has failed
DTC: P0132 **1T CCM, MIL: Yes** **1996** **Models:** Passport **Engines:** 2.6L VIN E **Transmissions:** M/T	**HO2S-11 (Bank 1 Sensor 1) Circuit High Input** DTC P0106, P0107, P0108, P0112, P0113, P0117, P0118, P0121, P0122, P0123, P0171, P0172 and P0300 not set, no Device Control or Intrusive tests active, engine running in closed loop with the A/F ratio at 14.5-14.8:1, throttle angle from 3-19% for 5 seconds, and the PCM detected the HO2S-11 signal was less than 952 mv for 77 seconds over a 90 second period; or the HO2S-11 signal was more than 500 mv during Decel Fuel Cutoff mode for 3 seconds. **Possible Causes:** • HO2S signal circuit shorted to system power (oil in connector) • HO2S contaminated by water or wrong fuel • HO2S is damaged or has failed • PCM has failed

DTC	Trouble Code Title, Conditions & Possible Causes
DTC: P0132 **1T CCM, MIL: Yes** **1996, 1997, 1998, 1999, 2000, 2001, 2002** **Models:** Passport **Engines:** 3.2L VIN V, 3.2L VIN W **Transmissions:** A/T, M/T	**HO2S-11 (Bank 1 Sensor 1) Circuit High Input** DTC P0106, P0107, P0108, P0112, P0113, P0117, P0118, P0121, P0122, P0123, P0171, P0172 and P0300, P0301-P0306 not set, engine running in closed loop with the A/F ratio from 14.5-14.8:1, throttle angle from 3-19% for 5 seconds, the PCM detected the HO2S-11 signal was less than 952 mv for 77 seconds over a 90 second period; or the HO2S-11 signal was more than 500 mv during Decel Fuel Cutoff mode for 3 seconds during the CCM test. **Possible Causes:** • HO2S signal circuit shorted to system power (oil in connector) • HO2S contaminated by water or wrong fuel • HO2S is damaged or has failed • PCM has failed
DTC: P0133 **1T O2S1, MIL: Yes** **1995** **Models:** Accord **Engines:** 2.7L VIN CE6 **Transmissions:** A/T	**HO2S-11 (Bank 1 Sensor 1) Circuit Slow Response** Engine running in closed loop in D4 position at over 55 mph at steady speed, and the PCM detected the HO2S response time to switch between 300-600 mv was too slow, or that the rich to lean or lean to rich switch time was too slow. **Possible Causes:** • Exhaust leak present in the exhaust manifold or exhaust pipes • O2S element fuel contamination • O2S element has deteriorated • PCM has failed
DTC: P0133 **1T O2S1, MIL: Yes** **1995** **Models:** Accord **Engines:** 2.7L VIN CE6 **Transmissions:** M/T	**HO2S-11 (Bank 1 Sensor 1) Circuit Slow Response** Engine running in closed loop in 4th gear at over 55 mph at steady speed, and the PCM detected the HO2S response time to switch between 300-600 mv was too slow, or that the rich to lean or lean to rich switch time was too slow. **Possible Causes:** • Exhaust leak present in the exhaust manifold or exhaust pipes • O2S element fuel contamination • O2S element has deteriorated • PCM has failed
DTC: P0133 **1T O2S1, MIL: Yes** **1996, 1997, 1998, 1999, 2000, 2001, 2002, 2003, 2004, 2005, 2006** **Models:** Accord, Civic, Civic del Sol, CR-V, Odyssey, Pilot, Prelude, Ridgeline, S2000 **Engines:** All **Transmissions:** M/T	**HO2S-11 (Bank 1 Sensor 1) Circuit Slow Response** Engine running in closed loop in 4th or 6th gear at over 55 mph at steady speed, and the PCM detected the HO2S response time to switch between 300-600 mv was too slow, or that the rich to lean or lean to rich switch time was too slow. **Possible Causes:** • Exhaust leak present in the exhaust manifold or exhaust pipes • O2S element fuel contamination • O2S element has deteriorated • PCM has failed
DTC: P0133 **1T O2S1, MIL: Yes** **2003, 2004, 2005, 2006** **Models:** Element **Engines:** All **Transmissions:** All	**Air/Fuel Sensor (Sensor 1) Circuit Slow Response** Engine running in closed loop in 4th or 6th gear at over 55 mph at steady speed, and the PCM detected the HO2S response time to switch between 300-600 mv was too slow, or that the rich to lean or lean to rich switch time was too slow. **Possible Causes:** • Exhaust leak present in the exhaust manifold or exhaust pipes • O2S element fuel contamination • O2S element has deteriorated • PCM has failed
DTC: P0133 **2T O2S2, MIL: Yes** **1996** **Models:** Passport **Engines:** 2.6L VIN E **Transmissions:** A/T, M/T	**HO2S-11 (Bank 1 Sensor 1) Slow Response** DTC P0106, P0107, P0108, P0112, P0113, P0117, P0118, P0121, P0122, P0123, P0171, P0172 and P0300 not set, engine runtime over 1 minute, ECT sensor more than 122°F, engine speed from 1500-3000 rpm, Calculated airflow from 9-42 g/sec, conditions met for 3 seconds, then 90 seconds after entering closed loop, the PCM detected the HO2S-11 lean-to-rich average transition time was over 85 ms, or the rich-to-lean average transition time was 100 ms during the Oxygen Sensor Monitor test. **Possible Causes:** • Exhaust leak present in the exhaust manifold or exhaust pipes • HO2S element fuel contamination • HO2S element has deteriorated • PCM has failed

DTC	Trouble Code Title, Conditions & Possible Causes
DTC: P0133 **2T O2S2, MIL: Yes** **1996, 1997, 1998, 1999, 2000,** **2001, 2002** **Models:** Passport **Engines:** 3.2L VIN V, 3.2L VIN W **Transmissions:** A/T, M/T	**HO2S-11 (Bank 1 Sensor 1) Slow Response** DTC P0106, P0107, P0108, P0112, P0113, P0117, P0118, P0121, P0122, P0123, P0171, P0172 and P0300, P0301-P0306 not set, engine runtime 1 minute in closed loop, ECT sensor more than 122°F, engine speed from 1500-3000 rpm, MAF sensor from 9-42 g/sec, Purge duty cycle over 1%, conditions met for 3 seconds, then 90 seconds after entering closed loop, the PCM detected the HO2S-11 lean-to-rich average transition time was over 94 ms, or the rich-to-lean average transition time was over 105 ms during the test. **Possible Causes:** • Exhaust leak present in the exhaust manifold or exhaust pipes • HO2S element fuel contamination • HO2S element has deteriorated • PCM has failed
DTC: P0134 **1T O2S2, MIL: Yes** **1996** **Models:** Passport **Engines:** 2.6L VIN E **Transmissions:** M/T	**HO2S-11 (Bank 1 Sensor 1) Insufficient Activity Detected** DTC P0106, P0107, P0108, P0112, P0113, P0117, P0118, P0121, P0122, P0123, P0171, P0172 and P0300 not set, system voltage from 10-16v, engine runtime over 90 seconds, then after the PCM determined the Oxygen Sensor Heater test passed, it detected the HO2S-11 signal remained from 400-500 mv for 77 seconds over a 90 second period during the Oxygen Sensor Monitor test. **Possible Causes:** • Exhaust leak present in exhaust manifold or exhaust pipes • HO2S element fuel contamination or has deteriorated • HO2S signal circuit or the ground circuit has high resistance • HO2S heater element has failed, or the heater circuit is open • PCM has failed
DTC: P0134 **1T O2S2, MIL: Yes** **1996, 1997, 1998, 1999, 2000,** **2001, 2002** **Models:** Passport **Engines:** 3.2L VIN V, 3.2L VIN W **Transmissions:** A/T, M/T	**HO2S-11 (Bank 1 Sensor 1) Insufficient Activity Detected** DTC P0106, P0107, P0108, P0112, P0113, P0117, P0118, P0121, P0122, P0123, P0171, P0172 and P0300, P0301-306 not set, system voltage from 11-16v, engine runtime over 40 seconds, then after the PCM determined the Oxygen Sensor Heater test passed, it detected the HO2S-11 signal remained from 400-500 mv for 77 seconds over a 90 second period in the Oxygen Sensor Monitor test. **Possible Causes:** • Exhaust leak present in exhaust manifold or exhaust pipes • HO2S element fuel contamination or has deteriorated • HO2S signal circuit or the ground circuit has high resistance • HO2S heater element has failed, or the heater circuit is open • PCM has failed
DTC: P0134 **1T O2S HTR1, MIL: Yes** **2003, 2004, 2005, 2006** **Models:** Element **Engines:** All **Transmissions:** All	**Air/Fuel Sensor (Sensor 1) Heater Circuit Malfunction** Engine runtime over 80 seconds, and the PCM detected an incorrect signal value at the HO2S heater circuit during the test period. **Possible Causes:** • Main relay output (power) circuit to the heater is open • HO2S element fuel contamination or has deteriorated • HO2S signal circuit or the ground circuit has high resistance • HO2S heater element has failed, or the heater circuit is open • PCM has failed
DTC: P0135 **1T O2S HTR1, MIL: Yes** **1995** **Models:** Accord **Engines:** 2.7L VIN CE6 **Transmissions:** A/T, M/T	**HO2S-11 (Bank 1 Sensor 1) Heater Circuit Malfunction** Engine runtime over 80 seconds, and the PCM detected an incorrect signal value at the HO2S heater circuit during the test period. **Possible Causes:** • Main relay output (power) circuit to the heater is open • O2S heater ground circuit is open • O2S heater element has high resistance • O2S heater element has an open condition • O2S heater element has a shorted condition • PCM has failed
DTC: P0135 **1T O2S HTR1, MIL: Yes** **1996, 1997, 1998, 1999, 2000,** **2001, 2002, 2003, 2004, 2005,** **2006** **Models:** Accord, Civic, Civic del Sol, CR-V, Odyssey, Pilot, Prelude, Ridgeline, S2000 **Engines:** All **Transmissions:** A/T, M/T	**HO2S-11 (Bank 1 Sensor 1) Heater Circuit Malfunction** Engine runtime over 80 seconds, and the PCM detected an incorrect signal value at the HO2S heater circuit during the test period. **Possible Causes:** • Main relay output (power) circuit to the heater is open • O2S heater ground circuit is open • O2S heater element has high resistance • O2S heater element has an open condition • O2S heater element has a shorted condition • PCM has failed

DTC	Trouble Code Title, Conditions & Possible Causes
DTC: P0135 **2T O2S HTR2, MIL: Yes** 1996 **Models:** Passport **Engines:** 2.6L VIN E **Transmissions:** M/T	**HO2S-11 (Bank 1 Sensor 1) Heater Circuit Malfunction** No HO2S-11 codes set, ECT and IAT sensors less than 90°F and within 14°F at startup, engine running, system voltage from 10-16v, then with the average Calculated airflow less than 15 g/sec during the test period, the PCM detected the HO2S-11 signal did not vary more than 150 mv from the bias voltage of 400 to 500 mv after 150 seconds during the Oxygen Sensor Heater Monitor test. **Possible Causes:** • HO2S heater power circuit is open (check the 20A heater fuse) • HO2S heater ground circuit is open • HO2S heater element has high resistance or has failed • PCM has failed
DTC: P0135 **2T O2S HTR2, MIL: Yes** 1996, 1997, 1998, 1999, 2000, 2001, 2002 **Models:** Passport **Engines:** 3.2L VIN V, 3.2L VIN W **Transmissions:** A/T, M/T	**HO2S-11 (Bank 1 Sensor 1) Heater Circuit Malfunction** Engine started cold, ECT and IAT sensors less than 90°F and within 14°F at startup, engine running, system voltage from 11-16v, then with the average Calculated airflow less than 15 g/sec during the test period, the PCM detected the HO2S-11 signal did not vary more than 150 mv from the bias voltage of 400 to 500 mv for up to 150 seconds during the Oxygen Sensor Heater Monitor test. **Possible Causes:** • HO2S heater power circuit is open (check the 20A heater fuse) • HO2S heater ground circuit is open • HO2S heater element has high resistance or has failed • PCM has failed
DTC: P0135 **1T O2S HTR1, MIL: Yes** 2003, 2004, 2005, 2006 **Models:** Element **Engines:** All **Transmissions:** All	**Air/Fuel Sensor (Sensor 1) Heater Circuit Malfunction** Engine runtime over 80 seconds, and the PCM detected an incorrect signal value at the HO2S heater circuit during the test period. **Possible Causes:** • Main relay output (power) circuit to the heater is open • HO2S element fuel contamination or has deteriorated • HO2S signal circuit or the ground circuit has high resistance • HO2S heater element has failed, or the heater circuit is open • PCM has failed
DTC: P0137 **1T CCM, MIL: Yes** 1995 **Models:** Accord **Engines:** 2.7L VIN CE6 **Transmissions:** A/T	**HO2S-12 (Bank 1 Sensor 2) Circuit Low Input** Engine running in closed loop in D4 position at cruise speed, and the PCM detected the HO2S signal was fixed at less than 0.30v. **Note: The actual value where the code sets is in the PCM memory.** **Possible Causes:** • HO2S signal circuit is open • HO2S signal circuit is shorted to ground • HO2S ground circuit is open • HO2S may be contaminated or may have failed • PCM has failed
DTC: P0137 **1T CCM, MIL: Yes** 1995 **Models:** Accord **Engines:** 2.7L VIN CE6 **Transmissions:** M/T	**HO2S-12 (Bank 1 Sensor 2) Circuit Low Input** Engine running in closed loop in 4th gear at cruise speed, and the PCM detected the HO2S signal was fixed at less than 0.30v. **Note: The actual value where the code sets is in the PCM memory.** **Possible Causes:** • HO2S signal circuit is open • HO2S signal circuit is shorted to ground • HO2S ground circuit is open • HO2S may be contaminated or may have failed • PCM has failed
DTC: P0137 **1T CCM, MIL: Yes** 1996, 1997, 1998, 1999, 2000, 2001, 2002, 2003, 2004, 2005, 2006 **Models:** Accord, Civic, Civic del Sol, CR-V, Insight, Odyssey, Pilot, Prelude, Ridgeline, S2000 **Engines:** All **Transmissions:** A/T	**HO2S-12 (Bank 1 Sensor 2) Circuit Low Input** Engine running in closed loop in D4 position at cruise speed, and the PCM detected the HO2S signal was fixed at less than 0.30v. **Note: The actual value where the code sets is in the PCM memory.** **Possible Causes:** • HO2S signal circuit is open • HO2S signal circuit is shorted to ground • HO2S ground circuit is open • HO2S may be contaminated or may have failed • PCM has failed

DTC	Trouble Code Title, Conditions & Possible Causes
DTC: P0137 **1T CCM, MIL: Yes** **2003, 2004, 2005, 2006** **Models:** Element **Engines:** All **Transmissions:** All	**HO2S-12 (Bank 1 Sensor 2) Circuit Low Input** Engine running in closed loop in D4 position at cruise speed, and the PCM detected the HO2S signal was fixed at less than 0.30v. **Note: The actual value where the code sets is in the PCM memory.** **Possible Causes:** • HO2S signal circuit is open • HO2S signal circuit is shorted to ground • HO2S ground circuit is open • HO2S may be contaminated or may have failed • PCM has failed
DTC: P0137 **1T CCM, MIL: Yes** **1996, 1997, 1998, 1999, 2000, 2001, 2002, 2003, 2004, 2005, 2006** **Models:** Accord, Civic, Civic del Sol, CR-V, Odyssey, Pilot, Prelude, Ridgeline, S2000 **Engines:** All **Transmissions:** M/T	**HO2S-12 (Bank 1 Sensor 2) Circuit Low Input** Engine running in closed loop in 4th or 6th gear at cruise speed, and the PCM detected the HO2S signal was fixed at less than 0.30v. **Note: The actual value where the code sets is in the PCM memory.** **Possible Causes:** • HO2S signal circuit is open • HO2S signal circuit is shorted to ground • HO2S ground circuit is open • HO2S may be contaminated or may have failed • PCM has failed
DTC: P0137 **1T CCM, MIL: Yes** **1996** **Models:** Passport **Engines:** 2.6L VIN E **Transmissions:** M/T	**HO2S-12 (Bank 1 Sensor 2) Circuit Low Input** DTC P0106, P0107, P0108, P0112, P0113, P0117, P0118, P0121, P0122, P0123, P0171, P0172 and P0300 not set, no Intrusive or Device Control tests active, engine started, engine running in closed loop with the A/F ratio command at 14.5-14.8:1, throttle angle from 3-19% for 5 seconds, the PCM detected the HO2S-12 signal was less than 26 mv for 106 seconds over a 125 second period, or it detected the HO2S-12 signal remained below 400 mv during Power Enrichment mode for 5 seconds during the CCM Rationality test. **Possible Causes:** • HO2S signal circuit is open or shorted to ground • HO2S contaminated by water or wrong fuel, or it has failed • PCM has failed
DTC: P0137 **1T CCM, MIL: Yes** **1996, 1997, 1998, 1999, 2000, 2001, 2002** **Models:** Passport **Engines:** 3.2L VIN V, 3.2L VIN W **Transmissions:** A/T, M/T	**HO2S-12 (Bank 1 Sensor 2) Circuit Low Input** DTC P0106, P0107, P0108, P0112, P0113, P0117, P0118, P0121, P0122, P0123, P0171, P0172 and P0300, P0301-P0306 not set, engine running in closed loop with the A/F ratio from 14.5-14.8:1, ECT sensor more than 140°F, throttle angle from 3-19% for 5 seconds, the PCM detected the HO2S-12 signal was less than 26 mv for 106 seconds over a 125 second period during the CCM test. **Possible Causes:** • HO2S signal circuit is open or shorted to ground • HO2S contaminated by water or wrong fuel, or it has failed • PCM has failed
DTC: P0138 **1T CCM, MIL: Yes** **1995** **Models:** Accord **Engines:** 2.7L VIN CE6 **Transmissions:** A/T	**HO2S-12 (Bank 1 Sensor 2) Circuit High Input** Engine running in closed loop in D4 position at cruise speed, and the PCM detected the HO2S signal was fixed at more than 0.60v **Note: The actual value where the code sets is in the PCM memory.** **Possible Causes:** • HO2S signal tracking (wet/oily) in connector causing a short between the signal circuit and heater power circuit • HO2S signal circuit is open, or the ground circuit is open • HO2S heater supply circuit is open • PCM has failed
DTC: P0138 **1T CCM, MIL: Yes** **1995** **Models:** Accord **Engines:** 2.7L VIN CE6 **Transmissions:** M/T	**HO2S-12 (Bank 1 Sensor 2) Circuit High Input** Engine running in closed loop in 4th gear at cruise speed, and the PCM detected the HO2S signal was fixed at more than 0.60v. **Note: The actual value where the code sets is in the PCM memory.** **Possible Causes:** • HO2S signal tracking (wet/oily) in connector causing a short between the signal circuit and heater power circuit • HO2S signal circuit is open, or the ground circuit is open • HO2S heater supply circuit is open • PCM has failed

DTC	Trouble Code Title, Conditions & Possible Causes
DTC: P0138 **1T CCM, MIL: Yes** **1996, 1997, 1998, 1999, 2000, 2001, 2002, 2003, 2004, 2005, 2006** **Models:** Accord, Civic, Civic del Sol, CR-V, Insight, Odyssey, Pilot, Prelude, Ridgeline, S2000 **Engines:** All **Transmissions:** A/T	**HO2S-12 (Bank 1 Sensor 2) Circuit High Input** Engine running in closed loop in D4 position at cruise speed, and the PCM detected the HO2S signal was fixed at more than 0.60v **Note: The actual value where the code sets is in the PCM memory.** **Possible Causes:** • HO2S signal tracking (wet/oily) in connector causing a short between the signal circuit and heater power circuit • HO2S signal circuit is open, or the ground circuit is open • HO2S heater supply circuit is open • PCM has failed
DTC: P0138 **1T CCM, MIL: Yes** **2003, 2004, 2005, 2006** **Models:** Element **Engines:** All **Transmissions:** All	**HO2S-12 (Bank 1 Sensor 2) Circuit High Input** Engine running in closed loop in D4 position at cruise speed, and the PCM detected the HO2S signal was fixed at more than 0.60v **Note: The actual value where the code sets is in the PCM memory.** **Possible Causes:** • HO2S signal tracking (wet/oily) in connector causing a short between the signal circuit and heater power circuit • HO2S signal circuit is open, or the ground circuit is open • HO2S heater supply circuit is open • PCM has failed
DTC: P0138 **1T CCM, MIL: Yes** **1996, 1997, 1998, 1999, 2000, 2001, 2002, 2003, 2004, 2005, 2006** **Models:** Accord, Civic, Civic del Sol, CR-V, Odyssey, Pilot, Prelude, Ridgeline, S2000 **Engines:** All **Transmissions:** M/T	**HO2S-12 (Bank 1 Sensor 2) Circuit High Input** Engine running in closed loop in 4th or 6th gear at cruise speed, and the PCM detected the HO2S signal was fixed at more than 0.60v. **Note: The actual value where the code sets is in the PCM memory.** **Possible Causes:** • HO2S signal tracking (wet/oily) in connector causing a short between the signal circuit and heater power circuit • HO2S signal circuit is open, or the ground circuit is open • HO2S heater supply circuit is open • PCM has failed
DTC: P0138 **1T CCM, MIL: Yes** **1996** **Models:** Passport **Engines:** 2.6L VIN E **Transmissions:** M/T	**HO2S-12 (Bank 1 Sensor 2) Circuit High Input** DTC P0106, P0107, P0108, P0112, P0113, P0117, P0118, P0121, P0122, P0123, P0171, P0172 and P0300 not set, no Intrusive or Device Control tests active, engine running in closed loop with the A/F ratio command at 14.5-14.8:1, throttle angle from 3-19% for 5 seconds, and the PCM detected the HO2S-12 signal was less than 952 mv for 106 seconds over a 125 second period, or it detected the HO2S-12 signal was above 500 mv in Decel Fuel Cutoff (DFCO) mode for up to 5 seconds during the CCM Rationality test. **Possible Causes:** • HO2S signal circuit shorted to system power (oil in connector) • HO2S contaminated by water or wrong fuel, damaged or it has failed • PCM has failed
DTC: P0138 **1T CCM, MIL: Yes** **1996, 1997, 1998, 1999, 2000, 2001, 2002** **Models:** Passport **Engines:** 3.2L VIN V, 3.2L VIN W **Transmissions:** A/T, M/T	**HO2S-12 (Bank 1 Sensor 2) Circuit High Input** DTC P0106, P0107, P0108, P0112, P0113, P0117, P0118, P0121, P0122, P0123, P0171, P0172 and P0300, P0301-P0306 not set, engine running in closed loop with the A/F ratio command at 14.5-14.8:1, throttle angle from 3-19% for 5 seconds, the PCM detected the HO2S-12 signal was less than 952 mv for 106 seconds over a 125 second period; or the HO2S-12 signal was more than 500 mv during Decel Fuel Cutoff mode for 3 seconds during the CCM test. **Possible Causes:** • HO2S signal circuit shorted to system power (oil in connector) • HO2S contaminated by water or wrong fuel • HO2S is damaged or has failed • PCM has failed
DTC: P0139 **1T O2S1, MIL: Yes** **1995** **Models:** Accord **Engines:** 2.7L VIN CE6 **Transmissions:** A/T	**HO2S-12 (Bank 1 Sensor 2) Circuit Slow Response** Engine running in closed loop in D4 position at over 55 mph at steady speed, and the PCM detected the HO2S response time to switch between 300-600 mv was too slow, or that the rich to lean or lean to rich switch time was too slow. **Possible Causes:** • Exhaust leak present in the exhaust manifold or exhaust pipes • HO2S element fuel contamination • HO2S element has deteriorated • PCM has failed

DTC	Trouble Code Title, Conditions & Possible Causes
DTC: P0139 **1T O2S1, MIL: Yes** **1995** **Models:** Accord **Engines:** 2.7L VIN CE6 **Transmissions:** M/T	**HO2S-12 (Bank 1 Sensor 2) Circuit Slow Response** Engine running in closed loop in 4th gear at over 55 mph at steady speed, and the PCM detected the HO2S response time to switch between 300-600 mv was too slow, or that the rich to lean or lean to rich switch time was too slow. **Possible Causes:** • Exhaust leak present in the exhaust manifold or exhaust pipes • HO2S element fuel contamination • HO2S element has deteriorated • PCM has failed
DTC: P0139 **2T O2S1, MIL: Yes** **1996, 1997, 1998, 1999, 2000,** **2001, 2002, 2003, 2004, 2005,** **2006** **Models:** Accord, Civic, Civic del Sol, CR-V, Insight, Odyssey, Pilot, Prelude, Ridgeline, S2000 **Engines:** All **Transmissions:** A/T	**HO2S-12 (Bank 1 Sensor 2) Circuit Slow Response** Engine running in closed loop in D4 position at over 55 mph at steady speed, and the PCM detected the HO2S response time to switch between 300-600 mv was too slow, or that the rich to lean or lean to rich switch time was too slow. **Possible Causes:** • Exhaust leak present in the exhaust manifold or exhaust pipes • HO2S element fuel contamination • HO2S element has deteriorated • PCM has failed
DTC: P0139 **2T O2S1, MIL: Yes** **2003, 2004, 2005, 2006** **Models:** Element **Engines:** All **Transmissions:** All	**HO2S-12 (Bank 1 Sensor 2) Circuit Slow Response** Engine running in closed loop in D4 position at over 55 mph at steady speed, and the PCM detected the HO2S response time to switch between 300-600 mv was too slow, or that the rich to lean or lean to rich switch time was too slow. **Possible Causes:** • Exhaust leak present in the exhaust manifold or exhaust pipes • HO2S element fuel contamination • HO2S element has deteriorated • PCM has failed
DTC: P0139 **2T O2S1, MIL: Yes** **1996, 1997, 1998, 1999, 2000,** **2001, 2002, 2003, 2004, 2005,** **2006** **Models:** Accord, Civic, Civic del Sol, CR-V, Odyssey, Pilot, Prelude, Ridgeline, S2000 **Engines:** All **Transmissions:** M/T	**HO2S-12 (Bank 1 Sensor 2) Circuit Slow Response** Engine running in closed loop in 4th or 6th gear at over 55 mph at steady speed, and the PCM detected the HO2S response time to switch between 300-600 mv was too slow, or that the rich to lean or lean to rich switch time was too slow. **Possible Causes:** • Exhaust leak present in the exhaust manifold or exhaust pipes • HO2S element fuel contamination • HO2S element has deteriorated • PCM has failed
DTC: P0140 **1T O2S2, MIL: Yes** **1996** **Models:** Passport **Engines:** 2.6L VIN E **Transmissions:** M/T	**HO2S-12 (Bank 1 Sensor 2) Insufficient Activity Detected** DTC P0106, P0107, P0108, P0112, P0113, P0117, P0118, P0121, P0122, P0123, P0171, P0172 and P0300 not set, system voltage from 10-16v, engine runtime over 90 seconds, then after the PCM determined the Oxygen Sensor Heater test passed, it detected the HO2S-12 signal remained from 426-474 mv for 105 seconds over a 125 second period during the Oxygen Sensor Monitor test. **Possible Causes:** • Exhaust leak present in exhaust manifold or exhaust pipes • HO2S element fuel contamination or has deteriorated • HO2S signal circuit or the ground circuit has high resistance • HO2S heater element has failed, or the heater circuit is open • PCM has failed
DTC: P0140 **1T O2S2, MIL: Yes** **1996, 1997, 1998, 1999, 2000,** **2001, 2002** **Models:** Passport **Engines:** 3.2L VIN V, 3.2L VIN W **Transmissions:** A/T, M/T	**HO2S-12 (Bank 1 Sensor 2) Insufficient Activity Detected** DTC P0106, P0107, P0108, P0112, P0113, P0117, P0118, P0121, P0122, P0123, P0171, P0172 and P0300, P0301-306 not set, system voltage from 11-16v, engine runtime over 40 seconds, then after the PCM determined the Oxygen Sensor Heater test passed, it detected the HO2S-12 signal remained from 426-474 mv for 105 seconds of a 125 second period in the Oxygen Sensor Monitor test. **Possible Causes:** • Exhaust leak present in exhaust manifold or exhaust pipes • HO2S element fuel contamination or has deteriorated • HO2S signal circuit or the ground circuit has high resistance • HO2S heater element has failed, or the heater circuit is open • PCM has failed

DTC	Trouble Code Title, Conditions & Possible Causes
DTC: P0141 **1T O2S HTR1, MIL: Yes** **1995** **Models:** Accord **Engines:** 2.7L VIN CE6 **Transmissions:** A/T, M/T	**HO2S-12 (Bank 1 Sensor 2) Heater Circuit Malfunction** Engine runtime over 80 seconds, and the PCM detected an incorrect signal value at the HO2S heater circuit during the test period. **Possible Causes:** • Main relay output (power) circuit to the heater is open • HO2S heater ground circuit is open • HO2S heater element is open or it has high resistance • HO2S heater element has a shorted condition • PCM has failed
DTC: P0141 **1T O2S HTR1, MIL: Yes** **1996, 1997, 1998, 1999, 2000,** **2001, 2002, 2003, 2004, 2005,** **2006** **Models:** Accord, Civic, Civic del Sol, CR-V, Insight, Odyssey, Pilot, Prelude, Ridgeline, S2000 **Engines:** All **Transmissions:** A/T, M/T	**HO2S-12 (Bank 1 Sensor 2) Heater Circuit Malfunction** Engine runtime over 80 seconds, and the PCM detected an incorrect signal value at the HO2S heater circuit during the test period. **Possible Causes:** • Main relay output (power) circuit to the heater is open • HO2S heater ground circuit is open • HO2S heater element has high resistance • HO2S heater element has an open condition • HO2S heater element has a shorted condition • PCM has failed
DTC: P0141 **1T O2S HTR1, MIL: Yes** **2003, 2004, 2005, 2006** **Models:** Element **Engines:** All **Transmissions:** All	**HO2S-12 (Bank 1 Sensor 2) Heater Circuit Malfunction** Engine runtime over 80 seconds, and the PCM detected an incorrect signal value at the HO2S heater circuit during the test period. **Possible Causes:** • Main relay output (power) circuit to the heater is open • HO2S heater ground circuit is open • HO2S heater element has high resistance • HO2S heater element has an open condition • HO2S heater element has a shorted condition • PCM has failed
DTC: P0141 **2T O2S HTR2, MIL: Yes** **1996** **Models:** Passport **Engines:** 2.6L VIN E **Transmissions:** M/T	**HO2S-12 (Bank 1 Sensor 2) Heater Circuit Malfunction** Engine started cold, ECT and IAT sensors less than 90°F and within 14°F at startup, engine running, system voltage from 10-16v, then with the average Calculated airflow less than 15 g/sec during the test period, the PCM detected the HO2S-11 signal did not vary more than 150 mv from the bias voltage of 400 to 500 mv for up to 150 seconds during the Oxygen Sensor Heater Monitor test. **Possible Causes:** • HO2S heater power circuit is open (check the 20A heater fuse) • HO2S heater ground circuit is open • HO2S heater element has high resistance or has failed • PCM has failed
DTC: P0141 **2T O2S HTR2, MIL: Yes** **1996, 1997, 1998, 1999, 2000,** **2001, 2002** **Models:** Passport **Engines:** 3.2L VIN V, 3.2L VIN W **Transmissions:** A/T, M/T	**HO2S-12 (Bank 1 Sensor 2) Heater Circuit Malfunction** Engine started cold, ECT and IAT sensors less than 90°F and within 11°F at startup, engine running, system voltage from 10-16v, then with the average Calculated airflow less than 23 g/sec during the test period, the PCM detected the HO2S-12 signal did not vary more than 150 mv from the bias voltage of 400 to 500 mv for up to 300 seconds during the Oxygen Sensor Heater Monitor test. **Possible Causes:** • HO2S heater power circuit is open (check the 20A heater fuse) • HO2S heater ground circuit is open • HO2S heater element has high resistance or has failed • PCM has failed
DTC: P0143 **1T CCM, MIL: Yes** **1996, 1997** **Models:** Passport **Engines:** 3.2L VIN V **Transmissions:** M/T	**HO2S-13 (Bank 1 Sensor 3) Circuit Low Input** DTC P0106, P0107, P0108, P0112, P0113, P0117, P0118, P0121, P0122, P0123, P0171, P0172 and P0300, P0301-P0306 not set, engine running in closed loop with the A/F ratio from 14.5-14.8:1, ECT sensor more than 140°F, throttle angle from 3-19% for 5 seconds, the PCM detected the HO2S-13 signal was less than 22 mv for 106 seconds over a 125 second period during the CCM test. **Possible Causes:** • HO2S signal circuit is open or shorted to ground • HO2S contaminated by water or wrong fuel, or it has failed • PCM has failed

DTC	Trouble Code Title, Conditions & Possible Causes
DTC: P0144 **1T CCM, MIL: Yes** **1996, 1997** **Models:** Passport **Engines:** 3.2L VIN V **Transmissions:** M/T	**HO2S-13 (Bank 1 Sensor 3) Circuit High Input** DTC P0106, P0107, P0108, P0112, P0113, P0117, P0118, P0121, P0122, P0123, P0171, P0172 and P0300, P0301-P0306 not set, engine running in closed loop with the A/F ratio from 14.5-14.8:1, ECT sensor more than 140°F, throttle angle from 3-19% for 5 seconds, the PCM detected the HO2S-13 signal was more than 952 mv for 106 seconds over a 125 second period during the CCM test. **Possible Causes:** • HO2S signal shorted to system power in harness (intermittent) • HO2S signal circuit shorted to system power (oil in connector) • HO2S contaminated by water or wrong fuel, damaged or failed • PCM has failed
DTC: P0146 **1T O2S2, MIL: Yes** **1996, 1997** **Models:** Passport **Engines:** 3.2L VIN V **Transmissions:** M/T	**HO2S-13 (Bank 1 Sensor 3) Insufficient Activity Detected** DTC P0106, P0107, P0108, P0112, P0113, P0117, P0118, P0121, P0122, P0123, P0171, P0172 and P0300, P0301-306 not set, system voltage from 11-16v, engine runtime over 40 seconds, then after the PCM determined the Oxygen Sensor Heater test passed, it detected the HO2S-13 signal remained from 426-474 mv for 105 seconds of a 125 second period in the Oxygen Sensor Monitor test. **Possible Causes:** • Exhaust leak present in exhaust manifold or exhaust pipes • HO2S element fuel contamination or has deteriorated • HO2S signal circuit or the ground circuit has high resistance • HO2S heater element has failed, or the heater circuit is open • PCM has failed
DTC: P0147 **2T O2S HTR2, MIL: Yes** **1996, 1997** **Models:** Passport **Engines:** 3.2L VIN V **Transmissions:** M/T	**HO2S-13 (Bank 1 Sensor 3) Heater Circuit Malfunction** Engine started cold, ECT and IAT sensors less than 90°F and within 11°F at startup, engine running, system voltage from 11-16v, then with the average Calculated airflow less than 23 g/sec during the test period, the PCM detected the HO2S-13 signal did not vary more than 150 mv from the bias voltage of 400 to 500 mv for up to 300 seconds during the Oxygen Sensor Heater Monitor test. **Possible Causes:** • HO2S power circuit is open (check the O2S heater fuse) • HO2S heater ground circuit is open • HO2S heater element has high resistance • HO2S heater element has failed (open or shorted) • PCM has failed
DTC: P0151 **1T CCM, MIL: Yes** **1996, 1997, 1998, 1999, 2000, 2001, 2002, 2003, 2004, 2005, 2006** **Models:** Accord, Odyssey, Pilot, Ridgeline **Engines:** 2.7L VIN CE6, 3.0L VIN CG1, 3.0L VIN CG2, 3.5L VIN RL1 **Transmissions:** A/T	**HO2S-21 (Bank 2 Sensor 1) Circuit Low Input** Engine running in closed loop in D4 position at cruise speed, and the PCM detected the HO2S signal was fixed at less than 0.10v. **Note: The actual value where the code sets is in the PCM memory.** **Possible Causes:** • HO2S signal circuit is open • HO2S signal circuit is shorted to ground • HO2S ground circuit is open • HO2S may be contaminated or may have failed • PCM has failed
DTC: P0151 **1T CCM, MIL: Yes** **1996, 1997, 1998, 1999, 2000, 2001, 2002** **Models:** Passport **Engines:** 3.2L VIN V, 3.2L VIN W **Transmissions:** A/T, M/T	**HO2S-21 (Bank 2 Sensor 1) Circuit Low Input** DTC P0106, P0107, P0108, P0112, P0113, P0117, P0118, P0121, P0122, P0123, P0171, P0172 and P0300, P0301-P0306 not set, engine running in closed loop with the A/F ratio from 14.5-14.8:1, ECT sensor more than 140°F, throttle angle from 3-19% for 5 seconds, the PCM detected the HO2S-21 signal was less than 22 mv for 77 seconds over a 90 second period during the CCM test. **Possible Causes:** • HO2S signal circuit is open or shorted to ground • HO2S contaminated by water or wrong fuel • HO2S is damaged or has failed • PCM has failed
DTC: P0152 **1T CCM, MIL: Yes** **1996, 1997, 1998, 1999, 2000, 2001, 2002, 2003, 2004, 2005, 2006** **Models:** Accord, Odyssey, Pilot, Ridgeline **Engines:** 2.7L VIN CE6, 3.0L VIN CG1, 3.0L VIN CG2, 3.5L VIN RL1 **Transmissions:** A/T	**HO2S-21 (Bank 2 Sensor 1) Circuit High Input** Engine running in closed loop in D4 position at cruise speed, and the PCM detected the HO2S signal was fixed at more than 0.90v **Note: The actual value where the code sets is in the PCM memory.** **Possible Causes:** • HO2S signal tracking (wet/oily) in connector causing a short between the signal circuit and heater power circuit • HO2S signal circuit is open, or the ground circuit is open • HO2S heater supply circuit is open • PCM has failed

DTC	Trouble Code Title, Conditions & Possible Causes
DTC: P0152 **1T CCM, MIL: Yes** **1996, 1997, 1998, 1999, 2000,** **2001, 2002** **Models:** Passport **Engines:** 3.2L VIN V, 3.2L VIN W **Transmissions:** A/T, M/T	**HO2S-21 (Bank 2 Sensor 1) Circuit High Input** DTC P0106, P0107, P0108, P0112, P0113, P0117, P0118, P0121, P0122, P0123, P0171, P0172 and P0300, P0301-P0306 not set, engine running in closed loop with the A/F ratio command at 14.5-14.8:1, throttle angle from 3-19% for 5 seconds, the PCM detected the HO2S-21 signal was less than 952 mv for 77 seconds over a 90 second period; or the HO2S-21 signal was more than 500 mv during Decel Fuel Cutoff mode for 3 seconds during the CCM test. **Possible Causes:** • HO2S signal circuit shorted to system power (oil in connector) • HO2S contaminated by water or wrong fuel • HO2S is damaged or has failed • PCM has failed
DTC: P0153 **2T O2S1, MIL: Yes** **1996, 1997, 1998, 1999, 2000,** **2001, 2002, 2003, 2004, 2005,** **2006** **Models:** Accord, Odyssey, Pilot, Ridgeline **Engines:** 2.7L VIN CE6, 3.0L VIN CG1, 3.0L VIN CG2, 3.5L VIN RL1 **Transmissions:** A/T	**HO2S-21 (Bank 2 Sensor 1) Circuit Slow Response** Engine running in closed loop in D4 position at over 55 mph at steady speed, and the PCM detected the HO2S response time to switch between 300-600 mv was too slow, or that the rich to lean or lean to rich switch time was too slow. **Possible Causes:** • Exhaust leak present in the exhaust manifold or exhaust pipes • HO2S element fuel contamination • HO2S element has deteriorated • PCM has failed
DTC: P0153 **2T O2S2, MIL: Yes** **1996, 1997, 1998, 1999, 2000,** **2001, 2002** **Models:** Passport **Engines:** 3.2L VIN V, 3.2L VIN W **Transmissions:** A/T, M/T	**HO2S-21 (Bank 2 Sensor 1) Slow Response** DTC P0106, P0107, P0108, P0112, P0113, P0117, P0118, P0121, P0122, P0123, P0171, P0172 and P0300, P0301-P0306 not set, engine runtime 1 minute in closed loop, ECT sensor more than 122°F, engine speed from 1500-3000 rpm, MAF sensor from 9-42 g/sec, Purge duty cycle over 1%, conditions met for 3 seconds, then 90 seconds after entering closed loop, the PCM detected the HO2S-21 lean-to-rich average transition time was over 94 ms, or the rich-to-lean average transition time was over 105 ms during the Oxygen Sensor Monitor test. **Possible Causes:** • Exhaust leak present in the exhaust manifold or exhaust pipes • HO2S element fuel contamination • HO2S element has deteriorated • PCM has failed
DTC: P0154 **1T O2S2, MIL: Yes** **1996, 1997, 1998, 1999, 2000,** **2001, 2002** **Models:** Passport **Engines:** 3.2L VIN V, 3.2L VIN W **Transmissions:** A/T, M/T	**HO2S-21 (Bank 2 Sensor 1) Insufficient Activity Detected** DTC P0106, P0107, P0108, P0112, P0113, P0117, P0118, P0121, P0122, P0123, P0171, P0172 and P0300, P0301-306 not set, system voltage from 11-16v, engine runtime over 40 seconds, then after the PCM determined the Oxygen Sensor Heater test passed, it detected the HO2S-21 signal remained from 400-500 mv for 77 seconds over a 90 second period in the Oxygen Sensor Monitor test. **Possible Causes:** • Exhaust leak present in exhaust manifold or exhaust pipes • HO2S element fuel contamination or has deteriorated • HO2S signal circuit or the ground circuit has high resistance • HO2S heater element has failed, or the heater circuit is open • PCM has failed
DTC: P0155 **1T O2S HTR1, MIL: Yes** **1996, 1997, 1998, 1999, 2000,** **2001, 2002, 2003, 2004, 2005,** **2006** **Models:** Accord, Odyssey, Pilot, Ridgeline **Engines:** 2.7L VIN CE6, 3.0L VIN CG1, 3.0L VIN CG2, 3.5L VIN RL1 **Transmissions:** A/T, M/T	**HO2S-21 (Bank 2 Sensor 1) Heater Circuit Malfunction** Engine runtime over 80 seconds, and the PCM detected an incorrect signal value at the HO2S heater circuit during the test period. **Possible Causes:** • Main relay output (power) circuit to the heater is open • O2S heater ground circuit is open • O2S heater element has high resistance • O2S heater element has an open condition • O2S heater element has a shorted condition • PCM has failed
DTC: P0155 **2T O2S HTR2, MIL: Yes** **1996, 1997, 1998, 1999, 2000,** **2001, 2002** **Models:** Passport **Engines:** 3.2L VIN V, 3.2L VIN W **Transmissions:** A/T, M/T	**HO2S-21 (Bank 2 Sensor 1) Heater Circuit Malfunction** DTC P0151, P0152, P0153 and P0154 not set, ECT and IAT sensor signals less than 90°F, and within 11°F at startup, system voltage from 11-16v, throttle angle under 40%, average Calculated airflow less than 18 g/sec in the sample period, and the PCM detected the HO2S-21 signal did not vary more than 150 mv from the bias voltage (400-500 mv) for too long a period (maximum time is 120 seconds). **Possible Causes:** • HO2S power circuit is open (from the O2S heater fuse) • HO2S heater ground circuit is open • HO2S heater element has high resistance • HO2S heater element has failed (open or shorted) • PCM has failed

DTC	Trouble Code Title, Conditions & Possible Causes
DTC: P0157 **1T CCM, MIL: Yes** **1996, 1997, 1998, 1999, 2000,** **2001, 2002, 2003, 2004, 2005,** **2006** **Models:** Accord, Odyssey, Pilot, Ridgeline **Engines:** 2.7L VIN CE6, 3.0L VIN CG1, 3.0L VIN CG2, 3.5L VIN RL1 **Transmissions:** A/T	**HO2S-22 (Bank 2 Sensor 2) Circuit Low Input** Engine running in closed loop in D4 position at cruise speed, and the PCM detected the HO2S signal was fixed at less than 0.30v. **Note: The actual value where the code sets is in the PCM memory.** **Possible Causes:** • HO2S signal circuit is open • HO2S signal circuit is shorted to ground • HO2S ground circuit is open • HO2S may be contaminated or may have failed • PCM has failed
DTC: P0157 **1T CCM, MIL: Yes** **1996, 1997, 1998, 1999, 2000,** **2001, 2002** **Models:** Passport **Engines:** 3.2L VIN V, 3.2L VIN W **Transmissions:** A/T, M/T	**HO2S-22 (Bank 2 Sensor 2) Circuit Low Input** DTC P0161 not set, ECT sensor more than 140°F, engine running in closed loop with the A/F ratio at 14.5-14.8:1, throttle angle from 3-19%, and the PCM detected the HO2S-22 signal was less than 26 mv for 106 seconds of a 125 second period; or that it was more than 400 mv in Power Enrichment Mode during the test. **Possible Causes:** • HO2S signal circuit is open or shorted to ground • HO2S contaminated by water or wrong fuel, or it has failed • PCM has failed
DTC: P0158 **1T CCM, MIL: Yes** **1996, 1997, 1998, 1999, 2000,** **2001, 2002, 2003, 2004, 2005,** **2006** **Models:** Accord, Odyssey, Pilot, Ridgeline **Engines:** 2.7L VIN CE6, 3.0L VIN CG1, 3.0L VIN CG2, 3.5L VIN RL1 **Transmissions:** M/T	**HO2S-22 (Bank 2 Sensor 2) Circuit High Input** Engine running in closed loop in 4th or 6th gear at cruise speed, and the PCM detected the HO2S signal was fixed at more than 0.60v. **Note: The actual value where the code sets is in the PCM memory.** **Possible Causes:** • HO2S signal tracking (wet/oily) in connector causing a short between the signal circuit and heater power circuit • HO2S signal circuit is open, or the ground circuit is open • HO2S heater supply circuit is open • PCM has failed
DTC: P0158 **1T CCM, MIL: Yes** **1996, 1997, 1998, 1999, 2000,** **2001, 2002** **Models:** Passport **Engines:** 3.2L VIN V, 3.2L VIN W **Transmissions:** A/T, M/T	**HO2S-22 (Bank 2 Sensor 2) Circuit High Input** DTC P0106, P0107, P0108, P0112, P0113, P0117, P0118, P0121, P0122, P0123, P0171, P0172 and P0300, P0301-P0306 not set, engine running in closed loop with the A/F ratio command at 14.5-14.8:1, throttle angle from 3-19% for 5 seconds, the PCM detected the HO2S-22 signal was less than 952 mv for 106 seconds over a 125 second period; or the HO2S-12 signal was more than 500 mv during Decel Fuel Cutoff mode for 3 seconds during the CCM test. **Possible Causes:** • HO2S signal shorted to system power in harness (intermittent) • HO2S signal circuit shorted to system power (oil in connector) • HO2S contaminated by water or wrong fuel, damaged or failed • PCM has failed
DTC: P0159 **2T O2S1, MIL: Yes** **1996, 1997, 1998, 1999, 2000,** **2001, 2002, 2003, 2004, 2005,** **2006** **Models:** Accord, Odyssey, Pilot, Ridgeline **Engines:** 2.7L VIN CE6, 3.0L VIN CG1, 3.0L VIN CG2, 3.5L VIN RL1 **Transmissions:** A/T	**HO2S-22 (Bank 2 Sensor 2) Circuit Slow Response** Engine running in closed loop in D4 position at over 55 mph at steady speed, and the PCM detected the HO2S response time to switch between 300-600 mv was too slow, or that the rich to lean or lean to rich switch time was too slow. **Possible Causes:** • Exhaust leak present in the exhaust manifold or exhaust pipes • O2S element fuel contamination • O2S element has deteriorated • PCM has failed
DTC: P0160 **1T O2S2, MIL: Yes** **1996, 1997, 1998, 1999, 2000,** **2001, 2002** **Models:** Passport **Engines:** 3.2L VIN V, 3.2L VIN W **Transmissions:** A/T, M/T	**HO2S-22 (Bank 2 Sensor 2) Insufficient Activity Detected** DTC P0106, P0107, P0108, P0112, P0113, P0117, P0118, P0121, P0122, P0123, P0171, P0172 and P0300, P0301-P0306 not set, system voltage from 11-16v, engine runtime over 40 seconds, then after the PCM determined the Oxygen Sensor Heater test passed, it detected the HO2S-22 signal remained from 426-474 mv for 105 seconds of a 125 second period in the Oxygen Sensor Monitor test. **Possible Causes:** • Exhaust leak present in exhaust manifold or exhaust pipes • HO2S element has fuel contamination or has deteriorated • HO2S signal circuit or the ground circuit has high resistance • PCM has failed

DTC	Trouble Code Title, Conditions & Possible Causes
DTC: P0161 **1T O2S HTR1, MIL: Yes** **1996, 1997, 1998, 1999, 2000, 2001, 2002, 2003, 2004, 2005, 2006** **Models:** Accord, Odyssey, Pilot, Ridgeline **Engines:** 2.7L VIN CE6, 3.0L VIN CG1, 3.0L VIN CG2, 3.5L VIN RL1 **Transmissions:** A/T, M/T	**HO2S-22 (Bank 2 Sensor 2) Heater Circuit Malfunction** Engine runtime over 80 seconds, and the PCM detected an incorrect signal value at the HO2S heater circuit during the test period. **Possible Causes:** • Main relay output (power) circuit to the heater is open • O2S heater ground circuit is open • O2S heater element has high resistance • O2S heater element has an open condition • O2S heater element has a shorted condition • PCM has failed
DTC: P0161 **2T O2S HTR2, MIL: Yes** **1996, 1997, 1998, 1999, 2000, 2001, 2002** **Models:** Passport **Engines:** 3.2L VIN V, 3.2L VIN W **Transmissions:** A/T, M/T	**HO2S-22 (Bank 2 Sensor 2) Heater Circuit Malfunction** Cold engine startup, ECT and IAT sensor less than 90°F, and within 11°F at startup, system voltage at 11-16v, average Calculated airflow during the test period less than 23 g/sec, and the PCM detected the HO2S-22 signal changed less than 150 mv from the bias voltage of 400-500 mv for up to 300 seconds during the test. **Possible Causes:** • HO2S power circuit is open (check the O2S heater fuse) • HO2S heater element has failed (it may be open or shorted) • PCM has failed
DTC: P0171 **1T FUEL, MIL: Yes** **1995** **Models:** Accord **Engines:** 2.7L VIN CE6 **Transmissions:** A/T, M/T	**Fuel System Too Lean (Bank 1)** DTC P0107, P0108, P0135, P0137, P0138, P0141, P0151, P0152 and P0155 not set, engine running in closed loop, and the PCM detected the LONGFT value exceeded the calibrated lean limit value **Possible Causes:** • Air leaks in intake manifold, exhaust pipes or exhaust manifold • One or more injectors restricted or pressure regulator has failed • Air is being drawn in from leaks in gaskets or other seals • O2S element is deteriorated or has failed • A "fuel control" sensor is out of calibration (ECT, IAT or MAP) • PCM has failed
DTC: P0171 **2T FUEL, MIL: Yes** **1996, 1997, 1998, 1999, 2000, 2001, 2002, 2003, 2004, 2005, 2006** **Models:** Accord, Civic, Civic del Sol, CR-V, Insight, Odyssey, Pilot, Prelude, Ridgeline, S2000 **Engines:** All **Transmissions:** A/T, M/T	**Fuel System Too Lean (Bank 1)** DTC P0107, P0108, P0135, P0137, P0138, P0141, P1128, P1129 and P1259 not set, engine running in closed loop, and the PCM detected the LONGFT value exceeded the calibrated lean limit value **Possible Causes:** • Air leaks in intake manifold, exhaust pipes or exhaust manifold • One or more injectors restricted or pressure regulator has failed • Air is being drawn in from leaks in gaskets or other seals • O2S element is deteriorated or has failed • A "fuel control" sensor is out of calibration (ECT, IAT or MAP) • PCM has failed
DTC: P0171 **2T FUEL, MIL: Yes** **2003, 2004, 2005, 2006** **Models:** Element **Engines:** All **Transmissions:** All	**Fuel System Too Lean (Bank 1)** DTC P0107, P0108, P0135, P0137, P0138, P0141, P1128, P1129 and P1259 not set, engine running in closed loop, and the PCM detected the LONGFT value exceeded the calibrated lean limit value **Possible Causes:** • Air leaks in intake manifold, exhaust pipes or exhaust manifold • One or more injectors restricted or pressure regulator has failed • Air is being drawn in from leaks in gaskets or other seals • O2S element is deteriorated or has failed • A "fuel control" sensor is out of calibration (ECT, IAT or MAP) • PCM has failed

DTC	Trouble Code Title, Conditions & Possible Causes
DTC: P0171 **2T FUEL, MIL: Yes** **1996** **Models:** Passport **Engines:** 2.6L VIN E **Transmissions:** M/T	**Fuel System Too Lean (Bank 1)** DTC P0106, P0107, P0108, P0112, P0113, P0117, P0118, P0121, P0122, P0123, P0131, P0132, P0133, P0134, P0135, P0137, P0138, P0201-204, P0300, P0325, P0336, P0341, P0342, P0401, P0502, P0503, P0506, P0507, P1406 and P1441 not set, engine running in closed loop, BARO sensor over 72.5 kPa, ECT sensor signal from 77-212°F, IAT sensor signal from -40 to 248°F, MAP sensor signal at 24-99 kPa, throttle angle less than 95%, VSS under 85 mph, engine speed from 400-6000 rpm, MAF sensor signal from 2-20 g/sec, Purge duty cycle over 0% if "on", and the PCM detected the average of the Long Term fuel trim values was more than +20%. **Possible Causes:** • Air leaks after the MAF sensor, or in the EGR or PCV system • Base engine "mechanical" fault affecting one or more cylinders • Exhaust leaks before or near where the front HO2S is mounted • Fuel control sensor is out of calibration (i.e., ECT, IAT or MAP) • Fuel delivery system supplying too little fuel during cruise or idle periods (e.g., faulty fuel pump or dirty, restricted fuel filter) • Fuel injector (one or more) dirty or pressure regulator has failed • HO2S is contaminated, deteriorated or it has failed • Vehicle driven low on fuel or until it ran out of fuel
DTC: P0171 **2T FUEL, MIL: Yes** **1996, 1997, 1998, 1999, 2000, 2001, 2002** **Models:** Passport **Engines:** 3.2L VIN V, 3.2L VIN W **Transmissions:** A/T, M/T	**Fuel System Too Lean (Bank 1)** DTC P0106, P0107, P0108, P0112, P0113, P0117, P0118, P0121, P0122, P0123, P0131, P0132, P0133, P0134, P0135, P0137, P0138, P0201-206, P0300, P0301=P0306, P0401, P0502, P0503, P0506, P0507, P1406 and P1441 not set, engine running in closed loop, system voltage from 11-16v, BARO sensor over 72.5 kPa, ECT sensor from 77-212°F, IAT sensor from -40 to 248°F, MAP sensor from 24-99 kPa, throttle angle less than 95%, VSS under 85 mph, engine speed from 400-6000 rpm, MAF sensor from 2-20 g/sec, Purge duty cycle over 0%, and the PCM detected the average of the Long Term fuel trim values was more than +20%. **Possible Causes:** • Air leaks after the MAF sensor, or in the EGR or PCV system • Base engine "mechanical" fault affecting one or more cylinders • Exhaust leaks before or near where the front HO2S is mounted • Fuel control sensor is out of calibration (i.e., ECT, IAT or MAP) • Fuel delivery system supplying too little fuel during cruise or idle periods (e.g., faulty fuel pump or dirty, restricted fuel filter) • Fuel injector (one or more) dirty or pressure regulator has failed • HO2S is contaminated, deteriorated or it has failed • Vehicle driven low on fuel or until it ran out of fuel
DTC: P0172 **1T FUEL, MIL: Yes** **1995** **Models:** Accord **Engines:** 2.7L VIN CE6 **Transmissions:** A/T, M/T	**Fuel System Too Rich (Bank 1)** DTC P0107, P0108, P0135, P0137, P0138, P0141, P0151, P0152 and P0155 not set, engine running in closed loop, and the PCM detected the LONGFT value exceeded the calibrated rich limit. **Note: A high MAP sensor signal at idle can cause this code to set.** **Possible Causes:** • Leaking/contaminated fuel injector(s) or fuel pressure regulator • HO2S element may be contaminated with water or alcohol • EVAP vapor recovery system has failed (pulling vacuum) • Engine oil overfill condition • Base engine fault (i.e., cam timing incorrect, engine oil too high
DTC: P0172 **2T FUEL, MIL: Yes** **1996, 1997, 1998, 1999, 2000, 2001, 2002, 2003, 2004, 2005, 2006** **Models:** Accord, Civic, Civic del Sol, CR-V, Insight, Odyssey, Pilot, Prelude, Ridgeline, S2000 **Engines:** All **Transmissions:** A/T, M/T	**Fuel System Too Rich (Bank 1)** DTC P0107, P0108, P0135, P0137, P0138, P0141, P1128, P1129 and P1259 not set, engine running in closed loop, and the PCM detected the LONGFT value exceeded the calibrated rich limit. **Note: A high MAP sensor signal at idle can cause this code to set.** **Possible Causes:** • Leaking/contaminated fuel injector(s) or fuel pressure regulator • HO2S element may be contaminated with water or alcohol • EVAP vapor recovery system has failed (pulling vacuum) • Base engine fault (i.e., cam timing incorrect, engine oil too high
DTC: P0172 **2T FUEL, MIL: Yes** **2003, 2004, 2005, 2006** **Models:** Element **Engines:** All **Transmissions:** All	**Fuel System Too Rich (Bank 1)** DTC P0107, P0108, P0135, P0137, P0138, P0141, P1128, P1129 and P1259 not set, engine running in closed loop, and the PCM detected the LONGFT value exceeded the calibrated rich limit. **Note: A high MAP sensor signal at idle can cause this code to set.** **Possible Causes:** • Leaking/contaminated fuel injector(s) or fuel pressure regulator • HO2S element may be contaminated with water or alcohol • EVAP vapor recovery system has failed (pulling vacuum) • Base engine fault (i.e., cam timing incorrect, engine oil too high

DTC	Trouble Code Title, Conditions & Possible Causes
DTC: P0172 **2T FUEL, MIL:** Yes **1996** **Models:** Passport **Engines:** 2.6L VIN E **Transmissions:** M/T	**Fuel System Too Rich (Bank 1)** DTC P0106, P0107, P0108, P0112, P0113, P0117, P0118, P0121, P0122, P0123, P0131, P0132, P0133, P0134, P0135, P0137, P0138, P0201-204, P0300, P0325, P0336, P0341, P0342, P0401, P0502, P0503, P0506, P0507, P1406 and P1441 not set, engine running in closed loop, BARO sensor over 72.5 kPa, ECT sensor from 77-212°F, IAT sensor from -40 to 248°F, MAP sensor from 24-99 kPa, throttle angle less than 95%, VSS under 85 mph, engine speed from 400-6000 rpm, MAF sensor signal from 2-20 g/sec, Purge duty cycle over 0% if "on", and the PCM detected the average of the Long Term fuel trim values was more than -14%. **Possible Causes:** • Base engine "mechanical" fault affecting one or more cylinders • EVAP system component has failed or canister fuel saturated • Fuel control sensor is out of calibration (i.e., ECT, IAT or MAP) • Fuel delivery system supplying too much fuel during cruise or idle periods (e.g., faulty fuel pump, or faulty pressure regulator) • Fuel injector(s) is leaking or stuck partially open (one or more) • HO2S is contaminated, deteriorated or it has failed
DTC: P0172 **2T FUEL, MIL:** Yes **1996, 1997, 1998, 1999, 2000, 2001, 2002** **Models:** Passport **Engines:** 3.2L VIN V, 3.2L VIN W **Transmissions:** A/T, M/T	**Fuel System Too Rich (Bank 1)** DTC P0106, P0107, P0108, P0112, P0113, P0117, P0118, P0121, P0122, P0123, P0131, P0132, P0133, P0134, P0135, P0137, P0138, P0201-206, P0300, P0301=P0306, P0401, P0502, P0503, P0506, P0507, P1406 and P1441 not set, engine running in closed loop, BARO sensor over 72.5 kPa, ECT sensor from 77-212°F, IAT sensor from -40 to 248°F, MAP sensor from 24-99 kPa, throttle angle less than 95%, VSS under 85 mph, engine speed from 400-6000 rpm, MAF sensor from 2-20 g/sec, Purge duty cycle over 0%, and the PCM detected the average of the Long Term fuel trim values was more than -14% during the Fuel System Monitor test. **Possible Causes:** • Base engine "mechanical" fault affecting one or more cylinders • EVAP system component has failed or canister fuel saturated • Fuel control sensor is out of calibration (i.e., ECT, IAT or MAP) • Fuel delivery system supplying too much fuel during cruise or idle periods (e.g., faulty fuel pump, or faulty pressure regulator) • Fuel injector(s) is leaking or stuck partially open (one or more) • HO2S is contaminated, deteriorated or it has failed
DTC: P0174 **2T FUEL, MIL:** Yes **1996, 1997, 1998, 1999, 2000, 2001, 2002, 2003, 2004, 2005, 2006** **Models:** Accord, Odyssey, Pilot, Ridgeline **Engines:** 2.7L VIN CE6, 3.0L VIN CG1, 3.0L VIN CG2, 3.5L VIN RL1 **Transmissions:** A/T	**Fuel System Too Lean (Bank 2)** DTC P0107, P0108, P0135, P0137, P0138, P0141, P1128, P1129 and P1259 not set, engine running in closed loop, and the PCM detected the LONGFT value exceeded the calibrated lean limit value **Possible Causes:** • Air leaks in intake manifold, exhaust pipes or exhaust manifold • One or more injectors restricted or pressure regulator has failed • Air is being drawn in from leaks in gaskets or other seals • O2S element is deteriorated or has failed • A "fuel control" sensor is out of calibration (ECT, IAT or MAP)
DTC: P0174 **2T FUEL, MIL:** Yes **1996, 1997, 1998, 1999, 2000, 2001, 2002** **Models:** Passport **Engines:** 3.2L VIN V, 3.2L VIN W **Transmissions:** A/T, M/T	**Fuel System Too Lean (Bank 2)** DTC P0106, P0107, P0108, P0112, P0113, P0117, P0118, P0121, P0122, P0123, P0131, P0132, P0133, P0134, P0135, P0137, P0138, P0201-206, P0300, P0301=P0306, P0401, P0502, P0503, P0506, P0507, P1406 and P1441 not set, engine running in closed loop, system voltage from 11-16v, BARO sensor over 72.5 kPa, ECT sensor from 77-212°F, IAT sensor from -40 to 248°F, MAP sensor from 24-99 kPa, throttle angle steady at less than 95%, VSS under 85 mph, engine speed from 400-6000 rpm, MAF sensor from 2-20 g/sec, Purge duty cycle over 0%, and the PCM detected the average of the Long Term fuel trim values was more than +20%. **Possible Causes:** • Air leaks after the MAF sensor, or in the EGR or PCV system • Base engine "mechanical" fault affecting one or more cylinders • Exhaust leaks before or near where the front HO2S is mounted • Fuel control sensor is out of calibration (i.e., ECT, IAT or MAP) • Fuel delivery system supplying too little fuel during cruise or idle periods (e.g., faulty fuel pump or dirty, restricted fuel filter) • Fuel injector (one or more) dirty or pressure regulator has failed • HO2S is contaminated, deteriorated or it has failed • Vehicle driven low on fuel or until it ran out of fuel

DTC	Trouble Code Title, Conditions & Possible Causes
DTC: P0175 **2T FUEL, MIL: Yes** **1996, 1997, 1998, 1999, 2000, 2001, 2002, 2003, 2004, 2005, 2006** **Models:** Accord, Odyssey, Pilot, Ridgeline **Engines:** 2.7L VIN CE6, 3.0L VIN CG1, 3.0L VIN CG2, 3.5L VIN RL1 **Transmissions:** A/T	**Fuel System Too Rich (Bank 2)** DTC P0107, P0108, P0135, P0137, P0138, P0141, P1128, P1129 and P1259 not set, engine running in closed loop, and the PCM detected the LONGFT value exceeded the calibrated rich limit. **Note: A high MAP sensor signal at idle can cause this code to set.** **Possible Causes:** • Leaking/contaminated fuel injector(s) or fuel pressure regulator • HO2S element may be contaminated with water or alcohol • EVAP vapor recovery system has failed (pulling vacuum) • A "fuel control" sensor is out of calibration (ECT, IAT or MAP) • Base engine fault (i.e., cam timing incorrect, engine oil too high
DTC: P0175 **2T FUEL, MIL: Yes** **1996, 1997, 1998, 1999, 2000, 2001, 2002** **Models:** Passport **Engines:** 3.2L VIN V, 3.2L VIN W **Transmissions:** A/T, M/T	**Fuel System Too Rich (Bank 2)** DTC P0106, P0107, P0108, P0112, P0113, P0117, P0118, P0121, P0122, P0123, P0131, P0132, P0133, P0134, P0135, P0137, P0138, P0201-206, P0300, P0301=P0306, P0401, P0502, P0503, P0506, P0507, P1406 and P1441 not set, engine running in closed loop, BARO sensor over 72.5 kPa, ECT sensor from 77-212ºF, IAT sensor from -40 to 248ºF, MAP sensor from 24-99 kPa, throttle angle less than 95%, VSS under 85 mph, engine speed from 400-6000 rpm, MAF sensor from 2-20 g/sec, Purge duty cycle over 0%, and the PCM detected the average of the Long Term fuel trim values was more than -14% during the Fuel System Monitor test. **Possible Causes:** • Base engine "mechanical" fault affecting one or more cylinders • EVAP system component has failed or canister fuel saturated • Fuel control sensor is out of calibration (i.e., ECT, IAT or MAP) • Fuel delivery system supplying too much fuel during cruise or idle periods (e.g., faulty fuel pump, or faulty pressure regulator) • Fuel injector(s) is leaking or stuck partially open (one or more) • HO2S is contaminated, deteriorated or it has failed
DTC: P0191 **1T CCM, MIL: Yes** **2001, 2002, 2003, 2004, 2005, 2006** **Models:** Civic **Engines:** 1.7L VIN EN2 **Transmissions:** A/T	**CNG Fuel Pressure Sensor Range/Performance** Engine running at hot idle speed, and the PCM detected the Fuel Pressure sensor signal indicated a fuel pressure out of specification. **Note: The fuel pressure sensor PID reading a hot idle is 2.33-2.93v.** **Possible Causes:** • Fuel leaks somewhere in the system • Fuel pressure regulator is damaged or has failed • Fuel pressure sensor is out of calibration or has failed • PCM has failed
DTC: P0192 **1T CCM, MIL: Yes** **2001, 2002, 2003, 2004, 2005, 2006** **Models:** Civic **Engines:** 1.7L VIN EN2 **Transmissions:** A/T	**CNG Fuel Pressure Sensor Circuit Low Input** Key on or engine running, and the PCM detected an unexpected low voltage condition on the Fuel Pressure sensor signal circuit. **Note: The fuel pressure sensor PID reading a hot idle is 2.33-2.93v.** **Possible Causes:** • Fuel pressure sensor signal circuit is shorted to ground • Fuel pressure sensor is damaged or has failed • PCM has failed
DTC: P0193 **1T CCM, MIL: Yes** **2001, 2002, 2003, 2004, 2005, 2006** **Models:** Civic **Engines:** 1.7L VIN EN2 **Transmissions:** A/T	**CNG Fuel Pressure Sensor Circuit High Input** Key on or engine running, and the PCM detected an unexpected high voltage condition on the Fuel Pressure sensor signal circuit. **Note: The fuel pressure sensor PID reading a hot idle is 2.33-2.93v.** **Possible Causes:** • Fuel pressure sensor signal circuit is open • Fuel pressure sensor signal circuit is shorted to VREF • Fuel pressure sensor is damaged or has failed • PCM has failed
DTC: P0201 **1T CCM, MIL: Yes** **1996** **Models:** Passport **Engines:** 2.6L VIN E **Transmissions:** M/T	**Fuel Injector Circuit 1 Malfunction** Engine running, system voltage over 9v, and PCM detected the injector voltage for Cylinder 1 did not equal the ignition voltage with the injector commanded "off", or that the injector voltage did not equal zero (0) volts with the injector commanded "on". **Possible Causes:** • Fuel injector control circuit is open or shorted to ground • Fuel injector power circuit is open between injector and relay • Fuel Injector has failed • PCM has failed (injector driver circuit may be open or shorted)

DTC	Trouble Code Title, Conditions & Possible Causes
DTC: P0201 **1T CCM, MIL: Yes** **1996, 1997, 1998, 1999, 2000,** **2001, 2002** **Models:** Passport **Engines:** 3.2L VIN V, 3.2L VIN W **Transmissions:** A/T, M/T	**Fuel Injector Circuit 1 Malfunction** Engine started, engine running, system voltage over 9v, and PCM detected the injector voltage for Cylinder 1 did not equal the system voltage with the injector commanded "off", or that the injector voltage did not equal zero (0) volts with the injector commanded "on". **Possible Causes:** • Fuel injector control circuit is open or shorted to ground • Fuel injector power circuit is open between injector and relay • Fuel Injector has failed • PCM has failed (injector driver circuit may be open or shorted)
DTC: P0202 **1T CCM, MIL: Yes** **1996** **Models:** Passport **Engines:** 2.6L VIN E **Transmissions:** M/T	**Fuel Injector Circuit 2 Malfunction** Engine running, system voltage over 9v, and PCM detected the injector voltage for Cylinder 2 did not equal the ignition voltage with the injector commanded "off", or that the injector voltage did not equal zero (0) volts with the injector commanded "on". **Possible Causes:** • Fuel injector control circuit is open or shorted to ground • Fuel injector power circuit open between injector and ECM fuse • Fuel Injector has failed • PCM has failed (injector driver circuit may be open or shorted)
DTC: P0202 **1T CCM, MIL: Yes** **1996, 1997, 1998, 1999, 2000,** **2001, 2002** **Models:** Passport **Engines:** 3.2L VIN V, 3.2L VIN W **Transmissions:** A/T, M/T	**Fuel Injector Circuit 2 Malfunction** Engine started, engine running, system voltage over 9v, and PCM detected the injector voltage for Cylinder 2 did not equal the system voltage with the injector commanded "off", or that the injector voltage did not equal zero (0) volts with the injector commanded "on". **Possible Causes:** • Fuel injector control circuit is open or shorted to ground • Fuel injector power circuit open between injector and ECM fuse • Fuel Injector has failed • PCM has failed (injector driver circuit may be open or shorted)
DTC: P0203 **1T CCM, MIL: Yes** **1996** **Models:** Passport **Engines:** 2.6L VIN E **Transmissions:** M/T	**Fuel Injector Circuit 3 Malfunction** Engine running, system voltage over 9v, and PCM detected the injector voltage for Cylinder 3 did not equal the ignition voltage with the injector commanded "off", or that the injector voltage did not equal zero (0) volts with the injector commanded "on". **Possible Causes:** • Fuel injector control circuit is open or shorted to ground • Fuel injector power circuit open between injector and ECM fuse • Fuel Injector has failed • PCM has failed (injector driver circuit may be open or shorted)
DTC: P0203 **1T CCM, MIL: Yes** **1996, 1997, 1998, 1999, 2000,** **2001, 2002** **Models:** Passport **Engines:** 3.2L VIN V, 3.2L VIN W **Transmissions:** A/T, M/T	**Fuel Injector Circuit 3 Malfunction** Engine started, engine running, system voltage over 9v, and PCM detected the injector voltage for Cylinder 3 did not equal the system voltage with the injector commanded "off", or that the injector voltage did not equal zero (0) volts with the injector commanded "on". **Possible Causes:** • Fuel injector control circuit is open or shorted to ground • Fuel injector power circuit open between injector and ECM fuse • Fuel Injector has failed • PCM has failed (injector driver circuit may be open or shorted)
DTC: P0204 **1T CCM, MIL: Yes** **1996** **Models:** Passport **Engines:** 2.6L VIN E **Transmissions:** M/T	**Fuel Injector Circuit 4 Malfunction** Engine running, system voltage over 9v, and PCM detected the injector voltage for Cylinder 4 did not equal the ignition voltage with the injector commanded "off", or that the injector voltage did not equal zero (0) volts with the injector commanded "on". **Possible Causes:** • Fuel injector control circuit is open or shorted to ground • Fuel injector power circuit open between injector and ECM fuse • Fuel Injector has failed • PCM has failed (injector driver circuit may be open or shorted)
DTC: P0204 **1T CCM, MIL: Yes** **1996, 1997, 1998, 1999, 2000,** **2001, 2002** **Models:** Passport **Engines:** 3.2L VIN V, 3.2L VIN W **Transmissions:** A/T, M/T	**Fuel Injector Circuit 4 Malfunction** Engine started, engine running, system voltage over 9v, and PCM detected the injector voltage for Cylinder 4 did not equal the system voltage with the injector commanded "off", or that the injector voltage did not equal zero (0) volts with the injector commanded "on". **Possible Causes:** • Fuel injector control circuit is open or shorted to ground • Fuel injector power circuit open between injector and ECM fuse • Fuel Injector has failed • PCM has failed (injector driver circuit may be open or shorted)

DTC	Trouble Code Title, Conditions & Possible Causes
DTC: P0205 **1T CCM, MIL: Yes** **1996, 1997, 1998, 1999, 2000, 2001, 2002** **Models:** Passport **Engines:** 3.2L VIN V, 3.2L VIN W **Transmissions:** A/T, M/T	**Fuel Injector Circuit 5 Malfunction** Engine started, engine running, system voltage over 9v, and PCM detected the injector voltage for Cylinder 5 did not equal the system voltage with the injector commanded "off", or that the injector voltage did not equal zero (0) volts with the injector commanded "on". **Possible Causes:** • Fuel injector circuit is open, shorted to ground or to power • Fuel Injector power circuit is open or the injector has failed • PCM has failed (injector driver circuit may be open or shorted)
DTC: P0206 **1T CCM, MIL: Yes** **1996, 1997, 1998, 1999, 2000, 2001, 2002** **Models:** Passport **Engines:** 3.2L VIN V, 3.2L VIN W **Transmissions:** A/T, M/T	**Fuel Injector Circuit 6 Malfunction** Engine started, engine running, system voltage over 9v, and PCM detected the injector voltage for Cylinder 6 did not equal the system voltage with the injector commanded "off", or that the injector voltage did not equal zero (0) volts with the injector commanded "on". **Possible Causes:** • Fuel injector circuit is open, shorted to ground or to power • Fuel Injector power circuit is open or the injector has failed • PCM has failed (injector driver circuit may be open or shorted)
DTC: P0218 **1T CCM, MIL: No** **1998, 1999, 2000, 2001, 2002** **Models:** Passport **Engines:** 3.2L VIN W **Transmissions:** A/T	**Transmission Fluid Over-Temperature Malfunction** DTC P0712 and P0713 not set, and the PCM detected the Transmission Fluid Temperature (TFT) sensor signal indicated more than 275°F, condition met for 21 seconds during the CCM test. **Note: The CHECK TRANS lamp is "on" if the TFT sensor signal exceeds 293°F.** **Possible Causes:** • TFT sensor signal circuit is shorted to ground • TFT sensor is out-of-calibration (skewed), or it has failed • Torque converter stator is damaged or has failed • PCM has failed
DTC: P0300 **2T MISFIRE, MIL: Yes** **1996, 1997, 1998, 1999, 2000, 2001, 2002, 2003, 2004, 2005, 2006** **Models:** Accord, Civic, Civic del Sol, CR-V, Element, Insight, Odyssey, Pilot, Prelude, Ridgeline, S2000 **Engines:** All **Transmissions:** A/T, M/T	**Multiple Misfire Detected** DTC P0107, P0108, P0131, P0132, P0171, P0172, P1128, P0335, P0336, P0505, P1128, P1129, P1259, P1361, P1362, P1366, P1367 and P1519 not set, engine running under positive torque conditions, and the PCM detected a misfire in 2 or more cylinders. **Note: If the misfire is severe, the MIL will flash on/off on the 1st trip!** **Possible Causes:** • CKP or CMP sensor problem affecting more than one cylinder • Fuel system problem affecting more than one cylinder • Ignition system problem affecting more than one cylinder • Base engine mechanical fault affecting more than 1 cylinder
DTC: P0300 **2T MISFIRE, MIL: Yes** **1996** **Models:** Passport **Engines:** 2.6L VIN E **Transmissions:** M/T	**Random Misfire Detected** DTC P0106, P0107, P0108, P0117, P0118, P0121, P0122, P0123, P0336, P0341, P0342, P0502, P0503, P1390, P1391, P392 and P1393 not set, ECT sensor from 20-248°F, system voltage from 11-16v, engine speed from 800-5500 rpm, throttle angle stable with any change less than 3% within 125 ms, and the PCM detected a deceleration in crankshaft speed in more than one cylinder characteristic of a misfire during the Misfire Monitor Detection test. Note: If the misfire is severe, the MIL will flash on/off on the 1st trip! **Possible Causes:** • Base engine mechanical fault that affects one or more cylinders • Fuel metering fault that affects more than one cylinder • Fuel pressure too low or too high, fuel supply contaminated • EVAP system problem or the EVAP canister is fuel saturated • EGR valve is stuck open or the PCV system has a vacuum leak • IC control circuit is shorted to ground (an intermittent fault) • Ignition system fault (a coil) that affects more than one cylinder • MAF sensor contamination (it can cause a very lean condition)

DTC	Trouble Code Title, Conditions & Possible Causes
DTC: P0300 **2T MISFIRE, MIL: Yes** **1996, 1997, 1998, 1999, 2000,** **2001, 2002** **Models:** Passport **Engines:** 3.2L VIN V, 3.2L VIN W **Transmissions:** A/T, M/T	**Random Misfire Detected** DTC P0101, P0102, P0103, P0106, P0107, P0108, P0117, P0118, P0121, P0122, P0123, P0336, P0341, P0342, P0502 and P0503 not set, ECT sensor from 20-248°F, system voltage from 11-16v, engine speed from 800-5500 rpm, throttle angle stable (± 3%), and the PCM detected a crankshaft speed variation in one or more cylinders characteristic of a misfire condition during the Misfire Monitor test. Note: If the misfire is severe, the MIL will flash on/off on the 1st trip! **Possible Causes:** • Base engine mechanical fault that affects one or more cylinders • Fuel metering fault that affects more than one cylinder • Fuel pressure too low or too high, fuel supply contaminated • EVAP system problem or the EVAP canister is fuel saturated • EGR valve is stuck open or the PCV system has a vacuum leak • IC control circuit is shorted to ground (an intermittent fault) • Ignition system fault (a coil) that affects more than one cylinder • MAF sensor contamination (it can cause a very lean condition)
DTC: P0301 **1T MISFIRE, MIL: Yes** **1995** **Models:** Accord **Engines:** 2.7L VIN CE6 **Transmissions:** A/T, M/T	**Cylinder 1 Misfire Detected** DTC P0107, P0108, P0131, P0132, P0171, P0172, P1128, P0335, P0336, P0505, P1128, P1129, P1259, P1361, P1362, P1366, P1367 and P1519 not set, engine running under positive torque conditions, and the PCM detected a misfire condition in one cylinder. **Note: If the misfire is severe, the MIL will flash on/off on the 1st trip!** **Possible Causes:** • Fuel system problem affecting only Cylinder 1 • Ignition system problem affecting Cylinder 1 • Base engine (mechanical) problem affecting only Cylinder 1
DTC: P0301 **2T MISFIRE, MIL: Yes** **1996, 1997, 1998, 1999, 2000,** **2001, 2002, 2003, 2004, 2005,** **2006** **Models:** Accord, Civic, Civic del Sol, CR-V, Element, Insight, Odyssey, Pilot, Prelude, Ridgeline, S2000 **Engines:** All **Transmissions:** A/T, M/T	**Cylinder 1 Misfire Detected** DTC P0107, P0108, P0131, P0132, P0171, P0172, P1128, P0335, P0336, P0505, P1128, P1129, P1259, P1361, P1362, P1366, P1367 and P1519 not set, engine running under positive torque conditions, and the PCM detected a misfire condition in one cylinder. **Note: If the misfire is severe, the MIL will flash on/off on the 1st trip!** **Possible Causes:** • Fuel system problem affecting only Cylinder 1 • Ignition system problem affecting Cylinder 1 • Base engine (mechanical) problem affecting only Cylinder 1
DTC: P0301 **2T MISFIRE, MIL: Yes** **1996, 1997, 1998, 1999, 2000,** **2001, 2002** **Models:** Passport **Engines:** 3.2L VIN V, 3.2L VIN W **Transmissions:** A/T, M/T	**Cylinder 1 Misfire Detected** DTC P0101, P0102, P0103, P0106, P0107, P0108, P0117, P0118, P0121, P0122, P0123, P0336, P0341, P0342, P0502 and P0503 not set, ECT sensor from 20-248°F, system voltage from 11-16v, engine speed from 800-5500 rpm, throttle angle stable (± 3%), and the PCM detected a crankshaft speed variation in one cylinder characteristic of a misfire condition during the Misfire Diagnostic Monitor test. Note: If the misfire is severe, the MIL will flash on/off on the 1st trip! **Possible Causes:** • Base engine mechanical fault that affects only one cylinder • Fuel metering fault that affects only one cylinder • IC control circuit is shorted to ground on only one cylinder • Ignition system fault (a coil) that affects only one cylinder
DTC: P0302 **1T MISFIRE, MIL: Yes** **1995** **Models:** Accord **Engines:** 2.7L VIN CE6 **Transmissions:** A/T, M/T	**Cylinder 2 Misfire Detected** DTC P0107, P0108, P0131, P0132, P0171, P0172, P1128, P0335, P0336, P0505, P1128, P1129, P1259, P1361, P1362, P1366, P1367 and P1519 not set, engine running under positive torque conditions, and the PCM detected a misfire condition in one cylinder. **Note: If the misfire is severe, the MIL will flash on/off on the 1st trip!** **Possible Causes:** • Fuel system problem affecting only Cylinder 2 • Ignition system problem affecting Cylinder 2 • Base engine (mechanical) problem affecting only Cylinder 2

DTC	Trouble Code Title, Conditions & Possible Causes
DTC: P0302 **2T MISFIRE, MIL: Yes** **1996, 1997, 1998, 1999, 2000, 2001, 2002, 2003, 2004, 2005, 2006** **Models:** Accord, Civic, Civic del Sol, CR-V, Element, Insight, Odyssey, Pilot, Prelude, Ridgeline, S2000 **Engines:** All **Transmissions:** A/T, M/T	**Cylinder 2 Misfire Detected** DTC P0107, P0108, P0131, P0132, P0171, P0172, P1128, P0335, P0336, P0505, P1128, P1129, P1259, P1361, P1362, P1366, P1367 and P1519 not set, engine running under positive torque conditions, and the PCM detected a misfire condition in one cylinder. **Note: If the misfire is severe, the MIL will flash on/off on the 1st trip!** **Possible Causes:** • Fuel system problem affecting only Cylinder 2 • Ignition system problem affecting Cylinder 2 • Base engine (mechanical) problem affecting only Cylinder 2
DTC: P0302 **2T MISFIRE, MIL: Yes** **1996, 1997, 1998, 1999, 2000, 2001, 2002** **Models:** Passport **Engines:** 3.2L VIN V, 3.2L VIN W **Transmissions:** A/T, M/T	**Cylinder 2 Misfire Detected** DTC P0101, P0102, P0103, P0106, P0107, P0108, P0117, P0118, P0121, P0122, P0123, P0336, P0341, P0342, P0502 and P0503 not set, ECT sensor from 20-248°F, system voltage from 11-16v, engine speed from 800-5500 rpm, throttle angle stable (± 3%), and the PCM detected a crankshaft speed variation in one cylinder characteristic of a misfire condition during the Misfire Diagnostic Monitor test. Note: If the misfire is severe, the MIL will flash on/off on the 1st trip! **Possible Causes:** • Base engine mechanical fault that affects only one cylinder • Fuel metering fault that affects only one cylinder • IC control circuit is shorted to ground on only one cylinder • Ignition system fault (a coil) that affects only one cylinder
DTC: P0303 **1T MISFIRE, MIL: Yes** **1995** **Models:** Accord **Engines:** 2.7L VIN CE6 **Transmissions:** A/T, M/T	**Cylinder 3 Misfire Detected** DTC P0107, P0108, P0131, P0132, P0171, P0172, P1128, P0335, P0336, P0505, P1128, P1129, P1259, P1361, P1362, P1366, P1367 and P1519 not set, engine running under positive torque conditions, and the PCM detected a misfire condition in one cylinder. **Note: If the misfire is severe, the MIL will flash on/off on the 1st trip!** **Possible Causes:** • Fuel system problem affecting only Cylinder 3 • Ignition system problem affecting Cylinder 3 • Base engine (mechanical) problem affecting only Cylinder 3
DTC: P0303 **2T MISFIRE, MIL: Yes** **1996, 1997, 1998, 1999, 2000, 2001, 2002, 2003, 2004, 2005, 2006** **Models:** Accord, Civic, Civic del Sol, CR-V, Element, Insight, Odyssey, Pilot, Prelude, Ridgeline, S2000 **Engines:** All **Transmissions:** A/T, M/T	**Cylinder 3 Misfire Detected** DTC P0107, P0108, P0131, P0132, P0171, P0172, P1128, P0335, P0336, P0505, P1128, P1129, P1259, P1361, P1362, P1366, P1367 and P1519 not set, engine running under positive torque conditions, and the PCM detected a misfire condition in one cylinder. **Note: If the misfire is severe, the MIL will flash on/off on the 1st trip!** **Possible Causes:** • Fuel system problem affecting only Cylinder 3 • Ignition system problem affecting Cylinder 3 • Base engine (mechanical) problem affecting only Cylinder 3
DTC: P0303 **2T MISFIRE, MIL: Yes** **1996, 1997, 1998, 1999, 2000, 2001, 2002** **Models:** Passport **Engines:** 3.2L VIN V, 3.2L VIN W **Transmissions:** A/T, M/T	**Cylinder 3 Misfire Detected** DTC P0101, P0102, P0103, P0106, P0107, P0108, P0117, P0118, P0121, P0122, P0123, P0336, P0341, P0342, P0502 and P0503 not set, ECT sensor from 20-248°F, system voltage from 11-16v, engine speed from 800-5500 rpm, throttle angle stable (± 3%), and the PCM detected a crankshaft speed variation in one cylinder characteristic of a misfire condition during the Misfire Diagnostic Monitor test. Note: If the misfire is severe, the MIL will flash on/off on the 1st trip! **Possible Causes:** • Base engine mechanical fault that affects only one cylinder • Fuel metering fault that affects only one cylinder • IC control circuit is shorted to ground on only one cylinder • Ignition system fault (a coil) that affects only one cylinder
DTC: P0304 **1T MISFIRE, MIL: Yes** **1995** **Models:** Accord **Engines:** 2.7L VIN CE6 **Transmissions:** A/T, M/T	**Cylinder 4 Misfire Detected** DTC P0107, P0108, P0131, P0132, P0171, P0172, P1128, P0335, P0336, P0505, P1128, P1129, P1259, P1361, P1362, P1366, P1367 and P1519 not set, engine running under positive torque conditions, and the PCM detected a misfire condition in one cylinder. **Note: If the misfire is severe, the MIL will flash on/off on the 1st trip!** **Possible Causes:** • Fuel system problem affecting only Cylinder 4 • Ignition system problem affecting Cylinder 4 • Base engine (mechanical) problem affecting only Cylinder 4

DTC	Trouble Code Title, Conditions & Possible Causes
DTC: P0304 **2T MISFIRE, MIL: Yes** **1996, 1997, 1998, 1999, 2000, 2001, 2002, 2003, 2004, 2005, 2006** **Models:** Accord, Civic, Civic del Sol, CR-V, Element, Insight, Odyssey, Pilot, Prelude, Ridgeline, S2000 **Engines:** All **Transmissions:** A/T, M/T	**Cylinder 4 Misfire Detected** DTC P0107, P0108, P0131, P0132, P0171, P0172, P1128, P0335, P0336, P0505, P1128, P1129, P1259, P1361, P1362, P1366, P1367 and P1519 not set, engine running under positive torque conditions, and the PCM detected a misfire condition in one cylinder. **Note: If the misfire is severe, the MIL will flash on/off on the 1st trip!** **Possible Causes:** • Fuel system problem affecting only Cylinder 4 • Ignition system problem affecting Cylinder 4 • Base engine (mechanical) problem affecting only Cylinder 4
DTC: P0304 **2T MISFIRE, MIL: Yes** **1996, 1997, 1998, 1999, 2000, 2001, 2002** **Models:** Passport **Engines:** 3.2L VIN V, 3.2L VIN W **Transmissions:** A/T, M/T	**Cylinder 4 Misfire Detected** DTC P0101, P0102, P0103, P0106, P0107, P0108, P0117, P0118, P0121, P0122, P0123, P0336, P0341, P0342, P0502 and P0503 not set, ECT sensor from 20-248°F, system voltage from 11-16v, engine speed from 800-5500 rpm, throttle angle stable (± 3%), and the PCM detected a crankshaft speed variation in one cylinder characteristic of a misfire condition during the Misfire Diagnostic Monitor test. Note: If the misfire is severe, the MIL will flash on/off on the 1st trip! **Possible Causes:** • Base engine mechanical fault that affects only one cylinder • Fuel metering fault that affects only one cylinder • IC control circuit is shorted to ground on only one cylinder • Ignition system fault (a coil) that affects only one cylinder
DTC: P0305 **1T MISFIRE, MIL: Yes** **1995** **Models:** Accord **Engines:** 2.7L VIN CE6 **Transmissions:** A/T, M/T	**Cylinder 5 Misfire Detected** DTC P0107, P0108, P0131, P0132, P0171, P0172, P1128, P0335, P0336, P0505, P1128, P1129, P1259, P1361, P1362, P1366, P1367 and P1519 not set, engine running under positive torque conditions, and the PCM detected a misfire condition in one cylinder. **Note: If the misfire is severe, the MIL will flash on/off on the 1st trip!** **Possible Causes:** • Fuel system problem affecting only Cylinder 5 • Ignition system problem affecting Cylinder 5 • Base engine (mechanical) problem affecting only Cylinder 5
DTC: P0305 **2T MISFIRE, MIL: Yes** **1996, 1997, 1998, 1999, 2000, 2001, 2002, 2003, 2004, 2005, 2006** **Models:** Accord, Odyssey, Pilot, Ridgeline **Engines:** 2.7L VIN CE6, 3.0L VIN CG1, 3.0L VIN CG2, 3.5L VIN RL1 **Transmissions:** A/T, M/T	**Cylinder 5 Misfire Detected** DTC P0107, P0108, P0131, P0132, P0171, P0172, P1128, P0335, P0336, P0505, P1128, P1129, P1259, P1361, P1362, P1366, P1367 and P1519 not set, engine running under positive torque conditions, and the PCM detected a misfire condition in one cylinder. Note: If the misfire is severe, the MIL will flash on/off on the 1st trip! **Possible Causes:** • Fuel system problem affecting only Cylinder 5 • Ignition system problem affecting Cylinder 5 • Base engine (mechanical) problem affecting only Cylinder 5
DTC: P0305 **2T MISFIRE, MIL: Yes** **1996, 1997, 1998, 1999, 2000, 2001, 2002** **Models:** Passport **Engines:** 3.2L VIN V, 3.2L VIN W **Transmissions:** A/T, M/T	**Cylinder 5 Misfire Detected** DTC P0101, P0102, P0103, P0106, P0107, P0108, P0117, P0118, P0121, P0122, P0123, P0336, P0341, P0342, P0502 and P0503 not set, ECT sensor from 20-248°F, system voltage from 11-16v, engine speed from 800-5500 rpm, throttle angle stable (± 3%), and the PCM detected a crankshaft speed variation in one cylinder characteristic of a misfire condition during the Misfire Diagnostic Monitor test. Note: If the misfire is severe, the MIL will flash on/off on the 1st trip! **Possible Causes:** • Base engine mechanical fault that affects only one cylinder • Fuel metering fault that affects only one cylinder • IC control circuit is shorted to ground on only one cylinder • Ignition system fault (a coil) that affects only one cylinder
DTC: P0306 **1T MISFIRE, MIL: Yes** **1995** **Models:** Accord **Engines:** 2.7L VIN CE6 **Transmissions:** A/T, M/T	**Cylinder 6 Misfire Detected** DTC P0107, P0108, P0131, P0132, P0171, P0172, P1128, P0335, P0336, P0505, P1128, P1129, P1259, P1361, P1362, P1366, P1367 and P1519 not set, engine running under positive torque conditions, and the PCM detected a misfire condition in one cylinder. **Note: If the misfire is severe, the MIL will flash on/off on the 1st trip!** **Possible Causes:** • Fuel system problem affecting only Cylinder 6 • Ignition system problem affecting Cylinder 6 • Base engine (mechanical) problem affecting only Cylinder 6

DTC	Trouble Code Title, Conditions & Possible Causes

DTC: P0306
2T MISFIRE, MIL: Yes
1996, 1997, 1998, 1999, 2000, 2001, 2002, 2003, 2004, 2005, 2006
Models: Accord, Odyssey, Pilot, Ridgeline
Engines: 2.7L VIN CE6, 3.0L VIN CG1, 3.0L VIN CG2, 3.5L VIN RL1
Transmissions: A/T, M/T
	Cylinder 6 Misfire Detected
	DTC P0107, P0108, P0131, P0132, P0171, P0172, P1128, P0335, P0336, P0505, P1128, P1129, P1259, P1361, P1362, P1366, P1367 and P1519 not set, engine running under positive torque conditions, and the PCM detected a misfire condition in one cylinder.
	Note: If the misfire is severe, the MIL will flash on/off on the 1st trip!
	Possible Causes:
	   • Fuel system problem affecting only Cylinder 6
	   • Ignition system problem affecting Cylinder 6
	   • Base engine (mechanical) problem affecting only Cylinder 6

DTC: P0306
2T MISFIRE, MIL: Yes
1996, 1997, 1998, 1999, 2000, 2001, 2002
Models: Passport
Engines: 3.2L VIN V, 3.2L VIN W
Transmissions: A/T, M/T
	Cylinder 6 Misfire Detected
	DTC P0101, P0102, P0103, P0106, P0107, P0108, P0117, P0118, P0121, P0122, P0123, P0336, P0341, P0342, P0502 and P0503 not set, ECT sensor from 20-248°F, system voltage from 11-16v, engine speed from 800-5500 rpm, throttle angle stable (± 3%), and the PCM detected a crankshaft speed variation in one cylinder characteristic of a misfire condition during the Misfire Diagnostic Monitor test.
	Note: If the misfire is severe, the MIL will flash on/off on the 1st trip!
	Possible Causes:
	   • Base engine mechanical fault that affects only one cylinder
	   • Fuel metering fault that affects only one cylinder
	   • IC control circuit is shorted to ground on only one cylinder
	   • Ignition system fault (a coil) that affects only one cylinder

DTC: P0325
2T CCM, MIL: Yes
1996, 1997, 1998, 1999, 2000, 2001, 2002, 2003, 2004, 2005, 2006
Models: Accord, Civic, Civic del Sol, CR-V, Element, Insight, Odyssey, Pilot, Prelude, Ridgeline, S2000
Engines: All
Transmissions: A/T, M/T
	Knock Sensor (Rear) Circuit Malfunction
	Engine running for over 1 minute, and the PCM detected an incorrect signal at the rear Knock Sensor (KS) circuit during the test.
	Possible Causes:
	   • Knock sensor signal circuit is open (rear bank of engine)
	   • Knock sensor signal circuit is grounded (rear bank of engine)
	   • Knock sensor not tightened properly
	   • Knock sensor damaged or has failed (it may be open internally)
	   • PCM has failed

DTC: P0325
2T CCM, MIL: Yes
1996, 1997, 1998, 1999, 2000, 2001, 2002
Models: Passport
Engines: 3.2L VIN V, 3.2L VIN W
Transmissions: A/T, M/T
	Knock Sensor Module Range/Performance
	DTC P0327 not set, engine started, system voltage from 11-16v, engine runtime over 120 seconds, and the PCM detected the Knock Sensor (KS) signal was present for over 5 seconds in the CCM test.
	Possible Causes:
	   • KS signal circuit is open or shorted to ground
	   • KS signal circuit is shorted to VREF or system power (B+)
	   • Knock Sensor is damaged or has failed
	   • PCM has failed

DTC: P0327
2T CCM, MIL: Yes
1996, 1997, 1998, 1999
Models: Passport
Engines: 3.2L VIN V, 3.2L VIN W
Transmissions: A/T, M/T
	Knock Sensor Circuit Low Input
	Engine started, engine runtime over 10 seconds, system voltage from 11-16v, ECT sensor more than 140°F, engine speed from 2000-4000 rpm, throttle angle over 5%, and the PCM detected the Knock Sensor (KS) signal indicated less than 0.20v, or indicated ore than 4.8v for over 15 seconds during the CCM test.
	Possible Causes:
	   • KS signal circuit is open or shorted to ground
	   • KS signal circuit is shorted to VREF or system power (B+)
	   • Knock Sensor is damaged or has failed
	   • PCM has failed

DTC: P0330
2T CCM, MIL: Yes
1996, 1997, 1998, 1999, 2000, 2001, 2002, 2003, 2004, 2005, 2006
Models: Accord, Civic, Civic del Sol, CR-V, Odyssey, Pilot, Prelude, Ridgeline, S2000
Engines: All
Transmissions: A/T, M/T
	Knock Sensor (Front) Circuit Malfunction
	Engine running for over 1 minute, and the PCM detected an incorrect signal at the front Knock Sensor (KS) circuit during the test.
	Possible Causes:
	   • Knock sensor signal circuit is open (front bank of engine)
	   • Knock sensor signal circuit is grounded (front bank of engine)
	   • Knock sensor not tightened properly
	   • Knock sensor damaged or has failed (it may be open internally)
	   • PCM has failed

DTC	Trouble Code Title, Conditions & Possible Causes
DTC: P0335 **1T CCM, MIL: Yes** **1995** **Models:** Accord **Engines:** 2.7L VIN CE6 **Transmissions:** A/T, M/T	**CKP Sensor 'A' Circuit Malfunction (No Signal)** Engine running, and the PCM did not detect any signals from the Crankshaft Position (CKP) Sensor 'A' during the test period. **Note: The engine will crank for a longer period of time, may buck or jerk, but it will start and run without the CKP sensor signal present.** **Possible Causes:** • CKP Sensor 'A' signal circuit is open or shorted to ground • CKP Sensor 'A' signal circuit shorted to VREF or system power • CKP Sensor 'A' is damaged or has failed • PCM has failed
DTC: P0335 **2T CCM, MIL: Yes** **1996, 1997, 1998, 1999, 2000, 2001, 2002, 2003, 2004, 2005, 2006** **Models:** Accord, Civic, Civic del Sol, CR-V, Odyssey, Pilot, Prelude, Ridgeline, S2000 **Engines:** All **Transmissions:** A/T, M/T	**CKP Sensor 'A' Circuit Malfunction (No Signal)** Engine running, and the PCM did not detect any signals from the Crankshaft Position (CKP) Sensor 'A' during the test period. **Note: The engine will crank for a longer period of time, may buck or jerk, but it will start and run without the CKP sensor signal present.** **Possible Causes:** • CKP Sensor 'A' signal circuit is open or shorted to ground • CKP Sensor 'A' signal circuit shorted to VREF or system power • CKP Sensor 'A' is damaged or has failed • PCM has failed
DTC: P0336 **1T CCM, MIL: Yes** **1995** **Models:** Accord **Engines:** 2.7L VIN CE6 **Transmissions:** A/T, M/T	**CKP Sensor 'A' Circuit Range/Performance** Engine running, and the PCM detected the Crankshaft Position (CKP) Sensor 'A' signal was missing for a short period of time. **Note: This trouble code is usually caused by an intermittent fault.** **Possible Causes:** • CKP Sensor 'A' signal circuit is open or shorted to ground • CKP Sensor 'A' signal circuit shorted to VREF or system power • CKP Sensor 'A' is damaged or has failed • PCM has failed
DTC: P0336 **2T CCM, MIL: Yes** **1996, 1997, 1998, 1999, 2000, 2001, 2002, 2003, 2004, 2005, 2006** **Models:** Accord, Civic, Civic del Sol, CR-V, Element, Insight, Odyssey, Pilot, Prelude, Ridgeline, S2000 **Engines:** All **Transmissions:** A/T, M/T	**CKP Sensor 'A' Circuit Range/Performance** Engine running, and the PCM detected the Crankshaft Position (CKP) Sensor 'A' signal was missing for a short period of time. **Note: This trouble code is usually caused by an intermittent fault.** **Possible Causes:** • CKP Sensor 'A' signal circuit is open or shorted to ground • CKP Sensor 'A' signal circuit shorted to VREF or system power • CKP Sensor 'A' is damaged or has failed • PCM has failed
DTC: P0336 **2T CCM, MIL: Yes** **1996** **Models:** Passport **Engines:** 2.6L VIN E **Transmissions:** M/T	**Crankshaft Position 58X Sensor Circuit Malfunction** Engine started, and the PCM detected extra or missing pulses between consecutive Crankshaft Position (CKP) 58X sensor signals during 10 out of 100 revolutions during the CCM test. **Possible Causes:** • CKP sensor 58X signal circuit is open or shorted to ground • CKP sensor ground circuit is open or has high resistance • CKP sensor power circuit is open between the sensor and PCM • CKP sensor is damaged, or the reluctor wheel is damaged • PCM has failed (the Ignition module function is inside the PCM)
DTC: P0336 **2T CCM, MIL: Yes** **1996, 1997, 1998, 1999, 2000, 2001, 2002** **Models:** Passport **Engines:** 3.2L VIN V, 3.2L VIN W **Transmissions:** A/T, M/T	**Crankshaft Position 58X Sensor Circuit Malfunction** Engine started, and the PCM detected extra or missing pulses between consecutive Crankshaft Position (CKP) 58X sensor signals during 10 out of 100 revolutions during the CCM test. **Possible Causes:** • CKP sensor 58X signal circuit is open or shorted to ground • CKP sensor ground circuit is open or has high resistance • CKP sensor power circuit is open between the sensor and PCM • CKP sensor is damaged, or the reluctor wheel is damaged • PCM has failed (the Ignition module function is inside the PCM)

DTC	Trouble Code Title, Conditions & Possible Causes
DTC: P0337 **2T CCM, MIL: Yes** **1996** **Models:** Passport **Engines:** 2.6L VIN E **Transmissions:** M/T	**Crankshaft Position 58X Sensor Circuit Low Input** DTC P0341 and P0342 not set, engine started, and the PCM did not detect any Crankshaft Position (CKP) 58X sensor signals present between two (2) CMP sensor pulses, or it did not detect any CKP sensor signals over a 24 second period during the CCM test. **Possible Causes:** • CKP sensor 58X signal circuit is open or shorted to ground • CKP sensor ground circuit is open or has high resistance • CKP sensor power circuit is open between the sensor and PCM • CKP sensor is damaged, or the reluctor wheel is damaged • PCM has failed (the Ignition module function is inside the PCM)
DTC: P0337 **2T CCM, MIL: Yes** **1996, 1997, 1998, 1999, 2000,** **2001, 2002** **Models:** Passport **Engines:** 3.2L VIN V, 3.2L VIN W **Transmissions:** A/T, M/T	**Crankshaft Position 58X Sensor Circuit Low Input** DTC P0341 and P0342 not set, engine started, and the PCM did not detect any Crankshaft Position (CKP) 58X sensor pulses present between two (2) CMP sensor pulses, or it did not detect any CKP sensor pulses within 8 CMP sensor pulses during the CCM test. **Possible Causes:** • CKP sensor 58X signal circuit is open or shorted to ground • CKP sensor ground circuit is open or has high resistance • CKP sensor power circuit is open between the sensor and PCM • CKP sensor is damaged, or the reluctor wheel is damaged • PCM has failed (the Ignition module function is inside the PCM)
DTC: P0339 **2T CCM, MIL: Yes** **2003, 2004, 2005, 2006** **Models:** Element **Engines:** All **Transmissions:** All	**Camshaft Position Sensor Intermittent** Engine started, engine running, and the PCM the Camshaft Position (CMP) sensor signals were not detected at the correct time interval after detecting for 100 occurrences in a 200-test sample in the test. **Possible Causes:** • CMP sensor circuit is open or shorted to ground (intermittent) • CMP sensor ground circuit is open (an intermittent fault) • CMP sensor is damaged or has failed • PCM has failed
DTC: P0340 **2T CCM, MIL: Yes** **2003, 2004, 2005, 2006** **Models:** Element **Engines:** All **Transmissions:** All	**Camshaft Position Sensor A No Signal** Engine started, engine running, and the PCM the Camshaft Position (CMP) sensor signals were not detected. **Possible Causes:** • CMP sensor circuit is open or shorted to ground (intermittent) • CMP sensor ground circuit is open (an intermittent fault) • CMP sensor is damaged or has failed • PCM has failed
DTC: P0341 **2T CCM, MIL: Yes** **2003, 2004, 2005, 2006** **Models:** Element **Engines:** All **Transmissions:** All	**Camshaft Position Sensor A and Crankshaft Position Sensor Incorrect Phase** Engine started, engine running, and the PCM detected that the Camshaft and Crankshaft position sensors are out of phase. **Possible Causes:** • VTEC oil control solenoid faulty • Timing chain auto tensioner faulty • Timing chain worn or damaged • VTC Actuator failed
DTC: P0341 **2T CCM, MIL: Yes** **1996, 1997** **Models:** Passport **Engines:** 2.6L VIN E **Transmissions:** M/T	**Camshaft Position Sensor Range/Performance** Engine started, engine running, and the PCM the Camshaft Position (CMP) sensor signals were not detected at the correct time interval after detecting for 100 occurrences in a 200-test sample in the test. **Possible Causes:** • CMP sensor circuit is open or shorted to ground (intermittent) • CMP sensor ground circuit is open (an intermittent fault) • CMP sensor is damaged or has failed • PCM has failed
DTC: P0341 **2T CCM, MIL: Yes** **1996, 1997, 1998, 1999, 2000** **Models:** Passport **Engines:** 3.2L VIN V, 3.2L VIN W **Transmissions:** A/T, M/T	**Camshaft Position Sensor Range/Performance** Engine started, engine running with CMP sensor pulses received, and the PCM detected an incorrect number of CMP signals were received during 10 tests over a 100-test sample period (that lasts 15.6 ms) during the CCM Rationality test. **Note: If a CKP sensor code is also set, check the common ground circuit between the CKP and CMP sensors for an open condition. If a fuel injector code is also set, check the power feed circuit as it connects to the fuel injectors and to the CMP sensor.** **Possible Causes:** • CMP sensor circuit is open or shorted to ground (intermittent) • CMP sensor ground circuit is open (an intermittent fault) • CMP sensor is damaged or has failed

DTC	Trouble Code Title, Conditions & Possible Causes
DTC: P0342 **2T CCM, MIL: Yes** **1996** **Models:** Passport **Engines:** 2.6L VIN E **Transmissions:** M/T	**Camshaft Position Sensor Circuit Low Input** Engine started, engine running, and PCM did not detect a Camshaft Position (CMP) sensor pulse at least once for every four (4) rotations of the crankshaft in a 10 second period during the CCM test. **Possible Causes:** • CMP sensor signal circuit is open • CMP sensor signal circuit is shorted to ground • CMP sensor power circuit open between the sensor and PCM • CMP sensor ground circuit is open • CMP sensor is damaged or has failed
DTC: P0342 **2T CCM, MIL: Yes** **1996, 1997, 1998, 1999, 2000** **Models:** Passport **Engines:** 3.2L VIN V, 3.2L VIN W **Transmissions:** A/T, M/T	**Camshaft Position Sensor Circuit Low Input** Engine started, engine running, and PCM did not detect a Camshaft Position (CMP) sensor pulse at least once for every six (6) rotations of the crankshaft in a 10 second period during the CCM test. **Possible Causes:** • CMP sensor signal circuit is open • CMP sensor signal circuit is shorted to ground • CMP sensor power circuit open between the sensor and PCM • CMP sensor ground circuit is open • CMP sensor is damaged or has failed
DTC: P0344 **2T CCM, MIL: Yes** **2003, 2004, 2005, 2006** **Models:** Element **Engines:** All **Transmissions:** All	**CMP Sensor A Circuit Intermittent Interruption** Engine started, engine running, and PCM did not detect a Camshaft Position (CMP) sensor pulse at least once for every six (6) rotations of the crankshaft in a 10 second period during the CCM test. **Possible Causes:** • CMP sensor signal circuit is open • CMP sensor signal circuit is shorted to ground • CMP sensor power circuit open between the sensor and PCM • CMP sensor ground circuit is open • CMP sensor is damaged or has failed
DTC: P0351 **2T CCM, MIL: Yes** **1996** **Models:** Passport **Engines:** 2.6L VIN E **Transmissions:** M/T	**Ignition Coil Primary Circuit Malfunction** Engine running, and the PCM detected the IC output signal did not equal 5v with the output commanded "on", or it did not equal 0v with the output commanded "off" in 20 tests over a 40 sample period. **Possible Causes:** • Ignition Coil control circuit is open or shorted to ground • Ignition Coil control circuit is shorted to system power (B+) • Ignition Module ground circuit is open • Ignition Coil power circuit is open (check the COIL 15A fuse) • Ignition control module is damaged or has failed
DTC: P0351 **1T CCM, MIL: Yes** **1996, 1997, 1998, 1999, 2000, 2001, 2002** **Models:** Passport **Engines:** 3.2L VIN V, 3.2L VIN W **Transmissions:** A/T, M/T	**Ignition Control Module Circuit 1 Malfunction** Engine running with CKP 58X signals received, and the PCM detected the IC output signal did not equal 5v with the output commanded "on", or it did not equal 0v with the output commanded "off" in 20 tests over a 40 sample period during the CCM test. **Possible Causes:** • EST signal circuit is open between module and coil or the PCM • EST signal circuit shorted between module and coil or the PCM • ION module is damaged or has failed • PCM has failed
DTC: P0352 **1T CCM, MIL: Yes** **1996, 1997, 1998, 1999, 2000, 2001, 2002** **Models:** Passport **Engines:** 3.2L VIN V, 3.2L VIN W **Transmissions:** A/T, M/T	**Ignition Control Module Circuit 2 Malfunction** Engine running with CKP 58X signals received, and the PCM detected the IC output signal did not equal 5v with the output commanded "on", or it did not equal 0v with the output commanded "off" in 20 tests over a 40 sample period during the CCM test. **Possible Causes:** • EST signal circuit is open between module and coil or the PCM • EST signal circuit shorted between module and coil or the PCM • ION module is damaged or has failed • PCM has failed
DTC: P0353 **1T CCM, MIL: Yes** **1996, 1997, 1998, 1999, 2000, 2001, 2002** **Models:** Passport **Engines:** 3.2L VIN V, 3.2L VIN W **Transmissions:** A/T, M/T	**Ignition Control Module Circuit 3 Malfunction** Engine running with CKP 58X signals received, and the PCM detected the IC output signal did not equal 5v with the output commanded "on", or it did not equal 0v with the output commanded "off" in 20 tests over a 40 sample period during the CCM test. **Possible Causes:** • EST signal circuit is open between module and coil or the PCM • EST signal circuit shorted between module and coil or the PCM • ION module is damaged or has failed • PCM has failed

DTC	Trouble Code Title, Conditions & Possible Causes
DTC: P0354 **1T CCM, MIL: Yes** **1996, 1997, 1998, 1999, 2000,** **2001, 2002** **Models:** Passport **Engines:** 3.2L VIN V, 3.2L VIN W **Transmissions:** A/T, M/T	**Ignition Control Module Circuit 4 Malfunction** Engine running with CKP 58X signals received, and the PCM detected the IC output signal did not equal 5v with the output commanded "on", or it did not equal 0v with the output commanded "off" in 20 tests over a 40 sample period during the CCM test. **Possible Causes:** • EST signal circuit is open between module and coil or the PCM • EST signal circuit shorted between module and coil or the PCM • ION module is damaged or has failed • PCM has failed
DTC: P0355 **1T CCM, MIL: Yes** **1996, 1997, 1998, 1999, 2000,** **2001, 2002** **Models:** Passport **Engines:** 3.2L VIN V, 3.2L VIN W **Transmissions:** A/T, M/T	**Ignition Control Module Circuit 5 Malfunction** Engine running with CKP 58X signals received, and the PCM detected the IC output signal did not equal 5v with the output commanded "on", or it did not equal 0v with the output commanded "off" in 20 tests over a 40 sample period during the CCM test. **Possible Causes:** • EST signal circuit is open between module and coil or the PCM • EST signal circuit shorted between module and coil or the PCM • ION module is damaged or has failed • PCM has failed
DTC: P0356 **1T CCM, MIL: Yes** **1996, 1997, 1998, 1999, 2000,** **2001, 2002** **Models:** Passport **Engines:** 3.2L VIN V, 3.2L VIN W **Transmissions:** A/T, M/T	**Ignition Control Module Circuit 6 Malfunction** Engine running with CKP 58X signals received, and the PCM detected the IC output signal did not equal 5v with the output commanded "on", or it did not equal 0v with the output commanded "off" in 20 tests over a 40 sample period during the CCM test. **Possible Causes:** • EST signal circuit is open between module and coil or the PCM • EST signal circuit shorted between module and coil or the PCM • ION module is damaged or has failed • PCM has failed
DTC: P0365 **2T CCM, MIL: Yes** **2003, 2004, 2005, 2006** **Models:** Element **Engines:** All **Transmissions:** All	**Camshaft Position Sensor B No Signal** Engine started, engine running, and the PCM the Camshaft Position (CMP) sensor signals were not detected. **Possible Causes:** • CMP sensor circuit is open or shorted to ground (intermittent) • CMP sensor ground circuit is open (an intermittent fault) • CMP sensor is damaged or has failed • PCM has failed
DTC: P0369 **2T CCM, MIL: Yes** **2003, 2004, 2005, 2006** **Models:** Element **Engines:** All **Transmissions:** All	**CMP Sensor B Circuit Intermittent Interruption** Engine started, engine running, and PCM did not detect a Camshaft Position (CMP) sensor pulse at least once for every six (6) rotations of the crankshaft in a 10 second period during the CCM test. **Possible Causes:** • CMP sensor signal circuit is open • CMP sensor signal circuit is shorted to ground • CMP sensor power circuit open between the sensor and PCM • CMP sensor ground circuit is open • CMP sensor is damaged or has failed
DTC: P0401 **2T EGR1, MIL: Yes** **1995** **Models:** Accord **Engines:** 2.7L VIN CE6 **Transmissions:** A/T, M/T	**EGR System Insufficient Flow Detected** Cold engine startup requirement met (ECT sensor less than 76°F at startup), engine running in closed loop at 40-55 mph for 2 minutes in Drive (D4), followed by a deceleration period back to 35 mph with the throttle closed, and the PCM detected a signal from the EGR position sensor that indicated insufficient EGR flow during the test. **Possible Causes:** • EGR valve source vacuum supply line open or restricted • EGR intake or exhaust manifold passages are restricted • EGR valve assembly or solenoid valve damaged or has failed • EGR constant vacuum control (CVC) valve is dirty or damaged • PCM has failed
DTC: P0401 **2T EGR1, MIL: Yes** **1996, 1997, 1998, 1999, 2000,** **2001, 2002, 2003, 2004, 2005,** **2006** **Models:** Accord, Civic, Civic del Sol, CR-V, Insight, Odyssey, Pilot, Prelude, Ridgeline, S2000 **Engines:** All **Transmissions:** A/T, M/T	**EGR System Insufficient Flow Detected** Cold engine startup requirement met (ECT sensor less than 76°F at startup), engine running in closed loop at 40-55 mph for 2 minutes in Drive (D4), followed by a deceleration period back to 35 mph with the throttle closed, and the PCM detected a signal from the EGR position sensor that indicated insufficient EGR flow during the test. **Possible Causes:** • EGR valve source vacuum supply line open or restricted • EGR intake or exhaust manifold passages are restricted • EGR valve assembly or solenoid valve damaged or has failed • EGR constant vacuum control (CVC) valve is dirty or damaged • PCM has failed

DTC	Trouble Code Title, Conditions & Possible Causes
DTC: P0401 **2T EGR2, MIL: Yes** 1996 **Models:** Passport **Engines:** 2.6L VIN E **Transmissions:** M/T	**Insufficient EGR System Flow Detected** No ECT, EGR Pintle Position sensor, EVAP, IAC, IAT, MAP, Misfire, TP or VSS codes set, engine running, system voltage from 11-16v, ECT sensor more than 140°F, BARO sensor more than 75 kPa, VSS more than 15 mph, IAC position stable with any change less than 10 counts, engine speed from 1100-2000 rpm, A/C clutch status stable, then with the throttle angle under 1%, EGR pintle position less than 1%, and during a gradual deceleration period with the EGR valve command over 95%, any engine speed change less than 100 rpm and any vehicle speed change less than 5 mph, the PCM detected the compensated MAP sensor signal indicated a value 10-40 kPa. **Possible Causes:** • EGR valve "low" circuit is open or shorted to ground • EGR valve "low" circuit is shorted to system power (B+) • EGR valve VREF (5-volt) is open between sensor and the PCM • EGR valve feedback circuit is open or shorted to ground • EGR valve is stuck closed, or partially open during the test • EGR exhaust flow path may be restricted • EGR valve is damaged, or has failed • PCM has failed
DTC: P0401 **2T EGR2, MIL: Yes** 1996, 1997, 1998, 1999, 2000, 2001, 2002 **Models:** Passport **Engines:** 3.2L VIN V, 3.2L VIN W **Transmissions:** A/T, M/T	**Insufficient EGR System Flow Detected** No ECT, EGR Pintle Position, EVAP, IAC, IAT, MAP, Misfire, TP or VSS codes set, system voltage from 11-16v, ECT sensor more than 140°F, BARO sensor over 75 kPa, IAC position stable (±10 counts), A/C Clutch and TCC status unchanged, and VSS over 15 mph, then with the throttle closed (TP angle under 1%), EGR duty cycle under 1%, MAP sensor from 10-40 kPa (±2 kPa), engine speed from 1100-2000 rpm, the PCM detected the compensated MAP sensor signal indicated a value from 10.3-49.8 kPa during the EGR System test. **Possible Causes:** • EGR valve "low" circuit is open, shorted to ground or to power • EGR valve VREF (5-volt) is open between sensor and the PCM • EGR valve feedback circuit is open or shorted to ground • EGR valve is stuck closed, or partially open during the test • EGR exhaust flow path may be restricted • EGR valve is damaged, or has failed • PCM has failed
DTC: P0402 **2T EGR2, MIL: Yes** 1998, 1999, 2000, 2001, 2002 **Models:** Passport **Engines:** 3.2L VIN W **Transmissions:** A/T, M/T	**Excessive EGR System Excessive Flow Detected** Engine started, IAT sensor more than 38°F, engine running, system voltage at 11-16v, and the PCM detected the EGR position sensor signal indicated more than 21% over a 625 ms period during the EGR System flow test right after engine startup. **Possible Causes:** • Linear EGR valve control circuit is shorted to ground • EGR valve is stuck partially open during the initial startup test • EGR valve is damaged, or has excessive carbon buildup • PCM has failed
DTC: P0404 **2T CCM, MIL: Yes** 1998, 1999, 2000, 2001, 2002 **Models:** Passport **Engines:** 3.2L VIN W **Transmissions:** A/T, M/T	**EGR Pintle Position Sensor Circuit Range/Performance** Engine started, IAT sensor more than 38°F, engine speed less than 600 rpm, system voltage at 11-16v, then with the Desired EGR position at over 0%, the PCM detected the difference between the Actual and Desired EGR position was more than 15% for over 15 seconds. This fault must occur 3 times in a single trip to set a code. **Possible Causes:** • Linear EGR valve control circuit is shorted to ground • EGR valve is stuck partially open during the initial startup test • EGR valve is damaged, or has excessive carbon buildup • PCM has failed
DTC: P0405 **2T CCM, MIL: Yes** 1998, 1999, 2000, 2001, 2002 **Models:** Passport **Engines:** 3.2L VIN W **Transmissions:** A/T, M/T	**EGR Pintle Position Sensor Circuit Low Input** Key on or engine running, system voltage from 11-16v, IAT sensor more than 140°F, and the PCM detected the EGR position sensor indicated less than 0.10v for 10 seconds during the CCM test. **Possible Causes:** • Linear EGR valve control circuit is shorted to ground • EGR valve is stuck partially open during the initial startup test • EGR valve is damaged, or has excessive carbon buildup • PCM has failed
DTC: P0406 **1T CCM, MIL: Yes** 1998, 1999, 2000, 2001, 2002 **Models:** Passport **Engines:** 3.2L VIN W **Transmissions:** A/T, M/T	**EGR Pintle Position Sensor Circuit High Input** Engine started, engine running, system voltage from 11-16v, IAT sensor more than 41°F, and the PCM detected the EGR position sensor indicated more than 4.80v for 10 seconds during the CCM test. **Possible Causes:** • EGR feedback signal circuit is open between sensor and PCM • EGR sensor is damaged or has failed • PCM has failed

DTC	Trouble Code Title, Conditions & Possible Causes
DTC: P0410 **2T AIR, MIL: Yes** **2000, 2001, 2002, 2003, 2004, 2005, 2006** **Models:** S2000 **Engines:** All **Transmissions:** M/T	**Secondary Air Pump Circuit Malfunction** Cold startup requirement met (ECT sensor from 32-158°F at engine startup), engine running, and the PCM detected an unexpected voltage condition on the Air Pump control circuit during the test. **Possible Causes:** • Secondary air pump control circuit is open or shorted to ground • Secondary air pump power (B+) circuit is open • Secondary air pump solenoid is damaged or has failed • PCM has failed
DTC: P0411 **2T AIR, MIL: Yes** **2000, 2001, 2002, 2003, 2004, 2005, 2006** **Models:** S2000 **Engines:** All **Transmissions:** M/T	**Secondary Air System Incorrect Flow** Cold startup requirement met (ECT sensor from 32-158°F at engine startup), engine running at idle speed, and the PCM detected an incorrect amount of airflow from the Air Injection system in the test. **Possible Causes:** • AIR solenoid air injection valve is damaged or has failed • AIR solenoid source vacuum hoses loose or disconnected • AIR solenoid air injection tube is restricted or clogged • PCM has failed
DTC: P0420 **2T CAT1, MIL: Yes** **1995** **Models:** Accord **Engines:** 2.7L VIN CE6 **Transmissions:** A/T, M/T	**Catalyst Efficiency Below Threshold (Bank 1)** DTC P0137, P0138 and P0141 not set, engine running in closed loop at 40-55 mph for 2 minutes, followed by a deceleration period to 35 mph at closed throttle, and the PCM detected excessive activity in the Catalyst oxygen sensor (rear HO2S) during the test period. **Possible Causes:** • Air leaks at the exhaust manifold or in the exhaust pipes • Catalytic converter damaged or has failed (deteriorated) • Front HO2S older (aged) than the rear HO2S (HO2S is lazy) • Front HO2S or rear HO2S is contaminated with fuel or moisture
DTC: P0420 **2T CAT1, MIL: Yes** **1996, 1997, 1998, 1999, 2000, 2001, 2002, 2003, 2004, 2005, 2006** **Models:** Accord, Civic, Civic del Sol, CR-V, Element, Insight, Odyssey, Pilot, Prelude, Ridgeline, S2000 **Engines:** All **Transmissions:** A/T, M/T	**Catalyst Efficiency Below Threshold (Bank 1)** DTC P0137, P0138 and P0141 not set, engine running in closed loop at 40-55 mph for 2 minutes, followed by a deceleration period to 35 mph at closed throttle, and the PCM detected excessive activity in the Catalyst oxygen sensor (rear HO2S) during the test period. **Possible Causes:** • Air leaks at the exhaust manifold or in the exhaust pipes • Catalytic converter damaged or has failed (deteriorated) • Front HO2S older (aged) than the rear HO2S (HO2S is lazy) • Front HO2S or rear HO2S is contaminated with fuel or moisture
DTC: P0420 **1T CAT2, MIL: Yes** **1996, 1997, 1998, 1999, 2000, 2001, 2002** **Models:** Passport **Engines:** 3.2L VIN V, 3.2L VIN W **Transmissions:** A/T, M/T	**Catalyst Efficiency Below Normal (Bank 1)** DTC P0106, P0107, P0108, P0112, P0113, P0117, P0118, P0121, P0122, P0123, P0131, P0132, P0133, P0134, P0137, P0138, P0140, P0141, P0171, P0172, P0300, P0301-P0306, P0336, P0341, P0342, P0401, P0502, P0506 and P0507 not set, engine speed less than 3500 rpm in closed loop, ECT sensor more than 140°F, MAF sensor from 8-50 g/sec, engine load less than 99% (±8%), predicted Catalyst temperature over 750°F, vehicle speed from 16-75 mph, and the PCM detected the catalyst oxygen storage capacity was below an acceptable threshold during the Catalyst test. **Possible Causes:** • Air leaks at the exhaust manifold or in the exhaust pipes • Front HO2S or rear HO2S is contaminated with fuel or moisture • Front HO2S older (aged) than the rear HO2S (HO2S is lazy) • Front HO2S and/or the rear HO2S is loose in the mounting hole • Catalytic converter is damaged or has failed
DTC: P0420 **1T CAT2, MIL: Yes** **1996, 1997** **Models:** Passport **Engines:** 2.6L VIN E **Transmissions:** M/T	**Catalyst Efficiency Below Normal (Bank 1)** DTC P0106, P0107, P0108, P0112, P0113, P0117, P0118, P0121, P0122, P0123, P0131, P0132, P0133, P0134, P0137, P0138, P0140, P0141, P0171, P0172, P0300, P0336, P0341, P0342, P0401, P0502, P0506 and P0507 not set, ECT sensor more than 140°F, engine speed less than 3500 rpm in closed loop, Calculated airflow from 7-41 g/sec, any change in engine load less than 8%, predicted Catalyst temperature over 750°F, vehicle speed from 16-75 mph, and the PCM detected the catalyst oxygen storage capacity was below an acceptable threshold during the Catalyst Monitor test. **Possible Causes:** • Air leaks at the exhaust manifold or in the exhaust pipes • Front HO2S or rear HO2S is contaminated with fuel or moisture • Front HO2S older (aged) than the rear HO2S (HO2S is lazy) • Front HO2S and/or the rear HO2S is loose in the mounting hole • Catalytic converter is damaged or has failed

DTC	Trouble Code Title, Conditions & Possible Causes
DTC: P0430 **1T CAT2, MIL: Yes** **1996, 1997, 1998, 1999, 2000,** **2001, 2002** **Models:** Passport **Engines:** 3.2L VIN V, 3.2L VIN W **Transmissions:** A/T, M/T	**Catalyst Efficiency Below Normal (Bank 2)** DTC P0106, P0107, P0108, P0112, P0113, P0117, P0118, P0121, P0122, P0123, P0131, P0132, P0133, P0134, P0137, P0138, P0140, P0141, P0171, P0172, P0300, P0301-P0306, P0336, P0341, P0342, P0401, P0502, P0506 and P0507 not set, engine speed less than 3500 rpm in closed loop, ECT sensor more than 140°F, MAF sensor from 8-50 g/sec, engine load less than 99% (±8%), predicted Catalyst temperature over 750°F, vehicle speed from 16-75 mph, and the PCM detected the catalyst oxygen storage capacity was below an acceptable threshold during the Catalyst test. **Possible Causes:** • Air leaks at the exhaust manifold or in the exhaust pipes • Front HO2S or rear HO2S is contaminated with fuel or moisture • Front HO2S older (aged) than the rear HO2S (HO2S is lazy) • Front HO2S and/or the rear HO2S is loose in the mounting hole • Catalytic converter is damaged or has failed
DTC: P0440 **2T EVADTC: P2, MIL: Yes** **1996** **Models:** Passport **Engines:** 3.2L VIN V **Transmissions:** A/T, M/T	**EVAP System Performance** DTC P0106, P0107, P0108, P0112, P0113, P0117, P0118, P0121, P0122, P0123, P1640 ad P1650 not set, ECT and IAT sensors less than 90°F and within 13°F at startup, ECT sensor over 39°F at startup, IAT sensor over 4°F at startup, system voltage from 11-16v, BARO sensor over 75 kPa, throttle angle from 7-30%, fuel level from 15-85% with minimum fuel slosh, vehicle speed under 75 mph, and the PCM determined it was unable to achieve or maintain vacuum in the system for 60-180 seconds during the EVAP Monitor leak test. **Possible Causes:** • Charcoal canister is loaded with fuel or moisture • ECT, IAT, MAP, VSS or TP sensor signals out-of-calibration • Fuel filler cap loose, cross-threaded, incorrect part or damaged • Fuel tank pressure sensor is damaged or has failed • Fuel tank or fuel tank sender assembly 'O' ring is leaking • Fuel tank vapor line(s) block, damaged or disconnected • Purge or Vent solenoid control circuit open or shorted to ground • Purge or Vent solenoid power circuit is open (check the fuse)
DTC: P0440 **2T EVADTC: P2, MIL: Yes** **1998, 1999, 2000, 2001, 2002** **Models:** Passport **Engines:** 3.2L VIN W **Transmissions:** A/T, M/T	**EVAP System Performance** DTC P0106, P0107, P0108, P0112, P0113, P0117, P0118, P0121, P0122, P0123, P1640 ad P1650 not set, ECT and IAT sensors less than 90°F and within 13°F at startup, ECT sensor over 39°F at startup, IAT sensor over 4°F at startup, system voltage from 11-16v, BARO sensor over 75 kPa, throttle angle from 7-30%, fuel level from 15-85% with minimum fuel slosh, vehicle speed under 75 mph, and the PCM determined it was unable to achieve or maintain vacuum in the system for 60-180 seconds during the EVAP Monitor leak test. **Possible Causes:** • Charcoal canister is loaded with fuel or moisture • ECT, IAT, MAP, VSS or TP sensor signals out-of-calibration • Fuel filler cap loose, cross-threaded, incorrect part or damaged • Fuel tank pressure sensor is damaged or has failed • Fuel tank or fuel tank sender assembly 'O' ring is leaking • Fuel tank vapor line(s) block, damaged or disconnected • Purge or Vent solenoid control circuit open or shorted to ground • Purge or Vent solenoid power circuit is open (check the fuse)
DTC: P0441 **2T EVADTC: P1, MIL: Yes** **1995** **Models:** Accord **Engines:** 2.7L VIN CE6 **Transmissions:** A/T, M/T	**EVAP System Incorrect Purge Flow** Cold startup requirement met (ECT sensor signal more than 154°F and IAT sensor signal more than 14°F at engine startup), engine runtime from 3-5 minutes, followed by an acceleration period to 50-60 mph at steady throttle, and the PCM did not detect enough purge flow through the EVAP system during the test period. **Possible Causes:** • EVAP purge control diaphragm valve hose loose/disconnected • EVAP purge cutoff or purge control solenoid valve is damaged • EVAP purge control diaphragm valve is damaged or has failed • PCM has failed
DTC: P0441 **2T EVADTC: P1, MIL: Yes** **1996, 1997, 1998, 1999, 2000,** **2001, 2002, 2003, 2004, 2005,** **2006** **Models:** Accord, Civic, Civic del Sol, CR-V, Odyssey, Pilot, Prelude, Ridgeline, S2000 **Engines:** All **Transmissions:** A/T, M/T	**EVAP System Incorrect Purge Flow** Cold engine startup (ECT sensor signal less than 154°F and IAT sensor signal more than 14°F at startup), engine runtime from 3-5 minutes, followed by an acceleration period to 50-60 mph at steady throttle, and the PCM did not detect enough purge flow through the EVAP system during the EVAP Monitor flow test. **Possible Causes:** • EVAP purge control diaphragm valve hose loose/disconnected • EVAP purge cutoff or purge control solenoid valve is damaged • EVAP purge control diaphragm valve is damaged or has failed • PCM has failed

DTC	Trouble Code Title, Conditions & Possible Causes
DTC: P0441 **2T EVADTC: P2, MIL: Yes** 1996 **Models:** Passport **Engines:** 2.6L VIN E **Transmissions:** A/T, M/T	**EVAP System No Flow During Purge Detected** DTC P0106, P0107, P0108, P0112, P0113, P0117, P0118, P0121, P0122, P0123 and P1442 not set, ECT and IAT sensors more than 41°F and within 45°F at startup, ECT sensor less than 158°F during testing, BARO sensor over 85 kPa, Calculated manifold pressure over 10 kPa, throttle angle over 14%, engine speed from 800-6000 rpm, system voltage from 11-16v, Purge duty cycle over 95%, and the PCM detected the EVAP vacuum switch indicated a "closed" position (with Purge enabled) for 3 seconds in the EVAP Purge test. **Possible Causes:** • Charcoal canister is damaged, clogged or restricted • Purge solenoid circuit is open or shored to ground (intermittent) • Purge valve vacuum line is clogged, restricted or disconnected • Purge vacuum switch is damaged or has failed • PCM has failed
DTC: P0442 **2T EVADTC: P2, MIL: Yes** 1996, 1997, 1998, 1999, 2000, 2001, 2002 **Models:** Passport **Engines:** 3.2L VIN V, 3.2L VIN W **Transmissions:** A/T, M/T	**EVAP System Small Leak (0.040") Detected** DTC P0106, P0107, P0108, P0112, P0113, P0117, P0118, P0121, P0122, P0123, P0440, P0446, P0562, P0563, P1640 and P1650 not set, ECT and IAT sensors less than 90°F and within 13°F at startup, ECT sensor over 39°F at startup, IAT sensor over 4°F at startup, system voltage from 11-16v, BARO sensor over 75 kPa, throttle angle from 7-30%, fuel level from 15-85% with minimum fuel slosh, vehicle speed under 75 mph, and the PCM detected a vacuum decaying condition characteristic of a small leak (0.040") in the system during the EVAP Monitor leak test. **Possible Causes:** • Charcoal canister is loaded with fuel or moisture • ECT, IAT, MAP, VSS or TP sensor signals out-of-calibration • Fuel filler cap loose, cross-threaded, incorrect part or damaged • Fuel tank pressure sensor is damaged or has failed • Fuel tank or fuel tank sender assembly 'O' ring is leaking • Fuel tank vapor line(s) block, damaged or disconnected • Purge or Vent solenoid control circuit open or shorted to ground • Purge or Vent solenoid power circuit is open (check the fuse)
DTC: P0442 **2T EVADTC: P2, MIL: Yes** 2003, 2004, 2005, 2006 **Models:** Element **Engines:** All **Transmissions:** All	**EVAP System Small Leak (0.040") Detected** DTC P0106, P0107, P0108, P0112, P0113, P0117, P0118, P0121, P0122, P0123, P0440, P0446, P0562, P0563, P1640 and P1650 not set, ECT and IAT sensors less than 90°F and within 13°F at startup, ECT sensor over 39°F at startup, IAT sensor over 4°F at startup, system voltage from 11-16v, BARO sensor over 75 kPa, throttle angle from 7-30%, fuel level from 15-85% with minimum fuel slosh, vehicle speed under 75 mph, and the PCM detected a vacuum decaying condition characteristic of a small leak (0.040") in the system during the EVAP Monitor leak test. **Possible Causes:** • Charcoal canister is loaded with fuel or moisture • ECT, IAT, MAP, VSS or TP sensor signals out-of-calibration • Fuel filler cap loose, cross-threaded, incorrect part or damaged • Fuel tank pressure sensor is damaged or has failed • Fuel tank or fuel tank sender assembly 'O' ring is leaking • Fuel tank vapor line(s) block, damaged or disconnected • Purge or Vent solenoid control circuit open or shorted to ground • Purge or Vent solenoid power circuit is open (check the fuse)
DTC: P0443 **1T CCM, MIL: No** 2003, 2004, 2005, 2006 **Models:** Element **Engines:** All **Transmissions:** All	**EVAP Purge Solenoid Circuit Malfunction (Open)** Engine started, engine running, system voltage from 11-16v, and the PCM detected an unexpected voltage condition on the EVAP Purge solenoid control circuit. **Possible Causes:** • Purge solenoid control circuit is open between device and PCM • Purge solenoid control circuit is shorted to system power (B+) • Purge solenoid is damaged or has failed • PCM has failed (the solenoid driver circuit is damaged)
DTC: P0444 **1T CCM, MIL: No** 2001, 2002 **Models:** Passport **Engines:** 3.2L VIN W **Transmissions:** A/T, M/T	**EVAP Canister Purge Valve Circuit Malfunction** Engine started, engine running, system voltage from 11-16v, and the PCM detected an unexpected "high" voltage condition on the EVAP Purge solenoid control circuit with the solenoid commanded "on". **Possible Causes:** • Purge solenoid control circuit is open between device and PCM • Purge solenoid control circuit is shorted to system power (B+) • Purge solenoid is damaged or has failed • PCM has failed (the solenoid driver circuit is damaged)

DTC	Trouble Code Title, Conditions & Possible Causes
DTC: P0445 **1T CCM, MIL: No** **2001, 2002** **Models:** Passport **Engines:** 3.2L VIN W **Transmissions:** A/T, M/T	**EVAP Purge Solenoid Circuit Malfunction (Shorted)** Engine started, engine running, system voltage from 11-16v, and the PCM detected an unexpected "low" voltage condition on the EVAP Purge solenoid control circuit with the solenoid commanded "off". **Possible Causes:** • Purge solenoid control circuit is shorted to chassis ground • Purge solenoid power circuit is open (check the ENGINE fuse) • Purge solenoid is damaged or has failed • PCM has failed (the solenoid driver circuit is damaged)
DTC: P0446 **1T EVADTC: P2, MIL: Yes** **1996, 1997, 1998, 1999, 2000, 2001, 2002** **Models:** Passport **Engines:** 3.2L VIN V, 3.2L VIN W **Transmissions:** A/T, M/T	**EVAP Vent Control System Performance** DTC P0106, P0107, P0108, P0112, P0113, P0121, P0122, P0123, P1640 and P1650 not set, ECT and IAT sensors from 39-86°F at startup, ECT signal within 12°F of the IAT signal at startup, and the IAT signal within 2°F of the ECT signal at startup, system voltage from 11-16v, BARO sensor more than 72 kPa, fuel tank level from 12-87%, Purge duty cycle over 50%, and the PCM detected the FTP sensor did no indicate close to -10" H2O under normal purge conditions with the Canister Vent solenoid "open", or it detected the FTP sensor did not indicate about -1.5 to +1.5" H2O at key "on". **Possible Causes:** • Charcoal canister is clogged, plugged or restricted • EVAP vent control solenoid control circuit is shorted to ground • EVAP vent control solenoid hose is bent, kinked or plugged • EVAP vent control solenoid is damaged or has failed • PCM has failed (the solenoid driver circuit may be shorted)
DTC: P0447 **1T CCM, MIL: No** **2001, 2002** **Models:** Passport **Engines:** 3.2L VIN W **Transmissions:** A/T, M/T	**EVAP Vent Solenoid Circuit Malfunction (Open)** Engine started, engine running, system voltage from 11-16v, and the PCM detected an unexpected "high" voltage condition on the EVAP Vent solenoid control circuit with the solenoid commanded "on". **Possible Causes:** • Vent solenoid control circuit is open between device and PCM • Vent solenoid control circuit is shorted to system power (B+) • Vent solenoid is damaged or has failed • PCM has failed (the solenoid driver circuit is damaged)
DTC: P0448 **1T CCM, MIL: No** **2001, 2002** **Models:** Passport **Engines:** 3.2L VIN W **Transmissions:** A/T, M/T	**EVAP Vent Solenoid Circuit Malfunction (Shorted)** Engine started, engine running, system voltage from 11-16v, and the PCM detected an unexpected "low" voltage condition on the EVAP Vent solenoid control circuit with the solenoid commanded "off". **Possible Causes:** • Vent solenoid control circuit is shorted to chassis ground • Vent solenoid power circuit is open (check the ENGINE fuse) • Vent solenoid is damaged or has failed • PCM has failed (the solenoid driver circuit is damaged)
DTC: P0451 **2T CCM, MIL: Yes** **1998, 1999, 2000, 2001, 2002, 2003, 2004, 2005, 2006** **Models:** Accord, Civic, CR-V, Element, Insight, Odyssey, Pilot, Prelude, Ridgeline, S2000 **Engines:** All **Transmissions:** A/T, M/T	**Fuel Tank Pressure Sensor Range/Performance** Engine running, and the PCM detected the fuel tank pressure (FTP) sensor signal was less than the allowable range stored in the PCM memory (a calibrated range adjusted to current conditions). **Note: The FTP sensor PID should be near 2.5v with the fuel cap off.** **Possible Causes:** • FTP sensor vacuum lines loose, damaged or disconnected • Fuel tank pressure sensor is damaged or has failed • PCM has failed
DTC: P0452 **1T CCM, MIL: Yes** **1998, 1999, 2000, 2001, 2002, 2003, 2004, 2005, 2006** **Models:** Accord, Civic, CR-V, Element, Insight, Odyssey, Pilot, Prelude, Ridgeline, S2000 **Engines:** All **Transmissions:** A/T, M/T	**Fuel Tank Pressure Sensor Circuit Low Input** Key on or engine running, and the PCM detected the fuel tank pressure (FTP) sensor signal was less than 0.16v during the test. **Note: The FTP sensor PID should be near 2.5v with the fuel cap off.** **Possible Causes:** • FTP sensor signal circuit is shorted to ground • FTP sensor vacuum lines loose, damaged or disconnected • Fuel tank pressure sensor is damaged or has failed • PCM has failed
DTC: P0452 **1T CCM, MIL: Yes** **1998, 1999, 2000, 2001, 2002** **Models:** Passport **Engines:** 3.2L VIN W **Transmissions:** A/T, M/T	**Fuel Tank Pressure Sensor Circuit Low Input** Key on or engine running, and the PCM detected the Fuel Tank Pressure (FTP) sensor signal was less than 0.20v for 12.5 seconds, 100 test failures over a 200 test period during the CCM test. **Possible Causes:** • FTP sensor signal circuit is shorted to ground • FTP sensor power circuit is open between sensor and the PCM • FTP sensor is damaged or has failed • PCM has failed

DTC	Trouble Code Title, Conditions & Possible Causes
DTC: P0453 **1T CCM, MIL: Yes** **1998, 1999, 2000, 2001, 2002,** **2003, 2004, 2005, 2006** **Models:** Accord, Civic, CR-V, Element, Insight, Odyssey, Pilot, Prelude, Ridgeline, S2000 **Engines:** All **Transmissions:** A/T, M/T	**Fuel Tank Pressure Sensor Circuit High Input** Key on or engine running, and the PCM detected the fuel tank pressure (FTP) sensor signal was more than 4.90v during the test. **Note: The FTP sensor PID should be near 2.5v with the fuel cap off.** **Possible Causes:** • FTP sensor signal circuit is shorted to VREF or power (B+) • FTP sensor ground circuit is open • FTP sensor vacuum lines loose, damaged or disconnected • Fuel tank pressure sensor is damaged or has failed • PCM has failed
DTC: P0453 **1T CCM, MIL: Yes** **1998, 1999, 2000, 2001, 2002** **Models:** Passport **Engines:** 3.2L VIN W **Transmissions:** A/T, M/T	**Fuel Tank Pressure Sensor Circuit High Input** Key on or engine running, and the PCM detected the Fuel Tank Pressure (FTP) sensor signal was more than 4.98v for 12.5 seconds, 100 test failures over a 200 test period in the CCM test. **Possible Causes:** • FTP sensor signal circuit is open between sensor and the PCM • FTP sensor ground circuit open between the sensor and PCM • FTP sensor signal circuit is shorted to VREF or system power • FTP sensor is damaged or has failed • PCM has failed
DTC: P0456 **2T EVADTC: P2, MIL: Yes** **2000, 2001, 2002** **Models:** Passport **Engines:** 3.2L VIN W **Transmissions:** A/T, M/T	**EVAP System Very Small Leak (0.020") Detected** DTC P0106, P0107, P0108, P0112, P0113, P0117, P0118, P0121, P0122, P0123, P0452 and P0453 codes set, BARO sensor more than 70 kPa, system voltage at 11-16v, engine speed under 1200 rpm, VSS indicating less than 5 mph, fuel tank level from 40-80%, throttle angle below 1.1%, VSS over 65 mph, and the PCM detected a vacuum decay condition that indicated the presence of a very small leak (under 0.020" in diameter) somewhere in the system. **Possible Causes:** • Canister Purge solenoid is damaged, leaking or has failed • EVAP purge or vent solenoid is damaged or has failed • Fuel filler cap loose, cross-threaded, incorrect part or damaged • Fuel tank is cracked (leaking), or a leak exists in the 'O' ring • Fuel tank pressure sensor is damaged or has failed • Fuel vapor line(s), fuel pipes or hoses damaged or leaking • PCM has failed
DTC: P0457 **2T EVADTC: P2, MIL: Yes** **2003, 2004, 2005, 2006** **Models:** Element **Engines:** All **Transmissions:** All	**EVAP System Leak Detected/Fuel Fill Cap Loose or Missing** BARO sensor more than 70 kPa, system voltage at 11-16v, engine speed under 1200 rpm, VSS indicating less than 5 mph, fuel tank level from 40-80%, throttle angle below 1.1%, VSS over 65 mph, and the PCM detected a vacuum decay condition that indicated the presence of a very large leak somewhere in the system. **Possible Causes:** • Fuel filler cap loose, cross-threaded, incorrect part or damaged • Missing or damaged fuel cap seal • Fuel tank is cracked (leaking), or a leak exists in the 'O' ring • Fuel tank pressure sensor is damaged or has failed • Fuel vapor line(s), fuel pipes or hoses damaged or leaking • EVAP canister vent shut valve damaged
DTC: P0456 **2T EVADTC: P2, MIL: Yes** **2003, 2004, 2005, 2006** **Models:** Element **Engines:** All **Transmissions:** All	**EVAP System Very Small Leak (0.020") Detected** DTC P0106, P0107, P0108, P0112, P0113, P0117, P0118, P0121, P0122, P0123, P0452 and P0453 codes set, BARO sensor more than 70 kPa, system voltage at 11-16v, engine speed under 1200 rpm, VSS indicating less than 5 mph, fuel tank level from 40-80%, throttle angle below 1.1%, VSS over 65 mph, and the PCM detected a vacuum decay condition that indicated the presence of a very small leak (under 0.020" in diameter) somewhere in the system. **Possible Causes:** • Canister Purge solenoid is damaged, leaking or has failed • EVAP purge or vent solenoid is damaged or has failed • Fuel filler cap loose, cross-threaded, incorrect part or damaged • Fuel tank is cracked (leaking), or a leak exists in the 'O' ring • Fuel tank pressure sensor is damaged or has failed • Fuel vapor line(s), fuel pipes or hoses damaged or leaking • PCM has failed

DTC	Trouble Code Title, Conditions & Possible Causes
DTC: P0461 **1T CCM, MIL: Yes** **1996, 1997, 1998, 1999, 2000, 2001, 2002** **Models:** Passport **Engines:** 3.2L VIN V, 3.2L VIN W **Transmissions:** A/T, M/T	**Fuel Level Sensor Circuit Range/Performance** Engine started, engine running, fuel tank level from 15-85%, Fuel Tank Level Slosh Test and Fuel Level Main Test both completed, fuel tank data is valid, and the PCM detected the fuel level did not change over a distance of 62.2 miles during the CCM test. **Note: For additional help with this code, view the Failure Records.** **Possible Causes:** • Fuel level signal circuit is open, shorted to ground or to power • Fuel tank empty or overfull (fuel sender is stuck mechanically) • Wrong fuel gauge is installed, or instrument panel is damaged • Fuel gauge sender unit is damaged or has failed • PCM has failed
DTC: P0462 **1T CCM, MIL: Yes** **1996, 1997, 1998, 1999, 2000, 2002** **Models:** Passport **Engines:** 3.2L VIN V, 3.2L VIN W **Transmissions:** A/T, M/T	**Fuel Level Sensor Circuit Low Input** Key on or engine running, and the PCM detected the Fuel Level sensor signal was less than 0.39v for 20 seconds during the test. **Note: For additional help with this code, view the Failure Records.** **Possible Causes:** • Fuel level signal circuit is shorted to ground • Wrong fuel gauge is installed, or instrument panel is damaged • Fuel gauge sender unit is damaged or has failed • PCM has failed
DTC: P0463 **1T CCM, MIL: Yes** **1996, 1997, 1998, 1999, 2000, 2001, 2002** **Models:** Passport **Engines:** 3.2L VIN V, 3.2L VIN W **Transmissions:** A/T, M/T	**Fuel Level Sensor Circuit High input** Key on or engine running, and the PCM detected the Fuel Level sensor signal was less than 2.90v for 20 seconds during the test. **Note: For additional help with this code, view the Failure Records.** **Possible Causes:** • Fuel level signal circuit is open or shorted to system power (B+) • Wrong fuel gauge is installed, or instrument panel is damaged • Fuel gauge sender unit is damaged or has failed • PCM has failed
DTC: P0464 **1T CCM, MIL: No** **2001, 2002** **Models:** Passport **Engines:** All **Transmissions:** A/T, M/T	**Fuel Level Sensor Circuit Malfunction (Noisy)** DTC P0461, P0462 and P0463 not set, fuel level over 3%, IAT sensor signal more than 40°F, engine running, and the PCM detected an unexpected "noisy" condition on the Fuel level sensor circuit during the CCM Rationality test. **Possible Causes:** • Fuel level sensor signal circuit open (an intermittent fault) • Fuel level sensor signal circuit shorted to power (intermittent) • Fuel level sensor damaged or has failed (an intermittent fault) • PCM has failed
DTC: P0496 **1T CCM, MIL: No** **2003, 2004, 2005, 2006** **Models:** Element **Engines:** All **Transmissions:** All	**EVAP System High Purge Flow** The PCM detected an unexpected purge flow condition during the CCM Rationality test. **Possible Causes:** • Poor connections or loose terminals at the FTP sensor, the EVAP canister purge valve, the EVAP canister vent shut valve, or the ECM/PCM • Faulty EVAP canister purge valve
DTC: P0497 **1T CCM, MIL: No** **2003, 2004, 2005, 2006** **Models:** Element **Engines:** All **Transmissions:** All	**EVAP System Low Purge Flow** The PCM detected an unexpected voltage condition during the CCM Rationality test. **Possible Causes:** • Poor connections or loose terminals at the FTP sensor, the EVAP canister purge valve, the EVAP canister vent shut valve, or the ECM/PCM • Loose, blocked, or damaged EVAP canister purge line between the intake manifold and the EVAP canister purge valve. • Faulty EVAP canister purge valve
DTC: P0498 **1T CCM, MIL: No** **2003, 2004, 2005, 2006** **Models:** Element **Engines:** All **Transmissions:** All	**EVAP Canister Vent Shut Valve Circuit Low Voltage** The PCM detected an unexpected voltage condition during the CCM Rationality test. **Possible Causes:** • Poor connections or loose terminals at the EVAP canister vent shut valve, or the ECM/PCM • Faulty EVAP canister Vent Shut valve

DTC	Trouble Code Title, Conditions & Possible Causes
DTC: P0499 **1T CCM, MIL: No** **2003, 2004, 2005, 2006** **Models:** Element **Engines:** All **Transmissions:** All	**EVAP Canister Vent Shut Valve Circuit High Voltage** The PCM detected an unexpected purge flow condition during the CCM Rationality test. **Possible Causes:** • Poor connections or loose terminals at the EVAP canister vent shut valve, or the ECM/PCM • Faulty EVAP canister Vent Shut valve
DTC: P0500 **1T CCM, MIL: Yes** **1995** **Models:** Accord **Engines:** 2.7L VIN CE6 **Transmissions:** A/T, M/T	**Vehicle Speed Sensor Circuit Low Input** Engine at idle, then accelerated to 4000 rpm in 2nd gear, followed by a deceleration period to 1500 rpm at closed throttle, and the PCM did not detect any VSS signals during the CCM test. **Note: The VSS signal should pulse from 0-5v as the vehicle moves.** **Possible Causes:** • VSS signal circuit is open or shorted to ground • VSS signal circuit is shorted to VREF or system power (B+) • VSS is damaged or has failed
DTC: P0500 **1T CCM, MIL: Yes** **1996, 1997, 1998, 1999, 2000,** **2001, 2002, 2003, 2004, 2005,** **2006** **Models:** Accord, Civic, Civic del Sol, Insight, Odyssey, Pilot, Prelude, Ridgeline, S2000 **Engines:** All **Transmissions:** A/T, M/T	**Vehicle Speed Sensor Circuit Low Input** Engine running, then the vehicle was accelerated to 4000 rpm in 2nd gear, followed by a deceleration period to 1500 rpm at closed throttle and the PCM did not detect any VSS signal during the CCM test. **Note: The VSS signal should pulse from 0-5v as the vehicle moves.** **Possible Causes:** • VSS signal circuit is open • VSS signal circuit is shorted to ground • VSS signal circuit is shorted to VREF or system power (B+) • VSS is damaged or has failed
DTC: P0501 **2T CCM, MIL: Yes** **1996, 1997, 1998, 1999, 2000,** **2001, 2002, 2003, 2004, 2005,** **2006** **Models:** Accord, Civic, Civic del Sol, Odyssey, Pilot, Prelude, Ridgeline, S2000 **Engines:** All **Transmissions:** A/T, M/T	**Vehicle Speed Sensor Circuit Performance** Engine running at Cruise speed under road load conditions, and the PCM detected the VSS signal was erratic or too low during the test. Note: The VSS signal should pulse from 0-5v as the vehicle moves. **Possible Causes:** • VSS signal circuit is open • VSS signal circuit is shorted to ground • VSS signal circuit is shorted to VREF or system power (B+) • VSS is damaged or has failed
DTC: P0502 **1T CCM, MIL: Yes** **1996** **Models:** Passport **Engines:** 2.6L VIN E **Transmissions:** A/T, M/T	**Vehicle Speed Sensor Circuit Low Input** Engine started, engine speed from 1800-2500, system voltage from 10-16v, ECT sensor more than 140°F, throttle angle from 10-40%, engine load over 40%, conditions met for 5 seconds, and the PCM did not detect any vehicle speed sensor signals during the CCM test. **Possible Causes:** • VSS signal circuit is open or shorted to ground • VSS power circuit is open (check the Meter 15A fuse) • VSS ground circuit is open between the sensor and ground • VSS is damaged or has failed, or the VSS rotor is cracked • PCM has failed
DTC: P0502 **1T CCM, MIL: Yes** **1996, 1997, 1998, 1999, 2000,** **2001, 2002** **Models:** Passport **Engines:** 3.2L VIN V, 3.2L VIN W **Transmissions:** A/T, M/T	**Vehicle Speed Sensor Circuit Low Input** Engine started, engine speed from 1800-2500, system voltage from 11-16v, ECT sensor more than 140°F, throttle angle from 10-40%, engine load over 50%, and the PCM did not detect any VSS signals for 12.5 seconds over a 15 second period during the CCM test. **Possible Causes:** • VSS signal circuit is open or shorted to ground • VSS power circuit is open (check the Meter 10A fuse) • VSS ground circuit is open between the sensor and ground • VSS is damaged or has failed, or the VSS rotor is cracked • PCM has failed

DTC	Trouble Code Title, Conditions & Possible Causes
DTC: P0505 **1T CCM, MIL: Yes** **1995** **Models:** Accord **Engines:** 2.7L VIN CE6 **Transmissions:** A/T, M/T	**Idle Speed Control System** DTC P1519 not set, engine running at hot idle speed, and the PCM detected the Actual and Target idle speed were too far apart. **Possible Causes:** • IAC valve circuit open, shorted to ground or to power (B+) • IAC valve is damaged or has failed • Fast idle thermo valve is damaged or has failed (some models) • Throttle body is dirty or full of sludge • PCM has failed
DTC: P0505 **1T CCM, MIL: Yes** **1996, 1997, 1998, 1999, 2000,** **2001, 2002, 2003, 2004, 2005,** **2006** **Models:** Accord, Civic, Civic del Sol, Insight, Odyssey, Pilot, Prelude, Ridgeline, S2000 **Engines:** All **Transmissions:** A/T, M/T	**Idle Speed Control System** DTC P1519 not set, engine running at hot idle speed, and the PCM detected the Actual and Target idle speed values were too far apart. **Possible Causes:** • IAC valve circuit open, shorted to ground or to power (B+) • IAC valve is damaged or has failed • Fast idle thermo valve is damaged or has failed (some models) • Throttle body is dirty or full of sludge • PCM has failed
DTC: P0506 **1T CCM, MIL: Yes** **1996, 1997, 1998, 1999, 2000,** **2001, 2002** **Models:** Passport **Engines:** 3.2L VIN V, 3.2L VIN W **Transmissions:** A/T, M/T	**Idle Air Control System Low RPM** DTC P0106, P0107, P0108, P0112, P0113, P0117, P0118, P0121, P0122, P0123, P0125, P0131, P0132, P0133, P0134, P0200-206, P0300, P0301-P306, P0335, P0341, P0342, P0404, P0405, P0440, P0442, P0446, P0452, P0453, P0502, P0507, P0601, P0602, P0705, P1133, P1404 and P1441 not set, engine runtime over 125 seconds, system voltage from 11-16v, ECT sensor more than 122°F, IAT sensor more than -40°F, MAP sensor less than 40 kPa, Purge duty cycle over 10%, BARO sensor over 75 kPa, vehicle speed less than 2 mph with the throttle closed, and the PCM detected the Actual speed was 100-200 rpm below the Desired idle speed for 10 seconds based on the current engine coolant temperature. **Possible Causes:** • High resistance between the IAC 'A' high or low control circuits • Short to ground between the IAC 'B' high or low control circuits • IAC valve is damaged, dirty, sticking or has failed • The throttle plate is carbon fouled (it may need to be cleaned)
DTC: P0506 **2T CCM, MIL: Yes** **1996** **Models:** Passport **Engines:** 2.6L VIN E **Transmissions:** A/T, M/T	**Idle Air Control System Low RPM** DTC P0106, P0107, P0108, P112, P113, P0117, P0118, P0121, P0122, P0123, P0171, P0172, P0300, P0441, P0502, P0562 and P0563 not set, engine running at idle speed with the throttle closed, system voltage from 10-16v, ECT sensor more than 118°F, IAT sensor more than -40°F, MAP sensor less than 40 kPa, Purge duty cycle over 10%, BARO sensor over 75kPa, vehicle speed less than 2 mph, then if the non-intrusive test fails, the PCM runs the Intrusive test, and the PCM detected the Actual idle speed was 100-200 rpm below the Desired idle speed based on the coolant temperature. **Possible Causes:** • High resistance between the IAC 'A' high or low control circuits • Short to ground between the IAC 'B' high or low control circuits • IAC valve is damaged, dirty, sticking or has failed • The throttle plate is carbon fouled (it may need to be cleaned)
DTC: P0506 **1T CCM, MIL: Yes** **2003, 2004, 2005, 2006** **Models:** Element **Engines:** All **Transmissions:** All	**Idle Air Control System Low RPM** DTC P0511 not set, engine runtime over 125 seconds, system voltage from 11-16v, ECT sensor more than 122°F, IAT sensor more than -40°F, MAP sensor less than 40 kPa, Purge duty cycle over 10%, BARO sensor over 75 kPa, vehicle speed less than 2 mph with the throttle closed, and the PCM detected the Actual speed was 100-200 rpm below the Desired idle speed for 10 seconds based on the current engine coolant temperature. **Possible Causes:** • High resistance between the IAC 'A' high or low control circuits • Short to ground between the IAC 'B' high or low control circuits • IAC valve is damaged, dirty, sticking or has failed • The throttle plate is carbon fouled (it may need to be cleaned)

DTC	Trouble Code Title, Conditions & Possible Causes
DTC: P0507 **1T CCM, MIL: Yes** **1996, 1997, 1998, 1999, 2000, 2001, 2002** **Models:** Passport **Engines:** 3.2L VIN V, 3.2L VIN W **Transmissions:** A/T, M/T	**Idle Air Control System High RPM** DTC P0106, P0107, P0108, P0112, P0113, P0117, P0118, P0121, P0122, P0123, P0125, P0131, P0132, P0133, P0134, P0201-206, P0300, P0301-P0306, P0335, P0341, P0342, P0404, P0405, P0440, P0442, P0446, P0452, P0453, P0502, P0507, P0601, P0602, P0705, P1133, P1404 and P1441 not set, engine runtime over 125 seconds, system voltage from 11-16v, ECT sensor more than 122°F, IAT sensor more than -40°F, MAP sensor less than 40 kPa, Purge duty cycle over 10%, BARO sensor over 75 kPa, vehicle speed under 2 mph with the throttle closed, and the PCM detected the Actual speed was 100-200 rpm above the Desired idle speed for 10 seconds based on the current engine coolant temperature. **Possible Causes:** • High resistance between the IAC 'A' high or low control circuits • Short to ground between the IAC 'B' high or low control circuits • IAC valve is damaged, dirty, sticking or has failed • The throttle plate is carbon fouled (it may need to be cleaned)
DTC: P0507 **2T CCM, MIL: Yes** **1996** **Models:** Passport **Engines:** 2.6L VIN E **Transmissions:** A/T, M/T	**Idle Air Control System High RPM** DTC P0106, P0107, P0108, P112, P113, P0117, P0118, P0121, P0122, P0123, P0171, P0172, P0300, P0441, P0502, P0562 and P0563 not set, engine running at idle speed with the throttle closed, system voltage from 10-16v, ECT sensor more than 118°F, IAT sensor more than -40°F, MAP sensor less than 40 kPa, Purge duty cycle over 10%, BARO sensor over 75kPa, vehicle speed less than 2 mph, then if the non-intrusive test fails, the PCM runs the Intrusive test, and the PCM detected the Actual idle speed was 100-200 rpm above the Desired idle speed based on the coolant temperature. **Possible Causes:** • High resistance between the IAC 'A' high or low control circuits • Short to ground between the IAC 'B' high or low control circuits • IAC valve is damaged, dirty, sticking or has failed • The throttle plate is carbon fouled (it may need to be cleaned)
DTC: P0507 **1T CCM, MIL: Yes** **2003, 2004, 2005, 2006** **Models:** Element **Engines:** All **Transmissions:** All	**Idle Air Control System High RPM** DTC P0511 not set, engine runtime over 125 seconds, system voltage from 11-16v, ECT sensor more than 122°F, IAT sensor more than -40°F, MAP sensor less than 40 kPa, Purge duty cycle over 10%, BARO sensor over 75 kPa, vehicle speed under 2 mph with the throttle closed, and the PCM detected the Actual speed was 100-200 rpm above the Desired idle speed for 10 seconds based on the current engine coolant temperature. **Possible Causes:** • High resistance between the IAC 'A' high or low control circuits • Short to ground between the IAC 'B' high or low control circuits • IAC valve is damaged, dirty, sticking or has failed • The throttle plate is carbon fouled (it may need to be cleaned)
DTC: P0511 **1T CCM, MIL: Yes** **2003, 2004, 2005, 2006** **Models:** Element **Engines:** All **Transmissions:** All	**IAC Valve Circuit Malfunction** Engine runtime over 125 seconds, system voltage from 11-16v, ECT sensor more than 122°F, IAT sensor more than -40°F, MAP sensor less than 40 kPa, Purge duty cycle over 10%, BARO sensor over 75 kPa, vehicle speed under 2 mph with the throttle closed, and the PCM is unable to control the idle speed **Possible Causes:** • High resistance between the IAC 'A' high or low control circuits • Short to ground between the IAC 'B' high or low control circuits • IAC valve is damaged, dirty, sticking or has failed • The throttle plate is carbon fouled (it may need to be cleaned)
DTC: P0560 **1T CCM, MIL: Yes** **2000, 2001, 2002, 2003, 2004, 2005, 2006** **Models:** Accord, Civic, CR-V, Odyssey, Pilot, Prelude, Ridgeline **Engines:** All **Transmissions:** A/T, M/T	**PCM Backup Circuit Low Voltage** Key on or engine running, and the PCM detected a low voltage condition on the PCM Backup circuit. Note: This circuit is connected to the Backup/Radio 7.5 amp fuse. **Possible Causes:** • PCM backup circuit is open • PCM backup circuit is shorted to ground • PCM backup circuit has high resistance • PCM has failed
DTC: P0562 **1T CCM, MIL: No** **1996, 1997, 1998, 1999, 2000, 2001, 2002** **Models:** Passport **Engines:** All **Transmissions:** A/T, M/T	**System Voltage Low Input** Engine started, engine runtime over 15 minutes, and the PCM detected the system voltage was less than 11.5v in the CCM test. **Note: For additional help with this code, view the Failure Records.** **Possible Causes:** • Check the drive belt for excessive wear and the proper tension • Check for high resistance at the battery connections or at the starter solenoid connection that connects to PCM power circuit • Check the generator output and the battery condition

DTC	Trouble Code Title, Conditions & Possible Causes
DTC: P0563 **1T CCM, MIL: No** 1996, 1997, 1998, 1999, 2000, 2001, 2002 **Models:** Passport **Engines:** All **Transmissions:** A/T, M/T	**System Voltage High Input** Engine started, engine runtime over 15 minutes, and the PCM detected the system voltage was more than 16.0v in the CCM test. **Note: For additional help with this code, view the Failure Records.** **Possible Causes:** • Check the generator output and the battery condition
DTC: P0563 **1T CCM, MIL: No** 2003, 2004, 2005, 2006 **Models:** Element **Engines:** All **Transmissions:** All	**ECM/PCM Power Source Circuit Unexpected Voltage** Engine off and the PCM is unable to shut down system power **Possible Causes:** • Short to ground between PCM and PGM-FI Relay • Faulty PGM-FI relay
DTC: P0565 **1T CCM, MIL: No** 2000, 2001, 2002 **Models:** Passport **Engines:** 3.2L VIN W **Transmissions:** A/T, M/T	**Cruise Control Main Switch Circuit Malfunction** Engine started, engine running, system voltage from 11-16v, and the PCM detected noise from Cruise Control (C/C) main switch contacts 60 times within one second, or the C/C main switch remained "on" for 15 seconds during the CCM Rationality test. **Possible Causes:** • C/C main switch signal circuit is shorted to ground • C/C main switch signal circuit is shorted to another circuit • C/C main switch is damaged or has failed
DTC: P0566 **1T CCM, MIL: No** 2000, 2001, 2002 **Models:** Passport **Engines:** 3.2L VIN W **Transmissions:** A/T, M/T	**Cruise Control Cancel Switch Circuit Malfunction** Engine started, engine running, system voltage from 11-16v, and the PCM detected noise from Cruise Control (C/C) cancel switch contacts 100 times within 1.6 seconds, or the C/C cancel switch remained "on" for 40 seconds during the CCM Rationality test. **Possible Causes:** • C/C cancel switch signal circuit is shorted to ground • C/C cancel switch signal circuit is shorted to another circuit • C/C cancel switch is damaged or has failed
DTC: P0567 **1T CCM, MIL: No** 2000, 2001, 2002 **Models:** Passport **Engines:** 3.2L VIN W **Transmissions:** A/T, M/T	**Cruise Control Resume Switch Circuit Malfunction** Engine started, engine running, system voltage from 11-16v, and the PCM detected noise from Cruise Control (C/C) resume switch contacts 100 times within 1.6 seconds, or the C/C resume switch remained "on" for 50 seconds during the CCM Rationality test. **Possible Causes:** • C/C resume switch signal circuit is shorted to ground • C/C resume switch signal circuit is shorted to another circuit • C/C resume switch is damaged or has failed
DTC: P0568 **1T CCM, MIL: No** 2000, 2001, 2002 **Models:** Passport **Engines:** 3.2L VIN W **Transmissions:** A/T, M/T	**Cruise Control Set Switch Circuit Malfunction** Engine started, engine running, system voltage from 11-16v, and the PCM detected noise from Cruise Control (C/C) set switch contacts 100 times within 1.6 seconds, or the C/C set switch remained "on" for 120 seconds during the CCM Rationality test. **Possible Causes:** • C/C set switch signal circuit is shorted to ground • C/C set switch signal circuit is shorted to another circuit • C/C set switch is damaged or has failed
DTC: P0571 **1T CCM, MIL: No** 1998, 1999, 2000, 2001, 2002 **Models:** Passport **Engines:** 3.2L VIN W **Transmissions:** A/T, M/T	**Brake Light Switch Circuit Malfunction** DTC P0502 not set, engine started, then driven to a speed of over 12.5 mph, and the PCM detected that two (2) brake switch signals did not agree with the brake switch status during the CCM test. **Possible Causes:** • Brake switch signal circuit is open between switch and the PCM • Brake switch signal circuit is shorted to ground • Brake switch power circuit open between switch and 15A fuse • Check the battery condition (it may have failed)
DTC: P0600 **1T CCM, MIL: No** 2003, 2004, 2005, 2006 **Models:** Element **Engines:** All **Transmissions:** All	**Serial Communication Link Malfunction** Necessary information is communicated through one signal line between the engine control module (ECM)/powertrain control module (PCM), the multiplex control unit, and the gauge assembly. If communication between the ECM/PCM, the multiplex control unit, and the gauge assembly is interrupted for more than the specified time, a malfunction is detected and a DTC is stored. **Possible Causes:** • This DTC is stored when there is a problem in the body electrical system. Check for body electrical DTCs

DTC	Trouble Code Title, Conditions & Possible Causes

DTC: P0601
1T DTC: PCM, MIL: No
1996, 1997, 1998, 1999, 2000, 2001, 2002
Models: Passport
Engines: All
Transmissions: A/T, M/T
	PCM Internal Check Sum Error
	Key on, and the PCM detected a check sum error had occurred during its initial self-test.
	Note: For additional help with this code, view the Failure Records.
	Possible Causes:
	    • The contents of the EEPROM have changed
	    • PCM needs to be replaced and reprogrammed to repair this trouble code

DTC: P0602 **1T DTC: PCM, MIL: No** 1998, 1999, 2000, 2001, 2002 **Models:** Passport **Engines:** 3.2L VIN W **Transmissions:** A/T, M/T	**PCM Programming Error** Key on or engine cranking, and the PCM detected that it was not programmed properly. **Note: A failure record is stored when this trouble code is set.** **Possible Causes:** • PCM must be replaced and then reprogrammed to repair this trouble code
DTC: P0603 **1T DTC: PCM, MIL: Yes** 2003, 2004, 2005, 2006 **Models:** Element **Engines:** All **Transmissions:** All	**ECM/PCM Internal Control Module Keep Alive Memory (KAM) Error** If something is wrong in the engine control module (ECM)/powertrain control module (PCM), and the data read from EEPROM is abnormal, the ECM/PCM detects a malfunction and a DTC is stored **Note: A failure record is stored when this trouble code is set.** **Possible Causes:** • PCM must be replaced and then reprogrammed to repair this trouble code
DTC: P0604 **1T DTC: PCM, MIL: No** 1998, 1999, 2000, 2001, 2002 **Models:** Passport **Engines:** 3.2L VIN W **Transmissions:** A/T, M/T	**PCM Random Access Memory Error** Key on, and the PCM detected inconsistencies between the Main CPU and the Watchdog CPU software calibration **Note: A failure record is stored when this trouble code is set.** **Possible Causes:** • PCM must be replaced and then reprogrammed to repair this trouble code
DTC: P0606 **1T DTC: PCM, MIL: No** 1998, 1999, 2000, 2001, 2002 **Models:** Passport **Engines:** 3.2L VIN W **Transmissions:** A/T, M/T	**PCM Internal Performance Error** Key on, and the PCM detected inconsistencies between the Main CPU and the Watchdog CPU software calibration **Note: A failure record is stored when this trouble code is set.** **Possible Causes:** • PCM must be replaced and then reprogrammed to repair this trouble code
DTC: P0606 **1T DTC: PCM, MIL: Yes** 2003, 2004, 2005, 2006 **Models:** Element **Engines:** All **Transmissions:** All	**PCM Internal Performance Error** If something is wrong in the engine control module (ECM)/powertrain control module (PCM), and the monitor signal from the digital knock system (DKS) CPU is not received for a set period of time, or a signal communication error remains for a set period time, the ECM/PCM detects a malfunction and a DTC is stored **Possible Causes:** • PCM must be replaced and then reprogrammed to repair this trouble code
DTC: P0685 **1T DTC: PCM, MIL: Yes** 2003, 2004, 2005, 2006 **Models:** Element **Engines:** All **Transmissions:** All	**ECM/PCM Power Control Circuit Malfunction** If something is wrong in the engine control module (ECM)/powertrain control module (PCM) that causes the PGM-FI main relay to turn OFF before the ignition switch is turned OFF, the ECM/PCM detects a malfunction and a DTC is stored. **Possible Causes:** • PCM must be replaced and then reprogrammed to repair this trouble code
DTC: P0700 **1T CCM, MIL: Yes** 1996, 1997, 1998, 1999, 2000, 2001, 2002, 2003, 2004, 2005, 2006 **Models:** Accord, Civic, CR-V, Odyssey, Pilot, Prelude, Ridgeline **Engines:** All **Transmissions:** A/T	**Automatic Transaxle** Engine running and the PCM detected an Automatic Transaxle fault. Note: DTC P0700 sets along with several other TCM trouble codes. **Possible Causes:** • Check for other A/T related trouble codes, and then refer to the Possible Causes for these trouble codes for more information.

DTC	Trouble Code Title, Conditions & Possible Causes
DTC: P0705 **1T CCM, MIL: No** **1998, 1999, 2000, 2001, 2002** **Models:** Passport **Engines:** 3.2L VIN W **Transmissions:** A/T	**Transmission Range Switch Illegal Position Malfunction** Engine started, then driven to a speed of over 8 mph, and the PCM detected "illegal" TR Range or Mode switch signals for 5 seconds. **Note: For additional help with this code, view the Failure Records.** **Possible Causes:** • TR range switch signal is open • TR range switch signal shorted to another switch position signal • TR range switch is damaged or has failed • PCM has failed
DTC: P0705 **1T CCM, MIL: No** **2003, 2004, 2005, 2006** **Models:** Element **Engines:** All **Transmissions:** All	**Short in Transmission Range Switch Circuit** PCM detected multiple TR Range or Mode switch signals for 5 seconds. **Possible Causes:** • TR range switch signal shorted to another switch position signal • TR range switch is damaged or has failed • PCM has failed
DTC: P0706 **1T CCM, MIL: No** **1998, 1999, 2000, 2001, 2002** **Models:** Passport **Engines:** 3.2L VIN W **Transmissions:** A/T	**Transmission Range Switch Circuit Performance** DTC P0122, P0123, P0722 and P0723 not set, engine started, then driven with the output speed over 3200 rpm, and the PCM detected the TR Switch indicated Reverse position, or with the Output speed under 3000 rpm and the throttle angle over 20%, it detected the TR switch indicated Park or Neutral position for 4 seconds in the test. **Note: For additional help with this code, view the Failure Records.** **Possible Causes:** • TR switch signal is open • TR switch signal shorted to another switch position signal • TR switch is damaged or has failed • PCM has failed
DTC: P0706 **1T CCM, MIL: No** **2003, 2004, 2005, 2006** **Models:** Element **Engines:** All **Transmissions:** All	**Open in Transmission Range Switch Circuit** PCM detected no TR Range or Mode switch signals while vehicle driving **Possible Causes:** • TR range switch signal is open • TR range switch is damaged or has failed • PCM has failed
DTC: P0711 **1T CCM, MIL: No** **1998, 1999, 2000, 2001, 2002** **Models:** Passport **Engines:** 3.2L VIN W **Transmissions:** A/T	**Transmission Fluid Temperature Sensor Performance** DTC P0722, P0723 and P1870 not set, engine started, system voltage from 11-16v, TFT sensor from -40°F to 69.8°F at startup, ECT sensor more than 150°F and has changed more than 90°F since startup, vehicle speed over 5 mph with the TCC slip speed over 120 rpm for 410 seconds, and the PCM detected the TFT sensor changed less than 2 counts since startup, or that its delta change was over 36°F at least 14 times during a 7 second period. **Possible Causes:** • TFT signal or ground circuit has a high resistance condition • TFT sensor is out-of-calibration (it may be skewed) • TFT sensor is damaged or has failed • PCM has failed
DTC: P0711 **1T CCM, MIL: No** **2003, 2004, 2005, 2006** **Models:** Element **Engines:** All **Transmissions:** All	**Problem in ATF Temperature Sensor Circuit** When the ATF temperature is low, the sensor resistance increases and the PCM detects a high signal voltage. As the ATF temperature rises, the sensor resistance decreases and the PCM detects a low signal voltage. If the ATF temperature sensor signal does not change, the PCM detects a malfunction and a DTC is stored. **Possible Causes:** • TFT signal or ground circuit has a high resistance condition • TFT sensor is out-of-calibration (it may be skewed) • TFT sensor is damaged or has failed • PCM has failed
DTC: P0712 **1T CCM, MIL: No** **1998, 1999, 2000, 2001, 2002** **Models:** Passport **Engines:** 3.2L VIN W **Transmissions:** A/T	**Transmission Fluid Temperature Sensor Low Input** Engine started, system voltage from 11-16v, and the PCM detected the TFT sensor was less than 0.40v for 20 seconds during the test. **Note: For additional help with this code, view the Failure Records.** **Possible Causes:** • TFT sensor signal circuit is shorted to sensor ground • TFT sensor signal circuit is shorted to chassis ground • TFT sensor is damaged (it may be shorted internally) • PCM has failed

DTC	Trouble Code Title, Conditions & Possible Causes
DTC: P0712 **1T CCM, MIL: No** **2003, 2004, 2005, 2006** **Models:** Element **Engines:** All **Transmissions:** All	**Transmission Fluid Temperature Sensor Low Input** Engine started, system voltage from 11-16v, and the PCM detected the TFT sensor was less than 0.40v for 20 seconds during the test. **Note: For additional help with this code, view the Failure Records.** **Possible Causes:** • TFT sensor signal circuit is shorted to sensor ground • TFT sensor signal circuit is shorted to chassis ground • TFT sensor is damaged (it may be shorted internally) • PCM has failed
DTC: P0713 **1T CCM, MIL: No** **2003, 2004, 2005, 2006** **Models:** Element **Engines:** All **Transmissions:** All	**Transmission Fluid Temperature Sensor High Input** Engine started, system voltage from 11-16v, and the PCM detected the TFT sensor was more than 4.86v for 20 seconds during the test. **Note: For additional help with this code, view the Failure Records.** **Possible Causes:** • TFT sensor signal circuit is open between the sensor and PCM • TFT sensor ground circuit is open between sensor and ground • TFT sensor signal circuit is shorted to VREF or system power • TFT sensor is damaged (it may be open internally) • PCM has failed
DTC: P0713 **1T CCM, MIL: No** **1998, 1999, 2000, 2001, 2002** **Models:** Passport **Engines:** 3.2L VIN W **Transmissions:** A/T	**Transmission Fluid Temperature Sensor High Input** Engine started, system voltage from 11-16v, and the PCM detected the TFT sensor was more than 4.86v for 20 seconds during the test. **Note: For additional help with this code, view the Failure Records.** **Possible Causes:** • TFT sensor signal circuit is open between the sensor and PCM • TFT sensor ground circuit is open between sensor and ground • TFT sensor signal circuit is shorted to VREF or system power • TFT sensor is damaged (it may be open internally) • PCM has failed
DTC: P0715 **1T CCM, MIL: Yes** **1995** **Models:** Accord **Engines:** 2.7L VIN CE6 **Transmissions:** A/T	**TCM A/T Mainshaft Speed Sensor Circuit Malfunction** Engine running with VSS inputs received, and the PCM detected an unexpected voltage condition on the Mainshaft speed sensor circuit. **Possible Causes:** • Mainshaft speed sensor circuit is open or shorted to ground • Mainshaft speed sensor circuit is shorted to VREF or power • Mainshaft speed sensor is damaged or has failed • PCM has failed
DTC: P0715 **1T CCM, MIL: Yes** **1996, 1997, 1998, 1999, 2000, 2001, 2002, 2003, 2004, 2005, 2006** **Models:** Accord, Civic, CR-V, Odyssey, Pilot, Prelude, Ridgeline **Engines:** All **Transmissions:** A/T	**TCM A/T Mainshaft Speed Sensor Circuit Malfunction** Engine running with VSS inputs received, and the PCM detected an unexpected voltage condition on the Mainshaft speed sensor circuit. **Possible Causes:** • Mainshaft speed sensor circuit is open or shorted to ground • Mainshaft speed sensor circuit is shorted to VREF or power • Mainshaft speed sensor is damaged or has failed • PCM has failed
DTC: P0716 **1T CCM, MIL: Yes** **2003, 2004, 2005, 2006** **Models:** Element **Engines:** All **Transmissions:** All	**Problem in Input Shaft (Mainshaft) Speed Sensor Circuit** Engine running with VSS inputs received, and the PCM detected no inputs from the Mainshaft speed sensor circuit. **Possible Causes:** • Mainshaft speed sensor circuit is open or shorted to ground • Mainshaft speed sensor circuit is shorted to VREF or power • Mainshaft speed sensor is damaged or has failed • PCM has failed
DTC: P0717 **1T CCM, MIL: Yes** **2003, 2004, 2005, 2006** **Models:** Element **Engines:** All **Transmissions:** All	**Problem in Input Shaft (Mainshaft) Speed Sensor Circuit** Engine running with VSS inputs received, and the PCM detected no inputs from the Mainshaft speed sensor circuit. **Possible Causes:** • Mainshaft speed sensor circuit is open or shorted to ground • Mainshaft speed sensor circuit is shorted to VREF or power • Mainshaft speed sensor is damaged or has failed • PCM has failed

DTC	Trouble Code Title, Conditions & Possible Causes
DTC: P0718 **1T CCM, MIL: Yes** **2003, 2004, 2005, 2006** **Models:** Element **Engines:** All **Transmissions:** All	**Input Shaft (Mainshaft) Speed Sensor Intermittent Failure** Engine running with VSS inputs received, and the PCM detected erratic inputs from the Mainshaft speed sensor circuit. **Possible Causes:** • Mainshaft speed sensor circuit is open or shorted to ground • Mainshaft speed sensor circuit is shorted to VREF or power • Mainshaft speed sensor is damaged or has failed • PCM has failed
DTC: P0719 **1T CCM, MIL: No** **1998, 1999, 2000, 2001, 2002** **Models:** Passport **Engines:** 3.2L VIN W **Transmissions:** A/T	**TCC Brake Switch Circuit High (Stuck On)** DTC P0722 and P0723 not set, engine started, then driven to a speed over 5 mph, then driven to over 20 mph for 5 seconds, then back to a speed of 5-20 mph for 4 seconds, and the PCM detected an "open" Brake switch condition for 15 minutes with no change in its status, condition occurred at least 7 times during the CCM test. **Possible Causes:** • TCC brake switch signal circuit is open or shorted to ground • TCC brake switch is damaged (it may be open internally) • PCM has failed
DTC: P0720 **1T CCM, MIL: Yes** **1995** **Models:** Accord **Engines:** 2.7L VIN CE6 **Transmissions:** A/T	**TCM A/T Countershaft Speed Sensor Circuit Malfunction** Engine running with VSS inputs received, and the PCM detected an unexpected voltage on the Countershaft Speed Sensor circuit. **Possible Causes:** • Countershaft speed sensor circuit is open or shorted to ground • Countershaft speed sensor circuit is shorted to VREF or power • Countershaft speed sensor is damaged or has failed • PCM has failed
DTC: P0720 **1T CCM, MIL: Yes** **1996, 1997, 1998, 1999, 2000, 2001, 2002, 2003, 2004, 2005, 2006** **Models:** Accord, Civic, CR-V, Odyssey, Pilot, Prelude, Ridgeline **Engines:** All **Transmissions:** A/T	**TCM A/T Countershaft Speed Sensor Circuit Malfunction** Engine running with VSS inputs received, and the PCM detected an unexpected voltage on the Countershaft Speed Sensor circuit. **Possible Causes:** • Countershaft speed sensor circuit is open or shorted to ground • Countershaft speed sensor circuit is shorted to VREF or power • Countershaft speed sensor is damaged or has failed • PCM has failed
DTC: P0721 **2T CCM, MIL: Yes** **2003, 2004, 2005, 2006** **Models:** Element **Engines:** All **Transmissions:** All	**Output Speed Sensor Low Input** TR switch indicating other than Park or Neutral position, throttle angle over 10%, engine vacuum from 0-70 kPa, engine speed from 3000-5000 rpm, and the PCM detected the Output Speed Sensor (OSS) signal indicated zero (0) rpm, condition met for 5 seconds. **Possible Causes:** • OSS (+) signal circuit is open or shorted to ground • OSS (-) signal circuit is open or shorted to ground • OSS is damaged or has failed • PCM has failed
DTC: P0722 **2T CCM, MIL: Yes** **1998, 1999, 2000, 2001, 2002** **Models:** Passport **Engines:** 3.2L VIN W **Transmissions:** A/T	**Output Speed Sensor Low Input** DTC P0106, P0107, P0108, P0122, P0123, P1106 and P1107 not set, TR switch indicating other than Park or Neutral position, throttle angle over 10%, engine vacuum from 0-70 kPa, engine speed from 3000-5000 rpm, and the PCM detected the Output Speed Sensor (OSS) signal indicated zero (0) rpm, condition met for 5 seconds. **Possible Causes:** • OSS (+) signal circuit is open or shorted to ground • OSS (-) signal circuit is open or shorted to ground • OSS is damaged or has failed • PCM has failed
DTC: P0722 **2T CCM, MIL: Yes** **2003, 2004, 2005, 2006** **Models:** Element **Engines:** All **Transmissions:** All	**Output Speed Sensor Low Input** TR switch indicating other than Park or Neutral position, throttle angle over 10%, engine vacuum from 0-70 kPa, engine speed from 3000-5000 rpm, and the PCM detected the Output Speed Sensor (OSS) signal indicated zero (0) rpm, condition met for 5 seconds. **Possible Causes:** • OSS (+) signal circuit is open or shorted to ground • OSS (-) signal circuit is open or shorted to ground • OSS is damaged or has failed • PCM has failed

DTC	Trouble Code Title, Conditions & Possible Causes

DTC: P0723
2T CCM, MIL: Yes
1998, 1999, 2000, 2001, 2002
Models: Passport
Engines: 3.2L VIN W
Transmissions: A/T
	Output Speed Sensor Signal Malfunction (Intermittent)
	Engine started, engine running, TR switch indicating not in Park or Neutral, OSS signal over 1000 rpm for 2 seconds, then OSS signal over 512 rpm for 2 seconds, engine vacuum less than 20 kPa, then OSS signal over 1380 for 1 second, then the NORAW-NOLAST was less than 200 rpm for 2-6 seconds, and the PCM determined the OSS signal did not vary during the CCM Rationality test.
	Note: NORAW indicates the latest OSS raw data and NOLAST indicated the filtered previous data from the OSS.
	Possible Causes:
	    • OSS (+) signal circuit is open or shorted to ground (intermittent)
	    • OSS (-) signal circuit is open or shorted to ground (intermittent)
	    • OSS is damaged or has failed (an intermittent fault)
	    • PCM has failed

DTC: P0723
2T CCM, MIL: Yes
2003, 2004, 2005, 2006
Models: Element
Engines: All
Transmissions: All
	Output Speed Sensor Signal Malfunction (Intermittent)
	Engine started, engine running, TR switch indicating not in Park or Neutral, OSS signal over 1000 rpm for 2 seconds, then OSS signal over 512 rpm for 2 seconds, engine vacuum less than 20 kPa, then OSS signal over 1380 for 1 second, then the NORAW-NOLAST was less than 200 rpm for 2-6 seconds, and the PCM determined the OSS signal did not vary during the CCM Rationality test.
	Possible Causes:
	    • OSS (+) signal circuit is open or shorted to ground (intermittent)
	    • OSS (-) signal circuit is open or shorted to ground (intermittent)
	    • OSS is damaged or has failed (an intermittent fault)
	    • PCM has failed

DTC: P0724
1T CCM, MIL: No
1998, 1999, 2000, 2001, 2002
Models: Passport
Engines: 3.2L VIN W
Transmissions: A/T
	TCC Brake Switch Circuit High (Stuck Off)
	DTC P0722 and P0723 not set, engine started, then driven to a speed of over 5 mph, then driven to over 20 mph for 5 seconds, then back to a speed of 5-20 mph for 4 seconds, and the PCM detected a "closed" Brake switch condition for 15 minutes without it changing its status, conditions occurred at least 7 times during the CCM test.
	Possible Causes:
	    • TCC brake switch signal circuit is shorted to system power (B+)
	    • TCC brake switch out of adjustment or mechanically closed
	    • TCC brake switch is damaged (it may be shorted internally)
	    • PCM has failed

DTC: P0725 **1T CCM, MIL: No** **1995** **Models:** Accord **Engines:** 2.7L VIN CE6 **Transmissions:** A/T	**Automatic Transaxle Malfunction** Engine running and the PCM detected an Automatic Transaxle fault. Note: This trouble code sets along with several other Automatic Transaxle related trouble codes. **Possible Causes:** • Check for other A/T related trouble codes, and then refer to the Possible Causes for these trouble codes for more information.
DTC: P0725 **1T CCM, MIL: Yes** **1996, 1997, 1998, 1999, 2000, 2001, 2002, 2003, 2004, 2005, 2006** **Models:** Accord, Civic, CR-V, Odyssey, Pilot, Prelude, Ridgeline **Engines:** All **Transmissions:** A/T	**Automatic Transaxle** Engine running and the PCM detected an Automatic Transaxle fault. Note: This trouble code sets along with several other Automatic Transaxle related trouble codes. **Possible Causes:** • Check for other A/T related trouble codes, and then refer to the Possible Causes for these trouble codes for more information.
DTC: P0730 **1T CCM, MIL: Yes** **1995** **Models:** Accord **Engines:** 2.7L VIN CE6 **Transmissions:** A/T	**TCM A/T Shift Control System** No other A/T trouble codes set, engine running at cruise speed with VSS inputs received, and the PCM detected the lockup clutch did not lock or unlock correctly. **Possible Causes:** • Refer to the repair instructions in a transmission repair manual or the information in other electronic media to repair this code.
DTC: P0730 **1T CCM, MIL: Yes** **1996, 1997, 1998, 1999, 2000, 2001, 2002, 2003, 2004, 2005, 2006** **Models:** Accord, Civic, CR-V, Odyssey, Pilot, Prelude, Ridgeline **Engines:** All **Transmissions:** A/T	**TCM A/T Shift Control System** No other A/T trouble codes set, engine running at cruise speed with VSS inputs received, and the PCM detected the lockup clutch did not lock or unlock correctly. **Possible Causes:** • Refer to the repair instructions in a transmission repair manual or the information in other electronic media to repair this code.

DTC	Trouble Code Title, Conditions & Possible Causes
DTC: P0730 **1T CCM, MIL:** No **1998, 1999** **Models:** Passport **Engines:** 3.2L VIN W **Transmissions:** A/T	**Transmission Incorrect Gear Ratio** DTC P0722 and P0723 not set, gear selector not in Park, Neutral or Reverse position, engine speed more than 3500 rpm, 3 seconds have passed since an Upshift event, and the PCM detected a slip value over 753 rpm in 1st gear, a slip value over 713 rpm in 2nd gear, a slip value over 694 rpm in 3rd gear, or a slip value over 685 rpm in 4th gear, condition met for 5.5 seconds during the CCM test. **Possible Causes:** • OSS signal is open or shorted to ground (an intermittent fault) • OSS is out-of-calibration (i.e., the tire size or rear axle ratio) • OSS is damaged or has failed • PCM has failed
DTC: P0730 **1T CCM, MIL:** No **2000, 2001, 2002** **Models:** Passport **Engines:** 3.2L VIN W **Transmissions:** A/T	**Transmission Gear Error Without Input Speed** DTC P0705, P0706, P0722 and P0723 not set, gear selector not in Park, Neutral or Reverse position, engine speed more than 3500 rpm, 3 seconds have passed since an Upshift event, and the PCM detected a slip value over 720 rpm in 1st gear, a slip value over 680 rpm in 2nd gear, a slip value over 660 rpm in 3rd gear, or a slip value over 650 rpm in 4th gear for 5.5 seconds during the CCM test. **Possible Causes:** • A/T fluid contaminated, tool or high (causing slipping condition) • OSS signal is open or shorted to ground (an intermittent fault) • OSS is damaged or has failed • PCM has failed
DTC: P0731 **1T CCM, MIL:** No **2003, 2004, 2005, 2006** **Models:** Element **Engines:** All **Transmissions:** All	**Problem in 1st Clutch and 1st Clutch Hydraulic Circuit** With 1st gear selected, the powertrain control module (PCM) computes the ratio of the mainshaft speed to the countershaft speed. When the ratio is not the 1st gear ratio, it is detected as a malfunction of the hydraulic circuit or the 1st clutch, and a DTC is stored. **Possible Causes:** • A/T fluid contaminated, tool or high (causing slipping condition) • OSS signal is open or shorted to ground (an intermittent fault) • OSS is damaged or has failed • PCM has failed
DTC: P0732 **1T CCM, MIL:** No **2003, 2004, 2005, 2006** **Models:** Element **Engines:** All **Transmissions:** All	**Problem in 2nd Clutch and 2nd Clutch Hydraulic Circuit** With 2nd gear selected, the powertrain control module (PCM) computes the ratio of the mainshaft speed to the countershaft speed. When the ratio is not the 2nd gear ratio, it is detected as a malfunction of the hydraulic circuit or the 2nd clutch, and a DTC is stored. **Possible Causes:** • A/T fluid contaminated, tool or high (causing slipping condition) • OSS signal is open or shorted to ground (an intermittent fault) • OSS is damaged or has failed • PCM has failed
DTC: P0733 **1T CCM, MIL:** No **2003, 2004, 2005, 2006** **Models:** Element **Engines:** All **Transmissions:** All	**Problem in 3rd Clutch and 3rd Clutch Hydraulic Circuit** With 3rd gear selected, the powertrain control module (PCM) computes the ratio of the mainshaft speed to the countershaft speed. When the ratio is not the 3rd gear ratio, it is detected as a malfunction of the hydraulic circuit or the 3rd clutch, and a DTC is stored. **Possible Causes:** • A/T fluid contaminated, tool or high (causing slipping condition) • OSS signal is open or shorted to ground (an intermittent fault) • OSS is damaged or has failed • PCM has failed
DTC: P0734 **1T CCM, MIL:** No **2003, 2004, 2005, 2006** **Models:** Element **Engines:** All **Transmissions:** All	**Problem in 4th Clutch and 4th Clutch Hydraulic Circuit** With 4th gear selected, the powertrain control module (PCM) computes the ratio of the mainshaft speed to the countershaft speed. When the ratio is not the 4th gear ratio, it is detected as a malfunction of the hydraulic circuit or the 4th clutch, and a DTC is stored. **Possible Causes:** • A/T fluid contaminated, tool or high (causing slipping condition) • OSS signal is open or shorted to ground (an intermittent fault) • OSS is damaged or has failed • PCM has failed

DTC	Trouble Code Title, Conditions & Possible Causes
DTC: P0741 **1T CCM, MIL: No** **2003, 2004, 2005, 2006** **Models:** Element **Engines:** All **Transmissions:** All	**Torque Converter Clutch Hydraulic Circuit Stuck OFF** DTCs P0116, P0117, P0118, P0122, P0123, P0705, P0706, P0716, P0717, P0718, P0721, P0722, P0723, P0747, P0752, P0761, P0771, P0776, P0777, P0780, P0796, P0797, P0962, P0963, P0966, P0967, P0970, P0971, P0973, P0974, P0976, P0977, P0979, P0980, P0985, P0986, P1121, P1122, P1731, P1732, P1735, P1736 not set and the ratio of the engine revolution to the transmission input pulse does not reach about 100% for at least 20 seconds. If the ratio of engine speed and mainshaft speed is not about 1:1 while the PCM is issuing the command to turn shift solenoid valve E and A/T clutch pressure control solenoid valve A ON, the PCM detects a faulty lock-up control system and stores a DTC. **Possible Causes:** • Pressure control solenoid control circuit is shorted to ground • Pressure control solenoid valve E is damaged or has failed • PCM has failed
DTC: P0747 **1T CCM, MIL: Yes** **2003, 2004, 2005, 2006** **Models:** Element **Engines:** All **Transmissions:** All	**A/T Clutch Pressure Control Solenoid Valve A Stuck ON** DTCs P0122, P0123, P0705, P0706, P0716, P0717, P0718, P0721, P0722, P0723, P0752, P0761, P0771, P0776, P0777, P0780, P0796, P0797, P0962, P0963, P0966, P0967, P0970, P0971, P0973, P0974, P0976, P0977, P0979, P0980, P0985, P0986, P1121, P1122, P1731, P1732, P1735, P1736 not set. The powertrain control module (PCM) computes the actual ratio of mainshaft and countershaft revolutions. If a difference between the actual ratio and the commanded gear occurs when shifting to each gear position, a malfunction in A/T clutch pressure control solenoid valve A or the hydraulic system is detected and a DTC is stored. **Possible Causes:** • Pressure control solenoid control circuit is shorted to ground • Pressure control solenoid valve A is damaged or has failed • PCM has failed
DTC: P0748 **1T CCM, MIL: No** **1998, 1999, 2000, 2001, 2002** **Models:** Passport **Engines:** 3.2L VIN W **Transmissions:** A/T	**A/T Pressure Control Solenoid Circuit Malfunction** Engine started, engine running, and the PCM detected the difference between the Actual and Commanded Pressure Control solenoid (PCS) current level was more than 200 mA during the CCM test. **Note: For additional help with this code, view the Failure Records.** **Possible Causes:** • Pressure control solenoid control circuit is shorted to ground • Pressure control solenoid is damaged or has failed • PCM has failed
DTC: P0751 **2T CCM, MIL: Yes** **1998, 1999, 2000, 2001, 2002** **Models:** Passport **Engines:** 3.2L VIN W **Transmissions:** A/T	**Shift Solenoid 'A' Performance Without Input Speed** DTC P0122, P0123, P0722, P0723, P0742, P0753, P0758 and P1860 not set, vehicle driven in D4 Gear at over 6.25 mph, TFT sensor signal at 68-257°F, then during a 1-2 Shift, TP angle at 10-60% (± 3%), VSS at 11-31 mph, the PCM detected the engine speed in 2nd Gear was 100 rpm more than it was in 1st Gear (1); or during a 2-3 Shift, TP angle at 13-60% (± 5%), VSS at 20-45 mph, the engine speed in 3rd Gear was 64 rpm less than it was in 2nd Gear (2); or during a 3-4 Shift, TP at 7-60% (± 5%), VSS at 25-87 mph, the engine speed in 4th gear was 60 rpm more than it was in 3rd Gear (3); or while in 4th Gear, TP angle at 13-60% (± 5%), speed ratio at 0.85 to 1.2, the TCC slip speed was 100-2000 rpm for 3 seconds (4); or while in 4th Gear with TCC "on", speed ratio at 0.5-0.85, the TCC slip speed was -50 to +500 for 3 seconds (5). **Note: This code is set if the conditions in (1), (2), (3) or (4) are met, or if the conditions in (1), (2), (3) or (5) are met twice in a row.** **Possible Causes:** • Shift solenoid 'A' is damaged or has failed mechanically (on) • Other internal transmission concerns can cause this problem
DTC: P0752 **2T CCM, MIL: Yes** **1998, 1999, 2000, 2001, 2002** **Models:** Passport **Engines:** 3.2L VIN W **Transmissions:** A/T	**Shift Solenoid 'A' Performance (Stuck Off)** DTC P0705, P0706, P0722, P0723, P0742, P0753, P0758, P1860 and P1870 not set, engine started, then driven in D4 with the output speed over 375 rpm, TFT sensor signal from 68-266°F, and with 1st Gear commanded "on" for 1 second, the PCM detected: 40 < engine torque < 400 Nm, speed ratio > 0.3, throttle angle > 10%, 800 < TCC slip < 8000 rpm, transmission output speed > 375 rpm, 0.62 < modeled ratio < 2.4 for 0.687 seconds; or with 4th Gear commanded "on" for 1 second, the PCM detected: 40 < engine torque < 400 Nm, throttle angle > 10%, 800 < TCC slip < 8000 rpm, speed ratio > 0.6, 0.92 < modeled ratio < 1.5 for 7 seconds. **Note: This fault must occur twice in one trip to set this trouble code.** **Possible Causes:** • Shift solenoid 'A' is damaged or has failed mechanically (off) • Other internal transmission concerns can cause this fault

DTC	Trouble Code Title, Conditions & Possible Causes
DTC: P0752 **2T CCM, MIL: Yes** **2003, 2004, 2005, 2006** **Models:** Element **Engines:** All **Transmissions:** All	**Shift Solenoid Valve A Stuck ON** DTC P0122, P0123, P0705, P0706, P0716, P0717, P0718, P0721, P0722, P0723, P0747, P0761, P0771, P0776, P0777, P0780, P0796, P0797, P0962, P0963, P0966, P0967, P0970, P0971, P0973, P0974, P0976, P0977, P0979, P0980, P0985, P0986, P1121, P1122, P1731, P1732, P1735, P1736 not set. The PCM monitors the mainshaft speed and the countershaft speed at the gear change determined by the shift schedule. When an improper gear ratio is output compared to the predetermined gear change mode, a shift solenoid valve A ON failure is detected and a DTC is stored. **Possible Causes:** • Shift solenoid 'A' is damaged or has failed mechanically • Other internal transmission concerns can cause this fault
DTC: P0753 **1T CCM, MIL: Yes** **1995** **Models:** Accord **Engines:** 2.7L VIN CE6 **Transmissions:** A/T	**TCM A/T Lockup Solenoid 'A' Circuit Malfunction** Engine running with VSS inputs, and the PCM detected an unexpected voltage condition on the Solenoid Valve 'A' circuit. **Possible Causes:** • TCM Solenoid 'A' control circuit is open or shorted to ground • TCM Solenoid 'A' control circuit is shorted to system power • TCM Solenoid 'A' is damaged or has failed • TCM or PCM has failed
DTC: P0753 **1T CCM, MIL: Yes** **1996, 1997, 1998, 1999, 2000, 2001, 2002, 2003, 2004, 2005, 2006** **Models:** Accord, Civic, CR-V, Odyssey, Pilot, Prelude, Ridgeline **Engines:** All **Transmissions:** A/T	**TCM A/T Lockup Solenoid 'A' Circuit Malfunction** Engine running with VSS inputs, and the PCM detected an unexpected voltage condition on the Solenoid Valve 'A' circuit. **Possible Causes:** • TCM Solenoid 'A' control circuit is open or shorted to ground • TCM Solenoid 'A' control circuit is shorted to system power • TCM Solenoid 'A' is damaged or has failed • TCM or PCM has failed
DTC: P0753 **2T CCM, MIL: Yes** **1998, 1999, 2000, 2001, 2002** **Models:** Passport **Engines:** 3.2L VIN W **Transmissions:** A/T	**Shift Solenoid 'A' Circuit Malfunction** Engine started, engine running, then with Shift Solenoid 'A' (SSA) commanded "on", and PCM detected the solenoid control signal was 12v, or with SSA commanded "off", the solenoid control signal was near 0v, either condition met for 0.84 to 1.0 seconds during the test. **Possible Causes:** • SSA control circuit is open or shorted to ground • SSA control circuit is shorted to system power (B+) • SSA power circuit (from the PCM) is open • SSA is damaged or has failed • PCM has failed
DTC: P0756 **2T CCM, MIL: Yes** **1998, 1999, 2000, 2001, 2002** **Models:** Passport **Engines:** 3.2L VIN W **Transmissions:** A/T	**Shift Solenoid 'B' Performance Without Input Speed** DTC P0122, P0123, P705, P706, P0722, P0723, P0742, P0753, P0758, P1106, P1107 and P1860 not set, vehicle driven in D4 Gear to over 6.25 mph at under 8000 rpm, MAP at 0-70 kPa, TFT signal at 68-257°F, the TCC "off", TP angle over 4%, 1st Gear "on", speed ratio at 0.5-2.65, TSS signal at 320-2000 rpm, the TCC slip speed was -200 to -4000 rpm for 1.8 seconds (1), or during a 2-3 Shift, TP angle at 10-60% (± 5%), VSS at 20-45 mph, the 3rd Gear speed was 64 rpm less than it was in 2nd gear (2), or during a 3-4 shift, TP angle at 7-60% (± 5%), VSS at 25-87 mph, the 4th Gear speed was 60 rpm less than it was in 3rd Gear (3); or in 4th Gear, TP angle at 13-60, speed ratio at 0.5-1.20, the TCC slip speed was 100-2000 rpm (4), or in 4th Gear, TP angle at 13-60, speed ratio at 0.5-0.85, the TCC slip speed was -50 to -500 rpm (5). **Note: This code is set if the conditions in (1), (3) or (4) are met twice (stuck on), or if the conditions in (1) and (3) are met twice (stuck off).** **Possible Causes:** • Shift solenoid 'B' is damaged or has failed mechanically (on) • Other internal transmission concerns can cause this fault
DTC: P0757 **2T CCM, MIL: Yes** **1998, 1999, 2000, 2001, 2002** **Models:** Passport **Engines:** 3.2L VIN W **Transmissions:** A/T	**Shift Solenoid 'B' Performance (Stuck On)** DTC P0705, P0706, P0722, P0723, P0742, P0753, P0758, P1860 and P1870 not set, engine started, then driven in D4 with the output speed over 375 rpm, TFT sensor signal from 68-266°F, and with 1st Gear commanded "on" for 1 second, the PCM detected: 40 < engine torque < 400 Nm, speed ratio > 0.6, transmission output speed > 375 rpm, throttle angle > 10%, 1.44 < modeled ratio < 2.4.49 for 41 seconds; or with 4th Gear commanded "on" for 1 second, the PCM detected: 15 < engine torque < 400 Nm, throttle angle > 10%, -8000 < TCC slip < 8000 rpm, speed ratio > 0.6, 2.75 < modeled ratio < 3.2 for 2 seconds during the CCM Rationality test. **Note: This fault must occur twice in one trip to set this trouble code.** **Possible Causes:** • Shift solenoid 'B' is damaged or has failed mechanically (off) • Other internal transmission concerns can cause this fault

DTC	Trouble Code Title, Conditions & Possible Causes
DTC: P0758 **1T CCM, MIL: Yes** **1995** **Models:** Accord **Engines:** 2.7L VIN CE6 **Transmissions:** A/T	**TCM A/T Lockup Solenoid 'B' Circuit Malfunction** Engine running with VSS inputs, and the PCM detected an unexpected voltage condition on the Solenoid Valve 'B' circuit. **Possible Causes:** • TCM Solenoid 'B' control circuit is open or shorted to ground • TCM Solenoid 'B' control circuit is shorted to system power • TCM Solenoid 'B' is damaged or has failed • TCM or PCM has failed
DTC: P0758 **1T CCM, MIL: Yes** **1996, 1997, 1998, 1999, 2000, 2001, 2002, 2003, 2004, 2005, 2006** **Models:** Accord, Civic, CR-V, Odyssey, Pilot, Prelude, Ridgeline **Engines:** All **Transmissions:** A/T	**TCM A/T Lockup Solenoid 'B' Circuit Malfunction** Engine running with VSS inputs, and the PCM detected an unexpected voltage condition on the Solenoid Valve 'B' circuit. **Possible Causes:** • TCM Solenoid 'B' control circuit is open or shorted to ground • TCM Solenoid 'B' control circuit is shorted to system power • TCM Solenoid 'B' is damaged or has failed • TCM or PCM has failed
DTC: P0758 **1T CCM, MIL: Yes** **1998, 1999, 2000, 2001, 2002** **Models:** Passport **Engines:** 3.2L VIN W **Transmissions:** A/T	**Shift Solenoid 'B' Circuit Malfunction** Engine started, engine running, then with Shift Solenoid 'B' (SSB) commanded "on", and PCM detected the solenoid control signal was 12v, or with SSB commanded "off", the solenoid control signal was near 0v, either condition met for 0.84 to 1.0 seconds during the test. **Possible Causes:** • SSB control circuit is open or shorted to ground • SSB control circuit is shorted to system power (B+) • SSB power circuit (from the PCM) is open • SSB is damaged or has failed • PCM has failed
DTC: P0761 **1T CCM, MIL: Yes** **2003, 2004, 2005, 2006** **Models:** Element **Engines:** All **Transmissions:** All	**Shift Solenoid Valve C Stuck OFF** DTCs P0122, P0123, P0705, P0706, P0716, P0717, P0718, P0721, P0722, P0723, P0747, P0752, P0771, P0776, P0777, P0780, P0796, P0797, P0962, P0963, P0966, P0967, P0970, P0971, P0973, P0974, P0976, P0977, P0979, P0980, P0985, P0986, P1121, P1122, P1731, P1732, P1735, P1736 not set. The PCM monitors the mainshaft speed and the countershaft speed at the gear change determined by the shift schedule. When an improper gear ratio is output compared to the predetermined gear change mode, a shift solenoid valve C OFF failure is detected and a DTC is stored.. **Possible Causes:** • Pressure control solenoid control circuit is shorted to ground • Pressure control solenoid valve C is damaged or has failed • PCM has failed
DTC: P0763 **1T CCM, MIL: Yes** **1996, 1997, 1998, 1999, 2000, 2001, 2002, 2003, 2004, 2005, 2006** **Models:** Accord, Civic, CR-V, Odyssey, Pilot, Prelude, Ridgeline **Engines:** All **Transmissions:** A/T	**TCM A/T Control Unit or Related Circuit Malfunction** Engine running with VSS inputs received, and the PCM detected a fault in the TCM A/T Control Unit or one of its related circuits. **Possible Causes:** • Refer to the repair instructions in a transmission repair manual or the information in other electronic media to repair this code.
DTC: P0771 **1T CCM, MIL: Yes** **2003, 2004, 2005, 2006** **Models:** Element **Engines:** All **Transmissions:** A/T	**Shift Solenoid Valve E Stuck OFF** DTCs P0122, P0123, P0705, P0706, P0716, P0717, P0718, P0721, P0722, P0723, P0747, P0752, P0771, P0776, P0777, P0780, P0796, P0797, P0962, P0963, P0966, P0967, P0970, P0971, P0973, P0974, P0976, P0977, P0979, P0980, P0985, P0986, P1121, P1122, P1731, P1732, P1735, P1736 not set. The PCM monitors the mainshaft speed and the countershaft speed at the gear change determined by the shift schedule. When an improper gear ratio is output compared to the predetermined gear change mode, a shift solenoid valve E OFF failure is detected and a DTC is stored.. **Possible Causes:** • Pressure control solenoid control circuit is shorted to ground • Pressure control solenoid valve E is damaged or has failed • PCM has failed

DTC	Trouble Code Title, Conditions & Possible Causes
DTC: P0776 **1T CCM, MIL: Yes** **2003, 2004, 2005, 2006** **Models:** Element **Engines:** All **Transmissions:** A/T	**A/T Clutch Pressure Control Solenoid Valve B Stuck OFF** DTCs P0122, P0123, P0705, P0706, P0716, P0717, P0718, P0721, P0722, P0723, P0752, P0761, P0771, P0776, P0777, P0780, P0796, P0797, P0962, P0963, P0966, P0967, P0970, P0971, P0973, P0974, P0976, P0977, P0979, P0980, P0985, P0986, P1121, P1122, P1731, P1732, P1735, P1736 not set. The powertrain control module (PCM) computes the actual ratio of mainshaft and countershaft revolutions. If a difference between the actual ratio and the commanded gear occurs when shifting to each gear position, a malfunction in A/T clutch pressure control solenoid valve B or the hydraulic system is detected and a DTC is stored. **Possible Causes:** • Pressure control solenoid control circuit is shorted to ground • Pressure control solenoid valve B is damaged or has failed • PCM has failed
DTC: P0777 **1T CCM, MIL: Yes** **2003, 2004, 2005, 2006** **Models:** Element **Engines:** All **Transmissions:** A/T	**A/T Clutch Pressure Control Solenoid Valve B Stuck ON** DTCs P0122, P0123, P0705, P0706, P0716, P0717, P0718, P0721, P0722, P0723, P0752, P0761, P0771, P0776, P0777, P0780, P0796, P0797, P0962, P0963, P0966, P0967, P0970, P0971, P0973, P0974, P0976, P0977, P0979, P0980, P0985, P0986, P1121, P1122, P1731, P1732, P1735, P1736 not set. The powertrain control module (PCM) computes the actual ratio of mainshaft and countershaft revolutions. If a difference between the actual ratio and the commanded gear occurs when shifting to each gear position, a malfunction in A/T clutch pressure control solenoid valve A or the hydraulic system is detected and a DTC is stored. **Possible Causes:** • Pressure control solenoid control circuit is shorted to ground • Pressure control solenoid valve B is damaged or has failed • PCM has failed
DTC: P0780 **1T CCM, MIL: Yes** **1996, 1997, 1998, 1999, 2000, 2001, 2002, 2003, 2004, 2005, 2006** **Models:** Accord, Civic, CR-V, Odyssey, Pilot, Prelude, Ridgeline **Engines:** All **Transmissions:** A/T	**Automatic Transaxle Malfunction** Engine running and the PCM detected an Automatic Transaxle fault. Note: This trouble code (P0780) sets with along with several TCM related trouble codes. **Possible Causes:** • Refer to the repair instructions in a transmission repair manual or the information in other electronic media to repair this code.
DTC: P0780 **1T CCM, MIL: Yes** **2003, 2004, 2005, 2006** **Models:** Element **Engines:** All **Transmissions:** A/T	**Shift Control System** This fault code is a general (specified by SAE) DTC that is stored at a time any of the following DTC codes (P1731, P01732, P1735 or P1736) are detected. **Possible Causes:** • Refer to the repair instructions for DTC P1731, P01732, P1735 or P1736
DTC: P0796 **1T CCM, MIL: Yes** **2003, 2004, 2005, 2006** **Models:** Element **Engines:** All **Transmissions:** A/T	**A/T Clutch Pressure Control Solenoid Valve C Stuck OFF** DTCs P0122, P0123, P0705, P0706, P0716, P0717, P0718, P0721, P0722, P0723, P0747, P0752, P0761, P0771, P0776, P0777, P0780, P0797, P0962, P0963, P0966, P0967, P0970, P0971, P0973, P0974, P0976, P0977, P0979, P0980, P0985, P0986, P1121, P1122, P1731, P1732, P1735, P1736 not set. The powertrain control module (PCM) computes the actual ratio of mainshaft and countershaft revolutions. If a difference between the actual ratio and the commanded gear occurs when shifting to each gear position, a malfunction in A/T clutch pressure control solenoid valve C or the hydraulic system is detected and a DTC is stored. **Possible Causes:** • Failed A/T clutch pressure control solenoid valve C • Failing automatic transmission
DTC: P0797 **1T CCM, MIL: Yes** **2003, 2004, 2005, 2006** **Models:** Element **Engines:** All **Transmissions:** A/T	**A/T Clutch Pressure Control Solenoid Valve C Stuck ON** DTCs P0122, P0123, P0705, P0706, P0716, P0717, P0718, P0721, P0722, P0723, P0752, P0761, P0771, P0776, P0777, P0780, P0796, P0797, P0962, P0963, P0966, P0967, P0970, P0971, P0973, P0974, P0976, P0977, P0979, P0980, P0985, P0986, P1121, P1122, P1731, P1732, P1735, P1736 not set. The powertrain control module (PCM) computes the actual ratio of mainshaft and countershaft revolutions. If a difference between the actual ratio and the commanded gear occurs when shifting to each gear position, a malfunction in A/T clutch pressure control solenoid valve B or the hydraulic system is detected and a DTC is stored. **Possible Causes:** • Pressure control solenoid control circuit is shorted to ground • Pressure control solenoid valve C is damaged or has failed • PCM has failed

DTC	Trouble Code Title, Conditions & Possible Causes
DTC: P0812 **2T CCM, MIL: No** **2003, 2004, 2005, 2006** **Models:** Element **Engines:** All **Transmissions:** A/T	**Open in Transmission Range Switch ATP RVS Switch Circuit** If the Reverse switch is OPEN with the shift lever in R position while shifting between the P, R, and N positions, the PCM detects a switch OPEN failure and a DTC is stored. **Possible Causes:** • Open circuit between the transmission range switch and the PCM • Failed Automatic Transmission Range switch • PCM has failed
DTC: P0842 **1T CCM, MIL: Yes** **2003, 2004, 2005, 2006** **Models:** Element **Engines:** All **Transmissions:** A/T	**Short in 2nd Clutch Transmission Fluid Pressure Switch Circuit, or 2nd Clutch Transmission Fluid Pressure Switch Stuck ON** If the 2nd clutch transmission fluid pressure switch is ON while driving the vehicle with the speed ratio of the countershaft to mainshaft other than 2nd (the ratio is Neutral or 4th), the PCM detects a 2nd clutch transmission fluid pressure switch failure and a DTC is stored. **Possible Causes:** • Failed 2nd clutch transmission fluid pressure switch • Faulty wiring between PCM and fluid pressure switch • PCM has failed
DTC: P0843 **1T CCM, MIL: Yes** **2003, 2004, 2005, 2006** **Models:** Element **Engines:** All **Transmissions:** A/T	**Open in 2nd Clutch Transmission Fluid Pressure Switch Circuit, or 2nd Clutch Transmission Fluid Pressure Switch Stuck OFF** If the 2nd clutch transmission fluid pressure switch is OFF while driving with the rotation speed ratio of the input/output pulses in 2nd gear, the PCM detects a malfunction in the 2nd clutch transmission fluid pressure switch and stores a DTC. **Possible Causes:** • Failed 2nd clutch transmission fluid pressure switch • Faulty wiring between PCM and fluid pressure switch • PCM has failed
DTC: P0847 **1T CCM, MIL: Yes** **2003, 2004, 2005, 2006** **Models:** Element **Engines:** All **Transmissions:** A/T	**Short in 3rd Clutch Transmission Fluid Pressure Switch Circuit, or 3rd Clutch Transmission Fluid Pressure Switch Stuck ON** If the 3rd clutch transmission fluid pressure switch is ON while driving the vehicle with the speed ratio of the countershaft to mainshaft other than 3rd (the ratio is Neutral or 4th), the PCM detects a 3rd clutch transmission fluid pressure switch failure and a DTC is stored. **Possible Causes:** • Failed 3rd clutch transmission fluid pressure switch • Faulty wiring between PCM and fluid pressure switch • PCM has failed
DTC: P0848 **1T CCM, MIL: Yes** **2003, 2004, 2005, 2006** **Models:** Element **Engines:** All **Transmissions:** A/T	**Open in 3rd Clutch Transmission Fluid Pressure Switch Circuit, or 3rd Clutch Transmission Fluid Pressure Switch Stuck OFF** If the 3rd clutch transmission fluid pressure switch is OFF while driving with the rotation speed ratio of the input/output pulses in 3rd gear, the PCM detects a malfunction in the 3rd clutch transmission fluid pressure switch and stores a DTC. **Possible Causes:** • Failed 3rd clutch transmission fluid pressure switch • Faulty wiring between PCM and fluid pressure switch • PCM has failed

OBD II Trouble Code List (P1xxx Codes)

DTC	Trouble Code Title, Conditions & Possible Causes
DTC: P1106 **1T CCM, MIL: Yes** **1996, 1997, 1998, 1999, 2000, 2001, 2002, 2003, 2004, 2005, 2006** **Models:** Accord, Civic, Civic del Sol, CR-V, Insight, Odyssey, Pilot, Prelude, Ridgeline, S2000 **Engines:** All **Transmissions:** A/T, M/T	**BARO Pressure Sensor Performance** Engine running in 4th gear, then accelerated to WOT, and the PCM detected the BARO sensor input did not change sufficiently within a specified period of time. **Possible Causes:** • BARO sensor signal circuit is open or shorted to ground • BARO sensor ground circuit has high resistance • BARO sensor is damaged or it may be out of calibration • PCM has failed
DTC: P1106 **1T CCM, MIL: No** **1996** **Models:** Passport **Engines:** 2.6L VIN E **Transmissions:** A/T, M/T	**MAP Sensor Circuit High Input (Intermittent)** DTC P0121, P0122 and P0123 not set, engine speed less than 1000 rpm with the throttle angle less than 3%, or with the engine speed over 1000 rpm with the throttle angle less than 10%, and the PCM detected the MAP sensor circuit was interrupted (e.g., the MAP input indicated over 80 kPa for 5 seconds over a 16 second period). **Note: For additional help with this code, view the Failure Records.** **Possible Causes:** • MAP sensor signal circuit is open (an intermittent fault) • MAP sensor signal circuit shorted to ground (intermittent fault) • MAP sensor VREF circuit is open or shorted to ground • MAP sensor is damaged or has failed
DTC: P1106 **1T CCM, MIL: No** **1996, 1997, 1998, 1999, 2000, 2001, 2002** **Models:** Passport **Engines:** 3.2L VIN V, 3.2L VIN W **Transmissions:** A/T, M/T	**MAP Sensor Circuit High Input (Intermittent)** DTC P0121, P0122 and P0123 not set, engine runtime 10 seconds, engine speed less than 1000 rpm with throttle angle less than 3%, or the engine speed is more than 1000 rpm with the throttle angle less than 10%, and the PCM detected an unexpected high value (over 80 kPa) on the MAP sensor circuit for 5 seconds of a 16 second period. **Note: For additional help with this code, view the Failure Records.** **Possible Causes:** • MAP sensor signal circuit is open (an intermittent fault) • MAP sensor signal circuit shorted to ground (intermittent fault) • MAP sensor VREF circuit is open or shorted to ground • MAP sensor is damaged or has failed
DTC: P1107 **1T CCM, MIL: Yes** **1995** **Models:** Accord **Engines:** 2.7L VIN CE6 **Transmissions:** A/T	**BARO Pressure Sensor Circuit Low Input** Key on or engine running, and the PCM detected the BARO sensor signal was less than a value in stored in backup memory. **Possible Causes:** • BARO sensor signal circuit is shorted to signal ground • BARO sensor signal circuit is shorted to chassis ground • BARO sensor is damaged (it may be shorted internally) • BARO sensor signal circuit to the TCM is open or grounded • TCM or the PCM has failed
DTC: P1107 **1T CCM, MIL: Yes** **1996, 1997, 1998, 1999, 2000, 2001, 2002, 2003, 2004, 2005, 2006** **Models:** Accord, Civic, Civic del Sol, CR-V, Insight, Odyssey, Pilot, Prelude, Ridgeline, S2000 **Engines:** All **Transmissions:** A/T, M/T	**BARO Pressure Sensor Circuit Low Input** Key on or engine running, and the PCM detected the BARO sensor signal was less than a value in stored in backup memory. **Possible Causes:** • BARO sensor signal circuit is shorted to signal ground • BARO sensor signal circuit is shorted to chassis ground • BARO sensor is damaged (it may be shorted internally) • BARO sensor signal circuit to the TCM is open or grounded • TCM or the PCM has failed
DTC: P1107 **1T CCM, MIL: No** **1996** **Models:** Passport **Engines:** 2.6L VIN E **Transmissions:** A/T, M/T	**MAP Sensor Circuit Low Input (Intermittent)** DTC P0121, P0122 and P0123 not set, engine started, engine speed less than 1000 rpm with the throttle angle less than 3%, or with the engine speed over 1000 rpm and the throttle angle over 2%, the PCM detected a sudden low voltage condition on the MAP sensor circuit on the MAP sensor circuit (i.e., the MAP sensor signal was less 11 kPa for 5 seconds over a 16 second period). **Note: For additional help with this code, view the Failure Records.** **Possible Causes:** • MAP sensor signal circuit is open (an intermittent fault) • MAP sensor signal circuit shorted to ground (intermittent fault) • MAP sensor VREF circuit is open or shorted to ground • MAP sensor is damaged or has failed

DTC	Trouble Code Title, Conditions & Possible Causes
DTC: P1107 **1T CCM, MIL: No** **1996, 1997, 1998, 1999, 2000, 2001, 2002** **Models:** Passport **Engines:** 3.2L VIN V, 3.2L VIN W **Transmissions:** A/T, M/T	**MAP Sensor Circuit Low Input (Intermittent)** DTC P0121, P0122, P0123 not set, engine running, engine speed less than 1000 rpm and throttle angle over 1%, or the engine speed more than 1000 rpm and throttle angle more than 2%, and the PCM detected an unexpected low value (below 11 kPa) on the MAP sensor circuit for 5 seconds of a 16 second period. **Note: For additional help with this code, view the Failure Records.** **Possible Causes:** • MAP sensor signal circuit is open (an intermittent fault) • MAP sensor signal circuit shorted to ground (intermittent fault) • MAP sensor VREF circuit is open or shorted to ground • MAP sensor is damaged or has failed
DTC: P1108 **1T CCM, MIL: Yes** **1995** **Models:** Accord **Engines:** 2.7L VIN CE6 **Transmissions:** A/T	**BARO Pressure Sensor Circuit High Input** Key on or engine running, and the PCM detected the BARO sensor signal was more than a value in stored in backup memory. **Possible Causes:** • BARO sensor signal circuit shorted to VREF • BARO sensor signal circuit is shorted to system power (B+) • BARO sensor is damaged (it may be open internally) • BARO sensor signal circuit to the TCM is shorted to power • TCM or the PCM has failed
DTC: P1108 **1T CCM, MIL: Yes** **1996, 1997, 1998, 1999, 2000, 2001, 2002, 2003, 2004, 2005, 2006** **Models:** Accord, Civic, Civic del Sol, CR-V, Insight, Odyssey, Pilot, Prelude, Ridgeline, S2000 **Engines:** All **Transmissions:** A/T, M/T	**BARO Pressure Sensor Circuit High Input** Key on or engine running, and the PCM detected the BARO sensor signal was more than a value in stored in backup memory. **Possible Causes:** • BARO sensor signal circuit shorted to VREF • BARO sensor signal circuit is shorted to system power (B+) • BARO sensor is damaged (it may be open internally) • BARO sensor signal circuit to the TCM is shorted to power • TCM or the PCM has failed
DTC: P1109 **1T CCM, MIL: Yes** **2003, 2004, 2005, 2006** **Models:** Element **Engines:** All **Transmissions:** A/T	**BARO Pressure Sensor Circuit High Input** Key on or engine running, and the PCM detected the BARO sensor signal was more than a value in stored in backup memory. **Possible Causes:** • BARO sensor signal circuit shorted to VREF • BARO sensor signal circuit is shorted to system power (B+) • BARO sensor is damaged (it may be open internally) • BARO sensor signal circuit to the TCM is shorted to power • TCM or the PCM has failed
DTC: P1111 **1T CCM, MIL: No** **1996** **Models:** Passport **Engines:** 2.6L VIN E **Transmissions:** A/T, M/T	**IAT Sensor Circuit High Input (Intermittent)** Engine started, engine runtime over 4 minutes, ECT sensor more than 140°F, vehicle driven to a speed of over 20 mph, Calculated airflow less than 20 g/sec, and the PCM detected an unexpected high voltage condition (over 4.90v) on the IAT sensor circuit (Scan Tool reads -38°F) for 2.5 seconds over a 25 second period. **Note: For additional help with this code, view the Failure Records.** **Possible Causes:** • IAT sensor signal circuit is open (an intermittent fault) • IAT sensor is damaged or has failed (an intermittent fault) • PCM has failed
DTC: P1111 **1T CCM, MIL: No** **1996, 1997, 1998, 1999, 2000, 2001, 2002** **Models:** Passport **Engines:** 3.2L VIN V, 3.2L VIN W **Transmissions:** A/T, M/T	**IAT Sensor Circuit High Input (Intermittent)** Engine started, engine runtime over 4 minutes, ECT sensor more than 140°F, vehicle speed under 20 mph, MAF sensor less than 20 g/sec, and the PCM detected an unexpected high signal of over 4.90v (Scan Tool reads -38°F) on the IAT sensor circuit for 2.5 seconds over a 25 second period during the CCM test. **Note: For additional help with this code, view the Failure Records.** **Possible Causes:** • IAT sensor signal circuit is open (an intermittent fault) • IAT sensor is damaged or has failed (an intermittent fault) • PCM has failed

DTC	Trouble Code Title, Conditions & Possible Causes
DTC: P1112 **1T CCM, MIL: No** **1996** **Models:** Passport **Engines:** 2.6L VIN E **Transmissions:** A/T, M/T	**IAT Sensor Circuit Intermittent Low Input Conditions:** Engine started, engine runtime over 2 minutes, ECT sensor more than 140°F, vehicle speed more than 30 mph, MAF sensor more than 20 g/sec, and the PCM detected an unexpected low signal of under 0.10v [Scan Tool reads 298°F]) on the IAT sensor circuit for 2.5 seconds over a 25 second time period. **Note: For additional help with this code, view the Failure Records.** **Possible Causes:** • IAT sensor signal circuit is shorted to ground (intermittent fault) • IAT sensor is damaged or has failed (an intermittent fault) • PCM has failed
DTC: P1112 **1T CCM, MIL: No** **1996, 1997, 1998, 1999, 2000,** **2001, 2002** **Models:** Passport **Engines:** 3.2L VIN V, 3.2L VIN W **Transmissions:** A/T, M/T	**IAT Sensor Circuit Low Input (Intermittent)** Engine started, engine runtime over 4 minutes, ECT sensor more than 140°F, vehicle driven to a speed of over 20 mph, Calculated airflow less than 20 g/sec, and the PCM detected an unexpected low voltage condition of less than 0.10v (Scan Tool reads 298°F) on the IAT sensor circuit for 2.5 seconds over a 25 second period. **Note: For additional help with this code, view the Failure Records.** **Possible Causes:** • IAT sensor signal circuit is shorted to ground (intermittent fault) • IAT sensor is damaged or has failed (an intermittent fault) • PCM has failed
DTC: P1114 **1T CCM, MIL: No** **1996** **Models:** Passport **Engines:** 2.6L VIN E **Transmissions:** A/T, M/T	**ECT Sensor Circuit Low Input (Intermittent)** Engine started, engine runtime more than 1 minute, and the PCM detected an unexpected low voltage condition of under 0.10v (Scan Tool reads 302°F) on the ECT sensor circuit for 10 seconds of a 100 second period. **Note: For additional help with this code, view the Failure Records.** **Possible Causes:** • ECT sensor signal circuit shorted to ground (intermittent fault) • ECT sensor is damaged or has failed • PCM has failed
DTC: P1114 **1T CCM, MIL: No** **1996, 1997, 1998, 1999, 2000,** **2001, 2002** **Models:** Passport **Engines:** 3.2L VIN V, 3.2L VIN W **Transmissions:** A/T, M/T	**ECT Sensor Circuit Low Input (Intermittent)** Engine started, engine runtime more than 1 minute, and the PCM detected an unexpected low voltage condition of less than 0.10v (Scan Tool read 302°F) on the ECT sensor circuit for 10 seconds of a 100 second period. **Note: For additional help with this code, view the Failure Records.** **Possible Causes:** • ECT sensor signal circuit shorted to ground (intermittent fault) • ECT sensor is damaged or has failed • PCM has failed
DTC: P1115 **1T CCM, MIL: No** **1996** **Models:** Passport **Engines:** 2.6L VIN E **Transmissions:** A/T, M/T	**ECT Sensor Circuit High Input (Intermittent)** Engine started, engine running for 1 minute, and the PCM detected an unexpected high voltage of over 4.90v (Scan Tool reads -38°F) on the ECT sensor circuit for 10 seconds of a 100 second period. **Note: For additional help with this code, view the Failure Records.** **Possible Causes:** • ECT sensor signal circuit is open (an intermittent fault) • ECT sensor is damaged or has failed (an intermittent fault) • PCM has failed
DTC: P1115 **1T CCM, MIL: No** **1996, 1997, 1998, 1999, 2000,** **2001, 2002** **Models:** Passport **Engines:** 3.2L VIN V, 3.2L VIN W **Transmissions:** A/T, M/T	**ECT Sensor Circuit High Input (Intermittent)** Engine started, engine running for 1 minute, and the PCM detected an unexpected high voltage of over 4.90v (Scan Tool reads -38°F)] on the ECT sensor circuit for 10 seconds of a 100 second period. **Note: For additional help with this code, view the Failure Records.** **Possible Causes:** • ECT sensor signal circuit is open (an intermittent fault) • ECT sensor is damaged or has failed (an intermittent fault) • PCM has failed
DTC: P1120 **2T CCM, MIL: Yes** **2000, 2001, 2002** **Models:** Passport **Engines:** 3.2L VIN W **Transmissions:** A/T, M/T	**Throttle Position Sensor 1 Circuit Malfunction** Key on or engine running, and the PCM detected the TP Sensor 1 (TP1) circuit indicated less than 2.5%, or more than 97.5% for 93.6 ms or for 18 tests out of 500 test samples during the CCM test. **Possible Causes:** • TP1 sensor signal circuit is open between the sensor and PCM • TP1 sensor signal circuit is grounded between sensor and PCM • TP1 sensor VREF circuit is open between sensor and the PCM • TP1 sensor is damaged or has failed

DTC	Trouble Code Title, Conditions & Possible Causes
DTC: P1121 **1T CCM, MIL: Yes** **1996, 1997, 1998, 1999, 2000, 2001, 2002, 2003, 2004, 2005, 2006** **Models:** Accord, Civic, Civic del Sol, CR-V, Element, Insight, Odyssey, Pilot, Prelude, Ridgeline, S2000 **Engines:** All **Transmissions:** A/T, M/T	**TP Sensor Input Lower Than Expected** Engine running, and the PCM detected the TP sensor input was lower than an expected value with the throttle wide open (<13.7%). Note: This trouble code sets if this circuit fails the rationality test. **Possible Causes:** • Throttle plate is dirty, clogged, or it is binding • TP sensor circuit open or shorted to ground between the PCM and the TCM • TP sensor is damaged or has failed • PCM has failed
DTC: P1121 **1T CCM, MIL: No** **1996** **Models:** Passport **Engines:** 2.6L VIN E **Transmissions:** A/T, M/T	**Throttle Position Sensor Intermittent High Input** Engine running, and the PCM detected an unexpected high signal of over 4.90v on the TP sensor circuit for 0.15 seconds over a 1.5 second period during the CCM test. **Note: For additional help with this code, view the Failure Records.** **Possible Causes:** • TP sensor signal circuit is open (may be an intermittent fault) • TP sensor ground circuit is open (may be an intermittent fault) • TP sensor signal circuit is shorted to VREF (intermittent fault) • TP sensor is damaged or has failed
DTC: P1121 **1T CCM, MIL: No** **1996, 1997, 1998, 1999** **Models:** Passport **Engines:** 3.2L VIN V, 3.2L VIN W **Transmissions:** A/T, M/T	**Throttle Position Sensor Intermittent High Input** Engine started, engine running for 5 seconds, and the PCM detected an unexpected high signal of over 4.90v on the TP sensor circuit for 0.15 seconds of a 1.5 second period during the CCM test. **Note: For additional help with this code, view the Failure Records.** **Possible Causes:** • TP sensor signal circuit is open between the sensor and PCM • TP sensor ground circuit is open between the sensor and PCM • TP sensor signal circuit is shorted to VREF or system power • TP sensor is damaged or has failed
DTC: P1122 **1T CCM, MIL: Yes** **1996, 1997, 1998, 1999, 2000, 2001, 2002, 2003, 2004, 2005, 2006** **Models:** Accord, Civic, Civic del Sol, CR-V, Element, Insight, Odyssey, Pilot, Prelude, Ridgeline, S2000 **Engines:** All **Transmissions:** A/T, M/T	**TP Sensor Input Higher Than Expected** Engine running, and the PCM detected that the TP sensor input higher than the expected value with the throttle closed (>16.9%). Note: This trouble code sets if this circuit fails the rationality test. **Possible Causes:** • Throttle plate is dirty, clogged, or it is binding • TP sensor signal circuit shorted to VREF or it is open between the PCM and the TCM • TP sensor is damaged or has failed • PCM has failed
DTC: P1122 **1T CCM, MIL: No** **1996** **Models:** Passport **Engines:** 2.6L VIN E **Transmissions:** A/T, M/T	**Throttle Position Sensor Intermittent Low Input** Engine running, and the PCM detected an unexpected low signal of less than 0.22v on the TP sensor circuit for 0.15 seconds over a 1.5 second period during the CCM test. **Note: For additional help with this code, view the Failure Records.** **Possible Causes:** • TP sensor signal circuit is shorted to ground (intermittent fault) • TP sensor VREF circuit shorted to ground (test other sensors) • TP sensor is damaged or has failed
DTC: P1122 **1T CCM, MIL: No** **1996, 1997, 1998, 1999** **Models:** Passport **Engines:** 3.2L VIN V, 3.2L VIN W **Transmissions:** A/T, M/T	**Throttle Position Sensor Intermittent Low Input** Engine running, and the PCM detected an unexpected low signal of less than 0.22v on the TP sensor circuit for 0.15 seconds over a 1.5 second period during the CCM test. **Note: For additional help with this code, view the Failure Records.** **Possible Causes:** • TP sensor signal circuit is shorted to ground • TP sensor VREF circuit shorted to ground (test other sensors) • TP sensor is damaged or has failed

DTC	Trouble Code Title, Conditions & Possible Causes
DTC: P1125 **1T CCM, MIL: Yes** **2000, 2001, 2002** **Models:** Passport **Engines:** 3.2L VIN W **Transmissions:** A/T, M/T	**Electronic Throttle Control Limit Performance Mode** Key on or engine running, and the PCM detected the TP Sensor 1 (TP1) circuit indicated less than 2.5%, or more than 97.5% for 18 counts out of 500 test samples during the CCM Rationality test. **Possible Causes:** • TP sensor signal circuit is open between the sensor and PCM • TP sensor signal circuit is grounded between sensor and PCM • TP sensor VREF circuit is open between sensor and the PCM • TP sensor is damaged or has failed
DTC: P1128 **1T CCM, MIL: Yes** **1996, 1997, 1998, 1999, 2000, 2001, 2002, 2003, 2004, 2005, 2006** **Models:** Accord, Civic, Civic del Sol, CR-V, Element, Insight, Odyssey, Pilot, Prelude, Ridgeline, S2000 **Engines:** All **Transmissions:** A/T, M/T	**MAP Sensor Value Less Than Expected** Engine running at cruise speed, then back to idle speed, and the PCM detected a MAP sensor signal lower than the expected value. Note: This trouble code sets if this circuit fails the rationality test. **Possible Causes:** • MAP sensor signal circuit shorted to ground (intermittent fault) • MAP sensor vacuum line bent or plugged at intake manifold • MAP sensor is damaged or it is out-of-calibration • PCM has failed
DTC: P1129 **1T CCM, MIL: Yes** **1996, 1997, 1998, 1999, 2000, 2001, 2002, 2003, 2004, 2005, 2006** **Models:** Accord, Civic, Civic del Sol, CR-V, Element, Insight, Odyssey, Pilot, Prelude, Ridgeline, S2000 **Engines:** All **Transmissions:** A/T, M/T	**MAP Sensor Value Higher Than Expected** Engine running at cruise speed, then back to idle speed, and the PCM detected a MAP sensor signal higher than the expected value. Note: This trouble code sets if this circuit fails the rationality test. **Possible Causes:** • MAP sensor signal circuit shorted to VREF (intermittent fault) • MAP sensor ground circuit has high resistance • MAP sensor is damaged or it is out-of-calibration • PCM has failed
DTC: P1133 **2T O2S2, MIL: Yes** **1996** **Models:** Passport **Engines:** 2.6L VIN E **Transmissions:** M/T	**HO2S-11 (Bank 1 Sensor 1) Insufficient Switching** DTC P0106, P0107, P0108, P0112, P0113, P0117, P0118, P0121, P0122, P0123, P0131, P0132, P0133, P0134, P0135, P0300, P0441 and P1441 not set, system voltage from 10-16v, engine speed from 1500-3000 rpm in closed loop for 1 minute, ECT sensor more than 122°F, Calculated airflow from 9-42 g/sec, Purge duty cycle over 2%, conditions met for 3 seconds, then 90 seconds after entering closed loop control, the PCM detected less than 18 rich-to-lean or lean-to-rich switches from the front HO2S-11 during the Oxygen Sensor Monitor test. **Possible Causes:** • Air leaks after the MAF sensor, or air leaks in the PCV system • Air leaks at the EGR gasket, or at the EGR valve diaphragm • Exhaust leaks before or near where the front HO2S is mounted • Fuel control sensor is out of calibration (i.e., ECT, IAT or MAP) • Fuel delivery system supplying too much or too little fuel during cruise or idle periods (e.g., faulty fuel pump, or dirty fuel filter) • Fuel injector (one or more) dirty, leaking or sticking • Fuel pressure regulator leaking, damaged or has failed • HO2S is contaminated, deteriorated or it has failed
DTC: P1133 **2T O2S2, MIL: Yes** **1996, 1997, 1998, 1999, 2000, 2001, 2002** **Models:** Passport **Engines:** 3.2L VIN V, 3.2L VIN W **Transmissions:** A/T, M/T	**HO2S-11 (Bank 1 Sensor 1) Insufficient Switching** DTC P0101, P0102, P0103, P0106, P0107, P0108, P0117, P0118, P0121, P0122, P0123, P0131, P0132, P0133, P0134, P0135, P0300, P0301-P0306, P0441 and P1441 not set, engine speed from 1500-3000 rpm in closed loop for 1 minute, system voltage from 11-16v, ECT sensor more than 122°F, MAF sensor from 9-42 g/sec, Purge duty cycle over 2%, conditions met for 3 seconds, and the PCM detected less than 23 rich-to-lean or lean-to-rich switches on the HO2S-11 signal circuit. **Possible Causes:** • Air leaks after the MAF sensor, or in the EGR or PCV system • Exhaust leaks before or near where the front HO2S is mounted • Fuel control sensor is out of calibration (i.e., ECT, IAT or MAP) • Fuel delivery system supplying too much or too little fuel during cruise or idle periods (e.g., faulty fuel pump, or dirty fuel filter) • Fuel injector (one or more) dirty, leaking or sticking • Fuel pressure regulator leaking, damaged or has failed • HO2S is contaminated, deteriorated or it has failed

DTC	Trouble Code Title, Conditions & Possible Causes
DTC: P1134 **2T O2S2, MIL: Yes** **1996, 1997** **Models:** Passport **Engines:** 2.6L VIN E **Transmissions:** M/T	**HO2S-11 (Bank 1 Sensor 1) Transition Time Ratio Error** DTC P0106, P0107, P0108, P0112, P0113, P0117, P0118, P0121, P0122, P0123, P0131, P0132, P0133, P0134, P0135, P0300, P0441 and P1441 not set, system voltage from 10-16v, engine speed from 1500-3000 rpm in closed loop for 1 minute, ECT sensor more than 122°F, Calculated airflow from 9-42 g/sec, Purge duty cycle over 2%, conditions met for 3 seconds, then 90 seconds after entering closed loop control, the PCM detected the transition time ratio to switch from lean-to-rich or rich to lean from the HO2S-11 was below 0.8 or over 3.8 during the Oxygen Sensor Monitor test. **Possible Causes:** • Air leaks at the exhaust manifold or exhaust pipes • HO2S signal circuit is open or shorted to ground (intermittent) • HO2S heater power circuit is open, or the heater has failed • HO2S contaminated with wrong fuel, has deteriorated or failed
DTC: P1134 **2T O2S2, MIL: Yes** **1996, 1997, 1998, 1999, 2000,** **2001, 2002** **Models:** Passport **Engines:** 3.2L VIN V, 3.2L VIN W **Transmissions:** A/T, M/T	**HO2S-11 (Bank 1 Sensor 1) Transition Time Ratio Error** DTC P0101, P0102, P0103, P0106, P0107, P0108, P0117, P0118, P0121, P0122, P0123, P0131, P0132, P0133, P0134, P0135, P0300, P0301-P0306, P0441 and P1441 not set, engine speed from 1500-3000 rpm in closed loop for 1 minute, system voltage from 11-16v, ECT sensor more than 122-167°F, MAF sensor at 18-42 g/sec, Purge duty cycle over 2%, conditions met for 3 seconds, and the PCM detected the HO2S-11 transition ratio from lean-to-rich and rich-to-lean was less than 0.44 or more than 3.8 during the test. **Possible Causes:** • Air leaks after the MAF sensor, or in the EGR or PCV system • Exhaust leaks before or near where the front HO2S is mounted • Fuel control sensor is out of calibration (i.e., ECT, IAT or MAP) • Fuel delivery system supplying too much or too little fuel during cruise or idle periods (e.g., faulty fuel pump, or dirty fuel filter) • Fuel injector (one or more) dirty, leaking or sticking • Fuel pressure regulator leaking, damaged or has failed • HO2S is contaminated, deteriorated or it has failed
DTC: P1149 **1T O2S1, MIL: Yes** **1996, 1997, 1998, 1999, 2000,** **2001, 2002, 2003, 2004, 2005,** **2006** **Models:** Accord, Civic, Civic del Sol, CR-V, Odyssey, Pilot, Prelude, Ridgeline **Engines:** All **Transmissions:** A/T	**HO2S-11 (Bank 1 Sensor 1) Performance** Vehicle driven at 55 mph in closed loop at a steady throttle, then back to idle speed, and the PCM detected the front HO2S response time was too slow between 300-600 mv, or it detected the rich to lean or lean to rich switch rate was too slow. **Possible Causes:** • Exhaust leak present in the exhaust manifold or exhaust pipes • O2S element fuel contamination • O2S element has deteriorated • PCM has failed
DTC: P1153 **2T CCM, MIL: Yes** **1996, 1997, 1998, 1999, 2000,** **2001, 2002** **Models:** Passport **Engines:** 3.2L VIN V, 3.2L VIN W **Transmissions:** A/T, M/T	**HO2S-21 (Bank 2 Sensor 1) Insufficient Switching** DTC P0101, P0102, P0103, P0106, P0107, P0108, P0117, P0118, P0121, P0122, P0123, P0131, P0132, P0133, P0134, P0135, P0300, P0301-P0306, P0441 and P1441 not set, system voltage from 11-16v, engine speed from 1500-3000 rpm in closed loop for 1 minute, ECT sensor more than 122-167°F, MAF sensor from 9-42 g/sec, Purge duty cycle over 2%, conditions met for 3 seconds, and the PCM detected less than 27 rich-to-lean or lean-to-rich switches on the HO2S-21 signal circuit in the Oxygen Sensor Monitor test. **Possible Causes:** • Air leaks after the MAF sensor, or in the EGR or PCV system • Exhaust leaks before or near where the front HO2S is mounted • Fuel control sensor is out of calibration (i.e., ECT, IAT or MAP) • Fuel delivery system supplying too much or too little fuel during cruise or idle periods (e.g., faulty fuel pump, or dirty fuel filter) • Fuel injector (one or more) dirty, leaking or sticking • Fuel pressure regulator leaking, damaged or has failed • HO2S is contaminated, deteriorated or it has failed
DTC: P1154 **2T O2S2, MIL: Yes** **1996, 1997, 1998, 1999, 2000,** **2001, 2002** **Models:** Passport **Engines:** 3.2L VIN V, 3.2L VIN W **Transmissions:** A/T, M/T	**HO2S-21 (Bank 2 Sensor 1) Transition Time Ratio** DTC P0101, P0102, P0103, P0106, P0107, P0108, P0117, P0118, P0121, P0122, P0123, P0131, P0132, P0133, P0134, P0135, P0300, P0301-P0306, P0441 and P1441 not set, system voltage from 10-16v, engine speed from 1500-3000 rpm in closed loop for 1 minute, ECT sensor more than 122-167°F, MAF sensor from 9-42 g/sec, Purge duty cycle over 2%, conditions met for 3 seconds, then 90 seconds after entering closed loop control, the PCM detected the transition time ratio to switch from lean-to-rich or rich to lean from the HO2S-21 was less than 0.44 or more than 3.8 during the test. **Possible Causes:** • Air leaks at the exhaust manifold or exhaust pipes • HO2S signal circuit is open or shorted to ground (intermittent) • HO2S heater power circuit is open, or the heater has failed • HO2S contaminated with wrong fuel, has deteriorated or failed

DTC	Trouble Code Title, Conditions & Possible Causes
DTC: P1157 **1T CCM, MIL: Yes** **2003, 2004, 2005, 2006** **Models:** Element **Engines:** All **Transmissions:** A/T	**Air-Fuel Ratio (A-F) Sensor (Sensor 1) AFS Line High Voltage** If the element is not activated for a set time period when power is drawn by the A/F sensor (Sensor 1) heater, a malfunction is detected and a DTC is stored. **Possible Causes:** • Open circuit between A/F Sensor and PCM • Failed A/F Sensor heater • PCM has failed
DTC: P1162 **1T CCM, MIL: Yes** **1998, 1999, 2000, 2001, 2002,** **2003, 2004, 2005, 2006** **Models:** Accord **Engines:** 2.3L VIN CG3, 2.3L VIN CG6 **Transmissions:** A/T	**Lean A/F Sensor (Bank 1 Sensor 1) Circuit Malfunction** DTC P0131, P0132, P0133, P1163 not set, vehicle driven while in closed loop at over 55 mph in D4 for 1-2 minutes, and the PCM detected an unexpected voltage condition on the LAF circuit during the HO2S Monitor test. **Possible Causes:** • HO2S may be contaminated or may have failed • Fuel supply system is too lean (exhaust leaks in front of HO2S) • Fuel supply system is too rich (fuel filter is clogged or dirty) • PCM has failed
DTC: P1162 **1T CCM, MIL: Yes** **1998, 1999, 2000, 2001, 2002,** **2003, 2004, 2005, 2006** **Models:** Accord **Engines:** 2.3L VIN CG3, 2.3L VIN CG6 **Transmissions:** M/T	**Lean A/F Sensor (Bank 1 Sensor 1) Circuit Malfunction** DTC P0131, P0132, P0133, P1163 not set, vehicle driven while in closed loop at over 55 mph for 1-2 minutes in 5th gear, and the PCM detected an unexpected voltage condition on the front Lean Air Fuel Sensor (LAF) circuit during the HO2S Monitor test. **Possible Causes:** • LAF sensor may be contaminated or may have failed • Fuel supply system is too lean (exhaust leaks in front of LAF) • Fuel supply system is too rich (fuel filter is clogged or dirty) • PCM has failed
DTC: P1162 **1T CCM, MIL: Yes** **1996, 1997, 1998, 1999** **Models:** Civic **Engines:** 1.6L VIN EJ7 **Transmissions:** A/T, M/T	**Lean A/F Sensor (Bank 1 Sensor 1) Circuit Malfunction** DTC P0131, P0132, P0133, P1163 not set, vehicle driven while in closed loop at over 55 mph in D4 for 1-2 minutes, and the PCM detected an unexpected voltage condition on the LAF circuit during the HO2S Monitor test. **Possible Causes:** • HO2S may be contaminated or may have failed • Fuel supply system is too lean (exhaust leaks in front of HO2S) • Fuel supply system is too rich (fuel filter is clogged or dirty) • PCM has failed
DTC: P1162 **1T CCM, MIL: Yes** **2000, 2001, 2002, 2003, 2004,** **2005, 2006** **Models:** Insight **Engines:** All **Transmissions:** A/T, M/T	**HO2S-11 (Bank 1 Sensor 1) Circuit Malfunction** Vehicle driven for over 2 minutes, engine at hot idle speed, and the PCM detected an unusual voltage condition on the front HO2S. **Possible Causes:** • LAF sensor signal circuit open or grounded (intermittent fault) • LAF sensor signal circuit shorted to heater power (intermittent) • LAF sensor may be contaminated or may have failed • PCM has failed
DTC: P1163 **2T O2S1, MIL: Yes** **1996, 1997, 1998, 1999, 2000,** **2001, 2002, 2003, 2004, 2005,** **2006** **Models:** All **Engines:** All **Transmissions:** All	**Lean A/F Sensor (Bank 1 Sensor 1) Slow Response** Engine at idle speed, then accelerated to 55 mph for 5 seconds, then back to idle speed for 5 seconds, and the PCM detected the LAF sensor response time was too slow, or that the R-L or L-R switch rate was too slow during the HO2S Monitor test period. **Possible Causes:** • LAF sensor may be contaminated or may have failed • Fuel supply system is too lean (exhaust leaks in front of LAF) • Fuel supply system is too rich (fuel filter is clogged or dirty) • PCM has failed
DTC: P1163 **1T O2S1, MIL: Yes** **1996, 1997, 1998, 1999, 2000,** **2001, 2002, 2003, 2004, 2005,** **2006** **Models:** All **Engines:** All **Transmissions:** All	**Lean A/F Sensor (Bank 1 Sensor 1) Slow Response** Engine at idle speed, then accelerated to 55 mph for 5 seconds, then back to idle speed for 5 seconds, and the PCM detected the LAF sensor response time was too slow, or that the R-L or L-R switch rate was too slow during the HO2S Monitor test period. **Possible Causes:** • LAF sensor may be contaminated or may have failed • Fuel supply system is too lean (exhaust leaks in front of LAF) • Fuel supply system is too rich (fuel filter is clogged or dirty) • PCM has failed

DTC	Trouble Code Title, Conditions & Possible Causes
DTC: P1163 **2T O2S1, MIL: Yes** **1996, 1997, 1998, 1999, 2000, 2001, 2002, 2003, 2004, 2005, 2006** **Models:** All **Engines:** All **Transmissions:** All	**HO2S (Bank 1 Sensor 1) Signal Slow Response** Engine at idle speed, then accelerated to over 55 mph for 5 seconds at a steady throttle, then back to idle speed for 5 seconds, and the PCM detected the front HO2S response time was too slow, or that the R-L or L-R switch rate was too slow during the HO2S test period. **Possible Causes:** • HO2S may be contaminated or may have failed • Fuel supply system is too lean (exhaust leaks in front of LAF) • Fuel supply system is too rich (fuel filter is clogged or dirty) • PCM has failed
DTC: P1164 **2T O2S1, MIL: Yes** **1996, 1997, 1998, 1999, 2000, 2001, 2002, 2003, 2004, 2005, 2006** **Models:** All **Engines:** All **Transmissions:** All	**Lean A/F Sensor (Bank 1 Sensor 1) Range/Performance** Engine speed over 1500 rpm in 4th gear in closed loop, then a quick acceleration to WOT, followed by a 5 second deceleration period with the throttle closed, and the PCM detected a problem in the LAF sensor response time, or that the rich-to lean or lean-to rich switch rate was too slow during the test. **Possible Causes:** • LAF sensor may be contaminated or may have failed • Fuel supply system is too lean (exhaust leaks in front of LAF) • Fuel supply system is too rich (fuel filter is clogged or dirty) • PCM has failed
DTC: P1164 **1T O2S1, MIL: Yes** **1996, 1997, 1998, 1999, 2000, 2001, 2002, 2003, 2004, 2005, 2006** **Models:** All **Engines:** All **Transmissions:** All	**Lean A/F Sensor (Bank 1 Sensor 1) Range/Performance** Engine speed over 1500 rpm in 4th gear in closed loop, then a quick acceleration to WOT, followed by a 5 second deceleration period with the throttle closed, and the PCM detected a problem in the LAF sensor response time, or that the rich-to lean or lean-to rich switch rate was too slow during the test. **Possible Causes:** • LAF sensor may be contaminated or may have failed • Fuel supply system is too lean (exhaust leaks in front of LAF) • Fuel supply system is too rich (fuel filter is clogged or dirty) • PCM has failed
DTC: P1164 **2T O2S1, MIL: Yes** **1996, 1997, 1998, 1999, 2000, 2001, 2002, 2003, 2004, 2005, 2006** **Models:** All **Engines:** All **Transmissions:** All	**HO2S-11 (Bank 1 Sensor 1) Signal Range/Performance** Vehicle driven to over 1500 rpm in high gear in closed loop, then a quick acceleration to WOT, followed by a 5 second deceleration period with the throttle closed, and the PCM detected the front HO2S response time or the R-L or L-R switch rate was too slow. **Possible Causes:** • HO2S may be contaminated or may have failed • Fuel supply system is too lean (exhaust leaks in front of LAF) • Fuel supply system is too rich (fuel filter is clogged or dirty) • PCM has failed
DTC: P1165 **2T CCM, MIL: Yes** **1996, 1997, 1998, 1999, 2000, 2001, 2002, 2003, 2004, 2005, 2006** **Models:** All **Engines:** All **Transmissions:** All	**Lean A/F-11 (Bank 1 Sensor 1) Range/Performance** Vehicle driven at an engine speed of 1500-2500 rpm at 45-60 mph for 2-3 minutes, followed by a deceleration period of 3 seconds back to idle speed with the throttle closed, and the PCM detected the LAF sensor signal was too high or too low during the HO2S Monitor test. **Possible Causes:** • LAF signal circuit is open or grounded (intermittent fault) • LAF signal circuit is shorted to heater power (intermittent) • LAF may be contaminated or may have failed • PCM has failed
DTC: P1165 **1T CCM, MIL: Yes** **1996, 1997, 1998, 1999, 2000, 2001, 2002, 2003, 2004, 2005, 2006** **Models:** All **Engines:** All **Transmissions:** All	**Lean A/F-11 (Bank 1 Sensor 1) Circuit Malfunction** Vehicle driven at an engine speed of 1500-2500 rpm at 45-60 mph for 2-3 minutes, followed by a deceleration period of 3 seconds back to idle speed with the throttle closed, and the PCM detected the LAF sensor signal was too high or too low during the HO2S Monitor test. **Possible Causes:** • LAF signal circuit is open or grounded (intermittent fault) • LAF signal circuit is shorted to heater power (intermittent) • LAF may be contaminated or may have failed • PCM has failed
DTC: P1165 **2T O2S1, MIL: Yes** **2000, 2001, 2002, 2003, 2004, 2005, 2006** **Models:** Insight **Engines:** All **Transmissions:** A/T, M/T	**HO2S-11 (Bank 1 Sensor 1) Signal Range/Performance** Vehicle driven at an engine speed of 1500-2500 rpm at 45-60 mph for 2-3 minutes, followed by a deceleration period of 3 seconds back to idle speed with the throttle closed, and the PCM detected the HO2S signal was too high or too low during the HO2S Monitor test. **Possible Causes:** • HO2S signal circuit is open or grounded (intermittent fault) • HO2S signal circuit is shorted to heater power (intermittent) • HO2S may be contaminated or may have failed • PCM has failed

DTC	Trouble Code Title, Conditions & Possible Causes
DTC: P1166 **1T O2S HTR1, MIL: Yes** 1998, 1999, 2000, 2001, 2002, 2003, 2004, 2005, 2006 **Models:** Accord **Engines:** 2.3L VIN CG3, 2.3L VIN CG6 **Transmissions:** A/T, M/T	**Lean A/F-11 (Bank 1 Sensor 1) Heater Circuit Malfunction** Engine runtime over 80 seconds, and the PCM detected an unexpected voltage condition on the LAF sensor heater circuit. **Possible Causes:** • LAF sensor heater circuit is open or shorted to ground • LAF sensor heater is damaged or has failed • PCM has failed
DTC: P1166 **1T O2S HTR1, MIL: Yes** 1996, 1997, 1998, 1999 **Models:** Civic **Engines:** 1.6L VIN EJ7 **Transmissions:** A/T, M/T	**Lean A/F-11 (Bank 1 Sensor 1) Heater Circuit Malfunction** Engine runtime over 80 seconds, and the PCM detected an unexpected voltage condition on the LAF sensor heater circuit. **Possible Causes:** • LAF sensor heater circuit is open or shorted to ground • LAF sensor heater is damaged or has failed • PCM has failed
DTC: P1166 **1T O2S HTR1, MIL: Yes** 2000, 2001, 2002, 2003, 2004, 2005, 2006 **Models:** Insight **Engines:** All **Transmissions:** A/T, M/T	**HO2S-11 (Bank 1 Sensor 1) Heater Circuit Malfunction** Engine runtime over 80 seconds, and the PCM detected an unexpected voltage condition on the front HO2S heater circuit. **Possible Causes:** • HO2S heater circuit is open or shorted to ground • HO2S heater circuit is shorted to system power (B+) • HO2S heater is damaged or has failed • PCM has failed
DTC: P1167 **1T O2S HTR1, MIL: Yes** 1998, 1999, 2000, 2001, 2002, 2003, 2004, 2005, 2006 **Models:** Accord **Engines:** 2.3L VIN CG3, 2.3L VIN CG6 **Transmissions:** A/T, M/T	**Lean A/F-11 (Bank 1 Sensor 1) Heater Circuit Malfunction** Engine runtime over 80 seconds, and the PCM detected an unexpected voltage condition on the LAF sensor Heater circuit. **Possible Causes:** • LAF sensor heater power supply circuit is open (check the fuse) • LAF sensor heater circuit is shorted to system power (B+) • LAF sensor heater is damaged or has failed • PCM has failed
DTC: P1167 **1T O2S HTR1, MIL: Yes** 1996, 1997, 1998, 1999 **Models:** Civic **Engines:** 1.6L VIN EJ7 **Transmissions:** A/T, M/T	**Lean A/F-11 (Bank 1 Sensor 1) Heater Circuit Malfunction** Engine runtime over 80 seconds, and the PCM detected an unexpected voltage condition on the LAF sensor heater circuit. **Possible Causes:** • LAF sensor heater circuit is open or shorted to ground • LAF sensor heater is damaged or has failed • PCM has failed
DTC: P1167 **1T O2S HTR1, MIL: Yes** 2000, 2001, 2002, 2003, 2004, 2005, 2006 **Models:** Insight **Engines:** All **Transmissions:** A/T, M/T	**HO2S-11 (Bank 1 Sensor 1) Heater Circuit Malfunction** Engine runtime over 80 seconds, and the PCM detected an unexpected voltage condition on the front HO2S heater circuit. **Possible Causes:** • HO2S heater circuit is open or shorted to ground • HO2S heater circuit is shorted to system power (B+) • HO2S heater is damaged or has failed • PCM has failed
DTC: P1167 **1T FUEL, MIL: Yes** 2000, 2001, 2002 **Models:** Passport **Engines:** 3.2L VIN W **Transmissions:** A/T, M/T	**Fuel System Rich During Decel Fuel Cutoff (Bank 1)** No related codes set, ECT sensor signal more than 140°F, engine running in Power Enrichment mode in closed loop for 3 seconds, and the PCM detected the HO2S-11 signal was more than 600 mv in a Decel Fuel Cutoff period during the Fuel System Monitor test. **Possible Causes:** • Base engine "mechanical" fault affecting one or more cylinders • EVAP system component has failed or canister fuel saturated • Exhaust leaks before or near where the front HO2S is mounted • Fuel control sensor is out of calibration (i.e., ECT, IAT or MAP) • Fuel delivery system supplying too much fuel during cruise or idle periods (e.g., faulty fuel pump, or faulty pressure regulator) • Fuel injectors (one or more) leaking, pressure regulator leaking • HO2S is contaminated, deteriorated or it has failed
DTC: P1168 **1T CCM, MIL: Yes** 1996, 1997, 1998, 1999 **Models:** Civic **Engines:** 1.6L VIN EJ7 **Transmissions:** A/T, M/T	**Lean A/F Sensor (Bank 1 Sensor 1) Label Circuit Low Input** Engine runtime over 2 minutes at idle speed, and the PCM detected the LAF sensor signal remained at less than the low threshold limit for too long a period of time during the HO2S Monitor test period. **Possible Causes:** • LAF sensor signal circuit is open or it is shorted to ground • LAF sensor is damaged or has failed (it may be contaminated) • PCM has failed

DTC	Trouble Code Title, Conditions & Possible Causes
DTC: P1168 **1T CCM, MIL: Yes** **2000, 2001, 2002, 2003, 2004, 2005, 2006** **Models:** Insight **Engines:** All **Transmissions:** A/T, M/T	**HO2S-11 (Bank 1 Sensor 1) Label Circuit Low Input** Engine runtime over 2 minutes at idle speed, and the PCM detected the front HO2S signal remained at less than the low threshold limit for too long a period of time during the HO2S Monitor test period. **Possible Causes:** • HO2S signal circuit is open or it is shorted to ground • HO2S is damaged or has failed (it may be contaminated) • PCM has failed
DTC: P1169 **2T CCM, MIL: Yes** **1996, 1997, 1998, 1999** **Models:** Civic **Engines:** 1.6L VIN EJ7 **Transmissions:** A/T, M/T	**Lean A/F Sensor (Bank 1 Sensor 1) Label Circuit High Input** Engine running for 2 minutes, and the PCM detected the LAF sensor signal remained at more than the high threshold limit for too long a period of time during the HO2S Monitor test period. **Possible Causes:** • LAF sensor signal circuit is shorted to system power (B+) • LAF sensor is damaged or has failed (it may be contaminated) • PCM has failed
DTC: P1169 **2T CCM, MIL: Yes** **2000, 2001, 2002, 2003, 2004, 2005, 2006** **Models:** Insight **Engines:** All **Transmissions:** A/T, M/T	**HO2S-11 (Bank 1 Sensor 1) Label Circuit High Input** Engine running for 2 minutes, and the PCM detected the front HO2S signal remained at more than the high threshold limit for too long a period of time during the HO2S Monitor test period. **Possible Causes:** • HO2S signal circuit is shorted to system power (B+) • HO2S is damaged or has failed (it may be contaminated) • PCM has failed
DTC: P1169 **1T FUEL, MIL: No** **2000, 2001, 2002** **Models:** Passport **Engines:** 3.2L VIN W **Transmissions:** A/T, M/T	**Fuel System Rich During Decel Fuel Cutoff (Bank 2)** No related codes set, ECT sensor signal more than 140°F, engine running in closed loop, Power Enrichment mode "on" for 3 seconds, and the PCM detected the Bank 2 HO2S-21 signal remained above 600 mv in a Decel Fuel Cutoff period during the Fuel System Test. **Possible Causes:** • Base engine "mechanical" fault affecting one or more cylinders • EVAP system component has failed or canister fuel saturated • Exhaust leaks before or near where the front HO2S is mounted • Fuel control sensor is out of calibration (i.e., ECT, IAT or MAP) • Fuel delivery system supplying too much fuel during cruise or idle periods (e.g., faulty fuel pump, or faulty pressure regulator) • Fuel injectors (one or more) leaking, pressure regulator leaking • HO2S is contaminated, deteriorated or it has failed
DTC: P1171 **1T FUEL, MIL: Yes** **1996** **Models:** Passport **Engines:** 2.6L VIN E **Transmissions:** A/T, M/T	**Fuel System Lean During Acceleration Detected** DTC P0131, P0132, P0133, P0134 and P1133 not set, ECT sensor signal more than 140°F, engine running in closed loop mode, Power Enrichment mode enabled, and the PCM detected the HO2S-11 signal indicated less than 400 mv for 5 seconds during the test. **Possible Causes:** • Air intake leaks in the engine, or in the PCV system (valve) • Air leaks at the EGR gasket, or at the EGR valve diaphragm • Base engine "mechanical" fault affecting one or more cylinders • Exhaust leaks before or near where the front HO2S is mounted • Fuel injectors (one or more) restricted (allowing too little fuel) • Fuel delivery system supplying too little fuel during acceleration periods (e.g., faulty fuel pump, dirty or restricted fuel filter) • Fuel control sensor out of calibration (i.e., IAT, MAF or MAP) • HO2S is contaminated, deteriorated or it has failed • Vehicle driven low on fuel or until it ran out of fuel
DTC: P1171 **1T FUEL, MIL: Yes** **1996, 1997, 1998, 1999, 2000, 2001, 2002** **Models:** Passport **Engines:** 3.2L VIN V, 3.2L VIN W **Transmissions:** A/T	**Fuel System Lean During Acceleration Detected** DTC P0131, P0132, P0133, P0134 and P1133 not set, ECT sensor more than 140°F, engine running in Power Enrichment mode in closed loop, and the PCM detected the HO2S-11 signal indicated less than 400 mv for 5 seconds during the Fuel System Monitor test. **Possible Causes:** • Air intake leaks in the engine, or in the PCV system (valve) • Air leaks at the EGR gasket, or at the EGR valve diaphragm • Base engine "mechanical" fault affecting one or more cylinders • Exhaust leaks before or near where the front HO2S is mounted • Fuel injectors (one or more) restricted (allowing too little fuel) • Fuel delivery system supplying too little fuel during acceleration periods (e.g., faulty fuel pump, dirty or restricted fuel filter) • Fuel control sensor out of calibration (i.e., IAT, MAF or MAP) • HO2S is contaminated, deteriorated or it has failed • Vehicle driven low on fuel or until it ran out of fuel

DTC	Trouble Code Title, Conditions & Possible Causes
DTC: P1182 **1T CCM, MIL: Yes** 2001, 2002, 2003, 2004, 2005, 2006 **Models:** Civic **Engines:** 1.7L VIN EN2 **Transmissions:** A/T	**CNG Fuel Temperature Sensor Circuit Low Input** Key on or engine running, and the PCM detected an unexpected low voltage condition on the Fuel Temperature sensor signal circuit. **Possible Causes:** • Fuel temperature sensor signal circuit is shorted to ground • Fuel temperature sensor is damaged or has failed • PCM has failed
DTC: P1183 **1T CCM, MIL: Yes** 2001, 2002, 2003, 2004, 2005, 2006 **Models:** Civic **Engines:** 1.7L VIN EN2 **Transmissions:** A/T	**CNG Fuel Temperature Sensor Circuit High Input** Key on or engine running, and the PCM detected an unexpected high voltage condition on the Fuel Temperature sensor signal circuit. **Possible Causes:** • Fuel temperature sensor signal circuit is open • Fuel temperature sensor signal is shorted to VREF • Fuel temperature sensor is damaged or has failed • PCM has failed
DTC: P1187 **1T CCM, MIL: Yes** 2001, 2002, 2003, 2004, 2005, 2006 **Models:** Civic **Engines:** 1.7L VIN EN2 **Transmissions:** A/T	**CNG Fuel Tank Temperature Sensor Circuit Low Input** Key on or engine running, and the PCM detected an unexpected low voltage condition on the Fuel Tank Temperature sensor signal circuit. **Possible Causes:** • Fuel tank temperature sensor signal circuit is shorted to ground • Fuel tank temperature sensor is damaged or has failed • PCM has failed
DTC: P1188 **1T CCM, MIL: Yes** 2001, 2002, 2003, 2004, 2005, 2006 **Models:** Civic **Engines:** 1.7L VIN EN2 **Transmissions:** A/T	**CNG Fuel Tank Temperature Sensor Circuit High Input** Key on or engine running, and the PCM detected an unexpected high voltage condition on the Fuel Tank Temperature sensor signal circuit. **Possible Causes:** • Fuel tank temperature sensor signal circuit is open • Fuel tank temperature sensor signal circuit is shorted to ground • Fuel tank temperature sensor is damaged or has failed • PCM has failed
DTC: P1192 **1T CCM, MIL: Yes** 2001, 2002, 2003, 2004, 2005, 2006 **Models:** Civic **Engines:** 1.7L VIN EN2 **Transmissions:** A/T	**CNG Fuel Tank Pressure Sensor Circuit Low Input** Key on or engine running, and the PCM detected an unexpected low voltage condition on the Fuel Tank Pressure sensor signal circuit. **Possible Causes:** • Fuel tank pressure sensor signal circuit is shorted to ground • Fuel tank pressure sensor is damaged or has failed • PCM has failed
DTC: P1193 **1T CCM, MIL: Yes** 2001, 2002, 2003, 2004, 2005, 2006 **Models:** Civic **Engines:** 1.7L VIN EN2 **Transmissions:** A/T	**CNG Fuel Tank Pressure Sensor Circuit High Input** Key on or engine running, and the PCM detected an unexpected high voltage condition on the Fuel Tank Pressure sensor signal circuit. **Possible Causes:** • Fuel tank pressure sensor signal circuit is open • Fuel tank pressure sensor signal circuit is shorted to VREF • Fuel tank pressure sensor is damaged or has failed • PCM has failed
DTC: P1220 **1T CCM, MIL: Yes** 2000, 2001, 2002 **Models:** Passport **Engines:** 3.2L VIN W **Transmissions:** A/T, M/T	**Throttle Position Sensor 2 Circuit Malfunction** Key on or engine running, and the PCM detected the TP Sensor 2 (TP2) circuit was less than 2.5% or over 97.5% of VREF (5v), condition met for 18 counts within 500 test samples over 15.6 ms. **Possible Causes:** • TP2 sensor signal circuit is open or shorted to ground • TP2 sensor signal circuit is shorted to VREF or system power • TP2 sensor power circuit is open between sensor and the PCM • TP2 sensor is damaged or has failed • PCM has failed
DTC: P1221 **1T CCM, MIL: Yes** 2000, 2001, 2002 **Models:** Passport **Engines:** 3.2L VIN W **Transmissions:** A/T, M/T	**Throttle Position Sensor 1-2 Circuit Performance** Key on or engine running, and the PCM detected the TP Sensor 2 (TP2) signal did not correlate with the TP Sensor 1 (TP1) signal. **Possible Causes:** • Check for excessive deposits in the ETC passage or the spring • Check for excessive deposits in throttle bore and throttle valve • Check for objects blocking the DC motor or throttle bore • ETC DC motor circuit(s) has high resistance • TP2 sensor signal circuit has high resistance

DTC	Trouble Code Title, Conditions & Possible Causes
DTC: P1253 **1T CCM, MIL: Yes** **1996** **Models:** Accord **Engines:** 2.2L VIN CD5 **Transmissions:** A/T, M/T	**VTEC Solenoid Circuit Malfunction** Engine running and the PCM detected an unexpected voltage condition in the Variable Timing Electronic Control (VTEC) solenoid circuit during the CCM test. **Possible Causes:** • VTEC solenoid control circuit is open or shorted to ground • VTEC solenoid control circuit is shorted to system power (B+) • VTEC solenoid is damaged or has failed • PCM has failed
DTC: P1257 **1T CCM, MIL: Yes** **1999** **Models:** Accord, Civic **Engines:** All **Transmissions:** A/T, M/T	**VTEC System Malfunction** Engine running under hard acceleration, than back to road load, and the PCM detected a fault in the operation of the VTEC system. **Possible Causes:** • VTEC solenoid is damaged or has failed • VTEC switch is damaged or has failed • PCM has failed
DTC: P1258 **1T CCM, MIL: Yes** **1999** **Models:** Accord, Civic **Engines:** All **Transmissions:** A/T, M/T	**VTEC System Malfunction** Engine running under hard acceleration, than back to road load, and the PCM detected a fault in the operation of the VTEC system. **Possible Causes:** • VTEC solenoid is damaged or has failed • VTEC switch is damaged or has failed • PCM has failed
DTC: P1259 **1T CCM, MIL: Yes** **1996, 1997, 1998, 1999, 2000, 2001, 2002, 2003, 2004, 2005, 2006** **Models:** Accord, Civic, Civic del Sol, Insight, Odyssey, Pilot, Prelude, Ridgeline, S2000 **Engines:** All **Transmissions:** A/T, M/T	**VTEC System Malfunction (Bank 1)** Engine running in closed loop, then accelerated in 1st gear to over 6000 rpm for 2 seconds, and the PCM detected a fault in the VTEC solenoid or the VTEC switch. **Possible Causes:** • VTEC solenoid is damaged or has failed • VTEC switch is damaged or has failed • PCM has failed
DTC: P1271 **1T CCM, MIL: Yes** **2000, 2001, 2002** **Models:** Passport **Engines:** 3.2L VIN W **Transmissions:** A/T, M/T	**Throttle Position Sensor 1-2 Correlation Error** Key on or engine running, and the PCM detected the difference in the angle of the accelerator pedal for ASP1 and ASP2 was less than 4.5% for 50 counts within 50 test samples over a 15.6 time period. **Possible Causes:** • Check for excessive deposits in the ETC passage or the spring • Check for excessive deposits in throttle bore and throttle valve • Check for objects blocking the DC motor or throttle bore • ETC DC motor circuit(s) has high resistance • TP2 sensor signal circuit has high resistance • TSB 01-047 (8/14/01) contains a repair procedure for this code
DTC: P1272 **1T CCM, MIL: Yes** **2000, 2001, 2002** **Models:** Passport **Engines:** 3.2L VIN W **Transmissions:** A/T, M/T	**Acceleration Position Sensor 2-3 Correlation Error** Key on or engine running, and the PCM detected the difference in the angle of the accelerator pedal for ASP2 and ASP3 was less than 4.5% for 50 counts within 50 test samples over a 15.6 time period. **Possible Causes:** • Check for excessive deposits in the ETC passage or the spring • Check for excessive deposits in throttle bore and throttle valve • Check for objects blocking the DC motor or throttle bore • ETC DC motor circuit(s) has high resistance • TP2 sensor signal circuit has high resistance
DTC: P1273 **1T CCM, MIL: Yes** **2000, 2001, 2002** **Models:** Passport **Engines:** 3.2L VIN W **Transmissions:** A/T, M/T	**Acceleration Position Sensor 1-3 Correlation Error** Key on or engine running, and the PCM detected the difference in the angle of the accelerator pedal for ASP1 and ASP3 was less than 4.5% for 50 counts within 50 test samples over a 15.6 time period. **Possible Causes:** • Check for excessive deposits in the ETC passage or the spring • Check for excessive deposits in throttle bore and throttle valve • Check for objects blocking the DC motor or throttle bore • ETC DC motor circuit(s) has high resistance • TP2 sensor signal circuit has high resistance • TSB 01-047 (8/14/01) contains a repair procedure for this code

DTC	Trouble Code Title, Conditions & Possible Causes
DTC: P1275 **1T CCM, MIL:** Yes **2000, 2001, 2002** **Models:** Passport **Engines:** 3.2L VIN W **Transmissions:** A/T, M/T	**Acceleration Position Sensor 1 Circuit Malfunction** Key on or engine running, and the PCM detected the Acceleration Position Sensor 1 (APS1) circuit was less than 2.5% or over 97.5% of VREF (5v) for 12 counts within 500 test samples over 15.6 ms. **Possible Causes:** • APS1 sensor signal circuit is open or shorted to ground • APS1 sensor signal circuit is shorted to VREF or system power • APS1 sensor power circuit is open between sensor and PCM • APS1 sensor is damaged or has failed • PCM has failed • TSB 01-047 (8/14/01) contains a repair procedure for this code
DTC: P1280 **1T CCM, MIL:** Yes **2000, 2001, 2002** **Models:** Passport **Engines:** 3.2L VIN W **Transmissions:** A/T, M/T	**Acceleration Position Sensor 2 Circuit Malfunction** Key on or engine running, and the PCM detected the Acceleration Position Sensor 2 (APS2) circuit was less than 2.5% or over 97.5% of VREF (5v) for 12 counts within 500 test samples over 15.6 ms. **Possible Causes:** • APS2 sensor signal circuit is open or shorted to ground • APS2 sensor signal circuit is shorted to VREF or system power • APS2 sensor power circuit is open between sensor and PCM • APS2 sensor is damaged or has failed • PCM has failed • TSB 01-047 (8/14/01) contains a repair procedure for this code
DTC: P1285 **1T CCM, MIL:** Yes **2000, 2001, 2002** **Models:** Passport **Engines:** 3.2L VIN W **Transmissions:** A/T, M/T	**Acceleration Position Sensor 3 Circuit Malfunction** Key on or engine running, and the PCM detected the Acceleration Position Sensor 3 (APS3) circuit was less than 2.5% or over 97.5% of VREF (5v) for 12 counts within 500 test samples over 15.6 ms. **Possible Causes:** • APS3 sensor signal circuit is open or shorted to ground • APS3 sensor signal circuit is shorted to VREF or system power • APS3 sensor power circuit is open between sensor and PCM • APS3 sensor is damaged or has failed • PCM has failed • TSB 01-047 (8/14/01) contains a repair procedure for this code
DTC: P1290 **1T CCM, MIL:** Yes **2000, 2001, 2002** **Models:** Passport **Engines:** 3.2L VIN W **Transmissions:** A/T, M/T	**Electronic Throttle Control Forced Idle Mode** Key on or engine running, and the PCM detected the Electronic Throttle Control (ETC) system had entered Forced Idle Mode. **Possible Causes:** • ETC module has detected a problem in the system • Forced Idle Mode is "active" • ETC module is damaged or has failed
DTC: P1295 **1T CCM, MIL:** Yes **2000, 2001, 2002** **Models:** Passport **Engines:** 3.2L VIN W **Transmissions:** A/T, M/T	**Electronic Throttle Control Power Management Mode** Key on or engine running, and the PCM detected the Electronic Throttle Control (ETC) system had entered Power Management Mode and is operating under "failsafe" mode conditions. **Possible Causes:** • ETC module has detected a problem in the system • Power Management Mode is "active" • ETC module is damaged or has failed
DTC: P1297 **1T CCM, MIL:** Yes **1996, 1997, 1998, 1999, 2000,** **2001, 2002** **Models:** Accord, Civic, Civic del Sol, CR-V, Element, Insight, Odyssey, Pilot, Prelude, Ridgeline, S2000 **Engines:** All **Transmissions:** A/T, M/T	**Electrical Load Detector Circuit Low Input** Engine running at hot idle speed or at cruise speed, headlights "on", and the PCM detected the ELD signal was less than a stored value. **Possible Causes:** • ELD sensor signal circuit is open or shorted to ground • ELD sensor power circuit is open or shorted to ground • ELD sensor is damaged or has failed • PCM has failed

DTC	Trouble Code Title, Conditions & Possible Causes
DTC: P1298 **1T CCM, MIL: Yes** **1996, 1997, 1998, 1999, 2000, 2001, 2002** **Models:** Accord, Civic, Civic del Sol, CR-V, Element, Insight, Odyssey, Pilot, Prelude, Ridgeline, S2000 **Engines:** All **Transmissions:** A/T, M/T	**Electrical Load Detector Circuit High Input** Engine running at hot idle speed or at cruise speed, headlights "on", and the PCM detected the ELD signal was more than a stored value. **Possible Causes:** • ELD sensor signal circuit is shorted to VREF • ELD sensor signal circuit is shorted to system power (B+) • ELD sensor is damaged or has failed • PCM has failed
DTC: P1299 **1T CCM, MIL: Yes** **2000, 2001, 2002** **Models:** Passport **Engines:** 3.2L VIN W **Transmissions:** A/T, M/T	**Electronic Throttle Control Forced Engine Shutdown Mode** Key on or engine running, and the PCM detected the Electronic Throttle Control (ETC) system had entered Forced Engine Shutdown Mode and is operating under "failsafe" mode conditions. **Possible Causes:** • ETC module has detected a problem in the system • ETC module is damaged or has failed
DTC: P1310 **1T CCM, MIL: Yes** **2000, 2001, 2002** **Models:** Passport **Engines:** 3.2L VIN W **Transmissions:** A/T, M/T	**ION Sensing Module Diagnosis** No CKP or System Voltage codes set, engine started, engine speed from 650-6500 rpm, system voltage from 10-16v, MAF sensor signal from 26-100 g/sec, fuel level indicating more than 10%, and the PCM detected "missing" Combustion Quality (CQ) pulses, multiple CQ pulses or CQ pulsewidth calculation errors in the CQ quality test. **Possible Causes:** • Secondary Line 1 circuit is open or shorted to ground • ION sensing module is damaged or has failed
DTC: P1311 **1T CCM, MIL: Yes** **2000, 2001, 2002** **Models:** Passport **Engines:** 3.2L VIN W **Transmissions:** A/T, M/T	**ION Sensing Module Secondary Line 1 Circuit Malfunction** Engine started, engine speed from 650-6500 rpm, system voltage from 10-16v, MAF sensor signal from 26-100 g/sec, fuel level indicating more than 10%, and the PCM detected an unexpected voltage condition on the Secondary Line 1 circuit to the ION Module. **Possible Causes:** • Secondary Line 1 circuit is open or shorted to ground • Secondary Line 1 circuit is shorted to system power (B+) • ION sensing module is damaged or has failed
DTC: P1312 **1T CCM, MIL: Yes** **1T CCM, MIL: Yes** **2000, 2001, 2002** **Models:** Passport **Engines:** 3.2L VIN W **Transmissions:** A/T, M/T	**ION Sensing Module Secondary Line 2 Circuit Malfunction** Engine started, engine speed from 650-6500 rpm, system voltage from 10-16v, MAF sensor signal from 26-100 g/sec, fuel level indicating more than 10%, and the PCM detected an unexpected voltage condition on the Secondary Line 2 circuit to the ION Module. **Possible Causes:** • Secondary Line 2 circuit is open or shorted to ground • Secondary Line 2 circuit is shorted to system power (B+) • ION sensing module is damaged or has failed
DTC: P1324 **1T CCM, MIL: Yes** **2001, 2002, 2003, 2004, 2005, 2006** **Models:** Civic **Engines:** 1.7L VIN ES1, 1.7L VIN ES2 **Transmissions:** A/T, M/T	**Knock Sensor Power Source Circuit Low Input** Key on or engine running, and the PCM detected a low input on the Knock Sensor Power Source circuit during the CCM test. **Possible Causes:** • Knock sensor power source circuit is open • Knock sensor power source circuit is shorted to ground • PCM has failed
DTC: P1326 **1T CCM, MIL: Yes** **2000, 2001, 2002** **Models:** Passport **Engines:** 3.2L VIN W **Transmissions:** A/T, M/T	**ION Sensing Module Combustion Quality** No CKP or CMP Sensor codes set, engine started, system voltage from 10-16v, and the PCM detected an unexpected voltage condition on the Combustion Quality (CQ) line circuit, faults in the ION Module or one or more faults in the analog input signals to the PCM. **Possible Causes:** • CQ line circuit is open or shorted to ground • One or more analog inputs to the PCM has failed or is missing • ION Sensing Module is damaged or has failed

DTC	Trouble Code Title, Conditions & Possible Causes
DTC: P1336 **1T CCM, MIL: Yes** **1996, 1997, 1998, 1999, 2000** **Models:** Civic, Civic del Sol, CR-V **Engines:** All **Transmissions:** A/T, M/T	**Crankshaft Speed Fluctuation Sensor Circuit Malfunction** Engine running, and PCM detected an unexpected or intermittent interruption of the crankshaft speed fluctuation (CSF) sensor circuit. **Possible Causes:** • CSF signal circuit is open or shorted to ground • CSF signal circuit is shorted to VREF or system power (B+) • CSF is damaged or has failed • CSF pickup assembly or its pulse rotor is damaged • PCM has failed
DTC: P1336 **1T CCM, MIL: Yes** **2001, 2002, 2003, 2004, 2005, 2006** **Models:** CR-V **Engines:** All **Transmissions:** A/T, M/T	**Engine Speed Fluctuation Sensor Circuit Malfunction** Engine running, and PCM detected an unexpected or intermittent interruption of the engine speed fluctuation (ESF) sensor signal. **Possible Causes:** • ESF signal circuit is open or shorted to ground • ESF signal circuit is shorted to VREF or system power (B+) • ESF is damaged or has failed • ESF pickup assembly or its pulse rotor is damaged • PCM has failed
DTC: P1336 **1T CCM, MIL: Yes** **2001, 2002, 2003, 2004, 2005, 2006** **Models:** CR-V **Engines:** All **Transmissions:** A/T, M/T	**Engine Speed Fluctuation Sensor Circuit Malfunction** Engine running, and PCM detected an unexpected or intermittent interruption of the engine speed fluctuation (ESF) sensor signal. **Possible Causes:** • ESF signal circuit is open or shorted to ground • ESF signal circuit is shorted to VREF or system power (B+) • ESF is damaged or has failed • ESF pickup assembly or its pulse rotor is damaged • PCM has failed
DTC: P1337 **1T CCM, MIL: Yes** **1996, 1997, 1998, 1999, 2000** **Models:** Civic, Civic del Sol, CR-V **Engines:** All **Transmissions:** A/T, M/T	**Crankshaft Speed Fluctuation Sensor No Signal** Engine running, and the PCM detected that it did not receive any signals from the Crankshaft Speed Fluctuation (CSF) sensor. **Possible Causes:** • CSF signal circuit is open or shorted to ground • CSF signal circuit is shorted to VREF or system power (B+) • CSF is damaged or has failed • CSF pickup assembly or its pulse rotor is damaged • PCM has failed
DTC: P1340 **1T CCM, MIL: Yes** **2000, 2001, 2002** **Models:** Passport **Engines:** 3.2L VIN W **Transmissions:** A/T, M/T	**ION Sensing Module Cylinder ID Malfunction** No ECT, Fuel Trim, Injector, Misfire or System Voltage codes set, engine started, system voltage from 10-16v, and the PCM detected that the "cylinder synchronization" routine had not been completed after a predetermined number of events occurred after startup. **Possible Causes:** • Knock sensor and Combustion Line connectors are swapped • One or more spark plugs (or boots) are shorted to ground • ION Sensing Module is damaged or has failed • PCM related hardware is damaged or has failed
DTC: P1359 **1T CCM, MIL: Yes** **1996, 1997, 1998, 1999, 2000, 2001, 2002, 2003, 2004, 2005, 2006** **Models:** Accord, CR-V, Prelude **Engines:** All **Transmissions:** A/T, M/T	**CKP/TDC Sensor Circuit Malfunction** Engine running, and the PCM detected an unexpected voltage condition on the CKP/TDC sensor circuit during the CCM test. **Possible Causes:** • TDC signal circuit is open or shorted to ground • TDC signal circuit is shorted to VREF or system power (B+) • TDC pickup assembly or its pulse rotor is damaged • PCM has failed
DTC: P1359 **1T CCM, MIL: Yes** **1T CCM, MIL: Yes** **1996, 1997, 1998, 1999, 2000** **Models:** Civic, Civic del Sol **Engines:** All **Transmissions:** A/T, M/T	**CKP/TDC Sensor Circuit Malfunction** Engine running, and the PCM detected an unexpected voltage condition on the CKP/TDC sensor circuit during the CCM test. **Possible Causes:** • TDC signal circuit is open or shorted to ground • TDC signal circuit is shorted to VREF or system power (B+) • TDC pickup assembly or its pulse rotor is damaged • PCM has failed

DTC	Trouble Code Title, Conditions & Possible Causes
DTC: P1359 **1T CCM, MIL: Yes** **1996, 1997, 1998, 1999** **Models:** Odyssey, Pilot, Ridgeline **Engines:** 2.2L VIN RA1, 2.3L VIN RA3 **Transmissions:** A/T, M/T	**CKP/TDC Sensor Circuit Malfunction** Engine running, and the PCM detected an unexpected voltage condition on the CKP/TDC sensor circuit during the CCM test. **Possible Causes:** • TDC signal circuit is open or shorted to ground • TDC signal circuit is shorted to VREF or system power (B+) • TDC pickup assembly or its pulse rotor is damaged • PCM has failed
DTC: P1361 **2T CCM, MIL: Yes** **1995** **Models:** Accord **Engines:** 2.7L VIN CE6 **Transmissions:** A/T, M/T	**Top Dead Center Sensor Intermittent Signal** Engine running, and the PCM detected an unexpected or intermittent interruption of the Top Dead Center (TDC) sensor signal. **Possible Causes:** • TDC signal circuit is open or shorted to ground • TDC signal circuit is shorted to VREF or system power • TDC pickup assembly or its pulse rotor is damaged • PCM has failed
DTC: P1361 **1T CCM, MIL: Yes** **1996, 1997, 1998, 1999, 2000, 2001, 2002, 2003, 2004, 2005, 2006** **Models:** Accord, Civic, Civic del Sol, CR-V, Insight, Odyssey, Pilot, Prelude, Ridgeline, S2000 **Engines:** All **Transmissions:** A/T, M/T	**Top Dead Center Sensor 1 Circuit Intermittent Signal** Engine running, and the PCM detected an unexpected or intermittent interruption of the Top Dead Center 1 (TDC1) sensor signal. **Possible Causes:** • TDC1 signal circuit is open or shorted to ground • TDC1 signal circuit is shorted to VREF or system power • TDC1 pickup assembly or its pulse rotor is damaged • TDC1 is damaged or has failed • PCM has failed
DTC: P1362 **2T CCM, MIL: Yes** **1995** **Models:** Accord **Engines:** 2.7L VIN CE6 **Transmissions:** A/T, M/T	**Top Dead Center Sensor No Signal** Engine cranking or running, and the PCM did not receive any signals from the Top Dead Center (TDC) sensor during the CCM test. **Note: The engine will start and run without the TDC sensor 1 signal.** **Possible Causes:** • TDC signal circuit is open or shorted to ground • TDC pickup assembly or its pulse rotor is damaged • PCM has failed
DTC: P1362 **1T CCM, MIL: Yes** **1996, 1997, 1998, 1999, 2000, 2001, 2002, 2003, 2004, 2005, 2006** **Models:** Accord, Civic, Civic del Sol, CR-V, Insight, Odyssey, Pilot, Prelude, Ridgeline, S2000 **Engines:** All **Transmissions:** A/T, M/T	**Top Dead Center Sensor 1 No Signal** Engine cranking or running, and the PCM did not receive any signals from the Top Dead Center 1 (TDC1) sensor during the CCM test. **Note: The engine will start and run without the TDC sensor 1 signal.** **Possible Causes:** • TDC1 signal circuit is open or shorted to ground • TDC1 pickup assembly or its pulse rotor is damaged • TDC1 is damaged or has failed • PCM has failed
DTC: P1366 **1T CCM, MIL: Yes** **1998, 1999, 2000, 2001, 2002, 2003, 2004, 2005, 2006** **Models:** Accord **Engines:** 3.0L VIN CG1, 3.0L VIN CG2 **Transmissions:** A/T, M/T	**Top Dead Center Sensor 2 Circuit Intermittent Signal** Engine running and the PCM detected an unexpected or intermittent interruption of the Top Dead Center Sensor 2 (TDC2) signal. **Possible Causes:** • TDC2 signal circuit is open or shorted to ground • TDC2 signal circuit is shorted to VREF or system power • TDC2 pickup assembly or its pulse rotor is damaged • TDC2 is damaged or has failed • PCM has failed
DTC: P1366 **1T CCM, MIL: Yes** **1998, 1999, 2000, 2001, 2002, 2003, 2004, 2005, 2006** **Models:** Insight, Odyssey, Pilot, Ridgeline, S2000 **Engines:** All **Transmissions:** A/T, M/T	**Top Dead Center Sensor 2 Circuit Intermittent Signal** Engine running and the PCM detected an unexpected or intermittent interruption of the Top Dead Center Sensor 2 (TDC2) signal. **Possible Causes:** • TDC2 signal circuit is open or shorted to ground • TDC2 signal circuit is shorted to VREF or system power • TDC2 pickup assembly or its pulse rotor is damaged • TDC2 is damaged or has failed • PCM has failed

DTC	Trouble Code Title, Conditions & Possible Causes
DTC: P1367 **1T CCM, MIL: Yes** **1998, 1999, 2000, 2001, 2002,** **2003, 2004, 2005, 2006** **Models:** Accord **Engines:** 3.0L VIN CG1, 3.0L VIN CG2 **Transmissions:** A/T, M/T	**Top Dead Sensor 2 No Signals** Engine cranking or running, and the PCM did not detect any signals from the Top Dead Center Sensor 2 (TDC2) during the CCM test. **Note: The engine will start and run without the TDC sensor 2 signal.** **Possible Causes:** • TDC2 signal circuit is open or shorted to ground • TDC2 pickup assembly or its pulse rotor is damaged • TDC2 is damaged or has failed • PCM has failed
DTC: P1367 **1T CCM, MIL: Yes** **1998, 1999, 2000, 2001, 2002,** **2003, 2004, 2005, 2006** **Models:** Insight, Odyssey, Pilot, Ridgeline, S2000 **Engines:** All **Transmissions:** A/T, M/T	**Top Dead Sensor 2 No Signals** Engine cranking or running, and the PCM did not detect any signals from the Top Dead Center Sensor 2 (TDC2) during the CCM test. **Note: The engine will start and run without the TDC sensor 2 signal.** **Possible Causes:** • TDC2 signal circuit is open or shorted to ground • TDC2 pickup assembly or its pulse rotor is damaged • TDC2 is damaged or has failed • PCM has failed
DTC: P1381 **2T CCM, MIL: Yes** **1995, 1996, 1997** **Models:** Accord **Engines:** 2.7L VIN CE6 **Transmissions:** A/T, M/T	**Camshaft Position Sensor 1 Circuit Malfunction** Engine running and the PCM detected an unexpected or intermittent interruption of the Camshaft Position (CMP) sensor 'A' signal. **Possible Causes:** • CMP signal circuit is open or shorted to ground • CMP signal circuit is shorted to VREF or system power • CMP pickup assembly or CMP sensor is damaged or has failed • PCM has failed
DTC: P1381 **1T CCM, MIL: Yes** **1996, 1997, 1998, 1999, 2000,** **2001, 2002, 2003, 2004, 2005,** **2006** **Models:** Accord **Engines:** 2.2L VIN CD5, 2.2L VIN CD7, 2.2L VIN CE1, 2.3L VIN CF8, 2.3L VIN CG3, 2.3L VIN CG5, 2.3L VIN CG6 **Transmissions:** A/T, M/T	**Camshaft Position Sensor 1 Circuit Malfunction** Engine running and the PCM detected an unexpected or intermittent interruption of the Camshaft Position (CMP) sensor 1 signal. **Possible Causes:** • CMP signal circuit is open or shorted to ground • CMP signal circuit is shorted to VREF or system power • CMP pickup assembly or CMP sensor is damaged or has failed • PCM has failed
DTC: P1381 **1T CCM, MIL: Yes** **1996, 1997, 1998, 1999, 2000** **Models:** Civic, Civic del Sol **Engines:** All **Transmissions:** A/T, M/T	**Camshaft Position Sensor 1 Circuit Malfunction** Engine running and the PCM detected an unexpected or intermittent interruption of the Camshaft Position (CMP) sensor 'A' signal. **Possible Causes:** • CMP signal circuit is open or shorted to ground • CMP signal circuit is shorted to VREF or system power • CMP pickup assembly or CMP sensor is damaged or has failed • PCM has failed
DTC: P1381 **1T CCM, MIL: Yes** **1996, 1997, 1998, 1999** **Models:** Odyssey, Pilot, Ridgeline **Engines:** 2.2L VIN RA1, 2.3L VIN RA3 **Transmissions:** A/T, M/T	**Camshaft Position Sensor 1 Circuit Malfunction** Engine running and the PCM detected an unexpected or intermittent interruption of the Camshaft Position (CMP) sensor 1 signal. **Possible Causes:** • CMP signal circuit is open or shorted to ground • CMP signal circuit is shorted to VREF or system power • CMP pickup assembly or CMP sensor is damaged or has failed • PCM has failed
DTC: P1381 **1T CCM, MIL: Yes** **1996, 1997, 1998, 1999, 2000,** **2001, 2002, 2003, 2004, 2005,** **2006** **Models:** CR-V, Prelude **Engines:** All **Transmissions:** A/T, M/T	**Camshaft Position Sensor 1 Circuit Malfunction** Engine running and the PCM detected an unexpected or intermittent interruption of the Camshaft Position (CMP) sensor 1 signal. **Possible Causes:** • CMP signal circuit is open or shorted to ground • CMP signal circuit is shorted to VREF or system power • CMP pickup assembly or CMP sensor is damaged or has failed • PCM has failed

DTC	Trouble Code Title, Conditions & Possible Causes
DTC: P1382 **2T CCM, MIL: Yes** **1995, 1996, 1997** **Models:** Accord **Engines:** 2.7L VIN CE6 **Transmissions:** A/T, M/T	**Camshaft Position Sensor 1 No Signal** Engine cranking or running, and the PCM did not detect any signals from the Camshaft Position (CMP) sensor 1 during the CCM test. **Note: The engine will start and run without the CMP sensor 1 signal.** **Possible Causes:** • CMP signal circuit is open or shorted to ground • CMP pickup assembly or CMP sensor is damaged or has failed • PCM has failed
DTC: P1382 **1T CCM, MIL: Yes** **1996, 1997, 1998, 1999, 2000,** **2001, 2002, 2003, 2004, 2005,** **2006** **Models:** Accord **Engines:** 2.2L VIN CD5, 2.2L VIN CD7, 2.2L VIN CE1, 2.3L VIN CF8, 2.3L VIN CG3, 2.3L VIN CG5, 2.3L VIN CG6 **Transmissions:** A/T, M/T	**Camshaft Position Sensor 1 No Signal** Engine cranking or running, and the PCM did not detect any signals from the Camshaft Position (CMP) sensor 1 during the CCM test. **Note: The engine will start and run without the CMP sensor 1 signal.** **Possible Causes:** • CMP signal circuit is open or shorted to ground • CMP pickup assembly or CMP sensor is damaged or has failed • PCM has failed
DTC: P1382 **1T CCM, MIL: Yes** **1996, 1997, 1998, 1999, 2000** **Models:** Civic, Civic del Sol **Engines:** All **Transmissions:** A/T, M/T	**Camshaft Position Sensor 1 No Signal** Engine cranking or running, and the PCM did not detect any signals from the Camshaft Position (CMP) sensor 1 during the CCM test. **Note: The engine will start and run without the CMP sensor 1 signal.** **Possible Causes:** • CMP signal circuit is open or shorted to ground • CMP pickup assembly or CMP sensor is damaged or has failed • PCM has failed
DTC: P1382 **1T CCM, MIL: Yes** **1996, 1997, 1998, 1999** **Models:** Odyssey, Pilot, Ridgeline **Engines:** 2.2L VIN RA1, 2.3L VIN RA3 **Transmissions:** A/T, M/T	**Camshaft Position Sensor 1 No Signal** Engine cranking or running, and the PCM did not detect any signals from the Camshaft Position (CMP) sensor 1 during the CCM test. **Note: The engine will start and run without the CMP sensor 1 signal.** **Possible Causes:** • CMP signal circuit is open or shorted to ground • CMP pickup assembly or CMP sensor is damaged or has failed • PCM has failed
DTC: P1382 **1T CCM, MIL: Yes** **1996, 1997, 1998, 1999, 2000,** **2001, 2002, 2003, 2004, 2005,** **2006** **Models:** CR-V, Prelude **Engines:** All **Transmissions:** A/T, M/T	**Camshaft Position Sensor 1 No Signal** Engine cranking or running, and the PCM did not detect any signals from the Camshaft Position (CMP) sensor 1 during the CCM test. **Note: The engine will start and run without the CMP sensor 1 signal.** **Possible Causes:** • CMP signal circuit is open or shorted to ground • CMP pickup assembly or CMP sensor is damaged or has failed • PCM has failed
DTC: P1390 **1T CCM, MIL: No** **1996, 1997** **Models:** Passport **Engines:** 2.6L VIN E, 3.2L VIN V **Transmissions:** A/T, M/T	**Acceleration 'G' Sensor Intermittent Low Input** Engine started, engine running, and the PCM detected an unexpected low signal of less than 0.5v on the 'G' Sensor circuit over a 6.25 second period. The 'G' sensor is used to sense vertical acceleration due to road vibration during Misfire diagnostics. **Note: For additional help with this code, view the Failure Records.** **Possible Causes:** • 'G' sensor signal circuit is shorted to ground • 'G' sensor is damaged or has failed • PCM has failed

DTC	Trouble Code Title, Conditions & Possible Causes
DTC: P1391 **1T CCM, MIL: No** **1996, 1997** **Models:** Passport **Engines:** 2.6L VIN E, 3.2L VIN V **Transmissions:** A/T, M/T	**Acceleration 'G' Sensor Range/Performance** Engine started, engine running, vehicle speed indicating (0) mph, and the PCM detected the G-Sensor signal was more than 2.5v, or it was less than 1.5v for 1 minute, or with vehicle speed at 30-80 mph, the G-Sensor signal indicated a change of 0.0002v each 10 ms. **Note: For additional help with this code, view the Failure Records.** **Possible Causes:** • 'G' sensor signal circuit is shorted to ground (intermittent fault) • 'G' sensor VREF circuit is open or shorted to ground (this circuit is also connected to the MAP sensor - check the MAP sensor) • 'G' sensor ground circuit is open (an intermittent fault) • 'G' sensor is damaged or has failed • PCM has failed
DTC: P1392 **1T CCM, MIL: Yes** **1996, 1997** **Models:** Passport **Engines:** 2.6L VIN E, 3.2L VIN V **Transmissions:** A/T, M/T	**Acceleration 'G' Sensor Circuit Low Input** Engine started, engine running, and the PCM detected the 'G' Sensor signal indicated less than 0.5v for 12.5 seconds over a 25 second period of time during the CCM test. **Possible Causes:** • 'G' sensor signal circuit is shorted to ground • 'G' sensor VREF circuit is open or shorted to ground (this circuit is also connected to the MAP sensor - check the MAP sensor) • 'G' sensor is damaged or has failed • PCM has failed
DTC: P1393 **1T CCM, MIL: Yes** **1996, 1997** **Models:** Passport **Engines:** 2.6L VIN E, 3.2L VIN V **Transmissions:** A/T, M/T	**Acceleration 'G' Sensor Circuit High Input** Engine started, engine running, and the PCM detected the 'G' Sensor signal indicated more than 4.5v for 12.5 seconds over a 25 second period of time during the CCM test. **Possible Causes:** • 'G' sensor signal circuit is open between the sensor and PCM • 'G' sensor ground circuit is open between sensor and ground • 'G' sensor is damaged or has failed • PCM has failed
DTC: P1394 **1T CCM, MIL: No** **1996, 1997** **Models:** Passport **Engines:** 2.6L VIN E, 3.2L VIN V **Transmissions:** A/T, M/T	**Acceleration 'G' Sensor Intermittent High Input** Engine started, engine running, and the PCM detected the 'G' Sensor signal indicated an intermittent high signal (4.8v) for 12.5 seconds over a 25 second time period. This sensor is used to sense vertical acceleration due to road vibration during Misfire diagnostics. **Possible Causes:** • 'G' sensor signal circuit is open (an intermittent fault) • 'G' sensor ground circuit is open (an intermittent fault) • 'G' sensor is damaged or has failed (an intermittent fault) • PCM has failed
DTC: P1404 **2T EGR2, MIL: Yes** **1998, 1999, 2000, 2001, 2002** **Models:** Passport **Engines:** 3.2L VIN W **Transmissions:** A/T, M/T	**EGR Valve Stuck Closed Malfunction** Engine started, engine running, system voltage from 11-16v, ECT sensor from 176-248, IAT sensor less than 212°F, Desired EGR valve position is 0%, and the PCM detected the difference between the Actual and Desired EGR pintle position was over 30% for 5 seconds, test must fail three times in one trip to set this code. **Possible Causes:** • EGR valve sticking or binding (check for deposits on the valve) • EGR valve is damaged or has failed (if the valve shows signs of excessive heat, check the converter and pipes for a restriction) • PCM has failed
DTC: P1406 **2T CCM, MIL: Yes** **1996, 1997** **Models:** Passport **Engines:** 2.6L VIN E, 3.2L VIN V **Transmissions:** A/T, M/T	**EGR Valve Pintle Position Sensor Circuit Malfunction** Engine started, system voltage over 12v, engine running with the EGR valve commanded to 0%, and the PCM detected the Actual EGR position was 0.20v more or less than the EGR closed valve position for 5 seconds; or with the EGR valve commanded to more than 0%, the PCM detected the Actual EGR position was more than 15% greater than, or 15% less than the Desired EGR position for 5 seconds; or the PCM detected the Actual EGR position was less than 0.10v for over 5 seconds at any time during the CCM test. **Possible Causes:** • EGR position sensor signal circuit is open (intermittent fault) • EGR valve sticking or binding (check for moisture or deposits on the EGR valve or on the valve seat - the valve may lock up) • EGR valve is damaged or has failed (if the valve shows signs of excessive heat, check the converter and pipes for a restriction) • PCM has failed

DTC	Trouble Code Title, Conditions & Possible Causes
DTC: P1410 **2T AIR, MIL: Yes** 2000, 2001, 2002, 2003, 2004, 2005, 2006 **Models:** S2000 **Engines:** All **Transmissions:** M/T	**Air Pump System Malfunction** Cold startup completed (ECT sensor from 32-158°F at startup), and then with engine running for 10 seconds at idle speed under no load conditions after startup, the PCM detected the amount of airflow was incorrect with the Air System air pump commanded "on". **Possible Causes:** • Secondary AIR system component problem (check the air pump, air injection hoses, or AIR relay or AIR solenoid • PCM has failed
DTC: P1415 **1T CCM, MIL: Yes** 2000, 2001, 2002, 2003, 2004, 2005, 2006 **Models:** S2000 **Engines:** All **Transmissions:** M/T	**AIR Pump Electric Current Sensor Signal Low Input** Key on or engine running, and the PCM detected a low voltage condition on the Secondary Air Pump control circuit during the test. **Possible Causes:** • AIR pump electric current sensor circuit is open • AIR pump electric current sensor circuit is shorted to ground • AIR pump electric current sensor power circuit is open • PCM has failed
DTC: P1416 **2T CCM, MIL: Yes** 2000, 2001, 2002, 2003, 2004, 2005, 2006 **Models:** S2000 **Engines:** All **Transmissions:** M/T	**AIR Pump Electric Current Sensor Signal High Input** Key on or engine running, and the PCM detected a high voltage condition on the Secondary Air Pump control circuit during the test. **Possible Causes:** • Air pump electric current sensor power circuit is open • Air pump electric current sensor circuit shorted to system power • PCM has failed
DTC: P1441 **2T CCM, MIL: Yes** 1996, 1997 **Models:** Passport **Engines:** 2.6L VIN E **Transmissions:** A/T, M/T	**EVAP Vacuum Switch Circuit High Input** No IAT, MAP or TP sensor codes set, system voltage at 11-16v, ECT and IAT sensor signals more than 41°F, startup ECT sensor signal less than 122°F, key on (prior to engine cranking), and the PCM detected the EVAP Purge Switch indicated 12v (open circuit) at least 4 out of the last 8 times the ignition switch was turned "on". **Possible Causes:** • Vacuum switch signal circuit is open between switch and PCM • Vacuum switch ground circuit open between switch and ground • Vacuum switch is damaged or has failed • PCM has failed
DTC: P1441 **2T CCM, MIL: Yes** 1996, 1997, 1998, 1999, 2000, 2001, 2002 **Models:** Passport **Engines:** 3.2L VIN V, 3.2L VIN W **Transmissions:** A/T, M/T	**EVAP Vacuum Switch Circuit High Input** DTC P0106, P0107, P0108, P0112, P0113, P0121, P0122, P0123, P1640 and P1650 not set, system voltage at 11-16v, IAT sensor more than 32°F, fuel level from 15-85%, and the PCM detected a continuous "open" purge condition during the EVAP Monitor test. **Possible Causes:** • Vacuum switch signal circuit is open between switch and PCM • Vacuum switch ground circuit open between switch and ground • Vacuum switch is damaged or has failed • PCM has failed
DTC: P1454 **2T EVADTC: P1, MIL: Yes** 2003, 2004, 2005, 2006 **Models:** Element **Engines:** All **Transmissions:** All	**Fuel Tank Pressure Sensor Range/Performance Problem** Cold startup completed (IAT sensor signal from 32-86°F at engine startup), vehicle driven at over 5 mph for over 2 minutes, then with ECT sensor signal more than 154°F and the EVAP Control and Vent solenoids enabled, the PCM detected the fuel tank pressure was incorrect due to a leak in the fuel tank area during the Leak Test. **Possible Causes:** • Fuel tank cap damaged, loose or the wrong part number • Fuel tank leaks at the fuel fill pipe or at the fuel tank seals • Fuel vapor control valve is damaged or has failed • Fuel tank vapor recirculation valve or vapor tube is damaged • Fuel tank vapor control vent tube is damaged or has failed
DTC: P1456 **2T EVADTC: P1, MIL: Yes** 1998, 1999, 2000, 2001, 2002, 2003, 2004, 2005, 2006 **Models:** Accord, Civic, CR-V, Insight, Odyssey, Pilot, Prelude, Ridgeline, S2000 **Engines:** All **Transmissions:** A/T, M/T	**EVAP System Leak Detected (Fuel Tank Area)** Cold startup completed (IAT sensor signal from 32-86°F at engine startup), vehicle driven at over 5 mph for over 2 minutes, then with ECT sensor signal more than 154°F and the EVAP Control and Vent solenoids enabled, the PCM detected the fuel tank pressure was incorrect due to a leak in the fuel tank area during the Leak Test. **Possible Causes:** • Fuel tank cap damaged, loose or the wrong part number • Fuel tank leaks at the fuel fill pipe or at the fuel tank seals • Fuel vapor control valve is damaged or has failed • Fuel tank vapor recirculation valve or vapor tube is damaged • Fuel tank vapor control vent tube is damaged or has failed

DTC	Trouble Code Title, Conditions & Possible Causes
DTC: P1457 **2T EVADTC: P1, MIL: Yes** **1998, 1999, 2000, 2001, 2002, 2003, 2004, 2005, 2006** **Models:** Accord, Civic, CR-V, Insight, Odyssey, Pilot, Prelude, Ridgeline, S2000 **Engines:** All **Transmissions:** A/T, M/T	**EVAP System Leak Detected (Canister Area)** Cold startup completed (IAT sensor signal from 32-86°F at engine startup), vehicle driven at over 5 mph for over 2 minutes, then with ECT sensor signal more than 154°F and the EVAP Control and Vent solenoids enabled, the PCM detected the fuel tank pressure was incorrect due to a leak in the canister area during the Leak Test. **Possible Causes:** • EVAP canister is leaking, damaged or full of water • EVAP canister purge line is loose, damaged or blocked • EVAP two-way valve or ORVR vent shut valve is damaged • EVAP fuel tank vapor control valve is damaged or has failed • PCM has failed
DTC: P1459 **1T CCM, MIL: Yes** **1995, 1996, 1997** **Models:** Accord **Engines:** 2.7L VIN CE6 **Transmissions:** A/T, M/T	**EVAP Purge Flow Switch Circuit Malfunction** Engine running at cruise speed and than back to idle speed, ECT sensor signal more than 154°F, and the PCM detected an unexpected voltage condition on Purge Flow switch circuit. **Possible Causes:** • Purge flow switch is open, shorted to ground or to power (B+) • Purge flow switch vacuum line is loose, bent or damaged • Purge flow switch is damaged or has failed • PCM has failed
DTC: P1486 **2T ECT, MIL: Yes** **2000, 2001, 2002, 2003, 2004, 2005, 2006** **Models:** Accord **Engines:** 2.3L VIN CF8, 2.3L VIN CG3, 2.3L VIN CG5, 2.3L VIN CG6 **Transmissions:** A/T, M/T	**Thermostat Range/Performance Malfunction** DTC P0107, P0108, P0112, P0113, P0116, P0117, P0118, P0300, P0301, P0302, P0303, P0304, P0305, P0306, P0335, P0336, P0401, P0500, P0505, P1106, P1107, P1108, P1128, P1129, P1253, P1257, P1258, P1259, P1359, P1399, P1491, P1498 and P1519 not set, vehicle driven for over 10 minutes, and the PCM detected the ECT signal did not reach the correct closed loop value. **Note: This trouble code can set if the engine remains under hot idle conditions with the hood open for an extended period of time.** **Possible Causes:** • ECT sensor is out of calibration • Check for low coolant level or incorrect coolant mixture • Cooling system component failure (thermostat is stuck open) • TSB 01-064 (9/11/01) contains a repair procedure for this code
DTC: P1486 **2T ECT, MIL: Yes** **2000, 2001, 2002, 2003, 2004, 2005, 2006** **Models:** Insight **Engines:** All **Transmissions:** A/T, M/T	**Thermostat Range/Performance Malfunction** DTC P0107, P0108, P0112, P0113, P0116, P0117, P0118, P0300, P0301, P0302, P0303, P0304, P0305, P0306, P0335, P0336, P0401, P0500, P0505, P1106, P1107, P1108, P1128, P1129, P1253, P1257, P1258, P1259, P1359, P1399, P1491, P1498 and P1519 not set, vehicle driven for over 10 minutes, and the PCM detected the ECT signal did not reach the correct closed loop value. **Note: This trouble code can set if the engine remains under hot idle conditions with the hood open for an extended period of time.** **Possible Causes:** • ECT sensor is out of calibration • Check for low coolant level or incorrect coolant mixture • Cooling system component failure (thermostat is stuck open)
DTC: P1491 **2T CCM, MIL: Yes** **1995** **Models:** Accord **Engines:** 2.7L VIN CE6 **Transmissions:** A/T, M/T	**EGR Valve Lift Sensor Insufficient Flow Detected** Vehicle driven in closed loop at 1700-2500 rpm for over 10 minutes, and the PCM detected the EGR valve lift sensor (EGRV) signal indicated insufficient EGR flow during the EGR Monitor test. **Possible Causes:** • EGR valve lift sensor is stuck, damaged or has failed • EGR control solenoid circuit is open or shorted to ground • EGR control solenoid valve is damaged or has failed • PCM has failed
DTC: P1491 **2T EGR1, MIL: Yes** **1996, 1997, 1998, 1999, 2000, 2001, 2002, 2003, 2004, 2005, 2006** **Models:** Accord, Insight, Odyssey, Pilot, Prelude, Ridgeline **Engines:** All **Transmissions:** A/T, M/T	**EGR Valve Lift Sensor Insufficient Flow Detected** Vehicle driven in closed loop at 1700-2500 rpm for over 10 minutes, and the PCM detected the EGR valve lift sensor (EGRV) signal indicated insufficient EGR flow during the EGR Monitor test. **Possible Causes:** • EGR valve lift sensor is stuck, damaged or has failed • EGR control solenoid circuit is open or shorted to ground • EGR control solenoid valve is damaged or has failed • PCM has failed

DTC	Trouble Code Title, Conditions & Possible Causes
DTC: P1491 **2T EGR1, MIL: Yes** **1996, 1997, 1998, 1999, 2000, 2001, 2002, 2003, 2004, 2005, 2006** **Models:** Civic **Engines:** 1.6L VIN EJ7, 1.7L VIN ES1, 1.7L VIN ES2 **Transmissions:** A/T, M/T	**EGR Valve Lift Sensor Insufficient Flow Detected** Vehicle driven in closed loop at 1700-2500 rpm for over 10 minutes, and the PCM detected the EGR valve lift sensor (EGRV) signal indicated insufficient EGR flow during the EGR Monitor test. **Possible Causes:** • EGR valve lift sensor is stuck, damaged or has failed • EGR control solenoid circuit is open or shorted to ground • EGR control solenoid valve is damaged or has failed • PCM has failed
DTC: P1498 **1T CCM, MIL: Yes** **1995** **Models:** Accord **Engines:** 2.7L VIN CE6 **Transmissions:** A/T, M/T	**EGR Valve Lift Sensor High Input** Key on or engine running, and the PCM detected the EGR Valve Lift sensor signal was more than an allowable range stored in memory. **Possible Causes:** • EGR valve lift sensor circuit is open or shorted to power • EGR valve lift sensor is shorted to VREF or system power (B+) • EGR valve lift sensor is stuck, damaged or has failed • PCM has failed
DTC: P1498 **1T CCM, MIL: Yes** **1996, 1997, 1998, 1999, 2000, 2001, 2002, 2003, 2004, 2005, 2006** **Models:** Accord, Insight, Odyssey, Pilot, Prelude, Ridgeline **Engines:** All **Transmissions:** A/T, M/T	**EGR Valve Lift Sensor High Input** Key on or engine running, and the PCM detected the EGR Valve Lift sensor signal was more than an allowable range stored in memory. **Possible Causes:** • EGR valve lift sensor circuit is open or shorted to power • EGR valve lift sensor is shorted to VREF or system power (B+) • EGR valve lift sensor is stuck, damaged or has failed • PCM has failed
DTC: P1498 **1T CCM, MIL: Yes** **1996, 1997, 1998, 1999, 2000, 2001, 2002, 2003, 2004, 2005, 2006** **Models:** Civic **Engines:** 1.6L VIN EJ7, 1.7L VIN ES1, 1.7L VIN ES2 **Transmissions:** A/T, M/T	**EGR Valve Lift Sensor High Input** Key on or engine running, and the PCM detected the EGR Valve Lift sensor signal was more than an allowable range in memory. **Possible Causes:** • EGR valve lift sensor circuit is open or shorted to power • EGR valve lift sensor is shorted to VREF or system power (B+) • EGR valve lift sensor is stuck, damaged or has failed • PCM has failed
DTC: P1508 **1T CCM, MIL: Yes** **1995** **Models:** Accord **Engines:** 2.7L VIN CE6 **Transmissions:** A/T, M/T	**Idle Air Control Valve Circuit Malfunction** Key on or engine running, and the PCM detected an unexpected voltage condition on the IAC valve control circuit during the test. **Possible Causes:** • IAC valve control circuit is open or shorted to ground • IAC valve control circuit is shorted to system power (B+) • IAC valve power circuit is open or shorted to ground • IAC valve is damaged or has failed • PCM has failed
DTC: P1508 **1T CCM, MIL: Yes** **1996, 1997** **Models:** Accord, Civic del Sol, Odyssey, Pilot, Ridgeline **Engines:** All **Transmissions:** A/T, M/T	**Idle Air Control Valve Circuit Malfunction** Key on or engine running, and the PCM detected an unexpected voltage condition on the IAC valve control circuit during the test. **Possible Causes:** • IAC valve control circuit is open or shorted to ground • IAC valve control circuit is shorted to system power (B+) • IAC valve power circuit is open or shorted to ground • IAC valve is damaged or has failed • PCM has failed
DTC: P1508 **1T CCM, MIL: Yes**	**Idle Air Control System Low RPM** Engine running at idle speed while in closed loop, and the PCM detected the Actual idle speed was more than 100 rpm below the Target idle speed during the CCM test. **Possible Causes:** • Air inlet dirty, restricted or the IAC valve is stuck partially closed • IAC valve is damaged or has failed • The throttle plate is carbon fouled (it may need to be cleaned) • PCM has failed

DTC	Trouble Code Title, Conditions & Possible Causes
DTC: P1508 **1T CCM, MIL: Yes** 1996, 1997, 1998, 1999, 2000, 2001, 2002, 2003, 2004, 2005, 2006 **Models:** CR-V, Prelude **Engines:** All **Transmissions:** A/T, M/T	**Idle Air Control System Low RPM** Engine running at idle speed while in closed loop, and the PCM detected the Actual idle speed was more than 100 rpm below the Target idle speed during the CCM test. **Possible Causes:** • Air inlet dirty, restricted or the IAC valve is stuck partially closed • IAC valve is damaged or has failed • The throttle plate is carbon fouled (it may need to be cleaned) • PCM has failed
DTC: P1508 **2T CCM, MIL: Yes** 1996, 1997, 1998, 1999 **Models:** Passport **Engines:** 3.2L VIN W **Transmissions:** A/T, M/T	**Idle Speed Control System Low RPM** DTC P0101, P0102, P0103, P0106, P0107, P0108, P0112, P0113, P0117, P0118, P0121, P0122, P0123, P0171, P0172, P0201-P0206, P0351-P0356, P0401, P0440, engine runtime 120 seconds, system voltage from 11-16v, ECT sensor over 120°F, BARO sensor over 75 kPa, Scan Tool tests all "off", vehicle speed under 1 mph with the throttle closed, and the PCM detected the Actual engine speed was 100-200 rpm less than the Desired idle speed for 5 seconds (Desired idle speed is based on ECT signal during the test). **Possible Causes:** • Fuel delivery system is too lean or too rich during the test • Inspect the engine mounts for damage to the mounts • Inspect the throttle linkage adjustment and tension • Inspect the throttle body bore for dirt or foreign material • Inspect for any air intake system leaks in the engine or hoses • IAC valve may be damaged or have failed • IAC motor control circuit is open or shorted to ground • Perform an IAC Reset function with the Scan Tool
DTC: P1509 **1T CCM, MIL: Yes** 1996, 1997 **Models:** Accord, Civic del Sol, Odyssey, Pilot, Ridgeline **Engines:** All **Transmissions:** A/T, M/T	**Idle Air Control System High RPM** Engine running at idle speed while in closed loop, and the PCM detected the Actual idle speed was more than 100 rpm above the Target idle speed during the CCM test. **Possible Causes:** • Air inlet dirty, restricted or the IAC valve is stuck partially open • IAC valve is damaged or has failed • The throttle plate is carbon fouled (it may need to be cleaned) • PCM has failed
DTC: P1509 **1T CCM, MIL: Yes** 1996, 1997, 1998, 1999, 2000, 2001, 2002, 2003, 2004, 2005, 2006 **Models:** CR-V, Prelude **Engines:** All **Transmissions:** A/T, M/T	**Idle Air Control System High RPM** Engine running at idle speed while in closed loop, and the PCM detected the Actual idle speed was more than 100 rpm below the Target idle speed during the CCM test. **Possible Causes:** • Air inlet dirty, restricted or the IAC valve is stuck partially closed • IAC valve is damaged or has failed • The throttle plate is carbon fouled (it may need to be cleaned) • PCM has failed
DTC: P1509 **2T CCM, MIL: Yes** 1998, 1999, 2000, 2001, 2002 **Models:** Passport **Engines:** 3.2L VIN W **Transmissions:** A/T, M/T	**Idle Speed Control System High RPM** DTC P0101, P0102, P0103, P0106, P0107, P0108, P0112, P0113, P0117, P0118, P0121, P0122, P0123, P0171, P0172, P0201-P0206, P0351-P0356, P0401, P0440, engine runtime 120 seconds, system voltage from 11-16v, ECT sensor over 120°F, BARO sensor over 75 kPa, Scan Tool tests all "off", vehicle speed under 1 mph with the throttle closed, and the PCM detected the Actual engine speed was 100-200 rpm more than the Desired idle speed for 5 seconds (Desired idle speed is based on ECT signal during the test). **Possible Causes:** • Fuel delivery system is too lean or too rich during the test • Inspect the engine mounts for damage to the mounts • Inspect the throttle linkage adjustment and tension • Inspect the throttle body bore for dirt or foreign material • Inspect for any air intake system leaks in the engine or hoses • IAC valve may be damaged or have failed • IAC motor control circuit is open or shorted to ground • Perform an IAC Reset function with the Scan Tool
DTC: P1519 **1T CCM, MIL: Yes** 1998, 1999, 2000, 2001, 2002, 2003, 2004, 2005, 2006 **Models:** Accord, Insight, Odyssey, Pilot, Ridgeline, S2000 **Engines:** All **Transmissions:** A/T, M/T	**Idle Air Control Valve Circuit Malfunction** Key on or engine running, and the PCM detected an unexpected voltage condition on the Idle Air Control (IAC) valve control circuit. **Possible Causes:** • IAC valve control circuit is open or shorted to power • IAC valve control circuit is shorted to system power (B+) • IAC valve power circuit is open or shorted to ground • IAC valve is damaged or has failed • PCM is damaged

DTC	Trouble Code Title, Conditions & Possible Causes
DTC: P1519 **1T CCM, MIL: Yes** 2001, 2002, 2003, 2004, 2005, 2006 **Models:** Civic **Engines:** All **Transmissions:** A/T, M/T	**Idle Air Control Valve Circuit Malfunction** Key on or engine running, and the PCM detected an unexpected voltage condition on the Idle Air Control (IAC) valve control circuit. **Possible Causes:** • IAC valve control circuit is open or shorted to power • IAC valve control circuit is shorted to system power (B+) • IAC valve power circuit is open or shorted to ground • IAC valve is damaged or has failed • PCM is damaged
DTC: P1522 **1T CCM, MIL: Yes** 2000, 2001, 2002, 2003, 2004, 2005, 2006 **Models:** Insight **Engines:** All **Transmissions:** A/T, M/T	**Master Power Vacuum Sensor Low Input** Engine running at hot idle speed, and the PCM detected the Master Power Vacuum Sensor signal remained in a low state. **Possible Causes:** • Master power vacuum sensor power circuit is open • Master power vacuum sensor ground circuit open • Master power vacuum sensor signal circuit shorted to ground • Master power vacuum sensor is damaged or has failed • PCM has failed
DTC: P1523 **1T CCM, MIL: Yes** 2000, 2001, 2002, 2003, 2004, 2005, 2006 **Models:** Insight **Engines:** All **Transmissions:** A/T, M/T	**Master Power Vacuum Sensor High Input** Engine running at hot idle speed, and the PCM detected the Master Power Vacuum Sensor signal remained in a high state. **Possible Causes:** • Master power vacuum sensor signal circuit shorted to VREF • Master power vacuum sensor signal circuit shorted to power • Master power vacuum sensor is damaged or has failed • PCM has failed
DTC: P1541 **1T CCM, MIL: Yes** 2000, 2001, 2002, 2003, 2004, 2005, 2006 **Models:** Insight **Engines:** All **Transmissions:** A/T, M/T	**HTRS Passenger Heater Standby Circuit Low Input** Key on or engine running, and the PCM detected the HTRS Passenger Compartment Heater Standby signal was in a low state. **Possible Causes:** • HTRS passenger heater standby circuit is open • HTRS passenger heater standby circuit is shorted to ground • HTRS passenger heater is damaged or has failed • PCM has failed
DTC: P1542 **1T CCM, MIL: Yes** 2000, 2001, 2002, 2003, 2004, 2005, 2006 **Models:** Insight **Engines:** All **Transmissions:** A/T, M/T	**HTRS Passenger Heater Standby Circuit High Input** Key on or engine running, and the PCM detected the HTRS Passenger Compartment Heater Standby signal was in a high state. **Possible Causes:** • HTRS passenger heater standby circuit is shorted to VREF • HTRS passenger heater standby circuit is shorted to power • HTRS passenger heater is damaged or has failed • PCM has failed
DTC: P1607 **1T DTC: PCM, MIL: Yes** 1995 **Models:** Accord **Engines:** 2.7L VIN CE6 **Transmissions:** A/T, M/T	**PCM Internal Circuit 'A' Malfunction** Key on, and the PCM detected an Internal Fault 'A' condition. **Note: This trouble code indicates an internal failure in the PCM. The OEM repair procedure recommends replacing the original PCM with a "known good" PCM and then verify the code does not reset.** **Possible Causes:** • PCM is damaged or has failed
DTC: P1607 **1T DTC: PCM, MIL: Yes** 1996, 1997, 1998, 1999, 2000, 2001, 2002, 2003, 2004, 2005, 2006 **Models:** Accord, Civic, Civic del Sol, CR-V, Insight, Odyssey, Pilot, Prelude, Ridgeline, S2000 **Engines:** All **Transmissions:** A/T, M/T	**PCM Internal Circuit 'A' Malfunction** Key on, and the PCM detected an Internal Fault 'A' condition. **Note: This trouble code indicates an internal failure in the PCM. The OEM repair procedure recommends replacing the original PCM with a "known good" PCM and then verify the code does not reset.** **Possible Causes:** • PCM is damaged or has failed
DTC: P1618 **1T DTC: PCM, MIL: No** 1996, 1997, 1998, 1999, 2000, 2001, 2002 **Models:** Passport **Engines:** 3.2L VIN V, 3.2L VIN W **Transmissions:** A/T, M/T	**Serial Peripheral Interface Communication** Key on for 2 seconds, system voltage over 9v, and the PCM detected an internal program fault (the check sum of the data communications error) for 3 out of 6 seconds with no TCM resets during the test period of 2 seconds. **Note: For additional help with this code, view the Failure Records.** **Possible Causes:** • Check the PCM calibration to verify it is the latest calibration • Recalibrate the PCM as required • PCM may need to be replaced and reprogrammed

DTC	Trouble Code Title, Conditions & Possible Causes
DTC: P1635 **1T CCM, MIL: Yes** **2000, 2001, 2002** **Models:** Passport **Engines:** All **Transmissions:** A/T, M/T	**5-Volt Reference Voltage 'A' Circuit Malfunction** Engine started, engine running, system voltage more than 6.3v, and the PCM detected the 5v VREF 'A' was less than 4.88v or more than 5.12v for 5 seconds within a 10 second sample during the CCM test. **Possible Causes:** • VREF circuit is open (an internal fault inside the PCM) • VREF circuit is shorted to ground (an internal fault in the PCM) • VREF circuit shorted to ground between a sensor and the PCM • PCM has failed
DTC: P1639 **1T CCM, MIL: Yes** **2000, 2001, 2002** **Models:** Passport **Engines:** All **Transmissions:** A/T, M/T	**5-Volt Reference Voltage 'B' Circuit Malfunction** Engine started, engine running, system voltage more than 6.3v, and the PCM detected the 5v VREF 'B' was less than 4.88v or more than 5.12v for 5 seconds within a 10 second sample during the CCM test. **Possible Causes:** • VREF circuit is open (an internal fault inside the PCM) • VREF circuit is shorted to ground (an internal fault in the PCM) • VREF circuit shorted to ground between a sensor and the PCM • PCM has failed
DTC: P1640 **1T CCM, MIL: Yes** **2000, 2001, 2002, 2003, 2004,** **2005, 2006** **Models:** Insight **Engines:** All **Transmissions:** A/T, M/T	**ACTRQ Motor Torque Signal Circuit Low Input** Key on or engine running, and the PCM detected the ACTRQ Motor Torque Signal circuit remained in a low state. **Possible Causes:** • ACTRQ motor torque signal circuit is open • ACTRQ motor torque signal circuit is shorted to ground • ACTRQ motor is damaged or has failed • PCM has failed
DTC: P1640 **1T CCM, MIL: No** **1996, 1997** **Models:** Passport **Engines:** 2.6L VIN E **Transmissions:** M/T	**Output Driver Module Circuit Malfunction** DTC P1618 not set, engine started, engine running, system voltage over 13.2v for 4 seconds, and the PCM detected a "high" voltage condition on the Output Driver Module (ODM) circuit with the device "on", or a "low" voltage condition on the ODM circuit with the device "off", condition met for 1 second during the CCM Rationality test. **Note: For additional help with this code, view the Failure Records.** **Possible Causes:** • One or more output device driver circuits connected to ODM has an open circuit condition • One or more output device driver circuits connected to ODM has a short-to-voltage condition • Check for an open power circuit to the related output devices • Disconnect the A/C clutch relay and Purge solenoid to find fault
DTC: P1640 **1T CCM, MIL: No** **1998, 1999, 2000, 2001, 2002** **Models:** Passport **Engines:** 3.2L VIN W **Transmissions:** A/T, M/T	**Output Driver Module 'A' Circuit Malfunction** DTC P1618 not set, engine started, engine running, system voltage over 13.2v for 4 seconds, and the PCM detected an open circuit condition and an unexpected high voltage condition on the Output Driver Module circuit (A/C Clutch Relay or Purge Solenoid) with the device "on" for 2.5 seconds during the CCM Rationality test. **Note: For additional help with this code, view the Failure Records.** **Possible Causes:** • One or more output device driver circuits connected to ODM 'A' has an open circuit condition • One or more output device driver circuits connected to ODM 'A' has a short-to-voltage condition • Check for an open power circuit to the related output devices • Disconnect the A/C clutch relay and Purge solenoid to find fault
DTC: P1641 **1T CCM, MIL: Yes** **2000, 2001, 2002, 2003, 2004,** **2005, 2006** **Models:** Insight **Engines:** All **Transmissions:** A/T, M/T	**ACTRQ Motor Torque Signal Circuit High Input** Key on or engine running, and the PCM detected the ACTRQ Motor Torque Signal circuit remained in a high state. **Possible Causes:** • ACTRQ motor torque signal circuit is shorted to VREF • ACTRQ motor torque signal circuit is shorted to power • ACTRQ motor is damaged or has failed • PCM has failed
DTC: P1642 **1T CCM, MIL: Yes** **2000, 2001, 2002, 2003, 2004,** **2005, 2006** **Models:** Insight **Engines:** All **Transmissions:** A/T, M/T	**QBATT Battery Signal Circuit Low Input** Key on or engine running, and the PCM detected that the QBATT Battery Signal circuit remained in a low state. **Possible Causes:** • QBATT battery signal circuit is open • QBATT battery signal circuit is shorted to ground • QBATT battery is damaged or has failed • PCM has failed

DTC	Trouble Code Title, Conditions & Possible Causes

DTC: P1643
1T CCM, MIL: Yes
2000, 2001, 2002, 2003, 2004,
2005, 2006
Models: Insight
Engines: All
Transmissions: A/T, M/T
	QBATT Battery Signal Circuit High Input
	Key on or engine running, and the PCM detected the QBATT Battery Signal circuit remained in a high state.
	Possible Causes:
	   • QBATT battery signal circuit is shorted to VREF
	   • QBATT battery signal circuit is shorted to power
	   • QBATT battery is damaged or has failed
	   • PCM has failed

DTC: P1644
1T CCM, MIL: Yes
2000, 2001, 2002, 2003, 2004,
2005, 2006
Models: Insight
Engines: All
Transmissions: A/T, M/T
	MOTFSA Signal Circuit Malfunction
	Key on or engine running, and the PCM detected an unexpected voltage condition on the MOTFSA signal circuit during the CCM test.
	Possible Causes:
	   • MOTFSA signal circuit is shorted to VREF or system power
	   • MOTFSA signal circuit is open or shorted to ground
	   • MOTFSA is damaged or has failed
	   • PCM has failed

DTC: P1645
1T CCM, MIL: Yes
2000, 2001, 2002, 2003, 2004,
2005, 2006
Models: Insight
Engines: All
Transmissions: A/T, M/T
	MOTFSA Signal Circuit Malfunction
	Key on or engine running, and the PCM detected an unexpected voltage condition on the MOTFSA signal circuit during the CCM test.
	Possible Causes:
	   • MOTFSA signal circuit is shorted to VREF or system power
	   • MOTFSA signal circuit is open or shorted to ground
	   • MOTFSA is damaged or has failed
	   • PCM has failed

DTC: P1646
1T CCM, MIL: Yes
2000, 2001, 2002, 2003, 2004,
2005, 2006
Models: Insight
Engines: All
Transmissions: A/T, M/T
	MOTFSA Signal Circuit Malfunction
	Key on or engine running, and the PCM detected an unexpected voltage condition on the MOTFSA signal circuit during the CCM test.
	Possible Causes:
	   • MOTFSA signal circuit is shorted to VREF or system power
	   • MOTFSA signal circuit is open or shorted to ground
	   • MOTFSA is damaged or has failed
	   • PCM has failed

DTC: P1650
1T CCM, MIL: No
1998, 1999, 2000, 2001, 2002
Models: Passport
Engines: 3.2L VIN W
Transmissions: A/T, M/T
	Output Driver Module Circuit Malfunction
	DTC P1618 not set, engine started, engine running, system voltage over 13.2v for 4 seconds, and the PCM detected the voltage on the Output Driver Module (ODM) circuit did not indicate less than 1.0 volt with the device commanded "on" for 0.5 seconds in the CCM test.
	Note: For additional help with this code, view the Failure Records.
	Possible Causes:
	   • One or more output device driver circuits connected to ODM has an open circuit condition or has a short-to-voltage condition
	   • Check for an open power circuit to the related output devices
	   • Disconnect the A/C clutch relay and Purge solenoid to find fault

DTC: P1655
1T CCM, MIL: Yes
1996, 1997, 1998, 1999, 2000
Models: Civic
Engines: 1.6L VIN EJ7
Transmissions: A/T
	TMA or TMB Signal Line Circuit Malfunction
	Engine running, and the PCM detected an unexpected voltage condition on the TMA or TMB circuit during the CCM test.
	Note: This trouble code is for vehicles with a CVT style transaxle.
	Possible Causes:
	   • TMA or TMB signal line is open or shorted to ground
	   • TMA or TMB signal line is shorted to VREF or system power
	   • TCM or PCM has failed

DTC: P1655
1T CCM, MIL: Yes
1997, 1998, 1999, 2000, 2001
Models: Prelude
Engines: All
Transmissions: A/T
	SEFA or SEAF Signal Line Circuit Malfunction
	Engine running, and the PCM detected an unexpected voltage condition on the TCM SEFA or SEAF circuit during the CCM test.
	Possible Causes:
	   • SEFA or SEAF signal line is open or shorted to ground
	   • SEFA or SEAF signal line is VREF or shorted to power
	   • TCM or PCM has failed

DTC: P1660
1T CCM, MIL: Yes
1995, 1996, 1997
Models: Accord
Engines: 2.7L VIN CE6
Transmissions: A/T
	TCM A/T FI Data Line Circuit Malfunction
	Key on or engine running, and the PCM detected an unexpected voltage condition on the TCM FI Data Line circuit during the test.
	Possible Causes:
	   • A/T FI data line is open or shorted to ground
	   • A/T FI data line is shorted to VREF or system power
	   • TCM or PCM has failed

DTC	Trouble Code Title, Conditions & Possible Causes
DTC: P1671 **1T CCM, MIL: Yes** **1996, 1997, 1998, 1999, 2000, 2001, 2002, 2003, 2004, 2005, 2006** **Models:** Accord, Civic, Civic del Sol, CR-V, Odyssey, Pilot, Prelude, Ridgeline **Engines:** All **Transmissions:** A/T	**TCM A/T FI Data Line No Signal** Key on or engine running, and the PCM detected a fault in the TCM FI Data Line circuit during the CCM test. **Possible Causes:** • A/T FI data line is open or shorted to ground • A/T FI data line is shorted to VREF or system power • TCM or PCM has failed
DTC: P1672 **1T CCM, MIL: Yes** **1996, 1997, 1998, 1999, 2000, 2001, 2002, 2003, 2004, 2005, 2006** **Models:** Accord, Civic, Civic del Sol, CR-V, Odyssey, Pilot, Prelude, Ridgeline **Engines:** All **Transmissions:** A/T	**TCM A/T FI Data Line Circuit Malfunction** Key on or engine running, and the PCM detected an unexpected voltage condition on the TCM FI Data Line circuit during the test. **Possible Causes:** • A/T FI data line is open or shorted to ground • A/T FI data line is shorted to VREF or system power • TCM or PCM has failed
DTC: P1676 **1T CCM, MIL: Yes** **1996, 1997, 1998, 1999, 2000, 2001, 2002, 2003, 2004, 2005, 2006** **Models:** Accord, Civic, Civic del Sol, CR-V, Odyssey, Pilot, Prelude, Ridgeline **Engines:** All **Transmissions:** A/T	**TCM A/T FI Data Line Circuit Malfunction** Key on or engine running, and the PCM detected an unexpected voltage condition on the TCM FI Data Line circuit during the test. **Possible Causes:** • A/T FI data line is open or shorted to ground • A/T FI data line is shorted to VREF or system power • TCM or PCM has failed
DTC: P1677 **1T CCM, MIL: Yes** **1996, 1997, 1998, 1999, 2000, 2001, 2002, 2003, 2004, 2005, 2006** **Models:** Accord, Civic, Civic del Sol, CR-V, Odyssey, Pilot, Prelude, Ridgeline **Engines:** All **Transmissions:** A/T	**TCM A/T FI Data Line Circuit Malfunction** Key on or engine running, and the PCM detected an unexpected voltage condition on the TCM FI Data Line circuit during the test. **Possible Causes:** • A/T FI data line is open or shorted to ground • A/T FI data line is shorted to VREF or system power • TCM or PCM has failed
DTC: P1678 **1T CCM, MIL: Yes** **1996, 1997, 1998, 1999, 2000, 2001, 2002, 2003, 2004, 2005, 2006** **Models:** Accord, Civic, Civic del Sol, CR-V, Odyssey, Pilot, Prelude, Ridgeline **Engines:** All **Transmissions:** A/T	**TCM A/T FPTDR Signal Line Circuit Malfunction** Key on or engine running, and the PCM detected an unexpected voltage condition on the TCM FPTDR Line circuit during the test. **Possible Causes:** • A/T FTPDR data line is open or shorted to ground • A/T FTPDR data line is shorted to VREF or system power • TCM or PCM has failed
DTC: P1681 **1T CCM, MIL: Yes** **1995, 1996, 1997** **Models:** Accord **Engines:** 2.7L VIN CE6 **Transmissions:** A/T	**TCM A/T FI Signal 'A' Low Input** Key on or engine running, and the PCM detected a "low input" condition on the TCM A/T FI Signal 'A' circuit during the CCM test. **Possible Causes:** • A/T FI data line is open • A/T FI data line is shorted to ground • TCM or PCM has failed

DTC	Trouble Code Title, Conditions & Possible Causes
DTC: P1682 **1T CCM, MIL: Yes** **1995, 1996, 1997** **Models:** Accord **Engines:** 2.7L VIN CE6 **Transmissions:** A/T	**TCM A/T FI Signal 'A' High Input** Key on or engine running, and the PCM detected a "high input" condition on the TCM A/T FI Signal 'A' circuit during the CCM test. **Possible Causes:** • A/T FI data line is open • A/T FI data line is shorted to ground • TCM or PCM has failed
DTC: P1683 **1T CCM, MIL: Yes** **1995, 1996, 1997** **Models:** Accord **Engines:** 2.7L VIN CE6 **Transmissions:** A/T	**TCM A/T FI Signal 'B' Low Input** Key on or engine running, and the PCM detected a "low input" condition on the TCM A/T FI Signal 'B' circuit during the CCM test. **Possible Causes:** • A/T FI data line is open • A/T FI data line is shorted to ground • TCM or PCM has failed
DTC: P1684 **1T CCM, MIL: Yes** **1995, 1996, 1997** **Models:** Accord **Engines:** 2.7L VIN CE6 **Transmissions:** A/T	TCM A/T FI Signal 'B' High Input Key on or engine running, and the PCM detected a "high input" condition on the TCM A/T FI Signal 'B' circuit during the CCM test. **Possible Causes:** • A/T FI data line is open • A/T FI data line is shorted to ground • TCM or PCM has failed
DTC: P1705 **1T CCM, MIL: Yes** **1995, 1996, 1997** **Models:** Accord **Engines:** 2.7L VIN CE6 **Transmissions:** A/T	**TCM A/T Gear Position Switch Low Input** Engine running, and the PCM detected a "low input" condition in the Gear Position Switch. **Note: The transaxle has no lockup function when this code is set.** **Possible Causes:** • A/T gear position switch signal circuit is shorted to ground • A/T gear position switch signal circuit is shorted to another wire • A/T gear position switch is damaged or has failed
DTC: P1705 **1T CCM, MIL: Yes** **1996, 1997, 1998, 1999, 2000,** **2001, 2002, 2003, 2004, 2005,** **2006** **Models:** Accord, Civic, Civic del Sol, CR-V, Odyssey, Pilot, Prelude, Ridgeline **Engines:** All **Transmissions:** A/T	**TCM A/T Gear Position Switch Low Input** Engine running, and the PCM detected a "low input" condition in the Gear Position Switch. **Note: The transaxle has no lockup function when this code is set.** **Possible Causes:** • A/T gear position switch signal circuit is shorted to ground • A/T gear position switch signal circuit is shorted to another wire • A/T gear position switch is damaged or has failed
DTC: P1706 **1T CCM, MIL: Yes** **1995, 1996, 1997** **Models:** Accord **Engines:** 2.7L VIN CE6 **Transmissions:** A/T	**TCM A/T Gear Position Switch High Input** Engine running, and the PCM detected a "high input" condition in the Gear Position Switch. **Note: The transaxle has no lockup function when this code is set.** **Possible Causes:** • A/T gear position switch signal circuit is open • A/T gear position switch is damaged or has failed
DTC: P1706 **1T CCM, MIL: Yes** **1996, 1997, 1998, 1999, 2000,** **2001, 2002, 2003, 2004, 2005,** **2006** **Models:** Accord, Civic, Civic del Sol, CR-V, Odyssey, Pilot, Prelude, Ridgeline **Engines:** All **Transmissions:** A/T	**TCM A/T Gear Position Switch High Input** Key on or engine running, and the PCM detected a "high input" condition in the Gear Position Switch. **Note: The transaxle has no lockup function when this code is set.** **Possible Causes:** • A/T gear position switch signal circuit is open • A/T gear position switch is damaged or has failed
DTC: P1709 **1T CCM, MIL: Yes** **1996, 1997, 1998, 1999, 2000,** **2001, 2002, 2003, 2004, 2005,** **2006** **Models:** Odyssey, Pilot, Prelude, Ridgeline **Engines:** All **Transmissions:** A/T	**TCM A/T Mode Switch Circuit Malfunction** Key on or engine running, and the PCM detected an unexpected voltage condition on the A/T Mode switch circuit during the test. **Possible Causes:** • A/T mode switch signal circuit is open • A/T mode switch signal circuit is shorted to ground • A/T mode switch signal circuit is shorted to system power • A/T mode switch is damaged or has failed • TCM or PCM has failed

DTC	Trouble Code Title, Conditions & Possible Causes
DTC: P1731 **1T CCM, MIL: Yes** **2003, 2004, 2005, 2006** **Models:** Element **Engines:** All **Transmissions:** A/T	**Problem in Shift Control System.** Vehicle driven through 1st, 2nd, 3rd and 4th gears, and the PCM detected a mechanical problem in the A/T Control system. **Possible Causes:** • Shift Solenoid Valve E Stuck ON • Shift Valve E Stuck • A/T Clutch Pressure Control Solenoid Valve A Stuck OFF • TCM or PCM has failed
DTC: P1732 **1T CCM, MIL: Yes** **2003, 2004, 2005, 2006** **Models:** Element **Engines:** All **Transmissions:** A/T	**Problem in Shift Control System.** Vehicle driven through 1st, 2nd, 3rd and 4th gears, and the PCM detected a mechanical problem in the A/T Control system. **Possible Causes:** • Shift Solenoid Valves B and C Stuck ON • Shift Valves B and C Stuck
DTC: P1735 **1T CCM, MIL: Yes** **2003, 2004, 2005, 2006** **Models:** Element **Engines:** All **Transmissions:** A/T	**Problem in Shift Control System.** Vehicle driven through 1st, 2nd, 3rd and 4th gears, and the PCM detected a mechanical problem in the A/T Control system. **Possible Causes:** • Shift Solenoid Valves B and C Stuck OFF • Shift Solenoid Valve E Stuck ON • Shift Valves B, C, and E Stuck • A/T Clutch Pressure Control Solenoid Valve A Stuck OFF
DTC: P1736 **1T CCM, MIL: Yes** **2003, 2004, 2005, 2006** **Models:** Element **Engines:** All **Transmissions:** A/T	**Problem in Shift Control System.** Vehicle driven through 1st, 2nd, 3rd and 4th gears, and the PCM detected a mechanical problem in the A/T Control system. **Possible Causes:** • Shift Solenoid Valve B Stuck OFF • Shift Solenoid Valve E Stuck ON • Shift Valves B and E Stuck • A/T Clutch Pressure Control Solenoid Valve A Stuck OFF
DTC: P1738 **1T CCM, MIL: No** **1996, 1997, 1998, 1999, 2000,** **2001, 2002, 2003, 2004, 2005,** **2006** **Models:** Accord, Odyssey, Pilot, Prelude, Ridgeline **Engines:** All **Transmissions:** A/T	**TCM A/T 2nd Pressure Switch Circuit Malfunction** Engine running in gear, and the PCM detected an unexpected voltage condition on the 2nd pressure switch circuit during the test. **Possible Causes:** • A/T 2nd pressure switch signal circuit is open • A/T 2nd pressure switch signal circuit is shorted to ground • A/T 2nd pressure switch signal circuit shorted to system power • A/T 2nd pressure switch is damaged or has failed • TCM or PCM has failed
DTC: P1739 **1T CCM, MIL: No** **1996, 1997, 1998, 1999, 2000,** **2001, 2002, 2003, 2004, 2005,** **2006** **Models:** Accord, Odyssey, Pilot, Prelude, Ridgeline **Engines:** All **Transmissions:** A/T	**TCM A/T 3rd Pressure Switch Circuit Malfunction** Engine running in gear, and the PCM detected an unexpected voltage condition on the 3rd pressure switch circuit during the test. **Possible Causes:** • A/T 3rd pressure switch signal circuit is open • A/T 3rd pressure switch signal circuit is shorted to ground • A/T 3rd pressure switch signal circuit shorted to system power • A/T 3rd pressure switch is damaged or has failed • TCM or PCM has failed
DTC: P1750 **1T CCM, MIL: Yes** **2000, 2001, 2002, 2003, 2004,** **2005, 2006** **Models:** Accord **Engines:** All **Transmissions:** A/T	**TCM A/T System Mechanical Malfunction** Vehicle driven through 1st, 2nd, 3rd and 4th gears, and the PCM detected a mechanical problem in the A/T Control system. **Possible Causes:** • Clutch pressure control solenoid valve 'A' is damaged or failed • Clutch pressure control solenoid valve 'B' is damaged or failed • A/T hydraulic control system mechanical problem • TCM or PCM has failed
DTC: P1751 **1T CCM, MIL: Yes** **2000, 2001, 2002, 2003, 2004,** **2005, 2006** **Models:** Accord **Engines:** All **Transmissions:** A/T	**TCM A/T System Mechanical Malfunction** Vehicle driven through 1st, 2nd, 3rd and 4th gears, and the PCM detected a mechanical problem in the A/T Control system. **Possible Causes:** • Shift solenoid valve 'B' is damaged or has failed • Clutch pressure control solenoid valve 'A' is damaged or failed • Clutch pressure control solenoid valve 'B' is damaged or failed • A/T hydraulic control system mechanical problem • TCM or PCM has failed

DTC	Trouble Code Title, Conditions & Possible Causes
DTC: P1753 **1T CCM, MIL: Yes** **1995** **Models:** Accord **Engines:** 2.7L VIN CE6 **Transmissions:** A/T	**TCM A/T Lockup Solenoid Valve 'A' Circuit Malfunction** Vehicle driven in 1st, 2nd, 3rd and 4th gears, and the PCM detected an unexpected voltage condition on the Solenoid 'A' circuit during the CCM test. **Note: The D4 lamp on the dash will blink when this code is set.** **Possible Causes:** • Lockup solenoid 'A' control circuit is open or shorted to ground • Lockup solenoid 'A' circuit is shorted to system power • Lockup solenoid 'A' is damaged or has failed • TCM or PCM has failed
DTC: P1753 **1T CCM, MIL: Yes** **1996, 1997, 1998, 1999, 2000, 2001, 2002, 2003, 2004, 2005, 2006** **Models:** Accord, Civic, Civic del Sol, CR-V, Odyssey, Pilot, Prelude, Ridgeline **Engines:** All **Transmissions:** A/T	**TCM A/T Lockup Solenoid Valve 'A' Circuit Malfunction** Vehicle driven in 1st, 2nd, 3rd and 4th gears, and the PCM detected an unexpected voltage condition on the Solenoid 'A' circuit during the CCM test. **Note: The D4 lamp on the dash will blink when this code is set.** **Possible Causes:** • Lockup solenoid 'A' control circuit is open or shorted to ground • Lockup solenoid 'A' circuit is shorted to system power • Lockup solenoid 'A' is damaged or has failed • TCM or PCM has failed
DTC: P1758 **1T CCM, MIL: Yes** **1995** **Models:** Accord **Engines:** 2.7L VIN CE6 **Transmissions:** A/T	**TCM A/T Lockup Solenoid Valve 'B' Circuit Malfunction** Vehicle driven in 1st, 2nd, 3rd and 4th gears, and the PCM detected an unexpected voltage condition on the Solenoid 'B' circuit during the CCM test. **Note: The D4 lamp on the dash will blink when this code is set.** **Possible Causes:** • Lockup solenoid 'B' control circuit is open or shorted to ground • Lockup solenoid 'B' circuit is shorted to system power • Lockup solenoid 'B' is damaged or has failed • TCM or PCM has failed
DTC: P1758 **1T CCM, MIL: Yes** **1996, 1997, 1998, 1999, 2000, 2001, 2002, 2003, 2004, 2005, 2006** **Models:** Accord, Civic, Civic del Sol, CR-V, Odyssey, Pilot, Prelude, Ridgeline **Engines:** All **Transmissions:** A/T	**TCM A/T Lockup Solenoid Valve 'B' Circuit Malfunction** Vehicle driven in 1st, 2nd, 3rd and 4th gears, and the PCM detected an unexpected voltage condition on the Solenoid 'B' circuit during the CCM test. **Note: The D4 lamp on the dash will blink when this code is set.** **Possible Causes:** • Lockup solenoid 'B' control circuit is open or shorted to ground • Lockup solenoid 'B' circuit is shorted to system power • Lockup solenoid 'B' is damaged or has failed • TCM or PCM has failed
DTC: P1768 **1T CCM, MIL: Yes** **1995** **Models:** Accord **Engines:** 2.7L VIN CE6 **Transmissions:** A/T	**TCM A/T Clutch Pressure Solenoid 'A' Circuit Malfunction** Vehicle driven in 1st, 2nd, 3rd and 4th gears, and the PCM detected an unexpected voltage condition on the Clutch Pressure Solenoid 'A' during the CCM test. **Note: The D4 lamp on the dash will blink when this code is set.** **Possible Causes:** • Clutch pressure solenoid 'A' circuit is open or shorted to ground • Clutch solenoid 'A' control circuit is shorted to system power • Clutch solenoid valve 'A' is damaged or has failed • TCM or PCM has failed
DTC: P1768 **1T CCM, MIL: Yes** **1996, 1997, 1998, 1999, 2000, 2001, 2002, 2003, 2004, 2005, 2006** **Models:** Accord, Civic, Civic del Sol, CR-V, Odyssey, Pilot, Prelude, Ridgeline **Engines:** All **Transmissions:** A/T	**TCM A/T Clutch Pressure Solenoid 'A' Circuit Malfunction** Vehicle driven in 1st, 2nd, 3rd and 4th gears, and the PCM detected an unexpected voltage condition on the Clutch Pressure Solenoid 'A' during the CCM test. **Note: The D4 lamp on the dash will blink when this code is set.** **Possible Causes:** • Clutch pressure solenoid 'A' circuit is open or shorted to ground • Clutch solenoid 'A' control circuit is shorted to system power • Clutch solenoid valve 'A' is damaged or has failed • TCM or PCM has failed

DTC	Trouble Code Title, Conditions & Possible Causes
DTC: P1773 **1T CCM, MIL: Yes** 1996, 1997, 1998, 1999, 2000, 2001, 2002, 2003, 2004, 2005, 2006 **Models:** Accord, Civic, Civic del Sol, CR-V, Odyssey, Pilot, Prelude, Ridgeline **Engines:** All **Transmissions:** A/T	**TCM A/T Clutch Pressure Solenoid 'B' Circuit Malfunction** Vehicle driven in 1st, 2nd, 3rd and 4th gears, and the PCM detected an unexpected voltage condition on the Clutch Pressure Solenoid 'B' during the CCM test. **Note: The D4 lamp on the dash will blink when this code is set.** **Possible Causes:** • Clutch pressure solenoid 'B' circuit is open or shorted to ground • Clutch solenoid 'B' control circuit is shorted to system power • Clutch solenoid valve 'B' is damaged or has failed • TCM or PCM has failed
DTC: P1785 **1T CCM, MIL: Yes** 1999, 2000, 2001, 2002, 2003, 2004, 2005, 2006 **Models:** Accord, Civic, Prelude **Engines:** All **Transmissions:** A/T	**TCM A/T System Malfunction** Vehicle driven in 1st, 2nd, 3rd and 4th gears, and the PCM detected a fault somewhere in the A/T system during the CCM test. **Note: The D4 lamp on the dash will blink when this code is set.** **Possible Causes:** • A/T hydraulic system problem in the transmission • TCM or PCM has failed
DTC: P1786 **1T CCM, MIL: Yes** 1995, 1996, 1997 **Models:** Accord **Engines:** 2.7L VIN CE6 **Transmissions:** A/T	**TCM A/T FAS Signal Circuit Malfunction** Key on or engine running, and the PCM detected an unexpected voltage condition on the A/T FAS signal circuit to the TCM. **Possible Causes:** • A/T FAS signal circuit to the TCM is open • A/T FAS signal circuit to the TCM is shorted to ground • A/T FAS signal circuit to the TCM is shorted to VREF • TCM or PCM has failed
DTC: P1790 **1T CCM, MIL: Yes** 1995, 1996, 1997 **Models:** Accord **Engines:** 2.7L VIN CE6 **Transmissions:** A/T	**TCM A/T Throttle Position Sensor Circuit Malfunction** Key on or engine running, and the PCM detected an unexpected voltage condition on the TP sensor signal circuit to the TCM. **Possible Causes:** • TP sensor signal circuit to the TCM is open • TP sensor signal circuit to the TCM is shorted to ground • TP sensor signal circuit to the TCM is shorted to VREF • TCM or PCM has failed
DTC: P1791 **1T CCM, MIL: Yes** 1995, 1996, 1997 **Models:** Accord **Engines:** 2.7L VIN CE6 **Transmissions:** A/T	**TCM A/T Vehicle Speed Sensor Circuit Malfunction** Vehicle driven for 15 seconds, and the PCM detected an unexpected voltage condition on the vehicle speed signal circuit to the TCM. **Possible Causes:** • VSS signal circuit to the TCM is open • VSS signal circuit to the TCM is shorted to ground • VSS signal circuit to the TCM is shorted to VREF • TCM or PCM has failed
DTC: P1794 **1T CCM, MIL: Yes** 1995, 1996, 1997 **Models:** Accord **Engines:** 2.7L VIN CE6 **Transmissions:** A/T	**TCM A/T Barometric Pressure Sensor Circuit Malfunction** Key on or engine running, and the PCM detected an unexpected voltage condition on the BARO sensor signal circuit to the TCM. **Possible Causes:** • BARO signal circuit to the TCM is open • BARO signal circuit to the TCM is shorted to ground • BARO signal circuit to the TCM is shorted to VREF • TCM or PCM has failed
DTC: P1850 **1T CCM, MIL: No** 1998, 1999, 2000, 2001, 2002 **Models:** Passport **Engines:** 3.2L VIN W **Transmissions:** A/T	**A/T Brake Band Apply Solenoid Circuit Malfunction** Engine started, engine running, then with A/T Brake Band Apply solenoid commanded "on", and PCM detected the solenoid control signal was 12v, or with A/T Brake Band Apply solenoid commanded "off", the solenoid control signal was 1.34 to 1.56 seconds. **Note: The PCM controls this solenoid with a pulsewidth modulated (PWM) control signal.** **Possible Causes:** • A/T Brake band solenoid circuit is open or shorted to ground • A/T Brake band solenoid High circuit open or shorted to ground • A/T brake band apply solenoid is damaged or has failed • PCM has failed

DTC	Trouble Code Title, Conditions & Possible Causes
DTC: P1860 **2T CCM, MIL: Yes** **1998, 1999, 2000, 2001, 2002** **Models:** Passport **Engines:** 3.2L VIN W **Transmissions:** A/T	**Torque Converter Clutch Solenoid Circuit Malfunction** DTC P0751, P0752, P0753, P0756, P0757 and P0758 not set, engine started, engine running, then with the TCC PWM solenoid commanded "on", the PCM detected the solenoid control circuit signal was 12v, or with the TCC PWM solenoid commanded "off", it detected the TCC control circuit signal was 0v for 1.25 seconds. **Possible Causes:** • TCC PWM solenoid control circuit open or shorted to ground • TCC PWM solenoid control circuit is shorted to power • TCC PWM solenoid is damaged or has failed • PCM has failed
DTC: P1870 **2T CCM, MIL: Yes** **1998, 1999, 2000, 2001, 2002** **Models:** Passport **Engines:** 3.2L VIN W **Transmissions:** A/T	**Transmission Component Slipping Malfunction** DTC P0722, P0723, P0742, P0751, P0752, P0753, P0756, P0757, P0758, P1860 and P1870 not set, engine started, then driven to a speed of 15-58 mph, engine speed from 1000-3500 rpm, TP sensor signal from 15-99%, MAP sensor signal from 0-70 kPa, 50 < Engine Torque < 300 Nm, gear selector in D4, TFT sensor signal from 68-302°F, speed ratio at 0.6-0.95, and the PCM detected the TCC slip speed was 250-800 rpm (event occurred 3 times within 7 seconds). **Possible Causes:** • Engine speed signal circuit open or shorted (intermittent fault) • Internal transmission component problem • TCC PWM or Shift Solenoids have failed (mechanical fault) • TR switch is damaged, out-of-adjustment or has failed
DTC: P1882 **1T CCM, MIL: Yes** **1996, 1997, 1998, 1999, 2000** **Models:** Civic **Engines:** 1.6L VIN EJ7 **Transmissions:** A/T	**A/T Inhibitor Solenoid Circuit Malfunction** Engine running, and the PCM detected an unexpected voltage condition on the A/T Inhibitor solenoid circuit during the CCM test. **Note: This trouble code is for vehicles with a CVT style transaxle.** **Possible Causes:** • A/T inhibitor solenoid control circuit open or shorted to ground • A/T inhibitor solenoid is shorted to VREF or system power • A/T inhibitor solenoid is damaged or has failed • TCM or PCM has failed
DTC: P1885 **1T CCM, MIL: Yes** **1996, 1997, 1998, 1999, 2000** **Models:** Civic **Engines:** 1.6L VIN EJ7 **Transmissions:** A/T	**A/T Drive Pulley Speed Sensor Circuit Malfunction** Engine running, and the PCM detected an unexpected voltage condition on the A/T Drive Pulley Speed sensor circuit in the test. The drive pulley sensor resistance is f360-600 ohms at 68°F. **Note: This trouble code is for vehicles with a CVT style transaxle.** **Possible Causes:** • A/T drive pulley speed sensor circuit open or shorted to ground • A/T drive pulley speed sensor is shorted to VREF • A/T drive pulley speed sensor is damaged or has failed • TCM or PCM has failed
DTC: P1886 **1T CCM, MIL: Yes** **1996, 1997, 1998, 1999, 2000** **Models:** Civic **Engines:** 1.6L VIN EJ7 **Transmissions:** A/T	**A/T Driven Pulley Speed Sensor Circuit Malfunction** Engine running, and the PCM detected an unexpected voltage condition on the A/T Driven Pulley Speed sensor circuit in the test. The drive pulley sensor resistance is f360-600 ohms at 68°F. **Note: This trouble code is for vehicles with a CVT style transaxle.** **Possible Causes:** • A/T driven pulley speed sensor circuit is open • A/T driven pulley speed sensor circuit is shorted to ground • A/T driven pulley speed sensor is shorted to VREF • A/T driven pulley speed sensor is damaged or has failed • TCM or PCM has failed
DTC: P1888 **1T CCM, MIL: Yes** **1996, 1997, 1998, 1999, 2000** **Models:** Civic **Engines:** 1.6L VIN EJ7 **Transmissions:** A/T	**A/T Second Gear Shaft Speed Sensor Circuit Malfunction** Engine running, and the PCM detected an unexpected voltage condition on the A/T 2nd Gear Shaft Speed sensor circuit in the test. **Note: This trouble code is for vehicles with a CVT style transaxle.** **Possible Causes:** • A/T 2nd gear shaft speed sensor circuit is open • A/T 2nd gear shaft speed sensor circuit is shorted to ground • A/T 2nd gear shaft speed sensor is shorted to VREF • A/T 2nd gear shaft speed sensor is damaged or has failed • TCM or PCM has failed

DTC	Trouble Code Title, Conditions & Possible Causes

DTC: P1890
1T CCM, MIL: Yes
1996, 1997, 1998, 1999, 2000
Models: Civic
Engines: 1.6L VIN EJ7
Transmissions: A/T
	A/T Shift Control System Malfunction
	Vehicle driven through several gears, and the PCM detected a problem somewhere in the A/T Shift Control system during the test.
	Note: This trouble code is for vehicles with a CVT style transaxle.
	Possible Causes:
	    • A/T shift control system is damaged
	    • A/T shift control system has failed
	    • TCM or PCM has failed

DTC: P1891
1T CCM, MIL: Yes
1996, 1997, 1998, 1999, 2000
Models: Civic
Engines: 1.6L VIN EJ7
Transmissions: A/T
	A/T Start Clutch Control System Malfunction
	Vehicle driven through several gears, and the PCM detected a problem somewhere in the Start Clutch Control system in the test.
	Note: This trouble code is for vehicles with a CVT style transaxle.
	Possible Causes:
	    • A/T start clutch control system is damaged
	    • A/T start clutch control system has failed
	    • TCM or PCM has failed

HONDA
COMPONENT TESTING

TABLE OF CONTENTS

COMPONENT TESTING

General Testing

INTERMITTENT TESTS

Intermittent problems can be difficult to diagnose because they may or may not be present while the vehicle is available to the repair technician. An intermittent fault can appear and leave so quickly that the vehicle computer does not detect a fault and store a trouble code in memory.

A computer will have to see the fault for a specific period of time before a fault is detected and a code set. While intermittent problems may appear to be occasional by nature, they usually occur under specific conditions. Therefore, the technician must identify and duplicate these conditions. Since an intermittent fault is difficult to duplicate, it is important to follow a logical, systematic routine (or checklist) when attempting to find the faulty system, circuit or component.

Some intermittent faults can be due to a loose connection, a wiring problem or a warped circuit board. Some of these problems are caused by poor test procedures that can damage the male or female ends of a connector. To test for loose or damaged connection, take the male end of a used connector from a wiring harness and carefully push it into the female terminal to verify the opening is tight. There should be some resistance felt when inserting the male into the female part of the terminal.

The Wiggle Test

A wiggle test can be used to locate some intermittent faults. To connect to the circuits, backprobe the actuator, sensor or switch and the PCM.

To perform this test, wiggle the suspect wiring, connector, and component while watching for a change on the DVOM. If it is available, use the Min/Max record mode on the DVOM to capture faults.

This is an excellent method to use to attempt to identify a condition that could cause an intermittent fault to occur.

Intermittents

This test procedure contains instructions for isolating an intermittent fault while using a Scan Tool, Breakout Box, fuel pressure gauge, vacuum gauge and a DVOM. Actual values from the vehicle under test can be compared to a typical set of values taken from the Reference Values.

PRELIMINARY CHECKS

All preliminary checks listed below are required prior to performing this test for intermittent faults. Preliminary checks include:

- All related electrical connections
- For vacuum leaks in related components or mounting hardware
- Fuel level - both quantity and quality
- Ignition wiring connections
- Air intake system filters, tubes and gaskets
- Basic engine components (compression, valve and ignition timing, etc.)
- Any aftermarket add-on devices

Locating and Repairing Intermittent Faults

Intermittent faults are generally associated with circuit problems. In order to pinpoint the fault area, a particular component or its wiring and connectors must be thoroughly inspected and tested. Prior to starting your test sequence, turn off all accessories and vehicle lighting. Also, verify that the battery and vehicle charging system are free of problems as these areas can disguise or mask a problem. Several of these tests are discussed next.

Change Input and Verify Output Response

The purpose of this type of test is to monitor how the PCM and its output devices respond to changes in sensor or switch inputs.

- Connect the Honda Test Harness or BOB.
- Record any pin voltages that related to an intermittent code or symptom
- Create a condition to cause the selected input condition to change
- Monitor the change in the pin voltage for a particular actuator signal on the DVOM (i.e., increase the throttle angle under engine load, and watch the IAC and TP Sensor pin voltage changes)

Actuator "Click" Testing

The purpose of this type of test is to monitor a particular PCM controlled relay or solenoid while watching and listening for a change of state

- Turn the key on or start the engine to actuate the device or switch
- If necessary, remove the device and test or actuate it on the bench
- Listen to verify that certain relays (PGM-FI, A/C) actually click on and off. If a Breakout Box is connected to the PCM, measure the control circuit while turning the outputs on and then off. A voltage change to close to battery voltage should occur during the on and off transition.

Test for Open Conditions in a Harness

The purpose of this type of test is to check a suspect wiring harness for an open circuit condition.

- Turn the key off and install the Honda Test Harness or BOB
- Remove the "suspect" component from the vehicle wiring harness
- Use a DVOM (ohmmeter function on the low range) to measure from one end of the suspect circuit at the test harness or BOB to the other end of the circuit at the particular component connector pin.
- The continuity test result should read under 5 ohms (If the test leads are okay and the ohmmeter is zeroed, the actual test results will read near 0.1 ohms).

Test for Shorted Conditions in a Harness

The purpose of this type of test is to check a suspect wiring harness for a short-to-power or short-to-ground condition.

- Turn the key off and disconnect the wiring harness from the PCM
- Remove the "suspect" component from its wiring harness connector.
- Use a DVOM (ohmmeter function on the high range) at the sensor harness connector to measure between the "suspect" circuit and signal return (ground) and to vehicle power or the voltage reference circuit.
- The continuity test result read more than 10,000 ohms (Infinity, >> or OL on some meters). If the actual reading is less than this amount, the two circuits are making contact somewhere in the wiring harness.

REFERENCE CHARTS

➡ **These example charts were designed for use with the Honda Test Harness or a Breakout Box and a DVOM. Readings were collected from a "known good" vehicle at Key On, Engine Off unless otherwise noted.**

ECT Sensor Conversion Chart

Degrees F	Resistance	Voltage
0°F	14,000-20,000 ohms	4.0v
100°F	900-1200 ohms	2-3v
180°F	250-350 ohms	0.5-0.6v

<GNAM "29130_HOND_C0001">

IAT Sensor Conversion Chart

Degrees F	Resistance	Voltage
0°F	14,000-20,000 ohms	4.0v
100°F	900-1200 ohms	2-3v
192°F	150-350 ohms	0.49-0.55v

<GNAM "29130_HOND_C0002">

Atmospheric Pressure Sensor Chart

BARO Pressure (inches Hg)	Voltage
0" Hg	2.8-3.0v
5" Hg	2.3-2.5v
10" Hg	1.8-2.0v
15" Hg	1.3-1.5v
20"Hg	0.8-1.0v
25" Hg	0.3-0.5v

<GNAM "29130_ACUR_C0003">

MAP Sensor Chart

MAP Pressure (inches Hg)	Voltage
0" Hg	2.8-3.0v
5" Hg	2.3-2.5v
10" Hg	1.8-2.0v
15" Hg	1.3-1.5v
20"Hg	0.8-1.0v
25" Hg	0.3-0.5v

<GNAM "29130_ACUR_C0004">

Fuel Tank Pressure Chart

Fuel Tank Pressure (in. Hg)	Voltage
0" Hg	2.8-3.0v
5" Hg	2.3-2.5v
10" Hg	1.8-2.0v
15" Hg	1.3-1.5v
20"Hg	0.8-1.0v
25" Hg	0.3-0.5v

<GNAM "29130_ACUR_C0005">

EGR Lift Sensor Chart

EGR Valve Position	Voltage
Valve Fully Closed (idle)	1.0-1.2v
Valve Fully Open (off-idle)	4.5v

<GNAM "29130_ACUR_C0005">

Component Locations

ACCORD

4 Cylinder Engines

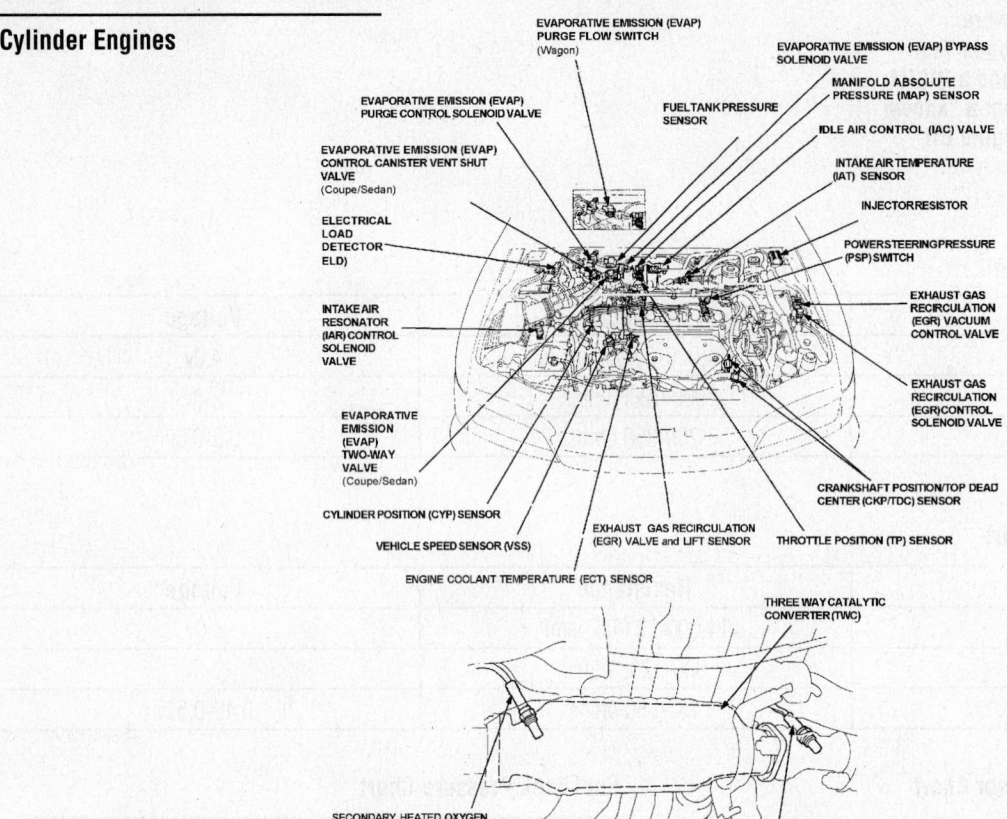

EVAPORATIVE EMISSION (EVAP)
PURGE FLOW SWITCH
(Wagon)

EVAPORATIVE EMISSION (EVAP)
PURGE CONTROL SOLENOID VALVE

FUEL TANK PRESSURE
SENSOR

EVAPORATIVE EMISSION (EVAP) BYPASS
SOLENOID VALVE

MANIFOLD ABSOLUTE
PRESSURE (MAP) SENSOR

EVAPORATIVE EMISSION (EVAP)
CONTROL CANISTER VENT SHUT
VALVE
(Coupe/Sedan)

IDLE AIR CONTROL (IAC) VALVE

INTAKE AIR TEMPERATURE
(IAT) SENSOR

INJECTOR RESISTOR

ELECTRICAL
LOAD
DETECTOR
ELD)

POWER STEERING PRESSURE
(PSP) SWITCH

INTAKE AIR
RESONATOR
(IAR) CONTROL
SOLENOID
VALVE

EXHAUST GAS
RECIRCULATION
(EGR) VACUUM
CONTROL VALVE

EXHAUST GAS
RECIRCULATION
(EGR) CONTROL
SOLENOID VALVE

EVAPORATIVE
EMISSION
(EVAP)
TWO-WAY
VALVE
(Coupe/Sedan)

CRANKSHAFT POSITION/TOP DEAD
CENTER (CKP/TDC) SENSOR

CYLINDER POSITION (CYP) SENSOR

EXHAUST GAS RECIRCULATION
(EGR) VALVE and LIFT SENSOR

THROTTLE POSITION (TP) SENSOR

VEHICLE SPEED SENSOR (VSS)

ENGINE COOLANT TEMPERATURE (ECT) SENSOR

THREE WAY CATALYTIC
CONVERTER (TWC)

SECONDARY HEATED OXYGEN
SENSOR
(SECONDARY H02S) (SENSOR 2)

PRIMARY HEATED OXYGEN SENSOR
(PRIMARY H02S) (SENSOR 1)

29130_HOND_G0001

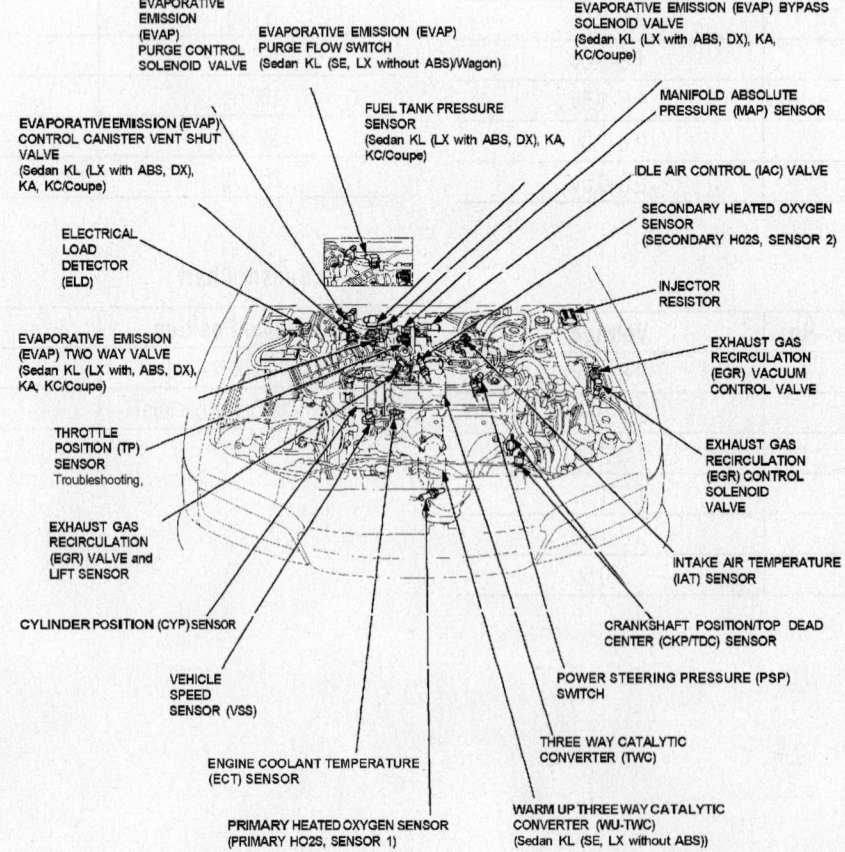

EVAPORATIVE
EMISSION
(EVAP)
PURGE CONTROL
SOLENOID VALVE

EVAPORATIVE EMISSION (EVAP)
PURGE FLOW SWITCH
(Sedan KL (SE, LX without ABS)/Wagon)

EVAPORATIVE EMISSION (EVAP) BYPASS
SOLENOID VALVE
(Sedan KL (LX with ABS, DX), KA,
KC/Coupe)

MANIFOLD ABSOLUTE
PRESSURE (MAP) SENSOR

EVAPORATIVE EMISSION (EVAP)
CONTROL CANISTER VENT SHUT
VALVE
(Sedan KL (LX with ABS, DX),
KA, KC/Coupe)

FUEL TANK PRESSURE
SENSOR
(Sedan KL (LX with ABS, DX), KA,
KC/Coupe)

IDLE AIR CONTROL (IAC) VALVE

SECONDARY HEATED OXYGEN
SENSOR
(SECONDARY H02S, SENSOR 2)

ELECTRICAL
LOAD
DETECTOR
(ELD)

INJECTOR
RESISTOR

EVAPORATIVE EMISSION
(EVAP) TWO WAY VALVE
(Sedan KL (LX with, ABS, DX),
KA, KC/Coupe)

EXHAUST GAS
RECIRCULATION
(EGR) VACUUM
CONTROL VALVE

THROTTLE
POSITION (TP)
SENSOR
Troubleshooting,

EXHAUST GAS
RECIRCULATION
(EGR) CONTROL
SOLENOID
VALVE

EXHAUST GAS
RECIRCULATION
(EGR) VALVE and
LIFT SENSOR

INTAKE AIR TEMPERATURE
(IAT) SENSOR

CYLINDER POSITION (CYP) SENSOR

CRANKSHAFT POSITION/TOP DEAD
CENTER (CKP/TDC) SENSOR

VEHICLE
SPEED
SENSOR (VSS)

POWER STEERING PRESSURE (PSP)
SWITCH

THREE WAY CATALYTIC
CONVERTER (TWC)

ENGINE COOLANT TEMPERATURE
(ECT) SENSOR

WARM UP THREE WAY CATALYTIC
CONVERTER (WU-TWC)
(Sedan KL (SE, LX without ABS))

PRIMARY HEATED OXYGEN SENSOR
(PRIMARY H02S, SENSOR 1)

29130_HOND_G002

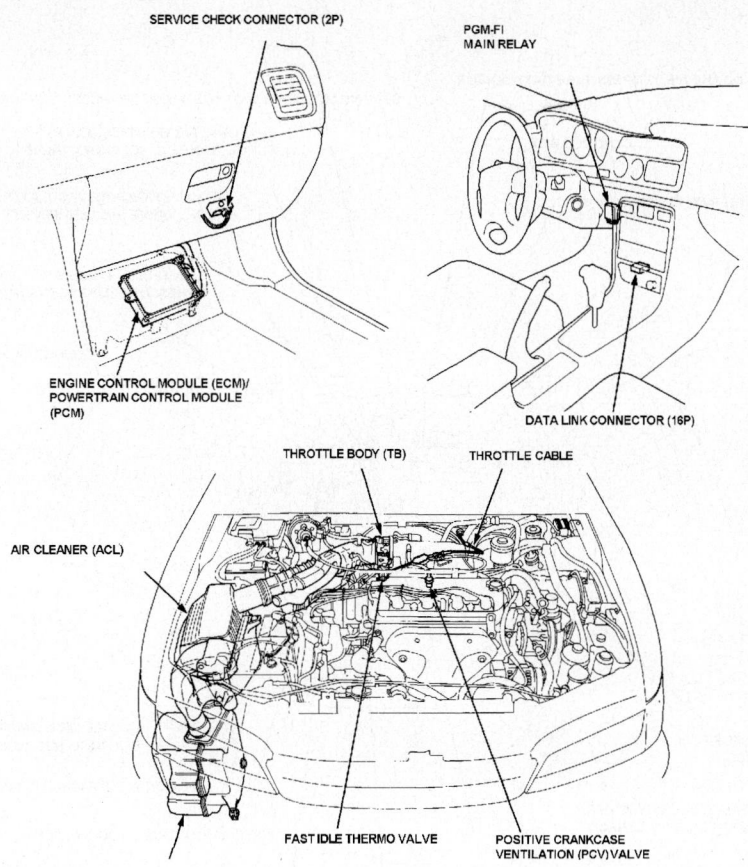

SERVICE CHECK CONNECTOR (2P)

PGM-FI MAIN RELAY

ENGINE CONTROL MODULE (ECM)/ POWERTRAIN CONTROL MODULE (PCM)

DATA LINK CONNECTOR (16P)

THROTTLE BODY (TB)

THROTTLE CABLE

AIR CLEANER (ACL)

FAST IDLE THERMO VALVE

POSITIVE CRANKCASE VENTILATION (PCV) VALVE

29130_HOND_G0003

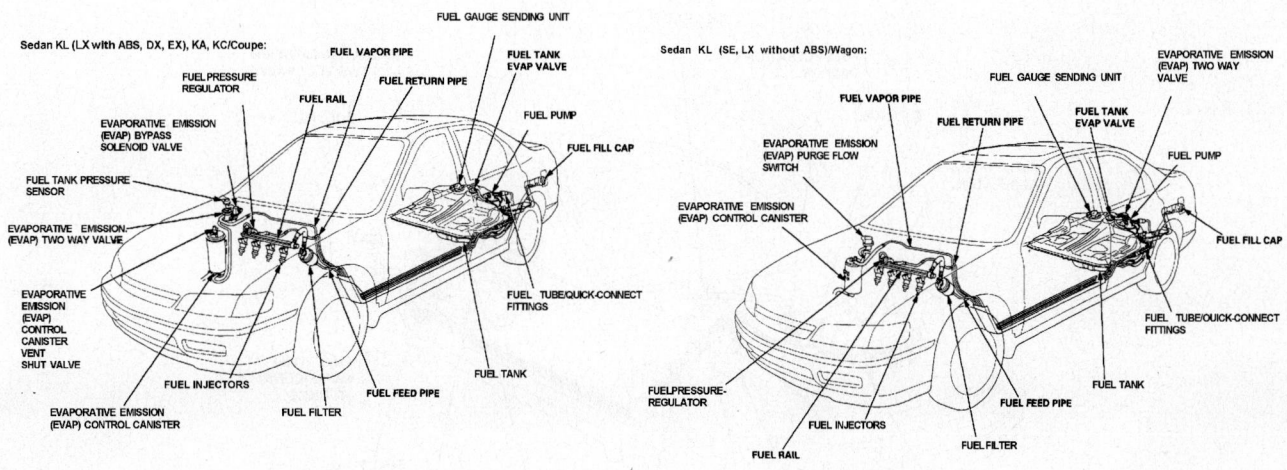

Sedan KL (LX with ABS, DX, EX), KA, KC/Coupe:

FUEL GAUGE SENDING UNIT

FUEL VAPOR PIPE

FUEL TANK EVAP VALVE

FUEL PRESSURE REGULATOR

FUEL RETURN PIPE

FUEL RAIL

FUEL PUMP

EVAPORATIVE EMISSION (EVAP) BYPASS SOLENOID VALVE

FUEL FILL CAP

FUEL TANK PRESSURE SENSOR

EVAPORATIVE EMISSION (EVAP) TWO WAY VALVE

FUEL TUBE/QUICK-CONNECT FITTINGS

EVAPORATIVE EMISSION (EVAP) CONTROL CANISTER VENT SHUT VALVE

FUEL TANK

FUEL INJECTORS

FUEL FEED PIPE

EVAPORATIVE EMISSION (EVAP) CONTROL CANISTER

FUEL FILTER

Sedan KL (SE, LX without ABS)/Wagon:

FUEL GAUGE SENDING UNIT

EVAPORATIVE EMISSION (EVAP) TWO WAY VALVE

FUEL VAPOR PIPE

FUEL RETURN PIPE

FUEL TANK EVAP VALVE

FUEL PUMP

EVAPORATIVE EMISSION (EVAP) PURGE FLOW SWITCH

EVAPORATIVE EMISSION (EVAP) CONTROL CANISTER

FUEL FILL CAP

FUEL TUBE/QUICK-CONNECT FITTINGS

FUEL PRESSURE REGULATOR

FUEL TANK

FUEL INJECTORS

FUEL FEED PIPE

FUEL FILTER

FUEL RAIL

29130_HOND_G0004

6 Cylinder Engines

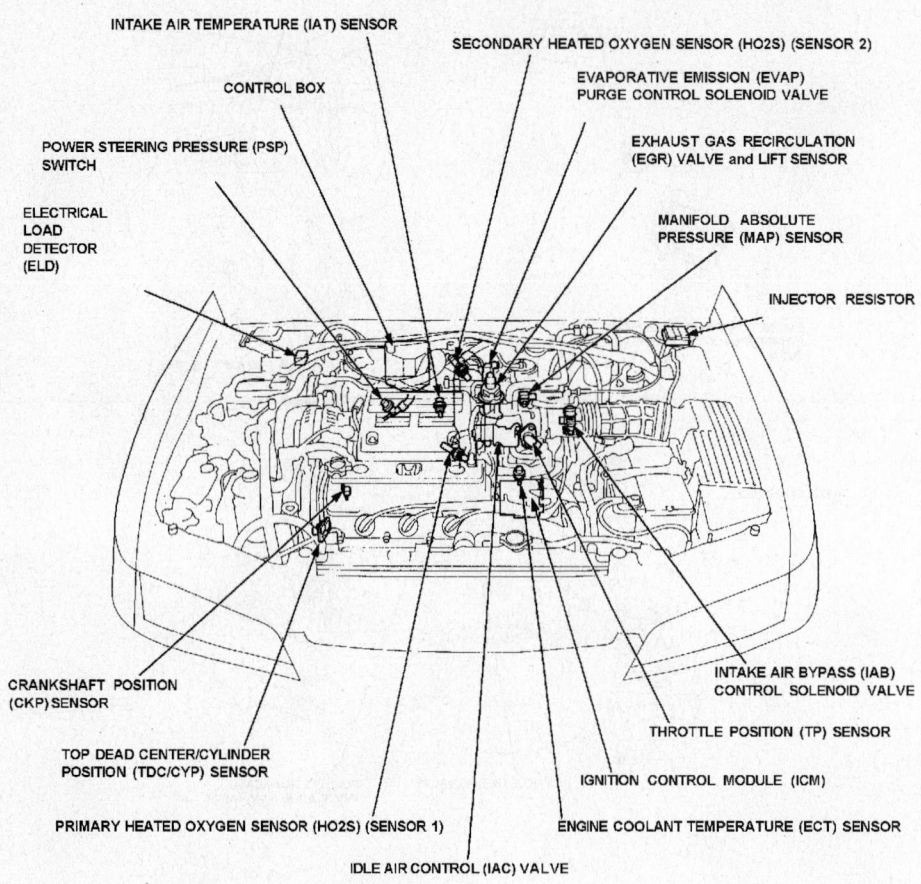

INTAKE AIR TEMPERATURE (IAT) SENSOR

SECONDARY HEATED OXYGEN SENSOR (HO2S) (SENSOR 2)

CONTROL BOX

EVAPORATIVE EMISSION (EVAP) PURGE CONTROL SOLENOID VALVE

POWER STEERING PRESSURE (PSP) SWITCH

EXHAUST GAS RECIRCULATION (EGR) VALVE and LIFT SENSOR

ELECTRICAL LOAD DETECTOR (ELD)

MANIFOLD ABSOLUTE PRESSURE (MAP) SENSOR

INJECTOR RESISTOR

CRANKSHAFT POSITION (CKP) SENSOR

INTAKE AIR BYPASS (IAB) CONTROL SOLENOID VALVE

TOP DEAD CENTER/CYLINDER POSITION (TDC/CYP) SENSOR

THROTTLE POSITION (TP) SENSOR

IGNITION CONTROL MODULE (ICM)

PRIMARY HEATED OXYGEN SENSOR (HO2S) (SENSOR 1)

ENGINE COOLANT TEMPERATURE (ECT) SENSOR

IDLE AIR CONTROL (IAC) VALVE

29130_HOND_G0005

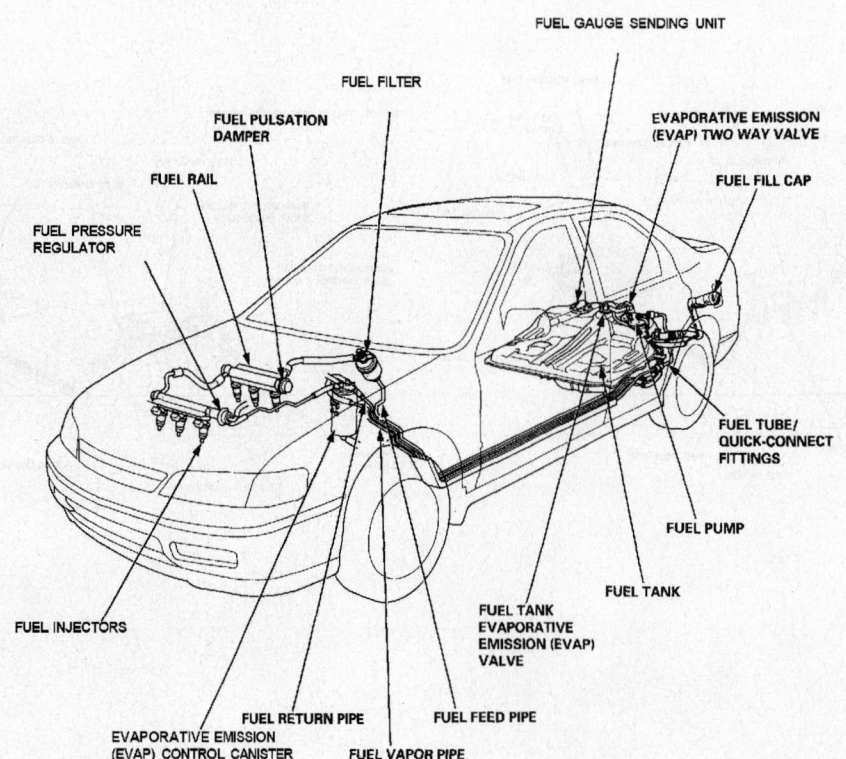

FUEL GAUGE SENDING UNIT

FUEL FILTER

FUEL PULSATION DAMPER

EVAPORATIVE EMISSION (EVAP) TWO WAY VALVE

FUEL RAIL

FUEL FILL CAP

FUEL PRESSURE REGULATOR

FUEL TUBE/ QUICK-CONNECT FITTINGS

FUEL PUMP

FUEL TANK

FUEL INJECTORS

FUEL TANK EVAPORATIVE EMISSION (EVAP) VALVE

FUEL RETURN PIPE

FUEL FEED PIPE

EVAPORATIVE EMISSION (EVAP) CONTROL CANISTER

FUEL VAPOR PIPE

29130_HOND_G0007

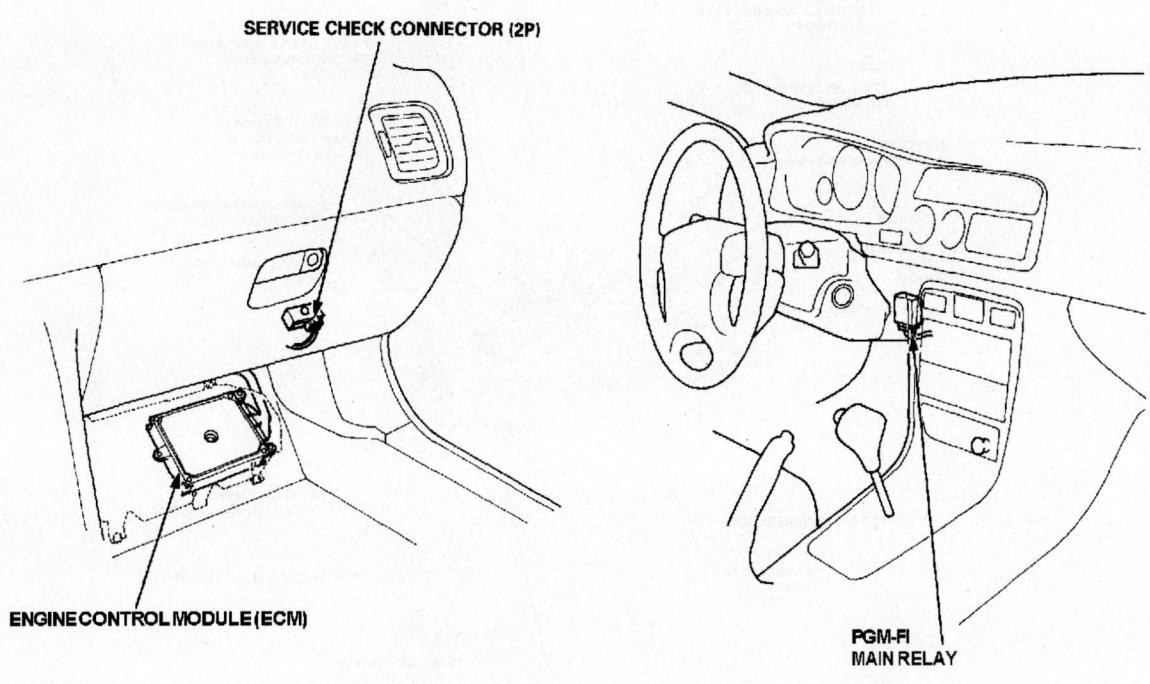

SERVICE CHECK CONNECTOR (2P)

ENGINE CONTROL MODULE (ECM)

PGM-FI
MAIN RELAY

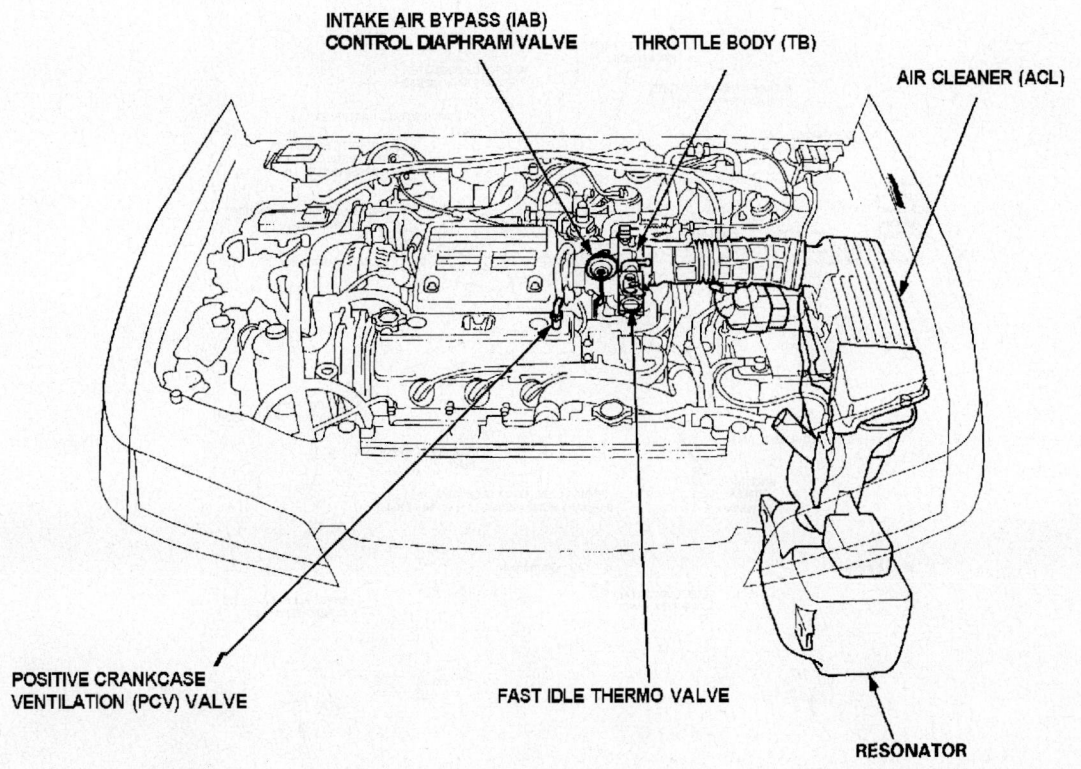

INTAKE AIR BYPASS (IAB)
CONTROL DIAPHRAM VALVE

THROTTLE BODY (TB)

AIR CLEANER (ACL)

POSITIVE CRANKCASE
VENTILATION (PCV) VALVE

FAST IDLE THERMO VALVE

RESONATOR

CIVIC

4 Cylinder Engines

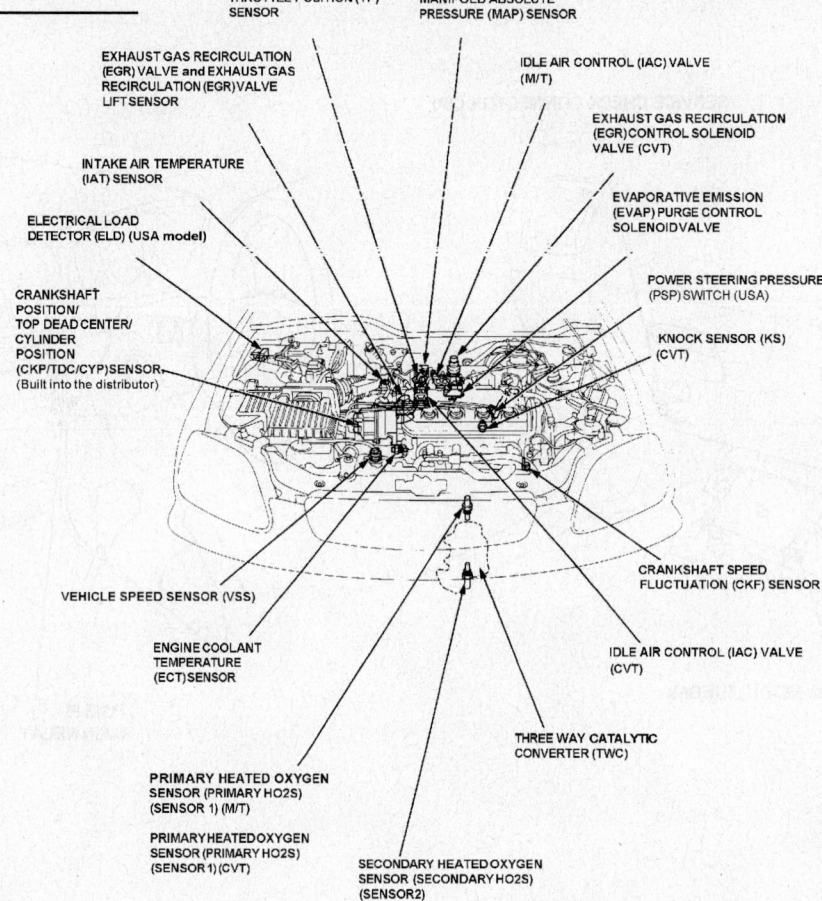

THROTTLE POSITION (TP) SENSOR

MANIFOLD ABSOLUTE PRESSURE (MAP) SENSOR

EXHAUST GAS RECIRCULATION (EGR) VALVE and EXHAUST GAS RECIRCULATION (EGR) VALVE LIFT SENSOR

IDLE AIR CONTROL (IAC) VALVE (M/T)

INTAKE AIR TEMPERATURE (IAT) SENSOR

EXHAUST GAS RECIRCULATION (EGR) CONTROL SOLENOID VALVE (CVT)

ELECTRICAL LOAD DETECTOR (ELD) (USA model)

EVAPORATIVE EMISSION (EVAP) PURGE CONTROL SOLENOID VALVE

CRANKSHAFT POSITION/ TOP DEAD CENTER/ CYLINDER POSITION (CKP/TDC/CYP)SENSOR (Built into the distributor)

POWER STEERING PRESSURE (PSP) SWITCH (USA)

KNOCK SENSOR (KS) (CVT)

VEHICLE SPEED SENSOR (VSS)

CRANKSHAFT SPEED FLUCTUATION (CKF) SENSOR

ENGINE COOLANT TEMPERATURE (ECT) SENSOR

IDLE AIR CONTROL (IAC) VALVE (CVT)

THREE WAY CATALYTIC CONVERTER (TWC)

PRIMARY HEATED OXYGEN SENSOR (PRIMARY HO2S) (SENSOR 1) (M/T)

PRIMARY HEATED OXYGEN SENSOR (PRIMARY HO2S) (SENSOR 1) (CVT)

SECONDARY HEATED OXYGEN SENSOR (SECONDARY HO2S) (SENSOR2)

29130_HOND_G0008

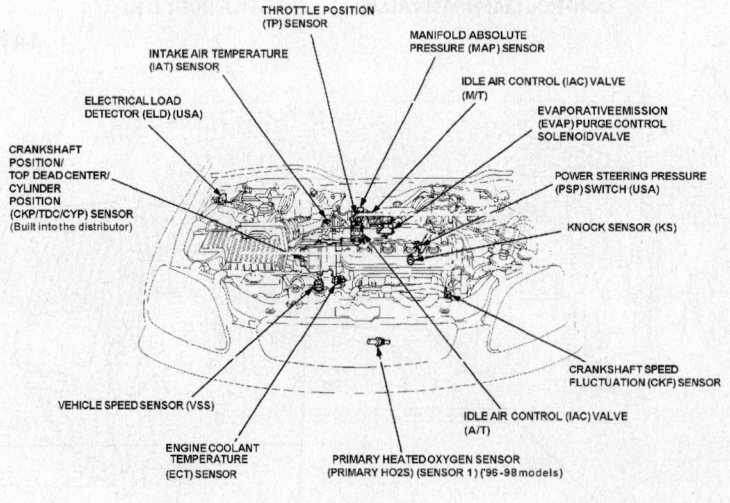

THROTTLE POSITION (TP) SENSOR

INTAKE AIR TEMPERATURE (IAT) SENSOR

MANIFOLD ABSOLUTE PRESSURE (MAP) SENSOR

IDLE AIR CONTROL (IAC) VALVE (M/T)

ELECTRICAL LOAD DETECTOR (ELD) (USA)

EVAPORATIVE EMISSION (EVAP) PURGE CONTROL SOLENOID VALVE

CRANKSHAFT POSITION/ TOP DEAD CENTER/ CYLINDER POSITION (CKP/TDC/CYP) SENSOR (Built into the distributor)

POWER STEERING PRESSURE (PSP) SWITCH (USA)

KNOCK SENSOR (KS)

VEHICLE SPEED SENSOR (VSS)

CRANKSHAFT SPEED FLUCTUATION (CKF) SENSOR

ENGINE COOLANT TEMPERATURE (ECT) SENSOR

IDLE AIR CONTROL (IAC) VALVE (A/T)

PRIMARY HEATED OXYGEN SENSOR (PRIMARY HO2S) (SENSOR 1) ('96-98 models)

'96 - 98 models:

'99 - 00 models:

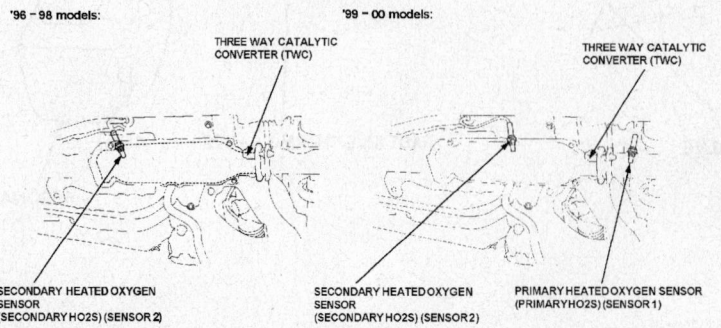

THREE WAY CATALYTIC CONVERTER (TWC)

THREE WAY CATALYTIC CONVERTER (TWC)

SECONDARY HEATED OXYGEN SENSOR (SECONDARY HO2S) (SENSOR 2)

SECONDARY HEATED OXYGEN SENSOR (SECONDARY HO2S) (SENSOR2)

PRIMARY HEATED OXYGEN SENSOR (PRIMARY HO2S) (SENSOR 1)

29130_HOND_G0009

THROTTLE POSITION (TP) SENSOR

MANIFOLD ABSOLUTE PRESSURE (MAP) SENSOR

VEHICLE SPEED SENSOR (VSS)

INTAKE AIR TEMPERATURE (IAT)SENSOR

ELECTRICAL LOAD DETECTOR (ELD) (USA)

POWER STEERING PRESSURE (PSP) SWITCH (USA)

CRANKSHAFT POSITION/ TOP DEAD CENTER/ CYLINDER POSITION (CKP/TDC/CYP)SENSOR (Built into the distributor)

IDLE AIR CONTROL (IAC) VALVE

CRANKSHAFTSPEED FLUCTUATION (CKF) SENSOR

ENGINE COOLANT TEMPERATURE (ECT) SENSOR

PRIMARY HEATED OXYGEN SENSOR (PRIMARY H02S) (SENSOR 1)

EVAPORATIVE EMISSION (EVAP) PURGE CONTROL SOLENOID VALVE

SECONDARY HEATED OXYGEN SENSOR(SECONDARYH02S) (SENSOR2)

THREE WAY CATALYTIC CONVERTER (TWC)

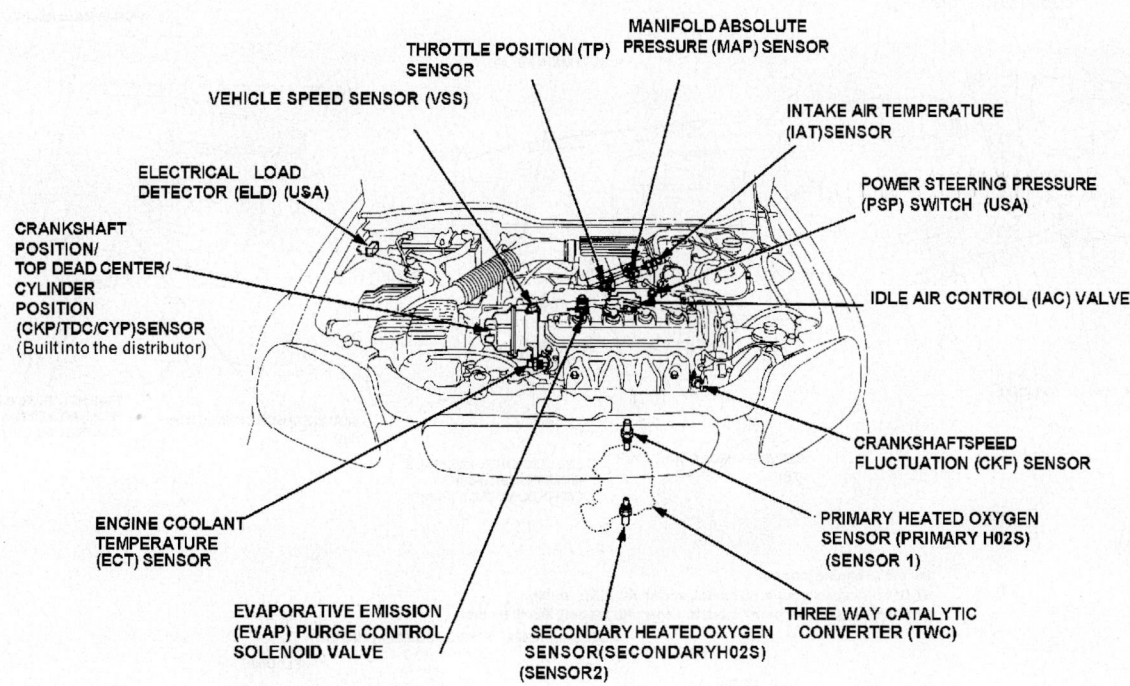

29130_HOND_G0010

INTAKE AIR TEMPERATURE (IAT)SENSOR

MANIFOLD ABSOLUTE PRESSURE (MAP) SENSOR

IDLE AIR CONTROL (IAC) VALVE

ELECTRICAL LOAD DETECTOR (ELD) (USA)

THROTTLE POSITION (TP) SENSOR

POWER STEERING PRESSUR (PSP) SWITCH (USA)

CRANKSHAFT POSITION/ TOP DEAD CENTER/ CYLINDER POSITION (CKP/TDC/CYP) SENSOR (Built into the distributor)

VEHICLE SPEED SENSOR (VSS)

ENGINECOOLANT TEMPERATURE (ECT) SENSOR

KNOCK SENSOR (KS)

CRANKSHAFT SPEED FLUCTUATION (CKF) SENSOR

THREE WAY CATALYTIC CONVERTER (TWC)

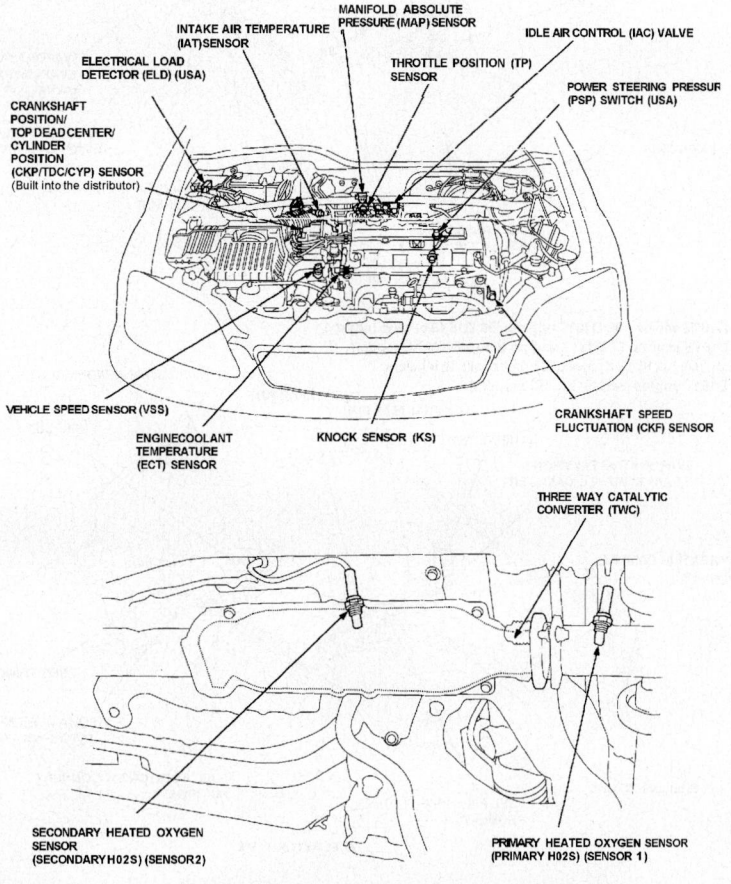

SECONDARY HEATED OXYGEN SENSOR (SECONDARYH02S) (SENSOR2)

PRIMARY HEATED OXYGEN SENSOR (PRIMARY H02S) (SENSOR 1)

29130_HOND_G0011

CIVIC

4 Cylinder Engines, cont'd

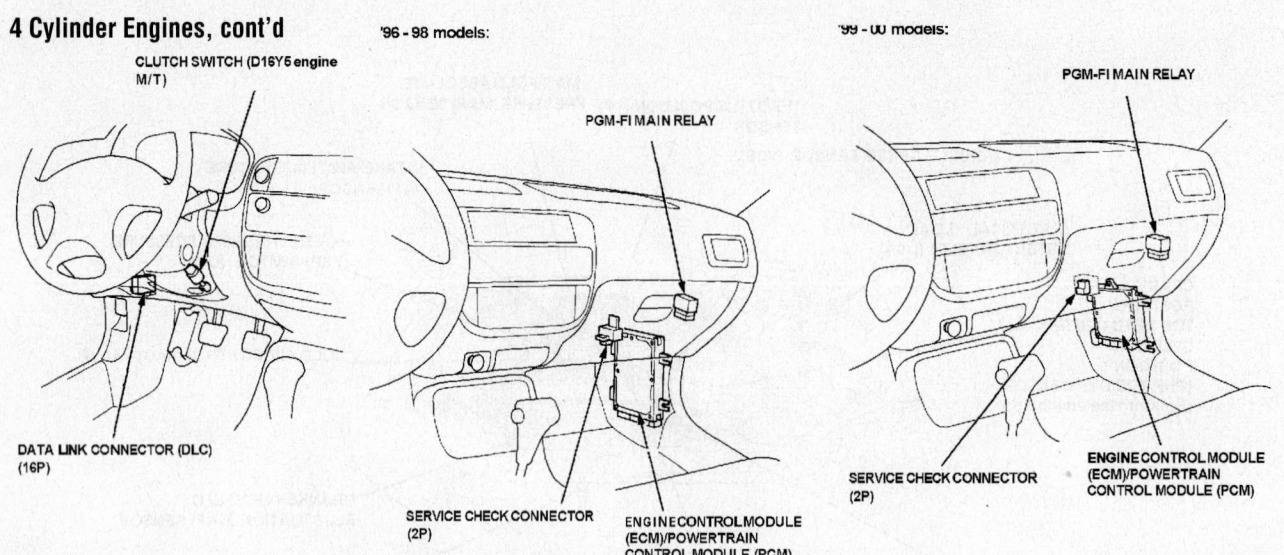

'96 - 98 models:

'99 - 00 models:

CLUTCH SWITCH (D16Y5 engine M/T)

PGM-FI MAIN RELAY

PGM-FI MAIN RELAY

DATA LINK CONNECTOR (DLC) (16P)

SERVICE CHECK CONNECTOR (2P)

ENGINE CONTROL MODULE (ECM)/POWERTRAIN CONTROL MODULE (PCM)

SERVICE CHECK CONNECTOR (2P)

ENGINE CONTROL MODULE (ECM)/POWERTRAIN CONTROL MODULE (PCM)

29130_HOND_G0012

'96 D16Y8 engine (coupe),
'97 D16Y7 engine (coupe: KL model, sedan: KL (LX) model),
'97 D16Y8 engine (coupe: all models, sedan: KL model), '98-all models:

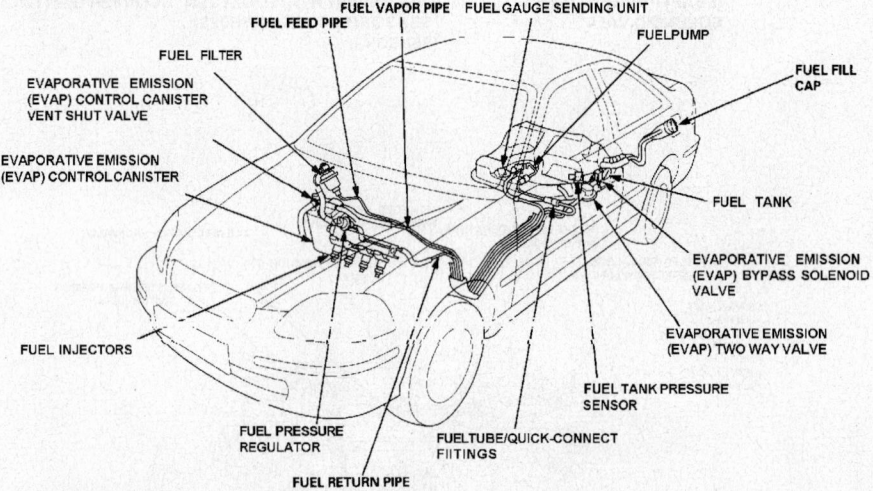

FUEL VAPOR PIPE FUEL GAUGE SENDING UNIT

FUEL FEED PIPE

FUEL FILTER

FUELPUMP

EVAPORATIVE EMISSION (EVAP) CONTROL CANISTER VENT SHUT VALVE

FUEL FILL CAP

EVAPORATIVE EMISSION (EVAP) CONTROL CANISTER

FUEL TANK

EVAPORATIVE EMISSION (EVAP) BYPASS SOLENOID VALVE

EVAPORATIVE EMISSION (EVAP) TWO WAY VALVE

FUEL INJECTORS

FUEL TANK PRESSURE SENSOR

FUEL PRESSURE REGULATOR

FUEL TUBE/QUICK-CONNECT FIITINGS

FUEL RETURN PIPE

'96 D16Y5 engine, '96 D16Y7 engine, '96 D16Y8 engine (sedan),
'97 D16Y5 engine, '97 D16Y7 engine (coupe: KA, KC models,
sedan: KA, KC, KL (DX) models, hatchback: all models),
'97 D16Y8 engine (sedan: KA, KC models):

FUEL VAPOR PIPE

FUELGAUGE SENDING UNIT

FUEL FEED PIPE

FUELFILTER

FUELPUMP

EVAPORATIVE EMISSION (EVAP) CONTROL CANISTER

FUEL PULSATION DAMPER (D16Y5 engine)

FUEL FILL CAP

FUEL TANK

EVAPORATIVE EMISSION (EVAP) TWO WAY VALVE

FUEL INJECTORS

FUEL TUBE/QUICK-CONNECT FITTINGS

FUEL PRESSURE REGULATOR

FUEL RETURN PIPE

29130_HOND_G0013

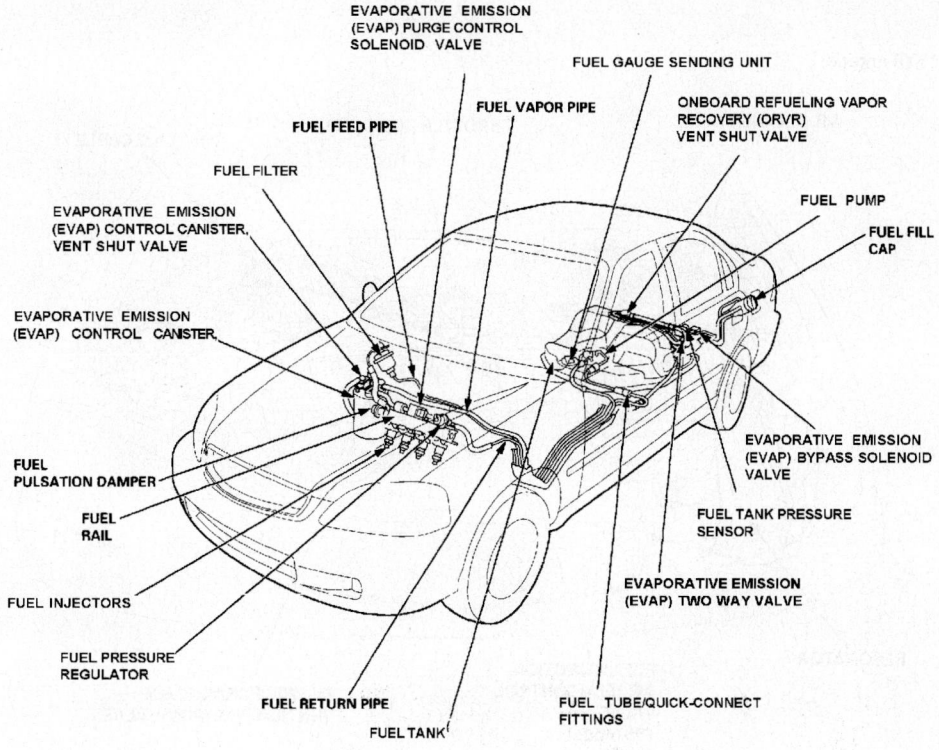

EVAPORATIVE EMISSION
(EVAP) PURGE CONTROL
SOLENOID VALVE

FUEL GAUGE SENDING UNIT

FUEL VAPOR PIPE

ONBOARD REFUELING VAPOR
RECOVERY (ORVR)
VENT SHUT VALVE

FUEL FEED PIPE

FUEL FILTER

FUEL PUMP

FUEL FILL
CAP

EVAPORATIVE EMISSION
(EVAP) CONTROL CANISTER,
VENT SHUT VALVE

EVAPORATIVE EMISSION
(EVAP) CONTROL CANISTER

EVAPORATIVE EMISSION
(EVAP) BYPASS SOLENOID
VALVE

FUEL
PULSATION DAMPER

FUEL TANK PRESSURE
SENSOR

FUEL
RAIL

EVAPORATIVE EMISSION
(EVAP) TWO WAY VALVE

FUEL INJECTORS

FUEL PRESSURE
REGULATOR

FUEL RETURN PIPE

FUEL TUBE/QUICK-CONNECT
FITTINGS

FUEL TANK

29130_HOND_G0014

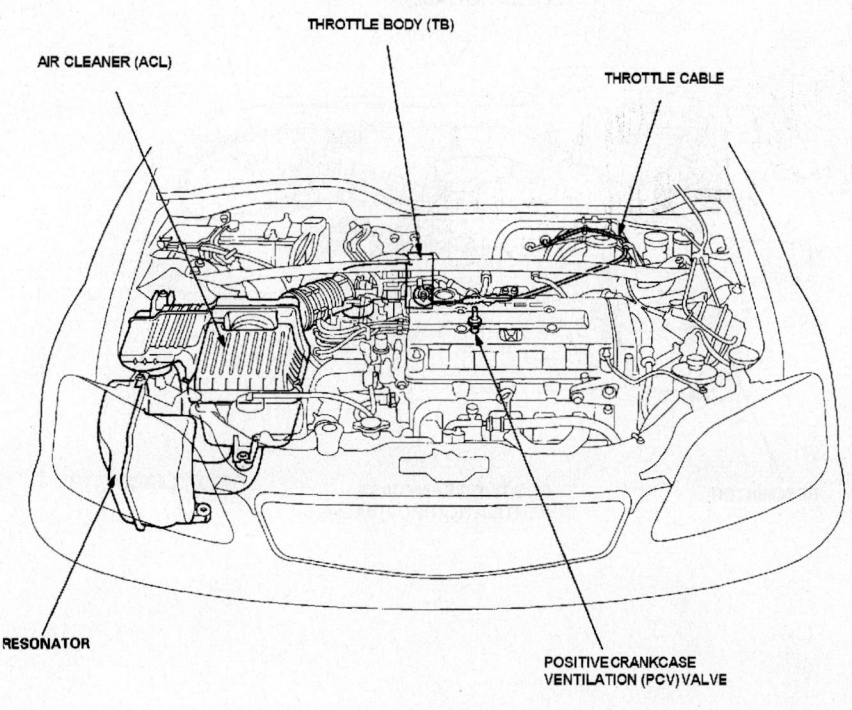

THROTTLE BODY (TB)

AIR CLEANER (ACL)

THROTTLE CABLE

RESONATOR

POSITIVE CRANKCASE
VENTILATION (PCV) VALVE

29130_HOND_G0016

D16Y5, D16Y8 engine:

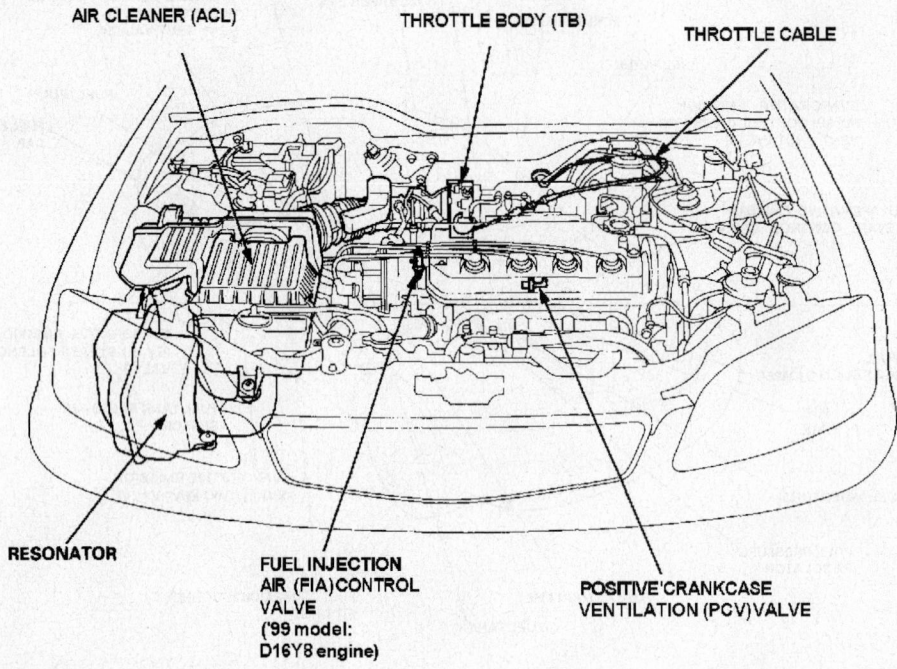

AIR CLEANER (ACL)

THROTTLE BODY (TB)

THROTTLE CABLE

RESONATOR

FUEL INJECTION
AIR (FIA) CONTROL
VALVE
('99 model:
D16Y8 engine)

POSITIVE CRANKCASE
VENTILATION (PCV) VALVE

D16Y7 engine:

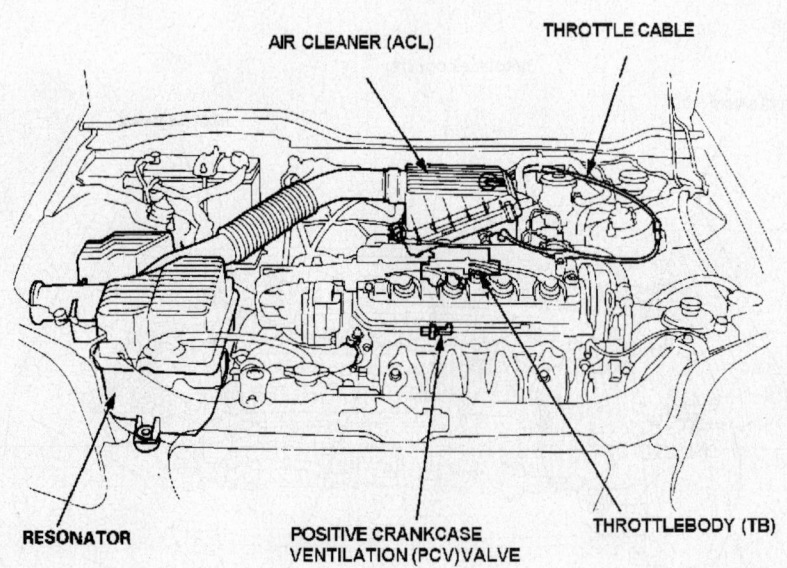

AIR CLEANER (ACL)

THROTTLE CABLE

RESONATOR

POSITIVE CRANKCASE
VENTILATION (PCV) VALVE

THROTTLE BODY (TB)

CR-V

4 Cylinder Engines

MANIFOLD ABSOLUTE
PRESSURE (MAP) SENSOR

IDLE AIR CONTROL (IAC) VALVE

THROTTLE POSITION (TP) SENSOR

EVAPORATIVE EMISSION (EVAP)
CONTROL CANISTER VENT SHUT
VALVE ('98 model)

INTAKE AIR TEMPERATURE
(IAT) SENSOR

ELECTRICAL LOAD
DETECTOR (ELD) (USA model)

POWER STEERING PRESSURE
(PSP) SWITCH

VEHICLE SPEED
SENSOR (VSS)

CRANKSHAFT
POSITION/
TOP DEAD CENTER/
CYLINDER POSITION
(CKP/TDC/CYP) SENSOR
(Built into the distributor)

CRANKSHAFT SPEED
FLUCTUATION (CKF) SENSOR

ENGINE COOLANT
TEMPERATURE
(ECT) SENSOR

EVAPORATIVE EMISSION
(EVAP) PURGE CONTROL
SOLENOID VALVE

29130_HOND_G0017

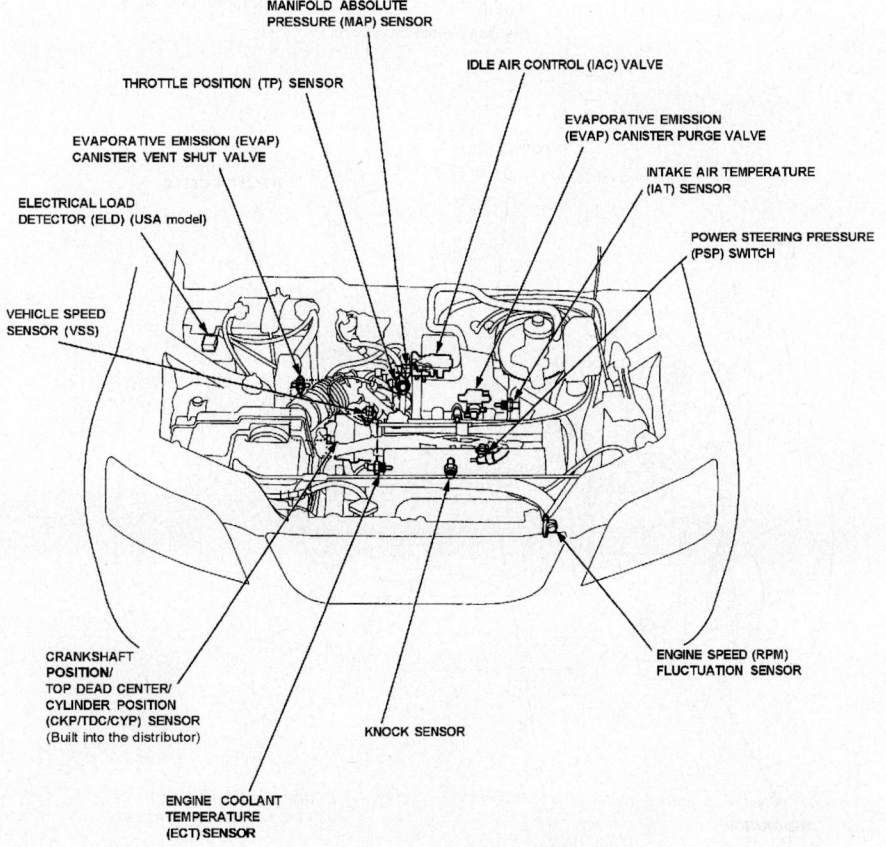

MANIFOLD ABSOLUTE
PRESSURE (MAP) SENSOR

THROTTLE POSITION (TP) SENSOR

IDLE AIR CONTROL (IAC) VALVE

EVAPORATIVE EMISSION (EVAP)
CANISTER VENT SHUT VALVE

EVAPORATIVE EMISSION
(EVAP) CANISTER PURGE VALVE

ELECTRICAL LOAD
DETECTOR (ELD) (USA model)

INTAKE AIR TEMPERATURE
(IAT) SENSOR

POWER STEERING PRESSURE
(PSP) SWITCH

VEHICLE SPEED
SENSOR (VSS)

CRANKSHAFT
POSITION/
TOP DEAD CENTER/
CYLINDER POSITION
(CKP/TDC/CYP) SENSOR
(Built into the distributor)

ENGINE SPEED (RPM)
FLUCTUATION SENSOR

KNOCK SENSOR

ENGINE COOLANT
TEMPERATURE
(ECT) SENSOR

29130_HOND_G0018

CR-V

4 Cylinder Engines, cont'd

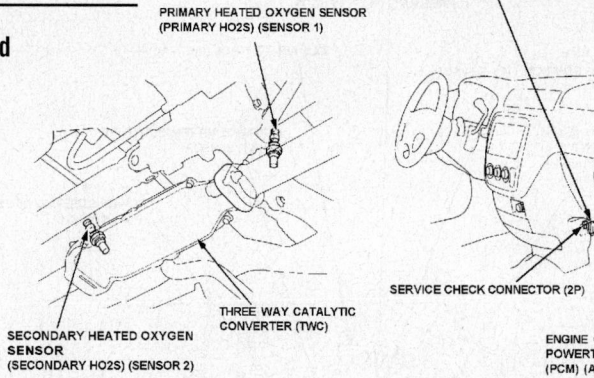

PRIMARY HEATED OXYGEN SENSOR (PRIMARY HO2S) (SENSOR 1)

DATA LINK CONNECTOR (DLC) (16P)

PGM-FI MAIN RELAY

SERVICE CHECK CONNECTOR (2P)

ENGINE CONTROL MODULE (ECM) (M/T)/ POWERTRAIN CONTROL MODULE (PCM) (A/T)

THREE WAY CATALYTIC CONVERTER (TWC)

SECONDARY HEATED OXYGEN SENSOR (SECONDARY HO2S) (SENSOR 2)

'97 - 98 models:

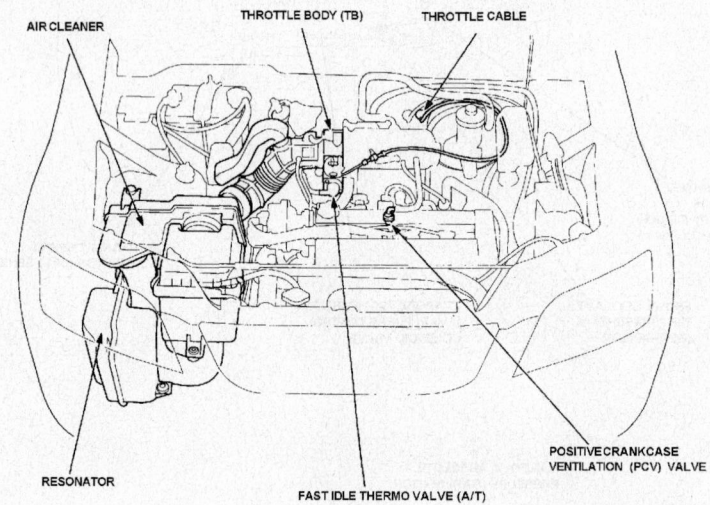

AIR CLEANER

THROTTLE BODY (TB)

THROTTLE CABLE

RESONATOR

FAST IDLE THERMO VALVE (A/T)

POSITIVE CRANKCASE VENTILATION (PCV) VALVE

29130_HOND_G0019

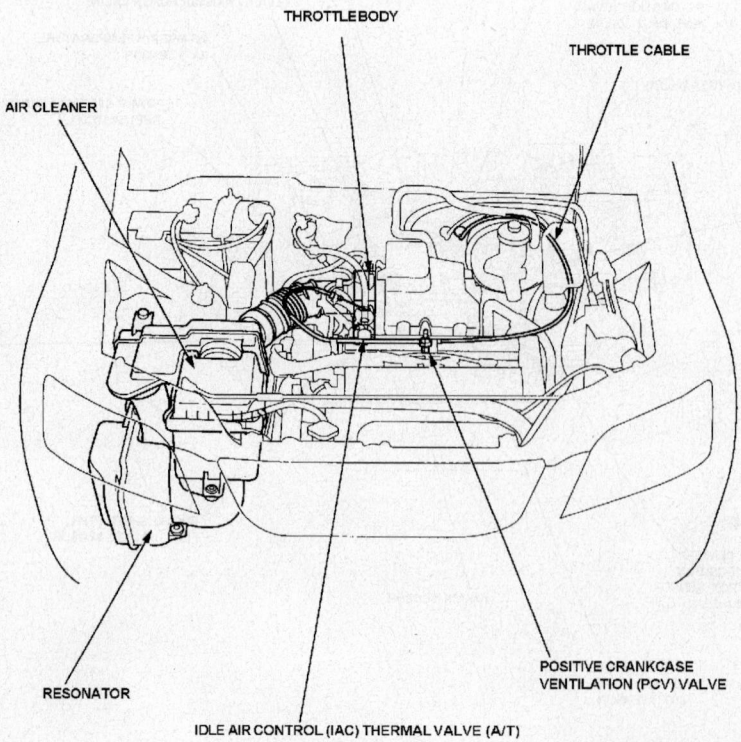

THROTTLE BODY

THROTTLE CABLE

AIR CLEANER

RESONATOR

IDLE AIR CONTROL (IAC) THERMAL VALVE (A/T)

POSITIVE CRANKCASE VENTILATION (PCV) VALVE

29130_HOND_G0020

'97 model:

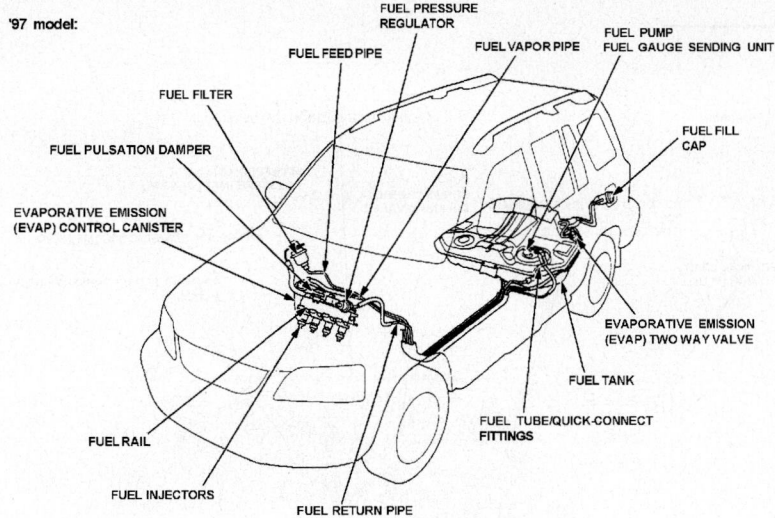

'98 model:

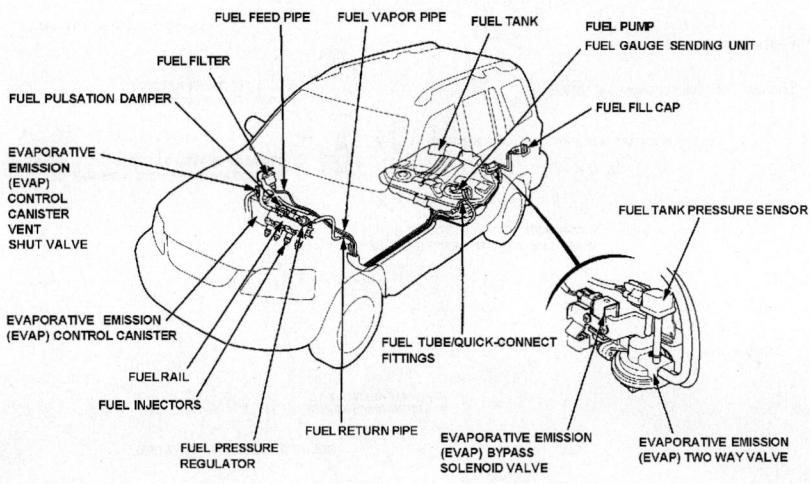

29130_HOND_G0021

'99 - 01 models:

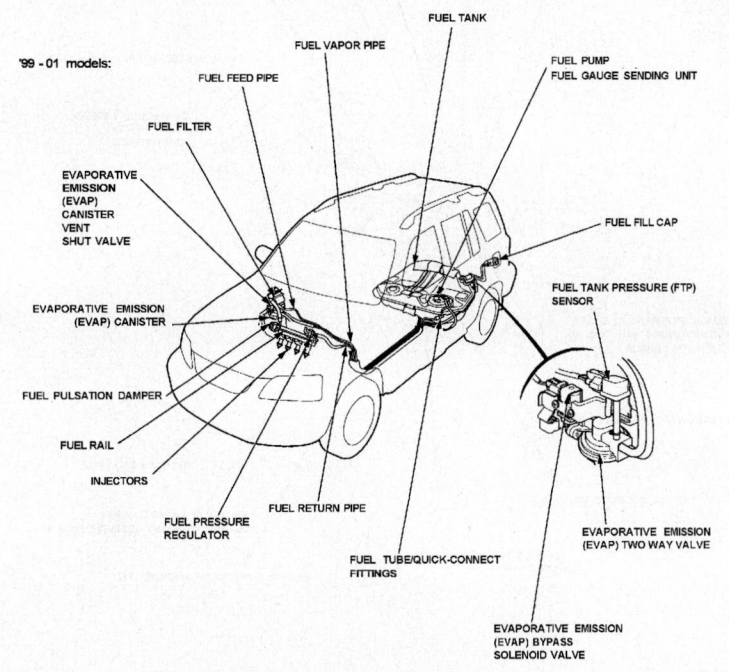

29130_HOND_G0022

DEL SOL

4 Cylinder Engines

D16Y7 engine:

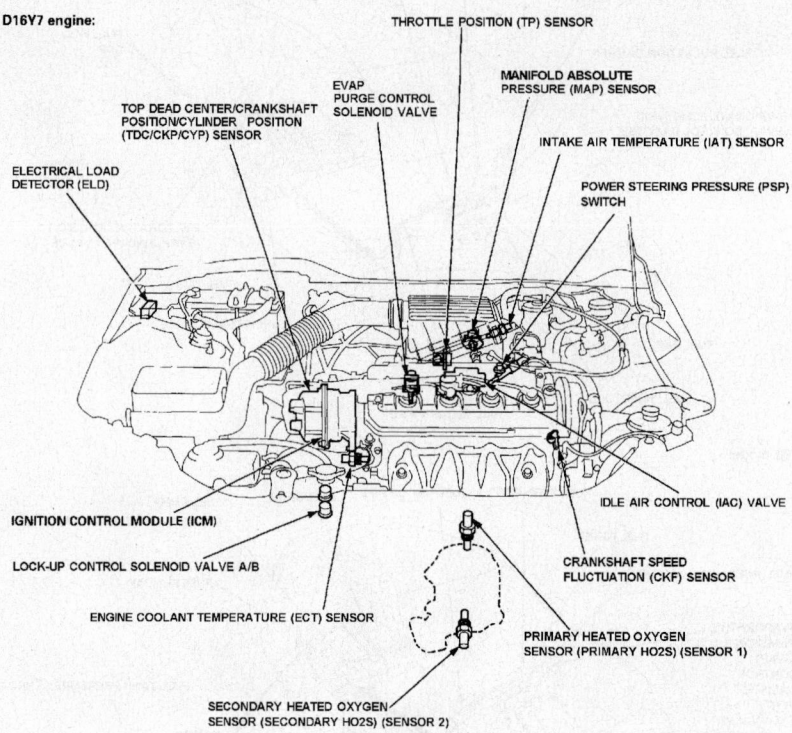

THROTTLE POSITION (TP) SENSOR

MANIFOLD ABSOLUTE
PRESSURE (MAP) SENSOR

EVAP
PURGE CONTROL
SOLENOID VALVE

INTAKE AIR TEMPERATURE (IAT) SENSOR

TOP DEAD CENTER/CRANKSHAFT
POSITION/CYLINDER POSITION
(TDC/CKP/CYP) SENSOR

POWER STEERING PRESSURE (PSP)
SWITCH

ELECTRICAL LOAD
DETECTOR (ELD)

IDLE AIR CONTROL (IAC) VALVE

IGNITION CONTROL MODULE (ICM)

CRANKSHAFT SPEED
FLUCTUATION (CKF) SENSOR

LOCK-UP CONTROL SOLENOID VALVE A/B

ENGINE COOLANT TEMPERATURE (ECT) SENSOR

PRIMARY HEATED OXYGEN
SENSOR (PRIMARY HO2S) (SENSOR 1)

SECONDARY HEATED OXYGEN
SENSOR (SECONDARY HO2S) (SENSOR 2)

29130_HOND_G0023

D16Y8 engine:

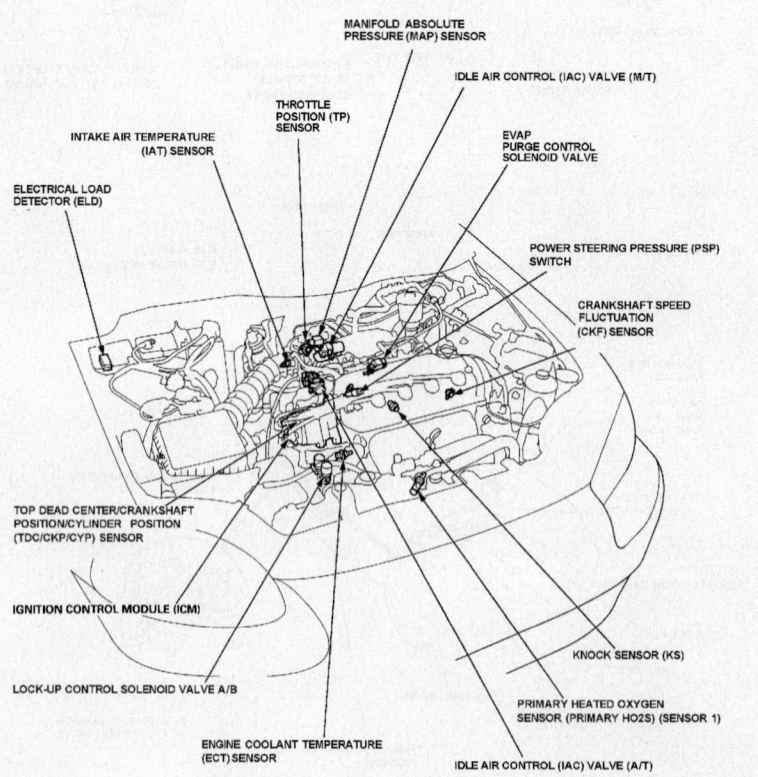

MANIFOLD ABSOLUTE
PRESSURE (MAP) SENSOR

IDLE AIR CONTROL (IAC) VALVE (M/T)

THROTTLE
POSITION (TP)
SENSOR

EVAP
PURGE CONTROL
SOLENOID VALVE

INTAKE AIR TEMPERATURE
(IAT) SENSOR

POWER STEERING PRESSURE (PSP)
SWITCH

ELECTRICAL LOAD
DETECTOR (ELD)

CRANKSHAFT SPEED
FLUCTUATION
(CKF) SENSOR

TOP DEAD CENTER/CRANKSHAFT
POSITION/CYLINDER POSITION
(TDC/CKP/CYP) SENSOR

IGNITION CONTROL MODULE (ICM)

KNOCK SENSOR (KS)

LOCK-UP CONTROL SOLENOID VALVE A/B

PRIMARY HEATED OXYGEN
SENSOR (PRIMARY HO2S) (SENSOR 1)

ENGINE COOLANT TEMPERATURE
(ECT) SENSOR

IDLE AIR CONTROL (IAC) VALVE (A/T)

29130_HOND_G0024

B16A2 engine:

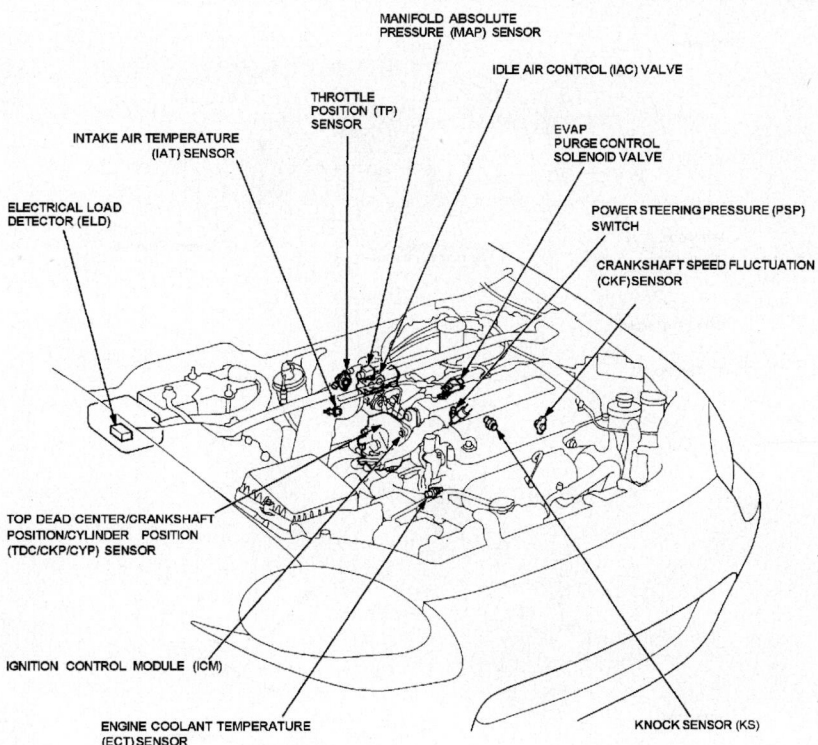

MANIFOLD ABSOLUTE
PRESSURE (MAP) SENSOR

IDLE AIR CONTROL (IAC) VALVE

THROTTLE
POSITION (TP)
SENSOR

EVAP
PURGE CONTROL
SOLENOID VALVE

INTAKE AIR TEMPERATURE
(IAT) SENSOR

POWER STEERING PRESSURE (PSP)
SWITCH

ELECTRICAL LOAD
DETECTOR (ELD)

CRANKSHAFT SPEED FLUCTUATION
(CKF) SENSOR

TOP DEAD CENTER/CRANKSHAFT
POSITION/CYLINDER POSITION
(TDC/CKP/CYP) SENSOR

IGNITION CONTROL MODULE (ICM)

ENGINE COOLANT TEMPERATURE
(ECT) SENSOR

KNOCK SENSOR (KS)

29130_HOND_G0025

D16Y8, B16A2 engine:

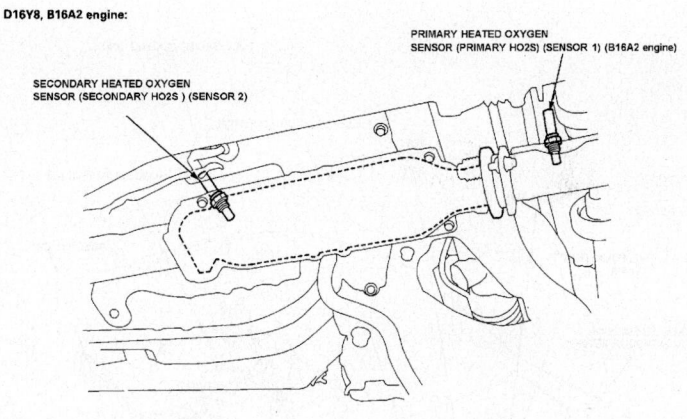

PRIMARY HEATED OXYGEN
SENSOR (PRIMARY HO2S) (SENSOR 1) (B16A2 engine)

SECONDARY HEATED OXYGEN
SENSOR (SECONDARY HO2S) (SENSOR 2)

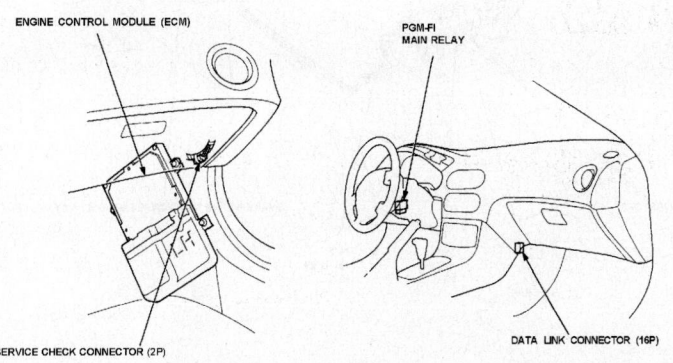

ENGINE CONTROL MODULE (ECM)

PGM-FI
MAIN RELAY

SERVICE CHECK CONNECTOR (2P)

DATA LINK CONNECTOR (16P)

29130_HOND_G0026

DEL SOL

4 Cylinder Engines, cont'd

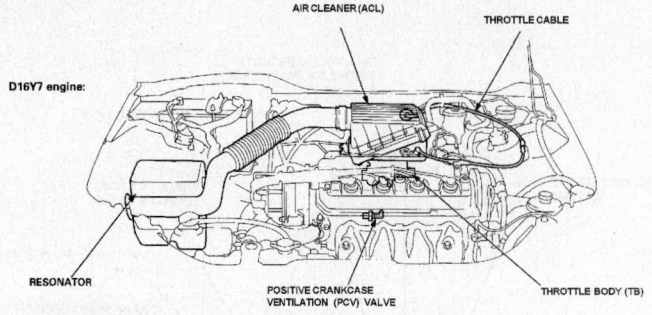

D16Y7 engine:

AIR CLEANER (ACL)

THROTTLE CABLE

RESONATOR

POSITIVE CRANKCASE VENTILATION (PCV) VALVE

THROTTLE BODY (TB)

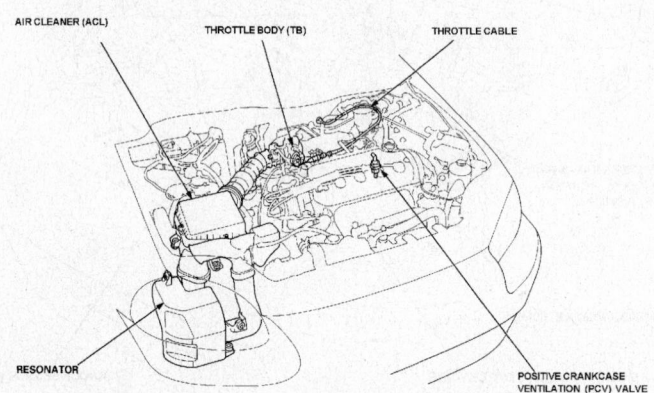

D16Y8, B16A2 engine:

AIR CLEANER (ACL)

THROTTLE BODY (TB)

THROTTLE CABLE

RESONATOR

POSITIVE CRANKCASE VENTILATION (PCV) VALVE

29130_HOND_G0027

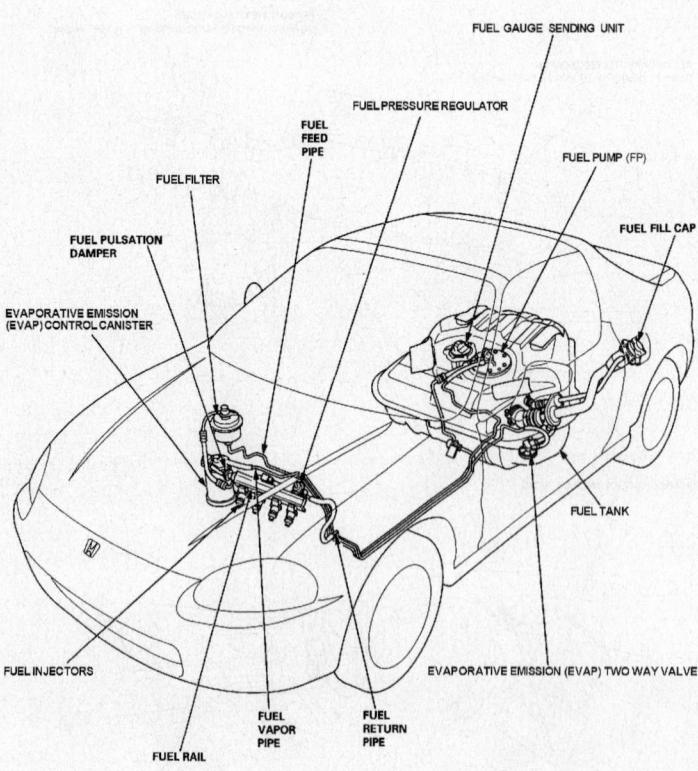

FUEL GAUGE SENDING UNIT

FUEL PRESSURE REGULATOR

FUEL FEED PIPE

FUEL PUMP (FP)

FUEL FILTER

FUEL FILL CAP

FUEL PULSATION DAMPER

EVAPORATIVE EMISSION (EVAP) CONTROL CANISTER

FUEL TANK

FUEL INJECTORS

EVAPORATIVE EMISSION (EVAP) TWO WAY VALVE

FUEL RAIL

FUEL VAPOR PIPE

FUEL RETURN PIPE

29130_HOND_G0028

ELEMENT

4 Cylinder Engines

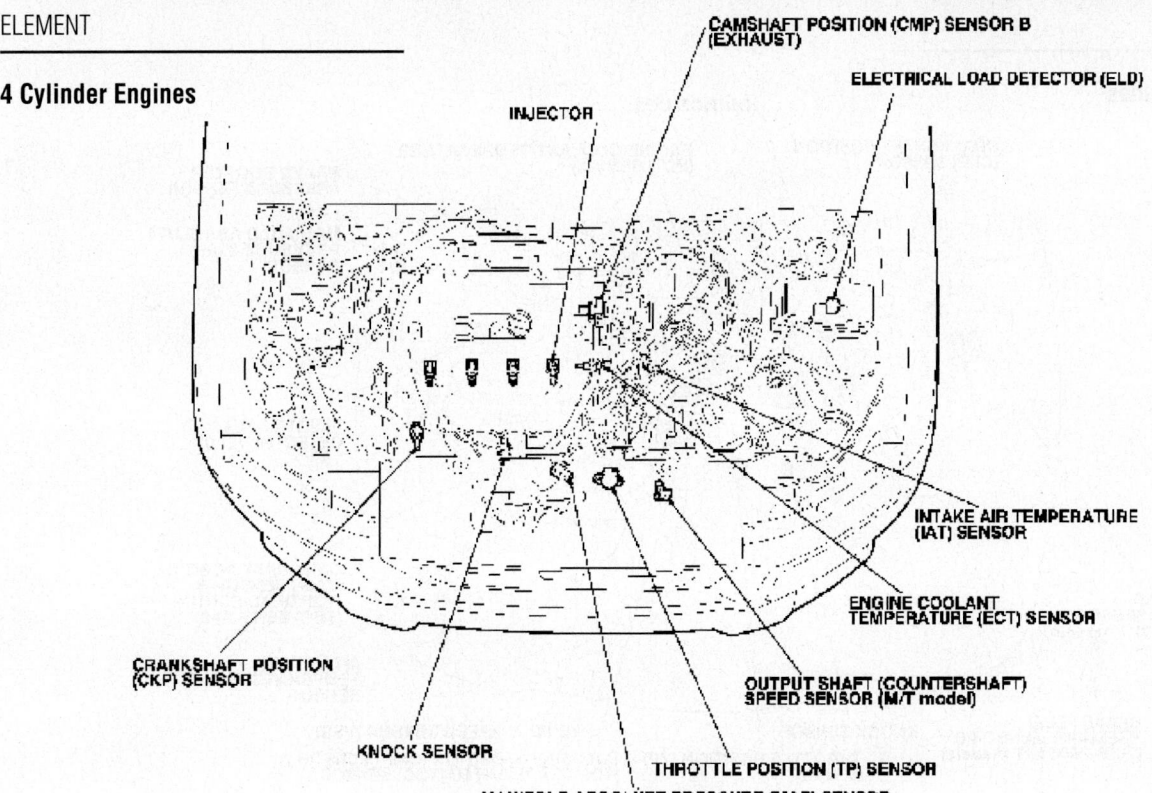

CAMSHAFT POSITION (CMP) SENSOR B
(EXHAUST)

ELECTRICAL LOAD DETECTOR (ELD)

INJECTOR

INTAKE AIR TEMPERATURE
(IAT) SENSOR

ENGINE COOLANT
TEMPERATURE (ECT) SENSOR

OUTPUT SHAFT (COUNTERSHAFT)
SPEED SENSOR (M/T model)

CRANKSHAFT POSITION
(CKP) SENSOR

KNOCK SENSOR

THROTTLE POSITION (TP) SENSOR

MANIFOLD ABSOLUTE PRESSURE (MAP) SENSOR

29130_HOND_G0029

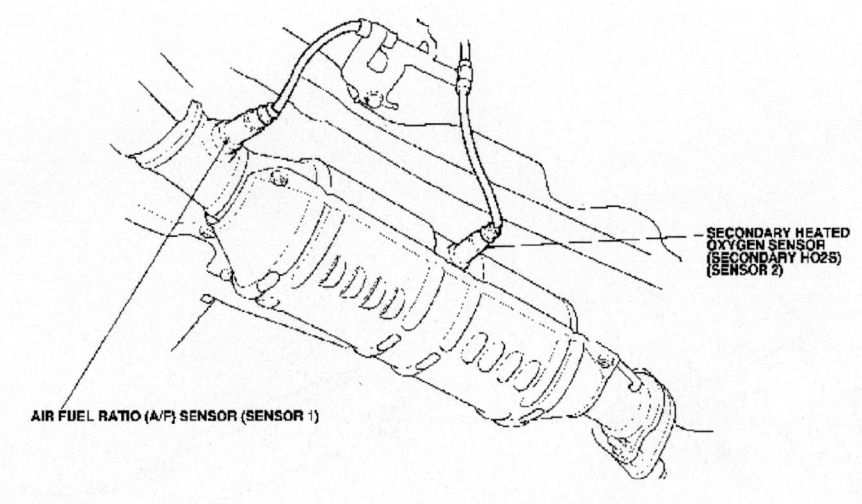

SECONDARY HEATED
OXYGEN SENSOR
(SECONDARY HO2S)
(SENSOR 2)

AIR FUEL RATIO (A/F) SENSOR (SENSOR 1)

29130_HOND_G0030

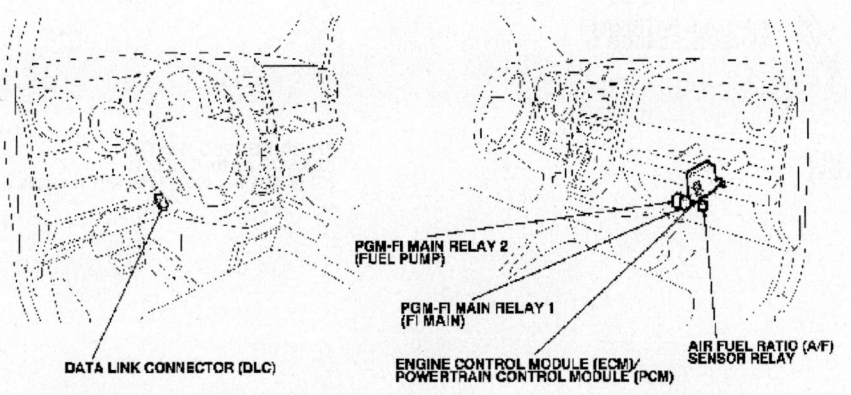

PGM-FI MAIN RELAY 2
(FUEL PUMP)

PGM-FI MAIN RELAY 1
(FI MAIN)

AIR FUEL RATIO (A/F)
SENSOR RELAY

DATA LINK CONNECTOR (DLC)

ENGINE CONTROL MODULE (ECM)/
POWERTRAIN CONTROL MODULE (PCM)

29130_HOND_G0031

INSIGHT

4 Cylinder Engines

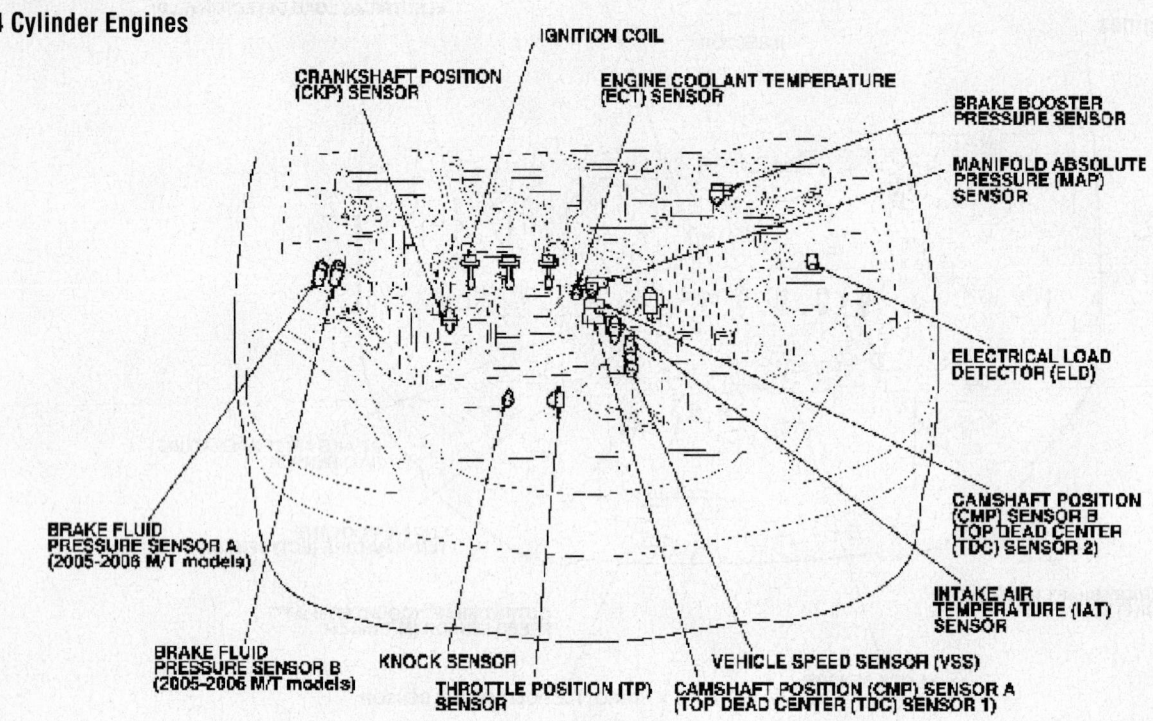

CRANKSHAFT POSITION (CKP) SENSOR

IGNITION COIL

ENGINE COOLANT TEMPERATURE (ECT) SENSOR

BRAKE BOOSTER PRESSURE SENSOR

MANIFOLD ABSOLUTE PRESSURE (MAP) SENSOR

ELECTRICAL LOAD DETECTOR (ELD)

CAMSHAFT POSITION (CMP) SENSOR B (TOP DEAD CENTER (TDC) SENSOR 2)

INTAKE AIR TEMPERATURE (IAT) SENSOR

BRAKE FLUID PRESSURE SENSOR A (2005-2006 M/T models)

BRAKE FLUID PRESSURE SENSOR B (2005-2006 M/T models)

KNOCK SENSOR

THROTTLE POSITION (TP) SENSOR

VEHICLE SPEED SENSOR (VSS)

CAMSHAFT POSITION (CMP) SENSOR A (TOP DEAD CENTER (TDC) SENSOR 1)

29130_HOND_G0032

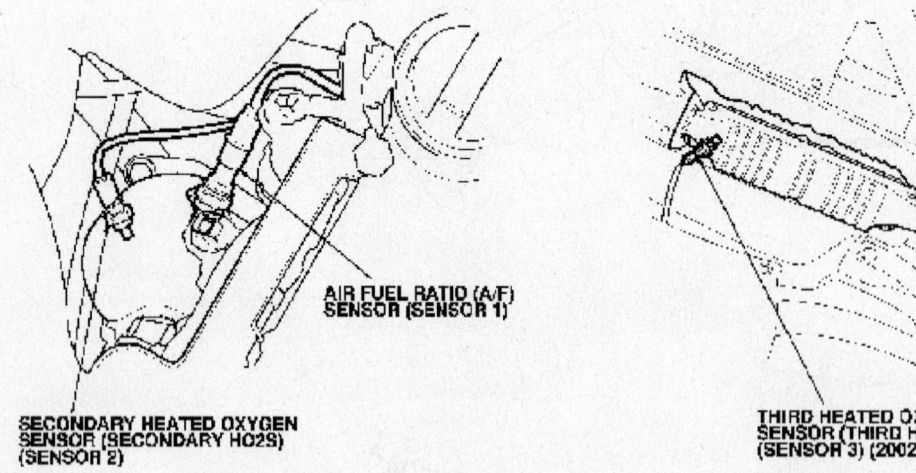

AIR FUEL RATIO (A/F) SENSOR (SENSOR 1)

SECONDARY HEATED OXYGEN SENSOR (SECONDARY HO2S) (SENSOR 2)

THIRD HEATED OXYGEN SENSOR (THIRD HO2S) (SENSOR 3) (2002-2006 M/T models only)

29130_HOND_G0033

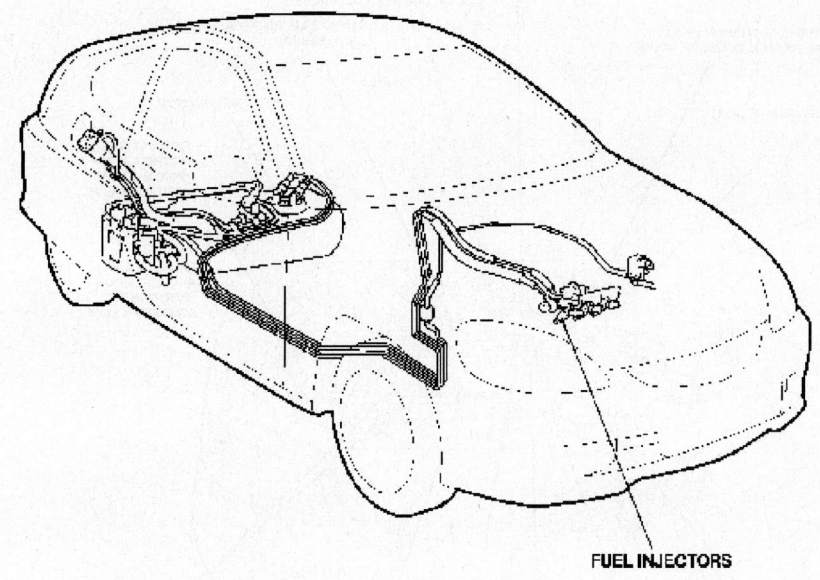

FUEL INJECTORS

29130_HOND_G0034

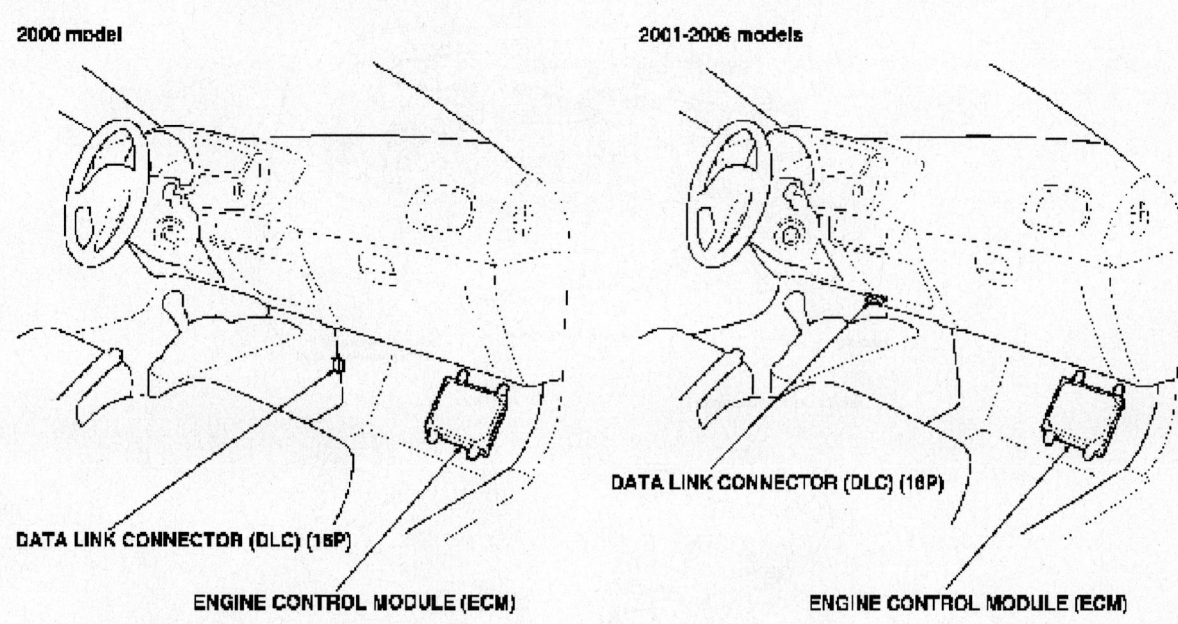

2000 model

2001-2006 models

DATA LINK CONNECTOR (DLC) (16P)

DATA LINK CONNECTOR (DLC) (16P)

ENGINE CONTROL MODULE (ECM)

ENGINE CONTROL MODULE (ECM)

29130_HOND_G0035

ODYSSEY

4 Cylinder Engines

MANIFOLD ABSOLUTE
PRESSURE (MAP) SENSOR

EVAPORATIVE EMISSION (EVAP)
PURGE FLOW SWITCH

IDLE AIR CONTROL (IAC) VALVE

EVAPORATIVE EMISSION (EVAP)
PURGE CONTROL SOLENOID VALVE

INTAKE AIR TEMPERATURE
(IAT) SENSOR

VEHICLE SPEED SENSOR (VSS)

INJECTOR RESISTOR

ELECTRICAL
LOAD
DETECTOR
(ELD)

EXHAUST GAS
RECIRCULATION
(EGR) VACUUM
CONTROL VALVE

EXHAUST GAS
RECIRCULATION
(EGR) CONTROL
SOLENOID VALVE

CRANKSHAFT POSITION/TOP DEAD
CENTER (CKP/TDC) SENSOR

CYLINDER POSITION (CYP) SENSOR
(Built into the distributor)

POWER STEERING PRESSURE
(PSP) SWITCH

ENGINE COOLANT TEMPERATURE (ECT) SENSOR

THROTTLE POSITION (TP) SENSOR

EXHAUST GAS RECIRCULATION
(EGR) VALVE and LIFT SENSOR

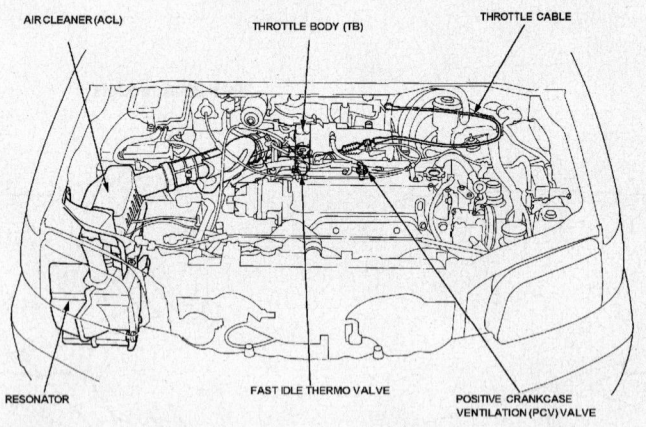

29130_HOND_G0036

AIR CLEANER (ACL)

THROTTLE BODY (TB)

THROTTLE CABLE

RESONATOR

FAST IDLE THERMO VALVE

POSITIVE CRANKCASE
VENTILATION (PCV) VALVE

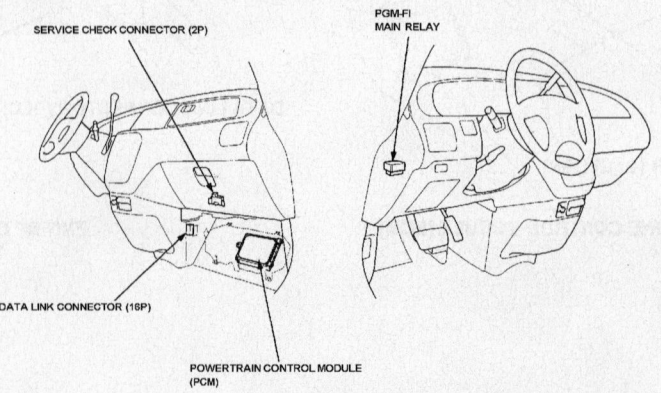

SERVICE CHECK CONNECTOR (2P)

PGM-FI
MAIN RELAY

DATA LINK CONNECTOR (16P)

POWERTRAIN CONTROL MODULE
(PCM)

29130_HOND_G0037

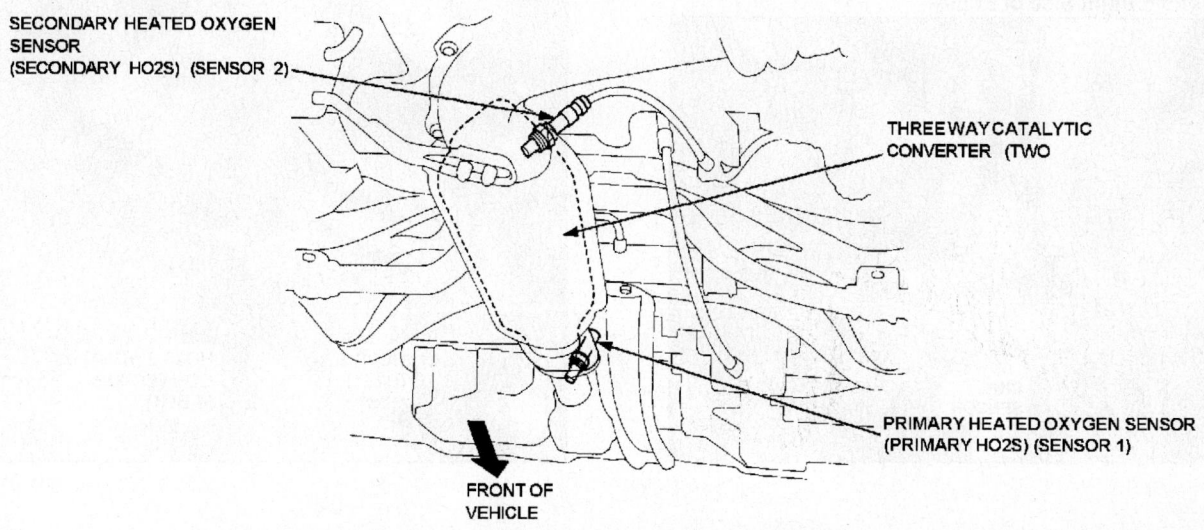

SECONDARY HEATED OXYGEN
SENSOR
(SECONDARY HO2S) (SENSOR 2)

THREE WAY CATALYTIC
CONVERTER (TWO

PRIMARY HEATED OXYGEN SENSOR
(PRIMARY HO2S) (SENSOR 1)

FRONT OF
VEHICLE

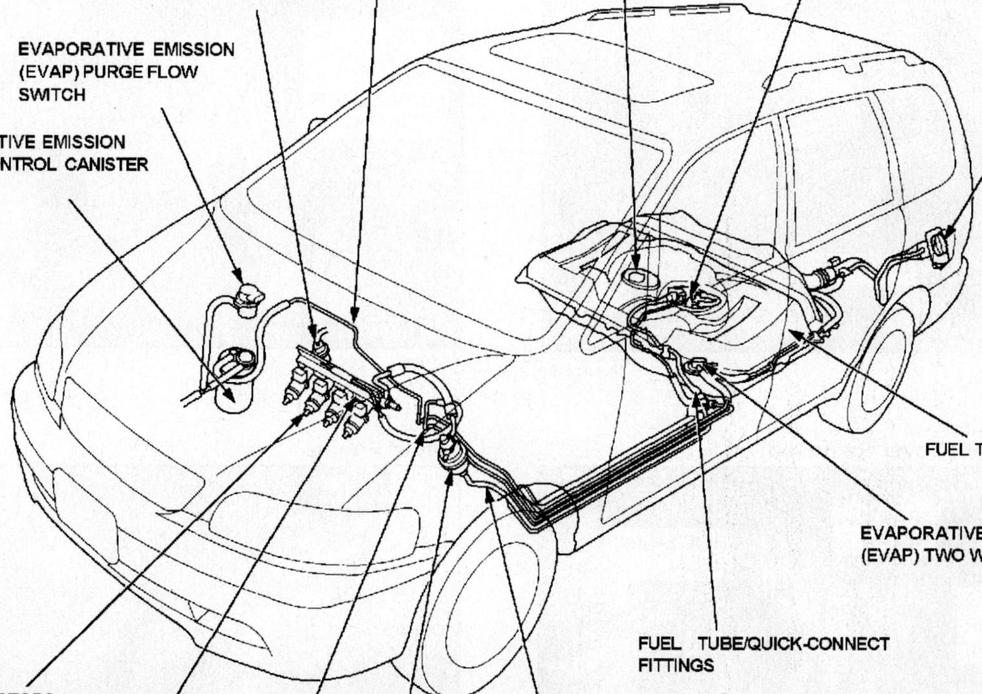

FUEL VAPOR PIPE

FUEL GAUGE SENDING UNIT

FUEL PUMP

FUEL
PRESSURE REGULATOR

EVAPORATIVE EMISSION
(EVAP) PURGE FLOW
SWITCH

EVAPORATIVE EMISSION
(EVAP) CONTROL CANISTER

FUEL FILL CAP

FUEL TANK

EVAPORATIVE EMISSION
(EVAP) TWO WAY VALVE

FUEL TUBE/QUICK-CONNECT
FITTINGS

FUEL INJECTORS

FUEL RAIL

FUEL FILTER

FUEL FEED PIPE

FUEL RETURN PIPE

29130_HOND_G0038

PASSPORT

6 Cylinder Engines

Bottom Right Side of Engine

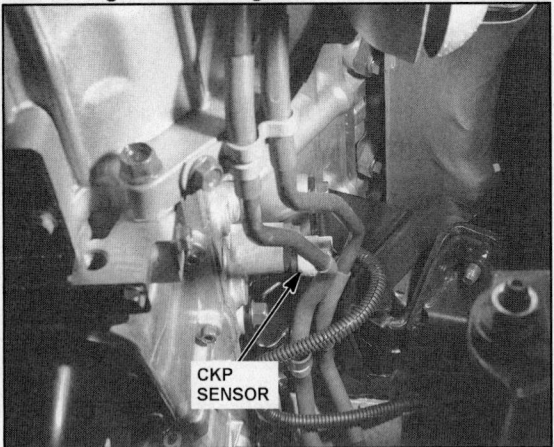

CKP SENSOR

29130_HOND_P0001

Left Front Exhaust Downpipe

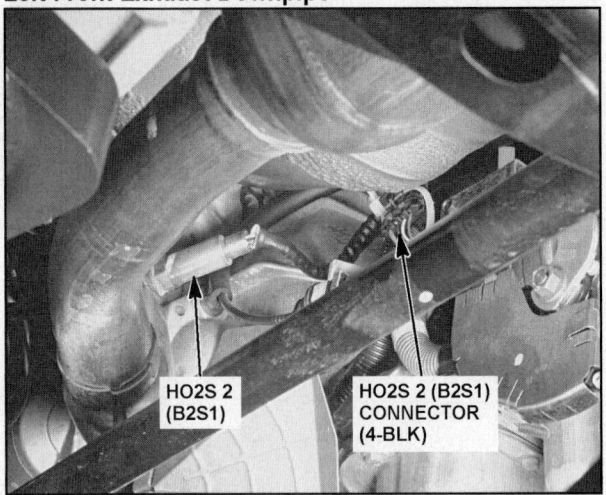

HO2S 2 (B2S1)

HO2S 2 (B2S1) CONNECTOR (4-BLK)

29130_HOND_P0004

Front of Engine

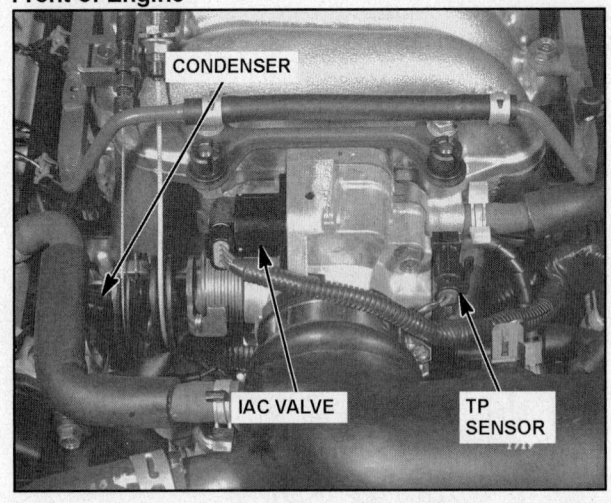

CONDENSER

IAC VALVE

TP SENSOR

29130_HOND_P0002

Left Front of Engine

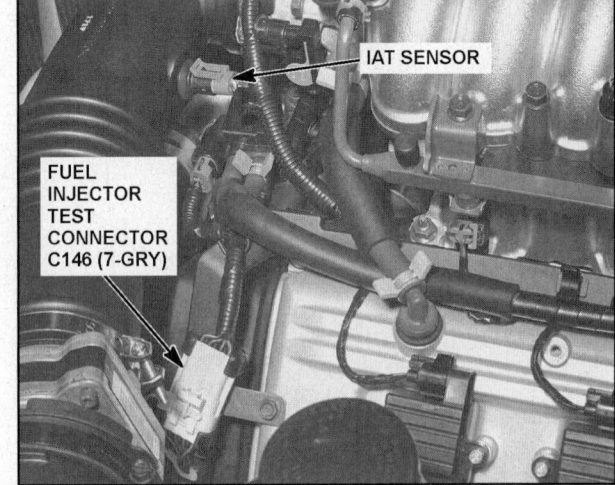

IAT SENSOR

FUEL INJECTOR TEST CONNECTOR C146 (7-GRY)

29130_HOND_P0005

Fuse/Relay Box (Cover Removed)

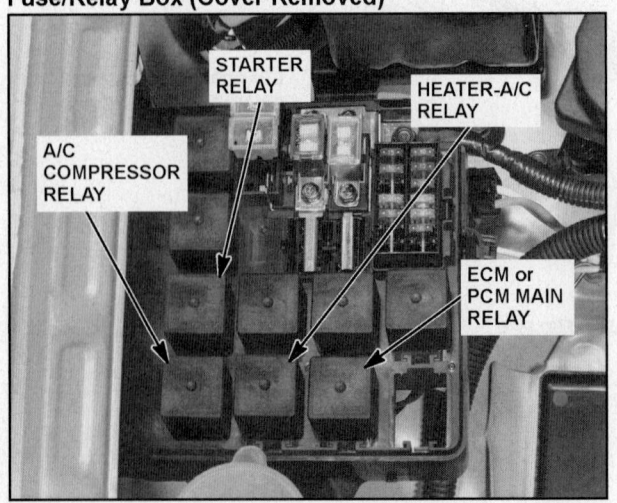

STARTER RELAY

HEATER-A/C RELAY

A/C COMPRESSOR RELAY

ECM or PCM MAIN RELAY

29130_HOND_P0003

Left of Engine

MAF SENSOR

29130_HOND_P0006

Left Side of Dash

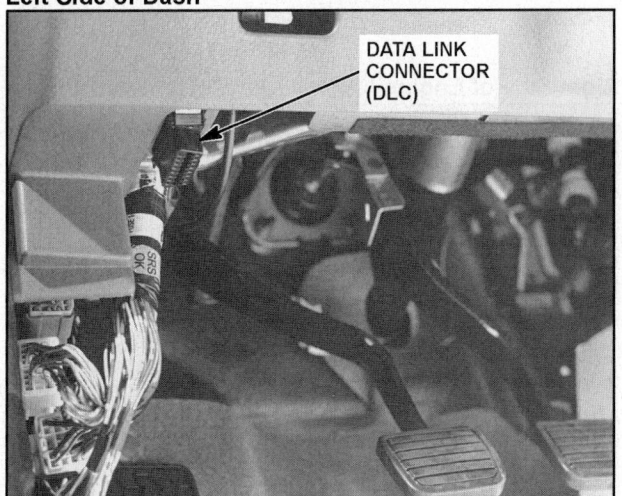

DATA LINK
CONNECTOR
(DLC)

29130_HOND_P0007

Rear of Left Catalytic Converter

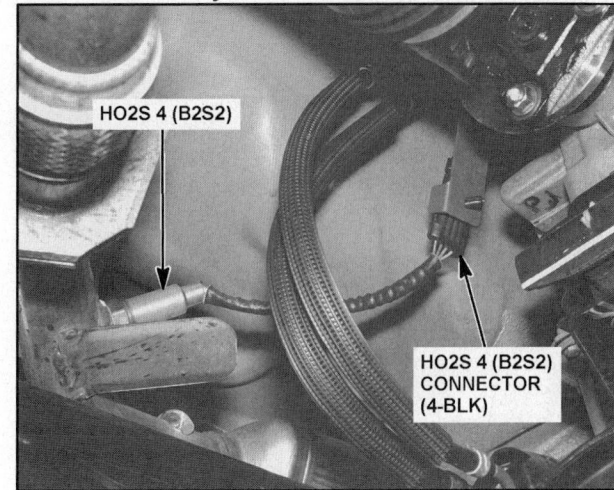

HO2S 4 (B2S2)

HO2S 4 (B2S2)
CONNECTOR
(4-BLK)

29130_HOND_P0010

Left Side of Engine

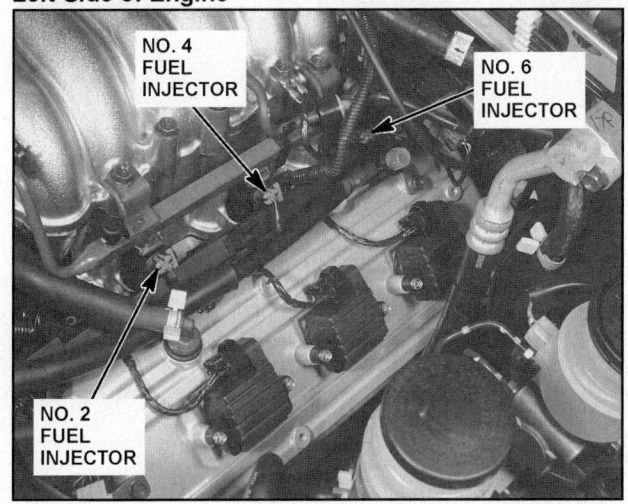

NO. 4
FUEL
INJECTOR

NO. 6
FUEL
INJECTOR

NO. 2
FUEL
INJECTOR

29130_HOND_P0008

Rear of Right Catalytic Converter

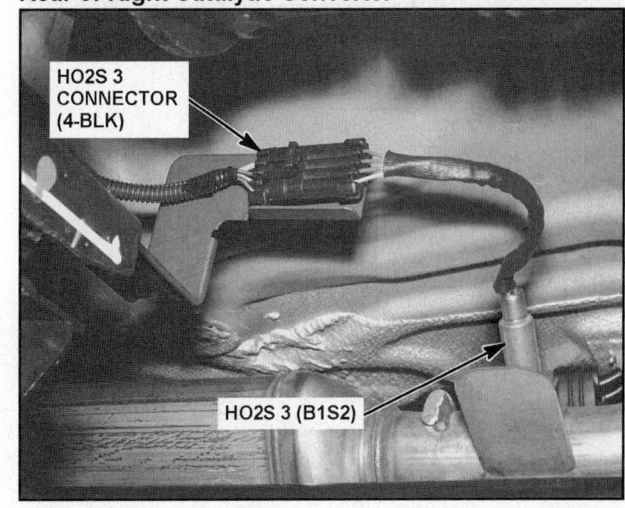

HO2S 3
CONNECTOR
(4-BLK)

HO2S 3 (B1S2)

29130_HOND_P0011

Beneath Center of Vehicle, on Transfer Case (A/T)

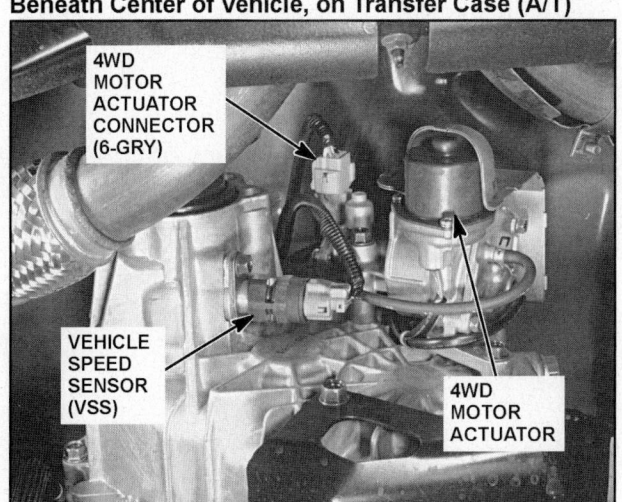

4WD
MOTOR
ACTUATOR
CONNECTOR
(6-GRY)

VEHICLE
SPEED
SENSOR
(VSS)

4WD
MOTOR
ACTUATOR

29130_HOND_P0009

Right Front Exhaust Downpipe

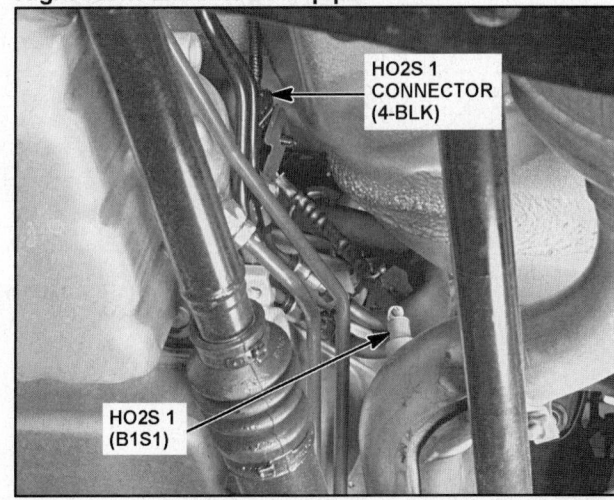

HO2S 1
CONNECTOR
(4-BLK)

HO2S 1
(B1S1)

29130_HOND_P0012

PASSPORT

6 Cylinder Engines, cont'd

Right Side of Engine

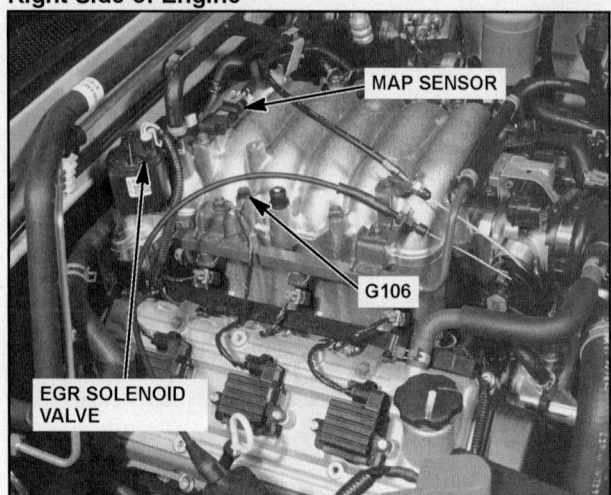

29130_HOND_P0013

Right Side of Engine

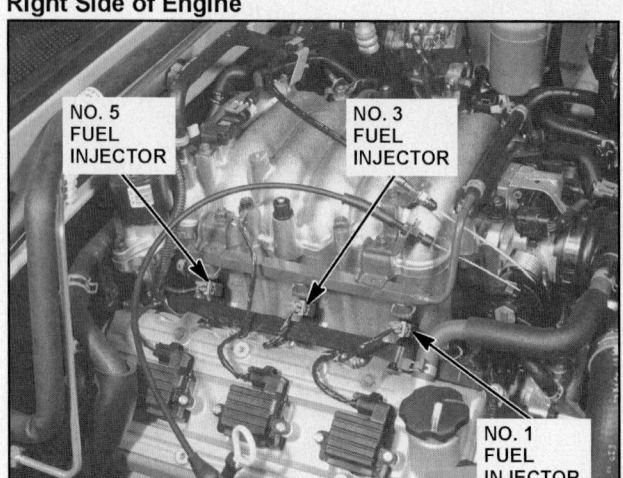

29130_HOND_P0014

PILOT

6 Cylinder Engines

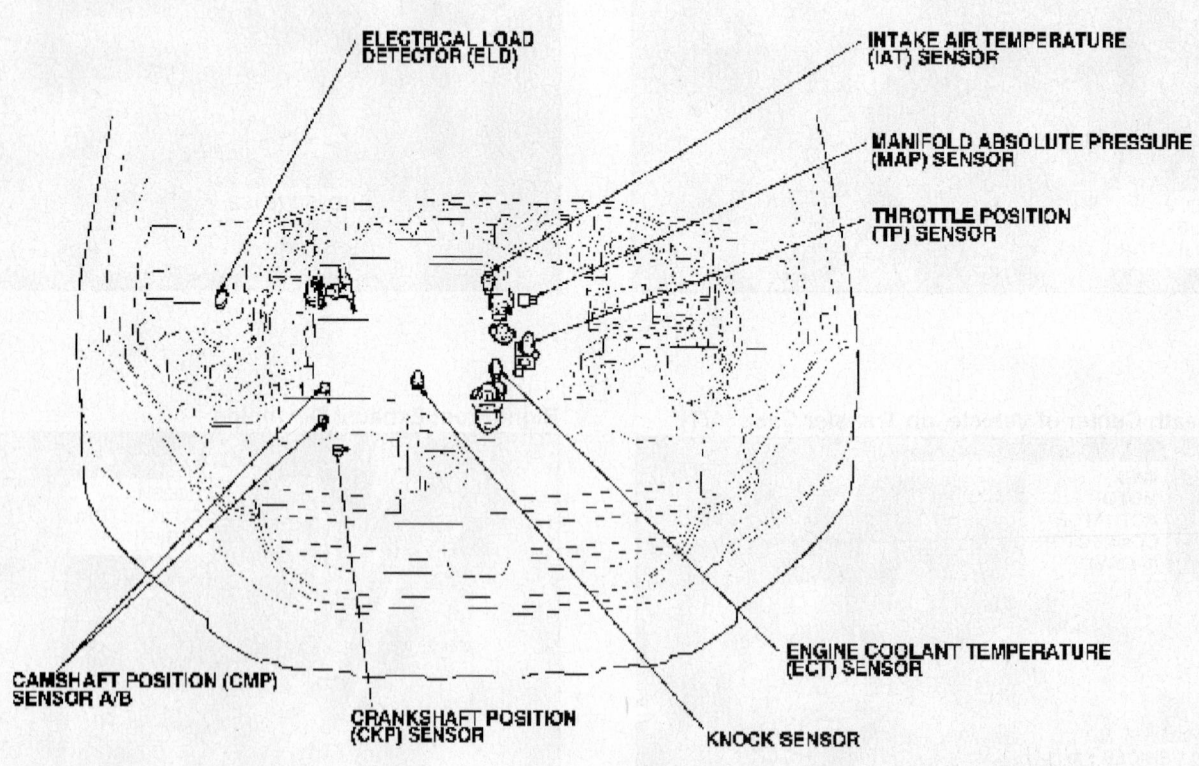

29130_HOND_G0039

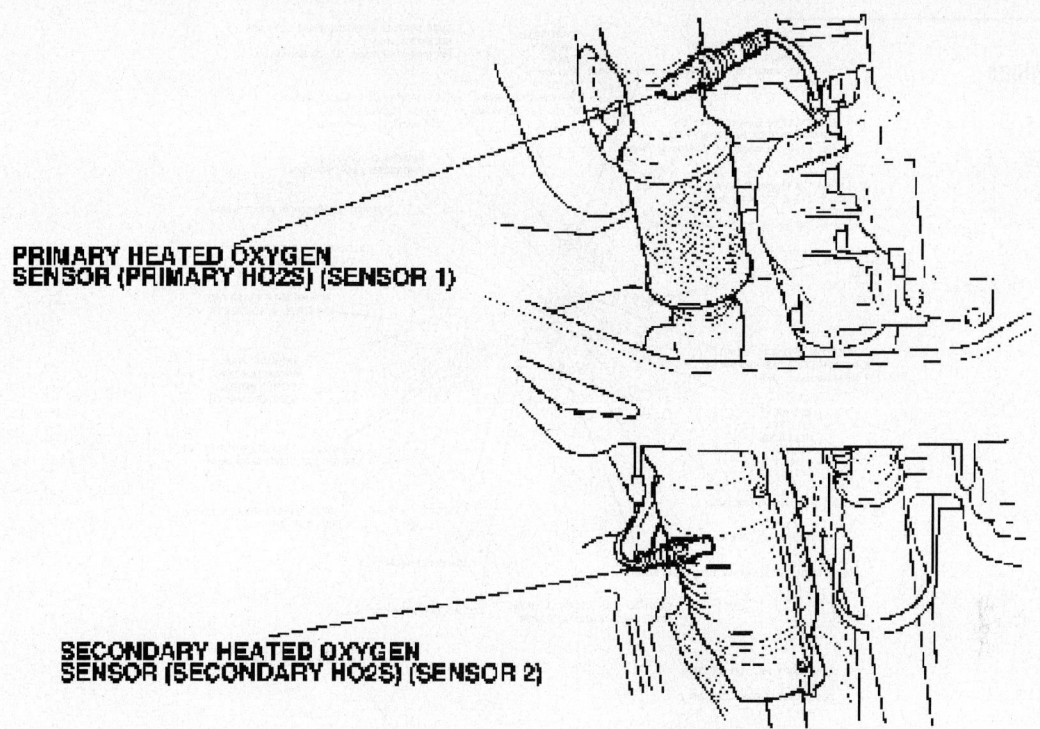

PRIMARY HEATED OXYGEN
SENSOR (PRIMARY HO2S) (SENSOR 1)

SECONDARY HEATED OXYGEN
SENSOR (SECONDARY HO2S) (SENSOR 2)

29130_HOND_G0040

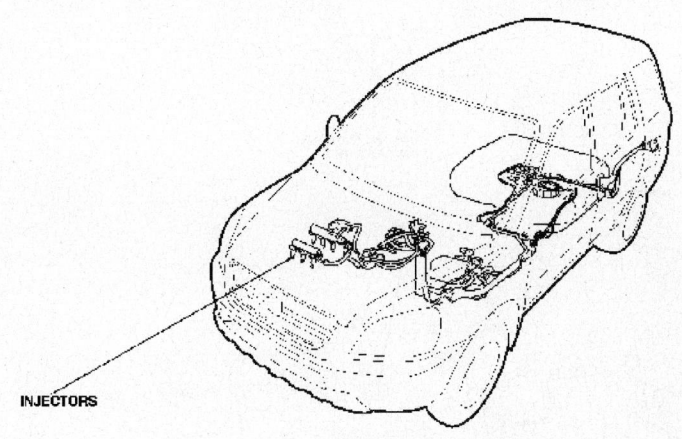

INJECTORS

29130_HOND_G0041

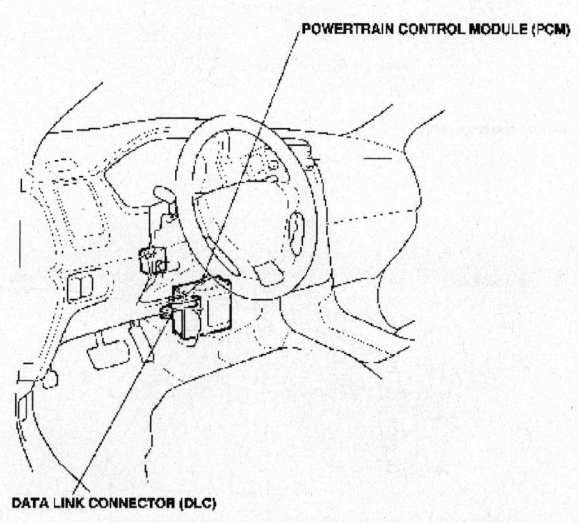

POWERTRAIN CONTROL MODULE (PCM)

DATA LINK CONNECTOR (DLC)

29130_HOND_G0042

PRELUDE

4 Cylinder Engines

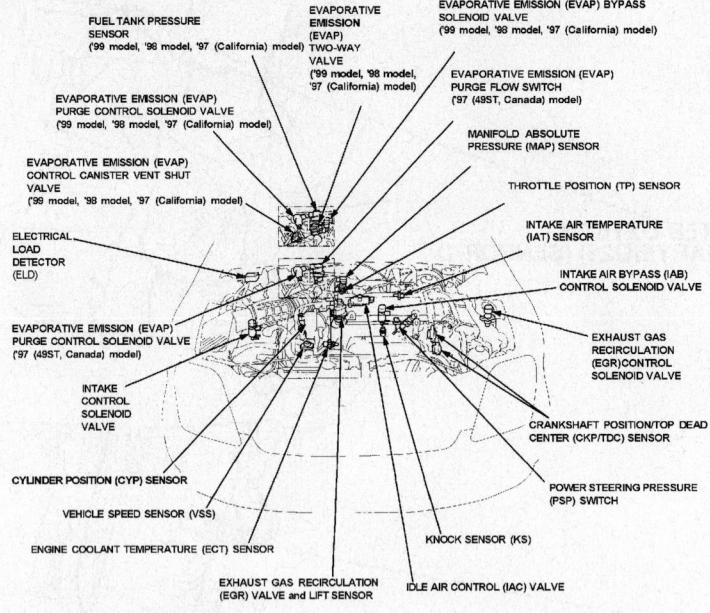

FUEL TANK PRESSURE SENSOR
('99 model, '98 model, '97 (California) model)

EVAPORATIVE EMISSION (EVAP) TWO-WAY VALVE
('99 model, '98 model, '97 (California) model)

EVAPORATIVE EMISSION (EVAP) BYPASS SOLENOID VALVE
('99 model, '98 model, '97 (California) model)

EVAPORATIVE EMISSION (EVAP) PURGE CONTROL SOLENOID VALVE
('99 model, '98 model, '97 (California) model)

EVAPORATIVE EMISSION (EVAP) PURGE FLOW SWITCH
('97 (49ST, Canada) model)

EVAPORATIVE EMISSION (EVAP) CONTROL CANISTER VENT SHUT VALVE
('99 model, '98 model, '97 (California) model)

MANIFOLD ABSOLUTE PRESSURE (MAP) SENSOR

THROTTLE POSITION (TP) SENSOR

ELECTRICAL LOAD DETECTOR (ELD)

INTAKE AIR TEMPERATURE (IAT) SENSOR

INTAKE AIR BYPASS (IAB) CONTROL SOLENOID VALVE

EVAPORATIVE EMISSION (EVAP) PURGE CONTROL SOLENOID VALVE
('97 (49ST, Canada) model)

EXHAUST GAS RECIRCULATION (EGR) CONTROL SOLENOID VALVE

INTAKE CONTROL SOLENOID VALVE

CRANKSHAFT POSITION/TOP DEAD CENTER (CKP/TDC) SENSOR

CYLINDER POSITION (CYP) SENSOR

POWER STEERING PRESSURE (PSP) SWITCH

VEHICLE SPEED SENSOR (VSS)

KNOCK SENSOR (KS)

ENGINE COOLANT TEMPERATURE (ECT) SENSOR

EXHAUST GAS RECIRCULATION (EGR) VALVE and LIFT SENSOR

IDLE AIR CONTROL (IAC) VALVE

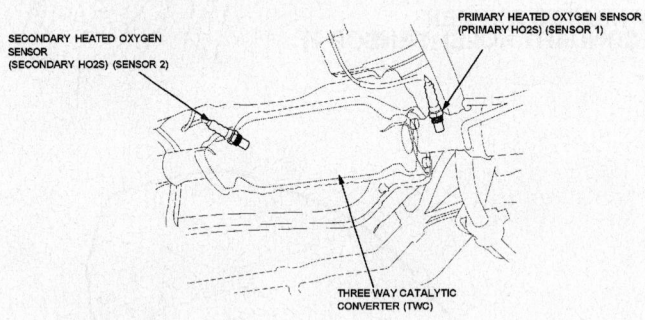

SECONDARY HEATED OXYGEN SENSOR (SECONDARY HO2S) (SENSOR 2)

PRIMARY HEATED OXYGEN SENSOR (PRIMARY HO2S) (SENSOR 1)

THREE WAY CATALYTIC CONVERTER (TWC)

29130_HOND_G0043

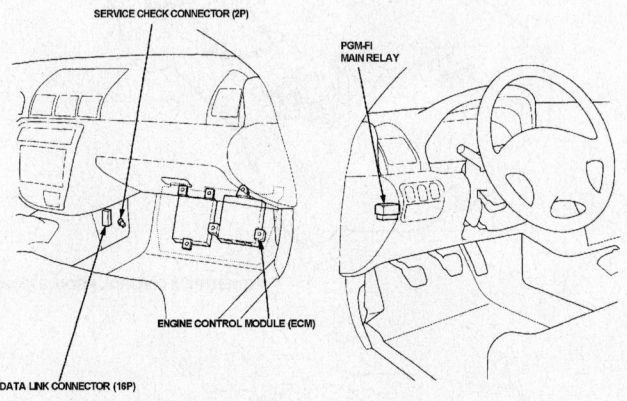

SERVICE CHECK CONNECTOR (2P)

PGM-FI MAIN RELAY

ENGINE CONTROL MODULE (ECM)

DATA LINK CONNECTOR (16P)

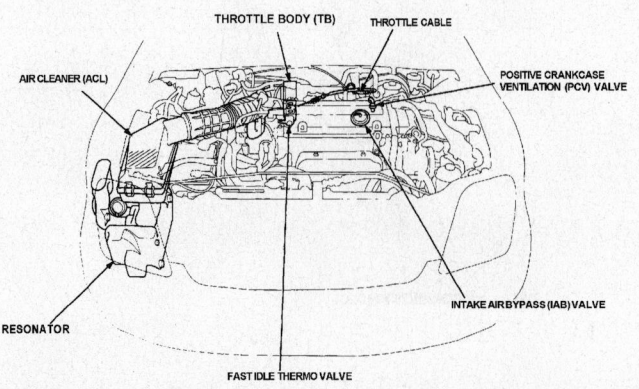

THROTTLE BODY (TB)

THROTTLE CABLE

AIR CLEANER (ACL)

POSITIVE CRANKCASE VENTILATION (PCV) VALVE

RESONATOR

INTAKE AIR BYPASS (IAB) VALVE

FAST IDLE THERMO VALVE

29130_HOND_G0044

'98 model, '97 (California) model:

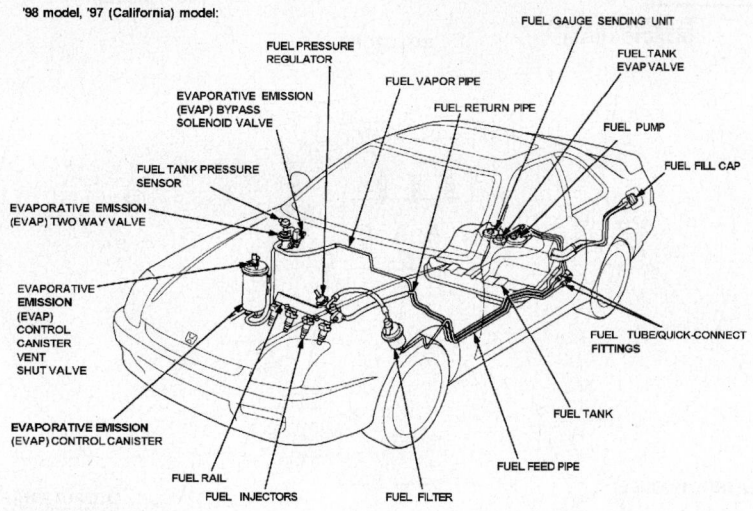

'97 (49ST, Canada) model:

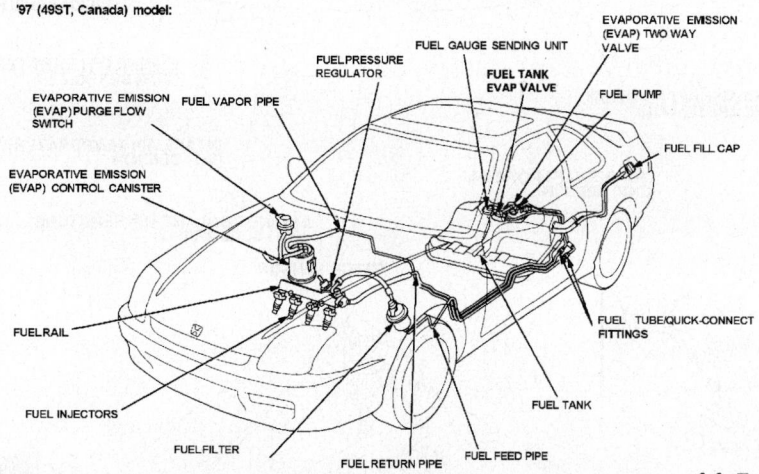

11-5

29130_HOND_G0045

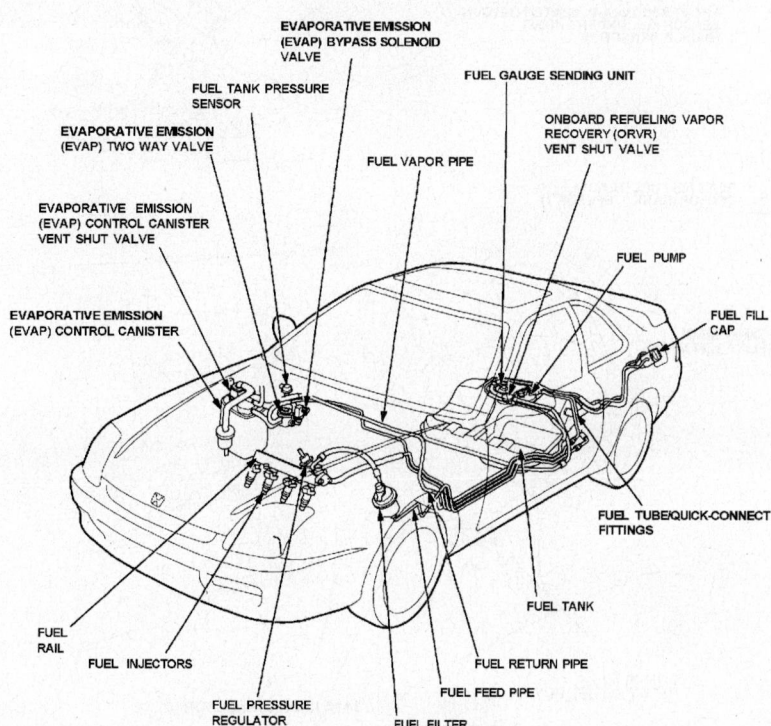

RIDGELINE

6 Cylinder Engines

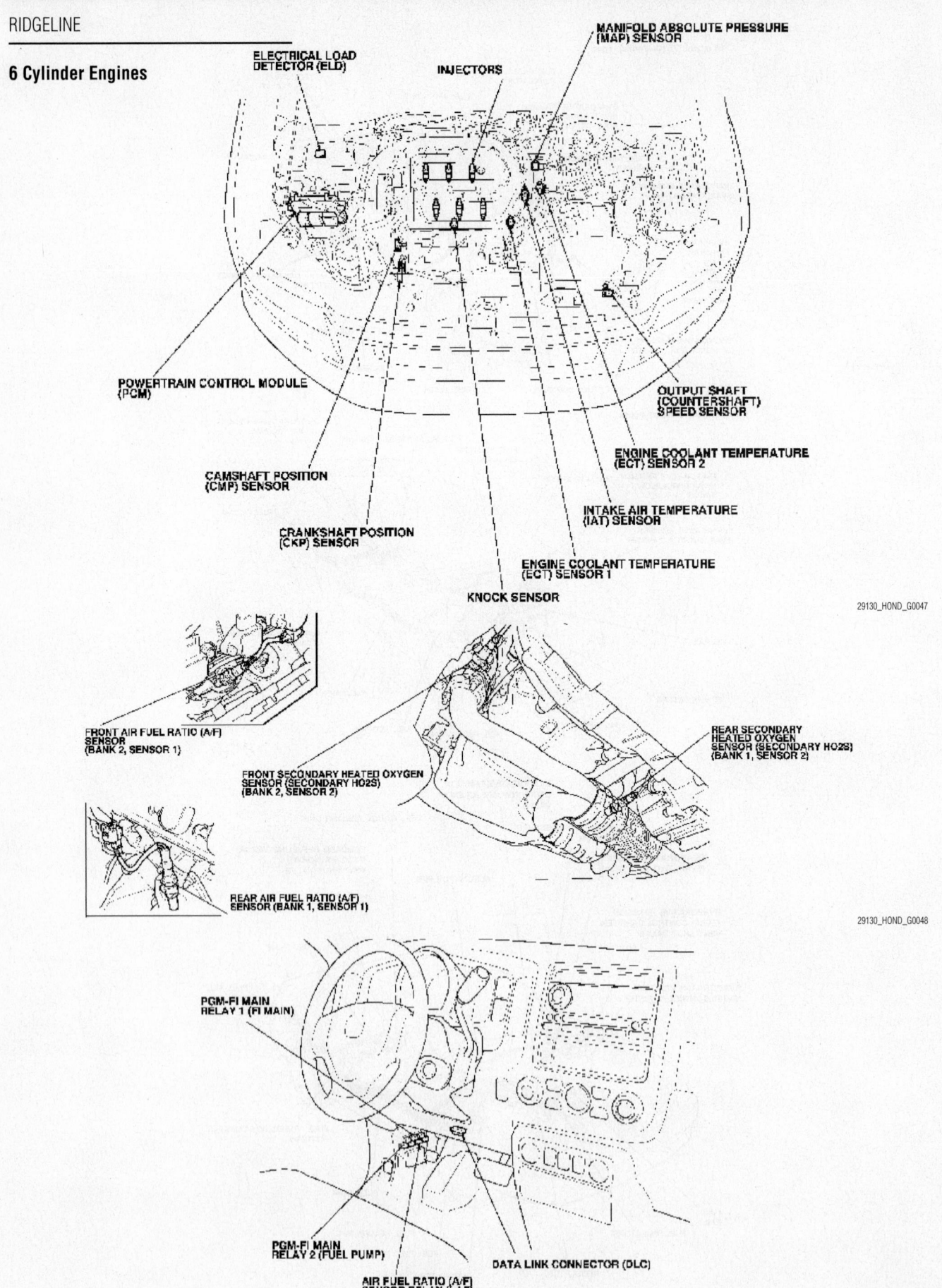

ELECTRICAL LOAD
DETECTOR (ELD)

MANIFOLD ABSOLUTE PRESSURE
(MAP) SENSOR

INJECTORS

POWERTRAIN CONTROL MODULE
(PCM)

OUTPUT SHAFT
(COUNTERSHAFT)
SPEED SENSOR

CAMSHAFT POSITION
(CMP) SENSOR

ENGINE COOLANT TEMPERATURE
(ECT) SENSOR 2

CRANKSHAFT POSITION
(CKP) SENSOR

INTAKE AIR TEMPERATURE
(IAT) SENSOR

ENGINE COOLANT TEMPERATURE
(ECT) SENSOR 1

KNOCK SENSOR

29130_HOND_G0047

FRONT AIR FUEL RATIO (A/F)
SENSOR
(BANK 2, SENSOR 1)

REAR SECONDARY
HEATED OXYGEN
SENSOR (SECONDARY HO2S)
(BANK 1, SENSOR 2)

FRONT SECONDARY HEATED OXYGEN
SENSOR (SECONDARY HO2S)
(BANK 2, SENSOR 2)

REAR AIR FUEL RATIO (A/F)
SENSOR (BANK 1, SENSOR 1)

29130_HOND_G0048

PGM-FI MAIN
RELAY 1 (FI MAIN)

PGM-FI MAIN
RELAY 2 (FUEL PUMP)

DATA LINK CONNECTOR (DLC)

AIR FUEL RATIO (A/F)
SENSOR RELAY (LAF)

29130_HOND_G0049

S2000

4 Cylinder Engines

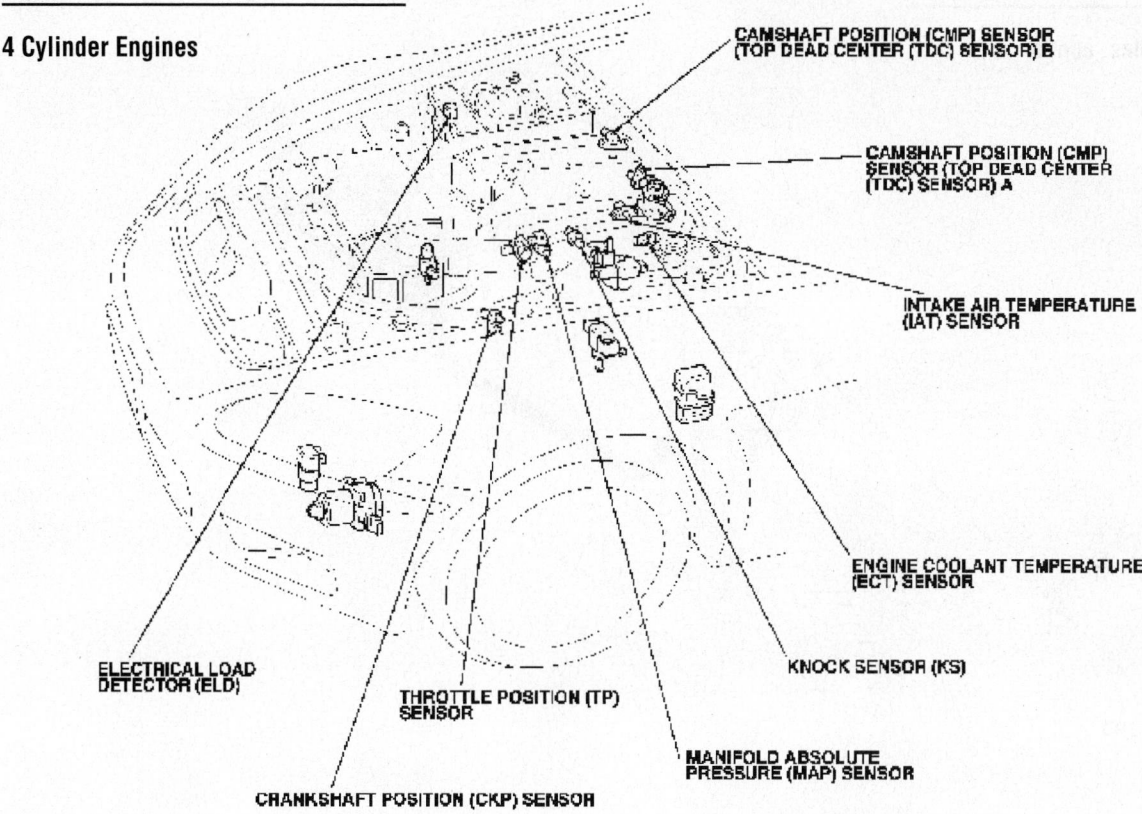

CAMSHAFT POSITION (CMP) SENSOR
(TOP DEAD CENTER (TDC) SENSOR) B

CAMSHAFT POSITION (CMP)
SENSOR (TOP DEAD CENTER
(TDC) SENSOR) A

INTAKE AIR TEMPERATURE
(IAT) SENSOR

ENGINE COOLANT TEMPERATURE
(ECT) SENSOR

KNOCK SENSOR (KS)

ELECTRICAL LOAD
DETECTOR (ELD)

THROTTLE POSITION (TP)
SENSOR

MANIFOLD ABSOLUTE
PRESSURE (MAP) SENSOR

CRANKSHAFT POSITION (CKP) SENSOR

29130_HOND_G0050

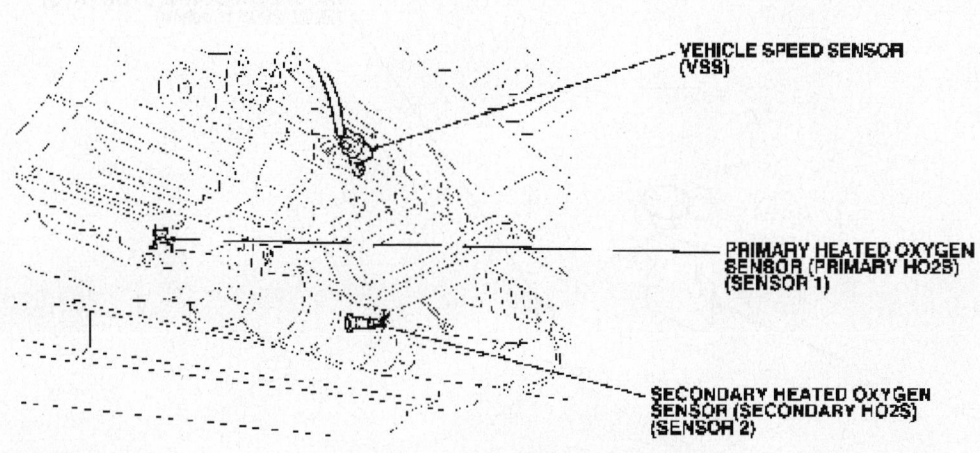

VEHICLE SPEED SENSOR
(VSS)

PRIMARY HEATED OXYGEN
SENSOR (PRIMARY HO2S)
(SENSOR 1)

SECONDARY HEATED OXYGEN
SENSOR (SECONDARY HO2S)
(SENSOR 2)

29130_HOND_G0051

S2000

4 Cylinder Engines, cont'd

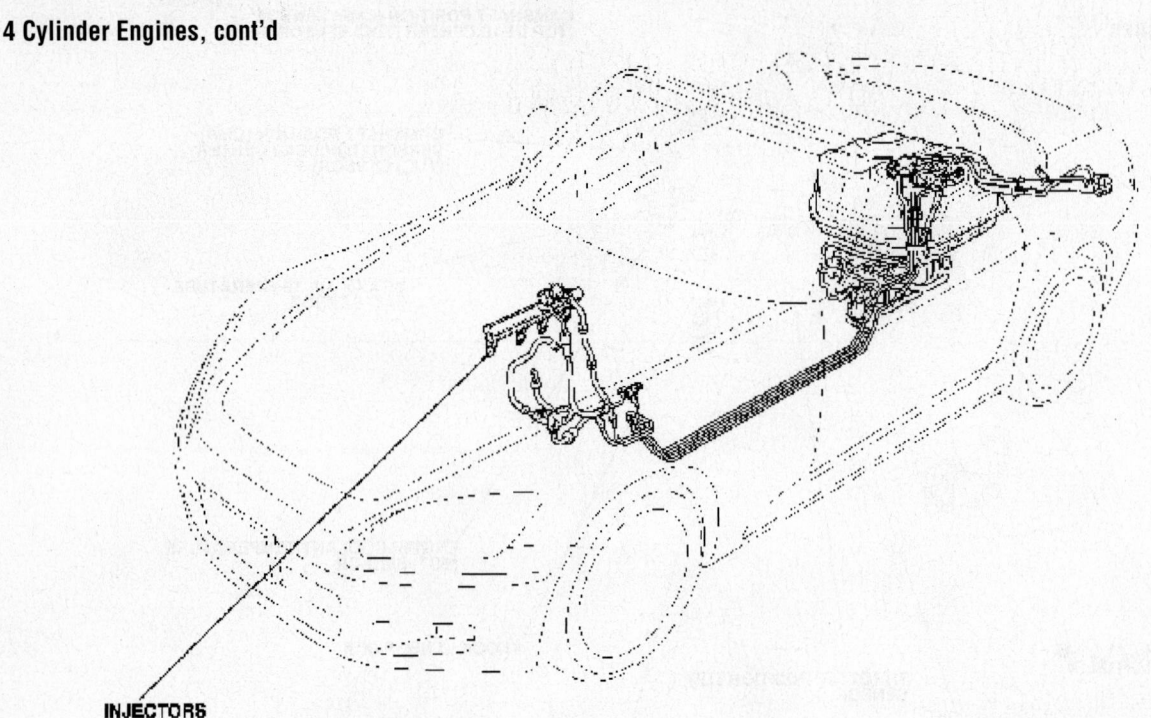

INJECTORS

29130_HOND_G0052

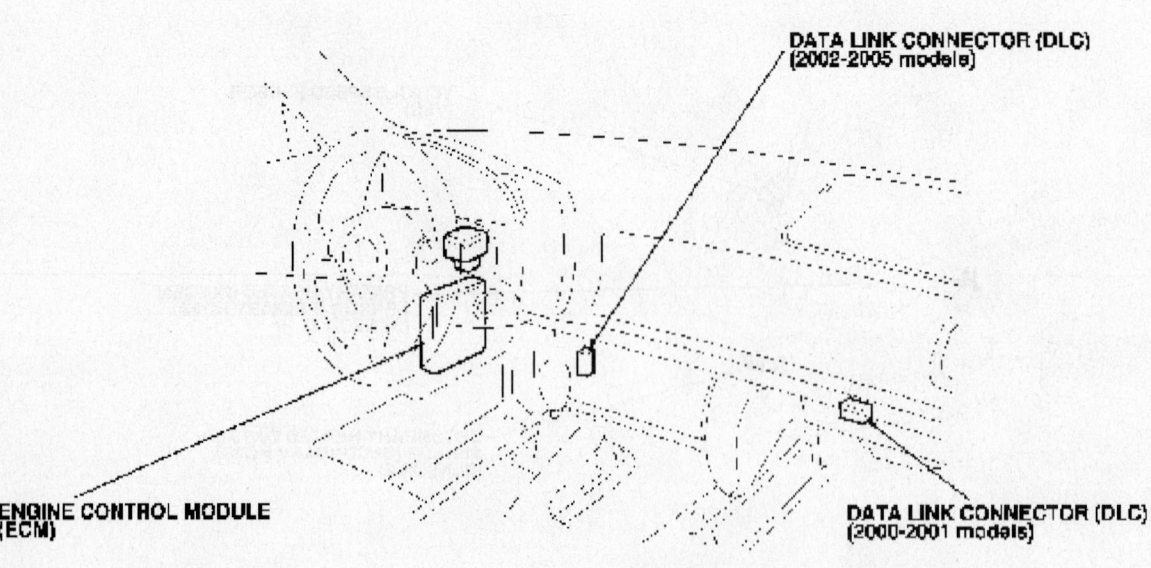

DATA LINK CONNECTOR (DLC)
(2002-2005 models)

ENGINE CONTROL MODULE
(ECM)

DATA LINK CONNECTOR (DLC)
(2000-2001 models)

29130_HOND_G0053

HONDA
PIN CHARTS

TABLE OF CONTENTS

PIN CHARTS

Introduction

A Pin Voltage Table is a term used to describe a table that identifies PCM pins, wire colors of the PCM circuits, circuit descriptions and "known good" values for devices that connect to the PCM. These tables include the following information:

- Signals from various sensors (ECT, IAT, MAP, TPS, etc.)
- Signals from various switches (PNP, PSP, WOT, etc.)
- Signals from oxygen sensors (O2S, HO2S)
- Signals from output devices (IAC, INJ, TCC, etc.)
- Power & ground signals

Pin Voltage Tables

Information contained within the Pin Voltage Tables can be used to:

- Test circuits for open, short to power or short to ground faults
- Check the operation of a component before or after a repair
- Check the operation of a component or system by viewing signals on PCM input/output circuits with a DVOM or Lab Scope

Using a Breakout Box

There are several Breakout Box (BOB) designs available for use to test the PCM and its input and output circuits. However, all of them require removal of the wire harness to the PCM so that the BOB can be installed between the PCM and wire harness connector. Several breakout boxes require the use of overlays in order to allow the tool to be used on more than one year or engine type. Always verify that the correct adapter and overlays are used to prevent connection to the wrong circuits and a misdiagnosis.

Power and Ground Circuit Checks

Measurements made at the BOB are accomplished via test leads and probes from the DVOM or a Lab Scope. If any of the terminals on the PCM or BOB are damaged or loose, test measurements made at the Breakout Box will be inaccurate. To verify the PCM battery power and ground circuits are normal (correct) at the BOB, test the condition of the circuit between the battery negative (-) post and these circuits prior to starting a test sequence.

Diagnosis with Pin Voltage Tables

See Figure 1.

Once an actual PCM pin voltage reading is recorded, it can be compared to an example from a vehicle with "known good" values. In the example shown the Value at Hot Idle for the EVP sensor signal (0.4v) is the "known good" value.

Wire Color Changes

Every effort has been made to obtain and list the correct circuit wire colors for all vehicles. However, running changes from the vehicle manufacturer can cause the wrong colors to be listed.

PCM Pin #	W/Color	Circuit Description (60-Pin)	Value at Hot Idle
27	BN/LG	EVP Sensor Signal	0.4v

Fig. 1 Example

ACCORD PIN CHARTS

1990-93 Coupe 2.2L I4 MFI VIN CB7 [All] 26P 'A' Connector

PCM Pin #	W/Color	Circuit Description (26-Pin)	Value at Hot Idle
A1	BRN	Injector 1 Control	2.0-3.3 ms
A2	YEL	Injector 4 Control	2.0-3.3 ms
A3	RED	Injector 2 Control	2.0-3.3 ms
A4	---	Not Used	---
A5	BLU	Injector 3 Control	2.0-3.3 ms
A6	ORN	HO2S Heater Control	Relay Off: 12v, On: 1v
A7	GRN/BLK	Fuel Pump Relay	Relay Off: 12v, On: 1v
A8	GRN/BLK	Fuel Pump Relay	Relay Off: 12v, On: 1v
A9	BLK/BLU	Electronic Air Control Valve	Pulse Signals
A10	---	Not Used	---
A11	PNK	EGR Control Solenoid	1.2v
A12	BLU	Radiator Fan Relay	Relay Off: 12v, On: 1v
A13	GRN/RED	Check Engine Light	MIL Off: 12v, On: 1v
A14	---	Not Used	---
A15	RED/BLU	A/C Clutch Relay	Relay Off: 12v, On: 1v
A16	WHT/GRN	Alternator Charging Signal	Lights Off: 12v, On: 0v
A17	---	Not Used	---
A18	BRN/WHT	A/T: Control Unit	---
A19	WHT	Intake Air Regulator Control	Solenoid Off: 12v, On: 1v
A20	RED/GRN	EVAP Purge Solenoid	Solenoid Off: 12v, On: 1v
A21	YEL/GRN	Igniter Control	Hot idle 10% d/cycle
A22	YEL/GRN	Igniter Control	Hot idle 10% d/cycle
A23	BLK	Power Ground	<0.1v
A24	BLK	Power Ground	<0.1v
A25	YEL/BLK	Ignition Power	12v
A26	BLK/RED	Chassis Ground	<0.050v

1990-93 Coupe 2.2L I4 MFI VIN CB7 [All] 16P 'B' Connector

PCM Pin #	W/Color	Circuit Description (16-Pin)	Value at Hot Idle
B1	YEL/BLK	Main Relay Power (B+)	12-14v
B2	BRN/BLK	Power Ground	<0.1v
B3	WHT/RED	A/T: Control Unit	Digital Signals
B4	GRN	A/T: Control Unit	Digital Signals
B5	BLU/BLK	A/C Switch Signal	Switch Off: 12v, On: 1v
B6	---	Not Used	---
B7	LT GRN	A/T: Neutral Position Switch	In N: 0v, Others: 12v
B8	RED	PSP Switch	Straight: 0v, Turning: 11v
B9	BLU/RED	A/T: Start Signal	Cranking: 9-11v
B9	BLK/GRN	M/T: Start Signal	Cranking: 9-11v
B10	ORN	Vehicle Speed Sensor	Moving: pulse signals
B11	ORN	CYP Sensor Signal	AC pulse signals
B12	WHT	CYP Sensor Ground	<0.050v
B13	ORN/BLU	TDC Sensor Signal	AC pulse signals
B14	WHT/BLU	TDC Sensor Ground	<0.050v
B15	BLU/GRN	CKP Sensor Signal	AC pulse signals
B16	BLU/YEL	CKP Sensor Ground	<0.050v

1990-93 Coupe 2.2L I4 MFI VIN CB7 [All] 22P 'D' Connector

PCM Pin #	W/Color	Circuit Description (22-Pin)	Value at Hot Idle
D1	WHT/YEL	Keep Alive Power (VBU)	12-14v
D2	GRN/WHT	Brake Switch Signal	Brake Off: 0v, On: 12v
D3	---	Not Used	---
D4	ORN/RED	Service Check Connector	SCS Open: 4.80v
D5-D8	---	Not Used	---
D9	WHT/RED	Alternator 'FR' Signal	Digital Signals: 0-5-0v
D10	GRN/RED	Electric Load Detector	Varies: 0.5-4.5v
D11	RED/BLK	TP Sensor Signal	0.5-0.6v
D12	WHT/BLK	A/T: EGR Lift Sensor Signal	1.1-1.2v
D13	YEL/GRN	ECT Sensor Signal	At 180°F: 0.52v
D14	WHT	Oxygen Sensor	0.1-1.1v
D15	RED/YEL	IAT Sensor Signal	Varies w/temp. (0.5-4.9v)
D16	---	Not Used	---
D17	WHT/BLU	MAP Sensor Signal	0.8-0.9v
D18	GRN/BLK	A/T: Control Unit	---
D19	RED/WHT	MAP Sensor VREF	4.9-5.1v
D20	YEL/WHT	Sensor VREF	4.9-5.1v
D21	BLU/WHT	Sensor Ground	<0.050v
D22	GRN/WHT	Sensor Ground	<0.050v

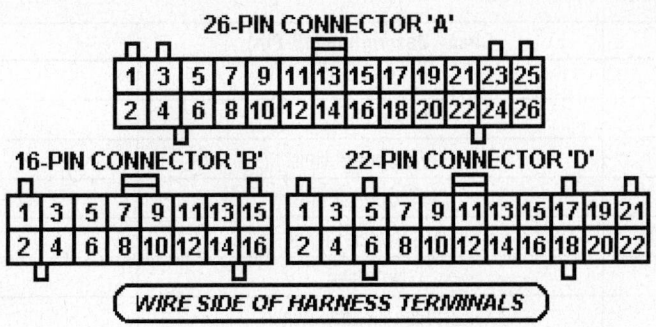

05533_ADIA_G597

Pin Connector Graphic

1994-95 Coupe 2.2L I4 VIN CD7 [All] 26P 'A' Connector

PCM Pin #	W/Color	Circuit Description (26-Pin)	Value at Hot Idle
A1	BRN	Injector 1 Control	2.0-3.3 ms
A2	YEL	Injector 4 Control	2.0-3.3 ms
A3	RED	Injector 2 Control	2.0-3.3 ms
A4	---	Not Used	---
A5	BLU	Injector 3 Control	2.0-3.3 ms
A6	ORN/BLK	HO2S Heater Control	Relay Off: 12v, On: 1v
A7	GRN/BLK	Fuel Pump Relay	Relay Off: 12v, On: 1v
A8	GRN/BLK	Fuel Pump Relay	Relay Off: 12v, On: 1v
A9	BLK/BLU	Intake Air Control Solenoid	Solenoid Off: 12v, On: 1v
A10	GRN/WHT	Engine Mount Solenoid	Solenoid Off: 12v, On: 1v
A11	RED	EGR Control Solenoid	Solenoid Off: 12v, On: 1v
A12	GRN	Radiator Fan Relay	Relay Off: 12v, On: 1v
A13	GRN/RED	Check Engine Light	MIL Off: 12v, On: 1v
A14	---	Not Used	---
A15	RED/BLU	A/C Clutch Relay	Relay Off: 12v, On: 1v
A16	WHT/GRN	Alternator Charging Signal	Lights Off: 12v, On: 0v
A17	---	Not Used	---
A18	BRN/WHT	A/T: Control Unit	---
A19	ORN/GRN	Intake Air Regulator Control	Solenoid Off: 12v, On: 1v
A20	RED/YEL	EVAP Purge Solenoid	Solenoid Off: 12v, On: 1v
A21	YEL/GRN	Igniter Control	Digital Signals: 0-12-0v
A22	YEL/GRN	Igniter Control	Digital Signals: 0-12-0v
A23	BLK	Power Ground	<0.1v
A24	BLK	Power Ground	<0.1v
A25	YEL/BLK	Ignition Power	12v
A26	BLK/RED	Chassis Ground	<0.050v

1994-95 Coupe 2.2L I4 VIN CD7 [All] 16P 'B' Connector

PCM Pin #	W/Color	Circuit Description (16-Pin)	Value at Hot Idle
B1	YEL/BLK	Ignition Power	12v
B2	BRN/BLK	Power Ground	<0.1v
B3	WHT/RED	A/T: Control Unit	Digital Signals
B4	GRN	A/T: Control Unit	Digital Signals
B5	RED/WHT	A/C Switch Signal	Switch Off: 12v, On: 1v
B6	---	Not Used	---
B7	LT GRN	A/T: Park/Neutral Switch	In P/N: 0v, Others: 11v
B8	GRN	PSP Switch Signal	Straight: 0v, Turning: 11v
B9	BLU/RED	A/T: Start Signal	Cranking: 9-11v
B9	BLU/RED	M/T: Start Signal	Cranking: 9-11v
B10	ORN	Vehicle Speed Sensor	Moving: pulse signals
B11	ORN	CYP Sensor Signal	AC pulse signals
B12	WHT	CYP Sensor Ground	<0.050v
B13	ORN/BLU	TDC Sensor Signal	AC pulse signals
B14	WHT/BLU	TDC Sensor Ground	<0.050v
B15	BLU/GRN	CKP Sensor Signal	AC pulse signals
B16	BLU/YEL	CKP Sensor Ground	<0.050v

1994-95 Coupe 2.2L I4 VIN CD7 [All] 22P 'D' Connector

PCM Pin #	W/Color	Circuit Description (22-Pin)	Value at Hot Idle
D1	WHT/YEL	Keep Alive Power (VBU)	12-14v
D2	GRN/WHT	Brake Switch Signal	Brake Off: 0v, On: 12v
D3	---	Not Used	---
D4	RED	Service Check Connector	SCS Open: 4.80v
D5	BLU/WHT	A/T: TCM Signal	Digital Signals
D6	---	Not Used	---
D7	GRN/RED	Data Link Connector	5v
D8	---	Not Used	---
D9	WHT/RED	Alternator 'FR' Signal	Digital Signals: 0-5-0v
D10	GRN/RED	Electric Load Detector	Varies: 0.5-4.5v
D11	RED/BLK	TP Sensor Signal	0.5-0.6v
D12	WHT/BLK	A/T: EGR Lift Sensor Signal	1.1-1.2v
D13	RED/WHT	ECT Sensor Signal	At 180°F: 0.52v
D14	WHT/RED	HO2S Signal	0.1-1.1v
D15	RED/YEL	IAT Sensor Signal	Varies w/temp. (0.5-4.9v)
D16	---	Not Used	---
D17	WHT/YEL	MAP Sensor Signal	0.8-0.9v
D18	GRN/BLK	A/T: Control Unit VREF	4.9-5.1v
D19	YEL/WHT	MAP Sensor VREF	4.9-5.1v
D20	YEL/BLU	Sensor VREF	4.9-5.1v
D21	GRN/WHT	MAP Sensor Ground	<0.050v
D22	GRN/BLU	Sensor Ground	<0.050v

26-PIN CONNECTOR 'A'

1	3	5	7	9	11	13	15	17	19	21	23	25
2	4	6	8	10	12	14	16	18	20	22	24	26

16-PIN CONNECTOR 'B'   **22-PIN CONNECTOR 'D'**

1	3	5	7	9	11	13	15
2	4	6	8	10	12	14	16

1	3	5	7	9	11	13	15	17	19	21
2	4	6	8	10	12	14	16	18	20	22

WIRE SIDE OF HARNESS TERMINALS

05533_ADIA_G597

Pin Connector Graphic
Standard Colors and Abbreviations

Abbreviation	Color	Abbreviation	Color	Abbreviation	Color
BLK	Black	LT BLU	Lt. Blue	TAN	Tan
BLU	Blue	LT GRN	Lt. Green	VIO	Violet
BRN	Brown	ORN	Orange	WHT	White
GRY	Gray	PNK	Pink	YEL	Yellow
GRN	Green	PPL	Purple		

1996-97 Coupe 2.2L I4 VTEC VIN CD7 [All] 32P 'A' Connector

PCM Pin #	W/Color	Circuit Description (32-Pin)	Value at Hot Idle
A1	YEL	Injector 4 Control	2.0-3.3 ms
A2	BLU	Injector 3 Control	2.0-3.3 ms
A3	RED	Injector 2 Control	2.0-3.3 ms
A4	BLU	Injector 1 Control	2.0-3.3 ms
A5	ORN/BLU	HO2S-12 (B1 S2) Heater	Digital Signals: 0-12-0v
A6	ORN/BLK	HO2S-11 (B1 S1) Heater	Digital Signals: 0-12-0v
A7	RED	EGR Solenoid	Solenoid Off: 12v, On: 1v
A8	GRN/YEL	VTEC Solenoid Control	0v, Hi-Speed: 12v
A9	BRN/BLK	Sensor Ground	<0.050v
A10	BLK	Power Ground	<0.1v
A11	YEL/BLK	Main Relay Power (B+)	12-14v
A12	BLK/BLU	Idle Air Control Valve Signal	Pulse Signals
A13	GRN/WHT	Engine Mount Solenoid	Solenoid Off: 12v, On: 1v
A14, 21	---	Not Used	---
A15	RED/YEL	EVAP Purge Solenoid	Solenoid Off: 12v, On: 1v
A16	GRN/BLK	Fuel Pump Relay	Relay Off: 12v, On: 1v
A17	RED/BLU	A/C Clutch Relay	Relay Off: 12v, On: 1v
A18	GRN/RED	Check Engine Light	MIL Off: 12v, On: 1v
A19	WHT/GRN	Alternator Charging Signal	Lights Off: 12v, On: 0v
A20	YEL/GRN	Igniter Control	Digital Signals: 0-12-0v
A22	BRN/BLK	Sensor Ground	<0.050v
A23	BLK	Power Ground	<0.1v
A24	YEL/BLK	Main Relay Power (B+)	12-14v
A25	ORN/GRN	Intake Air Resonator Control	Solenoid Off: 12v, On: 1v
A26	---	Not Used	---
A27	GRN	Radiator Fan Relay	Relay Off: 12v, On: 1v
A28	GRN/WHT	EVAP Bypass Solenoid	Solenoid Off: 12v, On: 1v
A29	ORN/GRN	EVAP Vent Solenoid	Solenoid Off: 12v, On: 1v
A30-32	---	Not Used	---

1996-97 Coupe 2.2L I4 VTEC VIN CD7 [All] 25P 'B' Connector

PCM Pin #	W/Color	Circuit Description (25-Pin)	Value at Hot Idle
B3	BLU/YEL	A/T: Shift Solenoid 'A'	D2-3: 12v, D1-4: 0v
B4	GRN/BLK	A/T: Lockup Solenoid 'B'	LSB On: 12v, Off: 0v
B5	YEL	A/T: Lockup Solenoid 'A'	LSA On: 12v, Off: 0v
B8	GRN/BLU	A/T: Gear Position Switch	In D3: 0v, Others: 12v
B11	GRN/WHT	A/T: Shift Solenoid 'B'	D1-2: 12v, D3-4: 0v
B12	WHT/GRN	Interlock Control Unit Signal	Key & Brake On: 12v
B13	BLU/RED	A/T: D4 Indicator Light Switch	D4 On: 0v, Off: 12v
B14	WHT/BLU	Mainshaft Speed Ground	<0.050v
B15	ORN/BLU	Mainshaft Speed Signal	Moving: AC pulses
B16	GRN/RED	A/T: Gear Position Switch	In 'R': 0v, Others: 12v
B17	GRN/YEL	A/T: Gear Position Switch	In D2: 0v, Others: 12v
B18	GRN/WHT	A/T: Gear Position Switch	In D1: 0v, Others: 12v
B22	BLU/YEL	Sensor Ground	<0.050v
B23	BLU/GRN	Countershaft Speed Sensor	Moving: AC pulses
B24	GRN/BLK	A/T: Gear Position Switch	In D4: 0v, Others: 12v
B25	LT GRN	A/T: Gear Position Switch	In P/N: 0v, Others: 12v

1996-97 Coupe 2.2L I4 VTEC VIN CD7 [All] 31P 'C' Connector

PCM Pin #	W/Color	Circuit Description (31-Pin)	Value at Hot Idle
C2	BLU	CYP Sensor Signal	AC pulse signals
C3	GRN	TDC Sensor Signal	AC pulse signals
C4	YEL	CKP Sensor Signal	AC pulse signals
C5	RED/WHT	A/C Switch Signal	Switch Off: 12v, On: 1v
C6	BLU/RED	Start Signal	Cranking: 9-11v
C7	RED	Service Check Connector	SCS Open: 4.80v
C8	LT GRN	K-Line Signal	12v
C10	WHT/YEL	Keep Alive Power (VBU)	12-14v
C12	WHT	CKP Sensor Ground	<0.050v
C13	RED	TDC Sensor Ground	<0.050v
C14	BLK	CYP Sensor Ground	<0.050v
C16	GRN	PSP Switch	Straight: 0v, Turning: 11v
C17	WHT/RED	Alternator 'FR' Signal	Digital Signals: 0-5-0v
C18	ORN	Vehicle Speed Sensor	Moving: pulse signals
C20	BRN	EVAP Purge Flow Switch	Switch on: 0v, off: 5v
C21-31	---	---	---

1996-97 Coupe 2.2L I4 VTEC VIN CD7 [All] 16P 'D' Connector

PCM Pin #	W/Color	Circuit Description (16-Pin)	Value at Hot Idle
D1	RED/BLK	TP Sensor Signal	0.5-0.6v
D2	RED/WHT	ECT Sensor Signal	At 180°F: 0.52v
D3	WHT/YEL	MAP Sensor Signal	0.8-0.9v
D4	YEL/WHT	MAP Sensor VREF	4.9-5.1v
D5	GRN/WHT	Brake Switch Signal	Brake Off: 0v, On: 12v
D7	WHT/RED	HO2S-11 (B1 S1) Signal	0.1-1.1v
D8	RED/YEL	IAT Sensor Signal	Varies w/temp. (0.5-4.9v)
D9	WHT/BLK	EGR Valve Lift Sensor	1.2v
D10	YEL/BLU	Sensor VREF	4.9-5.1v
D11	GRN/BLU	Sensor Ground	<0.050v
D12	GRN/WHT	MAP Sensor Ground	<0.050v
D13	RED/WHT	HO2S-12 Ground	<0.050v
D14	WHT/RED	HO2S-12 (B1 S2) Signal	0.1-1.1v
D15	GRN/RED	Fuel Tank Pressure Sensor	Fuel Cap off: 2.5v
D16	GRN/RED	Electric Load Detector	Varies: 0.5-4.5v

Pin Connector Graphic

05533_ADIA_G598

1996-97 Coupe 2.2L I4 MFI VIN CD7 [All] 32P 'A' Connector

PCM Pin #	W/Color	Circuit Description (32-Pin)	Value at Hot Idle
A1	YEL	Injector 4 Control	2.0-3.3 ms
A2	BLU	Injector 3 Control	2.0-3.3 ms
A3	RED	Injector 2 Control	2.0-3.3 ms
A4	BLU	Injector 1 Control	2.0-3.3 ms
A5	ORN/BLU	HO2S-12 (B1 S2) Heater	Digital Signals: 0-12-0v
A6	ORN/BLK	HO2S-11 (B1 S1) Heater	Digital Signals: 0-12-0v
A7	RED	EGR Solenoid Control	Solenoid Off: 12v, On: 1v
A8	---	Not Used	---
A9	BRN/BLK	Sensor Ground	<0.050v
A10	BLK	Power Ground	<0.1v
A11	YEL/BLK	Main Relay Power (B+)	12-14v
A12	BLK/BLU	Idle Air Control Valve Signal	Pulse Signals
A13	GRN/WHT	Engine Mount Solenoid	Solenoid Off: 12v, On: 1v
A14	---	Not Used	---
A15	RED/YEL	EVAP Purge Solenoid	Solenoid Off: 12v, On: 1v
A16	GRN/BLK	Fuel Pump Relay	Relay Off: 12v, On: 1v
A17	RED/BLU	A/C Clutch Relay	Relay Off: 12v, On: 1v
A18	GRN/RED	Check Engine Light	MIL Off: 12v, On: 1v
A19	WHT/GRN	Alternator Charging Signal	Lights Off: 12v, On: 0v
A20	YEL/GRN	Igniter Control	Digital Signals: 0-12-0v
A21, A26	---	Not Used	---
A22	BRN/BLK	Sensor Ground	<0.050v
A23	BLK	Power Ground	<0.1v
A24	YEL/BLK	Main Relay Power (B+)	12-14v
A25	ORN/GRN	Intake Air Resonator Control	Solenoid Off: 12v, On: 1v
A27	GRN	Radiator Fan Relay	Relay Off: 12v, On: 1v
A28	GRN/WHT	EVAP Bypass Solenoid	Solenoid Off: 12v, On: 1v
A29	ORN/GRN	EVAP Vent Solenoid	Solenoid Off: 12v, On: 1v
A30-32	---	Not Used	---

1996-97 Coupe 2.2L I4 MFI VIN CD7 [All] 25P 'B' Connector

PCM Pin #	W/Color	Circuit Description (25-Pin)	Value at Hot Idle
B3	BLU/YEL	A/T: Shift Solenoid 'A'	D2-3: 12v, D1-4: 0v
B4	GRN/BLK	A/T: Lockup Solenoid 'B'	LSB On: 12v, Off: 0v
B5	YEL	A/T: Lockup Solenoid 'A'	LSA On: 12v, Off: 0v
B8	GRN/BLU	A/T: Gear Position Switch	In D3: 0v, Others: 12v
B11	GRN/WHT	A/T: Shift Solenoid 'B'	D1-2: 12v, D3-4: 0v
B12	WHT/GRN	Interlock Control Unit Signal	Key & Brake On: 12v
B13	BLU/RED	A/T: D4 Indicator Light Switch	D4 On: 0v, Off: 12v
B14	WHT/BLU	Mainshaft Speed Ground 'N'	<0.050v
B15	ORN/BLU	Mainshaft Speed Signal 'P'	Moving: AC pulses
B16	GRN/RED	A/T: Gear Position Switch	In 'R': 0v, Others: 12v
B17	GRN/YEL	A/T: Gear Position Switch	In D2: 0v, Others: 12v
B18	GRN/WHT	A/T: Gear Position Switch	In D2: 0v, Others: 12v
B22	BLU/YEL	Countershaft Speed Sensor N	<0.050v
B23	BLU/GRN	Countershaft Speed Sensor	Moving: AC pulses
B24	GRN/BLK	A/T: Gear Position Switch	In D4: 0v, Others: 12v
B25	LT GRN	A/T: Gear Position Switch	In P/N: 0v, Others: 12v

1996-97 Coupe 2.2L I4 MFI VIN CD7 [All] 31P 'C' Connector

PCM Pin #	W/Color	Circuit Description (31-Pin)	Value at Hot Idle
C2	BLU	CYP Sensor Signal	AC pulse signals
C3	GRN	TDC Sensor Signal	AC pulse signals
C4	YEL	CKP Sensor Signal	AC pulse signals
C5	RED/WHT	A/C Switch Signal	Switch Off: 12v, On: 1v
C6	BLU/RED	Start Signal	Cranking: 9-11v
C7	RED	Service Check Connector	SCS Open: 4.80v
C8	LT GRN	K-Line Signal	12v
C10	WHT/YEL	Keep Alive Power (VBU)	12-14v
C12	WHT	CKP Sensor Ground	<0.050v
C13	RED	TDC Sensor Ground	<0.050v
C14	BLK	CYP Sensor Ground	<0.050v
C16	GRN	PSP Switch	Straight: 0v, Turning: 11v
C17	WHT/RED	Alternator 'FR' Signal	Digital Signals: 0-5-0v
C18	ORN	Vehicle Speed Sensor	Moving: pulse signals
C20	BRN	EVAP Purge Flow Switch	Switch on: 0v, off: 5v
C21-31	---	---	---

1996-97 Coupe 2.2L I4 MFI VIN CD7 [All] 16P 'D' Connector

PCM Pin #	W/Color	Circuit Description (16-Pin)	Value at Hot Idle
D1	RED/BLK	TP Sensor Signal	0.5-0.6v
D2	RED/WHT	ECT Sensor Signal	At 180°F: 0.52v
D3	WHT/YEL	MAP Sensor Signal	0.8-0.9v
D4	YEL/WHT	MAP Sensor VREF	4.9-5.1v
D5	GRN/WHT	Brake Switch Signal	Brake Off: 0v, On: 12v
D7	WHT/RED	HO2S-11 (B1 S1) Signal	0.1-1.1v
D8	RED/YEL	IAT Sensor Signal	Varies w/temp. (0.5-4.9v)
D9	WHT/BLK	EGR Valve Lift Sensor	1.2v
D10	YEL/BLU	Sensor VREF	4.9-5.1v
D11	GRN/BLU	Sensor Ground	<0.050v
D12	GRN/WHT	MAP Sensor Ground	<0.050v
D13	RED/WHT	HO2S-12 Ground	<0.050v
D14	WHT/RED	HO2S-12 (B1 S2) Signal	0.1-1.1v
D15	GRN/RED	Fuel Tank Pressure Sensor	Fuel Cap off: 2.5v
D16	GRN/RED	Electric Load Detector	Varies: 0.5-4.5v

Pin Connector Graphic

1998-99 Coupe 2.3L I4 VTEC VIN CG3 [All] 32P 'A' Connector

PCM Pin #	W/Color	Circuit Description (32-Pin)	Value at Hot Idle
A2	GRN/WHT	Engine Mount Solenoid	Solenoid Off: 12v, On: 1v
A3	BLU	EVAP Bypass Solenoid	Solenoid Off: 12v, On: 1v
A4	GRN/WHT	EVAP Vent Solenoid	Solenoid Off: 12v, On: 1v
A5	BLU/GRN	Cruise Control Signal	C/C On: pulse signals
A6	RED/YEL	EVAP Purge Solenoid	Solenoid Off: 12v, On: 1v
A8	BLK/WHT	HO2S-12 (B1 S2) Heater	Digital Signals: 0-12-0v
A9	BLU/WHT	A/T: Vehicle Speed Output	Moving: pulse signals
A10	BRN	Service Check Connector	SCS Open: 4.80v
A12	PNK	Immobilizer Indicator Lamp	Lamp Off: 12v, On: 1v
A13, 25	BLU, RED	Immobilizer Enable, Code	Digital Signals
A14	GRN/BLK	A/T: D4 Light Switch	D4 On: 0v, Off: 12v
A15	GRN/YEL	Immobilizer Fuel Pump Relay	Relay Off: 12v, On: 1v
A17	RED	A/C Clutch Relay Control	Relay Off: 12v, On: 1v
A18	GRN/ORN	Malfunction Indicator Light	MIL Off: 12v, On: 1v
A19	BLU	Engine Speed Pulse (NEP)	Digital Signals

1998-99 Coupe 2.3L I4 VTEC VIN CG3 [All] 32P 'A' Connector, *continued*

A20	GRN	Radiator Fan Relay Control	Relay Off: 12v, On: 1v
A21	GRY	K-Line Signal	12v
A23	WHT/RED	HO2S-12 (B1 S2) Signal	0.1-1.1v
A24	BLU/ORN	Starter Switch Signal	Cranking: 9-11v
A26	GRN	PSP Switch Signal	Straight: 0v, Turning: 11v
A27	BLU/RED	A/C Switch Signal	Switch Off: 12v, On: 1v
A28	WHT/RED	Interlock Control Unit Signal	Key & Brake On: 12v
A29	LT GRN	Fuel Tank Pressure Sensor	Fuel Cap off: 2.5v
A30	GRN/RED	Electric Load Detector	Varies: 0.5-4.5v
A32	WHT/BLK	Brake Switch Signal	Brake Off: 0v, On: 12v

1998-99 Coupe 2.3L I4 VTEC VIN CG3 [All] 16P 'D' Connector

PCM Pin #	W/Color	Circuit Description (16-Pin)	Value at Hot Idle
D1	YEL	A/T: Lockup Control Solenoid	LUS On: 12v, Off: 0v
D2	GRN/WHT	A/T: Shift Solenoid 'B'	SSB Off: 0v, On: 12v
D3	GRN	A/T: Shift Solenoid 'C'	SSC on: 12v, off: 0v
D5	BLK/YEL	A/T: Solenoid Feed (B+)	12-14v
D6	WHT	A/T: Gear Position Switch	In 'R': 0v, Others: 12v
D7	BLU/YEL	A/T: Shift Solenoid 'A'	SSA Off: 0v, On: 12v
D8	PNK	A/T: Gear Position Switch	In D3: 0v, Others: 12v
D9	YEL	A/T: Gear Position Switch	In D4: 0v, Others: 12v
D10	BLU	Countershaft Speed Sensor P	Moving: AC pulses
D11	RED	Mainshaft Speed Sensor 'P'	AC pulse signals
D12	WHT	Mainshaft Speed Sensor 'N'	<0.050v
D13	BLU/WHT	A/T: Gear Position Switch	In P/N: 0v, Others: 12v
D14, 15	BLU, BRN	A/T: Gear Position Switches	In D2: 0v, In D1: 0v
D16	GRN	Countershaft Speed Sensor N	<0.050v

Pin Connector Graphic

05533_ADIA_G599

1998-99 Coupe 2.3L I4 VTEC VIN CG3 [All] 25P 'B' Connector

PCM Pin #	W/Color	Circuit Description (25-Pin)	Value at Hot Idle
B1, B9	YEL/BLK	Main Relay Power (B+)	12-14v
B2, B10	BLK	Power Ground	<0.1v
B3, B4	RED, BLU	Injector 2, Injector 3 Control	2.0-3.3 ms
B5, B11	YEL, BRN	Injector 4, Injector 1 Control	2.0-3.3 ms
B7	PNK	E-EGR Control Solenoid	Digital Signals: 0-12-0v
B8, B17	WHT, RED	A/T: Clutch Solenoid 'A-', 'A+'	Pulse Signals
B12	GRN/YEL	VTEC Control Solenoid	0v, Hi-Speed: 12v
B13	YEL/GRN	Ignition Control Signal	Digital Signals: 0-12-0v
B14	BLU/BLK	A/T: 2nd Clutch Pressure Sw.	Open: 12v, Closed: 1v
B16	GRN/RED	HO2S-11 Heater Relay	Relay Off: 12v, On: 1v
B18, B25	GRN, ORN	A/T: Clutch Solenoid 'B-', 'B+'	Pulse Signals
B19	BLK/WHT	HO2S-11 (B1 S1) Heater	Digital Signals: 0-12-0v
B20, B22	BRN/BLK	Logic Ground	<0.050v
B21	WHT/YEL	Keep Alive Power (VBU)	12-14v
B23	BLK/BLU	Idle Air Control Valve	Pulse Signals
B24	BLU/WHT	A/T: 3rd Clutch Pressure Sw.	Open: 12v, Closed: 1v

1998-99 Coupe 2.3L I4 VTEC VIN CG3 [All] 31P 'C' Connector

PCM Pin #	W/Color	Circuit Description (31-Pin)	Value at Hot Idle
C2	WHT/GRN	Alternator Charging Signal	Lights Off: 12v, On: 0v
C3	RED/BLU	Knock Sensor Signal	No Detonation: 18mv AC
C5	WHT/RED	Alternator 'FR' Signal	Digital Signals: 0-5-0v
C6	WHT/BLK	EGR Valve Lift Sensor	1.2v
C7	GRN	MAP Sensor Ground	<0.050v
C8	BLU	CKP Sensor Signal	AC pulse signals
C9	WHT	CKP Sensor Ground	<0.050v
C10	BLU/BLK	VTEC Pressure Switch	0v, Hi-Speed: 12v
C13	WHT	HO2S-11 Heater Relay	Relay Off: 12v, On: 1v
C14	RED	HO2S-11 (B1 S1) Signal	0.1-1.1v
C15	BLU	HO2S-11 Ground	<0.050v
C16	WHT	HO2S-11 (B1 S1) Signal	0.1-1.1v
C17	RED/GRN	MAP Sensor Signal	0.8-0.9v
C18	GRN	Sensor Ground	<0.050v
C19, C28	YEL, YEL	MAP VREF, Sensor VREF	4.9-5.1v
C20	GRN	TDC Sensor Signal	AC pulse signals
C21	RED	TDC Sensor Ground	<0.050v
C23	BLU/WHT	M/T: VSS Input Signal	Moving: 0-5-0v
C24	YEL/GRN	ECT Sensor Signal to TCM	Digital Signals
C25	RED/YEL	IAT Sensor Signal	Varies w/temp (0.5-4.9v)
C26	RED/WHT	ECT Sensor Signal	At 180°F: 0.5-0.6v
C27	RED/BLK	TP Sensor Signal	0.5-0.6v
C29, C30	YEL, BLK	CYP Sensor Signal, Ground	AC pulse signals

1	2		3	4	5		6	7	8	1	2	3	4		5	6	7		8	9	10
9	10	11	12	13	14	15	16	17	18	11	12	13	14	15	16	17	18	19	20	21	22
19	20	21	22		23	24	25			23	24	25		26	27	28		29	30	31	

25-PIN 'B'   *WIRE SIDE OF HARNESS TERMINALS*   31-PIN 'C'

05533_ADIA_G600

Pin Connector Graphic
2000-03 Coupe 2.3L I4 VTEC VIN CG3 [All] 32P 'A' Connector

PCM Pin #	W/Color	Circuit Description (32-Pin)	Value at Hot Idle
A2	GRN/WHT	Engine Mount Solenoid	Solenoid Off: 12v, On: 1v
A3	BLU	EVAP Bypass Solenoid	Solenoid Off: 12v, On: 1v
A4	GRN/WHT	EVAP Vent Solenoid	Solenoid Off: 12v, On: 1v
A5	BLU/GRN	Cruise Control Signal	C/C On: pulse signals
A6	RED/YEL	EVAP Purge Solenoid	Solenoid Off: 12v, On: 1v
A8	BLK/WHT	HO2S-12 (B1 S2) Heater	Digital Signals: 0-12-0v
A9	BLU/WHT	A/T: Vehicle Speed Output	Moving: pulse signals
A10	BRN	Service Check Connector	SCS Open: 4.80v
A12	PNK	Immobilizer Indicator Lamp	Lamp Off: 12v, On: 1v
A13, 25	BLU, RED	Immobilizer Enable, Code	Digital Signals
A14	GRN/BLK	A/T: D4 Light Switch	D4 On: 0v, Off: 12v
A15	GRN/YEL	Immobilizer Fuel Pump Relay	Relay Off: 12v, On: 1v
A17	RED	A/C Clutch Relay Control	Relay Off: 12v, On: 1v
A18	GRN/ORN	Malfunction Indicator Light	MIL Off: 12v, On: 1v
A19	BLU	Engine Speed Pulse (NEP)	Digital Signals
A20	GRN	Radiator Fan Relay Control	Relay Off: 12v, On: 1v
A21	GRY	K-Line Signal	12v
A23	WHT/RED	HO2S-12 (B1 S2) Signal	0.1-1.1v
A24	BLU/ORN	Starter Switch Signal	Cranking: 9-11v
A26	GRN	PSP Switch Signal	Straight: 0v, Turning: 11v
A27	BLU/RED	A/C Switch Signal	Switch Off: 12v, On: 1v
A28	WHT/RED	Interlock Control Unit Signal	Key & Brake On: 12v
A29	LT GRN	Fuel Tank Pressure Sensor	Fuel Cap off: 2.5v
A30	GRN/RED	Electric Load Detector	Varies: 0.5-4.5v
A32	WHT/BLK	Brake Switch Signal	Brake Off: 0v, On: 12v

2000-03 Coupe 2.3L I4 VTEC VIN CG3 [All] 16P 'D' Connector

PCM Pin #	W/Color	Circuit Description (16-Pin)	Value at Hot Idle
D1	YEL	A/T: Lockup Control Solenoid	LUS On: 12v, Off: 0v
D2	GRN/WHT	A/T: Shift Solenoid 'B'	SSB Off: 0v, On: 12v
D3	GRN	A/T: Shift Solenoid 'C'	SSC on: 12v, off: 0v
D5	BLK/YEL	A/T: Solenoid Feed (B+)	12-14v
D6	WHT	A/T: Gear Position Switch	In 'R': 0v, Others: 12v
D7	BLU/YEL	A/T: Shift Solenoid 'A'	SSA Off: 0v, On: 12v
D8	PNK	A/T: Gear Position Switch	In D3: 0v, Others: 12v
D9	YEL	A/T: Gear Position Switch	In D4: 0v, Others: 12v
D10, D16	BLU, GRN	A/T Countershaft Speed P, N	Moving: AC pulses
D11	RED	A/T: Mainshaft Speed 'P'	AC pulse signals
D12	WHT	A/T: Mainshaft Speed 'N'	<0.050v
D13	BLU/WHT	A/T: Gear Position Switch	In P/N: 0v, Others: 12v
D14	BLU	A/T: Gear Position Switch	In D2: 0v, Others: 12v
D15	BRN	A/T: Gear Position Switch	In D1: 0v, Others: 12v

WIRE SIDE OF HARNESS TERMINALS

05533_ADIA_G599

Pin Connector Graphic

2000-03 Coupe 2.3L I4 VTEC VIN CG3 [All] 25P 'B' Connector

PCM Pin #	W/Color	Circuit Description (25-Pin)	Value at Hot Idle
B1, B9	YEL/BLK	Main Relay Power (B+)	12-14v
B2, B10	BLK	Power Ground	<0.1v
B3, B4	RED, BLU	Injector 2, Injector 3 Control	2.0-3.3 ms
B5, B11	YEL, BRN	Injector 4, Injector 1 Control	2.0-3.3 ms
B7	PNK	E-EGR Control Solenoid	Digital Signals: 0-12-0v
B8, B17	WHT, RED	A/T: Clutch Solenoid 'A-', 'A+'	Pulse Signals
B12	GRN/YEL	VTEC Control Solenoid	0v, Hi-Speed: 12v
B13	YEL/GRN	Ignition Control Signal	Digital Signals: 0-12-0v
B14	BLU/BLK	A/T: 2nd Clutch Pressure Sw.	Open: 12v, Closed: 1v
B16	GRN/RED	HO2S-11 Heater Relay	Relay Off: 12v, On: 1v
B18, B25	GRN, ORN	A/T: Clutch Solenoid 'B-', 'B+'	Pulse Signals
B20, B22	BRN/BLK	Logic Ground	<0.050v
B21	WHT/YEL	Keep Alive Power (VBU)	12-14v
B23	BLK/BLU	Idle Air Control Valve	Pulse Signals
B24	BLU/WHT	A/T: 3rd Clutch Pressure Sw.	Open: 12v, Closed: 1v

2000-03 Coupe 2.3L I4 VTEC VIN CG3 [All] 31P 'C' Connector

PCM Pin #	W/Color	Circuit Description (31-Pin)	Value at Hot Idle
C1	BLK/WHT	HO2S-11 (B1 S1) Heater	Digital Signals: 0-12-0v
C2	WHT/GRN	Alternator Charging Signal	Lights Off: 12v, On: 0v
C3	RED/BLU	Knock Sensor Signal	No Detonation: 18mv AC
C5	WHT/RED	Alternator 'FR' Signal	Digital Signals: 0-5-0v
C6	WHT/BLK	EGR Valve Lift Sensor	1.2v
C7	GRN	MAP Sensor Ground	<0.050v
C8	BLU	CKP Sensor Signal	AC pulse signals
C9	WHT	CKP Sensor Ground	<0.050v
C10	BLU/BLK	VTEC Pressure Switch	0v, Hi-Speed: 12v
C13	WHT	HO2S-11 Heater Relay	Relay Off: 12v, On: 1v
C14	RED	HO2S-11 (B1 S1) Signal	0.1-1.1v
C15	BLU	HO2S-11 Ground	<0.050v
C16	WHT	HO2S-11 (B1 S1) Signal	0.1-1.1v
C17	RED/GRN	MAP Sensor Signal	0.8-0.9v
C18	GRN	Sensor Ground	<0.050v
C19, C28	YEL; YEL	MAP VREF, Sensor VREF	4.9-5.1v
C20	GRN	TDC Sensor Signal	AC pulse signals
C21	RED	TDC Sensor Ground	<0.050v
C23	BLU/WHT	M/T: VSS Input Signal	Moving: 0-5-0v
C24	YEL/GRN	ECT Sensor Signal to TCM	Digital Signals
C25	RED/YEL	IAT Sensor Signal	Varies w/temp (0.5-4.9v)
C26	RED/WHT	ECT Sensor Signal	At 180°F: 0.5-0.6v
C27	RED/BLK	TP Sensor Signal	0.5-0.6v
C29, C30	YEL, BLK	CYP Sensor Signal, Ground	AC pulse signals

1	2		3	4	5		6	7	8		1	2	3	4		5	6	7		8	9	10
9	10	11	12	13	14	15	16	17	18	11	12	13	14	15	16	17	18	19	20	21	22
19	20	21	22		23	24	25			23	24	25		26	27	28		29	30	31

25-PIN 'B'   *WIRE SIDE OF HARNESS TERMINALS*   **31-PIN 'C'**

05533_ADIA_G600

Pin Connector Graphic

1998-99 Coupe 3.0L V6 VTEC VIN CG2 [All] 32P 'A' Connector

PCM Pin #	W/Color	Circuit Description (32-Pin)	Value at Hot Idle
A2	GRN/WHT	Engine Mount Solenoid	Solenoid Off: 12v, On: 1v
A3	BLU	EVAP Bypass Solenoid	Solenoid Off: 12v, On: 1v
A4	GRN/WHT	EVAP Vent Solenoid	Solenoid Off: 12v, On: 1v
A5	BLU/GRN	Cruise Control Signal	C/C On: pulse signals
A6	RED/YEL	EVAP Purge Solenoid	Solenoid Off: 12v, On: 1v
A8	BLK/WHT	HO2S-12 (B1 S2) Heater	Digital Signals: 0-12-0v
A9	BLU/WHT	A/T: Vehicle Speed Output	Moving: pulse signals
A10	BRN	Service Check Connector	SCS Open: 4.80v
A12	PNK	Immobilizer Indicator Lamp	Lamp Off: 12v, On: 1v
A13, A25	BLU, RED	Immobilizer Enable, Code	Digital Signals
A14	GRN/BLK	A/T: D4 Light Switch	D4 On: 0v, Off: 12v
A15	GRN/YEL	Immobilizer Fuel Pump Relay	Relay Off: 12v, On: 1v
A17	RED	A/C Clutch Relay Control	Relay Off: 12v, On: 1v
A18	GRN/ORN	Malfunction Indicator Light	MIL Off: 12v, On: 1v
A19	BLU	Engine Speed Pulse (NEP)	AC Pulse Signals
A20	GRN	Radiator Fan Relay Control	Relay Off: 12v, On: 1v
A21	GRY	K-Line Signal	12v
A23	WHT/RED	HO2S-12 (B1 S2) Signal	0.1-1.1v
A24	BLU/ORN	Starter Switch Signal	Cranking: 9-11v
A26	GRN	PSP Switch	Straight: 0v, Turning: 11v
A27	BLU/RED	A/C Switch Signal	Switch Off: 12v, On: 1v
A28	WHT/RED	Interlock Control Unit Signal	Key & Brake On: 12v
A29	LT GRN	Fuel Tank Pressure Sensor	Fuel Cap off: 2.5v
A30	GRN/RED	Electric Load Detector	Varies: 0.5-4.5v
A32	WHT/BLK	Brake Switch Signal	Brake Off: 0v, On: 12v

1998-99 Coupe 3.0L V6 VTEC VIN CG2 [All] 16P 'D' Connector

PCM Pin #	W/Color	Circuit Description (16-Pin)	Value at Hot Idle
D1	YEL	Lockup Control Solenoid	LUS On: 12v, Off: 0v
D2	GRN/WHT	A/T: Shift Solenoid 'B'	SSB in 3rd, 4th Gear: 0v
D3	GRN	A/T: Shift Solenoid 'C'	SSC in 2nd, 4th Gear: 0v
D5	BLK/YEL	VB Solenoid Feed (B+)	Sol. on: 12v, off: 0v
D6	WHT	A/T: Gear Position Switch	In 'R': 0v, Others: 12v
D7	BLU/YEL	A/T: Shift Solenoid 'A'	SSA in 1st, 4th Gear: 0v
D8	PNK	A/T: Gear Position Switch	In D3: 0v, Others: 12v
D9	YEL	A/T: Gear Position Switch	In D4: 0v, Others: 12v
D10	BLU	Countershaft Speed Sensor P	Moving: AC pulses
D11	RED	Mainshaft Speed Sensor 'P'	Moving: AC pulses
D12	WHT	Mainshaft Speed Sensor 'N'	Moving: AC pulses
D14	BLU	A/T: Gear Position Switch	In D2: 0v, Others: 12v
D15	BRN	A/T: Gear Position Switch	In D1: 0v, Others: 12v
D16	GRN	Countershaft Speed Sensor N	<0.050v

```
┌──┬──┬──┬──┐ ┌──┬──┬──┐ ┌──┬──┬──┬──┐  ┌──┐ ┌──┬──┬──┐ ┌──────┐
│ 1│ 2│ 3│ 4│ │ 5│ 6│ 7│ │ 8│ 9│10│11│  │ 1│ │ 2│ 3│ 4│ │  5   │
├──┼──┼──┼──┼─┼──┼──┼──┤ │  │  │  │  │  ├──┼─┼──┼──┼──┼─┤      │
│12│13│14│15│16│17│18│19│20│21│22│ 23 │ 24 │ 6│ 7│ 8│ 9│10│11│ 12   │
├──┼──┼──┼──┤ └──┴──┴──┘ └──┴──┴──────┘  ├──┼──┼──┼──┤ └──────┘
│25│26│27│ │28│29│30│31│ │   32   │   │13│14│15│ │16│ 16-PIN
32-PIN                                            'D'
  'A'
```

(WIRE SIDE OF HARNESS TERMINALS)

05533_ADIA_G599

Pin Connector Graphic
1998-99 Coupe 3.0L V6 VTEC VIN CG2 [All] 25P 'B' Connector

PCM Pin #	W/Color	Circuit Description (25-Pin)	Value at Hot Idle
B1, B9	YEL/BLK	Main Relay Power (B+)	12-14v
B2, B10	BLK	Power Ground	<0.1v
B3	BLK/RED	Injector 5 Control	2.0-3.3 ms
B4, B5	YEL, RED	Injector 4, Injector 2 Control	2.0-3.3 ms
B6	WHT/BLU	Injector 6 Control	2.0-3.3 ms
B7	PNK	E-EGR Solenoid Control	Digital Signals: 0-12-0v
B8	WHT	A/T: Clutch Solenoid 'A-'	AC Pulse Signals
B11	BRN	Injector 1 Control	2.0-3.3 ms
B12	GRN/YEL	VTEC Control Solenoid	0v, Hi-Speed: 12v
B13	YEL/GRN	Ignition Control Signal	Digital Signals: 0-12-0v
B14	BLU/BLK	A/T: 2nd Clutch Pressure Sw.	Open: 12v, Closed: 1v
B15	BLU	Injector 3 Control	2.0-3.3 ms
B17	RED	A/T: Clutch Solenoid 'A+'	AC Pulse Signals
B18	GRN	A/T: Clutch Solenoid 'B-'	Pulse Signals
B20, B22	BRN/BLK	Logic Ground	<0.050v
B21	WHT/YEL	Keep Alive Power (VBU)	12-14v
B23	BLK/BLU	Idle Air Control Valve	Pulse Signals
B24	BLU/WHT	A/T: 3rd Clutch Pressure Sw.	Open: 12v, Closed: 1v
B25	ORN	A/T: Clutch Solenoid 'B+'	Pulse Signals

1998-99 Coupe 3.0L V6 VTEC VIN CG2 [All] 31P 'C' Connector

PCM Pin #	W/Color	Circuit Description (31-Pin)	Value at Hot Idle
C1	BLK/WHT	HO2S-11 (B1 S1) Heater	Digital Signals: 0-12-0v
C2	WHT/GRN	Alternator Charging Signal	Lights Off: 12v, On: 0v
C3	WHT/BLU	Alternator Charging Signal	Lights Off: 12v, On: 0v
C5	WHT/RED	Alternator 'FR' Signal	Digital Signals: 0-5-0v
C6	WHT/BLK	EGR Valve Lift Sensor	1.2v
C7, C18	GRN/WHT	MAP Sensor, Sensor Ground	<0.050v
C8	BLU	CKP Sensor Signal	AC pulse signals
C9	WHT	CKP Sensor Ground	<0.050v
C10	BLU/BLK	VTEC Pressure Switch	0v, Hi-Speed: 12v
C16	WHT	HO2S-11 (B1 S1) Signal	0.1-1.1v
C17	RED/GRN	MAP Sensor Signal	0.8-0.9v
C19, C28	YEL/RED	MAP VREF, Sensor VREF	4.9-5.1v
C20	GRN	TDC1 Sensor Signal	AC pulse signals
C21	RED	TDC1 Sensor Ground	<0.050v
C25	RED/YEL	IAT Sensor Signal	Varies w/temp. (0.5-4.9v)
C26	RED/WHT	ECT Sensor Signal	At 180°F: 0.5-0.6v
C27	RED/BLK	TP Sensor Signal	0.5-0.6v
C29	YEL	TDC2 Sensor Signal	AC pulse signals
C30	BLK	TDC2 Sensor Ground	<0.050v

25-PIN 'B'   WIRE SIDE OF HARNESS TERMINALS   31-PIN 'C'

05533_ADIA_G601

Pin Connector Graphic

2000-03 Coupe 3.0L V6 VTEC VIN CG2 [A/T] 32P 'A' Connector

PCM Pin #	W/Color	Circuit Description (32-Pin)	Value at Hot Idle
A1	YEL/GRN	ECT Signal to TCM	Digital Signals
A2	GRN/WHT	A/T: Engine Mount Solenoid	Solenoid Off: 12v, On: 1v
A3	BLU	EVAP Bypass Solenoid	Solenoid Off: 12v, On: 1v
A4	GRN/WHT	EVAP Vent Solenoid	Solenoid Off: 12v, On: 1v
A5	BLU/GRN	Cruise Control Signal	C/C On: pulse signals
A6	RED/YEL	EVAP Purge Solenoid	Solenoid Off: 12v, On: 1v
A7, 11, 13	WT,GN,BL	ABS-TCS Signals	Digital Signals
A8	BLK/WHT	HO2S-12 (B1 S2) Heater	Digital Signals: 0-12-0v
A9	BLU/WHT	A/T: Vehicle Speed Output	Moving: pulse signals
A10	BRN	Service Check Connector	SCS Open: 4.80v
A12	PNK	Immobilizer Indicator Lamp	Lamp Off: 12v, On: 1v
A14	GRN/BLK	A/T: D4 Light Switch	D4 On: 0v, Off: 12v
A15	GRN/YEL	Immobilizer Fuel Pump Relay	Relay Off: 12v, On: 1v
A16, A25	BLU, RED	Immobilizer Enable, Code	Digital Signals
A17	RED	A/C Clutch Relay Control	Relay Off: 12v, On: 1v
A18	GRN/ORN	Malfunction Indicator Light	MIL Off: 12v, On: 1v

2000-03 Coupe 3.0L V6 VTEC VIN CG2 [A/T] 32P 'A' Connector, *continued*

A19	BLU	Engine Speed Pulse (NEP)	AC Pulse Signals
A20	GRN	Radiator Fan Relay Control	Relay Off: 12v, On: 1v
A21	GRY	K-Line Signal	12v
A22, A31	PNK, RED	ABS-TCS Signals	Digital Signals
A23	WHT/RED	HO2S-12 (B1 S2) Signal	0.1-1.1v
A24	BLU/ORN	Starter Switch Signal	Cranking: 9-11v
A26	GRN	PSP Switch	Straight: 0v, Turning: 11v
A27	BLU/RED	A/C Switch Signal	Switch Off: 12v, On: 1v
A28	WHT/RED	Interlock Control Unit Signal	Key & Brake On: 12v
A29	LT GRN	Fuel Tank Pressure Sensor	Fuel Cap off: 2.5v
A30	GRN/RED	Electric Load Detector	Varies: 0.5-4.5v
A32	WHT/BLK	Brake Switch Signal	Brake Off: 0v, On: 12v

2000-03 Coupe 3.0L V6 VTEC VIN CG2 [A/T] 16P 'D' Connector

PCM Pin #	W/Color	Circuit Description (16-Pin)	Value at Hot Idle
D1	YEL	Lockup Control Solenoid	LUS On: 12v, Off: 0v
D2	GRN/WHT	A/T: Shift Solenoid 'B'	SSB in 3rd, 4th Gear: 0v
D3	GRN	A/T: Shift Solenoid 'C'	SSC in 2nd, 4th Gear: 0v
D5	BLK/YEL	VB Solenoid Feed (B+)	Sol. on: 12v, off: 0v
D6	WHT	A/T: Gear Position Switch	In 'R': 0v, Others: 12v
D7	BLU/YEL	A/T: Shift Solenoid 'A'	SSA in 1st, 4th Gear: 0v
D8, D9	PNK, YEL	A/T: Gear Position Switches	In D3: 0v, In D4: 0v
D10, GRN	BLU, GRN	A/T: Countershaft Speed P, N	Moving: AC pulses
D11, D12	RED, WHT	Mainshaft Speed Sensor P, N	Moving: AC pulses
D13	BLU/BLK	A/T: 2nd Clutch Pressure Sw.	Open: 12v, Closed: 1v
D14, 15	BLU, BRN	A/T: Gear Position Switch	D2: 0v, D1: 0v

Pin Connector Graphic

05533_ADIA_G599

2000-03 Coupe 3.0L V6 VTEC VIN CG2 [A/T] 25P 'B' Connector

PCM Pin #	W/Color	Circuit Description (25-Pin)	Value at Hot Idle
B1, B9	YEL/BLK	Main Relay Power (B+)	12-14v
B2, B10	BLK	Power Ground	<0.1v
B3	BLK/RED	Injector 5 Control	2.0-3.3 ms
B4, B5	YEL, RED	Injector 4, Injector 2 Control	2.0-3.3 ms
B6	WHT/BLU	Injector 6 Control	2.0-3.3 ms
B7	PNK	E-EGR Solenoid Control	Digital Signals: 0-12-0v
B8	WHT	A/T: Clutch Solenoid 'A-'	AC Pulse Signals
B11	BRN	Injector 1 Control	2.0-3.3 ms
B12	GRN/YEL	VTEC Control Solenoid	0v, Hi-Speed: 12v
B14	BLU/WHT	A/T: Gear Position Switch	In P/N: 0v, Others: 12v
B15	BLU	Injector 3 Control	2.0-3.3 ms
B17	RED	A/T: Clutch Solenoid 'A+'	AC Pulse Signals
B18	GRN	A/T: Clutch Solenoid 'B-'	Pulse Signals
B20, B22	BRN/BLK	Logic Ground (LG1, LG2)	<0.050v
B21	WHT/YEL	Keep Alive Power (VBU)	12-14v
B23	BLK/BLU	Idle Air Control Valve	Pulse Signals
B24	BLU/WHT	A/T: 3rd Clutch Pressure Sw.	Open: 12v, Closed: 1v
B25	ORN	A/T: Clutch Solenoid 'B+'	Pulse Signals

2000-03 Coupe 3.0L V6 VTEC VIN CG2 [A/T] 31P 'C' Connector

PCM Pin #	W/Color	Circuit Description (31-Pin)	Value at Hot Idle
C1	BLK/WHT	HO2S-11 (B1 S1) Heater	Digital Signals: 0-12-0v
C2	WHT/GRN	Alternator Charging Signal	Lights Off: 12v, On: 0v
C3	WHT/BLU	Ignition Coil 3 Control	Digital Signals: 0-12-0v
C4	YEL/GRN	Ignition Coil 1 Control	Digital Signals: 0-12-0v
C5	WHT/RED	Alternator 'FR' Signal	Digital Signals: 0-5-0v
C6	WHT/BLK	EGR Valve Lift Sensor	1.2v
C7, C18	GRN/WHT	MAP Sensor, Sensor Ground	<0.050v
C8, C9	BLU, WHT	CKP Sensor Signal, Ground	AC pulse signals
C10	BLU/BLK	VTEC Pressure Switch	0v, Hi-Speed: 12v
C12	BLK/RED	Ignition Coil 5 Control	Digital Signals: 0-12-0v
C13	BRN	Ignition Coil 4 Control	Digital Signals: 0-12-0v
C14	BLU/RED	Ignition Coil 2 Control	Digital Signals: 0-12-0v
C16	WHT	HO2S-11 (B1 S1) Signal	0.1-1.1v
C17	RED/GRN	MAP Sensor Signal	0.8-0.9v
C19, C28	YEL/RED	MAP VREF, Sensor VREF	4.9-5.1v
C20, C21	GRN, RED	TDC1 Sensor Signal, Ground	AC pulse signals
C23	BRN/WHT	Ignition Coil 6 Control	Digital Signals: 0-12-0v
C25	RED/YEL	IAT Sensor Signal	Varies w/temp. (0.5-4.9v)
C26	RED/WHT	ECT Sensor Signal	At 180°F: 0.5-0.6v
C27	RED/BLK	TP Sensor Signal	0.5-0.6v
C29, C30	YEL, BLK	TDC2 Sensor Signal, Ground	AC pulse signals

1	2		3	4	5		6	7	8	1	2	3	4		5	6	7		8	9	10
9	10	11	12	13	14	15	16	17	18	11	12	13	14	15	16	17	18	19	20	21	22
19	20	21	22		23	24	25			23	24	25		26	27	28		29	30	31	

25-PIN 'B' ⟨ *WIRE SIDE OF HARNESS TERMINALS* ⟩ **31-PIN 'C'**

05533_ADIA_G601

Pin Connector Graphic
1990-93 Sedan 2.2L I4 MFI VIN CB7 [All] 26P 'A' Connector

PCM Pin #	W/Color	Circuit Description (26-Pin)	Value at Hot Idle
A1	BRN	Injector 1 Control	2.0-3.3 ms
A2	YEL	Injector 4 Control	2.0-3.3 ms
A3	RED	Injector 2 Control	2.0-3.3 ms
A4	---	Not Used	---
A5	BLU	Injector 3 Control	2.0-3.3 ms
A6	ORN/BLK	HO2S Heater Control	Relay Off: 12v, On: 1v
A7	GRN/BLK	Fuel Pump Relay	Relay Off: 12v, On: 1v
A8	GRN/BLK	Fuel Pump Relay	Relay Off: 12v, On: 1v
A9	BLK/BLU	Electronic Air Control Valve	Pulse Signals
A10	---	Not Used	---
A11	PNK	EGR Control Solenoid	1.2v
A12	BLU	Radiator Fan Relay	Relay Off: 12v, On: 1v
A13	GRN/RED	Check Engine Light	MIL Off: 12v, On: 1v
A14	---	Not Used	---
A15	RED/BLU	A/C Clutch Relay	Relay Off: 12v, On: 1v
A16	WHT/GRN	Alternator Charging Signal	Lights Off: 12v, On: 0v
A17	---	Not Used	---
A18	BRN/WHT	A/T: Control Unit	---
A19	WHT	Intake Air Regulator Control	Solenoid Off: 12v, On: 1v
A20	RED/GRN	EVAP Purge Solenoid	Solenoid Off: 12v, On: 1v
A21	YEL/GRN	Igniter Control	Hot idle 10% d/cycle
A22	YEL/GRN	Igniter Control	Hot idle 10% d/cycle
A23	BLK	Power Ground	<0.1v
A24	BLK	Power Ground	<0.1v
A25	YEL/BLK	Ignition Power	12-14v
A26	BLK/RED	Chassis Ground	<0.050v

1990-93 Sedan 2.2L I4 MFI VIN CB7 [All] 16P 'B' Connector

PCM Pin #	W/Color	Circuit Description (16-Pin)	Value at Hot Idle
B1	YEL/BLK	Main Relay Power (B+)	12-14v
B2	BRN/BLK	Power Ground	<0.1v
B3	WHT/RED	A/T: Control Unit	---
B4	GRN	A/T: Control Unit	---
B5	BLU/BLK	A/C Switch Signal	Switch Off: 12v, On: 1v
B6	---	Not Used	---
B7	LT GRN	A/T: Neutral Position Switch	In N: 0v, Others: 12v
B8	RED	PSP Switch	Straight: 0v, Turning: 11v
B9	BLU/RED	A/T: Start Signal	Cranking: 9-11v
B9	BLK/GRN	M/T: Start Signal	Cranking: 9-11v
B10	ORN	Vehicle Speed Sensor	Moving: pulse signals
B11	ORN	CYP Sensor Signal	AC pulse signals
B12	WHT	CYP Sensor Ground	<0.050v
B13	ORN/BLU	TDC Sensor Signal	AC pulse signals
B14	WHT/BLU	TDC Sensor Ground	<0.050v
B15	BLU/GRN	CKP Sensor Signal	AC pulse signals
B16	BLU/YEL	CKP Sensor Ground	<0.050v

1990-93 Sedan 2.2L I4 MFI VIN CB7 [All] 22P 'D' Connector

PCM Pin #	W/Color	Circuit Description (22-Pin)	Value at Hot Idle
D1	WHT/YEL	Keep Alive Power (VBU)	12-14v
D2	GRN/WHT	Brake Switch Signal	Brake Off: 0v, On: 12v
D3	---	Not Used	---
D4	ORN/RED	Service Check Connector	SCS Open: 4.80v
D5-D8	---	Not Used	---
D9	WHT/RED	Alternator 'FR' Signal	Digital Signals: 0-5-0v
D10	GRN/RED	Electric Load Detector	Varies: 0.5-4.5v
D11	RED/BLK	TP Sensor Signal	0.5-0.6v
D12	WHT/BLK	A/T: EGR Lift Sensor Signal	1.1-1.2v
D13	YEL/GRN	ECT Sensor Signal	At 180°F: 0.52v
D14	WHT	Oxygen Sensor	0.1-1.1v
D15	RED/YEL	IAT Sensor Signal	Varies w/temp. (0.5-4.9v)
D16	---	Not Used	---
D17	WHT/BLU	MAP Sensor Signal	0.8-0.9v
D18	GRN/BLK	A/T: Control Unit VREF	4.9-5.1v
D19	RED/WHT	MAP Sensor VREF	4.9-5.1v
D20	YEL/WHT	Sensor VREF	4.9-5.1v
D21	BLU/WHT	Sensor Ground	<0.050v
D22	GRN/WHT	Sensor Ground	<0.050v

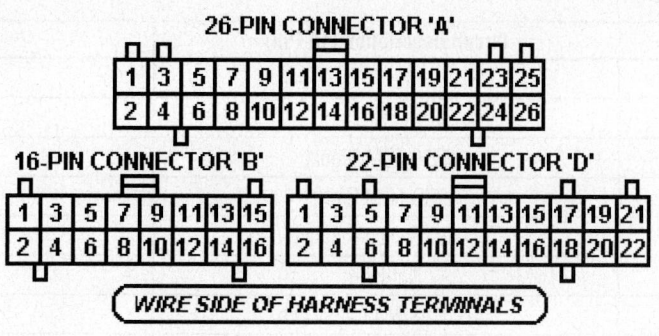

26-PIN CONNECTOR 'A'

16-PIN CONNECTOR 'B'   22-PIN CONNECTOR 'D'

WIRE SIDE OF HARNESS TERMINALS

05533_ADIA_G597

Pin Connector Graphic

1994 Sedan 2.2L I4 VTEC VIN CD5 [All] 26P 'A' Connector

PCM Pin #	W/Color	Circuit Description (26-Pin)	Value at Hot Idle
A1	BRN	Injector 1 Control	2.0-3.3 ms
A2	YEL	Injector 4 Control	2.0-3.3 ms
A3	RED	Injector 2 Control	2.0-3.3 ms
A4	GRN/YEL	VTEC Control Solenoid	0v, Hi-Speed: 12v
A5	BLU	Injector 3 Control	2.0-3.3 ms
A6	RED	EGR Control Solenoid	Solenoid Off: 12v, On: 1v
A7	GRN/BLK	Fuel Pump Relay	Relay Off: 12v, On: 1v
A8	---	Not Used	---
A9	BLK/BLU	Intake Air Regulator Control	Solenoid Off: 12v, On: 1v
A10	GRN/WHT	Engine Mount Solenoid	Solenoid Off: 12v, On: 1v
A11	GRN/BLK	HO2S Heater Control	Relay Off: 12v, On: 1v
A12	GRN	Radiator Fan Relay Control	Relay Off: 12v, On: 1v
A13	GRN/RED	Check Engine Light	MIL Off: 12v, On: 1v
A14	WHT/YEL	FIA Control Solenoid	Solenoid Off: 12v, On: 1v
A15	RED/BLU	A/C Clutch Relay	Relay Off: 12v, On: 1v
A16	WHT/GRN	Alternator Charging Signal	Lights Off: 12v, On: 0v
A17	---	Not Used	---
A18	BRN/WHT	FAS TCM Signal	Digital Signals
A19	ORN/GRN	Intake Air Regulator Control	Solenoid Off: 12v, On: 1v
A20	RED/YEL	EVAP Purge Solenoid	Solenoid Off: 12v, On: 1v
A21, 22	YEL/GRN	Igniter Control	Digital Signals: 0-12-0v
A23, 24	BLK, BLK	Power Ground	<0.1v
A25	YEL/BLK	Main Relay Power (B+)	12-14v
A26	BLK/RED	Logic Ground	<0.050v

1994 Sedan 2.2L I4 VTEC VIN CD5 [All] 16P 'B' Connector

PCM Pin #	W/Color	Circuit Description (16-Pin)	Value at Hot Idle
B1	YEL/BLK	Main Relay Power (B+)	12-14v
B2	BRN/BLK	Logic Ground	<0.050v
B3	WHT/RED	AFSA TCM Signal	Digital Signals
B4	GRN	AFSB TCM Signal	Digital Signals
B5	RED/WHT	A/C Switch Signal	Switch Off: 12v, On: 1v
B6	---	Not Used	---
B7	LT GRN	A/T: Park/Neutral Switch	In P/N: 0v, Others: 11v
B8	GRN	PSP Switch	Straight: 0v, Turning: 11v
B9	BLU/RED	A/T: Start Signal	Cranking: 9-11v
B9	BLU/RED	M/T: Start Signal	Cranking: 9-11v
B10	ORN	Vehicle Speed Sensor	Moving: pulse signals
B11	ORN	CYP Sensor Signal	AC pulse signals
B12	WHT	CYP Sensor Ground	<0.050v
B13	ORN/BLU	TDC Sensor Signal	AC pulse signals
B14	WHT/BLU	TDC Sensor Ground	<0.050v
B15	BLU/GRN	CKP Sensor Signal	AC pulse signals
B16	BLU/YEL	CKP Sensor Ground	<0.050v

1994 Sedan 2.2L I4 VTEC VIN CD5 [All] 22P 'D' Connector

PCM Pin #	W/Color	Circuit Description (22-Pin)	Value at Hot Idle
D1	WHT/YEL	Keep Alive Power (VBU)	12-14v
D2	GRN/WHT	Brake Switch Signal	Brake Off: 0v, On: 12v
D3	---	Not Used	---
D4	ORN/RED	Service Check Connector	SCS Open: 4.80v
D5	BLU/WHT	A/T: TCM BARO Signal	3v (at sea level)
D6	BLU/BLK	VTEC Pressure Switch	0v, Hi-Speed: 12v
D7	GRN/RED	Data Link Connector	5v
D8	---	Not Used	---
D9	WHT/RED	Alternator 'FR' Signal	Digital Signals: 0-5-0v
D10	GRN/RED	Electric Load Detector	Varies: 0.5-4.5v
D11	RED/BLK	TP Sensor Signal	0.5-0.6v
D12	WHT/BLK	EGR Lift Sensor	1.1-1.2v
D13	RED/WHT	ECT Sensor Signal	At 180°F: 0.51v
D14	WHT/RED	HO2S Signal	0.1-1.1v
D15	RED/YEL	IAT Sensor Signal	Varies w/temp. (0.5-4.9v)
D16	---	Not Used	---
D17	WHT/YEL	MAP Sensor Signal	0.8-0.9v
D18	GRN/BLK	VREF To TCM Signal	4.9-5.1v
D19	YEL/WHT	MAP Sensor VREF	4.9-5.1v
D20	YEL/BLU	Sensor VREF	4.9-5.1v
D21	GRN/WHT	MAP Sensor Ground	<0.050v
D22	GRN/BLU	Sensor Ground	<0.050v

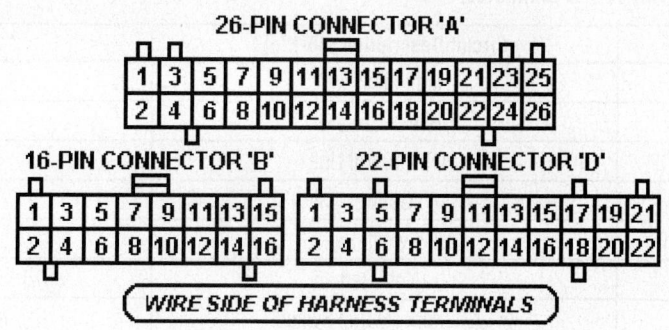

26-PIN CONNECTOR 'A'

16-PIN CONNECTOR 'B'   22-PIN CONNECTOR 'D'

WIRE SIDE OF HARNESS TERMINALS

05533_ADIA_G597

Pin Connector Graphic
1994 Sedan 2.2L I4 MFI VIN CD5 [All] 26P 'A' Connector

PCM Pin #	W/Color	Circuit Description (26-Pin)	Value at Hot Idle
A1	BRN	Injector 1 Control	2.0-3.3 ms
A2	YEL	Injector 4 Control	2.0-3.3 ms
A3	RED	Injector 2 Control	2.0-3.3 ms
A4	---	Not Used	---
A5	BLU	Injector 3 Control	2.0-3.3 ms
A6	RED	EGR Control Solenoid	Solenoid Off: 12v, On: 1v
A7-8	GRN/BLK	Fuel Pump Relay	Relay Off: 12v, On: 1v
A9	BLK/BLU	Idle Air Control Motor	Pulse Signals
A10	GRN/WHT	Engine Mount Solenoid	Solenoid Off: 12v, On: 1v
A11	GRN/BLK	HO2S Heater Control	Relay Off: 12v, On: 1v
A12	GRN	Radiator Fan Relay	Relay Off: 12v, On: 1v
A13	GRN/RED	Check Engine Light	MIL Off: 12v, On: 1v
A14	---	Not Used	---
A15	RED/BLU	A/C Clutch Relay	Relay Off: 12v, On: 1v
A16	WHT/GRN	Alternator Charging Signal	Lights Off: 12v, On: 0v
A17	---	Not Used	---
A18	BRN/WHT	A/T: Control Unit VREF	5v
A19	ORN/GRN	Intake Air Regulator Control	Solenoid Off: 12v, On: 1v
A20	RED/YEL	EVAP Purge Solenoid	Solenoid Off: 12v, On: 1v
A21	YEL/GRN	Igniter Control	Digital Signals: 0-12-0v
A22	YEL/GRN	Igniter Control	Digital Signals: 0-12-0v
A23	BLK	Power Ground	<0.1v
A24	BLK	Power Ground	<0.1v
A25	YEL/BLK	Ignition Power	12-14v
A26	BLK/RED	Chassis Ground	<0.050v

1994 Sedan 2.2L I4 MFI VIN CD5 [All] 16P 'B' Connector

PCM Pin #	W/Color	Circuit Description (16-Pin)	Value at Hot Idle
B1	YEL/BLK	Ignition Power	12-14v
B2	BRN/BLK	Power Ground	<0.1v
B3	WHT/RED	A/T: Control Unit	Digital Signals
B4	GRN	A/T: Control Unit	Digital Signals
B5	RED/WHT	A/C Switch Signal	Switch Off: 12v, On: 1v
B6	---	Not Used	---
B7	LT GRN	A/T: Park Neutral Switch	In P/N: 0v, Others: 11v
B8	GRN	PSP Switch	Straight: 0v, Turning: 11v
B9	BLU/RED	A/T: Start Signal	Cranking: 9-11v
B9	BLU/RED	M/T: Start Signal	Cranking: 9-11v
B10	ORN	Vehicle Speed Sensor	Moving: pulse signals
B11	ORN	CYP Sensor Signal	AC pulse signals
B12	WHT	CYP Sensor Ground	<0.050v
B13	ORN/BLU	TDC Sensor Signal	AC pulse signals
B14	WHT/BLU	TDC Sensor Ground	<0.050v
B15	BLU/GRN	CKP Sensor Signal	AC pulse signals
B16	BLU/YEL	CKP Sensor Ground	<0.050v

1994 Sedan 2.2L I4 MFI VIN CD5 [All] 22P 'D' Connector

PCM Pin #	W/Color	Circuit Description (22-Pin)	Value at Hot Idle
D1	WHT/YEL	Keep Alive Power (VBU)	12-14v
D2	GRN/WHT	Brake Switch Signal	Brake Off: 0v, On: 12v
D3	---	Not Used	---
D4	ORN/RED	Service Check Connector	SCS Open: 4.80v
D5	BLU/WHT	A/T TCM Signal	Digital Signals
D6	---	Not Used	---
D7	GRN/RED	Data Link Connector	5v
D8	---	Not Used	---
D9	WHT/RED	Alternator 'FR' Signal	Digital Signals: 0-5-0v
D10	GRN/RED	Electric Load Detector	Varies: 0.5-4.5v
D11	RED/BLK	TP Sensor Signal	0.5-0.6v
D12	WHT/BLK	A/T: EGR Lift Sensor Signal	1.1-1.2v
D13	RED/WHT	ECT Sensor Signal	At 180°F: 0.52v
D14	WHT/RED	HO2S Signal	0.1-1.1v
D15	RED/YEL	IAT Sensor Signal	Varies w/temp. (0.5-4.9v)
D16	---	Not Used	---
D17	WHT/YEL	MAP Sensor Signal	0.8-0.9v
D18	GRN/BLK	A/T: Control Unit VREF	5v
D19	YEL/WHT	MAP Sensor VREF	4.9-5.1v
D20	YEL/BLU	Sensor VREF	4.9-5.1v
D21	GRN/WHT	MAP Sensor Ground	<0.050v
D22	GRN/BLU	Sensor Ground	<0.050v

26-PIN CONNECTOR 'A'

| 1 | 3 | 5 | 7 | 9 | 11 | 13 | 15 | 17 | 19 | 21 | 23 | 25 |
| 2 | 4 | 6 | 8 | 10 | 12 | 14 | 16 | 18 | 20 | 22 | 24 | 26 |

16-PIN CONNECTOR 'B'

| 1 | 3 | 5 | 7 | 9 | 11 | 13 | 15 |
| 2 | 4 | 6 | 8 | 10 | 12 | 14 | 16 |

22-PIN CONNECTOR 'D'

| 1 | 3 | 5 | 7 | 9 | 11 | 13 | 15 | 17 | 19 | 21 |
| 2 | 4 | 6 | 8 | 10 | 12 | 14 | 16 | 18 | 20 | 22 |

WIRE SIDE OF HARNESS TERMINALS

05533_ADIA_G597

Pin Connector Graphic
1995 Sedan 2.2L I4 VTEC VIN CD5 [All] 26P 'A' Connector

PCM Pin #	W/Color	Circuit Description (26-Pin)	Value at Hot Idle
A1	BRN	Injector 1 Control	2.0-3.3 ms
A2	YEL	Injector 4 Control	2.0-3.3 ms
A3	RED	Injector 2 Control	2.0-3.3 ms
A4	GRN/YEL	VTEC Solenoid Control	0v, Hi-Speed: 12v
A5	BLU	Injector 3 Control	2.0-3.3 ms
A6	ORN/BLK	HO2S Heater Control	Relay Off: 12v, On: 1v
A7	GRN/BLK	Fuel Pump Relay	Relay Off: 12v, On: 1v
A8	---	Not Used	---
A9	BLK/BLU	Idle Air Control Valve Signal	Pulse Signals
A10	---	Not Used	---
A11	RED	EGR Solenoid	Solenoid Off: 12v, On: 1v
A12	GRN	Radiator Fan Relay	Relay Off: 12v, On: 1v
A13	GRN/RED	Check Engine Light	MIL Off: 12v, On: 1v
A14	---	Not Used	---
A15	RED/BLU	A/C Clutch Relay	Relay Off: 12v, On: 1v
A16	WHT/GRN	Alternator Charging Signal	Lights Off: 12v, On: 0v
A17	---	Not Used	---
A18	BRN/WHT	A/T: Control Unit	Digital Signals
A19	WHT	Intake Air Regulator Control	Solenoid Off: 12v, On: 1v
A20	RED/YEL	EVAP Purge Solenoid	Solenoid Off: 12v, On: 1v
A21	YEL/GRN	Igniter Control	Digital Signals: 0-12-0v
A22	---	Not Used	---
A23	BLK	Power Ground	<0.1v
A24	BLK	Power Ground	<0.1v
A25	YEL/BLK	Ignition Power	12-14v
A26	BLK/RED	Chassis Ground	<0.050v

1995 Sedan 2.2L I4 VTEC VIN CD5 [All] 16P 'B' Connector

PCM Pin #	W/Color	Circuit Description (16-Pin)	Value at Hot Idle
B1	YEL/BLK	Ignition Power	12-14v
B2	BRN/BLK	Power Ground	<0.1v
B3	WHT/RED	A/T: Control Unit	Digital Signals
B4	GRN	A/T: Control Unit	Digital Signals
B5	RED/WHT	A/C Switch Signal	Switch Off: 12v, On: 1v
B6	---	Not Used	---
B7	LT GRN	A/T: Park/Neutral Switch	In P/N: 0v, Others: 11v
B8	RED	PSP Switch	Straight: 0v, Turning: 11v
B9	BLU/RED	A/T: Start Signal	Cranking: 9-11v
B9	BLU/RED	MT: Start Signal	Cranking: 9-11v
B10	ORN	Vehicle Speed Sensor	Moving: pulse signals
B11	ORN	CYP Sensor Signal	AC pulse signals
B12	WHT	CYP Sensor Ground	<0.050v
B13	ORN/BLU	TDC Sensor Signal	AC pulse signals
B14	WHT/BLU	TDC Sensor Ground	<0.050v
B15	BLU/GRN	CKP Sensor Signal	AC pulse signals
B16	BLU/YEL	CKP Sensor Ground	<0.050v

1995 Sedan 2.2L I4 VTEC VIN CD5 [All] 22P 'D' Connector

PCM Pin #	W/Color	Circuit Description (22-Pin)	Value at Hot Idle
D1	WHT/YEL	Keep Alive Power (VBU)	12-14v
D2	GRN/WHT	Brake Switch Signal	Brake Off: 0v, On: 12v
D3	---	Not Used	---
D4	RED	Service Check Connector	SCS Open; 4.80v
D5	BLU/WHT	A/T TCM Signal	---
D6	BLU/BLK	VTEC Pressure Switch	0v, Hi-Speed: 12v
D7	GRN/RED	Data Link Connector	5v
D3	---	Not Used	---
D9	WHT/RED	Alternator 'FR' Signal	Digital Signals: 0-5-0v
D10	GRN/RED	Electric Load Detector	Varies: 0.5-4.5v
D11	RED/BLK	TP Sensor Signal	0.5-0.6v
D12	WHT/BLK	A/T: EGR Lift Sensor Signal	1.1-1.2v
D13	RED/WHT	ECT Sensor Signal	At 180°F: 0.52v
D14	WHT/RED	HO2S Signal	0.1-1.1v
D15	RED/YEL	IAT Sensor Signal	Varies w/temp. (0.5-4.9v)
D16	---	Not Used	---
D17	WHT/YEL	MAP Sensor Signal	0.8-0.9v
D18	GRN/BLK	A/T: Control Unit VREF	5v
D19	YEL/WHT	MAP Sensor VREF	4.9-5.1v
D20	YEL/BLU	Sensor VREF	4.9-5.1v
D21	GRN/WHT	MAP Sensor Ground	<0.050v
D22	GRN/BLU	Sensor Ground	<0.050v

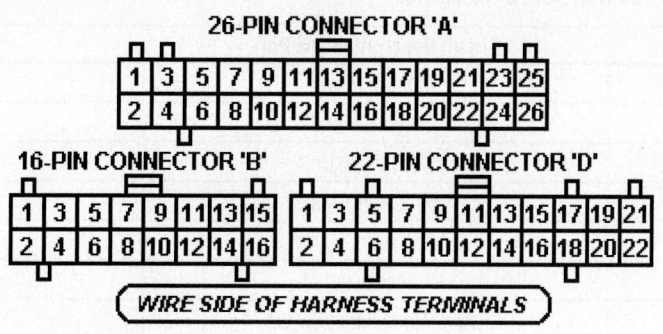

26-PIN CONNECTOR 'A'

16-PIN CONNECTOR 'B'   **22-PIN CONNECTOR 'D'**

WIRE SIDE OF HARNESS TERMINALS

05533_ADIA_G597

Pin Connector Graphic
1995 Sedan 2.2L I4 MFI VIN CD5, CD6 [All] 26P 'A' Connector

PCM Pin #	W/Color	Circuit Description (26-Pin)	Value at Hot Idle
A1, 2	BRN, YEL	Injector 1 Control, 4	2.0-3.3 ms
A3, 5	RED, BLU	Injector 2 Control, 3	2.0-3.3 ms
A4	---	Not Used	---
A6	ORN/BLK	HO2S Heater Control	Relay Off: 12v, On: 1v
A7	GRN/BLK	Fuel Pump Relay	Relay Off: 12v, On: 1v
A8	---	Not Used	---
A9	BLK/BLU	Intake Air Control Solenoid	Solenoid Off: 12v, On: 1v
A10	---	Not Used	---
A11	RED	EGR Control Solenoid	Solenoid Off: 12v, On: 1v
A12	GRN	Radiator Fan Relay	Relay Off: 12v, On: 1v
A13	GRN/RED	Check Engine Light	MIL Off: 12v, On: 1v
A14	---	Not Used	---
A15	RED/BLU	A/C Clutch Relay	Relay Off: 12v, On: 1v
A16	WHT/GRN	Alternator Charging Signal	Lights Off: 12v, On: 0v
A17	---	Not Used	---
A18	BRN/WHT	A/T: Control Unit	---
A19	WHT	Intake Air Regulator Control	Solenoid Off: 12v, On: 1v
A20	RED/YEL	EVAP Purge Solenoid	Solenoid Off: 12v, On: 1v
A21	YEL/GRN	Igniter Control	Digital Signals: 0-12-0v
A22	---	Not Used	---
A23	BLK	Power Ground	<0.1v
A24	BLK	Power Ground	<0.1v
A25	YEL/BLK	Ignition Power	12-14v
A26	BLK/RED	Chassis Ground	<0.050v

1995 Sedan 2.2L I4 MFI VIN CD5, CD6 [All] 16P 'B' Connector

PCM Pin #	W/Color	Circuit Description (16-Pin)	Value at Hot Idle
B1	YEL/BLK	Ignition Power	12v
B2	BRN/BLK	Power Ground	<0.1v
B3	WHT/RED	A/T: Control Unit	---
B4	GRN	A/T: Control Unit	---
B5	RED/WHT	A/C Switch Signal	Switch Off: 12v, On: 1v
B6	---	Not Used	---
B7	LT GRN	A/T: Park/Neutral Switch	In P/N: 0v, Others: 11v
B8	RED	PSP Switch	Straight: 0v, Turning: 11v
B9	BLU/RED	A/T: Start Signal	Cranking: 9-11v
B9	BLU/RED	MT: Start Signal	Cranking: 9-11v
B10	ORN	Vehicle Speed Sensor	Moving: pulse signals
B11	ORN	CYP Sensor Signal	AC pulse signals
B12	WHT	CYP Sensor Ground	<0.050v
B13	ORN/BLU	TDC Sensor Signal	AC pulse signals
B14	WHT/BLU	TDC Sensor Ground	<0.050v
B15	BLU/GRN	CKP Sensor Signal	AC pulse signals
B16	BLU/YEL	CKP Sensor Ground	<0.050v

1995 Sedan 2.2L I4 MFI VIN CD5, CD6 [All] 22P 'D' Connector

PCM Pin #	W/Color	Circuit Description (22-Pin)	Value at Hot Idle
D1	WHT/YEL	Keep Alive Power (VBU)	12-14v
D2	GRN/WHT	Brake Switch Signal	Brake Off: 0v, On: 12v
D3	---	Not Used	---
D4	RED	Service Check Connector	SCS Open: 4.80v
D5	BLU/WHT	A/T TCM Signal	Digital Signals
D6	---	Not Used	---
D7	GRN/RED	Data Link Connector	5v
D8	---	Not Used	---
D9	WHT/RED	Alternator 'FR' Signal	Digital Signals: 0-5-0v
D10	GRN/RED	Electric Load Detector	Varies: 0.5-4.5v
D11	RED/BLK	TP Sensor Signal	0.5-0.6v
D12	WHT/BLK	A/T: EGR Lift Sensor Signal	1.1-1.2v
D13	RED/WHT	ECT Sensor Signal	At 180°F: 0.52v
D14	WHT/RED	HO2S Signal	0.1-1.1v
D15	RED/YEL	IAT Sensor Signal	Varies w/temp. (0.5-4.9v)
D16	---	Not Used	---
D17	WHT/YEL	MAP Sensor Signal	0.8-0.9v
D18	GRN/BLK	A/T: Control Unit	---
D19	YEL/WHT	MAP Sensor VREF	4.9-5.1v
D20	YEL/BLU	Sensor VREF	4.9-5.1v
D21	GRN/WHT	MAP Sensor Ground	<0.050v
D22	GRN/BLU	Sensor Ground	<0.050v

26-PIN CONNECTOR 'A'

1	3	5	7	9	11	13	15	17	19	21	23	25
2	4	6	8	10	12	14	16	18	20	22	24	26

16-PIN CONNECTOR 'B'   **22-PIN CONNECTOR 'D'**

1	3	5	7	9	11	13	15
2	4	6	8	10	12	14	16

1	3	5	7	9	11	13	15	17	19	21
2	4	6	8	10	12	14	16	18	20	22

WIRE SIDE OF HARNESS TERMINALS

05533_ADIA_G597

Pin Connector Graphic
Standard Colors and Abbreviations

Abbreviation	Color	Abbreviation	Color	Abbreviation	Color
BLK	Black	LT BLU	Light Blue	TAN	Tan
BLU	Blue	LT GRN	Lt. Green	VIO	Violet
BRN	Brown	ORN	Orange	WHT	White
GRY	Gray	PNK	Pink	YEL	Yellow
GRN	Green	PPL	Purple		

1996-97 Sedan 2.2L I4 VTEC VIN CD5 [All] 32P 'A' Connector

PCM Pin #	W/Color	Circuit Description (32-Pin)	Value at Hot Idle
A1	YEL	Injector 4 Control	2.0-3.3 ms
A2	BLU	Injector 3 Control	2.0-3.3 ms
A3	RED	Injector 2 Control	2.0-3.3 ms
A4	BRN	Injector 1 Control	2.0-3.3 ms
A5	ORN/BLU	HO2S-12 (B1 S2) Heater	Digital Signals: 0-12-0v
A6	ORN/BLK	HO2S-11 (B1 S1) Heater	Digital Signals: 0-12-0v
A7	RED	EGR Solenoid	Solenoid Off: 12v, On: 1v
A8	GRN/YEL	VTEC Solenoid	0v, Hi-Speed: 12v
A9, A22	BRN/BLK	Sensor Ground	<0.050v
A10, 23	BLK	Power Ground	<0.1v
A11, A24	YEL/BLK	Main Relay Power (B+)	12-14v
A12	BLK/BLU	Idle Air Control Valve Signal	Pulse Signals
A13	GRN/WHT	Engine Mount Solenoid	Solenoid Off: 12v, On: 1v
A15	RED/YEL	EVAP Purge Solenoid	Solenoid Off: 12v, On: 1v
A16	GRN/BLK	Fuel Pump Relay	Relay Off: 12v, On: 1v
A17	RED/BLU	A/C Clutch Relay	Relay Off: 12v, On: 1v
A18	GRN/RED	Check Engine Light	MIL Off: 12v, On: 1v
A19	WHT/GRN	Alternator Charging Signal	Lights Off: 12v, On: 0v
A20	YEL/GRN	Igniter Control	Digital Signals: 0-12-0v
A25	ORN/GRN	Intake Air Resonator Control	Solenoid Off: 12v, On: 1v
A27	GRN	Radiator Fan Relay	Relay Off: 12v, On: 1v
A28	GRN/WHT	EVAP Bypass Solenoid	Solenoid Off: 12v, On: 1v
A29	ORN/GRN	EVAP Vent Solenoid	Solenoid Off: 12v, On: 1v

1996-97 Sedan 2.2L I4 VTEC VIN CD5 [All] 25P 'B' Connector

PCM Pin #	W/Color	Circuit Description (25-Pin)	Value at Hot Idle
B3	BLU/YEL	Shift Solenoid 'A'	In D2-3: 12v, D1-4: 0v
B4	GRN/BLK	Lockup Solenoid 'B'	LSB On: 12v, Off: 0v
B5	YEL	Lockup Solenoid 'A'	LSA On: 12v, Off: 0v
B8	GRN/BLU	A/T: Gear Position Switch	In D3: 0v, Others: 12v
B11	BLU/YEL	Shift Solenoid 'B'	D2/D3: 12v, D1/D4: 0v
B12	WHT/GRN	Interlock Control Unit Signal	Key & Brake On: 12v
B13	BLU/RED	A/T: D4 Indicator Light Switch	D4 On: 0v, Off: 12v
B14	WHT/BLU	Mainshaft Speed Sensor 'N'	<0.050v
B15	ORN/BLU	Mainshaft Speed Sensor 'P'	Moving: AC pulses
B16	GRN/RED	A/T: Gear Position Switch	In 'R': 0v, Others: 12v
B17	GRN/YEL	A/T: Gear Position Switch	In D2: 0v, Others: 12v
B18	GRN/WHT	A/T: Gear Position Switch	In D1: 0v, Others: 12v
B22	BLU/YEL	Countershaft Speed Sensor N	<0.050v
B23	BLU/GRN	Countershaft Speed Sensor P	Moving: AC pulses
B24	PNK/GRN	A/T: Gear Position Switch	In D4: 0v, Others: 12v
B25	LT GRN	A/T: Gear Position Switch	In P/N: 0v, Others: 12v

32-PIN 'A'   WIRE SIDE OF HARNESS TERMINALS   25-PIN 'B'

05533_ADIA_G602

Pin Connector Graphic

1996-97 Sedan 2.2L I4 VTEC VIN CD5 [All] 31P 'C' Connector

PCM Pin #	W/Color	Circuit Description (31-Pin)	Value at Hot Idle
C1	---	Not Used	---
C2	BLU	CKP Sensor Signal	AC pulse signals
C3	GRN	TDC Sensor Signal	AC pulse signals
C4	YEL	CYP Sensor Signal	AC pulse signals
C5	RED/WHT	A/C Switch Signal	Switch Off: 12v, On: 1v
C6	BLU/RED	Start Signal	Cranking: 9-11v
C7	RED	Service Check Connector	SCS Open: 4.80v
C8	LT GRN	K-Line Signal	12v
C10	WHT/YEL	Keep Alive Power (VBU)	12-14v
C12	WHT	CKP Sensor Ground	<0.050v
C13	RED	TDC Sensor Ground	<0.050v
C14	BLK	CYP Sensor Ground	<0.050v
C16	GRN	PSP Switch	Straight: 0v, Turning: 11v
C17	WHT/RED	Alternator 'FR' Signal	Digital Signals: 0-5-0v
C18	ORN	Vehicle Speed Sensor	Moving: pulse signals
C20	BRN	EVAP Purge Flow Switch	Switch on: 0v, off: 5v
C21-31	---	Not Used	---

1996-97 Sedan 2.2L I4 VTEC VIN CD5 [All] 16P 'D' Connector

PCM Pin #	W/Color	Circuit Description (16-Pin)	Value at Hot Idle
D1	RED/BLK	TP Sensor Signal	0.5-0.6v
D2	RED/WHT	ECT Sensor Signal	At 180°F: 0.52v
D3	WHT/YEL	MAP Sensor Signal	0.8-0.9v
D4	YEL/WHT	MAP Sensor VREF	4.9-5.1v
D5	GRN/WHT	Brake Switch Signal	Brake Off: 0v, On: 12v
D7	WHT/RED	HO2S-11 (B1 S1) Signal	0.1-1.1v
D8	RED/YEL	IAT Sensor Signal	Varies w/temp. (0.5-4.9v)
D9	WHT/BLK	EGR Valve Lift Sensor	1.2v
D10	YEL/BLU	Sensor VREF	4.9-5.1v
D11	GRN/BLU	Sensor Ground	<0.050v
D12	GRN/WHT	MAP Sensor Ground	<0.050v
D13	RED/WHT	HO2S-12 Ground	<0.050v
D14	WHT/RED	HO2S-12 (B1 S2) Signal	0.1-1.1v
D15	GRN/RED	Fuel Tank Pressure Sensor	Fuel Cap off: 2.5v
D16	GRN/RED	Electric Load Detector	Varies: 0.5-4.5v

31-PIN CONNECTOR 'C'   **16-PIN CONNECTOR 'D'**

1	2	3	4		5	6	7		8	9	10		1		2	3	4		5
11	12	13	14	15	16	17	18	19	20	21	22		6	7	8	9	10	11	12
	23	24	25		26	27	28		29	30	31		13	14	15		16		

WIRE SIDE OF HARNESS TERMINALS

05533_ADIA_G603

Pin Connector Graphic

1996-97 Sedan 2.2L I4 MFI VIN CD5 [All] 32P 'A' Connector

PCM Pin #	W/Color	Circuit Description (32-Pin)	Value at Hot Idle
A1	YEL	Injector 4 Control	2.0-3.3 ms
A2	BLU	Injector 3 Control	2.0-3.3 ms
A3	RED	Injector 2 Control	2.0-3.3 ms
A4	BRN	Injector 1 Control	2.0-3.3 ms
A5	ORN/BLU	HO2S-12 (B1 S2) Heater	Digital Signals: 0-12-0v
A6	ORN/BLK	HO2S-11 (B1 S1) Heater	Digital Signals: 0-12-0v
A7	RED	EGR Solenoid	Solenoid Off: 12v, On: 1v
A9, A22	BRN/BLK	Sensor Ground	<0.050v
A10, 23	BLK	Power Ground	<0.1v
A11, 24	YEL/BLK	Main Relay Power (B+)	12-14v
A12	BLK/BLU	Idle Air Control Valve Signal	Pulse Signals
A13	GRN/WHT	Engine Mount Solenoid	Solenoid Off: 12v, On: 1v
A15	RED/YEL	EVAP Purge Solenoid	Solenoid Off: 12v, On: 1v
A16	GRN/BLK	Fuel Pump Relay	Relay Off: 12v, On: 1v
A17	RED/BLU	A/C Clutch Relay	Relay Off: 12v, On: 1v
A18	GRN/RED	Check Engine Light	MIL Off: 12v, On: 1v
A19	WHT/GRN	Alternator Charging Signal	Lights Off: 12v, On: 0v
A20	YEL/GRN	Igniter Control	Digital Signals: 0-12-0v
A25	ORN/GRN	Intake Air Resonator Control	Solenoid Off: 12v, On: 1v
A27	GRN	Radiator Fan Relay	Relay Off: 12v, On: 1v
A28	GRN/WHT	EVAP Bypass Solenoid	Solenoid Off: 12v, On: 1v
A29	ORN/GRN	EVAP Vent Solenoid	Solenoid Off: 12v, On: 1v

1996-97 Sedan 2.2L I4 MFI VIN CD5 [All] 25P 'B' Connector

PCM Pin #	W/Color	Circuit Description (25-Pin)	Value at Hot Idle
B3	BLU/YEL	Shift Solenoid 'A'	In D2-3: 12v, D1-4: 0v
B4	GRN/BLK	Lockup Solenoid 'B'	LSB On: 12v, Off: 0v
B5	YEL	Lockup Solenoid 'A'	LSA On: 12v, Off: 0v
B8	GRN/BLU	A/T: Gear Position Switch	In D3: 0v, Others: 12v
B11	BLU/YEL	Shift Solenoid 'B'	D2/D3: 12v, D1/D4: 0v
B12	WHT/GRN	Interlock Control Unit Signal	Key & Brake On: 12v
B13	BLU/RED	A/T: D4 Indicator Light Switch	D4 On: 0v, Off: 12v
B14	WHT/BLU	Mainshaft Speed Sensor 'N'	<0.050v
B15	ORN/BLU	Mainshaft Speed Sensor 'P'	Moving: AC pulses
B16	GRN/RED	A/T: Gear Position Switch	In 'R': 0v, Others: 12v
B17	GRN/YEL	A/T: Gear Position Switch	In D2: 0v, Others: 12v
B18	GRN/WHT	A/T: Gear Position Switch	In D1: 0v, Others: 12v
B22	BLU/YEL	Countershaft Speed Sensor P	Moving: AC pulses
B23	BLU/GRN	Countershaft Speed Sensor N	<0.050v
B24	PNK/GRN	A/T: Gear Position Switch	In D4: 0v, Others: 12v
B25	LT GRN	A/T: Gear Position Switch	In P/N: 0v, Others: 12v

| 1 | 2 | 3 | 4 | | 5 | 6 | 7 | | 8 | 9 | 10 | | 11 | | | 1 | | 2 | | | 3 | 4 | 5 | | 6 | 7 | 8 |

WIRE SIDE OF HARNESS TERMINALS

32-PIN 'A'   25-PIN 'B'

05533_ADIA_G602

Pin Connector Graphic
1996-97 Sedan 2.2L I4 MFI VIN CD5 [All] 31P 'C' Connector

PCM Pin #	W/Color	Circuit Description (31-Pin)	Value at Hot Idle
C1	---	Not Used	---
C2	BLU	CYP Sensor Signal	AC pulse signals
C3	GRN	TDC Sensor Signal	AC pulse signals
C4	YEL	CKP Sensor Signal	AC pulse signals
C5	RED/WHT	A/C Switch Signal	Switch Off: 12v, On: 1v
C6	BLU/RED	Start Signal	Cranking: 9-11v
C7	RED	Service Check Connector	SCS Open: 4.80v
C8	LT GRN	K-Line Signal	12v
C10	WHT/YEL	Keep Alive Power (VBU)	12-14v
C12	WHT	CKP Sensor Ground	<0.050v
C13	RED	TDC Sensor Ground	<0.050v
C14	BLK	CYP Sensor Ground	<0.050v
C16	GRN	PSP Switch	Straight: 0v, Turning: 11v
C17	WHT/RED	Alternator 'FR' Signal	Digital Signals: 0-5-0v
C18	ORN	Vehicle Speed Sensor	Moving: pulse signals
C20	BRN	EVAP Purge Flow Switch	Switch on: 0v, off: 5v

1996-97 Sedan 2.2L I4 MFI VIN CD5 [All] 16P 'D' Connector

PCM Pin #	W/Color	Circuit Description (16-Pin)	Value at Hot Idle
D1	RED/BLK	TP Sensor Signal	0.5-0.6v
D2	RED/WHT	ECT Sensor Signal	At 180°F: 0.52v
D3	WHT/YEL	MAP Sensor Signal	0.8-0.9v
D4	YEL/WHT	MAP Sensor VREF	4.9-5.1v
D5	GRN/WHT	Brake Switch Signal	Brake Off: 0v, On: 12v
D7	WHT/RED	HO2S-11 (B1 S1) Signal	0.1-1.1v
D8	RED/YEL	IAT Sensor Signal	Varies w/temp. (0.5-4.9v)
D9	WHT/BLK	EGR Valve Lift Sensor	1.2v
D10	YEL/BLU	Sensor VREF	4.9-5.1v
D11	GRN/BLU	Sensor Ground	<0.050v
D12	GRN/WHT	MAP Sensor Ground	<0.050v
D13	RED/WHT	HO2S-12 Ground	<0.050v
D14	WHT/RED	HO2S-12 (B1 S2) Signal	0.1-1.1v
D15	GRN/RED	Fuel Tank Pressure Sensor	Fuel Cap off: 2.5v
D16	GRN/RED	Electric Load Detector	Varies: 0.5-4.5v

31-PIN CONNECTOR 'C'   16-PIN CONNECTOR 'D'

1	2	3	4		5	6	7		8	9	10		1		2	3	4		5
11	12	13	14	15	16	17	18	19	20	21	22		6	7	8	9	10	11	12
	23	24	25		26	27	28		29	30	31		13	14	15		16	

WIRE SIDE OF HARNESS TERMINALS

05533_ADIA_G603

Pin Connector Graphic

1998-99 Sedan 2.3L I4 MFI VIN CF8 [All] 32P 'A' Connector

PCM Pin #	W/Color	Circuit Description (32-Pin)	Value at Hot Idle
A2	GRN/WHT	A/T: Engine Mount Solenoid	Solenoid Off: 12v, On: 1v
A3	BLU	EVAP Bypass Solenoid	Solenoid Off: 12v, On: 1v
A4	GRN/WHT	EVAP Vent Solenoid	Solenoid Off: 12v, On: 1v
A5	BLU/GRN	Cruise Control Signal	C/C On: pulse signals
A6	RED/YEL	EVAP Purge Solenoid	Solenoid Off: 12v, On: 1v
A8	BLK/WHT	HO2S-12 (B1 S2) Heater	Digital Signals: 0-12-0v
A9	BLU/WHT	A/T: VSS Output Signal	Moving: pulse signals
A10	BRN	Service Check Connector	SCS Open: 4.80v
A12	PNK	Immobilizer Indicator Lamp	Lamp Off: 12v, On: 1v
A13, 25	BLU, RED	Immobilizer Enable, Code	Digital Signals
A14	GRN/BLK	A/T: D4 Light Switch	D4 On: 0v, Off: 12v
A15	GRN/YEL	Immobilizer Fuel Pump Relay	Relay Off: 12v, On: 1v
A17	RED	A/C Clutch Relay Control	Relay Off: 12v, On: 1v
A18	GRN/ORN	Malfunction Indicator Light	MIL Off: 12v, On: 1v
A19	BLU	Engine Speed Pulse (NEP)	Pulse Signals
A20	GRN	Radiator Fan Relay	Relay Off: 12v, On: 1v
A21	GRY	K-Line Signal	12v
A23	WHT/RED	HO2S-12 (B1 S2) Signal	0.1-1.1v
A24	BLU/ORN	Starter Switch Signal	Cranking: 9-11v
A26	GRN	PSP Switch	Straight: 0v, Turning: 11v
A27	BLU/RED	A/C Switch Signal	Switch Off: 12v, On: 1v
A28	WHT/RED	A/T: Interlock Control Unit	Key & Brake On: 12v
A29	LT GRN	Fuel Tank Pressure Sensor	Fuel Cap off: 2.5v
A30	GRN/RED	Electric Load Detector	Varies: 0.5-3.5v
A32	WHT/BLK	Brake Switch Signal	Brake Off: 0v, On: 12v

1998-99 Sedan 2.3L I4 MFI VIN CF8 [All] 16P 'D' Connector

PCM Pin #	W/Color	Circuit Description (16-Pin)	Value at Hot Idle
D1	YEL	A/T: Lockup Control Solenoid	LCS Off: 0v, On: 12v
D2	GRN/WHT	A/T: Shift Solenoid 'B'	SSB Off: 0v, On: 12v
D3	GRN	A/T: Shift Solenoid 'C'	SSC Off: 12v, On: 1v
D5	BLK/YEL	A/T: VB Solenoid Feed (B+)	12-14v
D6	WHT	A/T: Gear Position Switch	In 'R': 0v, Others: 12v
D7	BLU/YEL	Shift Solenoid 'A'	SSA Off: 0v, On: 12v
D8	PNK	A/T: Gear Position Switch	In D3: 0v, Others: 12v
D9	YEL	A/T: Gear Position Switch	In D4: 0v, Others: 12v
D10, D16	BLU, GRN	A/T: Countershaft Speed P, N	Moving: AC pulses
D11	RED	Mainshaft Speed Sensor 'P'	Moving: AC pulses
D12	WHT	Mainshaft Speed Sensor 'N'	<0.050v
D13	BLU/WHT	A/T: Gear Position Switch	In P/N: 0v, Others: 12v
D14	BLU	A/T: Gear Position Switch	In D2: 0v, Others: 12v
D15	BRN	A/T: Gear Position Switch	In D1: 0v, Others: 12v

Pin Connector Graphic

05533_ADIA_G599

1998-99 Sedan 2.3L I4 MFI VIN CF8 [All] 25P 'B' Connector

PCM Pin #	W/Color	Circuit Description (25-Pin)	Value at Hot Idle
B1, B9	YEL/BLK	Main Relay Power (B+)	12-14v
B2, B10	BLK	Power Ground	<0.1v
B3	RED	Injector 2 Control	2.0-3.3 ms
B4	BLU	Injector 3 Control	2.0-3.3 ms
B5	YEL	Injector 4 Control	2.0-3.3 ms
B7	PNK	E-EGR Solenoid Control	Digital Signals: 0-12-0v
B8	WHT	A/T: Clutch Solenoid 'A-'	LSA On: pulse signals
B11	BRN	Injector 1 Control	2.0-3.3 ms
B13	YEL/GRN	Ignition Control Signal	Digital Signals: 0-12-0v
B14	BLU/BLK	A/T: 2nd Clutch Pressure Sw.	Open: 12v, Closed: 1v
B16, B19	---	Not Used	---
B17	RED	A/T: Clutch Solenoid 'A+'	LSA On: pulse signals
B18	GRN	A/T: Clutch Solenoid 'B-'	Pulse Signals
B20, 22	BRN/BLK	Logic Ground (LG1, LG2)	<0.050v
B21	WHT/YEL	Keep Alive Power (VBU)	12-14v
B23	BLK/BLU	Idle Air Control Valve	Pulse Signals
B24	BLU/BLK	A/T: 3rd Clutch Pressure Sw.	Open: 12v, Closed: 1v
B25	ORN	A/T: Clutch Solenoid 'B+'	Pulse Signals

1998-99 Sedan 2.3L I4 MFI VIN CF8 [All] 31P 'C' Connector

PCM Pin #	W/Color	Circuit Description (31-Pin)	Value at Hot Idle
C1	BLK/WHT	HO2S-11 (B1 S1) Heater	Digital Signals: 0-12-0v
C2	WHT/GRN	Alternator Charging Signal	Lights Off: 12v, On: 0v
C3	RED/BLU	Knock Sensor Signal	No Detonation: 18mv AC
C5	WHT/RED	Alternator 'FR' Signal	Digital Signals: 0-5-0v
C6	WHT/BLK	EGR Valve Lift Sensor	1.2v
C7	GRN/WHT	MAP Sensor Ground (SG1)	<0.050v
C8	BLU	CKP Sensor Signal	AC pulse signals
C9	WHT	CKP Sensor Ground	<0.050v
C16	WHT	HO2S-11 (B1 S1) Signal	0.1-1.1v
C17	RED/GRN	MAP Sensor Signal	0.8-0.9v
C18	GRN/BLK	Sensor Ground (SG2)	<0.050v
C19	YEL/RED	MAP Sensor VREF (VCC1)	4.9-5.1v
C20	GRN	TDC Sensor Signal	AC pulse signals
C21	RED	TDC Sensor Ground	<0.050v
C23	BLU/WHT	M/T: VSS Input Signal	Moving: 0-5-0v
C25	RED/YEL	IAT Sensor Signal	Varies w/temp. (0.5-4.9v)
C26	RED/WHT	ECT Sensor Signal	At 180°F: 0.5-0.6v
C27	RED/BLK	TP Sensor Signal	0.5-0.6v
C28	YEL/BLU	Sensor VREF (VCC2)	4.9-5.1v
C29	YEL	CYP Sensor Signal	AC pulse signals
C30	BLK	CYP Sensor Ground	<0.050v

05533_ADIA_G600

Pin Connector Graphic

2000-03 Sedan 2.3L I4 MFI VIN CF8 [All] 32P 'A' Connector

PCM Pin #	W/Color	Circuit Description (32-Pin)	Value at Hot Idle
A2	GRN/WHT	A/T: Engine Mount Solenoid	Solenoid Off: 12v, On: 1v
A3	BLU	EVAP Bypass Solenoid	Solenoid Off: 12v, On: 1v
A4	GRN/WHT	EVAP Vent Solenoid	Solenoid Off: 12v, On: 1v
A5	BLU/GRN	Cruise Control Signal	C/C On: pulse signals
A6	RED/YEL	EVAP Purge Solenoid	Solenoid Off: 12v, On: 1v
A8	BLK/WHT	HO2S-12 (B1 S2) Heater	Digital Signals: 0-12-0v
A9	BLU/WHT	A/T: VSS Output Signal	Moving: pulse signals
A10	BRN	Service Check Connector	SCS Open: 4.80v
A12	PNK	Immobilizer Indicator Lamp	Lamp Off: 12v, On: 1v
A13, 25	BLU, RED	Immobilizer Enable, Code	Digital Signals
A14	GRN/BLK	A/T: D4 Light Switch	D4 On: 0v, Off: 12v
A15	GRN/YEL	Immobilizer Fuel Pump Relay	Relay Off: 12v, On: 1v
A17	RED	A/C Clutch Relay Control	Relay Off: 12v, On: 1v
A18	GRN/ORN	Malfunction Indicator Light	MIL Off: 12v, On: 1v
A19	BLU	Engine Speed Pulse (NEP)	Pulse Signals
A20	GRN	Radiator Fan Relay	Relay Off: 12v, On: 1v
A21	GRY	K-Line Signal	12v
A23	WHT/RED	HO2S-12 (B1 S2) Signal	0.1-1.1v
A24	BLU/ORN	Starter Switch Signal	Cranking: 9-11v
A26	GRN	PSP Switch	Straight: 0v, Turning: 11v
A27	BLU/RED	A/C Switch Signal	Switch Off: 12v, On: 1v
A28	WHT/RED	A/T: Interlock Control Unit	Key & Brake On: 12v
A29	LT GRN	Fuel Tank Pressure Sensor	Fuel Cap off: 2.5v
A30	GRN/RED	Electric Load Detector	Varies: 0.5-3.5v
A32	WHT/BLK	Brake Switch Signal	Brake Off: 0v, On: 12v

2000-03 Sedan 2.3L I4 MFI VIN CF8 [All] 16P 'D' Connector

PCM Pin #	W/Color	Circuit Description (16-Pin)	Value at Hot Idle
D1	YEL	A/T: Lockup Control Solenoid	LCS Off: 0v, On: 12v
D2	GRN/WHT	A/T: Shift Solenoid 'B'	SSB Off: 0v, On: 12v
D3	GRN	A/T: Shift Solenoid 'C'	SSC Off: 12v, On: 1v
D5	BLK/YEL	A/T: VB Solenoid Feed (B+)	12-14v
D6	WHT	A/T: Gear Position Switch	In 'R': 0v, Others: 12v
D7	BLU/YEL	Shift Solenoid 'A'	SSA Off: 0v, On: 12v
D8	PNK	A/T: Gear Position Switch	In D3: 0v, Others: 12v
D9	YEL	A/T: Gear Position Switch	In D4: 0v, Others: 12v
D10, D16	BLU, GRN	A/T: Countershaft Speed P, N	Moving: AC pulses
D11	RED	Mainshaft Speed Sensor 'P'	Moving: AC pulses
D12	WHT	Mainshaft Speed Sensor 'N'	<0.050v
D13	BLU/WHT	A/T: Gear Position Switch	In P/N: 0v, Others: 12v
D14	BLU	A/T: Gear Position Switch	In D2: 0v, Others: 12v
D15	BRN	A/T: Gear Position Switch	In D1: 0v, Others: 12v

```
 1  2  3  4     5  6  7     8  9  10   11    1    2  3  4     5
12 13 14 15 16 17 18 19 20 21 22  23   24    6  7  8  9 10 11  12
32-PIN 25 26 27   28 29 30 31    32       13 14 15   16  16-PIN
  'A'                                              'D'
        WIRE SIDE OF HARNESS TERMINALS
```

05533_ADIA_G599

Pin Connector Graphic
2000-03 Sedan 2.3L I4 MFI VIN CF8 [All] 25P 'B' Connector

PCM Pin #	W/Color	Circuit Description (25-Pin)	Value at Hot Idle
B1, B9	YEL/BLK	Main Relay Power (B+)	12-14v
B2, B10	BLK	Power Ground	<0.1v
B3	RED	Injector 2 Control	2.0-3.3 ms
B4	BLU	Injector 3 Control	2.0-3.3 ms
B5	YEL	Injector 4 Control	2.0-3.3 ms
B7	PNK	E-EGR Solenoid Control	Digital Signals: 0-12-0v
B8	WHT	A/T: Clutch Solenoid 'A-'	LSA On: pulse signals
B11	BRN	Injector 1 Control	2.0-3.3 ms
B13	YEL/GRN	Ignition Control Signal	Digital Signals: 0-12-0v
B14	BLU/BLK	A/T: 2nd Clutch Pressure Sw.	Open: 12v, Closed: 1v
B16, B19	---	Not Used	---
B17	RED	A/T: Clutch Solenoid 'A+'	LSA On: pulse signals
B18, B25	GRN, ORN	A/T: Clutch Solenoid 'B-', 'B+'	Pulse Signals
B20, 22	BRN/BLK	Logic Ground	<0.050v
B21	WHT/YEL	Keep Alive Power (VBU)	12-14v
B23	BLK/BLU	Idle Air Control Valve	Pulse Signals
B24	BLU/BLK	A/T: 3rd Clutch Pressure Sw.	Open: 12v, Closed: 1v

2000-03 Sedan 2.3L I4 MFI VIN CF8 [All] 31P 'C' Connector

PCM Pin #	W/Color	Circuit Description (31-Pin)	Value at Hot Idle
C1	BLK/WHT	HO2S-11 (B1 S1) Heater	Digital Signals: 0-12-0v
C2	WHT/GRN	Alternator Charging Signal	Lights Off: 12v, On: 0v
C3	RED/BLU	Knock Sensor Signal	No Detonation: 18mv AC
C5	WHT/RED	Alternator 'FR' Signal	Digital Signals: 0-5-0v
C6	WHT/BLK	EGR Valve Lift Sensor	1.2v
C7	GRN/WHT	MAP Sensor Ground (SG1)	<0.050v
C8	BLU	CKP Sensor Signal	AC pulse signals
C9	WHT	CKP Sensor Ground	<0.050v
C16	WHT	HO2S-11 (B1 S1) Signal	0.1-1.1v
C17	RED/GRN	MAP Sensor Signal	0.8-0.9v
C18	GRN/BLK	Sensor Ground (SG2)	<0.050v
C19	YEL/RED	MAP Sensor VREF (VCC1)	4.9-5.1v
C20	GRN	TDC Sensor Signal	AC pulse signals
C21	RED	TDC Sensor Ground	<0.050v
C23	BLU/WHT	M/T: VSS Input Signal	Moving: 0-5-0v

2000-03 Sedan 2.3L I4 MFI VIN CF8 [All] 31P 'C' Connector, *continued*

C24	YEL/GRN	ECT Signal to TCM	Digital Signals
C25	RED/YEL	IAT Sensor Signal	Varies w/temp. (0.5-4.9v)
C26	RED/WHT	ECT Sensor Signal	At 180°F: 0.5-0.6v
C27	RED/BLK	TP Sensor Signal	0.5-0.6v
C28	YEL/BLU	Sensor VREF (VCC2)	4.9-5.1v
C29	YEL	CYP Sensor Signal	AC pulse signals
C30	BLK	CYP Sensor Ground	<0.050v

05533_ADIA_G600

Pin Connector Graphic

1998-99 Sedan 2.3L I4 VTEC VIN CG5, CG6 [All] 32P A Connector

PCM Pin #	W/Color	Circuit Description (32-Pin)	Value at Hot Idle
A2	GRN/WHT	A/T: Engine Mount Solenoid	Solenoid Off: 12v, On: 1v
A3	BLU	EVAP Bypass Solenoid	Solenoid Off: 12v, On: 1v
A4	GRN/WHT	EVAP Vent Solenoid	Solenoid Off: 12v, On: 1v
A5	BLU/GRN	Cruise Control Signal	C/C On: pulse signals
A6	RED/YEL	EVAP Purge Solenoid	Solenoid Off: 12v, On: 1v
A8	BLK/WHT	HO2S-12 (B1 S2) Heater	Digital Signals: 0-12-0v
A9	BLU/WHT	A/T: Vehicle Speed Output	Moving: pulse signals
A10	BRN	Service Check Connector	SCS Open: 4.80v
A12	PNK	Immobilizer Indicator Lamp	Lamp Off: 12v, On: 1v
A13, 25	BLU, RED	Immobilizer Enable, Code	Digital Signals
A14	GRN/BLK	A/T: D4 Light Switch	D4 On: 0v, Off: 12v
A15	GRN/YEL	Immobilizer Fuel Pump Relay	Relay Off: 12v, On: 1v
A17	RED	A/C Clutch Relay Control	Relay Off: 12v, On: 1v
A18	GRN/ORN	Malfunction Indicator Light	MIL Off: 12v, On: 1v
A19	BLU	Engine Speed Pulse (NEP)	Digital Signals
A20	GRN	Radiator Fan Relay Control	Relay Off: 12v, On: 1v
A21	GRY	K-Line Signal	12v
A23	WHT/RED	HO2S-12 (B1 S2) Signal	0.1-1.1v
A24	BLU/ORN	Starter Switch Signal	Cranking: 9-11v
A26	GRN	PSP Switch Signal	Straight: 0v, Turning: 11v
A27	BLU/RED	A/C Switch Signal	Switch Off: 12v, On: 1v
A28	WHT/RED	Interlock Control Unit Signal	Key & Brake On: 12v
A29	LT GRN	Fuel Tank Pressure Sensor	Fuel Cap off: 2.5v
A30	GRN/RED	Electric Load Detector	Varies: 0.5-4.5v
A32	WHT/BLK	Brake Switch Signal	Brake Off: 0v, On: 12v

1998-99 Sedan 2.3L I4 VTEC VIN CG5, CG6 [All] 16P D Connector

PCM Pin #	W/Color	Circuit Description (16-Pin)	Value at Hot Idle
D1	YEL	A/T: Lockup Control Solenoid	LUS On: 12v, Off: 0v
D2	GRN/WHT	A/T: Shift Solenoid 'B'	SSB Off: 0v, On: 12v
D3	GRN	A/T: Shift Solenoid 'C'	SSC on: 12v, off: 0v
D5	BLK/YEL	A/T: Solenoid Feed (B+)	12-14v
D6	WHT	A/T: Gear Position Switch	In 'R': 0v, Others: 12v
D7	BLU/YEL	A/T: Shift Solenoid 'A'	SSA Off: 0v, On: 12v
D8	PNK	A/T: Gear Position Switch	In D3: 0v, Others: 12v
D9	YEL	A/T: Gear Position Switch	In D4: 0v, Others: 12v
Q	BLU, GRN	A/T: Countershaft Speed P, N	Moving: AC pulses
D11	RED	Mainshaft Speed Sensor 'P'	AC pulse signals
D12	WHT	Mainshaft Speed Sensor 'N'	<0.050v
D13	BLU/WHT	A/T: Gear Position Switch	In P/N: 0v, Others: 12v
D14	BLU	A/T: Gear Position Switch	In D2: 0v, Others: 12v
D15	BRN	A/T: Gear Position Switch	In D1: 0v, Others: 12v

05533_ADIA_G599

Pin Connector Graphic

1998-99 Sedan 2.3L I4 VTEC VIN CG5, CG6 [All] 25P B Connector

PCM Pin #	W/Color	Circuit Description (25-Pin)	Value at Hot Idle
B1, B9	YEL/BLK	Main Relay Power (B+)	12-14v
B2, B10	BLK	Power Ground	<0.1v
B3, B4	RED, BLU	Injector 2, Injector 3 Control	2.0-3.3 ms
B5, B11	YEL, BRN	Injector 4, Injector 1 Control	2.0-3.3 ms
B7	PNK	E-EGR Control Solenoid	Digital Signals: 0-12-0v
B8, B17	WHT, RED	A/T: Clutch Solenoid 'A-', 'A+'	Pulse Signals
B12	GRN/YEL	VTEC Control Solenoid	0v, Hi-Speed: 12v
B13	YEL/GRN	Ignition Control Signal	Digital Signals: 0-12-0v
B14	BLU/BLK	A/T: 2nd Clutch Pressure Sw.	Open: 12v, Closed: 1v
B16	GRN/RED	HO2S-11 Heater Relay	Relay Off: 12v, On: 1v
B18, B25	GRN, ORN	A/T: Clutch Solenoid 'B-', 'B+'	Pulse Signals
B19	BLK/WHT	HO2S-11 (B1 S1) Heater	Digital Signals: 0-12-0v
B20, B22	BRN/BLK	Logic Ground	<0.050v
B21	WHT/YEL	Keep Alive Power (VBU)	12-14v
B23	BLK/BLU	Idle Air Control Valve	Pulse Signals
B24	BLU/WHT	A/T: 3rd Clutch Pressure Sw.	Open: 12v, Closed: 1v

1998-99 Sedan 2.3L I4 VTEC VIN CG5, CG6 [All] 31P C Connector

PCM Pin #	W/Color	Circuit Description (31-Pin)	Value at Hot Idle
C2	WHT/GRN	Alternator Charging Signal	Lights Off: 12v, On: 0v
C3	RED/BLU	Knock Sensor Signal	No Detonation: 18mv AC
C5	WHT/RED	Alternator 'FR' Signal	Digital Signals: 0-5-0v
C6	WHT/BLK	EGR Valve Lift Sensor	1.2v
C7	GRN	MAP Sensor Ground	<0.050v
C8	BLU	CKP Sensor Signal	AC pulse signals
C9	WHT	CKP Sensor Ground	<0.050v
C10	BLU/BLK	VTEC Pressure Switch	0v, Hi-Speed: 12v
C13	WHT	HO2S-11 Heater Relay	Relay Off: 12v, On: 1v
C14	RED	HO2S-11 (B1 S1) Signal	0.1-1.1v
C15	BLU	HO2S-11 Ground	<0.050v
C16	WHT	HO2S-11 (B1 S1) Signal	0.1-1.1v
C17	RED/GRN	MAP Sensor Signal	0.8-0.9v
C18	GRN	Sensor Ground	<0.050v
C19, C28	YEL, YEL	MAP VREF, Sensor VREF	4.9-5.1v
C20	GRN	TDC Sensor Signal	AC pulse signals
C21	RED	TDC Sensor Ground	<0.050v
C23	BLU/WHT	M/T: VSS Input Signal	Moving: 0-5-0v
C24	YEL/GRN	ECT Sensor Signal to TCM	Digital Signals
C25	RED/YEL	IAT Sensor Signal	Varies w/temp (0.5-4.9v)
C26	RED/WHT	ECT Sensor Signal	At 180°F: 0.5-0.6v
C27	RED/BLK	TP Sensor Signal	0.5-0.6v
C29, C30	YEL, BLK	CYP Sensor Signal, Ground	AC pulse signals

Pin Connector Graphic

05533_ADIA_G600

2000-03 Sedan 2.3L I4 VTEC VIN CG5, CG6 [All] 32P A Connector

PCM Pin #	W/Color	Circuit Description (32-Pin)	Value at Hot Idle
A2	GRN/WHT	A/T: Engine Mount Solenoid	Solenoid Off: 12v, On: 1v
A3	BLU	EVAP Bypass Solenoid	Solenoid Off: 12v, On: 1v
A4	GRN/WHT	EVAP Vent Solenoid	Solenoid Off: 12v, On: 1v
A5	BLU/GRN	Cruise Control Signal	C/C On: pulse signals
A6	RED/YEL	EVAP Purge Solenoid	Solenoid Off: 12v, On: 1v
A8	BLK/WHT	HO2S-12 (B1 S2) Heater	Digital Signals: 0-12-0v
A9	BLU/WHT	A/T: Vehicle Speed Output	Moving: pulse signals
A10	BRN	Service Check Connector	SCS Open: 4.80v
A12	PNK	Immobilizer Indicator Lamp	Lamp Off: 12v, On: 1v
A13, 25	BLU, RED	Immobilizer Enable, Code	Digital Signals
A14	GRN/BLK	A/T: D4 Light Switch	D4 On: 0v, Off: 12v
A15	GRN/YEL	Immobilizer Fuel Pump Relay	Relay Off: 12v, On: 1v
A17	RED	A/C Clutch Relay Control	Relay Off: 12v, On: 1v
A18	GRN/ORN	Malfunction Indicator Light	MIL Off: 12v, On: 1v
A19	BLU	Engine Speed Pulse (NEP)	Digital Signals
A20	GRN	Radiator Fan Relay Control	Relay Off: 12v, On: 1v
A21	GRY	K-Line Signal	12v
A23	WHT/RED	HO2S-12 (B1 S2) Signal	0.1-1.1v
A24	BLU/ORN	Starter Switch Signal	Cranking: 9-11v
A26	GRN	PSP Switch Signal	Straight: 0v, Turning: 11v
A27	BLU/RED	A/C Switch Signal	Switch Off: 12v, On: 1v
A28	WHT/RED	Interlock Control Unit Signal	Key & Brake On: 12v
A29	LT GRN	Fuel Tank Pressure Sensor	Fuel Cap off: 2.5v
A30	GRN/RED	Electric Load Detector	Varies: 0.5-4.5v
A32	WHT/BLK	Brake Switch Signal	Brake Off: 0v, On: 12v

2000-03 Sedan 2.3L I4 VTEC VIN CG5, CG6 [All] 16P D Connector

PCM Pin #	W/Color	Circuit Description (16-Pin)	Value at Hot Idle
D1	YEL	A/T: Lockup Control Solenoid	LUS On: 12v, Off: 0v
D2	GRN/WHT	A/T: Shift Solenoid 'B'	SSB Off: 0v, On: 12v
D3	GRN	A/T: Shift Solenoid 'C'	SSC on: 12v, off: 0v
D5	BLK/YEL	A/T: Solenoid Feed (B+)	12-14v
D6	WHT	A/T: Gear Position Switch	In 'R': 0v, Others: 12v
D7	BLU/YEL	A/T: Shift Solenoid 'A'	SSA Off: 0v, On: 12v
D8	PNK	A/T: Gear Position Switch	In D3: 0v, Others: 12v
D9	YEL	A/T: Gear Position Switch	In D4: 0v, Others: 12v
D10, D16	BLU, GRN	A/T: Countershaft Speed P, N	Moving: AC pulses
D11	RED	Mainshaft Speed Sensor 'P'	AC pulse signals
D12	WHT	Mainshaft Speed Sensor 'N'	<0.050v
D13	BLU/WHT	A/T: Gear Position Switch	In P/N: 0v, Others: 12v
D14	BLU	A/T: Gear Position Switch	In D2: 0v, Others: 12v
D15	BRN	A/T: Gear Position Switch	In D1: 0v, Others: 12v

WIRE SIDE OF HARNESS TERMINALS

05533_ADIA_G599

Pin Connector Graphic

2000-03 Sedan 2.3L I4 VTEC VIN CG5, CG6 [All] 25P B Connector

PCM Pin #	W/Color	Circuit Description (25-Pin)	Value at Hot Idle
B1, B9	YEL/BLK	Main Relay Power (B+)	12-14v
B2, B10	BLK	Power Ground	<0.1v
B3, B4	RED, BLU	Injector 2, Injector 3 Control	2.0-3.3 ms
B5, B11	YEL, BRN	Injector 4, Injector 1 Control	2.0-3.3 ms
B7	PNK	E-EGR Control Solenoid	Digital Signals: 0-12-0v
B8, B17	WHT, RED	A/T: Clutch Solenoid 'A-', 'A+'	Pulse Signals
B12	GRN/YEL	VTEC Control Solenoid	0v, Hi-Speed: 12v
B13	YEL/GRN	Ignition Control Signal	Digital Signals: 0-12-0v
B14	BLU/BLK	A/T: 2nd Clutch Pressure Sw.	Open: 12v, Closed: 1v
B16	GRN/RED	HO2S-11 Heater Relay	Relay Off: 12v, On: 1v
B18, B25	GRN, ORN	A/T: Clutch Solenoid 'B-', 'B+'	Pulse Signals
B20, B22	BRN/BLK	Logic Ground	<0.050v
B21	WHT/YEL	Keep Alive Power (VBU)	12-14v
B23	BLK/BLU	Idle Air Control Valve	Pulse Signals
B24	BLU/WHT	A/T: 3rd Clutch Pressure Sw.	Open: 12v, Closed: 1v

2000-03 Sedan 2.3L I4 VTEC VIN CG5, CG6 [All] 31P C Connector

PCM Pin #	W/Color	Circuit Description (31-Pin)	Value at Hot Idle
C1	BLK/WHT	HO2S-11 (B1 S1) Heater	Digital Signals: 0-12-0v
C2	WHT/GRN	Alternator Charging Signal	Lights Off: 12v, On: 0v
C3	RED/BLU	Knock Sensor Signal	No Detonation: 18mv AC
C5	WHT/RED	Alternator 'FR' Signal	Digital Signals: 0-5-0v
C6	WHT/BLK	EGR Valve Lift Sensor	1.2v
C7	GRN	MAP Sensor Ground	<0.050v
C8	BLU	CKP Sensor Signal	AC pulse signals
C9	WHT	CKP Sensor Ground	<0.050v
C10	BLU/BLK	VTEC Pressure Switch	0v, Hi-Speed: 12v
C13	WHT	HO2S-11 Heater Relay	Relay Off: 12v, On: 1v
C14	RED	HO2S-11 (B1 S1) Signal	0.1-1.1v
C15	BLU	HO2S-11 Ground	<0.050v
C16	WHT	HO2S-11 (B1 S1) Signal	0.1-1.1v
C17	RED/GRN	MAP Sensor Signal	0.8-0.9v
C18	GRN	Sensor Ground	<0.050v
C19, C28	YEL, YEL	MAP VREF, Sensor VREF	4.9-5.1v
C20	GRN	TDC Sensor Signal	AC pulse signals
C21	RED	TDC Sensor Ground	<0.050v
C23	BLU/WHT	M/T: VSS Input Signal	Moving: 0-5-0v
C24	YEL/GRN	ECT Sensor Signal to TCM	Digital Signals
C25	RED/YEL	IAT Sensor Signal	Varies w/temp (0.5-4.9v)
C26	RED/WHT	ECT Sensor Signal	At 180°F: 0.5-0.6v
C27	RED/BLK	TP Sensor Signal	0.5-0.6v
C29, C30	YEL, BLK	CYP Sensor Signal, Ground	AC pulse signals

05533_ADIA_G600

Pin Connector Graphic

1995 Sedan 2.7L V6 MFI VIN CE6 [A/T] 26P 'A' Connector

PCM Pin #	W/Color	Circuit Description (26-Pin)	Value at Hot Idle
A1	BRN	Injector 1 Control	2.0-3.3 ms
A2	RED	Injector 2 Control	2.0-3.3 ms
A3	BLU	Injector 3 Control	2.0-3.3 ms
A4	GRN/BLK	Fuel Pump Relay	Relay Off: 12v, On: 1v
A5	BLK/BLU	Idle Air Control Valve	Pulse Signals
A6	ORN/BLK	Right HO2S Heater Control	Digital Signals: 0-12-0v
A7	GRN/RED	Check Engine Light	C/E on: <1v, of: 12v
A8	RED/BLU	A/C Clutch Relay	Relay Off: 12v, On: 1v
A9	RED	EGR Control Solenoid	Solenoid Off: 12v, On: 1v
A10	PNK/BLU	Intake Air Bypass Solenoid	Solenoid Off: 12v, On: 1v
A11	YEL/GRN	Igniter Control	Digital Signals: 0-12-0v

1995 Sedan 2.7L V6 MFI VIN CE6 [A/T] 26P 'A' Connector, *continued*

A12	BLK	Power Ground	<0.1v
A13	YEL/BLK	Ignition Power	12-14v
A14	YEL	Injector 4 Control	2.0-3.3 ms
A15	BLK/RED	Injector 5 Control	2.0-3.3 ms
A16	WHT/BLU	Injector 6 Control	2.0-3.3 ms
A17	ORN/BLU	Left HO2S Heater Control	Digital Signals: 0-12-0v
A18	GRN/WHT	Engine Mount Solenoid	Solenoid Off: 12v, On: 1v
A19	GRN	Radiator Fan Relay Control	On: <1v, Off: 12v
A21	WHT/GRN	Alternator Charging Signal	Lights on: 12v, off: 0v
A22	BRN/WHT	A/T: TCM Control Unit	N/A
A23	RED/YEL	EVAP Purge Solenoid	Solenoid Off: 12v, On: 1v
A24	BRN	EGR Solenoid	Solenoid Off: 12v, On: 1v
A25	BLK	Power Ground	<0.050v
A26	BRN/BLK	Chassis Ground	<0.050v

1995 Sedan 2.7L V6 MFI VIN CE6 [All] 16P 'B' Connector

PCM Pin #	W/Color	Circuit Description (16-Pin)	Value at Hot Idle
B1	YEL/BLK	Ignition Power	12v
B2	WHT/RED	A/T: TCM Control Unit	N/A
B3	RED/WHT	A/C Switch Signal	Switch Off: 12v, On: 1v
B4	LT GRN	A/T: Park Neutral Position	In P/N: 0v, Others: 11v
B5	BLK/GRN	Starter Switch Signal	Cranking: 9-11v
B6	YEL	CYP Sensor Signal	AC pulse signals
B7	GRN	TDC Sensor Signal	AC pulse signals
B8	BLU	CKP Sensor Signal	AC pulse signals
B9	BRN/BLK	Chassis Ground	<0.050v
B10	GRN	A/T: TCM Control Unit	N/A
B11	---	Not Used	---
B12	GRN	PSP Switch	Straight: 0v, Turning: 11v
B13	ORN	Vehicle Speed Sensor	Moving: pulse signals
B14	BLK	CYP Sensor Ground	<0.050v
B15	RED	TDC Sensor Ground	<0.050v
B16	WHT	CKP Sensor Ground	<0.050v

1995 Sedan 2.7L V6 MFI VIN CE6 [All] 22P 'D' Connector

PCM Pin #	W/Color	Circuit Description (22-Pin)	Value at Hot Idle
D1	WHT/YEL	Keep Alive Power (VBU)	12-14v
D2	RED/WHT	Rear HO2S Ground	<0.050v
D3	BLU/WHT	A/T TCM Signal	Digital Signals
D4	LT GRN	Data Link Connector	5v
D5	WHT/RED	Alternator 'FR' Signal	Digital Signals: 0-5-0v
D6	RED/BLK	TP Sensor Signal	0.5-0.6v
D7	RED/WHT	ECT Sensor Signal	At 180°F: 0.52v
D8	RED/YEL	IAT Sensor Signal	Varies w/temp. (0.5-4.9v)
D9	WHT/YEL	MAP Sensor Signal	0.8-0.9v
D10	YEL/WHT	MAP Sensor VREF	4.9-5.1v
D11	GRN/WHT	MAP Sensor Ground	<0.050v
D12	GRN/WHT	Brake Switch Signal	Brake Off: 0v, On: 12v
D13	RED	Service Check Connector	SCS Open: 4.80v
D14	RED/GRN	A/T TCM Signal	Digital Signals
D15	---	Not Used	---
D16	GRN/RED	Electric Load Detector	Varies: 0.5-4.5v
D17	WHT/BLK	A/T: EGR Lift Sensor Signal	1.2v
D18	WHT/RED	HO2S-11 (B1 S1) Signal	0.1-1.1v
D19	WHT/RED	HO2S-12 (B1 S2) Signal	0.1-1.1v
D20	GRN/BLK	Sensor VREF	4.9-5.1v
D21	YEL/BLU	A/T: EGR Lift Sensor Signal	1.1-1.2v
D22	GRN/BLU	Sensor Ground	<0.050v

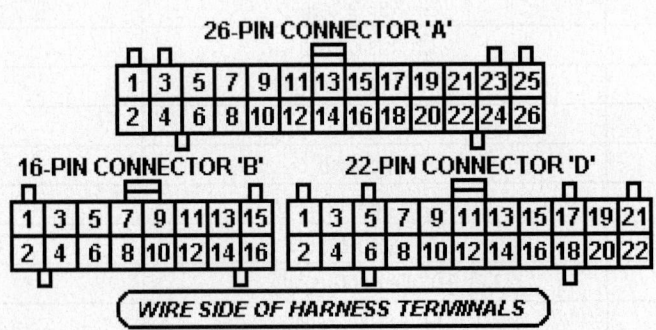

Pin Connector Graphic

Standard Colors and Abbreviations

Abbreviation	Color	Abbreviation	Color	Abbreviation	Color
BLK	Black	LT BLU	Lt. Blue	TAN	Tan
BLU	Blue	LT GRN	Lt. Green	VIO	Violet
BRN	Brown	ORN	Orange	WHT	White
GRY	Gray	PNK	Pink	YEL	Yellow
GRN	Green	PPL	Purple		

05533_ADIA_G597

1996-97 Sedan 2.7L V6 MFI VIN CE6 [A/T] 26P 'A' Connector

PCM Pin #	W/Color	Circuit Description (26-Pin)	Value at Hot Idle
A1	BRN	Injector 1 Control	2.0-3.3 ms
A2	RED	Injector 2 Control	2.0-3.3 ms
A3	BLU	Injector 3 Control	2.0-3.3 ms
A4	GRN/BLK	Fuel Pump Relay	Relay Off: 12v, On: 1v
A5	BLK/BLU	Idle Air Control Valve	Pulse Signals
A6	ORN/BLK	HO2S-11 (B1 S1) Heater	Digital Signals: 0-12-0v
A7	GRN/RED	Check Engine Light	MIL Off: 12v, On: 1v
A8	RED/BLU	A/C Clutch Relay	Relay Off: 12v, On: 1v
A9	RED	EGR Solenoid	Solenoid Off: 12v, On: 1v
A10	PNK/BLU	Intake Air Bypass Solenoid	Solenoid Off: 12v, On: 1v
A11	YEL/GRN	Igniter Control	Digital Signals: 0-12-0v
A12	BLK	Power Ground	<0.1v
A13	YEL/BLK	Main Relay Power (B+)	12-14v
A14	YEL	Injector 4 Control	2.0-3.3 ms
A15	BLK/RED	Injector 5 Control	2.0-3.3 ms
A16	WHT/BLU	Injector 6 Control	2.0-3.3 ms
A17	ORN/BLU	HO2S-12 (B1 S2) Heater	Digital Signals: 0-12-0v
A18	GRN/WHT	Engine Mount Solenoid	Solenoid Off: 12v, On: 1v
A19	GRN	Radiator Fan Relay	Relay Off: 12v, On: 1v
A20	---	Not Used	---
A21	WHT/GRN	Alternator Charging Signal	Lights Off: 12v, On: 0v
A22	BRN/WHT	A/T TCM Signal	Digital Signals
A23	RED/YEL	EVAP Purge Solenoid	Solenoid Off: 12v, On: 1v
A24	BRN	EVAP Purge Flow Switch	Switch on: 0v, off: 5v
A25	BLK	Power Ground	<0.1v
A26	BRN/BLK	Chassis Ground	<0.050v

1996-97 Sedan 2.7L V6 MFI VIN CE6 [A/T] 16P 'B' Connector

PCM Pin #	W/Color	Circuit Description (16-Pin)	Value at Hot Idle
B1	YEL/BLK	Main Relay Power (B+)	12-14v
B2	WHT/RED	A/T TCM Signal	Digital Signals
B3	RED/WHT	A/C Switch Signal	Switch Off: 12v, On: 1v
B4	LT GRN	A/T: Gear Position Indicator	In P/N: 0v, Others: 11v
B5	BLK/GRN	Starter Switch Signal	Cranking: 9-11v
B6	YEL	CYP Sensor Signal	AC pulse signals
B7	GRN	TDC Sensor Signal	AC pulse signals
B8	BLU	CKP Sensor Signal	AC pulse signals
B9	BRN/BLK	Chassis Ground	<0.050v
B10	GRN	A/T: TCM Signal	5v
B11	---	Not Used	---
B12	GRN	PSP Switch	Straight: 0v, Turning: 11v
B13	ORN	Vehicle Speed Sensor	Moving: pulse signals
B14	BLK	CYP Sensor Ground	<0.050v
B15	RED	TDC Sensor Ground	<0.050v
B16	WHT	CKP Sensor Ground	<0.050v

1996-97 Sedan 2.7L V6 MFI VIN CE6 [A/T] 22P 'D' Connector

PCM Pin #	W/Color	Circuit Description (22-Pin)	Value at Hot Idle
D1	WHT/YEL	Keep Alive Power (VBU)	12-14v
D2	RED/WHT	Rear HO2S Sensor Ground	<0.050v
D3	BLU/WHT	A/T: TCM Signal	5v
D4	LT GRN	Data Link Connector	5v
D5	WHT/RED	Alternator 'FR' Signal	Digital Signals: 0-5-0v
D6	RED/BLK	TP Sensor Signal	0.5-0.6v
D7	RED/WHT	ECT Sensor Signal	At 180°F: 0.52v
D8	RED/YEL	IAT Sensor Signal	Varies w/temp. (0.5-4.9v)
D9	WHT/YEL	MAP Sensor Signal	0.8-0.9v
D10	YEL/WHT	MAP Sensor VREF	4.9-5.1v
D11	GRN/WHT	MAP Sensor Ground	<0.050v
D12	GRN/WHT	Brake Switch Signal	Brake Off: 0v, On: 12v
D13	RED	Service Check Connector	SCS Open: 4.80v
D14	RED/GRN	A/T: TCM Signal	5v
D15	---	Not Used	---
D16	GRN/RED	Electric Load Detector	Varies: 0.5-4.5v
D17	WHT/BLK	A/T: EGR Lift Sensor Signal	1.2v
D18	WHT/RED	HO2S-11 (B1 S1) Signal	0.1-1.1v
D19	WHT/RED	HO2S-12 (B1 S2) Signal	0.1-1.1v
D20	GRN/BLK	A/T: TCM Signal	5v
D21	YEL/BLU	Sensor VREF	4.9-5.1v
D22	GRN/BLU	Sensor Ground	<0.050v

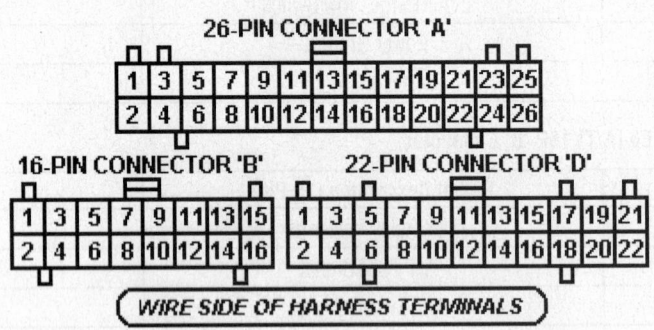

05533_ADIA_G597

Pin Connector Graphic

1998-99 Sedan 3.0L V6 VTEC VIN CG1 [A/T] 32P 'A' Connector

PCM Pin #	W/Color	Circuit Description (32-Pin)	Value at Hot Idle
A2	GRN/WHT	Engine Mount Solenoid	Solenoid Off: 12v, On: 1v
A3	BLU	EVAP Bypass Solenoid	Solenoid Off: 12v, On: 1v
A4	GRN/WHT	EVAP Vent Solenoid	Solenoid Off: 12v, On: 1v
A5	BLU/GRN	Cruise Control Signal	C/C On: pulse signals
A6	RED/YEL	EVAP Purge Solenoid	Solenoid Off: 12v, On: 1v
A8	BLK/WHT	HO2S-12 (B1 S2) Heater	Digital Signals: 0-12-0v
A9	BLU/WHT	A/T: Vehicle Speed Output	Moving: pulse signals
A10	BRN	Service Check Connector	SCS Open: 4.80v

1998-99 Sedan 3.0L V6 VTEC VIN CG1 [A/T] 32P 'A' Connector, *continued*

A12	PNK	Immobilizer Indicator Lamp	Lamp Off: 12v, On: 1v
A13, A25	BLU, RED	Immobilizer Enable, Code	Digital Signals
A14	GRN/BLK	A/T: D4 Light Switch	D4 On: 0v, Off: 12v
A15	GRN/YEL	Immobilizer Fuel Pump Relay	Relay Off: 12v, On: 1v
A17	RED	A/C Clutch Relay Control	Relay Off: 12v, On: 1v
A18	GRN/ORN	Malfunction Indicator Light	MIL Off: 12v, On: 1v
A19	BLU	Engine Speed Pulse (NEP)	AC Pulse Signals
A20	GRN	Radiator Fan Relay Control	Relay Off: 12v, On: 1v
A21	GRY	K-Line Signal	12v
A23	WHT/RED	HO2S-12 (B1 S2) Signal	0.1-1.1v
A24	BLU/ORN	Starter Switch Signal	Cranking: 9-11v
A26	GRN	PSP Switch	Straight: 0v, Turning: 11v
A27	BLU/RED	A/C Switch Signal	Switch Off: 12v, On: 1v
A28	WHT/RED	Interlock Control Unit Signal	Key & Brake On: 12v
A29	LT GRN	Fuel Tank Pressure Sensor	Fuel Cap off: 2.5v
A30	GRN/RED	Electric Load Detector	Varies: 0.5-4.5v
A32	WHT/BLK	Brake Switch Signal	Brake Off: 0v, On: 12v

1998-99 Sedan 3.0L V6 VTEC VIN CG1 [A/T] 16P 'D' Connector

PCM Pin #	W/Color	Circuit Description (16-Pin)	Value at Hot Idle
D1	YEL	Lockup Control Solenoid	LUS On: 12v, Off: 0v
D2	GRN/WHT	A/T: Shift Solenoid 'B'	SSB in 3rd, 4th Gear: 0v
D3	GRN	A/T: Shift Solenoid 'C'	SSC in 2nd, 4th Gear: 0v
D5	BLK/YEL	VB Solenoid Feed (B+)	Sol. on: 12v, off: 0v
D6	WHT	A/T: Gear Position Switch	In 'R': 0v, Others: 12v
D7	BLU/YEL	A/T: Shift Solenoid 'A'	SSA in 1st, 4th Gear: 0v
D8	PNK	A/T: Gear Position Switch	In D3: 0v, Others: 12v
D9	YEL	A/T: Gear Position Switch	In D4: 0v, Others: 12v
D10	BLU	Countershaft Speed Sensor P	Moving: AC pulses
D11	RED	Mainshaft Speed Sensor 'P'	Moving: AC pulses
D12	WHT	Mainshaft Speed Sensor 'N'	Moving: AC pulses
D14	BLU	A/T: Gear Position Switch	In D2: 0v, Others: 12v
D15	BRN	A/T: Gear Position Switch	In D1: 0v, Others: 12v
D16	GRN	Countershaft Speed Sensor N	<0.050v

```
1  2  3  4    5  6  7    8  9  10  11      1    2  3  4      5
12 13 14 15 16 17 18 19 20 21 22  23   24  6  7  8  9 10 11  12
32-PIN 25 26 27    28 29 30 31    32      13 14 15    16  16-PIN
  'A'                                                        'D'
        WIRE SIDE OF HARNESS TERMINALS
```

Pin Connector Graphic

1998-99 Sedan 3.0L V6 VTEC VIN CG1 [A/T] 25P 'B' Connector

PCM Pin #	W/Color	Circuit Description (25-Pin)	Value at Hot Idle
B1, B9	YEL/BLK	Main Relay Power (B+)	12-14v
B2, B10	BLK	Power Ground	<0.1v
B3	BLK/RED	Injector 5 Control	2.0-3.3 ms
B4, B5	YEL, RED	Injector 4, Injector 2 Control	2.0-3.3 ms
B6	WHT/BLU	Injector 6 Control	2.0-3.3 ms
B7	PNK	E-EGR Solenoid Control	Digital Signals: 0-12-0v
B8	WHT	A/T: Clutch Solenoid 'A-'	AC Pulse Signals
B11	BRN	Injector 1 Control	2.0-3.3 ms
B12	GRN/YEL	VTEC Control Solenoid	0v, Hi-Speed: 12v
B13	YEL/GRN	Ignition Control Signal	Digital Signals: 0-12-0v
B14	BLU/BLK	A/T: 2nd Clutch Pressure Sw.	Open: 12v, Closed: 1v
B15	BLU	Injector 3 Control	2.0-3.3 ms
B17	RED	A/T: Clutch Solenoid 'A+'	AC Pulse Signals
B18	GRN	A/T: Clutch Solenoid 'B-'	Pulse Signals
B20, B22	BRN/BLK	Logic Ground	<0.050v
B21	WHT/YEL	Keep Alive Power (VBU)	12-14v
B23	BLK/BLU	Idle Air Control Valve	Pulse Signals
B24	BLU/WHT	A/T: 3rd Clutch Pressure Sw.	Open: 12v, Closed: 1v
B25	ORN	A/T: Clutch Solenoid 'B+'	Pulse Signals

1998-99 Sedan 3.0L V6 VTEC VIN CG1 [A/T] 31P 'C' Connector

PCM Pin #	W/Color	Circuit Description (31-Pin)	Value at Hot Idle
C1	BLK/WHT	HO2S-11 (B1 S1) Heater	Digital Signals: 0-12-0v
C2	WHT/GRN	Alternator Charging Signal	Lights Off: 12v, On: 0v
C3	WHT/BLU	Alternator Charging Signal	Lights Off: 12v, On: 0v
C5	WHT/RED	Alternator 'FR' Signal	Digital Signals: 0-5-0v
C6	WHT/BLK	EGR Valve Lift Sensor	1.2v
C7, C18	GRN/WHT	MAP Sensor, Sensor Ground	<0.050v
C8	BLU	CKP Sensor Signal	AC pulse signals
C9	WHT	CKP Sensor Ground	<0.050v
C10	BLU/BLK	VTEC Pressure Switch	0v, Hi-Speed: 12v
C16	WHT	HO2S-11 (B1 S1) Signal	0.1-1.1v
C17	RED/GRN	MAP Sensor Signal	0.8-0.9v
C19, C28	YEL/RED	MAP VREF, Sensor VREF	4.9-5.1v
C20	GRN	TDC1 Sensor Signal	AC pulse signals
C21	RED	TDC1 Sensor Ground	<0.050v
C25	RED/YEL	IAT Sensor Signal	Varies w/temp. (0.5-4.9v)
C26	RED/WHT	ECT Sensor Signal	At 180°F: 0.5-0.6v
C27	RED/BLK	TP Sensor Signal	0.5-0.6v
C29	YEL	TDC2 Sensor Signal	AC pulse signals
C30	BLK	TDC2 Sensor Ground	<0.050v

Pin Connector Graphic

2000-03 Sedan 3.0L V6 VTEC VIN CG1 [A/T] 32P 'A' Connector

PCM Pin #	W/Color	Circuit Description (32-Pin)	Value at Hot Idle
A1	YEL/GRN	ECT Signal to TCM	Digital Signals
A2	GRN/WHT	A/T: Engine Mount Solenoid	Solenoid Off: 12v, On: 1v
A3	BLU	EVAP Bypass Solenoid	Solenoid Off: 12v, On: 1v
A4	GRN/WHT	EVAP Vent Solenoid	Solenoid Off: 12v, On: 1v
A5	BLU/GRN	Cruise Control Signal	C/C On: pulse signals
A6	RED/YEL	EVAP Purge Solenoid	Solenoid Off: 12v, On: 1v
A7, 11, 13	WT,GN,BL	ABS-TCS Signals	Digital Signals
A8	BLK/WHT	HO2S-12 (B1 S2) Heater	Digital Signals: 0-12-0v
A9	BLU/WHT	A/T: Vehicle Speed Output	Moving: pulse signals
A10	BRN	Service Check Connector	SCS Open: 4.80v
A12	PNK	Immobilizer Indicator Lamp	Lamp Off: 12v, On: 1v
A14	GRN/BLK	A/T: D4 Light Switch	D4 On: 0v, Off: 12v
A15	GRN/YEL	Immobilizer Fuel Pump Relay	Relay Off: 12v, On: 1v
A16, A25	BLU, RED	Immobilizer Enable, Code	Digital Signals
A17	RED	A/C Clutch Relay Control	Relay Off: 12v, On: 1v
A18	GRN/ORN	Malfunction Indicator Light	MIL Off: 12v, On: 1v
A19	BLU	Engine Speed Pulse (NEP)	AC Pulse Signals
A20	GRN	Radiator Fan Relay Control	Relay Off: 12v, On: 1v
A21	GRY	K-Line Signal	12v
A22, A31	PNK, RED	ABS-TCS Signals	Digital Signals
A23	WHT/RED	HO2S-12 (B1 S2) Signal	0.1-1.1v
A24	BLU/ORN	Starter Switch Signal	Cranking: 9-11v
A26	GRN	PSP Switch	Straight: 0v, Turning: 11v
A27	BLU/RED	A/C Switch Signal	Switch Off: 12v, On: 1v
A28	WHT/RED	Interlock Control Unit Signal	Key & Brake On: 12v
A29	LT GRN	Fuel Tank Pressure Sensor	Fuel Cap off: 2.5v
A30	GRN/RED	Electric Load Detector	Varies: 0.5-4.5v
A32	WHT/BLK	Brake Switch Signal	Brake Off: 0v, On: 12v

2000-03 Sedan 3.0L V6 VTEC VIN CG1 [A/T] 16P 'D' Connector

PCM Pin #	W/Color	Circuit Description (16-Pin)	Value at Hot Idle
D1	YEL	Lockup Control Solenoid	LUS On: 12v, Off: 0v
D2	GRN/WHT	A/T: Shift Solenoid 'B'	SSB in 3rd, 4th Gear: 0v
D3	GRN	A/T: Shift Solenoid 'C'	SSC in 2nd, 4th Gear: 0v
D5	BLK/YEL	VB Solenoid Feed (B+)	Sol. on: 12v, off: 0v
D6	WHT	A/T: Gear Position Switch	In 'R': 0v, Others: 12v
D7	BLU/YEL	A/T: Shift Solenoid 'A'	SSA in 1st, 4th Gear: 0v
D8, D9	PNK, YEL	A/T: Gear Position Switches	In D3: 0v, In D4: 0v
D10, D16	BLU, GRN	A/T: Countershaft Speed P, N	Moving: AC pulses
D11, D12	RED, WHT	Mainshaft Speed Sensor P, N	Moving: AC pulses
D13	BLU/BLK	A/T: 2nd Clutch Pressure Sw.	Open: 12v, Closed: 1v
D14, 15	BLU, BRN	A/T: Gear Position Switch	D2: 0v, D1: 0v

05533_ADIA_G599

Pin Connector Graphic

2000-03 Sedan 3.0L V6 VTEC VIN CG1 [A/T] 25P 'B' Connector

PCM Pin #	W/Color	Circuit Description (25-Pin)	Value at Hot Idle
B1, B9	YEL/BLK	Main Relay Power (B+)	12-14v
B2, B10	BLK	Power Ground	<0.1v
B3	BLK/RED	Injector 5 Control	2.0-3.3 ms
B4, B5	YEL, RED	Injector 4, Injector 2 Control	2.0-3.3 ms
B6	WHT/BLU	Injector 6 Control	2.0-3.3 ms
B7	PNK	E-EGR Solenoid Control	Digital Signals: 0-12-0v
B8	WHT	A/T: Clutch Solenoid 'A-'	AC Pulse Signals
B11	BRN	Injector 1 Control	2.0-3.3 ms
B12	GRN/YEL	VTEC Control Solenoid	0v, Hi-Speed: 12v
B14	BLU/BLK	A/T: Gear Position Switch	In P/N: 0v, Others: 12v
B15	BLU	Injector 3 Control	2.0-3.3 ms
B17	RED	A/T: Clutch Solenoid 'A+'	AC Pulse Signals
B18	GRN	A/T: Clutch Solenoid 'B-'	Pulse Signals
B20, B22	BRN/BLK	Logic Ground (LG1, LG2)	<0.050v
B21	WHT/YEL	Keep Alive Power (VBU)	12-14v
B23	BLK/BLU	Idle Air Control Valve	Pulse Signals
B24	BLU/WHT	A/T: 3rd Clutch Pressure Sw.	Open: 12v, Closed: 1v
B25	ORN	A/T: Clutch Solenoid 'B+'	Pulse Signals

2000-03 Sedan 3.0L V6 VTEC VIN CG1 [A/T] 31P 'C' Connector

PCM Pin #	W/Color	Circuit Description (31-Pin)	Value at Hot Idle
C1	BLK/WHT	HO2S-11 (B1 S1) Heater	Digital Signals: 0-12-0v
C2	WHT/GRN	Alternator Charging Signal	Lights Off: 12v, On: 0v
C3	WHT/BLU	Ignition Coil 3 Control	Digital Signals: 0-12-0v
C4	YEL/GRN	Ignition Coil 1 Control	Digital Signals: 0-12-0v
C5	WHT/RED	Alternator 'FR' Signal	Digital Signals: 0-5-0v
C6	WHT/BLK	EGR Valve Lift Sensor	1.2v
C7, C18	GRN/WHT	MAP Sensor, Sensor Ground	<0.050v
C8, C9	BLU, WHT	CKP Sensor Signal, Ground	AC pulse signals
C10	BLU/BLK	VTEC Pressure Switch	0v, Hi-Speed: 12v
C12	BLK/RED	Ignition Coil 5 Control	Digital Signals: 0-12-0v
C13	BRN	Ignition Coil 4 Control	Digital Signals: 0-12-0v
C14	BLU/RED	Ignition Coil 2 Control	Digital Signals: 0-12-0v
C16	WHT	HO2S-11 (B1 S1) Signal	0.1-1.1v
C17	RED/GRN	MAP Sensor Signal	0.8-0.9v
C19, C28	YEL/RED	MAP VREF, Sensor VREF	4.9-5.1v
C20, C21	GRN, RED	TDC1 Sensor Signal, Ground	AC pulse signals
C23	BRN/WHT	Ignition Coil 6 Control	Digital Signals: 0-12-0v
C25	RED/YEL	IAT Sensor Signal	Varies w/temp. (0.5-4.9v)
C26	RED/WHT	ECT Sensor Signal	At 180°F: 0.5-0.6v
C27	RED/BLK	TP Sensor Signal	0.5-0.6v
C29, C30	YEL, BLK	TDC2 Sensor Signal, Ground	AC pulse signals

1	2		3	4	5		6	7	8		1	2	3	4		5	6	7		8	9	10
9	10	11	12	13	14	15	16	17	18	11	12	13	14	15	16	17	18	19	20	21	22
19	20	21	22		23	24	25			23	24	25		26	27	28		29	30	31

25-PIN 'B'   *WIRE SIDE OF HARNESS TERMINALS*   **31-PIN 'C'**

05533_ADIA_G601

Pin Connector Graphic

1991-93 Wagon 2.2L I4 MFI VIN CB9 [All] 26P 'A' Connector

PCM Pin #	W/Color	Circuit Description (26-Pin)	Value at Hot Idle
A1	BRN	Injector 1 Control	2.0-3.3 ms
A2	YEL	Injector 4 Control	2.0-3.3 ms
A3	RED	Injector 2 Control	2.0-3.3 ms
A4	---	Not Used	---
A5	BLU	Injector 3 Control	2.0-3.3 ms
A6	ORN	HO2S Heater Control	Relay Off: 12v, On: 1v
A7	GRN/BLK	Fuel Pump Relay	Relay Off: 12v, On: 1v
A8	GRN/BLK	Fuel Pump Relay	Relay Off: 12v, On: 1v
A9	BLK/BLU	Electronic Air Control Valve	Pulse Signals
A10	---	Not Used	---
A11	PNK	EGR Control Solenoid	1.2v
A12	BLU	Radiator Fan Relay	Relay Off: 12v, On: 1v
A13	GRN/RED	Check Engine Light	MIL Off: 12v, On: 1v
A14, 17	---	Not Used	---
A15	RED/BLU	A/C Clutch Relay	Relay Off: 12v, On: 1v
A16	WHT/GRN	Alternator Charging Signal	Lights Off: 12v, On: 0v
A18	BRN/WHT	A/T: Control Unit	N/A
A19	WHT	Intake Air Regulator Control	Solenoid Off: 12v, On: 1v
A20	RED/GRN	EVAP Purge Solenoid	Solenoid Off: 12v, On: 1v
A21	YEL/GRN	Igniter Control	Hot idle 10% d/cycle
A22	YEL/GRN	Igniter Control	Hot idle 10% d/cycle
A23	BLK	Power Ground	<0.1v
A24	BLK	Power Ground	<0.1v
A25	YEL/BLK	Ignition Power	12-14v
A26	BLK/RED	Chassis Ground	<0.050v

1991-93 Wagon 2.2L I4 MFI VIN CB9 [All] 16P 'B' Connector

PCM Pin #	W/Color	Circuit Description (16-Pin)	Value at Hot Idle
B1	YEL/BLK	Main Relay Power (B+)	12-14v
B2	BRN/BLK	Power Ground	<0.1v
B3	WHT/RED	A/T: Control Unit	---
B4	GRN	A/T: Control Unit	---
B5	BLU/BLK	A/C Switch Signal	Switch Off: 12v, On: 1v
B6	---	Not Used	---
B7	LT GRN	A/T: Neutral Position Switch	In N: 0v, Others: 12v
B8	RED	PSP Switch	Straight: 0v, Turning: 11v
B9	BLU/RED	A/T: Start Signal	Cranking: 9-11v
B9	BLK/GRN	M/T: Start Signal	Cranking: 9-11v
B10	ORN	Vehicle Speed Sensor	Moving: pulse signals
B11	ORN	CYP Sensor Signal	AC pulse signals
B12	WHT	CYP Sensor Ground	<0.050v
B13	ORN/BLU	TDC Sensor Signal	AC pulse signals
B14	WHT/BLU	TDC Sensor Ground	<0.050v
B15	BLU/GRN	CKP Sensor Signal	<AC pulse signals
B16	BLU/YEL	CKP Sensor Ground	<0.050v

1991-93 Wagon 2.2L I4 MFI VIN CB9 [All] 22P 'D' Connector

PCM Pin #	W/Color	Circuit Description (22-Pin)	Value at Hot Idle
D1	WHT/YEL	Keep Alive Power (VBU)	12-14v
D2	GRN/WHT	Brake Switch Signal	Brake Off: 0v, On: 12v
D3	---	Not Used	---
D4	ORN/RED	Service Check Connector	SCS Open: 4.80v
D5-D8	---	Not Used	---
D9	WHT/RED	Alternator 'FR' Signal	Digital Signals: 0-5-0v
D10	GRN/RED	Electric Load Detector	Varies: 0.5-4.5v
D11	RED/BLK	TP Sensor Signal	0.5-0.6v
D12	WHT/BLK	A/T: EGR Lift Sensor Signal	1.1-1.2v
D13	YEL/GRN	ECT Sensor Signal	At 180°F: 0.52v
D14	WHT	Oxygen Sensor	0.1-1.1v
D15	RED/YEL	IAT Sensor Signal	Varies w/temp. (0.5-4.9v)
D16	---	Not Used	---
D17	WHT/BLU	MAP Sensor Signal	0.8-0.9v
D18	GRN/BLK	A/T: Control Unit	---
D19	RED/WHT	MAP Sensor VREF	4.9-5.1v
D20	YEL/WHT	Sensor VREF	4.9-5.1v
D21	BLU/WHT	Sensor Ground	<0.050v
D22	GRN/WHT	Sensor Ground	<0.050v

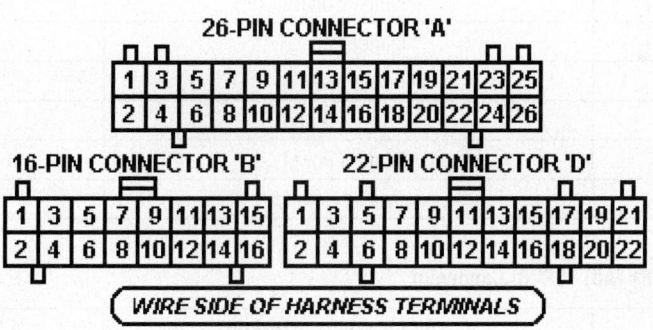

Pin Connector Graphic

05533_ADIA_G597

1994-95 Wagon 2.2L I4 MFI VIN CE1 [All] 26P 'A' Connector

PCM Pin #	W/Color	Circuit Description (26-Pin)	Value at Hot Idle
A1	BRN	Injector 1 Control	2.0-3.3 ms
A2	YEL	Injector 4 Control	2.0-3.3 ms
A3	RED	Injector 2 Control	2.0-3.3 ms
A4	---	Not Used	---
A5	BLU	Injector 3 Control	2.0-3.3 ms
A6	RED	EGR Control Solenoid	Solenoid Off: 12v, On: 1v
A7	GRN/BLK	Fuel Pump Relay	Relay Off: 12v, On: 1v
A8	GRN/BLK	Fuel Pump Relay	Relay Off: 12v, On: 1v
A9	BLK/BLU	Idle Air Control Motor	Pulse Signals
A10	GRN/WHT	Engine Mount Solenoid	Solenoid Off: 12v, On: 1v
A11	GRN/BLK	HO2S Heater Control	Relay Off: 12v, On: 1v
A12	GRN	Radiator Fan Relay	Relay Off: 12v, On: 1v
A13	GRN/RED	Check Engine Light	MIL Off: 12v, On: 1v
A14	---	Not Used	---
A15	RED/BLU	A/C Clutch Relay	Relay Off: 12v, On: 1v
A16	WHT/GRN	Alternator Charging Signal	Lights Off: 12v, On: 0v
A17	---	Not Used	---
A18	BRN/WHT	A/T: Control Unit	---
A19	ORN/GRN	Intake Air Regulator Control	Solenoid Off: 12v, On: 1v
A20	RED/YEL	EVAP Purge Solenoid	Solenoid Off: 12v, On: 1v
A21	YEL/GRN	Igniter Control	Digital Signals: 0-12-0v
A22	YEL/GRN	Igniter Control	Digital Signals: 0-12-0v
A23	BLK	Power Ground	<0.1v
A24	BLK	Power Ground	<0.1v
A25	YEL/BLK	Ignition Power	12v
A26	BLK/RED	Chassis Ground	<0.050v

1994-95 Wagon 2.2L I4 MFI VIN CE1 [All] 16P 'B' Connector

PCM Pin #	W/Color	Circuit Description (16-Pin)	Value at Hot Idle
B1	YEL/BLK	Ignition Power	12-14v
B2	BRN/BLK	Power Ground	<0.1v
B3	WHT/RED	A/T: Control Unit	---
B4	GRN	A/T: Control Unit	---
B5	RED/WHT	A/C Switch Signal	Switch Off: 12v, On: 1v
B6	---	Not Used	---
B7	LT GRN	A/T: Park Neutral Switch	In P/N: 0v, Others: 11v
B8	GRN	PSP Switch	Straight: 0v, Turning: 11v
B9	BLU/RED	A/T: Start Signal	Cranking: 9-11v

1994-95 Wagon 2.2L I4 MFI VIN CE1 [All] 16P 'B' Connector, *continued*

B9	BLU/RED	M/T: Start Signal	Cranking: 9-11v
B10	ORN	Vehicle Speed Sensor	Moving: pulse signals
B11	ORN	CYP Sensor Signal	AC pulse signals
B12	WHT	CYP Sensor Ground	<0.050v
B13	ORN/BLU	TDC Sensor Signal	AC pulse signals
B14	WHT/BLU	TDC Sensor Ground	<0.050v
B15	BLU/GRN	CKP Sensor Signal	<AC pulse signals
B16	BLU/YEL	CKP Sensor Ground	<0.050v

1994-95 Wagon 2.2L I4 MFI VIN CE1 [All] 22P 'D' Connector

PCM Pin #	W/Color	Circuit Description (22-Pin)	Value at Hot Idle
D1	WHT/YEL	Keep Alive Power (VBU)	12-14v
D2	GRN/WHT	Brake Switch Signal	Brake Off: 0v, On: 12v
D3	---	Not Used	---
D4	ORN/RED	Service Check Connector	SCS Open: 4.80v
D5	BLU/WHT	A/T: TCM Signal	Digital Signals
D6	---	Not Used	---
D7	GRN/RED	Data Link Connector	5v
D8	---	Not Used	---
D9	WHT/RED	Alternator 'FR' Signal	Digital Signals: 0-5-0v
D10	GRN/RED	Electric Load Detector	Varies: 0.5-4.5v
D11	RED/BLK	TP Sensor Signal	0.5-0.6v
D12	WHT/BLK	A/T: EGR Lift Sensor Signal	1.1-1.2v
D13	RED/WHT	ECT Sensor Signal	At 180°F: 0.52v
D14	WHT/RED	HO2S Signal	0.1-1.1v
D15	RED/YEL	IAT Sensor Signal	Varies w/temp. (0.5-4.9v)
D16	---	Not Used	---
D17	WHT/YEL	MAP Sensor Signal	0.8-0.9v
D18	GRN/BLK	A/T: Control Unit	---
D19	YEL/WHT	MAP Sensor VREF	4.9-5.1v
D20	YEL/BLU	Sensor VREF	4.9-5.1v
D21	GRN/WHT	MAP Sensor Ground	<0.050v
D22	GRN/BLU	Sensor Ground	<0.050v

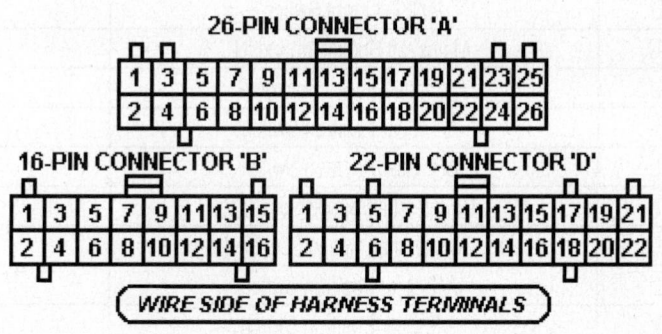

Pin Connector Graphic

1996-97 Wagon 2.2L I4 MFI VIN CE1 [All] 32P 'A' Connector

PCM Pin #	W/Color	Circuit Description (32-Pin)	Value at Hot Idle
A1	YEL	Injector 4 Control	2.0-3.3 ms
A2	BLU	Injector 3 Control	2.0-3.3 ms
A3	RED	Injector 2 Control	2.0-3.3 ms
A4	BRN	Injector 1 Control	2.0-3.3 ms
A5	ORN/BLU	HO2S-12 (B1 S2) Heater	Digital Signals: 0-12-0v
A6	ORN/BLK	HO2S-11 (B1 S1) Heater	Digital Signals: 0-12-0v
A7	RED	EGR Solenoid	Solenoid on: 1v, off: 12v
A9, 22	BRN/BLK	Sensor Ground	<0.050v
A10	BLK, BLK	Power Ground	<0.1v
A11	YEL/BLK	Main Relay Power (B+)	12-14v
A12	BLK/BLU	Idle Air Control Valve Signal	Pulse Signals
A13	GRN/WHT	Engine Mount Solenoid	Solenoid Off: 12v, On: 1v
A15	RED/YEL	EVAP Purge Solenoid	Solenoid Off: 12v, On: 1v
A16	GRN/BLK	Fuel Pump Relay	Relay Off: 12v, On: 1v
A17	RED/BLU	A/C Clutch Relay Control	Relay Off: 12v, On: 1v
A18	GRN/RED	Check Engine Light	MIL Off: 12v, On: 1v
A19	WHT/GRN	Alternator Charging Signal	Lights Off: 12v, On: 0v
A20	YEL/GRN	Igniter Control	At idle: 5 dwell
A23	BLK	Power Ground	<0.1v
A24	YEL/BLK	Main Relay Power (B+)	12-14v
A25	ORN/GRN	Intake Air Resonator Control	Solenoid Off: 12v, On: 1v
A27	GRN	Radiator Fan Relay Control	Relay Off: 12v, On: 1v
A28	GRN/WHT	EVAP Bypass Solenoid	Solenoid Off: 12v, On: 1v
A29	ORN/GRN	EVAP Vent Solenoid	Solenoid Off: 12v, On: 1v

1996-97 Wagon 2.2L I4 MFI VIN CE1 [All] 25P 'B' Connector

PCM Pin #	W/Color	Circuit Description (25-Pin)	Value at Hot Idle
B3	BLU/YEL	A/T: Shift Control Solenoid 'A'	SSA on: 1v, off: 12v
B4	GRN/BLK	A/T: Lockup solenoid 'B'	LSB On: 12v, Off: 0v
B5	YEL	A/T: Lockup solenoid 'A'	LSA On: 12v, Off: 0v
B8	GRN/BLU	A/T: Gear Position Switch	In D3: 0v, Others: 12v
B11	BLU/YEL	Shift Control Solenoid 'B'	SSB on: 1v, off: 12v
B12	WHT/GRN	Interlock Control Unit Signal	Key & Brake On: 12v
B13	BLU/RED	A/T: D4 Light Switch	In D4: 0v, Others: 12v
B14	WHT/BLU	Mainshaft Speed Ground 'N'	<0.050v
B15	ORN/BLU	Mainshaft Speed Sensor 'P'	Moving: AC pulses
B16	GRN/RED	A/T: Gear Position Switch	In 'R': 0v, Others: 12v
B17, 18	GRN/YEL	A/T: Gear Position Switch	In D2, D1: 0v
B22	BLU/YEL	Countershaft Speed Sensor P	Moving: AC pulses
B23	BLU/YEL	Countershaft Speed Sensor N	<0.050v
B24	GRN/BLU	A/T: Gear Position Switch	In D4: 0v
B25	LT GRN	A/T: Gear Position Switch	In P/N: 0v, Others: 12v

1	2	3	4		5	6	7		8	9	10		11		1		2		3	4	5		6	7	8
12	13	14	15	16	17	18	19	20	21	22	23		24		9		10	11	12	13	14	15	16	17	18
	25	26	27		28	29	30	31		32					19		20	21	22		23	24	25

32-PIN 'A'   *WIRE SIDE OF HARNESS TERMINALS*   **25-PIN 'B'**

05533_ADIA_G602

Pin Connector Graphic
1996-97 Wagon 2.2L I4 MFI VIN CE1 [All] 31P 'C' Connector

PCM Pin #	W/Color	Circuit Description (31-Pin)	Value at Hot Idle
C1	---	Not Used	---
C2	BLU	CYP Sensor	AC pulse signals
C3	GRN	TDC Sensor	AC pulse signals
C4	YEL	CKP Sensor	AC pulse signals
C5	RED/WHT	A/C Switch Signal	Switch Off: 12v, On: 1v
C6	BLU/RED	Starter Switch Signal	Cranking: 9-11v
C7	RED	Service Check Connector	SCS Open: 4.80v
C8	LT GRN	K-Line Signal	12v
C9	---	Not Used	---
C10	WHT/YEL	Keep Alive Power (VBU)	12-14v
C11	---	Not Used	---
C12	WHT	CKP Sensor Ground	<0.050v
C13	RED	TDC Sensor Ground	<0.050v
C14	BLK	CYP Sensor Ground	<0.050v
C15	---	Not Used	---
C16	GRN	PSP Switch	Straight: 0v, Turning: 11v
C17	WHT/RED	Alternator 'FR' Signal	Digital Signals: 0-5-0v
C18	ORN	Vehicle Speed Sensor	Moving: pulse signals
C19	---	Not Used	---
C20	BRN	EVAP Purge Flow Switch	Switch on: 0v, off: 5v
C21-31	---	Not Used	---

1996-97 Wagon 2.2L I4 MFI VIN CE1 [All] 16P 'D' Connector

PCM Pin #	W/Color	Circuit Description (16-Pin)	Value at Hot Idle
D1	RED/BLK	TP Sensor Signal	0.5-0.6v
D2	RED/WHT	ECT Sensor Signal	At 180°F: 0.52v
D3	WHT/YEL	MAP Sensor Signal	0.8-0.9v
D4	YEL/WHT	MAP Sensor VREF	4.9-5.1v
D5	GRN/WHT	Brake Switch Signal	Brake Off: 0v, On: 12v
D7	WHT/RED	HO2S-11 (B1 S1) Signal	0.1-1.1v
D8	RED/YEL	IAT Sensor Signal	Varies w/temp. (0.5-4.9v)
D9	WHT/BLK	EGR Valve Lift Sensor	1.2v
D10	YEL/BLU	Sensor VREF	4.9-5.1v
D11	GRN/BLU	Sensor Ground	<0.050v
D12	GRN/WHT	MAP Sensor Ground	<0.050v
D13	RED/WHT	HO2S-12 Ground	<0.050v
D14	WHT/RED	HO2S-12 (B1 S2) Signal	0.1-1.1v
D15	GRN/RED	Fuel Tank Pressure Sensor	Fuel Cap off: 2.5v
D16	GRN/RED	Electric Load Detector	Varies: 0.5-4.5v

31-PIN CONNECTOR 'C'   **16-PIN CONNECTOR 'D'**

1	2	3	4		5	6	7		8	9	10		1		2	3	4		5
11	12	13	14	15	16	17	18	19	20	21	22		6	7	8	9	10	11	12
	23	24	25		26	27	28		29	30	31		13	14	15		16		

WIRE SIDE OF HARNESS TERMINALS

05533_ADIA_G603

Pin Connector Graphic

1996-97 Wagon 2.2L I4 VTEC VIN CE1 [All] 32P 'A' Connector

PCM Pin #	W/Color	Circuit Description (32-Pin)	Value at Hot Idle
A1	YEL	Injector 4 Control	2.0-3.3 ms
A2	BLU	Injector 3 Control	2.0-3.3 ms
A3	RED	Injector 2 Control	2.0-3.3 ms
A4	BRN	Injector 1 Control	2.0-3.3 ms
A5	ORN/BLU	HO2S-12 (B1 S2) Heater	Digital Signals: 0-12-0v
A6	ORN/BLK	HO2S-11 (B1 S1) Heater	Digital Signals: 0-12-0v
A7	RED	EGR Solenoid	Solenoid Off: 12v, On: 1v
A8	GRN/YEL	VTEC Control Solenoid	0v, Hi-Speed: 12v
A9, 22	BRN/BLK	Sensor Ground	<0.050v
A10, 23	BLK, BLK	Power Ground	<0.1v
A11, 24	YEL/BLK	Main Relay Power (B+)	12-14v
A12	BLK/BLU	Idle Air Control Valve Signal	Pulse Signals
A13	GRN/WHT	Engine Mount Solenoid	Solenoid Off: 12v, On: 1v
A15	RED/YEL	EVAP Purge Solenoid	Solenoid Off: 12v, On: 1v
A16	GRN/BLK	Fuel Pump Relay	Relay Off: 12v, On: 1v
A17	RED/BLU	A/C Clutch Relay Control	Relay Off: 12v, On: 1v
A18	GRN/RED	Check Engine Light	MIL Off: 12v, On: 1v
A19	WHT/GRN	Alternator Charging Signal	Lights Off: 12v, On: 0v
A20	YEL/GRN	Igniter Control	At idle: 5 dwell
A25	ORN/GRN	Intake Air Resonator Control	Solenoid Off: 12v, On: 1v
A27	GRN	Radiator Fan Relay Control	Relay Off: 12v, On: 1v
A28	GRN/WHT	EVAP Bypass Solenoid	Solenoid Off: 12v, On: 1v
A29	ORN/GRN	EVAP Vent Solenoid	Solenoid Off: 12v, On: 1v

1996-97 Wagon 2.2L I4 VTEC VIN CE1 [All] 25P 'B' Connector

PCM Pin #	W/Color	Circuit Description (25-Pin)	Value at Hot Idle
B3	BLU/YEL	Shift Control Solenoid 'A'	SSA on: 1v, off: 12v
B4	GRN/BLK	Lockup Control Solenoid 'B'	SSB on: 1v, off: 12v
B5	YEL	Lockup Control Solenoid 'A'	LSA on: 1v, off: 12v
B8, 24	GRN/BLU	A/T: Gear Position Switch	In D3: 0v, in D4: 0v
B11	GRN/WHT	Shift Control Solenoid 'B'	Solenoid Off: 12v, On: 1v
B12	WHT/GRN	Interlock Control Unit Signal	Key & Brake On: 12v
B13	BLU/RED	A/T: D4 Light Switch	In D4: 0v, Off: 5v
B14	WHT/BLU	Mainshaft Speed Ground 'N'	<0.050v
B15	ORN/BLU	Mainshaft Speed Signal 'P'	Moving: AC pulses
B16	GRN/RED	A/T: Gear Position Switch	In 'R': 0v, Others: 12v
B17	GRN/YEL	A/T: Gear Position Switch	In D2: 0v, Others: 12v

1996-97 Wagon 2.2L I4 VTEC VIN CE1 [All] 25P 'B' Connector, *continued*

B18	GRN/YEL	A/T: Gear Position Switch	In D1: 0v, Others: 12v
B22	BLU/YEL	Countershaft Speed Sensor P	Moving: AC pulses
B23	BLU/GRN	Countershaft Speed Sensor N	<0.050v
B25	LT GRN	A/T: Gear Position Switch	In P/N: 0v, Others: 12v

05533_ADIA_G602

Pin Connector Graphic

1996-97 Wagon 2.2L I4 VTEC VIN CE1 [All] 31P 'C' Connector

PCM Pin #	W/Color	Circuit Description (31-Pin)	Value at Hot Idle
C2	BLU	CYP Sensor Signal	AC pulse signals
C3	GRN	TDC Sensor Signal	AC pulse signals
C4	YEL	CKP Sensor Signal	AC pulse signals
C5	RED/WHT	A/C Switch Signal	Switch Off: 12v, On: 1v
C6	BLU/RED	Starter Switch Signal	Cranking: 9-11v
C7	RED	Service Check Connector	SCS Open: 4.80v
C8	LT GRN	K-Line Signal	12v
C10	WHT/YEL	Keep Alive Power (VBU)	12-14v
C12	WHT	CKP Sensor Ground	<0.050v
C13	RED	TDC Sensor Ground	<0.050v
C14	BLK	CYP Sensor Ground	<0.050v
C16	GRN	PSP Switch	Straight: 0v, Turning: 11v
C17	WHT/RED	Alternator 'FR' Signal	Digital Signals: 0-5-0v
C18	ORN	Vehicle Speed Sensor	Moving: pulse signals
C20	BRN	EVAP Purge Flow Switch	Switch on: 0v, off: 5v

1996-97 Wagon 2.2L I4 VTEC VIN CE1 [All] 16P 'D' Connector

PCM Pin #	W/Color	Circuit Description (16-Pin)	Value at Hot Idle
D1	RED/BLK	TP Sensor Signal	0.5-0.6v
D2	RED/WHT	ECT Sensor Signal	At 180°F: 0.52v
D3	WHT/YEL	MAP Sensor Signal	0.8-0.9v
D4	YEL/WHT	MAP Sensor VREF	4.9-5.1v
D5	GRN/WHT	Brake Switch Signal	Brake Off: 0v, On: 12v
D7	WHT/RED	HO2S-11 (B1 S1) Signal	0.1-1.1v
D8	RED/YEL	IAT Sensor Signal	Varies w/temp. (0.5-4.9v)
D9	WHT/BLK	EGR Valve Lift Sensor	1.2v
D10	YEL/BLU	Sensor VREF	4.9-5.1v
D11	GRN/BLU	Sensor Ground	<0.050v
D12	GRN/WHT	MAP Sensor Ground	<0.050v
D13	RED/WHT	HO2S-12 Ground	<0.050v
D14	WHT/RED	HO2S-12 (B1 S2) Signal	0.1-1.1v
D15	GRN/RED	Fuel Tank Pressure Sensor	Fuel Cap off: 2.5v
D16	GRN/RED	Electric Load Detector	Varies: 0.5-4.5v

31-PIN CONNECTOR 'C'   16-PIN CONNECTOR 'D'

1	2	3	4		5	6	7		8	9	10		1		2	3	4		5
11	12	13	14	15	16	17	18	19	20	21	22	6	7	8	9	10	11	12
	23	24	25		26	27	28		29	30	31	13	14	15		16		

WIRE SIDE OF HARNESS TERMINALS

05533_ADIA_G603

Pin Connector Graphic

CIVIC PIN CHARTS

1994-95 Coupe 1.6L I4 VTEC VIN EJ1 26P 'A' Connector

PCM Pin #	W/Color	Circuit Description (26-Pin)	Value at Hot Idle
A1	BRN	Injector 1 Control	2.0-3.3 ms
A2	YEL	Injector 4 Control	2.0-3.3 ms
A3	RED	Injector 2 Control	2.0-3.3 ms
A4	ORN/WHT	VTEC Solenoid	0v, Hi-Speed: 12v
A5	BLU	Injector 3 Control	2.0-3.3 ms
A6	ORN/BLK	HO2S Heater Control	Relay Off: 12v, On: 1v
A7	GRN/YEL	Fuel Pump Relay	Relay Off: 12v, On: 1v
A8	---	Not Used	---
A9	GRN/WHT	Idle Air Control Valve	Pulse Signals
A10-11	---	Not Used	---
A12	YEL/GRN	Radiator Fan Relay	Relay Off: 12v, On: 1v
A13	GRN/ORN	Check Engine Light	MIL Off: 12v, On: 1v
A14	---	Not Used	---
A15	BLK/RED	A/C Clutch Relay	Relay Off: 12v, On: 1v
A16	WHT/YEL	Alternator Charging Signal	Lights Off: 12v, On: 0v
A17	GRN/BLK	A/T: Lockup Control Solenoid	Solenoid Off: 12v, On: 1v
A18	---	Not Used	---
A19	YEL	A/T: Lockup Control Solenoid	Solenoid Off: 12v, On: 1v
A20	RED	EVAP Purge Solenoid	Solenoid Off: 12v, On: 1v
A21	RED/GRN	Igniter Control	Digital Signals: 0-12-0v
A22	---	Not Used	---
A23	BLK	Power Ground	<0.1v
A24	BLK	Power Ground	<0.1v
A25	YEL/BLK	Main Relay Power (B+)	12-14v
A26	BLK/RED	Power Ground	<0.1v

1994-95 Coupe 1.6L I4 VTEC VIN EJ1 16P 'B' Connector

PCM Pin #	W/Color	Circuit Description (16-Pin)	Value at Hot Idle
B1	WHT/GRN	Main Relay Power (B+)	12-14v
B2	BRN/BLK	Logic Ground	<0.050v
B3	GRN/BLU	A/T: Shift Selector Signal	KOEO: 12-14v
B4	GRN/BLK	A/T: Shift Selector Signal	KOEO: 12-14v
B5	BLU/RED	A/C Switch Signal	Switch Off: 12v, On: 1v
B6	---	Not Used	---
B7	GRN	A/T Park/Neutral Switch	In P/N: 0v, Others: 11v
B8	BRN/RED	PSP Switch	Straight: 0v, Turning: 11v
B9	BLU/WHT	Starter Switch Signal	Cranking: 9-11v
B10	YEL/BLU	Vehicle Speed Sensor	Moving: pulse signals
B11	ORN	CYP Sensor Signal	AC pulse signals
B12	WHT	CYP Sensor Ground	<0.050v
B13	ORN/BLU	TDC Sensor Signal	AC pulse signals
B14	WHT/BLU	TDC Sensor Ground	<0.050v
B15	BLU/GRN	CKP Sensor Signal	AC pulse signals
B16	BLU/YEL	CKP Sensor Ground	<0.050v

1994-95 Coupe 1.6L I4 VTEC VIN EJ1 22P 'D' Connector

PCM Pin #	W/Color	Circuit Description (22-Pin)	Value at Hot Idle
D1	WHT/BLU	Keep Alive Power (VBU)	12-14v
D2	GRN/WHT	Brake Switch Signal	Brake Off: 0v, On: 12v
D3	RED/BLU	Knock Sensor Signal	No Detonation: 18mv AC
D4	BRN	Service Check Connector	SCS Open: 4.80v
D5	---	Not Used	---
D6	ORN/BLU	VTEC Pressure Switch	0v, Hi-Speed: 12v
D7	LT BLU	Data Link Connector	5v
D8	---	Not Used	---
D9	PNK	Alternator 'FR' Signal	Digital Signals: 0-5-0v
D10	GRN/RED	Electric Load Detector	Varies: 0.5-4.5v
D11	PNK/BLK	TP Sensor Signal	0.5-0.6v
D12	---	Not Used	---
D13	RED/WHT	ECT Sensor Signal	At 180°F: 0.51v
D14	WHT	HO2S-11 (B1 S1) Signal	0.1-1.1v
D15	RED/YEL	IAT Sensor Signal	Varies w/temp. (0.5-4.9v)
D16	---	Not Used	---
D17	WHT	MAP Sensor Signal	0.8-0.9v
D18	PNK/GRN	A/T: Interlock Control Unit	Key & Brake On: 12v
D19	YEL/GRN	MAP Sensor VREF	4.9-5.1v
D20	YEL/WHT	Sensor VREF	4.9-5.1v
D21	GRN/BLU	MAP Sensor GND	<0.050v
D22	GRN/WHT	Sensor Ground	<0.050v

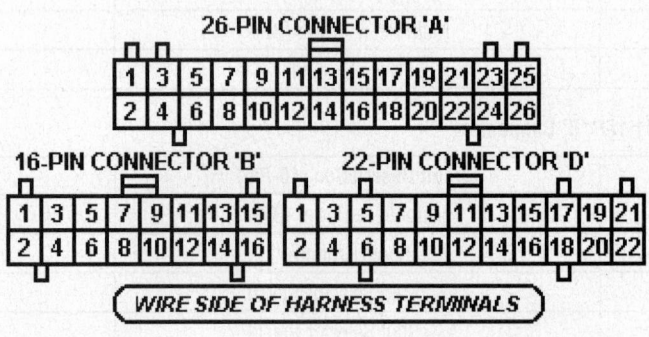

05533_ADIA_G597

Pin Connector Graphic

Standard Colors and Abbreviations

Abbreviation	Color	Abbreviation	Color	Abbreviation	Color
BLK	Black	LT BLU	Lt. Blue	TAN	Tan
BLU	Blue	LT GRN	Lt. Green	VIO	Violet
BRN	Brown	ORN	Orange	WHT	White
GRY	Gray	PNK	Pink	YEL	Yellow
GRN	Green	PPL	Purple		

1994-95 Coupe 1.6L I4 MFI VIN EJ2 [All] 26P 'A' Connector

PCM Pin #	W/Color	Circuit Description (26-Pin)	Value at Hot Idle
A1	BRN	Injector 1 Control	2.0-3.3 ms
A2	YEL	Injector 4 Control	2.0-3.3 ms
A3	RED	Injector 2 Control	2.0-3.3 ms
A4	---	Not Used	---
A5	BLU	Injector 3 Control	2.0-3.3 ms
A6	ORN/BLK	HO2S Heater Control	Relay Off: 12v, On: 1v
A7	GRN/YEL	Fuel Pump Relay	Relay Off: 12v, On: 1v
A8	---	Not Used	---
A9	GRN/WHT	Idle Air Control Valve	Pulse Signals
A10-11	---	Not Used	---
A12	YEL/GRN	Radiator Fan Relay	Relay Off: 12v, On: 1v
A13	GRN/ORN	Check Engine Light	MIL Off: 12v, On: 1v
A14	---	Not Used	---
A15	BLK/RED	A/C Clutch Relay	Relay Off: 12v, On: 1v
A16	WHT/YEL	Alternator Charging Signal	Lights Off: 12v, On: 0v
A17	GRN/BLK	A/T: Lockup Control Solenoid	Solenoid Off: 12v, On: 1v
A18	---	Not Used	---
A19	YEL	A/T: Lockup Control Solenoid	Solenoid Off: 12v, On: 1v
A20	RED	EVAP Purge Solenoid	Solenoid Off: 12v, On: 1v
A21	RED/GRN	Igniter Control	Digital Signals: 0-12-0v
A22	---	Not Used	---
A23	BLK	Power Ground	<0.1v
A24	BLK	Power Ground	<0.1v
A25	YEL/BLK	Main Relay Power (B+)	12-14v
A26	BLK/RED	Power Ground	<0.1v

1994-95 Coupe 1.6L I4 MFI VIN EJ2 [All] 16P 'B' Connector

PCM Pin #	W/Color	Circuit Description (16-Pin)	Value at Hot Idle
B1	WHT/GRN	Main Relay Power (B+)	12-14v
B2	BRN/BLK	Logic Ground	<0.050v
B3	GRN/BLU	A/T: Shift Selector Signal	KOEO: 12-14v
B4	GRN/BLK	A/T: Shift Selector Signal	KOEO: 12-14v
B5	BLU/RED	A/C Switch Signal	Switch Off: 12v, On: 1v
B6	---	Not Used	---
B7	GRN	A/T Park/Neutral Switch	In P/N: 0v, Others: 11v
B8	BRN/RED	PSP Switch	Straight: 0v, Turning: 11v
B9	BLU/WHT	Starter Switch Signal	Cranking: 9-11v
B10	YEL/BLU	Vehicle Speed Sensor	Moving: pulse signals
B11	ORN	CYP Sensor Signal	AC pulse signals
B12	WHT	CYP Sensor Ground	<0.050v
B13	ORN/BLU	TDC Sensor Signal	AC pulse signals
B14	WHT/BLU	TDC Sensor Ground	<0.050v
B15	BLU/GRN	CKP Sensor Signal	AC pulse signals
B16	BLU/YEL	CKP Sensor Ground	<0.050v

1994-95 Coupe 1.6L I4 MFI VIN EJ2 [All] 22P 'D' Connector

PCM Pin #	W/Color	Circuit Description (22-Pin)	Value at Hot Idle
D1	WHT/BLU	Keep Alive Power (VBU)	12-14v
D2	GRN/WHT	Brake Switch Signal	Brake Off: 0v, On: 12v
D3	RED/BLU	Knock Sensor Signal	No Detonation: 18mv AC
D4	BRN	Service Check Connector	SCS Open: 4.80v
D5-6	---	Not Used	---
D7	LT BLU	Data Link Connector	5v
D8	---	Not Used	---
D9	PNK	Alternator 'FR' Signal	Digital Signals: 0-5-0v
D10	GRN/RED	Electric Load Detector	Varies: 0.5-4.5v
D11	PNK/BLK	TP Sensor Signal	0.5-0.6v
D12	---	Not Used	---
D13	RED/WHT	ECT Sensor Signal	At 180°F: 0.51v
D14	WHT	HO2S-11 (B1 S1) Signal	0.1-1.1v
D15	RED/YEL	IAT Sensor Signal	Varies w/temp. (0.5-4.9v)
D16	---	Not Used	---
D17	WHT	MAP Sensor Signal	0.8-0.9v
D18	PNK/GRN	A/T: Interlock Control Unit	Key & Brake On: 12v
D19	YEL/GRN	MAP Sensor VREF	4.9-5.1v
D20	YEL/WHT	Sensor VREF	4.9-5.1v
D21	GRN/BLU	MAP Sensor GND	<0.050v
D22	GRN/WHT	Sensor Ground	<0.050v

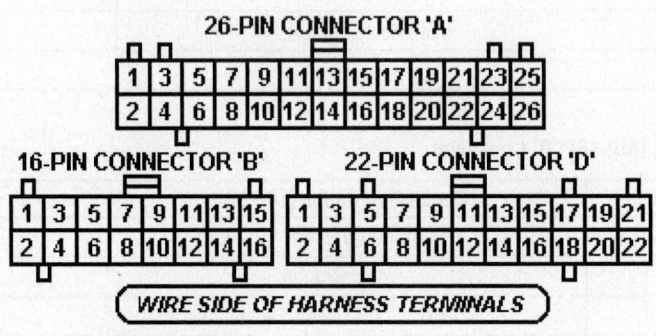

05533_ADIA_G597

Pin Connector Graphic

Standard Colors and Abbreviations

Abbreviation	Color	Abbreviation	Color	Abbreviation	Color
BLK	Black	LT BLU	Lt. Blue	TAN	Tan
BLU	Blue	LT GRN	Lt. Green	VIO	Violet
BRN	Brown	ORN	Orange	WHT	White
GRY	Gray	PNK	Pink	YEL	Yellow
GRN	Green	PPL	Purple		

1996 Coupe 1.6L I4 MFI VIN EJ6 [All] 32P 'A' Connector

PCM Pin #	W/Color	Circuit Description (32-Pin)	Value at Hot Idle
A1	YEL	Injector 4 Control	2.0-3.3 ms
A2	BLU	Injector 4 Control	2.0-3.3 ms
A3	RED	Injector 2 Control	2.0-3.3 ms
A4	BRN	Injector 1 Control	2.0-3.3 ms
A5	BLK/WHT	HO2S-12 (B1 S2) Heater	Digital Signals: 0-12-0v
A6	BLK/WHT	HO2S-11 (B1 S1) Heater	Digital Signals: 0-12-0v
A9, A22	BRN/BLK	Logic Ground	<0.050v
A10, 23	BLK	Power Ground	<0.1v
A11, 24	YEL/BLK	Main Relay Power (B+)	12-14v
A12	BLK/BLU	M/T: Idle Air Control Valve	Pulse Signals
A13	ORN	A/T: Idle Air Control Valve 'N'	Pulse Signals
A14	BLK/BLU	A/T: Idle Air Control Valve 'P'	Pulse Signals
A15	RED/YEL	EVAP Purge Solenoid	Solenoid Off: 12v, On: 1v
A16	GRN/YEL	Fuel Pump Relay	Relay Off: 12v, On: 1v
A17	BLK/RED	A/C Clutch Relay	Relay Off: 12v, On: 1v
A18	GRN/ORN	Check Engine Light	MIL Off: 12v, On: 1v
A19	WHT/GRN	Alternator Charging Signal	Lights Off: 12v, On: 0v
A20	YEL/GRN	Igniter Control	Digital Signals: 0-12-0v
A27	GRN	Radiator Fan Relay	Relay Off: 12v, On: 1v

1996 Coupe 1.6L I4 MFI VIN EJ6 [All] 25P 'B' Connector

PCM Pin #	W/Color	Circuit Description (25-Pin)	Value at Hot Idle
B1	WHT	A/T: Linear Solenoid (-)	Pulse Signals
B2	RED	A/T: Linear Solenoid (+)	Pulse Signals
B3	BLU/YEL	A/T: Shift Solenoid 'A'	SSA on: 1v: off: 12v
B4	GRN/BLK	A/T: Lockup Solenoid 'B'	LSB On: 12v, Off: 0v
B5	YEL	A/T: Lockup Solenoid 'A'	LSA On: 12v, Off: 0v
B8, 17	PNK, BLU	A/T: Gear Position Switch	In D3: 0v, Others: 12v
B11	GRN/WHT	A/T: Shift Solenoid 'B'	SSB on: 1v: off: 12v
B12	WHT/RED	Interlock Control Unit Signal	Key & Brake On: 12v
B13	GRN/BLK	A/T: D4 Indicator Light	In D4 on: 12v, others: 0v
B14	WHT	Mainshaft Speed Sensor 'N'	AC Pulse Signals
B15	RED	Mainshaft Speed Sensor 'P'	AC Pulse Signals
B16	WHT	A/T: Gear Position Switch	In 'R': 0v, Others: 12v
B17	BLU	A/T" Gear Position Switch	In D2: 0v, Others: 12v
B22, 23	GRN	Countershaft Speed Sensor N	Moving: AC pulses
B24	YEL	A/T: Gear Position Switch	In D4: 0v, Others: 12v
B25	LT GRN	A/T: Gear Position Switch	In P/N: 0v, Others: 12v

```
         32-PIN CONNECTOR 'A'              25-PIN CONNECTOR 'B'
    ┌──┬──┬──┬──┐ ┌──┬──┬──┐ ┌──┬──┬──┬──┐ ┌──┐ ┌──┐ ┌──┬──┬──┐ ┌──┬──┬──┬──┐
    │ 1│ 2│ 3│ 4│ │ 5│ 6│ 7│ │ 8│ 9│10│  │11│ │ 1│ │ 2│ │ 3│ 4│ 5│ │ 6│ 7│ 8│
    ├──┼──┼──┼──┼─┼──┼──┼──┼─┼──┼──┤  │  ├──┼──┼──┼──┼──┼──┼──┼──┼──┼──┼──┼──┤
    │12│13│14│15│16│17│18│19│20│21│22│ 23 │ 24 │ 9│ 10│11│12│13│14│15│16│17│18│
    ├──┼──┼──┼──┼──┼──┼──┼──┼──┼──┼──┼──┼──┼──┼──┼──┼──┼──┼──┤
    │25│26│27│ │28│29│30│31│ │   32   │ │ 19 │ 20 │ │21│22│ │23│24│25│
    └──┴──┴──┘ └──┴──┴──┴──┘ └────────┘ └────┴────┘ └──┴──┘ └──┴──┴──┘
```

WIRE SIDE OF HARNESS TERMINALS

05533_ADIA_G604

Pin Connector Graphic
1996 Coupe 1.6L I4 MFI VIN EJ6 [All] 31P 'C' Connector

PCM Pin #	W/Color	Circuit Description (31-Pin)	Value at Hot Idle
C1	BLU/RED	CKF Sensor Signal	AC pulse signals
C2	BLU	CKP Sensor Signal	AC pulse signals
C3	GRN	TDC Sensor Signal	AC pulse signals
C4	YEL	CYP Sensor Signal	AC pulse signals
C5	BLU/RED	A/C Switch Signal	Switch Off: 12v, On: 1v
C6	BLU/ORN	Starter Switch Signal	Cranking: 9-11v
C7	BRN	Service Check Connector	SCS Open: 4.80v
C8	LT BLU	K-Line Signal	12v
C10	WHT/BLU	Keep Alive Power (VBU)	12-14v
C11	WHT/RED	CKF Sensor Ground	<0.050v
C12	WHT	CKP Sensor Ground	<0.050v
C13	RED	TDC Sensor Ground	<0.050v
C14	BLK	CYP Sensor Ground	<0.050v
C16	GRN	PSP Switch	Straight: 0v, Turning: 11v
C17	WHT/RED	Alternator 'FR' Signal	Digital Signals: 0-5-0v
C18	BLU/WHT	Vehicle Speed Sensor	Moving: pulse signals
C30	PNK	A/T: CVT TMB Signal	Pulse Signals

1996 Coupe 1.6L I4 MFI VIN EJ6 [All] 16P 'D' Connector

PCM Pin #	W/Color	Circuit Description (16-Pin)	Value at Hot Idle
D1	RED/BLK	TP Sensor Signal	0.5-0.6v
D2	RED/WHT	ECT Sensor Signal	At 180°F: 0.51v
D3	RED/GRN	MAP Sensor Signal	0.8-0.9v
D4	YEL/RED	MAP Sensor VREF	4.9-5.1v
D5	GRN/WHT	Brake Switch Signal	Brake Off: 0v, On: 12v
D7	WHT	HO2S-11 (B1 S1) Signal	0.1-1.1v
D8	RED/YEL	IAT Sensor Signal	Varies w/temp. (0.5-4.9v)
D9	---	Not Used	---
D10	YEL/BLU	Sensor VREF	4.9-5.1v
D11	GRN/BLK	Sensor Ground	<0.050v
D12	GRN/WHT	MAP Sensor Ground	<0.050v
D13	RED/YEL	HO2S-12 Ground	<0.050v
D14	WHT/RED	HO2S-12 (B1 S2) Signal	0.1-1.1v
D15	---	Not Used	---
D16	GRN/RED	Electric Load Detector	Varies: 0.5-4.5v

31-PIN CONNECTOR 'C'   **16-PIN CONNECTOR 'D'**

```
| 1 | 2 | 3 | 4 |  | 5 | 6 | 7 |  | 8 | 9 |10|  | 1 |  | 2 | 3 | 4 |  | 5 |
|11|12|13|14|15|  |16|17|18|19|20|21|22|  | 6 | 7 | 8 | 9 |10|11|  |12|
   |23|24|25|  |26|27|28|  |29|30|31|  |13|14|15|  |16|
```

WIRE SIDE OF HARNESS TERMINALS

05533_ADIA_G605

Pin Connector Graphic

1997-99 Coupe 1.6L MFI VIN EJ6 [All] 32P 'A' Connector

PCM Pin #	W/Color	Circuit Description (32-Pin)	Value at Hot Idle
A1, A2	YEL, BLU	Injector 4, Injector 3 Control	2.0-3.3 ms
A3, A4	RED, BRN	Injector 2, Injector 1 Control	2.0-3.3 ms
A5	BLK/WHT	HO2S-12 (B1 S2) Heater	Digital Signals: 0-12-0v
A6	BLK/WHT	HO2S-11 (B1 S1) Heater	Digital Signals: 0-12-0v
A7	RED	A/T: EGR Solenoid Control	Solenoid Off: 12v, On: 1v
A7	PNK	M/T: E-EGR Solenoid	Digital Signals: 0-12-0v
A9, A22	BRN/BLK	Logic Ground (LG1, LG2)	<0.050v
A10, 23	BLK	Power Ground (PG1, PG2)	<0.1v
A11, 24	YEL/BLK	Main Relay Power (B+)	12-14v
A12	BLK/BLU	M/T: Idle Air Control Valve	Pulse Signals
A13	ORN	A/T: Idle Air Control Valve 'N'	Pulse Signals
A14	BLK/BLU	A/T: Idle Air Control Valve 'P'	Pulse Signals
A15	RED/YEL	EVAP Purge Solenoid	Solenoid Off: 12v, On: 1v
A16	GRN/YEL	Fuel Pump Relay	Relay Off: 12v, On: 1v
A17	BLK/RED	A/C Clutch Relay	Relay Off: 12v, On: 1v
A18	GRN/ORN	Malfunction Indicator Light	MIL Off: 12v, On: 1v
A19	WHT/GRN	Alternator Charging Signal	Lights Off: 12v, On: 0v
A20	YEL/GRN	Igniter Control	Digital Signals: 0-12-0v
A27	GRN	Radiator Fan Relay Control	Relay Off: 12v, On: 1v
A28	BLU	EVAP Bypass Solenoid	Solenoid Off: 12v, On: 1v
A29	WHT/RED	EVAP Vent Solenoid	Solenoid Off: 12v, On: 1v

1997-99 Coupe 1.6L MFI VIN EJ6 [All] 25P 'B' Connector

PCM Pin #	W/Color	Circuit Description (25-Pin)	Value at Hot Idle
B1	WHT	A/T: Linear Solenoid (-)	Pulse Signals
B2	RED	A/T: Linear Solenoid (+)	Pulse Signals
B3	BLU/YEL	A/T: Shift Solenoid 'A'	SSA On: 12v: Off: 0v
B4	GRN/BLK	A/T: Lockup Solenoid 'B'	LSB On: 12v: Off: 0v
B5	YEL	A/T: Lockup Solenoid 'A'	LSA On: 12v, Off: 0v
B8	PNK	A/T: Gear Position Switch	In D3: 0v, Others: 12v
B11	GRN/WHT	A/T: Shift Solenoid 'B'	SSB Off: 0v, On: 12v
B12	WHT/RED	Interlock Control Unit Signal	Key & Brake On: 12v
B13	GRN/BLK	A/T: D4 Indicator Light	D4 On: 12v, Off: 0v
B14	WHT	Mainshaft Speed Sensor 'P'	AC Pulse Signals
B15	RED	Mainshaft Speed Sensor 'N'	AC Pulse Signals
B16	WHT	A/T: Gear Position Switch	In 'R': 0v, Others: 12v
B17	BLU	A/T: Gear Position Switch	In D2: 0v, Others: 12v
B22	GRN	Countershaft Speed Sensor N	Moving: AC pulses
B23	BLU	Countershaft Speed Sensor P	Moving: AC pulses
B24	YEL	A/T: Gear Position Switch	In D4: 0v, Others: 12v
B25	LT GRN	A/T: Gear Position Switch	In P/N: 0v, Others: 12v

32-Pin 'A'   *WIRE SIDE OF HARNESS TERMINALS*   25-Pin 'B'

05533_ADIA_G606

Pin Connector Graphic

1997-99 Coupe 1.6L MFI VIN EJ6 [All] 31P 'C' Connector

PCM Pin #	W/Color	Circuit Description (31-Pin)	Value at Hot Idle
C1	BLU/RED	CKF Sensor Signal	AC pulse signals
C2	BLU	CKP Sensor Signal	AC pulse signals
C3	GRN	TDC Sensor Signal	AC pulse signals
C4	YEL	CYP Sensor Signal	AC pulse signals
C5	BLU/RED	A/C Switch Signal	Switch Off: 12v, On: 1v
C6	BLU/ORN	Starter Switch Signal	Cranking: 9-11v
C7	BRN	Service Check Connector	SCS Open: 4.80v
C8	LT BLU	K-Line Signal	12v
C10	WHT/BLU	Keep Alive Power (VBU)	12-14v
C11	WHT/RED	CKF Sensor Ground	<0.050v
C12	WHT	CKP Sensor Ground	<0.050v
C13	RED	TDC Sensor Ground	<0.050v
C14	BLK	CYP Sensor Ground	<0.050v
C16	GRN	PSP Switch Signal	Straight: 0v, Turning: 11v
C17	WHT/RED	Alternator 'FR' Signal	Digital Signals: 0-5-0v
C18	BLU/WHT	Vehicle Speed Sensor	Moving: pulse signals
C23	BLK	LAF Pump Cell (IP+)	0.5-3.5v
C24	RED	LAF Common (IP-, VS-)	2.6-2.8v
C25	WHT	LAF VS Cell Voltage (VS+)	7v
C29	LT GRN	A/T: Gear Position Switch	In P/N: 0v, Others: 12v
C29	RED	M/T: Clutch Switch Signal	Clutch In: 0v, Out: 5v
C30	PNK	CVT TMB Signal	Pulse Signals

1997-99 Coupe 1.6L MFI VIN EJ6 [All] 16P 'D' Connector

PCM Pin #	W/Color	Circuit Description (16-Pin)	Value at Hot Idle
D1	RED/BLK	TP Sensor Signal	0.5-0.6v
D2	RED/WHT	ECT Sensor Signal	At 180°F: 0.51v
D3	RED/GRN	MAP Sensor Signal	0.8-0.9v
D4	YEL/RED	MAP Sensor VREF	4.9-5.1v
D5	GRN/WHT	Brake Switch Signal	Brake Off: 0v, On: 12v
D6	RED/BLU	Knock Sensor Signal	No detonation: 18mv AC
D7	WHT	HO2S-11 (B1 S1) Signal	0.1-1.1v
D8	RED/YEL	IAT Sensor Signal	Varies w/temp. (0.5-4.9v)
D9	WHT/BLK	EGR Valve Lift Sensor	1.2v
D10	YEL/RED	Sensor VREF	4.9-5.1v
D11	GRN/BLK	Sensor Ground	<0.050v
D12	GRN/WHT	MAP Sensor Ground	<0.050v
D13	RED/YEL	HO2S-12 Ground	<0.050v
D14	WHT/RED	HO2S-12 (B1 S2) Signal	0.1-1.1v
D15	LT GRN	Fuel Tank Pressure Sensor	Fuel Cap off: 2.5v
D16	GRN/RED	Electric Load Detector	Varies: 0.5-4.5v

31-Pin 'C'   WIRE SIDE OF HARNESS TERMINALS   16-Pin 'D'

05533_ADIA_G607

Pin Connector Graphic

2000 Coupe 1.6L MFI VIN EJ6 [All] 32P 'A' Connector

PCM Pin #	W/Color	Circuit Description (32-Pin)	Value at Hot Idle
A1, A2	YEL, BLU	Injector 4, Injector 3 Control	2.0-3.3 ms
A3, A4	RED, BRN	Injector 2, Injector 1 Control	2.0-3.3 ms
A5	BLK/WHT	HO2S-12 (B1 S2) Heater	Digital Signals: 0-12-0v
A6	BLK/WHT	HO2S-11 (B1 S1) Heater	Digital Signals: 0-12-0v
A7	RED	A/T: EGR Solenoid Control	Solenoid Off: 12v, On: 1v
A7	PNK	M/T: E-EGR Solenoid	Digital Signals: 0-12-0v
A9, A22	BRN/BLK	Logic Ground (LG1, LG2)	<0.050v
A10, 23	BLK	Power Ground (PG1, PG2)	<0.1v
A11, 24	YEL/BLK	Main Relay Power (B+)	12-14v
A12	BLK/BLU	M/T: Idle Air Control Valve	Pulse Signals
A13	ORN	A/T: Idle Air Control Valve 'N'	Pulse Signals
A14	BLK/BLU	A/T: Idle Air Control Valve 'P'	Pulse Signals
A15	RED/YEL	EVAP Purge Solenoid	Solenoid Off: 12v, On: 1v
A16	GRN/YEL	Fuel Pump Relay	Relay Off: 12v, On: 1v
A17	BLK/RED	A/C Clutch Relay	Relay Off: 12v, On: 1v
A18	GRN/ORN	Malfunction Indicator Light	MIL Off: 12v, On: 1v
A19	WHT/GRN	Alternator Charging Signal	Lights Off: 12v, On: 0v
A20	YEL/GRN	Igniter Control	Digital Signals: 0-12-0v
A27	GRN	Radiator Fan Relay Control	Relay Off: 12v, On: 1v
A28	BLU	EVAP Bypass Solenoid	Solenoid Off: 12v, On: 1v
A29	WHT/RED	EVAP Vent Solenoid	Solenoid Off: 12v, On: 1v

2000 Coupe 1.6L MFI VIN EJ6 [All] 25P 'B' Connector

PCM Pin #	W/Color	Circuit Description (25-Pin)	Value at Hot Idle
B1	WHT	A/T: Linear Solenoid (-)	Pulse Signals
B2	RED	A/T: Linear Solenoid (+)	Pulse Signals
B3	BLU/YEL	A/T: Shift Solenoid 'A'	SSA On: 12v: Off: 0v
B4	GRN/BLK	A/T: Lockup Solenoid 'B'	LSB On: 12v: Off: 0v
B5	YEL	A/T: Lockup Solenoid 'A'	LSA On: 12v: Off: 0v
B8	PNK	A/T: Gear Position Switch	In D3: 0v, Others: 12v
B11	GRN/WHT	A/T: Shift Solenoid 'B'	SSB Off: 0v, On: 12v
B12	WHT/RED	Interlock Control Unit Signal	Key & Brake On: 12v
B13	GRN/BLK	A/T: D4 Indicator Light	D4 On: 12v, Off: 0v
B14	WHT	Mainshaft Speed Sensor 'P'	AC Pulse Signals
B15	RED	Mainshaft Speed Sensor 'N'	AC Pulse Signals
B16	WHT	A/T: Gear Position Switch	In 'R': 0v, Others: 12v
B17	BLU	A/T: Gear Position Switch	In D2: 0v, Others: 12v
B22	GRN	Countershaft Speed Sensor N	Moving: AC pulses
B23	BLU	Countershaft Speed Sensor P	Moving: AC pulses
B24	YEL	A/T: Gear Position Switch	In D4: 0v, Others: 12v
B25	LT GRN	A/T: Gear Position Switch	In P/N: 0v, Others: 12v

32-Pin 'A'   *WIRE SIDE OF HARNESS TERMINALS*   25-Pin 'B'

05533_ADIA_G606

Pin Connector Graphic

2000 Coupe 1.6L MFI VIN EJ6 [All] 31P 'C' Connector

PCM Pin #	W/Color	Circuit Description (31-Pin)	Value at Hot Idle
C1	BLU/RED	CKF Sensor Signal	AC pulse signals
C2	BLU	CKP Sensor Signal	AC pulse signals
C3	GRN	TDC Sensor Signal	AC pulse signals
C4	YEL	CYP Sensor Signal	AC pulse signals
C5	BLU/RED	A/C Switch Signal	Switch Off: 12v, On: 1v
C6	BLU/ORN	Starter Switch Signal	Cranking: 9-11v
C7	BRN	Service Check Connector	SCS Open: 4.80v
C8	LT BLU	K-Line Signal	12v
C10	WHT/BLU	Keep Alive Power (VBU)	12-14v
C11	WHT/RED	CKF Sensor Ground	<0.050v
C12	WHT	CKP Sensor Ground	<0.050v
C13	RED	TDC Sensor Ground	<0.050v
C14	BLK	CYP Sensor Ground	<0.050v
C16	GRN	PSP Switch Signal	Straight: 0v, Turning: 11v
C17	WHT/RED	Alternator 'FR' Signal	Digital Signals: 0-5-0v
C18	BLU/WHT	Vehicle Speed Sensor	Moving: pulse signals
C23	BLK	LAF Pump Cell (IP+)	0.5-3.5v
C24	RED	LAF Common (IP-, VS-)	2.6-2.8v
C25	WHT	LAF VS Cell Voltage (VS+)	7v
C29	LT GRN	A/T: Gear Position Switch	In P/N: 0v, Others: 12v
C29	RED	M/T: Clutch Switch Signal	Clutch In: 0v, Out: 5v
C30	PNK	CVT TMB Signal	Pulse Signals

2000 Coupe 1.6L MFI VIN EJ6 [All] 16P 'D' Connector

PCM Pin #	W/Color	Circuit Description (16-Pin)	Value at Hot Idle
D1	RED/BLK	TP Sensor Signal	0.5-0.6v
D2	RED/WHT	ECT Sensor Signal	At 180°F: 0.51v
D3	RED/GRN	MAP Sensor Signal	0.8-0.9v
D4	YEL/RED	MAP Sensor VREF	4.9-5.1v
D5	GRN/WHT	Brake Switch Signal	Brake Off: 0v, On: 12v
D6	RED/BLU	Knock Sensor Signal	No detonation: 18mv AC
D7	WHT	HO2S-11 (B1 S1) Signal	0.1-1.1v
D8	RED/YEL	IAT Sensor Signal	Varies w/temp. (0.5-4.9v)
D9	WHT/BLK	EGR Valve Lift Sensor	1.2v
D10	YEL/RED	Sensor VREF	4.9-5.1v
D11	GRN/BLK	Sensor Ground	<0.050v
D12	GRN/WHT	MAP Sensor Ground	<0.050v
D13	RED/YEL	HO2S-12 Ground	<0.050v
D14	WHT/RED	HO2S-12 (B1 S2) Signal	0.1-1.1v
D15	LT GRN	Fuel Tank Pressure Sensor	Fuel Cap off: 2.5v
D16	GRN/RED	Electric Load Detector	Varies: 0.5-4.5v

1	2	3	4		5	6	7		8	9	10
11	12	13	14	15	16	17	18	19	20	21	22
	23	24	25		26	27	28		29	30	31

31-Pin 'C'

1		2	3	4		5	
6	7	8	9	10	11	12	
13	14	15		16			

(WIRE SIDE OF HARNESS TERMINALS)   **16-Pin 'D'**

05533_ADIA_G607

Pin Connector Graphic
1996 Coupe 1.6L I4 VTEC-E VIN EJ7 [All] 32P 'A' Connector

PCM Pin #	W/Color	Circuit Description (32-Pin)	Value at Hot Idle
A1	YEL	Injector 4 Control	2.0-3.3 ms
A2	BLU	Injector 3 Control	2.0-3.3 ms
A3	RED	Injector 2 Control	2.0-3.3 ms
A4	BRN	Injector 1 Control	2.0-3.3 ms
A5	BLK/WHT	HO2S-12 (B1 S2) Heater	Digital Signals: 0-12-0v
A6	BLK/WHT	HO2S-11 (B1 S1) Heater	Digital Signals: 0-12-0v
A7	RED	A/T CVT: EGR Solenoid	Solenoid Off: 12v, On: 1v
A8	GRN/YEL	VTEC Solenoid Control	0v, Hi-Speed: 12v
A9, A22	BRN/BLK	Logic Ground	<0.050v
A10, 23	BLK	Power Ground	<0.1v
A11, 24	YEL/BLK	Main Relay Power (B+)	12-14v
A12	BLK/BLU	M/T: Idle Air Control Valve	Pulse Signals
A13	ORN	Idle Air Control Valve 'N'	Pulse Signals
A14	BLK/BLU	Idle Air Control Valve 'P'	Pulse Signals
A15	RED/YEL	EVAP Purge Solenoid	Solenoid Off: 12v, On: 1v
A16	GRN/YEL	Fuel Pump Relay	Relay Off: 12v, On: 1v
A17	BLK/RED	A/C Clutch Relay	Relay Off: 12v, On: 1v
A18	GRN/ORN	Check Engine Light	MIL Off: 12v, On: 1v
A19	WHT/GRN	Alternator Charging Signal	Lights Off: 12v, On: 0v
A20	YEL/GRN	Igniter Control	Digital Signals: 0-12-0v
A27	GRN	Radiator Fan Relay	Relay Off: 12v, On: 1v
A30	WHT/RED	Interlock Control Unit Signal	Key & Brake On: 12v

1996 Coupe 1.6L I4 VTEC-E VIN EJ7 [All] 16P 'D' Connector

PCM Pin #	W/Color	Circuit Description (16-Pin)	Value at Hot Idle
D1	RED/BLK	TP Sensor Signal	0.5-0.6v
D2	RED/WHT	ECT Sensor Signal	At 180°F: 0.51v
D3	RED/GRN	MAP Sensor Signal	0.8-0.9v
D4	YEL/RED	MAP VREF	4.9-5.1v
D5	GRN/WHT	Brake Switch Signal	Brake Off: 0v, On: 12v
D6	RED/BLU	Knock Sensor Signal	No Detonation: 18mv AC
D7	WHT	HO2S-11 (B1 S1) Signal	0.1-1.1v
D8	RED/YEL	IAT Sensor Signal	Varies w/temp. (0.5-4.9v)
D9	WHT/BLK	EGR Valve Lift Sensor	1.2v
D10	YEL/RED	Sensor VREF	4.9-5.1v
D11, 12	GRN/BLK	MAP Sensor, Sensor Ground	<0.050v
D13	GRN/BLK	HO2S-12 Ground	<0.050v
D14	WHT/RED	HO2S-12 (B1 S2) Signal	0.1-1.1v
D16	GRN/RED	Electric Load Detector	Varies: 0.5-4.5v

05533_ADIA_G608

Pin Connector Graphic

1996 Coupe 1.6L I4 VTEC-E VIN EJ7 [All] 31P 'C' Connector

PCM Pin #	W/Color	Circuit Description (31-Pin)	Value at Hot Idle
C1	BLU/RED	CKF Sensor Signal	AC pulse signals
C2	BLU	CKP Sensor Signal	AC pulse signals
C3	GRN	TDC Sensor Signal	AC pulse signals
C4	YEL	CYP Sensor Signal	AC pulse signals
C5	BLU/RED	A/C Switch Signal	Switch Off: 12v, On: 1v
C6	BLU/ORN	Starter Switch Signal	Cranking: 9-11v
C7	BRN	Service Check Connector	SCS Open: 4.80v
C8	LT BLU	K-Line Signal	12v
C9	GRY	TMA Signal	Pulse Signals
C10	WHT/BLU	Keep Alive Power (VBU)	12-14v
C11	WHT/RED	CKF Sensor Ground	<0.050v
C12	WHT	CKP Sensor Ground	<0.050v
C13	RED	TDC Sensor Ground	<0.050v
C14	BLK	CYP Sensor Ground	<0.050v
C15	BLU/BLK	VTEC Pressure Switch	0v, Hi-Speed: 12v
C16	GRN	PSP Switch	Straight: 0v, Turning: 11v
C17	WHT/RED	Alternator 'FR' Signal	Digital Signals: 0-5-0v
C18	BLU/WHT	Vehicle Speed Sensor	Moving: pulse signals

1996 Coupe 1.6L I4 VTEC-E VIN EJ7 [All] 31P 'C' Connector, *continued*

C23	BLK	M/T: LAF Pump Cell (IP+)	0.5-5.3v
C24	RED	M/T: LAF IP-, VS-	2.6-2.8v
C25	WHT	M/T: LAF VS Cell VS+ Volts	Key on: 6.5-7.5v
C29	LT GRN	A/T CVT Gear Position Switch	In P/N: 0v, Others: 12v
C29	RED	M/T: Clutch Switch Signal	Clutch In: 0v, Out: 5v
C30	PNK	A/T: CVT TMB Signal	Pulse Signals

31-PIN CONNECTOR 'C'

WIRE SIDE OF HARNESS TERMINALS

05533_ADIA_G609

Pin Connector Graphic

1997-98 Coupe 1.6L VTEC-E VIN EJ7 [All] 32P 'A' Connector

PCM Pin #	W/Color	Circuit Description (32-Pin)	Value at Hot Idle
A1	YEL	Injector 4 Control	2.0-3.3 ms
A2	BLU	Injector 3 Control	2.0-3.3 ms
A3, A4	RED, BRN	Injector 2, Injector 1 Control	2.0-3.3 ms
A5	BLK/WHT	HO2S-12 (B1 S2) Heater	Digital Signals: 0-12-0v
A6	BLK/WHT	HO2S-11 (B1 S1) Heater	Digital Signals: 0-12-0v
A7	RED	A/T: EGR Solenoid	Solenoid on: 1v: off: 12v
A8	GRN/YEL	VTEC Solenoid	0v, Hi-Speed: 12v
A9, A22	BRN/BLK	Logic Ground	<0.050v
A10, 23	BLK	Power Ground	<0.1v
A11, 24	YEL/BLK	Main Relay Power (B+)	12-14v
A13	ORN	Idle Air Control Valve 'N'	Pulse Signals
A14	BLK/BLU	Idle Air Control Valve 'P'	Pulse Signals
A15	RED/YEL	EVAP Purge Solenoid	Solenoid Off: 12v, On: 1v
A16	GRN/YEL	Fuel Pump Relay	Relay Off: 12v, On: 1v
A17	BLK/RED	A/C Clutch Relay	Relay Off: 12v, On: 1v
A18	GRN/ORN	Check Engine Light	MIL Off: 12v, On: 1v
A19	WHT/GRN	Alternator Charging Signal	Lights Off: 12v, On: 0v
A20	YEL/GRN	Igniter Control	Digital Signals: 0-12-0v
A27	GRN	Radiator Fan Relay	Relay Off: 12v, On: 1v
A28	BLU	EVAP Bypass Solenoid	Solenoid Off: 12v, On: 1v
A29	GRN/WHT	EVAP Vent Solenoid	Solenoid Off: 12v, On: 1v
A30	WHT/RED	CVT Interlock Control Unit	Key & Brake On: 12v

1997-98 Coupe 1.6L VTEC-E VIN EJ7 [All] 16P 'D' Connector

PCM Pin #	W/Color	Circuit Description (16-Pin)	Value at Hot Idle
D1	RED/BLK	TP Sensor Signal	0.5-0.6v
D2	RED/WHT	ECT Sensor Signal	At 180°F: 0.51v
D3	RED/GRN	MAP Sensor Signal	0.8-0.9v
D4	YEL/RED	Sensor VREF	4.9-5.1v
D5	GRN/WHT	Brake Switch Signal	Brake Off: 0v, On: 12v
D6	RED/BLU	Knock Sensor Signal	No Detonation: 18mv AC
D7	WHT	HO2S-11 (B1 S1) Signal	0.1-1.1v
D8	RED/YEL	IAT Sensor Signal	Varies w/temp. (0.5-4.9v)
D9	WHT/BLK	CVT EGR Valve Lift Sensor	1.2v
D10	YEL/BLU	MAP Sensor VREF	4.9-5.1v
D11, D12	GRN/BLK	Sensor Ground, MAP Ground	<0.050v
D13	GRN	HO2S-12 (B1 S2) Signal	0.1-1.1v
D14	BLK	HO2S-12 Ground	<0.050v
D15	LT GRN	Fuel Tank Pressure Sensor	Fuel Cap off: 2.5v
D16	GRN/RED	Electric Load Detector	Varies: 0.5-4.5v

05533_ADIA_G610

Pin Connector Graphic

1997-98 Coupe 1.6L VTEC-E VIN EJ7 [All] 31P 'C' Connector

PCM Pin #	W/Color	Circuit Description (31-Pin)	Value at Hot Idle
C1	BLU/RED	CKF Sensor Signal	AC pulse signals
C2	BLU	CKP Sensor Signal	AC pulse signals
C3	GRN	TDC Sensor Signal	AC pulse signals
C4	YEL	CYP Sensor Signal	AC pulse signals
C5	BLU/RED	A/C Switch Signal	Switch Off: 12v, On: 1v
C6	BLU/ORN	Starter Switch Signal	Cranking: 9-11v
C7	BRN	Service Check Connector	SCS Open: 4.80v
C8	LT BLU	K-Line Signal	12v
C9	GRY	TMA Signal	Pulse Signals
C10	WHT/BLU	Keep Alive Power (VBU)	12-14v
C11	WHT/RED	CKF Sensor Ground	<0.050v
C12	WHT	CKP Sensor Ground	<0.050v
C13	RED	TDC Sensor Ground	<0.050v
C14	BLK	CYP Sensor Ground	<0.050v
C15	BLU/BLK	VTEC Pressure Switch	0v, Hi-Speed: 12v
C16	GRN	PSP Switch	Straight: 0v, Turning: 11v
C17	WHT/RED	Alternator 'FR' Signal	Digital Signals: 0-5-0v
C18	BLU/WHT	Vehicle Speed Sensor	Moving: pulse signals
C23	BLK	M/T HO2S Pump Cell (IP+)	Key on: 0.5-5.3v

1997-98 Coupe 1.6L VTEC-E VIN EJ7 [All] 31P 'C' Connector, *continued*

C24	RED	M/T HO2S IP-, VS-	Hot Engine: 2.6-2.8v
C25	WHT	M/T VS Cell Voltage VS+	Key on: 6.5-7.5v
C29	LT GRN	CVT Gear Position Switch	In P/N: 0v, Others: 12v
C29	RED	M/T Clutch Switch Signal	Clutch In: 0v, Out: 5v
C30	PNK	CVT TMB Signal	Pulse Signals

31-PIN CONNECTOR 'C'

1	2	3	4		5	6	7		8	9	10
11	12	13	14	15	16	17	18	19	20	21	22
	23	24	25		26	27	28		29	30	31

WIRE SIDE OF HARNESS TERMINALS

05533_ADIA_G609

Pin Connector Graphic

1999-2000 Coupe 1.6L VTEC-E VIN EJ7 32P 'A' Connector

PCM Pin #	W/Color	Circuit Description (32-Pin)	Value at Hot Idle
A1	YEL	Injector 4 Control	2.0-3.3 ms
A2	BLU	Injector 3 Control	2.0-3.3 ms
A3, A4	RED, BRN	Injector 2, Injector 1 Control	2.0-3.3 ms
A5	BLK/WHT	HO2S-12 (B1 S2) Heater	Digital Signals: 0-12-0v
A6	BLK/WHT	HO2S-11 (B1 S1) Heater	Digital Signals: 0-12-0v
A7	RED	A/T: EGR Solenoid	Digital Signals: 0-12-0v
A7	PNK	M/T: E-EGR Solenoid	Digital Signals: 0-12-0v
A8	GRN/YEL	VTEC Solenoid Control	0v, Hi-Speed: 12v
A9, A22	BRN/BLK	Logic Ground	<0.050v
A10, A23	BLK	Power Ground	<0.1v
A11, A24	YEL/BLK	Main Relay Power (B+)	12-14v
A13	ORN	Idle Air Control Valve 'N'	Pulse Signals
A14	BLK/BLU	Idle Air Control Valve 'P'	Pulse Signals
A15	RED/YEL	EVAP Purge Solenoid	Solenoid Off: 12v, On: 1v
A16	GRN/YEL	Fuel Pump Relay	Relay Off: 12v, On: 1v
A17	BLK/RED	A/C Clutch Relay	Relay Off: 12v, On: 1v
A18	GRN/ORN	Malfunction Indicator Light	MIL Off: 12v, On: 1v
A19	WHT/GRN	Alternator Charging Signal	Lights Off: 12v, On: 0v
A20	YEL/GRN	Igniter Control	Digital Signals: 0-12-0v
A27	GRN	Radiator Fan Relay	Relay Off: 12v, On: 1v
A28	BLU	EVAP Bypass Solenoid	Solenoid Off: 12v, On: 1v
A29	GRN/WHT	EVAP Vent Solenoid	Solenoid Off: 12v, On: 1v
A30	WHT/RED	CVT Interlock Control Unit	Key & Brake On: 12v

1999-2000 Coupe 1.6L VTEC-E VIN EJ7 16P 'D' Connector

PCM Pin #	W/Color	Circuit Description (16-Pin)	Value at Hot Idle
D1	RED/BLK	TP Sensor Signal	0.5-0.6v
D2	RED/WHT	ECT Sensor Signal	At 180°F: 0.51v
D3	RED/GRN	MAP Sensor Signal	0.8-0.9v
D4	YEL/RED	Sensor VREF	4.9-5.1v
D5	GRN/WHT	Brake Switch Signal	Brake Off: 0v, On: 12v
D6	RED/BLU	Knock Sensor Signal	No Detonation: 18mv AC
D7	WHT	LAF Sensor Label Signal	0.3-4.9v
D8	RED/YEL	IAT Sensor Signal	Varies w/temp. (0.5-4.9v)
D9	WHT/BLK	EGR Valve Lift Sensor	1.2v
D10	YEL/BLU	MAP Sensor VREF	4.9-5.1v
D11, D12	GRN/BLK	Sensor Ground, MAP Ground	<0.050v
D13	GRN/BLK	HO2S-12 (B1 S2) Signal	0.1-1.1v
D14	BLK	HO2S-12 Ground	<0.050v
D15	LT GRN	Fuel Tank Pressure Sensor	Fuel Cap off: 2.5v
D16	GRN/RED	Electric Load Detector	Varies: 0.5-4.5v

```
1  2  3  4   5  6  7   8  9  10   11    1    2  3  4    5
12 13 14 15 16 17 18 19 20 21 22  23   24   6  7  8  9 10 11  12
32-PIN 25 26 27  28 29 30 31   32      13 14 15   16  16-PIN
  'A'                                              'D'
```

WIRE SIDE OF HARNESS TERMINALS

05533_ADIA_G610

Pin Connector Graphic

1999-2000 Coupe 1.6L VTEC-E VIN EJ7 25P 'B' Connector

PCM Pin #	W/Color	Circuit Description (25-Pin)	Value at Hot Idle
B1	WHT	A/T: Linear Solenoid LS-	Pulse Signals
B2	RED	A/T: Linear Solenoid LS+	Pulse Signals
B3	BLU/YEL	A/T: Shift Solenoid 'A'	SSA Off: 0v, On: 12v
B4	GRN/BLK	A/T: Lockup Solenoid 'B'	LUS On: 12v, Off: 0v
B5	YEL	A/T: Lockup Solenoid 'A'	LUS On: 12v, Off: 0v
B8	PNK	A/T: Gear Position Switch	In D3: 0v, Others: 12v
B11	GRN/WHT	A/T: Shift Solenoid 'B'	SSB Off: 0v, On: 12v
B12	WHT/RED	Interlock Control Unit Signal	Key & Brake On: 12v
B13	GRN/BLK	A/T: D4 Indicator Light Driver	D4 On: 12v, Off: 0v
B14, B15	WHT, RED	Mainshaft Speed Sensor 'N', 'P'	Pulse Signals
B16	WHT	A/T: Gear Position Switch	In 'R': 0v, Others: 12v
B17	BLU	A/T: Gear Position Switch	In D2: 0v, Others: 12v
B22, B23	GRN, BLU	Countershaft speed sensor N, P	<0.050v
B24	YEL	A/T: Gear Position Switch	In D4: 0v, Others: 12v
B25	LT GRN	A/T: Gear Position Switch	In P/N: 0v, Others: 12v

1999-2000 Coupe 1.6L VTEC-E VIN EJ7 31P 'C' Connector

PCM Pin #	W/Color	Circuit Description (31-Pin)	Value at Hot Idle
C1	BLU/RED	CKF Sensor Signal	AC pulse signals
C2	BLU	CKP Sensor Signal	AC pulse signals
C3	GRN	TDC Sensor Signal	AC pulse signals
C4	YEL	CYP Sensor Signal	AC pulse signals
C5	BLU/RED	A/C Switch Signal	Switch Off: 12v, On: 1v
C6	BLU/ORN	Starter Switch Signal	Cranking: 9-11v
C7	BRN	Service Check Connector	SCS Open: 4.80v
C8	LT BLU	K-Line Signal	12v
C9	GRY	ECM Communication to TCM	Digital Signals
C10	WHT/BLU	Keep Alive Power (VBU)	12-14v
C11	WHT/RED	CKF Sensor Ground	<0.050v
C12	WHT	CKP Sensor Ground	<0.050v
C13	RED	TDC Sensor Ground	<0.050v
C14	BLK	CYP Sensor Ground	<0.050v
C15	BLU/BLK	VTEC Pressure Switch	0v, Hi-Speed: 12v
C16	GRN	PSP Switch	Straight: 0v, Turning: 11v
C17	WHT/RED	Alternator 'FR' Signal	Digital Signals: 0-5-0v
C18	BLU/WHT	Vehicle Speed Sensor	Moving: pulse signals
C23	BLK	M/T: LAF Pump Cell (IP+)	0.5-5.3v
C24	RED	M/T: LAF Sensor IP-, VS-	Hot Engine: 2.6-2.8v
C25	WHT	M/T: VS Cell Voltage VS+	Key on: 6.5-7.5v
C29	LT GRN	A/T: Gear Position Switch	In P/N: 0v, Others: 12v
C29	RED	M/T Clutch Switch Signal	Clutch In: 0v, Out: 5v
C30	PNK	ECM Communication to TCM	Digital Signals

05533_ADIA_G600

Pin Connector Graphic

1996 Coupe 1.6L I4 VTEC VIN EJ8 [All] 32P 'A' Connector

PCM Pin #	W/Color	Circuit Description (32-Pin)	Value at Hot Idle
A1, A2	YEL, BLU	Injector 4 Control	2.0-3.3 ms
A3, A4	RED, BRN	Injector 2 Control	2.0-3.3 ms
A5	BLK/WHT	HO2S-12 (B1 S2) Heater	Digital Signals: 0-12-0v
A6	BLK/WHT	HO2S-11 (B1 S1) Heater	Digital Signals: 0-12-0v
A8	GRN/YEL	VTEC Solenoid	0v, Hi-Speed: 12v
A9, A22	BRN/BLK	Logic Ground	<0.050v
A10, 23	BLK	Power Ground	<0.1v
A11, 24	YEL/BLK	Main Relay Power (B+)	12-14v
A12	BLK/BLU	M/T: Idle Air Control Valve	Pulse Signals
A13	ORN	A/T: IAC Valve Control 'N'	Pulse Signals
A14	BLK/BLU	A/T: IAC Valve Control 'P'	Pulse Signals
A15	RED/YEL	EVAP Purge Solenoid	Solenoid Off: 12v, On: 1v
A16	GRN/YEL	Fuel Pump Relay	Relay Off: 12v, On: 1v
A17	BLK/RED	A/C Clutch Relay	Relay Off: 12v, On: 1v
A18	GRN/ORN	Check Engine Light	MIL Off: 12v, On: 1v
A19	WHT/GRN	Alternator Charging Signal	Lights Off: 12v, On: 0v
A20	YEL/GRN	Igniter Control	Digital Signals: 0-12-0v
A25	WHT/RED	A/T: TCM VREF Signal	4.9-5.1v
A27	GRN	Radiator Fan Relay	Relay Off: 12v, On: 1v
A28	BLU	EVAP Bypass Solenoid	Solenoid Off: 12v, On: 1v
A29	GRN/WHT	Canister Vent Solenoid	Solenoid Off: 12v, On: 1v

1996 Coupe 1.6L I4 VTEC VIN EJ8 [All] 25P 'B' Connector

PCM Pin #	W/Color	Circuit Description (25-Pin)	Value at Hot Idle
B1	WHT	A/T: Linear Solenoid (-)	Pulse Signals
B2	RED	A/T: Linear Solenoid (+)	Pulse Signals
B3	BLU/YEL	A/T: Shift Solenoid 'A'	SSA Off: 0v, On: 12v
B4	GRN/BLK	Lockup Solenoid 'B'	LSB On: 12v, Off: 0v
B5	YEL	Lockup Solenoid ''A'	LSA On: 12v, Off: 0v
B8, 17	PNK, BLU	A/T: Gear Position Switch	In D3: 0v, in D2: 0v
B11	GRN/WHT	A/T: Shift Solenoid 'B'	SSB Off: 0v, On: 12v
B12	WHT/RED	Interlock Control Unit Signal	Key & Brake On: 12v
B13	GRN/BLK	A/T: D4 Indicator Light	In D4: 12v, others: 0v
B14	WHT	Mainshaft Speed Sensor 'N'	<0.050v
B15	RED	Mainshaft Speed Sensor 'P'	AC Pulse Signals
B16, B25	WHT-GRN	A/T: Gear Position Switch	In 'R': 0v, Others: 12v
B22	GRN	Countershaft Speed Sensor N	<0.050v
B23	BLU	Countershaft Speed Sensor P	AC Pulse Signals
B24	YEL	A/T: Gear Position Switch	In D4: 0v, Others: 12v

32-PIN CONNECTOR 'A'   **25-PIN CONNECTOR 'B'**

1	2	3	4		5	6	7		8	9	10		11		1		2		3	4	5		6	7	8
12	13	14	15	16	17	18	19	20	21	22	23		24		9		10	11	12	13	14	15	16	17	18
	25	26	27		28	29	30	31		32				19		20		21	22		23	24	25

WIRE SIDE OF HARNESS TERMINALS

05533_ADIA_G604

Pin Connector Graphic
1996 Coupe 1.6L I4 VTEC VIN EJ8 [All] 31P 'D' Connector

PCM Pin #	W/Color	Circuit Description (31-Pin)	Value at Hot Idle
C1	BLU/RED	CKF Sensor Signal	AC pulse signals
C2	BLU, WHT	CKP Sensor Signal	AC pulse signals
C3	GRN	TDC Sensor Signal	AC pulse signals
C4	YEL	CYP Sensor Signal	AC pulse signals
C5	BLU/RED	A/C Switch Signal	Switch Off: 12v, On: 1v
C6	BLU/ORN	Starter Switch Signal	Cranking: 9-11v
C7	BRN	Service Check Connector	SCS Open: 4.80v
C8	LT BLU	K-Line Signal	12v
C9	---	Not Used	---
C10	WHT/BLU	Keep Alive Power (VBU)	12-14v
C11	WHT/RED	CKF Sensor Ground	<0.050v
C12	WHT	CKP Sensor Ground	<0.050v
C13	RED	TDC Sensor Ground	<0.050v
C14	BLK	CYP Sensor Ground	<0.050v
C15	BLU/BLK	VTEC Pressure Switch	0v, Hi-Speed: 12v
C16	GRN	PSP Switch	Straight: 0v, Turning: 11v
C17	WHT/RED	Alternator 'FR' Signal	Digital Signals: 0-5-0v
C18	BLU/WHT	Vehicle Speed Sensor	Moving: pulse signals
C19-31	---	Not Used	---

1996 Coupe 1.6L I4 VTEC VIN EJ8 [All] 16P 'D' Connector

PCM Pin #	W/Color	Circuit Description (16-Pin)	Value at Hot Idle
D1	RED/BLK	TP Sensor Signal	0.5-0.6v
D2	RED/WHT	ECT Sensor Signal	At 180°F: 0.51v
D3	RED/GRN	MAP Sensor Signal	0.8-0.9v
D4	YEL/RED	MAP Sensor VREF	4.9-5.1v
D5	GRN/WHT	Brake Switch Signal	Brake Off: 0v, On: 12v
D6	RED/BLU	Knock Sensor Signal	No Detonation: 18mv AC
D7	WHT	HO2S-11 (B1 S1) Signal	0.1-1.1v
D8	RED/YEL	IAT Sensor Signal	Varies w/temp. (0.5-4.9v)
D9	---	Not Used	---
D10	YEL/BLU	Sensor VREF	4.9-5.1v
D11	GRN/BLK	Sensor Ground	<0.050v
D12	GRN/WHT	MAP Sensor Ground	<0.050v
D13	GRN/BLK	HO2S-12 Ground	<0.050v
D14	WHT/RED	HO2S-12 (B1 S2) Signal	0.1-1.1v
D15	LT GRN	Fuel Tank Pressure	Fuel Cap off: 2.5v
D16	GRN/RED	Electric Load Detector	Varies: 0.5-4.5v

31-PIN CONNECTOR 'C'   16-PIN CONNECTOR 'D'

WIRE SIDE OF HARNESS TERMINALS

05533_ADIA_G605

Pin Connector Graphic

1997-99 Coupe 1.6L VTEC VIN EJ8 [All] 32P 'A' Connector

PCM Pin #	W/Color	Circuit Description (32-Pin)	Value at Hot Idle
A1, A2	YEL, BLU	Injector 4, Injector 3 Control	2.0-3.3 ms
A3, A4	RED, BRN	Injector 2, Injector 1 Control	2.0-3.3 ms
A5	BLK/WHT	HO2S-12 (B1 S2) Heater	Digital Signals: 0-12-0v
A6	BLK/WHT	HO2S-11 (B1 S1) Heater	Digital Signals: 0-12-0v
A8	GRN/YEL	VTEC Solenoid	0v, Hi-Speed: 12v
A9, A22	BRN/BLK	Logic Ground	<0.050v
A10, 23	BLK	Power Ground	<0.1v
A11	YEL/BLK	Main Relay Power (B+)	12-14v
A13	ORN	Idle Air Control Valve 'N'	Pulse Signals
A14	BLK/BLU	Idle Air Control Valve 'P'	Pulse Signals
A15	RED/YEL	EVAP Purge Solenoid	Solenoid Off: 12v, On: 1v
A16	GRN/YEL	Fuel Pump Relay	Relay Off: 12v, On: 1v
A17	BLK/RED	A/C Clutch Relay	Relay Off: 12v, On: 1v
A18	GRN/ORN	Check Engine Light	MIL Off: 12v, On: 1v
A19	WHT/GRN	Alternator Charging Signal	Lights Off: 12v, On: 0v
A20	YEL/GRN	Igniter Control	Digital Signals: 0-12-0v
A24	YEL/BLK	Main Relay Power (B+)	12-14v

1997-99 Coupe 1.6L VTEC VIN EJ8 [All] 32P 'A' Connector, *continued*

A25	WHT/RED	A/T: TCM VREF Signal	4.9-5.1v
A27	GRN	Radiator Fan Relay	Relay Off: 12v, On: 1v
A28	BLU	EVAP Bypass Solenoid	Solenoid Off: 12v, On: 1v
A29	GRN/WHT	EVAP Vent Solenoid	Solenoid Off: 12v, On: 1v

1997-99 Coupe 1.6L I4 VTEC VIN EJ8 [All] 25P 'B' Connector

PCM Pin #	W/Color	Circuit Description (25-Pin)	Value at Hot Idle
B1	WHT	A/T: Linear Solenoid (-)	Pulse Signals
B2	RED	A/T: Linear Solenoid (+)	Pulse Signals
B3	BLU/YEL	A/T: Shift Solenoid 'A'	SSA Off: 0v, On: 12v
B4	GRN/BLK	A/T: Lockup Solenoid 'B'	LSB On: 12v, Off: 0v
B5	YEL	A/T: Lockup Solenoid 'A'	LSA On: 12v, Off: 0v
B8	PNK	A/T: Gear Position Switch	In D3: 0v, Others: 12v
B11	GRN/WHT	A/T: Shift Solenoid 'B'	SSA Off: 0v, On: 12v
B12	WHT/RED	Interlock Control Unit Signal	Key & Brake On: 12v
B13	GRN/BLK	A/T: D4 Indicator Light	D4 On: 12v, Off: 0v
B14	WHT	Mainshaft Speed Sensor 'N'	<0.050v
B15	RED	Mainshaft Speed Sensor 'P'	AC Pulse Signals
B16	WHT	A/T: Gear Position Switch	In 'R': 0v, Others: 12v
B17	BLU	A/T: Gear Position Switch	In D2: 0v, Others: 12v
B22	GRN	Countershaft Speed Sensor N	<0.050v
B23	BLU	Countershaft Speed Sensor P	Moving: AC pulses
B24	YEL	A/T: Gear Position Switch	In D4: 0v, Others: 12v
B25	LT GRN	A/T: Gear Position Switch	In P/N: 0v, Others: 12v

```
┌─┬─┬─┬─┐ ┌─┬─┬─┐ ┌──┬─┬──┬────┐   ┌────┬────┐ ┌──┬─┬─┐ ┌──┬─┬─┐
│1│2│3│4│ │5│6│7│ │8 │9│10│ 11 │   │ 1  │ 2  │ │3 │4│5│ │6 │7│8│
├─┼─┼─┼─┤ ├─┼─┼─┤ ├──┼─┼──┼────┤   ├────┼────┤ ├──┼─┼─┤ ├──┼─┼─┤
│12│13│14│15│16│17│18│19│20│21│22│ 23 │ 24 │ │ 9 │ 10 │11│12│13│14│15│16│17│18│
├──┼──┼──┤ ├──┼──┼──┼──┤ ├────┤   ├────┼────┤ ├──┼─┤ ├──┼─┬─┐
│25│26│27│ │28│29│30│31│ │ 32 │   │ 19 │ 20 │ │21│22│ │23│24│25│
└──┴──┴──┘ └──┴──┴──┴──┘ └────┘   └────┴────┘ └──┴─┘ └──┴─┴─┘
    32-Pin 'A'      ⟨ WIRE SIDE OF HARNESS TERMINALS ⟩    25-Pin 'B'
```

05533_ADIA_G606

Pin Connector Graphic

1997-99 Coupe 1.6L VTEC VIN EJ8 [All] 31P 'C' Connector

PCM Pin #	W/Color	Circuit Description (31-Pin)	Value at Hot Idle
C1	BLU/RED	CKF Sensor Signal	AC pulse signals
C2	BLU	CKP Sensor Signal	AC pulse signals
C3	GRN	TDC Sensor Signal	AC pulse signals
C4	YEL	CYP Sensor Signal	AC pulse signals
C5	BLU/RED	A/C Switch Signal	Switch Off: 12v, On: 1v
C6	BLU/ORN	Starter Switch Signal	Cranking: 9-11v
C7	BRN	Service Check Connector	SCS Open: 4.80v
C8	LT BLU	K-Line Signal	12v
C9	GRY	ECM Communication to TCM	Digital Signals
C10	WHT/BLU	Keep Alive Power (VBU)	12-14v
C11	WHT/RED	CKF Sensor Ground	<0.050v
C12	WHT	CKP Sensor Ground	<0.050v
C13	RED	TDC Sensor Ground	<0.050v
C14	BLK	CYP Sensor Ground	<0.050v
C15	BLU/BLK	VTEC Pressure Switch	0v, Hi-Speed: 12v
C16	GRN	PSP Switch	Straight: 0v, Turning: 11v
C17	WHT/RED	Alternator 'FR' Signal	Digital Signals: 0-5-0v
C18	BLU/WHT	Vehicle Speed Sensor	Moving: pulse signals
C23	BLK	LAF Pump Cell (IP+)	0.5-3.5v
C24	RED	LAF Common (IP-, VS-)	2.6-2.8v
C25	WHT	LAF VS Cell Voltage (VS+)	7v
C29	LT GRN	A/T: Gear Position Switch	In P/N: 0v, Others: 12v
C29	RED	M/T: Clutch Switch Signal	Clutch In: 0v, Out: 5v
C30	PNK	A/T: CVT TMB Signal	Pulse Signals

1997-99 Coupe 1.6L VTEC VIN EJ8 [All] 16P 'D' Connector

PCM Pin #	W/Color	Circuit Description (16-Pin)	Value at Hot Idle
D1	RED/BLK	TP Sensor Signal	0.5-0.6v
D2	RED/WHT	ECT Sensor Signal	At 180°F: 0.51v
D3	RED/GRN	MAP Sensor Signal	0.8-0.9v
D4	YEL/RED	MAP Sensor VREF	4.9-5.1v
D5	GRN/WHT	Brake Switch Signal	Brake Off: 0v, On: 12v
D6	RED/BLU	Knock Sensor Signal	No Detonation: 18mv AC
D7	WHT	HO2S-11 (B1 S1) Signal	0.1-1.1v
D8	RED/YEL	IAT Sensor Signal	Varies w/temp. (0.5-4.9v)
D10	YEL/RED	Sensor VREF	4.9-5.1v
D11	GRN/BLK	Sensor Ground	<0.050v
D12	GRN/BLK	MAP Sensor Ground	<0.050v
D13	GRN/BLK	HO2S-12 Ground	<0.050v
D14	WHT/RED	HO2S-12 (B1 S2) Signal	0.1-1.1v
D15	LT GRN	Fuel Tank Pressure Sensor	Fuel Cap off: 2.5v
D16	GRN/RED	Electric Load Detector	Varies: 0.5-4.5v

```
1  2  3  4    5  6  7    8  9 10        1    2  3  4      5
11 12 13 14 15 16 17 18 19 20 21 22    6  7  8  9 10 11  12
      23 24 25    26 27 28    29 30 31  13 14 15    16
```

31-Pin 'C'   *WIRE SIDE OF HARNESS TERMINALS*   **16-Pin 'D'**

05533_ADIA_G607

Pin Connector Graphic
2000 Coupe 1.6L VTEC VIN EJ8 [All] 32P 'A' Connector

PCM Pin #	W/Color	Circuit Description (32-Pin)	Value at Hot Idle
A1, A2	YEL, BLU	Injector 4, Injector 3 Control	2.0-3.3 ms
A3, A4	RED, BRN	Injector 2, Injector 1 Control	2.0-3.3 ms
A5	BLK/WHT	HO2S-12 (B1 S2) Heater	Digital Signals: 0-12-0v
A6	BLK/WHT	HO2S-11 (B1 S1) Heater	Digital Signals: 0-12-0v
A8	GRN/YEL	VTEC Solenoid	0v, Hi-Speed: 12v
A9, A22	BRN/BLK	Logic Ground	<0.050v
A10, 23	BLK	Power Ground	<0.1v
A11	YEL/BLK	Main Relay Power (B+)	12-14v
A13	ORN	Idle Air Control Valve 'N'	Pulse Signals
A14	BLK/BLU	Idle Air Control Valve 'P'	Pulse Signals
A15	RED/YEL	EVAP Purge Solenoid	Solenoid Off: 12v, On: 1v
A16	GRN/YEL	Fuel Pump Relay	Relay Off: 12v, On: 1v
A17	BLK/RED	A/C Clutch Relay	Relay Off: 12v, On: 1v
A18	GRN/ORN	Check Engine Light	MIL Off: 12v, On: 1v
A19	WHT/GRN	Alternator Charging Signal	Lights Off: 12v, On: 0v
A20	YEL/GRN	Igniter Control	Digital Signals: 0-12-0v
A24	YEL/BLK	Main Relay Power (B+)	12-14v
A25	WHT/RED	A/T: TCM VREF Signal	4.9-5.1v
A27	GRN	Radiator Fan Relay	Relay Off: 12v, On: 1v
A28	BLU	EVAP Bypass Solenoid	Solenoid Off: 12v, On: 1v
A29	GRN/WHT	EVAP Vent Solenoid	Solenoid Off: 12v, On: 1v

2000 Coupe 1.6L I4 VTEC VIN EJ8 [All] 25P 'B' Connector

PCM Pin #	W/Color	Circuit Description (25-Pin)	Value at Hot Idle
B1	WHT	A/T: Linear Solenoid (-)	Pulse Signals
B2	RED	A/T: Linear Solenoid (+)	Pulse Signals
B3	BLU/YEL	A/T: Shift Solenoid 'A'	SSA Off: 0v, On: 12v
B4	GRN/BLK	A/T: Lockup Solenoid 'B'	LSB On: 12v, Off: 0v
B5	YEL	A/T: Lockup Solenoid 'A'	LSA On: 12v, Off: 0v
B8	PNK	A/T: Gear Position Switch	In D3: 0v, Others: 12v
B11	GRN/WHT	A/T: Shift Solenoid 'B'	SSA Off: 0v, On: 12v
B12	WHT/RED	Interlock Control Unit Signal	Key & Brake On: 12v
B13	GRN/BLK	A/T: D4 Indicator Light	D4 On: 12v, Off: 0v
B14	WHT	Mainshaft Speed Sensor 'N'	<0.050v
B15	RED	Mainshaft Speed Sensor 'P'	AC Pulse Signals
B16	WHT	A/T: Gear Position Switch	In 'R': 0v, Others: 12v
B17	BLU	A/T: Gear Position Switch	In D2: 0v, Others: 12v
B22	GRN	Countershaft Speed Sensor N	<0.050v
B23	BLU	Countershaft Speed Sensor P	Moving: AC pulses
B24	YEL	A/T: Gear Position Switch	In D4: 0v, Others: 12v
B25	LT GRN	A/T: Gear Position Switch	In P/N: 0v, Others: 12v

32-Pin 'A'   *WIRE SIDE OF HARNESS TERMINALS*   25-Pin 'B'

05533_ADIA_G606

Pin Connector Graphic

2000 Coupe 1.6L VTEC VIN EJ8 [All] 31P 'C' Connector

PCM Pin #	W/Color	Circuit Description (31-Pin)	Value at Hot Idle
C1	BLU/RED	CKF Sensor Signal	AC pulse signals
C2	BLU	CKP Sensor Signal	AC pulse signals
C3	GRN	TDC Sensor Signal	AC pulse signals
C4	YEL	CYP Sensor Signal	AC pulse signals
C5	BLU/RED	A/C Switch Signal	Switch Off: 12v, On: 1v
C6	BLU/ORN	Starter Switch Signal	Cranking: 9-11v
C7	BRN	Service Check Connector	SCS Open: 4.80v
C8	LT BLU	K-Line Signal	12v
C9	GRY	ECM Communication to TCM	Digital Signals
C10	WHT/BLU	Keep Alive Power (VBU)	12-14v
C11	WHT/RED	CKF Sensor Ground	<0.050v
C12	WHT	CKP Sensor Ground	<0.050v
C13	RED	TDC Sensor Ground	<0.050v
C14	BLK	CYP Sensor Ground	<0.050v
C15	BLU/BLK	VTEC Pressure Switch	0v, Hi-Speed: 12v
C16	GRN	PSP Switch	Straight: 0v, Turning: 11v

2000 Coupe 1.6L VTEC VIN EJ8 [All] 31P 'C' Connector, *continued*

C17	WHT/RED	Alternator 'FR' Signal	Digital Signals: 0-5-0v
C18	BLU/WHT	Vehicle Speed Sensor	Moving: pulse signals
C23	BLK	LAF Pump Cell (IP+)	0.5-3.5v
C24	RED	LAF Common (IP-, VS-)	2.6-2.8v
C25	WHT	LAF VS Cell Voltage (VS+)	7v
C29	LT GRN	A/T: Gear Position Switch	In P/N: 0v, Others: 12v
C29	RED	M/T: Clutch Switch Signal	Clutch In: 0v, Out: 5v
C30	PNK	A/T: CVT TMB Signal	Pulse Signals

2000 Coupe 1.6L VTEC VIN EJ8 [All] 16P 'D' Connector

PCM Pin #	W/Color	Circuit Description (16-Pin)	Value at Hot Idle
D1	RED/BLK	TP Sensor Signal	0.5-0.6v
D2	RED/WHT	ECT Sensor Signal	At 180°F: 0.51v
D3	RED/GRN	MAP Sensor Signal	0.8-0.9v
D4	YEL/RED	MAP Sensor VREF	4.9-5.1v
D5	GRN/WHT	Brake Switch Signal	Brake Off: 0v, On: 12v
D6	RED/BLU	Knock Sensor Signal	No Detonation: 18mv AC
D7	WHT	HO2S-11 (B1 S1) Signal	0.1-1.1v
D8	RED/YEL	IAT Sensor Signal	Varies w/temp. (0.5-4.9v)
D10	YEL/RED	Sensor VREF	4.9-5.1v
D11	GRN/BLK	Sensor Ground	<0.050v
D12	GRN/BLK	MAP Sensor Ground	<0.050v
D13	GRN/BLK	HO2S-12 Ground	<0.050v
D14	WHT/RED	HO2S-12 (B1 S2) Signal	0.1-1.1v
D15	LT GRN	Fuel Tank Pressure Sensor	Fuel Cap off: 2.5v
D16	GRN/RED	Electric Load Detector	Varies: 0.5-4.5v

```
 1  2  3  4   5  6  7   8  9 10      1     2  3  4     5
11 12 13 14 15 16 17 18 19 20 21 22   6  7  8  9 10 11  12
   23 24 25   26 27 28   29 30 31    13 14 15    16
    31-Pin 'C'    ( WIRE SIDE OF HARNESS TERMINALS )   16-Pin 'D'
```

05533_ADIA_G607

Pin Connector Graphic

2001-03 Coupe 1.7L MFI VIN ES1 [All] 31P Gray 'A' Connector

PCM Pin #	W/Color	Circuit Description (31-Pin)	Value at Hot Idle
A1	BLK/WHT	HO2S-11 (B1 S1) Heater	Digital Signals: 0-12-0v
A2, A3	YEL/BLK	Main Relay Power (B+)	12-14v
A4, A5	BLK	Power Ground (PG2, PG1)	<0.1v
A6	WHT	HO2S-11 (B1 S1) Signal	0.1-1.1v
A7	BLU	CKP Sensor Signal	Digital Signals
A8	YEL	Sensor VREF (VCCR)	4.9-5.1v
A9	RED/BLU	Knock Sensor Signal	No detonation: 18mv AC
A10	GRN/YEL	Sensor Ground (SG2)	<0.050v
A11	GRN/WHT	Sensor Ground (SG1)	<0.050v
A12	BLK/RED	Idle Air Control Valve	Pulse Signals
A13	WHT/BLK	EGR Valve Position Sensor	0.6-1.1v
A14	BLK/WHT	HO2S-12 (B1 S2) Heater	Digital Signals: 0-12-0v
A15	RED/BLK	TP Sensor Signal	0.5-0.6v
A17	BRN/WHT	Injector Mode Signal	Digital Signals
A18	WHT/GRN	Vehicle Speed Signal	Moving: 0-5-0v
A19	GRN/RED	MAP Sensor Signal	0.8-0.9v
A20, A21	YEL/BLU	Sensor VREF (VCC2, VCC1)	4.9-5.1v
A23, A24	BRN/YEL	Logic Ground (LG2, LG1)	<0.1v
A25	WHT/RED	HO2S-12 (B1 S2) Signal	0.1-1.1v
A26	GRN	TDC Sensor Signal	AC pulse signals
A27	BRN	Coil 4 Driver Control	Digital Signals
A28	WHT/BLU	Coil 3 Driver Control	Digital Signals
A29	BLU/RED	Coil 2 Driver Control	Digital Signals
A30	YEL/GRN	Coil 1 Driver Control	Digital Signals

2001-03 Coupe 1.7L MFI VIN ES1 [All] 24P White 'B' Connector

PCM Pin #	W/Color	Circuit Description (24-Pin)	Value at Hot Idle
B2	YEL	Injector 4 Control	2.0-3.3 ms
B3	BLU	Injector 3 Control	2.0-3.3 ms
B4	RED	Injector 2 Control	2.0-3.3 ms
B5	BRN	Injector 1 Control	2.0-3.3 ms
B6	GRN	Fan Relay Control	Relay Off: 12v, On: 1v
B7	RED/BLK	A/T: Linear Solenoid 'A+'	LSA Off: 0v, On: 12v
B8	RED/WHT	ECT Sensor Signal	At 180°F: 0.51v
B10	WHT/BLU	Alternator Load Signal	Lights Off: 12v, On: 0v
B13	WHT/RED	Alternator 'FR' Signal	Digital Signals: 0-5-0v
B14	BLU/RED	EGR Solenoid Control	Solenoid Off: 12v, On: 1v
B16	BLK/RED	A/T: Lockup Solenoid 'B'	LSB On: 12v: Off: 0v
B17	RED/YEL	IAT Sensor Signal	Varies w/temp. (0.5-4.9v)
B18	WHT/GRN	Alternator Charging Signal	Digital Signals: 0-12-0v
B21	YEL/BLU	EVAP Purge Solenoid	Solenoid Off: 12v, On: 1v

31-Pin 'A'   *WIRE SIDE OF HARNESS TERMINALS*   24-Pin 'B'

05533_ADIA_G611

Pin Connector Graphic
2001-03 Coupe 1.7L MFI VIN ES1 [W/O CVT] 22P C Black Connector

PCM Pin #	W/Color	Circuit Description (22-Pin)	Value at Hot Idle
C1	WHT/BLK	A/T: Linear Solenoid 'A–'	LSA Off: 12v, On: 1v
C2	YEL/BLU	A/T: TCC Solenoid	Solenoid Off: 12v, On: 1v
C4	GRN/WHT	A/T: Shift Solenoid 'B'	SSB Off: 0v, On: 12v
C6	BLU/BLK	A/T: Shift Solenoid 'A'	SSA On: 12v: Off: 0v
C7	WHT/RED	Mainshaft Speed Sensor 'P'	AC Pulse Signals
C8	BRN/WHT	A/T: Lockup Solenoid 'B–'	LSB On: 12v: Off: 0v
C9	RED	A/T: Gear Position Switch	In D3: 0v, Others: 12v
C10	WHT	A/T: Gear Position Switch	In 'R': 0v, Others: 12v
C11	BLU	A/T: Gear Position Switch	In D2: 0v, Others: 12v
C12	BLU/WHT	A/T: Gear Position Switch	In P/N: 0v, others: 12v
C14	GRN	Countershaft Speed Sensor N	Moving: AC pulses
C15	BLU	Countershaft Speed Sensor P	Moving: AC pulses
C18	BLY/REL	A/T: Gear Position Switch	Forward: 0v, Others: 12v
C20	YEL	A/T: Gear Position Switch	In 'D': 0v, Others: 12v
C21	WHT/GRN	Mainshaft Speed Sensor 'N'	AC Pulse Signals

2001-03 Coupe 1.7L MFI VIN ES1 [All] 31P White 'E' Connector

PCM Pin #	W/Color	Circuit Description (31-Pin)	Value at Hot Idle
E1	GRN/YEL	Immobilizer Fuel Pump Relay	Relay Off: 12v, On: 1v
E3	BRN/YEL	Logic Ground (LG3)	<0.050v
E4	PNK	Sensor Ground (SG3)	<0.050v
E5	YEL/BLU	Sensor VREF (VCC3)	4.9-5.1v
E7	RED/YEL	Main Relay Control	Relay Off: 12v, On: 1v
E9	YEL/BLK	Main Relay Power (B+)	12-14v
E12	BLU/ORN	Cruise Control Unit Signal	Digital Signals
E13	WHT/BLU	Multiplex Control Unit	Digital Signals
E14	LT GRN	Fuel Tank Pressure Sensor	Fuel Cap off: 2.5v
E15	GRN/RED	Electric Load Detector	Varies: 0.5-4.5v
E16	GRN/BLK	PSP Switch Signal	Straight: 0v, Turning: 11v
E18	RED	A/C Clutch Relay	Relay Off: 12v, On: 1v
E20	BLU/RED	EVAP Bypass Solenoid	Solenoid Off: 12v, On: 1v
E21	GRN/RED	EVAP Vent Solenoid	Solenoid Off: 12v, On: 1v
E22	WHT/BLK	Brake Switch Signal	Brake Off: 0v, On: 12v
E23	LT BLU	K-Line Signal	12v
E24	YEL	SEFMJ Signal (Multiplex Unit)	Digital Signals
E25	BLU/WHT	VSS Out Signal	Pulse Signals
E26	BLU	Engine Speed Pulse (NEP)	Digital Signals
E27	RED/BLU	Immobilizer Code Signal	Digital Signals
E29	BRN	Service Check Connector	SCS Open: 4.80v
E30	RED/WHT	WEN Terminal in DLC	0v
E31	GRN/ORN	Malfunction Indicator Lamp	MIL Off: 12v, On: 1v

24-Pin 'C'

WIRE SIDE OF HARNESS TERMINALS

31-Pin 'E'

05533_ADIA_G612

Pin Connector Graphic

2001-03 Coupe 1.7L VTEC VIN ES2 [All] 31P Gray 'A' Connector

PCM Pin #	W/Color	Circuit Description (31-Pin)	Value at Hot Idle
A1	BLK/WHT	HO2S-11 (B1 S1) Heater	Digital Signals: 0-12-0v
A2, A3	YEL/BLK	Main Relay Power (B+)	12-14v
A4, A5	BLK	Power Ground (PG2, PG1)	<0.1v
A6	WHT	HO2S-11 (B1 S1) Signal	0.1-1.1v
A7	BLU	CKP Sensor Signal	Digital Signals
A8	YEL	Sensor VREF (VCCR)	4.9-5.1v
A9	RED/BLU	Knock Sensor Signal	No detonation: 18mv AC
A10	GRN/YEL	Sensor Ground (SG2)	<0.050v
A11	GRN/WHT	Sensor Ground (SG1)	<0.050v
A12	BLK/RED	Idle Air Control Valve	Pulse Signals
A13	WHT/BLK	EGR Valve Position Sensor	0.6-1.1v

2001-03 Coupe 1.7L VTEC VIN ES2 [All] 31P Gray 'A' Connector, *continued*

A14	BLK/WHT	HO2S-12 (B1 S2) Heater	Digital Signals: 0-12-0v
A15	RED/BLK	TP Sensor Signal	0.5-0.6v
A17	BRN/WHT	Injector Mode Signal	Digital Signals
A18	WHT/GRN	Vehicle Speed Signal	Moving: 0-5-0v
A19	GRN/RED	MAP Sensor Signal	0.8-0.9v
A20, A21	YEL/BLU	Sensor VREF (VCC2, VCC1)	4.9-5.1v
A23, A24	BRN/YEL	Logic Ground (LG2, LG1)	<0.1v
A25	WHT/RED	HO2S-12 (B1 S2) Signal	0.1-1.1v
A26	GRN	TDC Sensor Signal	AC pulse signals
A27	BRN	Coil 4 Driver Control	Digital Signals
A28	WHT/BLU	Coil 3 Driver Control	Digital Signals
A29	BLU/RED	Coil 2 Driver Control	Digital Signals
A30	YEL/GRN	Coil 1 Driver Control	Digital Signals

2001-03 Coupe 1.7L VTEC VIN ES2 [All] 24P White 'B' Connector

PCM Pin #	W/Color	Circuit Description (24-Pin)	Value at Hot Idle
B2, B3	YEL, BLU	Injector 4, Injector 3 Control	2.0-3.3 ms
B4, B5	RED, BRN	Injector 2, Injector 1 Control	2.0-3.3 ms
B6	GRN	Fan Relay Control	Relay Off: 12v, On: 1v
B7	RED/BLK	A/T: Linear Solenoid 'A+'	LSA Off: 0v, On: 12v
B8	RED/WHT	ECT Sensor Signal	At 180°F: 0.51v
B9	BLU/BLK	VTEC Pressure Switch	0v, Hi-Speed: 12v
B10	WHT/BLU	Alternator Load Signal	Lights Off: 12v, On: 0v
B13	WHT/RED	Alternator 'FR' Signal	Digital Signals: 0-5-0v
B14	BLU/RED	EGR Solenoid Control	Solenoid Off: 12v, On: 1v
B15	GRN/YEL	VTEC Solenoid Control	0v, Hi-Speed: 12v
B16	BLK/RED	A/T: Lockup Solenoid 'B'	LSB On: 12v: Off: 0v
B17	RED/YEL	IAT Sensor Signal	Varies w/temp. (0.5-4.9v)
B18	WHT/GRN	Alternator Charging Signal	Digital Signals: 0-12-0v
B21	YEL/BLU	EVAP Purge Solenoid	Solenoid Off: 12v, On: 1v

31-Pin 'A'   *WIRE SIDE OF HARNESS TERMINALS*   24-Pin 'B'

05533_ADIA_G611

Pin Connector Graphic

2001-03 Coupe 1.7L VTEC VIN ES2 [W/O CVT] 22P 'C' Connector

PCM Pin #	W/Color	Circuit Description (22-Pin)	Value at Hot Idle
C1	WHT/BLK	A/T: Linear Solenoid 'A-'	LSA Off: 12v, On: 1v
C2	YEL/BLU	A/T: TCC Solenoid	Solenoid Off: 12v, On: 1v
C4	GRN/WHT	A/T: Shift Solenoid 'B'	SSB Off: 0v, On: 12v
C6	BLU/BLK	A/T: Shift Solenoid 'A'	SSA On: 12v, Off: 0v
C7	WHT/RED	Mainshaft Speed Sensor 'P'	AC Pulse Signals
C8	BRN/WHT	A/T: Lockup Solenoid 'B-'	LSB On: 12v, Off: 0v
C9	RED	A/T: Gear Position Switch	In D3: 0v, Others: 12v
C10	WHT	A/T: Gear Position Switch	In 'R': 0v, Others: 12v
C11	BLU	A/T: Gear Position Switch	In D2: 0v, Others: 12v
C12	BLU/WHT	A/T: Gear Position Switch	In P/N: 0v, others: 12v
C14	GRN	Countershaft Speed Sensor N	Moving: AC pulses
C15	BLU	Countershaft Speed Sensor P	Moving: AC pulses
C18	BLY/REL	A/T: Gear Position Switch	Forward: 0v, Others: 12v
C20	YEL	A/T: Gear Position Switch	In 'D': 0v, Others: 12v
C21	WHT/GRN	Mainshaft Speed Sensor 'N'	AC Pulse Signals

2001-03Coupe 1.7L VTEC VIN ES2 [All] 31P White 'E' Connector

PCM Pin #	W/Color	Circuit Description (31-Pin)	Value at Hot Idle
E1	GRN/YEL	Immobilizer Fuel Pump Relay	Relay Off: 12v, On: 1v
E3	BRN/YEL	Logic Ground (LG3)	<0.050v
E4	PNK	Sensor Ground (SG3)	<0.050v
E5	YEL/BLU	Sensor VREF (VCC3)	4.9-5.1v
E7	RED/YEL	Main Relay Control	Relay Off: 12v, On: 1v
E9	YEL/BLK	Main Relay Power (B+)	12-14v
E12	BLU/ORN	Cruise Control Unit Signal	Digital Signals
E13	WHT/BLU	Multiplex Control Unit	Digital Signals
E14	LT GRN	Fuel Tank Pressure Sensor	Fuel Cap off: 2.5v
E15	GRN/RED	Electric Load Detector	Varies: 0.5-4.5v
E16	GRN/BLK	PSP Switch Signal	Straight: 0v, Turning: 11v
E18	RED	A/C Clutch Relay	Relay Off: 12v, On: 1v
E20	BLU/RED	EVAP Bypass Solenoid	Solenoid Off: 12v, On: 0v
E21	GRN/RED	EVAP Vent Solenoid	Solenoid Off: 12v, On: 1v
E22	WHT/BLK	Brake Switch Signal	Brake Off: 0v, On: 12v
E23	LT BLU	K-Line Signal	12v
E24	YEL	SEFMJ Signal (Multiplex Unit)	Digital Signals
E25	BLU/WHT	VSS Out Signal	Pulse Signals
E26	BLU	Engine Speed Pulse (NEP)	Digital Signals
E27	RED/BLU	Immobilizer Code Signal	Digital Signals
E29	BRN	Service Check Connector	SCS Open: 4.80v
E30	RED/WHT	WEN Terminal in DLC	0v
E31	GRN/ORN	Malfunction Indicator Lamp	MIL Off: 12v, On: 1v

24-Pin 'C'   *WIRE SIDE OF HARNESS TERMINALS*   31-Pin 'E'

05533_ADIA_G612

Pin Connector Graphic
1990 CRX Coupe 1.5L I4 MFI VIN ED7 [M/T] 18P 'A' Connector

PCM Pin #	W/Color	Circuit Description (18-Pin)	Value at Hot Idle
A1	BRN	Injector 1 Control	2.0-3.3 ms
A2	BLK	Power Ground	<0.1v
A3	RED	Injector 2 Control	2.0-3.3 ms
A4	BLK	Power Ground	<0.1v
A5	BLU	Injector 3 Control	2.0-3.3 ms
A6	GRN	EVAP Purge Solenoid	Solenoid Off: 12v, On: 1v
A7	YEL	Injector 4 Control	2.0-3.3 ms
A8-9	---	Not Used	---
A10	ORN	EGR Solenoid	Solenoid Off: 12v, On: 1v
A11	BLU/YEL	Electronic Air Control Valve	Pulse Signals
A12	GRN/BLK	Fuel Pump Relay	Relay Off: 12v, On: 1v
A13	YEL/BLK	Main Relay Power (B+)	12-14v
A14	GRN/BLK	Fuel Pump Relay	Relay Off: 12v, On: 1v
A15	YEL/BLK	Main Relay Power (B+)	12-14v
A16	BRN/BLK	Power Ground	<0.1v
A17	BLK/YEL	Ignition Power	12-14v
A18	BLK/RED	Power Ground	<0.1v

1990 CRX Coupe 1.5L I4 MFI VIN ED7 [M/T] 20P 'B' Connector

PCM Pin #	W/Color	Circuit Description (20-Pin)	Value at Hot Idle
B1	WHT/GRN	Keep Alive Power (VBU)	12-14v
B2	GRN/YEL	Upshift Indicator Light	Lamp Off: 12v, On: 1v
B3	YEL	A/C Clutch Relay	Relay Off: 12v, On: 1v
B4	YEL/GRN	Radiator Fan Relay	Relay Off: 12v, On: 1v
B5	WHT/YEL	Alternator Charging Signal	Lights Off: 12v, On: 0v
B6	GRN/ORN	Check Engine Light	MIL Off: 12v, On: 1v
B7	GRN	Heater Fan Switch	Fan Off: 0v, On: 12v
B8	BLU/RED	A/C Switch Signal	Switch Off: 12v, On: 1v
B9	GRN/BLK	Reverse Switch Signal	In 'R': 0v, Others: 12v
B10	ORN	CKP Sensor Signal	AC pulse signals
B11	GRN	Clutch Switch Signal	Clutch In: 0v, Out: 12v
B12	WHT	CKP Sensor Ground	<0.050v
B13	BLU/WHT	Start Signal	Cranking: 9-11v
B14	BLU	Alternator 'FR' Signal	Digital Signals: 0-5-0v
B15	WHT	Igniter Control	Digital Signals: 0-12-0v
B16	YEL/RED	Vehicle Speed Sensor	Moving: pulse signals
B17	WHT	Igniter Control	Digital Signals: 0-12-0v
B18	---	Not Used	---
B19	GRN/RED	Electric Load Detector	Varies: 0.5-4.5v
B20	BRN	Ignition Timing Adjuster	0.5-4.5v

1990 CRX Coupe 1.5L I4 MFI VIN ED7 [M/T] 16P 'C' Connector

PCM Pin #	W/Color	Circuit Description (16-Pin)	Value at Hot Idle
C1	BLU/GRN	CYP Sensor Signal	AC pulse signals
C2	BLU/YEL	CYP Sensor Ground	<0.050v
C3	ORN/BLU	TDC Sensor Signal	AC pulse signals
C4	WHT/BLU	TDC Sensor Ground	<0.050v
C5	RED/YEL	IAT Sensor Signal	Varies w/temp. (0.5-4.9v)
C6	RED/WHT	ECT Sensor Signal	At 180°F: 0.51v
C7	RED/BLU	Throttle Angle Sensor Signal	0.5-0.6v
C8	YEL	EGR Valve Lift Sensor	1.2v
C9	RED/WHT	Atmospheric Pressure Sensor	2.76-2.96v at sea level
C10	GRN/WHT	Brake Switch Signal	Brake Off: 0v, On: 12v
C11	WHT	MAP Sensor Signal	0.8-0.9v
C12	GRN/WHT	Sensor Ground	<0.050v
C13	YEL/WHT	Sensor VREF	4.9-5.1v
C14	GRN/WHT	MAP Sensor Ground	<0.050v
C15	YEL/RED	MAP Sensor VREF	4.9-5.1v
C16	WHT	Oxygen Sensor Signal	0.1-1.1v

18-PIN CONNECTOR 'A'

| 1 | 3 | 5 | 7 | | | 11 | 13 | 15 | 17 |
| 2 | 4 | 6 | 8 | | 10 | 12 | 14 | 16 | 18 |

20-PIN CONNECTOR 'B'   **16-PIN CONNECTOR 'C'**

| 1 | 3 | 5 | 7 | 9 | 11 | 13 | 15 | 17 | 19 |
| 2 | 4 | 6 | 8 | 10 | 12 | 14 | 16 | 18 | 20 |

| 1 | 3 | 5 | 7 | | 9 | 11 | 13 | 15 |
| 2 | 4 | 6 | 8 | | 10 | 12 | 14 | 16 |

WIRE SIDE OF HARNESS TERMINALS

05533_ADIA_G613

Pin Connector Graphic

1990-91 CRX Coupe 1.5L I4 MFI VIN ED8 [All] 18P 'A' Connector

PCM Pin #	W/Color	Circuit Description (18-Pin)	Value at Hot Idle
A1	BRN	Injector 1 Control	2.0-3.3 ms
A2	BLK	Power Ground	<0.1v
A3	RED	Injector 2 Control	2.0-3.3 ms
A4	BLK	Power Ground	<0.1v
A5	BLU	Injector 3 Control	2.0-3.3 ms
A6	GRN	EVAP Purge Solenoid	Solenoid Off: 12v, On: 1v
A7	YEL	Injector 4 Control	2.0-3.3 ms
A8	YEL	A/T: Lockup Solenoid 'A'	LSA On: 12v, Off: 0v
A9	---	Not Used	---
A10	RED	A/T: EGR Solenoid (Calif.)	Solenoid Off: 12v, On: 1v
A11	BLU/YEL	Electronic Air Control Valve	Pulse Signals
A12	GRN/BLK	Fuel Pump Relay	Relay Off: 12v, On: 1v
A13	YEL/BLK	Main Relay Power (B+)	12-14v
A14	GRN/BLK	Fuel Pump Relay	Relay Off: 12v, On: 1v
A15	YEL/BLK	Main Relay Power (B+)	12-14v
A16	BRN/BLK	Power Ground	<0.1v
A17	---	Not Used	---
A18	BLK/RED	Power Ground	<0.1v

1990-91 CRX Coupe 1.5L I4 MFI VIN ED8 [All] 20P 'B' Connector

PCM Pin #	W/Color	Circuit Description (20-Pin)	Value at Hot Idle
B1	WHT/GRN	Keep Alive Power (VBU)	12-14v
B2	ORN	Tandem Valve Solenoid	Solenoid Off: 12v, On: 1v
B3	YEL	A/C Clutch Relay	Relay Off: 12v, On: 1v
B4	YEL/GRN	Radiator Fan Relay	Relay Off: 12v, On: 1v
B5	WHT/YEL	Alternator Charging Signal	Lights Off: 12v, On: 0v
B6	GRN/ORN	Check Engine Light	MIL Off: 12v, On: 1v
B7	GRN	A/T: Park/Neutral Switch	In P/N: 0v, Others: 11v
B8	BLU/RED	A/C Switch Signal	Switch Off: 12v, On: 1v
B9	---	Not Used	---
B10	ORN	CYP Sensor Signal	AC pulse signals
B11	GRN/BLK	A/T: D4 Switch Signal	In D4: 0v, Others: 12v
B11	GRN	M/T: Clutch Switch Signal	Clutch In: 0v, Out: 12v
B12	WHT	CYP Sensor Ground	<0.050v
B13	BLU/WHT	Start Signal	Cranking: 9-11v
B14	BLU	Alternator 'FR' Signal	Digital Signals: 0-5-0v
B15	WHT	Igniter Control	Digital Signals: 0-12-0v
B16	YEL/RED	Vehicle Speed Sensor	Moving: pulse signals
B17	WHT	Igniter Control	Digital Signals: 0-12-0v
B18	---	Not Used	---
B19	GRN/RED	Electric Load Detector	Varies: 0.5-4.5v
B20	BRN	Ignition Timing Adjuster	0.5-4.5v

1990-91 CRX Coupe 1.5L I4 MFI VIN ED8 [All] 16P 'C' Connector

PCM Pin #	W/Color	Circuit Description (16-Pin)	Value at Hot Idle
C1	---	Not Used	---
C2	---	Not Used	---
C3	ORN/BLU	TDC Sensor Signal	AC pulse signals
C4	WHT/BLU	TDC Sensor Ground	<0.050v
C5	RED/YEL	IAT Sensor Signal	Varies w/temp. (0.5-4.9v)
C6	RED/WHT	ECT Sensor Signal	At 180°F: 0.51v
C7	RED/BLU	Throttle Angle Sensor Signal	0.5-0.6v
C8	YEL	EGRV Lift Sensor (California)	1.2v
C9	RED/WHT	Atmospheric Pressure Sensor	2.76-2.96v at sea level
C10	GRN/WHT	Brake Switch Signal	Brake Off: 0v, On: 12v
C11	WHT	MAP Sensor Signal	0.8-0.9v
C12	GRN/WHT	Sensor Ground	<0.050v
C13	YEL/WHT	Sensor VREF	4.9-5.1v
C14	GRN/WHT	MAP Sensor Ground	<0.050v
C15	YEL/RED	MAP Sensor VREF	4.9-5.1v
C16	WHT	Oxygen Sensor Signal	0.1-1.1v

18-PIN CONNECTOR 'A'

| 1 | 3 | 5 | 7 | | 11 | 13 | 15 | 17 |
| 2 | 4 | 6 | 8 | 10 | 12 | 14 | 16 | 18 |

20-PIN CONNECTOR 'B'   **16-PIN CONNECTOR 'C'**

| 1 | 3 | 5 | 7 | 9 | 11 | 13 | 15 | 17 | 19 |
| 2 | 4 | 6 | 8 | 10 | 12 | 14 | 16 | 18 | 20 |

| 1 | 3 | 5 | 7 | 9 | 11 | 13 | 15 |
| 2 | 4 | 6 | 8 | 10 | 12 | 14 | 16 |

WIRE SIDE OF HARNESS TERMINALS

05533_ADIA_G613

Pin Connector Graphic
1990-91 CRX Coupe 1.6L I4 MFI VIN ED9 [M/T] 18P 'A' Connector

PCM Pin #	W/Color	Circuit Description (18-Pin)	Value at Hot Idle
A1	BRN	Injector 1 Control	2.0-3.3 ms
A2	BLK	Power Ground	<0.1v
A3	RED	Injector 2 Control	2.0-3.3 ms
A4	BLK	Power Ground	<0.1v
A5	BLU	Injector 3 Control	2.0-3.3 ms
A6	GRN	EVAP Purge Solenoid	Solenoid Off: 12v, On: 1v
A7	YEL	Injector 4 Control	2.0-3.3 ms
A8-9	---	Not Used	---
A10	ORN	EGR Solenoid	Solenoid Off: 12v, On: 1v
A11	BLU/YEL	Electronic Air Control Valve	Pulse Signals
A12, 14	GRN/BLK	Fuel Pump Relay	Relay Off: 12v, On: 1v
A13, 15	YEL/BLK	Main Relay Power (B+)	12-14v
A16	BRN/BLK	Power Ground	<0.1v
A17	BLK/YEL	Ignition Power	12-14v
A18	BLK/RED	Power Ground	<0.1v

1990-91 CRX Coupe 1.6L I4 MFI VIN ED9 [M/T] 20P 'B' Connector

PCM Pin #	W/Color	Circuit Description (20-Pin)	Value at Hot Idle
B1	WHT/GRN	Keep Alive Power (VBU)	12-14v
B2	---	Not Used	---
B2	BLU	Fast Idle Control	Solenoid Off: 12v, On: 1v
B3	YEL	A/C Clutch Relay	Solenoid Off: 12v, On: 1v
B4	YEL/GRN	Radiator Fan Relay	Relay Off: 12v, On: 1v
B5	WHT/YEL	Alternator Charging Signal	Lights Off: 12v, On: 0v
B6	GRN/ORN	Check Engine Light	MIL Off: 12v, On: 1v
B7	---	Not Used	---
B8	BLU/RED	A/C Switch Signal	Switch Off: 12v, On: 1v
B9	---	Not Used	---
B10	ORN	CKP Sensor Signal	AC pulse signals
B11	---	Not Used	---
B12	WHT	CKP Sensor Ground	<0.050v
B13	BLU/WHT	Start Signal	Cranking: 9-11v
B14	BLU	Alternator 'FR' Signal	Digital Signals: 0-5-0v
B15	WHT	Igniter Control	Digital Signals: 0-12-0v
B16	YEL/RED	Vehicle Speed Sensor	Moving: pulse signals
B17	WHT	Igniter Control	Digital Signals: 0-12-0v
B18	---	Not Used	---
B19	GRN/RED	Electric Load Detector	Varies: 0.5-4.5v
B20	BRN	Ignition Timing Adjuster	0.5-4.5v

1990-91 CRX Coupe 1.6L I4 MFI VIN ED9 [M/T] 16P 'C' Connector

PCM Pin #	W/Color	Circuit Description (16-Pin)	Value at Hot Idle
C1	BLU/GRN	CYP Sensor Signal	AC pulse signals
C2	BLU/YEL	CYP Sensor Ground	<0.050v
C3	ORN/BLU	TDC Sensor Signal	AC pulse signals
C4	WHT/BLU	TDC Sensor Ground	<0.050v
C5	RED/YEL	IAT Sensor Signal	Varies w/temp. (0.5-4.9v)
C6	RED/WHT	ECT Sensor Signal	At 180°F: 0.51v
C7	RED/BLU	Throttle Angle Sensor Signal	0.5-0.6v
C8	---	Not Used	---
C9	RED/WHT	Atmospheric Pressure Sensor	2.76-2.96v at sea level
C10	GRN/WHT	Brake Switch Signal	Brake Off: 0v, On: 12v
C11	WHT	MAP Sensor Signal	0.8-0.9v
C12	GRN/WHT	Sensor Ground	<0.050v
C13	YEL/WHT	Sensor VREF	4.9-5.1v
C14	GRN/WHT	MAP Sensor Ground	<0.050v
C15	YEL/RED	MAP Sensor VREF	4.9-5.1v
C16	WHT	Oxygen Sensor Signal	0.1-1.1v

18-PIN CONNECTOR 'A'

| 1 | 3 | 5 | 7 | | 11 | 13 | 15 | 17 |
| 2 | 4 | 6 | 8 | 10 | 12 | 14 | 16 | 18 |

20-PIN CONNECTOR 'B'

| 1 | 3 | 5 | 7 | 9 | 11 | 13 | 15 | 17 | 19 |
| 2 | 4 | 6 | 8 | 10 | 12 | 14 | 16 | 18 | 20 |

16-PIN CONNECTOR 'C'

| 1 | 3 | 5 | 7 | 9 | 11 | 13 | 15 |
| 2 | 4 | 6 | 8 | 10 | 12 | 14 | 16 |

WIRE SIDE OF HARNESS TERMINALS

05533_ADIA_G613

Pin Connector Graphic
1990-91 Hatchback 1.5L I4 MFI VIN ED6 [All] 18P 'A' Connector

PCM Pin #	W/Color	Circuit Description (18-Pin)	Value at Hot Idle
A1	BRN	Injector 1 Control	2.0-3.3 ms
A2	BLK	Power Ground	<0.1v
A3	RED	Injector 2 Control	2.0-3.3 ms
A4	BLK	Power Ground	<0.1v
A5	BLU	Injector 3 Control	2.0-3.3 ms
A6	GRN	EVAP Purge Solenoid	Solenoid Off: 12v, On: 1v
A7	YEL	Injector 4 Control	2.0-3.3 ms
A8-9	---	Not Used	---
A10	ORN	EGR Solenoid	Solenoid Off: 12v, On: 1v
A11	BLU/YEL	Electronic Air Control Valve	Pulse Signals
A12	GRN/BLK	Fuel Pump Relay	Relay Off: 12v, On: 1v
A13	YEL/BLK	Main Relay Power (B+)	12-14v
A14	GRN/BLK	Fuel Pump Relay	Relay Off: 12v, On: 1v
A15	YEL/BLK	Main Relay Power (B+)	12-14v
A16	BRN/BLK	Power Ground	<0.1v
A17	BLK/YEL	Ignition Power	12-14v
A18	BLK/RED	Power Ground	<0.1v

1990-91 Hatchback 1.5L I4 MFI VIN ED6 [All] 20P 'B' Connector

PCM Pin #	W/Color	Circuit Description (20-Pin)	Value at Hot Idle
B1	WHT/GRN	Keep Alive Power (VBU)	12-14v
B2	GRN/YEL	Upshift Indicator Light	Lamp Off: 12v, On: 1v
B2	---	A/T: Not Used	---
B3	YEL	A/C Clutch Relay	Relay Off: 12v, On: 1v
B4	YEL/GRN	Radiator Fan Relay	Relay Off: 12v, On: 1v
B5	WHT/YEL	Alternator Charging Signal	Lights Off: 12v, On: 0v
B6	GRN/ORN	Check Engine Light	MIL Off: 12v, On: 1v
B7	GRN	Heater Fan Switch	Fan Off: 0v, On: 12v
B8	BLU/RED	A/C Switch Signal	Switch Off: 12v, On: 1v
B9	GRN/BLK	Reverse Switch	In 'R': 0v, Others: 12v
B10	ORN	CKP Sensor Signal	AC pulse signals
B11	GRN	M/T: Clutch Switch	Clutch In: 0v, Out: 5v
B12	WHT	CKP Sensor Ground	<0.050v
B13	BLU/WHT	Start Signal	Cranking: 9-11v
B14	BLU	Alternator 'FR' Signal	Digital Signals: 0-5-0v
B15	WHT	Igniter Control	Digital Signals: 0-12-0v
B16	YEL/RED	Vehicle Speed Sensor	Moving: pulse signals
B17	WHT	Igniter Control	Digital Signals: 0-12-0v
B18	---	Not Used	---
B19	GRN/RED	Electric Load Detector	Varies: 0.5-4.5v
B20	BRN	Ignition Timing Adjuster	0.5-4.5v

1990-91 Hatchback 1.5L I4 MFI VIN ED6 [All] 16P 'C' Connector

PCM Pin #	W/Color	Circuit Description (16-Pin)	Value at Hot Idle
C1	BLU/GRN	CYP Sensor Signal	AC pulse signals
C2	BLU/YEL	CYP Sensor Ground	<0.050v
C3	ORN/BLU	TDC Sensor Signal	AC pulse signals
C4	WHT/BLU	TDC Sensor Ground	<0.050v
C5	RED/YEL	IAT Sensor Signal	Varies w/temp. (0.5-4.9v)
C6	RED/WHT	ECT Sensor Signal	At 180°F: 0.51v
C7	RED/BLU	Throttle Angle Sensor Signal	0.5-0.6v
C8	YEL	EGR Valve Lift Sensor	1.2v
C9	RED/WHT	Atmospheric Pressure Sensor	2.76-2.96v at sea level
C10	GRN/WHT	Brake Switch Signal	Brake Off: 0v, On: 12v
C11	WHT	MAP Sensor Signal	0.8-0.9v
C12	GRN/WHT	Sensor Ground	<0.050v
C13	YEL/WHT	Sensor VREF	4.9-5.1v
C14	GRN/WHT	MAP Sensor Ground	<0.050v
C15	YEL/RED	MAP Sensor VREF	4.9-5.1v
C16	WHT	Oxygen Sensor Signal	0.1-1.1v

18-PIN CONNECTOR 'A'

| 1 | 3 | 5 | 7 | | 11 | 13 | 15 | 17 |
| 2 | 4 | 6 | 8 | 10 | 12 | 14 | 16 | 18 |

20-PIN CONNECTOR 'B'   **16-PIN CONNECTOR 'C'**

| 1 | 3 | 5 | 7 | 9 | 11 | 13 | 15 | 17 | 19 |
| 2 | 4 | 6 | 8 | 10 | 12 | 14 | 16 | 18 | 20 |

| 1 | 3 | 5 | 7 | 9 | 11 | 13 | 15 |
| 2 | 4 | 6 | 8 | 10 | 12 | 14 | 16 |

WIRE SIDE OF HARNESS TERMINALS

05533_ADIA_G613

Pin Connector Graphic
1992-93 Hatchback 1.5L MFI VIN EH2 [All] 26P 'A' Connector

PCM Pin #	W/Color	Circuit Description (26-Pin)	Value at Hot Idle
A1	YEL	Injector 1 Control	2.0-3.3 ms
A2	YEL	Injector 4 Control	2.0-3.3 ms
A3	RED	Injector 2 Control	2.0-3.3 ms
A4	---	Not Used	---
A5	BLU	Injector 3 Control	2.0-3.3 ms
A6	ORN/BLK	HO2S Heater Control	Heater on: 1v, Off: 12v
A7	GRN/YEL	Fuel Pump Relay	Relay Off: 12v, On: 1v
A8	GRN/YEL	Fuel Pump Relay	Relay Off: 12v, On: 1v
A9	GRN/WHT	Electronic Air Control Valve	Pulse Signals
A10-11	---	Not Used	---
A12	YEL/GRN	Radiator Fan Relay	Relay Off: 12v, On: 1v
A13	GRN/ORN	Check Engine Light	MIL Off: 12v, On: 1v
A14	---	Not Used	---
A15	BLK/RED	A/C Clutch Relay	Relay Off: 12v, On: 1v
A16	WHT/YEL	Alternator Charging Signal	Lights Off: 12v, On: 0v
A17	GRN/BLK	A/T: TCM Signal	Digital Signals
A18	---	Not Used	---
A19	YEL	A/T: TCM Signal	Digital Signals
A20	RED	EVAP Purge Solenoid	Solenoid Off: 12v, On: 1v
A21	RED/GRN	Igniter Control	Digital Signals: 0-12-0v
A22	RED/GRN	Igniter Control	Digital Signals: 0-12-0v
A23	BLK	Power Ground	<0.1v
A24	BLK	Power Ground	<0.1v
A25	YEL/BLK	Main Relay Power (B+)	12-14v
A26	BLK/RED	Logic Ground	<0.050v

1992-93 Hatchback 1.5L MFI VIN EH2 [All] 16P 'B' Connector

PCM Pin #	W/Color	Circuit Description (16-Pin)	Value at Hot Idle
B1	WHT/GRN	Main Relay Power (B+)	12-14v
B2	BRN/BLK	Logic Ground	<0.050v
B3	GRN/BLU	A/T: TCM Signal	In D3: 0v, Others: 5v
B4	GRN/BLK	A/T: TCM Signal	In D4: 0v, Others: 5v
B5	BLU/RED	A/C Switch Signal	Switch Off: 12v, On: 1v
B6	---	Not Used	---
B7	GRN	A/T: Park Neutral Switch	In P/N: 0v, Others: 11v
B8	BRN/RED	PSP Switch	Straight: 0v, Turning: 11v
B9	BLU/WHT	Starter Switch Signal	Cranking: 9-11v
B10	YEL/BLU	Vehicle Speed Sensor	Moving: pulse signals
B11	ORN	CYP Sensor Signal	AC pulse signals
B12	WHT	CYP Sensor Ground	<0.050v
B13	ORN/BLU	TDC Sensor Signal	AC pulse signals
B14	WHT/BLU	TDC Sensor Ground	<0.050v
B15	BLU/GRN	CKP Sensor Signal	AC pulse signals
B16	BLU/YEL	CKP Sensor Ground	<0.050v

1992-93 Hatchback 1.5L MFI VIN EH2 [All] 22P 'D' Connector

PCM Pin #	W/Color	Circuit Description (22-Pin)	Value at Hot Idle
D1	WHT/BLU	Keep Alive Power (VBU)	12-14v
D2	GRN/WHT	Brake Switch Signal	Brake Off: 0v, On: 12v
D3	---	Not Used	---
D4	BRN	Service Check Connector	SCS Open: 4.80v
D5-6	---	Not Used	---
D7	LT BLU	Data Link Connector	5v
D8	---	Not Used	---
D9	PNK	Alternator 'FR' Signal	Digital Signals: 0-5-0v
D10	GRN/RED	Electric Load Detector	Varies: 0.5-4.5v
D11	PNK/BLK	TP Sensor Signal	0.5-0.6v
D12	---	Not Used	---
D13	RED/WHT	ECT Sensor Signal	At 180°F: 0.51v
D14	ORN/BLU	HO2S-11 (B1 S1) Signal	0.1-1.1v
D14	WHT	HO2S-11 (B1 S1) Signal (CX)	0.1-1.1v
D15	RED/YEL	IAT Sensor Signal	Varies w/temp. (0.5-4.9v)
D16	---	Not Used	---
D17	WHT	MAP Sensor Signal	0.8-0.9v
D18	PNK/GRN	Economy Driving Indicator	Digital Signals
D19	YEL/GRN	MAP Sensor VREF	4.9-5.1v
D20	YEL/WHT	Sensor VREF	4.9-5.1v
D21	GRN/WHT	MAP Sensor Ground	<0.050v
D21	GRN/BLU	MAP Sensor Ground (CX)	<0.050v
D22	GRN/WHT	Sensor Ground	<0.050v

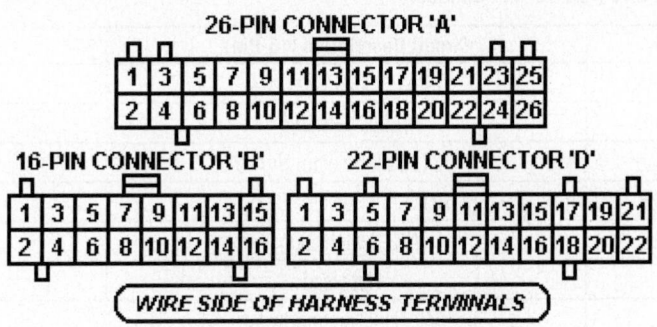

26-PIN CONNECTOR 'A'

16-PIN CONNECTOR 'B'   22-PIN CONNECTOR 'D'

WIRE SIDE OF HARNESS TERMINALS

05533_ADIA_G597

Pin Connector Graphic
1994-95 Hatchback 1.5L I4 MFI VIN EH2 [All] 26P 'A' Connector

PCM Pin #	W/Color	Circuit Description (26-Pin)	Value at Hot Idle
A1	BRN	Injector 1 Control	2.0-3.3 ms
A2	YEL	Injector 4 Control	2.0-3.3 ms
A3	RED	Injector 2 Control	2.0-3.3 ms
A4	---	Not Used	---
A5	BLU	Injector 3 Control	2.0-3.3 ms
A6	ORN/BLK	HO2S Heater Control	Relay Off: 12v, On: 1v
A7	GRN/YEL	Fuel Pump Relay	Relay Off: 12v, On: 1v
A8	---	Not Used	---
A9	GRN/WHT	Idle Air Control Valve	Pulse Signals
A10	---	Not Used	---
A11	---	Not Used	---
A12	YEL/GRN	Radiator Fan Relay	Relay Off: 12v, On: 1v
A13	GRN/ORN	Check Engine Light	MIL Off: 12v, On: 1v
A14	---	Not Used	---
A15	BLK/RED	A/C Clutch Relay	Relay Off: 12v, On: 1v
A16	WHT/YEL	Alternator Charging Signal	Lights Off: 12v, On: 0v
A17	GRN/BLK	A/T: Lockup Control Solenoid	Solenoid Off: 12v, On: 1v
A18	---	Not Used	---
A19	YEL	A/T: Lockup Control Solenoid	Solenoid Off: 12v, On: 1v
A20	RED	EVAP Purge Solenoid	Solenoid Off: 12v, On: 1v
A21	RED/GRN	Igniter Control	Digital Signals: 0-12-0v
A22	---	Not Used	---
A23	BLK	Power Ground	<0.1v
A24	BLK	Power Ground	<0.1v
A25	YEL/BLK	Main Relay Power (B+)	12-14v
A26	BLK/RED	Logic Ground	<0.050v

1994-95 Hatchback 1.5L I4 MFI VIN EH2 [All] 16P 'B' Connector

PCM Pin #	W/Color	Circuit Description (16-Pin)	Value at Hot Idle
B1	WHT/GRN	Main Relay Power (B+)	12-14v
B2	BRN/BLK	Logic Ground	<0.050v
B3	GRN/BLU	A/T: Shift Selector Signal	In P/N: 12v
B4	GRN/BLK	A/T: Shift Selector Signal	In P/N: 12v
B5	BLU/RED	A/C Switch Signal	Switch Off: 12v, On: 1v
B6	---	Not Used	---
B7	GRN	A/T: Park Neutral Switch	In P/N: 0v, Others: 11v
B7	GRN	M/T: Clutch Switch Signal	Clutch in: 11v
B8	BRN/RED	PSP Switch	Straight: 0v, Turning: 11v
B9	BLU/WHT	Starter Switch Signal	Cranking: 9-11v
B10	YEL/BLU	Vehicle Speed Sensor	Moving: pulse signals
B11	ORN	CYP Sensor Signal	AC pulse signals
B12	WHT	CYP Sensor Ground	<0.050v
B13	ORN/BLU	TDC Sensor Signal	AC pulse signals
B14	WHT/BLU	TDC Sensor Ground	<0.050v
B15	BLU/GRN	CKP Sensor Signal	AC pulse signals
B16	BLU/YEL	CKP Sensor Ground	<0.050v

1994-95 Hatchback 1.5L I4 MFI VIN EH2 [All] 22P 'D' Connector

PCM Pin #	W/Color	Circuit Description (22-Pin)	Value at Hot Idle
D1	WHT/BLU	Keep Alive Power (VBU)	12-14v
D2	GRN/WHT	Brake Switch Signal	Brake Off: 0v, On: 12v
D3	RED/BLU	Knock Sensor Signal	No Detonation: 18mv AC
D4	BRN	Service Check Connector	SCS Open: 4.80v
D5-6	---	Not Used	---
D7	LT BLU	Data Link Connector	5v
D8	---	Not Used	---
D9	PNK	Alternator 'FR' Signal	Digital Signals: 0-5-0v
D10	GRN/RED	Electric Load Detector	Varies: 0.5-4.5v
D11	PNK/BLK	TP Sensor Signal	0.5-0.6v
D12	---	Not Used	---
D13	RED/WHT	ECT Sensor Signal	At 180°F: 0.51v
D14	WHT	HO2S-11 (B1 S1) Signal	0.1-1.1v
D14	ORN/BLU	Oxygen Sensor Signal (CX)	0.1-1.1v
D15	RED/YEL	IAT Sensor Signal	Varies w/temp. (0.5-4.9v)
D16	---	Not Used	---
D17	WHT	MAP Sensor Signal	0.8-0.9v
D18	PNK/GRN	A/T: Interlock Control Unit	Key & Brake On: 12v
D19	YEL/GRN	MAP Sensor VREF	4.9-5.1v
D20	YEL/WHT	Sensor VREF	4.9-5.1v
D21	GRN/BLU	MAP Sensor Ground (CX, EX)	<0.050v
D21	GRN/WHT	MAP Sensor Ground (DC, LX)	<0.050v
D22	GRN/WHT	Sensor Ground	<0.050v

26-PIN CONNECTOR 'A'

| 1 | 3 | 5 | 7 | 9 | 11 | 13 | 15 | 17 | 19 | 21 | 23 | 25 |
| 2 | 4 | 6 | 8 | 10 | 12 | 14 | 16 | 18 | 20 | 22 | 24 | 26 |

16-PIN CONNECTOR 'B'   **22-PIN CONNECTOR 'D'**

| 1 | 3 | 5 | 7 | 9 | 11 | 13 | 15 |
| 2 | 4 | 6 | 8 | 10 | 12 | 14 | 16 |

| 1 | 3 | 5 | 7 | 9 | 11 | 13 | 15 | 17 | 19 | 21 |
| 2 | 4 | 6 | 8 | 10 | 12 | 14 | 16 | 18 | 20 | 22 |

WIRE SIDE OF HARNESS TERMINALS

05533_ADIA_G597

Pin Connector Graphic
1992-93 Hatchback 1.5L I4 VTEC VIN EH2 [All] 26P 'A' Connector

PCM Pin #	W/Color	Circuit Description (26-Pin)	Value at Hot Idle
A1	BRN	Injector 1 Control	2.0-3.3 ms
A2	YEL	Injector 4 Control	2.0-3.3 ms
A3	RED	Injector 2 Control	2.0-3.3 ms
A4	ORN/WHT	VTEC-E Solenoid	0v, Hi-Speed: 12v
A5	BLU	Injector 3 Control	2.0-3.3 ms
A6	ORN/BLK	HO2S Heater Control	Relay Off: 12v, On: 1v
A7, 8	GRN/YEL	Fuel Pump Relay	Relay Off: 12v, On: 1v
A9	GRN/WHT	Idle Air Control Valve	Pulse Signals
A10	---	Not Used	---
A11	ORN/BLU	EGR Valve Lift Sensor	1.2v
A12	YEL/GRN	Radiator Fan Relay	Relay Off: 12v, On: 1v
A13	GRN/ORN	Check Engine Light	MIL Off: 12v, On: 1v
A14	---	Not Used	---
A15	BLK/RED	A/C Clutch Relay	Relay Off: 12v, On: 1v
A16	WHT/YEL	Alternator Charging Signal	Lights Off: 12v, On: 0v
A17-19	---	Not Used	---
A20	RED	EVAP Purge Solenoid	Solenoid Off: 12v, On: 1v
A21	RED/GRN	Igniter Control	Digital Signals: 0-12-0v
A22	RED/GRN	Igniter Control	Digital Signals: 0-12-0v
A23	BLK	Power Ground	<0.1v
A24	BLK	Power Ground	<0.1v
A25	YEL/BLK	Main Relay Power (B+)	12-14v
A26	BLK/RED	Power Ground	<0.1v

1992-93 Hatchback 1.5L I4 VTEC VIN EH2 [All] 16P 'B' Connector

PCM Pin #	W/Color	Circuit Description (16-Pin)	Value at Hot Idle
B1	WHT/GRN	Main Relay Power (B+)	12-14v
B2	BRN/BLK	Logic Ground	<0.050v
B3-4	---	Not Used	---
B5	BLU/RED	A/C Switch Signal	Switch Off: 12v, On: 1v
B6	---	Not Used	---
B7	GRN	A/T: Park Neutral Switch	In P/N: 0v, Others: 11v
B8	BRN/RED	PSP Switch	Straight: 0v, Turning: 11v
B9	BLU/WHT	Starter Switch Signal	Cranking: 9-11v
B10	YEL/BLU	Vehicle Speed Sensor	Moving: pulse signals
B11	ORN	CYP Sensor Signal	AC pulse signals
B12	WHT	CYP Sensor Ground	<0.050v
B13	ORN/BLU	TDC Sensor Signal	AC pulse signals
B14	WHT/BLU	TDC Sensor Ground	<0.050v
B15	BLU/GRN	CKP Sensor Signal	AC pulse signals
B16	BLU/YEL	CKP Sensor Ground	<0.050v

1992-93 Hatchback 1.5L I4 VTEC VIN EH2 [All] 22P 'D' Connector

PCM Pin #	W/Color	Circuit Description (22-Pin)	Value at Hot Idle
D1	WHT/BLU	Keep Alive Power (VBU)	12-14v
D2	GRN/WHT	Brake Switch Signal	Brake Off: 0v, On: 12v
D3	BLU/YEL	LAF Signal	0.3-4.9v
D4	BRN	Service Check Connector	SCS Open: 4.80v
D6	ORN/BLU	VTEC-E Pressure Switch	0v, Hi-Speed: 12v
D7	LT BLU	Data Link Connector	5v
D8	WHT/BLU	LAF Sensor VS+ Signal	0.5-0.6v
D9	PNK	Alternator 'FR' Signal	Digital Signals: 0-5-0v
D10	GRN/RED	Electric Load Detector	Varies: 0.5-4.5v
D11	LT GRN	Throttle Angle Sensor Signal	0.5-0.6v
D12	WHT/BLK	EGR Valve Lift Sensor	1.2v
D13	RED/WHT	ECT Sensor Signal	At 180°F: 0.51v
D14	ORN/BLU	LAF Sensor IP+ Signal	0.5-5.3v
D15	RED/YEL	IAT Sensor Signal	Varies w/temp. (0.5-4.9v)
D16	BLU/GRN	LAF Sensor IP-, VS- Signal	2.6-2.8v
D17	WHT	MAP Sensor Signal	0.8-0.9v
D18	PNK/GRN	Economy Driving Indicator	---
D19	YEL/GRN	MAP Sensor VREF	4.9-5.1v
D20	YEL/WHT	Sensor VREF	4.9-5.1v
D21	GRN/WHT	MAP Sensor Ground	<0.050v
D22	GRN/WHT	Sensor Ground	<0.050v

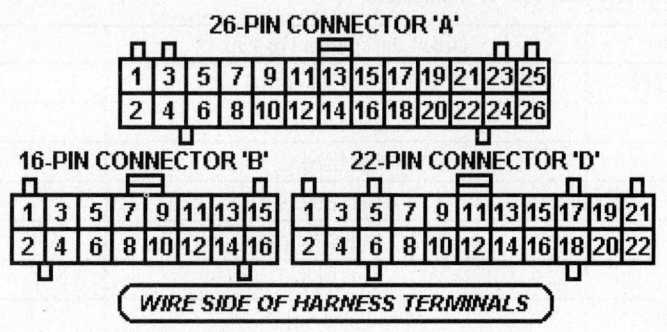

26-PIN CONNECTOR 'A'

16-PIN CONNECTOR 'B'   22-PIN CONNECTOR 'D'

WIRE SIDE OF HARNESS TERMINALS

05533_ADIA_G597

Pin Connector Graphic

1994-95 Hatchback 1.5L I4 VTEC VIN EH2 [All] 26P 'A' Connector

PCM Pin #	W/Color	Circuit Description (26-Pin)	Value at Hot Idle
A1	BRN	Injector 1 Control	2.0-3.3 ms
A2	YEL	Injector 4 Control	2.0-3.3 ms
A3	RED	Injector 2 Control	2.0-3.3 ms
A4	ORN/WHT	VTEC Solenoid	0v, Hi-Speed: 12v
A5	LT BLU	Injector 3 Control	2.0-3.3 ms
A6	ORN/BLK	HO2S Heater Control	Relay Off: 12v, On: 1v
A7	GRN/YEL	Fuel Pump Relay	Relay Off: 12v, On: 1v
A8	---	Not Used	---
A9	GRN/WHT	Idle Air Control Valve	Pulse Signals
A10	---	Not Used	---
A11	PNK/GRN	EGR Valve Lift Sensor	1.2v
A12	YEL/GRN	Radiator Fan Relay	Relay Off: 12v, On: 1v
A13	GRN/ORN	Check Engine Light	MIL Off: 12v, On: 1v
A14	---	Not Used	---
A15	BLK/RED	A/C Clutch Relay	Relay Off: 12v, On: 1v
A16	WHT/YEL	Alternator Charging Signal	Lights Off: 12v, On: 0v
A17	GRN/BLK	A/T: Lockup Control Solenoid	Solenoid Off: 12v, On: 1v
A18	---	Not Used	---
A19	YEL	A/T: Lockup Control Solenoid	Solenoid Off: 12v, On: 1v
A20	RED	EVAP Purge Solenoid	Solenoid Off: 12v, On: 1v
A21	RED/GRN	Igniter Control	Digital Signals: 0-12-0v
A22	---	Not Used	---
A23	BLK	Power Ground	<0.1v
A24	BLK	Power Ground	<0.1v
A25	YEL/BLK	Main Relay Power (B+)	12-14v
A26	BLK/RED	Power Ground	<0.1v

1994-95 Hatchback 1.5L I4 VTEC VIN EH2 [All] 16P 'B' Connector

PCM Pin #	W/Color	Circuit Description (16-Pin)	Value at Hot Idle
B1	WHT/GRN	Main Relay Power (B+)	12-14v
B2	BRN/BLK	Logic Ground	<0.050v
B3	---	Not Used	---
B4	---	Not Used	---
B5	BLU/RED	A/C Switch Signal	Switch Off: 12v, On: 1v
B6	---	Not Used	---
B7	GRN	M/T: Clutch Switch	Clutch In: 0v, Out: 5v
B8	BRN/RED	PSP Switch	Straight: 0v, Turning: 11v
B9	BLU/WHT	Starter Switch Signal	Cranking: 9-11v
B10	YEL/BLU	Vehicle Speed Sensor	Moving: pulse signals
B11	ORN	CYP Sensor Signal	AC pulse signals
B12	WHT	CYP Sensor Ground	<0.050v
B13	ORN/BLU	TDC Sensor Signal	AC pulse signals
B14	WHT/BLU	TDC Sensor Ground	<0.050v
B15	BLU/GRN	CKP Sensor Signal	AC pulse signals
B16	BLU/YEL	CKP Sensor Ground	<0.050v

1994-95 Hatchback 1.5L I4 VTEC VIN EH2 [All] 22P 'D' Connector

PCM Pin #	W/Color	Circuit Description (22-Pin)	Value at Hot Idle
D1	WHT/BLU	Keep Alive Power (VBU)	12-14v
D2	GRN/WHT	Brake Switch Signal	Brake Off: 0v, On: 12v
D3	BLU/YEL	LAF Sensor Signal	0.3-4.9v
D4	BRN	Service Check Connector	SCS Open: 4.80v
D6	ORN/BLU	VTEC Pressure Switch	0v, Hi-Speed: 12v
D7	LT BLU	Data Link Connector	5v
D8	WHT/BLU	LAF Sensor VS+ Signal	0.5-0.6v
D9	PNK	Alternator 'FR' Signal	Digital Signals: 0-5-0v
D10	GRN/RED	Electric Load Detector	Varies: 0.5-4.5v
D11	PNK/BLK	TP Sensor Signal	0.5-0.6v
D12	WHT/BLK	EGR Valve Lift Sensor	1.2v
D13	RED/WHT	ECT Sensor Signal	At 180°F: 0.51v
D14	ORN/BLU	LAF Sensor IP+ Signal	0.5-5.3v
D15	RED/YEL	IAT Sensor Signal	Varies w/temp. (0.5-4.9v)
D16	BLU/GRN	LAF Sensor IP-, VS- Signal	2.6-2.8v
D17	WHT	MAP Sensor Signal	0.8-0.9v
D18	PNK/GRN	Economy Driving Indicator	Digital Signals
D19	YEL/GRN	MAP Sensor VREF	4.9-5.1v
D20	YEL/WHT	Sensor VREF	4.9-5.1v
D21	GRN/WHT	MAP Sensor Ground	<0.050v
D22	GRN/WHT	Sensor Ground	<0.050v

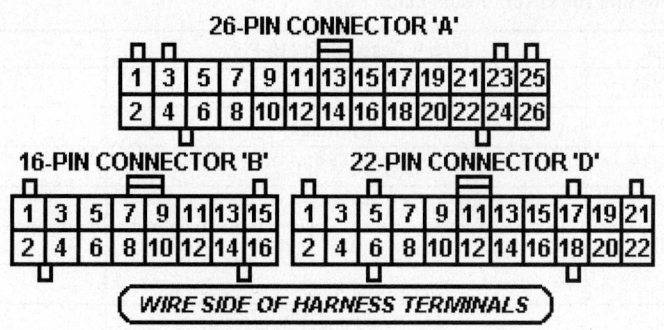

26-PIN CONNECTOR 'A'

16-PIN CONNECTOR 'B'   22-PIN CONNECTOR 'D'

WIRE SIDE OF HARNESS TERMINALS

05533_ADIA_G597

Pin Connector Graphic
1992-93 Hatchback 1.6L I4 VTEC VIN EH3 [M/T] 26P A Connector

PCM Pin #	W/Color	Circuit Description (26-Pin)	Value at Hot Idle
A1	BRN	Injector 1 Control	2.0-3.3 ms
A2	YEL	Injector 4 Control	2.0-3.3 ms
A3	RED	Injector 2 Control	2.0-3.3 ms
A4	ORN/WHT	VTEC Solenoid	Solenoid Off: 12v, On: 1v
A5	BLU	Injector 3 Control	2.0-3.3 ms
A6	ORN/BLK	HO2S Heater Control	Relay Off: 12v, On: 1v
A7, 8	GRN/YEL	Fuel Pump Relay	Relay Off: 12v, On: 1v
A9	GRN/WHT	Idle Air Control Valve	Pulse Signals
A10-11	---	Not Used	---
A12	YEL/GRN	Radiator Fan Relay	Relay Off: 12v, On: 1v
A13	GRN/ORN	Check Engine Light	MIL Off: 12v, On: 1v
A14	---	Not Used	---
A15	BLK/RED	A/C Clutch Relay	Relay Off: 12v, On: 1v
A16	WHT/YEL	Alternator Charging Signal	Lights Off: 12v, On: 0v
A17	LT GRN	A/T: TCM Signal	Digital Signals
A18	---	Not Used	---
A19	YEL	A/T: TCM Signal	Digital Signals
A20	RED	EVAP Purge Solenoid	Solenoid Off: 12v, On: 1v
A21	RED/GRN	Igniter Control	Digital Signals: 0-12-0v
A22	RED/GRN	Igniter Control	Digital Signals: 0-12-0v
A23	BLK	Power Ground	<0.1v
A24	BLK	Power Ground	<0.1v
A25	YEL/BLK	Main Relay Power (B+)	12-14v
A26	BLK/RED	Power Ground	<0.1v

1992-93 Hatchback 1.6L I4 VTEC VIN EH3 [M/T] 16P B Connector

PCM Pin #	W/Color	Circuit Description (16-Pin)	Value at Hot Idle
B1	WHT/GRN	Main Relay Power (B+)	12-14v
B2	BRN/BLK	Logic Ground	<0.050v
B3	GRN/BLU	A/T: TCM Signal	In D3: 0v, Others: 5v
B4	GRN/BLK	A/T: TCM Signal	In D4: 0v, Others: 5v
B5	BLU/RED	A/C Switch Signal	Switch Off: 12v, On: 1v
B6	---	Not Used	---
B7	GRN	A/T: Park Neutral Switch	In P/N: 0v, Others: 11v
B8	BRN/RED	PSP Switch	Straight: 0v, Turning: 11v
B9	BLU/WHT	Starter Switch Signal	Cranking: 9-11v
B10	YEL/BLU	Vehicle Speed Sensor	Moving: pulse signals
B11	ORN	CYP Sensor Signal	AC pulse signals
B12	WHT	CYP Sensor Ground	<0.050v
B13	ORN/BLU	TDC Sensor Signal	AC pulse signals
B14	WHT/BLU	TDC Sensor Ground	<0.050v
B15	BLU/GRN	CKP Sensor Signal	AC pulse signals
B16	BLU/YEL	CKP Sensor Ground	<0.050v

1992-93 Hatchback 1.6L I4 VTEC VIN EH3 [M/T] 22P D Connector

PCM Pin #	W/Color	Circuit Description (22-Pin)	Value at Hot Idle
D1	WHT/BLU	Keep Alive Power (VBU)	12-14v
D2	GRN/WHT	Brake Switch Signal	Brake Off: 0v, On: 12v
D4	BRN	Data Link Connector	5v
D6	ORN/BLU	VTEC Pressure Switch	0v, Hi-Speed: 12v
D7	LT BLU	Data Link Connector	5v
D9	PNK	Alternator 'FR' Signal	Digital Signals: 0-5-0v
D10	GRN/RED	Electric Load Detector	Varies: 0.5-4.5v
D11	LT GRN	Throttle Angle Sensor Signal	0.5-0.6v
D13	RED/WHT	ECT Sensor Signal	At 180°F: 0.51v
D14	ORN/BLU	HO2S Signal	0.1-1.1v
D15	RED/YEL	IAT Sensor Signal	Varies w/temp. (0.5-4.9v)
D17	WHT	MAP Sensor Signal	0.8-0.9v
D18	PNK/GRN	Economy Driving Indicator	Digital Signals
D19	YEL/GRN	MAP Sensor VREF	4.9-5.1v
D20	YEL/WHT	Sensor VREF	4.9-5.1v
D21	GRN/BLU	MAP Sensor Ground	<0.050v
D22	GRN/WHT	Sensor Ground	<0.050v

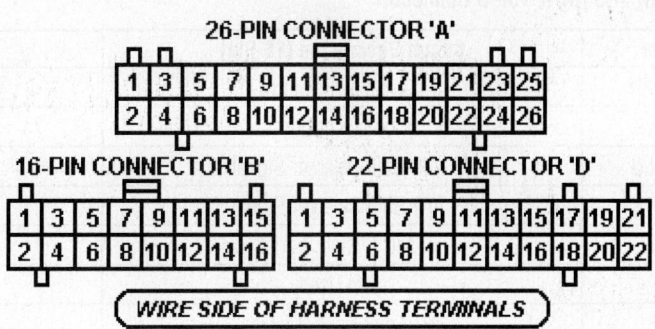

26-PIN CONNECTOR 'A'

16-PIN CONNECTOR 'B'   22-PIN CONNECTOR 'D'

WIRE SIDE OF HARNESS TERMINALS

05533_ADIA_G597

Pin Connector Graphic
Standard Colors and Abbreviations

Abbreviation	Color	Abbreviation	Color	Abbreviation	Color
BLK	Black	LT BLU	Lt. Blue	TAN	Tan
BLU	Blue	LT GRN	Lt. Green	VIO	Violet
BRN	Brown	ORN	Orange	WHT	White
GRY	Gray	PNK	Pink	YEL	Yellow
GRN	Green	PPL	Purple		

1994-95 Hatchback 1.6L I4 VTEC VIN EH3 [M/T] 26P A Connector

PCM Pin #	W/Color	Circuit Description (26-Pin)	Value at Hot Idle
A1	BRN	Injector 1 Control	2.0-3.3 ms
A2	YEL	Injector 4 Control	2.0-3.3 ms
A3	RED	Injector 2 Control	2.0-3.3 ms
A4	ORN/WHT	VTEC Solenoid	0v, Hi-Speed: 12v
A5	BLU	Injector 3 Control	2.0-3.3 ms
A6	ORN/BLK	HO2S Heater Control	Relay Off: 12v, On: 1v
A7	GRN/YEL	Fuel Pump Relay	Relay Off: 12v, On: 1v
A8	---	Not Used	---
A9	GRN/WHT	Idle Air Control Valve	Pulse Signals
A10-11	---	Not Used	---
A12	YEL/GRN	Radiator Fan Relay	Relay Off: 12v, On: 1v
A13	GRN/ORN	Check Engine Light	MIL Off: 12v, On: 1v
A14	---	Not Used	---
A15	BLK/RED	A/C Clutch Relay	Relay Off: 12v, On: 1v
A16	WHT/YEL	Alternator Charging Signal	Lights Off: 12v, On: 0v
A17	GRN/BLK	A/T: Lockup Control Solenoid	Solenoid Off: 12v, On: 1v
A18	---	Not Used	---
A19	YEL	A/T: Lockup Control Solenoid	Solenoid Off: 12v, On: 1v
A20	RED	EVAP Purge Solenoid	Solenoid Off: 12v, On: 1v
A21	RED/GRN	Igniter Control	Digital Signals: 0-12-0v
A22	---	Not Used	---
A23	BLK	Power Ground	<0.1v
A24	BLK	Power Ground	<0.1v
A25	YEL/BLK	Main Relay Power (B+)	12-14v
A26	BLK/RED	Power Ground	<0.1v

1994-95 Hatchback 1.6L I4 VTEC VIN EH3 [M/T] 16P B Connector

PCM Pin #	W/Color	Circuit Description (16-Pin)	Value at Hot Idle
B1	WHT/GRN	Main Relay Power (B+)	12-14v
B2	BRN/BLK	Logic Ground	<0.050v
B3	GRN/BLU	A/T: Shift Selector Signal	In D4: 0v, Others: 12v
B4	GRN/BLK	A/T: Shift Selector Signal	In D3: 0v, Others: 12v
B5	BLU/RED	A/C Switch Signal	Switch Off: 12v, On: 1v
B6	---	Not Used	---
B7	GRN	A/T: Park Neutral Switch	In P/N: 0v, Others: 12v
B8	BRN/RED	PSP Switch	Straight: 0v, Turning: 11v
B9	BLU/WHT	Starter Switch Signal	Cranking: 9-11v
B10	YEL/BLU	Vehicle Speed Sensor	Moving: pulse signals
B11	ORN	CYP Sensor Signal	AC pulse signals
B12	WHT	CYP Sensor Ground	<0.050v
B13	ORN/BLU	TDC Sensor Signal	AC pulse signals
B14	WHT/BLU	TDC Sensor Ground	<0.050v
B15	BLU/GRN	CKP Sensor Signal	AC pulse signals
B16	BLU/YEL	CKP Sensor Ground	<0.050v

1994-95 Hatchback 1.6L I4 VTEC VIN EH3 [M/T] 22P D Connector

PCM Pin #	W/Color	Circuit Description (22-Pin)	Value at Hot Idle
D1	WHT/BLU	Keep Alive Power (VBU)	12-14v
D2	GRN/WHT	Brake Switch Signal	Brake Off: 0v, On: 12v
D3	RED/BLU	Knock Sensor Signal	No Detonation: 18mv AC
D4	BRN	Service Check Connector	SCS Open: 4.80v
D5	---	Not Used	---
D6	ORN/BLU	VTEC Pressure Switch	0v, Hi-Speed: 12v
D7	LT BLU	Data Link Connector	5v
D8	---	Not Used	---
D9	PNK	Alternator 'FR' Signal	Digital Signals: 0-5-0v
D10	GRN/RED	Electric Load Detector	Varies: 0.5-4.5v
D11	PNK/BLK	TP Sensor Signal	0.5-0.6v
D12	---	Not Used	---
D13	RED/WHT	ECT Sensor Signal	At 180°F: 0.51v
D14	WHT	HO2S Signal	0.1-1.1v
D15	RED/YEL	IAT Sensor Signal	Varies w/temp. (0.5-4.9v)
D16	---	Not Used	---
D17	WHT	MAP Sensor Signal	0.8-0.9v
D18	PNK/GRN	A/T: Interlock Control Unit	Key & Brake On: 12v
D19	YEL/GRN	MAP Sensor VREF	4.9-5.1v
D20	YEL/WHT	Sensor VREF	4.9-5.1v
D21	GRN/BLU	MAP Sensor Ground	<0.050v
D22	GRN/WHT	Sensor Ground	<0.050v

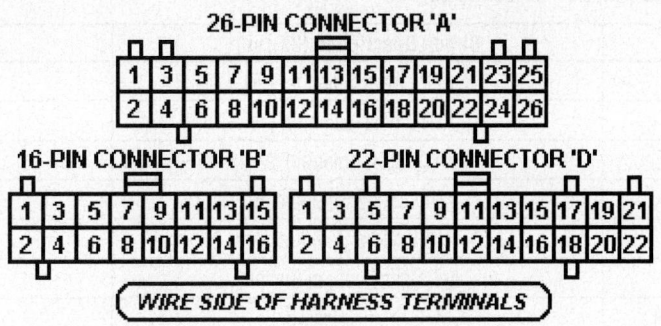

26-PIN CONNECTOR 'A'

16-PIN CONNECTOR 'B'   22-PIN CONNECTOR 'D'

WIRE SIDE OF HARNESS TERMINALS

05533_ADIA_G597

Pin Connector Graphic
1996 Hatchback 1.6L I4 MFI VIN EJ6 [All] 32P 'A' Connector

PCM Pin #	W/Color	Circuit Description (32-Pin)	Value at Hot Idle
A1	YEL	Injector 4 Control	2.0-3.3 ms
A2	BLU	Injector 3 Control	2.0-3.3 ms
A3	RED	Injector 2 Control	2.0-3.3 ms
A4	BRN	Injector 1 Control	2.0-3.3 ms
A5	BLK/WHT	HO2S-12 (B1 S2) Heater	Digital Signals: 0-12-0v
A6	BLK/WHT	HO2S-11 (B1 S1) Heater	Digital Signals: 0-12-0v
A9	BRN/BLK	Logic Ground	<0.050v
A10, 23	BLK	Power Ground	<0.1v
A11	YEL/BLK	Main Relay Power (B+)	12-14v
A12	BLK/BLU	M/T: Idle Air Control Valve	Pulse Signals
A13	ORN	Idle Air Control Valve 'N'	Pulse Signals
A14	BLK/BLU	Idle Air Control Valve 'P'	Pulse Signals
A15	RED/YEL	EVAP Purge Solenoid	Solenoid Off: 12v, On: 1v
A16	GRN/YEL	Fuel Pump Relay	Relay Off: 12v, On: 1v
A17	BLK/RED	A/C Clutch Relay	Relay Off: 12v, On: 1v
A18	GRN/ORN	Check Engine Light	MIL Off: 12v, On: 1v
A19	WHT/GRN	Alternator Charging Signal	Lights Off: 12v, On: 0v
A20	YEL/GRN	Igniter Control	Digital Signals: 0-12-0v
A22	BRN/BLK	Logic Ground	<0.050v
A23	BLK	Power Ground	<0.1v
A27	GRN	Radiator Fan Relay	Relay Off: 12v, On: 1v

1996 Hatchback 1.6L I4 MFI VIN EJ6 [All] 25P 'B' Connector

PCM Pin #	W/Color	Circuit Description (25-Pin)	Value at Hot Idle
B1	WHT	A/T: Linear Solenoid (-)	Pulse Signals
B2	RED	A/T: Linear Solenoid (+)	Pulse Signals
B3	BLU/YEL	A/T: Shift Solenoid 'A'	SSA Off: 0v, On: 12v
B4	GRN/BLK	A/T: Lockup Solenoid 'B'	LSB On: 12v, Off: 0v
B5	YEL	A/T: Lockup Solenoid 'A'	LSA On: 12v, Off: 0v
B8	PNK	A/T: Gear Position Switch	In D3: 0v, Others: 12v
B11	GRN/WHT	A/T: Shift Solenoid 'B'	SSB Off: 0v, On: 12v
B12	WHT/RED	Interlock Control Unit Signal	Key & Brake On: 12v
B13	GRN/BLK	A/T: D4 Indicator Light	D4 On: 12v, Off: 0v
B14	WHT	Mainshaft Speed Sensor 'N'	AC Pulse Signals
B15	RED	Mainshaft Speed Sensor 'P'	AC Pulse Signals
B16	WHT	A/T: Gear Position Switch	In 'R': 0v, Others: 12v
B17	BLU	A/T: Gear Position Switch	In D2: 0v, Others: 12v
B22	GRN	Countershaft Speed Sensor N	Moving: AC pulses
B23	ORN	Countershaft Speed Sensor P	Moving: AC pulses
B24	YEL	A/T: Gear Position Switch	In D4: 0v, Others: 12v
B25	LT GRN	A/T: Gear Position Switch	In P/N: 0v, Others: 12v

32-PIN CONNECTOR 'A'   **25-PIN CONNECTOR 'B'**

WIRE SIDE OF HARNESS TERMINALS

05533_ADIA_G604

Pin Connector Graphic

1996 Hatchback 1.6L I4 MFI VIN EJ6 [All] 31P 'C' Connector

PCM Pin #	W/Color	Circuit Description (31-Pin)	Value at Hot Idle
C1	BLU/RED	CKF Sensor Signal	AC pulse signals
C2	BLU	CKP Sensor Signal	AC pulse signals
C3	GRN	TDC Sensor Signal	AC pulse signals
C4	YEL	CYP Sensor Signal	AC pulse signals
C5	BLU/RED	A/C Switch Signal	Switch Off: 12v, On: 1v
C6	BLU/ORN	Starter Switch Signal	Cranking: 9-11v
C7	BRN	Service Check Connector	SCS Open: 4.80v
C8	LT BLU	K-Line Signal	12v
C9	---	Not Used	---
C10	WHT/BLU	Keep Alive Power (VBU)	12-14v
C11	WHT/RED	CKF Sensor Ground	<0.050v
C12	WHT	CKP Sensor Ground	<0.050v
C13	RED	TDC Sensor Ground	<0.050v
C14	BLK	CYP Sensor Ground	<0.050v

1996 Hatchback 1.6L I4 MFI VIN EJ6 [All] 31P 'C' Connector, *continued*

C16	GRN	PSP Switch	Straight: 0v, Turning: 11v
C17	WHT/RED	Alternator 'FR' Signal	Digital Signals: 0-5-0v
C18	BLU/WHT	Vehicle Speed Sensor	Moving: pulse signals
C19-29	---	Not Used	---
C30	PNK	A/T: CVT TMB Signal	Pulse Signals
C31	---	Not Used	---

1996 Hatchback 1.6L I4 MFI VIN EJ6 [All] 16P 'D' Connector

PCM Pin #	W/Color	Circuit Description (16-Pin)	Value at Hot Idle
D1	RED/BLK	TP Sensor Signal	0.5-0.6v
D2	RED/WHT	ECT Sensor Signal	At 180°F: 0.51v
D3	RED/GRN	MAP Sensor Signal	0.8-0.9v
D4	YEL/RED	MAP Sensor VREF	4.9-5.1v
D5	GRN/WHT	Brake Switch Signal	Brake Off: 0v, On: 12v
D7	WHT	HO2S-11 (B1 S1) Signal	0.1-1.1v
D8	RED/YEL	IAT Sensor Signal	Varies w/temp. (0.5-4.9v)
D9	---	Not Used	---
D10	YEL/BLU	Sensor VREF	4.9-5.1v
D11	GRN/BLK	Sensor Ground	<0.050v
D12	GRN/WHT	MAP Sensor Ground	<0.050v
D13	RED/YEL	HO2S-12 Ground	0.1-1.1v
D14	WHT/RED	HO2S-12 (B1 S2) Signal	0.1-1.1v
D16	GRN/RED	Electric Load Detector	Varies: 0.5-4.5v

31-PIN 'C'   *WIRE SIDE OF HARNESS TERMINALS*   16-PIN 'D'

05533_ADIA_G614

Pin Connector Graphic

1997-99 Hatchback 1.6L I4 MFI VIN EJ6 [All] 32P 'A' Connector

PCM Pin #	W/Color	Circuit Description (32-Pin)	Value at Hot Idle
A1, A2	YEL, BLU	Injector 4, Injector 3 Control	2.0-3.3 ms
A3, A4	RED, BRN	Injector 2, Injector 1 Control	2.0-3.3 ms
A5	BLK/WHT	HO2S-12 (B1 S2) Heater	Digital Signals: 0-12-0v
A6	BLK/WHT	HO2S-11 (B1 S1) Heater	Digital Signals: 0-12-0v
A7	RED	A/T: EGR Solenoid Control	Solenoid Off: 12v, On: 1v
A7	PNK	M/T: E-EGR Solenoid	Digital Signals: 0-12-0v
A9, A22	BRN/BLK	Logic Ground (LG1, LG2)	<0.050v
A10, 23	BLK	Power Ground (PG1, PG2)	<0.1v
A11, 24	YEL/BLK	Main Relay Power (B+)	12-14v
A12	BLK/BLU	M/T: Idle Air Control Valve	Pulse Signals
A13	ORN	A/T: Idle Air Control Valve 'N'	Pulse Signals
A14	BLK/BLU	A/T: Idle Air Control Valve 'P'	Pulse Signals
A15	RED/YEL	EVAP Purge Solenoid	Solenoid Off: 12v, On: 1v
A16	GRN/YEL	Fuel Pump Relay	Relay Off: 12v, On: 1v
A17	BLK/RED	A/C Clutch Relay	Relay Off: 12v, On: 1v
A18	GRN/ORN	Malfunction Indicator Light	MIL Off: 12v, On: 1v
A19	WHT/GRN	Alternator Charging Signal	Lights Off: 12v, On: 0v
A20	YEL/GRN	Igniter Control	Digital Signals: 0-12-0v
A27	GRN	Radiator Fan Relay Control	Relay Off: 12v, On: 1v
A28	BLU	EVAP Bypass Solenoid	Solenoid Off: 12v, On: 1v
A29	WHT/RED	EVAP Vent Solenoid	Solenoid Off: 12v, On: 1v

1997-99 Hatchback 1.6L I4 MFI VIN EJ6 [All] 25P 'B' Connector

PCM Pin #	W/Color	Circuit Description (25-Pin)	Value at Hot Idle
B1	WHT	A/T: Linear Solenoid (-)	Pulse Signals
B2	RED	A/T: Linear Solenoid (+)	Pulse Signals
B3	BLU/YEL	A/T: Shift Solenoid 'A'	SSA On: 12v: Off: 0v
B4	GRN/BLK	A/T: Lockup Solenoid 'B'	LSB On: 12v: Off: 0v
B5	YEL	A/T: Lockup Solenoid 'A'	LSA On: 12v: Off: 0v
B8	PNK	A/T: Gear Position Switch	In D3: 0v, Others: 12v
B11	GRN/WHT	A/T: Shift Solenoid 'B'	SSB Off: 0v, On: 12v
B12	WHT/RED	Interlock Control Unit Signal	Key & Brake On: 12v
B13	GRN/BLK	A/T: D4 Indicator Light	D4 On: 12v, Off: 0v
B14	WHT	Mainshaft Speed Sensor 'P'	AC Pulse Signals
B15	RED	Mainshaft Speed Sensor 'N'	AC Pulse Signals
B16	WHT	A/T: Gear Position Switch	In 'R': 0v, Others: 12v
B17	BLU	A/T: Gear Position Switch	In D2: 0v, Others: 12v
B22	GRN	Countershaft Speed Sensor N	Moving: AC pulses
B23	BLU	Countershaft Speed Sensor P	Moving: AC pulses
B24	YEL	A/T: Gear Position Switch	In D4: 0v, Others: 12v
B25	LT GRN	A/T: Gear Position Switch	In P/N: 0v, Others: 12v

| 1 | 2 | 3 | 4 | | 5 | 6 | 7 | | 8 | 9 | 10 | | 11 | | | 1 | | 2 | | 3 | 4 | 5 | | 6 | 7 | 8 |

(connector graphic)

32-Pin 'A'   **WIRE SIDE OF HARNESS TERMINALS**   25-Pin 'B'

05533_ADIA_G606

Pin Connector Graphic

1997-99 Hatchback 1.6L I4 MFI VIN EJ6 [All] 31P 'C' Connector

PCM Pin #	W/Color	Circuit Description (31-Pin)	Value at Hot Idle
C1	BLU/RED	CKF Sensor Signal	AC pulse signals
C2	BLU	CKP Sensor Signal	AC pulse signals
C3	GRN	TDC Sensor Signal	AC pulse signals
C4	YEL	CYP Sensor Signal	AC pulse signals
C5	BLU/RED	A/C Switch Signal	Switch Off: 12v, On: 1v
C6	BLU/ORN	Starter Switch Signal	Cranking: 9-11v
C7	BRN	Service Check Connector	SCS Open: 4.80v
C8	LT BLU	K-Line Signal	12v
C10	WHT/BLU	Keep Alive Power (VBU)	12-14v
C11	WHT/RED	CKF Sensor Ground	<0.050v
C12	WHT	CKP Sensor Ground	<0.050v
C13	RED	TDC Sensor Ground	<0.050v
C14	BLK	CYP Sensor Ground	<0.050v
C16	GRN	PSP Switch Signal	Straight: 0v, Turning: 11v
C17	WHT/RED	Alternator 'FR' Signal	Digital Signals: 0-5-0v
C18	BLU/WHT	Vehicle Speed Sensor	Moving: pulse signals
C23	BLK	LAF Pump Cell (IP+)	0.5-3.5v
C24	RED	LAF Common (IP-, VS-)	2.6-2.8v
C25	WHT	LAF VS Cell Voltage (VS+)	7v
C29	LT GRN	A/T: Gear Position Switch	In P/N: 0v, Others: 12v
C29	RED	M/T: Clutch Switch Signal	Clutch In: 0v, Out: 5v
C30	PNK	CVT TMB Signal	Pulse Signals

1997-99 Hatchback 1.6L I4 MFI VIN EJ6 16P 'D' Connector

PCM Pin #	W/Color	Circuit Description (16-Pin)	Value at Hot Idle
D1	RED/BLK	TP Sensor Signal	0.5-0.6v
D2	RED/WHT	ECT Sensor Signal	At 180°F: 0.51v
D3	RED/GRN	MAP Sensor Signal	0.8-0.9v
D4	YEL/RED	MAP Sensor VREF	4.9-5.1v
D5	GRN/WHT	Brake Switch Signal	Brake Off: 0v, On: 12v
D6	RED/BLU	Knock Sensor Signal	No detonation: 18mv AC
D7	WHT	HO2S-11 (B1 S1) Signal	0.1-1.1v
D8	RED/YEL	IAT Sensor Signal	Varies w/temp. (0.5-4.9v)
D9	WHT/BLK	EGR Valve Lift Sensor	1.2v
D10	YEL/RED	Sensor VREF	4.9-5.1v
D11	GRN/BLK	Sensor Ground	<0.050v
D12	GRN/WHT	MAP Sensor Ground	<0.050v
D13	RED/YEL	HO2S-12 Ground	<0.050v
D14	WHT/RED	HO2S-12 (B1 S2) Signal	0.1-1.1v
D15	LT GRN	Fuel Tank Pressure Sensor	Fuel Cap off: 2.5v
D16	GRN/RED	Electric Load Detector	Varies: 0.5-4.5v

31-Pin 'C'   WIRE SIDE OF HARNESS TERMINALS   16-Pin 'D'

Pin Connector Graphic

05533_ADIA_G607

2000 Hatchback 1.6L I4 MFI VIN EJ6 [All] 32P 'A' Connector

PCM Pin #	W/Color	Circuit Description (32-Pin)	Value at Hot Idle
A1, A2	YEL, BLU	Injector 4, Injector 3 Control	2.0-3.3 ms
A3, A4	RED, BRN	Injector 2, Injector 1 Control	2.0-3.3 ms
A5	BLK/WHT	HO2S-12 (B1 S2) Heater	Digital Signals: 0-12-0v
A6	BLK/WHT	HO2S-11 (B1 S1) Heater	Digital Signals: 0-12-0v
A7	RED	A/T: EGR Solenoid Control	Solenoid Off: 12v, On: 1v
A7	PNK	M/T: E-EGR Solenoid	Digital Signals: 0-12-0v
A9, A22	BRN/BLK	Logic Ground (LG1, LG2)	<0.050v
A10, 23	BLK	Power Ground (PG1, PG2)	<0.1v
A11, 24	YEL/BLK	Main Relay Power (B+)	12-14v
A12	BLK/BLU	M/T: Idle Air Control Valve	Pulse Signals
A13	ORN	A/T: Idle Air Control Valve 'N'	Pulse Signals
A14	BLK/BLU	A/T: Idle Air Control Valve 'P'	Pulse Signals
A15	RED/YEL	EVAP Purge Solenoid	Solenoid Off: 12v, On: 1v
A16	GRN/YEL	Fuel Pump Relay	Relay Off: 12v, On: 1v
A17	BLK/RED	A/C Clutch Relay	Relay Off: 12v, On: 1v
A18	GRN/ORN	Malfunction Indicator Light	MIL Off: 12v, On: 1v
A19	WHT/GRN	Alternator Charging Signal	Lights Off: 12v, On: 0v
A20	YEL/GRN	Igniter Control	Digital Signals: 0-12-0v
A27	GRN	Radiator Fan Relay Control	Relay Off: 12v, On: 1v
A28	BLU	EVAP Bypass Solenoid	Solenoid Off: 12v, On: 1v
A29	WHT/RED	EVAP Vent Solenoid	Solenoid Off: 12v, On: 1v

2000 Hatchback 1.6L I4 MFI VIN EJ6 [All] 25P 'B' Connector

PCM Pin #	W/Color	Circuit Description (25-Pin)	Value at Hot Idle
B1	WHT	A/T: Linear Solenoid (-)	Pulse Signals
B2	RED	A/T: Linear Solenoid (+)	Pulse Signals
B3	BLU/YEL	A/T: Shift Solenoid 'A'	SSA On: 12v: Off: 0v
B4	GRN/BLK	A/T: Lockup Solenoid 'B'	LSB On: 12v: Off: 0v
B5	YEL	A/T: Lockup Solenoid 'A'	LSA On: 12v, Off: 0v
B8	PNK	A/T: Gear Position Switch	In D3: 0v, Others: 12v
B11	GRN/WHT	A/T: Shift Solenoid 'B'	SSB Off: 0v, On: 12v
B12	WHT/RED	Interlock Control Unit Signal	Key & Brake On: 12v
B13	GRN/BLK	A/T: D4 Indicator Light	D4 On: 12v, Off: 0v
B14	WHT	Mainshaft Speed Sensor 'P'	AC Pulse Signals
B15	RED	Mainshaft Speed Sensor 'N'	AC Pulse Signals
B16	WHT	A/T: Gear Position Switch	In 'R': 0v, Others: 12v
B17	BLU	A/T: Gear Position Switch	In D2: 0v, Others: 12v
B22	GRN	Countershaft Speed Sensor N	Moving: AC pulses
B23	BLU	Countershaft Speed Sensor P	Moving: AC pulses
B24	YEL	A/T: Gear Position Switch	In D4: 0v, Others: 12v
B25	LT GRN	A/T: Gear Position Switch	In P/N: 0v, Others: 12v

32-Pin 'A'   WIRE SIDE OF HARNESS TERMINALS   25-Pin 'B'

05533_ADIA_G606

Pin Connector Graphic
2000 Hatchback 1.6L I4 MFI VIN EJ6 [All] 31P 'C' Connector

PCM Pin #	W/Color	Circuit Description (31-Pin)	Value at Hot Idle
C1	BLU/RED	CKF Sensor Signal	AC pulse signals
C2	BLU	CKP Sensor Signal	AC pulse signals
C3	GRN	TDC Sensor Signal	AC pulse signals
C4	YEL	CYP Sensor Signal	AC pulse signals
C5	BLU/RED	A/C Switch Signal	Switch Off: 12v, On: 1v
C6	BLU/ORN	Starter Switch Signal	Cranking: 9-11v
C7	BRN	Service Check Connector	SCS Open: 4.80v
C8	LT BLU	K-Line Signal	12v
C10	WHT/BLU	Keep Alive Power (VBU)	12-14v
C11	WHT/RED	CKF Sensor Ground	<0.050v
C12	WHT	CKP Sensor Ground	<0.050v
C13	RED	TDC Sensor Ground	<0.050v
C14	BLK	CYP Sensor Ground	<0.050v
C16	GRN	PSP Switch Signal	Straight: 0v, Turning: 11v
C17	WHT/RED	Alternator 'FR' Signal	Digital Signals: 0-5-0v
C18	BLU/WHT	Vehicle Speed Sensor	Moving: pulse signals
C23	BLK	LAF Pump Cell (IP+)	0.5-3.5v
C24	RED	LAF Common (IP-, VS-)	2.6-2.8v
C25	WHT	LAF VS Cell Voltage (VS+)	7v
C29	LT GRN	A/T: Gear Position Switch	In P/N: 0v, Others: 12v
C29	RED	M/T: Clutch Switch Signal	Clutch In: 0v, Out: 5v
C30	PNK	CVT TMB Signal	Pulse Signals

2000 Hatchback 1.6L I4 MFI VIN EJ6 [All] 16P 'D' Connector

PCM Pin #	W/Color	Circuit Description (16-Pin)	Value at Hot Idle
D1	RED/BLK	TP Sensor Signal	0.5-0.6v
D2	RED/WHT	ECT Sensor Signal	At 180°F: 0.51v
D3	RED/GRN	MAP Sensor Signal	0.8-0.9v
D4, D10	YEL/RED	MAP Sensor, Sensor VREF	4.9-5.1v
D5	GRN/WHT	Brake Switch Signal	Brake Off: 0v, On: 12v
D6	RED/BLU	Knock Sensor Signal	No detonation: 18mv AC
D7	WHT	HO2S-11 (B1 S1) Signal	0.1-1.1v
D8	RED/YEL	IAT Sensor Signal	Varies w/temp. (0.5-4.9v)
D9	WHT/BLK	EGR Valve Lift Sensor	1.2v

2000 Hatchback 1.6L I4 MFI VIN EJ6 [All] 16P 'D' Connector, *continued*

D11	GRN/BLK	Sensor Ground	<0.050v
D12	GRN/WHT	MAP Sensor Ground	<0.050v
D13	RED/YEL	HO2S-12 Ground	<0.050v
D14	WHT/RED	HO2S-12 (B1 S2) Signal	0.1-1.1v
D15	LT GRN	Fuel Tank Pressure Sensor	Fuel Cap off: 2.5v
D16	GRN/RED	Electric Load Detector	Varies: 0.5-4.5v

31-Pin 'C'   *WIRE SIDE OF HARNESS TERMINALS*   **16-Pin 'D'**

05533_ADIA_G607

Pin Connector Graphic

1990-91 Sedan 1.5L I4 MFI VIN ED3 [All] 18P 'A' Connector

PCM Pin #	W/Color	Circuit Description (18-Pin)	Value at Hot Idle
A1	BRN	Injector 1 Control	2.0-3.3 ms
A2	BLK	Power Ground	<0.1v
A3	RED	Injector 2 Control	2.0-3.3 ms
A4	BLK	Power Ground	<0.1v
A5	BLU	Injector 3 Control	2.0-3.3 ms
A6	GRN	EVAP Purge Solenoid	Solenoid Off: 12v, On: 1v
A7	YEL	Injector 4 Control	2.0-3.3 ms
A8-9	---	Not Used	---
A10	ORN	EGR Solenoid	Solenoid Off: 12v, On: 1v
A11	BLU/YEL	Electronic Air Control Valve	Pulse Signals
A12	GRN/BLK	Fuel Pump Relay	Relay Off: 12v, On: 1v
A13	YEL/BLK	Main Relay Power (B+)	12-14v
A14	GRN/BLK	Fuel Pump Relay	Relay Off: 12v, On: 1v
A15	YEL/BLK	Main Relay Power (B+)	12-14v
A16	BRN/BLK	Power Ground	<0.1v
A17	BLK/YEL	Ignition Power	12-14v
A18	BLK/RED	Power Ground	<0.1v

1990-91 Sedan 1.5L I4 MFI VIN ED3 [All] 20P 'B' Connector

PCM Pin #	W/Color	Circuit Description (20-Pin)	Value at Hot Idle
B1	WHT/GRN	Keep Alive Power (VBU)	12-14v
B2	GRN/YEL	Upshift Indicator Light	Lamp Off: 12v, On: 1v
B3	YEL	A/C Clutch Relay	Relay Off: 12v, On: 1v
B4	YEL/GRN	Radiator Fan Relay	Relay Off: 12v, On: 1v
B5	WHT/YEL	Alternator Charging Signal	Lights Off: 12v, On: 0v
B6	GRN/ORN	Check Engine Light	MIL Off: 12v, On: 1v
B7	GRN	Heater Fan Switch	Fan Off: 0v, On: 12v
B8	BLU/RED	A/C Switch Signal	Switch Off: 12v, On: 1v
B9	GRN/BLK	Reverse Switch	In 'R': 0v, Others: 12v
B10	ORN	CKP Sensor Signal	AC pulse signals
B11	GRN	Clutch Switch	Clutch In: 0v, Out: 5v
B12	WHT	CKP Sensor Ground	<0.050v
B13	BLU/WHT	Start Signal	Cranking: 9-11v
B14	BLU	Alternator 'FR' Signal	Digital Signals: 0-5-0v
B15	WHT	Igniter Control	Digital Signals: 0-12-0v
B16	YEL/RED	Vehicle Speed Sensor	Moving: pulse signals
B17	WHT	Igniter Control	Digital Signals: 0-12-0v
B18	---	Not Used	---
B19	GRN/RED	Electric Load Detector	Varies: 0.5-4.5v
B20	BRN	Ignition Timing Adjuster	0.5-4.5v

1990-91 Sedan 1.5L I4 MFI VIN ED3 [All] 16P 'C' Connector

PCM Pin #	W/Color	Circuit Description (16-Pin)	Value at Hot Idle
C1	BLU/GRN	CYP Sensor Signal	AC pulse signals
C2	BLU/YEL	CYP Sensor Ground	<0.050v
C3	ORN/BLU	TDC Sensor Signal	AC pulse signals
C4	WHT/BLU	TDC Sensor Ground	<0.050v
C5	RED/YEL	IAT Sensor Signal	Varies w/temp. (0.5-4.9v)
C6	RED/WHT	ECT Sensor Signal	At 180°F: 0.51v
C7	RED/BLU	Throttle Angle Sensor Signal	0.5-0.6v
C8	YEL	EGR Valve Lift Sensor	1.2v
C9	RED/WHT	Atmospheric Pressure Sensor	2.76-2.96v at sea level
C10	GRN/WHT	Brake Switch Signal	Brake Off: 0v, On: 12v
C11	WHT	MAP Sensor Signal	0.8-0.9v
C12	GRN/WHT	Sensor Ground	<0.050v
C13	YEL/WHT	Sensor VREF	4.9-5.1v
C14	GRN/WHT	MAP Sensor Ground	<0.050v
C15	YEL/RED	MAP Sensor VREF	4.9-5.1v
C16	WHT	Oxygen Sensor Signal	0.1-1.1v

18-PIN CONNECTOR 'A'

1	3	5	7			11	13	15	17
2	4	6	8		10	12	14	16	18

20-PIN CONNECTOR 'B'   **16-PIN CONNECTOR 'C'**

WIRE SIDE OF HARNESS TERMINALS

05533_ADIA_G613

Pin Connector Graphic
1992-93 Sedan 1.5L I4 MFI VIN EG8 [All] 26P 'A' Connector

PCM Pin #	W/Color	Circuit Description (26-Pin)	Value at Hot Idle
A1	BRN	Injector 1 Control	2.0-3.3 ms
A2	YEL	Injector 4 Control	2.0-3.3 ms
A3	RED	Injector 2 Control	2.0-3.3 ms
A4	BLU	Injector 2 Control	2.0-3.3 ms
A4	---	Not Used	---
A6	ORN/BLK	HO2S-11 (B1 S1) Heater	Heater Off: 12v, On: 1v
A7	GRN/YEL	Fuel Pump Relay	Relay Off: 12v, On: 1v
A8	GRN/YEL	Fuel Pump Relay	Relay Off: 12v, On: 1v
A9	GRN/WHT	Electronic Air Control Valve	Pulse Signals
A10-11	---	Not Used	---
A12	YEL/GRN	Radiator Fan Relay	Relay Off: 12v, On: 1v
A13	GRN/ORN	Check Engine Light	MIL Off: 12v, On: 1v
A14	---	Not Used	---
A15	BLK/RED	A/C Clutch Relay	Relay Off: 12v, On: 1v
A16	WHT/YEL	Alternator Charging Signal	Lights Off: 12v, On: 0v
A17	GRN/BLK	A/T: TCM Signal	In D3: 0v, Others: 5v
A19	YEL	A/T: TCM Signal	In D4: 0v, Others: 5v
A20	RED	EVAP Purge Solenoid	Solenoid Off: 12v, On: 1v
A21	RED/GRN	Igniter Control	Digital Signals: 0-12-0v
A22	RED/GRN	Igniter Control	Digital Signals: 0-12-0v
A23	BLK	Power Ground	<0.1v
A24	BLK	Power Ground	<0.1v
A25	YEL/BLK	Main Relay Power (B+)	12-14v
A26	BLK/RED	Power Ground	<0.1v

1992-93 Sedan 1.5L I4 MFI VIN EG8 [All] 16P 'B' Connector

PCM Pin #	W/Color	Circuit Description (16-Pin)	Value at Hot Idle
B1	WHT/GRN	Main Relay Power (B+)	12-14v
B2	BRN/BLK	Logic Ground	<0.050v
B3	GRN/BLU	A/T: TCM Signal	In D3: 0v, Others: 5v
B4	GRN/BLK	A/T: TCM Signal	In D4: 0v, Others: 5v
B5	BLU/RED	A/C Switch Signal	Switch Off: 12v, On: 1v
B6	---	Not Used	---
B7	GRN	A/T: Park Neutral Switch	In P/N: 0v, Others: 11v
B7	PNK/BLK	M/T: Clutch Pedal Switch	Clutch In: 0v, Out: 5v
B8	BRN/RED	PSP Switch	Straight: 0v, Turning: 11v
B9	BLU/WHT	Starter Switch Signal	Cranking: 9-11v
B10	YEL/BLU	Vehicle Speed Sensor	Moving: pulse signals
B11	ORN	CYP Sensor Signal	AC pulse signals
B12	WHT	CYP Sensor Ground	<0.050v
B13	ORN/BLU	TDC Sensor Signal	AC pulse signals
B14	WHT/BLU	TDC Sensor Ground	<0.050v
B15	BLU/GRN	CKP Sensor Signal	AC pulse signals
B16	BLU/YEL	CKP Sensor Ground	<0.050v

1992-93 Sedan 1.5L I4 MFI VIN EG8 [All] 22P 'D' Connector

PCM Pin #	W/Color	Circuit Description (22-Pin)	Value at Hot Idle
D1	WHT/BLU	Keep Alive Power (VBU)	12-14v
D2	GRN/WHT	Brake Switch Signal	Brake Off: 0v, On: 12v
D3	---	Not Used	---
D4	BRN	Service Check Connector	SCS Open: 4.80v
D5-6	---	Not Used	---
D7	LT BLU	Data Link Connector	5v
D8	---	Not Used	---
D9	PNK	Alternator 'FR' Signal	Digital Signals: 0-5-0v
D10	GRN/RED	Electric Load Detector	Varies: 0.5-4.5v
D11	LT GRN	Throttle Angle Sensor Signal	0.5-0.6v
D12	---	Not Used	---
D13	RED/WHT	ECT Sensor Signal	At 180°F: 0.51v
D14	WHT	O2S Signal (CX)	0.1-1.1v
D14	ORN/BLU	HO2S Signal ([All] others)	0.1-1.1v
D15	RED/YEL	IAT Sensor Signal	Varies w/temp. (0.5-4.9v)
D16	---	Not Used	---
D17	WHT	MAP Sensor Signal	0.8-0.9v
D18	PNK/GRN	Economy Driving Indicator	Digital Signals
D19	YEL/GRN	MAP Sensor VREF	4.9-5.1v
D20	YEL/WHT	Sensor VREF	4.9-5.1v
D21	GRN/WHT	MAP Sensor Ground (DX, LX)	<0.050v
D22	GRN/WHT	Sensor Ground	<0.050v

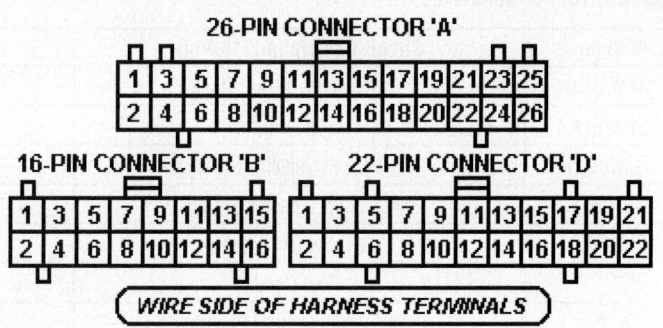

05533_ADIA_G597

Pin Connector Graphic

1994-95 Sedan 1.5L I4 MFI VIN EG8 [All] 26P 'A' Connector

PCM Pin #	W/Color	Circuit Description (26-Pin)	Value at Hot Idle
A1	BRN	Injector 1 Control	2.0-3.3 ms
A2	YEL	Injector 4 Control	2.0-3.3 ms
A3	RED	Injector 2 Control	2.0-3.3 ms
A4	---	Not Used	---
A5	BLU	Injector 3 Control	2.0-3.3 ms
A6	ORN/BLK	HO2S Heater Control	Relay Off: 12v, On: 1v
A7	GRN/YEL	Fuel Pump Relay	Relay Off: 12v, On: 1v
A8	---	Not Used	---
A9	GRN/WHT	Idle Air Control Valve	Pulse Signals
A10- 11	---	Not Used	---
A12	YEL/GRN	Radiator Fan Relay	Relay Off: 12v, On: 1v
A13	GRN/ORN	Check Engine Light	MIL Off: 12v, On: 1v
A14	---	Not Used	---
A15	BLK/RED	A/C Clutch Relay	Relay Off: 12v, On: 1v
A16	WHT/YEL	Alternator Charging Signal	Lights Off: 12v, On: 0v
A17	GRN/BLK	A/T: Lockup Control Solenoid	Solenoid Off: 12v, On: 1v
A18	---	Not Used	---
A19	YEL	A/T: Lockup Control Solenoid	Solenoid Off: 12v, On: 1v
A20	RED	EVAP Purge Solenoid	Solenoid Off: 12v, On: 1v
A21	RED/GRN	Igniter Control	Digital Signals: 0-12-0v
A22	---	Not Used	---
A23	BLK	Power Ground	<0.1v
A24	BLK	Power Ground	<0.1v
A25	YEL/BLK	Main Relay Power (B+)	12-14v
A26	BLK/RED	Logic Ground	<0.050v

1994-95 Sedan 1.5L I4 MFI VIN EG8 [All] 16P 'B' Connector

PCM Pin #	W/Color	Circuit Description (16-Pin)	Value at Hot Idle
B1	WHT/GRN	Main Relay Power (B+)	12-14v
B2	BRN/BLK	Logic Ground	<0.050v
B3	GRN/BLU	A/T: Shift Selector Signal	In D3: 0v, Others: 12v
B4	GRN/BLK	A/T: Shift Selector Signal	In D4: 0v, Others: 12v
B5	BLU/RED	A/C Switch Signal	Switch Off: 12v, On: 1v
B6	---	Not Used	---
B7	GRN	A/T: Park Neutral Switch	In P/N: 0v, Others: 11v
B7	GRN	M/T: Clutch Switch Signal	Clutch in: 11v
B8	BRN/RED	PSP Switch	Straight: 0v, Turning: 11v
B9	BLU/WHT	Starter Switch Signal	Cranking: 9-11v
B10	YEL/BLU	Vehicle Speed Sensor	Moving: pulse signals
B11	ORN	CYP Sensor Signal	AC pulse signals
B12	WHT	CYP Sensor Ground	<0.050v
B13	ORN/BLU	TDC Sensor Signal	AC pulse signals
B14	WHT/BLU	TDC Sensor Ground	<0.050v
B15	BLU/GRN	CKP Sensor Signal	AC pulse signals
B16	BLU/YEL	CKP Sensor Ground	<0.050v

1994-95 Sedan 1.5L I4 MFI VIN EG8 [All] 22P 'D' Connector

PCM Pin #	W/Color	Circuit Description (22-Pin)	Value at Hot Idle
D1	WHT/BLU	Keep Alive Power (VBU)	12-14v
D2	GRN/WHT	Brake Switch Signal	Brake Off: 0v, On: 12v
D3	RED/BLU	Knock Sensor Signal	No Detonation: 18mv AC
D4	BRN	Service Check Connector	SCS Open: 4.80v
D5-6	---	Not Used	---
D7	LT BLU	Data Link Connector	5v
D8	---	Not Used	---
D9	PNK	Alternator 'FR' Signal	Digital Signals: 0-5-0v
D10	GRN/RED	Electric Load Detector	Varies: 0.5-4.5v
D11	PNK/BLK	TP Sensor Signal	0.5-0.6v
D12	---	Not Used	---
D13	RED/WHT	ECT Sensor Signal	At 180°F: 0.51v
D14	WHT	HO2S Signal	0.1-1.1v
D15	RED/YEL	IAT Sensor Signal	Varies w/temp. (0.5-4.9v)
D16	---	Not Used	---
D17	WHT	MAP Sensor Signal	0.8-0.9v
D18	PNK/GRN	A/T: Interlock Control Unit	Key & Brake On: 12v
D19	YEL/GRN	MAP Sensor VREF	4.9-5.1v
D20	YEL/WHT	Sensor VREF	4.9-5.1v
D21	GRN/WHT	MAP Sensor GND	<0.050v
D22	GRN/WHT	Sensor Ground	<0.050v

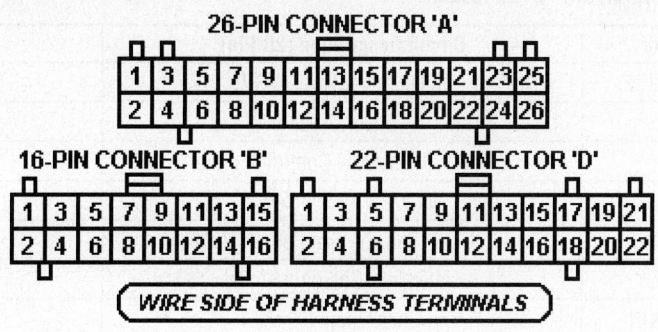

26-PIN CONNECTOR 'A'

16-PIN CONNECTOR 'B'   22-PIN CONNECTOR 'D'

WIRE SIDE OF HARNESS TERMINALS

05533_ADIA_G597

Pin Connector Graphic
Standard Colors and Abbreviations

Abbreviation	Color	Abbreviation	Color	Abbreviation	Color
BLK	Black	LT BLU	Lt. Blue	TAN	Tan
BLU	Blue	LT GRN	Lt. Green	VIO	Violet
BRN	Brown	ORN	Orange	WHT	White
GRY	Gray	PNK	Pink	YEL	Yellow
GRN	Green	PPL	Purple		

1990-91 Sedan 1.6L I4 MFI VIN ED4 [All] 18P 'A' Connector

PCM Pin #	W/Color	Circuit Description (18-Pin)	Value at Hot Idle
A1	BRN	Injector 1 Control	2.0-3.3 ms
A2	BLK	Power Ground	<0.1v
A3	RED	Injector 2 Control	2.0-3.3 ms
A4	BLK	Power Ground	<0.1v
A5	BLU	Injector 3 Control	2.0-3.3 ms
A6	GRN	EVAP Purge Solenoid	Solenoid Off: 12v, On: 1v
A7	YEL	Injector 4 Control	2.0-3.3 ms
A8-9	---	Not Used	---
A10	ORN	EGR Solenoid Control	Solenoid Off: 12v, On: 1v
A11	BLU/YEL	Electronic Air Control Valve	Pulse Signals
A12	GRN/BLK	Fuel Pump Relay	Relay Off: 12v, On: 1v
A13	YEL/BLK	Main Relay Power (B+)	12-14v
A14	GRN/BLK	Fuel Pump Relay	Relay Off: 12v, On: 1v
A15	YEL/BLK	Main Relay Power (B+)	12-14v
A16	BRN/BLK	Power Ground	<0.1v
A17	BLK/YEL	Ignition Power	12-14v
A18	BLK/RED	Power Ground	<0.1v

1990-91 Sedan 1.6L I4 MFI VIN ED4 [All] 20P 'B' Connector

PCM Pin #	W/Color	Circuit Description (20-Pin)	Value at Hot Idle
B1	WHT/GRN	Keep Alive Power (VBU)	12-14v
B2	---	Not Used	---
B2	BLU	Fast Idle Control	Solenoid Off: 12v, On: 1v
B3	YEL	A/C Clutch Relay	Solenoid Off: 12v, On: 1v
B4	YEL/GRN	Radiator Fan Relay	Relay Off: 12v, On: 1v
B5	WHT/YEL	Alternator Charging Signal	Lights Off: 12v, On: 0v
B6	GRN/ORN	Check Engine Light	MIL Off: 12v, On: 1v
B7	---	Not Used	---
B8	BLU/RED	A/C Switch Signal	Switch Off: 12v, On: 1v
B9	---	Not Used	---
B10	ORN	CKP Sensor Signal	AC pulse signals
B11	---	Not Used	---
B12	WHT	CKP Sensor Ground	<0.050v
B13	BLU/WHT	Start Signal	Cranking: 9-11v
B14	BLU	Alternator 'FR' Signal	Digital Signals: 0-5-0v
B15	WHT	Igniter Control	Digital Signals: 0-12-0v
B16	YEL/RED	Vehicle Speed Sensor	Moving: pulse signals
B17	WHT	Igniter Control	Digital Signals: 0-12-0v
B18	---	Not Used	---
B19	GRN/RED	Electric Load Detector	Varies: 0.5-4.5v
B20	BRN	Ignition Timing Adjuster	0.5-4.5v

1990-91 Sedan 1.6L I4 MFI VIN ED4 [All] 16P 'C' Connector

PCM Pin #	W/Color	Circuit Description (16-Pin)	Value at Hot Idle
C1	BLU/GRN	CYP Sensor Signal	AC pulse signals
C2	BLU/YEL	CYP Sensor Ground	<0.050v
C3	ORN/BLU	TDC Sensor Signal	AC pulse signals
C4	WHT/BLU	TDC Sensor Ground	<0.050v
C5	RED/YEL	IAT Sensor Signal	Varies w/temp. (0.5-4.9v)
C6	RED/WHT	ECT Sensor Signal	At 180°F: 0.51v
C7	RED/BLU	Throttle Angle Sensor Signal	0.5-0.6v
C8	---	Not Used	---
C9	RED/WHT	Atmospheric Pressure Sensor	2.76-2.96v at sea level
C10	GRN/WHT	Brake Switch Signal	Brake Off: 0v, On: 12v
C11	WHT	MAP Sensor Signal	0.8-0.9v
C12	GRN/WHT	Sensor Ground	<0.050v
C13	YEL/WHT	Sensor VREF	4.9-5.1v
C14	GRN/WHT	MAP Sensor Ground	<0.050v
C15	YEL/RED	MAP Sensor VREF	4.9-5.1v
C16	WHT	Oxygen Sensor Signal	0.1-1.1v

18-PIN CONNECTOR 'A'

| 1 | 3 | 5 | 7 | | 11 | 13 | 15 | 17 |
| 2 | 4 | 6 | 8 | 10 | 12 | 14 | 16 | 18 |

20-PIN CONNECTOR 'B'   **16-PIN CONNECTOR 'C'**

| 1 | 3 | 5 | 7 | 9 | 11 | 13 | 15 | 17 | 19 |
| 2 | 4 | 6 | 8 | 10 | 12 | 14 | 16 | 18 | 20 |

| 1 | 3 | 5 | 7 | 9 | 11 | 13 | 15 |
| 2 | 4 | 6 | 8 | 10 | 12 | 14 | 16 |

WIRE SIDE OF HARNESS TERMINALS

05533_ADIA_G613

Pin Connector Graphic
1992-93 Sedan 1.6L I4 VTEC VIN EH9 [All] 26P 'A' Connector

PCM Pin #	W/Color	Circuit Description (26-Pin)	Value at Hot Idle
A1	BRN	Injector 1 Control	2.0-3.3 ms
A2	YEL	Injector 4 Control	2.0-3.3 ms
A3	RED	Injector 2 Control	2.0-3.3 ms
A4	ORN/WHT	VTEC Solenoid	Solenoid Off: 12v, On: 1v
A5	BLU	Injector 3 Control	2.0-3.3 ms
A6	ORN/BLK	HO2S Heater Control	Relay Off: 12v, On: 1v
A7	GRN/YEL	Fuel Pump Relay	Relay Off: 12v, On: 1v
A8	GRN/YEL	Fuel Pump Relay	Relay Off: 12v, On: 1v
A9	GRN/WHT	Electronic Air Control Valve	Pulse Signals
A10-11	---	Not Used	---
A12	YEL/GRN	Radiator Fan Relay	Relay Off: 12v, On: 1v
A13	GRN/ORN	Check Engine Light	MIL Off: 12v, On: 1v
A14	---	Not Used	---
A15	BLK/RED	A/C Clutch Relay	Relay Off: 12v, On: 1v
A16	WHT/YEL	Alternator Charging Signal	Lights Off: 12v, On: 0v
A17	GRN/BLK	A/T: TCM Signal	Digital Signals
A18	---	Not Used	---
A19	YEL	A/T: TCM Signal	Digital Signals
A20	RED	EVAP Purge Solenoid	Solenoid Off: 12v, On: 1v
A21	RED/GRN	Igniter Control	Digital Signals: 0-12-0v
A22	RED/GRN	Igniter Control	Digital Signals: 0-12-0v
A23	BLK	Power Ground	<0.1v
A24	BLK	Power Ground	<0.1v
A25	YEL/BLK	Main Relay Power (B+)	12-14v
A26	BLK/RED	Power Ground	<0.1v

1992-93 Sedan 1.6L I4 VTEC VIN EH9 [All] 16P 'B' Connector

PCM Pin #	W/Color	Circuit Description (16-Pin)	Value at Hot Idle
B1	WHT/GRN	Main Relay Power (B+)	12-14v
B2	BRN/BLK	Logic Ground	<0.050v
B3	GRN/BLU	A/T: TCM Signal	In D3: 0v, Others: 5v
B4	GRN/BLK	A/T: TCM Signal	In D4: 0v, Others: 5v
B5	BLU/RED	A/C Switch Signal	Switch Off: 12v, On: 1v
B6	---	Not Used	---
B7	GRN	A/T: Park Neutral Switch	In P/N: 0v, Others: 11v
B8	BRN/RED	PSP Switch	Straight: 0v, Turning: 11v
B9	BLU/WHT	Starter Switch Signal	Cranking: 9-11v
B10	YEL/BLU	Vehicle Speed Sensor	Moving: pulse signals
B11	ORN	CYP Sensor Signal	AC pulse signals
B12	WHT	CYP Sensor Ground	<0.050v
B13	ORN/BLU	TDC Sensor Signal	AC pulse signals
B14	WHT/BLU	TDC Sensor Ground	<0.050v
B15	BLU/GRN	CKP Sensor Signal	AC pulse signals
B16	BLU/YEL	CKP Sensor Ground	<0.050v

1992-93 Sedan 1.6L I4 VTEC VIN EH9 [All] 22P 'D' Connector

PCM Pin #	W/Color	Circuit Description (22-Pin)	Value at Hot Idle
D1	WHT/BLU	Keep Alive Power (VBU)	12-14v
D2	GRN/WHT	Brake Switch Signal	Brake Off: 0v, On: 12v
D3	---	Not Used	---
D4	BRN	Service Check Connector	SCS Open: 4.80v
D5	---	Not Used	---
D6	ORN/BLU	VTEC Pressure Switch	0v, Hi-Speed: 12v
D7	LT BLU	Data Link Connector	5v
D8	---	Not Used	---
D9	PNK	Alternator 'FR' Signal	Digital Signals: 0-5-0v
D10	GRN/RED	Electric Load Detector	Varies: 0.5-4.5v
D11	LT GRN	Throttle Angle Sensor Signal	0.5-0.6v
D12	---	Not Used	---
D13	RED/WHT	ECT Sensor Signal	At 180°F: 0.51v
D14	ORN/BLU	HO2S Signal	0.1-1.1v
D15	RED/YEL	IAT Sensor Signal	Varies w/temp. (0.5-4.9v)
D16	---	Not Used	---
D17	WHT	MAP Sensor Signal	0.8-0.9v
D18	PNK/GRN	Economy Driving Indicator	---
D19	YEL/GRN	MAP Sensor VREF	4.9-5.1v
D20	YEL/WHT	Sensor VREF	4.9-5.1v
D21	GRN/BLU	MAP Sensor GND	<0.050v
D22	GRN/WHT	Sensor Ground	<0.050v

26-PIN CONNECTOR 'A'

1	3	5	7	9	11	13	15	17	19	21	23	25
2	4	6	8	10	12	14	16	18	20	22	24	26

16-PIN CONNECTOR 'B'

1	3	5	7	9	11	13	15
2	4	6	8	10	12	14	16

22-PIN CONNECTOR 'D'

1	3	5	7	9	11	13	15	17	19	21
2	4	6	8	10	12	14	16	18	20	22

WIRE SIDE OF HARNESS TERMINALS

05533_ADIA_G597

Pin Connector Graphic

1994-95 Sedan 1.6L I4 VTEC VIN EH9 [All] 26P 'A' Connector

PCM Pin #	W/Color	Circuit Description (26-Pin)	Value at Hot Idle
A1	BRN	Injector 1 Control	2.0-3.3 ms
A2	YEL	Injector 4 Control	2.0-3.3 ms
A3	RED	Injector 2 Control	2.0-3.3 ms
A4	ORN/WHT	VTEC Solenoid	0v, Hi-Speed: 12v
A5	LT BLU	Injector 3 Control	2.0-3.3 ms
A6	ORN/BLK	HO2S Heater Control	Relay Off: 12v, On: 1v
A7	GRN/YEL	Fuel Pump Relay	Relay Off: 12v, On: 1v
A8	---	Not Used	---
A9	GRN/WHT	Intake Air Control Solenoid	Solenoid Off: 12v, On: 1v
A10-11	---	Not Used	---
A12	YEL/GRN	Radiator Fan Relay	Relay Off: 12v, On: 1v
A13	GRN/ORN	Check Engine Light	MIL Off: 12v, On: 1v
A14	---	Not Used	---
A15	BLK/RED	A/C Clutch Relay	Relay Off: 12v, On: 1v
A16	WHT/YEL	Alternator Charging Signal	Lights Off: 12v, On: 0v
A17	GRN/BLK	A/T: Lockup Control Solenoid	Solenoid Off: 12v, On: 1v
A18	---	Not Used	---
A19	YEL	A/T: Lockup Control Solenoid	Solenoid Off: 12v, On: 1v
A20	RED	EVAP Purge Solenoid	Solenoid Off: 12v, On: 1v
A21	RED/GRN	Igniter Control	Digital Signals: 0-12-0v
A22	---	Not Used	---
A23	BLK	Power Ground	<0.1v
A24	BLK	Power Ground	<0.1v
A25	YEL/BLK	Main Relay Power (B+)	12-14v
A26	BLK/RED	Power Ground	<0.1v

1994-95 Sedan 1.6L I4 VTEC VIN EH9 [All] 16P 'B' Connector

PCM Pin #	W/Color	Circuit Description (16-Pin)	Value at Hot Idle
B1	WHT/GRN	Main Relay Power (B+)	12-14v
B2	BRN/BLK	Logic Ground	<0.050v
B3	GRN/BLU	A/T: Shift Selector Signal	In D3: 0v, Others: 5v
B4	GRN/BLK	A/T: Shift Selector Signal	In D4: 0v, Others: 5v
B5	BLU/RED	A/C Switch Signal	Switch Off: 12v, On: 1v
B6	---	Not Used	---
B7	GRN	A/T: Park Neutral Switch	In P/N: 0v, Others: 11v
B8	BRN/RED	PSP Switch	Straight: 0v, Turning: 11v
B9	BLU/WHT	Starter Switch Signal	Cranking: 9-11v
B10	YEL/BLU	Vehicle Speed Sensor	Moving: pulse signals
B11	ORN	CYP Sensor Signal	AC pulse signals
B12	WHT	CYP Sensor Ground	<0.050v
B13	ORN/BLU	TDC Sensor Signal	AC pulse signals
B14	WHT/BLU	TDC Sensor Ground	<0.050v
B15	BLU/GRN	CKP Sensor Signal	AC pulse signals
B16	BLU/YEL	CKP Sensor Ground	<0.050v

1994-95 Sedan 1.6L I4 VTEC VIN EH9 [All] 22P 'D' Connector

PCM Pin #	W/Color	Circuit Description (22-Pin)	Value at Hot Idle
D1	WHT/BLU	Keep Alive Power (VBU)	12-14v
D2	GRN/WHT	Brake Switch Signal	Brake Off: 0v, On: 12v
D3	RED/BLU	Knock Sensor Signal	No Detonation: 18mv AC
D4	BRN	Service Check Connector	SCS Open: 4.80v
D5	---	Not Used	---
D6	ORN/BLU	VTEC Pressure Switch	0v, Hi-Speed: 12v
D7	LT BLU	Data Link Connector	5v
D8	---	Not Used	---
D9	PNK	Alternator 'FR' Signal	Digital Signals: 0-5-0v
D10	GRN/RED	Electric Load Detector	Varies: 0.5-4.5v
D11	PNK/BLK	TP Sensor Signal	0.5-0.6v
D12	---	Not Used	---
D13	RED/WHT	ECT Sensor Signal	At 180°F: 0.51v
D14	WHT	HO2S Signal	0.1-1.1v
D15	RED/YEL	IAT Sensor Signal	Varies w/temp. (0.5-4.9v)
D16	---	Not Used	---
D17	WHT	MAP Sensor Signal	0.8-0.9v
D18	PNK/GRN	A/T: Interlock Control Unit	Key & Brake On: 12v
D19	YEL/GRN	MAP Sensor VREF	4.9-5.1v
D20	YEL/WHT	Sensor VREF	4.9-5.1v
D21	GRN/BLU	MAP Sensor Ground	<0.050v
D22	GRN/WHT	Sensor Ground	<0.050v

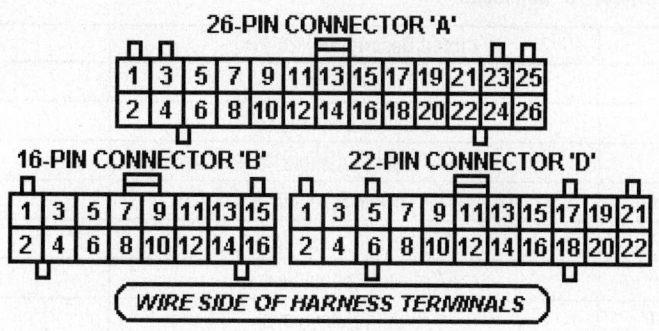

26-PIN CONNECTOR 'A'

16-PIN CONNECTOR 'B'   22-PIN CONNECTOR 'D'

WIRE SIDE OF HARNESS TERMINALS

05533_ADIA_G597

Pin Connector Graphic
1996 Sedan 1.6L I4 MFI VIN EJ6 [All] 32P 'A' Connector

PCM Pin #	W/Color	Circuit Description (32-Pin)	Value at Hot Idle
A1	YEL	Injector 4 Control	2.0-3.3 ms
A2	BLU	Injector 3 Control	2.0-3.3 ms
A3	RED	Injector 2 Control	2.0-3.3 ms
A4	BRN	Injector 1 Control	2.0-3.3 ms
A5	BLK/WHT	HO2S-12 (B1 S2) Heater	Digital Signals: 0-12-0v
A6	BLK/WHT	HO2S-11 (B1 S1) Heater	Digital Signals: 0-12-0v
A9, A22	BRN/BLK	Logic Ground	<0.050v
A10, A23	BLK	Power Ground	<0.1v
A11	YEL/BLK	Main Relay Power (B+)	12-14v
A12	BLK/BLU	M/T: Idle Air Control Valve	Pulse Signals
A13	ORN	A/T Idle Air Control Valve 'N'	Pulse Signals
A14	BLK/BLU	A/T Idle Air Control Valve 'P'	Pulse Signals
A15	RED/YEL	EVAP Purge Solenoid	Solenoid Off: 12v, On: 1v
A16	GRN/YEL	Fuel Pump Relay	Relay Off: 12v, On: 1v
A17	BLK/RED	A/C Clutch Relay	Relay Off: 12v, On: 1v
A18	GRN/ORN	Check Engine Light	MIL Off: 12v, On: 1v
A19	WHT/GRN	Alternator Charging Signal	Lights Off: 12v, On: 0v
A20	YEL/GRN	Igniter Control	Digital Signals: 0-12-0v
A24	YEL/BLK	Main Relay Power (B+)	12-14v
A27	GRN	Radiator Fan Relay	Relay Off: 12v, On: 1v

1996 Sedan 1.6L I4 MFI VIN EJ6 [All] 25P 'B' Connector

PCM Pin #	W/Color	Circuit Description (25-Pin)	Value at Hot Idle
B1	WHT	A/T: Linear Solenoid (-)	Pulse Signals
B2	RED	A/T: Linear Solenoid (+)	Pulse Signals
B3	BLU/YEL	A/T: Shift Solenoid 'A'	Solenoid Off: 12v, On: 1v
B4	GRN/BLK	A/T: Lockup Solenoid 'B'	LSB On: 12v, Off: 0v
B5	YEL	A/T: Lockup Solenoid 'A'	LSA On: 12v, Off: 0v
B8	PNK	A/T: Gear Position Switch	In D3: 0v, Others: 12v
B11	GRN/WHT	A/T: Shift Solenoid 'B'	Solenoid Off: 12v, On: 1v
B12	WHT/RED	Interlock Control Unit Signal	Key & Brake On: 12v
B13	GRN/BLK	A/T: D4 Indicator Light	D4 On: 12v, Off: 0v
B14	WHT	Mainshaft Speed Sensor 'N'	<0.050v
B15	RED	Mainshaft Speed Sensor 'P'	AC Pulse Signals
B16	WHT	A/T: Gear Position Switch	In 'R': 0v, Others: 12v
B17	BLU	A/T: Gear Position Switch	In D2: 0v, Others: 12v
B22	GRN	Countershaft Speed Sensor N	<0.050v
B23	LT GRN	Countershaft Speed Sensor P	Moving: AC pulses
B24	YEL	A/T: Gear Position Switch	In D4: 0v, Others: 12v
B25	LT GRN	A/T: Gear Position Switch	In P/N: 0v, Others: 12v

32-PIN CONNECTOR 'A'   25-PIN CONNECTOR 'B'

WIRE SIDE OF HARNESS TERMINALS

05533_ADIA_G604

Pin Connector Graphic

1996 Sedan 1.6L I4 MFI VIN EJ6 [All] 31P 'C' Connector

PCM Pin #	W/Color	Circuit Description (31-Pin)	Value at Hot Idle
C1	BLU/RED	CKF Sensor Signal	AC pulse signals
C2	BLU	CKP Sensor Signal	AC pulse signals
C3	GRN	TDC Sensor Signal	AC pulse signals
C4	YEL	CYP Sensor Signal	AC pulse signals
C5	BLU/RED	A/C Switch Signal	Switch Off: 12v, On: 1v
C6	BLU/ORN	Starter Switch Signal	Cranking: 9-11v
C7	BRN	Service Check Connector	SCS Open: 4.80v
C8	LT BLU	K-Line Signal	12v
C9	---	Not Used	---
C10	WHT/BLU	Keep Alive Power (VBU)	12-14v
C11	WHT/RED	CKF Sensor Ground	0.050v
C12	WHT	CKP Sensor Ground	0.050v
C13	RED	TDC Sensor Ground	0.050v
C14	BLK	CYP Sensor Ground	0.050v

1996 Sedan 1.6L I4 MFI VIN EJ6 [All] 31P 'C' Connector, *continued*

C16	GRN	PSP Switch	Straight: 0v, Turning: 11v
C17	WHT/RED	Alternator 'FR' Signal	Digital Signals: 0-5-0v
C18	BLU/WHT	Vehicle Speed Sensor	Moving: pulse signals
C19-29	---	Not Used	---
C30	PNK	A/T: CVT TMB Signal	Pulse Signals
C31	---	Not Used	---

1996 Sedan 1.6L I4 MFI VIN EJ6 [All] 16P 'D' Connector

PCM Pin #	W/Color	Circuit Description (16-Pin)	Value at Hot Idle
D1	RED/BLK	TP Sensor Signal	0.5-0.6v
D2	RED/WHT	ECT Sensor Signal	At 180°F: 0.51v
D3	RED/GRN	MAP Sensor Signal	0.8-0.9v
D4	YEL/RED	MAP Sensor VREF	4.9-5.1v
D5	GRN/WHT	Brake Switch Signal	Brake Off: 0v, On: 12v
D7	WHT	HO2S-11 (B1 S1) Signal	0.1-1.1v
D8	RED/YEL	IAT Sensor Signal	Varies w/temp. (0.5-4.9v)
D9	---	Not Used	---
D10	YEL/BLU	Sensor VREF	4.9-5.1v
D11	GRN/BLK	Sensor Ground	<0.050v
D12	GRN/WHT	MAP Sensor Ground	<0.050v
D13	RED/YEL	HO2S-12 Ground	<0.050v
D14	WHT/RED	HO2S-12 (B1 S2) Signal	0.1-1.1v
D15	---	Not Used	---
D16	GRN/RED	Electric Load Detector	Varies: 0.5-4.5v

31-PIN CONNECTOR 'C'   16-PIN CONNECTOR 'D'

1	2	3	4		5	6	7		8	9	10		1		2	3	4		5
11	12	13	14	15	16	17	18	19	20	21	22		6	7	8	9	10	11	12
	23	24	25		26	27	28		29	30	31		13	14	15		16		

WIRE SIDE OF HARNESS TERMINALS

05533_ADIA_G605

Pin Connector Graphic

1997-99 Sedan 1.6L I4 MFI VIN EJ6 [All] 32P 'A' Connector

PCM Pin #	W/Color	Circuit Description (32-Pin)	Value at Hot Idle
A1, A2	YEL, BLU	Injector 4, Injector 3 Control	2.0-3.3 ms
A3, A4	RED, BRN	Injector 2, Injector 1 Control	2.0-3.3 ms
A5	BLK/WHT	HO2S-12 (B1 S2) Heater	Digital Signals: 0-12-0v
A6	BLK/WHT	HO2S-11 (B1 S1) Heater	Digital Signals: 0-12-0v
A7	RED	A/T: EGR Solenoid Control	Solenoid Off: 12v, On: 1v
A7	PNK	M/T: E-EGR Solenoid	Digital Signals: 0-12-0v
A9, A22	BRN/BLK	Logic Ground (LG1, LG2)	<0.050v
A10, 23	BLK	Power Ground (PG1, PG2)	<0.1v
A11, 24	YEL/BLK	Main Relay Power (B+)	12-14v
A12	BLK/BLU	M/T: Idle Air Control Valve	Pulse Signals
A13	ORN	A/T: Idle Air Control Valve 'N'	Pulse Signals
A14	BLK/BLU	A/T: Idle Air Control Valve 'P'	Pulse Signals
A15	RED/YEL	EVAP Purge Solenoid	Solenoid Off: 12v, On: 1v
A16	GRN/YEL	Fuel Pump Relay	Relay Off: 12v, On: 1v
A17	BLK/RED	A/C Clutch Relay	Relay Off: 12v, On: 1v
A18	GRN/ORN	Malfunction Indicator Light	MIL Off: 12v, On: 1v
A19	WHT/GRN	Alternator Charging Signal	Lights Off: 12v, On: 0v
A20	YEL/GRN	Igniter Control	Digital Signals: 0-12-0v
A27	GRN	Radiator Fan Relay Control	Relay Off: 12v, On: 1v
A28	BLU	EVAP Bypass Solenoid	Solenoid Off: 12v, On: 1v
A29	WHT/RED	EVAP Vent Solenoid	Solenoid Off: 12v, On: 1v

1997-99 Sedan 1.6L I4 MFI VIN EJ6 [All] 25P 'B' Connector

PCM Pin #	W/Color	Circuit Description (25-Pin)	Value at Hot Idle
B1	WHT	A/T: Linear Solenoid (-)	Pulse Signals
B2	RED	A/T: Linear Solenoid (+)	Pulse Signals
B3	BLU/YEL	A/T: Shift Solenoid 'A'	SSA On: 12v: Off: 0v
B4	GRN/BLK	A/T: Lockup Solenoid 'B'	LSB On: 12v: Off: 0v
B5	YEL	A/T: Lockup Solenoid 'A'	LSA On: 12v: Off: 0v
B8	PNK	A/T: Gear Position Switch	In D3: 0v, Others: 12v
B11	GRN/WHT	A/T: Shift Solenoid 'B'	SSB Off: 0v, On: 12v
B12	WHT/RED	Interlock Control Unit Signal	Key & Brake On: 12v
B13	GRN/BLK	A/T: D4 Indicator Light	D4 On: 12v, Off: 0v
B14	WHT	Mainshaft Speed Sensor 'P'	AC Pulse Signals
B15	RED	Mainshaft Speed Sensor 'N'	AC Pulse Signals
B16	WHT	A/T: Gear Position Switch	In 'R': 0v, Others: 12v
B17	BLU	A/T: Gear Position Switch	In D2: 0v, Others: 12v
B22	GRN	Countershaft Speed Sensor N	Moving: AC pulses
B23	BLU	Countershaft Speed Sensor P	Moving: AC pulses
B24	YEL	A/T: Gear Position Switch	In D4: 0v, Others: 12v
B25	LT GRN	A/T: Gear Position Switch	In P/N: 0v, Others: 12v

```
 1  2  3  4    5  6  7    8  9  10    11      1      2     3  4  5    6  7  8
12 13 14 15 16 17 18 19 20 21 22  23   24    9     10   11 12 13 14 15 16 17 18
      25 26 27    28 29 30 31    32          19    20      21 22   23 24 25
```

32-Pin 'A'   **WIRE SIDE OF HARNESS TERMINALS**   **25-Pin 'B'**

05533_ADIA_G606

Pin Connector Graphic

1997-99 Sedan 1.6L I4 MFI VIN EJ6 [All] 31P 'C' Connector

PCM Pin #	W/Color	Circuit Description (31-Pin)	Value at Hot Idle
C1	BLU/RED	CKF Sensor Signal	AC pulse signals
C2	BLU	CKP Sensor Signal	AC pulse signals
C3	GRN	TDC Sensor Signal	AC pulse signals
C4	YEL	CYP Sensor Signal	AC pulse signals
C5	BLU/RED	A/C Switch Signal	Switch Off: 12v, On: 1v
C6	BLU/ORN	Starter Switch Signal	Cranking: 9-11v
C7	BRN	Service Check Connector	SCS Open: 4.80v
C8	LT BLU	K-Line Signal	12v
C10	WHT/BLU	Keep Alive Power (VBU)	12-14v
C11	WHT/RED	CKF Sensor Ground	<0.050v
C12	WHT	CKP Sensor Ground	<0.050v
C13	RED	TDC Sensor Ground	<0.050v
C14	BLK	CYP Sensor Ground	<0.050v
C16	GRN	PSP Switch Signal	Straight: 0v, Turning: 11v
C17	WHT/RED	Alternator 'FR' Signal	Digital Signals: 0-5-0v
C18	BLU/WHT	Vehicle Speed Sensor	Moving: pulse signals
C23	BLK	LAF Pump Cell (IP+)	0.5-3.5v
C24	RED	LAF Common (IP-, VS-)	2.6-2.8v
C25	WHT	LAF VS Cell Voltage (VS+)	7v
C29	LT GRN	A/T: Gear Position Switch	In P/N: 0v, Others: 12v
C29	RED	M/T: Clutch Switch Signal	Clutch In: 0v, Out: 5v
C30	PNK	CVT TMB Signal	Pulse Signals

1997-99 Sedan 1.6L I4 MFI VIN EJ6 [All] 16P 'D' Connector

PCM Pin #	W/Color	Circuit Description (16-Pin)	Value at Hot Idle
D1	RED/BLK	TP Sensor Signal	0.5-0.6v
D2	RED/WHT	ECT Sensor Signal	At 180°F: 0.51v
D3	RED/GRN	MAP Sensor Signal	0.8-0.9v
D4	YEL/RED	MAP Sensor VREF	4.9-5.1v
D5	GRN/WHT	Brake Switch Signal	Brake Off: 0v, On: 12v
D6	RED/BLU	Knock Sensor Signal	No detonation: 18mv AC
D7	WHT	HO2S-11 (B1 S1) Signal	0.1-1.1v
D8	RED/YEL	IAT Sensor Signal	Varies w/temp. (0.5-4.9v)
D9	WHT/BLK	EGR Valve Lift Sensor	1.2v
D10	YEL/RED	Sensor VREF	4.9-5.1v
D11	GRN/BLK	Sensor Ground	<0.050v
D12	GRN/WHT	MAP Sensor Ground	<0.050v
D13	RED/YEL	HO2S-12 Ground	<0.050v
D14	WHT/RED	HO2S-12 (B1 S2) Signal	0.1-1.1v
D15	LT GRN	Fuel Tank Pressure Sensor	Fuel Cap off: 2.5v
D16	GRN/RED	Electric Load Detector	Varies: 0.5-4.5v

31-Pin 'C'   *WIRE SIDE OF HARNESS TERMINALS*   16-Pin 'D'

05533_ADIA_G607

Pin Connector Graphic

2000 Sedan 1.6L I4 MFI VIN EJ6 [All] 32P 'A' Connector

PCM Pin #	W/Color	Circuit Description (32-Pin)	Value at Hot Idle
A1, A2	YEL, BLU	Injector 4, Injector 3 Control	2.0-3.3 ms
A3, A4	RED, BRN	Injector 2, Injector 1 Control	2.0-3.3 ms
A5	BLK/WHT	HO2S-12 (B1 S2) Heater	Digital Signals: 0-12-0v
A6	BLK/WHT	HO2S-11 (B1 S1) Heater	Digital Signals: 0-12-0v
A7	RED	A/T: EGR Solenoid Control	Solenoid Off: 12v, On: 1v
A7	PNK	M/T: E-EGR Solenoid	Digital Signals: 0-12-0v
A9, A22	BRN/BLK	Logic Ground (LG1, LG2)	<0.050v
A10, 23	BLK	Power Ground (PG1, PG2)	<0.1v
A11, 24	YEL/BLK	Main Relay Power (B+)	12-14v
A12	BLK/BLU	M/T: Idle Air Control Valve	Pulse Signals
A13	ORN	A/T: Idle Air Control Valve 'N'	Pulse Signals
A14	BLK/BLU	A/T: Idle Air Control Valve 'P'	Pulse Signals
A15	RED/YEL	EVAP Purge Solenoid	Solenoid Off: 12v, On: 1v
A16	GRN/YEL	Fuel Pump Relay	Relay Off: 12v, On: 1v
A17	BLK/RED	A/C Clutch Relay	Relay Off: 12v, On: 1v
A18	GRN/ORN	Malfunction Indicator Light	MIL Off: 12v, On: 1v
A19	WHT/GRN	Alternator Charging Signal	Lights Off: 12v, On: 0v
A20	YEL/GRN	Igniter Control	Digital Signals: 0-12-0v
A27	GRN	Radiator Fan Relay Control	Relay Off: 12v, On: 1v
A28	BLU	EVAP Bypass Solenoid	Solenoid Off: 12v, On: 1v
A29	WHT/RED	EVAP Vent Solenoid	Solenoid Off: 12v, On: 1v

2000 Sedan 1.6L I4 MFI VIN EJ6 [All] 25P 'B' Connector

PCM Pin #	W/Color	Circuit Description (25-Pin)	Value at Hot Idle
B1	WHT	A/T: Linear Solenoid (-)	Pulse Signals
B2	RED	A/T: Linear Solenoid (+)	Pulse Signals
B3	BLU/YEL	A/T: Shift Solenoid 'A'	SSA On: 12v: Off: 0v
B4	GRN/BLK	A/T: Lockup Solenoid 'B'	LSB On: 12v: Off: 0v
B5	YEL	A/T: Lockup Solenoid 'A'	LSA On: 12v, Off: 0v
B8	PNK	A/T: Gear Position Switch	In D3: 0v, Others: 12v
B11	GRN/WHT	A/T: Shift Solenoid 'B'	SSB Off: 0v, On: 12v
B12	WHT/RED	Interlock Control Unit Signal	Key & Brake On: 12v
B13	GRN/BLK	A/T: D4 Indicator Light	D4 On: 12v, Off: 0v
B14	WHT	Mainshaft Speed Sensor 'P'	AC Pulse Signals
B15	RED	Mainshaft Speed Sensor 'N'	AC Pulse Signals
B16	WHT	A/T: Gear Position Switch	In 'R': 0v, Others: 12v
B17	BLU	A/T: Gear Position Switch	In D2: 0v, Others: 12v
B22	GRN	Countershaft Speed Sensor N	Moving: AC pulses
B23	BLU	Countershaft Speed Sensor P	Moving: AC pulses
B24	YEL	A/T: Gear Position Switch	In D4: 0v, Others: 12v
B25	LT GRN	A/T: Gear Position Switch	In P/N: 0v, Others: 12v

```
 1  2  3  4     5  6  7     8  9  10    11      1       2      3  4  5     6  7  8
12 13 14 15 16 17 18 19 20 21 22  23     24      9      10   11 12 13 14 15 16 17 18
      25 26 27    28 29 30 31    32             19      20     21 22   23 24 25
       32-Pin 'A'          WIRE SIDE OF HARNESS TERMINALS        25-Pin 'B'
```

05533_ADIA_G606

Pin Connector Graphic
2000 Sedan 1.6L I4 MFI VIN EJ6 [All] 31P 'C' Connector

PCM Pin #	W/Color	Circuit Description (31-Pin)	Value at Hot Idle
C1	BLU/RED	CKF Sensor Signal	AC pulse signals
C2	BLU	CKP Sensor Signal	AC pulse signals
C3	GRN	TDC Sensor Signal	AC pulse signals
C4	YEL	CYP Sensor Signal	AC pulse signals
C5	BLU/RED	A/C Switch Signal	Switch Off: 12v, On: 1v
C6	BLU/ORN	Starter Switch Signal	Cranking: 9-11v
C7	BRN	Service Check Connector	SCS Open: 4.80v
C8	LT BLU	K-Line Signal	12v
C10	WHT/BLU	Keep Alive Power (VBU)	12-14v
C11	WHT/RED	CKF Sensor Ground	<0.050v
C12	WHT	CKP Sensor Ground	<0.050v
C13	RED	TDC Sensor Ground	<0.050v
C14	BLK	CYP Sensor Ground	<0.050v
C16	GRN	PSP Switch Signal	Straight: 0v, Turning: 11v
C17	WHT/RED	Alternator 'FR' Signal	Digital Signals: 0-5-0v
C18	BLU/WHT	Vehicle Speed Sensor	Moving: pulse signals
C23	BLK	LAF Pump Cell (IP+)	0.5-3.5v
C24	RED	LAF Common (IP-, VS-)	2.6-2.8v
C25	WHT	LAF VS Cell Voltage (VS+)	7v
C29	LT GRN	A/T: Gear Position Switch	In P/N: 0v, Others: 12v
C29	RED	M/T: Clutch Switch Signal	Clutch In: 0v, Out: 5v
C30	PNK	CVT TMB Signal	Pulse Signals

2000 Sedan 1.6L I4 MFI VIN EJ6 [All] 16P 'D' Connector

PCM Pin #	W/Color	Circuit Description (16-Pin)	Value at Hot Idle
D1	RED/BLK	TP Sensor Signal	0.5-0.6v
D2	RED/WHT	ECT Sensor Signal	At 180°F: 0.51v
D3	RED/GRN	MAP Sensor Signal	0.8-0.9v
D4	YEL/RED	MAP Sensor VREF	4.9-5.1v
D5	GRN/WHT	Brake Switch Signal	Brake Off: 0v, On: 12v
D6	RED/BLU	Knock Sensor Signal	No detonation: 18mv AC
D7	WHT	HO2S-11 (B1 S1) Signal	0.1-1.1v
D8	RED/YEL	IAT Sensor Signal	Varies w/temp. (0.5-4.9v)
D9	WHT/BLK	EGR Valve Lift Sensor	1.2v
D10	YEL/RED	Sensor VREF	4.9-5.1v
D11	GRN/BLK	Sensor Ground	<0.050v
D12	GRN/WHT	MAP Sensor Ground	<0.050v
D13	RED/YEL	HO2S-12 Ground	<0.050v
D14	WHT/RED	HO2S-12 (B1 S2) Signal	0.1-1.1v
D15	LT GRN	Fuel Tank Pressure Sensor	Fuel Cap off: 2.5v
D16	GRN/RED	Electric Load Detector	Varies: 0.5-4.5v

```
[1 2 3 4]  [5 6 7]  [8 9 10]      [1] [2 3 4] [ 5 ]
[11 12 13 14 15 16 17 18 19 20 21 22]  [6 7 8 9 10 11 12]
[23 24 25]  [26 27 28]  [29 30 31]    [13 14 15] [16]
```

31-Pin 'C'   *WIRE SIDE OF HARNESS TERMINALS*   **16-Pin 'D'**

05533_ADIA_G607

Pin Connector Graphic

1996 Sedan 1.6L I4 VTEC VIN EJ8 [All] 32P 'A' Connector

PCM Pin #	W/Color	Circuit Description (32-Pin)	Value at Hot Idle
A1, 2	YEL, BLU	Injector 4, Injector 3 Control	2.0-3.3 ms
A3, 4	RED, BRN	Injector 2, Injector 1 Control	2.0-3.3 ms
A5	BLK/WHT	HO2S-12 (B1 S2) Heater	Digital Signals: 0-12-0v
A6	BLK/WHT	HO2S-11 (B1 S1) Heater	Digital Signals: 0-12-0v
A8	GRN/YEL	VTEC Solenoid	0v, Hi-Speed: 12v
A9, A22	BRN/BLK	Logic Ground	<0.050v
A10, 23	BLK	Power Ground	<0.1v
A11, 24	YEL/BLK	Main Relay Power (B+)	12-14v
A12	BLK/BLU	M/T: Idle Air Control Valve	Pulse Signals
A13	ORN	Idle Air Control Valve 'N'	Pulse Signals
A14	BLK/BLU	Idle Air Control Valve 'P'	Pulse Signals
A15	RED/YEL	EVAP Purge Solenoid	Solenoid Off: 12v, On: 1v
A16	GRN/YEL	Fuel Pump Relay	Relay Off: 12v, On: 1v
A17	BLK/RED	A/C Clutch Relay	Relay Off: 12v, On: 1v
A18	GRN/ORN	Check Engine Light	MIL Off: 12v, On: 1v
A19	WHT/GRN	Alternator Charging Signal	Lights Off: 12v, On: 0v
A20	YEL/GRN	Igniter Control	Digital Signals: 0-12-0v
A25	WHT/RED	A/T: TCM VREF Signal	4.9-5.1v
A27	GRN	Radiator Fan Relay	Relay Off: 12v, On: 1v
A28	BLU	EVAP Bypass Solenoid	Solenoid Off: 12v, On: 1v
A29	GRN/WHT	EVAP Vent Solenoid	Solenoid Off: 12v, On: 1v

1996 Sedan 1.6L I4 VTEC VIN EJ8 [All] 25P 'B' Connector

PCM Pin #	W/Color	Circuit Description (25-Pin)	Value at Hot Idle
B1	WHT	A/T: Linear Solenoid (-)	Pulse Signals
B2	RED	A/T: Linear Solenoid (+)	Pulse Signals
B3	BLU/YEL	A/T: Shift Solenoid 'A'	SSA Off: 0v, On: 12v
B4	GRN/BLK	A/T: Lockup Solenoid 'B'	LSB On: 12v, Off: 0v
B5	YEL	A/T: Lockup Solenoid 'A'	LSA On: 12v, Off: 0v
B8	PNK	A/T: Gear Position Switch	In D3: 0v, Others: 12v
B11	GRN/WHT	A/T: Shift Solenoid 'B'	SSB Off: 0v, On: 12v
B12	WHT/RED	Interlock Control Unit Signal	Key & Brake On: 12v
B13	GRN/BLK	A/T: D4 Indicator Light	D4 On: 12v, Off: 0v
B14	WHT	Mainshaft Speed Sensor 'N'	<0.050v
B15	RED	Mainshaft Speed Sensor 'P'	AC Pulse Signals
B16, B25	WHT, YEL	A/T: Gear Position Switch	In 'R': 0v, Others: 12v
B17	BLU	A/T: Gear Position Switch	In D2: 0v, Others: 12v
B22	GRN	Countershaft Speed Sensor N	<0.050v
B23	BLU	Countershaft Speed Sensor P	Pulse Signals
B24	YEL	A/T: Gear Position Switch	In D4: 0v, Others: 12v

05533_ADIA_G604

Pin Connector Graphic

1996 Sedan 1.6L I4 VTEC VIN EJ8 [All] 31P 'D' Connector

PCM Pin #	W/Color	Circuit Description (31-Pin)	Value at Hot Idle
C1	BLU/RED	CKF Sensor Signal	AC pulse signals
C2	BLU	CKP Sensor Signal	AC pulse signals
C3	GRN	TDC Sensor Signal	AC pulse signals
C4	YEL	CYP Sensor Ground	AC pulse signals
C5	BLU/RED	A/C Switch Signal	Switch Off: 12v, On: 1v
C6	BLU/ORN	Starter Switch Signal	Cranking: 9-11v
C7	BRN	Service Check Connector	SCS Open: 4.80v
C8	LT BLU	K-Line Signal	12v
C9	---	Not Used	---
C10	WHT/BLU	Keep Alive Power (VBU)	12-14v
C11	WHT/RED	CKF Sensor Ground	<0.050v
C12	WHT	CKP Sensor Ground	<0.050v
C13	RED	TDC Sensor Ground	<0.050v
C14	BLK	CYP Sensor Ground	<0.050v
C15	BLU/BLK	VTEC Pressure Switch	0v, Hi-Speed: 12v
C16	GRN	PSP Switch	Straight: 0v, Turning: 11v
C17	WHT/RED	Alternator 'FR' Signal	Digital Signals: 0-5-0v
C18	BLU/WHT	Vehicle Speed Sensor	Moving: pulse signals
C19-31	---	Not Used	---

1996 Sedan 1.6L I4 VTEC VIN EJ8 [All] 16P 'D' Connector

PCM Pin #	W/Color	Circuit Description (16-Pin)	Value at Hot Idle
D1	RED/BLK	TP Sensor Signal	0.5-0.6v
D2	RED/WHT	ECT Sensor Signal	At 180°F: 0.51v
D3	RED/GRN	MAP Sensor Signal	0.8-0.9v
D4	YEL/RED	MAP Sensor VREF	4.9-5.1v
D5	GRN/WHT	Brake Switch Signal	Brake Off: 0v, On: 12v
D6	RED/BLU	Knock Sensor Signal	No Detonation: 18mv AC
D7	WHT	HO2S-11 (B1 S1) Signal	0.1-1.1v
D8	RED/YEL	IAT Sensor Signal	Varies w/temp. (0.5-4.9v)
D9	---	Not Used	---
D10	YEL/BLU	Sensor VREF	4.9-5.1v
D11	GRN/BLK	Sensor Ground	<0.050v
D12	GRN/WHT	MAP Sensor Ground	<0.050v
D13	GRN/BLK	HO2S-12 Ground	0.1-1.1v
D14	WHT/RED	HO2S-12 (B1 S2) Signal	0.1-1.1v
D15	LT GRN	Fuel Tank Pressure (Coupe)	Fuel Cap off: 2.5v
D16	GRN/RED	Electric Load Detector	Varies: 0.5-4.5v

31-PIN CONNECTOR 'C' **16-PIN CONNECTOR 'D'**

1	2	3	4		5	6	7		8	9	10		1		2	3	4		5	
11	12	13	14	15	16	17	18	19	20	21	22		6	7	8	9	10	11	12	
	23	24	25		26	27	28		29	30	31		13	14	15		16			

WIRE SIDE OF HARNESS TERMINALS

05533_ADIA_G605

Pin Connector Graphic

1997-99 Sedan 1.6L I4 VTEC VIN EJ8 [All] 32P 'A' Connector

PCM Pin #	W/Color	Circuit Description (32-Pin)	Value at Hot Idle
A3	BLU	EVAP Bypass Solenoid	Solenoid Off: 12v, On: 1v
A4	GRN/WHT	EVAP Vent Solenoid	Solenoid Off: 12v, On: 1v
A5	BLU/GRN	Cruise Control Signal	C/C On: pulse signals
A6	RED/YEL	EVAP Purge Solenoid	Solenoid Off: 12v, On: 1v
A8	BLK/WHT	HO2S-12 (B1 S2) Heater	Digital Signals: 0-12-0v
A9	LT GRN	A/T: Gear Position Switch	In P/N: 0v, Others: 12v
A10	BRN	Service Check Connector	SCS Open: 4.80v
A14	GRN/BLK	A/T: D4 Light Switch	D4 On: 0v, Off: 12v
A16	GRN/YEL	Fuel Pump Relay Control	Relay Off: 12v, On: 1v
A17	BLK/RED	A/C Clutch Relay Control	Relay Off: 12v, On: 1v
A18	GRN/ORN	Malfunction Indicator Light	MIL Off: 12v, On: 1v
A19	BLU	Engine Speed Pulse (NEP)	Digital Signals
A20	GRN	Radiator Fan Relay Control	Relay Off: 12v, On: 1v
A21	BLU/YEL	K-Line Signal	12v
A22	BLU	A/T: Gear Position Switch	In 'L': 0v, others: 12v
A23	WHT/RED	HO2S-12 (B1 S2) Signal	0.1-1.1v
A24	BLU/WHT	Starter Switch Signal	Cranking: 9-11v
A26	GRN	PSP Switch Signal	Straight: 0v, Turning: 11v
A27	BLU/RED	A/C Switch Signal	Switch Off: 12v, On: 1v
A28	WHT/RED	Interlock Control Unit Signal	Key & Brake On: 12v
A29	LT GRN	Fuel Tank Pressure Sensor	Fuel Cap off: 2.5v
A30	GRN/RED	Electric Load Detector	Varies: 0.5-4.5v
A32	WHT/BLK	Brake Switch Signal	Brake Off: 0v, On: 12v

1997-99 Sedan 1.6L I4 VTEC VIN EJ8 [All] 16P 'D' Connector

PCM Pin #	W/Color	Circuit Description (16-Pin)	Value at Hot Idle
D1	YEL	A/T: Lockup Control Solenoid	LUS On: 12v, Off: 0v
D2	GRN/WHT	A/T: Shift Solenoid 'B'	SSB Off: 0v, On: 12v
D3	GRN	A/T: Shift Solenoid 'C'	SSC on: 12v, off: 0v
D5	BLK/YEL	A/T: Solenoid Feed (B+)	12-14v
D6	WHT	A/T: Gear Position Switch	In 'R': 0v, Others: 12v
D7	BLU/YEL	A/T: Shift Solenoid 'A'	SSA Off: 0v, On: 12v
D8	PNK	A/T: Gear Position Switch	In D3: 0v, Others: 12v
D9	YEL	A/T: Gear Position Switch	In D4: 0v, Others: 12v
D10	BLU	Countershaft Speed Sensor P	Moving: AC pulses
D11	RED	Mainshaft Speed Sensor 'P'	Pulse Signals
D12	WHT	Mainshaft Speed Sensor 'N'	<0.050v
D13	BLU/WHT	A/T: Gear Position Switch	In P/N: 0v, Others: 12v
D14	GRN/BLK	A/T: D4 Indicator Light Driver	D4 on: 5v, off: 0v
D16	GRN	Countershaft Speed Sensor N	<0.050v

Pin Connector Graphic

05533_ADIA_G599

1997-99 Sedan 1.6L I4 VTEC VIN EJ8 [All] 25P 'B' Connector

PCM Pin #	W/Color	Circuit Description (25-Pin)	Value at Hot Idle
B1, B9	YEL/BLK	Main Relay Power (B+)	12-14v
B2, B10	BLK	Power Ground	<0.1v
B3, B4	RED, BLU	Injector 2, Injector 3 Control	2.0-3.3 ms
B5, B11	YEL, BRN	Injector 4, Injector 1 Control	2.0-3.3 ms
B7	RED	EGR Control Solenoid	Digital Signals: 0-12-0v
B8	PNK/BLK	A/T: Control Linear Solenoid 'N'	Pulse Signals
B12	GRN/YEL	VTEC Control Solenoid	0v, Hi-Speed: 12v
B13	YEL/GRN	Ignition Control Signal	Digital Signals: 0-12-0v
B15	ORN	Idle Air Control Valve 'N'	Pulse Signals
B17	GRN/WHT	A/T: Control Linear Solenoid 'P'	Pulse Signals
B18	PNK/BLU	A/T: Start Clutch Solenoid 'N'	Pulse Signals
B20, B22	BRN/BLK	Logic Ground (LG1, LG2)	<0.050v
B21	WHT/YEL	Keep Alive Power (VBU)	12-14v
B23	BLK/BLU	Idle Air Control Valve 'P'	Pulse Signals
B24	BLU/WHT	A/T: 3rd Clutch Pressure Sw.	Open: 12v, Closed: 1v
B25	YEL	A/T: Start Clutch Solenoid 'N'	Pulse Signals

1997-99 Sedan 1.6L I4 VTEC VIN EJ8 [All] 31P 'C' Connector

PCM Pin #	W/Color	Circuit Description (31-Pin)	Value at Hot Idle
C1	BLK/WHT	HO2S-11 (B1 S1) Heater	Digital Signals: 0-12-0v
C2	WHT/GRN	Alternator Charging Signal	Lights Off: 12v, On: 0v
C3	RED/BLU	Knock Sensor Signal	No Detonation: 18mv AC
C5	WHT/RED	Alternator 'FR' Signal	Digital Signals: 0-5-0v
C6	WHT/BLK	EGR Valve Lift Sensor	1.2v
C7	GRN/WHT	MAP Sensor Ground (SG1)	<0.050v
C8	BLU	CKP Sensor Signal	AC pulse signals
C9	WHT	CKP Sensor Ground	<0.050v
C10	BLU/BLK	VTEC Pressure Switch	0v, Hi-Speed: 12v
C16	WHT	HO2S-11 (B1 S1) Signal	0.1-1.1v
C17	RED/GRN	MAP Sensor Signal	0.8-0.9v
C18	GRN/BLK	Signal Ground (SG2)	<0.050v
C19, C28	YEL/RED	MAP Sensor, Sensor VREF	4.9-5.1v
C20	GRN	TDC Sensor Signal	AC pulse signals
C21	RED	TDC Sensor Ground	<0.050v
C22	BLU/RED	CKF Sensor Signal	AC pulse signals
C23	BLU/WHT	Vehicle Speed Sensor Signal	Moving: 0-5-0v
C25	RED/YEL	IAT Sensor Signal	Varies w/temp (0.5-4.9v)
C26	RED/WHT	ECT Sensor Signal	At 180°F: 0.5-0.6v
C27	RED/BLK	TP Sensor Signal	0.5-0.6v
C29	YEL	CYP Sensor Signal	AC pulse signals
C30	BLK	CYP Sensor Ground	<0.050v
C31	WHT/RED	CKF Sensor Ground	<0.050v

```
| 1 | 2 |   | 3 | 4 | 5 |   | 6 | 7 | 8 |   | 1 | 2 | 3 | 4 |   | 5 | 6 | 7 |   | 8 | 9 |10|
| 9 |10 |11 |12 |13 |14 |15 |16 |17 |18 |11 |12 |13 |14 |15 |16 |17 |18 |19 |20 |21 |22 |
|19 |20 |21 |22 |   |23 |24 |25 |   |23 |24 |25 |   |26 |27 |28 |   |29 |30 |31 |
     25-PIN 'B'      ( WIRE SIDE OF HARNESS TERMINALS )          31-PIN 'C'
```

05533_ADIA_G600

Pin Connector Graphic

2000 Sedan 1.6L I4 VTEC VIN EJ8 [All] 32P 'A' Connector

PCM Pin #	W/Color	Circuit Description (32-Pin)	Value at Hot Idle
A3	BLU	EVAP Bypass Solenoid	Solenoid Off: 12v, On: 1v
A4	GRN/WHT	EVAP Vent Solenoid	Solenoid Off: 12v, On: 1v
A5	BLU/GRN	Cruise Control Signal	C/C On: pulse signals
A6	RED/YEL	EVAP Purge Solenoid	Solenoid Off: 12v, On: 1v
A8	BLK/WHT	HO2S-12 (B1 S2) Heater	Digital Signals: 0-12-0v
A9	LT GRN	A/T: Gear Position Switch	In P/N: 0v, Others: 12v
A10	BRN	Service Check Connector	SCS Open: 4.80v
A14	GRN/BLK	A/T: D4 Light Switch	D4 On: 0v, Off: 12v
A16	GRN/YEL	Fuel Pump Relay Control	Relay Off: 12v, On: 1v
A17	BLK/RED	A/C Clutch Relay Control	Relay Off: 12v, On: 1v
A18	GRN/ORN	Malfunction Indicator Light	MIL Off: 12v, On: 1v
A19	BLU	Engine Speed Pulse (NEP)	Digital Signals
A20	GRN	Radiator Fan Relay Control	Relay Off: 12v, On: 1v
A21	BLU/YEL	K-Line Signal	12v
A22	BLU	A/T: Gear Position Switch	In 'L': 0v, others: 12v
A23	WHT/RED	HO2S-12 (B1 S2) Signal	0.1-1.1v
A24	BLU/WHT	Starter Switch Signal	Cranking: 9-11v
A26	GRN	PSP Switch Signal	Straight: 0v, Turning: 11v
A27	BLU/RED	A/C Switch Signal	Switch Off: 12v, On: 1v
A28	WHT/RED	Interlock Control Unit Signal	Key & Brake On: 12v
A29	LT GRN	Fuel Tank Pressure Sensor	Fuel Cap off: 2.5v
A30	GRN/RED	Electric Load Detector	Varies: 0.5-4.5v
A32	WHT/BLK	Brake Switch Signal	Brake Off: 0v, On: 12v

2000 Sedan 1.6L I4 VTEC VIN EJ8 [All] 16P 'D' Connector

PCM Pin #	W/Color	Circuit Description (16-Pin)	Value at Hot Idle
D1	YEL	A/T: Lockup Control Solenoid	LUS On: 12v, Off: 0v
D2	GRN/WHT	A/T: Shift Solenoid 'B'	SSB Off: 0v, On: 12v
D3	GRN	A/T: Shift Solenoid 'C'	SSC on: 12v, off: 0v
D5	BLK/YEL	A/T: Solenoid Feed (B+)	12-14v
D6	WHT	A/T: Gear Position Switch	In 'R': 0v, Others: 12v
D7	BLU/YEL	A/T: Shift Solenoid 'A'	SSA Off: 0v, On: 12v
D8	PNK	A/T: Gear Position Switch	In D3: 0v, Others: 12v
D9	YEL	A/T: Gear Position Switch	In D4: 0v, Others: 12v
D10	BLU	Countershaft Speed Sensor P	Moving: AC pulses
D11	RED	Mainshaft Speed Sensor 'P'	Pulse Signals
D12	WHT	Mainshaft Speed Sensor 'N'	<0.050v
D13	BLU/WHT	A/T: Gear Position Switch	In P/N: 0v, Others: 12v
D14	GRN/BLK	A/T: D4 Indicator Light Driver	D4 on: 5v, off: 0v
D16	GRN	Countershaft Speed Sensor N	<0.050v

```
1  2  3  4    5  6  7    8  9  10    11     1     2  3  4      5
12 13 14 15 16 17 18 19 20 21 22  23     24     6  7  8  9  10 11   12
32-PIN 25 26 27    28 29 30 31     32        13 14 15    16  16-PIN
  'A'                                                          'D'
          ( WIRE SIDE OF HARNESS TERMINALS )
```

05533_ADIA_G599

Pin Connector Graphic
2000 Sedan 1.6L I4 VTEC VIN EJ8 [All] 25P 'B' Connector

PCM Pin #	W/Color	Circuit Description (25-Pin)	Value at Hot Idle
B1, B9	YEL/BLK	Main Relay Power (B+)	12-14v
B2, B10	BLK	Power Ground	<0.1v
B3, B4	RED, BLU	Injector 2, Injector 3 Control	2.0-3.3 ms
B5, B11	YEL, BRN	Injector 4, Injector 1 Control	2.0-3.3 ms
B7	RED	EGR Control Solenoid	Digital Signals: 0-12-0v
B8	PNK/BLK	A/T: Control Linear Solenoid 'N'	Pulse Signals
B12	GRN/YEL	VTEC Control Solenoid	0v, Hi-Speed: 12v
B13	YEL/GRN	Ignition Control Signal	Digital Signals: 0-12-0v
B15	ORN	Idle Air Control Valve 'N'	Pulse Signals
B17	GRN/WHT	A/T: Control Linear Solenoid 'P'	Pulse Signals
B18	PNK/BLU	A/T: Start Clutch Solenoid 'N'	Pulse Signals
B20, B22	BRN/BLK	Logic Ground (LG1, LG2)	<0.050v
B21	WHT/YEL	Keep Alive Power (VBU)	12-14v
B23	BLK/BLU	Idle Air Control Valve 'P'	Pulse Signals
B24	BLU/WHT	A/T: 3rd Clutch Pressure Sw.	Open: 12v, Closed: 1v
B25	YEL	A/T: Start Clutch Solenoid 'N'	Pulse Signals

2000 Sedan 1.6L I4 VTEC VIN EJ8 [All] 31P 'C' Connector

PCM Pin #	W/Color	Circuit Description (31-Pin)	Value at Hot Idle
C1	BLK/WHT	HO2S-11 (B1 S1) Heater	Digital Signals: 0-12-0v
C2	WHT/GRN	Alternator Charging Signal	Lights Off: 12v, On: 0v
C3	RED/BLU	Knock Sensor Signal	No Detonation: 18mv AC
C5	WHT/RED	Alternator 'FR' Signal	Digital Signals: 0-5-0v
C6	WHT/BLK	EGR Valve Lift Sensor	1.2v
C7	GRN/WHT	MAP Sensor Ground (SG1)	<0.050v
C8	BLU	CKP Sensor Signal	AC pulse signals
C9	WHT	CKP Sensor Ground	<0.050v
C10	BLU/BLK	VTEC Pressure Switch	0v, Hi-Speed: 12v
C16	WHT	HO2S-11 (B1 S1) Signal	0.1-1.1v
C17	RED/GRN	MAP Sensor Signal	0.8-0.9v
C18	GRN/BLK	Signal Ground (SG2)	<0.050v
C19, C28	YEL/RED	MAP Sensor, Sensor VREF	4.9-5.1v
C20	GRN	TDC Sensor Signal	AC pulse signals
C21	RED	TDC Sensor Ground	<0.050v
C22	BLU/RED	CKF Sensor Signal	AC pulse signals
C23	BLU/WHT	Vehicle Speed Sensor Signal	Moving: 0-5-0v
C25	RED/YEL	IAT Sensor Signal	Varies w/temp (0.5-4.9v)
C26	RED/WHT	ECT Sensor Signal	At 180°F: 0.5-0.6v
C27	RED/BLK	TP Sensor Signal	0.5-0.6v
C29	YEL	CYP Sensor Signal	AC pulse signals
C30	BLK	CYP Sensor Ground	<0.050v
C31	WHT/RED	CKF Sensor Ground	<0.050v

WIRE SIDE OF HARNESS TERMINALS
25-PIN 'B'   31-PIN 'C'

05533_ADIA_G600

Pin Connector Graphic

2001-03 Sedan 1.7L CNG VIN EN2 [CVT] 31P Gray 'A' Connector

PCM Pin #	W/Color	Circuit Description (31-Pin)	Value at Hot Idle
A1	BLK/WHT	HO2S-11 (B1 S1) Heater	Digital Signals: 0-12-0v
A2, A3	YEL/BLK	Main Relay Power (B+)	12-14v
A4, A5	BLK	Power Ground (PG2, PG1)	<0.1v
A6	WHT	HO2S-11 (B1 S1) Signal	0.1-1.1v
A7	BLU	CKP Sensor Signal	Digital Signals
A10	GRN/YEL	Sensor Ground (SG2)	<0.050v
A11	GRN/WHT	Sensor Ground (SG1)	<0.050v
A12	BLK/RED	Idle Air Control Valve	Pulse Signals
A13	WHT/BLK	EGR Valve Position Sensor	0.6-1.1v
A14	BLK/WHT	HO2S-12 (B1 S2) Heater	Digital Signals: 0-12-0v
A15	RED/BLK	TP Sensor Signal	0.5-0.6v

2001-03 Sedan 1.7L CNG VIN EN2 [CVT] 31P Gray 'A' Connector, *continued*

A17	BRN/WHT	Injector Mode Signal	Digital Signals
A18	WHT/GRN	Vehicle Speed Signal	Moving: 0-5-0v
A19	GRN/RED	MAP Sensor Signal	0.8-0.9v
A20, A21	YEL/BLU	Sensor VREF (VCC2, VCC1)	4.9-5.1v
A23, A24	BRN/YEL	Logic Ground (LG2, LG1)	<0.1v
A25	WHT/RED	HO2S-12 (B1 S2) Signal	0.1-1.1v
A26	GRN	TDC Sensor Signal	AC pulse signals
A27	BRN	Coil 4 Driver Control	Digital Signals
A28	WHT/BLU	Coil 3 Driver Control	Digital Signals
A29	BLU/RED	Coil 2 Driver Control	Digital Signals
A30	YEL/GRN	Coil 1 Driver Control	Digital Signals

2001-03 Sedan 1.7L CNG VIN EN2 [CVT] 24P White 'B' Connector

PCM Pin #	W/Color	Circuit Description (24-Pin)	Value at Hot Idle
B2	YEL	Injector 4 Control	2.0-3.3 ms
B3	BLU	Injector 3 Control	2.0-3.3 ms
B4	RED	Injector 2 Control	2.0-3.3 ms
B5	BRN	Injector 1 Control	2.0-3.3 ms
B6	GRN	Fan Relay Control	Relay Off: 12v, On: 1v
B7	GRN/WHT	CVT: Pulley Control Valve	Pulse Signals
B8	RED/WHT	ECT Sensor Signal	At 180°F: 0.51v
B10	WHT/BLU	Alternator Load Signal	Lights Off: 12v, On: 0v
B13	WHT/RED	Alternator 'FR' Signal	Digital Signals: 0-5-0v
B14	BLU/RED	EGR Solenoid Control	Solenoid Off: 12v, On: 1v
B16	BLK/RED	CVT: Pressure Control Valve	Valve On: 12v: Off: 0v
B17	RED/YEL	IAT Sensor Signal	Varies w/temp. (0.5-4.9v)
B18	WHT/GRN	Alternator Charging Signal	Digital Signals: 0-12-0v

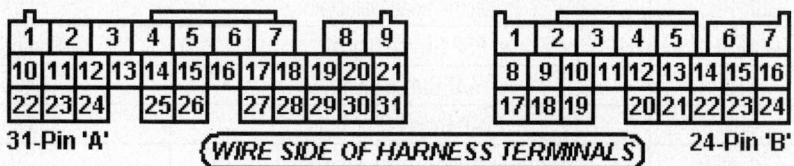

31-Pin 'A'   *WIRE SIDE OF HARNESS TERMINALS*   24-Pin 'B'

05533_ADIA_G611

Pin Connector Graphic

2001-03 Sedan 1.7L CNG VIN EN2 [CVT] 22P C Black Connector

PCM Pin #	W/Color	Circuit Description (22-Pin)	Value at Hot Idle
C1	BLK/RED	Pulley Press. Control Valve	Pulse Signals
C2-5	---	Not Used	---
C6	GRN/RED	Inhibitor Solenoid	INH On: 12v: Off: 0v
C7	RED/BLU	Drive Pulley Speed Sensor	Pulse Signals
C8	BLU/RED	Start Press. Control Valve	Pulse Signals
C9	RED	Gear Position Switch	In 2nd: 0v, Others: 12v
C10	WHT	Gear Position Switch	In 'R': 0v, Others: 12v
C11	BLU	Gear Position Switch	In 'L': 0v, Others: 12v
C12	BLU/WHT	Gear Position Switch	In P/N: 0v, others: 12v
C13-14	---	Not Used	---
C15	WHT/GRN	Driven Pulley Speed Sensor	Pulse Signals
C16	GRN/YEL	CVT: Speed Change Control	Pulse Signals
C18	BLU/YEL	Gear Position Switch	In FWD: 0v, Others: 12v
C20	YEL	Gear Position Switch	In 'D': 0v, Others: 12v
C21	WHT/RED	CVT Speed Sensor 1	Pulse Signals

2001-03 Sedan 1.7L CNG VIN EN2 [CVT] 31P White 'E' Connector

PCM Pin #	W/Color	Circuit Description (31-Pin)	Value at Hot Idle
E1	GRN/YEL	Immobilizer Fuel Pump Relay	Relay Off: 12v, On: 1v
E3	BRN/YEL	Logic Ground (LG3)	<0.050v
E4	PNK	Sensor Ground (SG3)	<0.050v
E5	YEL/BLU	Sensor VREF (VCC3)	4.9-5.1v
E7	RED/YEL	Main Relay Control	Relay Off: 12v, On: 1v
E9	YEL/BLK	Main Relay Power (B+)	12-14v
E12	BLU/ORN	Cruise Control Unit Signal	N/A
E13	WHT/BLU	Multiplex Control Unit	N/A
E14	LT GRN	Fuel Tank Pressure Sensor	Fuel Cap off: 2.5v
E15	GRN/RED	Electric Load Detector	Varies: 0.5-4.5v
E16	GRN/BLK	PSP Switch Signal	Straight: 0v, Turning: 11v
E18	RED	A/C Clutch Relay	Relay Off: 12v, On: 1v
E20	GRN	EVAP Bypass Solenoid	Solenoid Off: 12v, On: 1v
E21	GRN/RED	Fuel Tank Temp. Sensor	Varies: 0.1-4.9v
E22	WHT/BLK	Brake Switch Signal	Brake Off: 0v, On: 12v
E23	LT BLU	K-Line Signal	12v
E24	YEL	SEFMJ Signal (Multiplex Unit)	Digital Signals
E25	BLU/WHT	VSS Out Signal	Pulse Signals
E26	BLU	Engine Speed Pulse (NEP)	Digital Signals
E27	RED/BLU	Immobilizer Code Signal	Digital Signals
E29	BRN	Service Check Connector	SCS Open: 4.80v
E30	RED/WHT	WEN Terminal in DLC	0v
E31	GRN/ORN	Malfunction Indicator Lamp	MIL Off: 12v, On: 1v

24-Pin 'C'

WIRE SIDE OF HARNESS TERMINALS

31-Pin 'E'

05533_ADIA_G612

Pin Connector Graphic

2001-03 Sedan 1.7L MFI VIN ES1 [All] 31P Gray 'A' Connector

PCM Pin #	W/Color	Circuit Description (31-Pin)	Value at Hot Idle
A1	BLK/WHT	HO2S-11 (B1 S1) Heater	Digital Signals: 0-12-0v
A2, A3	YEL/BLK	Main Relay Power (B+)	12-14v
A4, A5	BLK	Power Ground (PG2, PG1)	<0.1v
A6	WHT	HO2S-11 (B1 S1) Signal	0.1-1.1v
A7	BLU	CKP Sensor Signal	Digital Signals
A8	YEL	Sensor VREF (VCCR)	4.9-5.1v
A9	RED/BLU	Knock Sensor Signal	No detonation: 18mv AC
A10	GRN/YEL	Sensor Ground (SG2)	<0.050v
A11	GRN/WHT	Sensor Ground (SG1)	<0.050v
A12	BLK/RED	Idle Air Control Valve	Pulse Signals
A13	WHT/BLK	EGR Valve Position Sensor	0.6-1.1v
A14	BLK/WHT	HO2S-12 (B1 S2) Heater	Digital Signals: 0-12-0v
A15	RED/BLK	TP Sensor Signal	0.5-0.6v
A17	BRN/WHT	Injector Mode Signal	Digital Signals
A18	WHT/GRN	Vehicle Speed Signal	Moving: 0-5-0v
A19	GRN/RED	MAP Sensor Signal	0.8-0.9v
A20, A21	YEL/BLU	Sensor VREF (VCC2, VCC1)	4.9-5.1v
A23, A24	BRN/YEL	Logic Ground (LG2, LG1)	<0.1v
A25	WHT/RED	HO2S-12 (B1 S2) Signal	0.1-1.1v
A26	GRN	TDC Sensor Signal	AC pulse signals
A27	BRN	Coil 4 Driver Control	Digital Signals
A28	WHT/BLU	Coil 3 Driver Control	Digital Signals
A29	BLU/RED	Coil 2 Driver Control	Digital Signals
A30	YEL/GRN	Coil 1 Driver Control	Digital Signals

2001-03 Sedan 1.7L MFI VIN ES1 [All] 24P White 'B' Connector

PCM Pin #	W/Color	Circuit Description (24-Pin)	Value at Hot Idle
B2	YEL	Injector 4 Control	2.0-3.3 ms
B3	BLU	Injector 3 Control	2.0-3.3 ms
B4	RED	Injector 2 Control •	2.0-3.3 ms
B5	BRN	Injector 1 Control	2.0-3.3 ms
B6	GRN	Fan Relay Control	Relay Off: 12v, On: 1v
B7	RED/BLK	A/T: Linear Solenoid 'A+'	LSA Off: 0v, On: 12v
B8	RED/WHT	ECT Sensor Signal	At 180°F: 0.51v
B10	WHT/BLU	Alternator Load Signal	Lights Off: 12v, On: 0v
B13	WHT/RED	Alternator 'FR' Signal	Digital Signals: 0-5-0v
B14	BLU/RED	EGR Solenoid Control	Solenoid Off: 12v, On: 1v
B16	BLK/RED	A/T: Lockup Solenoid 'B'	LSB On: 12v: Off: 0v
B17	RED/YEL	IAT Sensor Signal	Varies w/temp. (0.5-4.9v)
B18	WHT/GRN	Alternator Charging Signal	Digital Signals: 0-12-0v
B21	YEL/BLU	EVAP Purge Solenoid	Solenoid Off: 12v, On: 1v

Pin Connector Graphic

2001-03 Sedan 1.7L MFI VIN ES1 [W/O CVT] 22P C Black Connector

PCM Pin #	W/Color	Circuit Description (22-Pin)	Value at Hot Idle
C1	WHT/BLK	A/T: Linear Solenoid 'A-'	LSA Off: 12v, On: 1v
C2	YEL/BLU	A/T: TCC Solenoid	Solenoid Off: 12v, On: 1v
C4	GRN/WHT	A/T: Shift Solenoid 'B'	SSB Off: 0v, On: 12v
C6	BLU/BLK	A/T: Shift Solenoid 'A'	SSA On: 12v: Off: 0v
C7	WHT/RED	Mainshaft Speed Sensor 'P'	AC Pulse Signals
C8	BRN/WHT	A/T: Lockup Solenoid 'B-'	LSB On: 12v: Off: 0v
C9	RED	A/T: Gear Position Switch	In D3: 0v, Others: 12v
C10	WHT	A/T: Gear Position Switch	In 'R': 0v, Others: 12v
C11	BLU	A/T: Gear Position Switch	In D2: 0v, Others: 12v
C12	BLU/WHT	A/T: Gear Position Switch	In P/N: 0v, others: 12v
C14	GRN	Countershaft Speed Sensor N	Moving: AC pulses
C15	BLU	Countershaft Speed Sensor P	Moving: AC pulses
C18	BLY/REL	A/T: Gear Position Switch	Forward: 0v, Others: 12v
C20	YEL	A/T: Gear Position Switch	In 'D': 0v, Others: 12v
C21	WHT/GRN	Mainshaft Speed Sensor 'N'	AC Pulse Signals

2001-03 Coupe 1.7L I4 VIN ES1 [All] 31P White 'E' Connector

PCM Pin #	W/Color	Circuit Description (31-Pin)	Value at Hot Idle
E1	GRN/YEL	Immobilizer Fuel Pump Relay	Relay Off: 12v, On: 1v
E3	BRN/YEL	Logic Ground (LG3)	<0.050v
E4	PNK	Sensor Ground (SG3)	<0.050v
E5	YEL/BLU	Sensor VREF (VCC3)	4.9-5.1v
E7	RED/YEL	Main Relay Control	Relay Off: 12v, On: 1v
E9	YEL/BLK	Main Relay Power (B+)	12-14v
E12	BLU/ORN	Cruise Control Unit Signal	Digital Signals
E13	WHT/BLU	Multiplex Control Unit	Digital Signals
E14	LT GRN	Fuel Tank Pressure Sensor	Fuel Cap off: 2.5v
E15	GRN/RED	Electric Load Detector	Varies: 0.5-4.5v
E16	GRN/BLK	PSP Switch Signal	Straight: 0v, Turning: 11v
E18	RED	A/C Clutch Relay	Relay Off: 12v, On: 1v
E20	BLU/RED	EVAP Bypass Solenoid	Solenoid Off: 12v, On: 1v
E21	GRN/RED	EVAP Vent Solenoid	Solenoid Off: 12v, On: 1v
E22	WHT/BLK	Brake Switch Signal	Brake Off: 0v, On: 12v
E23	LT BLU	K-Line Signal	12v
E24	YEL	SEFMJ Signal (Multiplex Unit)	Digital Signals
E25	BLU/WHT	VSS Out Signal	Pulse Signals
E26	BLU	Engine Speed Pulse (NEP)	Digital Signals
E27	RED/BLU	Immobilizer Code Signal	Digital Signals
E29	BRN	Service Check Connector	SCS Open: 4.80v
E30	RED/WHT	WEN Terminal in DLC	0v
E31	GRN/ORN	Malfunction Indicator Lamp	MIL Off: 12v, On: 1v

24-Pin 'C'   WIRE SIDE OF HARNESS TERMINALS   31-Pin 'E'

05533_ADIA_G612

Pin Connector Graphic

2001-03 Sedan 1.7L VTEC VIN ES1 [All] 31P Gray 'A' Connector

PCM Pin #	W/Color	Circuit Description (31-Pin)	Value at Hot Idle
A1	BLK/WHT	HO2S-11 (B1 S1) Heater	Digital Signals: 0-12-0v
A2, A3	YEL/BLK	Main Relay Power (B+)	12-14v
A4, A5	BLK	Power Ground (PG2, PG1)	<0.1v
A6	WHT	HO2S-11 (B1 S1) Signal	0.1-1.1v
A7	BLU	CKP Sensor Signal	Digital Signals
A8	YEL	Sensor VREF (VCCR)	4.9-5.1v
A9	RED/BLU	Knock Sensor Signal	No detonation: 18mv AC
A10	GRN/YEL	Sensor Ground (SG2)	<0.050v
A11	GRN/WHT	Sensor Ground (SG1)	<0.050v
A12	BLK/RED	Idle Air Control Valve	Pulse Signals
A13	WHT/BLK	EGR Valve Position Sensor	0.6-1.1v
A14	BLK/WHT	HO2S-12 (B1 S2) Heater	Digital Signals: 0-12-0v
A15	RED/BLK	TP Sensor Signal	0.5-0.6v
A17	BRN/WHT	Injector Mode Signal	Digital Signals
A18	WHT/GRN	Vehicle Speed Signal	Moving: 0-5-0v
A19	GRN/RED	MAP Sensor Signal	0.8-0.9v
A20, A21	YEL/BLU	Sensor VREF (VCC2, VCC1)	4.9-5.1v
A23, A24	BRN/YEL	Logic Ground (LG2, LG1)	<0.1v
A25	WHT/RED	HO2S-12 (B1 S2) Signal	0.1-1.1v
A26	GRN	TDC Sensor Signal	AC pulse signals
A27	BRN	Coil 4 Driver Control	Digital Signals
A28	WHT/BLU	Coil 3 Driver Control	Digital Signals
A29	BLU/RED	Coil 2 Driver Control	Digital Signals
A30	YEL/GRN	Coil 1 Driver Control	Digital Signals

2001-03 Sedan 1.7L VTEC VIN ES1 [All] 24P White 'B' Connector

PCM Pin #	W/Color	Circuit Description (24-Pin)	Value at Hot Idle
B2, B3	YEL, BLU	Injector 4, Injector 3 Control	2.0-3.3 ms
B4, B5	RED, BRN	Injector 2, Injector 1 Control	2.0-3.3 ms
B6	GRN	Fan Relay Control	Relay Off: 12v, On: 1v
B7	RED/BLK	A/T: Linear Solenoid 'A+'	LSA Off: 0v, On: 12v
B8	RED/WHT	ECT Sensor Signal	At 180°F: 0.51v
B9	BLU/BLK	VTEC Pressure Switch	0v, Hi-Speed: 12v
B10	WHT/BLU	Alternator Load Signal	Lights Off: 12v, On: 0v
B13	WHT/RED	Alternator 'FR' Signal	Digital Signals: 0-5-0v
B14	BLU/RED	EGR Solenoid Control	Solenoid Off: 12v, On: 1v
B15	GRN/YEL	VTEC Solenoid Control	0v, Hi-Speed: 12v
B16	BLK/RED	A/T: Lockup Solenoid 'B'	LSB On: 12v: Off: 0v
B17	RED/YEL	IAT Sensor Signal	Varies w/temp. (0.5-4.9v)
B18	WHT/GRN	Alternator Charging Signal	Digital Signals: 0-12-0v
B21	YEL/BLU	EVAP Purge Solenoid	Solenoid Off: 12v, On: 1v

31-Pin 'A'   (WIRE SIDE OF HARNESS TERMINALS)   24-Pin 'B'

05533_ADIA_G615

Pin Connector Graphic

2001-03 Sedan 1.7L VTEC VIN ES1 [NO CVT] 22P C Black Connector

PCM Pin #	W/Color	Circuit Description (22-Pin)	Value at Hot Idle
C1	WHT/BLK	A/T: Linear Solenoid 'A-'	LSA Off: 12v, On: 1v
C2	YEL/BLU	A/T: TCC Solenoid	Solenoid Off: 12v, On: 1v
C4	GRN/WHT	A/T: Shift Solenoid 'B'	SSB Off: 0v, On: 12v
C6	BLU/BLK	A/T: Shift Solenoid 'A'	SSA On: 12v: Off: 0v
C7	WHT/RED	Mainshaft Speed Sensor 'P'	AC Pulse Signals
C8	BRN/WHT	A/T: Lockup Solenoid 'B-'	LSB On: 12v: Off: 0v
C9	RED	A/T: Gear Position Switch	In D3: 0v, Others: 12v
C10	WHT	A/T: Gear Position Switch	In 'R': 0v, Others: 12v
C11	BLU	A/T: Gear Position Switch	In D2: 0v, Others: 12v
C12	BLU/WHT	A/T: Gear Position Switch	In P/N: 0v, others: 12v
C14	GRN	Countershaft Speed Sensor N	Moving: AC pulses
C15	BLU	Countershaft Speed Sensor P	Moving: AC pulses
C18	BLY/REL	A/T: Gear Position Switch	Forward: 0v, Others: 12v
C20	YEL	A/T: Gear Position Switch	In 'D': 0v, Others: 12v
C21	WHT/GRN	Mainshaft Speed Sensor 'N'	AC Pulse Signals

2001-03 Sedan 1.7L VTEC VIN ES1 [All] 31P White 'E' Connector

PCM Pin #	W/Color	Circuit Description (31-Pin)	Value at Hot Idle
E1	GRN/YEL	Immobilizer Fuel Pump Relay	Relay Off: 12v, On: 1v
E3	BRN/YEL	Logic Ground (LG3)	<0.050v
E4	PNK	Sensor Ground (SG3)	<0.050v
E5	YEL/BLU	Sensor VREF (VCC3)	4.9-5.1v
E7	RED/YEL	Main Relay Control	Relay Off: 12v, On: 1v
E9	YEL/BLK	Main Relay Power (B+)	12-14v
E12	BLU/ORN	Cruise Control Unit Signal	N/A
E13	WHT/BLU	Multiplex Control Unit	N/A
E14	LT GRN	Fuel Tank Pressure Sensor	Fuel Cap off: 2.5v
E15	GRN/RED	Electric Load Detector	Varies: 0.5-4.5v
E16	GRN/BLK	PSP Switch Signal	Straight: 0v, Turning: 11v
E18	RED	A/C Clutch Relay	Relay Off: 12v, On: 1v
E20	BLU/RED	EVAP Bypass Solenoid	Solenoid Off: 12v, On: 1v
E21	GRN/RED	EVAP Vent Solenoid	Solenoid Off: 12v, On: 1v
E22	WHT/BLK	Brake Switch Signal	Brake Off: 0v, On: 12v
E23	LT BLU	K-Line Signal	12v
E24	YEL	SEFMJ Signal (Multiplex Unit)	Digital Signals
E25	BLU/WHT	VSS Out Signal	Pulse Signals
E26	BLU	Engine Speed Pulse (NEP)	Digital Signals
E27	RED/BLU	Immobilizer Code Signal	Digital Signals
E29	BRN	Service Check Connector	SCS Open: 4.80v
E30	RED/WHT	WEN Terminal in DLC	0v
E31	GRN/ORN	Malfunction Indicator Lamp	MIL Off: 12v, On: 1v

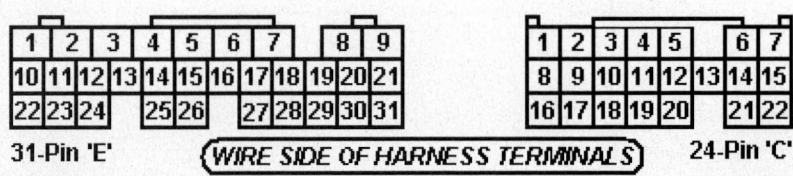

31-Pin 'E'   (WIRE SIDE OF HARNESS TERMINALS)   **24-Pin 'C'**

05533_ADIA_G616

Pin Connector Graphic

1990-91 Wagon 1.5L I4 MFI VIN EE2 [All] 18P 'A' Connector

PCM Pin #	W/Color	Circuit Description (18-Pin)	Value at Hot Idle
A1	BRN	Injector 1 Control	2.0-3.3 ms
A2	BLK	Power Ground	<0.1v
A3	RED	Injector 2 Control	2.0-3.3 ms
A4	BLK	Power Ground	<0.1v
A5	BLU	Injector 3 Control	2.0-3.3 ms
A6	GRN	EVAP Purge Solenoid	Solenoid Off: 12v, On: 1v
A7	YEL	Injector 4 Control	2.0-3.3 ms
A8	YEL/BLU	A/T: Control Unit Signal	Digital Signals
A9	---	Not Used	---
A10	ORN	EGR Solenoid	Solenoid Off: 12v, On: 1v
A11	BLU/YEL	Electronic Air Control Valve	Pulse Signals

1990-91 Wagon 1.5L I4 MFI VIN EE2 [All] 18P 'A' Connector, *continued*

A12	GRN/BLK	Fuel Pump Relay	Relay Off: 12v, On: 1v
A13	YEL/BLK	Main Relay Power (B+)	12-14v
A14	GRN/BLK	Fuel Pump Relay	Relay Off: 12v, On: 1v
A15	YEL/BLK	Main Relay Power (B+)	12-14v
A16	BRN/BLK	Power Ground	<0.1v
A17	---	Not Used	---
A18	BLK/RED	Power Ground	<0.1v

1990-91 Wagon 1.5L I4 MFI VIN EE2 [All] 20P 'B' Connector

PCM Pin #	W/Color	Circuit Description (20-Pin)	Value at Hot Idle
B1	WHT/GRN	Keep Alive Power (VBU)	12-14v
B2	GRN/YEL	Upshift Indicator Light	Lamp Off: 12v, On: 1v
B2	---	Not Used	---
B3	YEL	A/C Clutch Relay	Relay Off: 12v, On: 1v
B4	YEL/GRN	Radiator Fan Relay	Relay Off: 12v, On: 1v
B5	WHT/YEL	Alternator Charging Signal	Lights Off: 12v, On: 0v
B6	GRN/ORN	Check Engine Light	MIL Off: 12v, On: 1v
B7	GRN	A/T: P/N Switch (1989-91)	In P/N: 0v, Others: 11v
B8	BLU/RED	A/C Switch Signal	Switch Off: 12v, On: 1v
B9	GRN/BLK	Reverse Switch	In 'R': 0v, Others: 12v
B10	ORN	CKP Sensor Signal	AC pulse signals
B11	GRN	Clutch Switch	Clutch In: 0v, Out: 5v
B12	WHT	CKP Sensor Ground	0.050v
B13	BLU/WHT	Start Signal	Cranking: 9-11v
B14	BLU	Alternator 'FR' Signal	Digital Signals: 0-5-0v
B15	WHT	Igniter Control	Digital Signals: 0-12-0v
B16	YEL/RED	Vehicle Speed Sensor	Moving: pulse signals
B17	WHT	Igniter Control	Digital Signals: 0-12-0v
B18	---	Not Used	---
B19	GRN/RED	Electric Load Detector	Varies: 0.5-4.5v
B20	BRN	Ignition Timing Adjuster	0.5-4.5v

1990-91 Wagon 1.5L I4 MFI VIN EE2 [All] 16P 'C' Connector

PCM Pin #	W/Color	Circuit Description (16-Pin)	Value at Hot Idle
C1	BLU/GRN	CYP Sensor Signal	AC pulse signals
C2	BLU/YEL	CYP Sensor Ground	<0.050v
C3	ORN/BLU	TDC Sensor Signal	AC pulse signals
C4	WHT/BLU	TDC Sensor Ground	<0.050v
C5	RED/YEL	IAT Sensor Signal	Varies w/temp. (0.5-4.9v)
C6	RED/WHT	ECT Sensor Signal	At 180°F: 0.51v
C7	RED/BLU	Throttle Angle Sensor Signal	0.5-0.6v
C8	YEL	EGR Valve Lift Sensor	1.2v
C9	RED/WHT	Atmospheric Pressure Sensor	2.76-2.96v at sea level
C10	GRN/WHT	Brake Switch Signal	Brake Off: 0v, On: 12v
C11	WHT	MAP Sensor Signal	0.8-0.9v
C12	GRN/WHT	Sensor Ground	<0.050v
C13	YEL/WHT	Sensor VREF	4.9-5.1v
C14	GRN/WHT	MAP Sensor Ground	<0.050v
C15	YEL/RED	MAP Sensor VREF	4.9-5.1v
C16	WHT	Oxygen Sensor Signal	0.1-1.1v

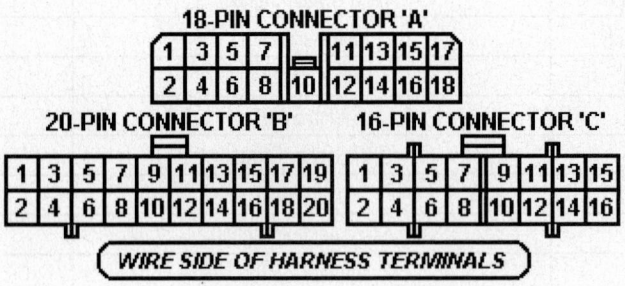

05533_ADIA_G613

Pin Connector Graphic

Standard Colors and Abbreviations

Abbreviation	Color	Abbreviation	Color	Abbreviation	Color
BLK	Black	LT BLU	Lt. Blue	TAN	Tan
BLU	Blue	LT GRN	Lt. Green	VIO	Violet
BRN	Brown	ORN	Orange	WHT	White
GRY	Gray	PNK	Pink	YEL	Yellow
GRN	Green	PPL	Purple		

1990-91 Wagon 1.6L I4 MFI VIN EE4 [All] 18P 'A' Connector

PCM Pin #	W/Color	Circuit Description (18-Pin)	Value at Hot Idle
A1	BRN	Injector 1 Control	2.0-3.3 ms
A2	BLK	Power Ground	<0.1v
A3	RED	Injector 2 Control	2.0-3.3 ms
A4	BLK	Power Ground	<0.1v
A5	BLU	Injector 3 Control	2.0-3.3 ms
A6	GRN	EVAP Purge Solenoid	Solenoid Off: 12v, On: 1v
A7	YEL	Injector 4 Control	2.0-3.3 ms
A8	YEL/BLU	A/T: Control Unit Signal	Digital Signals
A9-10	---	Not Used	---
A11	BLU/YEL	Electronic Air Control Valve	Pulse Signals
A12	GRN/BLK	Fuel Pump Relay	Relay Off: 12v, On: 1v
A13	YEL/BLK	Main Relay Power (B+)	12-14v
A14	GRN/BLK	Fuel Pump Relay	Relay Off: 12v, On: 1v
A15	YEL/BLK	Main Relay Power (B+)	12-14v
A16	BRN/BLK	Power Ground	<0.1v
A17	---	Not Used	---
A18	BLK/RED	Power Ground	<0.1v

1990-91 Wagon 1.6L I4 MFI VIN EE4 [All] 20P 'B' Connector

PCM Pin #	W/Color	Circuit Description (20-Pin)	Value at Hot Idle
B1	WHT/GRN	Keep Alive Power (VBU)	12-14v
B2	---	Not Used	---
B2	BLU	Fast Idle Control	Solenoid Off: 12v, On: 1v
B3	YEL	A/C Clutch Relay	Solenoid Off: 12v, On: 1v
B4	YEL/GRN	Radiator Fan Relay	Relay Off: 12v, On: 1v
B5	WHT/YEL	Alternator Charging Signal	Lights Off: 12v, On: 0v
B6	GRN/ORN	Check Engine Light	MIL Off: 12v, On: 1v
B7	---	Not Used	---
B7	GRN	A/T: P/N Switch	In P/N: 0v, Others: 12v
B8	BLU/RED	A/C Switch Signal	Switch Off: 12v, On: 1v
B9	---	Not Used	---
B10	ORN	CKP Sensor Signal	AC pulse signals
B11	---	Not Used	---
B12	WHT	CKP Sensor Ground	<0.050v
B13	BLU/WHT	Start Signal	Cranking: 9-11v
B14	BLU	Alternator 'FR' Signal	Digital Signals: 0-5-0v
B15	WHT	Igniter Control	Digital Signals: 0-12-0v
B16	YEL/RED	Vehicle Speed Sensor	Moving: pulse signals
B17	WHT	Igniter Control	Digital Signals: 0-12-0v
B18	---	Not Used	---
B19	GRN/RED	Electric Load Detector	Varies: 0.5-4.5v
B20	BRN	Ignition Timing Adjuster	0.5-4.5v

1990-91 Wagon 1.6L I4 MFI VIN EE4 [All] 16P 'C' Connector

PCM Pin #	W/Color	Circuit Description (16-Pin)	Value at Hot Idle
C1	BLU/GRN	CYP Sensor Signal	AC pulse signals
C2	BLU/YEL	CYP Sensor Ground	<0.050v
C3	ORN/BLU	TDC Sensor Signal	AC pulse signals
C4	WHT/BLU	TDC Sensor Ground	<0.050v
C5	RED/YEL	IAT Sensor Signal	Varies w/temp. (0.5-4.9v)
C6	RED/WHT	ECT Sensor Signal	At 180°F: 0.51v
C7	RED/BLU	Throttle Angle Sensor Signal	0.5-0.6v
C8	---	Not Used	---
C8	BLU/WHT	A/T: Control Unit Signal	Digital Signals
C9	RED/WHT	Atmospheric Pressure Sensor	2.76-2.96v at sea level
C10	GRN/WHT	Brake Switch Signal	Brake Off: 0v, On: 12v
C11	WHT	MAP Sensor Signal	0.8-0.9v
C12	GRN/WHT	Sensor Ground	<0.050v
C13	YEL/WHT	Sensor VREF	4.9-5.1v
C14	GRN/WHT	MAP Sensor Ground	<0.050v
C15	YEL/RED	MAP Sensor VREF	4.9-5.1v
C16	WHT	Oxygen Sensor Signal	0.1-1.1v

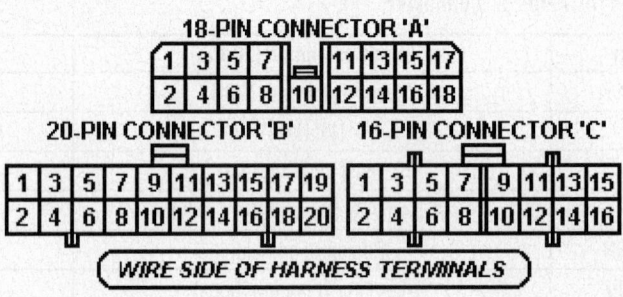

05533_ADIA_G613

Pin Connector Graphic

Standard Colors and Abbreviations

Abbreviation	Color	Abbreviation	Color	Abbreviation	Color
BLK	Black	LT BLU	Lt. Blue	TAN	Tan
BLU	Blue	LT GRN	Lt. Green	VIO	Violet
BRN	Brown	ORN	Orange	WHT	White
GRY	Gray	PNK	Pink	YEL	Yellow
GRN	Green	PPL	Purple		

CIVIC DEL SOL PIN CHARTS

1993 Coupe 1.5L I4 MFI VIN EG1 [All] 26P 'A' Connector

PCM Pin #	W/Color	Circuit Description (26-Pin)	Value at Hot Idle
A1	BRN	Injector 1 Control	2.0-3.3 ms
A2	YEL	Injector 1 Control	2.0-3.3 ms
A3	RED	Injector 2 Control	2.0-3.3 ms
A4	---	Not Used	---
A5	BLU	Injector 3 Control	2.0-3.3 ms
A6	ORN/BLK	HO2S-11 (B1 S1) Heater	Relay Off: 12v, On: 1v
A7	GRN/YEL	Fuel Pump Relay	Relay Off: 12v, On: 1v
A8	GRN/YEL	Fuel Pump Relay	Relay Off: 12v, On: 1v
A9	GRN/WHT	Idle Air Control Valve	Pulse Signals
A10-11	---	Not Used	---
A12	YEL/GRN	Radiator Fan Relay	Relay Off: 12v, On: 1v
A13	GRN/ORN	Check Engine Light	MIL Off: 12v, On: 1v
A14	---	Not Used	---
A15	BLK/RED	A/C Clutch Relay	Relay Off: 12v, On: 1v
A16	WHT/YEL	Alternator Charging Signal	Lights Off: 12v, On: 0v
A17	LT GRN	A/T: TCM Signal	Digital Signals
A18	---	Not Used	---
A19	YEL	A/T: TCM Signal	Digital Signals
A20	RED	EVAP Purge Solenoid	Solenoid Off: 12v, On: 1v
A21	RED/GRN	Igniter Control	Digital Signals: 0-12-0v
A22	RED/GRN	Igniter Control	Digital Signals: 0-12-0v
A23	BLK	Power Ground	<0.1v
A24	BLK	Power Ground	<0.1v
A25	YEL/BLK	Main Relay Power (B+)	12-14v
A26	BLK/RED	Logic Ground	<0.050v

1993 Coupe 1.5L I4 MFI VIN EG1 [All] 16P 'B' Connector

PCM Pin #	W/Color	Circuit Description (16-Pin)	Value at Hot Idle
B1	WHT/GRN	Main Relay Power (B+)	12-14v
B2	BRN/BLK	Logic Ground	<0.050v
B3	GRN/BLU	A/T: TCM Signal	In D3: 0v, Others: 5v
B4	GRN/BLK	A/T: TCM Signal	In D4: 0v, Others: 5v
B5	BLU/RED	A/C Switch Signal	Relay Off: 12v, On: 1v
B6	---	Not Used	---
B7	GRN	A/T: Park Neutral Switch	In P/N: 0v, Others: 12v
B8	BRN/RED	PSP Switch	Straight: 0v, Turning: 11v
B9	BLU/WHT	Starter Switch Signal	Cranking: 9-11v
B10	YEL/BLU	Vehicle Speed Sensor	Moving: pulse signals
B11	ORN	CYP Sensor Signal	AC pulse signals
B12	WHT	CYP Sensor Ground	<0.050v
B13	ORN/BLU	TDC Sensor Signal	AC pulse signals
B14	WHT/BLU	TDC Sensor Ground	<0.050v
B15	BLU/GRN	CKP Sensor Signal	AC pulse signals
B16	BLU/YEL	CKP Sensor Ground	<0.050v

1993 Coupe 1.5L I4 MFI VIN EG1 [All] 22P 'D' Connector

PCM Pin #	W/Color	Circuit Description (22-Pin)	Value at Hot Idle
D1	WHT/BLU	Keep Alive Power (VBU)	12-14v
D2	GRN/WHT	Brake Switch Signal	Brake Off: 0v, On: 12v
D3	---	Not Used	---
D4	BRN	Service Check Connector	SCS Open: 4.80v
D5	---	Not Used	---
D6	---	Not Used	---
D7	LT BLU	Data Link Connector	5v
D8	---	Not Used	---
D9	PNK	Alternator 'FR' Signal	Varies 1.5-3.5v
D10	GRN/RED	Electric Load Detector	Varies: 0.5-4.5v
D11	LT GRN	TP Sensor Signal	0.5-0.6v
D12	---	Not Used	---
D13	RED/WHT	ECT Sensor Signal	At 180°F: 0.52v
D14	ORN/BLU	HO2S-11 (B1 S1) Signal	0.1-1.1v
D15	RED/YEL	IAT Sensor Signal	Varies w/temp. (0.5-4.9v)
D16	---	Not Used	---
D17	WHT	MAP Sensor Signal	0.8-0.9v
D18	PNK/GRN	Economy Driving Indicator	Digital Signals
D19	YEL/GRN	MAP Sensor VREF	4.9-5.1v
D20	YEL/WHT	Sensor VREF	4.9-5.1v
D21	GRN/WHT	MAP Sensor Ground	<0.050v
D22	GRN/WHT	Sensor Ground	<0.050v

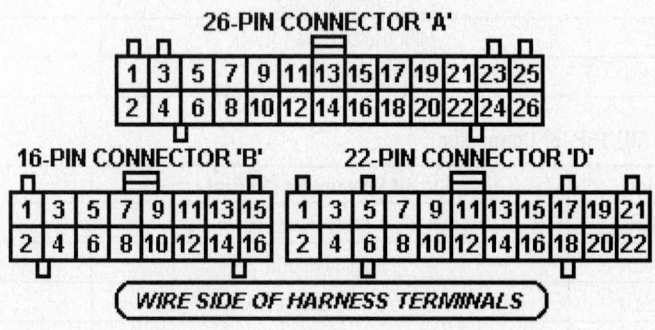

05533_ADIA_G617

Pin Connector Graphic

Standard Colors and Abbreviations

Abbreviation	Color	Abbreviation	Color	Abbreviation	Color
BLK	Black	LT BLU	Lt. Blue	TAN	Tan
BLU	Blue	LT GRN	Lt. Green	VIO	Violet
BRN	Brown	ORN	Orange	WHT	White
GRY	Gray	PNK	Pink	YEL	Yellow
GRN	Green	PPL	Purple		

1993 Coupe 1.6L I4 VTEC VIN EH6 [All] 26P 'A' Connector

PCM Pin #	W/Color	Circuit Description (26-Pin)	Value at Hot Idle
A1	BRN	Injector 1 Control	2.0-3.3 ms
A2	YEL	Injector 4 Control	2.0-3.3 ms
A3	RED	Injector 2 Control	2.0-3.3 ms
A4	ORN/WHT	VTEC Solenoid	0v, Hi-Speed: 12v
A5	BLU	Injector 3 Control	2.0-3.3 ms
A6	ORN/BLK	HO2S-1 Heater Control	Relay Off: 12v, On: 1v
A7	GRN/YEL	Fuel Pump Relay	Relay Off: 12v, On: 1v
A8	GRN/YEL	Fuel Pump Relay	Relay Off: 12v, On: 1v
A9	GRN/WHT	Idle Air Control Valve	Pulse Signals
A10-11	---	Not Used	---
A12	YEL/GRN	Radiator Fan Relay	Relay Off: 12v, On: 1v
A13	GRN/ORN	Check Engine Light	MIL Off: 12v, On: 1v
A14	---	Not Used	---
A15	BLK/RED	A/C Clutch Relay	Relay Off: 12v, On: 1v
A16	WHT/YEL	Alternator Charging Signal	Lights Off: 12v, On: 0v
A17	LT GRN	A/T TCM Signal	Digital Signal
A18	---	Not Used	---
A19	YEL	A/T TCM Signal	Digital Signals
A20	RED	EVAP Purge Solenoid	Solenoid Off: 12v, On: 1v
A21	RED/GRN	Igniter Control	Digital Signals: 0-12-0v
A22	RED/GRN	Igniter Control	Digital Signals: 0-12-0v
A23	BLK	Power Ground	<0.1v
A24	BLK	Power Ground	<0.1v
A25	YEL/BLK	Main Relay Power (B+)	12-14v
A26	BLK/RED	Logic Ground	<0.050v

1993 Coupe 1.6L I4 VTEC VIN EH6 [All] 16P 'B' Connector

PCM Pin #	W/Color	Circuit Description (16-Pin)	Value at Hot Idle
B1	WHT/GRN	Main Relay Power (B+)	12-14v
B2	BRN/BLK	Logic Ground	<0.050v
B3	GRN/BLU	A/T TCM Signal	In D3: 0v, Others: 5v
B4	GRN/BLK	A/T TCM Signal	In D4: 0v, Others: 5v
B5	BLU/RED	A/C Switch Signal	Relay Off: 12v, On: 1v
B6	---	Not Used	---
B7	GRN	A/T: Park Neutral Switch	In P/N: 0v, Others: 12v
B8	BRN/RED	PSP Switch	Straight: 0v, Turning: 11v
B9	BLU/WHT	Starter Switch Signal	Cranking: 9-11v
B10	YEL/BLU	Vehicle Speed Sensor	Moving: pulse signals
B11	ORN	CYP Sensor Signal	AC pulse signals
B12	WHT	CYP Sensor Ground	<0.050v
B13	ORN/BLU	TDC Sensor Signal	AC pulse signals
B14	WHT/BLU	TDC Sensor Ground	<0.050v
B15	BLU/GRN	CKP Sensor Signal	AC pulse signals
B16	BLU/YEL	CKP Sensor Ground	<0.050v

1993 Coupe 1.6L I4 VTEC VIN EH6 [All] 22P 'D' Connector

PCM Pin #	W/Color	Circuit Description (22-Pin)	Value at Hot Idle
D1	WHT/BLU	Keep Alive Power (VBU)	12-14v
D2	GRN/WHT	Brake Switch Signal	Brake Off: 0v, On: 12v
D4	BRN	Service Check Connector	SCS Open: 4.80v
D6	ORN/BLU	VTEC Pressure Switch	0v, Hi-Speed: 12v
D7	LT BLU	Data Link Connector	5v
D9	PNK	Alternator 'FR' Signal	Varies: 1.5-3.5v
D10	GRN/RED	Electric Load Detector	Varies: 0.5-4.5v
D11	LT GRN	TP Sensor Signal	0.5-0.6v
D13	RED/WHT	ECT Sensor Signal	At 180°F: 0.52v
D14	ORN/BLU	HO2S-11 (B1 S1) Signal	0.1-1.1v
D15	RED/YEL	IAT Sensor Signal	Varies w/temp. (0.5-4.9v)
D17	WHT	MAP Sensor Signal	0.8-0.9v
D18	PNK/GRN	Economy Driving Indicator	N/A
D19	YEL/GRN	MAP Sensor VREF	4.9-5.1v
D20	YEL/WHT	Sensor VREF	4.9-5.1v
D21	GRN/WHT	MAP Sensor Ground	<0.050v
D22	GRN/WHT	Sensor Ground	<0.050v

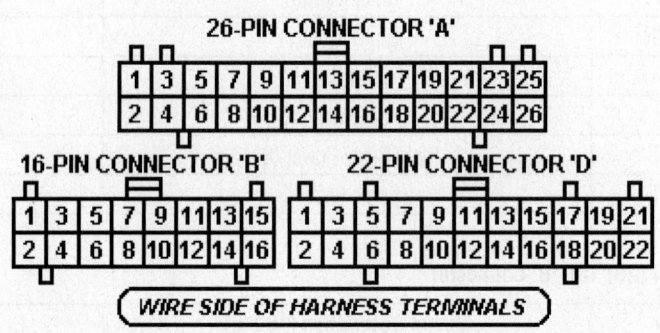

05533_ADIA_G617

Pin Connector Graphic

Standard Colors and Abbreviations

Abbreviation	Color	Abbreviation	Color	Abbreviation	Color
BLK	Black	LT BLU	Lt. Blue	TAN	Tan
BLU	Blue	LT GRN	Lt. Green	VIO	Violet
BRN	Brown	ORN	Orange	WHT	White
GRY	Gray	PNK	Pink	YEL	Yellow
GRN	Green	PPL	Purple		

1994-95 Coupe 1.5L I4 MFI VIN EG1 [All] 26P 'A' Connector

PCM Pin #	W/Color	Circuit Description (26-Pin)	Value at Hot Idle
A1	BRN	Injector 1 Control	2.0-3.3 ms
A2	YEL	Injector 4 Control	2.0-3.3 ms
A3	RED	Injector 2 Control	2.0-3.3 ms
A4	--	Not Used	---
A5	BLU	Injector 3 Control	2.0-3.3 ms
A6	ORN/BLK	HO2S Heater Control	Relay Off: 12v, On: 1v
A7	GRN/YEL	Fuel Pump Relay	Relay Off: 12v, On: 1v
A8	--	Not Used	---
A9	GRN/WHT	Idle Air Control Valve	Pulse Signals
A12	YEL/GRN	Radiator Fan Relay	Relay Off: 12v, On: 1v
A13	GRN/ORN	Check Engine Light	MIL Off: 12v, On: 1v
A15	BLK/RED	A/C Clutch Relay	Relay Off: 12v, On: 1v
A16	WHT/YEL	Alternator Charging Signal	Lights Off: 12v, On: 0v
A17	GRN/BLK	A/T: Lockup Solenoid 'B'	LSB On: 12v, Off: 0v
A19	YEL	A/T: Lockup Solenoid 'A'	LSA On: 12v, Off: 0v
A20	RED	EVAP Purge Solenoid	Solenoid Off: 12v, On: 1v
A21	RED/GRN	Igniter Control	Digital Signals: 0-12-0v
A22	---	Not Used	---
A23	BLK	Power Ground	<0.1v
A24	BLK	Power Ground	<0.1v
A25	YEL/BLK	Main Relay Power (B+)	12-14v
A26	BLK/RED	Logic Ground	<0.050v

1994-95 Coupe 1.5L I4 MFI VIN EG1 [All] 16P 'B' Connector

PCM Pin #	W/Color	Circuit Description (16-Pin)	Value at Hot Idle
B1	WHT/GRN	Main Relay Power (B+)	12-14v
B2	BRN/BLK	Logic Ground	<0.050v
B3	GRN/BLU	A/T: Shift Selector Signal	In D3: 0v, Others: 12v
B4	GRN/BLK	A/T: Shift Selector Signal	In D4: 0v, Others: 12v
B5	BLU/RED	A/C Switch Signal	Relay Off: 12v, On: 1v
B6	---	Not Used	---
B7	GRN	A/T: Park Neutral Switch	In P/N: 0v, Others: 12v
B8	BRN/RED	PSP Switch	Straight: 0v, Turning: 11v
B9	BLU/WHT	Starter Switch Signal	Cranking: 9-11v
B10	YEL/BLU	Vehicle Speed Sensor	Moving: pulse signals
B11	ORN	CYP Sensor Signal	AC pulse signals
B12	WHT	CYP Sensor Ground	<0.050v
B13	ORN/BLU	TDC Sensor Signal	AC pulse signals
B14	WHT/BLU	TDC Sensor Ground	<0.050v
B15	BLU/GRN	CKP Sensor Signal	AC pulse signals
B16	BLU/YEL	CKP Sensor Ground	<0.050v

1994-95 Coupe 1.5L I4 MFI VIN EG1 [All] 22P 'D' Connector

PCM Pin #	W/Color	Circuit Description (22-Pin)	Value at Hot Idle
D1	WHT/BLU	Keep Alive Power (VBU)	12-14v
D2	GRN/WHT	Brake Switch Signal	Brake Off: 0v, On: 12v
D3	RED/BLU	Knock Sensor Signal	No Detonation: 18mv AC
D4	BRN	Service Check Connector	SCS Open: 4.80v
D5-6, 8	---	Not Used	---
D7	LT BLU	Data Link Connector	5v
D9	PNK	Alternator 'FR' Signal	Varies: 1.5-3.5v
D10	GRN/RED	Electric Load Detector	Varies: 0.5-4.5v
D11	LT GRN	TP Sensor Signal	0.5-0.6v
D12	---	Not Used	---
D13	RED/WHT	ECT Sensor Signal	At 180°F: 0.52v
D14	WHT	HO2S-11 (B1 S1) Signal	0.1-1.1v
D15	RED/YEL	IAT Sensor Signal	Varies w/temp. (0.5-4.9v)
D16	---	Not Used	---
D17	WHT	MAP Sensor Signal	0.8-0.9v
D18	PNK/GRN	A/T: Interlock Control Unit	Key & Brake On: 12v
D19	YEL/GRN	MAP Sensor VREF	4.9-5.1v
D20	YEL/WHT	Sensor VREF	4.9-5.1v
D21	GRN/BLU	MAP Sensor Ground	<0.050v
D22	GRN/WHT	Sensor Ground	<0.050v

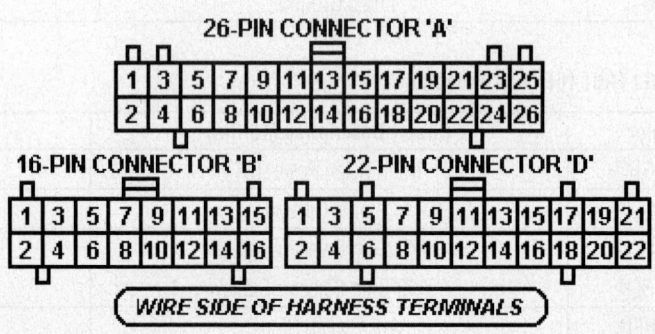

Pin Connector Graphic

Standard Colors and Abbreviations

Abbreviation	Color	Abbreviation	Color	Abbreviation	Color
BLK	Black	LT BLU	Lt. Blue	TAN	Tan
BLU	Blue	LT GRN	Lt. Green	VIO	Violet
BRN	Brown	ORN	Orange	WHT	White
GRY	Gray	PNK	Pink	YEL	Yellow
GRN	Green	PPL	Purple		

1994-95 Coupe 1.6L I4 VTEC VIN EH6 [All] 26P 'A' Connector

PCM Pin #	W/Color	Circuit Description (26-Pin)	Value at Hot Idle
A1	BRN	Injector 1 Control	2.0-3.3 ms
A2	YEL	Injector 4 Control	2.0-3.3 ms
A3	RED	Injector 2 Control	2.0-3.3 ms
A4	ORN/WHT	VTEC Solenoid	0v, Hi-Speed: 12v
A5	BLU	Injector 3 Control	2.0-3.3 ms
A6	ORN/BLK	HO2S Heater Control	Relay Off: 12v, On: 1v
A7	GRN/YEL	Fuel Pump Relay	Relay Off: 12v, On: 1v
A8	---	Not Used	---
A9	GRN/WHT	Idle Air Control Valve	Pulse Signals
A10-11	---	Not Used	---
A12	YEL/GRN	Radiator Fan Relay	Relay Off: 12v, On: 1v
A13	GRN/ORN	Check Engine Light	MIL Off: 12v, On: 1v
A14	---	Not Used	---
A15	BLK/RED	A/C Clutch Relay	Relay Off: 12v, On: 1v
A16	WHT/YEL	Alternator Charging Signal	Lights Off: 12v, On: 0v
A17	GRN/BLK	A/T: Lockup Solenoid 'B'	LSB On: 12v, Off: 0v
A18	---	Not Used	---
A19	YEL	A/T: Lockup Solenoid 'A'	LSA On: 12v, Off: 0v
A20	RED	EVAP Purge Solenoid	Solenoid Off: 12v, On: 1v
A21	RED/GRN	Igniter Control	Digital Signals: 0-12-0v
A22	---	Not Used	---
A23	BLK	Power Ground	<0.1v
A24	BLK	Power Ground	<0.1v
A25	YEL/BLK	Main Relay Power (B+)	12-14v
A26	BLK/RED	Logic Ground	<0.050v

1994-95 Coupe 1.6L I4 VTEC VIN EH6 [All] 16P 'B' Connector

PCM Pin #	W/Color	Circuit Description (16-Pin)	Value at Hot Idle
B1	WHT/GRN	Main Relay Power (B+)	12-14v
B2	BRN/BLK	Logic Ground	<0.050v
B3	GRN/BLU	A/T: Shift Selector Signal	In D3: 0v, Others: 12v
B4	GRN/BLK	A/T: Shift Selector Signal	In D4: 0v, Others: 12v
B5	BLU/RED	A/C Switch Signal	Relay Off: 12v, On: 1v
B7	GRN	A/T: Park Neutral Switch	In P/N: 0v, Others: 12v
B8	BRN/RED	PSP Switch	Straight: 0v, Turning: 11v
B9	BLU/WHT	Starter Switch Signal	Cranking: 9-11v
B10	YEL/BLU	Vehicle Speed Sensor	Moving: pulse signals
B11	ORN	CYP Sensor Signal	AC pulse signals
B12	WHT	CYP Sensor Ground	<0.050v
B13	ORN/BLU	TDC Sensor Signal	AC pulse signals
B14	WHT/BLU	TDC Sensor Ground	<0.050v
B15	BLU/GRN	CKP Sensor Signal	AC pulse signals
B16	BLU/YEL	CKP Sensor Ground	<0.050v

1994-95 Coupe 1.6L I4 VTEC VIN EH6 [All] 22P 'D' Connector

PCM Pin #	W/Color	Circuit Description (22-Pin)	Value at Hot Idle
D1	WHT/BLU	Keep Alive Power (VBU)	12-14v
D2	GRN/WHT	Brake Switch Signal	Brake Off: 0v, On: 12v
D3	RED/BLU	Knock Sensor Signal	No Detonation: 18mv AC
D4	BRN	Service Check Connector	SCS Open: 4.80v
D5	---	Not Used	---
D6	ORN/BLU	VTEC Pressure Switch	0v, Hi-Speed: 12v
D7	LT BLU	Data Link Connector	5v
D8	---	Not Used	---
D9	PNK	Alternator 'FR' Signal	Digital Signals: 0-5-0v
D10	GRN/RED	Electric Load Detector	Varies: 0.5-4.5v
D11	LT GRN	TP Sensor Signal	0.5-0.6v
D12	---	Not Used	---
D13	RED/WHT	ECT Sensor Signal	At 180°F: 0.52v
D14	WHT	HO2S Signal	0.1-1.1v
D15	RED/YEL	IAT Sensor Signal	Varies w/temp. (0.5-4.9v)
D16	---	Not Used	---
D17	WHT	MAP Sensor Signal	0.8-0.9v
D18	PNK/GRN	A/T: Interlock Control Unit	Key & Brake On: 12v
D19	YEL/GRN	MAP Sensor VREF	4.9-5.1v
D20	YEL/WHT	Sensor VREF	4.9-5.1v
D21	GRN/WHT	MAP Sensor Ground	<0.050v
D22	GRN/WHT	Sensor Ground	<0.050v

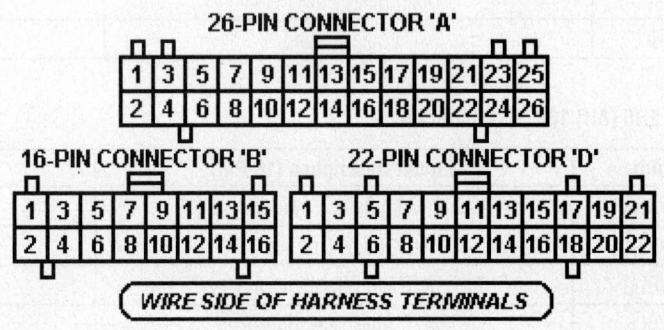

Pin Connector Graphic

Standard Colors and Abbreviations

Abbreviation	Color	Abbreviation	Color	Abbreviation	Color
BLK	Black	LT BLU	Lt. Blue	TAN	Tan
BLU	Blue	LT GRN	Lt. Green	VIO	Violet
BRN	Brown	ORN	Orange	WHT	White
GRY	Gray	PNK	Pink	YEL	Yellow
GRN	Green	PPL	Purple		

1994-95 Coupe 1.6L I4 VTEC VIN EG2 [M/T] 26P 'A' Connector

PCM Pin #	W/Color	Circuit Description (26-Pin)	Value at Hot Idle
A1	BRN	Injector 1 Control	2.0-3.3 ms
A2	YEL	Injector 4 Control	2.0-3.3 ms
A3	RED	Injector 2 Control	2.0-3.3 ms
A4	ORN/WHT	VTEC Solenoid	0v, Hi-Speed: 12v
A5	BLU	Injector 3 Control	2.0-3.3 ms
A6	ORN/BLK	HO2S Heater Control	Relay Off: 12v, On: 1v
A7	GRN/YEL	Fuel Pump Relay	Relay Off: 12v, On: 1v
A8	---	Not Used	---
A9	GRN/WHT	Idle Air Control Valve	Pulse Signals
A10-11	---	Not Used	---
A12	YEL/GRN	Radiator Fan Relay	Relay Off: 12v, On: 1v
A13	GRN/ORN	Check Engine Light	MIL Off: 12v, On: 1v
A14	---	Not Used	---
A15	BLK/RED	A/C Clutch Relay	Relay Off: 12v, On: 1v
A16	WHT/YEL	Alternator Charging Signal	Lights Off: 12v, On: 0v
A17-19	---	Not Used	---
A20	RED	EVAP Purge Solenoid	Solenoid Off: 12v, On: 1v
A21	RED/GRN	Igniter Control	Digital Signals: 0-12-0v
A22	---	Not Used	---
A23	BLK	Power Ground	<0.1v
A24	BLK	Power Ground	<0.1v
A25	YEL/BLK	Main Relay Power (B+)	12-14v
A26	BLK/RED	Logic Ground	<0.050v

1994-95 Coupe 1.6L I4 VTEC VIN EG2 [M/T] 16P 'B' Connector

PCM Pin #	W/Color	Circuit Description (16-Pin)	Value at Hot Idle
B1	WHT/GRN	Main Relay Power (B+)	12-14v
B2	BRN/BLK	Logic Ground	<0.050v
B3-4	---	Not Used	---
B5	BLU/RED	A/C Switch Signal	Relay Off: 12v, On: 1v
B6-7	---	Not Used	---
B8	BRN/RED	PSP Switch	Straight: 0v, Turning: 11v
B9	BLU/WHT	Starter Switch Signal	Cranking: 9-11v
B10	YEL/BLU	Vehicle Speed Sensor	Moving: pulse signals
B11	ORN	CYP Sensor Signal	AC pulse signals
B12	WHT	CYP Sensor Ground	<0.050v
B13	ORN/BLU	TDC Sensor Signal	AC pulse signals
B14	WHT/BLU	TDC Sensor Ground	<0.050v
B15	BLU/GRN	CKP Sensor Signal	AC pulse signals
B16	BLU/YEL	CKP Sensor Ground	<0.050v

1994-95 Coupe 1.6L I4 VTEC VIN EG2 [M/T] 22P 'D' Connector

PCM Pin #	W/Color	Circuit Description (22-Pin)	Value at Hot Idle
D1	WHT/BLU	Keep Alive Power (VBU)	12-14v
D2	GRN/WHT	Brake Switch Signal	Brake Off: 0v, On: 12v
D3	RED/BLU	Knock Sensor Signal	No Detonation: 18mv AC
D4	BRN	Service Check Connector	SCS Open: 4.80v
D5	---	Not Used	---
D6	ORN/BLU	VTEC Pressure Switch	0v, Hi-Speed: 12v
D7	LT BLU	Data Link Connector	5v
D8	---	Not Used	---
D9	PNK	Alternator 'FR' Signal	Digital Signals: 0-5-0v
D10	GRN/RED	Electric Load Detector	Varies: 0.5-4.5v
D11	LT GRN	TP Sensor Signal	0.5-0.6v
D12	---	Not Used	---
D13	RED/WHT	ECT Sensor Signal	At 180ºF: 0.52v
D14	WHT	HO2S-11 (B1 S1) Signal	0.1-1.1v
D15	RED/YEL	IAT Sensor Signal	Varies w/temp. (0.5-4.9v)
D16	---	Not Used	---
D17	WHT	MAP Sensor Signal	0.8-0.9v
D18	---	Not Used	---
D19	YEL/GRN	MAP Sensor VREF	4.9-5.1v
D20	YEL/WHT	Sensor VREF	4.9-5.1v
D21	GRN/WHT	MAP Sensor Ground	<0.050v
D22	GRN/WHT	Sensor Ground	<0.050v

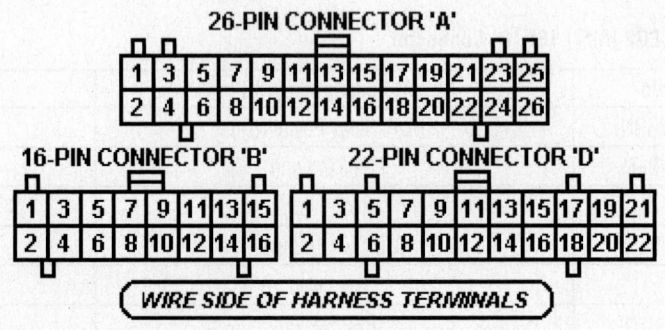

05533_ADIA_G617

Pin Connector Graphic

1996-97 Coupe 1.6L I4 VTEC VIN EG2 [M/T] 32P 'A' Connector

PCM Pin #	W/Color	Circuit Description (32-Pin)	Value at Hot Idle
A1	YEL	Injector 4 Control	2.0-3.3 ms
A2	BLU	Injector 3 Control	2.0-3.3 ms
A3	RED	Injector 2 Control	2.0-3.3 ms
A4	BRN	Injector 1 Control	2.0-3.3 ms
A5	BLK/WHT	HO2S-12 (B1 S2) Heater	Digital Signals: 0-12-0v
A6	ORN/BLK	HO2S-11 (B1 S1) Heater	Digital Signals: 0-12-0v
A7	---	Not Used	---
A8	ORN/WHT	VTEC Solenoid	0v, Hi-Speed: 12v

1996-97 Coupe 1.6L I4 VTEC VIN EG2 [M/T] 32P 'A' Connector, *continued*

A9	BLK/RED	Logic Ground	<0.050v
A10	BLK	Power Ground	<0.1v
A11	YEL/BLK	Main Relay Power (B+)	12-14v
A12	GRN/WHT	Idle Air Control Valve	Pulse Signals
A13-14	---	Not Used	---
A15	RED	EVAP Purge Solenoid	Solenoid Off: 12v, On: 1v
A16	GRN/YEL	Fuel Pump Relay	Relay Off: 12v, On: 1v
A17	BLK/RED	A/C Clutch Relay	Relay Off: 12v, On: 1v
A18	GRN/ORN	Check Engine Light	MIL Off: 12v, On: 1v
A19	WHT/YEL	Alternator Charging Signal	Lights Off: 12v, On: 0v
A20	RED/GRN	Igniter Control	Digital Signals: 0-12-0v
A21	---	Not Used	---
A22	BRN/BLK	Logic Ground	<0.050v
A23	BLK	Power Ground	<0.1v
A24	YEL/BLK	Main Relay Power (B+)	12-14v
A25-26	---	Not Used	---
A27	YEL/GRN	Radiator Fan Relay	Relay Off: 12v, On: 1v
A28-32	---	Not Used	---

Pin Connector Graphic

1996-97 Coupe 1.6L I4 VTEC VIN EG2 [M/T] 31P 'C' Connector

PCM Pin #	W/Color	Circuit Description (31-Pin)	Value at Hot Idle
C1	BLU/RED	CKF Sensor Signal	AC pulse signals
C2	BLU/GRN	CKP Sensor Signal	AC pulse signals
C3	ORN/BLU	TDC Sensor Signal	AC pulse signals
C4	ORN	CYP Sensor Signal	AC pulse signals
C5	BLU/RED	A/C Switch Signal	Relay Off: 12v, On: 1v
C6	BLU/WHT	Starter Switch Signal	Cranking: 9-11v
C7	BRN	Service Check Connector	SCS Open: 4.80v
C8	GRN/YEL	K-Line Signal	12v
C9	---	Not Used	---
C10	WHT/BLU	Keep Alive Power (VBU)	12-14v
C11	WHT/RED	CKF Sensor Ground	<0.050v
C12	BLU/YEL	CKP Sensor Ground	<0.050v
C13	WHT/BLU	TDC Sensor Ground	<0.050v
C14	WHT	CYP Sensor Ground	<0.050v
C15	ORN/BLU	VTEC Pressure Switch	0v, Hi-Speed: 12v
C16	BRN/RED	PSP Switch	Straight: 0v, Turning: 11v
C17	PNK	Alternator 'FR' Signal	Varies: 1.5-3.5v
C18	YEL/BLU	Vehicle Speed Sensor	Moving: pulse signals
C19-31	---	Not Used	---

1996-97 Coupe 1.6L I4 VTEC VIN EG2 [M/T] 16P 'D' Connector

PCM Pin #	W/Color	Circuit Description (16-Pin)	Value at Hot Idle
D1	RED/BLU	TP Sensor Signal	0.5-0.6v
D2	RED/WHT	ECT Sensor Signal	At 180°F: 0.52v
D3	WHT	MAP Sensor Signal	0.8-0.9v
D4	YEL/GRN	MAP Sensor VREF	4.9-5.1v
D5	GRN/WHT	Brake Switch Signal	Brake Off: 0v, On: 12v
D6	RED/BLU	Knock Sensor Signal	No Detonation: 18mv AC
D7	WHT	HO2S-11 (B1 S1) Signal	0.1-1.1v
D8	RED/YEL	IAT Sensor Signal	Varies w/temp. (0.5-4.9v)
D9	---	Not Used	---
D10	YEL/WHT	Sensor VREF	4.9-5.1v
D11	GRN/WHT	Sensor Ground	<0.050v
D12	GRN/WHT	MAP Sensor Ground	<0.050v
D13	RED/YEL	HO2S-12 Ground	<0.050v
D14	WHT/RED	HO2S-12 (B1 S2) Signal	0.1-1.1v
D15	---	Not Used	---
D16	GRN/RED	Electric Load Detector	Varies: 0.5-4.5v

1996-97 Coupe 1.6L I4 MFI VIN EH6 [All] 32P 'A' Connector

PCM Pin #	W/Color	Circuit Description (32-Pin)	Value at Hot Idle
A1	YEL	Injector 4 Control	2.0-3.3 ms
A2	BLU	Injector 3 Control	2.0-3.3 ms
A3	RED	Injector 2 Control	2.0-3.3 ms
A4	BRN	Injector 1 Control	2.0-3.3 ms
A5	BLK/WHT	HO2S-12 (B1 S2) Heater	Digital Signals: 0-12-0v
A6	ORN/BLK	HO2S-11 (B1 S1) Heater	Digital Signals: 0-12-0v
A7-8	---	Not Used	---
A9	BLK/RED	Logic Ground	<0.050v
A10	BLK	Power Ground	<0.1v
A11	YEL/BLK	Main Relay Power (B+)	12-14v
A12	---	Not Used	---
A13	ORN	A/T: Idle Air Control Valve 'N'	Pulse Signals
A14	BLK/BLU	A/T: Idle Air Control Valve 'P'	Pulse Signals
A15	RED	EVAP Purge Solenoid	Solenoid Off: 12v, On: 1v
A16	GRN/YEL	Fuel Pump Relay	Relay Off: 12v, On: 1v
A17	BLK/RED	A/C Clutch Relay	Relay Off: 12v, On: 1v
A18	GRN/ORN	Check Engine Light	MIL Off: 12v, On: 1v
A19	WHT/YEL	Alternator Charging Signal	Lights Off: 12v, On: 0v
A20	RED/GRN	Igniter Control	Digital Signals: 0-12-0v
A21	---	Not Used	---
A22	BRN/BLK	Logic Ground	<0.050v
A23	BLK	Power Ground	<0.1v
A24	YEL/BLK	Main Relay Power (B+)	12-14v
A25	GRN/BLK	A/T: Lockup Solenoid 'B'	LSB On: 12v, Off: 0v
A26	YEL	A/T: Lockup Solenoid 'A'	LSA On: 12v, Off: 0v
A27	YEL/GRN	Radiator Fan Relay	Relay Off: 12v, On: 1v
A28-29	---	Not Used	---
A30	WHT/RED	A/T Interlock Control Unit	Key & Brake On: 12v
A31-32	---	Not Used	---

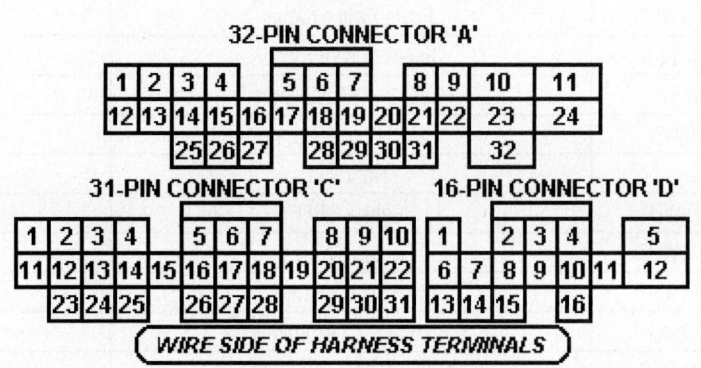

Pin Connector Graphic

1996-97 Coupe 1.6L I4 MFI VIN EH6 [All] 31P 'C' Connector

PCM Pin #	W/Color	Circuit Description (31-Pin)	Value at Hot Idle
C1	BLU/RED	CKF Sensor Signal	AC pulse signals
C2	BLU/GRN	CKP Sensor Signal	AC pulse signals
C3	ORN/BLU	TDC Sensor Signal	AC pulse signals
C4	ORN	CYP Sensor Signal	AC pulse signals
C5	BLU/RED	A/C Switch Signal	Relay Off: 12v, On: 1v
C6	BLU/WHT	Starter Switch Signal	Cranking: 9-11v
C7	BRN	Service Check Connector	SCS Open: 4.80v
C8	GRN/YEL	K-Line Signal	12v
C9	---	Not Used	---
C10	WHT/BLU	Keep Alive Power (VBU)	12-14v
C11	WHT/RED	CKF Sensor Ground	<0.050v
C12	BLU/YEL	CKP Sensor Ground	<0.050v
C13	WHT/BLU	TDC Sensor Ground	<0.050v
C14	WHT	CYP Sensor Ground	<0.050v
C15	---	Not Used	---
C16	BRN/RED	PSP Switch	Straight: 0v, Turning: 11v
C17	PNK	Alternator 'FR' Signal	Varies: 1.5-3.5v
C18	YEL/BLU	Vehicle Speed Sensor	Moving: pulse signals
C19-26	---	Not Used	---
C27	GRN/BLK	A/T: Gear Position Switch	In D4: 0v, Others: 12v
C28	GRN/BLU	A/T: Gear Position Switch	In D3: 0v, Others: 12v
C29	GRN	A/T: Gear Position Switch	In P/N: 0v, Others: 12v
C30-31	---	Not Used	---

1996-97 Coupe 1.6L I4 MFI VIN EH6 [All] 16P 'D' Connector

PCM Pin #	W/Color	Circuit Description (16-Pin)	Value at Hot Idle
D1	RED/BLU	TP Sensor Signal	0.5-0.6v
D2	RED/WHT	ECT Sensor Signal	At 180°F: 0.52v
D3	WHT	MAP Sensor Signal	0.8-0.9v
D4	YEL/GRN	MAP Sensor VREF	4.9-5.1v
D5	GRN/WHT	Brake Switch Signal	Brake Off: 0v, On: 12v
D6	RED/BLU	Knock Sensor Signal	No Detonation: 18mv AC
D7	WHT	HO2S-11 (B1 S1) Signal	0.1-1.1v
D8	RED/YEL	IAT Sensor Signal	Varies w/temp. (0.5-4.9v)
D9	---	Not Used	---
D10	YEL/WHT	Sensor VREF	4.9-5.1v
D11	GRN/WHT	Sensor Ground	<0.050v
D12	GRN/WHT	MAP Sensor Ground	<0.050v
D13	RED/YEL	HO2S-12 Ground	<0.050v
D14	WHT/RED	HO2S-12 (B1 S2) Signal	0.1-1.1v
D16	GRN/RED	Electric Load Detector	Varies: 0.5-4.5v

1996-97 Coupe 1.6L I4 VTEC VIN EH6 [A/T] 32P 'A' Connector

PCM Pin #	W/Color	Circuit Description (32-Pin)	Value at Hot Idle
A1	YEL	Injector 4 Control	2.0-3.3 ms
A2	BLU	Injector 3 Control	2.0-3.3 ms
A3	RED	Injector 2 Control	2.0-3.3 ms
A4	BRN	Injector 1 Control	2.0-3.3 ms
A5	BLK/WHT	HO2S-12 (B1 S2) Heater	Digital Signals: 0-12-0v
A6	ORN/BLK	HO2S-11 (B1 S1) Heater	Digital Signals: 0-12-0v
A8	ORN/WHT	VTEC Solenoid	0v, Hi-Speed: 12v
A9	BLK/RED	Logic Ground	<0.050v
A10	BLK	Power Ground	<0.1v
A11	YEL/BLK	Main Relay Power (B+)	12-14v
A12	GRN/WHT	M/T: Idle Air Control Valve	Pulse Signals
A13	ORN	A/T: Idle Air Control Valve 'N'	Pulse Signals
A14	BLK/BLU	A/T: Idle Air Control Valve 'P'	Pulse Signals
A15	RED	EVAP Purge Solenoid	Solenoid Off: 12v, On: 1v
A16	GRN/YEL	Fuel Pump Relay	Relay Off: 12v, On: 1v
A17	BLK/RED	A/C Clutch Relay	Relay Off: 12v, On: 1v
A18	GRN/ORN	Check Engine Light	MIL Off: 12v, On: 1v
A19	WHT/YEL	Alternator Charging Signal	Lights Off: 12v, On: 0v
A20	RED/GRN	Igniter Control	Digital Signals: 0-12-0v
A22	BRN/BLK	Logic Ground	<0.050v
A23	BLK	Power Ground	<0.1v
A24	YEL/BLK	Main Relay Power (B+)	12-14v
A25	GRN/BLK	A/T: Lockup Solenoid 'B'	LSB On: 12v, Off: 0v
A26	YEL	A/T: Lockup Solenoid 'A'	LSA On: 12v, Off: 0v
A27	YEL/GRN	Radiator Fan Relay	Relay Off: 12v, On: 1v
A28-29	---	Not Used	---
A30	WHT/RED	A/T Interlock Control Unit	Key & Brake On: 12v
A31-32	---	Not Used	---

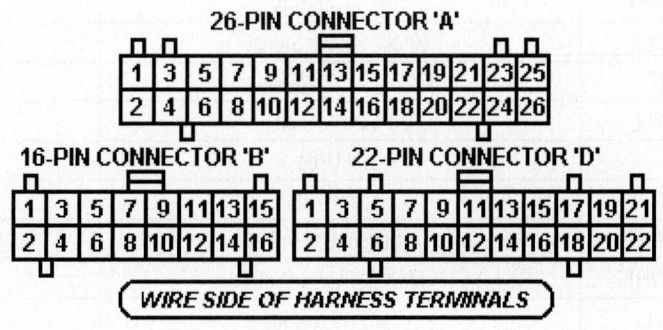

05533_ADIA_G597

Pin Connector Graphic

1996-97 Coupe 1.6L I4 VTEC VIN EH6 [A/T] 31P 'C' Connector

PCM Pin #	W/Color	Circuit Description (31-Pin)	Value at Hot Idle
C1	BLU/RED	CKF Sensor Signal	AC pulse signals
C2	BLU/GRN	CKP Sensor Signal	AC pulse signals
C3	ORN/BLU	TDC Sensor Signal	AC pulse signals
C4	ORN	CYP Sensor Signal	AC pulse signals
C5	BLU/RED	A/C Switch Signal	Relay Off: 12v, On: 1v
C6	BLU/WHT	Starter Switch Signal	Cranking: 9-11v
C7	BRN	Service Check Connector	SCS Open: 4.80v
C8	GRN/YEL	K-Line Signal	12v
C9	---	Not Used	---
C10	WHT/BLU	Keep Alive Power (VBU)	12-14v
C11	WHT/RED	CKF Sensor Ground	<0.050v
C12	BLU/YEL	CKP Sensor Ground	<0.050v
C13	WHT/BLU	TDC Sensor Ground	<0.050v
C14	WHT	CYP Sensor Ground	<0.050v
C15	ORN/BLU	VTEC Pressure Switch	0v, Hi-Speed: 12v
C16	BRN/RED	PSP Switch	Straight: 0v, Turning: 11v
C17	PNK	Alternator 'FR' Signal	Varies: 1.5-3.5v
C18	YEL/BLU	Vehicle Speed Sensor	Moving: pulse signals
C27	GRN/BLK	A/T: Gear Position Switch	In D4: 0v, Others: 12v
C28	GRN/BLU	A/T: Gear Position Switch	In D3: 0v, Others: 12v
C29	GRN	A/T: Gear Position Switch	In P/N: 0v, Others: 12v
C30-31	---	Not Used	---

1996-97 Coupe 1.6L I4 VTEC VIN EH6 [A/T] 16P 'D' Connector

PCM Pin #	W/Color	Circuit Description (16-Pin)	Value at Hot Idle
D1	RED/BLU	TP Sensor Signal	0.5-0.6v
D2	RED/WHT	ECT Sensor Signal	At 180°F: 0.52v
D3	WHT	MAP Sensor Signal	0.8-0.9v
D4	YEL/GRN	MAP Sensor VREF	4.9-5.1v
D5	GRN/WHT	Brake Switch Signal	Brake Off: 0v, On: 12v
D6	RED/BLU	Knock Sensor Signal	No Detonation: 18mv AC
D7	WHT	HO2S-11 (B1 S1) Signal	0.1-1.1v
D8	RED/YEL	IAT Sensor Signal	Varies w/temp. (0.5-4.9v)
D9	---	Not Used	---
D10	YEL/WHT	Sensor VREF	4.9-5.1v
D11	GRN/WHT	Sensor Ground	<0.050v
D12	GRN/WHT	MAP Sensor Ground	<0.050v
D13	RED/YEL	HO2S-12 Ground	<0.050v
D14	WHT/RED	HO2S-12 (B1 S2) Signal	0.1-1.1v
D16	GRN/RED	Electric Load Detector	Varies: 0.5-4.5v

1996-97 Coupe 1.6L I4 VTEC VIN EH6 [M/T] 32P 'A' Connector

PCM Pin #	W/Color	Circuit Description (32-Pin)	Value at Hot Idle
A1	YEL	Injector 4 Control	2.0-3.3 ms
A2	BLU	Injector 3 Control	2.0-3.3 ms
A3	RED	Injector 2 Control	2.0-3.3 ms
A4	BRN	Injector 1 Control	2.0-3.3 ms
A5	BLK/WHT	HO2S-12 (B1 S2) Heater	Digital Signals: 0-12-0v
A6	ORN/BLK	HO2S-11 (B1 S1) Heater	Digital Signals: 0-12-0v
A7	---	Not Used	---
A8	ORN/WHT	VTEC Solenoid	0v, Hi-Speed: 12v
A9	BLK/RED	Logic Ground	<0.050v
A10	BLK	Power Ground	<0.1v
A11	YEL/BLK	Main Relay Power (B+)	12-14v
A12	GRN/WHT	Idle Air Control Valve	Pulse Signals
A13-14	---	Not Used	---
A15	RED	EVAP Purge Solenoid	Solenoid Off: 12v, On: 1v
A16	GRN/YEL	Fuel Pump Relay	Relay Off: 12v, On: 1v
A17	BLK/RED	A/C Clutch Relay	Relay Off: 12v, On: 1v
A18	GRN/ORN	Check Engine Light	MIL Off: 12v, On: 1v
A19	WHT/YEL	Alternator Charging Signal	Lights Off: 12v, On: 0v
A20	RED/GRN	Igniter Control	Digital Signals: 0-12-0v
A21	---	Not Used	---
A22	BRN/BLK	Logic Ground	<0.050v
A23	BLK	Power Ground	<0.1v
A24	YEL/BLK	Main Relay Power (B+)	12-14v
A25-26	---	Not Used	---
A27	YEL/GRN	Radiator Fan Relay	Relay Off: 12v, On: 1v
A28-32	---	Not Used	---

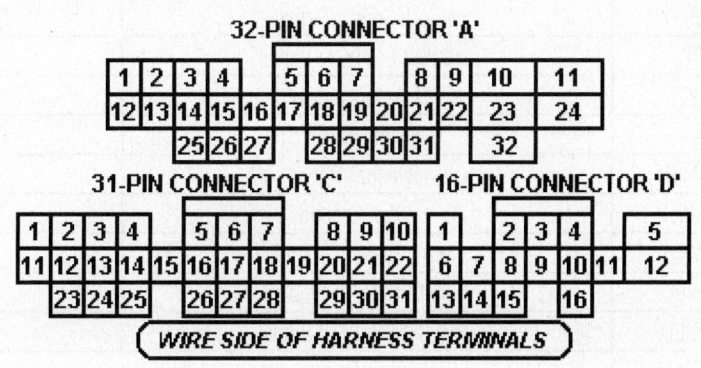

Pin Connector Graphic

1996-97 Coupe 1.6L I4 VTEC VIN EH6 [M/T] 31P 'C' Connector

PCM Pin #	W/Color	Circuit Description (31-Pin)	Value at Hot Idle
C1	BLU/RED	CKF Sensor Signal	AC pulse signals
C2	BLU/GRN	CKP Sensor Signal	AC pulse signals
C3	ORN/BLU	TDC Sensor Signal	AC pulse signals
C4	ORN	CYP Sensor Signal	AC pulse signals
C5	BLU/RED	A/C Switch Signal	Switch Off: 12v, On: 1v
C6	BLU/WHT	Starter Switch Signal	Cranking: 9-11v
C7	BRN	Service Check Connector	SCS Open: 4.80v
C8	GRN/YEL	K-Line Signal	12v
C9	---	Not Used	---
C10	WHT/BLU	Keep Alive Power (VBU)	12-14v
C11	WHT/RED	CKF Sensor Ground	<0.050v
C12	BLU/YEL	CKP Sensor Ground	<0.050v
C13	WHT/BLU	TDC Sensor Ground	<0.050v
C14	WHT	CYP Sensor Ground	<0.050v
C15	ORN/BLU	VTEC Pressure Switch	0v, Hi-Speed: 12v
C16	BRN/RED	PSP Switch	Straight: 0v, Turning: 11v
C17	PNK	Alternator 'FR' Signal	Varies: 1.5-3.5v
C18	YEL/BLU	Vehicle Speed Sensor	Moving: pulse signals
C19-31	---	Not Used	---

1996-97 Coupe 1.6L I4 VTEC VIN EH6 [M/T] 16P 'D' Connector

PCM Pin #	W/Color	Circuit Description (16-Pin)	Value at Hot Idle
D1	RED/BLU	TP Sensor Signal	0.5-0.6v
D2	RED/WHT	ECT Sensor Signal	At 180°F: 0.52v
D3	WHT	MAP Sensor Signal	0.8-0.9v
D4	YEL/GRN	MAP Sensor VREF	4.9-5.1v
D5	GRN/WHT	Brake Switch Signal	Brake Off: 0v, On: 12v
D6	RED/BLU	Knock Sensor Signal	No Detonation: 18mv AC
D7	WHT	HO2S-11 (B1 S1) Signal	0.1-1.1v
D8	RED/YEL	IAT Sensor Signal	Varies w/temp. (0.5-4.9v)
D9	---	Not Used	---
D10	YEL/WHT	Sensor VREF	4.9-5.1v
D11	GRN/WHT	Sensor Ground	<0.050v
D12	GRN/WHT	MAP Sensor Ground	<0.050v
D13	RED/YEL	HO2S-12 Ground	<0.050v
D14	WHT/RED	HO2S-12 (B1 S2) Signal	0.1-1.1v
D15	---	Not Used	---
D16	GRN/RED	Electric Load Detector	Varies: 0.5-4.5v

CR-V PIN CHARTS

1997 MPV 2.0L I4 MFI VIN RD1 [A/T] 32P 'A' Connector

PCM Pin #	W/Color	Circuit Description (32-Pin)	Value at Hot Idle
A1	YEL	Injector 4 Control	2.0-3.3 ms
A2	BLU	Injector 3 Control	2.0-3.3 ms
A3	RED	Injector 2 Control	2.0-3.3 ms
A4	BRN	Injector 1 Control	2.0-3.3 ms
A5	BLK/WHT	HO2S-12 (B1 S2) Heater	Digital Signals: 0-12-0v
A6	BLK/WHT	HO2S-11 (B1 S1) Heater	Digital Signals: 0-12-0v
A9, A22	BRN/BLK	Logic Ground	<0.050v
A10, 23	BLK	Power Ground	<0.1v
A11	YEL/BLK	Main Relay Power (B+)	12-14v
A12	BLK/BLU	Idle Air Control Valve	Pulse Signals
A15	RED/YEL	EVAP Purge Solenoid	Solenoid Off: 12v, On: 1v
A16	GRN/YEL	Fuel Pump Relay Control	Relay Off: 12v, On: 1v
A17	BLK/RED	A/C Clutch Relay Control	Relay Off: 12v, On: 1v
A18	GRN/ORN	Check Engine Light	MIL Off: 12v, On: 1v
A19	WHT/GRN	Alternator Charging Signal	Lights Off: 12v, On: 0v
A20	YEL/GRN	Igniter Control	At idle: 5 dwell
A24	YEL/BLK	Main Relay Power (B+)	12-14v
A27	GRN	Radiator Fan Relay Control	R/F on: <1v, off: 12v

1997 MPV 2.0L I4 MFI VIN RD1 [A/T] 25P 'B' Connector

PCM Pin #	W/Color	Circuit Description (25-Pin)	Value at Hot Idle
B1	WHT	A/T: Linear Solenoid (-)	Pulse Signals
B2	RED	A/T: Linear Solenoid (+)	Pulse Signals
B3	BLU/YEL	A/T: Shift Solenoid 'A'	SSA Off: 0v, On: 12v
B4	GRN/BLK	A/T: Lockup Solenoid 'B'	LSB On: 12v, Off: 0v
B5	YEL	A/T: Lockup Solenoid 'A'	LSA On: 12v, Off: 0v
B8	PNK	A/T: Gear Position Switch	In D3: 0v, others: 12v
B11	GRN/WHT	A/T: Shift Solenoid 'B'	SSB Off: 0v, On: 12v
B12	WHT/RED	Interlock Control Unit Signal	Key & Brake On: 12v
B13	GRN/BLK	A/T: D4 Indicator Light Switch	In D4: 12v, others: 0v
B14	WHT	Mainshaft Speed Sensor 'N'	<0.050v
B15	RED	Mainshaft Speed Sensor 'P'	AC Pulse Signals
B16	WHT	A/T: Gear Position Switch	In 'R': 0v, Others: 12v
B17	BLU	A/T: Gear Position Switch	In D2: 0v, Others: 12v
B18	BRN	A/T: Gear Position Switch	In D1: 0v, Others: 12v
B22	GRN	Countershaft Speed Sensor N	<0.050v
B23	BLU	Countershaft Speed Sensor P	Moving: pulse signals
B24	YEL	A/T: Gear Position Switch	In D4: 0v, others: 12v
B25	LT GRN	A/T: Gear Position Switch	In P/N: 0v, Others: 12v

```
        32-PIN CONNECTOR 'A'              25-PIN CONNECTOR 'B'
    ┌──┬──┬──┬──┬──┬──┬──┬──┬──┬──┬──┬──┐  ┌──┬──┬──┬──┬──┬──┬──┬──┐
    │ 1│ 2│ 3│ 4│ 5│ 6│ 7│ 8│ 9│10│11│  │ 1│ 2│ 3│ 4│ 5│ 6│ 7│ 8│ 24│
    ├──┼──┼──┼──┼──┼──┼──┼──┼──┼──┼──┤  ├──┼──┼──┼──┼──┼──┼──┼──┤
    │12│13│14│15│16│17│18│19│20│21│22│ 23│ 9│ 10│11│12│13│14│15│16│17│18│
    ├──┼──┼──┼──┼──┼──┼──┼──┼──┼──┼──┤  ├──┼──┼──┼──┼──┼──┼──┼──┤
    │25│26│27│28│29│30│31│ 32│ 19│ 20│21│22│23│24│25│
    └──┴──┴──┴──┴──┴──┴──┴──┴──┴──┴──┘  └──┴──┴──┴──┴──┴──┴──┴──┘

            WIRE SIDE OF HARNESS TERMINALS
                                              05533_ADIA_G619
```

Pin Connector Graphic
1997 MPV 2.0L I4 MFI VIN RD1 [A/T] 31P 'C' Connector

PCM Pin #	W/Color	Circuit Description (31-Pin)	Value at Hot Idle
C1	BLU/RED	CKF Sensor Signal	Idle: 0.900mv (AC)
C2	BLU	CKP Sensor Signal	Idle: 0.900mv (AC)
C3	GRN	TDC Sensor Signal	KOER: 1.00v (AC)
C4	YEL	CYP Sensor Signal	Idle: 0.250mv (AC)
C5	BLU/RED	A/C Switch Signal	Relay Off: 12v, On: 1v
C6	BLU/ORN	Starter Switch Signal	Cranking: 9-11v
C7	BRN	Service Check Connector	SCS Open: 4.80v
C8	LT BLU	K-Line Signal	12v
C9	---	Not Used	---
C10	WHT/BLU	Keep Alive Power (VBU)	12-14v
C11	WHT/RED	CKF Sensor Ground	<0.050v
C12	WHT	CKP Sensor Ground	<0.050v
C13	RED	TDC Sensor Ground	<0.050v
C14	BLK	CYP Sensor Ground	<0.050v
C16	GRN	PSP Switch	Wheel straight: 0v
C17	WHT/RED	Alternator 'FR' Signal	Varies: 1.5-3.5v
C18	BLU/WHT	Vehicle Speed Sensor	Moving: pulse signals
C19-31	---	Not Used	---

1997 MPV 2.0L I4 MFI VIN RD1 [A/T] 16P 'D' Connector

PCM Pin #	W/Color	Circuit Description (16-Pin)	Value at Hot Idle
D1	RED/BLK	TP Sensor Signal	At 0.5-0.6v
D2	RED/WHT	ECT Sensor Signal	At 180°F: 0.5-0.6v
D3	RED/GRN	MAP Sensor Signal	0.8-0.9v
D4	YEL/RED	MAP Sensor VREF	4.9-5.1v
D5	GRN/WHT	Brake Switch Signal	Brake Off: 0v, On: 12v
D7	WHT	HO2S-11 (B1 S1) Signal	0.1-1.1v
D8	RED/YEL	IAT Sensor Signal	Varies w/temp. (0.5-4.9v)
D9	---	Not Used	---
D10	YEL/BLU	Sensor VREF	4.9-5.1v
D11	GRN/BLK	Sensor Ground	<0.050v
D12	GRN/WHT	MAP Sensor Ground	<0.050v
D13	RED/YEL	HO2S-12 Ground	<0.050v
D14	WHT/RED	HO2S-12 (B1 S2) Signal	0.1-1.1v
D16	GRN/RED	Electronic Load Detector	Varies 2.5-3.5v

31-PIN CONNECTOR 'C'   **16-PIN CONNECTOR 'D'**

1	2	3	4		5	6	7		8	9	10		1		2	3	4		5
11	12	13	14	15	16	17	18	19	20	21	22		6	7	8	9	10	11	12
	23	24	25		26	27	28		29	30	31		13	14	15		16		

WIRE SIDE OF HARNESS TERMINALS

05533_ADIA_G620

Pin Connector Graphic

1998-03 MPV 2.0L I4 MFI VIN RD1, RD2 [All] 32P 'A' Connector

PCM Pin #	W/Color	Circuit Description (32-Pin)	Value at Hot Idle
A3	BLU	EVAP Bypass Solenoid	Solenoid Off: 12v, On: 1v
A4	GRN/WHT	EVAP Vent Solenoid	Solenoid Off: 12v, On: 1v
A5	BLU	Cruise Control Signal	C/C On: pulse signals
A6	RED/YEL	EVAP Purge Solenoid	Solenoid Off: 12v, On: 1v
A8	BLK/WHT	HO2S-12 (B1 S2) Heater	Digital Signals: 0-12-0v
A10	BRN	Service Check Connector	SCS Open: 4.80v
A14	GRN/BLK	A/T: D4 Light Switch	D4 On: 0v, Off: 12v
A16	GRN/YEL	Fuel Pump Relay Control	Relay Off: 12v, On: 1v
A17	BLK/RED	A/C Clutch Relay Control	Relay Off: 12v, On: 1v
A18	GRN/ORN	Malfunction Indicator Light	MIL Off: 12v, On: 1v
A19	BLU	Engine Speed Pulse (NEP)	Digital Signals
A20	GRN	Radiator Fan Relay Control	Relay Off: 12v, On: 1v
A21	BLU/YEL	K-Line Signal	12v
A23	WHT/RED	HO2S-12 (B1 S2) Signal	0.1-1.1v
A24	BLU/WHT	Starter Switch Signal	Cranking: 9-11v
A26	GRN	PSP Switch Signal	Straight: 0v, Turning: 11v
A27	BLU/RED	A/C Switch Signal	Switch Off: 12v, On: 1v
A28	WHT/RED	Interlock Control Unit Signal	Key & Brake On: 12v
A29	LT GRN	Fuel Tank Pressure Sensor	Fuel Cap off: 2.5v
A30	GRN/RED	Electric Load Detector	Varies: 0.5-4.5v
A32	WHT/BLK	Brake Switch Signal	Brake Off: 0v, On: 12v

1998-03 MPV 2.0L I4 MFI VIN RD1, RD2 [All] 16P 'D' Connector

PCM Pin #	W/Color	Circuit Description (16-Pin)	Value at Hot Idle
D1	YEL	A/T: Lockup Solenoid 'A'	LUS On: 12v, Off: 0v
D2	GRN/WHT	A/T: Shift Solenoid 'B'	SSB Off: 0v, On: 12v
D3	GRN/BLK	A/T: Lockup Solenoid 'B'	LUS On: 12v, Off: 0v
D5	BLK/YEL	A/T: Solenoid Feed (B+)	12-14v
D6	WHT	A/T: Gear Position Switch	In 'R': 0v, Others: 12v
D7	BLU/YEL	A/T: Shift Solenoid 'A'	SSA Off: 0v, On: 12v
D8	PNK	A/T: Gear Position Switch	In D3: 0v, Others: 12v
D8	PNK	M/T: Overdrive Switch	In O/D: 0v, others: 5v
D9	YEL	A/T: Gear Position Switch	In D4: 0v, Others: 12v
D10	BLU	Countershaft Speed Sensor P	Moving: AC pulses
D11	RED	Mainshaft Speed Sensor 'P'	Pulse Signals
D12	WHT	Mainshaft Speed Sensor 'N'	<0.050v
D13	LT GRN	A/T: Gear Position Switch	In P/N: 0v, Others: 12v
D14	BLU	A/T: D4 Indicator Light Driver	In D2: 0v, Off: 12v
D15	BRN	A/T: D4 Indicator Light Driver	In D12: 0v, Off: 12v
D16	GRN	Countershaft Speed Sensor N	<0.050v

05533_ADIA_G599

Pin Connector Graphic

1998-03 MPV 2.0L I4 MFI VIN RD1, RD2 [All] 25P 'B' Connector

PCM Pin #	W/Color	Circuit Description (25-Pin)	Value at Hot Idle
B1, B9	YEL/BLK	Main Relay Power (B+)	12-14v
B2, B10	BLK	Power Ground (PG1, PG2)	<0.1v
B3	RED	Injector 2 Control	2.0-3.3 ms
B4	BLU	Injector 3 Control	2.0-3.3 ms
B5	YEL	Injector 4 Control	2.0-3.3 ms
B8	WHT	A/T: Clutch Press. Solenoid 'N'	Pulse Signals
B11	BRN	Injector 1 Control	2.0-3.3 ms
B12	GRN/YEL	VTEC Control Solenoid	0v, Hi-Speed: 12v
B13	YEL/GRN	Ignition Control Signal	Digital Signals: 0-12-0v
B17	RED	A/T: Clutch Press. Solenoid 'P'	Pulse Signals
B18	PNK/BLU	A/T: Start Clutch Solenoid 'N'	Pulse Signals
B19	PNK	A/T: Overdrive Indicator Light	In O/D: 0v, Others: 12v
B20	BRN/BLK	Logic Ground (LG1)	<0.050v
B21	WHT/YEL	Keep Alive Power (VBU)	12-14v
B22	BRN/BLK	Logic Ground (LG2)	<0.050v
B23	BLK/BLU	Idle Air Control Valve	Pulse Signals
B24-25	---	Not Used	---

1998-03 MPV 2.0L I4 MFI VIN RD1, RD2 [All] 31P 'C' Connector

PCM Pin #	W/Color	Circuit Description (31-Pin)	Value at Hot Idle
C1	BLK/WHT	HO2S-11 (B1 S1) Heater	Digital Signals: 0-12-0v
C2	WHT/GRN	Alternator Charging Signal	Lights Off: 12v, On: 0v
C3	RED/BLU	Knock Sensor Signal	No Detonation: 18mv AC
C5	WHT/RED	Alternator 'FR' Signal	Digital Signals: 0-5-0v
C7	GRN/WHT	MAP Sensor Ground (SG1)	<0.050v
C8, C9	BLU, WHT	CKP Sensor Signal, Ground	AC pulse signals
C16	WHT	HO2S-11 (B1 S1) Signal	0.1-1.1v
C17	RED/GRN	MAP Sensor Signal	0.8-0.9v
C18	GRN/BLK	Sensor Ground (SG2)	<0.050v
C19	YEL/RED	Sensor VREF (VCC1)	4.9-5.1v
C20	GRN	TDC Sensor Signal	AC pulse signals
C21	RED	TDC Sensor Ground	<0.050v
C22	BLU/RED	CKF Sensor Signal	AC pulse signals
C23	BLU/WHT	Vehicle Speed Sensor Signal	Moving: 0-5-0v
C25	RED/YEL	IAT Sensor Signal	Varies w/temp (0.5-4.9v)
C26	RED/WHT	ECT Sensor Signal	At 180°F: 0.5-0.6v
C27	RED/BLK	TP Sensor Signal	0.5-0.6v
C28	YEL/BLU	Sensor VREF (VCC2)	4.9-5.1v
C29	YEL	CYP Sensor Signal	AC pulse signals
C30	BLK	CYP Sensor Ground	<0.050v
C31	WHT/RED	CKF Sensor Ground	<0.050v

Pin Connector Graphic

05533_ADIA_G600

INSIGHT PIN CHARTS

2000-03 Hatchback 1.0L I3 VTEC VIN ZE1 [M/T] 32P 'A' Connector

PCM Pin #	W/Color	Circuit Description (32-Pin)	Value at Hot Idle
A1	BLU/WHT	Starter Cut Relay	In Start: 12v
A2	BLK/YEL	Engine Ready Signal	Stop: 12v, No Stop: 0v
A3	BLU	EVAP Bypass Solenoid	Solenoid Off: 12v, On: 1v
A4	GRN/WHT	EVAP Vent Solenoid	Solenoid Off: 12v, On: 1v
A6	RED/YEL	EVAP Purge Solenoid	Solenoid Off: 12v, On: 1v
A7	RED/YEL	DC/DC Converter ECT Signal	Varies: 0.5-4.8v
A9	WHT/GRN	DC/DC Converter Control	Pulse Signals
A10	WHT/RED	Master Power Vacuum Motor	Varies: 1.0-3.0v
A12	PNK	Immobilizer Indicator Lamp	Lamp Off: 12v, On: 1v
A13	BLU/YEL	Immobilizer Enable Signal	Digital Signals
A15	GRN/YEL	Immobilizer Fuel Pump Relay	Relay Off: 12v, On: 1v
A17	RED	A/C Clutch Relay Control	Relay Off: 12v, On: 1v
A18	GRN/ORN	Malfunction Indicator Light	MIL Off: 12v, On: 1v
A19	BLU	Engine Speed Pulse (NEP)	Pulse Signals
A20	BLU/RED	Radiator Fan Relay Control	Relay Off: 12v, On: 1v
A21	GRY	K-Line Signal	12v
A23	BRN/YEL	Heater Standby Signal	Pulse Signals
A24	BLU/ORN	Starter Switch Signal	Cranking: 9-11v
A25	RED	Immobilizer Code Signal	Digital Signals
A27	BLU/BLK	A/C Switch Signal	Switch Off: 12v, On: 1v
A29	LT GRN	Fuel Tank Pressure Sensor	Fuel Cap off: 2.5v
A30	GRN/RED	Electric Load Detector	Varies: 0.5-4.5v
A32	GRN/WHT	Brake Switch Signal	Brake Off: 0v, On: 12v

2000-03 Hatchback 1.0L I3 VTEC VIN ZE1 [M/T] 16P 'D' Connector

PCM Pin #	W/Color	Circuit Description (16-Pin)	Value at Hot Idle
D1	BLU/WHT	Motor Control FSB Signal	Pulse Signals
D2	BLU/RED	Motor Control FSA Signal	Pulse Signals
D3	YEL/RED	Motor Control Standby Signal	Pulse Signals
D4	GRN	HO2S Pump Cell (+) Signal	0.5-3.5v
D5	BLK/WHT	HO2S-11 (B1 S1) Heater	Digital Signals: 0-12-0v
D6	RED/YEL	Motor Control Mode 1 Signal	Pulse Signals
D7	BLU	Engine Torque Signal	Pulse Signals
D8	BLU/BLK	Motor Power Signal	Pulse Signals
D9, D16	---	Not Used	---
D10	RED	HO2S Common IP- (VS-)	2.6-2.8v
D11	BLU	VS Cell Voltage (VS+)	7v
D12	WHT	LAF Label Signal	0.3-4.9v
D13	PNK	'Q' Battery Signal	Pulse Signals
D14	YEL	Motor Torque Signal	Pulse Signals
D15	WHT/RED	Motor Control Mode 2 Signal	Pulse Signals

25-Pin 'B'

WIRE SIDE OF HARNESS TERMINALS

31-Pin 'C'

05533_ADIA_G621

Pin Connector Graphic
2000-03 Hatchback 1.0L I3 VTEC VIN ZE1 [M/T] 25P 'B' Connector

PCM Pin #	W/Color	Circuit Description (25-Pin)	Value at Hot Idle
B1, B9	YEL/BLK	Main Relay Power (B+)	12-14v
B2, B10	BLK	Power Ground (G101)	<0.1v
B3, B4	RED, BLU	Injector 2, Injector 3 Control	2.0-3.3 ms
B5	GRN	TIM Signal to Gauge Assembly	Digital Signals
B7	PNK	E-EGR Solenoid Control	Digital Signals: 0-12-0v
B11	BRN	Injector 1 Control	2.0-3.3 ms
B12	GRN/YEL	VTEC Control Solenoid	0v, Hi-Speed: 12v
B14	RED	Clutch Switch Signal	Clutch In: 0v, Out: 5v
B15	RED/GRN	ECT Sensor Signal to Gauge	Digital Signals
B16	RED/BLK	Neutral Position Switch Signal	In 'N': 0v, others: 12v
B17	BRN	Service Check Connector	Open: 5v, closed: 0v
B20	BRN/BLK	Power Ground (G101)	<0.1v
B21	WHT/BLU	Keep Alive Power (VBU)	12-14v
B22	BRN/BLK	Power Ground (G101)	<0.1v
B23	BLK/BLU	Idle Air Control Valve	Pulse Signals
B24	GRN/BLK	Reverse Position Switch Signal	In 'R': 12v, Others: 0v

2000-03 Hatchback 1.0L I3 VTEC VIN ZE1 [M/T] 31P 'C' Connector

PCM Pin #	W/Color	Circuit Description (31-Pin)	Value at Hot Idle
C4	WHT	Ignition Coil 1 Control	Digital Signals: 0-12-0v
C5	BLU/WHT	Vehicle Speed Sensor Signal	Moving: 0-5-0v
C6	WHT/BLK	EGR Valve Lift Sensor	1.2v
C7, C18	GRN/WHT	Sensor Ground (SG1, SG2)	<0.050v
C8	BLU	CKP Sensor Signal	AC pulse signals
C9	WHT	CKP Sensor Ground	<0.050v
C10	BLU/BLK	VTEC Pressure Switch	0v, Hi-Speed: 12v
C12	BLK/WHT	HO2S-11 (B1 S1) Heater	Digital Signals: 0-12-0v
C13	WHT/GRN	Ignition Coil 2 Control	Digital Signals: 0-12-0v
C14	WHT/BLK	Ignition Coil 3 Control	Digital Signals: 0-12-0v
C15	WHT/RED	Shift Indicator Lamp Control	Lamp on: 1v, off: 12v
C17	RED/GRN	MAP Sensor Signal	0.8-0.9v
C19	YEL/RED	MAP Sensor VREF (VCC1)	4.9-5.1v
C20	GRN	TDC1 Sensor Signal	AC pulse signals
C21	RED	TDC1 Sensor Ground	<0.050v
C22	BLU/RED	Knock Sensor Signal	No Detonation: 18mv AC
C25	RED/YEL	IAT Sensor Signal	Varies w/temp (0.5-4.9v)
C26	RED/WHT	ECT Sensor Signal	At 180°F: 0.5-0.6v
C27	RED/BLK	TP Sensor Signal	0.5-0.6v
C28	YEL/BLU	Sensor VREF (VCC2)	4.9-5.1v
C29, C30	YEL, BLK	TDC2 Sensor Signal, Ground	AC pulse signals
C31	WHT/RED	HO2S-12 (B1 S2) Signal	0.1-1.1v

25-Pin 'B'   WIRE SIDE OF HARNESS TERMINALS   31-Pin 'C'

05533_ADIA_G621

Pin Connector Graphic

ODYSSEY PIN CHARTS

1995 Van 2.2L I4 MFI VIN RA1 [A/T] 26P 'A' Connector

PCM Pin #	W/Color	Circuit Description (26-Pin)	Value at Hot Idle
A1	BRN	Injector 1 Control	2.0-3.3 ms
A2	RED	Injector 2 Control	2.0-3.3 ms
A3	BLU	Injector 3 Control	2.0-3.3 ms
A4	GRN/BLK	Fuel Pump Relay	Relay Off: 12v, On: 1v
A5	BLK/BLU	Idle Air Control Valve	Pulse Signals
A6	ORN/BLK	HO2S-11 (B1 S1) Heater	Digital Signals: 0-12-0v
A7	BLU	Check Engine Light	MIL Off: 12v, On: 1v
A8	RED/BLU	A/C Clutch Relay	Relay Off: 12v, On: 1v
A11	YEL/GRN	Igniter Control	Digital Signals: 0-12-0v
A12	BLK	Power Ground	<0.1v
A13	YEL/BLK	Main Relay Power (B+)	12-14v
A14	YEL	Injector 4 Control	2.0-3.3 ms
A16	RED	EGR Solenoid	Solenoid Off: 12v, On: 1v
A18	GRN/WHT	Engine Mount Solenoid	Solenoid Off: 12v, On: 1v
A19	GRN	Radiator Fan Relay	Relay Off: 12v, On: 1v
A21	WHT/GRN	Alternator Charging Signal	Lights Off: 12v, On: 0v
A22	BRN/WHT	A/T: TCM FAS	N/A
A23	RED/YEL	EVAP Purge Solenoid	Solenoid Off: 12v, On: 1v
A25	BLK	Power Ground	<0.1v
A26	BRN/BLK	Logic Ground	<0.050v

1995 Van 2.2L I4 MFI VIN RA1 [A/T] 16P 'B' Connector

PCM Pin #	W/Color	Circuit Description (16-Pin)	Value at Hot Idle
B1	YEL/BLK	Main Relay Power (B+)	12-14v
B2	WHT/RED	A/T: TCM AFSA	---
B3	RED/WHT	A/C Switch Signal	Relay Off: 12v, On: 1v
B4	LT GRN	A/T: Gear Position Switch	P/N: 0v: others: 12v
B5	BLU/RED	Starter Switch Signal	Cranking: 9-11v
B6	ORN	CYP Sensor Signal	AC pulse signals
B7	ORN/BLU	TDC Sensor Signal	AC pulse signals
B8	BLU/GRN	CKP Sensor Signal	AC pulse signals
B9	BRN/BLK	Logic Ground	<0.050v
B10	GRN	A/T: TCM AFSB	N/A
B12	GRN	PSP Switch	Wheel straight: 0v
B13	ORN	Vehicle Speed Sensor	Moving: pulse signals
B14	WHT	CYP Sensor Ground	<0.050v
B15	WHT/BLU	TDC Sensor Ground	<0.050v
B16	BLU/YEL	CKP Sensor Ground	<0.050v

1995 Van 2.2L I4 MFI VIN RA1 [A/T] 22P 'D' Connector

PCM Pin #	W/Color	Circuit Description (22-Pin)	Value at Hot Idle
D1	WHT/YEL	Keep Alive Power (VBU)	12-14v
D3	BLU/WHT	A/T: TCM BARO Signal	3v (at sea level)
D4	GRN/RED	Data Link Connector	5v
D5	WHT/RED	Alternator 'FR' Signal	Digital Signals: 0-5-0v
D6	RED/BLK	TP Sensor Signal	0.5-0.6v
D7	RED/WHT	ECT Sensor Signal	At 180°F: 0.5-0.6v
D8	RED/YEL	IAT Sensor Signal	Varies w/temp. (0.5-4.9v)
D9	WHT/YEL	MAP Sensor Signal	0.8-0.9v
D10	YEL/WHT	MAP Sensor VREF	4.9-5.1v
D11	GRN/WHT	MAP Sensor Ground	<0.050v
D12	GRN/WHT	Brake Switch Signal	Brake Off: 0v, On: 12v
D13	RED	Service Check Connector	SCS Open: 4.80v
D16	GRN/RED	Electrical Load Detector	Varies: 0.5-4.5v
D17	WHT/BLK	EGR Valve Lift Sensor	1.2v
D18	WHT/RED	HO2S-11 (B1 S1) Signal	0.1-1.1v
D19	---	Not Used	---
D20	GRN/BLK	A/T: TCM VREF	4.9-5.1v
D21	YEL/BLU	Sensor VREF	4.9-5.1v
D22	GRN/BLU	Sensor Ground	<0.050v

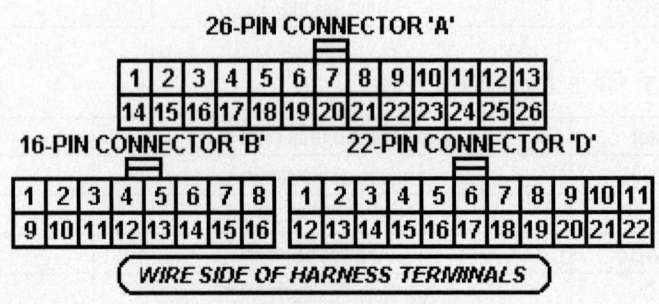

05533_ADIA_G622

Pin Connector Graphic

1996-98 Van 2.2L I4 MFI VIN RA1 [A/T] 32P 'A' Connector

PCM Pin #	W/Color	Circuit Description (32-Pin)	Value at Hot Idle
A1	YEL	Injector 4 Control	2.0-3.3 ms
A2	BLU	Injector 3 Control	2.0-3.3 ms
A3	RED	Injector 2 Control	2.0-3.3 ms
A4	BRN	Injector 1 Control	2.0-3.3 ms
A5	ORN/BLK	HO2S-12 (B1 S2) Heater	Digital Signals: 0-12-0v
A6	ORN/BLK	HO2S-11 (B1 S1) Heater	Digital Signals: 0-12-0v
A7	RED	EGR Solenoid	Solenoid Off: 12v, On: 1v
A9	BRN/BLK	Logic Ground	<0.050v
A10	BLK	Power Ground	<0.1v
A11	YEL/BLK	Main Relay Power (B+)	12-14v
A12	BLK/BLU	Idle Air Control Valve	Pulse Signals

1996-98 Van 2.2L I4 MFI VIN RA1 [A/T] 32P 'A' Connector, *continued*

A13	GRN/WHT	Engine Mount Solenoid	Solenoid Off: 12v, On: 1v
A15	RED/YEL	EVAP Purge Solenoid	Solenoid Off: 12v, On: 1v
A16	GRN/BLK	Fuel Pump Relay	Relay Off: 12v, On: 1v
A17	RED/BLU	A/C Clutch Relay	Relay Off: 12v, On: 1v
A18	BLU	Check Engine Light	MIL Off: 12v, On: 1v
A19	WHT/GRN	Alternator Charging Signal	Lights Off: 12v, On: 0v
A20	YEL/GRN	Igniter Control	Digital Signals: 0-12-0v
A22	BRN/BLK	Logic Ground	<0.050v
A23	BLK	Power Ground	<0.1v
A24	YEL/BLK	Main Relay Power (B+)	12-14v
A27	GRN	Radiator Fan Relay	Relay Off: 12v, On: 1v
A28-32	---	---	---

1996-98 Van 2.2L I4 MFI VIN RA1 [A/T] 25P 'B' Connector

PCM Pin #	W/Color	Circuit Description (25-Pin)	Value at Hot Idle
B1-2	---	Not Used	---
B3	BLU/YEL	A/T: Shift Solenoid 'A'	SSA Off: 0v, On: 12v
B4	GRN/BLK	A/T: Lockup Solenoid 'B'	LSB On: 12v, Off: 0v
B5	YEL	A/T: Lockup Solenoid 'A'	LSA On: 12v, Off: 0v
B8	GRN	A/T: Gear Position Switch	In D3: 0v, Others: 12v
B11	GRN/WHT	A/T: Shift Solenoid 'B'	SSB Off: 0v, On: 12v
B12	WHT/GRN	Interlock Control Unit Signal	Key & Brake On: 12v
B13	BLU/RED	A/T: D4 Indicator Light Switch	D4 On: 0v, Off: 12v
B14	WHT/BLU	Mainshaft Speed Sensor 'N'	<0.050v
B15	ORN/BLU	Mainshaft Speed Sensor 'P'	AC Pulse Signals
B16	WHT	A/T: Gear Position Switch	In 'R': 0v, Others: 12v
B17	BLU	A/T: Gear Position Switch	In D2: 0v, Others: 12v
B18	BRN	A/T: Gear Position Switch	In D1: 0v, Others: 12v
B22	BLU/YEL	Countershaft Speed Sensor N	<0.050v
B23	BLU/GRN	Countershaft Speed Sensor P	Moving: AC pulses
B24	YEL	A/T: Gear Position Switch	In D4: 0v, Others: 12v
B25	LT GRN	A/T: Gear Position Switch	In P/N: 0v, Others: 12v

1996-98 Van 2.2L I4 MFI VIN RA1 [A/T] 31P 'C' Connector

PCM Pin #	W/Color	Circuit Description (31-Pin)	Value at Hot Idle
C2	BLU	CKP Sensor Signal	AC pulse signals
C3	GRN	TDC Sensor Signal	AC pulse signals
C4	YEL	CYP Sensor Signal	AC pulse signals
C5	RED/WHT	A/C Switch Signal	Switch Off: 12v, On: 1v
C6	BLU/RED	Starter Switch Signal	Cranking: 9-11v
C7	RED	Service Check Connector	SCS Open: 4.80v
C8	LT GRN	K-Line Signal	12v
C9	---	Not Used	---
C10	WHT/YEL	Keep Alive Power (VBU)	12-14v
C12	BLU/YEL	CKP Sensor Ground	<0.050v
C13	WHT/BLU	TDC Sensor Ground	<0.050v
C14	WHT	CYP Sensor Ground	<0.050v
C16	GRN	PSP Switch	Wheels Turned: 11v
C17	WHT/RED	Alternator 'FR' Signal	Digital Signals: 0-5-0v
C18	ORN	Vehicle Speed Sensor	Moving: pulse signals
C19	---	Not Used	---
C20	BRN	EVAP Purge Solenoid	Solenoid Off: 12v, On: 1v
C21-31	---	Not Used	---

1996-98 Van 2.2L I4 MFI VIN RA1 [A/T] 16P 'D' Connector

PCM Pin #	W/Color	Circuit Description (16-Pin)	Value at Hot Idle
D1	RED/BLK	TP Sensor Signal	0.5-0.6v
D2	RED/WHT	ECT Sensor Signal	At 180°F: 0.5-0.6v
D3	WHT/YEL	MAP Sensor Signal	0.8-0.9v
D4	YEL/WHT	MAP Sensor VREF	4.9-5.1v
D5	GRN/WHT	Brake Switch Signal	Brake Off: 0v, On: 12v
D7	WHT/RED	HO2S-11 (B1 S1) Signal	0.1-1.1v
D8	RED/YEL	Intake Air Temperature	Varies w/temp. (0.5-4.9v)
D9	WHT/BLK	EGR Valve Lift Sensor	1.2v
D10	YEL/BLU	Sensor VREF	4.9-5.1v
D11	GRN/BLU	Sensor Ground	<0.050v
D12	GRN/WHT	MAP Sensor Ground	<0.050v
D13	RED/WHT	HO2S-12 Ground	<0.050v
D14	WHT/RED	HO2S-12 (B1 S2) Signal	0.1-1.1v
D16	GRN/RED	Electrical Load Detector	Varies: 0.5-4.5v

32-PIN CONNECTOR 'A'

1	2	3	4		5	6	7		8	9	10		11	
12	13	14	15	16	17	18	19	20	21	22	23		24	
	25	26	27		28	29	30	31		32				

31-PIN CONNECTOR 'C'

1	2	3	4		5	6	7		8	9	10
11	12	13	14	15	16	17	18	19	20	21	22
23	24	25		26	27	28		29	30	31	

25-PIN CONNECTOR 'B'

1	2		3	4	5		6	7	8
9	10	11	12	13	14	15	16	17	18
19	20		21	22		23	24	25	

16-PIN CONNECTOR 'D'

1		2	3	4		5
6	7	8	9	10	11	12
13	14	15		16		

WIRE SIDE OF HARNESS TERMINALS

05533_ADIA_G623

Pin Connector Graphic
1998 Van 2.3L I4 VTEC SOHC VIN RA3 [A/T] 32P 'A' Connector

PCM Pin #	W/Color	Circuit Description (32-Pin)	Value at Hot Idle
A2	GRN/WHT	Engine Mount Solenoid	Solenoid Off: 12v, On: 1v
A3	GRN/ORN	EVAP Bypass Solenoid	Solenoid Off: 12v, On: 1v
A4	ORN/GRN	EVAP Vent Solenoid	Solenoid Off: 12v, On: 1v
A5	BLU/GRN	Cruise Control Signal	C/C On: Pulse Signals
A6	RED/YEL	EVAP Purge Solenoid	Solenoid Off: 12v, On: 1v
A8	ORN/BLK	HO2S-12 (B1 S2) Heater	Digital Signals: 0-12-0v
A9	BLU/WHT	A/T: VSS Signal from CSS	Moving: pulse signals
A10	RED	Service Check Connector	SCS Open: 4.80v
A12	PNK	Immobilizer Indicator Lamp	Lamp Off: 12v, On: 1v
A13, 25	BLU, RED	Immobilizer Enable, Code	Digital Signals
A14	BLU/RED	A/T: D4 Position Indicator	D4 On: 0v, Off: 12v
A15	ORN/BLK	Immobilizer Fuel Pump Relay	Relay Off: 12v, On: 1v
A17	RED/BLU	A/C Clutch Relay Control	Relay Off: 12v, On: 1v
A18	BLU	Malfunction Indicator Light	MIL Off: 12v, On: 1v
A19	BLU	Engine Speed Pulse (NEP)	Pulse Signals
A20	GRN	Radiator Fan Relay Control	Relay Off: 12v, On: 1v
A21	LT GRN	K-Line Signal	12v
A23	WHT/RED	HO2S-12 (B1 S2) Signal	0.1-1.1v
A24	BLU/RED	Starter Switch Signal	Cranking: 9-11v
A26	GRN	PSP Switch Signal	Straight: 0v, Turning: 11v
A27	RED/WHT	A/C Switch Signal	Switch Off: 5v, On: 0v
A28	WHT/GRN	Interlock Control Unit Signal	Key & Brake On: 12v
A29	WHT/BLU	Fuel Tank Pressure Sensor	Fuel Cap off: 2.5v
A30	ORN/RED	Electric Load Detector	Varies: 0.5-4.5v
A32	GRN/WHT	Brake Switch Signal	Brake Off: 0v, On: 12v

1998 Van 2.3L I4 VTEC SOHC VIN RA3 [A/T] 16P 'D' Connector

PCM Pin #	W/Color	Circuit Description (16-Pin)	Value at Hot Idle
D1	YEL	A/T: Lockup Control Solenoid	LUS On: 12v, Off: 0v
D2	GRN/WHT	A/T: Shift Solenoid 'B'	SSB Off: 0v, On: 12v
D3	GRN	A/T: Shift Solenoid 'C'	SSC on: 12v, off: 0v
D5	BLK/YEL	A/T: Solenoid Feed (B+)	12-14v
D6	WHT	A/T: Gear Position Switch	In 'R': 0v, Others: 12v
D7	BLU/YEL	A/T: Shift Solenoid 'A'	SSA Off: 0v, On: 12v
D8	PNK	A/T: Gear Position Switch	In D3: 0v, Others: 12v
D9	YEL	A/T: Gear Position Switch	In D4: 0v, Others: 12v
D10, D16	BLU, GRN	Countershaft Speed 'P', 'N'	Moving: AC pulses
D11	RED	Mainshaft Speed Sensor 'P'	AC pulse signals
D12	WHT	Mainshaft Speed Sensor 'N'	<0.050v
D13	LT GRN	A/T: Gear Position Switch	In P/N: 0v, Others: 12v
D14	BLU	A/T: Gear Position Switch	In D2: 0v, Others: 12v
D15	BRN	A/T: Gear Position Switch	In D1: 0v, Others: 12v

```
 1  2  3  4    5  6  7    8  9  10   11      1     2  3  4      5
12 13 14 15 16 17 18 19 20 21 22   23   24    6  7  8  9 10 11   12
32-PIN 25 26 27    28 29 30 31     32        13 14 15    16  16-PIN
  'A'                                                        'D'
        WIRE SIDE OF HARNESS TERMINALS
```

05533_ADIA_G599

Pin Connector Graphic

1998 Van 2.3L I4 VTEC SOHC VIN RA3 [A/T] 25P 'B' Connector

PCM Pin #	W/Color	Circuit Description (25-Pin)	Value at Hot Idle
B1, B9	YEL/BLK	Main Relay Power (B+)	12-14v
B2, B10	BLK	Power Ground	<0.1v
B3, B4	RED, BLU	Injector 2, Injector 3 Control	2.0-3.3 ms
B5, B11	YEL, BRN	Injector 4, Injector 1 Control	2.0-3.3 ms
B7	PNK	E-EGR Control Solenoid	Digital Signals: 0-12-0v
B8	WHT	A/T: Clutch Solenoid 'A-'	Pulse Signals
B12	GRN/YEL	VTEC Control Solenoid	0v, Hi-Speed: 12v
B13	YEL/GRN	Ignition Control Signal	Digital Signals: 0-12-0v
B14	BLU/BLK	A/T: 2nd Clutch Pressure Sw.	Open: 12v, Closed: 1v
B16	GRN/RED	HO2S-11 Heater Relay	Relay Off: 12v, On: 1v
B17	RED	A/T: Clutch Solenoid 'A+'	Pulse Signals
B18	GRN	A/T: Clutch Solenoid 'B-'	Pulse Signals
B19	BLK/WHT	HO2S-11 (B1 S1) Heater	Digital Signals: 0-12-0v
B20, B22	BRN/BLK	Logic Ground (LG1, LG2)	<0.050v
B21	WHT/YEL	Keep Alive Power (VBU)	12-14v
B23	BLK/BLU	Idle Air Control Valve	Pulse Signals
B24	BLU/WHT	A/T: 3rd Clutch Pressure Sw.	Open: 12v, Closed: 1v
B25	ORN	A/T: Clutch Solenoid 'B+'	Pulse Signals

1998 Van 2.3L I4 VTEC SOHC VIN RA3 [A/T] 31P 'C' Connector

PCM Pin #	W/Color	Circuit Description (31-Pin)	Value at Hot Idle
C1	BLK/WHT	HO2S-11 (B1 S1) Heater	Digital Signals: 0-12-0v
C2	WHT/GRN	Alternator Charging Signal	Lights Off: 12v, On: 0v
C3	RED/BLU	Knock Sensor Signal	No Detonation: 18mv AC
C5	WHT/RED	Alternator 'FR' Signal	Digital Signals: 0-5-0v
C6	WHT/BLK	EGR Valve Lift Sensor	1.2v
C7	GRN/WHT	MAP Sensor Ground (SG1)	<0.050v
C8	BLU	CKP Sensor Signal	AC pulse signals
C9	WHT	CKP Sensor Ground	<0.050v
C10	BLU/BLK	VTEC Pressure Switch	0v, Hi-Speed: 12v
C16	WHT	HO2S-11 (B1 S1) Signal	0.1-1.1v
C17	RED/GRN	MAP Sensor Signal	0.8-0.9v
C18	GRN/BLK	Sensor Ground (SG2)	<0.050v
C19	YEL/RED	MAP Sensor VREF (VCC1)	4.9-5.1v
C20	GRN	TDC Sensor Signal	AC pulse signals
C21	RED	TDC Sensor Ground	<0.050v
C25	RED/YEL	IAT Sensor Signal	Varies w/temp (0.5-4.9v)
C26	RED/WHT	ECT Sensor Signal	At 180°F: 0.5-0.6v
C27	RED/BLK	TP Sensor Signal	0.5-0.6v
C28	YEL/BLU	Sensor VREF (VCC2)	4.9-5.1v
C29	YEL	CYP Sensor Signal	AC pulse signals
C30	BLK	CYP Sensor Ground	<0.050v

1	2		3	4	5		6	7	8	1	2	3	4		5	6	7		8	9	10
9	10	11	12	13	14	15	16	17	18	11	12	13	14	15	16	17	18	19	20	21	22
19	20	21	22		23	24	25				23	24	25		26	27	28		29	30	31

25-PIN 'B'   *WIRE SIDE OF HARNESS TERMINALS*   **31-PIN 'C'**

05533_ADIA_G600

Pin Connector Graphic

1999-2000 Van 3.5L V6 VTEC VIN RL1 [A/T] 32P 'A' Connector

PCM Pin #	W/Color	Circuit Description (32-Pin)	Value at Hot Idle
A2	GRN/WHT	Engine Mount Solenoid	Idle: 0v, off-idle: 12v
A3	BLU	EVAP Bypass Solenoid	Solenoid Off: 12v, On: 1v
A4	GRN/WHT	EVAP Vent Solenoid	Solenoid Off: 12v, On: 1v
A5	BLU/GRN	Cruise Control Signal	C/C On: pulse signals
A6	RED/YEL	EVAP Purge Solenoid	Solenoid Off: 12v, On: 1v
A7	WHT/RED	VREF to TCM (with TCS)	4.9-5.1v
A8	BLK/WHT	HO2S-12 (B1 S2) Heater	Digital Signals: 0-12-0v
A9	BLU/WHT	Vehicle Speed Output	Moving: pulse signals
A10	BRN	Service Check Connector	SCS Open: 4.80v
A11	LT GRN	Gear Position Signal (TCS)	In 'P': 4v
A12	PNK	Immobilizer Indicator Lamp	Lamp Off: 12v, On: 1v
A13, A25	BLU, RED	Immobilizer Enable, Code	Digital Signals
A14	GRN/BLK	D4 Indicator Light	D4 On: 0v, Off: 12v
A15	GRN/YEL	Fuel Pump Relay Control	Relay Off: 12v, On: 1v
A17	RED	A/C Clutch Relay Control	Relay Off: 12v, On: 1v
A18	GRN/ORN	Malfunction Indicator Light	MIL Off: 12v, On: 1v
A19	BLU	Engine Speed Pulse (NEP)	AC Pulse Signals
A20	BLU/RED	Radiator Fan Relay Control	Relay Off: 12v, On: 1v
A21	GRY	K-Line Signal	12v
A23	WHT/RED	HO2S-12 (B1 S2) Signal	0.1-1.1v
A24	BLU/ORN	Starter Switch Signal	Cranking: 9-11v
A26	GRN	PSP Switch Signal	Straight: 0v, Turning: 11v
A27	BLU/RED	A/C Switch Signal	Switch Off: 12v, On: 1v
A28	WHT/RED	Interlock Control Unit Signal	Key & Brake On: 12v
A29	LT GRN	Fuel Tank Pressure Sensor	Fuel Cap off: 2.5v
A30	GRN/RED	Electric Load Detector	Varies: 0.5-4.5v
A31	RED/BLK	TP Sensor Signal to TCS	At idle: 0.5v
A32	WHT/BLK	Brake Switch Signal	Brake Off: 0v, On: 12v

1999-2000 Van 3.5L V6 VTEC VIN RL1 [A/T] 16P 'D' Connector

PCM Pin #	W/Color	Circuit Description (16-Pin)	Value at Hot Idle
D1	YEL	Lockup Control Solenoid	LUS On: 12v, Off: 0v
D2	GRN/WHT	A/T: Shift Solenoid 'B'	SSB in 3rd, 4th Gear: 0v
D3	GRN	A/T: Shift Solenoid 'C'	SSC in 2nd, 4th Gear: 0v
D5	BLK/YEL	VB Solenoid Feed (B+)	12-14v
D6	WHT	A/T: Gear Position Switch	In 'R': 0v, Others: 12v
D7	BLU/YEL	A/T: Shift Solenoid 'A'	SSA in 1st, 4th Gear: 0v
D8, D9	PNK, YEL	A/T: Gear Position Switch	In D3: 0v, In D4: 0v
D10, D16	BLU, GRN	Countershaft Sensor 'P', 'N'	AC Pulses, <0.050v
D11, D12	RED, WHT	Mainshaft Speed Sensor P, N	Moving: AC pulses
D13	BLU/BLK	A/T: 2nd Clutch Pressure Sw.	Open: 12v, Closed: 1v
D14, 15	BLU, BRN	A/T: Gear Position Switch	D2: 0v, D1: 0v

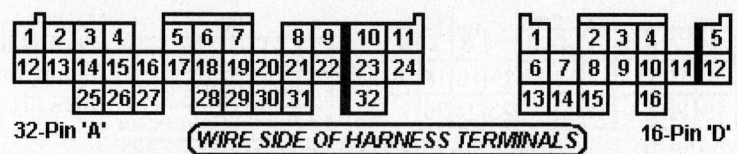

32-Pin 'A'

WIRE SIDE OF HARNESS TERMINALS

16-Pin 'D'

05533_ADIA_G624

Pin Connector Graphic

1999-2000 Van 3.5L V6 VTEC VIN RL1 [A/T] 25P 'B' Connector

PCM Pin #	W/Color	Circuit Description (25-Pin)	Value at Hot Idle
B1	YEL/BLK	Main Relay Power (B+)	12-14v
B2	BLK	Power Ground (PG1)	<0.1v
B3	BLK/RED	Injector 5 Control	2.0-3.3 ms
B4	YEL	Injector 2 Control	2.0-3.3 ms
B5	RED	Injector 2 Control	2.0-3.3 ms
B6	WHT/BLU	Injector 6 Control	2.0-3.3 ms
B7	PNK	E-EGR Solenoid Control	Digital Signals: 0-12-0v
B8	WHT	A/T: Clutch Solenoid 'A-'	AC Pulse Signals
B9	YEL/BLK	Main Relay Power (B+)	12-14v
B10	BLK	Power Ground (PG2)	<0.1v
B11	BRN	Injector 1 Control	2.0-3.3 ms
B12	GRN/YEL	VTEC Control Solenoid	0v, Hi-Speed: 12v
B14	BLU/BLK	A/T: Gear Position Switch	In P/N: 0v, Others: 12v
B15	BLU	Injector 3 Control	2.0-3.3 ms
B17	RED	A/T: Clutch Solenoid 'A+'	AC Pulse Signals
B18	GRN	A/T: Clutch Solenoid 'B-'	Pulse Signals
B20	BRN/BLK	Logic Ground (LG1)	<0.050v
B21	WHT/YEL	Keep Alive Power (VBU)	12-14v
B22	BRN/BLK	Logic Ground (LG2)	<0.050v
B23	BLK/BLU	Idle Air Control Valve	Pulse Signals
B24	WHT/RED	A/T: 3rd Clutch Pressure Sw.	Open: 12v, Closed: 1v
B25	ORN	A/T: Clutch Solenoid 'B+'	Pulse Signals

1999-2000 Van 3.5L V6 VTEC VIN RL1 [W/TCS] 12P 'E' Connector

PCM Pin #	W/Color	Circuit Description (12-Pin)	Value at Hot Idle
E1	PNK/BLK	PCM Torque Down Request	Digital Signals: 0-5-0v
E2	---	Not Used	---
E3	GRN/RED	TCS Operation Permission	Digital Signals: 0-5-0v
E4	---	Not Used	---
E5	---	Not Used	---
E6	---	Not Used	---
E7	---	Not Used	---
E8	---	Not Used	---
E9	---	Not Used	---
E10	---	Not Used	---
E11	---	Not Used	---
E12	---	Not Used	---

For Vehicles
Equipped with
ABS / TCS

WIRE SIDE OF HARNESS TERMINALS

05533_ADIA_G625

Pin Connector Graphic

1999-2000 Van 3.5L V6 VTEC VIN RL1 [A/T] 31P 'C' Connector

PCM Pin #	W/Color	Circuit Description (31-Pin)	Value at Hot Idle
C1	BLK/WHT	HO2S-11 (B1 S1) Heater	Digital Signals: 0-12-0v
C2	WHT/GRN	Alternator Charging Signal	Lights Off: 12v, On: 0v
C3	BLU	Ignition Coil 3 Control	Digital Signals: 0-12-0v
C4	YEL/GRN	Ignition Coil 1 Control	Digital Signals: 0-12-0v
C5	WHT/RED	Alternator 'FR' Signal	Digital Signals: 0-5-0v
C6	WHT/BLK	EGR Valve Lift Sensor	1.2v
C7	GRN/WHT	MAP Sensor Ground	<0.050v
C8	BLU	CKP Sensor Signal	AC pulse signals
C9	WHT	CKP Sensor Ground	<0.050v
C10	BLU/BLK	VTEC Pressure Switch	0v, Hi-Speed: 12v
C11	---	Not Used	----
C12	BLK/RED	Ignition Coil 5 Control	Digital Signals: 0-12-0v
C13	YEL	Ignition Coil 4 Control	Digital Signals: 0-12-0v
C14	RED	Ignition Coil 2 Control	Digital Signals: 0-12-0v
C15	---	Not Used	----
C16	WHT	HO2S-11 (B1 S1) Signal	0.1-1.1v
C17	RED/GRN	MAP Sensor Signal	0.8-0.9v
C18	GRN/WHT	Sensor Ground	<0.050v
C19	YEL/RED	MAP Sensor VREF (VCC1)	4.9-5.1v
C20	GRN	TDC1 Sensor Signal	AC pulse signals
C21	RED	TDC1 Sensor Ground	<0.050v
C22	RED/BLU	Knock Sensor Signal	No Detonation: 18mv AC
C23	WHT/BLU	Ignition Coil 6 Control	Digital Signals: 0-12-0v
C25	RED/YEL	IAT Sensor Signal	Varies w/temp. (0.5-4.9v)
C26	RED/WHT	ECT Sensor Signal	At 180°F: 0.5-0.6v
C27	RED/BLK	TP Sensor Signal	0.5-0.6v
C28	YEL/RED	Sensor VREF (VCC2)	4.9-5.1v
C29	YEL	TDC2 Sensor Signal	AC pulse signals
C30	BLK	TDC2 Sensor Ground	<0.050v
C31	---	Not Used	----

WIRE SIDE OF HARNESS TERMINALS

05533_ADIA_G626

Pin Connector Graphic

2001-03 Van 3.5L V6 VTEC VIN RL1 [A/T] 31P 'A' Connector

PCM Pin #	W/Color	Circuit Description (31-Pin)	Value at Hot Idle
A1	WHT/YEL	Keep Alive Power (VBU)	12-14v
A2, A3	YEL/BLK	Main Relay Power (B+)	12-14v
A4, A5	BLK	Power Ground (PG2, PG1)	<0.1v
A6	WHT	HO2S-11 (B1 S1) Signal	0.1-1.1v
A9	RED/BLU	Knock Sensor Signal	No detonation: 18mv AC
A10, A22	YEL, BLK	TDC2 Sensor Signal	AC pulse signals
A11, A23	GRN,RED	TDC1 Sensor Signal	AC pulse signals
A12, A24	BLU, WHT	CKP Sensor Signal	AC pulse signals
A13	RED/WHT	ECT Sensor Signal	At 180°F: 0.5-0.6v
A14	RED/YEL	IAT Sensor Signal	Varies w/temp. (0.5-4.9v)
A15	RED/BLK	TP Sensor Signal	0.5-0.6v
A16	WHT/BLK	EGR Valve Position Sensor	0.6-1.1v
A19	WHT/RED	Alternator 'FR' Signal	Digital Signals: 0-5-0v
A20	BLU/BLK	VTEC Pressure Switch	0v, Hi-Speed: 12v
A25, A26	BRN/BLK	Logic Ground (LG2, LG1)	<0.1v
A27	RED/GRN	MAP Sensor Signal	0.8-0.9v
A28, A29	GRN/BLK	Sensor Ground (SG2, SG1)	<0.050v
A30, A31	YEL/BLU	Sensor VREF (VCC2, VCC1)	4.9-5.1v

2001-03 Van 3.5L V6 VTEC VIN RL1 [A/T] 22P 'C' Connector

PCM Pin #	W/Color	Circuit Description (22-Pin)	Value at Hot Idle
C1	WHT	A/T: Pressure Solenoid 'A-'	AC pulse signals
C2	BLU/YEL	A/T: Shift Solenoid 'A'	SSA in 1st, 4th Gear: 0v
C4	PNK	A/T: Gear Position Switch	In D3: 0v, Others: 12v
D5	BLK/YEL	VB Solenoid Feed (B+)	12-14v
D6	WHT	A/T: Gear Position Switch	In 'R': 0v, Others: 12v
C6	WHT	Mainshaft Speed Sensor 'N'	<0.050v
C7	RED	Mainshaft Speed Sensor 'P'	Moving: AC pulses
C8	GRN	A/T: Pressure Solenoid 'B-'	AC pulse signals
C9	GRN/WHT	A/T: Shift Solenoid 'B'	SSB in 3rd, 4th Gear: 0v
C10	YEL	Lockup Control Solenoid	LUS On: 12v, Off: 0v
C11	BLU/BLK	A/T: 2nd Clutch Pressure Sw.	Open: 12v, Closed: 1v
C12	BLU	A/T: Gear Position Switch	In D2: 0v, Others: 12v
C13	YEL	A/T: Gear Position Switch	In D4: 0v, Others: 12v
C14	GRN	Countershaft Sensor 'N'	<0.050v
C15	BLU	Countershaft Sensor 'P'	AC pulse signals
C17	GRN	A/T: Shift Solenoid 'C'	SSC in 2nd, 4th Gear: 0v
C19	WHT/RED	A/T: 3rd Clutch Pressure Sw.	Open: 12v, Closed: 1v
C20	BRN	A/T: Gear Position Switch	In D1: 0v, Others: 12v
C21	WHT	A/T: Gear Position Switch	In 'R': 0v, Others: 12v
C22	BLU/WHT	A/T: Gear Position Switch	In P/N: 0v, Others: 12v

31-Pin 'A'   WIRE SIDE OF HARNESS TERMINALS   22-Pin 'C'

05533_ADIA_G627

Pin Connector Graphic

2001-03 Van 3.5L V6 VTEC VIN RL1 [A/T] 31P 'A' Connector

PCM Pin #	W/Color	Circuit Description (31-Pin)	Value at Hot Idle
B1	---	Not Used	---
B2	GRN/YEL	VTEC Control Solenoid	0v, Hi-Speed: 12v
B3	PNK	E-EGR Solenoid Control	Digital Signals: 0-12-0v
B4	---	Not Used	---
B5	BLK/WHT	HO2S-11 (B1 S1) Heater	Digital Signals: 0-12-0v
B6	WHT/GRN	Alternator Charging Signal	Lights Off: 12v, On: 0v
B7	RED	A/T: Pressure Solenoid 'A+'	AC pulse signals
B8	BLU	Coil 3 Driver Control	Digital Signals: 0-12-0v
B9	RED	Coil 2 Driver Control	Digital Signals: 0-12-0v
B10	YEL/GRN	Coil 1 Driver Control	Digital Signals: 0-12-0v
B11	---	Not Used	---
B12	BLU	Injector 3 Control	2.0-3.3 ms
B13	RED	Injector 2 Control	2.0-3.3 ms
B14	BRN	Injector 1 Control	2.0-3.3 ms
B15	---	Not Used	---
B16	ORN	A/T: Pressure Solenoid 'B+'	AC pulse signals
B17	WHT/BLU	Coil 6 Driver Control	Digital Signals: 0-12-0v
B18	BLK/RED	Coil 5 Driver Control	Digital Signals: 0-12-0v
B19	YEL	Coil 4 Driver Control	Digital Signals: 0-12-0v
B20	WHT/BLU	Injector 6 Control	2.0-3.3 ms
B21	BLK/RED	Injector 5 Control	2.0-3.3 ms
B22	YEL	Injector 4 Control	2.0-3.3 ms
B23	BLK/BLU	Idle Air Control Valve	Pulse Signals
B24	BLK/YEL	VB Solenoid Feed (B+)	12-14v

2001-03 Van 3.5L V6 VTEC VIN RL1 [A/T W/TCS] 17P 'D' Connector

PCM Pin #	W/Color	Circuit Description (17-Pin)	Value at Hot Idle
D1-2	---	Not Used	---
D3	PNK/BLK	TCM Retard Signal to PCM	Digital Signals: 0-5-0v
D4	---	Not Used	---
D5	GRN/RED	TCS Operation Permission	Digital Signals: 0-5-0v
D6-8	---	Not Used	---
D9	WHT/RED	ABS/TCS VREF	4.9-5.1v
D10	---	Not Used	---
D11	LT GRN	A/T Shift Position Signal	In 'P': near 4v
D12-15	---	Not Used	---
D16	RED/BLK	TPS Signal to ABS/TCS Unit	0.5-0.6v
D17	---	Not Used	---

```
 1  2  3  4  5     6  7
 8  9 10 11 12 13 14 15 16
17 18 19    20 21 22 23 24
24-Pin 'B'
```

```
 1  2  3  4  5  6
 7  8  9 10 11 12
13 14    15 16 17
17-Pin 'D'
```

WIRE SIDE OF HARNESS TERMINALS

jnew

05533_ADIA_G628

Pin Connector Graphic

2001-03 Van 3.5L V6 VTEC VIN RL1 [A/T] 31P 'E' Connector

PCM Pin #	W/Color	Circuit Description (31-Pin)	Value at Hot Idle
E1	GRN/YEL	Immobilizer Fuel Pump Relay	Relay Off: 12v, On: 1v
E2	BLK/WHT	HO2S-12 (B1 S2) Heater	Digital Signals: 0-12-0v
E3	BLU	Engine Speed (NEP) Signal	Pulse Signals
E4	GRN/OR	Malfunction Indicator Lamp	MIL Off: 12v, On: 1v
E5	GRN/BLK	A/T: Gear Position Switch	In D4: 0v, Others: 12v
E6	YEL/BLU	Sensor VREF (VCC3)	4.9-5.1v
E7	GRN/BLK	Sensor Ground (SG3)	<0.050v
E8	BLU/ORN	Starter Switch Signal	Cranking: 9-11v
E9	WHT/BLK	Brake Switch Signal	Brake Off; 0v, On: 12v
E10	---	Not Used	---
E11	RED/YEL	EVAP Purge Solenoid	Solenoid Off: 12v, On: 1v
E12	RED	A/C Clutch Relay Control	Relay Off: 12v, On: 1v
E13	BLU/RED	Radiator Fan Control Relay	Relay Off: 12v, On: 1v
E14	YEL/GRN	ECT Signal to TCM Unit	At 180°F: 0.5-0.6v
E15	WHT/RED	Illumination (Multiplex Unit)	Digital Signals
E16	BRN	Service Check Connector	SCS Open 5v
E17	BLU/GRN	Cruise Control Signal	C/C On: pulse signals
E18	WHT/RED	HO2S-12 (B1 S2) Signal	0.1-1.1v
E19	GRN/RED	Electric Load Detector	Varies: 0.5-4.5v
E20	GRN	PSP Switch Signal	Straight: 0v, Turning: 11v
E21	BLU/RED	A/C Switch Signal	Switch Off: 12v, On: 1v
E22	GRN/WHT	EVAP Vent Solenoid	Solenoid Off: 12v, On: 1v
E23	BLU	EVAP Bypass Solenoid	Solenoid Off: 12v, On: 1v
E24	GRN/WHT	Engine Mount Solenoid	Idle: 0v, off-idle: 12v
E25	---	Not Used	---
E26	BLU/WHT	VSS Signal to Speedometer	Digital Signals
E27	GRY	K-Line Signal to DCL	12v
E28	LT GRN	Fuel Tank Pressure Sensor	Fuel Cap off: 2.5v
E29	PNK	Immobilizer Indicator Lamp	Lamp Off: 12v, On: 1v
E30	BLU	Immobilizer Enable Signal	Digital Signals
E31	RED	Immobilizer Code Signal	Digital Signals

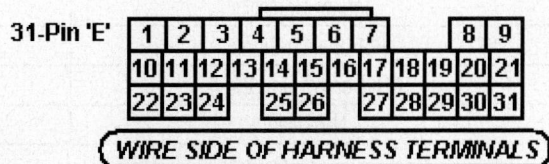

```
31-Pin 'E'   1  2  3  4  5  6  7        8  9
            10 11 12 13 14 15 16 17 18 19 20 21
            22 23 24    25 26    27 28 29 30 31
```

WIRE SIDE OF HARNESS TERMINALS

05533_ADIA_G629

Pin Connector Graphic

PRELUDE PIN CHARTS

1990-91 Coupe 2.0L I4 MFI VIN BA4 [All] 18P 'A' Connector

PCM Pin #	W/Color	Circuit Description (18-Pin)	Value at Hot Idle
A1	BRN	Injector 1 Control	2.0-3.3 ms
A2	BLK	Power Ground	<0.1v
A3	RED	Injector 2 Control	2.0-3.3 ms
A4	BLK	Power Ground	<0.1v
A5	BLU	Injector 3 Control	2.0-3.3 ms
A6	---	Not Used	---
A7	YEL	Injector 4 Control	2.0-3.3 ms
A8	YEL	Bypass Control Solenoid	Solenoid Off: 12v, On: 1v
A9	---	Not Used	---
A10	RED	EGR Solenoid Control	Solenoid Off: 12v, On: 1v
A11	BLU/RED	Electronic Air Control Valve	Pulse Signals
A12	GRN/BLK	Fuel Pump Relay Control	Relay Off: 12v, On: 1v
A13	YEL/BLK	Main Relay Power (B+)	12-14v
A14	GRN/BLK	Fuel Pump Relay Control	Relay Off: 12v, On: 1v
A15	YEL/BLK	Main Relay Power (B+)	12-14v
A16	BRN/BLK	Power Ground	<0.1v
A17	WHT/YEL	Keep Alive Power (VBU)	12-14v
A18	BLK/RED	Power Ground	<0.1v

1990-91 Coupe 2.0L I4 MFI VIN BA4 [All] 20P 'B' Connector

PCM Pin #	W/Color	Circuit Description (20-Pin)	Value at Hot Idle
B1	RED/BLU	A/C Clutch Relay Control	Relay Off: 12v, On: 1v
B2	YEL/BLK	EVAP Purge Solenoid	Solenoid Off: 12v, On: 1v
B3	---	Not Used	---
B4	---	Not Used	---
B5	BRN	Ignition Timing Adjuster	0.4-4.5v
B6	YEL/RED	Check Engine Light	C/E Off: 12v, On: 1v
B7	LT GRN	A/T: Park/Neutral Switch	In P/N: 0v, Others: 12v
B8	RED/GRN	A/C Switch Signal	Switch Off: 12v, On: 1v
B9	---	Not Used	---
B10	BLU/GRN	CKP Sensor Signal	AC pulse signals
B11	RED/BLU	Oxygen Sensor 'B' Signal	0.1-1.1v
B12	BLU/YEL	CKP Sensor Ground	<0.050v
B13	BLU/RED	Starter Switch Signal	Cranking: 9-11v
B14	WHT/RED	Alternator 'FR' Signal	Digital Signals: 0-5-0v
B15	WHT	Igniter Control	Digital Signals: 0-12-0v
B16	WHT/BLU	Vehicle Speed Sensor	Moving: pulse signals
B17	WHT	Igniter Control	Digital Signals: 0-12-0v
B18	BLU/RED	PSP Switch Signal	Straight: 0v, Turning: 11v
B19	YEL/WHT	A/T: Control Unit VREF	4.9-5.1v
B20	---	Not Used	---

1990-91 Coupe 2.0L I4 MFI VIN BA4 [All] 16P 'C' Connector

PCM Pin #	W/Color	Circuit Description (16-Pin)	Value at Hot Idle
C1	ORN	CYP Sensor Signal	AC pulse signals
C2	WHT	CYP Sensor Ground	<0.050v
C3	ORN/BLU	TDC Sensor Signal	AC pulse signals
C4	WHT/BLU	TDC Sensor Ground	<0.050v
C5	WHT/RED	IAT Sensor Signal	Varies w/temp. (0.5-4.9v)
C6	YEL/GRN	ECT Sensor Signal	At 180°F: 0.5-0.6v
C7	RED/YEL	Throttle Angle Sensor Signal	0.5-0.6v
C8	YEL	A/T: EGRV Lift Sensor Signal	1.2v
C9	RED	Atmospheric Pressure Sensor	2.76-2.96v at sea level
C10	---	Not Used	---
C11	WHT/BLU	MAP Sensor Signal	0.8-0.9v
C12	GRN/WHT	Sensor Ground	<0.050v
C13	YEL/WHT	Sensor VREF	4.9-5.1v
C14	BLU/WHT	Sensor Ground	<0.050v
C15	RED/WHT	Sensor VREF	4.9-5.1v
C16	WHT	Oxygen Sensor 'A' Signal	0.1-1.1v

05533_ADIA_G613

Pin Connector Graphic

Standard Colors and Abbreviations

Abbreviation	Color	Abbreviation	Color	Abbreviation	Color
BLK	Black	LT BLU	Lt. Blue	TAN	Tan
BLU	Blue	LT GRN	Lt. Green	VIO	Violet
BRN	Brown	ORN	Orange	WHT	White
GRY	Gray	PNK	Pink	YEL	Yellow
GRN	Green	PPL	Purple		

1990-91 Coupe 2.1L I4 MFI VIN BA4 [All] 26P 'A' Connector

PCM Pin #	W/Color	Circuit Description (26-Pin)	Value at Hot Idle
A1	BRN	Injector 1 Control	2.0-3.3 ms
A2	YEL	Injector 4 Control	2.0-3.3 ms
A3	RED	Injector 2 Control	2.0-3.3 ms
A4	---	Not Used	---
A5	BLU	Injector 3 Control	2.0-3.3 ms
A6	BLK	HO2S-11 (B1 S1) Heater	Relay Off: 12v, On: 1v
A7	GRN/BLK	Fuel Pump Relay Control	Relay Off: 12v, On: 1v
A8	GRN/BLK	Fuel Pump Relay Control	Relay Off: 12v, On: 1v
A9	BLU/RED	Electronic Air Control Valve	Pulse Signals
A10	---	Not Used	---
A11	RED	EGR Solenoid Control	1.2v
A12	---	Not Used	---
A13	YEL/RED	Check Engine Light	C/E Off: 12v, On: 1v
A14	---	Not Used	---
A15	RED/BLU	A/C Clutch Relay Control	Relay Off: 12v, On: 1v
A16-19	---	Not Used	---
A20	YEL/BLK	EVAP Purge Solenoid	Solenoid Off: 12v, On: 1v
A21	WHT	Igniter Control	Digital Signals: 0-12-0v
A22	WHT	Igniter Control	Digital Signals: 0-12-0v
A23	BLK	Power Ground	<0.1v
A24	BLK	Power Ground	<0.1v
A25	YEL/BLK	Main Relay Power (B+)	12-14v
A26	BLK/RED	Power Ground	<0.1v

1990-91 Coupe 2.1L I4 MFI VIN BA4 [All] 16P 'B' Connector

PCM Pin #	W/Color	Circuit Description (16-Pin)	Value at Hot Idle
B1	YEL/BLK	Main Relay Power (B+)	12-14v
B2	BRN/BLK	Power Ground	<0.1v
B3	---	Not Used	---
B4	---	Not Used	---
B5	RED/GRN	A/C Switch Signal	Switch Off: 12v, On: 1v
B6	---	Not Used	---
B7	LT GRN	A/T: Park/Neutral Switch	In P/N: 0v, Others: 12v
B8	BLU/RED	PSP Switch	Straight: 0v, Turning: 11v
B9	BLU/RED	Starter Switch Signal	Cranking: 9-11v
B10	WHT/BLU	Vehicle Speed Sensor	Moving: pulse signals
B11	ORN	CYP Sensor Signal	AC pulse signals
B12	WHT	CYP Sensor Ground	<0.050v
B13	ORN/BLU	TDC Sensor Signal	AC pulse signals
B14	WHT/BLU	TDC Sensor Ground	<0.050v
B15	BLU/GRN	CKP Sensor Signal	AC pulse signals
B16	BLU/YEL	CKP Sensor Ground	<0.050v

1990-91 Coupe 2.1L I4 MFI VIN BA4 [All] 22P 'D' Connector

PCM Pin #	W/Color	Circuit Description (22-Pin)	Value at Hot Idle
D1	WHT/YEL	Keep Alive Power (VBU)	12-14v
D2	---	Not Used	---
D3	---	Not Used	---
D4	BRN	Service Check Connector	SCS Open: 4.80v
D5-8	---	Not Used	---
D9	WHT/RED	Alternator 'FR' Signal	Digital Signals: 0-5-0v
D10	---	Not Used	---
D11	RED/YEL	Throttle Angle Sensor Signal	0.5-0.6v
D12	YEL	A/T: EGRV Lift Sensor Signal	1.2v
D13	YEL/GRN	ECT Sensor Signal	At 180°F: 0.5-0.6v
D14	WHT	HO2S-11 (B1 S1) Signal	0.1-1.1v
D15	WHT/RED	IAT Sensor Signal	Varies w/temp. (0.5-4.9v)
D16	---	Not Used	---
D17	WHT/BLU	MAP Sensor Signal	0.8-0.9v
D18	YEL/WHT	A/T: Control Unit VREF	4.9-5.1v
D19	RED/WHT	MAP Sensor VREF	4.9-5.1v
D20	YEL/WHT	Sensor VREF	4.9-5.1v
D21	BLU/WHT	MAP Sensor Ground	<0.050v
D22	RED/WHT	Sensor VREF	4.9-5.1v

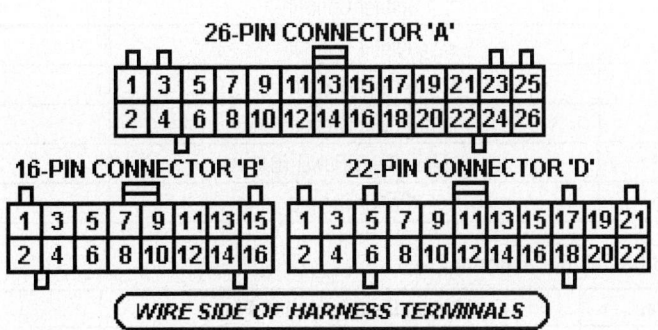

05533_ADIA_G597

Pin Connector Graphic

Standard Colors and Abbreviations

Abbreviation	Color	Abbreviation	Color	Abbreviation	Color
BLK	Black	LT BLU	Lt. Blue	TAN	Tan
BLU	Blue	LT GRN	Lt. Green	VIO	Violet
BRN	Brown	ORN	Orange	WHT	White
GRY	Gray	PNK	Pink	YEL	Yellow
GRN	Green	PPL	Purple		

1992 Coupe 2.2L I4 MFI VIN BA8 [All] 26P 'A' Connector

PCM Pin #	W/Color	Circuit Description (26-Pin)	Value at Hot Idle
A1	BRN	Injector 1 Control	2.0-3.3 ms
A2	YEL	Injector 4 Control	2.0-3.3 ms
A3	RED	Injector 2 Control	2.0-3.3 ms
A4	---	Not Used	---
A5	BLU	Injector 3 Control	2.0-3.3 ms
A6	ORN/WHT	HO2S Heater Control	Relay Off: 12v, On: 1v
A7	GRN/BLK	Fuel Pump Relay	Relay Off: 12v, On: 1v
A8	GRN/BLK	Fuel Pump Relay	Relay Off: 12v, On: 1v
A9	BLK/BLU	Electronic Air Control Valve	Pulse Signals
A10	---	Not Used	---
A11	RED	EGR Solenoid	Solenoid Off: 12v, On: 1v
A12	BLU/RED	Radiator Fan Relay	Relay Off: 12v, On: 1v
A13	BLU/WHT	Check Engine Light	MIL Off: 12v, On: 1v
A14	---	Not Used	---
A15	RED/BLU	A/C Compressor Unit	---
A16	WHT/GRN	Alternator Charging Signal	Lights Off: 12v, On: 0v
A17	PNK	Intake Air Bypass Solenoid	Solenoid Off: 12v, On: 1v
A18	ORN/RED	A/T: TCM (shift acknowledge)	N/A
A19	WHT	Intake Air Control Solenoid	Solenoid Off: 12v, On: 1v
A20	RED/GRN	EVAP Purge Solenoid	Solenoid Off: 12v, On: 1v
A21	YEL/GRN	Igniter Control	Digital Signals: 0-12-0v
A22	YEL/GRN	Igniter Control	Digital Signals: 0-12-0v
A23	BLK	Power Ground	<0.1v
A24	BLK	Power Ground	<0.1v
A25	YEL/BLK	Main Relay Power (B+)	12-14v
A26	BLK/RED	Power Ground	<0.1v

1992 Coupe 2.2L I4 MFI VIN BA8 [All] 16P 'B' Connector

PCM Pin #	W/Color	Circuit Description (16-Pin)	Value at Hot Idle
B1	YEL/BLK	Main Relay Power (B+)	12-14v
B2	BRN/BLK	Power Ground	<0.1v
B3	ORN	A/T: TCM Signal (up-shift)	Digital Signals
B4	PNK	A/T: TCM Signal (down-shift)	Digital Signals
B5	BLU/BLK	A/C Switch Signal	Switch Off: 12v, On: 1v
B6	---	Not Used	---
B7	LT GRN	A/T: Park/Neutral Switch	In N: 0v, Others: 12v
B8	RED/GRN	PSP Switch	Straight: 0v, Turning: 11v
B9	BLU/RED	Starter Switch Signal	Cranking: 9-11v
B10	ORN	Vehicle Speed Sensor	Moving: pulse signals
B11	ORN	CYP Sensor Signal	AC pulse signals
B12	WHT	CYP Sensor Ground	<0.050v
B13	ORN/BLU	TDC Sensor Signal	AC pulse signals
B14	WHT/BLU	TDC Sensor Ground	<0.050v
B15	BLU/GRN	CKP Sensor Signal	AC pulse signals
B16	BLU/YEL	CKP Sensor Ground	<0.050v

1992 Coupe 2.2L I4 MFI VIN BA8 [All] 22P 'D' Connector

PCM Pin #	W/Color	Circuit Description (22-Pin)	Value at Hot Idle
D1	WHT/YEL	Keep Alive Power (VBU)	12-14v
D2	GRN/WHT	Brake Switch Signal	Brake Off: 0v, On: 12v
D3	RED/BLU	Knock Sensor Signal Solenoid	Solenoid Off: 12v, On: 1v
D4	BRN/WHT	Service Check Connector	SCS Open: 4.80v
D5-6	---	Not Used	---
D7	GRN/RED	Data Link Connector	5v
D8	---	Not Used	---
D9	WHT/RED	Alternator 'FR' Signal	Digital Signals: 0-5-0v
D10	---	Not Used	---
D11	RED/BLK	Throttle Angle Sensor Signal	0.5-0.6v
D12	WHT/BLK	A/T: EGR Lift Sensor Signal	1.2v
D13	YEL/BLU	ECT Sensor Signal	At 180°F: 0.5-0.6v
D14	WHT	O2S-11 Signal	0.1-1.1v
D15	RED/YEL	IAT Sensor Signal	Varies w/temp. (0.5-4.9v)
D16	---	Not Used	---
D17	WHT/BLU	MAP Sensor Signal	0.8-0.9v
D18	ORN/BLK	A/T: Control Unit VREF	4.9-5.1v
D19	RED/WHT	Sensor VREF	4.9-5.1v
D20	YEL/WHT	Sensor VREF	4.9-5.1v
D21	BLU/WHT	Sensor Ground	<0.050v
D22	GRN/WHT	Sensor Ground	<0.050v

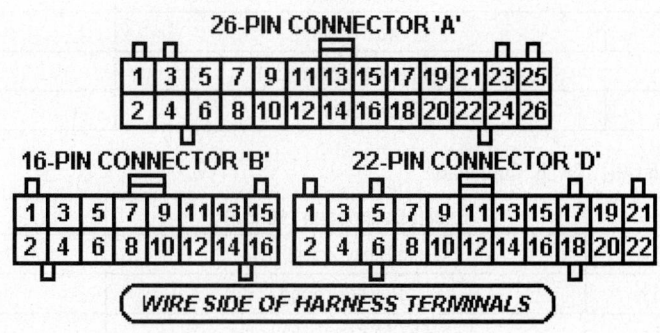

05533_ADIA_G597

Pin Connector Graphic

Standard Colors and Abbreviations

Abbreviation	Color	Abbreviation	Color	Abbreviation	Color
BLK	Black	LT BLU	Lt. Blue	TAN	Tan
BLU	Blue	LT GRN	Lt. Green	VIO	Violet
BRN	Brown	ORN	Orange	WHT	White
GRY	Gray	PNK	Pink	YEL	Yellow
GRN	Green	PPL	Purple		

1993-95 Coupe 2.2L I4 MFI VIN BA8 [All] 26P 'A' Connector

PCM Pin #	W/Color	Circuit Description (26-Pin)	Value at Hot Idle
A1	BRN	Injector 1 Control	2.0-3.3 ms
A2	YEL	Injector 4 Control	2.0-3.3 ms
A3	RED	Injector 2 Control	2.0-3.3 ms
A4	---	Not Used	---
A5	BLU	Injector 3 Control	2.0-3.3 ms
A6	ORN/WHT	HO2S Heater Control	Relay Off: 12v, On: 1v
A7	GRN/BLK	Fuel Pump Relay	Relay Off: 12v, On: 1v
A8	GRN/BLK	Fuel Pump Relay	Relay Off: 12v, On: 1v
A9	BLK/BLU	Idle Air Control Valve	Pulse Signals
A10	---	Not Used	---
A11	RED	EGR Solenoid	Solenoid Off: 12v, On: 1v
A12	BLU/RED	Radiator Fan Relay	Relay Off: 12v, On: 1v
A13	BLU/WHT	Check Engine Light	MIL Off: 12v, On: 1v
A14	---	Not Used	---
A15	RED/BLU	A/C Clutch Relay Control	Relay Off: 12v, On: 1v
A16	WHT/GRN	Alternator Charging Signal	Lights Off: 12v, On: 0v
A17	PNK	Intake Air Bypass Solenoid	Solenoid Off: 12v, On: 1v
A18	ORN/RED	A/T: TCM (shift acknowledge)	N/A
A19	WHT	Intake Control Solenoid	Solenoid Off: 12v, On: 1v
A20	RED/GRN	EVAP Purge Solenoid	Solenoid Off: 12v, On: 1v
A21	YEL/GRN	Igniter Control	Digital Signals: 0-12-0v
A22	YEL/GRN	Igniter Control	Digital Signals: 0-12-0v
A23	BLK	Power Ground	<0.1v
A24	BLK	Power Ground	<0.1v
A25	YEL/BLK	Main Relay Power (B+)	12-14v
A26	BLK/RED	Power Ground	<0.1v

1993-95 Coupe 2.2L I4 MFI VIN BA8 [All] 16P 'B' Connector

PCM Pin #	W/Color	Circuit Description (16-Pin)	Value at Hot Idle
B1	YEL/BLK	Main Relay Power (B+)	12-14v
B2	BRN/BLK	Power Ground	<0.1v
B3	ORN	A/T TCM Signal (Upshift)	Digital Signals
B4	PNK	A/T TCM Signal (Downshift)	Digital Signals
B5	BLU/BLK	A/C Switch Signal	Switch Off: 12v, On: 1v
B6	---	Not Used	---
B7	LT GRN	A/T: Park Neutral Switch	In P/N: 0v, Others: 5v
B8	RED/GRN	PSP Switch	Straight: 0v, Turning: 11v
B9	BLU/RED	Starter Switch Signal	Cranking: 9-11v
B10	ORN	Vehicle Speed Sensor	Moving: pulse signals
B11	ORN	CYP Sensor Signal	AC pulse signals
B12	WHT	CYP Sensor Ground	0.050v
B13	ORN/BLU	TDC Sensor	AC pulse signals
B14	WHT/BLU	TDC Sensor	0.050v
B15	BLU/GRN	CKP Sensor Signal	AC pulse signals
B16	BLU/YEL	CKP Sensor Ground	0.050v

1993-95 Coupe 2.2L I4 MFI VIN BA8 [All] 22P 'D' Connector

PCM Pin #	W/Color	Circuit Description (22-Pin)	Value at Hot Idle
D1	WHT/YEL	Keep Alive Power (VBU)	12-14v
D2	GRN/WHT	Brake Switch Signal	Brake Off: 0v, On: 12v
D3	RED/BLU	Knock Sensor Signal	No Detonation: 18mv AC
D4	BRN/WHT	Service Check Connector	SCS Open: 4.80v
D5	---	Not Used	---
D6	---	Not Used	---
D7	GRN/RED	Data Link Connector	5v
D8	---	Not Used	---
D9	WHT/RED	Alternator 'FR' Signal	Digital Signals: 0-5-0v
D10	GRN/RED	Electrical Load Detector	Varies: 0.5-4.5v
D11	RED/BLK	TP Sensor Signal	0.5-0.6v
D12	WHT/BLK	EGR Valve Lift Sensor	1.2v
D13	YEL/BLU	ECT Sensor Signal	At 180°F: 0.5-0.6v
D14	WHT	HO2S Signal	0.1-1.1v
D15	RED/YEL	IAT Sensor Signal	Varies w/temp. (0.5-4.9v)
D16	---	Not Used	---
D17	WHT/BLU	MAP Sensor Signal	0.8-0.9v
D18	GRN/BLK	A/T: Control Unit VREF	4.9-5.1v
D19	RED/WHT	Sensor VREF	4.9-5.1v
D20	YEL/WHT	Sensor VREF	4.9-5.1v
D21	BLU/WHT	Sensor Ground	<0.050v
D22	GRN/WHT	Sensor Ground	<0.050v

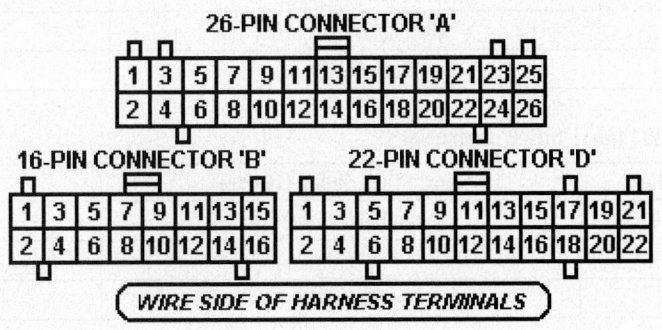

Pin Connector Graphic

Standard Colors and Abbreviations

Abbreviation	Color	Abbreviation	Color	Abbreviation	Color
BLK	Black	LT BLU	Lt. Blue	TAN	Tan
BLU	Blue	LT GRN	Lt. Green	VIO	Violet
BRN	Brown	ORN	Orange	WHT	White
GRY	Gray	PNK	Pink	YEL	Yellow
GRN	Green	PPL	Purple		

05533_ADIA_G597

1993-95 Coupe 2.2L I4 VTEC VIN BB1 [M/T] 26P 'A' Connector

PCM Pin #	W/Color	Circuit Description (26-Pin)	Value at Hot Idle
A1	BRN	Injector 1 Control	2.0-3.3 ms
A2	YEL	Injector 4 Control	2.0-3.3 ms
A3	RED	Injector 2 Control	2.0-3.3 ms
A4	GRN/YEL	VTEC Solenoid	0v, Hi-Speed: 12v
A5	BLU	Injector 3 Control	2.0-3.3 ms
A6	ORN/WHT	HO2S Heater Control	Relay Off: 12v, On: 1v
A7	GRN/BLK	Fuel Pump Relay	Relay Off: 12v, On: 1v
A8	GRN/BLK	Fuel Pump Relay	Relay Off: 12v, On: 1v
A9	BLK/BLU	Idle Air Control Valve	Pulse Signals
A10	---	Not Used	---
A11	RED	EGR Solenoid	Solenoid Off: 12v, On: 1v
A12	BLU/RED	Radiator Fan Relay Control	Relay Off: 12v, On: 1v
A13	BLU/WHT	Check Engine Light	MIL Off: 12v, On: 1v
A14	---	Not Used	---
A15	RED/BLU	A/C Clutch Relay Control	Relay Off: 12v, On: 1v
A16	WHT/GRN	Alternator Charging Signal	Lights Off: 12v, On: 0v
A17	PNK	Intake Air Bypass Solenoid	Solenoid Off: 12v, On: 1v
A18	---	Not Used	---
A19	WHT	Intake Control Solenoid	Solenoid Off: 12v, On: 1v
A20	RED/GRN	EVAP Purge Solenoid	Solenoid Off: 12v, On: 1v
A21	YEL/GRN	Igniter Control	Digital Signals: 0-12-0v
A22	YEL/GRN	Igniter Control	Digital Signals: 0-12-0v
A23	BLK	Power Ground	<0.1v
A24	BLK	Power Ground	<0.1v
A25	YEL/BLK	Main Relay Power (B+)	12-14v
A26	BLK/RED	Power Ground	<0.1v

1993-95 Coupe 2.2L I4 VTEC VIN BB1 [M/T] 16P 'B' Connector

PCM Pin #	W/Color	Circuit Description (16-Pin)	Value at Hot Idle
B1	YEL/BLK	Main Relay Power (B+)	12-14v
B2	BRN/BLK	Power Ground	<0.1v
B3	---	Not Used	---
B4	---	Not Used	---
B5	BLU/BLK	A/C Switch Signal	Switch Off: 12v, On: 1v
B6	---	Not Used	---
B7	---	Not Used	---
B8	RED/GRN	PSP Switch Signal	Straight: 0v, Turning: 11v
B9	BLU/RED	Starter Switch Signal	Cranking: 9-11v
B10	ORN	Vehicle Speed Sensor	Moving: pulse signals
B11	ORN	CYP Sensor Signal	AC pulse signals
B12	WHT	CYP Sensor Ground	<0.050v
B13	ORN/BLU	TDC Sensor Signal	AC pulse signals
B14	WHT/BLU	TDC Sensor Ground	<0.050v
B15	BLU/GRN	CKP Sensor Signal	AC pulse signals
B16	BLU/YEL	CKP Sensor Ground	<0.050v

1993-95 Coupe 2.2L I4 VTEC VIN BB1 [M/T] 22P 'D' Connector

PCM Pin #	W/Color	Circuit Description (26-Pin)	Value at Hot Idle
D1	WHT/YEL	Keep Alive Power (VBU)	12-14v
D2	GRN/WHT	Brake Switch Signal	Brake Off: 0v, On: 12v
D3	RED/BLU	Knock Sensor Signal	No Detonation: 18mv AC
D4	BRN/WHT	Service Check Connector	SCS Open: 4.80v
D5	---	Not Used	---
D6	LT BLU	VTEC Pressure Switch	0v, Hi-Speed: 12v
D7	GRN/RED	DLC TXD, RXD Signals	Digital Signals (0-5-0-5v)
D8	---	Not Used	---
D9	WHT/RED	Alternator 'FR' Signal	Digital Signals: 0-5-0v
D10	GRN/RED	Electrical Load Detector	Varies: 0.5-4.5v
D11	RED/BLK	TP Sensor Signal	0.5-0.6v
D12	WHT/BLK	EGR Valve Lift Sensor	1.2v
D13	YEL/BLU	ECT Sensor Signal	At 180°F: 0.5-0.6v
D14	WHT	HO2S Signal	0.1-1.1v
D15	RED/YEL	IAT Sensor Signal	Varies w/temp. (0.5-4.9v)
D16	---	---	---
D17	WHT/BLU	MAP Sensor Signal	0.8-0.9v
D18	---	Not Used	---
D19	RED/WHT	Sensor VREF	4.9-5.1v
D20	YEL/WHT	Sensor VREF	4.9-5.1v
D21	BLU/WHT	Sensor Ground	<0.050v
D22	GRN/WHT	Sensor Ground	<0.050v

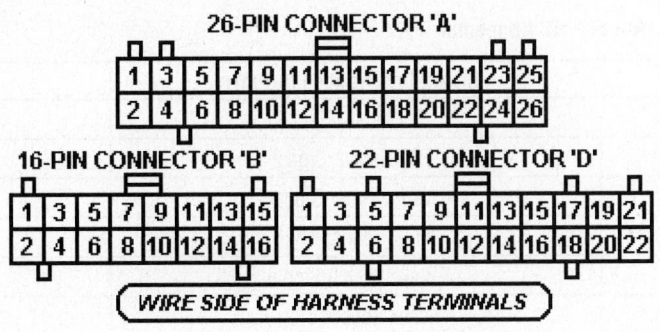

05533_ADIA_G597

Pin Connector Graphic

Standard Colors and Abbreviations

Abbreviation	Color	Abbreviation	Color	Abbreviation	Color
BLK	Black	LT BLU	Lt. Blue	TAN	Tan
BLU	Blue	LT GRN	Lt. Green	VIO	Violet
BRN	Brown	ORN	Orange	WHT	White
GRY	Gray	PNK	Pink	YEL	Yellow
GRN	Green	PPL	Purple		

1996 Coupe 2.2L I4 MFI VIN BA8 [All] 32P 'A' Connector

PCM Pin #	W/Color	Circuit Description (32-Pin)	Value at Hot Idle
A1	YEL	Injector 4 Control	2.0-3.3 ms
A2	BLU	Injector 3 Control	2.0-3.3 ms
A3	RED	Injector 2 Control	2.0-3.3 ms
A4	BRN	Injector 1 Control	2.0-3.3 ms
A5	ORN/BLK	HO2S-12 (B1 S2) Heater	Digital Signals: 0-12-0v
A6	ORN/WHT	HO2S-11 (B1 S1) Heater	Digital Signals: 0-12-0v
A7	RED	EGR Solenoid	Solenoid Off: 12v, On: 1v
A9	BLK/RED	Logic Ground	<0.050v
A10	BLK	Power Ground	<0.1v
A11	YEL/BLK	Main Relay Power (B+)	12-14v
A12	BLK/BLU	Idle Air Control Valve	Pulse Signals
A15	RED/YEL	EVAP Purge Solenoid	Solenoid Off: 12v, On: 1v
A16	GRN/BLK	Fuel Pump Relay	Relay Off: 12v, On: 1v
A17	RED/BLU	A/C Clutch Relay Control	Relay Off: 12v, On: 1v
A18	BLU/WHT	Check Engine Light	MIL Off: 12v, On: 1v
A19	WHT/GRN	Alternator Charging Signal	Lights Off: 12v, On: 0v
A20	YEL/GRN	Igniter Control	Digital Signals: 0-12-0v
A22	BRN/BLK	Logic Ground	<0.050v
A23	BLK	Power Ground	<0.1v
A24	YEL/BLK	Main Relay Power (B+)	12-14v
A25	WHT	Intake Air Control Solenoid	Solenoid Off: 12v, On: 1v
A26	PNK	Intake Air Bypass Solenoid	Solenoid Off: 12v, On: 1v
A27	BLU/RED	Radiator Fan Relay	Relay Off: 12v, On: 1v

1996 Prelude 2.2L I4 MFI VIN BA8 [All] 16P 'D' Connector

PCM Pin #	W/Color	Circuit Description (16-Pin)	Value at Hot Idle
D1	RED/BLK	TP Sensor Signal	0.5-0.6v
D2	YEL/BLU	ECT Sensor Signal	At 180°F: 0.5-0.6v
D3	WHT/YEL	MAP Sensor Signal	0.8-0.9v
D4	YEL/WHT	MAP Sensor VREF	4.9-5.1v
D5	GRN/WHT	Brake Switch Signal	Brake Off: 0v, On: 12v
D6	RED/BLU	Knock Sensor Signal	No Detonation: 18mv AC
D7	WHT	HO2S-11 (B1 S1) Signal	0.1-1.1v
D8	RED/YEL	IAT Sensor Signal	Varies w/temp. (0.5-4.9v)
D9	WHT/BLK	EGR Valve Lift Sensor	1.2v
D10	YEL/WHT	Sensor VREF	4.9-5.1v
D11	GRN/WHT	Sensor Ground	<0.050v
D12	GRN/WHT	MAP Sensor Ground	<0.050v
D13	GRN/BLU	HO2S-12 Ground	<0.050v
D14	WHT/RED	HO2S-12 (B1 S2) Signal	0.1-1.1v
D16	GRN/BLK	Electrical Load Detector	Varies: 0.5-4.5v

```
1  2  3  4    5  6  7    8  9  10    11    1     2  3  4    5
12 13 14 15 16 17 18 19 20 21 22  23    24   6  7  8  9 10 11   12
32-PIN 25 26 27   28 29 30 31   32        13 14 15    16  16-PIN
   'A'                                               'D'
```

WIRE SIDE OF HARNESS TERMINALS

05533_ADIA_G610

Pin Connector Graphic

1996 Coupe 2.2L I4 MFI VIN BA8 [All] 25P 'B' Connector

PCM Pin #	W/Color	Circuit Description (25-Pin)	Value at Hot Idle
B3	BLU/YEL	A/T: Shift Solenoid 'A'	SSA Off: 0v, On: 12v
B4	WHT/BLK	A/T: Lockup Solenoid 'B'	LSB On: 12v, Off: 0v
B5	RED/WHT	A/T: Lockup Solenoid 'A'	LSA On: 12v, Off: 0v
B8	GRN/BLU	A/T: Gear Position Switch	In D3: 0v, Others: 12v
B11	GRN/YEL	A/T: Shift Solenoid 'B'	SSB Off: 0v, On: 12v
B12	WHT/GRN	Interlock Control Unit Signal	Key & Brake On: 12v
B13	GRN/BLK	A/T: D4 Indicator Light Switch	D4 On: 0v, Off: 12v
B14	WHT/BLU	Mainshaft Speed Sensor 'N'	<0.050v
B15	ORN/BLU	Mainshaft Speed Sensor 'P'	AC Pulse Signals
B16	GRN/RED	A/T: Gear Position Switch	In 'R': 0v, Others: 12v
B17	GRN/YEL	A/T: Gear Position Switch	In D2: 0v, Others: 12v
B18	GRN/WHT	A/T: Gear Position Switch	In D1: 0v, Others: 12v
B22	BLU/YEL	Countershaft Speed Sensor N	<0.050v
B23	BLU/GRN	Countershaft Speed Sensor P	Moving: AC pulses
B24	GRN/BLK	A/T: Gear Position Switch	In D4: 0v, Others: 12v
B25	LT GRN	A/T: Gear Position Switch	In P/N: 0v, Others: 12v

1996 Coupe 2.2L I4 MFI VIN BA8 [All] 31P 'C' Connector

PCM Pin #	W/Color	Circuit Description (31-Pin)	Value at Hot Idle
C1	---	Not Used	---
C2	BLU	CKP Sensor Signal	AC pulse signals
C3	GRN	TDC Sensor Signal	AC pulse signals
C4	YEL	CYP Sensor Signal	AC pulse signals
C5	BLU/BLK	A/C Switch Signal	Switch Off: 12v, On: 1v
C6	BLU/RED	Starter Switch Signal	Cranking: 9-11v
C7	BRN/WHT	Service Check Connector	SCS Open: 4.80v
C8	LT GRN	K-Line Signal	12v
C9	---	Not Used	---
C10	WHT/YEL	Keep Alive Power (VBU)	12-14v
C12	WHT	CKP Sensor Ground	<0.050v
C13	RED	TDC Sensor Ground	<0.050v
C14	BLK	CYP Sensor Ground	<0.050v
C16	RED/GRN	PSP Switch	Straight: 0v, Turning: 11v
C17	WHT/RED	Alternator FR Signal	Digital Signals: 0-5-0v
C18	ORN	Vehicle Speed Sensor	Moving: pulse signals
C19	---	Not Used	---
C20	BRN	EVAP Purge Flow Switch	Switch on: 0v, off: 5v
C21-31	---	Not Used	---

1	2		3	4	5		6	7	8		1	2	3	4		5	6	7		8	9	10
9	10	11	12	13	14	15	16	17	18	11	12	13	14	15	16	17	18	19	20	21	22
19	20	21	22		23	24	25			23	24	25		26	27	28		29	30	31

25-PIN 'B' *WIRE SIDE OF HARNESS TERMINALS* 31-PIN 'C'

05533_ADIA_G630

Pin Connector Graphic

1996 Coupe 2.2L I4 VTEC VIN BB1 [All] 32P 'A' Connector

PCM Pin #	W/Color	Circuit Description (32-Pin)	Value at Hot Idle
A1	YEL	Injector 4 Control	2.0-3.3 ms
A2	BLU	Injector 3 Control	2.0-3.3 ms
A3	RED	Injector 2 Control	2.0-3.3 ms
A4	BRN	Injector 1 Control	2.0-3.3 ms
A5	ORN/BLK	HO2S-12 (B1 S2) Heater	Digital Signals: 0-12-0v
A6	ORN/WHT	HO2S-11 (B1 S1) Heater	Digital Signals: 0-12-0v
A7	RED	EGR Solenoid	Solenoid Off: 12v, On: 1v
A8	GRN/YEL	VTEC Solenoid	0v, Hi-Speed: 12v
A9	BLK/RED	Logic Ground	<0.050v
A10	BLK	Power Ground	<0.1v
A11	YEL/BLK	Main Relay Power (B+)	12-14v
A12	BLK/BLU	Idle Air Control Valve	Pulse Signals
A15	RED/YEL	EVAP Purge Solenoid	Solenoid Off: 12v, On: 1v
A16	GRN/BLK	Fuel Pump Relay	Relay Off: 12v, On: 1v
A17	RED/BLU	A/C Clutch Relay Control	Relay Off: 12v, On: 1v
A18	BLU/WHT	Check Engine Light	MIL Off: 12v, On: 1v
A19	WHT/GRN	Alternator Charging Signal	Lights Off: 12v, On: 0v
A20	YEL/GRN	Igniter Control	Digital Signals: 0-12-0v
A22	BLK/RED	Logic Ground	<0.050v
A23	BLK	Power Ground	<0.1v
A24	YEL/BLK	Main Relay Power (B+)	12-14v
A25	WHT	Intake Air Control Solenoid	Solenoid Off: 12v, On: 1v
A26	PNK	Intake Air Bypass Solenoid	Solenoid Off: 12v, On: 1v
A27	BLU/RED	Radiator Fan Relay Control	Relay Off: 12v, On: 1v

1996 Coupe 2.2L I4 VTEC VIN BB1 [All] 16P 'D' Connector

PCM Pin #	W/Color	Circuit Description (16-Pin)	Value at Hot Idle
D1	RED/BLK	TP Sensor Signal	0.5-0.6v
D2	YEL/BLU	ECT Sensor Signal	At 180°F: 0.5-0.6v
D3	WHT/YEL	MAP Sensor Signal	0.8-0.9v
D4	YEL/WHT	MAP Sensor VREF	4.9-5.1v
D5	GRN/WHT	Brake Switch Signal	Brake Off: 0v, On: 12v
D6	RED/BLU	Knock Sensor Signal	No Detonation: 18mv AC
D7	WHT	HO2S-11 (B1 S1) Signal	0.1-1.1v
D8	RED/YEL	IAT Sensor Signal	Varies w/temp. (0.5-4.9v)
D9	WHT/BLK	EGR Valve Lift Sensor	1.2v

1996 Coupe 2.2L I4 VTEC VIN BB1 [All] 16P 'D' Connector, *continued*

D10	YEL/WHT	Sensor VREF	4.9-5.1v
D11	GRN/WHT	Sensor Ground	<0.050v
D12	GRN/WHT	MAP Sensor Ground	<0.050v
D13	GRN/BLU	HO2S-12 Ground	<0.050v
D14	GRN/BLU	HO2S-12 (B1 S2) Signal	0.1-1.1v
D16	GRN/BLK	Electrical Load Detector	Varies: 0.5-4.5v

WIRE SIDE OF HARNESS TERMINALS

05533_ADIA_G610

Pin Connector Graphic

1996 Coupe 2.2L I4 VTEC VIN BB1 [All] 25P 'B' Connector

PCM Pin #	W/Color	Circuit Description (25-Pin)	Value at Hot Idle
B1-2	---	Not Used	---
B3	BLU/YEL	A/T: Shift Solenoid 'A'	SSA Off: 0v, On: 12v
B4	WHT/BLK	A/T: Lockup Solenoid 'B'	LSB On: 12v, Off: 0v
B5	RED/WHT	A/T: Lockup Solenoid 'A'	LSA On: 12v, Off: 0v
B6-7	---	Not Used	---
B8	GRN/BLU	A/T: Gear Position Switch	In D3: 0v, Others: 12v
B9-10	---	Not Used	---
B11	GRN/YEL	A/T: Shift Solenoid 'B'	SSB Off: 0v, On: 12v
B12	WHT/GRN	Interlock Control Unit Signal	Key & Brake On: 12v
B13	GRN/BLK	A/T: D4 Indicator Light Switch	D4 On: 0v, Off: 12v
B14	WHT/BLU	Mainshaft Speed Sensor 'N'	<0.050v
B15	ORN/BLU	Mainshaft Speed Sensor 'P'	AC Pulse Signals
B16	GRN/RED	A/T: Gear Position Switch	In 'R': 0v, Others: 12v
B17	GRN/YEL	A/T: Gear Position Switch	In D2: 0v, Others: 12v
B18	GRN/WHT	A/T: Gear Position Switch	In D1: 0v, Others: 12v
B19-21	---	Not Used	---
B22	BLU/YEL	Countershaft Speed Sensor N	<0.050v
B23	BLU/GRN	Countershaft Speed Sensor P	Moving: AC pulses
B24	GRN/BLK	A/T: Gear Position Switch	In D4: 0v, Others: 12v
B25	LT GRN	A/T: Gear Position Switch	In P/N: 0v, Others: 12v

1996 Coupe 2.2L I4 VTEC VIN BB1 [All] 31P 'C' Connector

PCM Pin #	W/Color	Circuit Description (31-Pin)	Value at Hot Idle
C2	BLU	CKP Sensor Signal	AC pulse signals
C3	GRN	TDC Sensor Signal	AC pulse signals
C4	YEL	CYP Sensor Signal	AC pulse signals
C5	BLU/BLK	A/C Switch Signal	Switch Off: 12v, On: 1v
C6	BLU/RED	Starter Switch Signal	Cranking: 9-11v
C7	BRN/WHT	Service Check Connector	SCS Open: 4.80v
C8	LT GRN	K-Line Signal	12v
C10	WHT/YEL	Keep Alive Power (VBU)	12-14v
C12	WHT	CKP Sensor Ground	<0.050v
C13	RED	TDC Sensor Ground	<0.050v
C14	BLK	CYP Sensor Ground	<0.050v
C15	LT BLU	VTEC Pressure Switch	0v, Hi-Speed: 12v
C16	RED/GRN	PSP Switch	Straight: 0v, Turning: 11v
C17	WHT/RED	Alternator 'FR' Signal	Digital Signals: 0-5-0v
C18	ORN	Vehicle Speed Sensor	Moving: pulse signals
C19	---	Not Used	---
C20	BRN	EVAP Purge Flow Switch	Switch on: 0v, off: 5v
C21-31	---	Not Used	---

05533_ADIA_G631

Pin Connector Graphic

1997-99 Coupe 2.2L I4 VTEC VIN BB6 [All] 32P 'A' Connector

PCM Pin #	W/Color	Circuit Description (32-Pin)	Value at Hot Idle
A1	YEL	Injector 4 Control	2.0-3.3 ms
A2	BLU	Injector 3 Control	2.0-3.3 ms
A3	RED	Injector 2 Control	2.0-3.3 ms
A4	BRN	Injector 1 Control	2.0-3.3 ms
A5	ORN/BLU	HO2S-12 (B1 S2) Heater	Digital Signals: 0-12-0v
A6	BLK/WHT	HO2S-11 (B1 S1) Heater	Digital Signals: 0-12-0v
A7	ORN	EGR Solenoid Control	Solenoid Off: 12v, On: 1v
A8	GRN/YEL	VTEC Solenoid Control	0v, Hi-Speed: 12v
A9	BRN/BLK	Logic Ground (LG1)	<0.050v
A10	BLK	Power Ground (PG1)	<0.1v
A11	YEL/BLK	Main Relay Power (IGP1)	12-14v
A12	BLK/BLU	Idle Air Control Valve	Pulse Signals
A13	---	Not Used	---
A14	---	Not Used	---
A15	RED/YEL	EVAP Purge Solenoid	Solenoid Off: 12v, On: 1v
A16	GRN/ORN	Fuel Pump Relay	Relay Off: 12v, On: 1v

1997-99 Coupe 2.2L I4 VTEC VIN BB6 [All] 32P 'A' Connector, *continued*

A17	PNK/BLU	A/C Clutch Relay Control	Relay Off: 12v, On: 1v
A18	GRY/RED	Malfunction Indicator Light	MIL Off: 12v, On: 1v
A19	WHT/GRN	Alternator Charging Signal	Lights Off: 12v, On: 0v
A20	YEL/GRN	Igniter Control	Digital Signals: 0-12-0v
A21	---	Not Used	---
A22	BRN/BLK	Logic Ground (LG2)	<0.050v
A23	BLK	Power Ground (LG1)	<0.1v
A24	YEL/BLK	Main Relay Power (IGP2)	12-14v
A25	WHT	Intake Air Control Solenoid	Solenoid Off: 12v, On: 1v
A26	RED/BLU	Intake Air Bypass Solenoid	Solenoid Off: 12v, On: 1v
A27	GRN	Radiator Fan Relay Control	Relay Off: 12v, On: 1v
A28	GRN/WHT	EVAP Bypass Solenoid	Solenoid Off: 12v, On: 1v
A29	ORN/GRN	EVAP Vent Solenoid	Solenoid Off: 12v, On: 1v
A30-32	---	Not Used	---

05533_ADIA_G632

Pin Connector Graphic

1997-99 Coupe 2.2L I4 VTEC VIN BB6 31P 'C' Connector

PCM Pin #	W/Color	Circuit Description (31-Pin)	Value at Hot Idle
C1	GRN/BLK	VREF Signal to TCM	4.9-5.1v
C2	BLU	CKP Sensor Signal	AC pulse signals
C3	GRN	TDC Sensor Signal	AC pulse signals
C4	YEL	CYP Sensor Signal	AC pulse signals
C5	BLU/ORN	A/C Switch Signal	Switch Off: 12v, On: 1v
C6	BLU/RED	Starter Switch Signal	Cranking: 9-11v
C7	RED/WHT	Service Check Connector	SCS Open: 4.80v
C8	LT GRN	K-Line Signal	12v
C9	---	Not Used	---
C10	WHT/YEL	Keep Alive Power (VBU)	12-14v
C11	---	Not Used	---
C12	WHT	CKP Sensor Ground	<0.050v
C13	RED	TDC Sensor Ground	<0.050v
C14	BLK	CYP Sensor Ground	<0.050v
C15	BLU/BLK	VTEC Pressure Switch	0v, Hi-Speed: 12v
C16	GRN	PSP Switch	Straight: 0v, Turning: 11v
C17	WHT/GRN	Alternator FR Signal	Digital Signals: 0-5-0v
C18	BLU/WHT	Vehicle Speed Sensor	Moving: pulse signals
C19	---	Not Used	---
C20 ('97)	BRN	EVAP Purge Flow Switch	Switch On: 0v, Off: 5v
C22	BRN/YEL	Immobilizer Code Signal	Digital Signals
C23-28	---	Not Used	---
C29	LT GRN	PNP Switch Signal	In P/N: 0v, Others: 12v
C30	GRN/BLU	A/T: TCM SEAF, FITX Signal	Pulse Signals
C31	GRN/YEL	A/T: TCM SEFA, FIRX Signal	Pulse Signals

1997-99 Coupe 2.2L I4 VTEC VIN BB6 16P 'D' Connector

PCM Pin #	W/Color	Circuit Description (16-Pin)	Value at Hot Idle
D1	RED/BLK	TP Sensor Signal	0.5-0.6v
D2	RED/WHT	ECT Sensor Signal	At 180°F: 0.5-0.6v
D3	RED/GRN	MAP Sensor Signal	0.8-0.9v
D4	YEL/RED	MAP Sensor VREF (VCC1)	4.9-5.1v
D5	WHT/BLK	Brake Switch Signal	Brake Off: 0v, On: 12v
D6	RED/BLU	Knock Sensor Signal	No Detonation: 18mv AC
D7	WHT	HO2S-11 (B1 S1) Signal	0.1-1.1v
D8	RED/YEL	IAT Sensor Signal	Varies w/temp. (0.5-4.9v)
D9	WHT/BLK	EGR Valve Lift Sensor	1.2v
D10	YEL/BLU	Sensor VREF	4.9-5.1v
D11	GRN/BLK	Sensor Ground (SG2)	<0.050v
D12	GRN/WHT	MAP Sensor Ground (SG1)	<0.050v
D14	WHT/RED	HO2S-12 (B1 S2) Signal	0.1-1.1v
D15	WHT/BLU	Fuel Tank Pressure Sensor	Fuel Cap off: 2.5v
D16	GRN/RED	Electrical Load Detector	Varies: 0.5-4.5v

2000-01 Coupe 2.2L I4 VTEC VIN BB6 [All] 32P 'A' Connector

PCM Pin #	W/Color	Circuit Description (32-Pin)	Value at Hot Idle
A1	YEL	Injector 4 Control	2.0-3.3 ms
A2	BLU	Injector 3 Control	2.0-3.3 ms
A3	RED	Injector 2 Control	2.0-3.3 ms
A4	BRN	Injector 1 Control	2.0-3.3 ms
A5	ORN/BLU	HO2S-12 (B1 S2) Heater	Digital Signals: 0-12-0v
A6	BLK/WHT	HO2S-11 (B1 S1) Heater	Digital Signals: 0-12-0v
A7	ORN	Electronic EGR Solenoid	Digital Signals: 0-12-0v
A8	GRN/YEL	VTEC Solenoid Control	0v, Hi-Speed: 12v
A9	BRN/BLK	Logic Ground (LG1)	<0.050v
A10	BLK	Power Ground (PG1)	<0.1v
A11	YEL/BLK	Main Relay Power (IGP1)	12-14v
A12	BLK/BLU	Idle Air Control Valve	Pulse Signals
A13	---	Not Used	---
A14	---	Not Used	---
A15	RED/YEL	EVAP Purge Solenoid	Solenoid Off: 12v, On: 1v
A16	GRN/ORN	Fuel Pump Relay	Relay Off: 12v, On: 1v
A17	PNK/BLU	A/C Clutch Relay Control	Relay Off: 12v, On: 1v
A18	GRY/RED	Malfunction Indicator Light	MIL Off: 12v, On: 1v
A19	WHT/GRN	Alternator Charging Signal	Lights Off: 12v, On: 0v
A20	YEL/GRN	Igniter Control	Digital Signals: 0-12-0v
A21	---	Not Used	---
A22	BRN/BLK	Logic Ground (LG2)	<0.050v
A23	BLK	Power Ground (LG1)	<0.1v
A24	YEL/BLK	Main Relay Power (IGP2)	12-14v
A25	WHT	Intake Air Control Solenoid	Solenoid Off: 12v, On: 1v
A26	RED/BLU	Intake Air Bypass Solenoid	Solenoid Off: 12v, On: 1v
A27	GRN	Radiator Fan Relay Control	Relay Off: 12v, On: 1v
A28	GRN/WHT	EVAP Bypass Solenoid	Solenoid Off: 12v, On: 1v
A29	ORN/GRN	EVAP Vent Solenoid	Solenoid Off: 12v, On: 1v
A30-32	---	Not Used	---

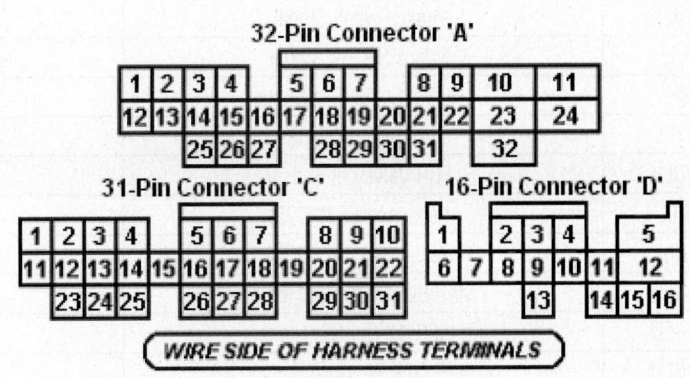

Pin Connector Graphic

2000-01 Coupe 2.2L I4 VTEC VIN BB6 [All] 31P 'C' Connector

PCM Pin #	W/Color	Circuit Description (31-Pin)	Value at Hot Idle
C1	GRN/BLK	VREF Signal to TCM	4.9-5.1v
C2	BLU	CKP Sensor Signal	AC pulse signals
C3	GRN	TDC Sensor Signal	AC pulse signals
C4	YEL	CYP Sensor Signal	AC pulse signals
C5	BLU/ORN	A/C Switch Signal	Switch Off: 12v, On: 1v
C6	BLU/RED	Starter Switch Signal	Cranking: 9-11v
C7	RED/WHT	Service Check Connector	SCS Open: 4.80v
C8	LT GRN	K-Line Signal	12v
C9	---	Not Used	---
C10	WHT/YEL	Keep Alive Power (VBU)	12-14v
C11	---	Not Used	---
C12	WHT	CKP Sensor Ground	<0.050v
C13	RED	TDC Sensor Ground	<0.050v
C14	BLK	CYP Sensor Ground	<0.050v
C15	BLU/BLK	VTEC Pressure Switch	0v, Hi-Speed: 12v
C16	GRN	PSP Switch	Straight: 0v, Turning: 11v
C17	WHT/GRN	Alternator FR Signal	Digital Signals: 0-5-0v
C18	BLU/WHT	Vehicle Speed Sensor	Moving: pulse signals
C19-20	---	Not Used	---
C22	BRN/YEL	Immobilizer Code Signal	Digital Signals
C23-28	---	Not Used	---
C29	LT GRN	A/T: PNP Switch Signal	In P/N: 0v, Others: 12v
C30	GRN/BLU	A/T: TCM SEAF, FITX Signal	Pulse Signals
C31	GRN/YEL	A/T: TCM SEFA, FIRX Signal	Pulse Signals

2000-01 Coupe 2.2L I4 VTEC VIN BB6 [All] 16P 'D' Connector

PCM Pin #	W/Color	Circuit Description (16-Pin)	Value at Hot Idle
D1	RED/BLK	TP Sensor Signal	0.5-0.6v
D2	RED/WHT	ECT Sensor Signal	At 180°F: 0.5-0.6v
D3	RED/GRN	MAP Sensor Signal	0.8-0.9v
D4	YEL/RED	MAP Sensor VREF (VCC1)	4.9-5.1v
D5	WHT/BLK	Brake Switch Signal	Brake Off: 0v, On: 12v
D6	RED/BLU	Knock Sensor Signal	No Detonation: 18mv AC
D7	WHT	HO2S-11 (B1 S1) Signal	0.1-1.1v
D8	RED/YEL	IAT Sensor Signal	Varies w/temp. (0.5-4.9v)
D9	WHT/BLK	EGR Valve Lift Sensor	1.2v
D10	YEL/BLU	Sensor VREF (VCC2)	4.9-5.1v
D11	GRN/BLK	Sensor Ground (SG2)	<0.050v
D12	GRN/WHT	MAP Sensor Ground (SG1)	<0.050v
D14	WHT/RED	HO2S-12 (B1 S2) Signal	0.1-1.1v
D15	WHT/BLU	Fuel Tank Pressure Sensor	Fuel Cap off: 2.5v
D16	GRN/RED	Electrical Load Detector	Varies: 0.5-4.5v

1992 Coupe 2.3L I4 MFI VIN BB2 [All] 26P 'A' Connector

PCM Pin #	W/Color	Circuit Description (26-Pin)	Value at Hot Idle
A1	BRN	Injector 1 Control	2.0-3.3 ms
A2	YEL	Injector 4 Control	2.0-3.3 ms
A3, 5	RED, BLU	Injector 2 Control, 3	2.0-3.3 ms
A6	ORN/WHT	HO2S Heater Control	Relay Off: 12v, On: 1v
A7, A8	GRN/BLK	Fuel Pump Relay	Relay Off: 12v, On: 1v
A9	BLK/BLU	Electronic Air Control Valve	Pulse Signals
A11	RED	EGR Solenoid	Solenoid Off: 12v, On: 1v
A12	BLU/RED	Radiator Fan Relay Control	Relay Off: 12v, On: 1v
A13	BLU/WHT	Check Engine Light	MIL Off: 12v, On: 1v
A15	RED/BLU	A/C Compressor Unit	---
A16	WHT/GRN	Alternator Charging Signal	Lights Off: 12v, On: 0v
A17	PNK	Intake Air Bypass Solenoid	Solenoid Off: 12v, On: 1v
A18	ORN/RED	A/T: TCM (shift acknowledge)	N/A
A19	WHT	Intake Air Control Solenoid	Solenoid Off: 12v, On: 1v
A20	RED/GRN	EVAP Purge Solenoid	Solenoid Off: 12v, On: 1v
A21, 22	YEL/GRN	Igniter Control	Digital Signals: 0-12-0v
A23, 24	BLK	Power Ground	<0.1v
A25	YEL/BLK	Main Relay Power (B+)	12-14v
A26	BLK/RED	Power Ground	<0.1v

1992 Coupe 2.3L I4 MFI VIN BB2 [All] 16P 'B' Connector

PCM Pin #	W/Color	Circuit Description (16-Pin)	Value at Hot Idle
B1	YEL/BLK	Main Relay Power (B+)	12-14v
B2	BRN/BLK	Power Ground	<0.1v
B3	ORN	A/T: TCM (Upshift compare)	N/A
B4	PNK	A/T: TCM (downshift)	N/A
B5	BLU/BLK	A/C Switch Signal	Switch Off: 12v, On: 1v
B7	LT GRN	A/T: Park/Neutral Switch	In P/N: 0v, Others: 12v
B8	RED/GRN	PSP Switch	Straight: 0v, Turning: 11v
B9	BLU/RED	Starter Switch Signal	Cranking: 9-11v
B10	ORN	Vehicle Speed Sensor	Moving: pulse signals
B11	ORN	CYP Sensor Signal	AC pulse signals
B12	WHT	CYP Sensor Ground	<0.050v
B13	ORN/BLU	TDC Sensor Signal	AC pulse signals
B14	WHT/BLU	TDC Sensor Ground	<0.050v
B15	BLU/GRN	CKP Sensor Signal	AC pulse signals
B16	BLU/YEL	CKP Sensor Ground	<0.050v

Standard Colors and Abbreviations

Abbreviation	Color	Abbreviation	Color	Abbreviation	Color
BLK	Black	LT BLU	Lt. Blue	TAN	Tan
BLU	Blue	LT GRN	Lt. Green	VIO	Violet
BRN	Brown	ORN	Orange	WHT	White
GRY	Gray	PNK	Pink	YEL	Yellow
GRN	Green	PPL	Purple		

1992 Coupe 2.3L I4 MFI VIN BB2 [All] 22P 'D' Connector

PCM Pin #	W/Color	Circuit Description (22-Pin)	Value at Hot Idle
D1	WHT/YEL	Keep Alive Power (VBU)	12-14v
D2	GRN/WHT	Brake Switch Signal	Brake Off: 0v, On: 12v
D3	RED/BLU	Knock Sensor Signal	No Detonation: 18mv AC
D4	BRN/WHT	Service Check Connector	SCS Open: 4.80v
D5	---	Not Used	---
D6	---	Not Used	---
D7	GRN/RED	Data Link Connector	5v
D8	---	Not Used	---
D9	WHT/RED	Alternator 'FR' Signal	Digital Signals: 0-5-0v
D10	---	Not Used	---
D11	RED/BLK	TP Sensor Signal	0.5-0.6v
D12	WHT/BLK	EGR Valve Lift Sensor	1.2v
D13	YEL/BLU	ECT Sensor Signal	At 180°F: 0.5-0.6v
D14	WHT	O2S-11 Signal	0.1-1.1v
D15	RED/YEL	IAT Sensor Signal	Varies w/temp. (0.5-4.9v)
D16	---	Not Used	---
D17	WHT/BLU	MAP Sensor Signal	0.8-0.9v
D18	GRN/BLK	A/T: Control Link VREF	4.9-5.1v
D19	RED/WHT	Sensor VREF	4.9-5.1v
D20	YEL/WHT	Sensor VREF	4.9-5.1v
D21	BLU/WHT	Sensor Ground	<0.050v
D22	GRN/WHT	Sensor Ground	<0.050v

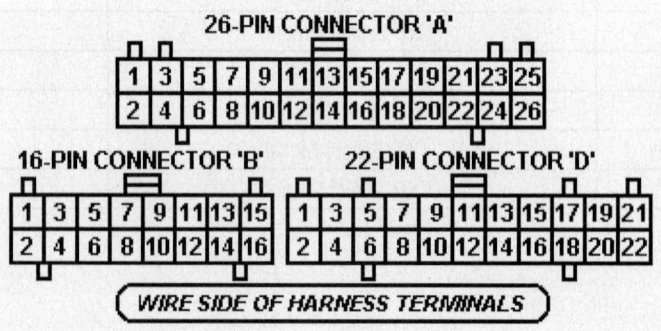

Pin Connector Graphic

05533_ADIA_G597

1993-95 Coupe 2.3L I4 MFI VIN BB2 [All] 26P 'A' Connector

PCM Pin #	W/Color	Circuit Description (26-Pin)	Value at Hot Idle
A1	BRN	Injector 1 Control	2.0-3.3 ms
A2	YEL	Injector 4 Control	2.0-3.3 ms
A3	RED	Injector 2 Control	2.0-3.3 ms
A4	---	Not Used	---
A5	BLU	Injector 3 Control	2.0-3.3 ms
A6	ORN/WHT	HO2S Heater Control	Relay Off: 12v, On: 1v
A7	GRN/BLK	Fuel Pump Relay	Relay Off: 12v, On: 1v
A8	GRN/BLK	Fuel Pump Relay	Relay Off: 12v, On: 1v
A9	BLK/BLU	Idle Air Control Valve	Pulse Signals
A10	---	Not Used	---
A11	RED	EGR Solenoid	Solenoid Off: 12v, On: 1v
A12	BLU/RED	Radiator Fan Relay Control	Relay Off: 12v, On: 1v
A13	BLU/WHT	Check Engine Light	MIL Off: 12v, On: 1v
A14	---	Not Used	---
A15	RED/BLU	A/C Compressor Unit	Solenoid Off: 12v, On: 1v
A16	WHT/GRN	Alternator Charging Signal	Lights Off: 12v, On: 0v
A17	PNK	Intake Air Bypass Solenoid	Solenoid Off: 12v, On: 1v
A18	ORN/RED	A/T: TCM Signal	Digital Signals
A19	WHT	Intake Control Solenoid	Solenoid Off: 12v, On: 1v
A20	RED/GRN	EVAP Purge Solenoid	Solenoid Off: 12v, On: 1v
A21	YEL/GRN	Igniter Control	Digital Signals: 0-12-0v
A22	YEL/GRN	Igniter Control	Digital Signals: 0-12-0v
A23	BLK	Power Ground	<0.1v
A24	BLK	Power Ground	<0.1v
A25	YEL/BLK	Main Relay Power (B+)	12-14v
A26	BLK/RED	Power Ground	<0.1v

1993-95 Coupe 2.3L I4 MFI VIN BB2 [All] 16P 'B' Connector

PCM Pin #	W/Color	Circuit Description (16-Pin)	Value at Hot Idle
B1	YEL/BLK	Main Relay Power (B+)	12-14v
B2	BRN/BLK	Power Ground	<0.1v
B3	ORN	A/T: TCM Signal (Upshift)	Digital Signals
B4	PNK	A/T: TCM Signal (downshift)	Digital Signals
B5	BLU/BLK	A/C Switch Signal	Switch Off: 12v, On: 1v
B6	---	Not Used	---
B7	LT GRN	A/T: Park Neutral Switch	P/N: 0v, others: 5v
B8	RED/GRN	PSP Switch	Straight: 0v, Turning: 11v
B9	BLU/RED	Starter Switch Signal	Cranking: 9-11v
B10	ORN	Vehicle Speed Sensor	Moving: pulse signals
B11	ORN	CYP Sensor Signal	AC pulse signals
B12	WHT	CYP Sensor Ground	<0.050v
B13	ORN/BLU	TDC Sensor Signal	AC pulse signals
B14	WHT/BLU	TDC Sensor Ground	<0.050v
B15	BLU/GRN	CKP Sensor Signal	AC pulse signals
B16	BLU/YEL	CKP Sensor Ground	<0.050v

1993-95 Coupe 2.3L I4 MFI VIN BB2 [All] 22P 'D' Connector

PCM Pin #	W/Color	Circuit Description (22-Pin)	Value at Hot Idle
D1	WHT/YEL	Keep Alive Power (VBU)	12-14v
D2	GRN/WHT	Brake Switch Signal	Brake Off: 0v, On: 12v
D3	RED/BLU	Knock Sensor Signal	No Detonation: 18mv AC
D4	BRN/WHT	Service Check Connector	SCS Open: 4.80v
D5-6	---	Not Used	---
D7	GRN/RED	Data Link Connector	5v
D8	---	Not Used	---
D9	WHT/RED	Alternator 'FR' Signal	Digital Signals: 0-5-0v
D10	GRN/RED	Electrical Load Detector	Varies: 0.5-4.5v
D11	RED/BLK	TP Sensor Signal	0.5-0.6v
D12	WHT/BLK	EGR Valve Lift Sensor	1.2v
D13	YEL/BLU	ECT Sensor Signal	At 180°F: 0.5-0.6v
D14	WHT	HO2S Signal	0.1-1.1v
D15	RED/YEL	IAT Sensor Signal	Varies w/temp. (0.5-4.9v)
D16	---	Not Used	---
D17	WHT/BLU	MAP Sensor Signal	0.8-0.9v
D18	GRN/BLK	A/T: Control Unit VRED	4.9-5.1v
D19	RED/WHT	Sensor VREF	4.9-5.1v
D20	YEL/WHT	Sensor VREF	4.9-5.1v
D21	BLU/WHT	Sensor Ground	<0.050v
D22	GRN/WHT	Sensor Ground	<0.050v

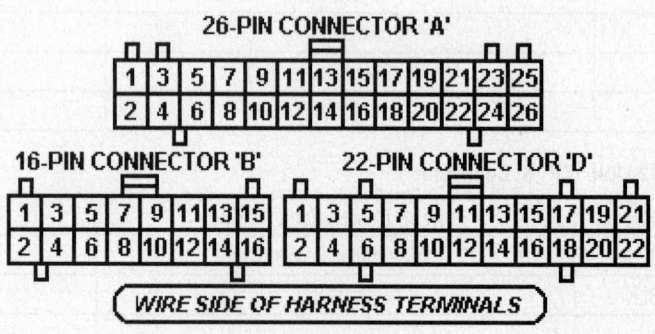

05533_ADIA_G597

Pin Connector Graphic

1996 Coupe 2.3L I4 MFI VIN BB2 [All] 32P 'A' Connector

PCM Pin #	W/Color	Circuit Description (32-Pin)	Value at Hot Idle
A1	YEL	Injector 1 Control	2.0-3.3 ms
A2	BLU	Injector 2 Control	2.0-3.3 ms
A3	RED	Injector 3 Control	2.0-3.3 ms
A4	BRN	Injector 4 Control	2.0-3.3 ms
A5	ORN/BLK	HO2S-12 (B1 S2) Heater	Digital Signals: 0-12-0v
A6	ORN/WHT	HO2S-11 (B1 S1) Heater	Digital Signals: 0-12-0v
A7	RED	EGR Solenoid	Solenoid Off: 12v, On: 1v
A9, A22	BLK, BRN	Logic Ground	<0.050v
A10, A23	BLK	Power Ground	<0.1v

1996 Coupe 2.3L I4 MFI VIN BB2 [All] 32P 'A' Connector, *continued*

A11	YEL/BLK	Main Relay Power (B+)	12-14v
A12	BLK/BLU	Idle Air Control Valve	Pulse Signals
A15	RED/YEL	EVAP Purge Solenoid	Solenoid Off: 12v, On: 1v
A16	GRN/BLK	Fuel Pump Relay	Relay Off: 12v, On: 1v
A17	RED/BLU	A/C Clutch Relay Control	Relay Off: 12v, On: 1v
A18	BLU/WHT	Check Engine Light	MIL Off: 12v, On: 1v
A19	WHT/GRN	Alternator Charging Signal	Lights Off: 12v, On: 0v
A20	YEL/GRN	Igniter Control	Digital Signals: 0-12-0v
A24	YEL/BLK	Main Relay Power (B+)	12-14v
A25	WHT	Intake Air Control Solenoid	Solenoid Off: 12v, On: 1v
A26	PNK	Intake Air Bypass Solenoid	Solenoid Off: 12v, On: 1v
A27	BLU/RED	Radiator Fan Relay Control	Relay Off: 12v, On: 1v
A28	PNK	Idle Air Bypass Solenoid	Solenoid Off: 12v, On: 1v

1996 Coupe 2.3L I4 MFI VIN BB2 [All] 25P 'B' Connector

PCM Pin #	W/Color	Circuit Description (25-Pin)	Value at Hot Idle
B3	BLU/YEL	A/T: Shift Solenoid 'A'	SSA Off: 0v, On: 12v
B4	WHT/BLK	A/T: Lockup Solenoid 'B'	LSB On: 12v, Off: 0v
B5	RED/WHT	A/T: Lockup Solenoid 'A'	LSA On: 12v, Off: 0v
B8	GRN/BLU	A/T: Gear Position Switch	In D3: 0v, Others: 12v
B11	GRN/YEL	A/T: Shift Solenoid 'B'	SSB Off: 0v, On: 12v
B12	WHT/GRN	Interlock Control Unit Signal	Key & Brake On: 12v
B13	GRN/BLK	A/T: D4 Indicator Light Switch	D3 on: 0v, off: 12v
B14	WHT/BLU	Mainshaft Speed Sensor 'N'	<0.050v
B15	ORN/BLU	Mainshaft Speed Sensor 'P'	Moving: AC pulses
B16	GRN/RED	A/T: Gear Position Switch	In 'R': 0v, Others: 12v
B17, 18	GRN/YEL	A/T: Gear Position Switch	In D2, D1: 0v
B22	BLU/YEL	Countershaft Speed Sensor N	<0.050v
B23	BLU/GRN	Countershaft Speed Sensor P	Moving: AC pulses
B24	GRN/BLK	A/T: Gear Position Switch	In D4: 0v, Others: 12v
B25	LT GRN	A/T: Gear Position Switch	In P/N: 0v, Others: 12v

32-PIN CONNECTOR 'A'   **25-PIN CONNECTOR 'B'**

1	2	3	4		5	6	7		8	9	10	11

WIRE SIDE OF HARNESS TERMINALS

05533_ADIA_G604

Pin Connector Graphic

1996 Coupe 2.3L I4 MFI VIN BB2 [All] 31P 'C' Connector

PCM Pin #	W/Color	Circuit Description (31-Pin)	Value at Hot Idle
C2	BLU	CKP Sensor Signal	AC pulse signals
C3	GRN	TDC Sensor Signal	AC pulse signals
C4	YEL	CYP Sensor Signal	AC pulse signals
C5	BLU/BLK	A/C Switch Signal	Switch Off: 12v, On: 1v
C6	BLU/RED	Starter Switch Signal	Cranking: 9-11v
C7	BRN/WHT	Service Check Connector	SCS Open: 4.80v
C8	LT GRN	K-Line Signal	12v
C10	WHT/YEL	Keep Alive Power (VBU)	12-14v
C12	WHT	CKP Sensor Ground	<0.050v
C13	RED	TDC Sensor Ground	<0.050v
C14	BLK	CYP Sensor Ground	<0.050v
C16	RED/GRN	PSP Switch	Straight: 0v, Turning: 11v
C17	WHT/RED	Alternator 'FR' Signal	Digital Signals: 0-5-0v
C18	ORN	Vehicle Speed Sensor	Moving: pulse signals
C20	BRN	EVAP Purge Flow Switch	Switch on: 0v, off: 5v

1996 Coupe 2.3L I4 MFI VIN BB2 [All] 16P 'D' Connector

PCM Pin #	W/Color	Circuit Description (16-Pin)	Value at Hot Idle
D1	RED/BLK	TP Sensor Signal	0.5-0.6v
D2	YEL/BLU	ECT Sensor Signal	At 180°F: 0.5-0.6v
D3	WHT/YEL	MAP Sensor Signal	0.8-0.9v
D4	YEL/WHT	MAP Sensor VREF	4.9-5.1v
D5	GRN/WHT	Brake Switch Signal	Brake Off: 0v, On: 12v
D6	RED/BLU	Knock Sensor Signal	No Detonation: 18mv AC
D7	WHT	HO2S-11 (B1 S1) Signal	0.1-1.1v
D8	RED/YEL	IAT Sensor Signal	Varies w/temp. (0.5-4.9v)
D9	WHT/BLK	EGR Valve Lift Sensor	1.2v
D10	YEL/WHT	Sensor VREF	4.9-5.1v
D11	GRN/WHT	Sensor Ground	<0.050v
D12	GRN/WHT	MAP Sensor Ground	<0.050v
D13	GRN/BLU	HO2S-12 Ground	<0.050v
D14	GRN/BLU	HO2S-12 (B1 S2) Signal	Accel: 0.5-1v
D16	GRN/BLK	Electrical Load Detector	Varies: 0.5-4.5v

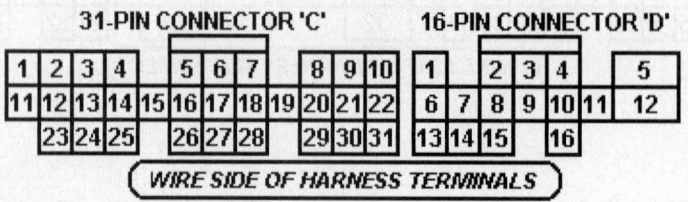

Pin Connector Graphic

05533_ADIA_G605

S2000 PIN CHARTS

2000-03 Convertible 2.0L I4 VTEC VIN API [M/T] 32P A Connector

PCM Pin #	W/Color	Circuit Description (32-Pin)	Value at Hot Idle
A1	YEL/GRN	ECT Sensor Signal to Gauge	Digital Signals
A2	RED	Air Control Solenoid Valve	Solenoid Off: 12v, On: 1v
A3	ORN	EVAP Bypass Solenoid	Solenoid Off: 12v, On: 1v
A4	GRN/WHT	EVAP Vent Solenoid	Solenoid Off: 12v, On: 1v
A5	---	Not Used	---
A6	RED/YEL	EVAP Purge Solenoid	Solenoid Off: 12v, On: 1v
A7-8	---	Not Used	---
A9	BLU/WHT	Vehicle Speed Output Signal	Digital Pulses
A10	BRN	Service Check Connector	SCS Open: 4.80v
A11	---	Not Used	---
A12	PNK	Immobilizer Indicator Lamp	Lamp Off: 12v, On: 1v
A13	PNK/BLU	Immobilizer Enable Signal	Digital Signals
A14	---	Not Used	---
A15	GRN/YEL	Immobilizer Fuel Pump Relay	2 sec's after startup: 12v
A16	---	Not Used	---
A17	RED	A/C Clutch Relay Control	Relay Off: 12v, On: 1v
A18	GRN/ORN	Malfunction Indicator Light	MIL Off: 12v, On: 1v
A19	BLU	Engine Speed Pulse (NEP)	Digital Signals
A20	GRN	Radiator Fan Relay Control	Relay Off: 12v, On: 1v
A21	GRY	K-Line Signal	12v
A22-23	---	Not Used	---
A24	BLU/ORN	Starter Switch Signal	Cranking: 9-11v
A25	RED/BLU	Immobilizer Code Signal	Digital Signals
A26	BLU/BLK	PSP Switch Signal	Straight: 0v, Turning: 11v
A27	BLU/RED	A/C Switch Signal	Switch Off: 12v, On: 1v
A28	BLU	Air Pump Relay	Relay Off: 12v, On: 1v
A29	LT GRN	Fuel Tank Pressure Sensor	Fuel Cap off: 2.5v
A30	GRN/RED	Electric Load Detector	Varies: 0.5-4.5v
A31	---	Not Used	---
A32	WHT/BLK	Brake Switch Signal	Brake Off: 0v, On: 12v

32-Pin 'A'

1	2	3	4		5	6	7		8	9	10	11
12	13	14	15	16	17	18	19	20	21	22	23	24
	25	26	27			28	29	30	31		32	

WIRE SIDE OF HARNESS TERMINALS

05533_ADIA_G634

Pin Connector Graphic

2000-03 Convertible 2.0L I4 VTEC VIN API [M/T] 25P B Connector

PCM Pin #	W/Color	Circuit Description (31-Pin)	Value at Hot Idle
B1	YEL/BLK	Main Relay Power (B+)	12-14v
B2	BLK	Power Ground (PG1)	<0.1v
B3	RED, BLU	Injector 2, Injector 3 Control	2.0-3.3 ms
B4	RED, BLU	Injector 2, Injector 3 Control	2.0-3.3 ms
B5	YEL	Injector 4 Control	2.0-3.3 ms
B7	---	Not Used	---
B8	---	Not Used	---
B9	YEL/BLK	Main Relay Power (B+)	12-14v
B10	BLK	Power Ground (PG1)	<0.1v
B11	BRN	Injector 1 Control	2.0-3.3 ms
B12	GRN/YEL	VTEC Control Solenoid	0v, Hi-Speed: 12v
B13	---	Not Used	---
B14	---	Not Used	---
B15	---	Not Used	---
B16	---	Not Used	---
B17	---	Not Used	---
B18	---	Not Used	---
B19	---	Not Used	---
B20	BRN/YEL	Logic Ground (LG1)	<0.050v
B21	WHT/RED	Keep Alive Power (VBU)	12-14v
B22	BRN/YEL	Logic Ground (LG1)	<0.050v
B23	BLK/RED	Idle Air Control Valve	Pulse Signals
B24	---	Not Used	---
B25	---	Not Used	---

25-Pin 'B'

WIRE SIDE OF HARNESS TERMINALS

05533_ADIA_G635

Pin Connector Graphic

2000-03 Convertible 2.0L I4 VTEC VIN API [M/T] 31P C Connector

PCM Pin #	W/Color	Circuit Description (31-Pin)	Value at Hot Idle
C1	BLK/WHT	HO2S-11 (B1 S1) Heater	Digital Signals: 0-12-0v
C2	WHT/GRN	Alternator Charging Signal	Lights Off: 12v, On: 0v
C4	WHT	Ignition Coil 1 Control	Digital Signals: 0-12-0v
C5	WHT/RED	Alternator 'FR' Signal	Digital Signals: 0-5-0v
C7	GRN/WHT	MAP Sensor Ground (SG1)	<0.050v
C8	BLU	CKP Sensor Signal	AC pulse signals
C9	WHT	CKP Sensor Ground	<0.050v
C10	BLU/BLK	VTEC Pressure Switch	0v, Hi-Speed: 12v
C11	BLK/WHT	HO2S-12 (B1 S2) Heater	Digital Signals: 0-12-0v
C12	WHT/GRN	Ignition Coil 2 Control	Digital Signals: 0-12-0v
C13	WHT/BLK	Ignition Coil 3 Control	Digital Signals: 0-12-0v
C14	WHT/BLU	Ignition Coil 4 Control	Digital Signals: 0-12-0v
C15	WHT/RED	HO2S-12 (B1 S2) Signal	0.1-1.1v
C16	WHT	HO2S-11 (B1 S1) Signal	0.1-1.1v
C17	GRN/RED	MAP Sensor Signal	0.8-0.9v
C18	GRN/YEL	Sensor Ground (SG2)	<0.050v
C19	YEL/RED	MAP Sensor VREF (VCC1)	4.9-5.1v
C20	GRN	TDC1 Sensor Signal	AC pulse signals
C21	RED	TDC1 Sensor Ground	<0.050v
C22	RED/BLU	Knock Sensor Signal	No Detonation: 18mv AC
C24	WHT/BLK	Air Pump Current Sensor	Pump on: 2-5v
C25	RED/YEL	IAT Sensor Signal	Varies w/temp (0.5-4.9v)
C26	RED/WHT	ECT Sensor Signal	At 180°F: 0.5-0.6v
C27	RED/BLK	TP Sensor Signal	0.3-0.4v
C28	YEL/BLU	Sensor VREF (VCC2)	4.9-5.1v
C29	YEL	TDC2 Sensor Signal	AC pulse signals
C30	BLK	TDC2 Sensor Ground	<0.050v
C31	---	Not Used	---

31-Pin 'C'

1	2	3	4		5	6	7		8	9	10
11	12	13	14	15	16	17	18	19	20	21	22
	23	24	25		26	27	28		29	30	31

WIRE SIDE OF HARNESS TERMINALS

05533_ADIA_G636

Pin Connector Graphic

PASSPORT PIN CHARTS

1994-95 DX 2.6L I4 MFI VIN E 2WD [M/T] "A28" 26P 'A' Connector

PCM Pin #	W/Color	Circuit Description (26-Pin)	Value at Hot Idle
A28-1	GRN/RED	B+ from Main Relay	12-14v
A28-2	BLK/GRN	Power Ground	<0.1v
A28-3	BLK/YEL	Data Link Connector	4.80v (no Scan Tool)
A28-4	BLU/YEL	Throttle Idle Switch	Idle: 12v, off-idle: 0.1v
A28-5	GRN/BLK	A/C Switch Signal	Switch on: 12v, off: 0v
A28-6	---	Not Used	---
A28-7	GRN/BLU	CKP Sensor Signal	Varies 2.0-3.0v
A28-8	GRN/YEL	Fuel Pump Relay Control	Relay on: 12v, off: 0v
A28-9	BRN/YEL	Air Management Solenoid	Sol. on: 1v. off: 12v
A28-10	LT GRN	EGR Duty Cycle Solenoid	Digital Signals: 0-12-0v
A28-11	---	Not Used	---
A28-12	WHT/BLK	Injector 2 Control	2.0-3.3 ms
Q	PUR/WHT	Injector 1 Control	2.0-3.3 ms
A28-14	GRN/BLK	B+ from Main Relay	12-14v
A28-15	---	Not Used	---
A28-16	---	Not Used	---
A28-17	---	Not Used	---
A28-18	GRN/ORN	Monitor Signals	Digital Signals
A28-19	GRN/YEL	Monitor Signals	Digital Signals
A28-20	WHT	CKP Sensor VREF	4.9-5.1v
A28-21	WHT/GRN	EGR Vacuum Control Valve	Sol on: 1v, off: 12v
A28-22	GRN	Malfunction Indicator Lamp	MIL Off: 12v, On: 1v
A28-23	GRN/WHT	EVAP Purge Solenoid	Digital Signals: 0-12-0v
A28-24	WHT/BLU	Injector 4 Control	2.0-3.3 ms
A28-25	WHT/RED	Injector 3 Control	2.0-3.3 ms
A28-26	BLK/GRN	Logic Ground	<0.050v

A28 26-PIN CONNECTOR

1	2	3	4	5	6	7	8	9	10	11	12	13
14	15	16	17	18	19	20	21	22	23	24	25	26

View is into back of Wire Harness Connector

05533_ADIA_G647

Pin Connector Graphic

Standard Colors and Abbreviations

Abbreviation	Color	Abbreviation	Color	Abbreviation	Color
BLK	Black	LT BLU	Lt. Blue	TAN	Tan
BLU	Blue	LT GRN	Lt. Green	VIO	Violet
BRN	Brown	ORN	Orange	WHT	White
GRY	Gray	PNK	Pink	YEL	Yellow
GRN	Green	PPL	Purple		

1994-95 DX 2.6L I4 MFI VIN E 2WD [M/T] "A27" 22P 'A' Connector

PCM Pin #	W/Color	Circuit Description (22-Pin)	Value at Hot Idle
A27-1	BLU	Engine Speed Sensor	Digital Signals
A27-2	BLK/RED	Power Ground	<0.1v
A27-3	GRN/ORN	MAP Sensor VREF	4.9-5.1v
A27-4	RED/WHT	Keep Alive Power	12-14v
A27-5	WHT/GRN	Vehicle Speed Sensor	Moving: 0-5-0v
A27-6	GRY	HO2S-11 (B1 S1) Ground	<0.050v
A27-7	BLK/WHT	ECT Sensor Ground	<0.050v
A27-8	BRN/WHT	ECT Sensor Signal	At 180F: 0.52v
A27-9	BLU/WHT	MAP Sensor Signal	0.6-1.3v
A27-10	---	Not Used	---
A27-11	BLK	Power Ground	<0.1v
A27-12	---	Not Used	---
A27-13	BLK/RED	Logic Ground	<0.050v
A27-14	RED	MAF Sensor Signal VREF	12-14v
A27-15	GRN/RED	Wide Open Throttle Switch	At WOT: 12-14v
A27-16	BLK/GRN	Starter Signal	KOEC: 9-11v
A27-17	GRY	MAF Sensor Shield Ground	<0.050v
A27-18	BLU/RED	MAP Sensor Ground	<0.050v
A27-19	BLK	MAF Sensor Ground	<0.050v
A27-20	RED	HO2S-11 (B1 S1) Signal	0.1-1.1v
A27-21	WHT	MAF Sensor Signal	7-9v
A27-22	BLK	Power Ground	<0.1v

A27 22-PIN CONNECTOR

1	2	3	4	5	6	7	8	9	10	11
12	13	14	15	16	17	18	19	20	21	22

View is into back of Wire Harness Connector

05533_ADIA_G648

Pin Connector Graphic

Standard Colors and Abbreviations

Abbreviation	Color	Abbreviation	Color	Abbreviation	Color
BLK	Black	LT BLU	Lt. Blue	TAN	Tan
BLU	Blue	LT GRN	Lt. Green	VIO	Violet
BRN	Brown	ORN	Orange	WHT	White
GRY	Gray	PNK	Pink	YEL	Yellow
GRN	Green	PPL	Purple		

1996-97 DX 2.6L I4 MFI VIN E 2WD [M/T] 32P Red 'A' Connector

PCM Pin #	W/Color	Circuit Description (32-Pin)	Value at Hot Idle
A1	GRY/BLU	Reference Voltage 'A'	4.9-5.1v
A2-3	---	Not Used	---
A4	RED/WHT	Keep Alive Power	12-14v
A5	BRN/BLK	Idle Air Control 'A' High	Pulse Signals
A6	BRN/WHT	Idle Air Control 'A' Low	Pulse Signals
A7	BRN/YEL	Idle Air Control 'B' Low	Pulse Signals
A8	GRN/BLU	Idle Air Control 'B' High	Pulse Signals
A9-12	---	Not Used	---
A13	GRN	Malfunction Indicator Lamp	MIL Off: 12v, On: 1v
A14, 16	---	Not Used	---
A15	GRN/WHT	EVAP Purge Solenoid	Purge on, 1v, off: 12-14v

1996-97 DX 2.6L I4 MFI VIN E 2WD [M/T] 32P Red 'B' Connector

PCM Pin #	W/Color	Circuit Description (32-Pin)	Value at Hot Idle
B1	BLU/ORN	Reference Voltage 'B'	4.9-5.1v
B2-6	---	Not Used	---
B7	GRY	Linear EGR Sensor Signal	0.5-0.6v
B8	GRY	IAT Sensor Signal	Varies w/temp: 0.5-4.8v
B9	---	Not Used	---
B10	GRN/WHT	Rough Road Sensor	2.5v
B11	GRN/RED	PSP Switch Signal	Straight: 12v, Turning: 0v
B12	GRN/YEL	Illuminated Switch Signal	Switch on: 0.1v, off: 12v
B13	ORN/BLK	Class 2 Serial Data Link	Varies: 0-7v
B14	GRY/RED	A/C Clutch Relay	Relay on: 12v, off: 0v
B15-16	---	Not Used	---

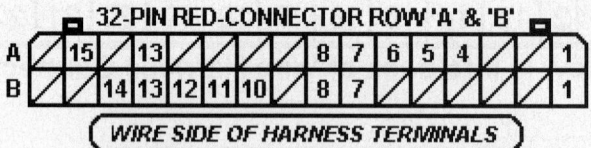

32-PIN RED-CONNECTOR ROW 'A' & 'B'

A | 15 | 13 | 8 7 6 5 4 | 1
B | 14 13 12 11 10 | 8 7 | 1

WIRE SIDE OF HARNESS TERMINALS

05533_ADIA_G649

Pin Connector Graphic

Standard Colors and Abbreviations

Abbreviation	Color	Abbreviation	Color	Abbreviation	Color
BLK	Black	LT BLU	Lt. Blue	TAN	Tan
BLU	Blue	LT GRN	Lt. Green	VIO	Violet
BRN	Brown	ORN	Orange	WHT	White
GRY	Gray	PNK	Pink	YEL	Yellow
GRN	Green	PPL	Purple		

1996-97 DX 2.6L MFI VIN E 2WD [M/T] 32P White 'C' Connector

PCM Pin #	W/Color	Circuit Description (32-Pin)	Value at Hot Idle
C1	BLU/YEL	Injector 2 Control	2.0-3.3 ms
C2-3	---	Not Used	---
C4	BLU	Igniter Control Signal	Digital Signals: 0-12-0v
C5	RED	CKP Sensor Signal	Digital Signals: 0-5-0v
C6	---	Not Used	---
C7	WHT	Power Ground	<0.1v
C8	BLK/RED	Power Ground	<0.1v
C9	BLK/BLU	Power Ground	<0.1v
C10	---	Not Used	---
C11	YEL/GRN	Upshift Lamp Control	Relay Off: 12v, On: 1v
C12	RED/WHT	Alternator Charge Signal	12-14v
C13	---	Not Used	---
C14	RED	HO2S-11 (B1 S1) Signal	0.1-1.1v
C15	GRN	HO2S-11 (B1 S1) Ground	<0.050v
C16	PNK	HO2S-12 (B1 S2) Signal	0.1-1.1v

1996-97 DX 2.6L MFI VIN E 2WD [M/T] 32P White 'D' Connector

PCM Pin #	W/Color	Circuit Description (32-Pin)	Value at Hot Idle
D1	BLU/RED	Injector 3 Control	2.0-3.3 ms
D2	---	Not Used	---
D3	BLU/BLK	Injector 1 Control	2.0-3.3 ms
D4	RED	Serial Data (8192 Baud)	Digital Signals
D5-6	---	Not Used	---
D7	WHT	Vehicle Speed Sensor	Moving: 0-12v
D8	GRN	Sensor Ground	<0.1v
D9	GRN	Sensor Ground	<0.1v
D10	---	Not Used	---
D11	WHT	CMP Sensor Signal	Digital Signals: 0-5-0v
D12-15	---	Not Used	---
D16	BLU	HO2S-12 (B1 S2) Ground	<0.050v

32-PIN WHITE-CONNECTOR ROW 'C' & 'D'

C | 16 | 15 | 14 | / | 12 | 11 | / | 9 | 8 | 7 | / | 5 | 4 | / | | 1
D | 16 | / | / | / | / | 11 | / | 9 | 8 | 7 | / | / | 4 | 3 | / | 1

WIRE SIDE OF HARNESS TERMINALS

05533_ADIA_G650

Pin Connector Graphic

Standard Colors and Abbreviations

Abbreviation	Color	Abbreviation	Color	Abbreviation	Color
BLK	Black	LT BLU	Lt. Blue	TAN	Tan
Q	Blue	LT GRN	Lt. Green	VIO	Violet
BRN	Brown	ORN	Orange	WHT	White
GRY	Gray	PNK	Pink	YEL	Yellow
GRN	Green	PPL	Purple		

1996-97 DX 2.6L I4 MFI VIN E 2WD [M/T] 32P Blue 'E' Connector

PCM Pin #	W/Color	Circuit Description (32-Pin)	Value at Hot Idle
E1-4	---	Not Used	---
E5	RED/GRN	Linear EGR Solenoid "High"	Pulse Signals
E6	GRY/RED	Linear EGR Solenoid "Low"	Pulse Signals
E7	---	Not Used	---
E8	BLU	TP Sensor Signal	0.5-0.6v
E9	GRY/BLK	ECT Sensor Signal	At 180F: 0.54v
E10	---	Not Used	---
E11	WHT/ORN	CKP Sensor VREF	4.9-5.1v
E12	---	Not Used	---
E13	PNK/WHT	Vehicle Speed Sensor	Moving: 0-5-0v
E14	---	Not Used	---
E15	GRN/BLK	A/C Switch Signal	Switch on: 12v, off: 0v
E16	RED/GRN	B+ from Main Relay	12-14v

1996-97 DX 2.6L I4 MFI VIN E 2WD [M/T] 32P Blue 'F' Connector

PCM Pin #	W/Color	Circuit Description (32-Pin)	Value at Hot Idle
F1-7	---	Not Used	---
F8	GRY/RED	MAP Sensor Signal	0.6-1.3v
F9	WHT/GRN	Vacuum Switch Signal	0v
F10-12		Not Used	
F13	BLU/PNK	Injector 4 Control	2.0-3.3 ms
F14-15	---	Not Used	---
F16	RED/GRN	B+ from Main Relay	12-14v

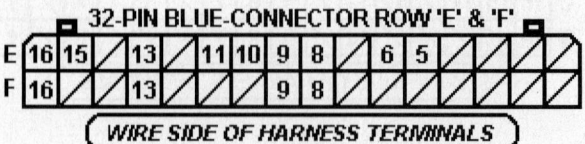

Pin Connector Graphic

Standard Colors and Abbreviations

Abbreviation	Color	Abbreviation	Color	Abbreviation	Color
BLK	Black	LT BLU	Lt. Blue	TAN	Tan
BLU	Blue	LT GRN	Lt. Green	VIO	Violet
BRN	Brown	ORN	Orange	WHT	White
GRY	Gray	PNK	Pink	YEL	Yellow
GRN	Green	PPL	Purple		

1994-95 LX 3.2L V6 MFI VIN V 2WD [All] 12P 'A' Connector

PCM Pin #	W/Color	Circuit Description (12-Pin)	Value at Hot Idle
A1	BLU/ORN	TP & MAP Sensor VREF	4.9-5.1v
A2	GRY/RED	A/C Clutch Relay	Relay Off: 12v, On: 1v
A3	---	Not Used	---
A4	BLUPNK	M/T: Shift Indicator Lamp	Lamp on: 1v, off: 12v
A5	GRN	Malfunction Indicator Lamp	MIL Off: 12v, On: 1v
A6	RED/GRN	B+ from Main Relay	12-14v
A7	WHT/RED	EVAP Purge Solenoid	Digital Signals: 0-12-0v
A8	WHT/GRN	EGR Solenoid Control	Sol. on: 1v, off: 12v
A9	YEL/BLU	TCM Signal ECT Input	At 180F: 0.54v
A10	---	Not Used	---
A11	GRY	Sensor Ground	<0.050v
A12	BLK/ORN	Power Ground	<0.1v

1994-95 LX 3.2L V6 MFI VIN V 2WD [All] 12P 'B' Connector

PCM Pin #	W/Color	Circuit Description (12-Pin)	Value at Hot Idle
B1	RED/WHT	Keep Alive Power	12-14v
B2	BLK/PNK	HO2S-11 (B1 S1) Ground	<0.050v
B3	RED	HO2S-11 (B1 S1) Signal	0.1-1.1v
B4	BRN/BLK	Idle Air Control 'A' High	Pulse Signals
B5	BRN/WHT	Idle Air Control 'A' Low	Pulse Signals
B6	BRN/YEL	Idle Air Control 'B' Low	Pulse Signals
B7	BRN/RED	Idle Air Control 'B' High	Pulse Signals
B8	YEL/VLT	Ignition VREF 'Lo'	<0.050v
B9	YEL/RED	Ignition VREF 'Hi'	Varies: 1.1-2.4v
B10-12	---	Not Used	---

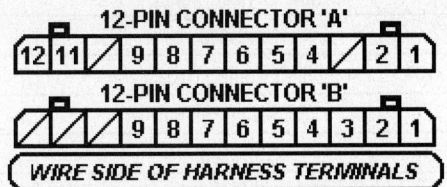

05533_ADIA_G652

Pin Connector Graphic

Standard Colors and Abbreviations

Abbreviation	Color	Abbreviation	Color	Abbreviation	Color
BLK	Black	LT BLU	Lt. Blue	TAN	Tan
BLU	Blue	LT GRN	Lt. Green	VIO	Violet
BRN	Brown	ORN	Orange	WHT	White
GRY	Gray	PNK	Pink	YEL	Yellow
GRN	Green	PPL	Purple		

1994-95 LX 3.2L V6 MFI VIN V 2WD [All] 16P 'C' Connector

PCM Pin #	W/Color	Circuit Description (16-Pin)	Value at Hot Idle
C1	---	Not Used	---
C2	ORN/BLK	Data Link Connector	Open: 4.80v
C3	PNK/WHT	Fuel Pump Relay Control	Relay on: 12v, off: 0v
C4	ORN/YEL	DLC Diagnostic Request	Digital Signals
C5	---	Not Used	---
C6	GRN/YEL	PSP Switch Signal	Straight: 12v, Turning: 0v
C7	ORN	A/T: PNP Switch Signal	In P/N: 0v, others: 12v
C8	---	Not Used	---
C9	YEL/BLK	A/C Switch Signal	Switch on: 12v, off: 0v
C10	---	Not Used	---
C11	BLU/YEL	Injector Pair 1 & 2	2.0-3.3 ms
C12	BLK/ORN	Power Ground	<0.1v
C13	---	Not Used	---
C14	BLU/WHT	Injector Pair 5 & 6	2.0-3.3 ms
C15	BLK/GRN	Power Ground	<0.1v
C16	RED/WHT	Keep Alive Power	12-14v

1994-95 LX 3.2L V6 MFI VIN V 2WD [All] 16P 'D' Connector

PCM Pin #	W/Color	Circuit Description (16-Pin)	Value at Hot Idle
D1	BLK/GRN	Power Ground	<0.1v
D2	BLK/WHT	Sensor Ground	<0.050v
D3-4	---	Not Used	---
D5	YEL	Igniter Control Signal	Digital Signals: 0-12-0v
D6	BLU/RED	TP Sensor Signal	0.5-0.6v
D7	BLU	IAT Sensor Signal	Varies w/temp: 0.5-4.8v
D8	WHT	Vehicle Speed Sensor	Moving: 0-5-0-5v
D9	GRY/RED	MAP Sensor Signal	0.6-1.1v
D10	---	Not Used	---
D11	YEL/GRN	Ignition Control Bypass	Hot idle: 4.9v
D12	GRY/BLK	ECT Sensor Signal	At 180F: 0.54v
D13-14	---	Not Used	---
D15	BLU/GRN	Injector Pair 3 & 4	2.0-3.3 ms
D16	BLU/GRN	Injector Pair 3 & 4	2.0-3.3 ms

16-PIN CONNECTOR 'C'

| 16 | 15 | 14 | / | 12 | 11 | / | 9 | / | 7 | 6 | / | 4 | 3 | 2 | / |

16-PIN CONNECTOR 'D'

| 16 | 15 | / | / | 12 | 11 | / | 9 | 8 | 7 | 6 | 5 | / | / | 2 | 1 |

WIRE SIDE OF HARNESS TERMINALS

05533_ADIA_G653

Pin Connector Graphic

1996-97 LX 3.2L V6 MFI VIN V 2WD [All] C209 Row 'A' Connector

PCM Pin #	W/Color	Circuit Description (16-Pin)	Value at Hot Idle
A1	RED	Reference Voltage 'A'	4.9-5.1v
A2	YEL	Knock Sensor Signal	No detonation: 18 mv AC
A3, A12	---	Not Used	---
A4	RED/WHT	Keep Alive Power	12-14v
A5	BLU	Idle Air Control 'A' High	Pulse Signals
A6	BLU/WHT	Idle Air Control 'A' Low	Pulse Signals
A7	BLU/BLK	Idle Air Control 'B' Low	Pulse Signals
A8	BLU/RED	Idle Air Control 'B' High	Pulse Signals
A9	ORN/BLK	Transmission Fluid Lamp	Lamp on: 1v.off: 12v
A10	PNK/GRN	Winter Lamp	Lamp on: 1v. off: 12v
A11	PNK/WHT	Power Lamp	Lamp on: 1v. off: 12v
A13	GRN	Malfunction Indicator Lamp	MIL Off: 12v, On: 1v
A14	PPL	Check Transmission Light	Lamp on: 1v, off: 12v
A15	RED/BLU	EVAP Purge Solenoid	Digital Signals: 0-12-0v
A16	BRN/YEL	A/T: Shift Solenoid Feed (B+)	12-14v

1996-97 LX 3.2L V6 MFI VIN V 2WD [All] C209 Row 'B' Connector

PCM Pin #	W/Color	Circuit Description (16-Pin)	Value at Hot Idle
B1	BLU/ORN	Reference Voltage 'B'	4.9-5.1v
B2	RED/WHT	Igniter 4 Control	Digital Signals: 0-12-0v
B3	RED/BLK	Igniter 2 Control	Digital Signals: 0-12-0v
B4	RED/GRN	Igniter 6 Control	Digital Signals: 0-12-0v
B5	ORN/GRN	Fuel Level Sensor	Tank empty: 1.8v
B6	GRY	Fuel Tank Pressure Sensor	With Fuel Cap Off: 2.5v
B7	YEL/RED	Linear EGR Sensor Signal	0.5-0.6v
B8	YEL/GRN	IAT Sensor Signal	Varies w/temp: 0.5-4.8v
B9	---	Not Used	---
B10	GRN/WHT	Rough Road Sensor	At idle: 2.5v
B11	GRN/YEL	PSP Switch Signal	Straight: 12v, Turning: 0v
B12	GRN/YEL	Headlights Switch Signal	Switch on: 0.1v, off: 12v
B13	ORN/BLK	Class 2 Serial Data Link	0v
B14	GRY/RED	A/C Clutch Relay	Relay Off: 12v, On: 1v
B15	PNK	Low Fuel Lamp	Lamp on: 1v, off: 12v
B16	BRN/WHT	EVAP Vent Solenoid	Sol. on: 1v, off: 12v

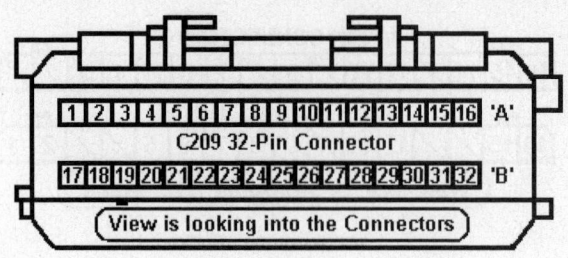

C209 32-Pin Connector

View is looking into the Connectors

05533_ADIA_G654

Pin Connector Graphic
1996-97 LX 3.2L MFI VIN V 2WD [All] C211 Row 'C' Connector

PCM Pin #	W/Color	Circuit Description (16-Pin)	Value at Hot Idle
C1	GRN/RED	Injector 4 Control	2.0-3.3 ms
C2	BRN/BLK	A/T: Shift Solenoid 'B'	SSB Off: 12v, On: 1v
C3	GRN/YEL	Injector 6 Control	2.0-3.3 ms
C4	RED	Igniter 1 Control	Digital Signals: 0-12-0v
C5	YEL	CKP Sensor Signal	Digital Signals: 0-5-0v
C7	YEL/BLK	Power Ground	<0.1v
C8	BLK/RED	Power Ground	<0.1v
C9	BLK/BLU	Power Ground	<0.1v
C10	BLK/RED	Tachometer Signal	Pulse Signals
C11	YEL/GRN	M/T: Upshift Lamp Control	Lamp on: 1v, off: 12v
C12	RED/WHT	Alternator Charge Signal	12-14v
C13	YEL/RED	Fuel Gauge PWM Signal	Duty Cycle: 0-100%
C14	WHT	HO2S-21 (B2 S1) Signal	0.1-1.1v
C15	RED	HO2S-21 (B2 S1) Ground	<0.050v
C16	RED	HO2S-22 (B2 S2) Signal	0.1-1.1v

1996-97 LX 3.2L V6 VIN V 2WD [All] C211 Row 'D' Connector

PCM Pin #	W/Color	Circuit Description (16-Pin)	Value at Hot Idle
D1	GRN/ORN	Injector 2 Control	2.0-3.3 ms
D2	RED/YEL	Torque Converter Clutch	TCC on: 1v, off: 12v
D3	GRN/WHT	Injector 1 Control	2.0-3.3 ms
D4	RED	Serial Data (8192 Baud)	Digital Signals
D5	RED/YEL	Igniter 5 Control	Digital Signals: 0-12-0v
D6	RED/BLU	Igniter 3 Control	Digital Signals: 0-12-0v
D7	WHT	Vehicle Speed Sensor	Moving: 0-5-0-5v
D8, D9	GRN	Sensor 'A', 'B' Ground	<0.050v
D10	YEL/BLU	MAF Sensor Signal	7-9v
D11	BLU	CMP Sensor Signal	Digital Signals: 0-5-0v
D12	BLU	HO2S-13 (B1 S3) Ground	<0.050v
D12	BLU	HO2S-12 (B1 S2) Ground	<0.050v
D13	GRN	HO2S-13 (B1 S3) Signal	0.1-1.1v
D13	GRN	HO2S-12 (B1 S2) Signal	0.1-1.1v
D14	BLU	HO2S-11 (B1 S1) Ground	<0.050v
D15	PNK	HO2S-11 (B1 S1) Signal	0.1-1.1v
D16	GRN	HO2S-22 (B2 S2) Ground	<0.050v

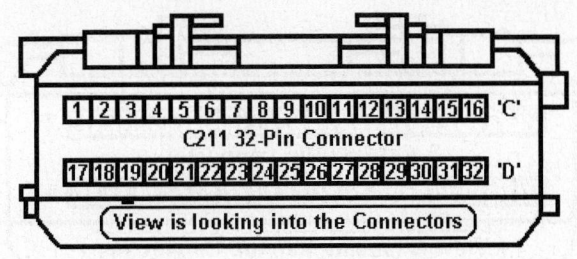

05533_ADIA_G655

Pin Connector Graphic

1996-97 LX 3.2L V6 MFI VIN V 2WD [All] C212 Row 'E' Connector

PCM Pin #	W/Color	Circuit Description (32-Pin)	Value at Hot Idle
E1	YEL	VSS "High" Signal	AC Pulse Signals
E2	BRN	VSS "Low" Signal	<0.050v
E3	RED/ORN	A/T: Pressure Control Low	Pulse Signals
E4	RED/BLK	A/T: Pressure Control High	Pulse Signals
E5	RED/GRN	Linear EGR Solenoid "High"	Pulse Signals
E6	YEL	Linear EGR Solenoid "Low"	Pulse Signals
E7	PNK	A/T: Range Signal 'B'	In P/N: 1v, others: 12v
E8	BLU	TP Sensor Signal	0.5-0.6v
E9	BLU/RED	ECT Sensor Signal	At 180F: 0.54v
E10	---	Not Used	---
E11	YEL/RED	CKP Sensor VREF	4.9-5.1v
E12	PNK/BLU	A/T: Range Signal 'A'	In P/N: 1v, others: 12v
E13	PNK/WHT	Fuel Pump Relay Control	Relay on: 12v, off: 0v
E14	BRN/WHT	A/T: Bank Apply Solenoid	Sol. on: 1v, off: 12v
E15	GRN/BLK	A/C Switch Signal	Switch on: 12v, off: 0v
E16	RED/BLU	B+ from Main Relay	12-14v

1996-97 LX 3.2L V6 MFI VIN V 2WD [All] C212 Row 'F' Connector

PCM Pin #	W/Color	Circuit Description (32-Pin)	Value at Hot Idle
F1, 9, 12	---	Not Used	---
F2	BLU/WHT	A/T: Range Signal 'C'	In P/N: 1v, others: 12v
F3	PNK/BLK	A/T: Range Signal 'P'	In P/N: 1v, others: 12v
F4	RED	Brake Switch Signal	On: 12v, Off: 0v
F5	PPL/RED	Power Switch Signal	On: 0.1v, off: 12v
F6	PPL/GRN	Winter Switch Signal	On: 0.1v, off: 12v
F7	GRN/RED	A/T: TOT Sensor Signal	Varies w/temp: 0.5-4.8v
F8	RED	MAP Sensor Signal	0.6-1.3v
F10	GRY/BLU	Cruise Control	On: 0.1v, off: 12v
F11	LT BLU	A/T: Kickdown Switch	On: 0.1v, off: 12v
F13	WHT/GRN	Injector 3 Control	2.0-3.3 ms
F14	YEL/GRN	A/T: Shift Solenoid 'A'	SSA Off: 12v, On: 1v
F15	GRN/BLK	Injector 5 Control	2.0-3.3 ms
F16	RED/BLU	B+ from Main Relay	12-14v

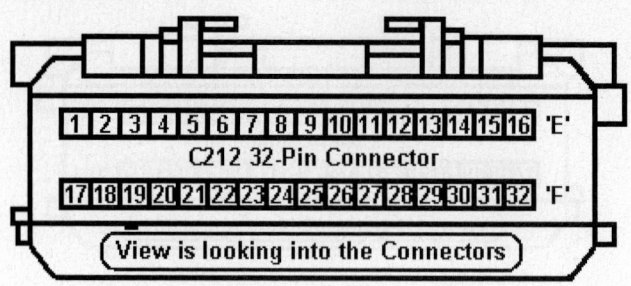

View is looking into the Connectors

05533_ADIA_G656

Pin Connector Graphic

1998-03 EX, LX 3.2L VIN W 2WD [A/T] 80P J21 Blue Connector

PCM Pin #	W/Color	Circuit Description (40-Pin)	Value at Hot Idle
1	BLK	Power Ground	<0.1v
2	GRN	Reference Voltage 'A'	4.9-5.1v
3	RED/WHT	Reference Voltage 'B'	4.9-5.1v
4	GRY/RED	A/C Clutch Relay Control	Relay Off: 12v, On: 1v
5	GRN/RED	A/T: Shift Solenoid 'A'	SSA Off: 12v, On: 1v
6	---	Not Used	---
7	BLK/RED	Tachometer Signal	Pulse Signals
8	---	Not Used	---
9	YEL	Knock Sensor 1 Signal	No detonation: 18 mv AC
10	RED	Knock Sensor 2 Signal	No detonation: 18 mv AC
11	---	Not Used	---
12	WHT/RED	CMP Sensor Signal	Digital Signals: 0-5-0v
13	PNK	To be done	---
14	BLK/BLU	MAF Sensor Signal	4.2v
15-16	---	Not Used	---
17	RED	HO2S-22 (B2 S2) Ground	0.1-1.1v
18	---	Not Used	---
19	RED/GRN	Ignition Power (Main Relay)	12-14v
20	RED/WHT	Keep Alive Power	12-14v
21	WHT	HO2S-11 (B1 S1) Ground	<0.050v
22	---	Not Used	---
23	BLU	HO2S-12 (B1 S2) Ground	<0.050v
24	---	Not Used	---
25	RED/BLU	Coil On Plug 3 Control	Digital Signals: 0-12-0v
26	RED/BLK	Coil On Plug 2 Control	Digital Signals: 0-12-0v
27	GRN	IAT Sensor Signal	Varies w/temp: 0.5-4.8v
28	---	Not Used	---
29	RED	MAP Sensor Signal	0.9v
30-36	---	Not Used	---
37	GRN/YEL	PSP Switch Signal	Straight: 12v, Turning: 0v
38	GRN/YEL	Headlights On Signal	Lights Off: 12v, On: 0v
39	RED	Brake Switch Signal	Brake Off: 0v, On: 12v
40	BLK/BLU	Power Ground	<0.1v

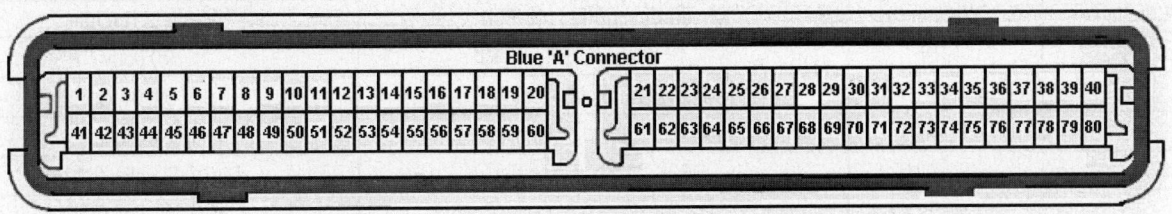

05533_ADIA_G657

Pin Connector Graphic

1998-03 EX, LX 3.2L VIN W 2WD [A/T] 80P J21 Blue Connector

PCM Pin #	W/Color	Circuit Description (40-Pin)	Value at Hot Idle
41	GRN	TP Sensor 1 Ground	<0.050v
42	PNK/WHT	Fuel Pump Relay Control	Relay Off: 0v, On: 12v
43	PPL/RED	Pressure Control Solenoid (-)	Pulse Signals
44	ORN/WHT	HO2S-21 (B2 S1) Heater	Digital Signals: 0-12-0v
45	GRN/BLK	A/C Switch Signal	A/C On: 12v, Off: 0v
46	BRN/RED	Brake Switch Signal	Brake Off: 12v, On: 0v
47	PPL/WHT	Pressure Control Solenoid (+)	Pulse Signals
48	BLU	Throttle Motor (-) Signal	Pulse Signals
49	BLK	Knock Sensor 1 Ground	<0.050v
50	ORN	HO2S-22 (B2 S2) Ground	<0.050v
51	---	Not Used	---
52	RED	HO2S-21 (B2 S1) Ground	<0.050v
53	BLK	ECT & EGR Sensor Ground	<0.050v
54	BLK	FTP Sensor Ground	<0.050v
55	---	Not Used	---
56	GRN/BLK	Injector 5 Control	2.0-3.3 ms
57	RED/WHT	Keep Alive Power	12-14v
58	ORN/BLK	Class 2 Serial Data Link	0v
59	---	Not Used	---
60	BLK	Knock Sensor 2 Ground	<0.050v
61	GRN/WHT	TP Sensor 2 Ground	<0.050v
62	GRN/ORN	Injector 2 Control	2.0-3.3 ms
63	RED	A/P Sensor 1 Ground	<0.1v
64	GRN/RED	Injector 4 Control	2.0-3.3 ms
65	BLU/WHT	TP Sensor 2 Signal	0.5-0.6v
66	GRN	Injector 3 Control	2.0-3.3 ms
67	YEL	PCM C.Q. Input (1-3-5)	Digital Signals
68	RED	Knock Sensor Input (2-4-6)	<0.050v
69	GRN/WHT	Injector 1 Control	2.0-3.3 ms
70	---	Not Used	---
71	GRN/WHT	HO2S-22 (B2 S2) Heater	Digital Signals: 0-12-0v
72	WHT/BLU	C/C Resume/Accel Input	Switch Off: 0v, On: 12v
73	WHT/BLU	CKP Sensor Signal	Digital Signals: 0-5-0v
74	BLU/RED	ECT Sensor Signal	At 180F: 0.54v
75	RED/GRN	Ignition Power (Main Relay)	12-14v
76	PNK/BLK	A/T: Range Signal 'P'	In 'P': 12v, Others: 0v
77	PNK/BLU	A/T: Range Signal 'A'	In 'N': 0v, Others: 12v
78	PNK/YEL	A/T: Range Signal 'C'	In 'C': 0v, Others: 12v
79	PNK	A/T: Range Signal 'B'	In 'B': 0v, Others: 12v
80	BLU	A. P. Sensor 2 Ground	<0.050v

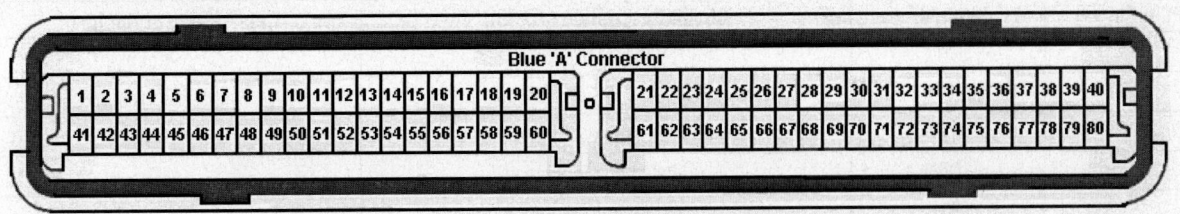

05533_ADIA_G657

Pin Connector Graphic

1998-03 EX, LX 3.2L VIN W 2WD [A/T] 80P J22 Red Connector

PCM Pin #	W/Color	Circuit Description (40-Pin)	Value at Hot Idle
1	GRN	Sensor Ground	<0.050v
2	BLK/ORN	HO2S-11 (B1 S1) Heater	Digital Signals: 0-12-0v
3	YEL	A/P Sensor 2 Signal	N/A
4	ORN	Reference Voltage	4.9-5.1v
5	WHT	CKP Sensor Reference	4.9-5.1v
6	GRY/GRN	Cruise Set Switch	Switch Off: 0v, On: 12v
7	RED/BLU	EVAP Purge Solenoid	Valve Off: 12v, On: 1v
8	GRN/WHT	Indicator Control	N/A
9	RED/GRN	Ignition Power (Main Relay)	12-14v
10	---	Not Used	---
11	WHT/GRN	Malfunction Indicator Lamp	MIL Off: 12v, On: 1v
12	GRN	Throttle Motor (+) Signal	Pulse Signals
13	RED/WHT	Coil On Plug 4 Control	Digital Signals: 0-12-0v
14	BLU	HO2S-12 (B1 S2) Ground	<0.050v
15	BLU	HO2S-22 (B2 S2) Ground	<0.050v
16	BLU	HO2S-22 (B2 S2) Ground	<0.050v
17	PNK	HO2S-21 (B1 S1) Ground	<0.050v
18	WHT	HO2S-11 (B1 S1) Ground	<0.050v
19	RED/YEL	Coil On Plug 5 Control	Digital Signals: 0-12-0v
20	YEL/GRN	Sensor Ground	<0.050v
21	BLU/YEL	Sensor Ground	<0.050v
22	BLU/YEL	OSS Sensor 'P' Signal	AC Pulses
24	RED	HO2S-22 (B2 S2) Ground	0.1-1.1v
25	---	Not Used	---
26	BLU/RED	EGR Position Solenoid	Digital Signals: 0-12-0v
27	BLK/YEL	Vehicle Speed Sensor	Moving: 0-5-0-5v
28	GRN/YEL	Injector 6 Control	2.0-3.3 ms
29	---	Not Used	---
30	WHT/GRN	Cruise Engage Switch	Switch Off: 0v, On: 12v
31	GRY/RED	FTP Sensor Reference	4.9-5.1v
32	RED/GRN	Ignition Power (Main Relay)	12-14v
33-34	---	Not Used	---
35	RED/GRN	Coil On Plug 6 Control	Digital Signals: 0-12-0v
36	BLK	A/P Sensor 1 Reference	4.9-5.1v
37	---	Not Used	---
38	RED	TP Sensor 1 Reference	4.9-5.1v
39-40	---	Not Used	---

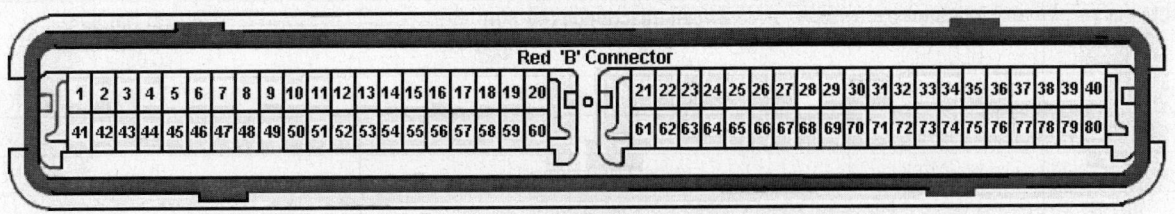

05533_ADIA_G658

Pin Connector Graphic

1998-03 EX, LX 3.2L VIN W 2WD [A/T] 80P J22 Red Connector

PCM Pin #	W/Color	Circuit Description (40-Pin)	Value at Hot Idle
41	BLU/ORN	A/P Sensor 3 Ground	<0.050v
42	WHT/RED	HO2S-12 (B1 S2) Heater	Digital Signals: 0-12-0v
43-46	---	Not Used	---
47	WHT/BLK	Water Gauge Signal	0-5v
48	---	Not Used	---
49	RED/WHT	Class 2 Data Line to DLC	0v
50	ORN/GRN	Fuel Level Sensor Signal	Duty Cycle Signal
51	---	Not Used	---
52	GRN	FTP Sensor Signal	With Fuel Cap Off: 2.5v
53	---	Not Used	---
54	BLU	HO2S-12 (B1 S2) Ground	<0.050v
55	BLU/GRN	OSS Sensor 'N' Signal	AC Pulses
56-57	---	Not Used	---
58	ORN/BLU	HO2S-11 (B1 S1) Signal	0.1-1.1v
59	LT GRN	Cruise Set Switch	Switch Off: 0v, On: 12v
60	BLU/YEL	Fuel Level Sensor Reference	12v
61	YEL/GRN	IAT Sensor Signal	Varies w/temp: 0.5-4.8v
62	RED	Coil On Plug 1 Control	Digital Signals: 0-12-0v
63	WHT	HO2S-21 (B2 S1) Signal	0.1-1.1v
64	PNK	HO2S-12 (B1 S2) Signal	0.1-1.1v
65	GRN	HO2S-22 (B2 S2) Signal	0.1-1.1v
66	---	Not Used	---
67	GRY/RED	EGR Position Sensor Signal	1.2-2.0v
68	WHT	A/P Sensor 1 Signal	N/A
69	GRN	Throttle Motor (+) Signal	Pulse Signals
70	---	Not Used	---
71	GRN/WHT	HO2S-22 (B2 S2) Heater	Digital Signals: 0-12-0v
72	RED/GRN	Ignition Power (Main Relay)	12-14v
73	GRN/WHT	Cruise Control Indicator	Lamp Off: 12v, On: 1v
74	YEL/BLK	Intake Air VSV Control	Valve Off: 12v, On: 1v
75	BRN/WHT	EVAP Purge Cut Solenoid	Valve Off: 12v, On: 1v
76	BLU	TP Sensor 1 Signal	0.5-0.8v
77	ORN	A/P Sensor 3 Reference	4.9-5.1v
78	YEL	A/P Sensor 2 Signal	N/A
79	BLU/GRN	A/P Sensor 3 Signal	N/A
80	GRN	MAP Sensor Reference	4.9-5.1v

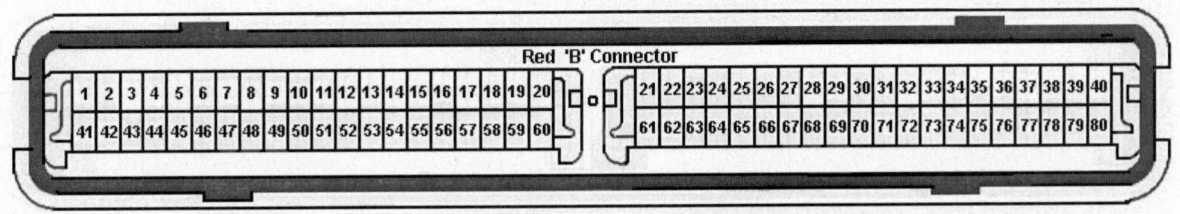

05533_ADIA_G658

Pin Connector Graphic

1998-03 EX, LX 3.2L VIN W 2WD [M/T] 80P J21 Blue Connector

PCM Pin #	W/Color	Circuit Description (40-Pin)	Value at Hot Idle
1	BLK	Power Ground	<0.1v
2	GRN	Reference Voltage 'A'	4.9-5.1v
3	RED/WHT	Reference Voltage 'B'	4.9-5.1v
4	GRY/RED	A/C Clutch Relay Control	Relay Off: 12v, On: 1v
5-6	---	Not Used	---
7	BLK/RED	Tachometer Signal	Pulse Signals
8	---	Not Used	---
9	YEL	Knock Sensor 1 Signal	No detonation: 18 mv AC
10	RED	Knock Sensor 2 Signal	No detonation: 18 mv AC
11	---	Not Used	---
12	WHT/RED	CMP Sensor Signal	Digital Signals: 0-5-0v
13	PNK	To be done	---
14	BLK/BLU	MAF Sensor Signal	4.2v
15-16	---	Not Used	---
17	RED	HO2S-22 (B2 S2) Ground	0.1-1.1v
18	---	Not Used	---
19	RED/GRN	Ignition Power (Main Relay)	12-14v
20	RED/WHT	Keep Alive Power	12-14v
21	WHT	HO2S-11 (B1 S1) Ground	<0.050v
22	---	Not Used	---
23	BLU	HO2S-12 (B1 S2) Ground	<0.050v
24	---	Not Used	---
25	RED/BLU	Coil On Plug 3 Control	Digital Signals: 0-12-0v
26	RED/BLK	Coil On Plug 2 Control	Digital Signals: 0-12-0v
27	GRN	IAT Sensor Signal	Varies w/temp: 0.5-4.8v
28	---	Not Used	---
29	RED	MAP Sensor Signal	0.9v
30-36	---	Not Used	---
37	GRN/YEL	PSP Switch Signal	Straight: 12v, Turning: 0v
38	GRN/YEL	Headlights On Signal	Lights Off: 12v, On: 0v
39	RED	Brake Switch Signal	Brake Off: 0v, On: 12v
40	BLK/BLU	Power Ground	<0.1v

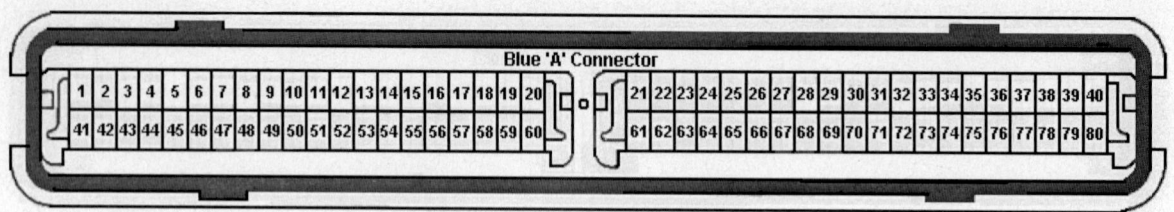

05533_ADIA_G657

Pin Connector Graphic

1998-03 EX, LX 3.2L VIN W 2WD [M/T] 80P J21 Blue Connector

PCM Pin #	W/Color	Circuit Description (40-Pin)	Value at Hot Idle
41	GRN	TP Sensor 1 Ground	<0.050v
42	PNK/WHT	Fuel Pump Relay Control	Relay Off: 0v, On: 12v
43	---	Not Used	---
44	ORN/WHT	HO2S-21 (B2 S1) Heater	Digital Signals: 0-12-0v
45	GRN/BLK	A/C Switch Signal	A/C On: 12v, Off: 0v
46	BRN/RED	Brake Switch Signal	Brake Off: 12v, On: 0v
47	---	Not Used	---
48	BLU	Throttle Motor (-) Signal	Pulse Signals
49	BLK	Knock Sensor 1 Ground	<0.050v
50	ORN	HO2S-22 (B2 S2) Ground	<0.050v
51	---	Not Used	---
52	RED	HO2S-21 (B2 S1) Ground	<0.050v
53	BLK	ECT & EGR Sensor Ground	<0.050v
54	BLK	FTP Sensor Ground	<0.050v
55	---	Not Used	---
56	GRN/BLK	Injector 5 Control	2.0-3.3 ms
57	RED/WHT	Keep Alive Power	12-14v
58	ORN/BLK	Class 2 Serial Data Link	0v
59	---	Not Used	---
60	BLK	Knock Sensor 2 Ground	<0.050v
61	GRN/WHT	TP Sensor 2 Ground	<0.050v
62	GRN/ORN	Injector 2 Control	2.0-3.3 ms
63	RED	A/P Sensor 1 Ground	<0.1v
64	GRN/RED	Injector 4 Control	2.0-3.3 ms
65	BLU/WHT	TP Sensor 2 Signal	0.5-0.6v
66	GRN	Injector 3 Control	2.0-3.3 ms
67	YEL	PCM C.Q. Input (1-3-5)	Digital Signals
68	RED	Knock Sensor Input (2-4-6)	<0.050v
69	GRN/WHT	Injector 1 Control	2.0-3.3 ms
70	---	Not Used	---
71	GRN/WHT	HO2S-22 (B2 S2) Heater	Digital Signals: 0-12-0v
72	WHT/BLU	C/C Resume/Accel Input	Switch Off: 0v, On: 12v

1998-03 EX, LX 3.2L VIN W 2WD [M/T] 80P J21 Blue Connector, *continued*

73	WHT/BLU	CKP Sensor Signal	Digital Signals: 0-5-0v
74	BLU/RED	ECT Sensor Signal	At 180F: 0.54v
75	RED/GRN	Ignition Power (Main Relay)	12-14v
76	PNK/BLK	Clutch Switch Signal	Clutch Out: 12v, In: 0v
77-79	---	Not Used	---
80	BLU	A. P. Sensor 2 Ground	<0.050v

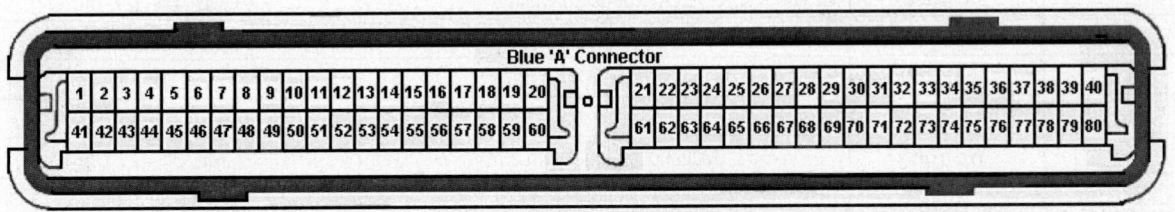

05533_ADIA_G657

Pin Connector Graphic

1998-03 EX, LX 3.2L VIN W 2WD [M/T] 80P J22 Red Connector

PCM Pin #	W/Color	Circuit Description (40-Pin)	Value at Hot Idle
1	GRN	Sensor Ground	<0.050v
2	BLK/ORN	HO2S-11 (B1 S1) Heater	Digital Signals: 0-12-0v
3	YEL	A/P Sensor 2 Signal	N/A
4	ORN	Reference Voltage	4.9-5.1v
5	WHT	CKP Sensor Reference	4.9-5.1v
6	GRY/GRN	Cruise Set Switch	Switch Off: 0v, On: 12v
7	RED/BLU	EVAP Purge Solenoid	Valve Off: 12v, On: 1v
8	GRN/WHT	Indicator Control	N/A
9	RED/GRN	Ignition Power (Main Relay)	12-14v
10	---	Not Used	---
11	WHT/GRN	Malfunction Indicator Lamp	MIL Off: 12v, On: 1v
12	GRN	Throttle Motor (+) Signal	Pulse Signals
13	RED/WHT	Coil On Plug 4 Control	Digital Signals: 0-12-0v
14	BLU	HO2S-12 (B1 S2) Ground	<0.050v
15	BLU	HO2S-22 (B2 S2) Ground	<0.050v
16	BLU	HO2S-22 (B2 S2) Ground	<0.050v
17	PNK	HO2S-21 (B1 S1) Ground	<0.050v
18	WHT	HO2S-11 (B1 S1) Ground	<0.050v
19	RED/YEL	Coil On Plug 5 Control	Digital Signals: 0-12-0v
20	YEL/GRN	Sensor Ground	<0.050v
21	BLU/YEL	Sensor Ground	<0.050v
22	---	Not Used	---
24	RED	HO2S-22 (B2 S2) Ground	0.1-1.1v
25	---	Not Used	---
26	BLU/RED	EGR Position Solenoid	Digital Signals: 0-12-0v
27	BLK/YEL	Vehicle Speed Sensor	Moving: 0-5-0-5v
28	GRN/YEL	Injector 6 Control	2.0-3.3 ms
29	---	Not Used	---
30	WHT/GRN	Cruise Engage Switch	Switch Off: 0v, On: 12v
31	GRY/RED	FTP Sensor Reference	4.9-5.1v
32	RED/GRN	Ignition Power (Main Relay)	12-14v
33-34	---	Not Used	---
35	RED/GRN	Coil On Plug 6 Control	Digital Signals: 0-12-0v
36	BLK	A/P Sensor 1 Reference	4.9-5.1v
37	---	Not Used	---
38	RED	TP Sensor 1 Reference	4.9-5.1v
39-40	---	Not Used	---

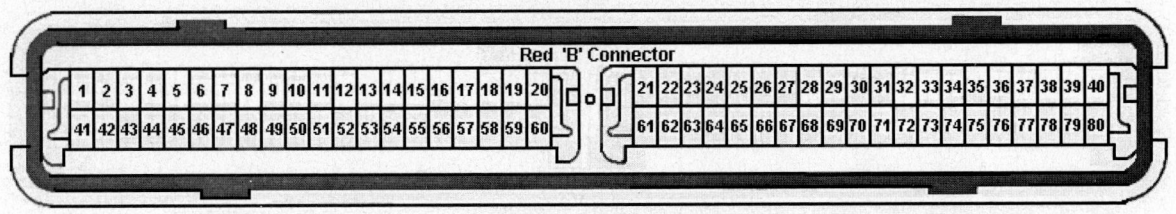

05533_ADIA_G658

Pin Connector Graphic
1998-03 EX, LX 3.2L VIN W 2WD [M/T] 80P J22 Red Connector

PCM Pin #	W/Color	Circuit Description (40-Pin)	Value at Hot Idle
41	BLU/ORN	A/P Sensor 3 Ground	<0.050v
42	WHT/RED	HO2S-12 (B1 S2) Heater	Digital Signals: 0-12-0v
43-46	---	Not Used	---
47	WHT/BLK	Water Gauge Signal	0-5v
48	---	Not Used	---
49	RED/WHT	Class 2 Data Line to DLC	0v
50	ORN/GRN	Fuel Level Sensor Signal	Duty Cycle Signal
51	---	Not Used	---
52	GRN	FTP Sensor Signal	With Fuel Cap Off: 2.5v
53	---	Not Used	---
54	BLU	HO2S-12 (B1 S2) Ground	<0.050v
55-57	---	Not Used	---
58	ORN/BLU	HO2S-11 (B1 S1) Signal	0.1-1.1v
59	LT GRN	Cruise Set Switch	Switch Off: 0v, On: 12v
60	BLU/YEL	Fuel Level Sensor Reference	12v
61	YEL/GRN	IAT Sensor Signal	Varies w/temp: 0.5-4.8v
62	RED	Coil On Plug 1 Control	Digital Signals: 0-12-0v
63	WHT	HO2S-21 (B2 S1) Signal	0.1-1.1v
64	PNK	HO2S-12 (B1 S2) Signal	0.1-1.1v
65	GRN	HO2S-22 (B2 S2) Signal	0.1-1.1v
66	---	Not Used	---
67	GRY/RED	EGR Position Sensor Signal	1.2-2.0v
68	WHT	A/P Sensor 1 Signal	N/A
69	GRN	Throttle Motor (+) Signal	Pulse Signals
70	---	Not Used	---
71	GRN/WHT	HO2S-22 (B2 S2) Heater	Digital Signals: 0-12-0v
72	RED/GRN	Ignition Power (Main Relay)	12-14v
73	GRN/WHT	Cruise Control Indicator	Lamp Off: 12v, On: 1v
74	YEL/BLK	Intake Air VSV Control	Valve Off: 12v, On: 1v
75	BRN/WHT	EVAP Purge Cut Solenoid	Valve Off: 12v, On: 1v
76	BLU	TP Sensor 1 Signal	0.5-0.8v
77	ORN	A/P Sensor 3 Reference	4.9-5.1v
78	YEL	A/P Sensor 2 Signal	N/A
79	BLU/GRN	A/P Sensor 3 Signal	N/A
80	GRN	MAP Sensor Reference	4.9-5.1v

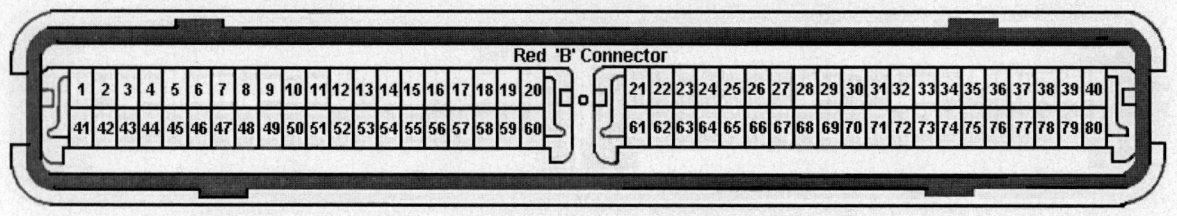

05533_ADIA_G658

Pin Connector Graphic

1994-95 EX, LX 3.2L V6 MFI 4WD VIN V [A/T] 12P 'A' Connector

PCM Pin #	W/Color	Circuit Description (12-Pin)	Value at Hot Idle
A1	BLU/ORN	TP & MAP VREF	4.9-5.1v
A2	GRY/YEL	A/C Clutch Relay	Relay Off: 12v, On: 1v
A3	---	Not Used	---
A4	BLUPNK	M/T: Shift Indicator Lamp	Lamp on: 1v, off: 12v
A5	GRN	Malfunction Indicator Lamp	MIL Off: 12v, On: 1v
A6	RED/GRN	B+ from Main Relay	12-14v
A7	WHT/RED	EVAP Purge Solenoid	Digital Signals: 0-12-0v
A8	WHT/GRN	EGR Solenoid Control	Sol. on: 1v, off: 12
A9	YEL/BLU	ECT Input to TCM	At 180F: 0.54v
A10	---	Not Used	---
A11	GRY	Sensor Ground	<0.050v
A12	BLK/ORN	Power Ground	<0.1v

1994-95 EX, LX 3.2L V6 MFI 4WD VIN V [A/T] 12P 'B' Connector

PCM Pin #	W/Color	Circuit Description (12-Pin)	Value at Hot Idle
B1	RED/WHT	Keep Alive Power	12-14v
B2	BLK/PNK	HO2S-11 (B1 S1) Ground	<0.050v
B3	RED	HO2S-11 (B1 S1) Signal	0.1-1.1v
B4	BRN/BLK	Idle Air Control 'A' High	Pulse Signals
B5	BRN/WHT	Idle Air Control 'A' Low	Pulse Signals
B6	BRN/YEL	Idle Air Control 'B' Low	Pulse Signals
B7	BRN/RED	Idle Air Control 'B' High	Pulse Signals
B8	YEL/VLT	Electronic Ignition VREF Low	<0.050v
B9	YEL/RED	Electronic Ignition VREF High	Varies: 1.1-2.4v
B10-12	---	Not Used	---

05533_ADIA_G652

Pin Connector Graphic

Standard Colors and Abbreviations

Abbreviation	Color	Abbreviation	Color	Abbreviation	Color
BLK	Black	LT BLU	Light Blue	TAN	Tan
BLU	Blue	LT GRN	Lt. Green	VIO	Violet
BRN	Brown	ORN	Orange	WHT	White
GRY	Gray	PNK	Pink	YEL	Yellow
GRN	Green	PPL	Purple		

1994-95 EX, LX 3.2L V6 MFI 4WD VIN V [A/T] 16P 'C' Connector

PCM Pin #	W/Color	Circuit Description (16-Pin)	Value at Hot Idle
C1	---	Not Used	---
C2	ORN/BLK	Data Link Connector	4.80v (no Scan Tool)
C3	PNK/WHT	Fuel Pump Relay Control	Relay on: 12v, off: 0v
C4	ORN/YEL	DLC Diagnostic Request	Digital Signals
C5	---	Not Used	---
C6	GRN/YEL	PSP Switch Signal	Straight: 12v, Turning: 0v
C7	ORN	PNP Switch Signal	In P/N: 0v, others: 12v
C8	---	Not Used	---
C9	YEL/BLK	A/C Switch Signal	Switch on: 12v, off: 0v
C10	---	Not Used	---
C11	BLU/YEL	Injector Pair 1 & 2	2.0-3.3 ms
C12	BLK/ORN	Power Ground	<0.1v
C13	---	Not Used	---
C14	BLU/WHT	Injector Pair 5 & 6	2.0-3.3 ms
C15	BLK/GRN	Power Ground	<0.1v
C16	RED/WHT	Keep Alive Power	12-14v

1994-95 EX, LX 3.2L V6 MFI 4WD VIN V [A/T] 16P 'D' Connector

PCM Pin #	W/Color	Circuit Description (16-Pin)	Value at Hot Idle
D1	BLK/GRN	Power Ground	<0.1v
D2	BLK/WHT	Sensor Ground	<0.050v
D3-4	---	Not Used	---
D5	YEL	Igniter Control Signal	Digital Signals: 0-12-0v
D6	BLU/RED	TP Sensor Signal	0.5-0.6v
D7	BLU	IAT Sensor Signal	Varies w/temp: 0.5-4.8v
D8	WHT	Vehicle Speed Sensor	Moving: 0-5-0-5v
D9	GRY/RED	MAP Sensor Signal	0.6-1.3v
D10	---	Not Used	---
D11	YEL/GRN	Ignition Control Bypass	Hot idle: 4.9v
D12	GRY/BLK	ECT Sensor Signal	At 180F: 0.54v
D13-14	---	Not Used	---
D15	BLU/GRN	Injector Pair 3 & 4	2.0-3.3 ms
D16	BLU/GRN	Injector Pair 3 & 4	2.0-3.3 ms

```
16-PIN CONNECTOR 'C'
16 15 14 / 12 11 / 9 / 7 6 / 4 3 2
16-PIN CONNECTOR 'D'
16 15 / / 12 11 / 9 8 7 6 5 / / 2 1
WIRE SIDE OF HARNESS TERMINALS
```

05533_ADIA_G653

Pin Connector Graphic

1994-95 EX, LX 3.2L V6 MFI 4WD VIN V [M/T] 12P 'A' Connector

PCM Pin #	W/Color	Circuit Description (12-Pin)	Value at Hot Idle
A1	BLU/ORN	TP & MAP Sensor VREF	4.9-5.1v
A2	GRY/YEL	A/C Clutch Relay	Relay Off: 12v, On: 1v
A3-4	---	Not Used	---
A5	GRN	Malfunction Indicator Lamp	MIL Off: 12v, On: 1v
A6	RED/GRN	B+ from Main Relay	12-14v
A7	WHT/RED	EVAP Purge Solenoid	Digital Signals: 0-12-0v
A8	WHT/GRN	EGR Solenoid Control	Sol. on: 1v, off: 12v
A9-10	---	Not Used	---
A11	GRY	Sensor Ground	<0.050v
A12	BLK/ORN	Power Ground	<0.1v

1994-95 EX, LX 3.2L V6 MFI 4WD VIN V [M/T] 12P 'B' Connector

PCM Pin #	W/Color	Circuit Description (12-Pin)	Value at Hot Idle
B1	RED/WHT	Keep Alive Power	12-14v
B2	BLK/PNK	HO2S-11 (B1 S1) Ground	<0.050v
B3	RED	HO2S-11 (B1 S1) Signal	0.1-1.1v
B4	BRN/BLK	Idle Air Control 'A' High	Pulse Signals
B5	BRN/WHT	Idle Air Control 'A' Low	Pulse Signals
B6	BRN/YEL	Idle Air Control 'B' Low	Pulse Signals
B7	BRN/RED	Idle Air Control 'B' High	Pulse Signals
B8	YEL/VLT	Electronic Ignition VREF Low	<0.050v
B9	YEL/RED	Electronic Ignition VREF High	Digital Pulses: 0-5-0-5v
B10-12	---	Not Used	---

```
12-PIN CONNECTOR 'A'
12 11 / / 8 7 6 5 / / 2 1
12-PIN CONNECTOR 'B'
/ / / 9 8 7 6 5 4 3 2 1
WIRE SIDE OF HARNESS TERMINALS
```

05533_ADIA_G659

Pin Connector Graphic

Standard Colors and Abbreviations

Abbreviation	Color	Abbreviation	Color	Abbreviation	Color
BLK	Black	LT BLU	Light Blue	TAN	Tan
BLU	Blue	LT GRN	Lt. Green	VIO	Violet
BRN	Brown	ORN	Orange	WHT	White
GRY	Gray	PNK	Pink	YEL	Yellow
GRN	Green	PPL	Purple		

1994-95 EX, LX 3.2L V6 MFI 4WD VIN V [M/T] 16P 'C' Connector

PCM Pin #	W/Color	Circuit Description (16-Pin)	Value at Hot Idle
C1	---	Not Used	---
C2	ORN/BLK	Data Link Connector	4.80v (no Scan Tool)
C3	PNK/WHT	Fuel Pump Relay Control	Relay on: 12v, off: 0v
C4	ORN/YEL	DLC Diagnostic Request	Digital Signals
C5	---	Not Used	---
C6	GRN/YEL	PSP Switch Signal	Straight: 12v, Turning: 0v
C7-8	---	Not Used	---
C9	YEL/BLK	A/C Switch Signal	Switch on: 12v, off: 0v
C10	---	Not Used	---
C11	BLU/YEL	Injector Pair 1 & 2	2.0-3.3 ms
C12	BLK/ORN	Power Ground	<0.1v
C13	---	Not Used	---
C14	BLU/WHT	Injector Pair 5 & 6	2.0-3.3 ms
C15	BLK/GRN	Power Ground	<0.1v
C16	RED/WHT	Keep Alive Power	12-14v

1994-95 EX, LX 3.2L V6 MFI 4WD VIN V [M/T] 16P 'D' Connector

PCM Pin #	W/Color	Circuit Description (16-Pin)	Value at Hot Idle
D1	BLK/GRN	Power Ground	<0.1v
D2	BLK/WHT	Sensor Ground	<0.050v
D3-4	---	Not Used	---
D5	YEL	Igniter Control Signal	Idle: 10% d/cycle
D6	BLU/RED	TP Sensor Signal	0.5-0.6v
D7	BLU	IAT Sensor Signal	Varies w/temp: 0.5-4.8v
D8	WHT	Vehicle Speed Sensor	Moving: 0-5-0-5v
D9	GRY/RED	MAP Sensor Signal	0.6-1.3v
D10	---	Not Used	---
D11	YEL/GRN	Ignition Control Bypass	Hot idle: 4.9v
D12	GRY/BLK	ECT Sensor Signal	At 180F: 0.54v
D13-14	---	Not Used	---
D15	BLU/GRN	Injector Pair 3 & 4	2.0-3.3 ms
D16	BLU/GRN	Injector Pair 3 & 4	2.0-3.3 ms

16-PIN CONNECTOR 'C'

| 16 | 15 | 14 | / | 12 | 11 | / | 9 | / | / | 6 | / | 4 | 3 | 2 | |

16-PIN CONNECTOR 'D'

| 16 | 15 | / | / | 12 | 11 | / | 9 | 8 | 7 | 6 | 5 | / | / | 2 | 1 |

WIRE SIDE OF HARNESS TERMINALS

05533_ADIA_G660

Pin Connector Graphic

1996-97 EX, LX 3.2L V6 4WD VIN V [A/T] C209 Red 'A' Connector

PCM Pin #	W/Color	Circuit Description (16-Pin)	Value at Hot Idle
A1	RED	Reference Voltage 'A'	4.9-5.1v
A2	YEL	Knock Sensor Signal	No detonation: 18 mv AC
A3, A12	---	Not Used	---
A4	RED/WHT	Keep Alive Power	12-14v
A5	BLU	Idle Air Control 'A' High	Pulse Signals
A6	BLU/WHT	Idle Air Control 'A' Low	Pulse Signals
A7	BLU/BLK	Idle Air Control 'B' Low	Pulse Signals
A8	BLU/RED	Idle Air Control 'B' High	Pulse Signals
A9	ORN/BLK	Transmission Fluid Lamp	Lamp on: 1v, off: 12v
A10	PNK/GRN	Winter Lamp	Lamp on: 1v. off: 12v
A11	PNK/WHT	Power Lamp	Lamp on: 1v. off: 12v
A13	GRN	Malfunction Indicator Lamp	MIL Off: 12v, On: 1v
A14	PPL	Check Transmission Light	Lamp on: 1v, off: 12v
A15	RED/BLU	EVAP Purge Solenoid	Digital Signals: 0-12-0v
A16	BRN/YEL	A/T: Shift Low Band Apply	Sol. on: 1v, off: 12v

1996-97 EX, LX 3.2L V6 4WD VIN V [A/T] C209 Red 'B' Connector

PCM Pin #	W/Color	Circuit Description (16-Pin)	Value at Hot Idle
B1	BLU/ORN	Reference Voltage 'B'	4.9-5.1v
B2	RED/WHT	Igniter 4 Control	Digital Signals: 0-12-0v
B3	RED/BLK	Igniter 2 Control	Digital Signals: 0-12-0v
B4	RED/GRN	Igniter 6 Control	Digital Signals: 0-12-0v
B5	ORN/GRN	Fuel Level Sensor	Tank empty: 1.8v
B6	GRY	Fuel Tank Pressure Sensor	With Fuel Cap Off: 2.5v
B7	YEL/RED	Linear EGR Sensor Signal	0.5-0.6v
B8	YEL/GRN	IAT Sensor Signal	Varies w/temp: 0.5-4.8v
B9	---	Not Used	---
B10	GRN/WHT	Rough Road Sensor	At idle: 2.5v
B11	GRN/YEL	PSP Switch Signal	Straight: 12v, Turning: 0v
B12	GRN/YEL	Headlight Switch Signal	Switch on: 0.1v, off: 12v
B13	ORN/BLK	Class 2 Serial Data Link	0v
B14	GRY/RED	A/C Clutch Relay	Relay Off: 12v, On: 1v
B15	PNK	Low Fuel Lamp	Lamp on: 1v, off: 12v
B16	BRN/WHT	EVAP Vent Solenoid	Sol. on: 1v, off: 12v

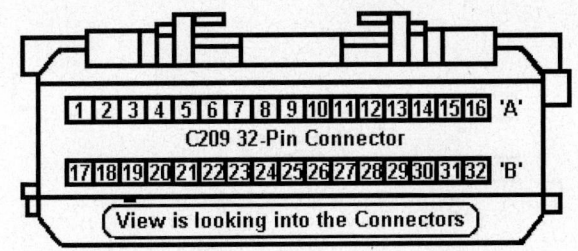

05533_ADIA_G654

Pin Connector Graphic
1996-97 EX, LX 3.2L 4WD VIN V [A/T] C211 White 'C' Connector

PCM Pin #	W/Color	Circuit Description (16-Pin)	Value at Hot Idle
C1	GRN/RED	Injector 4 Control	2.0-3.3 ms
C2	BRN/BLK	A/T: Shift Solenoid 'B'	SSB Off: 12v, On: 1v
C3	GRN/YEL	Injector 6 Control	2.0-3.3 ms
C4	RED	Igniter 1 Control	Digital Signals: 0-12-0v
C5	YEL	CKP Sensor Signal	Digital Signals: 0-5-0v
C7	YEL/BLK	Power Ground	<0.1v
C8	BLK/RED	Power Ground	<0.1v
C9	BLK/BLU	Power Ground	<0.1v
C10	BLK/RED	Tachometer Signal	Pulse Signals
C12	RED/WHT	Alternator Charge Signal	12-14v
C13	YEL/RED	Fuel Gauge PWM Signal	Duty cycle: 0-100%
C14	WHT	HO2S-21 (B2 S1) Signal	0.1-1.1v
C15	RED	HO2S-21 (B2 S1) Ground	<0.050v
C16	RED	HO2S-22 (B2 S2) Signal	0.1-1.1v

1996-97 EX, LX 3.2L 4WD VIN V [A/T] C211 White 'D' Connector

PCM Pin #	W/Color	Circuit Description (16-Pin)	Value at Hot Idle
D1	GRN/ORN	Injector 2 Control	2.0-3.3 ms
D2	RED/YEL	Torque Converter Clutch	TCC on: 1v, off: 12v
D3	GRN/WHT	Injector 1 Control	2.0-3.3 ms
D4	RED	Serial Data (8192 Baud)	Digital Signals
D5	RED/YEL	Igniter 5 Control	Digital Signals: 0-12-0v
D6	RED/BLU	Igniter 3 Control	Digital Signals: 0-12-0v
D7	WHT	Vehicle Speed Sensor	Moving: 0-5-0-5v
D8	GRN	Sensor 'A' Ground	<0.050v
D9	GRN	Sensor 'B' Ground	<0.050v
D10	YEL/BLU	MAF Sensor Signal	7-9v
D11	BLU	CMP Sensor Signal	Digital Signals: 0-5-0v
D12	BLU	HO2S-12 (B1 S2) Ground	<0.050v
D13	GRN	HO2S-12 (B1 S2) Signal	0.1-1.1v
D14	BLU	HO2S-11 (B1 S1) Ground	<0.050v
D15	PNK	HO2S-11 (B1 S1) Signal	0.1-1.1v
D16	GRN	HO2S-22 (B2 S2) Ground	<0.050v

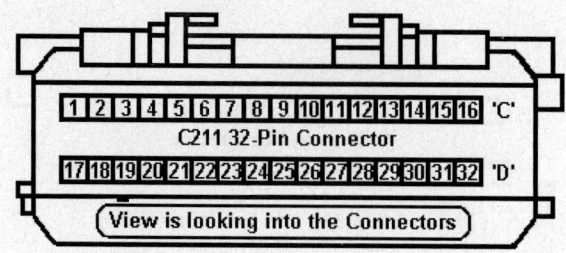

C211 32-Pin Connector

View is looking into the Connectors

05533_ADIA_G655

Pin Connector Graphic

1996-97 EX, LX 3.2L V6 4WD VIN V [A/T] C212 Blue 'E' Connector

PCM Pin #	W/Color	Circuit Description (16-Pin)	Value at Hot Idle
E1	YEL	VSS "High" Signal	AC Pulse Signals
E2	BRN	VSS "Low" Signal	<0.050v
E3	RED/ORN	A/T: Pressure Control 'N'	Pulse Signals
E4	RED/BLK	A/T: Pressure Control 'P'	Pulse Signals
E5	RED/GRN	Linear EGR Solenoid "High"	Pulse Signals
E6	YEL	Linear EGR Solenoid "Low"	Pulse Signals
E7	PNK	A/T: Range Signal 'B'	In P/N: 1v, others: 12v
E8	BLU	TP Sensor Signal	0.5-0.6v
E9	BLU/RED	ECT Sensor Signal	At 180F: 0.54v
E10	---	Not Used	---
E11	YEL/RED	CKP Sensor VREF	4.9-5.1v
E12	PNK/BLU	A/T: Range Signal 'A'	In P/N: 1v, others: 12v
E13	PNK/WHT	Fuel Pump Relay Control	Relay on: 12v, off: 0v
E14	BRN/WHT	A/T: Shift Solenoid Feed (B+)	12-14v
E15	GRN/BLK	A/C Switch Signal	Switch on: 12v, off: 0v
E16	RED/BLU	B+ from Main Relay	12-14v

1996-97 EX, LX 3.2L V6 4WD VIN V [A/T] C212 Blue 'F' Connector

PCM Pin #	W/Color	Circuit Description (16-Pin)	Value at Hot Idle
F1	---	Not Used	---
F2	BLU/WHT	A/T: Range Signal 'C'	In P/N: 1v, others: 12v
F3	PNK/BLK	A/T: Range Signal 'P'	In P/N: 1v, others: 12v
F4	RED	Brake Switch Signal	Brake on: 12v, Off: 0v
F5	PPL/RED	Power Switch Signal	Switch on: 0.1v, off: 12v
F6	PPL/GRN	Winter Switch Signal	Switch on: 0.1v, off: 12v
F7	GRN/RED	A/T: TOT Sensor Signal	Varies w/temp. (0.5-4.9v)
F8	RED	MAP Sensor Signal	0.6-1.3v
F9, F12	---	Not Used	---
F10	GRY/BLU	Cruise Control	C/C on: 0.1v, off: 12v
F11	LT BLU	A/T: Kickdown Switch	Switch on: 0.1v, off: 12v
F13	WHT/GRN	Injector 3 Control	2.0-3.3 ms
F14	YEL/GRN	A/T: Shift Solenoid 'A'	Sol. on: 1v, off: 12v
F15	GRN/BLK	Injector 5 Control	2.0-3.3 ms
F16	RED/BLU	B+ from Main Relay	12-14v

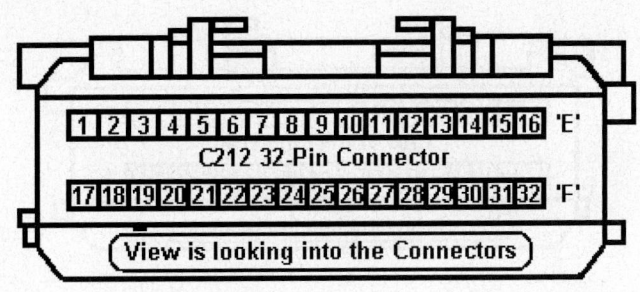

View is looking into the Connectors

05533_ADIA_G656

Pin Connector Graphic

1996-97 EX, LX 3.2L V6 4WD VIN V [M/T] C209 Red 'A' Connector

PCM Pin #	W/Color	Circuit Description (16-Pin)	Value at Hot Idle
A1	RED	Reference Voltage 'A'	4.9-5.1v
A2	YEL	Knock Sensor Signal	No detonation: 18 mv AC
A3	---	Not Used	---
A4	RED/WHT	Keep Alive Power	12-14v
A5	BLU	Idle Air Control 'A' High	Pulse Signals
A6	BLU/WHT	Idle Air Control 'A' Low	Pulse Signals
A7	BLU/BLK	Idle Air Control 'B' Low	Pulse Signals
A8	BLU/RED	Idle Air Control 'B' High	Pulse Signals
A9-12	---	Not Used	---
A13	GRN	Malfunction Indicator Lamp	MIL Off: 12v, On: 1v
A14	PPL	Check Transmission Light	Lamp on: 1v, off: 12v
A15	RED/BLU	EVAP Purge Solenoid	Digital Signals: 0-12-0v
A16	---	Not Used	---

1996-97 EX, LX 3.2L V6 4WD VIN V [M/T] C209 Red 'B' Connector

PCM Pin #	W/Color	Circuit Description (16-Pin)	Value at Hot Idle
B1	BLU/ORN	Reference Voltage 'B'	4.9-5.1v
B2	RED/WHT	Igniter 4 Control	Digital Signals: 0-12-0v
B3	RED/BLK	Igniter 2 Control	Digital Signals: 0-12-0v
B4	RED/GRN	Igniter 6 Control	Digital Signals: 0-12-0v
B5	ORN/GRN	Fuel Level Sensor	Tank empty: 1.8v
B6	GRY	Fuel Tank Pressure Sensor	With Fuel Cap Off: 2.5v
B7	YEL/RED	Linear EGR Sensor Signal	0.5-0.6v
B8	YEL/GRN	IAT Sensor Signal	Varies w/temp: 0.5-4.8v
B9	---	Not Used	---
B10	GRN/WHT	Rough Road Sensor	At idle: 2.5v
B11	GRN/YEL	PSP Switch Signal	Straight: 12v, Turning: 0v
B12	GRN/YEL	Headlight Switch Signal	Switch on: 0.1v, off: 12v
B13	ORN/BLK	Class 2 Serial Data Link	0v
B14	GRY/RED	A/C Clutch Relay	Relay Off: 12v, On: 1v
B15	PNK	Low Fuel Lamp	Lamp on: 1v, off: 12v
B16	BRN/WHT	EVAP Vent Solenoid	Sol. on: 1v, off: 12v

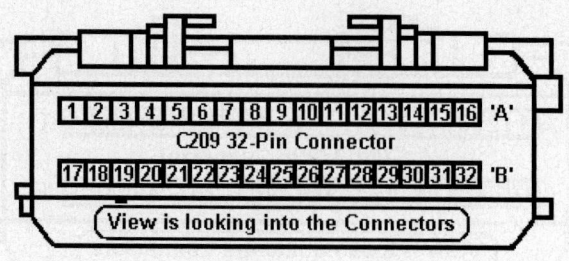

C209 32-Pin Connector

View is looking into the Connectors

05533_ADIA_G654

Pin Connector Graphic

1996-97 EX, LX 3.2L 4WD VIN V [M/T] C211 White 'C' Connector

PCM Pin #	W/Color	Circuit Description (16-Pin)	Value at Hot Idle
C1	GRN/RED	Injector 4 Control	2.0-3.3 ms
C2	---	Not Used	---
C3	GRN/YEL	Injector 6 Control	2.0-3.3 ms
C4	RED	Igniter 1 Control	Digital Signals: 0-12-0v
C5	YEL	CKP Sensor Signal	Digital Signals: 0-5-0v
C7	YEL/BLK	Power Ground	<0.1v
C8	BLK/RED	Power Ground	<0.1v
C9	BLK/BLU	Power Ground	<0.1v
C10	BLK/RED	Tachometer Signal	Pulse Signals
C11	YEL/GRN	Upshift Lamp Control	U/L on: 1v, off: 12
C12	RED/WHT	Alternator Charge Signal	12-14v
C13	YEL/RED	Fuel Gauge PWM Signal	Duty cycle: 0-100%
C14	WHT	HO2S-21 (B2 S1) Signal	0.1-1.1v
C15	RED	HO2S-21 (B2 S1) Ground	<0.050v
C16	RED	HO2S-22 (B2 S2) Signal	0.1-1.1v

1996-97 EX, LX 3.2L 4WD VIN V [M/T] C211 White 'D' Connector

PCM Pin #	W/Color	Circuit Description (16-Pin)	Value at Hot Idle
D1	GRN/ORN	Injector 2 Control	2.0-3.3 ms
D2	---	Not Used	---
D3	GRN/WHT	Injector 1 Control	2.0-3.3 ms
D4	RED	Serial Data (8192 Baud)	Digital Signals
D5	RED/YEL	Igniter 5 Control	Digital Signals: 0-12-0v
D6	RED/BLU	Igniter 3 Control	Digital Signals: 0-12-0v
D7	WHT	Vehicle Speed Sensor	Moving: 0-5-0-5v
D8	GRN	Sensor 'A' Ground	<0.050v
D9	GRN	Sensor 'B' Ground	<0.050v
D10	YEL/BLU	MAF Sensor Signal	7-9v
D11	BLU	CMP Sensor Signal	Digital Signals: 0-5-0v
D12	BLU	HO2S-13 (B1 S3) Ground	<0.050v
D13	GRN	HO2S-13 (B1 S3) Signal	0.1-1.1v
D14	BLU	HO2S-11 (B1 S1) Ground	<0.050v
D15	PNK	HO2S-11 (B1 S1) Signal	0.1-1.1v
D16	GRN	HO2S-22 (B2 S2) Ground	<0.050v

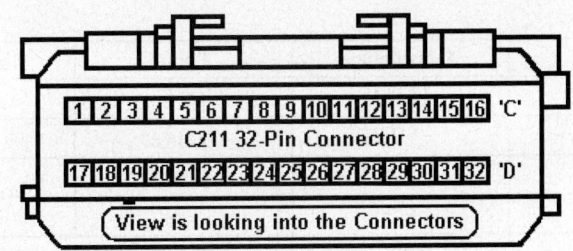

05533_ADIA_G655

Pin Connector Graphic
1996-97 EX, LX 3.2L V6 4WD VIN V [M/T] C212 Blue 'E' Connector

PCM Pin #	W/Color	Circuit Description (16-Pin)	Value at Hot Idle
E1-4	---	Not Used	---
E5	RED/GRN	Linear EGR Solenoid 'P'	Pulse Signals
E6	YEL	Linear EGR Solenoid 'N'	Pulse Signals
E7	---	Not Used	---
E8	BLU	TP Sensor Signal	0.5-0.6v
E9	BLU/RED	ECT Sensor Signal	At 180F: 0.54v
E10	---	Not Used	---
E11	YEL/RED	CKP Sensor VREF	4.9-5.1v
E12	---	Not Used	---
E13	PNK/WHT	Fuel Pump Relay Control	Relay on: 12v, off: 0v
E14	---	Not Used	---
E15	GRN/BLK	A/C Switch Signal	Switch on: 12v, off: 0v
E16	RED/BLU	B+ from Main Relay	12-14v

1996-97 EX, LX 3.2L V6 4WD VIN V [M/T] C212 Blue 'F' Connector

PCM Pin #	W/Color	Circuit Description (16-Pin)	Value at Hot Idle
F1-7	---	Not Used	---
F8	RED	MAP Sensor Signal	0.6-1.3v
F9-12	---	Not Used	---
F13	WHT/GRN	Injector 3 Control	2.0-3.3 ms
F14	---	Not Used	---
F15	GRN/BLK	Injector 5 Control	2.0-3.3 ms
F16	RED/BLU	B+ from Main Relay	12-14v

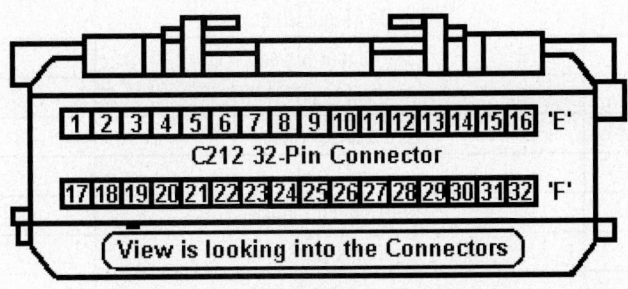

05533_ADIA_G656

Pin Connector Graphic

Standard Colors and Abbreviations

Abbreviation	Color	Abbreviation	Color	Abbreviation	Color
BLK	Black	LT BLU	Light Blue	TAN	Tan
BLU	Blue	LT GRN	Lt. Green	VIO	Violet
BRN	Brown	ORN	Orange	WHT	White
GRY	Gray	PNK	Pink	YEL	Yellow
GRN	Green	PPL	Purple		

1998-03 EX, LX 3.2L VIN W 4WD [A/T] 80P J21 Blue Connector

PCM Pin #	W/Color	Circuit Description (40-Pin)	Value at Hot Idle
1	BLK	Power Ground	<0.1v
2	GRN	Reference Voltage 'A'	4.9-5.1v
3	RED/WHT	Reference Voltage 'B'	4.9-5.1v
4	GRY/RED	A/C Clutch Relay Control	Relay Off: 12v, On: 1v
5	GRN/RED	A/T: Shift Solenoid 'A'	SSA Off: 12v, On: 1v
6	---	Not Used	---
7	BLK/RED	Tachometer Signal	Pulse Signals
8	---	Not Used	---
9	YEL	Knock Sensor 1 Signal	No detonation: 18 mv AC
10	RED	Knock Sensor 2 Signal	No detonation: 18 mv AC
11	---	Not Used	---
12	WHT/RED	CMP Sensor Signal	Digital Signals: 0-5-0v
13	PNK	To be done	---
14	BLK/BLU	MAF Sensor Signal	4.2v
15-16	---	Not Used	---
17	RED	HO2S-22 (B2 S2) Ground	0.1-1.1v
18	---	Not Used	---
19	RED/GRN	Ignition Power (Main Relay)	12-14v
20	RED/WHT	Keep Alive Power	12-14v
21	WHT	HO2S-11 (B1 S1) Ground	<0.050v
22	---	Not Used	---
23	BLU	HO2S-12 (B1 S2) Ground	<0.050v
24	---	Not Used	---
25	RED/BLU	Coil On Plug 3 Control	Digital Signals: 0-12-0v
26	RED/BLK	Coil On Plug 2 Control	Digital Signals: 0-12-0v
27	GRN	IAT Sensor Signal	Varies w/temp: 0.5-4.8v
28	---	Not Used	---
29	RED	MAP Sensor Signal	0.9v
30-36	---	Not Used	---
37	GRN/YEL	PSP Switch Signal	Straight: 12v, Turning: 0v
38	GRN/YEL	Headlights On Signal	Lights Off: 12v, On: 0v
39	RED	Brake Switch Signal	Brake Off: 0v, On: 12v
40	BLK/BLU	Power Ground	<0.1v

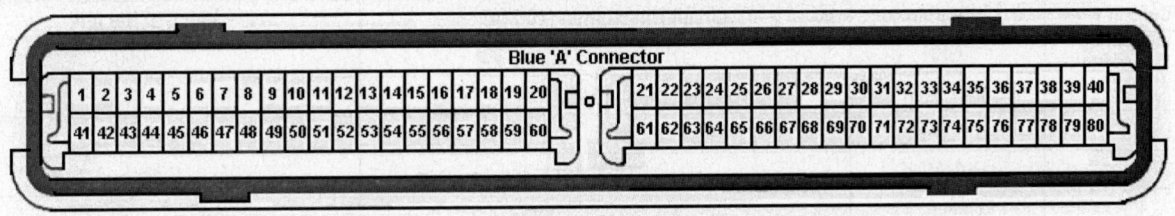

05533_ADIA_G657

Pin Connector Graphic

1998-03 EX, LX 3.2L VIN W 4WD [A/T] 80P J21 Blue Connector

PCM Pin #	W/Color	Circuit Description (40-Pin)	Value at Hot Idle
41	GRN	TP Sensor 1 Ground	<0.050v
42	PNK/WHT	Fuel Pump Relay Control	Relay Off: 0v, On: 12v
43	PPL/RED	Pressure Control Solenoid (-)	Pulse Signals
44	ORN/WHT	HO2S-21 (B2 S1) Heater	Digital Signals: 0-12-0v
45	GRN/BLK	A/C Switch Signal	A/C On: 12v, Off: 0v
46	BRN/RED	Brake Switch Signal	Brake Off: 12v, On: 0v
47	PPL/WHT	Pressure Control Solenoid (+)	Pulse Signals
48	BLU	Throttle Motor (-) Signal	Pulse Signals
49	BLK	Knock Sensor 1 Ground	<0.050v
50	ORN	HO2S-22 (B2 S2) Ground	<0.050v
51	---	Not Used	---
52	RED	HO2S-21 (B2 S1) Ground	<0.050v
53	BLK	ECT & EGR Sensor Ground	<0.050v
54	BLK	FTP Sensor Ground	<0.050v
55	---	Not Used	---
56	GRN/BLK	Injector 5 Control	2.0-3.3 ms
57	RED/WHT	Keep Alive Power	12-14v
58	ORN/BLK	Class 2 Serial Data Link	0v
59	---	Not Used	---
60	BLK	Knock Sensor 2 Ground	<0.050v
61	GRN/WHT	TP Sensor 2 Ground	<0.050v
62	GRN/ORN	Injector 2 Control	2.0-3.3 ms
63	RED	A/P Sensor 1 Ground	<0.1v
64	GRN/RED	Injector 4 Control	2.0-3.3 ms
65	BLU/WHT	TP Sensor 2 Signal	0.5-0.6v
66	GRN	Injector 3 Control	2.0-3.3 ms
67	YEL	PCM C.Q. Input (1-3-5)	N/A
68	RED	Knock Sensor Input (2-4-6)	<0.050v
69	GRN/WHT	Injector 1 Control	2.0-3.3 ms
70	---	Not Used	---
71	GRN/WHT	HO2S-22 (B2 S2) Heater	Digital Signals: 0-12-0v
72	WHT/BLU	C/C Resume/Accel Input	Switch Off: 0v, On: 12v
73	WHT/BLU	CKP Sensor Signal	Digital Signals: 0-5-0v
74	BLU/RED	ECT Sensor Signal	At 180F: 0.54v
75	RED/GRN	Ignition Power (Main Relay)	12-14v
76	PNK/BLK	A/T: Range Signal 'P'	In 'P': 12v, Others: 0v
77	PNK/BLU	A/T: Range Signal 'A'	In 'N': 0v, Others: 12v
78	PNK/YEL	A/T: Range Signal 'C'	In 'C': 0v, Others: 12v
79	PNK	A/T: Range Signal 'B'	In 'B': 0v, Others: 12v
80	BLU	A. P. Sensor 2 Ground	<0.050v

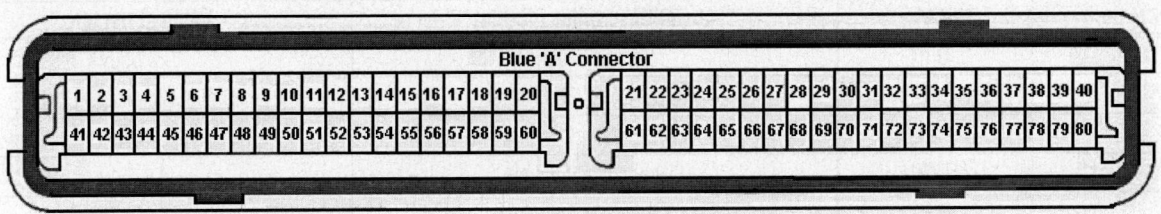

Blue 'A' Connector

05533_ADIA_G657

Pin Connector Graphic

1998-03 EX, LX 3.2L VIN W 4WD [A/T] 80P J22 Red Connector

PCM Pin #	W/Color	Circuit Description (40-Pin)	Value at Hot Idle
1	GRN	Sensor Ground	<0.050v
2	BLK/ORN	HO2S-11 (B1 S1) Heater	Digital Signals: 0-12-0v
3	YEL	A/P Sensor 2 Signal	N/A
4	ORN	Reference Voltage	4.9-5.1v
5	WHT	CKP Sensor Reference	4.9-5.1v
6	GRY/GRN	Cruise Set Switch	Switch Off: 0v, On: 12v
7	RED/BLU	EVAP Purge Solenoid	Valve Off: 12v, On: 1v
8	GRN/WHT	Indicator Control	N/A
9	RED/GRN	Ignition Power (Main Relay)	12-14v
10	---	Not Used	---
11	WHT/GRN	Malfunction Indicator Lamp	MIL Off: 12v, On: 1v
12	GRN	Throttle Motor (+) Signal	Pulse Signals
13	RED/WHT	Coil On Plug 4 Control	Digital Signals: 0-12-0v
14	BLU	HO2S-12 (B1 S2) Ground	<0.050v
15	BLU	HO2S-22 (B2 S2) Ground	<0.050v
16	BLU	HO2S-22 (B2 S2) Ground	<0.050v
17	PNK	HO2S-21 (B1 S1) Ground	<0.050v
18	WHT	HO2S-11 (B1 S1) Ground	<0.050v
19	RED/YEL	Coil On Plug 5 Control	Digital Signals: 0-12-0v
20	YEL/GRN	Sensor Ground	<0.050v
21	BLU/YEL	Sensor Ground	<0.050v
22	BLU/YEL	OSS Sensor 'P' Signal	AC Pulses
24	RED	HO2S-22 (B2 S2) Ground	0.1-1.1v
25	---	Not Used	---
26	BLU/RED	EGR Position Solenoid	Digital Signals: 0-12-0v
27	BLK/YEL	Vehicle Speed Sensor	Moving: 0-5-0-5v
28	GRN/YEL	Injector 6 Control	2.0-3.3 ms
29	---	Not Used	---
30	WHT/GRN	Cruise Engage Switch	Switch Off: 0v, On: 12v
31	GRY/RED	FTP Sensor Reference	4.9-5.1v
32	RED/GRN	Ignition Power (Main Relay)	12-14v
33-34	---	Not Used	---
35	RED/GRN	Coil On Plug 6 Control	Digital Signals: 0-12-0v
36	BLK	A/P Sensor 1 Reference	4.9-5.1v
37	---	Not Used	---
38	RED	TP Sensor 1 Reference	4.9-5.1v
39-40	---	Not Used	---

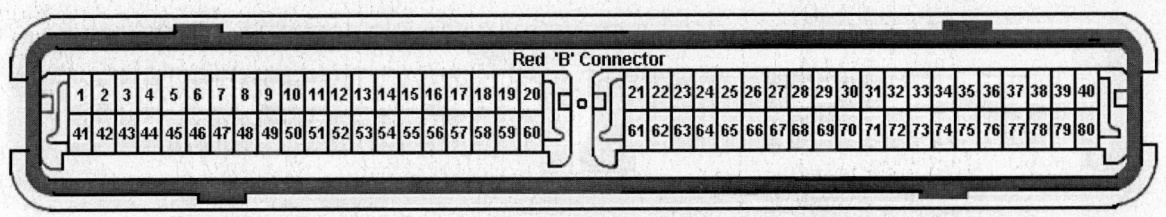

Red 'B' Connector

05533_ADIA_G658

Pin Connector Graphic
1998-03 EX, LX 3.2L VIN W 4WD [A/T] 80P J22 Red Connector

PCM Pin #	W/Color	Circuit Description (40-Pin)	Value at Hot Idle
41	BLU/ORN	A/P Sensor 3 Ground	<0.050v
42	WHT/RED	HO2S-12 (B1 S2) Heater	Digital Signals: 0-12-0v
43-46	---	Not Used	---
47	WHT/BLK	Water Gauge Signal	0-5v
48	---	Not Used	---
49	RED/WHT	Class 2 Data Line to DLC	0v
50	ORN/GRN	Fuel Level Sensor Signal	Duty Cycle Signal
51	---	Not Used	---
52	GRN	FTP Sensor Signal	With Fuel Cap Off: 2.5v
53	---	Not Used	---
54	BLU	HO2S-12 (B1 S2) Ground	<0.050v
55	BLU/GRN	OSS Sensor 'N' Signal	AC Pulses
56-57	---	Not Used	---
58	ORN/BLU	HO2S-11 (B1 S1) Signal	0.1-1.1v
59	LT GRN	Cruise Set Switch	Switch Off: 0v, On: 12v
60	BLU/YEL	Fuel Level Sensor Reference	12v
61	YEL/GRN	IAT Sensor Signal	Varies w/temp: 0.5-4.8v
62	RED	Coil On Plug 1 Control	Digital Signals: 0-12-0v
63	WHT	HO2S-21 (B2 S1) Signal	0.1-1.1v
64	PNK	HO2S-12 (B1 S2) Signal	0.1-1.1v
65	GRN	HO2S-22 (B2 S2) Signal	0.1-1.1v
66	---	Not Used	---
67	GRY/RED	EGR Position Sensor Signal	1.2-2.0v
68	WHT	A/P Sensor 1 Signal	N/A
69	GRN	Throttle Motor (+) Signal	Pulse Signals
70	---	Not Used	---
71	GRN/WHT	HO2S-22 (B2 S2) Heater	Digital Signals: 0-12-0v
72	RED/GRN	Ignition Power (Main Relay)	12-14v
73	GRN/WHT	Cruise Control Indicator	Lamp Off: 12v, On: 1v
74	YEL/BLK	Intake Air VSV Control	Valve Off: 12v, On: 1v
75	BRN/WHT	EVAP Purge Cut Solenoid	Valve Off: 12v, On: 1v
76	BLU	TP Sensor 1 Signal	0.5-0.8v
77	ORN	A/P Sensor 3 Reference	4.9-5.1v
78	YEL	A/P Sensor 2 Signal	N/A
79	BLU/GRN	A/P Sensor 3 Signal	N/A
80	GRN	MAP Sensor Reference	4.9-5.1v

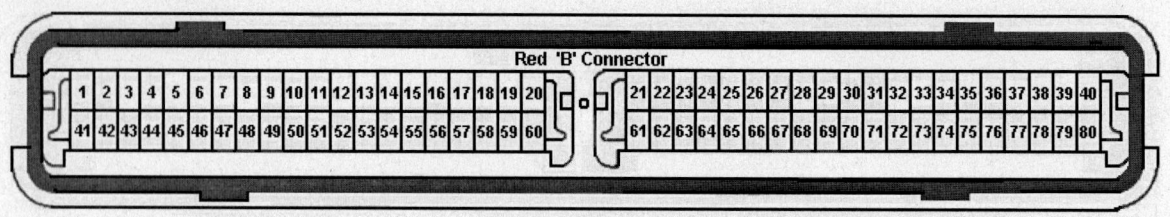

05533_ADIA_G658

Pin Connector Graphic

1998-03 EX, LX 3.2L VIN W 4WD [M/T] 80P J21 Blue Connector

PCM Pin #	W/Color	Circuit Description (40-Pin)	Value at Hot Idle
1	BLK	Power Ground	<0.1v
2	GRN	Reference Voltage 'A'	4.9-5.1v
3	RED/WHT	Reference Voltage 'B'	4.9-5.1v
4	GRY/RED	A/C Clutch Relay Control	Relay Off: 12v, On: 1v
5-6	---	Not Used	---
7	BLK/RED	Tachometer Signal	Pulse Signals
8	---	Not Used	---
9	YEL	Knock Sensor 1 Signal	No detonation: 18 mv AC
10	RED	Knock Sensor 2 Signal	No detonation: 18 mv AC
11	---	Not Used	---
12	WHT/RED	CMP Sensor Signal	Digital Signals: 0-5-0v
13	PNK	To be done	---
14	BLK/BLU	MAF Sensor Signal	4.2v
15-16	---	Not Used	---
17	RED	HO2S-22 (B2 S2) Ground	0.1-1.1v
18	---	Not Used	---
19	RED/GRN	Ignition Power (Main Relay)	12-14v
20	RED/WHT	Keep Alive Power	12-14v
21	WHT	HO2S-11 (B1 S1) Ground	<0.050v
22	---	Not Used	---
23	BLU	HO2S-12 (B1 S2) Ground	<0.050v
24	---	Not Used	---
25	RED/BLU	Coil On Plug 3 Control	Digital Signals: 0-12-0v
26	RED/BLK	Coil On Plug 2 Control	Digital Signals: 0-12-0v
27	GRN	IAT Sensor Signal	Varies w/temp: 0.5-4.8v
28	---	Not Used	---
29	RED	MAP Sensor Signal	0.9v
30-36	---	Not Used	---
37	GRN/YEL	PSP Switch Signal	Straight: 12v, Turning: 0v
38	GRN/YEL	Headlights On Signal	Lights Off: 12v, On: 0v
39	RED	Brake Switch Signal	Brake Off: 0v, On: 12v
40	BLK/BLU	Power Ground	<0.1v

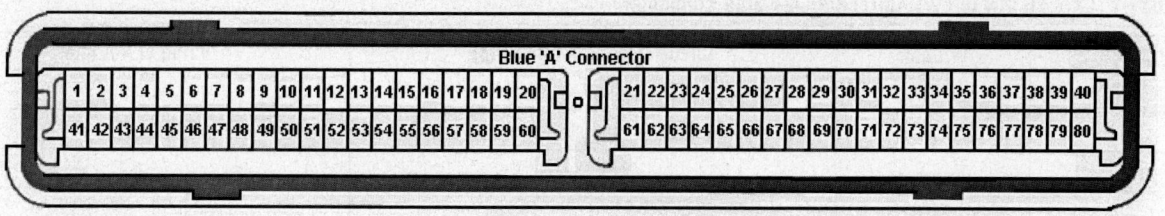

05533_ADIA_G657

Pin Connector Graphic

1998-03 EX, LX 3.2L VIN W 4WD [M/T] 80P J21 Blue Connector

PCM Pin #	W/Color	Circuit Description (40-Pin)	Value at Hot Idle
41	GRN	TP Sensor 1 Ground	<0.050v
42	PNK/WHT	Fuel Pump Relay Control	Relay Off: 0v, On: 12v
43	---	Not Used	---
44	ORN/WHT	HO2S-21 (B2 S1) Heater	Digital Signals: 0-12-0v
45	GRN/BLK	A/C Switch Signal	A/C On: 12v, Off: 0v
46	BRN/RED	Brake Switch Signal	Brake Off: 12v, On: 0v
47	---	Not Used	---
48	BLU	Throttle Motor (-) Signal	Pulse Signals
49	BLK	Knock Sensor 1 Ground	<0.050v
50	ORN	HO2S-22 (B2 S2) Ground	<0.050v
51	---	Not Used	---
52	RED	HO2S-21 (B2 S1) Ground	<0.050v
53	BLK	ECT & EGR Sensor Ground	<0.050v
54	BLK	FTP Sensor Ground	<0.050v
55	---	Not Used	---
56	GRN/BLK	Injector 5 Control	2.0-3.3 ms
57	RED/WHT	Keep Alive Power	12-14v
58	ORN/BLK	Class 2 Serial Data Link	0v
59	---	Not Used	---
60	BLK	Knock Sensor 2 Ground	<0.050v
61	GRN/WHT	TP Sensor 2 Ground	<0.050v
62	GRN/ORN	Injector 2 Control	2.0-3.3 ms
63	RED	A/P Sensor 1 Ground	<0.1v
64	GRN/RED	Injector 4 Control	2.0-3.3 ms
65	BLU/WHT	TP Sensor 2 Signal	0.5-0.6v
66	GRN	Injector 3 Control	2.0-3.3 ms
67	YEL	PCM C.Q. Input (1-3-5)	Digital Signals: 0-12-0v
68	RED	Knock Sensor Input (2-4-6)	<0.050v
69	GRN/WHT	Injector 1 Control	2.0-3.3 ms
70	---	Not Used	---
71	GRN/WHT	HO2S-22 (B2 S2) Heater	Digital Signals: 0-12-0v
72	WHT/BLU	C/C Resume/Accel Input	Switch Off: 0v, On: 12v
73	WHT/BLU	CKP Sensor Signal	Digital Signals: 0-5-0v
74	BLU/RED	ECT Sensor Signal	At 180F: 0.54v
75	RED/GRN	Ignition Power (Main Relay)	12-14v
76	PNK/BLK	Clutch Switch Signal	Clutch Out: 12v, In: 0v
77-79	---	Not Used	---
80	BLU	A. P. Sensor 2 Ground	<0.050v

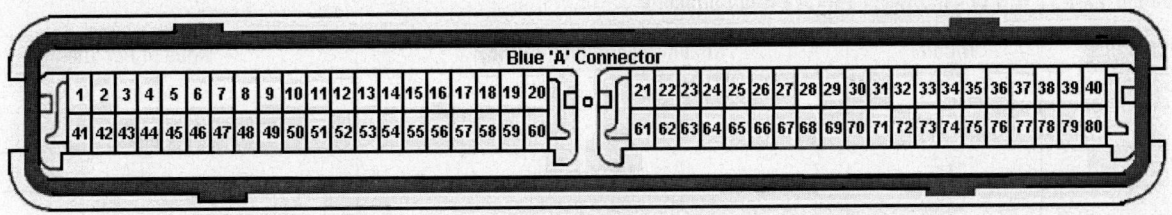

05533_ADIA_G657

Pin Connector Graphic

1998-03 EX, LX 3.2L VIN W 4WD [M/T] 80P J22 Red Connector

PCM Pin #	W/Color	Circuit Description (40-Pin)	Value at Hot Idle
1	GRN	Sensor Ground	<0.050v
2	BLK/ORN	HO2S-11 (B1 S1) Heater	Digital Signals: 0-12-0v
3	YEL	A/P Sensor 2 Signal	N/A
4	ORN	Reference Voltage	4.9-5.1v
5	WHT	CKP Sensor Reference	4.9-5.1v
6	GRY/GRN	Cruise Set Switch	Switch Off: 0v, On: 12v
7	RED/BLU	EVAP Purge Solenoid	Valve Off: 12v, On: 1v
8	GRN/WHT	Indicator Control	N/A
9	RED/GRN	Ignition Power (Main Relay)	12-14v
10	---	Not Used	---
11	WHT/GRN	Malfunction Indicator Lamp	MIL Off: 12v, On: 1v
12	GRN	Throttle Motor (+) Signal	Pulse Signals
13	RED/WHT	Coil On Plug 4 Control	Digital Signals: 0-12-0v
14	BLU	HO2S-12 (B1 S2) Ground	<0.050v
15	BLU	HO2S-22 (B2 S2) Ground	<0.050v
16	BLU	HO2S-22 (B2 S2) Ground	<0.050v
17	PNK	HO2S-21 (B1 S1) Ground	<0.050v
18	WHT	HO2S-11 (B1 S1) Ground	<0.050v
19	RED/YEL	Coil On Plug 5 Control	Digital Signals: 0-12-0v
20	YEL/GRN	Sensor Ground	<0.050v
21	BLU/YEL	Sensor Ground	<0.050v
22	---	Not Used	---
24	RED	HO2S-22 (B2 S2) Ground	0.1-1.1v
25	---	Not Used	---
26	BLU/RED	EGR Position Solenoid	Digital Signals: 0-12-0v
27	BLK/YEL	Vehicle Speed Sensor	Moving: 0-5-0-5v
28	GRN/YEL	Injector 6 Control	2.0-3.3 ms
29	---	Not Used	---
30	WHT/GRN	Cruise Engage Switch	Switch Off: 0v, On: 12v
31	GRY/RED	FTP Sensor Reference	4.9-5.1v
32	RED/GRN	Ignition Power (Main Relay)	12-14v
33-34	---	Not Used	---
35	RED/GRN	Coil On Plug 6 Control	Digital Signals: 0-12-0v
36	BLK	A/P Sensor 1 Reference	4.9-5.1v
37	---	Not Used	---
38	RED	TP Sensor 1 Reference	4.9-5.1v
39-40	---	Not Used	---

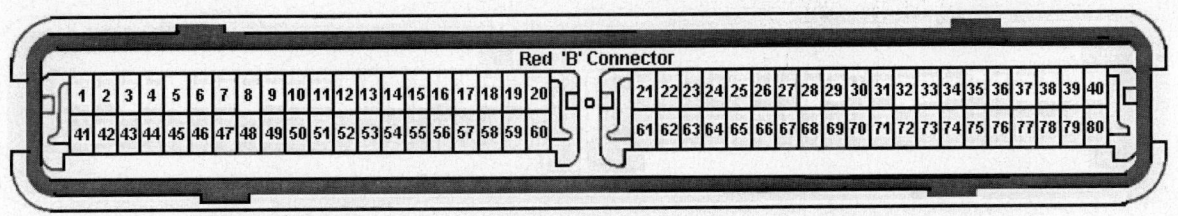

Red 'B' Connector

05533_ADIA_G658

Pin Connector Graphic

1998-03 EX, LX 3.2L VIN W 4WD [M/T] 80P J22 Red Connector

PCM Pin #	W/Color	Circuit Description (40-Pin)	Value at Hot Idle
41	BLU/ORN	A/P Sensor 3 Ground	<0.050v
42	WHT/RED	HO2S-12 (B1 S2) Heater	Digital Signals: 0-12-0v
43-46	---	Not Used	---
47	WHT/BLK	Water Gauge Signal	0-5v
48	---	Not Used	---
49	RED/WHT	Class 2 Data Line to DLC	0v
50	ORN/GRN	Fuel Level Sensor Signal	Duty Cycle Signal
51	---	Not Used	---
52	GRN	FTP Sensor Signal	With Fuel Cap Off: 2.5v
53	---	Not Used	---
54	BLU	HO2S-12 (B1 S2) Ground	<0.050v
55-57	---	Not Used	---
58	ORN/BLU	HO2S-11 (B1 S1) Signal	0.1-1.1v
59	LT GRN	Cruise Set Switch	Switch Off: 0v, On: 12v
60	BLU/YEL	Fuel Level Sensor Reference	12v
61	YEL/GRN	IAT Sensor Signal	Varies w/temp: 0.5-4.8v
62	RED	Coil On Plug 1 Control	Digital Signals: 0-12-0v
63	WHT	HO2S-21 (B2 S1) Signal	0.1-1.1v
64	PNK	HO2S-12 (B1 S2) Signal	0.1-1.1v
65	GRN	HO2S-22 (B2 S2) Signal	0.1-1.1v
66	---	Not Used	---
67	GRY/RED	EGR Position Sensor Signal	1.2-2.0v
68	WHT	A/P Sensor 1 Signal	N/A
69	GRN	Throttle Motor (+) Signal	Pulse Signals
70	---	Not Used	---
71	GRN/WHT	HO2S-22 (B2 S2) Heater	Digital Signals: 0-12-0v
72	RED/GRN	Ignition Power (Main Relay)	12-14v
73	GRN/WHT	Cruise Control Indicator	Lamp Off: 12v, On: 1v
74	YEL/BLK	Intake Air VSV Control	Valve Off: 12v, On: 1v
75	BRN/WHT	EVAP Purge Cut Solenoid	Valve Off: 12v, On: 1v
76	BLU	TP Sensor 1 Signal	0.5-0.8v
77	ORN	A/P Sensor 3 Reference	4.9-5.1v
78	YEL	A/P Sensor 2 Signal	N/A
79	BLU/GRN	A/P Sensor 3 Signal	N/A
80	GRN	MAP Sensor Reference	4.9-5.1v

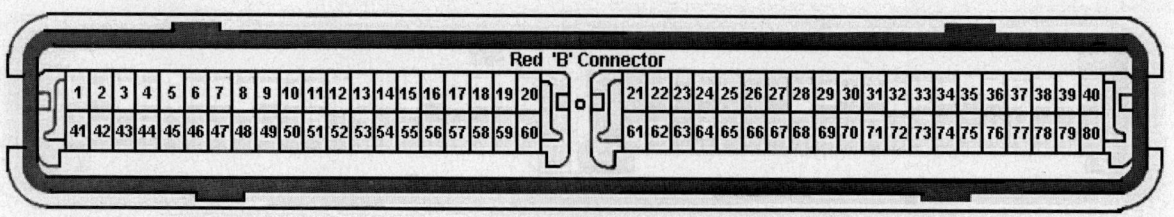

05533_ADIA_G658

Pin Connector Graphic

Chilton 2006 Labor Guide Manuals ISBN 1-4180-1688-8/Part No. 131688

We've added hundreds of new labor operations-including maintenance services and electronic system diagnosis to the Chilton 2006 Labor Guide Manuals. All labor times for the 1981 through 2005 (and available 2006) domestic and imported vehicles consider the real world environment in which technicians work. The parts terminology is more standardized across different OEMs to simplify reference. You can reference any of three labor times for many models: Chilton's Standard and Severe Service times, plus OEM warranty time. Each OEM is arranged alphabetically by section for easy reference, and improved indexing means easier access to today's repair industry standards. Chilton labor times are accepted by most insurance and extended warranty companies. Vehicle makes and models conform to current Automotive Aftermarket Industry Association standards. Make sure you have the latest edition of these manuals because our experts have updated hundreds of labor times for earlier models as well!

Labor Guide Manual Benefits:

- a total of 2,500 pages of updated Chilton labor times split into two volumes (Domestic and Imported vehicles)
- enjoy quicker referencing than ever by using separate, easier-to-handle, domestic and imported vehicle manuals
- find it fast by using tabs that display contents by manufacturer and model, two indexes — labor operations and systems — in each model group, and page numbering that includes manufacturer code so you know where you are in the book
- manufacturers are arranged alphabetically within each volume

Hardcover Manuals are 8 1/2" x 11", ©2006

Labor Guide CD-ROM Benefits:

- automatically calculates labor charges, taxes, and your parts as total job estimates
- creates professional estimates for your customers and worksheets for your technicians, allowing them to print them whenever needed
- keeps track of customers, prior estimates, and your own parts or package jobs

CD ISBN 1-4180-0605-X/Part No. 130605
©2006

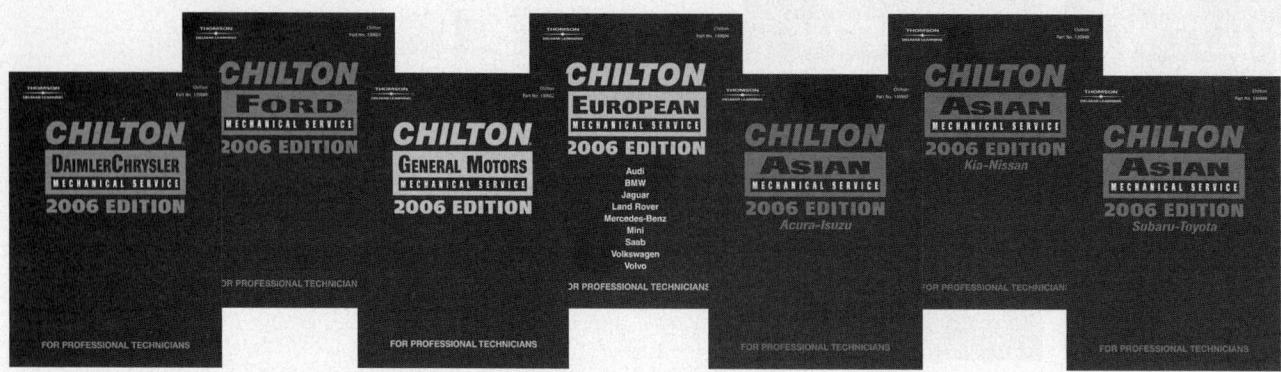

Chilton 2006 Mechanical Service Manuals

The *Chilton® 2006 Mechanical Service Manuals* provide updated coverage through 2005 models and even many 2006 models, as made available from original equipment manufacturers (OEMs). Chilton is still your reliable source for fast, accurate repairs and reassembly and it still provides the lowest-priced professional repair manuals on the market! These manuals are organized by make, model, and system so information gathering is easier. Now with even more illustrations and a streamlined index, it's no wonder more automotive professionals turn to Chilton Professional Manuals for their mechanical service and repair information.

Mechanical Service Manual Benefits:
* access up-to-date service and repair information covering model years 2002-2006, all logically arranged by manufacturer
* follow clear, step-by-step procedures—from drive train to chassis—to yield fast, accurate results
* service more mechanical systems, including brakes, engines, suspensions, steering, and related components
* know what special tools are required for specific jobs, as Chilton editors describe and illustrate them to make repair work go more smoothly

2006 Editions

Chilton 2006 DaimlerChrysler Mechanical Service Manual—ISBN 1-4180-0600-9/Part No. 130600
Chilton 2006 Ford Mechanical Service Manual—ISBN 1-4180-0601-7/Part No. 130601
Chilton 2006 General Motors Mechanical Service Manual—ISBN 1-4180-0602-5/Part No. 130602
Chilton 2006 Asian Mechanical Service Manual—Volume I—ISBN 1-4180-0947-4/Part No. 130947
Chilton 2006 Asian Mechanical Service Manual—Volume II—ISBN 1-4180-0948-2/Part No. 130948
Chilton 2006 Asian Mechanical Service Manual—Volume III—ISBN 1-4180-0949-0/Part No. 130949
Chilton 2006 Asian Mechanical Service Manual—3 Volume Set—ISBN 1-4180-0603-3/Part No. 130603
Chilton 2006 European Mechanical Service Manual—ISBN 1-4180-0604-1/Part No. 130604

Manuals are 8 1/2" x 11"

2005 Editions

Chilton 2005 General Motors Mechanical Service Manual—ISBN 1-4018-7146-1/Part No. 27146
Chilton 2005 Chrysler Mechanical Service Manual—ISBN 1-4018-6718-9/Part No. 26718
Chilton 2005 Ford Mechanical Service Manual—ISBN 1-4018-6719-7/Part No. 26719
Chilton 2005 European Mechanical Service Manual—ISBN 1-4018-6720-0/Part No. 126720
Chilton 2005 Asian Mechanical Service Manual – Volume I—(Acura-Mazda) ISBN 1-4018-6716-2/Part No. 26716
Chilton 2005 Asian Mechanical Service Manual – Volume II—(Mitsubishi-Toyota)
        ISBN 1-4018-6717-0/Part No. 26717
Chilton 2005 Asian Mechanical Service Manual – 2 Volume Set—ISBN 1-4018-7180-1/Part No. 27180

Manuals are 8 1/2" x 11", ©2005

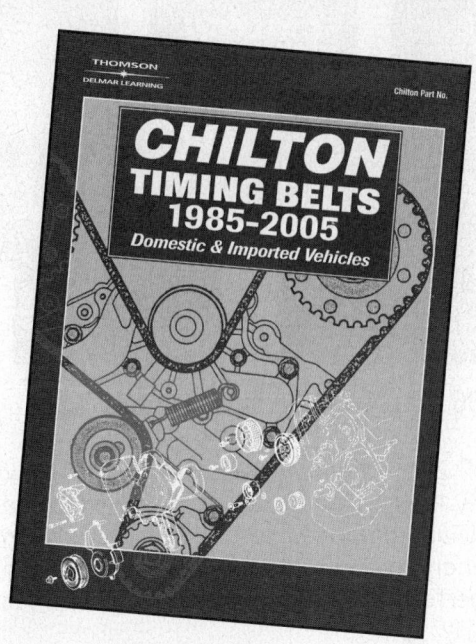

Chilton Timing Belts, 1985-2005

Chilton
ISBN 1-4018-9880-7/Part No. 129880

Timing belt procedures can represent increased profits for automotive repair shops and service stations, and this manual contains all the information automotive technicians need to properly service timing belts on domestic and imported cars, vans, and light trucks through 2005 models. Clear, straightforward procedures, illustrations, and specifications help to communicate 20 years of vehicle applications for fast, accurate inspection, replacement, and tensioning of timing belts. Readers will learn step-by-step how to perform key procedures both quickly and safely, while learning the correct labor time to charge for the service. OEM-recommended replacement intervals for proper maintenance of customer's vehicles are also featured. Professional technicians, trainers, industry professionals, and automotive enthusiasts will all benefit from this manual.

Benefits

- simplify your work - use illustrations showing camshaft and crankshaft timing alignment marks, exploded views of timing belt components, belt routings, special tools, and tensioning adjustments
- take advantage of manufacturer tips on performing correct procedures and adjustments
- prevent engine damage - identify interference or free-wheeling engines
- be confident - use updated manufacturer-recommended timing belt replacement procedures, intervals, and torque specifications
- provide accurate estimates - use trusted Chilton labor times for standard or severe service, including additional time for air conditioning, power steering, balance shaft belt, and water pump removal
- save time - quickly identify year, make, and model in the contents, turn to the start of each engine section, and confirm engine application or VIN code, labor times, and replacement intervals all in one table

544 pp, 8 1/2" x 11", softcover, ©2006

Chilton 2006 Diagnostic Service Manuals

The *Chilton® 2006 Diagnostic Service Manuals* provide technicians with the critical diagnostic information they need to accurately identify and solve engine performance problems. Clear explanations, specifications, and illustrations help technicians diagnose second generation on-board diagnostic (OBD-II) systems. *Chilton Diagnostic Service Manuals*, when used with an engine analyzer, scan tool, or lab scope, allow diagnosticians to understand functions of engine performance components and systems, simplify testing procedures, and diagnose trouble codes.

Diagnostic Service Manual Benefits:
- provide training information in addition to reference material
- explain engine performance components and system operation
- function as exceptional diagnostic companions when analyzing automotive drive-train performance problems
- provide a comprehensive list of trouble code titles, conditions, and possible causes
- reduce diagnostic and repair time using expert testing procedures and troubleshooting hints

2006 Editions

Chilton 2006 DaimlerChrysler Diagnostic Service Manual
 ISBN 1-4180-2118-0/Part No. 132118
Chilton 2006 Ford Diagnostic Service Manual
 ISBN 1-4180-2119-9/Part No. 132119
Chilton 2006 General Motors Diagnostic Service Manual
 ISBN 1-4180-2120-2/Part No. 132120
Chilton 2006 Asian Diagnostic Service Manual, Volume I
 ISBN 1-4180-2913-0/Part No. 132913
Chilton 2006 Asian Diagnostic Service Manual, Volume II
 ISBN 1-4180-2914-9/Part No. 132914
Chilton 2006 Asian Diagnostic Service Manual, Volume III
 ISBN 1-4180-2915-7/Part No. 132915
Chilton 2006 Asian Diagnostic Service Manual, 3 Volume Set
 ISBN 1-4180-2986-6/Part No. 132986
Chilton 2006 European Diagnostic Service Manual
 ISBN 1-4180-2924-6/Part No. 132924

Manuals are 8 1/2" x 11"

2005 Editions

Chilton 2005 General Motors Diagnostic Service Manual
 ISBN 1-4180-0552-5/Part No. 130552
Chilton 2005 Chrysler Diagnostic Service Manual
 ISBN 1-4180-0550-9/Part No. 130550
Chilton 2005 Ford Diagnostic Service Manual
 ISBN 1-4180-0551-7/Part No. 130551
Chilton 2005 Asian Diagnostic Service Manual
 ISBN 1-4180-0553-3/Part No. 130553

Manuals are 8 1/2" x 11", ©2005

Chilton Service Manuals - Perennial Editions

The *Chilton® Perennial Editions* contain repair and maintenance information for popular mechanical systems that may not be available elsewhere. They offer a wide range of repair information on cars, trucks, vans, and SUVs dating back to the early 1960s, and as current as 2002. Information for 1993 and later model years includes scheduled maintenance interval charts.

Benefits:
- covers the most common vehicle models found in the repair aftermarket today
- gain quick understanding of systems using exploded-view illustrations, diagrams, and charts
- simplify tough jobs with easy-to-follow removal and installation instructions for heater core and other components
- obtain complete coverage of repair procedures from drive train to chassis and associated components

Auto Repair Manual, 1998-2002, 1,426 pages
ISBN 0-8019-9362-8/Part No. 9362
Auto Repair Manual, 1993-1997, 2,064 pages
ISBN 0-8019-7919-6/Part No. 7919
Auto Repair Manual, 1988-1992, 1,284 pages
ISBN 0-8019-7906-4/Part No. 7906
Auto Repair Manual, 1980-1987, 1,344 pages
ISBN 0-8019-7670-7/Part No. 7670

Import Car Repair Manual, 1998-2002, 1,792 pps
ISBN 0-8019-9363-6/Part No. 9363
Import Car Repair Manual, 1993-1997, 2,080 pps
ISBN 0-8019-7920-X/Part No. 7920
Import Car Repair Manual, 1988-1992, 1,632 pages
ISBN 0-8019-7907-2/Part No. 7907
Import Car Repair Manual, 1980-1987, 1,488 pages
ISBN 0-8019-7672-3/Part No. 7672

Truck & Van Repair Manual, 1998-2002, 1,408 pages
ISBN 0-8019-9364-4/Part No. 9364
Truck & Van Repair Manual, 1993-1997, 2,096 pages
ISBN 0-8019-7921-8/Part No. 7921
Truck & Van Repair Manual, 1991-1995, 1,664 pages
ISBN 0-8019-7911-0/Part No. 7911
Truck & Van Repair Manual, 1986-1990, 1,536 pages
ISBN 0-8019-7902-1/Part No. 7902
Truck & Van Repair Manual, 1979-1986, 1,440 pages
ISBN 0-8019-7655-3/Part No. 7655

SUV Repair Manual, 1998-2002, 1,292 pages
ISBN 0-8019-9365-2/Part No. 9365

Hardcover manuals are 8 1/2" x 11"

Chilton Collector's Editions—*Reference Manuals for Vintage Vehicles*
Auto Repair Manual, 1964-1971, ISBN 0-8019-5974-8/Part No. 5974,
Truck & Van Repair Manual, 1971-1978, ISBN 0-8019-7012-1/Part No. 7012

ASE Test Preparation Series

Thomson Delmar Learning

ISBN 1-4180-3954-3
Part No. 133954

(Complete Set: A1-A8, L1, P2 X1, C1)

Thomson Delmar Learning has developed comprehensive ASE Test Preparation Manuals to help automotive technicians increase their success on these certification programs. The material covers the topics one might find during the test process. The booklets include many review questions and answers, as well as detailed descriptions of the repairs involved. Designed to look like the actual test, participants will feel more comfortable with practice, which will translate into greater success in taking the actual tests. The design of the Thomson Delmar Learning product also includes helpful test taking hints and student preparation ideas designed to enhance success.

Benefits
- history of the ASE
- test-taking strategies
- tasks lists and overview
- sample test questions
- ASE-style exams
- explanations to the answers (right and wrong)
- glossary of terms

(A1) Automotive Engine Repair, 4E

1-4180-3878-4
Part No. 133878

Includes the following topics: General Engine Diagnosis, Cylinder Head and Valve Train Diagnosis and Repair, Engine Block Diagnosis and Repair, Lubrication and Cooling Systems Diagnosis and Repair, and Fuel, Electrical, Ignition and Exhaust Systems Inspection and Service.

(A2) Automotive Transmissions and Transaxles, 4E

1-4180-3879-2
Part No. 133879

Includes the following topics: General Transmission/Transaxle Diagnosis (Mechanical/Hydraulic Systems and Electronic Systems), Transmission/Transaxle Maintenance and Adjustment, In-Vehicle Transmission/Transaxle Repair, Off-Vehicle Transmission/Transaxle Repair.

(A3) Automotive Manual Drive Trains and Axles, 4E

1-4180-3880-6
Part No.133880

Includes the following topics: Clutch Diagnosis and Repair, Transmission Diagnosis and Repair, Transaxle Diagnosis and Repair, Drive Shaft/Half Shaft and Universal Joint/Constant Velocity (CV) Joint Diagnosis and Repair (Front and Rear Wheel Drive), Rear Axle Diagnosis and Repair, Four Wheel Drive/All Wheel Drive Component Diagnosis and Repair.

(A4) Automotive Suspension and Steering, 4E

1-4180-3881-4
Part No. 133881

Includes the following topics: Steering Systems Diagnosis and Repair (Steering Columns and Manual Steering Gears, Power Assisted Steering Units, Steering Linkage), Suspension Systems Diagnosis and Repair (Front Suspensions, Rear Suspensions, Miscellaneous Services), Wheel Alignment Diagnosis, Adjustment and Repair, and Wheel and Tire Diagnosis and Repair.

(A5) Automotive Brakes, 4E

1-4180-3882-2
Part No. 133882

Iincludes the following topics: Hydraulic System Diagnosis and Repair, Drum Brake Diagnosis and Repair, Disc Brake Diagnosis and Repair, Power Assist Units Diagnosis and Repair, Miscellaneous Systems Diagnosis and Repair, Antilock Brake Systems (ABS) Diagnosis and Repair.

(A6) Automotive Electrical-Electronic Systems, 4E

1-4180-3883-0
Part No.133883

Includes the following topics: General Electrical/Electronic Systems Diagnosis, Battery Diagnosis and Service, Starting Systems Diagnosis and Repair, Charging Systems Diagnosis and Repair, Lighting Systems Diagnosis and Repair, Gauges, Warning Devices and Driver Information Systems Diagnosis and Repair, Horn and Wiper/Washer Diagnosis and Repair.

(A7) Automotive Heating and Air Conditioning, 4E

1-4180-3884-9
Part No. 133884

Includes the following topics: A/C System Diagnosis and Repair, Refrigeration System Component Diagnosis and Repair, Heating and Engine Cooling Systems Diagnosis and Repair, Operating Systems and Related Controls Diagnosis and Repair, Refrigerant Recovery, Recycling, Handling and Retrofit.

(A8) Automotive Engine Performance, 4E

1-4180-3885-7
Part No. 133885

Includes the following topics: General Engine Diagnosis, Ignition System Diagnosis and Repair, Fuel, Air Induction and Exhaust Systems Diagnosis and Repair, Emissions Control Systems Diagnosis and Repair (Including OBDII), Computerized Engine Controls Diagnosis and Repair (Including OBDII), Engine Electrical Systems Diagnosis and Repair.

(L1) Automotive Advanced Engine Performance, 4E

1-4180-3888-1
Part No. 133888

Includes the following topics: General Powertrain Diagnosis, Computerized Powertrain Controls Diagnosis (Including OBDII), Ignition System Diagnosis, Fuel Systems and Air Induction Systems Diagnosis, Emission Control Systems Diagnosis, I/M Failure Diagnosis.

(P2) Automobile Parts Specialist, 4E

1-4180-3887-3
Part No. 133887

Includes the following topics: General Operations, Customer Relations and Sales Skills, Vehicle Systems Knowledge, Vehicle Identification, Cataloging Skills, Inventory Management, Merchandising.

(X1) Exhaust Systems, 4E

1-4180-3886-5
Part No. 133886

Includes the following topics: Exhaust Systems Inspection and Repair, Emissions Systems Diagnosis, Exhaust System Fabrication, Exhaust System Installation, Exhaust System Repair Regulations.

(C1) Automotive Service Consultant, 4E

See next page for details

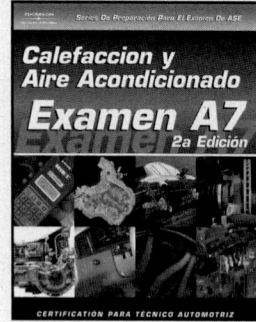

ASE Test Preparation Series in Español!

Thomson Delmar Learning

ISBN 1-4018-1530-8

(Complete Set: A1-A8, L1, P2, X1)

Now available in Español – the first of its kind for Spanish-speaking technicians! This comprehensive package of ASE test preparation booklets are intended for any Spanish-speaking automotive technician who is preparing to take an ASE examination. The series includes questions that relate to each competency required for certification by ASE. In addition to a multitude of questions, the reason why each answer is right or wrong is explained, along with task lists and overview, test-taking strategies, and more.

(A1) Reparación de Motores, 2A Edición
1-4018-1014-4/Part No. 21014

(A2) Transmision Automática/ Eje de Transmision Automática, 2A Edición
1-4018-1015-2/Part No. 21015

(A3) Tren de y Mando Ejes Manuales, 2A Edición
1-4018-1016-0/Part No. 21016

(A4) Suspensión y Dirección, 2A Edición
1-4018-1017-9/Part No. 21017

(A5) Frenos, 2A Edición
1-4018-1018-7/Part No. 21018

(A6) Sistemas Eléctricos/ Electrónicos, 2A Edición
1-4018-1019-5/Part No. 21019

(A7) Calefacción y Aire Acondicionado, 2A Edición
1-4018-1020-9/Part No. 21020

(A8) Funcionamiento de Motores, 2A Edición
1-4018-1021-7/Part No. 21021

(L1) Especialista en el Funciommiato Avansado de Motores, 2A Edición
1-4018-1022-5/Part No. 21022

(P2) Especialista en Partes de Automovil, 2A Edición
1-4018-1023-3/Part No. 21023

(X1) Sistemas de Escape, 2A Edición
1-4018-1024-1/Part No. 21024

ASE Test Preparation Manual—C1 Automotive Service Consultant, 4E
Thomson Delmar Learning
ISBN 1-4180-3889-X
Part No.133889

Prepare to pass the Service Consultant ASE Exam with help from this new test preparation booklet. The new C1 Exam is designed to measure systems knowledge and people skills of those who come in contact with the customer. It will contain questions on Communications, Product Knowledge, Sales Skills, and Shop Operations.

Benefits:
- the ASE task list is fully up-to-date, while current test prep questions reflect the most recent ASE task changes for the broadest knowledge possible
- hundreds of ASE-style exam questions adequately prepare readers to successfully pass the ASE exam
- readers are given multiple opportunities to check their understanding of critical concepts through sample problems, refresher materials, and competency-specific test questions
- overviews of each task provide a great reference point to help answer difficult ASE questions
- explanations for each answer help the user understand why the response is correct or incorrect

Softcover manual is
8 1/2" x 11", ©2006

ASE Test Preparation Manuals—M1 Engine Machinist
Thomson Delmar Learning
ISBN 0-7668-6283-6
Part No. 16283
(Complete Set: M1-M3)

With an abundance of quality content, Thomson Delmar Learning's ASE Test Preparation M-Series contains detailed information designed to help you pass the ASE exams. Each manual combines refresher materials with an abundance of sample test questions, as well as a wealth of information regarding test-taking strategies and the types of questions found in an ASE exam. In addition to the questions, thorough explanations are provided as to why each answer is correct or incorrect.

Benefits:
- the History section explains why the exams are important to the industry
- test-taking strategies help prepare technicians for the environment they will encounter during the actual exam

(M1) Cylinder Head Specialist
0-7668-6280-1/Part No. 16280

(M2) Cylinder Block Specialist
0-7668-6281-X/
Part No. 16281

(M3) Assembly Specialist
0-7668-6282-8/
Part No. 16282

Softcover manuals are
8 1/2" x 11", ©2002

ASE Test Preparation Manuals—B2-B6 Collison Repair and Refinishing
Thomson Delmar Learning
ISBN 1-4018-5120-7/Part No. 25120
(Complete Set: B2-B6)

This fully expanded third edition has been contains high quality ASE test preparation material designed to increase the test taking success of collision repair and refinish technicians. Each book in the series provides valuable preparation for technicians seeking certification in one or more of the ASE collision repair areas. Readers are afforded scores of opportunities to ascertain their knowledge of critical concepts through the extensive array of sample problems, ASE-style exams, and competency-specific test questions required for certification by ASE.

Benefits:
- current, job-related ASE-style exam questions reflecting the most recent ASE task changes test the skills that technicians need to know on the job
- each book contains a general knowledge pretest, a sample test, and additional practice learning that add up to the most real-test practice time available

(B2) Painting and Refinishing, 3E
1-4018-3664-X/Part No. 23664
(B3) Non-Structural Analysis and Damage Repair, 3E
1-4018-3665-8/Part No. 23665
(B4) Structural Analysis and Damage Repair, 3E
1-4018-3666-6/Part No. 23666
(B5) Mechanical and Electrical Components, 3E
1-4018-3667-4/Part No. 23667
(B6) Damage Analysis and Estimation, 3E
1-4018-3668-2/Part No. 23668

Softcover manuals are 8 1/2" x 11", ©2006

Technician Test Preparation— Automotive Bilingual Series
Thomson Delmar Learning

Now both English and Spanish speaking technicians seeking ASE certification can access online test preparation material with ease! The *TTP-Automotive (Bilingual)* series for automotive training and certification provides up-to-date technology and content for tests A1-A8, L1, P2, X1, and C1. An easy-to-use format combined with helpful remediation addresses the unique needs of technicians by clearly demonstrating text-based theory for enhanced learning and retention. Not only is *TTP-Automotive (Bilingual)* the ultimate in test preparation, but it is also an excellent learning tool!

Technician Test Preparation Benefits:
- maps to the latest ASE task lists to familiarize users with the actual work they should be able to do as technicians when taking the ASE tests
- well-illustrated remediation offered via digitized video clips, animations, and high impact graphics further explains key concepts for a more effective learning process
- practice questions provide helpful hints, insight into right and wrong answers, and links to further study specific task areas
- detailed reports provide accurate test results and instant feedback for selected test types so that users can pinpoint the task areas needing improvement
- switch between Spanish and English versions at the click of a button

Call Your Thomson Delmar Learning Sales Rep for Part Numbers & Pricing

Visit **www.techniciantestprep.com** to see the latest modules and a free demo!

The ASE "Passing Lane" Package
Thomson Delmar Learning

ISBN 0-7668-4338-6
(Complete Set: A1-A8, L1, P2)

The most comprehensive test preparation for Automotive Tests A1-A8, L1, and P2. Combining the most thorough ASE Test Preparation books with the latest in ASE videos, this package provides a program of self-study for the automotive ASE Tests.

Book Benefits:
- test-taking strategies
- tasks lists and overview
- sample test questions
- ASE-style exams
- explanations to the answers
- glossary of terms

Video Benefits:
- lively, easy to follow videos emphasize safety throughout
- covers major task areas and topics for each of the ASE exams
- accompanying Activity Sheets help users comprehend and retain information

(A1) Automotive Engine Repair Book/Video,
0-7668-4181-2
(A2) Automotive Transmissions and Transaxles Book/Video,
0-7668-4182-0
(A3) Automotive Manual Drive Trains and Axles Book/Video,
0-7668-4183-9
(A4) Automotive Suspension and Steering Book/Video,
0-7668-4184-7
(A5) Automotive Brakes Book/Video,
0-7668-4185-5
(A6) Automotive Electrical-Electronics Systems Book/Video,
0-7668-4186-3
(A7) Automotive Heating and Air Conditioning Book/Video,
0-7668-4187-1
(A8) Automotive Engine Performance Book/Video,
0-7668-4188-X
(L1) Automotive Advanced Engine Performance Book/Video,
0-7668-4189-8
(P2) Automobile Parts Specialist Book/Video,
0-7668-4190-1

Automotive Technician Certification Test Preparation Manual, 3E
Don Knowles

ISBN 1-4180-4926-3/ Part No. 134926

Filled with updated task list theory, practice tests, and abundant, demonstrative graphics, this revised edition provides all the latest information required to sufficiently prepare technicians to pass each of the A1-A8 and L1 ASE certification exams. Each chapter begins with a pretest that indicates the depth of preparation required to become familiar with the information in the chapter, followed by a description of each ASE task and the must-have information related to the task. ASE-type questions at the end of each chapter appear in the same format as on actual ASE tests to further prepare users to pass each exam.

Benefits
- current information provides practice questions which match the latest ASE task list
- answers to pretest questions and helpful analyses at the end of each chapter provide learners with faster access to accurate information
- supportive "Hints" throughout each chapter help users work through the process of determining the correct answers to the questions

CONTENTS

Chapter 1 - Engine Repair. Chapter 2 - Automatic Transmission/Transaxle. Chapter 3 - Manual Drive Train and Axles. Chapter 4 - Suspension and Steering. Chapter 5 - Brakes. Chapter 6 - Electrical/Electronic Systems. Chapter 7 - Heating, Ventilation, and Air Conditioning Systems. Chapter 8 - Engine Performance. Chapter 9 - Advanced Engine Performance.

656 pp, 8½" x 11", softcover, ©2007

ATC Challenge 3.0 CD-ROM
Thomson Delmar Learning

ISBN 0-7668-2982-0

These exciting interactive CD-ROMs have been designed to prepare technicians for successful completion of the Automotive ASE task areas (A1-A8, L1, P2, and F1). This multimedia software assesses strengths and weaknesses by identifying topics needing further study while allowing users to review ASE task areas at their own pace. Explanations, hints, notes, and a glossary aid the user in comprehension, critical thinking and retention. These CD-ROMs offer hundreds of ASE-style questions, a test taking strategy section and LAN compatibility.

CD-ROM, ©2001

Site License Available for Multiple Unit Purchases or Multiple Workstations for ATC Challenge 3.0:
User 1: Full Price (List or Net)
Users 2-5:
$80/workstation + Full Price
Users 6-10:
$70/workstation + Full Price
Users 11-20:
$60/workstation + Full Price
Users 21+:
$50/workstation + Full Price

ATC Challenge for P2
Thomson Delmar Learning
ISBN 0-7668-1827-6

This interactive CD-ROM contains material that will help prepare technicians for the Automotive Parts Specialist (P2) ASE certification exam.
CD-ROM, ©2000

NATEF Standards Job Sheets, 2E
Thomson Delmar Learning

ISBN 1-4180-2082-6

(Complete Set: A1-A8)

New from today's leading automotive education publisher, each of our eight NATEF (National Automotive Technicians Education Foundation) Standards Job Sheets workbooks has been thoughtfully designed to assist users in gaining valuable job preparedness skills and mastering specific technical competencies required for success as a professional automotive technician. Ideal for use as a stand-alone item, or with any comprehensive or topic-specific automotive text, the entire series is based on the 2005 NATEF tasks and consists of individual books for each of the following areas: Engine Repair, Automatic Transmissions/Transaxles, Manual Drive Trains and Axles, Suspension and Steering, Brakes, Electricity/Electronics, Heating and Air Conditioning, and Engine Performance.

Key Features
- manuals are not keyed to a specific text, making it easy to use one or more of them in any automotive training program in which NATEF coverage is desired
- NATEF tasks are addressed in each manual, providing students with a first-quality, comprehensive learning experience

JOB SHEETS AVAILABLE FOR:
(A1) **Automotive Engine Repair**, 1-4180-2074-5
(A2) **Automatic Transmissions and Transaxles**, 1-4180-2075-3
(A3) **Manual Drive Trains and Axles**, 1-4180-2076-1
(A4) **Automotive Suspension and Steering**, 1-4180-2077-X
(A5) **Automotive Brakes**, 1-4180-2078-8
(A6) **Automotive Electrical and Electronic Systems**, 1-4180-2079-6
(A7) **Automotive Heating and Air Conditioning**, 1-4180-2080-X
(A8) **Automotive Engine Performance**, 1-4180-2081-8

Softcover manuals are 8½" x 11", ©2006

The Service Consultant: Working in an Automotive Facility
Ronald A. Garner & C. William Garner

ISBN 1-4018-7990-X
Part No. 127990

This book examines the multi-faceted responsibilities of an automotive service consultant. It outlines task-oriented procedures for day-to day operations and provides an understanding of how service techniques are used to maximize customer satisfaction and profitability. Content follows the tasks identified by ASE for Automotive Service Consultant (C1). ASE terminology is used throughout to describe the people and businesses servicing the driving public. Coverage examines communications specific to customer relations and sales as well as internal communications, relations, and supervision. Customer delivery and follow up round out this thorough exploration of the functions of a successful automotive service consultant.

Key Features
- content correlating to the ASE tasks for Automotive Service Consultant (C1) helps readers better prepare for the ASE certification exam
- activities that require readers to interview local service facility managers provide opportunities for establishing important industry contacts for the future
- careful attention to the sequence of job duties for the service consultant gives readers a clear picture of the types of work and responsibilities that will be expected of them when they enter the workforce

224 pp, 7 3/8" x 9", hardcover, ©2005

Managing Automotive Businesses: Strategic Planning, Personnel and Finances
Ronald A. Garner & C. William Garner

ISBN 1-4018-9896-3
Part No. 129896

The success of any organization most often depends on the execution and management of such strategic issues as business development, personnel, and fiscal operations. This new book introduces readers to the duties and practices assigned to service managers in the successful operation of an automotive service facility. Coverage begins with a general discussion of the management structure and the service manager's role in facility operations. Consideration is then given to navigation of the personnel process from the recruitment of workers to supervision of their performance. The financial business practices of a service manager familiarizes readers with the importance of fiscal responsibility in the operation of a lucrative automotive service business.

Key Features
- fosters a thorough understanding of strategic planning from the owner's perspective to the service manager's level in the establishment of a productive service facility
- information on the recruitment and retention of employees and the establishment of rules and performance measures gives readers valuable insight into the "real world" of the automobile business
- the analysis of financial statements exposes readers to the basic activities and duties that are typical of an automotive service manager's position

272 pp, 7 3/8" x 9 1/4", hardcover, ©2006

This pioneering eight-book series offers automotive service shop owners and those wanting to be shop owners the necessary business and customer service skills to run a successful automotive service facility.

The series covers three main topic areas: personnel management, business management, and sales and marketing. Each book provides a framework to help technicians make consistent, high-quality, and productive service a part of every day shop operations. According to the author, "Great performance coupled with increased customer loyalty, trust, and operational excellence will almost always result in increased profits."

Automotive Service Management Series Benefits:

- real-world approach reflects author's experience as a fourth generation technician, a repair & service company owner, and an automotive industry trainer
- all-inclusive coverage spans from designing an automotive repair facility floor plan through financial management techniques, customer/staff relations, and more
- length of each book makes it easy to incorporate this series into workshops, seminars, and training/education courses
- information is available "as is" or for customization

Total Customer Relationship Management
ISBN 1-4018-2657-1/Part No. 22657
From Intent to Implementation
ISBN 1-4018-2658-X/Part No. 22658
Operational Excellence
ISBN 1-4018-2659-8/Part No. 22659
Building a Team
ISBN 1-4018-2660-1/Part No. 22660
The High Performance Shop
ISBN 1-4018-2661-X/Part No. 22661
Safety Communications
ISBN 1-4018-2662-8/Part No. 22662
Managing Dollars with Sense
ISBN 1-4018-2663-6/Part No. 22663
Operations Management
ISBN 1-4018-2665-2/Part No. 22665
Entire Set of 8 Books
ISBN 1-4018-2499-4/Part No. 22499

Softcover manuals are 8 1/2" x 11", ©2003

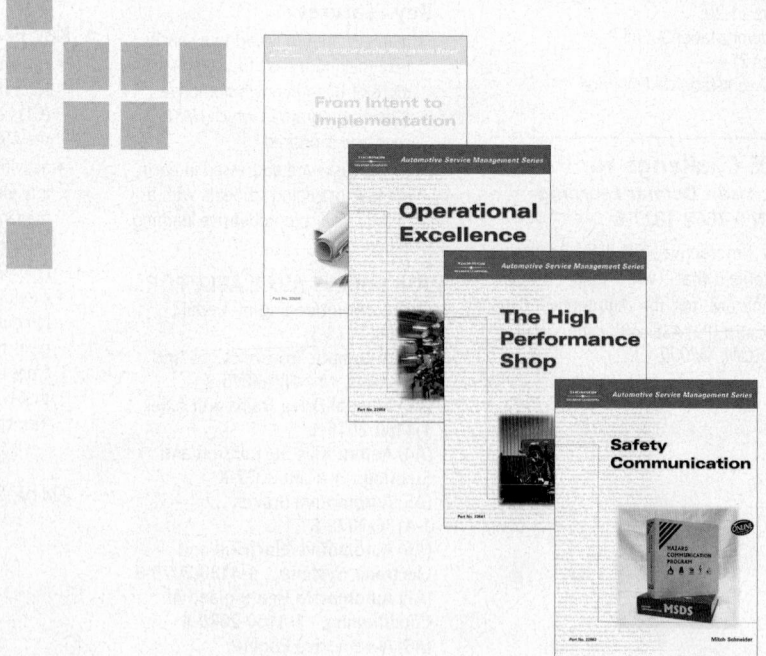

Introduction to OBDII
Roy Cox
ISBN 1-4180-1220-3
Part No. 131220

Here's an easy-to-understand, logical guide to the diagnosis and repair of today's complex and sophisticated automotive control systems! Introduction to On-Board Diagnostics (OBD II) readers will learn the fundamentals of how to perform diagnostic procedures, and be provided with valuable reference material for diagnosing and troubleshooting components and circuits. This book provides a simple, logical approach to explain the operation of the OBD II process and will teach the reader how to quickly spot problems and identify components that are not functioning correctly.

Benefits
- "quick hit" troubleshooting tricks teach readers how to diagnose problems when there is no stored OBDII trouble code, as well as how to handle situations where the trouble code is actually set by a basic mechanical problem rather than a failure of the indicated component
- information is useful for those who wish to expand their capabilities from more basic, mechanical repairs to complex electronics and drivability diagnosis and repair
- focuses on logical troubleshooting that can be done without expensive, complicated test equipment and special tools

CONTENTS
Chapter 1- Introduction, Chapter 2- Evolution of OBD, Chapter 3- OBDII Terminology, Chapter 4- System Operating Protocols, Chapter 5- System Monitors, Chapter 6- Drive Cycles, Chapter 7- Diagnostic Trouble Codes (DTC's), Chapter 8- Diagnostic Routines

256 pp, 8 1/2" x 11", softcover, ©2006

SUPPLEMENTS
Diagnostic Tool CD-ROM
1-4180-1221-1/Part No. 131221
Instructor's Guide 1-4180-1222-X

Professional Automotive Technician Training Series
Thomson Delmar Learning

This self-paced, interactive learning series contains must-have training for today's professional technicians. Each course in the series contains the most up-to-date content, reinforced using engaging graphics, animations, and user interactions. Section review questions, as well as the end of course review, are designed to reinforce user learning and progression.

These thought-provoking products combine theory, diagnosis, and repair information into one easy-to-use training tool! More than 8.5 hours of state-of-the-art instruction per course is provided. Available in both CD-ROM(CBT) and Web-based (WBT) formats, these training tools are ideal for all automotive technicians, and they have been developed to comply with both AICC and SCORM compliance standards.

Benefits:
- regular use of highly engaging animations and interactivity keeps users engaged throughout all the material
- bookmarking technology enables users to track their progress from beginning to end
- periodic progress checks and end-of-section reviews are integrated throughout to ensure the highest level of retention
- certificates of completion can be printed by users achieving a score of 80% or higher on the final course review
- all material is up-to-date to the latest ASE standards

Basic Automotive Service and Maintenance
    CBT: ISBN 1-4180-4100-9/Part No. 134100
    WBT: ISBN 1-4180-4101-7/Part No. 134101

Engine Performance
    CBT: ISBN 1-4180-4239-0/Part No. 134239
    WBT: ISBN 1-4180-4240-4/Part No. 134240

Brakes
    CBT: ISBN 1-4180-4235-8/Part No. 134235
    WBT: ISBN 1-4180-4236-6/Part No. 134236

Suspension and Steering
    CBT: ISBN 1-4180-4237-4/Part No. 134237
    WBT: ISBN 1-4180-4238-2/Part No. 134238

Electricity and Electronics
    CBT: ISBN 1-4180-4241-2/Part No. 134241
    WBT: ISBN 1-4180-4242-0/Part No. 134242

©2007

Comprehensive Skill Assessment Tool (CSAT) —Automotive
Thomson Delmar Learning

The Comprehensive Skill Assessment Tool for Automotive is an online skill gap analysis product, designed to help instructors and trainers implement targeted training programs. Strategic learning areas are measured to account for a technicians knowledge of theory, hands-on/application, and diagnostic knowledge across key automotive skill areas. While the pre-assessment will assist with the identification of areas needing improvement, the combined phases of education and training, and post-assessment allow trainers to track skill level growth and prove out return on investment against their training dollars invested. This is a must have tool for any training prganization.

Benefits
- a low-cost solution benefiting trainers and students
- individual users can take tests online to identify areas of strength and areas needing improvement
- account set-up that enables instructors to assess and track the results of individual users

Call Your
Thomson Delmar Learning
Sales Rep for
Part Numbers & Pricing

Visit **www.skillanalysis.com** to see the latest modules and a free demo!

Improving Fuel Economy
Roy Cox

ISBN 1-4018-8367-2/Part No. 28367

With gasoline prices on the rise, every driver will appreciate this complete, concise guide to fuel economy and how to improve it in today's automobiles! The only publication of its kind, *Fuel Economy* thoroughly discusses ways in which modern automotive systems work together and the interrelationships that affect fuel economy. Author Roy Cox uses simple, straightforward language to raise such important topics as conditions and components that cause specific problems, actions that can be taken to make your car use less fuel, and when to seek professional repairs.

"Drive slower," says Cox "Driving at 50 to 55 mph is measurably more economical. Drivers should also try to avoid braking and drive off-hours to as not to get caught in rush hour traffic jams." Fuel Economy details the beneficial effects of fuel economy maintenance. Cox includes step-by-step instructions for common do-it-yourself procedures. "Vehicle owners should drive less and drive better." Says Cox. "Regular maintenance and unclogging of air filters is important, too." This clever combination of "how it works" and "how to do the maintenance" approach fosters new insight into the key factors affecting fuel economy in cars, and what drivers can do to keep them performing at peak power and economy.

Benefits
- useful advice, such as what fuel choice to make, what motor oil and filter to buy, and what factors to consider when buying replacement parts, makes this a handy resource for new motorists and auto enthusiasts alike
- well illustrated charts provide valuable quick-reference information that can be stored in the glove box and consulted again and again
- useful guidance regarding what maintenance is appropriate to do oneself and what is best left to the professionals helps prevent consumers from getting in "over their heads"
- a beneficial glossary of common automotive acronyms and abbreviations helps readers sort out technical jargon and understand the reasons for recommended repairs and maintenance

CONTENTS
1. Factors that affect fuel economy, 2. Things you can do to make your car use less fuel, 3. When to seek professional repairs, 4. Keep the performance your car was born with: recipe for a long and happy engine life, Glossary of commonly used automotive abbreviations and acronyms

52 pp, 4 1/2" x 7 1/2", softcover, ©2005

Delmar's Automotive Dictionary
David W. South & Boyce Dwiggins

ISBN 0-8273-7405-4

This handy, ready-reference dictionary provides the automotive engineer, technician, mechanic, student, enthusiast or layperson with a single source for the most up-to-date definitions available of technical, professional and informal terminology used in today's automotive world. It is descriptive and covers the wide scope of terms pertinent to the automotive field. With multiple definitions and aids, and proper pronunciation of terms, this dictionary is a must for all!

Benefits
- over 3000 terms comprehensively covering more than 100 subject areas
- enhanced by a list of acronyms and abbreviations
- up-to-date definitions of today's automotive terminology
- aids for proper pronunciation
- each term has multiple definitions

281 pp, 6" x 9", softcover, ©1997

Math for the Automotive Trade, 3E
John C. Peterson & William deKryger

ISBN 0-8273-6712-0

Math for Automotive Trades, 3E provides excellent examples and problems that reflect technological requirements of workers in automotive technology. The text has three parts: review of basic mathematics skills, math applications to specific automotive situations, and an examination of measurement aspects beginning with angle and linear measurements and ending with an extensive look at measurement tools used in the automotive trade.

345 pp, 8½" x 11", softcover, ©1995
Instructor's Manual   **0-8273-6713-9**

Practical Problems in Mathematics for Automotive Technicians, 6E Text:
Larry Sformo, Todd Sformo, and George Moore

ISBN 1-4018-3999-1

Comprehensive and easy-to use, this updated edition covers every type of practical math problem that automotive technicians will face on the job. The subject matter is organized in a knowledge-building format that progresses from the basics of whole number operations into percentages, linear measurements, ratios, and the use of more complex formulas. Complete coverage of fundamentals, as well as more advanced computations make this book suitable for both beginning and advanced technicians. With a special section on graphs, scales, test meters, estimation, and invoices used in the workplace, this book is tailor-made for any automotive course of study!

Benefits
- new section on conversion of measurements makes English-to-metric conversions quick and easy
- proficiency using fractions is encouraged to assist with real world math applications
- step-by-step instructions, diagrams, charts, and examples cultivate problem-solving skills that are crucial to success on the job
- problems proceed from the simple to the more complex so that readers have the ample opportunities to develop their skills and confidence
- solid understanding of the practical applications for each mathematical process fosters a strong foundation in fundamental principles for beginners and advanced technicians

CONTENTS
Whole Numbers. Common Fractions. Decimal Fractions. Percent and Percentage. Measurement. Ratio and Proportion. Powers and Roots. Formulas. Graphs. Invoices.

288 pp, 7 7/8" x 9 1/4", softcover, ©2005
Instructor's Manual   **1-4018-4000-0**